# Amorphous and Nanocrystalline Silicon-Based Films—2003

MATERIALS RESEARCH SOCIETY
SYMPOSIUM PROCEEDINGS VOLUME 762

# Amorphous and Nanocrystalline Silicon-Based Films—2003

Symposium held April 22–25, 2003, San Francisco, California, U.S.A.

### EDITORS:

**John R. Abelson**
University of Illinois at Urbana-Champaign
Urbana, Illinois, U.S.A.

**Gautam Ganguly**
BP Solar
Toano, Virginia, U.S.A.

**Hideki Matsumura**
Japan Advanced Institute of Science and Technology
Ishikawa, Japan

**John Robertson**
Cambridge University
Cambridge, United Kingdom

**Eric A. Schiff**
Syracuse University
Syracuse, New York, U.S.A.

**Materials Research Society**
Warrendale, Pennsylvania

Single article reprints from this publication are available through
University Microfilms Inc., 300 North Zeeb Road, Ann Arbor, Michigan 48106

CODEN: MRSPDH

Copyright 2003 by Materials Research Society.
All rights reserved.

This book has been registered with Copyright Clearance Center, Inc. For further information, please contact the Copyright Clearance Center, Salem, Massachusetts.

Published by:

Materials Research Society
506 Keystone Drive
Warrendale, PA 15086
Telephone (724) 779-3003
Fax (724) 779-8313
Web site: http://www.mrs.org/

Manufactured in the United States of America

# CONTENTS

Preface ............................................................................................................. xix

Materials Research Society Symposium Proceedings ..................................... xx

## AMORPHOUS NETWORK STRUCTURES, ELECTRONIC METASTABILITY, DEFECTS AND PHOTOLUMINESCENCE

\* Numerical Studies of the Dynamics of Silicon: Relaxation,
Nucleation and Energy Landscape ............................................................... 3
    Normand Mousseau, Philippe Beaucage, and Francis Valiquette

Atomistic Character of Nanocrystalline and Mixed Phase Silicon ............... 15
    R. Biswas and B.C. Pan

Kinetics of Light-Induced Effects in Mixed-Phase Hydrogenated
Silicon Solar Cells ........................................................................................ 21
    Guozhen Yue, Baojie Yan, Jeffrey Yang, Kenneth Lord, and
    Subhendu Guha

Temperature Dependence of a Hydrogen Doublet Site in Light-
Soaked a-Si:H From $^1$H NMR ..................................................................... 27
    T. Su, P.C. Taylor, G. Ganguly, and D.E. Carlson

Evidence for Trap-Conversion Induced Instability in
Amorphous Silicon ....................................................................................... 33
    Vikram L. Dalal, Puneet Sharma, and Abdul Aziz

Evolution of Charged Gap States in a-Si:H Under Light Exposure ............. 39
    M. Zeman, V. Nádazdy, R.A.C.M.M. van Swaaij, R. Durny,
    and J.W. Metselaar

Metamict Transformation of Silica ............................................................... 45
    Ju-Yin Cheng, M.M.J. Treacy, and P.J. Keblinski

\* Metastable Defects in the Amorphous Silicon-Germanium Alloys ............ 51
    J. David Cohen

Temperature Dependence of the Decay of Optically Excited
Charge Carriers in Amorphous Silicon ......................................................... 63
    J. Whitaker and P.C. Taylor

\*Invited Paper

**Electron-Spin-Resonance Investigation of Laser Crystallized Polycrystalline Silicon** .................................................................................................... 69
    K. Brendel, N.H. Nickel, K. Lips, and W. Fuhs

**Time-Resolved Switching Studies in a-Si:H and Related Films** ........................................ 75
    P. Stradins, W.B. Jackson, H.M. Branz, J. Hu, C.L. Perkins, and Qi Wang

**Adsorption and Oxidation Effects in Microcrystalline Silicon** ............................................ 81
    T. Dylla, F. Finger, and R. Carius

**Silicon Nanostructured Films Formed by Pulsed-Laser Deposition in Inert Gas and Reactive Gas** ......................................................................... 87
    X.Y. Chen, Y.F. Lu, Y.H. Wu, B.J. Cho, and H. Hu

**Origin of the Low-Energy Photoluminescence in Microcrystalline Silicon Films** ............................................................................................. 93
    Joon-Yong Lee, Dong-Hyun Park, and Jong-Hwan Yoon

**Post-Transit Analysis of Transient Photocurrents From High-Deposition-Rate a-Si:H Samples** ............................................................................... 99
    Monica Brinza, W.M.M. Kessel, Arno H.M. Smets, M.C.M. van de Sanden, and Guy J. Adriaenssens

**Size Distribution of Embedded Nano-Crystallites in Polymorphous Silicon Studied by Raman Spectroscopy and Photoluminescence** ...................................... 105
    V. Tripathi, Y.N. Mohapatra, Md.N. Islam, V. Suendo, and P. Roca i Cabarrocas

**Thin Film Cavity Ringdown Spectroscopy and Second Harmonic Generation on Thin a-Si:H Films** ..................................................................... 111
    I.M.P. Aarts, B. Hoex, J.J.H. Gielis, C.M. Leewis, A.H.M. Smets, R. Engeln, M. Nesládek, W.M.M. Kessels, and M.C.M. van de Sanden

**High Frequency Electron Spin Resonance Study of Hydrogenated Microcrystalline Silicon** .................................................................................. 117
    Takashi Ehara

**Photoelectron Spectroscopic Investigations of Very Thin a-Si:H Layers** ............................................................................................................................ 125
    M. Schmidt, A. Schoepke, O. Milch, Th. Lussky, and W. Fuhs

Determination of Defect Densities by Constant Photocurrent
Method—Comparison of AC and DC Methods .................................................................131
    Charlie Main, Steve Reynolds, Ivica Zrinscak, and Amar Merazga

Depth Profiling of Light-Induced Defects in Hydrogenated
Amorphous Silicon by Transient Photocurrent Spectroscopy ...................................137
    Steve Reynolds, Charlie Main, and Rudi Brüggemann

A Study of Electronic Defects in Hydrogenated Amorphous Silicon
Prepared by the Expanding Thermal Plasma Technique ............................................143
    Steve Reynolds, Charlie Main, Ivica Zrinscak, Zdravka Aneva,
    and Diana Nesheva

## THIN FILM TRANSISTORS
## AND IMAGER ARRAYS

Performance of Thin-Film Silicon MEMS Resonators in Vacuum ...........................151
    J. Gaspar, V. Chu, and J.P. Conde

Area-Dependent Switching in Thin-Film Silicon Devices ...........................................157
    Jian Hu, Warren Jackson, Scott Ward, Pauls Stradins,
    Howard M. Branz, and Qi Wang

Properties of Silicon Nitride Films Prepared by Combination
of Catalytic-Nitridation and Catalytic-CVD .................................................................163
    A. Izumi, A. Kikkawa, K. Higashimine, and H. Matsumura

High-Rate (> 1nm/s) and Low-Temperature (< 400°C)
Deposition of Silicon Nitride Using an $N_2/SiH_4$ and $NH_3/SiH_4$
Expanding Thermal Plasma .............................................................................................169
    J. Hong, W.M.M. Kessels, and M.C.M. van de Sanden

Switch-On Transients and Static Characteristics of
Polymorphous and Amorphous Silicon Thin-Film Transistor ...................................175
    V. Tripathi, Y.N. Mohapatra, and P. Roca i Cabarrocas

Dependences of Structural Parameters on the Characteristics
of Poly-Si Thin-Film Transistors After Plasma Passivation .......................................181
    Cheng-Ming Yu, Tiao-Yuan Huang, Tan-Fu Lei, and
    Horng-Chih Lin

Above-Threshold Parameter Extraction Including Contact
Resistance Effects for a-Si:H TFTs on Glass and Plastic ............................................187
    Peyman Servati, Denis Striakhilev, and Arokia Nathan

Mechanical Stress and Process Integration of Direct X-ray
Detector and TFT in a-Si:H Technology .................................................................. 193
    Czang-Ho Lee, Isaac Chan, and Arokia Nathan

Stacked n-i-p-n-i-p Heterojunctions for Image Recognition ........................................ 199
    M. Vieira, A. Fantoni, M. Fernandes, P. Louro, and
    I. Rodrigues

Development of Vertically Integrated Imaging and Particle Sensors ........................ 205
    N. Wyrsch, C. Miazza, S. Dunand, A. Shah, N. Blanc,
    R. Kaufmann, L. Cavalier, G. Anelli, M. Despeisse, P. Jarron,
    D. Moraes, A.G. Sirvent, G. Dissertori, and G. Viertel

An Amorphous Silicon Photoconductor for UV Detection ........................................ 211
    Matthias Hillebrand, Frank Blecher, Jürgen Sterzel, and
    Markus Böhm

Correlation Between the Tunneling Oxide and I-V Curves of
MIS Photodiodes ........................................................................................................ 217
    H. Águas, L. Pereira, A. Goullet, R. Silva, E. Fortunato, and
    R. Martins

Numerical Simulation of the Influence of the Gap State of a-Si:H
on the Characteristics of a-Si:H p-i-n/OLED Coupling Device ................................ 223
    Chunya Wu, Yousu Chen, Juan Li, Guanghua Yang,
    Huidong Yang, Zhenhua Zhou, Ying Zhao, Zhiguo Meng,
    Xinhua Geng, Shaozheng Xiong, Lizhu Zhang, and Qi Wang

\* Low-Temperature Growth of Poly-Si and SiGe Thin Films by
Reactive Thermal CVD and Fabrication of High Mobility
TFTs Over 50 $cm^2/Vs$ ............................................................................................ 229
    Jun-ichi Hanna and Kousaku Shimizu

Improvement of Gate Oxide Integrity in Low Temperature
Poly Silicon TFT ........................................................................................................ 241
    Seok-Woo Lee, Dae Hyun Nam, Jin Mo Yoon, Hyun Sik Seo,
    Kyoung Moon Lim, and Chang-Dong Kim

Threshold Voltage Performance of a-Si:H TFTs for Analog
Applications ............................................................................................................... 247
    K.S. Karim, K. Sakariya, and A. Nathan

\*Invited Paper

**Characteristics of Bottom Gate Thin Film Transistors With Silicon Rich Poly-Si$_{1-x}$Ge$_x$ and Poly-Si Fabricated by Reactive Thermal Chemical Vapor Deposition** ...................253
    Kousaku Shimizu, Jian Jun Zhang, Jeong-Woo Lee, and Jun-ichi Hanna

**Optoelectronic Detection of DNA Molecules Using an Amorphous Silicon Photodetector** ...................259
    F. Fixe, D.M.F. Prazeres, V. Chu, and J.P. Conde

**Fabrication of Novel TFT LCD Panels With High Aperture Ratio Using a-SiCO:H Films as a Passivation Layer** ...................265
    W.S. Hong, K.W. Jung, B.K. Hwang, G. Cerny, S.H. Yang, J.H. Choi, and K. Chung

**Leakage Current Behavior In Common i-Layer a-Si:H p-i-n Photodiode Arrays** ...................271
    Jeremy A. Theil

**Enhanced Blue Sensitivity in ITO/a-SiN$_x$:H/a-Si:H MIS Photodetectors** ...................277
    S. Tao, Y. Vygranenko, and A. Nathan

## SOLAR CELLS

\* **Amorphous and Microcrystalline Silicon Based Solar Cells and Modules on Textured Zinc Oxide Coated Glass Substrates** ...................285
    Bernd Rech, Joachim Müller, Tobias Repmann, Oliver Kluth, Tobias Roschek, Jürgen Hüpkes, Helmut Stiebig, and Wolfgang Appenzeller

**Bandtail Limits to Solar Conversion Efficiencies in Amorphous Silicon Solar Cells** ...................297
    K. Zhu, J. Yang, W. Wang, E.A. Schiff, J. Liang, and S. Guha

**Carrier Transport and Recombination in a-Si:H p-i-n Solar Cells in Dark and Under Illumination** ...................303
    J. Deng, J.M. Pearce, V. Vlahos, R.J. Koval, R.W. Collins, and C.R. Wronski

\*Invited Paper

\* **Hydrogenated Microcrystalline Silicon Single-Junction and Multi-Junction Solar Cells**...............309
    Baojie Yan, Guozhen Yue, Jeffrey Yang, Arindam Banerjee,
    and Subhendu Guha

**Electronic Properties of Microcrystalline Silicon Investigated by Photoluminescence Spectroscopy on Films and Devices**...............321
    R. Carius, T. Merdzhanova, and F. Finger

\* **Localized States in Microcrystalline Silicon Photovoltaic Structures Studied by Post-Transit Time-of-Flight Spectroscopy**...............327
    Steve Reynolds, Vladimir Smirnov, Charlie Main,
    Reinhard Carius, and Friedhelm Finger

**Femtosecond Far-Infrared Studies of Carrier Dynamics in Hydrogenated Amorphous Silicon and Silicon-Germanium Alloys**...............333
    A.V.V. Nampoothiri, B.P. Nelson, and S.L. Dexheimer

**Micro-Raman Studies of Mixed-Phase Hydrogenated Silicon Solar Cells**...............339
    Jessica M. Owens, Daxing Han, Baojie Yan, Jeffrey Yang,
    Kenneth Lord, and Subhendu Guha

**Hole Drift-Mobility Measurements in Contemporary Amorphous Silicon**...............345
    S. Dinca, G. Ganguly, Z. Lu, E.A. Schiff, V. Vlahos,
    C.R. Wronski, and Q. Yuan

**Recombination in n-i-p (Substrate) a-Si:H Solar Cells With Silicon Carbide and Protocrystalline p-Layers**...............351
    V. Vlahos, J. Deng, J.M. Pearce, R.J. Koval, G.M. Ferreira,
    R.W. Collins, and C.R. Wronski

**Nanocrystalline Silicon (nc-Si) From Single Ion Beam Sputtering**...............357
    Z.B. Zhou, G.M. Hadi, R.Q. Cui, Z.M. Ding, and G. Li

**Correlation of Material Properties and Open-Circuit Voltage of Amorphous Silicon Based Solar Cells**...............363
    Baojie Yan, Jeffrey Yang, Guozhen Yue, and Subhendu Guha

\*Invited Paper

Simulations of Buffer Layers in a-Si:H Thin Film Solar Cells
Deposited With an Expanding Thermal Plasma .................................................................. 369
    A.M.H.N. Petit, M. Zeman, R.A.C.M.M. van Swaaij, and
    M.C.M. van de Sanden

Microcrystalline (Si,Ge):H Solar Cells ................................................................................ 375
    Jianhua Zhu and Vikram L. Dalal

Investigation of the Causes and Variation of Leakage Currents
in Amorphous Silicon p-i-n Diodes ..................................................................................... 381
    Todd R. Johnson, Gautam Ganguly, George S. Wood, and
    David E. Carlson

Deposition of Device Quality μc-Si Films and Solar Cells at
High Rates by HWCVD in a W Filament Regime Where W/Si
Formation is Minimal .......................................................................................................... 387
    E. Iwaniczko, A.H. Mahan, B. Yan, L.N. Gedvilas,
    D.L. Williamson, and B.P. Nelson

Room Temperature Recovery of Light Induced Changes in
Amorphous Silicon Solar Cells ........................................................................................... 393
    G. Ganguly, D.E. Carlson, M.S. Bennett, F. Willing, R.R. Arya,
    and P. Stradins

Toward Understanding the Degradation Without Light
Soaking in Hot-Wire a-Si:H Thin Films and Solar Cells ..................................................... 399
    Qi Wang, Keda Wang, and Daxing Han

Material Aspects of Reactively MF-Sputtered Zinc Oxide for
TCO Application in Silicon Thin-Film Solar Cells ............................................................. 405
    Jürgen Hüpkes, Bernd Rech, Oliver Kluth, Joachim Müller,
    Hilde Siekmann, Chitra Agashe, Hans P. Bochem, and Matthias Wuttig

## *GROWTH MECHANISMS, HOT FILAMENT CVD AND MICROCRYSTALLINE Si:H GROWTH*

\* Combinatorial Approach to Thin-Film Silicon Materials and Devices ...................... 413
    Qi Wang, Leandro R. Tessler, Helio Moutinho, Bobby To,
    John Perkins, Daxing Han, Dave Ginley, and Howard M. Branz

Calculations of $SiH_3$ Diffusion and Growth Processes on a-Si:H Surfaces ................. 425
    P. Vigneron, P.W. Peacock, K. Xiong, and J. Robertson

\*Invited Paper

The a-Si:H Growth Mechanism: Temperature Study of the
SiH$_3$ Surface Reactivity and the Surface Silicon Hydride
Composition During Film Growth .................................................................................... 431
    W.M.M. Kessels, Y. Barrell, P.J. van den Oever,
    J.P.M. Hoefnagels, and M.C.M. van de Sanden

Effects of Excitation Frequency and H$_2$ Dilution on Cluster
Generation in Silane High-Frequency Discharges ....................................................... 437
    Masaharu Shiratani, Kazunori Koga, Atsushi Harikai,
    Takanori Ogata, and Yukio Watanabe

*Application of Deposition Phase Diagrams for the
Optimization of a-Si:H-Based Materials and Solar Cells ........................................... 443
    R.W. Collins, A.S. Ferlauto, G.M. Ferreira, Joohyun Koh,
    Chi Chen, R.J. Koval, J.M. Pearce, C.R. Wronski,
    M.M. Al-Jassim, and K.M. Jones

Influence of Filament and Substrate Temperatures on
Structural and Optoelectronic Properties of Narrow Gap
a-SiGe:H Alloys Deposited by Hot-Wire CVD ............................................................ 455
    Yueqin Xu, Brent P. Nelson, D.L. Williamson,
    Lynn M. Gedvilas, and Robert C. Reedy

On the Role of Surface Diffusion and Its Relation to the
Hydrogen Incorporation During Hydrogenated Amorphous
Silicon Growth ...................................................................................................................... 461
    A.H.M. Smets, W.M.M. Kessels, and M.C.M. van de Sanden

Effect of Temperature and Temperature Uniformity on
Plasma and Device Stability ............................................................................................. 467
    G. Ganguly, M.S. Bennett, D.E. Carlson, and R.R. Arya

Hydrosilylation of Silicon Surfaces: Crystalline Versus Amorphous ........................ 473
    Andrea Lehner, Georg Steinhoff, Martin S. Brandt,
    Martin Eickhoff, and Martin Stutzmann

*Present Status of Hot Wire Chemical Vapor Deposition Technology ................... 479
    R.E.I. Schropp

Properties of High Quality p-Type Micro-Crystalline-Si
Prepared by Cat-CVD ........................................................................................................ 491
    Hideki Matsumura, Kouichi Katouno, Masaya Itoh, and
    Atsushi Masuda

*Invited Paper

**Investigations on the Real-Time Monitoring of the
Crystallinity of Hydrogenated Microcrystalline Silicon Films** .......................... 497
  Christoph Ross, Friedhelm Finger, and Reinhard Carius

**Towards Microcrystalline Silicon n-i-p Solar Cells With 10%
Conversion Efficiency** ................................................................................ 503
  L. Feitknecht, C. Droz, J. Bailat, X. Niquille, J. Guillet, and
  A. Shah

**Hydrogenated Amorphous Silicon Thin Films With
Nanocrystalline Silicon Inclusions** ............................................................. 509
  T.J. Belich, S. Thompson, C.R. Perrey, U. Kortshagen,
  C.B. Carter, and J. Kakalios

**Microstructure and Optical Functions of Transparent Conductors
and Their Impact on Collection in Amorphous Silicon Solar Cells** .......... 515
  G.M. Ferreira, Chi Chen, A.S. Ferlauto, P.I. Rovira, Ilsin An,
  C.R. Wronski, R.W. Collins, G. Ganguly, Joong Hwan Kwak,
  and Koeng Su Lim

**Reaction Control in Amorphous Silicon Film Deposition by
Hydrogen Chloride** ...................................................................................... 521
  Akihiro Takano, Takehito Wada, Shinji Fujikake,
  Takashi Yoshida, Tokio Ohto, and Eray S. Aydil

**Material Structure of Microcrystalline Silicon Deposited With
an Expanding Thermal Plasma** ................................................................. 527
  C. Smit, D.L. Williamson, M.C.M. van de Sanden, and
  R.A.C.M.M. van Swaaij

**Microcrystalline Silicon Thin Film Growth by Electron
Cyclotron Resonance Chemical Vapour Deposition at 80°C
for Plastic Application** ................................................................................ 533
  Ian Y.Y. Bu, A.J. Flewitt, J. Robertson, and W.I. Milne

**Evolution of Crystallinity in Mixed-Phase (a+μc)-Si:H as
Determined by Real Time Spectroscopic Ellipsometry** ........................... 539
  A.S. Ferlauto, G.M. Ferreira, R.J. Koval, J.M. Pearce,
  C.R. Wronski, R.W. Collins, M.M. Al-Jassim, and
  K.M. Jones

**Structural Characterization of Microcrystalline Silicon Solar
Cells Fabricated by Conventional RF-PECVD** ......................................... 545
  Liwei Li, Yuan-Min Li, J.A. Anna Selvan, Alan E. Delahoy,
  and Roland A. Levy

Hot-Wire Chemical Vapor Deposition for Epitaxial Silicon
Growth on Large-Grained Polycrystalline Silicon Templates ...................551
    M.S. Mason, C.M. Chen, and H.A. Atwater

## FILM GROWTH RELATED TO DEVICES

Helium Versus Hydrogen Dilution of Silane in the Deposition
of Polymorphous Silicon Films: Effects on the Structure and
the Transport Properties ...................559
    O. Saadane, S. Lebib, A.V. Kharchenko, V. Suendo,
    C. Longeaud, and P. Roca i Cabarrocas

Characterization of Nanocrystalline Silicon Film Grown by
LEPECVD for Photovoltaic Applications ...................565
    M. Bollani, S. Binetti, M. Acciarri, L. Fumagalli, A. Arcari,
    S. Pizzini, and H. von Känel

Influence of Substrate Temperature and Hydrogen Dilution
Ratio on the Properties of Nanocrystalline Silicon Thin Films
Grown by Hot-Wire Chemical Vapor Deposition ...................571
    H.R. Moutinho, C.-S. Jiang, B. Nelson, Y. Xu, J. Perkins,
    B. To, K.M. Jones, M.J. Romero, and M.M. Al-Jassim

Electrical Properties of Phosphorus-Doped and Boron-Doped
Nanocrystalline Germanium Thin-Films for p-i-n Devices ...................577
    William B. Jordan and Sigurd Wagner

p- and n-Type Microcrystalline SiC Fabricated by rf Plasma
CVD With Ethane Gas ...................583
    T. Toyama, Y. Nakano, T. Kosuge, A. Asano, and H. Okamoto

Polymorphous Silicon Films Produced in Large Area Reactors
by PECVD at 27.12 MHz and 13.56 MHz ...................589
    H. Águas, L. Raniero, L. Pereira, E. Fortunato, P. Roca i Cabarrocas,
    and R. Martins

Surface Roughness Study of Low-Temperature PECVD a-Si:H ...................595
    George T. Dalakos, Joel L. Plawsky, and Peter D. Persans

Hollow Electrode Enhanced RF Glow Plasma Generation and
Its Application to the Fast Deposition of Microcrystalline
Silicon Films ...................601
    Toshihiro Tabuchi, Masayuki Takashiri, Yasumasa Toyoshima,
    and Hiroyuki Mizukami

The Reliability of Measurements on Electron Energy
Distribution Function in Silane rf Glow Discharges .......................................................607
    Kuixun Lin, Xuanying Lin, Linfei Chi, Chuying Yu,
    Yunpeng Yu, and Shi Liu

High Rate Deposition of Stable Hydrogenated Amorphous Silicon
in Transition From Amorphous to Microcrystalline Silicon ...........................................613
    Guofu Hou, Xinhua Geng, Xiaodan Zhang, Ying Zhao,
    Junming Xue, Huizhi Ren, Jian Sun, Dekun Zhang, and
    Yueqin Xu

Properties of Nanocrystalline Germanium-Carbon Films and Devices ........................619
    X.J. Niu, Vikram L. Dalal, and Max Noack

Structural Evolution of Nanocrystalline Germanium Thin Films
With Film Thickness and Substrate Temperature ..........................................................625
    William B. Jordan, Eric D. Carlson, Todd R. Johnson, and
    Sigurd Wagner

Process Parameters for Poly-Silicon Deposition at a High
Growth Rate (1–7nm/s) by Hot-Wire Chemical Vapor
Deposition ...........................................................................................................................631
    J.K. Rath, A.J. Hardeman, C.H.M. van der Werf,
    P.A.T.T. van Veenendaal, M.Y.S. Rusche, and
    R.E.I. Schropp

High Temperature n- and p-Type Doped Microcrystalline Silicon
Layers Grown by VHF PECVD Layer-by-Layer Deposition .........................................637
    A. Gordijn, J.K. Rath, and R.E.I. Schropp

Deposition of Device Quality μc-Si Films and Solar Cells at
High Rates by HWCVD in a W Filament Regime Where W/Si
Formation is Minimal .........................................................................................................643
    E. Iwaniczko, A.H. Mahan, B. Yan, L.N. Gedvilas,
    D.L. Williamson, and B.P. Nelson

Influence of Hydrogen Dilution on Properties of Silicon Films
Prepared by D.C. Saddle-Field Glow-Discharge: Observation
of Microcrystallinity ..........................................................................................................649
    T. Allen, I. Milostnaya, D. Yeghikyan, K. Leong, F. Gaspari,
    N.P. Kherani, T. Kosteski, and S. Zukotynski

Reactive Pulsed Laser Deposition of Microcrystalline Ge-Based
Thin Films ...........................................................................................................................655
    Matthew R. Wills, Ruth Shinar, and Alan P. Constant

Electrical and Optical Properties of Amorphous and
Microcrystalline Hydrogenated Silicon Films Deposited
Using Saddle-Field Glow-Discharge .................................................................................. 661
    I. Milostnaya, T. Allen, F. Gaspari, N.P. Kherani,
    D. Yeghikyan, W.L. Roes, T. Kosteski, and S. Zukotynski

## CRYSTALLIZED FILMS

Stress Effects on Nanocrystal Formation by Ni-Induced
Crystallization of Amorphous Si .......................................................................................... 669
    Yaocheng Liu, Michael D. Deal, Mahmooda Sultana, and
    James D. Plummer

Field-Assisted Germanium Induced Crystallization of
Amorphous Silicon .............................................................................................................. 675
    J. Derakhshandeh, S. Mohajerzadeh, N. Golshani,
    E. Asl Soleimani, and M.D. Robertson

Prediction of the Interface Response Functions for Amorphous
and Crystalline Phases of Silicon and Germanium ............................................................. 681
    Erik J. Albenze, Laura A. Matejik, Nick F. Fynan, and
    Paulette Clancy

Advanced Lateral Crystal Growth of a-Si Thin Films by
Double-Pulsed Irradiation of All Solid-State Lasers ........................................................... 687
    Toshio Kudo, Koji Seike, Kazunori Yamazaki,
    Hirohito Komori, Sachi Yawaka, Shiro Hamada,
    and Cheng-Guo Jin

Influence of Laser Annealing on Hydrogen Bonding in
Disordered Silicon Thin Films ............................................................................................ 693
    N.H. Nickel and K. Brendel

Metal Containing Link Formed in Amorphous Silicon Metal-
to-Metal Antifuse ................................................................................................................ 699
    Frank Hawley, Farid Issaq, Jeewika Ranaweera, Roy Lambertson,
    and John McCollum

Direct Electrical Characterization of Metal Induced Lateral
Crystallization Regions by Spreading Resistance Probe Measurements ........................... 705
    Alexandre M. Myasnikov, Vincent M.C. Poon, Vincent T.C. Leung,
    Mansun Chan, and Lawrence C.F. Cheng

**Effect of SiO$_2$ Capping Layer on a Laser Crystallization of a-Si Thin Film** ............................................................................................................. 711
    Myung-Koo Kang, Hyun Jae Kim, Sook Young Kang,
    Su-Kyung Lee, Chi-Woo Kim, and Kyuha Chung

**Laser Interference Structuring of a-GeN for the Production of Optical Diffraction Gratings** ............................................................................ 717
    M. Mulato, A.R. Zanatta, D. Toet, and I.E. Chambouleyron

**Observation and Annealing of Incomplete Recrystallized Junction Defects Due to the Excimer Laser Beam Diffraction at the Gate Edge in Poly-Si TFT** ......................................................................... 723
    Woo-Jin Nam, Kee-Chan Park, Sang-Hoon Jung, Soo-Jeong Park,
    and Min-Koo Han

**2-Dimensional Controlled Large Lateral Grain Growth on the Floating Amorphous Silicon Film by Excimer Laser Recrystallization** ............................................................................................... 729
    In-Hyuk Song, Su-Hyuk Kang, Woo-Jin Nam, and Min-Koo Han

**Improved Electrical Properties in Nanocrystalline Si Formed by Metal Induced Growth** .............................................................................. 735
    Chunhai Ji and Wayne A. Anderson

**A Simple Lateral Grain Growth of Poly-Si by Single Excimer Laser Crystallization of Amorphous Silicon Film Deposited on Polygon Shaped Trench** ...................................................................................... 741
    Sang-Hoon Jung, Su-Hyuk Kang, Hee-Sun Shin, and
    Min-Koo Han

**Formation of Large, Orientation-Controlled, Nearly Single Crystalline Si Thin Films on SiO$_2$ Using Contact Printing of Rolled and Annealed Nickel Tapes** ................................................................. 747
    Hwang Huh and Jung H. Shin

**Lifetime Measurements of Stain Etched and Passivated Porous Silicon** ................... 753
    Ricardo Guerrero-Lemus, Fathi A. Ben-Hander, Cristoffer Ballif,
    Ali Kenanoglu, Dietmar Borchert, Cecilio Hernández-Rodríguez,
    Tomás Rodríguez, and José M. Martínez-Duart

**Anomalous Behavior of Stain Etched Porous Silicon Photoluminescence** ............................................................................................ 759
    Ricardo Guerrero-Lemus, Fathi A. Ben-Hander,
    Cecilio Hernández-Rodríguez, and José M. Martínez-Duart

**Scattering Rings in Birefringent Porous Silicon** .......................................................... 767
  Claudio J. Oton, Zeno Gaburro, Mher Ghulinyan,
  Nicola Daldosso, Lucio Pancheri, Paolo Bettotti,
  Luca Dal Negro, and Lorenzo Pavesi

**Light and Thermally Induced Metastabilities in
Nanocrystalline Silicon** ........................................................................................... 773
  N.P. Mandal and S.C. Agarwal

**Physicochemical Characterization of Porous Silicon Surfaces
Etched in Salt Solutions of Varying Compositions and pH** ................................. 779
  Mariem Rosario-Canales, Ana R. Guadalupe, Luis F. Fonseca,
  and Oscar Resto

**Author Index** ........................................................................................................... 785

**Subject Index** .......................................................................................................... 791

# PREFACE

Researchers from around the world met to attend and present papers at Symposium A, "Amorphous and Nanocrystalline Silicon-Based Films—2003," held April 22–25 at the 2003 MRS Spring Meeting in San Francisco, California. 170 scientists presented their work at this symposium, including a large component of younger researchers.

This field has been the subject of a symposium at the Materials Research Society every year since 1984. This remarkable longevity is due to the continuous emergence of new scientific questions and new technological challenges for Si thin films. In 2003, there was a strong emphasis on methods to achieve high deposition rates using plasma or hot wire chemical vapor deposition, and on the properties and applications of nanocrystalline Si films, which for example have been incorporated into stacked a-Si:H/nc-Si:H solar cells. There were many papers on the subjects of solar cells, imagers, thin film transistors, crystallization of a-Si:H, and defects.

The papers in Symposium A were selected from over 200 submitted abstracts by the five symposium organizers, together with five international advisors. The latter individuals were asked to participate because their internationally recognized expertise complemented that of the organizers. The advisors for this year's symposium were Greg Parsons, Reinhard Carius, J. David Cohen, Michio Kondo, and Philippe Fauchet. They contributed in the selection of invited speakers, ranking the submitted abstracts, chairing the sessions, and refereeing papers. The organizers wish to express their sincere thanks and appreciation to these individuals for all of their valuable input and effort.

The 120 papers appearing in this proceedings have been sorted under six chapter headings on the basis of subject matter. The pages are numbered serially; in addition, each paper has been identified with its program number from the meeting. Chapter I is concerned with amorphous network structures, electronic metastability, defects, and photoluminescence. Chapter II focuses on thin film transistors and imager arrays. Chapter III covers solar cells. Chapter IV is concerned with growth mechanisms, hot filament CVD, and nc-Si:H growth. Chapter V contains all remaining topics in film growth, especially those related to devices. Finally, Chapter VI focuses on crystallized film.

On behalf of all the participants, the organizers wish to thank the financial supporters of Symposium A: BP Solar, Inc., MVSystems, Inc., the National Renewable Energy Laboratory, Sanyo Electric Co., Ltd., United Solar Systems Corporation, Voltaix, Inc., and Xerox PARC. The organizing committee also particularly wishes to thank Mary Ann Woolf of the University of Utah for her invaluable help preparing this proceedings, as well as for all her other help that was so instrumental in making this year's Symposium A such a success.

John R. Abelson
Gautam Ganguly
Hideki Matsumura
John Robertson
Eric Schiff

August 2003

# MATERIALS RESEARCH SOCIETY SYMPOSIUM PROCEEDINGS

Volume 734— Polymer/Metal Interfaces and Defect Mediated Phenomena in Ordered Polymers, E.D. Manias, G.G. Malliaras, 2003, ISBN: 1-55899-671-0
Volume 735— Bioinspired Nanoscale Hybrid Systems, G. Schmid, U. Simon, S.J. Stranick, S.M. Arrivo, S. Hong, 2003, ISBN: 1-55899-672-9
Volume 736— Electronics on Unconventional Substrates—Electrotextiles and Giant-Area Flexible Circuits, M.S. Shur, P. Wilson, D. Urban, 2003, ISBN: 1-55899-673-7
Volume 737— Quantum Confined Semiconductor Nanostructures, J.M. Buriak, D.D.M. Wayner, F. Priolo, B. White, V. Klimov, L. Tsybeskov, 2003, ISBN: 1-55899-674-5
Volume 738— Spatially Resolved Characterization of Local Phenomena in Materials and Nanostructures, D.A. Bonnell, J. Piqueras, A.P. Shreve, F. Zypman, 2003, ISBN: 1-55899-675-3
Volume 739— Three-Dimensional Nanoengineered Assemblies, T.M. Orlando, L. Merhari, K. Ikuta, D.P. Taylor, 2003, ISBN: 1-55899-676-1
Volume 740— Nanomaterials for Structural Applications, C. Berndt, T.E. Fischer, I. Ovid'ko, G. Skandan, T. Tsakalakos, 2003, ISBN: 1-55899-677-X
Volume 741— Nano- and Microelectromechanical Systems (NEMS and MEMS) and Molecular Machines, A.A. Ayon, T. Buchheit, D.A. LaVan, M. Madou, 2003, ISBN: 1-55899-678-8
Volume 742— Silicon Carbide 2002—Materials, Processing and Devices, S.E. Saddow, D.J. Larkin, N.S. Saks, A. Schoener, 2003, ISBN: 1-55899-679-6
Volume 743— GaN and Related Alloys—2002, E.T. Yu, C.M. Wetzel, J.S. Speck, A. Rizzi, Y. Arakawa, 2003, ISBN: 1-55899-680-X
Volume 744— Progress in Semiconductors II—Electronic and Optoelectronic Applications, B.D. Weaver, M.O. Manasreh, C.C. Jagadish, S. Zollner, 2003, ISBN: 1-55899-681-8
Volume 745— Novel Materials and Processes for Advanced CMOS, M.I. Gardner, J-P. Maria, S. Stemmer, S. De Gendt, 2003, ISBN: 1-55899-682-6
Volume 746— Magnetoelectronics and Magnetic Materials—Novel Phenomena and Advanced Characterization, S. Zhang, W. Kuch, G. Guentherodt, C. Broholm, A. Kent, M.R. Fitzsimmons, I. Schuller, J.B. Kortright, T. Shinjo, Y. Zhu, 2003, ISBN: 1-55899-683-4
Volume 747— Crystalline Oxide-Silicon Heterostructures and Oxide Optoelectronics, D.S. Ginley, S. Guha, S. Carter, S.A. Chambers, R. Droopad, H. Hosono, D.C. Paine, D.G. Schlom, J. Tate, 2003, ISBN: 1-55899-684-2
Volume 748— Ferroelectric Thin Films XI, D. Kaufman, S. Hoffmann-Eifert, S.R. Gilbert, S. Aggarwal, M. Shimizu, 2003, ISBN: 1-55899-685-0
Volume 749— Morphological and Compositional Evolution of Thin Films, N. Bartelt, M.J. Aziz, I. Berbezier, J.B. Hannon, S. Hearne, 2003, ISBN: 1-55899-686-9
Volume 750— Surface Engineering 2002—Synthesis, Characterization and Applications, A. Kumar, W.J. Meng, Y-T. Cheng, J. Zabinski, G.L. Doll, S. Veprek, 2003, ISBN: 1-55899-687-7
Volume 751— Structure-Property Relationships of Oxide Surfaces and Interfaces II, X. Pan, K.B. Alexander, C.B. Carter, R.W. Grimes, T. Wood, 2003, ISBN: 1-55899-688-5
Volume 752— Membranes—Preparation, Properties and Applications, V.N. Burganos, R.D. Noble, M. Asaeda, A. Ayral, J.D. LeRoux, 2003, ISBN: 1-55899-689-3
Volume 753— Defect Properties and Related Phenomena in Intermetallic Alloys, E.P. George, H. Inui, M.J. Mills, G. Eggeler, 2003, ISBN: 1-55899-690-7
Volume 754— Supercooled Liquids, Glass Transition and Bulk Metallic Glasses, A.L. Greer, T. Egami, A. Inoue, S. Ranganathan, 2003, ISBN: 1-55899-691-5
Volume 755— Solid-State Chemistry of Inorganic Materials IV, M. Greenblatt, M.A. Alario-Franco, M.S. Whittingham, G. Rohrer, 2003, ISBN: 1-55899-692-3
Volume 756— Solid-State Ionics—2002, P. Knauth, J-M. Tarascon, E. Traversa, H.L. Tuller, 2003, ISBN: 1-55899-693-1
Volume 757— Scientific Basis for Nuclear Waste Management XXVI, R.J. Finch, D.B. Bullen, 2003, ISBN: 1-55899-694-X

# MATERIALS RESEARCH SOCIETY SYMPOSIUM PROCEEDINGS

Volume 758— Rapid Prototyping Technologies, A.S. Holmes, A. Piqué, D.B. Dimos, 2003, ISBN: 1-55899-695-8
Volume 759— Granular Material-Based Technologies, S. Sen, M.L. Hunt, A.J. Hurd, 2003, ISBN: 1-55899-696-6
Volume 760E—The Undergraduate Curriculum in Materials Science and Engineering, E.P. Douglas, O.D. Dubón Jr., J.A. Isaacs, W.B. Knowlton, M. Stanley Whittingham, 2003, ISBN: 1-55899-697-4
Volume 761E—Molecular Electronics, M-I. Baraton, E.L. Garfunkel, D.C. Martin, S.S.P. Parkin, 2003, ISBN: 1-55899-698-2
Volume 762— Amorphous and Nanocrystalline Silicon-Based Films—2003, J.R. Abelson, G. Ganguly, H. Matsumura, J. Robertson, E. Schiff, 2003, ISBN: 1-55899-699-0
Volume 763— Compound Semiconductor Photovoltaics, R. Noufi, D. Cahen, W. Shafarman, L. Stolt, 2003, ISBN: 1-55899-700-8
Volume 764— New Applications for Wide-Bandgap Semiconductors, S.J. Pearton, J. Han, A.G. Baca, J-I. Chyi, W.H. Chang, 2003, ISBN: 1-55899-701-6
Volume 765— CMOS Front-End Materials and Process Technology, T-J. King, B. Yu, R.J.P. Lander, S. Saito, 2003, ISBN: 1-55899-702-4
Volume 766— Materials, Technology and Reliability for Advanced Interconnects and Low-k Dielectrics—2003, A. McKerrow, J. Leu, O. Kraft, T. Kikkawa, 2003, ISBN: 1-55899-703-2
Volume 767— Chemical-Mechanical Planarization, M. Oliver, D. Boning, D. Stein, K. Devriendt, 2003, ISBN: 1-55899-704-0
Volume 768— Integration of Heterogeneous Thin-Film Materials and Devices, H.A. Atwater, M. Levy, M.I. Current, T. Sands, 2003, ISBN: 1-55899-705-9
Volume 769— Flexible Electronics—Materials and Device Technology, B.R. Chalamala, B.E. Gnade, N. Fruehauf, J. Jang, 2003, ISBN: 1-55899-706-7
Volume 770— Optoelectronics of Group-IV-Based Materials, T. Gregorkiewicz, R.G. Elliman, P.M. Fauchet, J.A. Hutchby, 2003, ISBN: 1-55899-707-5
Volume 771— Organic and Polymeric Materials and Devices, P.W.M. Blom, N.C. Greenham, C.D. Dimitrakopoulos, C.D. Frisbie, 2003, ISBN: 1-55899-708-3
Volume 772— Nanotube-Based Devices, P. Bernier, S. Roth, D. Carroll, G-T. Kim, 2003, ISBN: 1-55899-709-1
Volume 773— Biomicroelectromechanical Systems (BioMEMS), C. Ozkan, J. Santini, H. Gao, G. Bao, 2003, ISBN: 1-55899-710-5
Volume 774— Materials Inspired by Biology, J.L. Thomas, L. Gower, K.L. Kiick, 2003, ISBN: 1-55899-711-3
Volume 775— Self-Assembled Nanostructured Materials, C.J. Brinker, Y. Lu, M. Antonietti, C. Bai, 2003, ISBN: 1-55899-712-1
Volume 776— Unconventional Approaches to Nanostructures with Applications in Electronics, Photonics, Information Storage and Sensing, O.D. Velev, T.J. Bunning, Y. Xia, P. Yang, 2003, ISBN: 1-55899-713-X
Volume 777— Nanostructuring Materials with Energetic Beams, S. Roorda, H. Bernas, A. Meldrum, 2003, ISBN: 1-55899-714-8
Volume 778— Mechanical Properties Derived from Nanostructuring Materials, H. Kung, D.F. Bahr, N.R. Moody, K.J. Wahl, 2003, ISBN: 1-55899-715-6
Volume 779— Multiscale Phenomena in Materials—Experiments and Modeling Related to Mechanical Behavior, K.J. Hemker, D.H. Lassila, L.E. Levine, H.M. Zbib, 2003, ISBN: 1-55899-716-4
Volume 780— Advanced Optical Processing of Materials, I.W. Boyd, M. Dinescu, A.V. Rode, D.B. Chrisey, 2003, ISBN: 1-55899-717-2
Volume 781E—Mechanisms in Electrochemical Deposition and Corrosion, J.C. Barbour, R.M. Penner, P.C. Searson, 2003, ISBN: 1-55899-718-0

**Prior Materials Research Society Symposium Proceedings available by contacting Materials Research Society**

**Amorphous Network Structures,
Electronic Metastability, Defects
and Photoluminescence**

# Numerical Studies Of The Dynamics Of Silicon: Relaxation, Nucleation And Energy Landscape

Normand Mousseau* and Philippe Beaucage and Francis Valiquette

*Département de physique and Groupe de recherche en science et technologie des couches minces, Université de Montréal, C.P. 6128 succ. Centre-ville, Montréal (Québec), Canada, H3T 1J7*

## ABSTRACT

Using various simulation techniques, such as molecular dynamics and the activation-relaxation technique, we are slowly developing a consistent picture of the dynamical properties of amorphous silicon. For example, results of an extensive search for the activated events surrounding a single minimum, in a well-relaxed model represented by a modified Stillinger-Weber potential, confirm that barrier height at the transition point, for activated mechanisms, is determined essentially by the binding energy of a single bond and not the details of the mechanism. We will discuss these results in some detail as well as recent simulations of nucleation in liquid and amorphous silicon.

## INTRODUCTION

It is commonplace to affirm that the most studied element of all time, silicon, still deserves our attention. Amazingly, though, this affirmation holds true: although we have mapped in great details many of the properties of this material, a number of fundamental questions continue to elude us. This is particularly so of dynamical properties, which are difficult to study both experimentally and theoretically. In this paper, we present the results of simulations attempting to draw a first sketch of two dynamical processes that have received relatively little attention until now: the dynamics of nucleation from the liquid phase and the dynamics of relaxation and diffusion in the amorphous state.

Within the last five years, our group and collaborators have spent considerable efforts trying to obtain a reasonably clear picture of the structural and dynamical properties of the amorphous phase of silicon. For example, Barkema and one of us, NM, have developed various methods to generate high-quality $a$-Si models which can serve as a basis for further structural, dynamical or electronic studies. [1] One of these methods is an optimized version of the Wooten-Winer-Weaire bond-switching algorithm which generates an amorphous network starting from a perfectly 4-fold-coordinated random network. [2, 3] This is achieved using a rather artificial but remarkably satisfactory harmonic Keating potential. Although producing the best models available, this method is essentially static in nature, providing no information as to how real $a$-Si is produced or how defects relax and diffuse.

To look at these aspects of $a$-Si, we have used another method, the activation-relaxation

---

*Electronic address: `normand.mousseau@Umontreal.ca`

technique (ART and ART *nouveau*). [4, 5] This method, which was used with a slightly modified Stillinger-Weber potential [6, 7] to generate good quality models of a-Si, can also explore the energy landscape of complex materials, jumping literally over activation barriers of all heights and complexity. [8, 9] With ART nouveau, it is possible to characterize the energy landscape and the basic mechanisms responsible for relaxation and diffusion in a wide range of systems. In the next section, we discuss an application of this technique to sample the energy landscape surrounding a *single* energy minimum of a well-relaxed 1000-atom cell of a-Si.

Much less work has been done on another fundamental dynamical property of this material: the nucleation process. The study of this process can help us understand the origin of the amorphous phase as well as the rapid crystallization, from the amorphous state, that takes place at temperatures much below melting. In section III, we present some preliminary results on this process to indicate the feasibly of such a theoretical study. Almost everything remains to be done on this problem.

Together these two themes offer just a glimpse of the complexity of the dynamical processes taking place in such a simple material. Much more experimental and theoretical will be needed before we can claim to understand them.

## DIFFUSION AND RELAXATION IN AMORPHOUS SILICON

### Dynamics through the energy landscape

In the last ten years, the concept of energy landscape has gained considerable importance [10]. This places the dynamics of diffusion in complex systems such as disordered materials into a global point of view, where the whole configuration can be seen as jumping from one local minimum to another, through the energy landscape, instead of being just something happening somewhere in the lattice (see Fig. 1). This conceptual change has no impact on the mechanisms as such, of course, but help us understand diffusion and relaxation as the configuration moves from one macroscopic state to another.

The energy landscape representation has many advantages. This is particularly so when designing accelerated algorithms. The time scale associated with activated relaxation or diffusion in many materials, such as semiconductors, is many orders of magnitude larger than the typical phonon frequency, and might involve some rather complex atomistic rearrangements. This often goes beyond the limits of standard simulation techniques such as molecular dynamics or real-space Monte-Carlo — where atoms are moved one by one by a small amount.

To circumvent these limitations it is preferable to define moves directly in the energy landscape, where activation can be seen as a jump from one local energy minimum to a nearby one, going through a first-order saddle point. In this picture, the real-space complexity of the move, which might involve many tens or hundreds of atoms, is completely set aside, simplifying considerably the simulation. It is such an approach that we selected for our study of the energy landscape around a local energy minimum in a 1000-atom model of a-Si. We use the activation-relaxation technique (ART nouveau), which has been described in details in Ref. 5 and represents an improved version of the original algorithm of Barkema and Mousseau [4, 11].

But the energy landscape picture has its limits. In particular, it hides the fundamentally

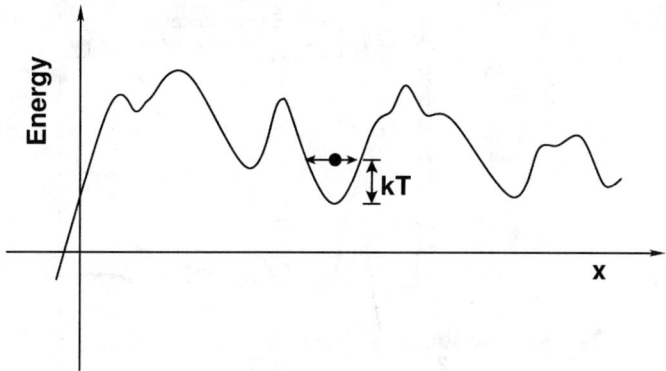

FIG. 1: The energy surface as a function of a generalized coordinate for a generic system. The circle represents a 3N-dimensional conformation oscillating in a local minimum at a temperature $T$.

localized nature of the dynamics in bulk materials: in these systems, all activated events always involve a finite number of atoms except at a phase transition. The 1+1 dimensional picture presented in Fig. 1 can therefore be very misleading. For example, much has been made recently of the barrier energy associated with higher-order saddle points implying some complex correlations between various levels of the energy landscape in disordered systems. [12] Analyzing these events in real space, we discover that these higher-order saddles correspond simply to a series of normal local events, each at the transition point and characterized by a direction of negative curvature in the energy landscape.

## Details of the simulation

We study the energy landscape around a well-relaxed 1000-atom $a$-Si unit cell prepared using ART nouveau. The configuration is described by a modified Stillinger-Weber potential, where the 3-body force constant has been increased by 50 % in order to lead to the correct amorphous structure. [4, 7] The radial distribution function for this initial configuration is in excellent agreement with a recently measured high-quality RDF, as shown in Fig. 2. The initial configuration has only 26 three-fold and 20 five-fold coordinated defects at a cut-off of 2.8 Å, with more than 95 % perfectly coordinated. The energy of the configuration is -4.000 eV/atom, compared to -4.137 for the crystalline state at zero pressure.

In previous work, our group and others studied the relaxation and diffusion process, as the configuration walked through the energy landscape, relaxing from a high-energy configuration to a well-relaxed one. [8, 9, 14] We are interested here in the characterization of the energy landscape in the *immediate surrounding* of a single minimum. Following a similar study of the landscape around local minimum in Lennard-Jones clusters, we want to evaluate, among others: (1) the structure and energy distributions of the barriers, (2) the number of escapes routes, (3) the bias of the sampling method (ART nouveau). We present

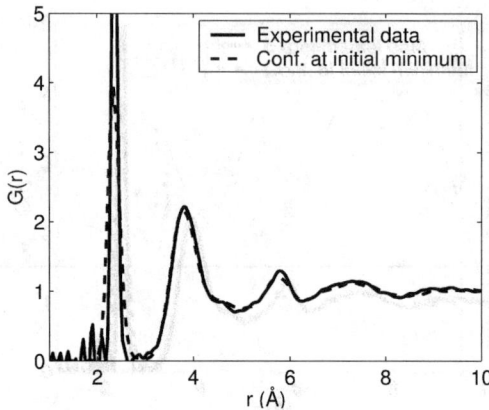

FIG. 2: Radial distribution function (RDF) of the 1000-atom $a$-Si configuration used as the initial minimum (solid line). The dashed represent the experimental RDF measured by Laaziri et al[13].

here results for a sampling of more than 42 000 events around the minimum described in the previous paragraph. Each event is started from the initial minimum by selecting an atom, $i$, at random and drawing a random direction in the subspace of the atom $i$ and its near neighbors. However, the activation to the saddle point is done without spatial restriction: all atoms are treated in exactly the same way. At this point, this technique represents the best way to ensure that no class of event is missed.[5] We have checked that these results are generic by sampling around a second minimum of similar structural quality and found that the results are independent of the fine details of the structure and topology of the network studied.

An event is characterized by an initial state and a final state, both representing local energy minima, and an intermediate state measured at the saddle point, defining the reaction state. As discussed in Ref. [5], these ART events are fully reversible, in the sense that a long-enough search at the final minimum will be able to find the saddle point and the original minimum. Although not guaranteeing detailed balance, this ensures that the search is ergodic.

## Properties of the energy landscape

By construction, the initial direction for leaving the minimum is selected uniformly throughout the configuration. However, because there is no limitation on the number of atoms involved in the event, we could expect that a few atoms or regions of the 1000-atom configuration dominate in terms of events. Fig. 3 shows that the probability for each atom to participate into an event is equivalent, within a factor three or so. The dynamics of relaxation and diffusion in a well-relaxed model of $a$-Si is therefore uniform, confirming its local nature.

Because the search for events is random, the same event can occur many times. This

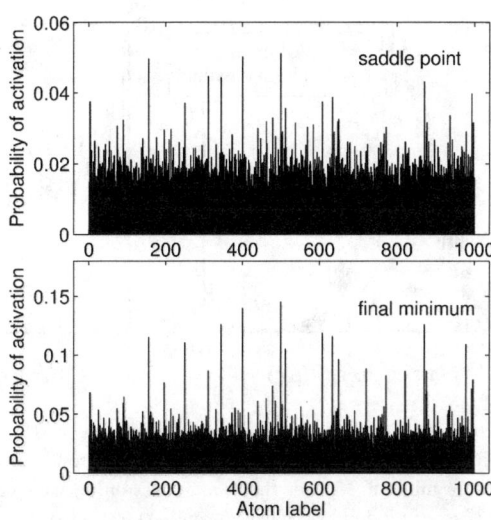

FIG. 3: Probability of participation to an event for each of the 1000 atoms in the model, computed over all generated events using displacement and change in coordination, as described in the text. The top panel shows that probability for at the saddle point and the bottom panel, at the final minimum. For this graph, an atom is involved in an event if it changes coordination or if its displacement is more than 0.1 Å. Results do not depend qualitatively on the exact value of the threshold selected.

helps us answer the fundamental question as to how many saddle points and minima surround any given local minimum in $a$-Si. Fig. 5shows the number of unique saddle points, minima and overall events as a function of the number of generated events; even 42 000 events are not sufficient to ensure a complete sampling of the landscape surrounding a local minimum in a 1000-atom cell. We find 6519 different minima, 8799 different saddles and 11014 different events, yet the curve do not seem to converge to any specific value. Because moves are defined in the continuous space, the identification of unique events is not perfect and the exact number depend on the criteria we use. Although the convergence criterion for the saddle point and the minimum can be fixed with any desired accuracy, it has a finite precision. To decrease the impact of this finite precision, we label an event using only atoms that have moved significantly: only atoms moving by at least $\delta r_{\text{threshold}} = 0.4$ and 0.2 Å, respectively, at the saddle point and the final minimum, are used to label a specific event. We consider that two events are identical if they have the same list of participating atoms and their respective energy barriers is the same within a precision of 0.2 eV. The exact value of these thresholds affects the quantitative number of unique events but not the overall conclusion.

Although the number of unique events as a function of trial event is not converged, we have enough information to try to assess the total number of activated paths around a

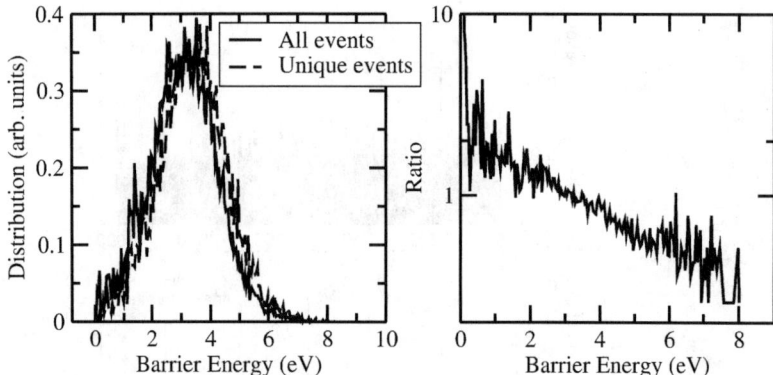

FIG. 4: Top: Normalized barrier energy distribution, as measured from the initial minimum, for all saddles (solid line) and unique saddles (dashed line). Bottom: Log-normal ratio of all saddles points generated from minimum 1 over the unique ones only. A histogram for both distributions is first constructed, as a function of the energy, and the ratio is taken over this histogram.

single minimum in configuration space. First, we need to identify the bias of ART nouveau in its selection of events. To do so, we compare the energy distribution of the all events to that of the unique events (top panel of Fig. 4). Similarly to what we had found for Lennard-Jones clusters, it seems that ART selects events with an exponential bias on the energy barrier. Here, we find that the bias is well fitted by $exp(-\Delta E/E_0)$ with $E_0 = 4.0$ eV. The origin of this bias is not understood at this point but it is likely related to the width of the valley leading to the saddle point.

We can now try to assess the total number of different events surrounding a given minimum in $a$-Si. If we suppose a random selection of events, drawn with an exponential preference for low energy events and taken from a distribution of energy such as that already calculated here, we find that the growth in the number of different events can be reproduced with a bias function $exp(-\Delta E/A)$ with $0.4 \leq A \leq 0.6$ and a total number of different events between 30 000 and 60 000, or 30 to 60 different events per atom.

Because of the approximation needed, the estimation of the total number of paths surrounding a minimum has to be taken as an order of magnitude. From the above result, this number is more than one order of magnitude greater than the number of degrees of freedom, but remains finite, as one could have expected. These results are not expected to be qualitatively affected by the use of better interaction potential or less relaxed configurations. [15] The main challenge now will be to connect these results with experiment.

## NUCLEATION OF SI

Amorphous silicon is known to re-crystallize well below the melting point of Si. The mechanisms underlying this transition are not clear, however, and we do not know what this implies regarding the structural and dynamical properties of Si. As a first step to-

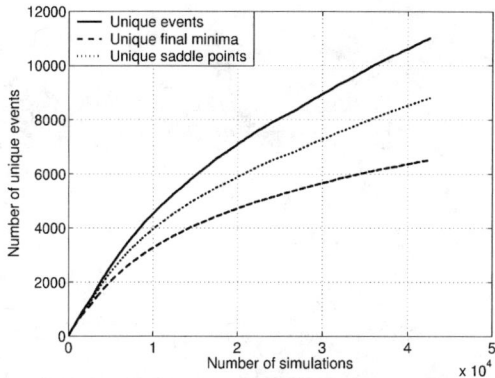

FIG. 5: Number of unique events (solid line), unique final minima (dashed line) and unique saddle points (dotted line) as a function of the number of events generated. The distribution remains the same if we recreate this curve by drawing events at random in the sequence generated. There is therefore no bias in the order of the events selected.

wards trying to answer this question, we consider the crystallization process from the liquid phase, which is more easily accessible in computer simulations as it happens at a higher temperature.

Numerical research on the nucleation process started with hard spheres and Lennard-Jones models [16–18]. For covalent material like silicon, crystallization was studied either from a liquid/crystal interface [19] or an amorphous/liquid interface [20]. In the first case, the interest was to the planar front crystallization was studied as a function of the orientation of the crystal. In the second case, the stability of amorphous clusters imbedded in a crystalline matrix was studied as a function of the size of these clusters.

In previous work, we have shown that it is possible to crystallize a 1000-atom cell of liquid silicon described by a modified version of the Stillinger-Weber potential used in the previous section without having to introduce seed clusters artificially. [21] The set of parameters used had been developed by Vink and collaborators [7] in order to obtain the right structural and vibrationnal properties in the amorphous phase. Although excellent in the amorphous phase, this potential, which includes a three-body term 50 percent larger than the original Stillinger-Weber potential, does not get the right density relation between the crystalline and the liquid silicon: contrary to experiment, this potential generates a $c$-Si denser than $l$-Si.

After toying with the potential, the results we present here are for a Stillinger-Weber potential with a three-body term strengthened by 12.5 % only, ensuring the crystallization of a 1000-atom cell while providing the proper phase diagram. This slight increase in the angular force with respect to the initial parameter set for the Stillinger-Weber potential seems to be necessary in order to crystallize a 1000-atom cell on a timescale reachable by molecular dynamics. Some problems still seem to remain much below the melting point and we are currently trying to address them. [22]

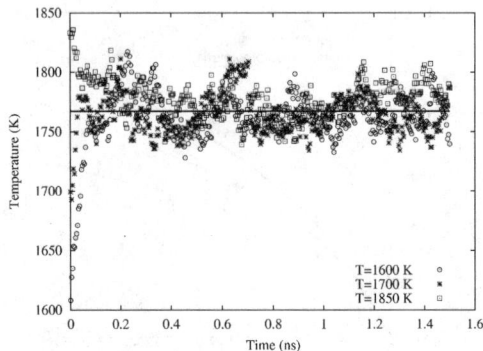

FIG. 6: Identification of the melting transition temperature using the procedure proposed by Keblinski and collaborators for the modified Stillinger-Weber used for crystallization [23].

## Melting temperature

Before discussing the nucleation process, we must establish the melting transition temperature with this set of parameters. Because crystal-liquid melting is a first-order, there is strong hysteresis and one cannot simply identify the temperature at which a cell is melting as the melting transition temperature. The standard procedure to measure this quantity is to use thermodynamical integration, a numerical method which can give the free energy as a function of temperature for both the crystalline and the liquid phase as a function of temperature. The intersection of these two curves is the melting transition temperature. Although well established, this procedure is heavy and requires extensive sets of simulations.

Here, we follow instead the elegant technique of Keblinski and collaborators, [23] which can identify the melting transition temperature with precision without having to resort to thermodynamical integration. A crystalline and a liquid slabs are first equilibrated near the melting temperature and then put in contact in an NPE MD simulation. As shown in Fig. 6, the crystalline or the liquid phase grow until equilibrium is reached, at the melting temperature. This procedure, which is much simpler than thermodynamical integration, gives the right melting transition temperature for the Stillinger-Weber potential and 1767 ± 16 K for the potential used here.

## Crystallization from the liquid phase

To limit size effects, we simulate the crystallization process in a 10648-atom cell with periodic boundary conditions. After equilibrating the liquid well above the melting temperature, we bring at once the system at a temperature 25 % below the melting temperature and pursue the simulation in the NVT ensemble, in a box corresponding to the crystalline density. This degree of undercooling is similar to what is used experimentally for the crystallization of a wide variety of materials [24] and also numerically for Lennard-Jones systems [17]. We select a NVT ensemble for numerical convenience. We have checked that crystallization also occurs in a NPT ensemble but found that it did so only after a

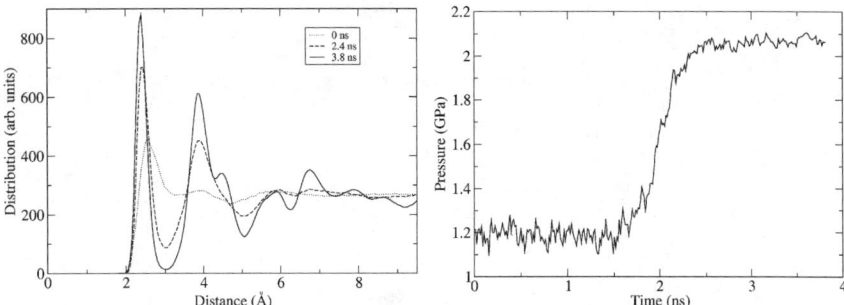

FIG. 7: Left panel: Radial distribution function at T=0, 2.4, 3.8 ns for the run described in the right panel. The sharp peaks correspond to crystalline distances. Right panel: Pressure as a function of time for the 10648-atom cell, during the crystallization process. The clock is started as the system is brought to 0.75 $T_m$.

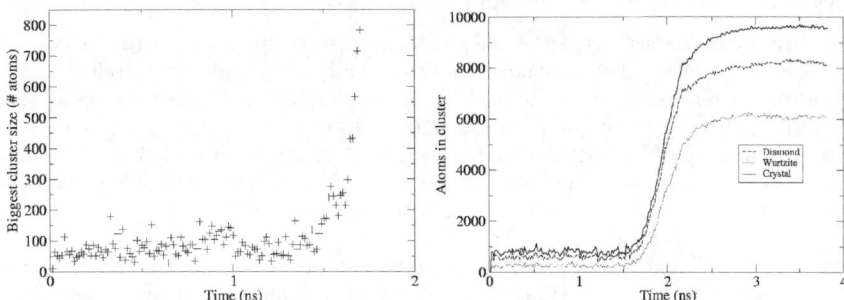

FIG. 8: Left panel: Evolution of the size of the largest cluster in the model as a function of time. Right: Number of atoms member of a Wurtzite, Diamond and both building block as a function of time. The total number of atoms part of a crystalline network is smaller than the sum of the two since some atoms might be member of both a Wurtzite and a Diamond local structure.

considerably longer simulation time. The analysis of the crystallization process does not show qualitative differences between the two trajectories so we use constant volume for the simulations.

In the NVT ensemble, the crystallization process of the 10648-atom cell starts typically between 1 and 3 ns from the time the system is brought at its final temperature. Comparing with smaller simulation cells, we find that this time seems to decrease with an increasing cell, in agreement with standard nucleation theory. For a 1000-atom cell, the typical nucleation time is between 5 to 15 ns. As can be seen in Fig. 7, once nucleation starts, the crystallization is rapid, overtaking the model in slightly less than 1 ns.

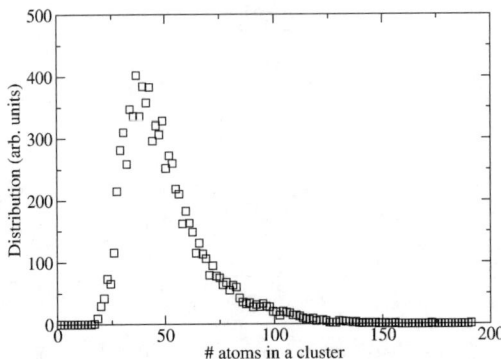

FIG. 9: Distribution of size for the largest cluster measured during the supercooled liquid phase. The critical size appears to be around 200 atoms.

## Characterization of the crystallization process

In order to characterize better the process of crystallization, we identify two basic blocks, representing the smallest rigid structures that can be extracted from tetrahedral structure. The first, counting 12 atoms, is made up of two 6-fold rings, placed on top of each other and connected by three bonds, is extracted from a Wurtzite (hexagonal) structure. The second, from the tetrahedral network, consists of 10 atoms assembled as four 6-fold rings back to back. Both structures appear with a very low incidence in high quality amorphous silicon networks and are therefore good indicators of crystallization.

Figure 8 shows the evolution of the nucleation process analyzed in terms of these building blocks. For simplicity, we plot only the evolution of the size of the largest crystallite as a function of time. We see that, in the metastable equilibrium of the supercooled liquid, the fluctuations are sizable. From another simulation that did not crystallize in 5 ns, we plot Fig. 9 the distribution of size for the largest cluster in the supercooled liquid region sampled every 1 ps. The distribution is smooth and asymmetrical, with a peak at about 40 atoms and a tail reaching 180. The critical size for nucleation, at a temperature 25 % below melting, seems therefore to be around 200. In Lennard-Jones systems, the critical size is 170 atoms [17].

At the end of the nucleation process about 95% of all atoms are part of at least one of the two building blocks. Given the size of the system the final configuration is clearly affected by the periodic boundary conditions: there is a single crystallite which forms grain boundaries with itself as it wraps around the box. These grain boundaries contain a significant fraction of the Wurtzite conformation, which does not correspond to the crystalline symmetry of silicon. At 0 K, the Stillinger-Weber potential is cut-off before the third shell of atoms and cannot, therefore, differentiate between the two phases. However, thermal fluctuation are sufficient and allow the potential to distinguish them. We expect, therefore, that the Wurtzite phase is only a result of the rapid crystallization process and that it would anneal into a tetrahedral conformation with further relaxation.

Much remains done in the analysis of the nucleation process in silicon. These results indicate however that this process is achievable on simulation time scale, which should allow us to make predictions that will need to be confirmed experimentally.

## CONCLUSIONS

In this short paper, we described two sets of simulations that attempt to describe some of the dynamical properties of silicon. In order to go much beyond these simulations, it will be necessary to establish ways to compare the results we obtain with experiment. This, in itself, represents a considerable challenge. Although the activation barrier distribution we generate with ART is in agreement with the barrier measured experimentally, we do not know if the wide distribution of mechanisms we see is occurring in real samples. Similarly, the nucleation process in silicon has been very little studied experimentally. It would be of great interest to have more experimental results on this transition.

## ACKNOWLEDGMENT

We thank Frédéric van Wijland for his help in analyzing the trend in the number of unique events as a function of sampled ones. This work is funded in part by NSERC (Canada) and NATEQ (Québec). Most of the simulations were run on the computers of the Réseau québécois de calcul de haute performance (RQCHP) whose support is gratefully acknowledged. NM is a Cottrell Scholar of the Research Corporation.

## REFERENCES

[1] S. Nakhmanson, P. Voyles, N. Mousseau, G. Barkema, and D. Drabold, Phys. Rev. B **63**, 235207 (2001).
[2] F. Wooten and D. Weaire, Solid State Physics **40**, 1 (1987).
[3] G. T. Barkema and N. Mousseau, Phys. Rev. B **62**, 4985 (2000).
[4] G. T. Barkema and N. Mousseau, Phys. Rev. Lett. **77**, 4358 (1996).
[5] R. Malek and N. Mousseau, Phys. Rev. E **62**, 7723 (2000).
[6] F. H. Stillinger and T. A. Weber, Phys. Rev. B **31**, 5262 (1985).
[7] R. L. C. Vink, G. T. Barkema, W. F. van der Weg, and N. Mousseau, J. Non-Cryst. Solids **282**(248) (2001).
[8] G. T. Barkema and N. Mousseau, Phys. Rev. Lett. **81**, 1865 (1998).
[9] N. Mousseau and G. T. Barkema, Phys. Rev. B **61**, 1898 (2000).
[10] D. Wales, *Energy landcapes* (Cambridge University Press, 2003), in press.
[11] N. Mousseau and G. T. Barkema, Phys. Rev. E **57**, 2419 (1998).
[12] A. Cavagan, I. Giardina, and G. Parisi, Phys. Rev. Lett. **83**, 108 (1999).
[13] K. Laaziri, S. Kycia, S. Roorda, M. Chicoine, J. L. Robertson, J. Wang, and S. C. Moss, Phys. Rev. Lett. **82**, 3460 (1999).
[14] L. J. Munro and D. J. Wales, Phys. Rev. B **59**, 3969 (1999).
[15] F. Valiquette and N. Mousseau, *Energy landscape of amorphous silicon* (2003), submitted.
[16] H. E. A. Huitema, J. P. van der Eerden, J. J. M. Janssen, and H. Human, Phys. Rev. B **62**, 14690 (2000).
[17] L. A. Baez and P. Clancy, J. Chem. Phys. **102**, 8138 (1995).

[18] W. C. Swope and H. C. Andersen, Phys. Rev. B **41**, 7042 (1990).
[19] U. Landman, W. D. Luetdke, M. W. Ribarsky, R. N. Barnett, and C. L. Cleveland, Phys. Rev. B **37**(4637-4646) (1988).
[20] J. K. Bording and J. Tafto, Phys. Rev. B **62**(8098-8103) (2000).
[21] S. Nakhmanson and N. Mousseau, J. Phys.: Condens. Matter **14**, 6627 (2002).
[22] P. Beaucage and N. Mousseau, *Crystallization of silicon* (2003), in preparation.
[23] P. Keblinski, M. Z. Bazant, R. K. Dash, and M. M. Treacy, Phys. Rev. B **66** (2002).
[24] K. A. Jackson, in *Nucleation phenomena: a symposium*, edited by A. S. Micheals (American Chemical Society, 1965), p. 37.

# ATOMISTIC CHARACTER OF NANOCRYSTALLINE AND MIXED PHASE SILICON

R. BISWAS* AND B. C. PAN*[§]
*Department of Physics and Astronomy, Microelectronics Research Center and Ames Laboratory-USDOE, Iowa State University, Ames, Iowa 50011
§Department of Physics, University of Science and Technology of China, Hefei 230026, People's Republic of China

## ABSTRACT

Materials grown close to the phase boundary of amorphous and microcrystalline growth have the best electronic properties for solar cells. Systematic molecular dynamics methods have generated such nano-crystalline silicon, consisting of a mixed phase of nano-crystallites in an amorphous matrix, using an embedding method. An excess density of H resides on the surface of the nano-crystallites. The structure of this heterogeneous phase will be characterized by atomic distribution functions and structure factors. The electronic band structure of smaller models of nanocrystalline silicon reveals no midgap states and is similar to a-Si:H. There is a highly strained region surrounding the crystallites. The presence of localized strain region may increase the stability of the material.

## INTRODUCTION

Mixed phase semiconductors grown near the amorphous and micro-crystalline phase boundary have emerged as an intense area of current research. These include two broad classes of new materials. One is microcrystalline silicon (µc), grown on the crystalline side of the phase boundary, containing a large volume fraction of crystalline grains. The coalesced crystalline grains are separated by amorphous tissue. Microcrystalline silicon is being employed as a stable low bandgap solar cell material [1,2] and is stable to light exposure.
The second class of material is grown on the amorphous side of the a-µc phase boundary, usually in a silane–hydrogen gas mixture with high H dilution ratio (R). Under H-dilution conditions small nano-crystallites with dimensions of a few nm appear in a background a-Si:H matrix [3,4,5]. As the film grows in thickness, the crystallites grow in size and eventually coalesce at a critical film thickness into the micro-crystalline phase [3,4]. Higher H-dilution decreases the onset of the film thickness where µc phase occurs. Most of the current a-Si:H solar cells employ thin layers of H-diluted materials grown well before the onset of microcrystallinity. Such H-diluted material is markedly more stable to light-induced degradation [6,7,8]. This material is closer in electronic properties and band gap to a-Si:H and has a crystallite volume fraction typically less than ~10%. Transmission electron microscopy has observed the presence of nm size crystallites dispersed in an amorphous matrix. In our present work we study this class of mixed phase material with molecular dynamics simulations. Among the key questions is why such material has improved stability to light-soaking, and whether the material has superior order over traditional a-Si:H. Another intriguing aspect is the dependence of the phase boundary on film thickness- with the transition to microcrystalline occurring faster for higher H-dilution.

**MODEL**

Molecular dynamics simulations have simulated bulk amorphous semiconductors, and growth processes on surfaces. However the simulation of mixed phases has received less attention so far [9]. Previous approaches to generating a-Si:H have involved quenching of a melt to generate an amorphous structure or disordering a crystalline structure with bond-switching rearrangements. We introduce here a new *embedding* approach to generating mixed phases. The basic idea is to utilize the existing bulk a-Si:H structures that have been most successful in describing metastability, structure and H-motion [10,11]. These structures are free of coordination defects and midgap electronic states. Starting from an a-Si:H model we remove a central portion of this structure. The central portion is then replaced with a crystalline seed with approximately the same volume and shape as the removed region. Different crystallite sizes can be introduced by varying the size of the removed region. We utilize crystallites with low energy [111] surfaces. A classical molecular dynamics scheme [12] was used for larger models and tight-binding for smaller (N<600 atom models).

The relaxation of this cell initially yields coordination defects. The mixed phase cell is then systematically annealed to remove these defects. MD simulations have intrinsic limitations in annealing because of the short simulation time scales. MD is still a valuable guide to understanding atomistic processes. We were able to remove weak silicon bonds by H-insertion. H is introduced to anneal the over-coordinated sites. Dangling bonds are individually saturated with H atoms. The addition of H atoms throughout the annealing phase is analogous to processes occurring in H-diluted growth. Resulting models (Fig. 1) have no coordination defects- which is essential if one is to realistically model device-quality material with dangling bond densities of $<10^{17}$ cm$^{-3}$ or less than 1 atom in $10^5$. We note that the background a-Si:H matrix does show zone-axes or remnants of crystalline rings. This is due to the a-Si:H structure being simulated by a disordering of a crystalline structure. However the bond-length and bond-angle distribution is typical for an amorphous network.

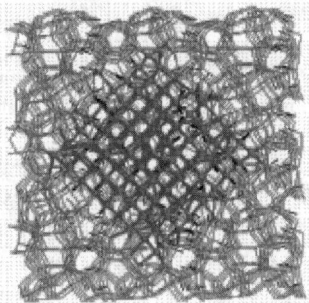

Fig. 1 A model of mixed phase nano-crystalline silicon. A smaller model with 1727 atoms is chosen for clarity. Nanocrystallite atoms are dark spheres. The bonds in the surrounding a-Si:H matrix are shown including the H atoms (small spheres). The crystallite contains 165 atoms. The H content is 14.6%.

## RESULTS AND ATOMISITIC STRUCTURE

The radial distribution function (Fig. 2) for the entire structure shows the characteristic sharp first neighbor peak expected for a-Si:H and a broader second neighbor peak. These features are very similar to the bulk amorphous structure because the volume fraction of the crystallite is quite small. Considerably more order is found for the atoms within the nano-crystallite (Fig. 2). Their radial distribution function reveals sharp peaks at the nearest neighbor distances corresponding to c-Si (Fig. 2). These sharp peaks are consistent with the medium range order (MRO) associated with mixed phase material although this are also the result of the simulation algorithm we used. The interior of the nanocrystallites are strongly ordered with the interior bond-lengths and bond-angles similar to the crystal. The surface atoms of the nanocrystallite are strongly distorted due to the matching of the crystallite with the amorphous matrix.

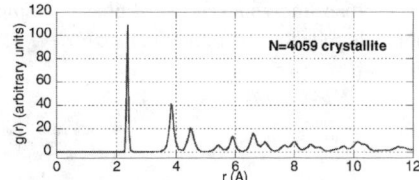 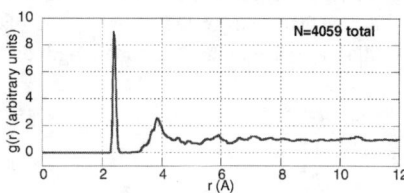

Fig. 2. Radial distribution functions (rdf) for the N=4059 atom model showing the total rdf compared to the rdf for only the crystallite. The crystallite has 455 atoms and the H content is 14.2%.

We have computed the structure factor $S(q)$ (Fig. 3) – a key quantity for diffraction measurements, that measures longer range order. $S(q)$ represents the Fourier transform of the radial distribution function. The peaks of $S(q)$ occur at the wave-vectors expected for bulk a-Si:H and compare well with older experiments [13]. The second peak in $S(q)$ is larger than the first peak in agreement with scattering experiments [13].

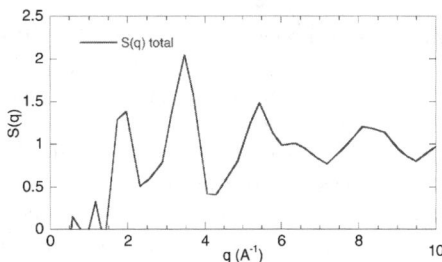 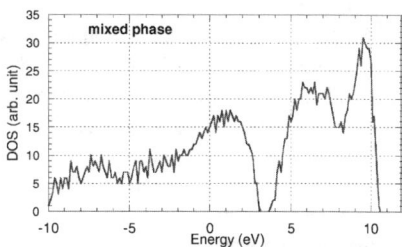

Fig. 3 Structure factor $S(q)$ for nano-crystalline mixed phase model.

Fig. 4 Electronic densities of states calculated with the tight binding method for the N=546 atom model.

The electronic densities of states was computed for a smaller model with N=546 atoms using the tight-binding method. The results for the larger model are expected to be similar. As expected there are no deep gap states since all the dangling bonds have been removed. The band tails and overall DOS are quite similar to those of bulk a-Si:H since the amorphous volume fraction is dominant. We learn that interfacial strained bonds do not generate gap states but have states in the band tails.

It is interesting to plot (Fig. 4) the rms bond angle in the mixed phase model averaged over all the atoms in the model, as a function of the cell size (or equivalently the number of atoms). We have separated the average bond angle into its contribution from the total cell (including the crystallite) and the portion of the angular distribution from the amorphous matrix (A). We find (Fig. 5) the small cells (cell size a <25 Å) are severely distorted with strained rms bond angle distributions exceeding 11.5°. The bond-angle strain reduces monotonically with cell size and is markedly reduced for larger cells. For the larger cells (a> 4nm) we find the average rms bond-angle variation in the amorphous region is below that (11.5°) of the homogenous a-Si:H network that we started with. This suggests beneficial ordering of the amorphous matrix, in the presence of the crystallite, a feature that is suggested by narrowing of the TO Raman line-width in mixed phase samples. A similar decrease in the rms bond length distribution is found as a function of cell size.

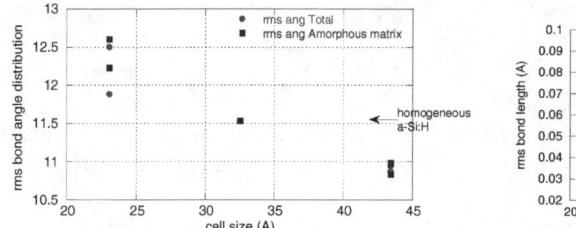
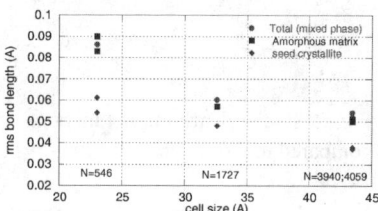

Fig. 5 Root-mean-square (rms) bond-angle values and rms bond-length values as a function of the cell size for the different mixed phase models. The rms angle for the homogeneous network is marked. The number of atoms N in the model is indicated in the bond-length plot.

## GRAIN BOUNDARY

Analysis of the individual atomic bond-angle and bond length variation reveals very interesting microscopic features. We have plotted the rms bond angle for each silicon atom as a function of its distance from the center of the crystallite. The distribution (Fig. 6) separates into three distinct regions. Upto a distance of ~1 nm the bond angles are very similar to crystalline values, indicating a well-ordered crystallite. Then there is a grain boundary region between the crystallite and amorphous matrix where the bond angles have very distorted rms values exceeding 15°, and even approaching 30-35°, also suggesting limitations of the relaxation algorithm. Finally there is the bulk amorphous phase, extending to the boundary of the simulation cell. We estimate grain boundary widths to be ~1 nm (or slightly less). The grain boundary arises because of the mismatch between the crystallite and the amorphous matrix, and results in severely strained bonds at the nano-crystallite surface. We expect that such strained grain boundaries are a universal feature of mixed phase material.

The strained grain boundary also contains severely strained silicon bonds, with elongated bond-lengths. We have also found the strained grain boundary to contain an excess density of H, with as much as twice the background H-density in the amorphous phase. This excess H passivates the large density of dangling bonds and weak bonds in the grain boundary region and may be the key influence of H-dilution in the growth process. The excess H at crystallite boundaries has been speculated by other investigators and may be responsible for the low-temperature H- evolution peak in mixed phase material.

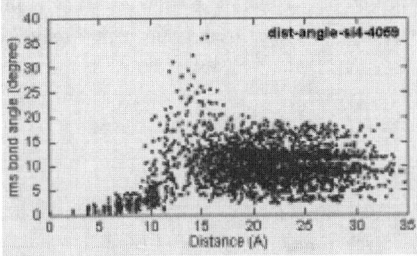

Fig. 6 Bond-angle distributions for individual silicon atoms as a function of distance from the center of the crystallite, showing the grain boundary. Results are for the N=4059 atom model.

The mixed phase material (presently used in a-Si:H solar cell i-layers) grown under H-dilution conditions is significantly more stable to light-induced degradation and also stabilizes more quickly under light soaking[4,6,7]. Furthermore the regime of micro-crystalline silicon (with a high volume fraction of crystallites is stable to light-soaking. Our present findings of a disordered grain boundary layer seem to be inconsistent with the known stability of the mixed phase. The region of strain at the grain boundary and the weak bonds in this strained region are counter-intuitive since such weak silicon bonds are expected to be unstable to light excitation and lead to dangling bond defects.

We propose a qualitatitive model to understand the stability. Light-soaking generates excited electrons and holes. The non-radiative recombination of such excited carriers breaks weak silicon bonds resulting in light-induced dangling bonds. Clearly the highly strained and elongated bonds in the grain boundary region will be preferentially broken. If the amorphous matrix is better ordered recombination may occur at the weak bonds in the grain boundary rather than the bulk amorphous matrix, since the weak bonds may generate tail states. From our previous study, Si-bond breaking results in a dangling bond accompanied by a neighboring floating bond. The mobile floating bonds are expected to diffuse within the grain boundary, since their migration barriers are lower within a more disordered region. Mobile floating bonds will then annihilate or recombine with the dangling bonds in the grain boundary and the net result will be no defect creation. This may lead to lower defect creation in heterogeneous silicon. Excess defect states with high optical cross-section at the crystallite-amorphous boundary have been found [14] from photocapacitance studies. More work is needed to quantify recombination in nanocrystalline silicon.

We also found that when two dangling bonds are created nearby in the strained shell surrounding the crystallites, these dangling bonds can recombine directly by rebonding of silicon network. We observed such rebonding processes with molecular dynamics simulations when dangling bonds are less than 5 Å apart in this region. Low energy barriers (<0.5 eV) are found for such recombination processes. We expect the grain boundary region to become more distorted after light-soaking. Qualitatively this would involve a decrease in volume of the crystallite on light-soaking. This would argue for an increase in the gap or an increase of open-circuit voltage (Voc) as found in recent measurements [15,16]. Spatially localizing the excess strain region may increase the stability of the material and reduce the density of light-induced defects.

In conclusion we have generated models of mixed phase or nano-crystalline silicon and analyzed the microscopic atomistic properties of the model. There is a strained grain boundary region between the crystallite and amorphous network. The localized strain region may lead to higher stability.

## ACKNOWLEDGEMENTS

We acknowledge support from NREL under contract ACQ-1-30619-08. Acknowledgement is made to the donors of the Petroleum Research Fund administered by the American Chemical Society for partial support of this research. BCP acknowledges support from the NSFC with grant No. 69876025. It is a pleasure to thank the National Amorphous Thin Film team for many stimulating discussions.

## REFERENCES

[1]   J. Meier, R. Fluckiger, H. Keppner, A. Shah, Appl. Phys. Lett. **65**, 860 (1994).
[2]   R.E.I. Schropp, Y. Xu, E. Iwanicko, G. Zacharia, and A. H. Mahan, MRS Symp. Proc. **715** 623 (2002).
[3]   J. Koh, Y. Lee, H. Fujiwara, C.R. Wronski, and R.W. Collins, Appl. Phys. Lett. **73**, 1526 (1998). A. S. Ferlauto, R. J. Koval, C. R. Wronski and R. W. Collins, Appl. Phys. Lett. **80**, 2666 (2002).
[4]   C. R. Wronski, J. M. Pearce, R. J. Koval, X. Niu, A. S. Ferlauto, J. Koh and R. W. Collins, Proceedings MRS **715**, 459 (2002).
[5]   D.V. Tsu, B.S. Chao, and S.R. Ovshinsky, S. Guha and J. Yang, Appl. Phys. Lett. **71**, 1317(1997).
[6]   T. Kamei, P.Stradins and A. Matsuda, Appl. Phys. Lett. **74**, 1707 (1999).
[7]   Y. Lubianiker and J. D. Cohen, H-C. Jin, and J. R. Abelson, Phys. Rev. B **60**, 4434 (1999).
[8]   D. Han, J. Baugh and G. Yue, Phys. Rev. B **62**, 7169 (2000).
[9]   C. R. S. da Silva and A. Fazzio, Phys. Rev. B **64**, 075301 (2001).
[10]  R. Biswas, B.C. Pan and Y. Y. Ye, Phys. Rev. Lett. **88**, 205502 (2002).
[11]  R. Biswas and B. C. Pan, Appl. Phys. Lett. **72**, 371 (1998).
[12]  U. Hansen and P. Vogl, Phys. Rev. B **57**, 13295 (1998).
[13]  T. A. Postol et al, Phys. Rev. Lett. **45**, 648 (1980).
[14]  D. Kwon, C._C. Chen. J. D. Cohen, H.C. Jin, E. Hollar, I. Robertson and J. R. Abelson, Phys. Rev. B **60**, 4442 (1999).
[15]  K. Lord, B. Yan, J. Yang and S. Guha, Appl. Phys. Lett. **79**, 3800 (2001).
[16]  J. Yang, K. Lord, B. Yan, A. Banerjee, S. Guha, D. Han and K. Wang, MRS **715**, 601 (2002).

## Kinetics of Light-induced Effects in Mixed-Phase Hydrogenated Silicon Solar Cells

Guozhen Yue, Baojie Yan, Jeffrey Yang, Kenneth Lord, and Subhendu Guha
United Solar Systems Corp., 1100 West Maple Rd., Troy, MI 48084, U.S.A.

## ABSTRACT

We have observed a significant light-induced increase in the open-circuit voltage ($V_{oc}$) of mixed-phase hydrogenated silicon solar cells. In this study, we investigate the kinetics of the light-induced effects. The results show that the cells with different initial $V_{oc}$ have different kinetic behavior. For the cells with a low initial $V_{oc}$ (less than 0.8 V), the increase in $V_{oc}$ is slow and does not saturate for light-soaking time of up to 16 hours. For the cells with medium initial $V_{oc}$ (0.8 ~ 0.95 V), the $V_{oc}$ increases rapidly and then saturates. Cells with high initial $V_{oc}$ (0.95 ~ 0.98 V) show an initial increase in $V_{oc}$, followed by a $V_{oc}$ decrease. All light-soaked cells exhibit a degradation in fill factor. The temperature dependence of the kinetics shows that light soaking at high temperatures causes $V_{oc}$ increase to saturate faster than at low temperatures. The observed results can be explained by our recently proposed two-diode equivalent-circuit model for mixed-phase solar cells.

## INTRODUCTION

Since the first observation of improved stability of hydrogenated amorphous silicon (a-Si:H) films prepared from a diluted mixture of silane and hydrogen [1], hydrogen dilution has been extensively used to improve the optoelectronic properties of a-Si:H materials and solar cells. An appropriate hydrogen dilution improves intermediate range order of the a-Si:H network, widens the optical band gap, decreases the defect density of the material, and consequently improves the solar cell performance. The best a-Si:H solar cell is made using a hydrogen dilution ratio just before the onset of microcrystallite formation [2]. Increasing the hydrogen dilution ratio beyond a threshold value leads to a heterogeneous material having a mixture of microcrystalline and amorphous phases. In a solar cell, a mixed-phase intrinsic (*i*) layer is characterized by a sharp drop in the open-circuit voltage ($V_{oc}$) [3]; the larger the microcrystalline volume fraction, the lower the $V_{oc}$.

The nature of the light-induced change for the mixed-phase solar cells is different from that of conventional a-Si:H solar cells. Our previous result [4] showed that $V_{oc}$ is enhanced after light soaking. The amount of the change depends on the initial $V_{oc}$ and light-soaking conditions. Further experiments [5] by applying a reverse bias to the cell during light soaking as well as applying a forward bias to the cell in the dark suggest that the $V_{oc}$ enhancement has the same origin as the Staebler-Wronski effect (SWE) [6], i.e. the recombination of photo-generated carrier is the driving force. However, when a fully amorphous silicon buffer layer (~ 500 Å) is inserted between the *i* and the *p* layers, the solar cells show high initial $V_{oc}$ and no light-induced enhancement. Based on these observations, we have recently proposed a two-diode equivalent-circuit model [7] to explain the $V_{oc}$ increase as being associated with the degradation of microcrystallite quality. In this paper, we investigate the kinetics of the light-induced change of the mixed-phase solar cells. The results provide an insight into the microscopic mechanism of light-induced change in the mixed-phase material; they can also be explained by the two-diode model.

## EXPERIMENTAL

Single-junction *nip* solar cells were deposited using radio frequency glow discharge on aluminum/zinc oxide coated stainless steel substrates. The deposition parameters were adjusted to obtain solar cells with a distribution of $V_{oc}$ from over 1.0 V to less than 0.5 V on the 4 cm × 4 cm deposition area. More than 40 indium tin oxide dots with areas of 0.05 cm$^2$ and 0.25 cm$^2$ were deposited on top of the cells. Current-voltage (J-V) characteristics were measured at 25 °C under an AM1.5 solar simulator. Light soaking for the kinetics was conducted at different temperatures under AM1.5 illumination. The parameters, $V_{oc}$, fill factor (FF), and short circuit current density ($J_{sc}$), were extracted from J-V characteristics.

## RESULTS

We first studied the kinetics of light-induced changes in cells having different initial values of $V_{oc}$. The J-V curves were recorded at predetermined intervals during light soaking at 25 °C. Figure 1 plots $V_{oc}$ and normalized $V_{oc}$ as a function of light-soaking time for the cells with initial $V_{oc}$ of (a) > 0.8 V and (b) < 0.8 V. For the cells with 0.80 V < $V_{oc}$ < 0.98 V, as shown in Fig.1(a), the $V_{oc}$ increases then saturates. The increase rate, saturation time, and saturation value depend critically on the initial $V_{oc}$. For higher initial $V_{oc}$, the increase rate and magnitude become smaller and the saturation time becomes shorter. In fact, for an initial $V_{oc}$ of 0.986 V, the $V_{oc}$ is almost constant for the first 3 hours, then decreases slightly, while the $V_{oc}$ of the cell with an initial value of 1.011 V decreases continuously during light soaking, as is typically observed in conventional a-Si:H solar cells.

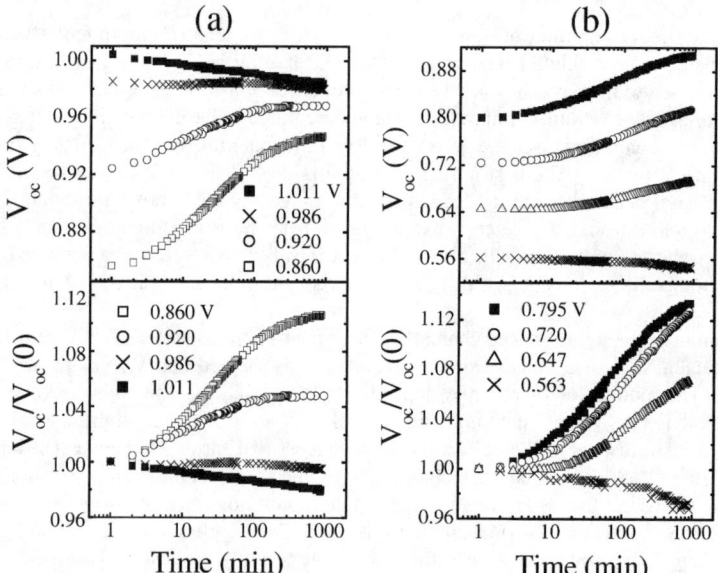

**Figure 1.** $V_{oc}$ and normalized $V_{oc}$ as a function of light-soaking time for cells with initial $V_{oc}$ (a) > 0.8 V and (b) < 0.8 V.

The kinetics of the light-induced changes in cells having initial $V_{oc} < 0.8$ V is shown in Fig.1(b). The $V_{oc}$ increases with time for cells having initial $V_{oc} > 0.57$ V, but decreases for cells with initial $V_{oc} < 0.57$ V. In contrast to the behavior of the cells in Fig. 1 (a), both the rate and the magnitude of the change increase with the increase in $V_{oc}$, there is no indication of saturation in 1000 min.

It is interesting to compare the details of the kinetics of $V_{oc}$ FF. Figure 2 shows the normalized $V_{oc}$ and FF versus light-soaking time for all the cells. First, $V_{oc}$ enhancement does not exhibit saturation in 1000 min. for cells with $V_{oc} < 0.8$ V but does for cells with $V_{oc} > 0.8$ V. FF shows a similar tendency. In addition, at the beginning of the light soaking, the increase rate of $V_{oc}$ is smaller for cells with $V_{oc} < 0.8$ V than that with $V_{oc} > 0.8$ V. After passing an intersection point, depending on the initial $V_{oc}$, the increase in $V_{oc}$ becomes faster than the cells with $V_{oc} > 0.8$ V.

In order to further investigate the above observations, light soaking was carried out on two solar cells for a much longer period. One has $V_{oc} > 0.8$ V and the other $V_{oc} < 0.8$ V. Figure 3 shows the $V_{oc}$ and FF light-soaking

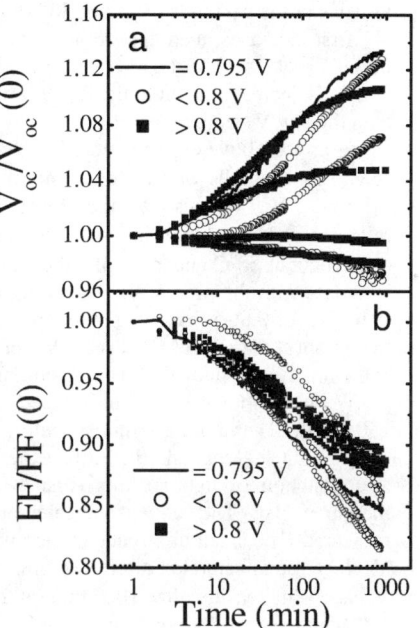

**Figure 2.** Comparison of the kinetics of the normalized $V_{oc}$ and FF for the cells with different initial $V_{oc}$.

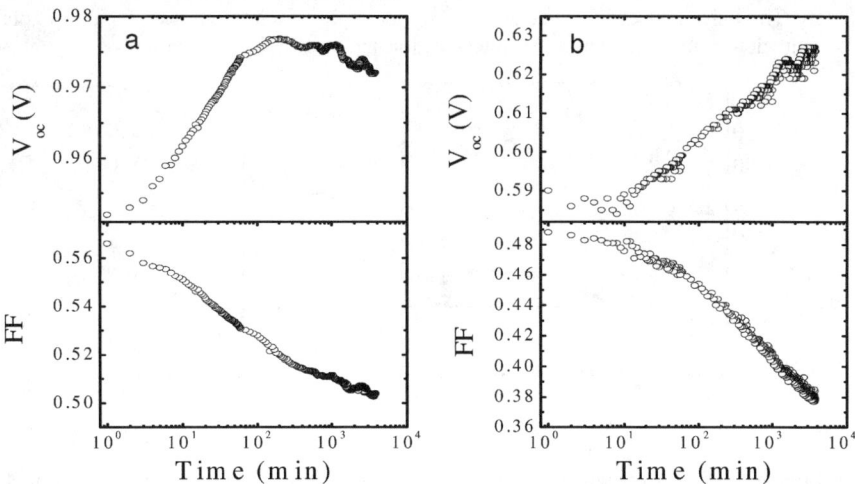

**Figure 3.** $V_{oc}$ and FF as a function of exposure time for the cells with initial $V_{oc}$ of (a) 0.952 V and (b) 0.590 V.

kinetics of the two cells over a period of 67 hours. For the cell with an initial $V_{oc}$ of 0.952V, the $V_{oc}$ first increases, then appears to saturate, and finally starts to decrease. For the cell with an initial $V_{oc}$ of 0.590 V, the $V_{oc}$ keeps increasing and has no indication of saturation. The FF of the two cells decreases monotonically with time, but the change for the high $V_{oc}$ cell is much slower than the low $V_{oc}$ cell.

From the above experimental data, we can see that the behavior of light-induced $V_{oc}$ change ($\Delta V_{oc}$) for the cells with different initial $V_{oc}$ is different. The rate of $\Delta V_{oc}$ depends not only on the initial $V_{oc}$, but also on the length of light-soaking time. The same phenomenon is also reflected on the relationship between $\Delta V_{oc}$ and the initial $V_{oc}$ for different light-soaking time carried out at 50°C under metal-halide lamp of intensity 100 mW/cm$^2$. As shown in Fig. 4, $\Delta V_{oc}$ increases with the time of light soaking. When the cells are exposed for 5 min., the cell with an initial $V_{oc}$ of 0.86 V has the maximum change. As the time of exposure increases to 40 hours, the maximum change is for a cell with $V_{oc}$ of 0.77 V, indicative of a faster initial $V_{oc}$ increase for the cell with higher initial $V_{oc}$. This is similar to the results described earlier.

We have shown that light-induced $V_{oc}$ enhancement can be annealed away in vacuum at 150°C [4]. Therefore, thermal annealing and light-induced change should occur simultaneously during light soaking, and they affect the $V_{oc}$ change in opposite directions. Figure 5 shows $V_{oc}$ as a function of light-soaking time at different temperatures. The kinetics at 100 °C was measured first. The cell was annealed at 150 °C for two hours before the kinetics at 25 °C was measured. The small difference in the initial $V_{oc}$ might be due to annealing. We can see that the $V_{oc}$ increase is saturated after 200 min. when the light-soaking temperature is 100 °C, whereas $V_{oc}$ keeps increasing after 1000 min. at 25 °C. $\Delta V_{oc}$ at 25 °C and 100 °C after light soaking for 1000 min. are ~ 60 mV and ~ 45 mV, respectively. Obviously, light soaking at higher temperatures causes the increase in $V_{oc}$ to saturate faster than at lower temperatures.

## DISCUSSION

The light-induced $V_{oc}$ increase in the mixed-phase solar cells was initially explained by a reduction in microcrystalline volume fraction induced by light soaking [3-5]. However, X-ray

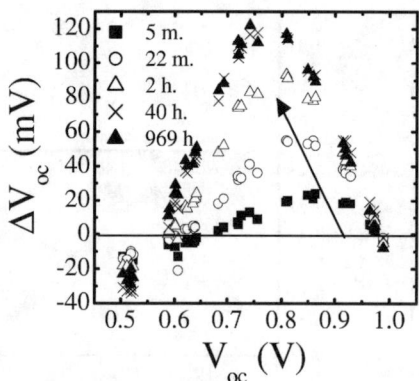

**Figure 4.** $\Delta V_{oc}$ as a function of the initial $V_{oc}$ for different exposure time. The arrow indicates the shift in the peak position.

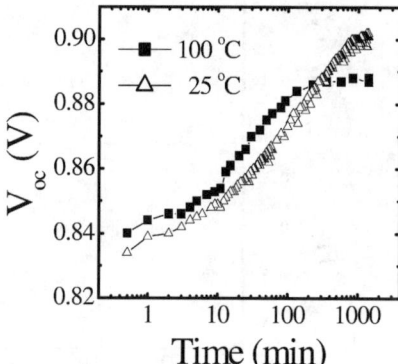

**Figure 5.** $V_{oc}$ as a function of light-soaking time at different temperatures.

diffraction and Raman spectroscopy did not find any light-induced structural change within detection limits [8, 9]. By analyzing all the experimental data, we recently proposed a two-diode equivalent model [7], and the key points are summarized below. A mixed-phase solar cell is equivalent to two diodes connected in parallel. One diode has the characteristics of an a-Si:H solar cell and the other of a μc-Si:H solar cell. When a forward bias, with a magnitude larger than the $V_{oc}$ of the μc-Si:H cell but smaller than that of the a-Si:H cell, is applied to the mixed-phase cell under illumination, the μc-Si:H cell is under forward current injection, while the a-Si:H cell collects negative photocurrent. $V_{oc}$ is reached when the forward injection current and the negative photocurrent are equal in magnitude. Therefore, the $V_{oc}$ of a mixed-phase cell depends not only on the microcrystalline volume fraction, but also on the forward current density of the μc-Si:H cell. To illustrate this point, we plot in Fig. 6 (a) the J-V characteristics of an a-Si:H cell and two μc-Si:H cells, and in Fig. 6 (b) the corresponding $V_{oc}$ versus microcrystalline volume fraction for the μc-Si:H cells. Upon light soaking, the $V_{oc}$ of a mixed-phase cell could increase due to a reduction of microcrystalline volume fraction, as indicated by arrow (1) or by the reduction of forward current density of the μc-Si:H cell, as indicated by arrow (2). In a real situation, the two processes might coexist. The area between the two curves represents an unstable region.

Based on the concept of the two-diode model, most of the kinetic results are understandable. First, the cells with an initial $V_{oc}$ around 0.6 - 0.8 V (depending on the cell structure and the light soaking condition) has a wider spread between the two curves of Fig. 6 (b), which means that any change in the microcrystalline volume fraction or forward current density would produce a large $\triangle V_{oc}$. Experimentally, the largest light-induced $V_{oc}$ increase appears at around 0.75 V with a bell-shaped curve on the $\triangle V_{oc}$ versus the initial $V_{oc}$ plot. Therefore, the increase rate for the cells with intermediate initial $V_{oc}$ is higher than those with smaller or larger initial $V_{oc}$. In addition, the cells with low initial $V_{oc}$ (~ 0.6 V) would move towards the unstable region as indicated by arrow (3), but the cells with high initial $V_{oc}$ would move out of the unstable region as indicated by arrow (4). Consequently, on the average, the cells with initial $V_{oc} < 0.8$ V show a

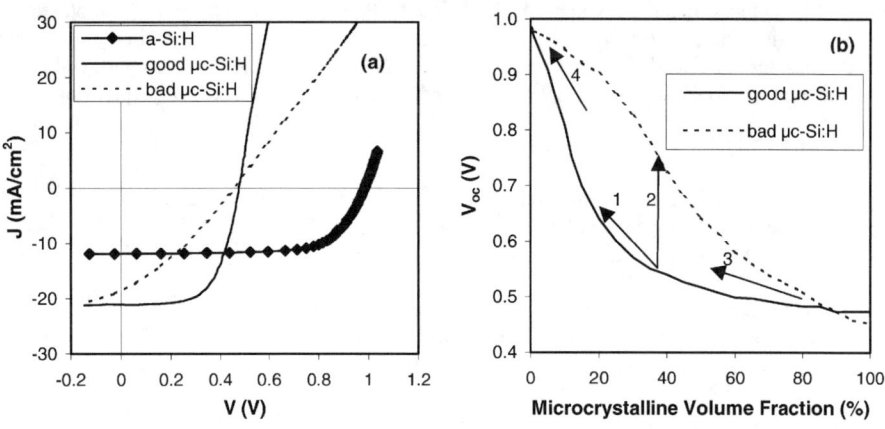

**Figure 6.** (a) J-V characteristics of an a-Si:H cell, a good μc-Si:H cell, and a poor μc-Si:H cell, and (b) estimated $V_{oc}$ as a function of crystal volume fraction in mixed phase solar cells by using the good and the bad μc-Si:H cell.

smaller increase rate and a slower tendency to saturate than cells with initial $V_{oc} > 0.8$ V, as shown in Fig. 2. As a result, the maximum on the $\triangle V_{oc}$ versus initial $V_{oc}$ plot shifts to a lower initial $V_{oc}$ as the light soaking time increases, (see Fig. 4). The saturation or decrease of $V_{oc}$ after a period of light soaking for the cells with a high initial $V_{oc}$ could be explained as follows. The decrease of forward current for the μc-Si:H cell due to a reduction of forward current density or reduction of microcrystalline volume fraction is offset by the reduction of current from the a-Si:H, due to the SWE that is more pronounced at voltages around 0.85 eV. We should point out that Fig. 6 is just an illustration. The choices of J-V curves for a-Si:H and μc-Si:H diodes are somewhat arbitrary. In a real mixed-phase cell, the characteristics of the equivalent a-Si:H and μc-Si:H are unknown. Therefore, quantitative predictions of the amount of light-induced $V_{oc}$ increase, and kinetic behavior cannot be made.

Previously, we have reported that the origin of light-induced $V_{oc}$ increase is the same as SWE [5]; the metastable defects are generated in μc-Si:H and a-Si:H phases. Therefore, it is not difficult to understand why the FF for all the cells decreases with light-soaking time. The earlier saturation in FF for the cells with $V_{oc} > 0.8$ V might be due to the better material quality in these cells than those with $V_{oc} < 0.8$V.

## SUMMARY

We have studied the kinetics of the light-induced change of mixed-phase solar cells. We find that the cells with different values of initial $V_{oc}$ exhibit different light-soaking kinetics. All the light-soaked cells show a decrease in FF. The temperature dependence of the kinetics shows that light soaking at high temperatures causes $\Delta V_{oc}$ to saturate faster than at low temperatures. The observed results can be explained by our recently proposed two-diode equivalent circuit model for mixed-phase solar cells.

## ACKNOWLEDGEMENTS

The authors thank E. Chen, A. Mohsin, T. Palmer, G. Pieka, and D. Wolf for sample preparations and A. Banerjee, K. Beernink, H. Fritzsche, and S. R. Ovshinsky for stimulating discussions. The work was supported in part by NREL under the thin-film partnership program Subcontract No. ZDJ-2-30630-19. We thank S. Sundquist for manuscript preparation.

## REFERENCES

1. S. Guha, K. L. Narasimhan, and S. M. Pietruszko, J. Appl. Phys. **52**, 859 (1981).
2. J. Yang, A. Banerjee, and S. Guha, Appl. Phys. Lett. **70**, 2975 (1997).
3. S. Guha, J. Yang, D.L. Williamson, Y. Lubianiker, J.D. Cohen, and A.H. Mahan, Appl. Phys. Lett. **74**, 1860 (1999).
4. K. Lord, B. Yan, J. Yang, and S. Guha, Appl. Phys. Lett. **79**, 3800 (2001).
5. J. Yang, K. Lord, B. Yan, A. Banerjee, and S. Guha, Proc. of 29th IEEE Photovoltaic Specialists Conference, New Orleans, 1094 (2002).
6. D.L. Staebler and C.R. Wronski, Appl. Phys. Lett. **31**, 292 (1977).
7. B. Yan, G. Yue, J. Yang, K. Lord, and S. Guha, submitted to Appl. Phys. Lett. (2003).
8. D. Han, private communication.
9. D. Williamson, *Solar Energy Materials and Solar Cells*, in press.

# TEMPERATURE DEPENDENCE OF A HYDROGEN DOUBLET SITE IN LIGHT-SOAKED a-Si:H FROM $^1$H NMR

T. Su and P. C. Taylor
Department of Physics, University of Utah
Salt Lake City, UT 84112, U.S.A.
G. Ganguly and D. E. Carlson
BP Solar
Toano, VA 23168, U.S.A.

## ABSTRACT

We investigate the temperature dependence of an additional $^1$H NMR signal that appears in a-Si:H at T = 7 K only after light soaking. This "doublet" signal is attributed to a pair of hydrogen atoms approximately 2.3 Å apart. We present the first lineshape of this doublet, and we examine the spin-lattice relaxation from 6.5 K to 20 K at a magnetic field of about 3.5 T. At low temperatures, the spin-lattice relaxation time of these paired hydrogen sites has a strong dependence on temperature, which is probably the result of some form of local motion.

## INTRODUCTION

We have previously reported an $^1$H nuclear magnetic resonance (NMR) signal in a-Si:H that occurs only after light soaking for 600 hours [1]. This signal, which is attributed to a pair of hydrogen atoms, exhibits similar annealing kinetics to that of the defects created during light-soaking, and the concentration of these sites is comparable to that of the defects measured by electron spin resonance (ESR). The distance between the two hydrogen atoms in the pair is about 2.3 Å. Preliminary data on the temperature dependence of the lineshape suggested that the pair may undergo some form of local motion as the temperature increases [1], but the evidence was not compelling. In addition, our earlier measurements could not exclude the possibility that the additional lineshape was due to a hydrogen triplet with the central peak masked by the narrow line from bonded hydrogen. In this paper we investigate the lineshape in further detail. In particular, we report the temperature dependence of $^1$H NMR due to this doublet at a higher magnetic field ($B_0$ = 3.5 T). The spin-lattice relaxation time ($T_1$) of the doublet has a strong temperature dependence, which may indicate a thermally activated motion. However, this dependence cannot be explained by a simple thermally activated, random diffusive process, such as that described by the Bloembergen, Purcell, and Pound model (BPP model). In particular, for temperatures between 6.5 and 20 K we find no evidence that the doublet lineshape is narrowed due to motional effects as would be expected if the simplest BPP model applied.

## EXPERIMENTAL DETAILS

The light-soaked, thin film of $a$-Si:H was made at BP Solar in a large area deposition system by DC plasma enhanced chemical vapor deposition (PECVD). The film was light-soaked for 600 hr with a solar simulator. Details of the sample preparation can be found elsewhere [1,2]. NMR measurements were performed on a typical pulsed-NMR spectrometer from Tecmag. Details can also be found elsewhere [1]. In order to improve the signal-to-noise ratios, the static magnetic field for the present experiments was increased from ~ 2 T to ~ 3.5 T, corresponding to an increase of the $^1$H NMR frequency from 86.556 MHz to 147.417 MHz. The sample was cooled using a gas-exchange cryostat from Oxford. The temperature drift during each measurement was about ±0.3 K.

To suppress the much stronger signal from hydrogen bonded to silicon atoms, a Jeener-Broekaert three-pulse sequence was used [3], and the NMR line shapes are obtained by recording the stimulated echo after the Jeener-Broekaert sequence. The echo intensity $I(\tau_1, \tau_2)$ is given by [4]

$$I(\tau_1,\tau_2) = I_0 \exp\{-\frac{2\tau_1}{T_2}\}\exp\{-(\frac{\tau_2}{T_{1D}})^{\frac{1}{2}}\} \quad , \tag{1}$$

where $\tau_1$ and $\tau_2$ are the separations between the first and second pulses, and the second and third pulses, respectively. $T_2$ and $T_{1D}$ are the spin-spin relaxation time and the dipolar spin lattice relaxation time, respectively. A typical pulse width for the 90° pulse is about 2 μs.

At a $^1$H NMR frequency $\omega_0 = 147.417$ MHz and at low temperatures, the values of $T_1$ for the hydrogen, which is bonded to silicon (bonded hydrogen), are too long for these hydrogen sites to fully recover using a practical pulse repetition rate. Therefore the signal from the bonded hydrogen is suppressed due to partial saturation. The dependence of the NMR signal $I$ on the delay time between two scans, $T_d$, is,

$$I = I_0(1 - e^{-T_d/T_1}) , \tag{2}$$

where $I_0$ is the amplitude of the fully relaxed signal. The inverse of $T_d$ is the rate at which the three-pulse sequence is repeated (repetition rate). When $T_d \ll T_1$, the relation becomes,

$$I = I_0 \frac{T_d}{T_1} . \tag{3}$$

**RESULTS**

Figure 1 shows the temperature dependence of the NMR lineshapes at $B_0 = 3.5$ T. The lineshape obtained earlier [1] at $B_0 = 2$ T and $T = 7$ K is also shown for comparison. [In all lineshapes shown in this paper the Zeeman frequency has been subtracted so that the peaks are centered about zero.] The composite lineshapes contain a narrow line (~ 4 kHz FWHM) and a broad line (~ 25 kHz FWHM) as is always seen in typical a-Si:H samples prepared by the PECVD technique. The narrow line is attributed to randomly occurring hydrogen atoms that are bonded to silicon, hereafter termed "dilute" hydrogen. The broad line is attributed to hydrogen atoms bonded to silicon in a clustered configuration, hereafter termed "clustered" hydrogen. In addition, a small doublet occurs with a splitting $\Delta v = 15$ kHz as indicated by the dashed lines. As can be seen from Fig. 1, $\Delta v$ is constant for $B_0 = 2$ T and $B_0 = 3.5$ T. At $B_0 = 3.5$ T, $\Delta v$ remains the same, within the experimental error, at $T = 7$ K and $T = 10$ K. The doublet is less obvious at $T = 6.5$ T and $T = 20$ K. As we will discuss below, this behavior is attributed to the different temperature dependences of the $T_1$'s for the bonded hydrogen and the doublet. A detailed comparison of the composite lineshape shows that no motional narrowing of the doublet occurs for temperatures up to 20 K.

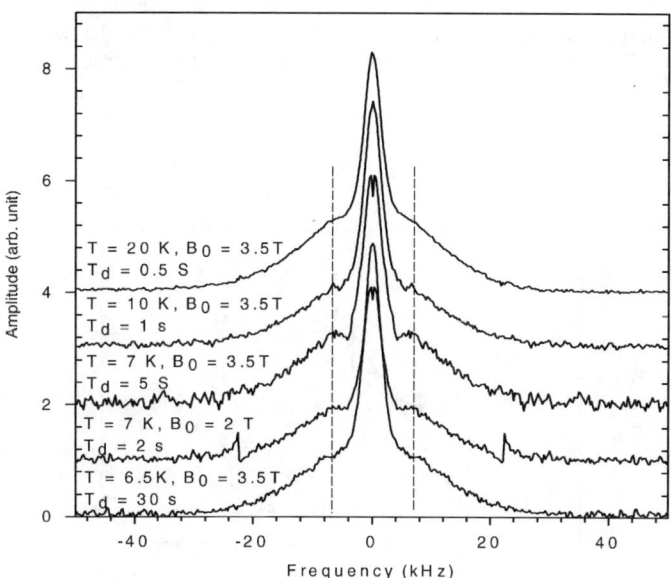

**Figure 1.** $^1$H NMR lineshapes from the light-soaked sample at different temperatures and at two magnetic fields, $B_0 = 2$ T and $B_0 = 3.5$ T. The two dashed lines indicate where the doublet appears.

Figure 2 shows the temperature dependence of $T_1$ for both the bonded hydrogen and the doublet, plotted as log($T_1$) against 1000/$T$. Open circles denote the data for bonded hydrogen. Solid circles denote the data for the doublet. The lines are aids to the eye. The value of $T_1$ for bonded hydrogen, hereafter denoted as $T_1(B)$, at 4.2 K is an extrapolation from the data at 92.5 MHz [3], assuming a field dependence of $T_1$ as $T_1 \propto \varpi^2$. As can be seen in Fig. 2, $T_1$ of the doublet, hereafter denoted as $T_1(D)$, increases much more rapidly than $T_1(B)$, as the temperature decreases. For temperatures below about 6 K, $T_1(D) > T_1(B)$. This relationship means that at the lowest temperatures the signal from the bonded hydrogen will recover faster than that of the doublet, resulting in an apparent suppression of the signal from the doublet. On the other hand, at $T = 20$ K, $T_1(B)$ is short enough compared to the pulse repetition rate so that most of the signal from bonded hydrogen fully recovers, and once again effectively reduces the signal from the doublet. In short, the doublet only becomes visible when the signal from bonded hydrogen is drastically reduced by the dipolar echo sequence (due to a much more rapid $T_{1d}$) and also reduced by partial saturation (due to a much longer $T_1$).

Figure 3 shows the comparison of the lineshape at 7 K, with a recovery time $T_d = 5$ s, and the lineshape at 20 K with $T_d = 0.5$ s, in which the signal from the bonded hydrogen has relaxed and the signal from the doublet is effectively suppressed. Solid circles and open circles represent the data at 7 K and 20 K, respectively. The solid line represents a subtraction of the 20 K data from the 7 K data, and the dashed line is a fit to this subtraction with a Pake doublet powder pattern that is broadened isotropically by a Gaussian whose width is approximately 10% of the splitting between the two divergences. The sharp negative spike near the center is an artifact due to small phase errors in the complex Fourier transforms. The splitting between the two divergences is about 16 kHz. This value is consistent with that previously estimated from measurements at a lower frequency [1].

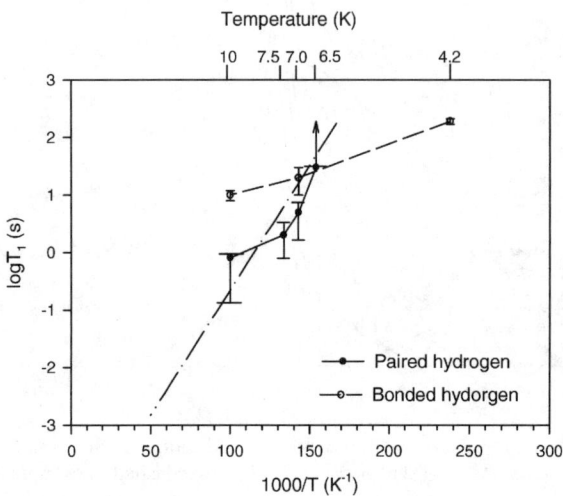

**Figure 2.** Temperature dependence of $T_1$ for the hydrogen doublet and the bonded hydrogen. The open circles represent the data for bonded hydrogen. The solid circles represent the data for the hydrogen doublet. The lines are aids to the eye. The dashed-dotted line is a straight line with a slope of 100 K. See text for details.

## DISCUSSION

The most interesting result of this study is the strong temperature dependence of the spin-lattice relaxation time of the doublet. In particular, for temperatures between 6.5 K and 20 K, the values of $T_1$ for the doublet are shorter than those of the bonded hydrogen. Since most of the intensity in the doublet line is well outside the width of the narrow line, the dipole-dipole coupling between the doublet and the dilute hydrogen should be rather weak. Therefore, if the relaxation mechanism were the same (spin diffusion of the energy to a molecular hydrogen relaxation center) the $T_1$ for the doublet should, if anything, be longer than that of the bonded hydrogen. Since the $T_1$ of the doublet is actually shorter, there must be a different relaxation mechanism. Specifically, there must be a relaxation center close to the hydrogen doublet that is more efficient than spin diffusion to molecular hydrogen. A paramagnetic "impurity", such as a silicon dangling bond, is one possibility, but in this case one must explain why this paramagnetic center does not also shorten the $T_{1d}$ of the doublet and why the doublets occur preferentially near such a relaxation center.

A second possibility is thermally activated local motion of the hydrogen pairs. The simplest case is the BPP-type relaxation, in which the atoms jump randomly between different sites due to thermal activation. The local dipolar field experienced by the nuclei is modulated, and therefore induces relaxation of the nuclei. If one assumes a thermally activated hopping of hydrogen atoms, the hopping time $\tau$ can be expressed as

$$\tau^{-1} = \tau_0^{-1} e^{-E_a/kT}, \tag{4}$$

where $E_a$ is the activation energy and $\tau_0$ is a pre-factor that is, on the order of a phonon frequency.

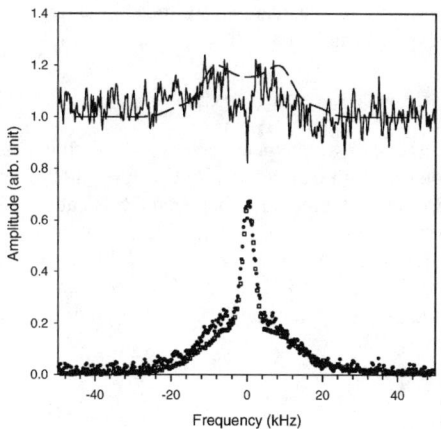

**Figure 3.** $^1$H NMR lineshapes from the light-soaked sample at different temperatures and at a magnetic field $B_0 = 3.5$ T. Solid circles represent data at 7 K taken with $T_d = 5$ s. Open circles represent data at 20 K taken with $T_d = 0.5$ s. The solid line represents the difference between the data at 7 K and 10 K. The dashed line is a fit using a Pake doublet powder pattern broadened isotropically by about 2 kHz, with a splitting between the two divergences of 16 kHz. See text for details.

In the BPP model $T_1$ is given as

$$T_1 \propto \left(\frac{\omega_0}{\Delta\omega}\right)^2 \tau_0 e^{E_a/kT}, \tag{5}$$

where $\Delta\omega$ is on the order of the dipolar linewidth and $\omega_0$ is the Zeeman frequency. In our case, eq. 5 does not appear to hold because the dependence of $T_1$ on $\omega_0$ is weaker than quadratic. In addition, for reasonable choices of $\Delta\omega$ and $\tau_0$ the observed values of $T_1$ imply that the doublet lineshape should be motionally narrowed, which is also not the case. Earlier works have suggested that if there exists a distribution of $\tau$, which is very likely in disordered materials, then one might be able to fit the observed temperature dependence of $T_1$ consistently [4]; however, even if this is the case, the observed lineshape will be a sum of the lineshapes with hydrogen atoms at different sites, and one would again expect a change in lineshape with temperature. Therefore, if the hydrogen pairs are undergoing local motion, it must be some form of correlated motion that maintains the distance between the two hydrogen atoms in a pair.

The second issue is the absence of motional narrowing of the doublet lineshape. From $T = 6.5$ K up to $T = 20$ K, the doublet lineshape remains unchanged. In the case of the random hopping of the atoms, motional narrowing of the NMR line occurs when $\tau = (\Delta v)^{-1}$, where $\Delta v$ is the linewidth. For the doublet, $\Delta v = 16$ kHz, which gives a value of $\tau$ between $10^{-5}$ and $10^{-4}$ s. Our calculation shows that the temperature at which the narrowing should occur is below 6 K, but we observe no narrowing between 6.5 K and 20 K.

The identification of a microscopic mechanism for the spin lattice relaxation of the hydrogen doublet will require further experimental data.

## SUMMARY

We have investigated the temperature dependence of the paired hydrogen site that is associated with light-soaking. We find that the spin-lattice relaxation time of these paired hydrogen atoms strongly depends on temperature between 6.5 K and 20 K. This dependence cannot be explained by a simple model assuming that hydrogen atoms are hopping randomly between different sites. For temperatures between 6.5 K and 20 K, we do not observe motional narrowing. A subtraction of the spectra at 7 K and 20 K provides an approximate lineshape for the site associated with light soaking and supports the attribution of this site to a pair of hydrogen atoms, which are approximately 2.3 Å apart.

## ACKNOWLEGEMENTS

The authors thank H. M. Branz, P. A. Fedders, and C. G. van de Walle for helpful discussions. The research at the University of Utah is supported by NREL under subcontract ADJ-2-30630-23 and by NSF under grant number DMR-0073004.

## REFERENCES

1. T. Su, P. C. Taylor, G. Ganguly, and D. E. Carlson, Phys. Rev. Lett. **89**, 015502 (2002).
2. D. E. Carlson, K. Rajan, and D. Bradley, in *Proc. 26th IEEE Photovoltaic Specialists Conf.* (IEEE, New York, 1997) pp. 595.
3. J. Jeener and P. Broekaert, Phys. Rev. 157, 232 (1967).
4. T. Su, S. Chen, R. S. Crandall, and A. H. Mahan), Phys. Rev. B **62**, 12849 (2000).
5. J. B. Boyce and S. E. Ready, Physica B **170**, 305 (1991); J. B. Boyce, S. E. Ready, M. Stutzmann, and R. E. Norberg, J. Non-Cryst. Solids, **114**, 211 (1989).
6. L.R. Lichty, J-W. Han, R. Ibanez-Meier, D. R. Torgeson, and R. G. Barnes, Phys. Rev. B **39**, 2012 (1989).

# Evidence For Trap-Conversion Induced Instability In Amorphous Silicon

Vikram L. Dalal, Puneet Sharma and Abdul Aziz
Dept. of Electrical and Computer Engr. and Microelectronics Research center
Iowa State University, Ames, Iowa 50011, USA

## ABSTRACT

It has been shown recently that there are two distinct types of recovery during annealing of amorphous Silicon after degradation due to light soaking. It has been postulated that the two different kinetics of annealing point to the existence of two different types of states, with perhaps one state being charged dangling bonds and the other state being neutral dangling bonds. To see if two kinds of states exist, in this paper, we study the kinetics of degradation within the first 100 seconds, and also study the entire absorption curve at all degradation times. An analytical model is derived for early time degradation based on the conversion of a $D^-$ state into a neutral dangling band by absorption of a light generated ( the trap-to-dangling bond conversion model of Adler) and the experimental data of degradation versus light intensity fit the predictions of the model very well. The model also predicts that the Adler-type negatively charged defect states, which have a negative correlation energy, upon conversion will transform into Do states at a higher energy, and therefore, there should be a decrease in absorption corresponding to states closer to the valence band, and an increase in absorption corresponding to states near the mid-gap. For the films where such $D^-$ states are deliberately introduced by using a small oxygen ( a donor atom) leak , we see strong evidence for such a behavior in absorption, with a decrease in the 1.3-1.4 eV photon energy range, and an increase in the 1.1 eV photon energy range. The increase in Do corresponds well with the decrease in photo-conductivity, even at the earliest times.

## INTRODUCTION

It has been known for some time that two types of states are created in a-Si:H upon degradation, the "fast" states and "slow" states, the designation being based on whether they anneal rapidly or slowly upon thermal annealing [1]. Recently, Heck et al[2], Stradins [3] and Wronski et al[4] have shown that indeed, the behavior of annealing of photo-conductivity and of fill factors of solar cells indicates that there are two distinct types of states in the material. In particular, one does not find a one-to-one inverse correlation between midgap defect density measured by absorption at ~1.2 eV below the conduction band and photo-conductivity, as the standard single defect model would suggest. It has been postulated that perhaps the fast states are charged defects. Adler proposed many years ago that there were charged defects present in the material because of structural inhomogeneities in the material, and that these charged defects were negatively correlated in energy, i.e. the $D^-$ state was below the Do or $D^+$ state [5]. He further proposed that the $D^-$ and $D^+$ states would convert into neutral Do dangling bonds, at a different energetic location, upon light excitation by absorbing a hole or an electrons respectively, and that this neutral bond would be a more effective recombination center for electrons, thus explaining the observed decrease in photo-conductivity. In a previous paper, Dalal showed that early behavior of photo-conductivity did not follow the predictions of the bond-breaking model, but could be explained by postulating the Adler model [6]. In this paper, we examine the kinetics of Adler's trap-to-dangling conversion model, and show that the predicted kinetics can be fitted very well with the experimental data on early degradation.

Another prediction of the Adler model would be that the shape of the subgap absorption curve would change upon degradation, with a decrease in the region where the D⁻ bonds exist, and an increase in the region where one finds Do bonds. We will show that our experimental data indeed show such a change in the shape of the absorption curve when the film contains an increased density of charged D⁻ bonds, created by deliberately introducing excess oxygen into the sample.

## EXPERIMENTAL TECHNIQUES

The samples used in this study were grown using the remote, low pressure ECR plasma technique previously described in detail[7]. All the samples were grown using either hydrogen or helium dilution. The substrate temperatures were about ~325 C. The growth rates were in the 1-1.5 A/s range, and film thicknesses were in the range of ~ 1 micrometer. To guard against false results arising from surface recombination effects, a novel film structure, consisting of a sandwich structure with a-Si:H film being sandwiched between two thin, larger gap a-(Si,C):H layers was used. The surface a-(Si,C):H layers, because of their larger bandgap, serve to drive electrons and holes away from the front and back surfaces. Subgap absorption coefficients were measured using a double-beam photo-conductivity technique. The photo-conductivity degradation was done using either ELH bulb or focused xenon light, filtered with an a-(Si,C):H film which only allowed red photons to pass through. The intensity of the filtered xenon source at the sample surface was adjusted to correspond to 2 x sunlight. The intensity of the ELH lamp used for kinetic studies was adjusted by using appropriate neutral density filters and by changing the geometrical arrangement of the source.

## RESULTS

In Fig. 1, we show the subgap absorption data in state A (annealed) for an a-Si:H film (2/6339) deposited without using any deliberate oxygen leak. The film had a very low of ppm B doping to compensate for any accidental oxygen doping[7]. The Urbach energy is ~45 meV. The Tauc gap was 1.71 eV, and E04 gap was 1.87 eV. The activation energy was 0.84 eV and the photo-conductivity was $8 \times 10^{-5}$ S/cm under AM1.5 illumination. The photo/dark conductivity ratio was ~4 x $10^5$. Then the film was subjected to light soaking under xenon lamp for different times, and the entire subgap absorption spectrum was remeasured at various times. In Fig. 1, we also show the changes in subgap absorption at various times. As expected, the absorption increases in the region of 1.1 eV below the conduction band upon light soaking even for 10 min light soaking.

Next, to introduce donors and D⁻ states, we deliberately made a film with an air leak but no ppm B compensation. The deliberate air leak results in a change in base pressure of the system from $3 \times 10^{-7}$ Torr to $2 \times 10^{-6}$ Torr. The resulting film was clearly more n-doped, as indicated by a smaller activation energy (0.64 eV) and a much higher photo-conductivity (3.5 x $10^{-4}$ S/cm). This is to be expected since oxygen is a donor state in a-Si:H. This film was then subjected to the same light soaking tests. The relevant absorption curves are shown in Fig.2 for the annealed and various light-soaked states. Very clearly, now, one can observe a decrease in absorption of ~ 40 % at ~1.3 eV and an increase at ~1.1 eV of ~70%. See Fig. 3. Thus, *there is change in the shape of the curve after light soaking*. The change in absorption implies a decrease in defect density at 1.3-1.4 eV below the conduction band and an increase at ~1.1 eV. Since the film was n-doped, clearly, D⁻ states were created during oxygen doping. From the data, one can infer that the D⁻ state may be at ~ 1.3-1.4 eV below the conduction band, and the Do state ( into which D⁻ converts upon trapping of a light-generated hole) is at ~1.1 eV below the conduction band. Thus, these D⁻ and Do states have a negative correlation energy, as postulated by Adler.

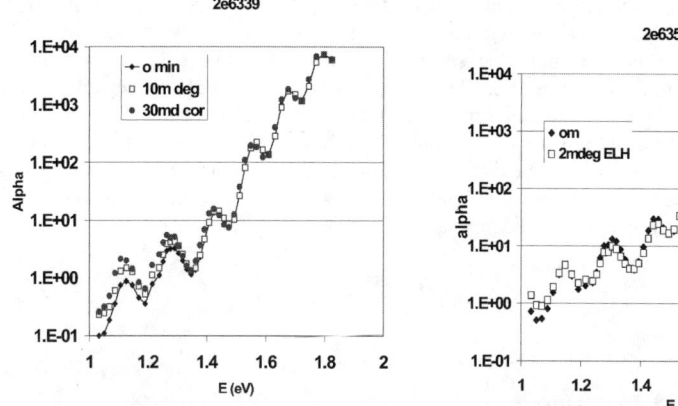

Fig. 1 Absorption of a-Si:H film without $O_2$ leak At different degradation times

Fig. 2 Absorption of a-Si:H film prepared with $O_2$ leak in annealed and 2 min degradation

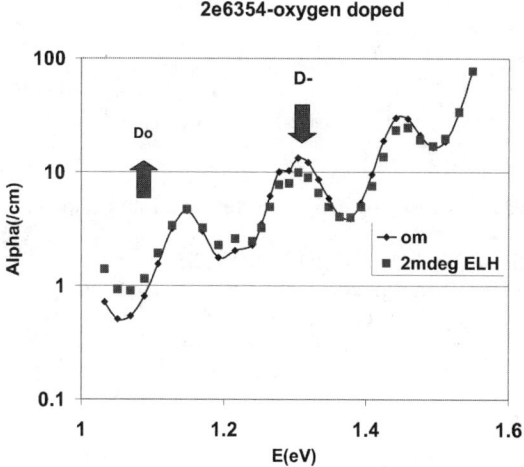

Fig. 3. Expanded view of Fig.2, clearly showing a decrease in absorption in 1.3-1.4 eV range and an increase in 1.1 eV range.

To show that as the density of D⁻ traps decreases, one sees less of an effect of trap conversion, a film (2/6353) was also grown with the oxygen leak, but with ppm B doping to compensate the oxygen donors. As expected, the Fermi level dropped to 0.79 eV below the conduction band, and the photo-conductivity decreased to $9 \times 10^{-5}$ S/cm. The Fermi level was higher than for the case of the film without any oxygen leak, and thus we expect that some D⁻ traps were present, though a smaller density than in film 2/6354. In Fig. 4, we show the absorption spectrum for this film (2/6353), and it clearly shows a hint of a negative change in absorption at ~1.3-1.4 eV photon energy, and an increase at ~ 1.1 eV photon energy. As expected, the decrease in the 1.3-1.4 eV range is less than for the film with higher n- doping (2/6354). Thus, the results from the three films confirm that introducing donor states leads to a trap-to-dangling bond conversion upon light soaking, with a decrease in deeper states (~ 1.3-1.4 eV below $E_c$) and an increase in midgap states (~1.1 eV below $E_c$).

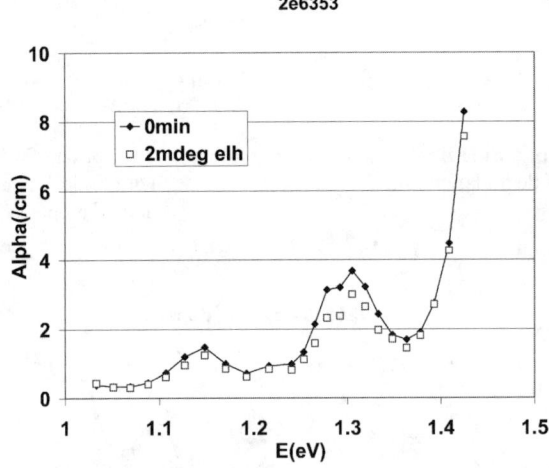

Fig. 4 Decrease in defect density upon light soaking in a film compensated with ppm levels of B

KINETICS OF THE TRAP CONVERSION PROCESS

One can develop a simple model for trap-to-dangling bond conversion as follows:
Let the density of Do states be D at any given time, and Do at time t=0
Let the density of D⁻ states be N at time t, and No at t=0.
Let the capture cross-section for holes of D⁻ state be σ- . Obviously, a D⁻ cannot capture an electron.
The photo-conductivity pc can be written as:
pc = qG $\mu_n \tau_n$ , where $\mu_n$ is the mobility, $\tau_n$ is the lifetime and G is the generation rate.
$\tau_n$ = 1/(vσD) where σ is the capture cross-section for electrons by D states.
Upon absorption of a hole, the rate at which the N states convert into D states is given by:
dN/dt = - cpN = -dD/dt ............... (1)
where c is a trap-to-dangling bond conversion rate constant.

Now, $p = G\tau_p$, and lifetime of holes, $\tau_p = 1/[v\sigma_p(D + aN)]$, where a is the ratio of capture-cross-sections between N and D states and $\sigma_p$ is the capture-cross-section of holes by Do states.
Thus, $dN/dt = -cNG/[v\sigma_p(D + aN)]$ \hfill (2)
But $D = Do + (No-N)$
Therefore, $dN/dt = -cNG/[v\sigma_p\{Do+No+(a-1)N\}]$ \hfill (3)
Using dimensionless variables, $d = Do/No$ and $n_t = N/No$, one gets:
$[(d+1)(1/n_t) +(a-1)](dn_t/dt) = -cG$ \hfill (4)
The solution is:
$(d+1)\ln(n_t-1) +(a-1)(n_t-1) = -cGt$ \hfill (5)
Notice from equation (5) that the time required to create a certain $n_t$, starting with a No ($n_t=1$), is <u>inversely</u> proportional to the generation rate G. If the neutral dangling bonds created by trap conversion dominate the Do created by bond breaking, which is clearly the case for oxygen doped samples at very early degradation times, then the decay in photo-conductivity should be controlled by how many $N_t$ states have converted into Do, i.e. by ($n_t$-1). Since, for a given ($n_t$-1), the time required to cause this change is inversely proportional to intensity, therefore, the time required to cause a given photo-conductivity decay should be inversely proportional to intensity.

In Fig. 5, we plot the experimental decay curves for the oxygen leak-doped sample for two intensities, 1 sun and 0.48 sun. Then, we replot the 0.48 sun data by invoking the reciprocity of time and intensity, and the result is the curve represented by the symbol Δ. The time-converted curve maps almost perfectly at early times ( when few additional dangling bonds have been generated by the bond-breaking model) onto the experimental curve for 1 sun degradation, thus proving the kinetic relationship postulated in Eqn. 5.

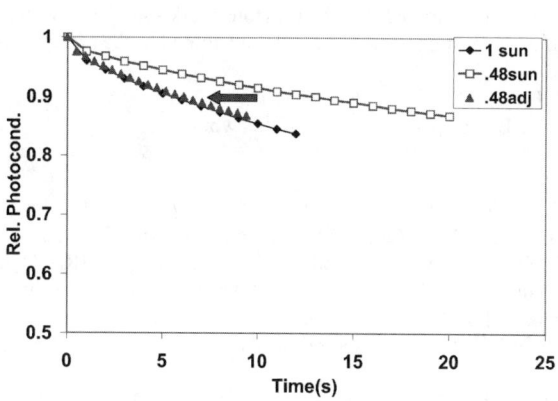

Fig. 5 Transient decay curves for a sample showing the inverse relationship between Intensity and time for a given amount of degradation. Equation(5) is verified.

Note that this is not the only sample where such behavior was observed. In Fig. 6, we show the kinetic data for another sample, this time deposited with He dilution, also with a slight oxygen leak, which also matches the time-intensity product behavior predicted by Eqn. 5. This sample also showed the decrease in subgap alpha at ~ 1.3 eV.

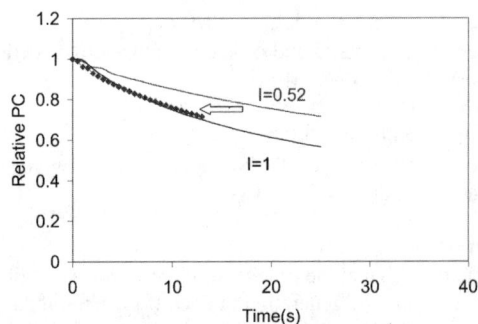

Fig. 6 Degradation kinetics of another a-Si:H sample made with He dilution measured under two intensities, showing the reciprocity of intensity and time to create a certain degradation.

CONCLUSIONS

From the experimental data, we conclude that there are two types of states present in a-Si:H, a charged dangling bond, and neutral dangling bond. An analytical model has been developed to explain the conversion of charged bonds into neutral bonds upon mono-molecular trapping, and the intensity behavior matches the analytical model very well. An increase in oxygen concentration, which creates more D⁻ states, leads to more pronounced decrease in absorption at ~1.3 eV, implying that the negatively charged, negatively correlated dangling bond at that energy has converted into a neutral dangling bond at ~ 1.1 eV below the conduction band. This result matches the predictions of Adler's model. Since oxygen is a common accidental donor in a-Si:H, it is not surprising that the previous work found a non-linear behavior between midgap alpha and inverse of photo-conductivity, attributed to the presence of a charged bond.

ACKNOWLEDGEMENTS

We thank NREL for partial support of this work.

REFERENCES
1. L.Yang and L. F. Chen. Proc. Of Mater. Res. Soc., 297,619(1993)
2. S. Heck and H. Branz, Proc. Of Mater.Res.Soc., 664,Paper 12.2(2001)
3. P. Stradins, S. Shimizu, M. Kondo and A. Matsuda, Proc. Of MRS, 664, Ppaer 12.1(2001)
4. J. Pearce, X. Niu, R. Koval, G. Ganguly, D. Carlson, R. W. Collins and C. Wronski, Proc. Of MRS, 664, Paper 12.3(2001)
5. D. Adler, in "Semiconductors and Semimetals", Ed. J. Pankove, (Academic Press)Vol. 21A,p. 291 (1984)
6. V. L. Dalal, Proc. of Amer. Inst. of Phys., 157, 249(1987)

## Evolution of Charged Gap States in *a*-Si:H Under Light Exposure

M. Zeman[1], V. Nádaždy[1], R.A.C.M.M. van Swaaij[1], R. Durný[2], J.W. Metselaar[1]
[1]Delft University of Technology – DIMES, P.O. Box 5053, 2600 GB Delft, The Netherlands
[2]Slovak University of Technology, Ilkovičova 3, 812 19 Bratislava, Slovakia

## ABSTRACT

The charge deep-level transient spectroscopy (Q-DLTS) experiments on undoped hydrogenated amorphous silicon (*a*-Si:H) demonstrate that during light soaking the states in the upper part of the gap disappear, while additional states around and below midgap are created. Since no direct correlation is observed in light-induced changes of the three groups of states that we identify from the Q-DLTS signal, we believe that we deal with three different types of defects. Positively charged states above midgap are related to a complex formed by a hydrogen molecule and a dangling bond. Negatively charged states below midgap are attributed to floating bonds. Various trends in the evolution of dark conductivity due to light soaking indicate that the kinetics of light-induced changes of the three gap-state components depend on their initial energy distributions and on the spectrum and intensity of light during exposure.

## INTRODUCTION

Inherent to hydrogenated amorphous silicon (*a*-Si:H) are the reversible changes in electronic properties of *a*-Si:H under light exposure. This is known today as the Staebler-Wronski effect (SWE) [1]. Since the observation of the SWE, a large effort has been put into obtaining understanding of the processes that cause the structural and opto-electronic light-induced changes in *a*-Si:H. It is generally accepted that light soaking leads to the creation of additional dangling-bond defects [2]. However, many unresolved issues regarding the SWE still remain, such as the exact role of hydrogen, the influence of local bonding configurations, and the presence of more than one metastable defect.

We present experimental results from the charge deep-level transient spectroscopy (Q-DLTS) [3] that reveal a surprising behavior of the gap states in *a*-Si:H during light soaking. The results of the Q-DLTS experiment show that prior to the creation of the defect states around midgap there is an initial decrease of the density of gap states located above midgap. This observation indicates that in the early stage of light soaking, annihilation of some type of defects takes place. Combining our results from the Q-DLTS experiments with recent results from the nuclear magnetic resonance (NMR) experiments [4], *ab initio* pseudopotential calculations [5,6], and molecular dynamics simulations (MDS) [7,8] we propose microscopic atomic configurations that introduce charged gap states in *a*-Si:H.

## Q-DLTS TECHNIQUE

The DLTS technique has proved to be a powerful and straightforward method to analyze gap states in *a*-Si:H [9,10]. The advantage of the DLTS technique in comparison to other techniques that are used to study the gap states in *a*-Si:H, such as electron-spin resonance or the constant photocurrent method is that the raw DLTS spectra give direct information about the energy distribution of gap states (EDOS). In addition, the DLTS signal is very sensitive to changes in the EDOS. A modification to the DLTS that can be applied to *undoped* a-Si:H samples is the Q-DLTS [11,12].

It was demonstrated that the Q-DLTS spectra of undoped a-Si:H changed as a function of the Fermi level position in a-Si:H [11]. The authors identified three components in the Q-DLTS spectra with peaks at about 320, 390 and 430 K. These components were related to three gap-state distributions with peaks at energies of 0.63, 0.82, and 1.25 eV with respect to the mobility edge of the conduction band, respectively. These gap-state distributions were related to positively charged, $D_h$, neutral, $D_z$, and negatively charged, $D_e$, defect-state distributions, respectively, as predicted by the defect-pool model (DPM) [13,14]. Recent comparison of the Q-DLTS and ESR measurements [12] has shown that the defect-state distribution with the peak at 0.82 eV corresponds to the ESR centers with g = 2.0055, which are associated with singly occupied dangling-bond defects.

## Q-DLTS SPECTRA OF UNDOPED a-Si:H UNDER LIGHT EXPOSURE

We carried out the Q-DLTS experiment on a 1-μm thick a-Si:H layer deposited on an n$^+$-type single crystalline Si wafer. The Q-DLTS directly detects the thermally emitted carriers by integrating the current in the external circuit. For successful Q-DLTS experiments on undoped a-Si:H a very thin insulating layer has to be created in the surface region of a-Si:H [11]. This insulating layer reduces the leakage current density of the test structure below $10^{-5}$ Acm$^{-2}$ even at a temperature of 450 K, which is negligible with respect to transient charge. A MOS test structure was prepared by evaporation of an Al semitransparent (≈35%) top electrode. Prior to each light soaking experiment we created an equilibrium EDOS in the a-Si:H by annealing the structure at a temperature of 500 K for 15 minutes. Light soaking of a-Si:H was carried out at room temperature through the semitransparent electrode using a He-Ne red laser ($\lambda$ = 633 nm) with 120 mW/cm$^2$. This wavelength was chosen to obtain an uniform absorption profile in the whole a-Si:H layer and to ensure that we observe degradation in the bulk. In one particular case we used white light with a high irradiation intensity of 4 W/cm$^2$ in order to measure degradation after heavy light soaking. A bias voltage $U_b$ = −3V, excitation pulses $\Delta U$ = 6V, and a rate window of 100 s$^{-1}$ were used in the Q-DLTS experiments.

Figure 1 shows a Q-DLTS signal for a-Si:H in the annealed state. A good matching between the measured and simulated Q-DLTS signal was obtained using the EDOS, which is shown in figure 2. The EDOS was calculated using the DPM [14] and the contributions to the total EDOS from the $D_h$, $D_z$, and $D_e$ states are included in figure 2. The DPM parameters used in the simulations are reported in figure 2. The simulations of the Q-DLTS signal clearly relate the low and high temperature part of the Q-DLTS spectra with the $D_h$ and $D_e$ states, respectively. Though we argue in this article that the charged $D_h$ and $D_e$ gap states do not represent dangling-bond defects as implied by the defect-pool models, we retain the notation of $D_h$ and $D_e$. This notation gives a satisfactory description of positively and negatively charged gap states.

The evolution of the Q-DLTS signal measured after light soaking is shown in figures 3 and 4. The arrows in figures 3 and 4 indicate the peak positions of $D_h$, $D_z$, and $D_e$ gap-state distributions in the Q-DLTS spectra. The Q-DLTS spectra measured after short exposure times (up to 30 s) clearly show that at low temperatures the signal decreases while the rest of the spectra remains unchanged (curves 2 and 3 in figure 3). The decay of the signal at low temperatures, which reflects the decrease of the $D_h$ states, means that particular defects are being annihilated.

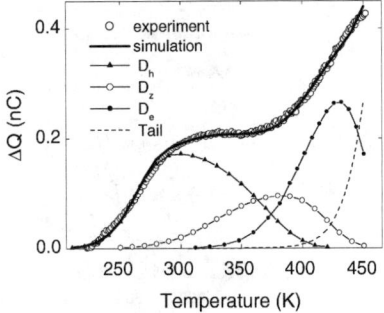
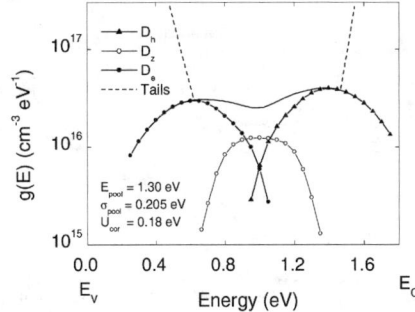

**Figure 1**. The Q-DLTS signal from $a$-Si:H in the annealed state. The contributions of the valence band tail and $D_h$, $D_z$, and $D_e$ states to the signal are included.

**Figure 2**. The EDOS in $a$-Si:H in the annealed state including the tails and the $D_h$, $D_z$, and $D_e$ gap-states distributions.

We refer to this part of light soaking as the first stage. We have to stress that only after the initial decay of the low-temperature part of the Q-DLTS spectra, the signal starts to increase at higher temperatures. In this second stage of light soaking we observe an enhancement of the Q-DLTS signal for both $D_z$ and $D_e$ components, while the low-temperature part continues to decrease, as demonstrated by curve 4 in figure 3. It should be noted that when a high light intensity is used in the experiment, the transition between the first two stages is difficult to distinguish. In the third stage of light soaking, the Q-DLTS spectra are dominated by an increase of the signal corresponding to the $D_z$ states, while the other two components do not change significantly (curves 2 to 4 in figure 4). Finally, we recognize a fourth stage of light soaking, in which one can observe an increase of the Q-DLTS signal for both $D_z$ and $D_e$ components and a shift of the peaks of $D_z$ ($D_e$) components towards a higher (lower) temperature (curve 5 in figure 4).

## ORIGIN OF CHARGED GAP STATES

The direct observation that the initial decrease of the $D_h$ component is not accompanied by an increase in the $D_z$ or $D_e$ components is remarkable and has serious implications for the understanding of the SWE. This finding suggests that the decrease of $D_h$ states is not a transition between two charged states of dangling-bond defects. Further, we do not observe a direct correlation in light-induced changes of all three components that we identify from the Q-DLTS spectra. Therefore, we believe that we deal with three different types of defects and the question arises which microscopic atomic configurations introduce the $D_h$ and $D_e$ states. We exclude that impurity atoms such as oxygen or carbon are the origin of the gap states as demonstrated by Kamei et al. [15] and consider only atomic configurations including Si and H atoms to be candidates for introducing the $D_h$ and $D_e$ gap states. We also rule out Si-H bonds as candidates for the $D_h$ and $D_e$ states since the Si-H bonds do not introduce states into the band gap.

An alternative approach is to consider a hydrogen molecule as a candidate in forming structural complexes that can introduce states into the band gap. This approach is supported by

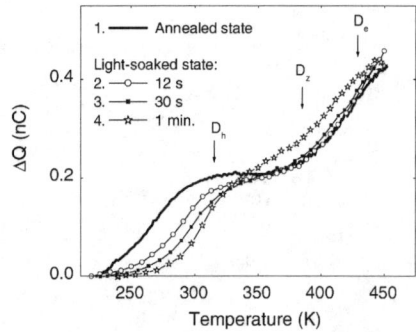 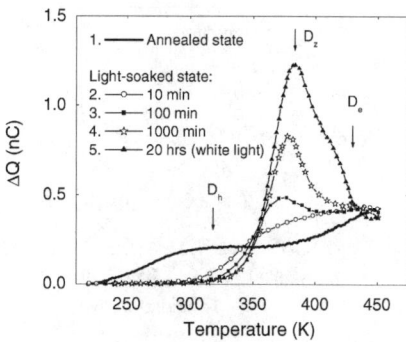

**Figure 3**. The Q-DLTS spectra of undoped a-Si:H for the annealed state (1) and after light soaking for three different times: 12s (2), 30s (3), 1 min (4).

**Figure 4**. The Q-DLTS spectra of undoped a-Si:H for the annealed state (1), after light soaking for three different times: 10 min (2), 100 min (3), 1000 min (4), 20 hrs (5).

recent proton NMR and spin echo double resonance experiments [4] that give convincing evidence that in high-quality a-Si:H nearly 40% of the incorporated hydrogen is in the form of hydrogen molecules. These hydrogen molecules are individually trapped in an amorphous equivalent to a tetragonal $T$ site. Further, *ab initio* pseudopotential calculations of hydrogen interaction with a-Si by Van de Walle and Tuttle [5] showed that the existence of an over-coordinated complex created with two hydrogen atoms placed in the dangling-bond region is possible. This complex introduces a deep level in the upper part of the band gap. The additional calculations by Van de Walle and Tuttle [5] indicate that also a configuration, in which a hydrogen molecule is placed near a dangling bond, is close in energy to the complex with two hydrogen atoms. Being aware of these facts we propose a microscopic configuration, in which a positively ionised hydrogen molecule forms a hybrid bond with a dangling bond, that introduces energy levels into the upper part of the band gap. These energy levels correspond to the positively charged $D_h$ states, which we measure as a low temperature Q-DLTS signal.

We propose that the defect states below midgap (the $D_e$ states observed by Q-DLTS) are induced by floating-bond defects, although the floating-bond defects have not been directly observed experimentally yet. Recent finite-temperature MDS of a supercell performed by Su and Pantelides, which demonstrate that in the annealed state the a-Si:H supercell contains both dangling bonds and floating bonds [7], support the existence of floating bonds in a-Si:H. Further, Peressi *et al.* carried out *ab initio* pseudopotetial calculations of the electronic structure in a-Si:H and clearly identified both dangling- and floating-bond defects, which both induced gap states [6]. However, Fedders showed recently that when using larger supercells in the MDS the number of defects generated in the supercell substantially decreased [16]. Biswas *et al.* performed the tight-binding MDS of network rebonding of a-Si:H under light soaking and showed that light exposure caused creation of floating bonds, which induced deep levels into the gap localized close to the valence band [8]. In contrast to Biswas *et al.*, who has not considered floating bonds in the network before light soaking, the Q-DLTS experiments suggest that floating bonds are present in the network already *before* light soaking, as indicated by the simulations [6,7].

# EVOLUTION OF DARK CONDUCTIVITY UNDER LIGHT EXPOSURE

The observed initial elimination of the $D_h$ states during light soaking can easily explain the initial drop in the dark conductivity, $\sigma_d$, of $a$-Si:H as reported e.g. by Staebler and Wronski [17]. Simulations with the ASA software package [18] confirmed that the decrease of the $D_h$ state density, while keeping the density of $D_z$ and $D_e$ states constant, naturally leads to a shift of the Fermi level towards the valence band mobility edge. We calculated the shift of the Fermi level due to light soaking for $a$-Si:H using the equilibrium EDOS as shown in figure 2. The light-soaked EDOS was obtained by decreasing the density of $D_h$ states by 20% without changing their energy distribution. This decrease was inferred from the difference in Q-DLTS signal intensity at 320 K between the spectra in the annealed state and after 30 s of light soaking (curve 3 in figure 3). The mobility gap of $a$-Si:H was taken 1.80 eV and the electron and hole extended-state mobility was 10 and 3 cm$^2$/Vs, respectively. The simulations show that the Fermi level drops from 0.78 eV to 0.85 eV with respect to the mobility edge of the conduction band for annealed and light-soaked state, respectively. Such a shift in the Fermi level corresponds to a decrease of the simulated $\sigma_d$ from $5\times10^{-11}$ ($\Omega$ cm)$^{-1}$ to $3\times10^{-12}$ ($\Omega$ cm)$^{-1}$ for the annealed and light-soaked state, respectively. It is not possible to achieve such a shift in the Fermi level by increasing only the density of neutral $D_z$ states. The simultaneous increase of both the $D_h$ and $D_z$ state density cannot shift the Fermi level more towards the valence band than in the case of increasing the $D_z$ state density alone. This finding is in contradiction with the attempts to explain the changes in opto-electronic properties of $a$-Si:H by an increase of the positively charged defect states above midgap [19] during the degradation.

The simulations also revealed that during light soaking various trends in the evolution of the $\sigma_d$ could be expected, even the non-monotonous behavior of the $\sigma_d$ as a function of light exposure time. The evolution of the $\sigma_d$ depends on the individual changes of $D_h$, $D_z$, and $D_e$ gap-state distributions during light soaking and their mutual proportions. Since these gap-state distributions are supposed to be introduced by different types of defects, the kinetics of changes in the $D_h$, $D_z$, and $D_e$ gap-state distributions are expected to be dependent on the light spectrum and intensity. In fact, we measured different behavior of the $\sigma_d$ of $a$-Si:H during light soaking with a semiconductor red laser ($\lambda$ = 670 nm) using two different intensities. Figure 5 shows the time evolution of the $\sigma_d$ for three $a$-Si:H samples, which differ in the annealed $\sigma_d$. Figure 5 demonstrates that the evolution of the $\sigma_d$ during light soaking depends not only on the light intensity but also on the annealed $\sigma_d$. In any case, we do not observe such a large change in the measured $\sigma_d$ in the initial stage of light soaking as compared to the simulated drop of $\sigma_d$, which was more than one order of magnitude. Closer evaluation of the Q-DLTS experiments revealed that the decrease of the density of $D_h$ states occurs asymmetric, which was not taken into account in the simulations.

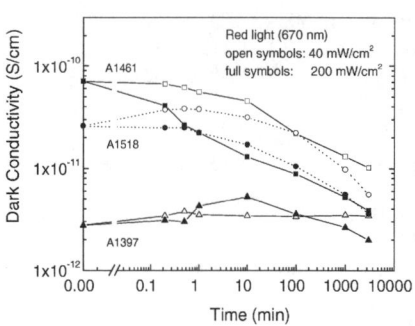

**Figure 5**. The measured $\sigma_d$ as function of exposure time for two different intensity levels of red-light illumination.

## CONCLUSIONS

We have presented the evolution of the Q-DLTS spectra during light soaking of $a$-Si:H. The Q-DLTS experiments demonstrate that during light soaking the states in the upper part of the gap disappear, while additional states around and below midgap are created. Since we have not observed a direct correlation in light-induced changes of the three groups of states that we identify from the Q-DLTS signal, we believe that we deal with three different types of defects. We propose that the positively charged defect states above midgap are mainly induced by "hydrogen molecule – dangling bond" complexes and the negatively charged defect states below midgap by floating-bond defects. Various trends in the evolution of $\sigma_d$ due to light soaking indicate that the kinetics of light-induced changes of the three gap-state components depend on their initial energy distributions and on the spectrum and intensity of light during exposure.

## ACKNOWLEDGEMENTS

The authors acknowledge the financial support of Delft University of Technology under the DIOC program and the VEGA grant agency of Slovak Republic under Project No. 2/1013/21.

## REFERENCES

1. D.L. Staebler and C.R. Wronski, *Appl. Phys. Lett.* **31**, 292 (1977).
2. H. Fritzsche, *Annu. Rev. Mater. Res.* **31**, 47 (2001).
3. I. Thurzo, V. Nádaždy, *Properties of amorphous silicon and its alloys*, ed. T.M. Searle (IEE Publishing, London, 1998).
4. P. A. Fedders, D. J. Leopold, P.H. Chan, R. Borzi, and R.E. Norberg, *Phys. Rev. Lett.* **85**, 401 (2000).
5. CH. G. Van de Walle and B. Tuttle, in *Amorphous and Heterogenous Silicon Thin Films: Fundamentals to Devices - 1999*, edited by H. Branz et al., (Mater. Res. Soc. Proc. No. **557**, Warrendale, 1999) p. 275.
6. M. Peressi, M. Fornari, S. De Gironcoli, L. De Santis and A. Baldereschi, *Philos. Mag. B* **80** (4), pp. 515-521 (2000).
7. Y.-S. Su and S.T. Pantelides, *Phys. Rev. Lett.* **88**, 165503 (2002).
8. R. Biswas, B.C. Pan, and Y.Y. Ye, *Phys. Rev. Lett.* **88**, 205502 (2002).
9. D.V. Lang, J.D. Cohen, and J.P. Harbison, *Phys. Rev. B* **25**, 5285 (1982).
10. J.D. Cohen, D.V. Lang, *Phys. Rev. B* **25**, 5321 (1982).
11. V. Nádaždy, R. Durný, and E. Pinčík, *Phys. Rev. Lett.* **78**, 1102 (1997).
12. V. Nádaždy, R. Durný, I. Thurzo, E. Pinčík, A. Nishida, J. Shimizu, M. Kumeda, T.Shimizu, *Phys. Rev. B* **66**, 195211 (2002).
13. M.J. Powell and S.C. Dean, *Phys. Rev. B* **48**, 10 815 (1993).
14. M.J. Powell and S.C. Dean, *Phys. Rev. B* **53**, 10 121 (1996).
15. T. Kamei, N. Hata, A. Matsuda, T. Uchiyama, T. Amano, K. Tsukamoto, Y. Yoshita, and T. Hirao, *Appl. Phys. Lett.* **68**, 2380 (1996).
16. P.A. Fedders, *Phys. Rev. B* **66**, 195308 (2002).
17. D.L. Staebler and C.R. Wronski, *J. Appl. Phys.* **51**, 3262 (1980).
18. M. Zeman, J.A. Willemen, L.L.A. Vosteen, G. Tao, and J.W. Metselaar, *Solar Energy Materials and Solar Cells* **46**, 81 (1997).
19. S. Heck and H.M. Branz, *Appl. Phys. Lett.* **79**, 3080 (2001).

## Metamict Transformation of Silica

Ju-Yin Cheng[1], M. M. J. Treacy[2] and P. J. Keblinski[1]
[1]Department of Material Science and Engineering, Rensselaer Polytechnic Institute, Troy, New York 12180
[2]NEC Research Institute, Inc., Princeton, New Jersey 08540

### ABSTRACT

Alpha quartz can be transformed to a metamict phase in a physical reaction such as irradiation with fast ions, neutrons and electrons. As truly inspired by this fact, we design a simple experiment to produce disordered silica and at the same time watch the transformation using a transmission electron microscope. This work simply presents electron-damaged silica in dark field at Bragg reflection. Since the transformation is not complete, disordered silica (at the sample edge) and retained quartz can be compared in each image. Not surprisingly, more speckles are found in the quartz than in the disordered structure. But it is too early to say no medium-range ordering in amorphous silica because the experiment only shows one Bragg diffraction. Systematic work to complete and measure the disordering again is being pursued.

### INTRODUCTION

Silica is a zeolite abundant in Earth. Its polymorphs (from low to high atmospheric temperatures) are quartz, tridymite and cristobalite. When heated up to 1723 °C, the melting point of cristobalite, the zeolite starts to melt. The liquid then transforms to a glassy phase by passing over crystallization under cooling. The glassy phase is called vitreous silica. In very early X-ray diffraction studies [1], vitreous silica showed a sharp peak around 10–30 $nm^{-1}$. This first sharp diffraction peak (FSDP) was interpreted as the evidence of intermediate-range structure in the glass. Later on, the microcrystallite theory was redefined in modeling the structure of vitreous silica [2].

On the other hand, silica has another glassy phase. This glassy phase is produced from irradiation of crystalline silica, so called "metamict" silica. Similar to vitreous silica, metamict silica shows both the glassy and crystalline characters in electron diffraction [3]. However, metamict and vitreous silicas presumably have different structures, since the former is caused by irradiation but the latter is due to melting. According to microcrystallite models [4, 5], the structure of vitreous silica is not built upon a low-temperature polymorph such as quartz. This is because in most cases vitreous silica is transformed from cristobalite, and the reverse is true (e.g. the product of devitrification is cristobalite alone) [6]. But quartz can be considered in the construction of structure models for metamict silica.

In our work, metamict silica is produced from alpha quartz crystal by electron irradiation. As the irradiation proceeds, we measure the diffracted intensity (at the first Bragg reflection) of the sample. Disordering is initiated at the surface. The intensity in dark field from disordered regions is much lower than that from retained quartz. Although the intermediate-range structure of the metamict phase is not largely seen in the chosen dark-field condition, a topological connection to quartz is still a good rationale. Finally, we propose a damage model and discuss the structure transformation.

## EXPERIMENTAL

Silica is high-purity quartz sands (IOTA, Inc.) of equal size about 300 μm in diameter. For TEM use, the powder was dispersed in acetone alcohol, smashed to a smaller size (less than 20 μm in diameter), and collected on the holey carbon film of a copper grid. To produce disordering in quartz, one selected crystal of average diameter about 0.5 μm was irradiated with a convergent electron beam at 200 keV at room temperature in a Hitachi 9000 microscope. The dose rate, measured from the Faraday cage inside the chamber, is about $2 \times 10^3$ e/Å$^2$ per second. While the crystal was being damaged, dark-field images at the Bragg condition $g = 210$ ($k = 6.21$ nm$^{-1}$) were acquired with a Gatan MSC 794 slow scan CCD camera. It is inferred from Z-contrast images ($k \approx 20$ nm$^{-1}$) that no artifacts such as mass loss were found in the irradiation, even at a higher dose rate ($\approx 6 \times 10^4$ e/Å$^2$ per second). The irradiation experiment was done for 45 minutes.

## RESULTS AND DISCUSSION

Alpha quartz has a hexagonal lattice. Figure 1 shows thickness fringes of the crystal in dark field at the first Bragg peak $g = 210$ ($k = 6.21$ nm$^{-1}$) viewed at the zone axis (0001) during the irradiation. According to the dynamic theory of contrast, the diffracted intensity is given by [7]

$$I_g = \frac{1}{1+w^2} \sin^2(\frac{\pi t}{\xi_g/\sqrt{1+w^2}}) \propto \sin^2[\pi(\frac{t}{\xi_g})_{\it{eff}}], \qquad (1)$$

where $w = s\xi_g$. $s$ is the deviation from the reciprocal lattice point, $\xi_g$ is the extinction distance and $t$ is the thickness of the crystal. In equation (1), the dark-field intensity is a sine-wave function of $(t/\xi_g)_{\it{eff}}$. As seen in figure (a), the breadth and amplitude of the sine wave change with thickness due to the absorption effect. In a few minutes of the damage, disordering begins at the edge and surface regions (including top and bottom) in the crystal, as shown in figures (b) and (c). Obviously, electron beams scatter with quartz far more strongly than with disordered SiO$_2$, so the damaged edge seems disappearing in dark field but in fact it is still there (see figure (f)). Here we neglect any diffraction from the carbon film. Near the finish of the irradiation, but not quite rendering the crystal to a complete disorder, no distinct thickness fringes are observed in figures (d) and (e). This is because of the decreasing $(t/\xi_g)_{\it{eff}}$ as well as the overlapping with disordered layers. As a result, it is very hard to obtain lattice information for the partially disordered quartz with a reduced thickness.

We are interested in two areas on these dark-field images: the sample edge, where speckles due to disordering evolve, and the first fringe (persistent evidence of the crystal). The irregular shape of the sample adds difficulties on image analysis.

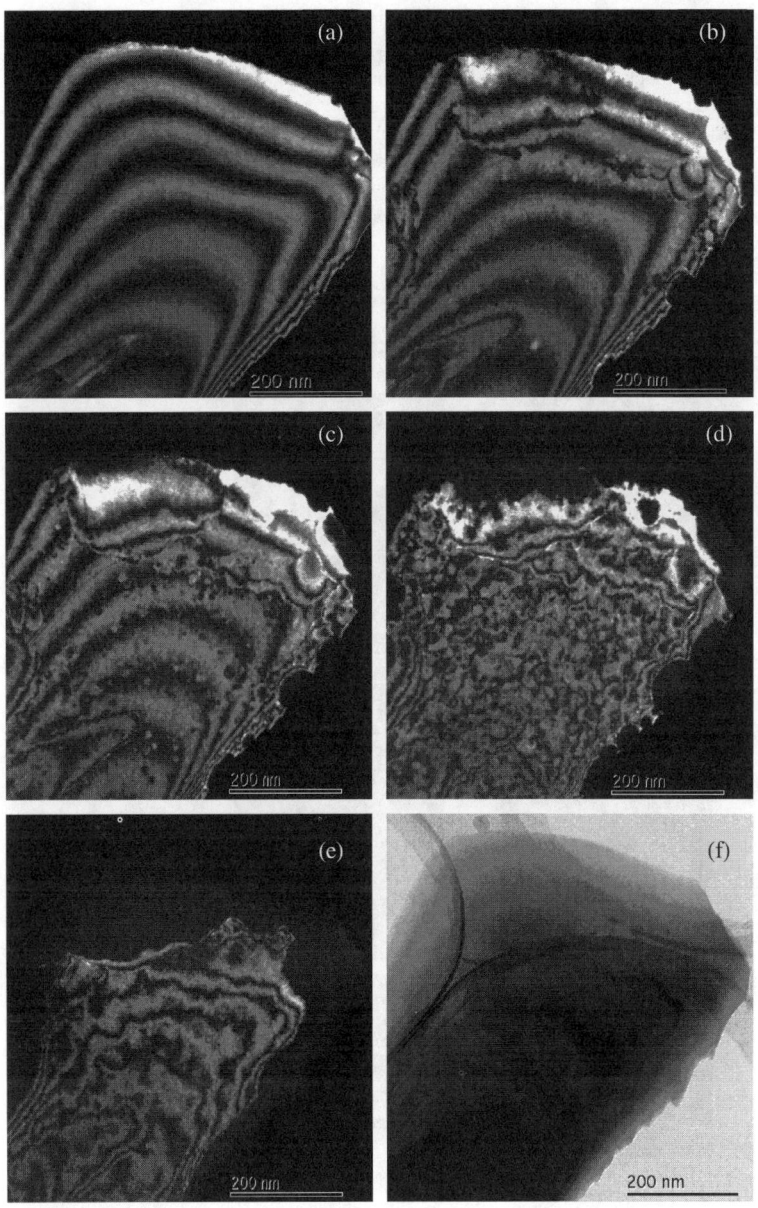

**Figure 1.** Dark field images ($k = 6.21$ nm$^{-1}$) during electron irradiation: (a) 2 min; (b) 11 min; (c) 24 min; (d) 32 min and (e) 45 min. (f) is the bright-field image after irradiation.

Nonetheless, using Digital Micrograph 3.3 our goals can be achieved. In this program, twenty-two raw images (image index is related to electron dose) are scaled up by 4, rotated counterclockwise by 9.5° and then scaled down by 4. Rectangles of 30 × 800 pixels are selected from these processed images, and arranged in two dimension after being averaged in intensity over each row (each column represents the increased thickness line). Accordingly, normalized variance for the averaged intensity can be obtained. The data are shown in Figure 2. Now we have a clearer view on how image contrast vary with thickness and dose.

Owing to the way of manipulating images, it is glooming (i.e. intensity drops but variance emerges) in the upper right-hand corner of the variance picture in figure 2. The beading curve on the variance picture, corresponding to the upper sharp edge in the intensity picture that indicates the interface between quartz and disordered silica, is also

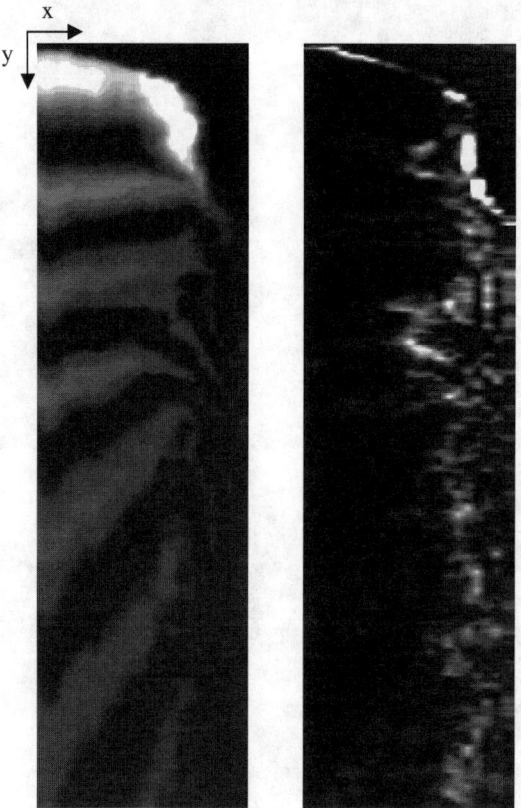

**Figure 2.** The variations of mean intensity (left) and normalized variance (right) with increased dose (x) and thickness (y).

an artifact caused by the analysis. It seems that disordered silica at the sample edge does not show as many speckles as we expected. In a contrast experiment where off-Bragg conditions are used, a very similar variance results. Remember that these data are measured under dynamical scattering conditions from a thickness up to 100 nm, for which any coherently diffracted beams from disordered regions can be extinct after passing through the sample.

To describe how quartz is transformed into metamict silica, we should understand the underlying mechanism of electron irradiation. Electron irradiation is a solid-phase reaction involving energy transfer (and perhaps momentum transfer, depending on incident electron energy) from incident electrons to atomic nuclei and electrons. In the case of damaging silica, the Si-O bond in a tetrahedron can be broken. The excess oxygen then joins neighboring ones to form a molecular unit. This event is known as radiolytic displacement. As a matter of single bond breakage, the network undergoes an awesome frustration, leading to topological disordering. The incipience of disordering usually occurs at the surface, where structure freedom is more than structure constraints. These indeed have been seen in our measurements.

What are the topologies of metamict silica is a curious question. It is found that the significant difference in geometry between crystalline polymorphs lies at the third-neighbor distance [3], so the metamict polymorph can be visualized using local clusters of medium-range sizes. Our simple model, as illustrated in figure 3, does not give a satisfatory answer to the key question but instead stimulates discussion about the structure. The investigation does not stop here. Indeed, we should find other supportive evidence for this model using more precise methods.

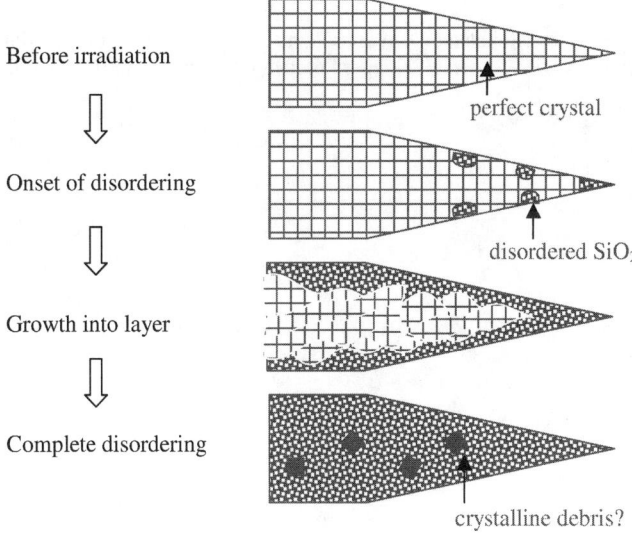

**Figure 3.** Aschematic diagram of metamict transformation in quartz.

## CONCLUSIONS

Alpha quartz is broken into disordered silica and a defective crystal in high-dose electron irradiation. The metamict transformation is monitored at a specific Bragg condition ($g = 210$ at $k = 6.21$ nm$^{-1}$). This dark field condition is set to seek any connection between the suspicious medium-range ordering in disordered silica and the parent crystal. From the images, we find incipient disordering at the surface. After a while when the irradiation has produced enough damage in quartz, the disordered regions at the sample edge reveal only a glimmer of contrast. In large part, the thickness fringes of the damaged quartz are the origin of intensity fluctuations. Whether or not the structure is topologically ordered is now under further investigations.

## ACKNOWLEDGMENTS

This work is supported by National Science Foundation under grant DMR 00-74273.

## REFERENCES

1. J. T. Randall, H. P. Rooksby, and B. S. Cooper, Nature **125**, 458 (1930).
2. N. Valenkov and E. A. Porai-Koshits, Nature **137**, 273 (1936).
3. L. W. Hobbs, J. Non-Cryst. Solids **182**, 27 (1995).
4. J. H. Konnert and J. Karle, Nature Phys. Sci. **236**, 92 (1972).
5. J. C. Phillips, Solid State Phys. **37**, 93 (1982).
6. R. B. Sosman, in *The Phases of Silica* (Rutgers University Press, New Jersey, 1965), p. 175-187.
7. P. Hirsch, A. Howie, R. B. Nicholson, D. W. Pashley, and M. J. Whelan, in *Electron Microscopy of Thin Crystals* (Robert E. Krieger Publishing, New York, 1977), p. 202.

# Metastable Defects in the Amorphous Silicon-Germanium Alloys

J. David Cohen
Department of Physics and Materials Science Institute, University of Oregon, Eugene, OR 97403

## ABSTRACT

This paper first briefly reviews a few of the early studies that established some of the salient features of light-induced degradation in a-Si,Ge:H. In particular, I discuss the fact that both Si and Ge metastable dangling bonds are involved. I then review some of the recent studies carried out by members of my laboratory concerning the details of degradation in the low Ge fraction alloys utilizing the modulated photocurrent method to monitor the individual changes in the Si and Ge deep defects. By relating the metastable creation and annealing behavior of these two types of defects, new insights into the fundamental properties of metastable defects have been obtained for amorphous silicon materials in general. I will conclude with a brief discussion of the microscopic mechanisms that may be responsible.

## INTRODUCTION

It is well known that the electronic properties of hydrogenated amorphous silicon (a-Si:H) are altered as it is exposed to light [1]. The details of how this affects transport properties, deep defect densities, and the properties of devices fabricated from this material have been the subject of literally thousands of studies. Although these effects in a-Si:H are not entirely understood, certain key aspects of this behavior are generally accepted, at least in the traditional or standard forms of the material. In contrast, the breadth of analogous studies in the amorphous silicon-germanium alloys (a-Si,Ge:H) is considerably less even though these alloys are widely employed in the fabrication of tandem and triple solar cells.

Initially it was thought that light-induced changes in the a-Si,Ge:H alloys were rather minor [2,3]. However, this was undoubtedly due to their initially quite poor electronic properties. In the latter half of the 1980's more systematic programs were undertaken to develop and understand the properties of the a-Si,Ge:H alloys [4,5,6,7], and these materials finally reached a high degree of optimization during the 1990's.[8,9,10] This can be identified by the fact that the Urbach energies for alloys, covering nearly the entire range of Ge fractions, could now be maintained at or below 50meV, essentially identical to those found in the best a-Si:H.[11] Likewise deep defect densities in such optimized alloys could be maintained at or below the mid $10^{16}$ cm$^{-3}$ level in the as-deposited state for Ge levels up to 50at.%. Thus, it is really only after this point (from the early 1990's onward) that the issue of light-induced degradation in the alloys became a truly relevant issue. This was at a time when much of the work in the field was shifting away from fundamental science and toward applications. This was somewhat unfortunate, particularly because of the insights such studies of the alloys can provide to the issue of degradation in unalloyed a-Si:H.

In the discussion below I will first review a few studies that established some of the key salient features of light-induced degradation in a-Si,Ge:H. I will then focus on the recent studies carried out by members of my own group that have been focusing on the details of degradation in the low Ge fraction alloys. These studies have indeed allowed us to address some fundamental aspects of degradation in amorphous silicon materials in general.

# THE GENERAL NATURE LIGHT INDUCED DEGRADATION IN THE ALLOYS

## Kinetics of defect creation and annealing in a-Si,Ge:H

The first careful attempts to examine the degradation kinetics in the a-Si,Ge:H alloys were published in two studies during 1992. One of these, carried out by G. Schumm et. al at the University of Stuttgart, remains one of the best such studies ever undertaken [12]. The degradation of photoconductivity was studied for four a-Si,Ge:H films with optical gaps varying between 1.66eV down to 1.45eV (Ge fractions up to 42at.%). To ensure the same optical exposure for the samples of varying gaps, a second piece of each sample was used as an optical filter to cut off photon energies significantly above the optical gap and thus ensure spatially uniform exposure. The intensities following the filtering were also adjusted to be the same. In this manner the carrier generation rates were reasonably well matched (near $2 \times 10^{22}$ cm$^{-3}$s$^{-1}$) for the entire set of samples. The results of this study indicated that the photoconductivity degraded roughly as the cube root of exposure time at longer times, in good agreement with the degradation kinetics in pure amorphous silicon and thus indicating a bimolecular defect creation process. However, there was a faster component of degradation apparent at shorter times that appear to saturate earlier. This study also indicated a smaller absolute degradation rate as the optical gap was reduced, which agreed with indications from some earlier work [4]. A comparison was also made between the degradation rate in these alloy samples with some pure a-Si:H samples in which the optical gap was varied by about 0.1eV by modifying their hydrogen content. This led to the conclusion that the variation in degradation rates was due to changes in optical gaps rather than the Ge fraction directly.

This latter conclusion was confirmed quite convincingly by some much more recent studies by A. Terakawa [13] which examined the performance of p-i-n solar cell devices. The optical gaps of the i-layers were varied between 1.20eV and 1.58eV by changing the Ge alloy fraction and/or by changing the hydrogen content. For example, a 1.32eV optical gap alloy was achieved in four samples whose Ge content varied from 32at.% to 41at.% while varying the hydrogen content from 6at.% to 11at.%. This study found that the degradation rate, as determined by monitoring the normalized device efficiencies, was identical for all four samples. Thus, this clearly indicated that the optical gap rather than the Ge content itself controlled the degradation kinetics. This study also found that the annealing kinetics varied strongly with temperature but not with optical gap. This therefore seems to preclude any direct role of the carrier concentrations themselves in the annealing mechanism of the light-induced changes.

Also in 1992, N.W. Wang, et al at Princeton University reported an extensive study of metastable deep defect creation in the a-Si,Ge:H alloys as determined by the strength of the sub-band-gap absorption band measured by the constant photocurrent (CPM) method [14]. In agreement with the other studies mentioned above, they found no qualitative difference between the kinetics of deep defect creation, nor in the temperature dependence of saturation, between a-Si,Ge:H films with Tauc gaps varying between 1.51 and 1.62eV versus pure a-Si:H.

The above studies clearly established that degradation occurs in the a-Si,Ge:H alloys in a manner similar to pure a-Si:H. They suggest that the kinetics of deep defect creation and thermal annealing are also similar and that the systematic variation with increasing Ge content seems to be a result of varying optical gap, not the Ge content directly.

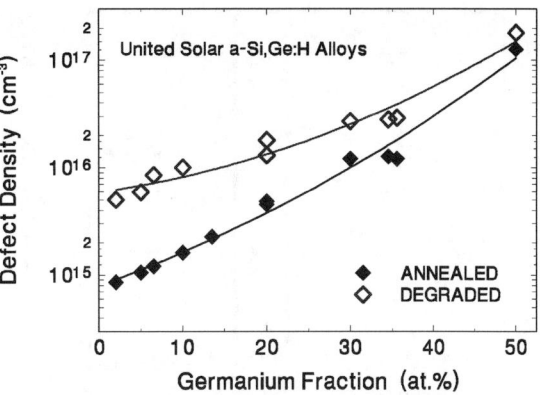

**Figure 1.** Initial and degraded deep defect densities of eleven United Solar glow discharge a-Si,Ge:H films as determined by the drive-level capacitance profiling method.[15] Note that although both the annealed and degraded state deep defect densities increase with increasing Ge fraction, the *relative* increase in deep defect density decreases.

**Decreased degradation in the higher Ge fraction alloys**

One important remaining question is whether deep defect creation actually persists to alloys with higher Ge fractions (lower gaps). For example, in Fig. 1 we have compiled data from a couple of studies [15] employing optimized United Solar a-Si,Ge:H films showing the deep defect densities before and after light induced degradation as determined by the drive-level capacitance profiling method [16]. These alloy samples all had Urbach energies at or below 50meV. These data clearly indicate that the factor of deep defect density increase becomes smaller as the Ge fraction increases. We note that the probable reason why the annealed defect density increases so markedly for alloys above 40at.% is that United Solar has optimized their deposition process for alloy fractions of 35at.% and below. Measurements on other optimized sources of material indicate that deep defect densities at or below $1 \times 10^{17}$ cm$^{-3}$ can be maintained even up to 100% Ge [17,18].

A couple of researchers have hypothesized that the production of metastable deep defects under light exposure might actually cease for optical gaps below a certain value. H. Fritzsche has pointed out that one has not observed light-induced deep defect creation in pure a-Ge:H even for cases of very low initial defect densities.[19] T. Unold suggested in 1994 that, based upon the data available at that time, it appeared that a-Si,Ge:H with Ge fractions above about 65at.% Ge (optical gaps below 1.4eV) would be stable.[20] If this turns out to be true it would suggest that there exists a certain energy threshold below which the recombination of electrons and holes does not lead to the creation of metastable deep defects.

**The Nature of the Metastable Defects**

ESR studies have shown that both Si and Ge dangling bonds exist in the a-Si,Ge:H alloys. Since germanium bonds are, in general, less strong than silicon bonds, one expects that Ge dangling bonds will dominate in alloys with appreciable fractions of Ge. This has indeed observed to be the case. Figure 2(a) shows some spectra from one study, and 2(b) summarizes results from 3 such studies [6,21,22]. Here we have plotted the ratio of Ge to Si dangling bonds *vs.* the Ge fraction in the film. We see that, for alloys with Ge fractions above 30at.%, the Ge dangling bonds dominate by at least an order of magnitude. The only cases for which the density of Si and Ge dangling bonds appear comparable is for Ge fractions below about 10at.%.

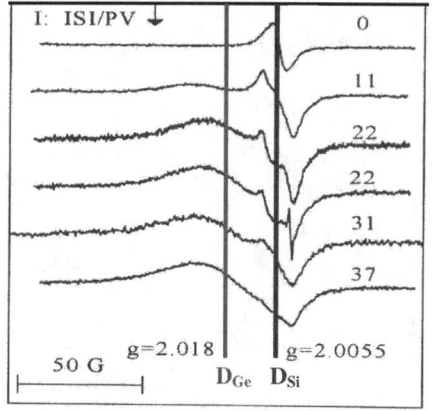
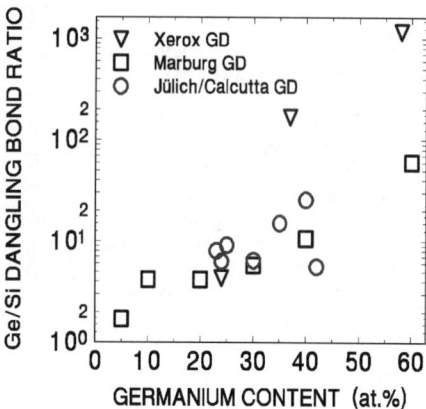

**Figure 2.** (a) Examples of ESR spectra for a series of a-Si,Ge:H alloy samples from a Jülich/Calcutta collaborative study [22] showing that Ge dangling bonds predominate once the Ge fraction exceeds about 20at.%. (b) Ratios of gemanium to silicon dangling bonds determined from ESR studies of a-Si,Ge:H samples at Xerox [6], Marburg [21], and from the Jülich/Calcutta collaboration [22]. Note the predominance of Ge dangling bonds when the Ge fraction becomes appreciable.

There have only been a couple ESR studies examining the effects of light soaking on the dangling bonds in the a-Si,Ge:H alloys. Although a study was published about 5 years ago [23] this study did not include any information about the optical gaps or Ge fractions of the samples examined. Therefore the best such study remains the early work carried out by Stutzmann et al.[6] In that study only the Si dangling bonds were observed to increase. However, since most of the films examined exhibited total defect densities of $10^{17}$ cm$^{-3}$ or higher, it is not clear that the behavior of such samples is very relevant to the alloy material currently being incorporated into devices. Moreover, due to the comparative difficulty in observing the much broader Ge dangling bond signal, optimized a-Si,Ge:H sample are really not very amenable to study by ESR where one is trying to detect changes in Ge dangling bond at the $10^{16}$ cm$^{-3}$ level or below.

## MODULATED PHOTOCURRENT SPECTROSCOPY

The difficulty in interpreting ESR to deduce types of dangling bonds led my own students to search for a more sensitive alternative method to identify the types of dangling bonds present in the optimized a-Si,Ge:H alloys. Since the 1990's various groups have been employing the modulated photocurrent (MPC) method [24,25] as a spectroscopic tool to monitor deep defect properties in a-Si:H. [26,27,28,29] A prominent band of electron carrier traps is disclosed by this technique in pure a-Si:H which releases its trapped electrons with an activation energy near 0.65eV (see Fig. 3). However, there has been considerable disagreement about whether this band of traps should be attributed to neutral or positively charged dangling bonds [27,29].

This question was answered to my satisfaction when Daewon Kwon of my group compared the ESR spin density and MPC defect band magnitude in both intrinsic and very lightly phosphorus doped films for a series of metastable states produced by sequential annealing of a strongly light soaked state in each sample.[30] For both types of samples, the ESR Si dangling bond signal and MPC disclosed defect band magnitudes tracked each other almost perfectly. However, because an n-type doped sample contains very few positive dangling bonds and, moreover, the ratio of the neutral to charged dangling bonds should actually change as such a sample is annealed, the close correspondence of the spectra clearly indicated that *neutral dangling bonds* must be responsible for the MPC defect band.

It is important to note that all of the MPC measurements carried out in my laboratory use a sandwich rather than a co-planar contact geometry, and use strongly absorbed light (530nm). In such measurements with the top contact at negative bias, the photocurrent signal arises from the electrons drifting through the sample. Moreover, the electric field in the normally undepleted part of the sample must reach a value that guarantees the removal of the photoelectrons at exactly the rate they are generated (ignoring the effect of carrier recombination which is nearly negligible under such conditions). This must happen because MPC is a steady-state measurement. This implies that the deduced trapping state density will actually be independent of the external

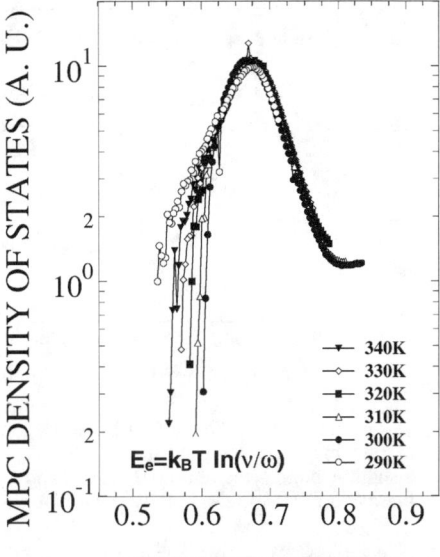

**Figure 3.** The deep defect band revealed using the MPC method on pure a-Si:H has an activation energy for the release of trapped electrons near 0.67eV. The density of states was obtained from the magnitude and phase shift of the photocurrent signal for a 570nm light source modulated between 3Hz and 10kHz using the analysis of R. Brüggeman, et al. [25]

bias applied across the sample (as long as the sample does not become fully depleted), a fact that has been verified experimentally in our measurements. Because of this, such MPC spectra are able to yield actual quantitative comparisons of the densities of electron trapping states, provided we maintain a constant carrier generation rate.

A few years ago my former student Kimon Palinginis began studying a-Si,Ge:H alloy samples with low Ge fractions, specifically a set of four samples with 2at.%, 5at.%, 7at.%, and 10at.% Ge fractions deposited at United Solar. A couple of the films were also deposited onto quartz substrates so that comparison ESR measurements could be carried out. In his attempts to observe two types of dangling bonds in such samples, he ultimately applied the MPC method to these samples.[31]. In Fig. 4 we see that, unlike the MPC spectra for pure a-Si:H, two distinct defect bands are disclosed for the 5at.% Ge alloy. The MPC spectra shown were obtained by starting with a fully light soaked state (state B) followed by--using a sequence of isochronal anneals of 15min each at increasing temperatures--a series of spectra for metastable states which end in a 470K annealed state (state A). In all cases one can fit these spectra to two Gaussian bands located near 0.68eV and 0.78eV, indicating bands of electron traps at those energies below the conduction band. We shall refer to these two bands as the shallow and deep bands,

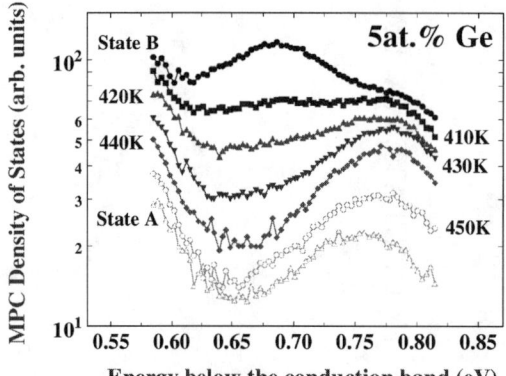
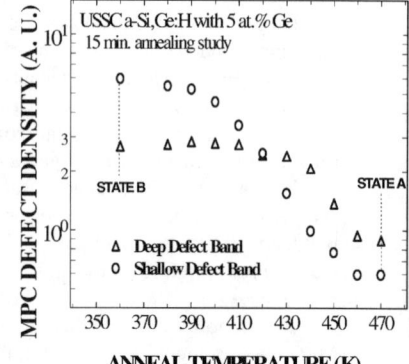

**Figure 4.** MPC spectra taken at 340K for a sequence of anneal states for the 5at.% a-Si,Ge:H sample. Two distinct bands of electron trapping states are observed: a shallow band lying at 0.68±0.01eV, and a deeper band at 0.78±0.015eV, both with similar widths.

**Figure 5.** Annealing kinetics of the shallow and deep defect bands for the 5at.% Ge film. The magnitudes were determined by fitting two Gaussians defect bands to the MPC spectra of Fig. 4.

respectively. We can see from this figure that the shallow band is dominant in State B while the deep band is dominant in State A. Qualitatively identical results were also obtained for the 2at.%, 7at.%, and 10at.% Ge fraction samples in this series. In each of the four samples he also found that the shallow defect density greatly exceeded the deeper defect density for state B, whereas the reverse is true for state A. Nevertheless, both types of defects were found to increase with light soaking. The annealing behavior of the magnitudes of these two defect bands for the 5at.% Ge sample deduced from Fig. 4 are shown in Fig. 5.

These shallow and deep bands of deep defects have been identified as originating from neutral Si and neutral Ge dangling bonds, respectively. The identification of the shallow band is quite solid: First, its energetic position and energy width closely matches the feature observed in the MPC spectra of pure a-Si:H (see Fig.3). Second, we carried out ESR measurements on a couple of matched samples and found quite good agreement between the factor of increase of the shallow MPC band with light soaking and the corresponding Si dangling bond density observed in ESR.[31] The ESR spectra also indicated the presence of a weak neutral Ge dangling bond signal whose magnitudes matched the deeper band, albeit with very large error bars. Moreover, if one rules out the alternative that the deeper band might originate from positively charged dangling bonds because of the similarity of its thermal emission prefactor to that of the shallow band, then the assignment of the deeper band to neutral Ge dangling bonds seems fairly certain. Thus the MPC spectra allow us to study the degradation and annealing kinetics of two types of deep defects within the same sample. More of these results are discussed below.

## INSIGHTS INTO THE GENERAL NATURE OF DEGRADATION

As discussed above, MPC measurements on a-Si,Ge:H films with low Ge fractions were found to exhibit two distinct bands of deep defects which appear to originate from neutral Si dangling bonds ($D^0_{Si}$) and neutral Ge dangling bonds ($D^0_{Ge}$). In addition, subsequent EXAFS studies made upon several United Solar a-Si,Ge:H films have indicated no evidence of Ge atom clustering in these alloys.[32] Thus, with a fair degree of confidence, we may assume for these samples that the the Ge atoms and dangling bonds are effectively isolated within these films and have only Si atom near neighbors.

As also mentioned above, one can fit the two bands with with Gaussians to determine their relative magnitudes. In Fig. 5 we displayed the annealing behavior of the 5at.% alloy for each of the two types of defects for a series of 15 minute isochronal annealing steps at increasing temperature. We clearly observe that the two types of defects anneal differently, with the shallow ($D^0_{Si}$) band appearing to anneal first followed, at the higher anneal temperatures, by the deeper, $D^0_{Ge}$ band. This strongly suggested two different activation energies for annealing and, if this were indeed the case, would argue for two independent processes such as the local rearrangement of atoms in the vicinity of these two types of dangling bonds. We thus tried to determine these two activation energies by repeating using isochronal anneals with times varying from 5 minutes to 450 minutes. This resulted in a shift of the crossing point of the two curves in Fig. 5 from 430K to 390K. However, much to our surprise, it did not change the relative positions of the annealing curves for the two defects.

In contrast, in Fig. 6, the individual defect densities are plotted as a function of the total deep defect density as determined by such MPC measurements.[31] This plot actually incorporates such an annealing series for *two different samples* where the defect magnitudes for the two samples have been normalized to be the same at the crossing point. One can see that the $D^0_{Ge}$ density actually remains fairly constant in the initial stages of the annealing (annealing proceeds from right to left), while the $D^0_{Si}$ drops significantly. However, once the $D^0_{Si}$ becomes appreciably smaller than the $D^0_{Ge}$ density, the latter starts to decrease as well. Quite similar results we found for all of the samples, with the only difference being the final ratio of the defect magnitudes at saturation. Moreover, when plotted in this manner, the behavior was found to be nearly independent of the times used for these isochronal anneals. Hence, these data indicated a *direct correlation* between the annealing processes of the two defects rather than different activation energies for the annealing of $D^0_{Ge}$ and $D^0_{Si}$.

The annealing behavior has been described by a simple model that leads to the fits shown in Fig. 6.[35] One assumes that every partial anneal reduces the total deep defect density by some $\Delta N_{tot}$. If there is a global reconfiguration process, then the Si and Ge defects will compete for their share of the drop in total defect density. Independent of possible underlying mechanisms

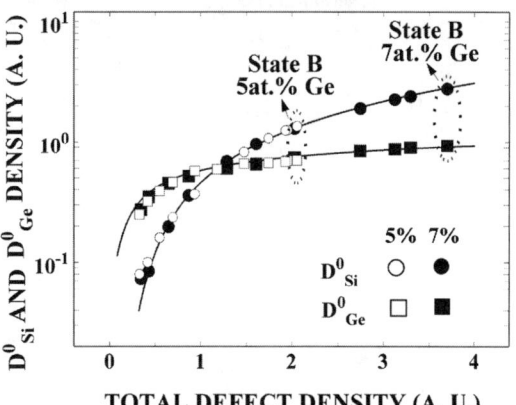

**Figure 6.** Estimated MPC densities of Si and Ge deep defects as a function of the total defect density during isochronal anneal sequences of 15 min duration for the 5 at.% Ge (open symbols) and the 7 at.% Ge (solid symbols) samples. The relative defect densities of the two samples have been rescaled to match each other at the crossing point. To fit the data we used the solution to Eq. (1) with $\eta \approx 0.25$. From K.C. Palinginis [35].

there will be a probability $k_{Si}N_{Si}$ that the $D^0_{Si}$ defects will be reduced and a probability $k_{Ge}N_{Ge}$ that the $D^0_{Ge}$ defects will be reduced. Therefore,

$$\frac{dN_i}{dN_{tot}} = \frac{k_i N_i}{k_{Si}N_{Si} + k_{Ge}N_{Ge}} \tag{1}$$

where i denotes Si or Ge, and $k_{Si}$ and $k_{Ge}$ are weighting factors or effective "capture coefficients" for the global entity responsible for the annealing process. The integration of Eq. (1) leads to the fits shown. Note that there is basically only one fitting parameter, namely $\eta = k_{Ge}/k_{Si} \approx 0.2$.

One possible microscopic scenario of such an annealing behavior might be long-range hydrogen diffusion. H gets released from some kind of reservoir, diffuses through the material and then gets trapped by a dangling bond. This interpretation would be along the lines of the recently proposed "hydrogen collision" model by H. Branz [33]. Other viable explanations include bond switching in conjunction with the release of network strain upon annealing, or propagation of floating bonds as recently proposed by R. Biswas [34].

A series of spectra for the degradation of the 7at.% Ge sample is shown in Fig. 7.[35] The change in the magnitudes of these two defect bands with exposure time is shown in Fig. 8. One can see that neither band follows the expected $t^{1/3}$ time dependence in all time regimes; however, the total deep defect density obtained from the sum of the two agrees with that expected time dependence reasonably well. Generally speaking, this means that the behavior of these alloy samples fits quite well to the predictions of the bimolecular recombination model developed for pure amorphous silicon if one generalizes the deep defect recombination to include both types of deep defects. This is explained in more detail below.

In Fig. 9 the saturation behavior of the two defect band is plotted as a function of the degradation temperature and for a couple values of light intensity.[36] The variation of the total actually agrees reasonably well with the results of earlier saturation studies [14]. However, it is quite striking that while the saturation value of the shallower (Si) defect band depends quite strongly on temperature, the deeper (Ge) defect band has a saturation value that is surprisingly

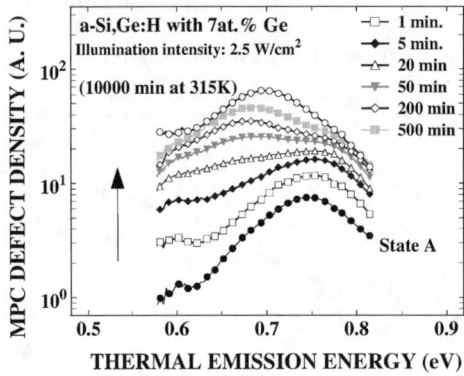

**Figure 7.** MPC spectra for a series of metastable states for a 7at.% Ge film produced by exposures of increasing duration as indicated at a light intensity of 2.5 W/cm² obtained from a red-filtered tungsten halogen light source [Ref. 35].

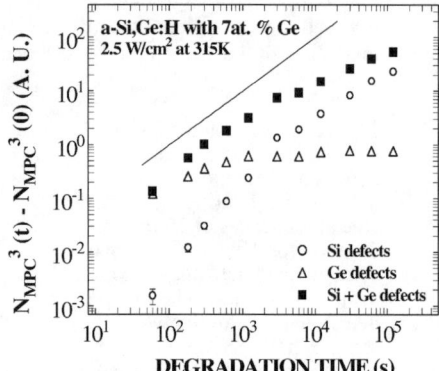

**Figure 8.** The cubes of the light-induced Si, Ge, and *total* defect densities of the a-Si,Ge:H sample with 7 at.% Ge determined from Fig. 7 as a function of exposure time. The line indicates a strictly $t^{1/3}$ dependence.

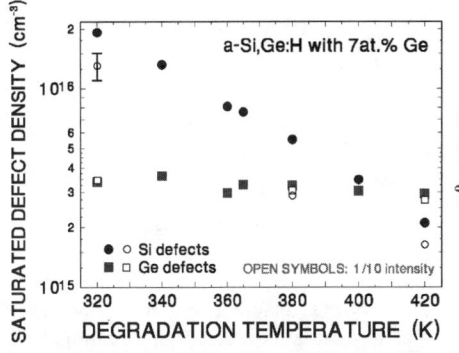
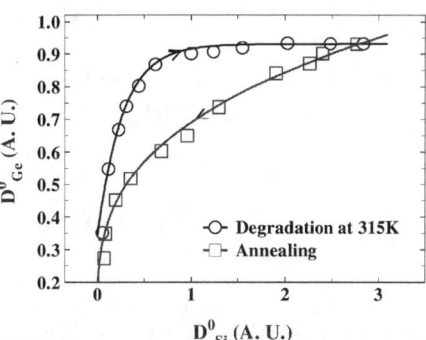

**Figure 9.** The densities of Si and Ge dangling bonds for long exposure times for the 7 at.% Ge sample as a function of the degradation temperature at two light intensities. The light intensity was 2.5 W/cm² for the solid symbols, and 250 mW/cm² for the open symbols.[Data from Ref. 36]

**Figure 10.** MPC determined Ge defect density plotted *vs.* the MPC determined Si defect density for an annealing sequence and a degradation sequence. The degradation fitted line was obtained from Eq. (4) with $\xi(315K) \approx 4$ and $N_{Ge}^P \approx 3.3 \times 10^{15}$ cm⁻³, while the annealing data were fitted via Eq. (1) with $\eta \approx 0.2$.

independent of temperature. These data thus strongly suggest a distinct and relatively low upper limit to the density of precursor sites for metastable Ge dangling bonds. In contrast, the saturation value of the Si dangling bonds varies greatly. Indeed, it depends upon the dynamical balance between light-induced defect creation and annealing, as has been previously well established for un-alloyed a-Si:H.[37]

Some additional insight into the degradation process in these alloys was obtained by directly examining the correlation of the Si and Ge defects during both the annealing and the degradation process. Figure 10 displays the degradation and annealing MPC densities of $D^0_{Ge}$ vs $D^0_{Si}$ for the 7at.% Ge alloy taken from Figs.6 and 8. One thus sees clearly that the degradation and annealing follow different paths. The annealing trajectory is correctly predicted from Eq. (1) to be $N_{Ge} \propto N_{Si}^\eta$. In contrast, the degradation behavior on this plot is found follow a simple exponential behavior which is correctly predicted using fairly general equations as discussed in reference 35. We reproduce this derivation here briefly as follows:

Defect creation in pure a-Si:H is usually described by the equation:

$$\frac{dN_{Si}}{dt} = c_{SW}^{Si} N_{Si}^P - A_{Si}(G,T) \qquad (2)$$

where $N_{Si}^P$ denotes the potential deep defect precursor sites and $A_{Si}(G,T)$ is a function describing the annealing terms which will depend on the generation rate G and degradation temperature T. The annealing terms are usually negligible prior to the onset of saturation after long exposure times. In 1985 Stutzmann, Jackson, and Tsai proposed a bi-molecular carrier recombination (bcr) mechanism to account for the observed $t^{1/3}$ time dependence.[38] In that analysis, the creation coefficient $c_{SW}^{Si}$ becomes $\alpha_{Si}G^2/N_{tot}^2$, where G is the optical carrier generation rate, $N_{tot}$ is the *total* deep defect density and $\alpha_{Si}$ is a constant of proportionality.

To apply such models to the a-Si,Ge:H alloys one extends the rate equations by incorporating a second type of dangling bond. However, from Fig. 9 it appears that there is a distinct and fairly low density of Ge dangling bond precursor sites. Thus, we should write:

$$\frac{dN_{Ge}}{dt} = c_{SW}^{Ge}(N_{Ge}^P - N_{Ge}) - A_{Ge}(G,T) \tag{3}$$

For the bcr mechanism the coefficient $c^{Ge}_{SW}$ would be given by $\alpha_{Ge}G^2/N_{tot}^2$, and again, prior to saturation, the second term can be ignored. Now simply dividing Eq. (3) by Eq. (2), we obtain

$$\frac{dN_{Ge}}{dN_{Si}} = \xi\left(1 - \frac{N_{Ge}}{N_{Ge}^P}\right), \text{ with } \xi = \frac{\alpha_{Ge}N_{Ge}^P}{\alpha_{Si}N_{Si}^P} \tag{4}$$

The solution of this rate equation yields a simple exponential, which gives the excellent fit to the $N_{Ge}$ vs $N_{Si}$ data shown in Fig. 10.

One of the most surprising conclusions from the above experiments on the low Ge fraction alloys has been the implication of a very low density of Ge dangling bond precursor sites. The evidence for this is really quite strong based upon the saturation studies shown in Fig. 9. To try to gain some more insight into this, we carried out experiments to specifically examine light induced annealing. These experiments followed procedures paralleling previous studies in pure a-Si:H.[37] To obtain the results displayed in Fig. 11, we degraded the 7at.% Ge film using full intensity (2.5W/cm$^2$) light exposure, holding the sample temperature at 365K. We then compared an *isothermal* dark anneal at 365K with an anneal at 365K during which time the sample was exposed to the degrading light, but at an intensity 100 times less intense.

One can clearly observe the effect of the light induced annealing in Fig. 11 by examining the time scales given at the top of each data set. In each case the total anneal time was over 200 hours; however, in neither case did we reach steady-state conditions. Nonetheless, the functional form of the dark isothermal anneal data in Fig. 11(a) agrees extremely well with the isochronal anneal data in Fig. 6 except that the defect density is plotted on a linear scale, and the state B silicon dangling bond density is a bit lower in Fig. 11 due to the higher temperature employed during degradation (365K *vs.* 315K).

In addition to the timescales, there are some other significant differences between the data in Figs 11(a) and 11(b). Most notably, and in contrast to the dark annealing process, $N_{Ge}$ appears to have reached a steady-state value after about 3 hours of annealing in the presence of the light. This implies that the creation and annealing of Ge dangling bonds comes into balance while the

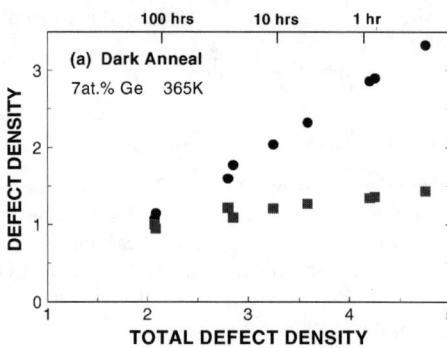 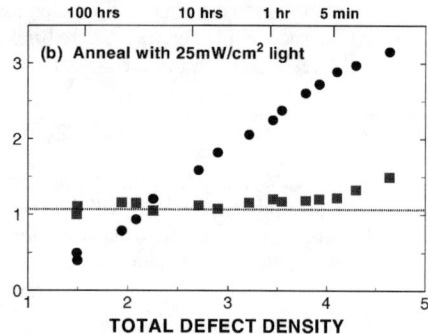

**Figure 11(a).** Isothermal anneal in the dark at 365K following degradation in red-filtered light intensity of 2.5W/cm$^2$ for 12 hours at 365K.

**Figure 11(b).** Isothermal anneal at 365K in the presence of 25mW/cm$^2$ light after the same degradation as for Fig. 11(a).

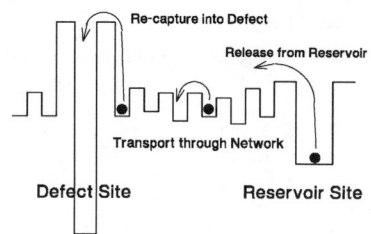

**Figure 12.** Schematic indicating possible annealing steps in a global mediated process. The observed annealing kinetics will be determined by the rate-limiting step of the three processes indicated.

Si dangling bonds continue to decrease. This must mean that the availability of the mediating entity is no longer the rate limiting step when weak light is present, in contrast to annealing in the dark. That is, the slower annealing rate for the Si defects must be limited by something else.

Figure 12 illustrates that, in general, there can be up to three steps in the types of global annealing process we have been envisioning. In the simplest process the rate-limiting step would be the release of the mediating entity from the reservoir site, and the *relative rates* of annealing between the two types of defects would remain the same. This appears to be the case for dark annealing, but not those assisted by the presence of the weak light. Instead, the rate limiting step for the Si defects appears to be the recapture of the mediating entity onto the defect site itself. For the Ge defects, on the other hand, it appears that optically induced defect creation (that is, the re-release of the mediating entity) is rapid enough to establish steady-state much more quickly.

## CONCLUSIONS

The above discussion has illustrated that light-induced degradation in general, and light-induced deep defect creation in particular, is an important aspect of the properties of the a-Si,Ge:H alloys. This has been very apparent since researchers were really able to optimize their electronic properties during the early part of the 1990's. We have discussed how light-induced deep defect creation in these alloys is even more complex than in pure a-Si:H.

This paper has focused on recent experimental results showing how the details of metastable behavior a-Si,Ge:H alloys has led to important insights into the metastable behavior in pure a-Si:H itself. This is due to the opportunity to examine correlations between two distinct types of metastable centers. Thus, based upon studies of the low Ge fraction alloys, it has been possible to establish clearly for the first time that a global mediating entity must be involved. Additional studies of light-induced annealing have indicated that it may be possible to actually take apart the different steps involved in the deep defect annealing process.

It is important to note that the low Ge fraction a-Si,Ge:H alloys undoubtedly represent the simplest case of these alloys, and that very fundamental issues remain nearly unexplored in materials with more significant Ge fractions. Hopefully, the technological importance of these alloy materials will continue long enough that many of these issues will be actively pursued.

## ACKNOWLEDGEMENTS

I gratefully acknowledge Fan Zhong, Chih-Chiang Chen, Daewon Kwon, Jennifer Heath, and particularly Kimon Palinginis as the scientists who actually carried out the fine experimental work on the a-Si,Ge:H alloys in my laboratory. I also wish to thank Jeffrey Yang and Subhendu Guha at United Solar Systems Corp. for providing the high quality alloy samples for our studies, and the National Renewable Energy Laboratory for financial support of the work at Oregon.

# REFERENCES

1. D.L. Staebler and C.R. Wronski, Appl. Phys. Lett. **31**, 292 (1977).
2. J. Bullot, M. Galin, M. Gauthier, and B. Bourdon, J. Phys. (Paris) **44**, 713 (1983).
3. G. Nakamara, K. Sato, and Y. Yukimoto, Solar Cells **9**, 75 (1983).
4. S. Aljishi, Z E. Smith, and S. Wagner, in *Amorphous Silicon and Related Materials*, ed. by H. Fritzsche (World Scientific, Sinapore, 1989), pp. 887-938.
5. S. Guha, J.S. Payson, S.C. Argawal, and S.R. Ovshinsky, J. Non-Cryst. Solids **97&98**, 1455 (1988).
6. M. Stutzmann., R.A. Street, C.C. Tsai, J.B. Boyce, and S.E. Ready, J. Appl. Phys. **66**, 569 (1989).
7. C.E. Nebel, H.C. Weller, and G.H. Bauer, Mat. Res. Soc. Symp. Proc. **118**, 507 (1988).
8. W. Paul, J.H. Chen, E.Z. Liu, A.E. Wetsel, and P. Wickboldt, J. Non-Cryst. Solids **164-166**, 1 (1993).
9. S. Guha, J. Yang, S.J. Jones, Y. Chen, and D.L. Williamson, Appl. Phys. Lett **61**, 1444 (1992).
10. T. Unold, J.D. Cohen, and C.M. Fortmann, Appl. Phys. Lett. **64**, 1714 (1994).
11. See J.D. Cohen in *Properties of Amorphous Silicon and its Alloys*, ed. by Tim Searle, EMIS Datareview Series No. 19 (INSPEC, London, 1998), pp. 180-187.
12. G. Schumm, C.D. Abel, and G.H. Bauer, Mat. Res. Soc. Symp. Proc. **258**, 505 (1992).
13. A. Terakawa, Ph.D. Thesis, Kyoto University, 1999.
14. N.W. Wang, P.A. Morin, V. Chu, and S. Wagner, Mat. Res. Soc. Symp. Proc. **258**, 589 (1992).
15. J.D. Cohen, NREL Reports: NREL/SR-520-25802 (1998) & NREL/SR-520-32535 (2002).
16. C. E. Michelson, A. V. Gelatos, and J. D. Cohen,, Appl. Phys. Lett. **47**, 412 (1985).
17. P. Wickboldt, D. Pang, W. Paul, J. H. Chen, F. Zhong, C.C. Chen, J.D. Cohen, and D.L. Williamson, J. Appl. Phys. **81**, 6252 (1997).
18. S.R. Sheng, R. Braunstein, and V.L. Dalal, Mat. Res. Soc. Symp. Proc. **664**, A8.4 (2001).
19. H. Fritzsche, P. Stradins, and G. Belomoin, Mat. Res. Soc. Symp. Proc. **420**, 563 (1996).
20. T. Unold, Mat. Res. Soc. Symp. Proc. **336**, 287 (1994).
21. W. Fuhs and F. Finger, J. Non-Cryst. Solids **114**, 1387 (1989).
22. C. Maltern, F. Finger, J. Fölsch, T. Kulessa, H. Wagner, S. Ray, A.R. Middya, and S. Hazra, Mat. Res. Soc. Symp. Proc. **377**, 559 (1995).
23. S. Hazra, A.R. Middya, and S. Ray, Philos. Mag. B**75**, 859 (1997).
24. H. Oheda, J. Appl. Phys. **52**, 6693 (1981).
25. R. Brüggemann, C. Main, J. Berkin, and S. Reynolds, Phil. Mag. B**62**, 29 (1990).
26. G. Schumm and G.H. Bauer, Phys. Rev. B**39**, 5311 (1989).
27. C. Longeaud and J.P. Kleider, Phys. Rev. B**53** 16133 (1996).
28. K. Hattori, Y. Niwano, H. Okamoto, Y. Hamakawa, J. Non-Cryst. Solids **137-138**, 363 (1991).
29. F. Zhong and J.D. Cohen, Mat. Res. Soc. Symp. Proc. **258**, 813 (1992).
30. J.D. Cohen and D. Kwon, J. Non-Cryst. Solids **227-230**, 348 (1998).
31. K.C. Palinginis, J.D. Cohen, J.C. Yang, and S. Guha, J. Non-Cryst. Solids **266**, 665 (2000).
32. B.D. Chapman, S.-W. Han, G.T. Seidler, E.A. Stern, J.D. Cohen, S. Guha, and J. Yang, J. Appl. Phys. **92**, 801 (2002).
33. H.M. Branz, Phys. Rev. B**59**, 5498 (1999).
34. R. Biswas, B.C. Pan, and Y. Ye, Mat. Res. Soc. Symp. Proc. **664**, 14.1 (2001).
35. K.C. Palinginis, J.D. Cohen, S. Guha, and J.C. Yang, Phys. Rev. B**63**, 201203(R) (2001).
36. J.D. Cohen, J. Heath, K.C. Palinginis, J.C. Yang, and S. Guha, J. Non-Cryst. Solids **299-302**, 449 (2002).
37. See, for example, H. Gleskova and S. Wagner, J. Non-Cryst. Solids **190**, 157 (1995).
38. M. Stutzmann, W. B. Jackson, C.-C, Tsai, Phys. Rev. B**32**, 23 (1985)

# TEMPERATURE DEPENDENCE OF THE DECAY OF OPTICALLY EXCITED CHARGE CARRIERS IN AMORPHOUS SILICON

J. Whitaker and P. C. Taylor
Department of Physics, University of Utah,
115 South 1400 East Room 201,
Salt Lake City UT 84112-0830

## ABSTRACT

We report the temperature dependence of the growth and decay of the optically induced electron spin resonance (LESR) on short and long time scales ($10^{-3}$ s < t < 2500 s). This range of times spans the region between previously published photoluminescence and the LESR data. In addition, we examine the steady-state density of optically excited charge carriers as a function of temperature. These measurements lead to a better understanding of the band tail structure of amorphous silicon as well as the kinetics of the excitation and recombination processes.

## INTRODUCTION

Modeling of the eventual recombination of optically excited electrons and holes in amorphous silicon at low temperatures (T < 40 K) involves two distinct processes, the hopping down in energy via tunneling of each individual carrier (diffusion) and recombination of electrons and holes via tunneling [1]. At short times the hopping down process dominates, and at long times the recombination process dominates. On intermediate time scales ($10^{-3}$ < t < 1 s) and at low temperatures the two processes compete: tunneling by a single carrier to a lower energy state, and recombination of an electron and a hole via tunneling. At finite temperature where excitations that increase the energy are possible, a third mechanism, namely variable range hopping of the charge carriers, must be considered. The optically induced electron spin resonance (LESR) signal of amorphous silicon at low temperatures has been well studied for time scales greater than about 1 s [2,3]. The photoluminescence (PL) signal, which is related to the LESR, in amorphous silicon has also been well studied for time scales less than about 1 ms [4]. In this paper we report measurements that extend the LESR results down to time scales approaching those probed in the PL experiments. We also examine the temperature dependence of the decay of optically excited carriers over a wide range of times from approximately $10^{-3}$ s to $10^3$ s. In particular, using LESR we examine the recombination kinetics of both short-lived and long-lived optically excited carriers in a-Si:H at finite temperatures using LESR.

The discovery in 1989 by Shklovskii et al. [1] that the simultaneous diffusion and recombination of electron-hole pairs in amorphous semiconductors is a universal property that does not depend on the densities of localized band-tail states has prompted renewed interest in the low temperature recombination processes in hydrogenated amorphous silicon (a-Si:H). At short times and low temperatures, hopping of the carriers downward in energy plays a major role in recombination, and at long times carriers become effectively trapped and recombination via tunneling is the only important process. At finite temperature where excitations that increase the energy are possible one would expect a more rapid diffusion of carriers and therefore enhanced recombination. Unlike the low temperature behavior, this diffusion, called variable range hopping, will depend on the details of the densities of localized band-tail states. Short times (t ≤ 1 ms) and low temperatures have been probed in photoluminescence (PL) or photoconductivity (PC) experiments [4] while long times (t ≥ 10 s) have been probed in optically excited electron spin resonance (LESR) experiments [2,3,5]. At temperatures up to 100 K, previous measurements have shown that there isn't single relaxation lifetime for the optically excited carriers, but rather the lifetimes vary by many orders of magnitude [2,5]. The previous LESR experiments concentrated on low-temperature, long-time decays where the theoretical interpretation is less complex. On shorter time scales there are two competing processes, tunneling by a single carrier to a lower energy state and recombination of an electron and a hole via tunneling. Even at these shorter times the decays are "universal" in the sense that they do not depend on the densities of localized band-tail states.

## EXPERIMENTAL DETAILS

These experiments employed thin film samples about 3 micrometers thick on quartz substrates. Five films were stacked together in order to increase the signal-to-noise ratios. The samples were the same as those used in a previous study of recombination kinetics of long and short lived carriers in a-Si:H at low temperatures [3,5,6]. The samples were irradiated at a wavelength of 632.8 nm using a Melles Griot Helium-Neon laser. The intensity of the light was approximately 1 mW/cm$^{-2}$ at the sample. The experiments were conducted with a Bruker Instruments ESR spectrometer. The second harmonic detection technique was the same as that used in ref. 5. All measurements were performed at microwave frequencies of about 9.5 GHz, and all spectra were taken with 0.4 mT modulation amplitude.

The magnetic field was set at the peak of the second harmonic LESR signal. This peak is mainly due to band-tail electrons, which exhibit a narrower line, but there is some contribution from the band-tail holes, which exhibit a broader line [7]. In this way, the peak LESR intensity was recorded as a function of time. Short time decay curves were taken for six different temperatures, between 20 and 100K. The sample was annealed above 150 K between each temperature scan. Initially the sample was cooled in the dark using a helium gas-exchange, Helitran transfer system from Advanced Research Products, and the temperature was controlled with an Air Products temperature controller. After cooling in the dark, the sample was irradiated for 140 milliseconds, after which the light was turned off and the data were acquired. In the short-time experiments, the total data acquisition time was 143 milliseconds. The laser pulse was obtained using an acousto-optic modulator. The modulator was switched by a TTL signal from a Hewlett-Packard function generator. The function generator also triggered the EPR spectrometer. The optics was aligned so that the first order deflected light from the acousto-optic modulator was used to irradiate the sample, and the zero-order light was blocked.

In a second series of experiments, the sample was annealed at 295 K and then cooled to various temperatures. Once at the desired temperature the sample was irradiated with a light pulse that was long enough to allow the density of optically excited electrons and holes to reach a steady state. Using similar triggering electronics, the data were acquired immediately after the light pulse was turned off. LESR data were acquired for six different temperatures over acquisition times up to 2600 seconds.

In the steady state experiments the first harmonic detection technique was used. This procedure allowed for a more accurate determination of the spin densities. The sample was irradiated with continuous light at 1 mW/cm$^2$. The microwave power was set below the saturation level for each temperature measured. The intensity of the signal was then compared to a standard weak pitch sample to estimate the spin density.

## RESULTS

Figure 1 shows the saturated, steady state spin density as a function of temperature. Although the initial light intensity remains unchanged at 1 mW/cm$^2$ the steady state number of spins decreases with temperature. The density decreases with increasing temperature because of the increased ability of the carriers to hop to higher energy states (variable range hopping) and therefore to recombine more readily. For this reason the number of steady state charge carriers decreases with temperature. This behavior has been known for many years.

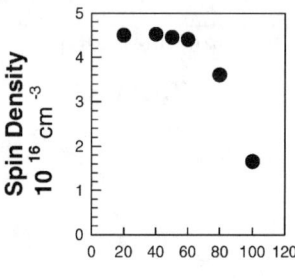

**Figure 1.** Temperature dependence of the saturated, steady-state LESR spin density in a-Si:H under irradiation at 632.8 nm with an intensity approximately 1 mW/cm$^2$.

Typical LESR decays are shown in Figure 2. As will be discussed in the next section, the two general features as temperature increases – a more rapid decay at short times and a slower decay at long times – are consistent with the expectations of the variable-range hopping process.

Figure 3 shows the decay of the charge carriers with time over a wide range of times (six decades). Although some of the details at short times are lost on the log time scale, the lower temperatures clearly exhibit a different shape than the higher temperatures. Data taken at 40 K and 50 K are not shown for clarity. The decays at 40 K and 50 K are very similar to that at 60 K. On the log time scale, the 100K curve appears to be almost flat, and there is a significant drop in the intensity at times shorter than those measured.

Figure 4 shows the long time decay curves of the LESR at 20 K and 100 K. The gray lines through the data are fits that will be described in the next section. These fits start at 10 seconds using a procedure described below.

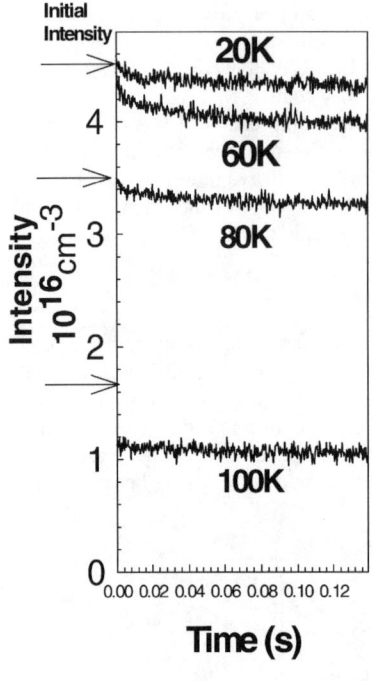 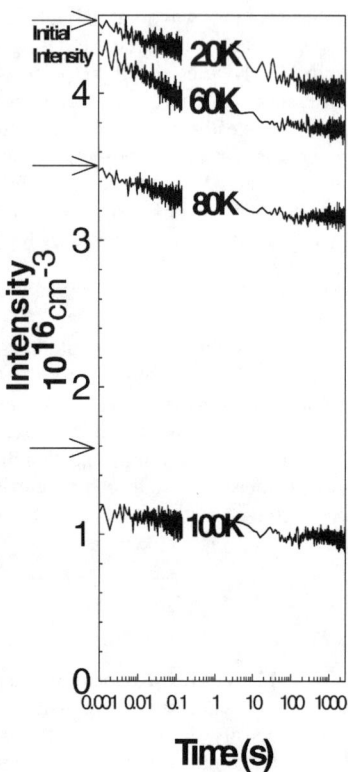

**Figure 2.** [Left] The short time decay of charge carriers in a-Si:H at 20, 60, 80 and 100 K. Arrows indicate the saturated, steady-state LESR spin densities. See text for details.

**Figure 3.** [Right] The decay of charge carriers in a-Si:H at 20, 60, 80 and 100 K on a log time scale. Arrows indicate the saturated, steady-state LESR spin densities. See text for details.

## DISCUSSION

The LESR intensities with the light on (Fig. 1) were obtained by comparing the first harmonic LESR signal at all temperatures with a calibrated standard run at 292 K. Due to the standard difficulties in measuring absolute intensities in ESR, these estimates are correct to within plus or minus a factor of approximately two. The relative errors are less than the sizes of the data points. As seen in Fig. 1, the density is essentially constant up to about 60 K above which it decays rapidly with temperature.

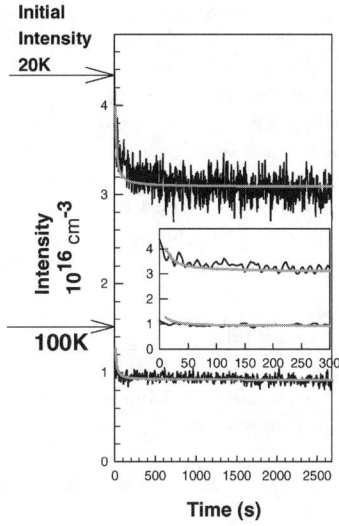

**Figure 4.** Decay of the LESR in a-Si:H as a function of time at low and high temperatures. The gray lines through the data are "zero" temperature fits as described in the text. The inset shows the shorter time data on an expanded scale.

In addition to the decrease in the saturated steady-state spin density with temperature, two qualitative features are discernable from the data in Figs. 2 and 3. As the temperature increases the short time decay becomes more rapid. This is particularly evident at 100 K where there is considerable decay at times shorter than the shortest measurement times (a few ms). This enhancement in the short time decay is consistent with the onset of variable range hopping because the increased diffusion of the carriers will lead to an increased probability for recombination. The second qualitative feature is that with increasing temperature the decay rates actually decrease at the longer times. This behavior is also particularly evident in the 100 K data of Figs. 2 and 3. This behavior is also a consequence of variable range hopping. Although diffusion of the carriers results initially in a more rapid decay, those few carriers that remain at longer times are trapped deeply enough that temperature no longer plays a significant role.

At "zero" temperature, where only hopping downward in energy is possible, and at the longest decay times the LESR intensities should converge on a common value, $n(t)$, which is independent of the initial intensity (independent of the generation rate). This asymptotic behavior, which is logarithmic in $t^{-3}$, and therefore highly non-exponential, is "universal" in the sense that it depends neither on the original excitation intensity nor on the shapes of the valence and conduction band tails of localized electronic states. The difficulty in estimating the time at which downward hopping is no longer important, the decay curves are usually fit to the appropriate functional form after a specific time, in our case 10 s. This fitting amounts to scaling the data for one of the decay curves to eliminate the uncertainties in the absolute value of the spin density. After fitting one decay curve, all others at different excitation intensities are obtained by using the initial spin densities, but with no further adjustable parameters.

In Fig. 4 this procedure has been applied to two curves, one at 20 K and one at 100 K. Clearly, the zero temperature approximation does not apply to 100 K, but the fit using only the steady-state value of the spin density is remarkable. This agreement confirms the comment made above that the only carriers remaining at long times are those trapped deeply enough such that temperature no longer plays a role. At the longest times, the only effect of finite temperature is to reduce the steady-state spin density at constant excitation intensity.

Although the simultaneous diffusion and recombination of optically excited carriers at low temperatures is a universal property of amorphous semiconductors, the shapes of the decay curves at elevated temperatures do depend on the densities of localized band tail states. Therefore, at higher temperatures one expects the decay curves to be different for different materials.

## SUMMARY

At an excitation intensity of approximately 1 mW/cm$^2$ the decay of the LESR in a-Si:H exhibits three qualitative features with increasing temperature. First, the saturated, steady-state densities decrease rapidly above approximately 60 K. Second, the long time decay (at times greater than a few seconds) at elevated temperatures can be fit using a low temperature approximation presumably because the carriers that remain are trapped deeply enough that temperature is no longer a factor in recombination. Third, the short time decay (at times less than a second) increases progressively with increasing temperature presumably because of the effects of variable range hopping.

## ACKNOWLEDGMENTS

This research was supported by NREL under subcontract ADJ-2-30630-23 and by NSF under grant number DMR-0073004. One of us (JW) acknowledges the support of a Graduate Research Fellowship from the University of Utah.

## REFERENCES

1. B. I. Shklovskii, H. Fritzche, and S. D. Baranovskii, Phys. Rev. Lett. **62**, 2989 (1989).
2. N. Shultz, B. Yan, A. L. Efros, and P.C. Taylor, J. Non-Cryst. Solids **266-269**, 372-375 (2000).
3. J. Whitaker, T. Su, and P.C. Taylor, Mat. Res. Soc. Symp. Proc. **715**, 275-281 (2002).
4. R.A. Street, *Hydrogenated Amorphous Silicon* (Cambridge University Press, Cambridge, 1991).
5. B. Yan, N. A. Schultz, A. L. Efros, and P. C. Taylor, Phys. Rev. Lett. **84**, 4180 (2000).
6. B. Yan and P. C. Taylor, MRS Symp. Proc. **507**, 805 (1998).
7. R. A. Street and D. K. Biegelsen, Solid State Commun. **33**, 1159 (1980); R. A. Street and D. K. Biegelsen, Solid State Commun. **44**, 501 (1982).

# Electron-Spin-Resonance Investigation of Laser Crystallized Polycrystalline Silicon

K. Brendel, N. H. Nickel, K. Lips and W. Fuhs
Hahn-Meitner-Institut Berlin, Silicon Photovoltaics
Kekuléstr. 5, D-12489 Berlin, Germany

## ABSTRACT

Doped and undoped laser crystallized polycrystalline silicon was investigated by electron-spin-resonance experiments. In P-doped samples two resonance are detected at g = 2.0053 and g = 1.998 which are due to silicon dangling bonds and conducting electrons, respectively. After crystallization a large amount of hydrogen remains in the samples. This residual hydrogen can be activated to reduce the spin density by passivating dangling-bonds. The temperature dependent investigation of the conducting electron resonance reveals that the susceptibility can be described by the sum of Pauli and Curie paramagnetism. The data are discussed in terms of models developed for single crystal and microcrystalline silicon.

## INTRODUCTION

Polycrystalline silicon (poly-Si) fabricated on low cost substrates has been subject of intensive research over the past years. Excimer laser crystallization of amorphous silicon ($a$-Si) is an attractive way to produce high quality poly-Si with grain sizes of up to 5 µm and a low dislocation density [1]. The dominant defects in poly-Si are silicon dangling-bonds (DB) that have been identified by electron spin resonance (ESR) spectroscopy [2]. To obtain device grade poly-Si these defects have to be passivated with hydrogen. Laser crystallization of hydrogenated amorphous silicon ($a$-Si:H) has the advantage that the resulting poly-Si film contains a high residual hydrogen concentration.

Poly-Si consists of single crystal grains that are separated by grain boundaries. On one hand, poly-Si shows characteristics known from amorphous silicon like band-tails which were first observed by Jackson et al.[3] in poly-Si and are located at grain boundaries. On the other hand poly-Si also shows characteristics only known from single crystal silicon (c-Si) like the Fano resonance in Raman scattering [4]. This effect has not been observed in amorphous or microcrystalline silicon (µc-Si:H) [5]. However, relatively little is known about the microscopic electronic states in this heterogeneous material.

In this paper we show a detailed ESR study on doped and undoped laser crystallized poly-Si. In order to understand the doping mechanism, we performed temperature dependent ESR measurements on P-doped samples. The resonance of conducting electrons (CE) at g= 1.998 appears for doping concentrations larger than $2 \times 10^{18}$ cm$^{-3}$. The temperature dependence of the susceptibility can be divided into a Curie and a Pauli paramagnetism. Comparing the data with results known from c-Si and µc-Si:H it will be deduced that the doping mechanism is not influenced by the presence of band-tails. Additionally, we will show that the residual hydrogen concentration after laser crystallization can be redistributed to decrease the defect density.

## EXPERIMENTAL

Polycrystalline silicon was prepared by laser crystallization of *a*-Si:H using a XeCl excimer laser. The amorphous starting material was deposited in a plasma enhanced chemical vapor deposition process at a substrate temperature of 230°C. The thickness of the specimens varied between 160 and 200 nm. The samples were deposited on quartz substrates. N-type doping was achieved by premixing silane with phosphine. The samples were crystallized applying a step-by-step crystallization procedure [6] to avoid explosive out-diffusion of hydrogen. Continuous wave ESR measurements were performed using an X-band spectrometer. The samples were cooled in a He gas flow cryostat. Doping and carrier concentrations were obtained from SIMS and Hall-effect measurements, respectively.

## RESULTS

Figure 1a shows ESR spectra of an undoped laser crystallized poly-Si sample before and after annealing at 400°C for 30 minutes. The resonance at a g-value of g = 2.0053 is characteristic of silicon dangling-bonds. Double integration of the spectrum yields a DB concentration of $N_S \approx 5 \times 10^{18}$ cm$^{-3}$ for the unheated sample. Laser crystallized poly-Si exhibits a high residual hydrogen concentration [7]. To activate hydrogen a post vacuum anneal was performed on the sample and the spin density was measured again (gray spectrum in Fig. 1). The spin density decreases to $N_S \approx 4 \times 10^{18}$ cm$^{-3}$. To get more insight into this important result the annealing kinetics were investigated. The open symbols in Fig. 1b show $N_S$ as a function of the annealing time for various temperatures. $N_S$ decreases with increasing annealing time. This decrease is faster for higher temperatures. Due to annealing $N_S$ can be reduced by a factor of 2. To investigate whether the hydrogen is responsible for the decrease of $N_S$ the annealing experiment was performed for poly-Si where unhydrogenated *a*-Si was used as the starting material (full squares in Figure 2b).

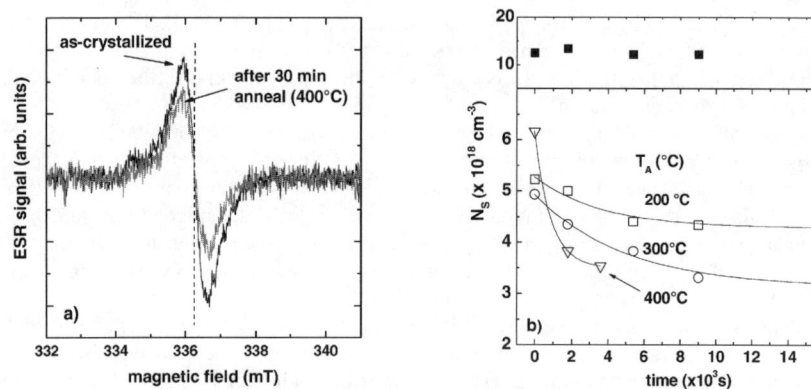

**Figure 1.** a) ESR spectra of as-crystallized poly-Si (black) and after annealing for 30 min. at 400°C (gray). b) Spin density, $N_S$, as a function of time for different annealing temperatures. The full squares represent the same experiment with unhydrogenated *a*-Si as the starting material.

$N_S$ does not decrease with time. This result clearly shows that the observed decrease of $N_S$ is due to hydrogen which is in the sample after crystallization.

In the following the influence of P-doping on the ESR spectra is investigated. Figure 2 shows ESR spectra of a P-doped sample measured at T = 5 and 60 K. Clearly a second resonance emerges with a g-value of g = 1.998 which is characteristic for the resonance of conducting electrons in c-Si [9]. The CE resonance is detectable for doping concentrations larger than $2 \times 10^{18}$ cm$^{-3}$. Because all samples were doped near or above the metal-isolator-transition no P-hyperfine splitting is detectable. The g-value decreases with increasing temperature and doping concentration. The ESR signal intensity, which is proportional to the susceptibility in cw-ESR experiments is strongly influenced by the temperature. The DB resonance is proportional to 1/T indicating a Boltzmann occupation of the Zeeman-states. This suggests that the dangling bonds are not interacting. The temperature dependence of the CE resonance intensity is more complicated. Figure 3 shows the susceptibility, $\chi$, normalized to unity at T = 5K as a function of temperature for samples with different doping concentrations. For low doping concentrations, $\chi$ decreases with increasing temperature. With increasing doping concentration the temperature dependence changes. For the sample with the highest doping concentration $\chi$ is constant for T < 40 K and decreases somewhat at higher temperatures. A similar behavior for the temperature dependence of $\chi$ of the CE resonance was found in c-Si [10]. In order to explain the temperature dependence the susceptibility can be divided into a Curie- and a Pauli- paramagnetic contribution. The Curie paramagnetism describes the magnetism of localized, non-interacting spins whose distribution can be described with the Maxwell-Boltzmann statistics. The susceptibility $\chi_{Curie}$ is proportional to the reciprocal temperature and is given by

$$\chi_{Curie} = N_{Curie} g^2 \mu_0 \mu_B^2 / 4 k_B T . \qquad (1)$$

$N_{Curie}$ is the number of spins that show Curie paramagnetism, $g$ is the g-value, $\mu_0$ the vacuum

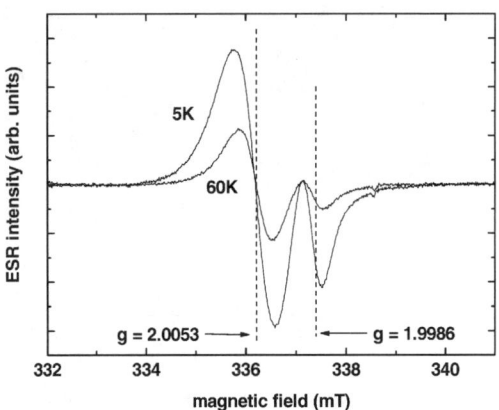

**Figure 2**. ESR spectra of a poly-Si sample doped with a P concentration of $8 \times 10^{18}$ cm$^{-3}$ for T = 5K and 60 K

permeability, $\mu_B$ the Bohr magneton, and $k_B$ the Boltzmann constant. For carriers in degenerate states like the conduction band eq. (1) can not be applied and the temperature dependence is described by the Pauli like paramagnetism for an electron gas

$$\chi_{Pauli}(T) = (N_{Pauli})^{1/3} \frac{g^2 \mu_0 \mu_B^2}{4k_B} \frac{1}{T} \frac{F'_{1/2}(\xi)}{F_{1/2}(\xi)} \text{ with } \xi = \frac{E_f}{k_B T}. \quad (2)$$

$N_{Pauli}$ is the number of electrons contributing to the Pauli paramagnetism, $F_{1/2}$ is the Fermi integral and $F'$ its derivative and $E_f$ is the Fermi energy. An example for the temperature dependence of $\chi_{Curie}$ and $\chi_{Pauli}$ is given in Fig. 4. In order to describe the temperature dependence of the susceptibility of poly-Si the data points of a sample with an electron concentration of $n = 7 \times 10^{18}$ cm$^{-3}$ are included in the picture and are fitted by a sum of $\chi_{Curie}$ and $\chi_{Pauli}$. In this procedure $N_{Pauli}$ and $N_{Curie}$ are the only fitting parameters. In Fig. 3 the fits are represented by the lines. They are in good agreement with the data. In Fig. 5 $N_{Pauli}$ and $N_{Curie}$ are plotted as a function of the doping concentration. With increasing doping concentration the contribution of carriers showing Pauli paramagnetism increases, while the contribution of carriers showing Curie paramagnetism decreases.

## DISCUSSION

As shown above, two resonance are observed in doped laser crystallized poly-Si that are attributed to a CE and a DB resonance. Previously it was shown that dangling bonds are predominately located at grain boundaries [8]. The spin density of as-crystallized poly-Si amounts to about $5 \times 10^{18}$ cm$^{-3}$. As mentioned above a CE resonance was observed for P

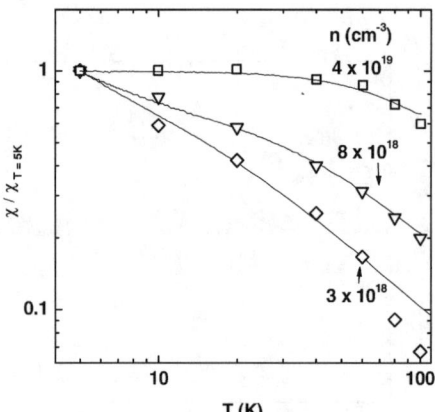

**Figure 3**. Normalized susceptibility as a function of temperature for samples with different doping concentrations. The lines are calculated. Details are given in the text

concentrations larger than about $2 \times 10^{18}$ cm$^{-3}$. Below this doping concentration nearly all carriers are trapped in the dangling bonds located at the grain boundaries. This might be also affected by the small grain size of about 300 nm. Normally a post hydrogen plasma treatment is performed to reduce the spin density [13]. In laser crystallized poly-Si a high residual hydrogen concentration of up to 20 at. % is still in the film [7]. A part of this hydrogen can be activated to passivate Si dangling-bonds. For an annealing temperature of 400°C $N_S$ decreases by about $3 \times 10^{18}$ cm$^{-3}$. Taken into account a residual hydrogen concentration of about $1 \times 10^{20}$ cm$^{-3}$ the amount of hydrogen necessary to passivate defects is about 30 times larger than the number of defects. On the other hand this value is about 10 for a post plasma treatment at T = 400°C [13]. One possible explanation of the lower passivation efficiency in laser crystallized poly-Si is that H is bound in other complexes due to laser annealing than in post plasma treated material.

The susceptibility of the CE resonance can be divided into a Pauli and a Curie paramagnetism. Similar observations were reported for c-Si and μc-Si:H [10, 11]. In c-Si the Pauli-paramagnetic contribution increases with increasing doping concentration indicating an increasing amount of carriers located in the defect or conduction band [10] while the Curie-paramagnetic contribution decreases. We observe a similar behavior in poly-Si. On the other hand, in μc-Si:H the increase of the Pauli contribution is also observed but the number of carriers showing Curie paramagnetism is constant independent of the doping concentration [11]. It is believed that this phenomenon is due to a constant amount of carriers localized in band-tail states. Microcrystalline silicon has smaller grains than poly-Si and exhibits a large amount of intra-grain defects [12]. The absence of a constant Curie contribution for high doping concentrations in poly-Si suggests that the density-of-states in the exponential tails in μc-Si:H is higher than in poly-Si.

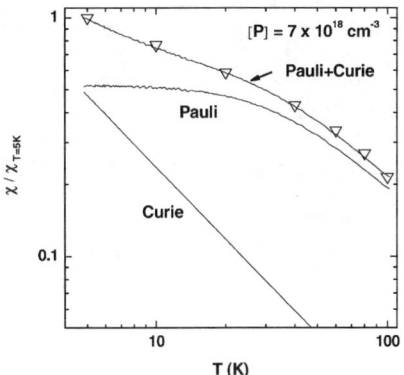

**Figure 4.** Temperature dependence of the Curie- and Pauli-paramagnetism and the sum of both. The data points are taken from a sample doped with $P = 7 \times 10^{18}$ cm$^{-3}$

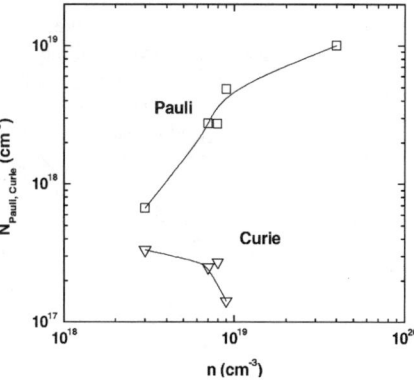

**Figure 5.** $N_{Curie}$ and $N_{Pauli}$ as a function of the carrier concentration.

## CONCLUSIONS

In summary, the spin density of laser crystallized poly-Si can be reduced in a low temperature vacuum anneal. The residual hydrogen in the film can be redistributed to allow a decrease in $N_S$ by a factor of 2. In P-doped samples a second resonance is observed that is attributed to conducting electrons. The temperature dependent susceptibility can be divided into a Pauli and a Curie paramagnetism. With increasing doping concentration the Pauli contribution to $\chi$ increases whereas the Curie contribution to $\chi$ decreases. This indicates that grain boundaries do not affect the CE states as was reported for µc-Si:H.

## ACKNOWLEDGEMENT

The authors are grateful to S. Brehme for Hall-effect measurements.

## REFERENCES

1. S. Christiansen, P. Lengsfeld, J. Krinke, M. Nerding, N. H. Nickel, and H. P. Strunk, J. Appl. Phys. **86** 5348 (2001).
2. N. M. Johnson, D. K. Biegelson, and M. D. Moyer, Appl. Phys. Lett. **40** 882 (1982).
3. W. B. Jackson, N. M. Johnson, and D. K. Biegelsen, Appl. Phys. Lett. **43** 195 (1983).
4. N. H. Nickel, P. Lengsfeld, and I. Sieber, Phys. Rev. B **61** 15558 (2000).
5. P. Lengsfeld, S. Brehme, K. Brendel, C. Genzel, and N. H. Nickel, phys. stat. sol. **235** 170 (2003).
6. P. Lengsfeld, N. H. Nickel, and W. Fuhs, Appl. Phys. Lett. **76** 1680 (2000).
7. N. H. Nickel and K. Brendel, Appl. Phys. Lett. **82** 3029 (2003).
8. K. Brendel and N. H. Nickel, in: T. Sameshima, T. Fukyuki, H. P. Strunk, and J. H. Werner (Ed.), Polycrystalline Semiconductors VII-Bulk Materials, Thin Films and Devices, Nara, Japan, Scitech Publ., Uettikon am See, Switzerland, 2002.
9. J. D. Quirt and J. R. Marko, Phys. Rev. B **5** 1716 (1972).
10. H. Ue and S. Maekawa, Phys. Rev. B **3** 4232 (1971).
11. K. Lips, P. Kanschat, S. Brehme, and W. Fuhs, J.of non-Cryst. solids **299 - 302** 350 (2002).
12. W. Fuhs, P. Kanschat, and K. Lips, J. Vac. Sci. Tech. **18** 1792 (2000).
13. N. H. Nickel, N. M. Johnson, and W. B. Jackson, Appl. Phys. Lett. **62** 3285 (1993).

## Time-resolved switching studies in a-Si:H and related films

**P. Stradins, W. B. Jackson[1], H. M. Branz, J. Hu, C. L. Perkins, and Qi Wang**

National Renewable Energy Laboratory, 1617 Cole Blvd. Golden, Colorado 80401, USA
[1] Hewlett Packard Laboratories, 1501 Page Mill Rd., Palo Alto, CA 94304, USA

## ABSTRACT

Switching in a-Si:H and a-Si:HN$_x$ layers is investigated by pulse current transient and Auger scanning microspectroscopy measurements. Switching in a-Si:H with Ag and Cr contacts exhibits 2 different regimes depending on the voltage pulse polarity. With a positive top Ag contact, switching occurs in nanoseconds after a certain latency time, which depends on voltage exponentially. For a negative Ag contact, there is no latency time provided the voltage exceeds a certain critical value. This might be related to interface effects on contact properties or field-assisted metal diffusion. Scanning Auger element micromaps reveal metallic filaments in the switched films. They contain both Ag and Cr throughout the film thickness. Two phases of the filament formation are suggested – a precursor phase and a post-switching phase characterized by local heating and atomic diffusion. Soft and hard switching are observed in a-Si:HN$_x$ films simultaneously and their rates depend strongly on the contact material and applied voltage. Soft switching might be related to the charge trapping in this wide bandgap material.

## INTRODUCTION

Switching in a-Si:H based devices is an interesting physical phenomenon and has a potential for electronic technology[1, 2]. Despite the recent progress, there is no satisfactory explanation of the switching mechanism, particularly in a-Si:H layers with metallic contacts. It is not clear whether the effect is caused by the avalanche-type electrical breakdown, current, thermal runaway, charge accumulation or their combination. The role of contact and interface properties also needs further understanding. There is experimental evidence that switching is accompanied by the formation of submicron diameter filaments extending from the electric contacts into the amorphous switching layer [3-5]. Metal atoms, Ag in particular, are assumed to migrate into these filaments and form permanent conductive paths[6]. Nevertheless, spatially resolved structure analysis has not yet been performed to establish the existence and composition of such metallic filaments on a microscopic level. Lastly, wide badgap materials such as a-Si:HN$_x$ are of interest as switches because of their very low conductivity and ability to accumulate charges at room temperature. The above problems are addressed in this work.

## EXPERIMENTAL

Devices studied in this work consisted of an amorphous switching layer sandwiched between two conducting electrodes. Both intrinsic and 2000ppm p-type a-Si:H switching layers were 100nm thick while a-Si:HN$_x$ was 30nm. These samples were prepared in our standard hot filament CVD reactor at a rate of several Å/s at a typical temperature of 160 °C. A Cr film on glass served as a bottom contact for a-Si:H switches. For a-Si:HN$_x$, the bottom conductive electrode was a multilayer structure on glass, of which the last, a-Si:H p$^+$-layer, contacted the a-Si:HN$_x$. Two types of top contacts were used: a pore structure and a touch contact. The former, prepared by photolithography, consisted of a (0.5mm)$^2$ evaporated an Ag pad with a well-defined square contact area in its center. Only the center area was in contact with the switching layer, while the rest of this pad was isolated by the underlying photoresist. The size of this square contact was typically 20μm. Alternatively, the delicate contact of the surface of the switch layer with a thin pure metal probe tip formed a roughly (50μm)² touch contact. Devices were subjected to voltage pulses from a pulse generator. Pulse rise times were typically 50ns. Series resistors were introduced between the sample and the pulse generator, and also to serve as loads for current transient measurements, which were recorded by a 1GHz digital oscilloscope.

Scanning Auger microspectroscopy of a-Si:H pore structure switches was performed after chemically removing the top Ag contact layer to avoid contamination during the ion milling. Elemental signal x-y maps at 30nm resolution were recorded at different depths of the switch layer, down to the bottom Cr contact.

## EXPERIMENTAL RESULTS

We first investigate the switching in a-Si:H layers with bottom Cr and top Ag contacts. Fig. 1 shows a typical transient response to a rectangular +19V voltage pulse starting at $t = 0$ applied to the p-type a-Si:H switch with a pore electrode. A 1kΩ series resistor limits the maximum current. The top Ag contact is positive with respect to the bottom Cr. Current through the device is low ( ≈ 0.2mA) during a certain latency time $t_L$. At $t = t_L$ the current increases to 17mA within a few 10ns, indicating switching to a low resistance state of 120 Ω. This transition is permanent and non-reversible. Similar behavior is observed using a positive Ag touch contact on intrinsic a-Si:H.

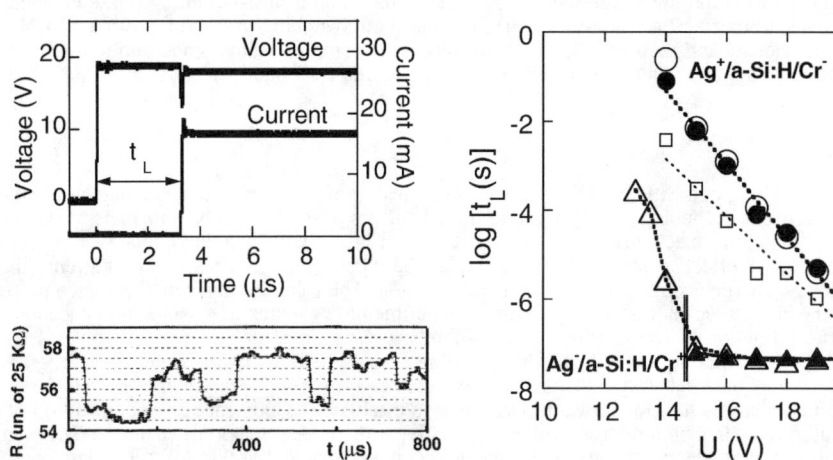

*Fig. 1.* Top - current transient through a p-type a-Si:H switch layer caused by a voltage step (top) applied via 1KΩ series resistor. Ag pore contact of the device is positively biased. Bottom – resistance jumps in the pre-switching state of a-Si:H.

*Fig.2* . Voltage dependences of the switching latency time $t_L$ in an a-Si:H layer with bottom Cr and top Ag contacts. Solid symbols – touch contact, intrinsic film; open symbols – pore contact, p-type film. Circles – top Ag contact positive, room T. Squares – Ag positive, 70°C. Triangles: Ag negative, room T.

The latency time $t_L$ strongly depends on the applied voltage $U$ and temperature, as shown in Fig.2. When the top Ag contact is positive, $t_L$ depends on $U$ exponentially. A change in $U$

from 14 to 20V lead to a decrease in $t_L$ from 100ms to 1µs. These dependences are almost identical for both the pore structure of p-type (open circles) and the intrinsic a-Si:H switch with touch Ag contact (solid circles). At 70°C a similar exponential behavior but faster switching times are observed.

Switching behavior becomes qualitatively different, however, when the switching bias has the opposite polarity (top Ag negative), see bottom curves of Fig.2. In this case, switching occurs after $t_L \approx$ 50ns, provided that applied voltage is above 15V. This small value of $t_L$ actually corresponds to the rise time of the voltage pulse, as confirmed by the transient measurements. Therefore, the actual $t_L$ is less than 50ns. Below 15V, switching becomes very slow or impossible. No switching was observed in intrinsic a-Si:H device with touch contact even for 1s long 14.5V pulses, as denoted by the borderline. In the p-type pore structure the switching was still observed at $U < $ 15V at longer $t_L$, however, this $U$-dependence is very steep. No switching could be induced by 1s pulses of 12.5V.

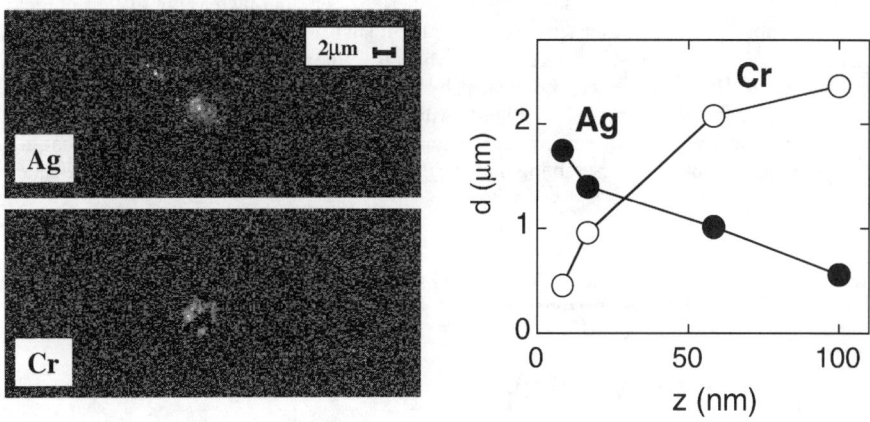

·Fig. 3. Left: scanning Auger x-y maps of Ag and Cr in a switched a-Si:H film after chemically removing the top Ag contact and subsequent ion-milling of 17nm top layer of a-Si:H. Right: Average diameters d of Ag and Cr filaments as functions of the distance z from the top of the a-Si:H film.

It should be noted that switching using both polarities with a 1KΩ series resistor typically resulted in a permanent, non-reversible low resistance state of few hundred ohms even when the pulse length was below 1µs. We have also attempted to establish the existence of a switching precursor states, perhaps formed by transient current filaments. Careful studies of the current transients just below the breakdown (switching) voltage with 100KΩ series resistor revealed that before switching, the current exhibits random jumps resembling telegraph noise (Fig. 1, bottom). The values of these jumps tend to cluster around half multiples of $e^2/\hbar = (25K\Omega)^{-1}$, a quantized conductance fluctuation. It is likely that when enough current is drawn through such a transient filament, a runaway process occurs drawing more current in the filament and causing permanent structural changes in surrounding area, accompanied by metal diffusion and forming a permanent metallic filament. To establish the existence of such a permanent filament, scanning Auger microspectroscopy was employed.

Fig. 3 shows the Auger maps for Ag and Cr in the x-y plane in permanently switched p-type a-Si:H. A 100μs, 18.5V bias pulse applied to a $(12\mu m)^2$ top Ag pore contact (positive) via a 120Ω series resistor resulted in a switching to 140Ω after $t_L$ = 60μs. Auger maps shown were recorded after milling away 17nm of the top of a-Si:H layer. The image brightness is proportional to the element concentration. We indeed observe a metallic region over 1 micron in diameter in which both Ag and Cr are present throughout the whole depth of the a-Si:H film. As shown by the graph on the right, the diameter of the Ag filament from its Auger map decreases from the top surface while that of the Cr increases. Such a metallic filament is caused by switching as confirmed by its absence in unswitched control samples. Typically, one filament per switch was found.

Switching behavior in a-Si:HN$_x$ is different from a-Si:H. Fig. 4 shows the time dependences of the current due to a 1s long positive voltage pulse applied to the touch Ag contact via a 1KΩ series resistor. Typically, there is a gradual increase in the current (curve A). This is not a short-term transient phenomenon but signifies soft switching from a high resistance (GΩ) state to a low resistance (KΩ) state, which is stable for several hours or days. A continuum of lower-resistance states can be created by varying the bias pulse length. The soft switching, however, may get interrupted by a hard switching similar to the one observed in a-Si:H. In the event shown by curve B, soft switching started and continued for 50ms. After that, a sudden current jump took place resembling that in Fig. 1, resulting in a stable low resistance state.

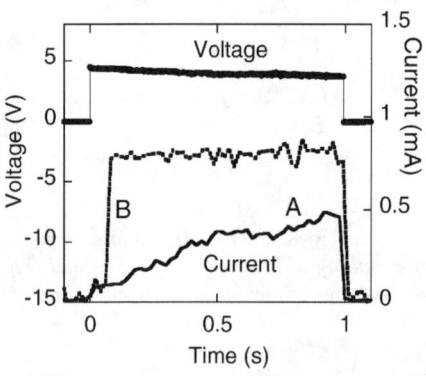
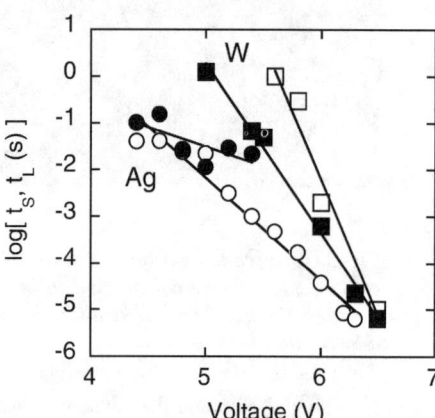

*Fig. 4.* Two types of switching in a-Si:HN$_x$ demonstrated by current transient A (soft switching) and B (hard switching). Switching is induced by a 1s voltage pulse (top).

*Fig. 5.* Average times of soft switching $t_S$ (solids) and hard switching $t_L$ (open) in a-Si:HN$_x$ biased using a positive touch contact tip of Ag (circles) and W (squares).

The characteristic times of the two processes vary statistically from one contact location to another. Their average values exhibit different voltage dependences as shown by Fig. 5. Here, we define soft switching time $t_S$ by ½ the time necessary to form the 5KΩ low-R state. The hard switching time $t_L$ is defined as in Fig.1. In Fig.5, $t_S$ are solid symbols and $t_L$ are open symbols. When Ag is a top positive contact we observe $t_S > t_L$. The hard switching time $t_L$ decreases exponentially with voltage. The changes of $t_S$ with voltage are much weaker. As a result, the soft switching in most cases is interrupted before its completion by the hard switching. This behavior is reversed, however, when the top positive Ag contact tip is replaced by a positive W tip. Here, both types of switching occur at higher voltages compared to the Ag tip. Surprisingly, $t_S < t_L$. Thus, for a W contact the soft switching dominates down to the times of 10µs.

## DISCUSSION

The exponentially decreasing latency time $t_L$ for a-Si:H layers with positively biased - top Ag contacts varies from 100ms to less than 1µs. Thus, the underlying physical process should have such wide timescales. Possible candidates may be slow charge accumulation, heating effects, or structural changes. We have tried to relate the observed $t_L$ to the current values prior to the switching. One might suggest that the exponential dependence of Fig. 2 comes from an exponential I-V dependence (diode behavior) of the device observed at lower voltages. At the high switching voltages, however, this diode behavior unlikely holds. When the voltage was changed from 14 to 20V, the pre-switching current via $(20µm)^2$ contact varied only between 23µA and 350µA, a factor much less than the changes in $t_L$. Also, The Fig.2 dependences of $t_L$ coincide for p-type and intrinsic a-Si:H while the current densities in these samples might be very different. The top curves of Fig.2 resemble the exponential dependences for current instabilities in a-Si:H $p^+$-n-i structures [1] , however, in our case permanent metallic filaments are formed.

The explanation for the drastic changes in the switching behavior with reversed polarity remains a challenging problem. For the p-type a-Si:H switch with a pore electrode, the initial low-voltage I-V curves are strongly asymmetric, with higher current for negative Ag. Because of the similar work functions for Ag and Cr (4.3 and 4.6eV, respectively), the asymmetric IV behavior might be related to a contaminated interface at one of the contacts (likely the top one), for example, an oxide or a residual photoresist layer. On the other hand, the I-V curves for the intrinsic film with touch Ag contacts are almost symmetric. Therefore, there seems to be no evident correlation with the current asymmetry from the contacts.

In our earlier experiments [6] we have observed polarity dependent switching in Cr/a-Si:H/Ag at very long $t_L$ (35s current ramp excitation). Switching with positive top Ag took place at lower current values distributed over a wide range, while higher switching currents and more abrupt transitions were observed with the Ag negative. Field-induced atomic motion of positively charged Ag ions was suggested. Fig.2 offers a somewhat different explanation. Indeed, it demonstrates that with positive Ag, switching is slow but it takes place at lower voltages. In contrast, reaching a relatively high threshold switching voltage is necessary for negative Ag, however, above it the switching is almost instantaneous ($t_L < 50$ns). This explains the easier switching for Ag positive at long $t_L$ in Ref. [6]. It, however, does not unambiguously point to the Ag motion as positive ions because of the observed very fast switching with Ag negative at higher voltages. Experiments with carefully prepared contact interfaces and micro-Auger measurements near the switching threshold might clarify the role of Ag or Cr motion.

The metallic filament (Fig. 3) is observed for the first time with fine spatial resolution. It is composed by both Ag and Cr. Taking into account the high current passed through the switch and the large diameter of the filament (actually, more resembling a disc), it is likely that after the switching, large local heating produces melting and metal diffusion from both contacts.

It seems likely that the first phase of the filament formation involves a local instability (metastable current filament). These transient filaments occur and disappear statistically resulting in telegraphic noise. If the transient filament exists long enough and draws enough current, it is heated and structurally modified by its own current. Due to a positive feedback, a runaway occurs, leading to high temperature and metal diffusion and electromigration from the contacts.

Soft switching in a-Si:$HN_x$ is likely a result of changes in the bulk of the film. Possible candidates may be defect formation or charge trapping. The latter plays an important role in this wide-bandgap material and has been used to explain even the photoinduced changes in subgap absorption [7]. To elucidate the switching mechanism, it is necessary to investigate the stability of the switched state against temperature and light exposure.

## SUMMARY

Time-resolved switching studies in a-Si:H with Ag and Cr reveal that, prior to switching, there is a telegraphic current noise resembling turning on and off of an individual transient current filament. Switching itself takes place in nanoseconds after a certain latency time. The latter varies from seconds to microseconds and depends exponentially on the applied voltage if the top Ag contact is biased positively. In contrast, switching times are in the nanosecond range for negatively biased Ag contacts above a threshold voltage value. Switching in our a-Si:H devices leads to the formation of permanent metallic filaments containing both Ag and Cr, as confirmed by scanning Auger microspectroscopy. Two phases of the filament formation are suggested – a precursor phase (metastable filaments) and a post-switching phase characterized by local heating and atomic diffusion. In a-Si:$HN_x$ films, two types of switching are observed: hard switching similar to that in a-Si:H and soft switching leading to a continuum of low resistance states. Both processes depend strongly on the contact material and applied voltage.

## ACKNOWLEDGEMENTS

The authors gratefully acknowledge the valuable help of Scott Ward and Anna Duda.

## REFERENCES

1. J. Hajto, A. E. Owen, A. J. Snell, P. G. Lecomber and M. J. Rose, in: Amorphous and Crystalline Semiconductor Devices, ed. by J. Kanicki (Artech House, 1992).
2. R. A. Street, Hydrogenated Amorphous Silicon (Cambridge Univ. Press, 1991).
3. P. G. Lecomber, A. E. Owen, W. E. Spear, J. Hajto, A. J. Snell, W. K. Choi, M. J. Rose and S. Reynolds, J. Non-Cryst. Solids 77-8 (1985) 1373.
4. M. Jafar and D. Haneman, Phys. Rev. B 49 (1994) 4605.
5. M. Jafar and D. Haneman, Phys. Rev. B 49 (1994) 13611.
6. J. Hu, H. M. Branz, R. S. Crandall, S.Ward, C.Perkins and Q.Wang, Mat. Res. Soc. Proc. 715 (2002) 763.
7. Y. Nakayama, P. Stradins and H. Fritzsche, J. Non-Cryst. Solids 166 (1993) 1061.

# ADSORPTION AND OXIDATION EFFECTS IN MICROCRYSTALLINE SILICON

T. Dylla[1,2], F. Finger[1], R. Carius[1]

[1] Institut für Photovoltaik, Forschungszentrum Jülich, 52425 Jülich, Germany
[2] Department of Physics, Syracuse University, Syracuse, NY 13244-1130 USA

## ABSTRACT

Electron spin resonance and conductivity measurements were used to study adsorption and oxidation effects on microcrystalline silicon with different structure compositions ranging from porous, highly crystalline to compact, mixed phase amorphous/crystalline. We found a correlation between active surface area and the magnitude of observed meta-stable and irreversible effects.

## INTRODUCTION

Microcrystalline silicon ($\mu$c-Si:H) is a widely studied material and successfully established as an absorber layer in thin film solar cells [1,2]. Especially the reports about the absence of light induced degradation, known as Staebler-Wronski effect, made this material popular [1]. However, the particular structure properties found in $\mu$c-Si:H suggest that instabilities and meta-stability phenomena may also occur in this material. Typically $\mu$c-Si:H consists of columnar clusters of coherent regions separated by crack like voids and disordered material. In the presence of such crack-like voids, impurities or atmospheric gases can easily diffuse along the column boundaries. Evidence for such cracks and diffusion comes from transmission electron microscopy [3,4], infrared spectroscopy [5,6] and hydrogen evolution [7] studies. In particular material grown at high hydrogen dilution, which yields the largest grain sizes and the highest crystalline volume fraction, shows pronounced porosity. The column boundaries are sensitive to chemical reactions leading to termination of surface states or formation of defects and thereby affect the electronic properties of the material. Increasing amorphous phase content seems to reduce the porosity and leads to a better termination of the column surfaces. If built-in into solar cells such compact material, which may contain some amorphous phase yields higher efficiencies [2,8]. Earlier investigations on highly crystalline material prepared with chemical transport deposition show that atmospheric gas adsorption and/or oxidation affects the surface states, electronic transport and the electron spin density [9]. Recently reports show similar effects on the electron spin resonance (ESR) signals in material prepared with PECVD and HW-CVD respectively [10,11]. However any conclusive description of the meta-stability in ESR signals in $\mu$c-Si:H and in particular the microscopic nature of these effects is missing. The most important defect in $\mu$c-Si:H which will contribute to the ESR signal is the Si dangling bond (DB). These DB can be in different environments of the material: the crystalline regions, the grain boundaries, the amorphous regions or in connection with impurity atoms like oxygen [12]. For typical high quality PECVD material the ESR spectrum shows a line, which can be de-convoluted into contributions at g-values of 2.0043 and

2.0052. While it has been suggested that the contributions are actually the contributions from an axially symmetric g-tensor of defects on grain surfaces [13], there are a number of indications that these are two independent states. The latter we also conclude from the present results.

In the present study we want to investigate and identify these instability effects caused by adsorption and oxidation in state of the art material. In particular we want to relate these effects to the structure composition of the material ranging from porous, highly crystalline to compact, mixed phase amorphous/crystalline.

## EXPERIMENT

The material was prepared with PECVD at 95 MHz at a substrate temperature of 200°C. The silane-hydrogen mixture, which was used as process gases, was varied resulting in material with structural compositions ranging from high to low crystalline volume fractions. To obtain highly conductive n-type material with Fermi level position close to the conduction band edge, 10 ppm phosphine was added to the process gas. Al and Mo foil and borosilicate glass were used as substrates. The powder used in ESR measurements was obtained from the Al foil by an HCl etch and sealed in He atmosphere after they were rinsed in de-ionized water and dried at ambient. From the Mo foil the material peeled off in flakes after bending the foil and could be sealed immediately without further treatment. The influence of different environments during storage or annealing was studied by removing the µc-Si:H from the He filled tubes and sealing them into Ar or $O_2$ atmosphere or treating them in HCl or water.

The Raman intensity ratio $I_C^{RS}$ was used as a semi-quantitative measure of the crystalline volume content [4] and will be used to classify the material structure.

ESR was measured in X-band on powdered samples or glass substrates at room temperature and at T=40K. To obtain the spin density for the superimposed resonance lines, a numerical fitting procedure was applied. The db-signals at g=2.0043 and g=2.0052 could be well approximated by Gaussian lines.

## RESULTS

In Fig.1 shows the influence of the etching (HCL + $H_2O$) and drying in ambient atmosphere at room temperature (RT), which is used to obtain powder material from Al foils is shown. Material with different structure composition, which was deposited on Mo foil was exposed to the same etching and drying process used for the Al foil. This process results in considerable increase of the spin density $N_S$ and a shift of the g-value to higher values. All spectra show the typical asymmetric µc-Si:H spectra with contributions at g=2.0043 and g=2.0052. A de-convolution into these two lines shows that changes only occur at g=2.0052. For the highly crystalline porous material with $I_C^{RS}$ = 0.82 the spin density at g=2.0052 increases by about $5 \times 10^{16}$ cm$^{-3}$. On the other hand the spin density of the resonance at g=2.0043 stays constant at values of about $1.5 \times 10^{16}$ cm$^{-3}$. The absolute changes in $N_S$ are considerably less for the more compact structure ($I_C^{RS}$=0.71), and in material at the transition between crystalline and amorphous growth ($I_C^{RS}$=0.47) and for completely amorphous material ($I_C^{RS}$=0). Both HCl/$H_2O$ and $H_2O$ treatments lead to the same increase of $N_S$. The final $N_S$ value after the treatment is very close to the values,

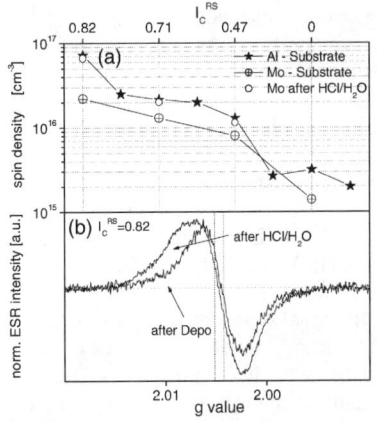

Fig.1 *(a) spin density $N_S$ for samples deposited on Al ($\star$) and Mo before ($\oplus$) and after (o) treatment in $HCl/H_2O$, and (b) ESR spectra before and after $H_2O$ treatment*

which have been reported for material prepared on Al foil (Fig. 1a) [14]. Finally also a simple storage of the samples in air leads to a similar increase of the ESR signal intensity.

The increase of the ESR signal upon contact with water or air at RT is reversible by low temperature annealing (80°C) in Ar. This is shown in Fig. 2a-c. With increasing amorphous content and therefore increasing compactness very little or no changes upon air break and annealing can be observed. For compact material with $I_C^{RS}=0.47$ we observe a little decrease in $N_S$. Again only spins at g=2.0052 are affected by the treatment and the spin density at g=2.0043 stays constant. The reversible changes in the ESR signal, meaning the air break/annealing cycles can be repeated many times without any sign of fatigue.

If on the other hand the material is annealed at 80°C in $O_2$ atmosphere, we observe an irreversible increase of the spin density. The corresponding resonance appears at still higher g-values and shows a different spectral shape compared to the db-signals at g=20052 and g=2.0043. For the $I_C^{RS}=0.82$ material this leads to additional spins of about $10^{17} cm^{-3}$ seen in ESR. With decreasing $I_C^{RS}$ this effect is less pronounced ($2 \times 10^{16} cm^{-3}$ additional spins for $I_C^{RS}=0.71$) and disappears completely for the $I_C^{RS}=0.41$ material.

The increase in $N_S$ upon annealing in $O_2$ or air is non-reversible by simply annealing in Ar but can be restored by an HF dip. This is shown in Fig. 3. For ease of sample handling we used material deposited on glass. The price for that is a considerably lower

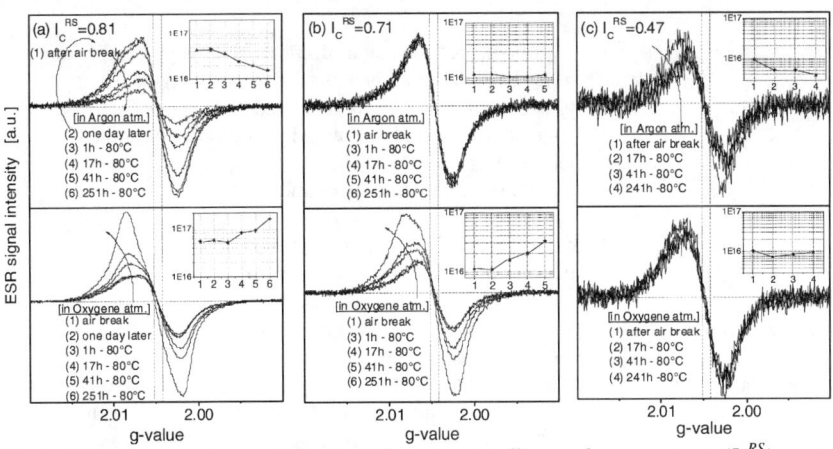

Fig.2 *ESR spectra of sample with different crystalline volume content ($I_C^{RS}$) upon annealing in Ar and $O_2$ atmosphere.*

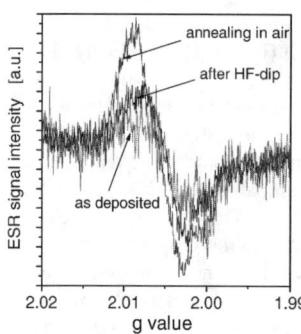

Fig.3 The *increase of $N_S$ due to annealing in oxygen can be restored to its initial value by an HF dip.*

signal to noise ratio due to the reduced sample volume. Still one can easily observe the increase in $N_S$ after annealing in air and the recovery caused by the HF dip.

Annealing steps have also been performed at elevated temperature of 160°C. Fig. 4 shows the results for annealing in Ar or $O_2$. The highly porous sample shown in Fig. 2a was first exposed to air at RT (lower trace Fig 4a, b). Upon annealing in Argon at 160°C a strong contribution of the conduction electron (CE) resonance at g=1.998 [15] and a decrease of the DB signal is observed. Again these changes are reversible and can be treated back to its initial state by exposure to air. Annealing in $O_2$ on the other hand, results in an irreversible increase of the $N_S$ at still higher g-values than g=2.0052.

Exposure and annealing experiments on n-type material with doping concentrations around $10^{17} cm^{-3}$ phosphorous atoms are shown in Fig. 5. The doping shifts the Fermi level towards the conduction band, resulting in occupation of conduction band tail states, which can be observed by the CE signal. Storing and annealing in $O_2$ atmosphere results in a decreasing CE resonance and an increasing dangling bond signal.

Conductivity measurements performed right after deposition in vacuum using a gap contact configuration show also reversible effects due to air break and annealing in Ar. High porous material show a decrease of up to three orders of magnitude due to storing in air. These changes can be restored to its initial value by simply annealing the sample in vacuum at temperatures of 180°C. This is in agreement with earlier investigations done by Veprek et al [9].

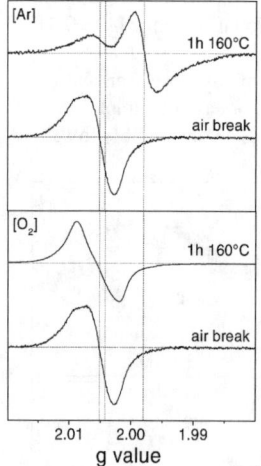

Fig.4 *Changes of the ESR spectra due to annealing at temperatures of T=160°C in Ar and $O_2$ atmosphere.*

## DISCUSSION

In- and out-diffusion of impurities and atmospheric gases leads to numerous instability and meta-stable phenomena on the ESR signal and the electronic conductivity in μc-Si:H already at or close to room temperature.

It appears that the amplitude of these changes is connected to the surface area of the material. $N_S$ changes due to adsorption are highly pronounced for $I_C^{RS}$=0.81 material which is known to be very porous. At lower $I_C^{RS}$ the effects are much smaller or absent. We propose that with increasing amorphous content the material is getting more compact, and the amorphous phase leads to a better termination of the crystal surfaces, which are therefore less susceptible for adsorption of impurities and atmospheric gases.

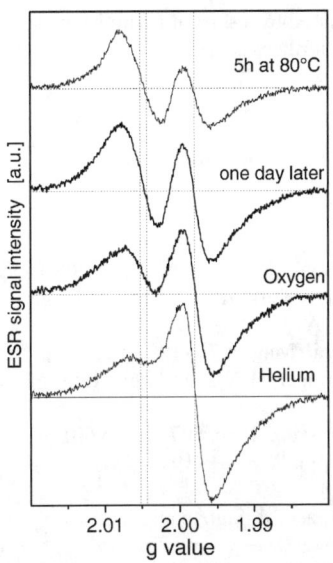

Fig.5 *ESR spectra of n-type doped μc-Si:H sample upon annealing in $O_2$*

It had already been suggested in the early investigations by Veprek and co-workers [9], that adsorption of oxygen or water could be the reason for the meta-stable effects in ESR and conductivity. Intentionally un-doped μc-Si:H generally shows n-type conductivity. The adsorption of an acceptor molecule like $O_2$ or $H_2O$ will create a depletion layer at the surface. The resulting shift of the Fermi level away from the conduction band will be observable as decreasing overall conductivity and also lead to a depopulation of states in the upper half of the gap. This could result in more DB states getting single occupied, which will then be observable in ESR. In particular for highly porous material these surface processes can play an important role or even dominate the electronic transport of the material. The increased amorphous content in the samples shown in Fig. 2b, c leads to a more compact structure and a reduced active surface and therefore adsorption effects are less pronounced and cannot be observed.

Compatible with the change in occupation of near-surface states is the effect of annealing at 160°C in Ar. In highly crystalline material this leads to the appearance of a strong CE signal (Fig. 4), resulting from a shift of the Fermi level into the conduction band tail. The observation is in agreement with the direction and magnitude of changes observed in $\sigma_D$ and indicates considerable n-type background doping of this nominally un-doped sample. The reason for this remains unknown.

The fact that the changes only occur at g=2.0052 supports an earlier assumption, that the related defect is separated from the defect resulting in the g=2.0043 line [12]. Probably the g=2.0052 defect is located at or near the grain boundaries. An alternative explanation for the increase of the resonance at g=2.0052 would relate the resonance with states of the adsorbed species. Further studies will be needed to conclusively decide between these explanations.

Unlike the reversible adsorption effects, the effects of oxidation on the ESR signals are not reversible by moderate temperature annealing in inert gas or long time storage in inert gas or in vacuum. This process seems to be thermally activated, starting already at room temperature but is considerably faster at 80°C and even faster at 160°C. The non-reversible occurrence of additional spins by annealing in oxygen is linked to this oxidation process. The effect is again closely related to the size of the surface area and the fact that the increase of $N_S$ after annealing in $O_2$ or air can be restored by an HF dip supports that the observed effect is a surface process. The depopulation of CE states due to creation of Si-O related defect states gives cause to the assumption that the energy of additional states observed in ESR is below the energy of conduction band tail states. If the additional ESR signal results from a reoccupation caused by a shift of Fermi level or from creation of new defects needs to be investigated in further studies.

## ACKNOWLEDGEMENTS

We are grateful to W. Beyer and M. Hülsbeck for their contributions to this work.

## REFERENCES

1. J. Meier, R. Flückiger, H. Keppner, and A. Shah, Appl. Phys. Lett. **65**, 860 (1994)
2. O. Vetterl, R. Carius, L. Houben, C. Scholten, M. Luysberg, A. Lambertz, F. Finger, and H. Wagner, Mater. Res. Soc. Symp. Proc. 609, A15.2 (2000).
3. M. Luysberg, P. Hapke, R. Carius and F. Finger, Phil. Mag. A **75** (1997), 31.
4. L. Houben, M. Luysberg, P. Hapke, R. Carius, F. Finger, H. Wagner, Phil. Mag. A **77** (1998), 1447.
5. W. Beyer and M.S. Abo Ghazala: Mater. Res. Soc. Symp. Proc. **507** (1998) 601.
6. M. Tzolov, F. Finger, R. Carius and P. Hapke, J. Appl. Phys. **81** (1997) 7376
7. W. Beyer (2002) private communication
8. S. Klein, J. Wolff, F. Finger, R. Carius, H. Wagner, and M. Stutzmann, Japanese Journal of Applied Physics, Part II: Letters **41,** L10 (2002)
9. S. Veprek, Z Iqpal, R. O. Kühne, P. Capezzuto, F.-A. Sarott, J. K. Gimzewski, J. Phys. C: **16**, 6241 (1983)
10. D.Will, C. Lerner, W.Fuhs and K. Lips, Mat. Res. Soc. Symp. Proc. Vol. 467 (1997) 361
11. P.Kanschat, K.Lips, R.Brueggemann, A.Hierzenberger, I. Sieber and W.Fuhs, Mat. Res. Soc. Symp. Proc. Vol. 507 (1998) 793
12. F.Finger, S.Klein, T.Dylla, A.L.Baia Neto, O. Vetterl and R. Carius, Mat. Res. Soc. Symp. Proc. Vol. 715 (2002) 123
13. M. Kondo, S. Yamasaki, A. Matsuda, J. Non-Cryst. Solids, **266-269** (2000) 544.
14. A. L. Baia Neto, A. Lambertz, R. Carius, and F. Finger, Phys. Status Solidi A **186,** R4 (2001)
15. F. Finger, J. Müller, C. Malten, R. Carius, H. Wagner, J. Non-Crystal. Solids 266–269, 511-518 (2000)

## Silicon Nanostructured Films Formed by Pulsed-Laser Deposition in Inert Gas and Reactive Gas

X. Y. Chen, Y. F. Lu, Y. H. Wu, B. J. Cho, H. Hu
Laser Microprocessing Laboratory and Silicon Nano Device Laboratory, Department of Electrical and Computer Engineering, National University of Singapore, 10 Kent Ridge Crescent, Singapore 119260

## ABSTRACT

We reported Si nanostructured films formed by pulsed-laser deposition (PLD) in both inert Ar gas and reactive $O_2$ gas. The as-deposited nanostructured films with visible photoluminescence (PL) show a transition from a film structure to a porous cauliflowerlike structure, as the ambient gas pressure increases from 1 mTorr to 1 Torr. The film consists of small crystals with size from 1 to 20 nm. The oxygen composition of $SiO_x$ increases with increasing $O_2$ gas pressure, while Si $2p$ peak of the Si dioxide also becomes dominate. At 100 mTorr $O_2$ gas, almost complete $SiO_2$ structure is formed. The PL at 1.8–2.1 eV is attributed to the quantum confinement effect (QCE) in Si nanocrystal core, while the PL band at 2.55 eV can be explained by the light emission from the localized surface states at $SiO_x$/Si interface. Laser annealing was applied to the as-deposited nanostructured films. The PL intensities are increased by about two to three times of magnitude after annealing. High laser fluence causes damages in the films and optimal laser fluence exists before film damages or laser ablation occur.

## INTRODUCTION

The observation of photoluminescence (PL) in Si nanostructures [1-3] has attracted great interest due to their potential applications in microelectronics for nanocrystal floating gate memory and in optoelectronics for light emitting devices. The area of Si nanocrystals is currently one of the most active frontiers in physics and chemistry. Si nanocrystals have been synthesized by several techniques. Fabricating size-, distribution- and surface- controlled Si nanocrystals with reproducibility is critical to their promising applications. Among the fabrication methods, pulsed-laser deposition (PLD) is one of the most flexible and promising techniques due to its ability of size distribution control and maintaining crystal purity in a cold-wall processing ambient [4]. In the PLD method, the size distribution of Si nanocrystals can be controlled by varying background gas species and pressure, laser fluence, target-to-substrate distance, and subsequent annealing or oxidation.

On the other hand, the properties and functions of the Si nanocrystals are greatly determined by surface condition and crystal structure. Annealing is necessary to achieve good crystallization. Laser annealing has been well developed to remove lattice damage and defects in crystals. Laser annealing can provide several advantages such as selected area processing, rapid crystallization without any change in intrinsic structure, and avoidance of surface contamination which plays a significant role in the strong visible luminescence from Si nanostructures. In this work, PLD was used to fabricate Si nanocrystals in inert Ar gas and reactive $O_2$ gas. Laser annealing was then applied to the as-deposited Si nanocrystal films. The purpose of this work is to study the effects of deposition conditions and laser annealing on the structures and properties of Si nanocrystals.

## EXPERIMENTAL DETAILS

The PLD system is shown in Fig. 1. The laser beam was directed by a mirror and then focused by a lens (50-cm focal length) onto the target at an incident angle of 45°. After laser ablation of the Si target made from compressed intrinsic Si powder, luminescent Si plasma plume perpendicular to the target surface was generated and expanded toward the substrates which were identical to the target. The ejected species (atoms, ions, clusters with a few atoms) in the plume were cooled down and condensed into nanocrystals in the ambient gas and deposited on Si(100) substrates. A pulsed KrF excimer laser (Lambda Physik 100, $\lambda$ = 248 nm, $\tau$ = 30 ns) was used as light source at a repetition rate of 10 Hz. The laser fluence was set at 3.0 J/cm$^2$ by focusing laser beam with an energy of 90 mJ to a laser spot size of 0.3 × 0.1 cm$^2$. The target was rotated constantly to provide each pulse a fresh surface. The substrates were not heated or cooled during deposition. The substrates were cleaned with acetone and ethanol ultrasonic baths before deposition. After base vacuum was pumped down to $1.0 \times 10^{-5}$ Torr, Ar (purity 99.999%) gas or $O_2$ (purity 99.999%) gas was introduced into the vacuum chamber and maintained at constant pressure during deposition. The deposition was carried out for 60 min at room temperature.

For PLD of Si in Ar or $O_2$ gas, $SiO_x$ (0<x≤2) is formed due to the strong reaction between Si and residual oxygen in the chamber during the expansion of the Si plasma plume. The as-deposited $SiO_x$ films were annealed by KrF excimer laser ($\lambda$ = 248 nm, $\tau$ = 30 ns) at room temperature. Single-pulse laser annealing was performed with laser fluences from 50 to 200 mJ/cm$^2$.

The deposited Si nanocrystals were characterized by several methods. The surface morphology was observed using a Hitachi S-4100 field-emission scanning electron microscope (SEM). The composition was determined by X-ray photoelectron spectroscopy (XPS) with a Physical Electronics Quantum 2000 Scanning ESCA microprobe. The PL spectra were recorded by PL spectroscopy (Renishaw Raman microscope) with an electrically cooled CCD detector at room temperature, using 514.5 nm Ar ion laser line and 325 nm HeCd laser line as excitation sources. The size distribution was investigated by a Digital Instruments atomic force microscope (AFM).

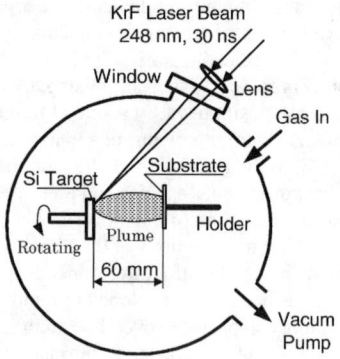

**Figure 1.** Schematic of PLD system.

## DISCUSSION

We deposited Si nanocrystals in Ar and $O_2$ gases at different pressures. Figure 2 shows the surface morphology of Si nanocrystals deposited in Ar observed by SEM. With increasing Ar pressure, there is a transition from a film structure to a porous cauliflowerlike structure. As shown in Fig. 2(a), at a gas pressure of 1 mTorr, SEM image shows a uniform background film as well as a big particle. The big particle is macroscopic droplet of target material that is deposited on substrates along with the film, which is a major drawback of PLD. As shown in Fig. 2(b), at a gas pressure of 100 mTorr, a film with undulating periodic crests and troughs is obtained. Figure 2(c) shows the image of Si nanocrystals deposited in 1 Torr. As shown in the left part of Fig. 2(c), separate crystals are deposited on the surface. They seem to agglomerate together, forming porous cauliflowerlike structure. From the enlarged image in the right part of Fig. 2(c), crystals with an average size of 10 nm can be observed. However, such crystals may be composed of much smaller crystals. The size distribution of Si nanocrystals with different gas pressures can be explained by collisions between the ejected species. The increase of gas pressure results in increasing collisions between the ejected species and the ambient gas. At a pressure of 1 mTorr, the mean free path of the ejected species is approximately 5 cm [5]. With our target-to-substrate distance of 6 cm, it means that there are only 1.2 collisions between the ejected species before they reach the substrate. As a result, solidified liquid droplets expelled from the target are predominant, while the vapor species are deposited as a background film. When the gas pressure increases, more collisions occur. The mean free path of the ejected species is 0.05 cm at a higher pressure of 100 mTorr [5]. The vapor species can undergo sufficient collisions to form small crystals and thus nucleation and growth take place during the Si plasma plume expands towards the substrates. On the other hand, the arrival of the liquid droplets at the substrates is hampered by the increased collisions and gas pressure. A similar transition of surface morphology from a film structure to a porous cauliflowerlike structure is also found in the $O_2$ pressure from 1 mTorr to 1 Torr. Therefore, it can be concluded that the size distribution of Si nanocrystals are greatly influenced by the ambient gas pressure.

The composition of Si nanocrystal films was examined by XPS. For the film deposited in 1 mTorr $O_2$, a peak of elemental Si ($Si^0$) at a binding energy of 99.8 eV and a peak of Si bonding to oxygen ($Si^{4+}$) at 103.5 eV [6] can be observed. When the $O_2$ pressure increases to 100 mTorr, no more elemental Si peak is found and the $Si^{4+}$ peak becomes stronger and shifts to 104.0 eV, which corresponds to $SiO_2$ [6]. Further XPS data analysis confirms that nearly $SiO_2$ stoichiometry is obtained at 100 mTorr $O_2$. Thus, it can be concluded that nearly $SiO_2$ is obtained above the $O_2$ pressure of 100 mTorr.

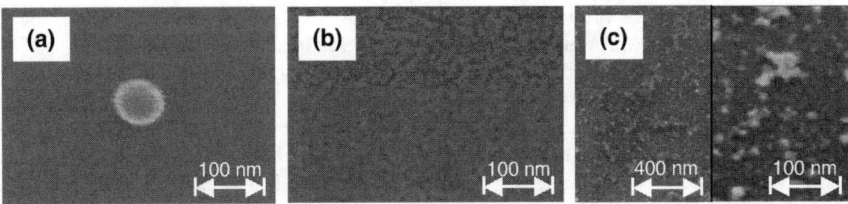

**Figure 2.** SEM images of Si nanocrystals deposited in Ar at a pressure of (a) 1 mTorr, (b) 100 mTorr, and (c) 1 Torr.

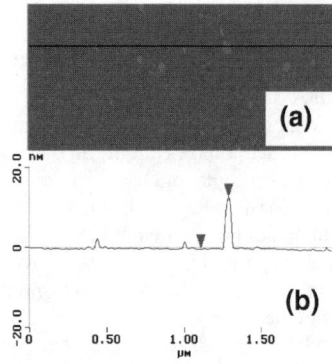 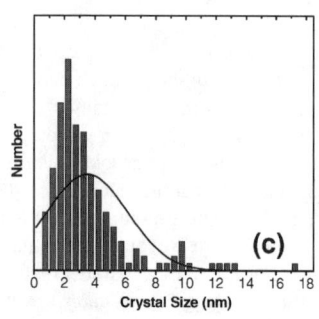

**Figure 3**. (a),(b) AFM image and (c) size distribution of Si nanocrystals deposited in 1 mTorr Ar.

AFM was used to characterize the size distribution of Si nanocrystals. As the Si nanocrystals are formed in the gas phase in the PLD process, the properties of the substrates are inferior. A fraction of a monolayer of clusters was deposited on a highly oriented pyrolitic graphite (HOPG) substrate, which has an atomically flat surface suitable for reliable AFM measurements of nanocluster size. Figure 3(a) presents the AFM image of Si nanocrystals deposited in 1 mTorr Ar gas with a laser fluence of 3.0 J/cm$^2$. It is known that the nanocluster size is presented by the height from AFM measurements, while their lateral dimensions are enlarged due to the tip-object convolution effect. Thus, the height was taken to be the crystal size. Figure 3(b) shows the section analysis and corresponding AFM image. Arrows in Fig. 3(b) indicate a cluster height of ~12 nm. The size distribution histogram from the AFM measurements is presented in Fig. 3(c). It is found that a wide distribution of crystal size from 1 to 20 nm is obtained. The mean crystal size is 3.5 nm while there are two peaks at ~ 2.5 and 10 nm. Theoretical calculations [7] and experimental data [8] have consistently shown that the optical band gap ranges from 1.6 to 2.3 eV for Si nanocrystals with an average size between 5.0 and 2.5 nm, which is in agreement with our results. From the AFM results, the size of the Si nanocrystals is not uniform. As mentioned above, the properties of the deposited species are greatly determined by the collisions between the ejected species. The non-uniformity is due to few collisions at 1 mTorr Ar.

The PL of Si nanocrystals deposited in Ar were measured using two excitation sources. During the measurements, particular attention was paid to keep the excitation power at a constant level. The PL spectra at long wavelength using Ar ion laser line as excitation source are presented in Fig. 4(a). A broad main peak is observed in all films. There is also a small shoulder for the film deposited in 1 mTorr Ar, which corresponds to 1.64 eV. The PL spectra are highly dependent on the Ar pressure. With increasing gas pressure from 0.1 to 1 mTorr, the peak is red-shifted from 2.03 to 1.81 eV while the PL intensity increases. Further increase of gas pressure from 1 mTorr to 1 Torr leads to a blue-shift of the peak position from 1.81 to 2.10 eV and a decrease of PL intensity. At 1 mTorr Ar, there is a turnaround of PL band shift and PL intensity. Figure 4(b) shows the PL spectra at short wavelength using HeCd laser line as excitation source. One single peak at 490 nm is observed in this range, which corresponds to 2.55 eV. The intensity increases with increasing Ar pressure while there is no shift in peak position. It is also found that the films deposited below 20 mTorr Ar show similar spectra as that deposited in 20 mTorr Ar, with a weak intensity PL peaked at 2.55 eV. From Fig. 4(a), it is clear that the PL band at 1.8–2.1 eV is highly dependent on the ambient gas pressure. Different gas pressure results in size variation of Si nanocrystals in the films as well as different surface morphology (different microstructure), which can be responsible for the PL band shift at 1.8–2.1 eV. Thus, the PL band at 1.8–2.1 eV is attributed to the quantum confinement effect (QCE) in Si nanocrystal core. In Fig. 4(b), the PL band at 2.55 eV is only apparent at high Ar pressure and the peak positions are fixed. The origin of this band can be explained by the light emission from the localized surface states at $SiO_x$/Si interface [9].

Single-pulse laser annealing was performed to the Si nanocrystal films deposited in 1 mTorr Ar. Figure 5(a) shows the SEM image after laser annealing at 100 mJ/cm$^2$. A large number of small crystals with sizes from 10 to 30 nm are formed. Figure 5(b) shows the PL spectra of the film after laser annealing at different laser fluences. It is found that the PL intensity increases with increasing laser fluence by about two to four times of magnitude, due to the re-crystallization of Si. At a high laser fluence of 200 mJ/cm$^2$, laser ablation in some local area is observed. The PL spectra measured from the film-unremoved regions show an increased PL intensity. However, the PL intensity in the damaged regions is very weak, since the film has been removed by the ablation. It means that a non-uniform PL is obtained after high fluence laser annealing. Thus, laser annealing should be carried out below the threshold fluence of laser ablation.

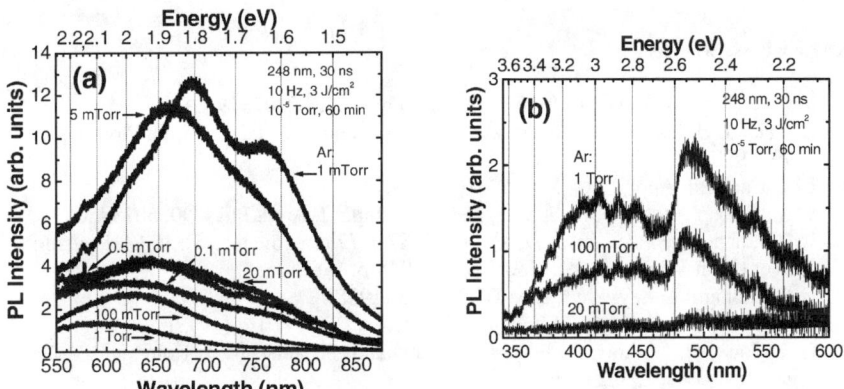

**Figure 4.** PL spectra of Si nanocrystals deposited in Ar at different pressures. (a) The long wavelength range and (b) The short wavelength range.

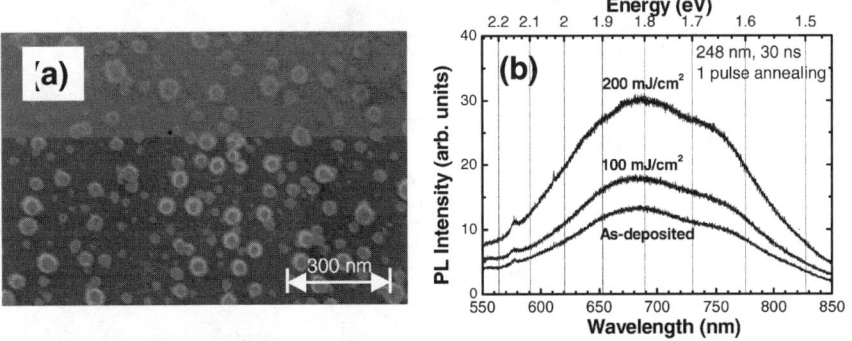

**Figure 5.** (a) SEM image and (b) PL spectra of Si nanocrystal films deposited in 1 mTorr Ar after single-pulse laser annealing.

## CONCLUSIONS

Si nanocrystals have been formed by PLD in Ar and $O_2$ gases. Laser annealing was applied to the Si nanocrystals. The as-deposited Si nanocrystals show a transition from a film structure to a porous cauliflowerlike structure with increasing ambient gas pressure, due to the collisions between the ejected species. The oxygen content of Si nanocrystals increases with increasing ambient pressure and nearly $SiO_2$ stoichiometry is obtained at higher $O_2$ pressure. PL bands peaked at 1.8–2.1 eV due to the QCE in Si nanocrystal core and 2.55 eV due to the localized surface states at $SiO_x$/Si interface were found from the Si nanocrystals. Single-pulse laser annealing improves the crystallinity in the films and enhances the PL intensity. Laser annealing is easy to induce damages and an optimal laser fluence exists for the best annealing effect before film damages or laser ablation occur.

## ACKNOWLEDGEMENTS

The authors would like to acknowledge the help of H. L. Koh and C. H. Soh for their technical support.

## REFERENCES

1. S. Furukawa and T. Miyasato, Jpn. J. Appl. Phys. **27**, L2207 (1988).
2. H. Takagi, H. Ogawa, Y. Yamazaki, A. Ishizaki, and T. Nakagiri, Appl. Phys. Lett. **56**, 2379 (1990).
3. L. T. Canham, Appl. Phys. Lett. **57**, 1046 (1990).
4. T. Makino, Y. Yamada, N. Suzuki, and T. Yoshida, J. Appl. Phys. **90**, 5075 (2001).
5. L. Y. Chen, in *Pulsed Laser Deposition of Thin Films*, edited by D. B. Chrisey and G. K. Hubler (John Wiley & Sons, New York, 1994), p. 186.
6. F. G. Bell and L. Ley, Phys. Rev. B **37**, 8383 (1988).
7. C. Delerue, G. Allan, and M. Lannoo, Phys. Rev. B **48**, 11024 (1993).
8. G. Ledoux, O. Guillois, D. Porterat, C. Reynaud, F. Huisken, B. Kohn, and V. Paillard, Phys. Rev. B **62**, 15942 (2000).
9. K. Kimura and S. Iwasaki, Jpn. J. Appl. Phys. **38**, 609 (1999).

# ORIGIN OF THE LOW-ENERGY PHOTOLUMINESCENCE IN MICROCRYSTALLINE SILICON FILMS

Joon-Yong Lee, Dong-Hyun Park, and Jong-Hwan Yoon
Department of Physics, College of Natural Science, Kangwon National University, Chuncheon, Kangwon-Do 200-701, Korea

## ABSTRACT

In this work we have investigated the low-energy photoluminescence (PL) band with a peak between 0.8 eV and 1.0 eV for microcrystalline silicon films ($\mu$c-Si:H) grown under various growth conditions. At least four subbands are observed, the peaks of which are located near 0.80 eV, 0.87 eV, 0.92 eV, and 0.97 eV, respectively. It is suggested that the low-energy PL band basically arises from a superposition of these subbands, whose intensities strongly depend on deposition conditions, and thus its peak is determined by the sum of these subband intensities. From the results, it is suggested that the subband centered at 0.92 eV originates from defect-related radiative recombination in the amorphous phase rather than radiative band tail-to-tail transitions in the grain boundaries.

## INTRODUCTION

It is well known that hydrogenated microcrystalline silicon films ($\mu$c-Si:H) reveal a low-energy photoluminescence (PL) band with a peak located between 0.8 eV and 1.0 eV, the peak of which strongly depends on growth conditions. Despite many studies on the low-energy PL, however, its origin is still controversy. Bhat et al.[1] observed two PL bands centered at 0.9 eV and 1.3 eV in the $\mu$c-Si:H films. They suggested that the low-energy PL band is due to the defect-related radiative recombination. Komuro et al.[2] observed several PL bands centered at energy between 0.76 eV and 1.24 eV, depending on deposition conditions. They suggested that the low-energy PL band is due to the defects created in the amorphous phase. Carius et al.[3] and Kalkan et al.[4] also reported the low-energy PL band, the peak position of which strongly depends on preparation conditions. They proposed that the low-energy PL in $\mu$c-Si:H arises from a superposition of different contributions from the defect-related recombination in the crystalline phase. Recently, on the other hand, Yue et al.[5, 6] have suggested that the low-energy PL originates from radiative tail-to-tail transitions in the grain-boundary region. Savchouk et al.[7] have also suggested that the 0.9 eV PL band observed in polycrystalline silicon films arises from the radiative transitions between conduction and valence band tails in the grain boundaries. In this work we present experimental results that the low-energy PL band basically arises from a superposition of several subbands and the 0.9 eV PL band arises from the defect-related recombination in amorphous region.

## EXPERIMENTAL DETAILS

The $\mu$c-Si:H films studied in this work were grown by a conventional plasma-enhanced chemical vapor deposition (PECVD) excited at 13.56 MHz using either hydrogen or argon diluted $SiH_4$ gas. The base pressure before deposition was less than $7 \times 10^{-7}$ torr while the pressure during deposition was kept constant at a value of 300 mtorr and 650 mtorr for hydrogen and argon diluted samples, respectively. The hydrogen and argon dilution ratios were varied in gas flow rates. The samples were grown on Corning 7059 glasses for electrical measurements while polished glasses were used for PL and Raman measurements. The microstructures of µc-Si:H films were analyzed using the Raman back scattering taken at room temperature using the 514.5 nm Ar-ion laser line. The deposition conditions and physical properties of the samples used in this work are described in Table I. The PL spectra were measured using a liquid nitrogen cooled Ge detector and lock-in technique. The 488 nm line of an Ar-ion laser was used as the excitation source and the PL spectra were corrected for the system response. The samples were mounted on a cold stage, whose temperature was varied from 10 K to 300 K using a closed-cycle helium refrigerator.

**Table I.** Deposition conditions and physical properties of µc-Si:H films studied. $E_a$, dark conductivity activation energy; $\sigma_d$, dark conductivity; $X_a$, amorphous phase volume fraction, $X_c$, crystalline volume fraction

| Sample | Gas Flow Rate (sccm) | | | $T_s$ (°C) | RF Power (Watt) | $E_a$ (eV) | $\sigma_d$ (S/cm) | $X_a$ | $X_c$ |
|---|---|---|---|---|---|---|---|---|---|
| | $SiH_4$ | $H_2$ | Ar | | | | | | |
| KM01 | 5 | 145 | 0 | 220 | 20 | 0.27 | $8.6 \times 10^{-5}$ | 17 | 83 |
| KM02 | 3 | 147 | 0 | 170 | 20 | 0.32 | $1.2 \times 10^{-5}$ | 33 | 67 |
| KM03 | 2 | 0 | 98 | 220 | 20 | 0.52 | $1.5 \times 10^{-4}$ | 18 | 82 |
| KM04 | 2 | 0 | 98 | 220 | 30 | 0.50 | $1.8 \times 10^{-4}$ | 18 | 82 |

## RESULTS AND DISCUSSION

### Characteristics of the low-energy PL band

First, characteristics of the low-energy PL band were examined by measuring the temperature dependence of the low-energy PL band for µc-Si:H films of high crystalline volume faction, $X_c$, grown from hydrogen and argon diluted silane. Figs. 1(a) and (b) show the temperature dependences of the low-energy PL band plotted as a function of emission energy for samples grown from $H_2$ (KM01) and Ar (KM03) dilution but with a similar value of $X_c$, respectively. For both the samples it is observed that the PL band centered at 1.2~1.4 eV, due to amorphous phase [8], is very small to be negligible. Fig. 1 shows that the characteristics of the low-energy PL band are significantly affected by deposition condition and temperature. At low temperatures below 30 K, the KM01 sample reveals the PL peak around 0.87 eV while the KM03 reveals the PL peak near 0.97 eV. As temperature increases, these PL peak intensities continuously decrease while other peaks appear to be more dominant. For the sample grown using $H_2$ dilution (KM01) it is observed that the band near 0.80 eV becomes dominant, while for the sample grown using Ar dilution (KM03) it is observed that the bands near 0.87 eV and 0.92 eV become dominant. Here, note that there is no distinct shift in the positions of the peaks. These results suggest that the low-energy PL band observed in μc-Si:H films may arise from a superposition of several subbands, the peaks of which are located near 0.80 eV, 0.87 eV, 0.92 eV, and 0.97 eV, due to either crystallites, grain boundaries, or amorphous phase.

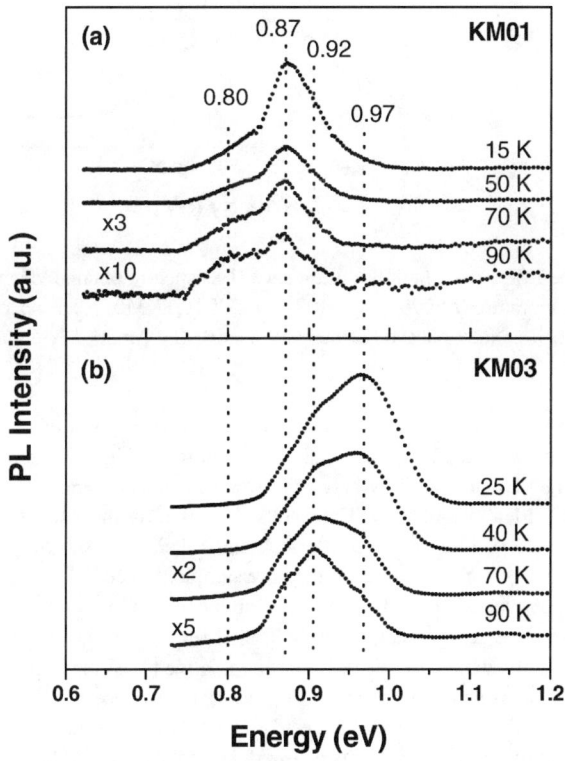

**Figure 1.** Temperature dependences of the low-energy photoluminescence bands obtained from μc-Si:H films of high crystalline volume fraction grown from (a) hydrogen and (b) argon dilution.

### The influence of defects on the low-energy PL band

In order to understand the origin of the subbands, the effect of defects on the low-energy PL band has been studied. Defect density was varied by using either low substrate temperature or high RF power. Fig. 2 represents the low-energy PL spectra obtained from two kinds of μc-Si:H films (KM01 and KM02) grown at different substrate temperatures using the same $H_2$ dilution.

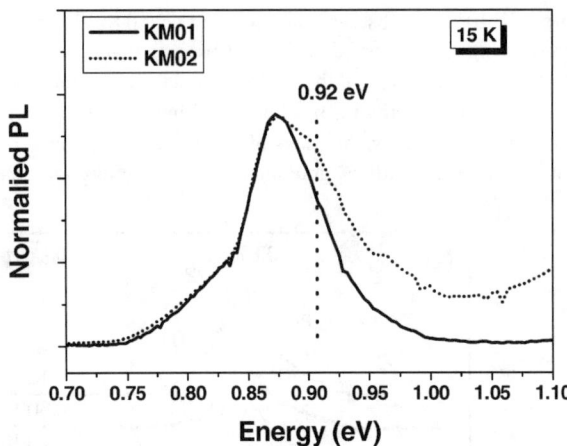

**Figure 2.** The low-energy photoluminescence (PL) spectra obtained from μc-Si:H films grown at substrate temperatures of 170 °C (KM02) and 220 °C (KM01). The spectra were normalized to the peak intensities. Note that the PL intensity near 0.92 eV markedly enhances in the film grown at 170 °C.

The PL spectra were measured at 15 K using an excitation intensity of 10 mW. They were normalized to the peak intensities near 0.87 eV.   Shown in Fig. 2 is there is no distinct change in the shapes of the PL below 0.87 eV between two samples, but there is a pronounced change in the shapes of the PL above 0.87 eV. The results show that for the low substrate temperature sample (KM02) the PL intensities near 0.92 eV and 1.3 eV ( not shown ) markedly increase, as compared to those of high substrate temperature sample. In general, a decrease in the substrate temperature below optimal condition results in an increase of amorphous volume fraction in the μc-Si:H, and an increase of defect density in a-Si:H films[9]. As a result, the increase in the PL intensity near 1.3 eV, which originates from the radiative transitions between the conduction and valence band tails, is due to the increase in the amorphous phase. On the other hand, the origin of the increase in the PL intensity near 0.92 eV is open to all possibilities except the radiative band tail-to-tail transitions in the amorphous phase: defect-related luminescence in amorphous phase; band tail- and defect-related luminescence in the grain and grain boundaries. However, the results in Fig. 1(b) show that the position of peak of 0.92 eV does not change as temperature increases. This is different from that observed in the band tail-related luminescence, accompanying the shift of peak position to low energy by rapid thermalization of band tail carries as temperature increases. As a result, the PL band centered at 0.92 eV shown in Fig. 2 is due to defect-related luminescence rather than band tail-related one. Furthermore, Raman spectrum of the low substrate temperature µc-Si:H film (KM02) shows that the intensity of shoulder near 480 cm$^{-1}$ significantly increases, which means an increase of amorphous phase and in turn, an increase of defect density. From these results it is suggested that the PL band centered at 0.92 eV may be due to the luminescence related to defects created in the amorphous phase rather than defects in the grain or grain boundaries.

The origin of the PL band near 0.92 eV was further examined for the µc-Si:H films grown at different RF powers using argon diluted silane gas. Fig. 3 shows the low-energy PL spectra obtained from two kinds of µc-Si:H films grown at RF powers of 20 W (KM03) and 30 W (KM04) using the same argon dilution.

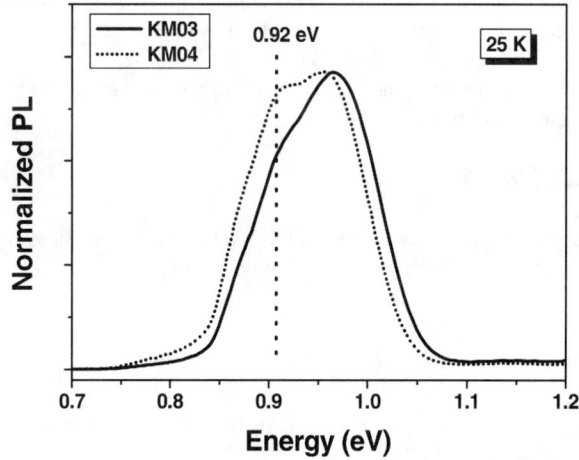

**Figure 3.** The low-energy photoluminescence (PL) spectra obtained from µc-Si:H films grown at RF powers of 20 W (KM03) and 30 W (KM04). The spectra were normalized to the peak intensities. Note that the PL intensity near 0.92 eV markedly enhances in the film grown at a power of 30 W.

The PL spectra were measured at 25 K using an excitation intensity of 10 mW. The PL spectra were normalized to the peak intensities near 0.97 eV. Fig. 3 shows a small red-shift in the PL band and an increase in the PL intensity near 0.92 eV with increasing RF power. No distinct change in the shapes of the Raman spectra between two samples is observed, indicating there are no changes in the crystalline, grain boundaries, and amorphous phase volume fractions. In general, it is known that argon dilution results in an increase of defect density by ion bombardment. In particular, the higher RF power is, the larger the defect density is[9]. These results indicate that the increase in the PL intensity near 0.92 eV as shown in Fig. 3 may be mainly due to an increase in the defect density in the amorphous phase. This suggestion can be also confirmed by the results in Figs. 1. Despite no change in the Raman spectra between H$_2$ and

Ar diluted samples, Fig. 1 shows that the PL near 0.92 eV appears in the Ar diluted sample while it is undetectable in the $H_2$ diluted one. The reason is that Ar dilution results in defect density larger than $H_2$ dilution. As a result, it is suggested that the PL band near 0.92 eV observed in μc-Si:H films may be due to the defects in the amorphous phase.

## SUMMARY

A systematic study has been made on the low-energy photoluminescence band with a peak between 0.8 eV and 1.0 eV in microcrystalline silicon. Temperature dependence of photoluminescence spectrum is examined for characterizing the low-energy photoluminescence band. And also substrate temperature, dilution gas, and RF power dependences of the low-energy photoluminescence band are examined for exploring the influence of defects on the low-energy photoluminescence. It has been suggested that the low-energy PL band basically arises from a superposition of subbands, centered at 0.80 eV, 0.87 eV, 0.92 eV, and 0.97 eV. It has been suggested that the PL subband centered at 0.92 eV originates from recombination at defect centers in the amorphous phase.

## ACKNOWLEDGMENT

This work was supported by Grant No. C1-2002-007-0-4 from the University Research Program supported by the Ministry of Information & Communication in Republic of Korea.

## REFERENCES

1. P. K. Bhat, G. Diprose, T. M. Searle, I. G. Austin, P. G. LeComber, and W. E. Spear, Physica B **117&118**, 917 (1983).
2. S. Komuro, Y. Aoyagi, Y. Segawa, and S. Namba, J. Appl. Phys. **58**, 943 (1985).
3. R. Carius, F. Finger, U. Backhausen, M. Luysberg, P. Hapke, L. Houben, and H. Overhof, Mater. Res. Soc. Symp. Proc. **467**, 283 (1997).
4. A. K. Kalkan, S. J. Fonash, and S. Cheng, Appl. Phys. Lett. **77**, 55 (2000).
5. G. Yue, J. D. Lorentzen, J. Lin, and D. Han, Appl. Phys. Lett. **75**, 492 (1999).
6. G. Yue, Daxing Han, L. E. McNeil, and Qi Wang, J. Appl. Phys. **88**, 4904 (2000).
7. A. U. Savchouk, S. Ostapenko, G. Nowak, and L. Jastrzebski, Appl. Phys. Lett. **67**, 82 (1995).
8. R. A. Street, J. C. Knights, and D. K. Biegelsen, Phys. Rev. B **18**, 1880 (1978).
9. D. K. Biegelsen, R. A. Street, C. C. Tsai, and J. C. Knights, Phys. Rev. B **20**, 4839 (1979).

## Post-transit Analysis of Transient Photocurrents from High-Deposition-Rate a-Si:H Samples

Monica Brinza[1], W.M.M. Kessel[2], Arno H.M Smets[2], M.C.M van de Sanden[2], Guy J. Adriaenssens[1]
[1]Halfgeleiderfysica, University of Leuven, Celestijnenlaan 200D, B-3001 Leuven, Belgium
[2] Department of Applied Physics, Eindhoven University of Technology, P.O. Box 513, 5600 MB Eindhoven, The Netherlands

### ABSTRACT

An interpretation of post-transit photocurrents in a time-of-flight experiment in terms of the underlying density of localized gap states in the sample is presented for the case of hydrogenated amorphous silicon cells prepared by the expanding thermal plasma technique. It is pointed out that part of the observed current is not generated by re-emission of trapped photo-generated charge and should, therefore, not be used for density-of-states calculations.

### INTRODUCTION

High growth rates as well as high hole drift mobilities make the expanding thermal plasma (ETP) technique an attractive method for the deposition of hydrogenated amorphous silicon [1]. Previous time-of-flight (TOF) measurements on ETP a-Si:H samples [2,3] showed an interesting behavior of the electron post-transit photocurrents. Above room temperature, the photocurrent decay changes from a $t^{-1}$ dependence to a much slower one. Converting the photocurrent into a density-of-states (DOS) curve, a huge bump in the DOS towards mid-gap is obtained. However, a thorough analysis of the data carried out for this study reveals the fact that the flattened part of the photocurrent is not due to the carriers being released from deep traps. Consequently, it should not be used when deriving the density of states. A correct analysis of the experimental data leads to gap state densities of several times $10^{16}$ cm$^{-3}$eV$^{-1}$. The likely origin of the excess non-emission current is also examined.

### EXPERIMENTAL DETAILS

Post-transit photocurrents were measured in a standard TOF set-up on samples with a sandwich contact configuration. A strongly absorbed laser pulse is used to create electron-hole pairs just behind the semi-transparent blocking electrode. An applied field of the appropriate polarity will drive one type of carriers through the sample. If the aim of the experiment is to measure transit times in order to calculate the carrier drift mobility, the voltage must be applied shortly before the light pulse such that a uniform field is ensured in the sample during the measurements. A HP214B voltage pulse generator with maximum pulse duration of 10ms is used for this purpose. In the post-transit regime, the uniform field is no longer an absolute requirement [4] and a DC voltage source is used to measure photocurrents beyond 10ms.

The transient photocurrents presented here were measured on ETP samples deposited at substrate temperatures between 250°C and 500°C (Table 1) at the Eindhoven University of Technology (The Netherlands) [1] with a growth rate of 6-7 nm/s. The samples listed in Table 1 are sandwiched between two Cr electrodes, the top one being semi-transparent. Apart from these, a sample provided with semi-transparent Mo contacts and deposited at 450°C substrate temperature was available. All samples were ~2.5μm thick.

## POST-TRANSIT PHOTOCURRENT ANALYSIS

During their transit through the sample the photo-generated carriers interact with the distribution of the localized states in the gap. In the framework of the multiple trapping transport model, the carriers that leave the sample at the transit time are those which sampled only the shallow part of DOS, while the carriers trapped deeper will be released to the transport path (and thus contribute to the photocurrent) at much longer times. On the assumption that retrapping in deep states is not a significant factor, standard post-transit photocurrent analysis (PTPA) derives a simple relationship between the measured photocurrent I(t) and the DOS g(E) [5]:

$$I(t)t = \frac{Q_0 t_0 v_0}{2g(0)} g(E), \qquad E = kT \ln(v_0 t), \tag{1}$$

Here $Q_0$ is the total photo-induced charge, $t_0$ the trap-free transit time, $v_0$ the attempt-to-escape frequency and $g(0)$ the DOS at the mobility edge. While $Q_0$ and $v_0$ can, in principle, be estimated from the post-transit photocurrent curves, values for $t_0$ and $g(0)$ have to be extracted from other experiments.

## EXPERIMENTAL RESULTS

Figure 1a shows transient electron photocurrents measured at different temperatures between room temperature and 85°C. After an initial bulge due to RC effects of the measurement circuit (region I), the photocurrent decay follows approximately a $t^{-1}$ law (region II) up to a time $t_1$. After this, the photocurrent decay is leveling off (region III). This behavior is observed for electrons in all Cr/a-Si:H/Cr samples, but the leveling-off of the photocurrent appears only at more elevated measurement temperatures for samples deposited at lower substrate temperatures. In the hole transients, the leveling-off seems to be related to the aging of the samples, being absent for relatively fresh samples.

The usual explanation for this behavior would be that the flattening of the photocurrent beginning at $t_1$ is due to a large number of carriers being released from a defect band deep in the gap. As the emission of trapped carriers is thermally assisted, the higher the temperature, the sooner the carriers will be released, with

$$t = v_0^{-1} \exp(E/kT) \tag{2}$$

representing the average release time of a carrier initially trapped at depth E.

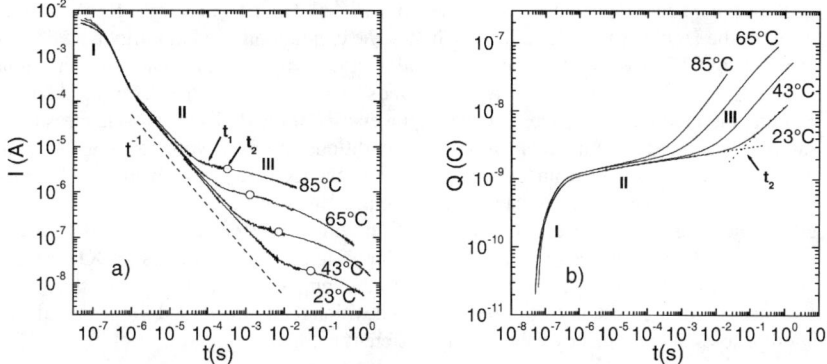

**Figure 1** a) Post-transit photocurrent transients measured in the sample RS604 at different temperatures; b) The photocharge curves obtained by integration of the photocurrents.

Instead of determining $t_1$ from I(t) curves, one can use the integrated photocurrent curves, Q(t) (Fig.1b), where the delimitation between region II and III is more distinctive. In practice, the data before and after the kink are fitted to two lines, and a time $t_2$ (with a similar meaning as $t_1$) is read at the intersection. It was found that $t_2$ is systematically somewhat larger than $t_1$, but that the current value $I(t_2)$ shows the same $t^{-1}$ dependence as $I(t_1)$.

The thermally activated behavior of $t_2$ is shown in Fig.2a for the samples of Table 1 with deposition temperatures of 325°C and higher. According to Eq.(2), this should allow us to interpret it as an "emission time" [6] of charge from a deep trap. The best fit to $t_2$ gives values of $v_0$ in the range $10^{13}$-$10^{15}$ s$^{-1}$ (Table 1). Such values are much higher than the commonly accepted $10^{12}$ s$^{-1}$ value for the attempt-to-escape frequency.

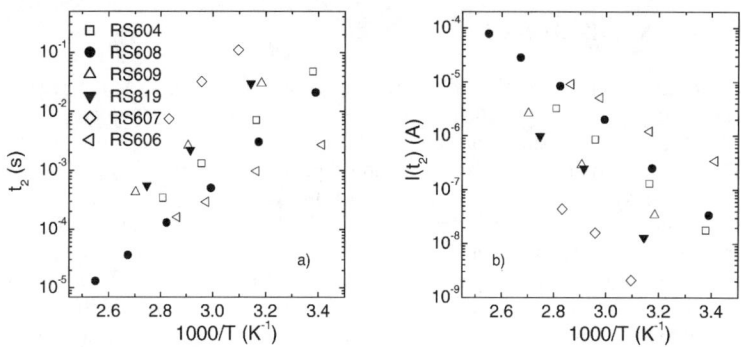

**Figure 2** Reciprocal temperature dependence of a) $t_2$ and b) $I(t_2)$ as measured in various samples.

Subsequent use of, for instance, the $v_0 = 5 \times 10^{14}$ s$^{-1}$ value from sample RS608 in Eq.(1) resolves the DOS between 0.5 and 1 eV below the conduction band mobility edge with a "defect band" below 0.75 eV. On the other hand, dark conductivity measurements give an activation energy $E_d = 0.73$ eV, which roughly corresponds to the position of the Fermi level below the conduction band mobility edge. The activation energy of the dark current was measured on samples deposited in comparable deposition conditions and fitted with coplanar electrodes, as well as on some samples from Table 1. As most of the localized states below the Fermi level are already filled with electrons in thermal equilibrium, they have a low probability of capturing excess electrons. As a consequence, no significant photoinduced electron emission is expected to be observed from states below 0.73 eV. Thus, in principle, no picture of the DOS can be obtained below this limit. To resolve this contradiction, we have to accept that the feature which we interpreted as emission from deep traps, and which then directly led to the unrealistically high attempt-to-escape frequency value, is of a different nature.

**Table 1.** Characteristic parameters of the samples, where $T_{dep}$ is the substrate temperature at deposition, $E_{a\_}t_2$ is the activation temperature of the time $t_2$ (as defined in the text), $E_a\_I(t_2)$ is the activation temperature of the photocurrent measured at $t_2$ and DOS is the average density of gap states calculated from curves as in Fig.3.

| Sample | $T_{dep}$(°C) | $v_0$ (s$^{-1}$) | $E_a\_t_2$(eV) | $E_a\_I(t_2)$(eV) | DOS(cm$^{-3}$eV$^{-1}$) |
|---|---|---|---|---|---|
| RS600 | 250 | - | - | - | 5 10$^{16}$ |
| RS607 | 325 | 6 10$^{14}$ | 0.88 | 0.95 | 3 10$^{16}$ |
| RS609 | 400 | 5 10$^{13}$ | 0.76 | 0.77 | 2 10$^{16}$ |
| RS819 | 400 | 2 10$^{15}$ | 0.87 | 0.95 | 2 10$^{16}$ |
| RS608 | 425 | 5 10$^{14}$ | 0.76 | 0.80 | 4 10$^{16}$ |
| RS604 | 450 | 1 10$^{14}$ | 0.74 | 0.78 | 4 10$^{16}$ |
| RS606 | 500 | 3 10$^{10}$ | 0.45 | 0.53 | 7 10$^{16}$ |

## ANALYSIS AND DISCUSSION

The re-examination of the PTPA data requires that a few corrections to the presumed DOS image be made:
(i) the final flattened part of the photocurrent traces cannot be used in deriving the DOS, as it is not due to thermal emission of trapped photogenerated carriers
(ii) by analogy, the estimated $Q_0$ value in Eq.(1) should be $Q(t_2)$

The maximum value of integrated photocurrent $Q_0 = 10^{-7}$ C (equivalent to 6 10$^{11}$ elementary charges) inferred from Fig.1b cannot be rejected as being too high by the simple comparison with the number of photons per light pulse. The laser specifications give 2 10$^{14}$ photons per pulse, but the number of photons reaching the a-Si:H layer is smaller and difficult to estimate directly due to losses in the dye cell, absorption in the semi-transparent metal contact and reflection.

Table 1 presents values for the gap DOS obtained when the pre-factors in Eq.(1) are chosen to be: $Q_0 = Q(t_2)$, $v_0 = 10^{12}$ s$^{-1}$, $g(0) = 10^{21}$ cm$^{-3}$eV$^{-1}$ and $\mu_0 = 8$ cm$^2$V$^{-1}$s$^{-1}$. The DOS is now resolved between 0.35 and 0.65 eV and shows a profile which is flat or slightly rising towards midgap (Fig.3). The new value for $Q_0$ does now correspond to similar values obtained from TOF measurements on a-Si:H samples produced by standard plasma-enhanced chemical vapor deposition techniques. On the basis of analogous considerations we deduce a DOS of 5 $10^{15}$ - 2 $10^{16}$ cm$^{-3}$eV$^{-1}$ on the valence band side of the gap from hole post-transit photocurrents.

Next, the origin of the final flattened part of the photocurrent will be investigated. The value of the current at $t_2$ is a good measure for the magnitude of the current in the region III immediately after the departure from the $t^{-1}$ decay. One interesting experimental finding was that the $I(t_2)$ shows a thermally activated behavior (Fig.2b) similar to the one of the dark current, with an activation energy (Table 1) somewhat higher but still comparable to the activation energy of the dark current (~0.73eV). This fact leads us to consider the leveling-off part of the photocurrent as a leakage current due to a modification of the contact barrier at the illuminated electrode with the incoming light. The current flow through a reverse biased Schottky metal/semiconductor contact is limited by the electrons having to overcome the potential barrier $\Phi_B$ equal (for the ideal case) to the difference between the extraction work of the metal $\Phi_M$ and the electron affinity $\Psi_S$ of the semiconductor. Without detailing the possible transport mechanisms through the barrier, we assume that a small lowering of this barrier due to, for example, light induced dipoles at the interface metal/semiconductor, will allow more of the thermally available carriers to flow through the sample. The barrier exposed at the laser pulse is expected to recover its equilibrium profile after a time interval equal to the dielectric relaxation time of the semiconductor. Experimentally it is indeed observed that the leakage current fades away for times longer that 10-100 ms.

Figure 4 depicts a simple model that can explain why the time $t_2$, whose interpretation as an emission time proved to be erroneous, shows a thermally activated behavior. We consider here $t_2$ as being the intersection point of a photocurrent which has a $t^{-1}$ decay $I_{Ph}=C\ t^{-1}$ and is basically temperature independent and a leakage current which we consider in the first approximation to be constant in time but showing a temperature activated behavior $I_L=I_0 \exp(-E_a/kT)$. From $I_{Ph}=I_L$ one gets $t_2 \sim \exp(E_a/kT)$. From Table 1 it is evident that even if the activation energy of $t_2$ is in general up to 0.07 eV lower than the activation energy of $I(t_2)$, the two sets of values are still consistent. On the premise of the model described above, no better agreement is expected.

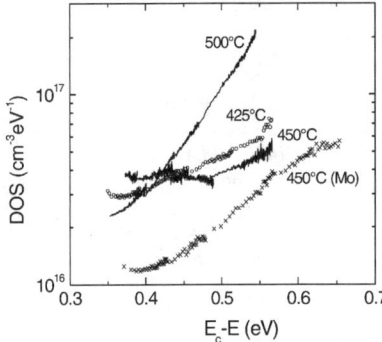

**Figure 3.** DOS profile for TOF samples deposited at different substrate temperatures. One sample has Mo contacts.

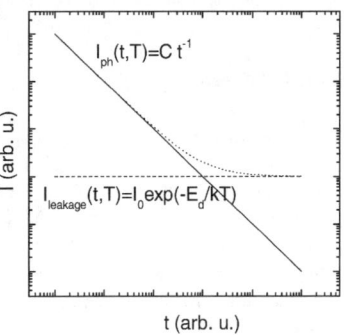

**Figure 4.** Schematic diagram showing the photocurrent as being made up of two components

The leakage current seems to appear only in the particular case of Cr/ETP a-Si:H contact. Post-transit photocurrents measured in our group on Cr/a-Si:H/Cr samples, with the a-Si:H layer deposited by different methods [4,5] have never shown such a peculiar behavior. Moreover, measurements performed on a test ETP sample sandwiched between semi-transparent Mo electrodes do not show a leakage current. By illuminating the sample through the substrate, electron post-transit photocurrents were measured up to 0.5 s and the resolved DOS profile is included in Fig.3 .

While helping to establish the role of the specific Cr/ETP Si barrier in generating the excess current, the Mo-sandwiched sample also illustrated the problems that can arise at the Schottky barriers. Indeed, in the case of Mo/a:Si:H/Mo sample, a new puzzle appears when the sample is illuminated through the top contact. In this situation, and in contrast to illumination through the bottom contact, the photocurrent measured when DC voltage is applied to the sample is almost one order of magnitude smaller than the photocurrent measured with the pulsed voltage. In the DC case the field in the sample is completely relaxed and confined to the region near the top contact while the rest of the sample is almost field free. The motion of carriers through such a non-uniform field can indeed produce a lower current in comparison with the pulsed voltage measurements when the field is uniform. A strong asymmetry of the top and bottom contacts can account for the differences in the post-transit photocurrents measured when either top or bottom contact is illuminated.

## CONCLUSIONS

Post-transit photocurrent analysis applied to ETP a-Si:H deposited at rates of 6 to 7 nm/s reveals a gap-state density of 2-4 $10^{16}$ $cm^{-3}eV^{-1}$ on the conduction band side for the deposition temperatures of 400 to 450°C that produce the best quality material at that deposition rate. A density of 5 $10^{15}$ - 2 $10^{16}$ $cm^{-3}eV^{-1}$ is found on the valence band side of the gap. To deduce this values, care has to be taken to avoid or disregard current contributions that are related to the Schottky barrier behavior rather than to the emission from deep a-Si:H traps.

## REFERENCES

1. W. M. M. Kessels, R. J. Severens, A. H. M. Smets, B. A. Korevaar, G. J. Adriaenssens, D. C. Schram and M. C. M. van de Sanden, *J. Appl. Phys.* **89**, 2404 (2001).
2. G. J. Adriaenssens, H.-Z. Song, V. I. Arkhipov, E. V. Emelianova, W. M. M. Kessels, A. H. M. Smets, B. A. Korevaar and M. C. M. van de Sanden, *J. Optoel. Adv. Mater.* **2**, 31 (2000).
3. M. Brinza, G. J. Adriaenssens, K. Iakoubovskii, A. Stesmans, W. M. M. Kessels, A. H. M. Smets and M. C. M. van de Sanden, *J. Non-Cryst. Solids* **299**, 420 (2002).
4. M. Nesládek, F. Schauer, P.Brada, G.J.Adriaenssens and L.M. Stals, , *J. Non-Cryst. Solids* **164**, 505 (1993)
5. G. F. Seynhaeve, R. P. Barclay, G. J. Adriaenssens and J. M. Marshall, *Phys. Rev. B* **39**, 10 196 (1989).
6. H. Antoniadis and E. A. Schiff, *Phys. Rev. B* **46**, 9482 (1992).

## Size Distribution Of Embedded Nano-Crystallites In Polymorphous Silicon Studied By Raman Spectroscopy And Photoluminescence

V. Tripathi, Y. N. Mohapatra
Department of Physics, IIT, and Kanpur-208016, INDIA E-mail: vibha@iitk.ac.in
Md. N. Islam
QAED/SRG, Space Application Centre (ISRO), Ahemdabad-380015 (India)
V. Suendo, P. Roca i Cabarrocas
Laboratoire de Physique des Interfaces et des Couches Minces (UMR 7647 du CNRS),
Ecole Polytechnique, 91128 Palaiseau Cedex, France

**Abstract.**

Polymorphous Silicon (pm-Si:H) deposited by Plasma Enhanced Chemical Vapour Deposition (PECVD) has emerged as an alternative material to amorphous silicon (a-Si:H). Deposition parameters of pm-Si:H are such that small crystallites get embedded in a relaxed amorphous silicon matrix, thus improving the optical and electrical properties. We study the size of crystallites and degree of order in pm-Si:H using Raman and photoluminescence (PL) spectra of pm-Si:H and a-Si:H. Raman Spectra of a variety of hydrogenated nanostructured silicon (pm-Si:H) and amorphous Silicon (a-Si:H) samples grown at different pressures were analyzed. Deconvolution of observed multiple peaks in photoluminescence spectra and fitting to Gaussian size distribution also yields particle size to be in the range of 2.3 to 3.5nm in agreement with Transmission Electron Microscopy and Raman results.

## INTRODUCTION

Hydrogenated amorphous Silicon (a-Si: H) is important to large area electronics in applications such as thin film transistors for flat panel displays, solar cells and photocopiers etc. Recently, it has been shown that optical and transport properties of such material can be improved by embedding small nanocrystallites in a-Si:H matrix. This material has been termed as hydrogenated polymorphous silicon (pm-Si: H). It is deposited by PECVD under conditions close to dust formation in the vacuum chamber. Presence of nanocrystallites in amorphous medium changes the properties of the material. pm-Si:H has been shown to have much lower density of states [2] and devices made by pm-Si:H have exhibited superior characteristics [3]. Understanding the role of crystallites in altering the properties of this material is crucial for its further improvement and application. Quantum confinement influences both phonon and electronic states. The influence of size effect on phonons can be traced to changes in lineshape of Raman lines corresponding to a particular mode. The size effect in nanoparticles of silicon can induce band to band transitions resulting in radiative recombination in an otherwise indirect band gap semiconductor. Photoluminescence due to nanostrucure of silicon has been widely studied in varied systems including isolated clusters, porous silicon, embedded nanocrystallites. Hence a large variety of information is already available. It has also been observed in PECVD grown a-Si:H which has nanocrystallites incorporated into it [4,5]. We seek to use this wealth of information to deduce particle size and their distribution in polymorphous silicon. The size dependence on bandgap is used as given by Proot et al[ 6]. Optical spectroscopic studies of such material can give dual benefit of being convenient characterization tool for nanoparticles and at the same time give insight into the electronic properties.

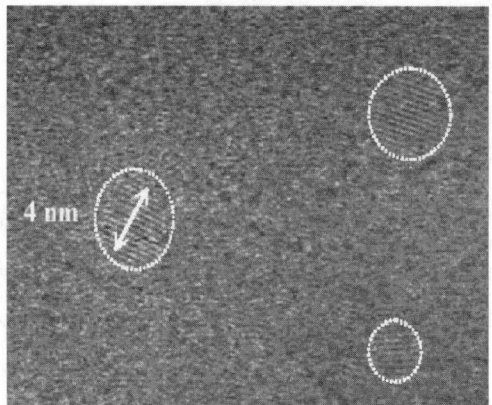

**Figure 1:** HRTEM picture of 10 nm thick polymorphous silicon film where nanocrystalline particles can be distinguished.

In this paper, we use Raman spectroscopy and photoluminescence spectroscopy(PL) to study the particle size distribution in PECVD grown pm-Si:H. We show that the incorporation of the Gaussian size distribution into phonon confinement models and size effect can be used to obtain mean size and distribution of nanoparticles in amorphous matrix.

**EXPERIMENT:**

A series of samples was deposited on corning glass by dissociation of strongly diluted (2-3%) silane in hydrogen ambient by rf (13.6MHz) glow discharge method[7]. Total gas pressure was varied between 0.6 torr to 1.7 torr. The resulting deposition rate was ~1Å/s and the sample thickness was ~ 300nm. Raman spectroscopy was performed on these samples using He-Ne laser and Ar-ion laser (514nm). The laser was operated at 100mW and the integration time was set to 100 sec in order to avoid crystallization during measurement. Photoluminescence experiments were performed using 270 nm line from a xenon lamp source.

**THEORY:**

The Raman intensity $I(\omega,L)$ given by the phonon confinement model for a spherical nanocrystal with diameter $L$ is [8,9,10]

$$I(\omega,L) \propto \int \frac{|C(q,L)|^2 d^3q}{(\omega-\omega(q))^2 + (\Gamma_0/2)^2} \quad (1)$$

where $\omega(q)$ is phonon dispersion relation with the phonon momentum $q$. $\Gamma_0$, the natural line width of bulk $c$-Si, and $C(q,L)$ is the Fourier coefficient of the phonon confinement function $W(r,L)$. Though there are different forms of $W(r,L)$, the Gaussian form is the most natural assuming unbiased randomness and is given by

$$W(r,L) = \exp(-\alpha r^2/L^2) \quad (2)$$

In order to include the effects of a random distribution of crystallite sizes in real samples, we incorporate the size dispersion in the Raman intensity profile by integrating Raman profile for a single nanocrystallite [Eq. (1)] over the appropriate crystal size distribution. Using Gaussian distribution for the crystallites and after reduction, the equation for intensity can be written as

$$I(\omega, L_0, \sigma) \propto \frac{f(q)\exp\left\{-\frac{q^2 L_0^2 f^2(q)}{2\alpha}\right\}}{(\omega - \omega(q))^2 + (\Gamma_0/2)^2} \quad (3)$$

where $f(q) = \left[1 + \frac{q^2 \sigma^2}{\alpha}\right]^{-1/2}$, $f(q)$ incorporates distribution-broadening parameter σ into the Raman line shape expression [11].

We have used an analytical model recently developed for interpretation of PL in case of nanoparticles of silicon[12]. The model incorporates features such as gap widening due to quantum confinement effect, the oscillator strength, exciton binding energy, and localized surface states and the surrounding medium. This phenomenological model has been shown to be able to predict PL peak position and line shape with respect to the mean crystallite size and its dispersion.

**RESULTS.**

Fig. 1 shows HRTEM picture of a 10 nm thick pm-Si:H deposited at 1.2 Torr. In this figure, a crystallite of 4nm in size is highlighted and several other crystallites can be deciphered on close inspection. Another discernible feature in this picture is the general improved ordering of the material[1]. This is due to the fact that ultra-small particles have fused into a-Si:H matrix, their surface energy being too small to retain their separate identity. There is also a definite size distribution in pm-Si:H. Fig. 2 shows experimental TO line for both type of samples along with fits to various models for comparison. As a consequence of presence of nano-crystallites, perfectly symmetric line shape of TO Raman signal found in a-Si:H becomes asymmetric.

For pm-Si:H, and in fact, this asymmetry is the first signal of the presence of

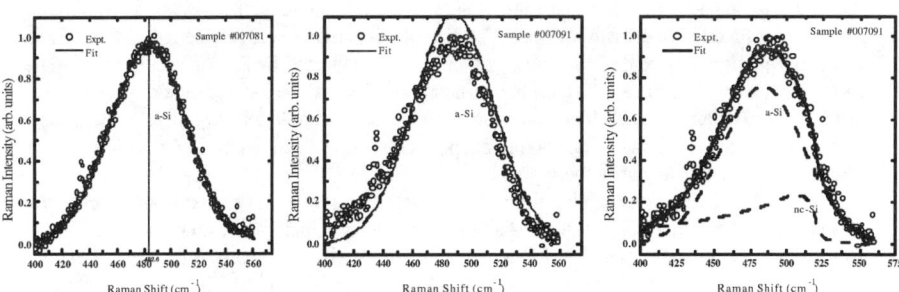

**Figure 2.** The left most graph shows Raman of a-Si:H sample, the second one shows a polymorphous sample with a-Si:H fit, and the third one is a pm-Si:H sample fitted with a-Si:H and crystalline phase.

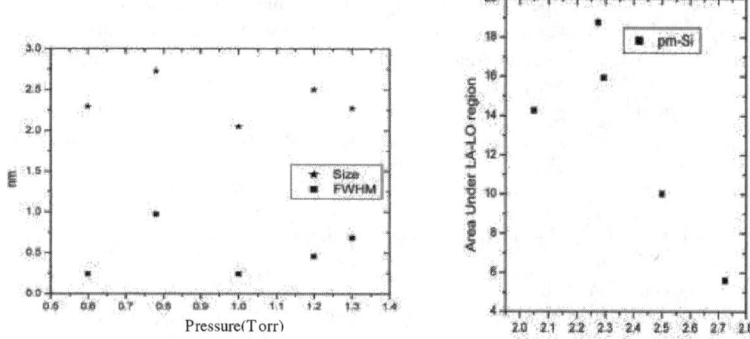

**Figure 3.** Variation of particle size(*) and FWHM of distribution (■) with pressure. Particle size is roughly same for all pm-Si:H samples

**Figure 4.** Plot of Area under LA-LO peak against Particle Size suggests a correlation between particle size and diminishing LA-LO shoulder.

nanocrystallites. Fig. 2(a) shows TO peak in Raman spectra of a-Si:H fitted with a gaussian line shape. Similar fit shown in Fig. 2(b) pm-Si:H gives higher standard error, which is also reflected in the quality of the fit in this graph. In Fig. 2(c), signal due to pm-Si:H has been fitted with our model, which includes amorphous silicon and nanocrystalline silicon; the particle size distribution is incorporated as described in the last section. As compared to other models, our model gives a larger downshift with crystallite size; this is a natural consequence of incorporating particle size distribution in the model. We note that whenever there is a distribution of particle size, the Raman peak position shifts towards lower wave numbers on decreasing the crystallite size. An increase in size dispersion influences the two sides of the peak asymmetrically. The response is weighted towards smaller crystallite size as larger numbers of phonons with smaller frequencies participate. This results in longer and stronger low-frequency Raman tail giving an asymmetric spectrum. Fig. 3 gives variation in particle size along with the FWHM as a function of growth pressure. The size of the particles seems to be roughly the same. The concentration may be different but that would not be detected by Raman spectroscopy. The variation with pressure in the polymorphous silicon samples may be masked either by the fact that in our model amorphous silicon and nanocrystalline silicon are highly co-related systems or indeed, there may not be much variation in the polymorphous silicon in terms of particle size. More work is needed to clarify these possibilities.

The LA-LO regions ranging from $300 cm^{-1}$ to $380 cm^{-1}$ and TA region from $160 cm^{-1}$ to $200 cm^{-1}$ also show interesting features. As fig. 4 demonstrates there seems to be definite correlation between area under LA-LO region and particle size. Comparison between TO and TA regions is shown in Fig. 5, where the Raman signal is between $160 cm^{-1}$ to $200 cm^{-1}$. pm-Si:H sample has comparatively much larger TO peak. Ratio between TA and TO peak has been considered to be a good indicator of medium range order in a given sample. Kshirsagar et al [14] and Voyles et al [13] have shown that $I_{TA}/I_{TO}$ ratio can provide a measure of medium range order. This ratio is similar for the three polymorphous silicon samples grown at different pressures indicating that the degree of medium range order is similar.

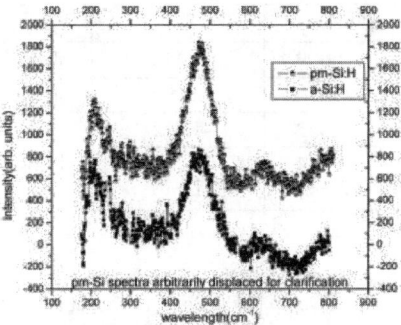

**Figure 5.** Untreated pm-Si and a-Si:H raman signals superimposed to show $I_{TA}/I_{TO}$ variation. Spectra are shifted along the y-axis for clarity

We have observed PL on a variety of samples. A typical spectra for a sample grown on a cold substrate at a pressure of 0.8 Torr is shown in fig. 6. Multiplicity of peaks in PL spectra in this range of wavelength are indicators of presence of nanocrystallites. The spectra were fitted to the model described elsewhere in detail [12]. The results of the parameters so obtained are listed in Table I. The table shows that the mean particle size varies between 2.1-3.5nm consistent with TEM and Raman spectroscopic results. Note that in contrast to line broadening in case of Raman spectroscopy, PL results in multiple peaks due to distinct sizes. The change of mean particle size for peaks in PL corresponds to approximately addition of one monolayer for particles as shown by number of layers in a particle listed in the last column of the Table I. Hence the distribution for luminescence purposes can be considered to be of consisting of discrete sizes in a narrow range of values of diameter. PL spectra of corning glass is also shown for reference.

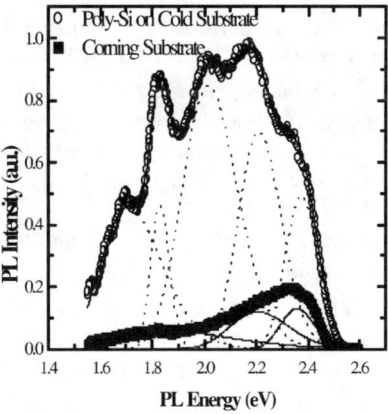

**Figure 6.** Photoluminescence spectra of pm-Si:H deposited at 800mT is fitted with particle size and size distribution. Details of the fit are listed in Table I. The reference spectra for the substrate i.e. corning glass alone is also shown.

| PeakPosition(eV) | Mean (Å) | Std.Dev (Å) | FWHM (Å) | L0/D111 |
|---|---|---|---|---|
| 1.7 | 34.090 | 4.009 | 9.440 | 10.857 |
| 1.85 | 31.618 | 0.976 | 2.298 | 10.069 |
| 2 | 26.962 | 2.052 | 4.832 | 8.587 |
| 2.2 | 23.990 | 1.184 | 2.788 | 7.640 |
| 2.38 | 21.977 | 0.727 | 1.712 | 6.999 |

**Table I:** PL Fitting Parameters: to model of ref. [11]. The last column is the ratio of the mean size to thickness of the monolayer in [111] plane. Note that last two peak positions are also present in Corning glass

## CONCLUSIONS:

Raman Spectra of five hydrogenated polymorphous silicon samples were analyzed in order to set a procedure that allows easy determination of local structure parameters of polymorphous silicon. By using a phenomenological model developed by us in the TO region of the spectra, we could predict particle size and particle size distribution- typical particle size being in the range of 3nm with width of the distribution ranging between 0.3 to 0.9nm. A deconvolution of multiple peaks in PL spectra also gives similar particle size distribution.

## References:

1. A. Fontecuberta i Morral, H. Hofmeister, P. Roca i Cabarrocas, J. Non Cryst. Solids (299-302),284,(200)
2. M. Meaudre, R. Meaudre, Butte´, S. Vignoli, C. Longeaud, J. P. Kleider, and P. Roca i. Cabarrocas, J. Appl. Phys. **86**, 946 (1999).
3. C. Voz, J. Puigdollers, A. Orpella, R. Alcubilla, A. Fontecuberta i. Morral, V. Tripathi, P. Roca i. Cabarrocas, *J. Non Cryst. Sol.* **299-302**,1345 (2002)
4. M.-B.Park and N.-H.Cho, *Appl. Surf. Sci.*, **190** 151-156, 2002.
5. G. Viera, S. Huet, E. Bertran, L. Boufendi , *J. Appl. Phys*, **90** ( 8 ) 4272 (2001).
6. J.P. Proot, C. Delerue, and G. Allan, *Appl. Phys. Lett.* **61**, 1948 (1992).
7. P. Roca i Cabarrocas, J. B. Chévrier, J. Huc, A. Lloret, J. Y. Parey, and J. P. M. Schmitt, *J. Vac. Sci. Technol. A* **9**, 2331 (1991)
8. H. Richter, Z. P. Wang, and L. Ley, *Solid State Commun.* **39**, 625 (1981).
9. P. A. M. Rodrigues, H. A. Cerdeira, and F. Cerdeira, *Int. J. Mod. Phys.B* **3**, 1167 (1989).
10. I. H. Campbell, and P. M. Fauchet, *Solid State Commun.* **58**, 739 (1986).
11. Md. N. Islam, and S. Kumar, *App. Phys. Lett.*, **78**(6) 715 (2001)
12. Md. N. Islam and Satyendra Kumar, *J. Appl. Phys.*, **93**(11)1753 (2003).
13. P. M. Voyles, N. Zotov; S. M. Nakhmanson, D. A. Drabold, J. M. Gibson, M. M. J. Treacy, and P. Keblinski, *J. Appl. Phys* **90**(9), 4438 (2001).
14. S. T. Kshirsagar, and J. S. Lannin, *Phys. Rev. B* **25**, 2196 (1982).

## Thin film cavity ringdown spectroscopy and second harmonic generation on thin a-Si:H films

I.M.P. Aarts, B. Hoex, J.J.H. Gielis, C.M. Leewis, A.H.M. Smets, R. Engeln, M. Nesládek[1], W.M.M. Kessels and M.C.M. van de Sanden
Department of Applied Physics, Eindhoven University of Technology, P.O. Box 513, 5600 MB Eindhoven, The Netherlands
[1]Institute for Materials Research, Limburgs Universitair Centrum, Wetenschapspark 1, B-3590 Diepenbeek, Belgium

## ABSTRACT

A set of 8 rf deposited a-Si:H thin films of various thickness (4-1031nm) have been used to explore the applicability of two optical techniques, thin film cavity ringdown spectroscopy (tf-CRDS) and second harmonic generation (SHG), for the measurement of small defect-related absorptions. In this paper we will give a first overview of the different aspects of these techniques, which are novel in the field of amorphous silicon materials. It is shown that tf-CRDS is capable of measuring defect-related absorptions (associated with dangling bonds) as small as $10^{-7}$ for a single measurement, without the need for elaborate calibration procedures. The results are compared with photothermal deflection spectroscopy (PDS) for a broad spectral range (0.7 – 1.7 eV) and show good agreement. Furthermore the existence of a defect-rich surface layer with a defect density of $1.1 \times 10^{12}$ cm$^{-2}$ has been proven. The absorption spectrum of a 4 nm thin film has revealed a different spectral signature than a bulk dominated (1031 nm) film. The SHG experiments on a-Si:H films have shown that the second harmonic signal arises from the surface states and polarization dependent studies have revealed that the surface states probed have an ∞m-symmetry. From this it can be deduced that the absorbing surface states are isotropically distributed. A spectral scan suggests that the second harmonic signal, whose origin has not been unrevealed yet, has a resonance at an incident photon energy of 1.22 eV.

## INTRODUCTION

Due to the continuous advancement in thin film technologies, there exists a constant need for improved and/or new film diagnostic techniques. For a-Si:H for example, a considerable number of techniques that give information on the defect density in the a-Si:H film is already available, however it is still believed that the research field can benefit from new diagnostics with promising perspectives such as: easy-to-apply without calibration procedures, ultrahigh sensitivity such that also very thin films can be probed, possibility to distinguish defects in the bulk and surface region of the films, applicable during light-soaking and during real-time film growth, etc. For this reason, we are currently looking into the application of two optical diagnostics in the field of a-Si:H defect spectroscopy.

The first technique, cavity ringdown spectroscopy (CRDS) is well known from gas phase and plasma absorption measurements [1,2]. CRDS is ultra-sensitive (minimum absorption sensitivity < $10^{-12}$ per pass) and gives directly the absorption value without a calibration procedure. This ultra-high sensitivity of CRDS, makes it also applicable in the field of a-Si:H defect measurements and it could be complementary to existing techniques such as photothermal deflection spectroscopy (PDS) and the constant photocurrent method (CPM). Recently, some studies have been performed [3-6] to introduce CRDS into thin solid films. In this paper we present results on the absorption spectrum of the sub-gap states of a-Si:H using a broad range of photon energies and we compare the results to those obtained with common techniques such as PDS and transmission spectroscopy [7,8].

The second technique is second harmonic generation (SHG). SHG is well known in surface science studies of dangling bonds on crystalline silicon surfaces [9]. In these studies the second harmonic signal is resonantly enhanced in the 1.1-1.5 eV photon energy range because these photon energies line up with optical transitions from the valence band to Si dangling bond surface states. Furthermore SHG is surface specific because SH generation is prohibited in bulk crystalline silicon due to symmetry considerations. Here we present SHG experiments on a-Si:H because it is expected that SHG might also reveal information about the (surface) dangling bond states in a-Si:H [10].

Both techniques have been applied to a set of 8 a-Si:H films deposited by an rf plasma (at the Delft University) under the same experimental condition but each with a different thickness (4-1031 nm). All samples are deposited on high quality quartz. The experiments have been performed *ex situ* such that a native oxide is present on the a-Si:H surface. Finally after presenting the results on both techniques, the future prospects of both techniques will be given such as the application during real time film growth of a-Si:H.

## THIN-FILM CAVITY RINGDOWN SPECTROSCOPY (tf-CRDS)

A conventional pulsed general CRDS measurement, see figure 1, uses a pulsed laser to excite a stable cavity that is formed by two highly reflective plano-concave mirrors. Due to finite reflection coefficient of the mirrors part of the electric field will leak through the mirrors, therefore the optical field inside the cavity will decay exponentially with a certain time constant, which is called the ringdown time.

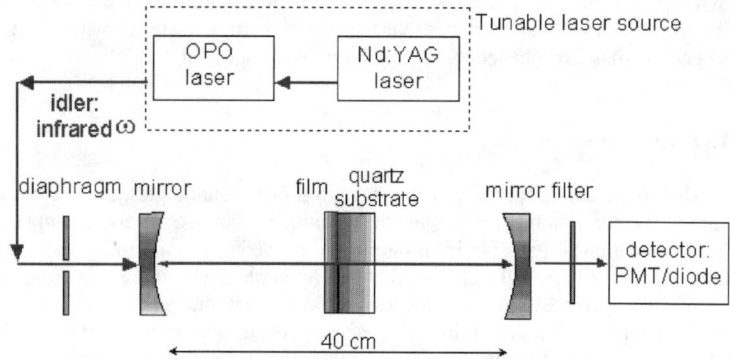

**Figure 1.** A general setup for tf-CRDS measurements. Two highly-reflective mirrors form the optical cavity. An optical parametric oscillator laser serves as monochromatic light source.

For low absorptions the Lambert-Beer law can be linearized and the ringdown time $\tau(\omega)$ is given by the relation

$$\tau(\omega) = \frac{t_r}{\sum_i L_i} = \frac{t_r}{L_0(\omega) + L_{abs}(\omega)}, \quad (1)$$

with $t_r$ the round-trip time of light propagating in the cavity, $L_0(\omega)$ the intrinsic losses of the cavity, and $L_{abs}(\omega)$ the absorption losses induced by the medium to be probed. The absorption losses $L_{abs}(\omega)$ in the a-Si:H can directly be determined by comparing the ringdown time for a cavity with a quartz substrate and a-Si:H film with the decay time of a cavity with only a quartz substrate. The sensitivity of CRDS is determined by the product of the intrinsic loss and relative uncertainty in the measured ringdown time: $(L_{abs})_{min} = L_0(\Delta \tau/\tau)_{min}$, and therefore the intrinsic loss should be as low as possible and the ringdown time should be measured with the highest accuracy. This has been achieved by using very high quality quartz as substrate material and a state of the art real-time data acquisition system. We have established that for our experimental configuration $(L_{abs})_{min} \approx 10^{-7}$ (single shot), which means that the tf-CRDS technique is indeed extremely sensitive. An extensive overview of the theory and applications of CRDS can be found in Ref. [2].

The absorption coefficient can now be determined from the measured absorption loss $L_{abs}(\omega)$ by correcting for interference effects that are present in the thin a-Si:H film. For this reason an *ab initio* model has been developed that corrects for these interference effects. Figure 2 shows the absorption coefficient as measured with tf-CRDS for a 1031 nm thick a-Si:H film as a function of photon energy. To prove the validity of the result we have performed transmission spectroscopy measurements as well as photothermal deflection spectroscopy (PDS) on the same sample. The results are also given in Fig. 2 and good agreement between the three techniques is observed. From Fig. 2 the average defect density can be calculated and is found to be $2.1 \times 10^{16}$ cm$^{-3}$ [11]. Figure 3 shows the average defect density for the complete sample set (4-1031 nm).

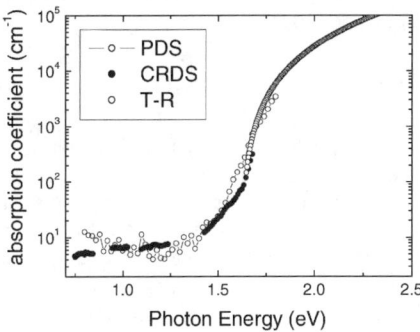

**Figure 2.** Absorption coefficient α as a function of photon energy for three different techniques: tf-CRDS, PDS and transmission spectroscopy.

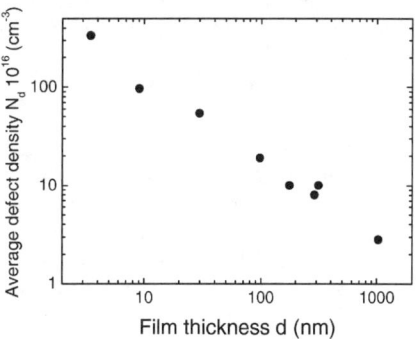

**Figure 3.** Average defect density in a-Si:H thin films as a function of film thickness d.

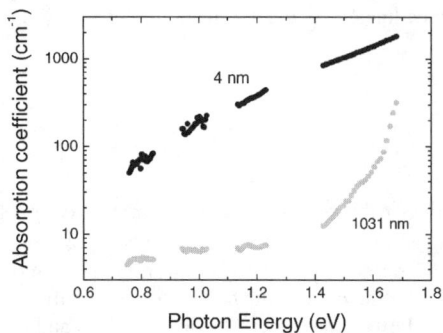

**Figure 4.** Absorption coefficient α as a function of photon energy for a 4 nm and 1031 nm a-Si:H film.

It can be seen that the defect density increases exponentially for decreasing film thickness, indicating the presence of a defect rich surface and/or interface layer, as reported before [12]. From a simple defect distribution model [13] we deduce a surface defect density for our samples of $1.1 \times 10^{12}$ cm$^{-2}$.

Because tf-CRDS is a highly-sensitive technique which directly gives the absorption values, the absorption spectrum of even very thin films (< 100 nm) could be measured easily. These very thin films show a remarkable difference in spectral signature of the absorption coefficient when compared to thicker films (>100 nm) as depicted in Fig. 4 where the absorption coefficient of the 4 nm film and the 1031 nm film are compared. The absorption coefficient at sub-gap energies decreases much more rapidly for the 4 nm film than it does for the 1031 nm film. It is likely that this spectral dissimilarity is caused by a different DOS, possibly caused by difference in the environment surrounding the defects, or by a difference in the defects probed. From the results, it can be concluded that tf-CRDS is an ultra-sensitive absorption technique that does not require any calibration. The technique is therefore very promising for a broad range of thin film research fields.

## SECOND HARMONIC GENERATION (SHG)

SHG is a nonlinear optical technique in which a laser beam with frequency $\omega$ incident on a surface generates light at the double frequency $2\omega$. Or in terms of electric fields, an electric field with frequency $\omega$ incident on a surface induces a polarization

$$\vec{P}(2\omega) = \varepsilon_0 \ddot{\chi}^{(2)} : \vec{E}(\omega)\vec{E}(\omega) \qquad (4)$$

oscillating with a frequency component at $2\omega$. $\chi^{(2)}$ is a third-rank tensor that represents the second order nonlinear susceptibility. Parity considerations show that $\chi^{(2)} \neq 0$ only in non-centrosymmetric media and at surfaces where the symmetry is broken [14,15]. This has the important implication that second harmonic signals are generated at surfaces for amorphous materials, which makes SHG a surface-specific technique. Furthermore, the second harmonic signal is resonantly enhanced whenever $\omega$ or $2\omega$ coincides with an optical transition from an initially occupied state to an initially unoccupied state. Such a transition can for example be an electronic or vibrational transition of species adsorbed at the surface. Information about the surface properties and/or surface species can be obtained by determining the individual elements of the tensor $\chi^{(2)}$. This can be done by measuring different combinations of the polarization states of the incoming and outgoing beams.

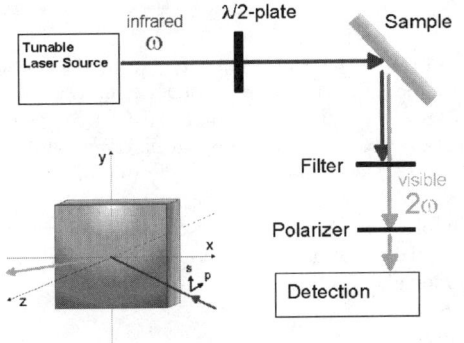

**Figure 5.** Schematic representation of the SHG setup. Inset shows the geometry of the experiment.

The measurements are performed on the same set of films as used in tf-CRDS. The setup is depicted schematically in Fig. 5. For incoming s-polarization and outgoing p-polarization (s-p configuration) we measured the SHG signal as a function of thickness of the a-Si:H samples. From Fig. 6 it can be seen that after correction for interference (using a simple model), the signal does not vary with film thickness. This indicates that the signal originates from the surface. To ascertain that the probed surface states are isotropically distributed the second harmonic signal has been measured (s-p configuration) while rotating the surface. No variation in the second harmonic signal is observed and therefore it is concluded that the surface states are isotropically distributed over the polar angle. To investigate the polarization dependence in more detail the incoming polarization is varied while the SH signal is detected under fixed polarization. Briefly, without going into details, from these measurements it can be concluded that the polarizability of the surface states behave as a isotropically distributed media over the polar angle $\varphi$ and azimuthal angle $\theta$ and resemble with the ∞m-symmetry tensor. To obtain information about the surface states generating the second harmonic signal, a spectral scan has been made on the 4 nm thin a-Si:H sample. Figure 7 gives the SH-signal as function of photon energy for the s-p configuration. A sharp rise in intensity at 1.22 eV is clearly visible and this indicates a resonance. Currently more studies are carried out to investigate whether this resonance originates from dangling bonds or from something such as the native oxide.

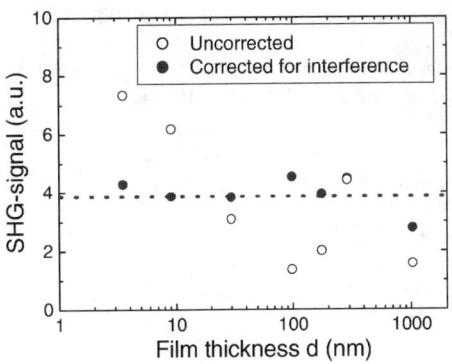

**Figure 6.** SHG signal as a function of film thickness for a-Si:H before and after applying a correction for interference determined from a simple model.

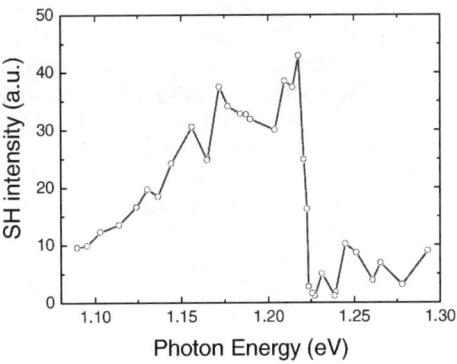

**Figure 7.** SH-signal as a function of the incident photon energy obtained for the s-p configuration.

## OUTLOOK

In the longer term, we plan to benefit from the advantages given by the two techniques to explore the role of dangling bonds on the a-Si:H surface during deposition. One particular advantage is that both methods can be used *in situ* during film growth. For tf-CRDS, the research goal is to monitor the dangling bond density in real time and to obtain information about the absolute defect density in the bulk and at the surface. This can reveal the role of the dangling bonds in film growth. SHG can also be very useful in this respect because it can give detailed information about the surface dynamics involved in dangling bonds creation and passivation. Both techniques are currently being implemented in our new UHV set-up GALAPAGOS, which is specifically designed to study the fundamentals of a-Si:H growth using well-defined radical sources.

## ACKNOWLEDGEMENTS

The authors would like to thank the skillful help of Ries van de Sande, Jo Janssen, Bertus Hüsken, Herman de Jong and Martijn Tijssen. This work is sponsored by the Foundation for Fundamental Research on Matter (FOM). The research of W.K. has been made possible by a fellowship of the Royal Netherlands Academy of Arts and Sciences (KNAW).

## REFERENCES

1. A. O'Keefe, and D.A.G. Deacon, *Rev. Sci. instrum.* **59**, 2544 (1988)
2. K.W. Bush and M.A. Bush (editors), *Cavity-ringdown spectroscopy: an ultratrace absorption measurement technique*, Americal chemical society, Washington DC, 1999
3. R. Engeln, G. von Helden, A. J. A. van Roij and G. Meijer, *J. Chem. Phys.* **110**, 2732 (1999)
4. S.L. Logunov, *Appl. Optics*, **40**, 1570 (2001)
5. A.H.M. Smets, J.H. van Helden, M. C. M. Sanden, *J. Non Cryst. Solids* **299-302**, 610 (2002)
6. A.C.R. Pipino, *Phys. Rev. Lett.* **83**, 2093 (1999)
7. W.B. Jackson, N. M. Armer, A. C. Boccara and D. Fournier, *Applied Optics* **20**, 1333 (1981)
8. M. Vanĕček, J. Kočka, A. Poruba, A. Fejfar, *J. Appl. Phys.* **78**, 6203 (1995)
9. U. Höfer, *Appl. Phys. A* **63**, 533-547 (1996)
10. S. Alexandrova, P. Danesh, I.A. Maslyanitsyn, *Phys. Rev. B.*, **61**, 11136 (2000)
11. W.B. Jackson and N. M. Amer, *Phys. Rev. B.* **25**, 5559 (1982)
12. A. Asano, M. Stuzmann, *J. Appl. Phys.*, **70**, 5025 (1991)
13. M Favre, A. V. Shah, *J. Non, Cryst. Solids* **97-98**, 731 (1987)
13. Y.R. Shen, Nature **337**, 519-525 (1989)
15. G.A. Reider and T.F. Heinz, *Photonic probes of surfaces*, ed. by P. Halevi, Elsevier, Amsterdam, 1995

# High Frequency Electron Spin Resonance Study of Hydrogenated Microcrystalline Silicon

Takashi Ehara
School of Science and Engineering, Ishinomaki Senshu University
1 Shinmoto, Minamisakai, Ishinomaki, Miyagi 986-8580, Japan

## ABSTRACT

Dangling bond defects (DB) in hydrogenated microcrystalline silicon (µc-Si:H) have been studied by X-band (9 GHz) Q-band (33 GHz) and W-band (90 GHz) electron spin resonance (ESR) spectroscopy. In X-band ESR spectra, all the samples showed asymmetric dangling bond defect signal at $g = 2.005 - 2.006$. The DB signal shape shows little dependence on substrate temperature in the X-band electron spin resonance (ESR) spectra. In the Q-band and W-band ESR spectra, existence of two centers in DB signals is clearly indicated by the shape of the spectra. The Q-band ESR spectra shape reviles that the peak of one center is at $g = 2.0055$ and the other is around at $g = 2.0060$. In addition, the DB signal showed dependence on substrate temperature. The dependence of the DB signals can be explained by difference of intensity ratio of the peaks by these two centers. The signal at $g = 2.0060$ is consistent with the asymmetric ESR signal observed in the microcrystalline silicon embedded in $SiO_2$. W-band ESR measurement indicates that the signal observed at $g = 2.0060$ is due to single inhomogeneous species and does not consist of plural species.

## INTRODUCTION

µc-Si:H has been recognized as promising materials for optical devices. In the µc-Si:H, DB are very important to control its electrical properties. The µc-Si:H is a heterogeneous material that consists of small crystallites with size in the order of 5 – 30 nm which are embedded in columns structure with a diameter of 50 – 200 nm [1,2]. In such mixed phase structure, ESR spectroscopy is an important method to study a particular electronic state, because the techniques are sensitive to the microscopic environment of this state. Previously, deconvolution of the ESR spectra of the DB in the μc-Si:H into two species has been reported [3]. This result supports the existence of plural kinds of the DB in the μc-Si:H. In addition, we have reported the high frequency ESR study of unhydrogenated μc-Si that showed the spectra could not be explained by single paramagnetic species [4].

In the present work, the DB in the μc-Si:H is studied by high resolution Q-band and W-band ESR spectroscopy. The details of the spectra are studied also by comparison with microcrystalline silicon embedded in $SiO_2$ [5].

## EXPERIMENTAL DETAILS

The μc-Si:H films were prepared by the plasma enhanced chemical vapor deposition (PECVD) method using hydrogen diluted $SiH_4$ gas at a concentration of 2 % and a feed rate of 50 sccm. The rf power and the deposition pressure were 100 W (13.56 MHz) and 133 Pa, respectively. Substrate temperatures were ranged from 70 to 270°C. The deposition time was 15 hours and thickness of the samples were about 1.0 μm. The microcrystalline embedded in $SiO_2$ samples were prepared by the rf co-sputtering of Si and $SiO_2$ followed by the thermal annealing. Twenty-four pieces of crystalline Si wafers cut into 0.5 mm x 5 mm x 15 mm section were placed on a $SiO_2$ target (10 cm in diameter) and co-sputtered with Ar gas at sputtering pressure of 2.66 Pa. The film depositions were carried out at an rf power of 200 W (13.56 MHz) for 3 hours. Thin films at a thickness of about 1.5 μm were deposited. The thermal annealing for 1 hour was carried out at 1000°C in a nitrogen atmosphere.

The ESR spectra were measured by using a Bruker ESP-300E X-band and Q-band ESR spectrometer and Bruker E-600 W-band ESR spectrometer at room temperature using a powdered sample. The amplitude and frequency for magnetic field modulation were 3 G and 100 kHz, respectively. The microwave power was 20 μW in X-band and Q-band, 60 μW in W-band. The *g* values were determined by using $Mn^{2+}$ (in MgO) and DPPH as the standard materials in X-band and Q-band.

## RESULTS AND DISCUSSION

The structures of the µc-Si:H prepared at substrate temperature of 70, 170 and 270°C have been confirmed by Raman scattering and X-ray diffraction. All the samples display a sharp and asymmetric peak at 515 – 520 cm$^{-1}$ in the Raman spectra. In the X-ray diffraction, the peaks of (111), (220) and (311) are observed in all samples. The results indicate the structure of microcrystalline silicon and are consistent with that in the previous reports [6]. Crystallite size determined from the peak width of the X-ray diffraction is 7 – 10 nm.

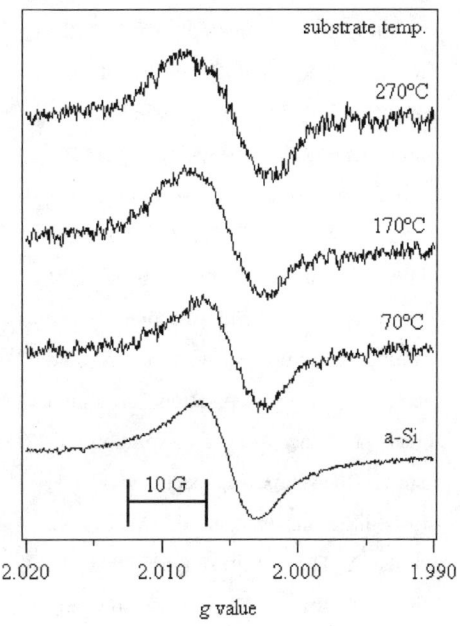

**Figure 1.** X-band ESR spectra of µc-Si:H prepared at various substrate temperature

Figure 1 shows the X-band ESR spectra of the µc-Si:H sample prepared at substrate temperature of 70°C, 170°C and 270°C. The ESR spectrum of unhydrogenated amorphous silicon (a-Si) is also shown for comparison. All the spectra show an asymmetric peak at $g$ = 2.005-6 with peak-to-peak width ($\Delta H_{pp}$) of 9 - 12 G. The $\Delta H_{pp}$ changes with substrate temperature, however, the difference of spectra shape between each samples are not clear in X-band spectra. The defect densities of the samples are $10^{17}$ - $10^{18}$ cm$^{-3}$, which decreased with substrate temperature in this temperature region.

In order to analyze detail of the ESR spectra, we measured Q-band ESR spectra. As shown in figure 2, well-resolved Q-band ESR spectra were obtained. The width of all the broad signals of μc-Si:H is about 35 G and is proportionally increased with the microwave frequency. The result indicates that the width of the signal is induced by inhomogeneous broadening. Similar increase of $\Delta H_{pp}$ is observed in the spectrum of the a-Si that has 19 G of width. The shape of the Q-band ESR spectra of the μc-Si:H becomes more asymmetric compared with the X-band spectra. This is separation of plural signals by increase of resolution. In contrast, the shape of spectrum of a-Si shows little dependence on microwave frequency. The result suggests that the ESR spectrum of the a-Si include only one signal, and that of μc-Si:H consist of plural signals. Dependence of spectra shape on the substrate temperature is also clearly observed in the Q-band spectra at low magnetic field region.

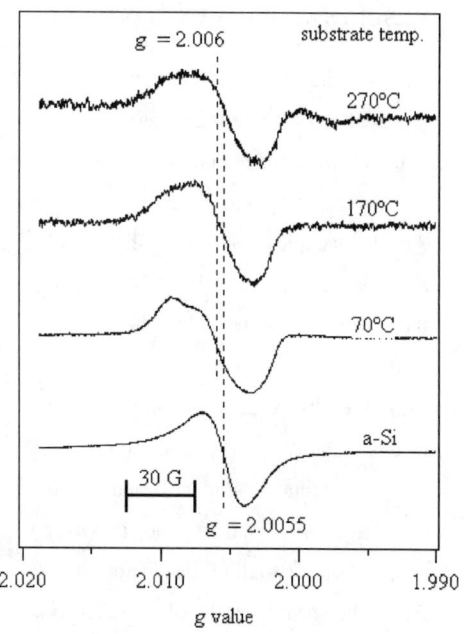

**Figure 2.** Q-band ESR spectra of μc-Si:H prepared at various substrate temperature

Above all, the spectrum of the sample prepared at the substrate temperature of 70°C clearly indicates the existence of plural signals. It can be speculated that the spectrum consists of at least two signals which has g value of about 2.005 – 2.006. One has width of 35 – 36 G and the other has width of about 20 G. Although the existence of plural species is not observed clearly in the sample prepared at 170 and 270°C, the difference of the spectra shape between each spectrum is ascribed to the difference of the intensity ratio of each signal. The width of the narrow signals (20 G) is very similar to that of the spectrum of a-Si. Due to the results, the narrow signals ascribed to be due to the defect in disordered fraction in the μc-Si:H, which might be amorphous structure.

The origin of the wide signal in the spectra of the μc-Si:H is tentatively assigned to the DB in crystalline fraction, because we assign the narrow one to the DB in disordered fraction. In order to study the origin of the wide signal, we have compared the Q-band ESR spectrum of μc-Si:H with that of the microcrystalline silicon embedded in SiO$_2$ (μc-Si in SiO$_2$). Figure 3 depicts the Q-band ESR spectra of μc-Si:H, a-Si, and μc-Si in SiO$_2$. In the μc-Si in SiO$_2$, a broad signal with ΔH$_{pp}$ of 35 G at $g$ = 2.0060 and a sharp signal at $g$ = 2.0026 are observed. However, the sharp one is due to carbon contaminations. The μc-Si in SiO$_2$ contains no silicon disorder fractions like μc-Si:H in it. Thus, it is possible to consider the defects in the crystallite fraction without superposition with the signal of the DB in the silicon disorder

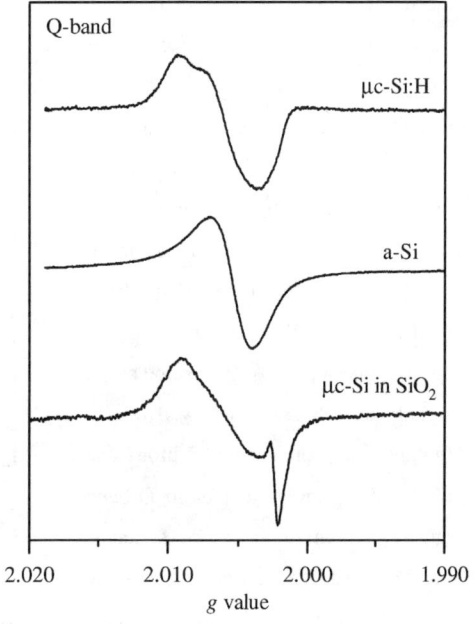

**Figure 3.** *Q-band ESR spectra of μc-Si:H prepared at substrate temperature of 70°C, a-Si, and μc-Si in SiO$_2$*

fraction. We do not deny the existence of defects in the SiO$_2$ fraction, however, the sputtered SiO$_2$ sample that annealed at 1000°C, prepared for comparison showed almost no peaks in ESR spectrum by the same measurement condition used in the spectrum of the μc-Si in SiO$_2$ shown in the figure 3. Thus, it can be concluded that the signal observed in the μc-Si in SiO$_2$ is only due to the defects in the crystalline fraction. The $g$ value and width of the broad signal is consistent with that of the wide signal in μc-Si:H. Thus, the ESR spectra of the μc-Si:H can be ascribed to be assigned as superposition of the signal of the DB in the a-Si with width of 19 G at $g$ = 2.0055 and defects in crystalline fraction that has same paramagnetic characteristics of the defects in μc-Si in SiO$_2$.

As shown in the figure 3, the broad signal at $g = 2.0060$ in the Q-band ESR spectrum of the μc-Si in $SiO_2$ is asymmetric. Two causes can be thought as the reason of the asymmetric signal shape. One is existence of plural peaks in the broad signal and the other is anisotropic structure of the defects, which induces the anisotropic distribution of g-tensor. In order to consider these two possibilities, measurement of the ESR spectroscopy using more higher microwave frequency (W-band, 90 GHz) is carried out. In figure 4, W-band ESR spectrum of the μc-Si in $SiO_2$ is shown. In the spectra, one sharp signal and one broad signal are observed as well as in Q-band. The signal shape of broad one observed is similar to that in Q-band and separation of the broad signal is not observed. The estimation of peak width of the broad peak is very difficult, however, it can be speculated

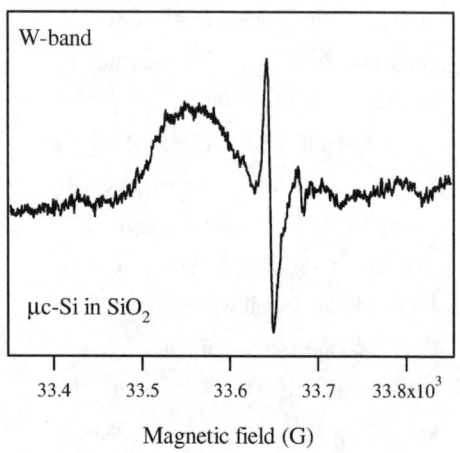

*Figure 4.* W-band ESR spectra of μc-Si in $SiO_2$

as about 90 G. The increase of width is proportional to microwave frequency. In addition, due to the signal shape of low magnetic field area (33500 – 33600 G), it can be concluded that the broad signal is consist of only one inhomogeneous species. The results suggest that the asymmetric ESR signal shape of the defects in crystalline fraction is caused by anisotropic defect structure [7], not by existence of plural species.

The present work suggests that the EPR spectra of μc-Si:H consist of only two species. One is a defect in disordered structure fraction that has amorphous structure and the other is the defect in the crystalline fraction. The signal of the defects in the crystalline fraction shows asymmetric shape, because of the anisotropic distribution of defect structure.

**REFERENCES**

1. M. Luysberg, P. Hapke, R. Carius, F. Finger, Phil. Mag. A 75, 31 (1997).

2. L. Houben, M. Luysberg, P. Haple, R. Carius, F. Finger and H. Wagner, Phil. Mag. A 77, 1447 (1998).

3. J. Müller, F. Finger, R. Carius and H. Wagner, Phys. Rev. B, 60, 11666 (1999).

4. T. Ehara, T. Ikoma, K. Akiyama, S. Tero-Kubota, J. Appl. Phys. 88, 1698 (2000).

5. For example, S. Hayashi, T. Nagareda, Y. Kanazawa and K. Yamamoto, *Jpn. J. Appl. Phys.* **32**, 3840 (1993).

6. For example, Z. Iqbal, S. Veprek, *J. Phys. C.* **15**, 377 (1982).

7. T. Ehara, T. Ikoma and S. Tero-Kubota, J. Non-Cryst. Solids, 266-269, 540 (2000).

## Photoelectron spectroscopic investigations of very thin a-Si:H layers

M. Schmidt, A. Schoepke, O. Milch, Th. Lussky, W. Fuhs
Hahn Meitner Institut Berlin, Abteilung Silizium-Photovoltaik,
Kekuléstr. 5, D-12489 Berlin, Germany

## ABSTRACT

We report on a detailed study on gap-state distribution in thin amorphous silicon layers (a-Si:H) with film thicknesses between 5 nm and 20 nm on c-Si wafers performed by UV excited photoelectron spectroscopy (UV-PES). We measured how the work function, the gap state density, the position of the Fermi-level and the Urbach-energy depend on the layer thickness and the doping level of the ultra thin a-Si:H(n) layers. It was found, that for phosphorous doping the position of the Fermi level saturates at $E_F - E_V = 1.47$ eV. This is achieved at a gas phase concentration of 10000 ppm $PH_3$ in the $SiH_4/H_2$ mixture which was used for the PECVD deposition process. The variation of the doping level from 0 to 20000 ppm $PH_3$ addition results in an increase of the Urbach energy from 65 meV to 101 meV and in an increase of the gap state density at midgap ($E_V - E_i = 0.86$ eV) from $3 \cdot 10^{18}$ to $2 \cdot 10^{19}$ cm$^{-3}$eV$^{-1}$.

## INTRODUCTION

The energetic distribution of gap states, of interface states and the band offsets play a crucial role in heterostructures. The detection of these quantities is possible by photo-excited electron emission [1,2] if the escape depth corresponds to the top layer thickness. The excitation with photon energies below 10 eV results in an increase of the escape depth up to 10nm [3]. Very thin amorphous silicon layers play a fundamental role in hetero emitter solar cells of the TCO/a-Si:H/c-Si type [4]. The thickness of such a-Si:H emitter layers amounts to about 5 nm [5]. They are doped with boron or phosphorus depending on the type of the c-Si substrate. Additionally, an undoped a-Si:H layer is often used for a better interface passivation of c-Si [4,5]. This layer is located among the doped amorphous emitter layer and the c-Si absorber. Therefore, the distribution of the gap state density of a thin a-Si:H layer which covers the c-Si surface, and their change with doping and layer thickness are of outmost interest.

The basic properties of doped and undoped a-Si:H layers are well known [6,7] and are based on many years research. But, the knowledge concerning the electronic properties of such very thin amorphous layers grown on crystalline silicon substrates is patchy.

## EXPERIMENTAL DETAILS

We used c-Si wafers (FZ) of p-type (75 Ohm cm) with a (111) surface orientation as substrates. Prior to deposition of the amorphous layer the wafers were cleaned by the standard RCA process followed by a HF-dip in 2% HF acid for 30 sec. The last step leads to a passivation of the Si surface by hydrogen (H-termination) which is stable for about 45 minutes. This allows to insert the samples into the deposition chamber without oxygen contamination of the silicon surface. The deposition of the thin amorphous silicon layers was realized by the conventional 13.56 MHz

plasma enhanced chemical vapor deposition (PECVD) technique in a high vacuum system ($10^{-7}$ mbar basic pressure) with a loadlock. Semiconductor grade silane (SiH$_4$) with gas flow rates of 2.5 sccm, phosphine diluted in hydrogen (0.5%) and pure hydrogen were used as source gases. The summarized gas flow rate amounts 7.5 sccm. This allowed a doping variation between 0 and 10000 ppm phosphine at nearly constant deposition conditions. The deposition was carried out at a chamber pressure of 0.5 mbar, rf power of 55 mW cm$^{-2}$, and a substrate temperature of 170 °C. The layer thicknesses range between 2 and 20 nm as measured by ellipsometry.
Immediately after deposition, the a-Si:H/c-Si samples were transferred into the UHV system where photoelectron spectroscopic investigations and annealing processes have been carried out.

## Gap state density analysis

The process of photoelectron emission can be divided into three steps which are assumed to be independent of each other [7]. The optical excitation of hot electrons (1), the transfer towards the surface (2) and the emission into the vacuum (3). The yield Y in dependence on the kinetic electron energy E$_K$ and photon energy hv is given by equations 1.

$$Y(E_K, h\nu) = C \cdot D(E) \cdot \delta(E_k - E) \cdot T(E, h\nu) \cdot P^2(h\nu) \cdot N_{occ}(E - h\nu) \cdot N_u(E) \quad (1)$$

where

$$T(E, h\nu) = \frac{\lambda_{imfp}(E)/\alpha^{-1}(h\nu)}{1 + \lambda_{imfp}(E)/\alpha^{-1}(h\nu)} \quad \text{and} \quad \begin{array}{ll} P^2(h\nu) = c & h\nu \leq 3.5\,\text{eV} \\ P^2(h\nu) \approx c \cdot h\nu^{-5} & h\nu > 3.5\,\text{eV} \end{array}$$

describes the probability of the electron transfer towards the surface depending on the generation depth given by the inverse optical absorption coefficient $\alpha^{-1}$(hv) and the inelastic mean free path $\lambda_{imfp}$(E). P$^2$(hv) is the energy dependent optical matrix element [9]. D(E) describes the energy dependent emission probability into the vacuum, $\delta$(E$_K$-E) denotes the analyzer detection function and C is an apparatus constant.

We want to determine the energetic distribution of the occupied gap states and upper valence band states, DOS, of the thin a-Si:H layer. Equation 1 shows that this is possible if all quantities are nearly constant except the occupied (initial) states N$_{occ}$(E-hv). This precondition is fulfilled for a-Si:H because the DOS of the conduction band N$_u$(E) (unoccupied final states) is constant [8] and the optical matrix element P$^2$(hv) remains nearly constant up to hv = 3.5 eV and decreases above this value as shown in detail in [9]. The electron interaction with phonons, electrons and plasmons is summarized by the energy dependent inelastic mean free path length $\lambda_{imfp}$(E) which determines the limit of the electron emission depth. For photon energies below 10 eV the emission depth increases strongly because the strong inelastic plasmon generation process becomes energetically impossible. Considering these arguments we can measure the DOS of the a-Si:H in the excitation energy range between the work function limit at 3.8 eV and about 7 eV. The corresponding emission depth ranges from 10 to 7 nm [3].

The photoelectrons were excited by strong monochromatic UV-light (4 –7 eV) generated by passing the light of a Xe lamp through a double-grating monochromator. In this low excitation energy range we are able to determine the number of absorbed photons using calibrated silicon diodes for the detection of the incoming and reflected photon beams. This procedure allows to measure the absolute photoelectron quantum yield at each photon energy, in two different modes. In the first mode, corresponding to standard UPS, we detected the kinetic energy distribution of the photoelectrons during excitation with a fixed photon energy [3]. The achieved energy resolution was 100 meV, using a SPECS EA10P energy analyzer. In the other measuring mode, the energy analyzer was operated at a fixed energy (final state energy) while changing the photon energy (4-7 eV), a technique called **C**onstant **F**inal **S**tate **Y**ield **S**pectroscopy (CFSYS) [10]. Additionally, we determined the total yield as a function of the photon energy, the so called **T**otal **Y**ield **S**pectrum ,TYS(hv). Here, all emitted electrons for each excitation energy are counted independent of their kinetic energy. The derivative of the TYS spectrum with respect to the photon energy hv represents the DOS, as shown in [1,2].

## RESULTS

Figure 1 shows the yield and the gap state distribution of a 20 nm thick a-Si:H(n) layer on c-Si measured by UPS, by CFSYS at two different analyzer energies and the derivative of the total yield (DTY).

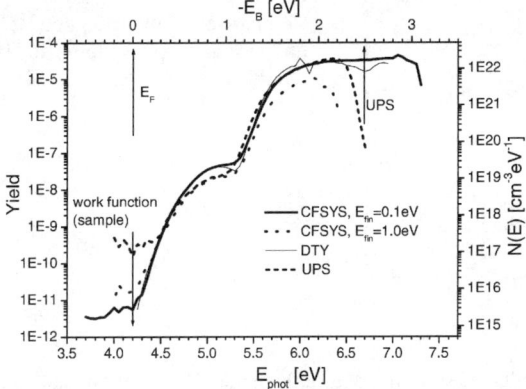

**Figure 1.**
Comparison of UPS results ($Y_{int}$ vs. $E_{kin}$) obtained at hv=6.5 eV excitation (top scale; $-E_B = E_{kin} - hv - q\phi_{Analyzer}$) with $dY_{tot}/dhv$ (DTY) and with CFSYS results vs. the photon energy (bottom scale). The CFSYS were performed at 0.1 and 1 eV above the vacuum energy. Measurements were performed on an a-Si:H(150 nm)/c-Si structure.

All spectra are in relatively good agreement but exhibit different low yield detection limits (dynamic ranges) as shown in figure1. The dynamic range of the CFSYS spectra decreases with increasing final state energy, and for UPS it is about two orders of magnitude lower than for DTY and CFSYS (0.1 eV final state energy).

The reason for the different dynamic ranges is not yet fully understood. The energies of the photons exciting electrons in the vicinity of the Fermi edge are different in UPS (6.5 eV) and CFSYS (about 4.5 eV). Therefore, the optical matrix element $P^2(hv)$ should be more than one order of magnitude lower in UPS compared to CFSYS (and DTY). In addition, the photon flux emitted from the Xe lamp decreases with rising photon energy. Both facts lead to a lower detection sensitivity for electrons from near the Fermi edge in UPS.

On the other hand, electrons emitted from the vicinity of the valence band edge are excited by photons of comparable energies in all detection modes and result in nearly identical yield values. Furthermore, scattering processes like electron phonon interaction might lead to a higher population of the low kinetic energy levels. The DTY method ignores this fact as long as the electrons are able to be emitted. The relatively good agreement between DTY and CFSYS (0.1 eV final state energy) clearly shows that this effect is not dominant.

Thus, the energy dependent optical matrix element and its influence on the supply function of the photoelectrons has to be proved. If this will be successfully performed, the detection of the photoelectrons as a function of excitation energy is of advantage over the selective measurement of the kinetic energy as used in UPS. Both methods DTY and CFSYS use this advantage. Work is in progress to solve this problem.

We determined the position of the Fermi level in reference to the valence band, the slope of the valence band tail states (Urbach energy $E_{ov}$) and the change in the gap state distribution in dependence on the doping level. The subsequently presented results are based on UPS measurements with an excitation energy of $h\nu = 6.5$ eV.

Figure 2 shows the comparison of the DOS for a nominally undoped a-Si:H(i) and an a-Si:H(n) layer doped with $10^4$ ppm phosphine. Three features can be clearly seen. The gap state distribution shows the exponential band tail states, the DOS in the gap region increases by nearly one order of magnitude and the position of the Fermi level shifts towards the conduction band edge for the phosphorous doped layer.

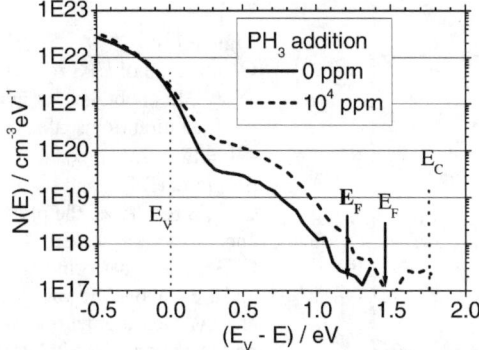

**Figure 2.**
Comparison of the DOS between an undoped and a phosphorus doped 10 nm thick a-Si:H layer deposited on a c-Si wafer (111).

From the sqrt Y(E) vs. E slope results the starting point of the parabolic density of states distribution at the valence band edge and allows to determine the valence band edge $E_V$. In [2,9] it was shown that the transition from the localized gap states to the extended valence band states takes place at a density of states of $N_{OV} \approx 2 \times 10^{21}$ cm$^{-3}$ eV$^{-1}$. This value was used to scale the photoemission yield with the DOS distribution. Furthermore, it is possible to determine the Urbach energy $E_{OV}$ of the exponentially distributed valence band tail states from the slope of a log-linear plot of N(E) according to equation 2.

$$N(E - E_V) = N_{ov} \cdot \exp(-(E - E_V)/E_{ov}) \quad (2)$$

The obtained value of $E_{OV}$ reflects the disorder broadening of the valence band. The value of $E_{OV}$ increases with disorder. The Urbach energy is $E_{OV} = 65.2$ meV for the undoped sample and increases to $E_{OV} = 101.3$ meV for the doped sample. The defect density is another measure of the disorder and also increases with the band tail slope as can be seen in Fig.2. These values are at or slightly above the upper limit compared to the values of thick films [6] and suggests a stronger disorder of such extremely thin films or non optimized preparation conditions. The influence of surface contamination like O or C can be ruled out as a reason as proved by XPS measurements. Figure 3 shows typical results of the influence of annealing at 350 °C whereby effusion of hydrogen takes place.

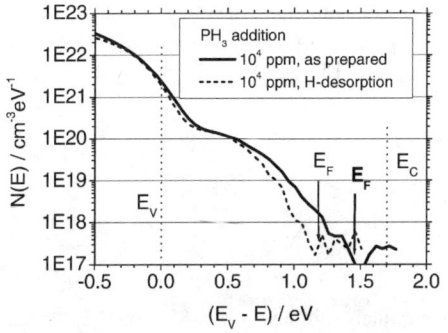

**Figure 3.**
Comparison of the DOS between a sample as prepared and after annealing for 10 min at 350 °C (H-desorption). The a-Si:H(n) layer on a c-Si (111) wafer was phosphorus doped and 10 nm thick.

Two features stand out in figure 3, the shift of the Fermi level by nearly 300 meV towards midgap after annealing and a decrease of the Urbach energy from $E_{OV} = 101.3$ meV to $E_{OV} = 89.6$ meV. This indicates that both the doping efficiency and the degree of disorder decrease during the annealing process (H-desorption (effusion)).

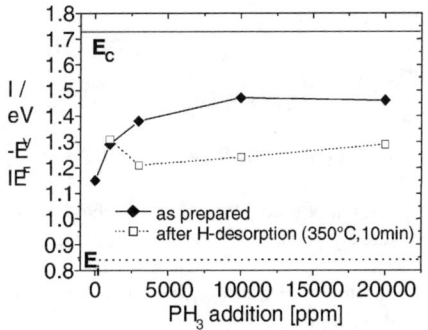

**Figure 4.**
Dependence of the Fermi level position on the phosphine addition during layer growth. Change of these values after annealing of the 10-15 nm thick samples at 350 ° for 10 min. The thickness of the samples varied between 10 and 15 nm.

The Fermi level shift saturates at about 1.5 eV above the valence band edge at a phosphine addition of 10000 ppm. This corresponds to a value of $E_F = 0.27$ eV below the conduction band edge, assuming a value of the gap width of 1.74 eV as determined for 100 nm thick films.
The saturation of the Fermi level is a self limiting process caused by the defect generation with energy levels located in the gap. The shift of the Fermi energy after annealing towards midgap is clearly seen in figure 4 . The Urbach energy shows a similar behavior cf. figure 5. This is not unexpected because the disorder and defect generation represents the background for this behavior in both cases. The assumption of a constant optical gap of the layers before and after annealing could not be proved. This will be done by spectral dependent ellipsometric measurements in the future.

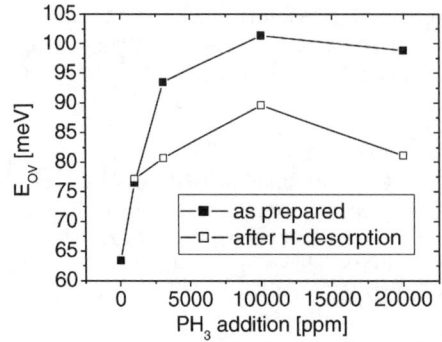

**Figure 5.**

Dependence of Urbach energy on the phosphine content of the silane during the layer growth. Change of these values after annealing at 350 °C for 10 min. The thicknesses of the samples varied between 10 and 15 nm.

## CONCLUSIONS

Our results lead to the conclusion that in such extremely thin layers of a-Si:H on c-Si the basic features like the electronic structure, the exponential tail states, the gap state distribution and the self limitation of the doping efficiency are preserved. Special attention has to be directed to the comparison of different measurement modes. The applied electron spectroscopic methods allow to determine the basic electronic properties like DOS and ($E_F$-$E_V$) of such extremely thin a-Si:H layers. The influence of energy dependent optical excitation cross sections and electron scattering processes on the estimated DOS and the change of band gap with thickness and/or post annealing processes are in our focus for future investigations.

## REFERENCES

[1] B.von Roedern, L.Ley, M.Cardona, F.W.Smith, Phil. Mag. B, **40,** 433 (1979).
[2] K.Wiener, L.Ley, Phys. Rev. B, **36**, 6072 (1987).
[3] M. Prutton, Introduction to Surface Physics p.24, Clarendon press, Oxford 1994, p. 23.
[4] M.Tanaka, M.Taguchi, T.Matsuyama, T.Sawada, S.Tsuda, S.Nakano, H.Hanafusa, Y. Kuwano, Jpn. J. Appl. Phys. **31**, 3518 (1992).
[5] M.L.Scherff, A.Froitzheim, A.Uljaschin, M.Schmidt, W.R.Fahrner, W.Fuhs, Proceedings European Photovoltaic Conference, Rome (2002) (in press).
[6] R.A. Street, Hydrogenated amorphous silicon, Cambridge university press, Cambridge 1991; Emis data review series No.19, Amorphous silicon and its alloys, ed. by T. Searl, INSPEC, London 1998.
[7] W.E. Spicer, Phys. Rev. **112**, 114 (1958).
[8] W.B.Jackson, S.M.Kelso, C.C.Tsai, J.W.Allen, S.-J.Oh, Phys. Rev. B, **31**, 5187 (1985).
[9] W.B.Jackson, S.-J.Oh, C.C.Tsai, J.W.Allen, Phys. Rev. Lett. **53**, 1481 (1984).
[10] M.Sebastiani, L.Di Gaspare, G.Capellini, C.Bittencourt, F.Evangelisti, Phys. Rev. Lett. **75,** 3352 (1995).

# Determination of Defect Densities by Constant Photocurrent Method – Comparison of AC and DC Methods

Charlie. Main, Steve Reynolds, Ivica. Zrinšćak and Amar Merazga[1],
University of Abertay Dundee, School of Computing and Advanced Technologies, Bell Street, Dundee DD1 1HG
[1] Faculté des Sciences et Sciences de l'ingénieur, Université Mohammed Khidir, Biskra, Algéria

## ABSTRACT

We report on discrepancies in the absorption spectrum of a-Si:H measured by DC and AC constant photocurrent methods (CPM). Our measurements reveal discrepancies in the absorption coefficient alpha, of up to an order of magnitude. DC measurement gives the higher value for alpha at photon energies below the Urbach tail. In this paper we examine free carrier generation paths in AC CPM, and the influence of these paths on the photocurrent frequency response to modulated sub-gap illumination. A simple kinetic model is used to attempt to explain quantitatively the differences in the photocurrent frequency response for sub- and super-gap excitation. At first sight the basic AC technique is of doubtful validity, even at exciting frequencies as low as 1 Hz, since maintaining a constant AC photocurrent does not guarantee a constant carrier lifetime. On the other hand, the DC method involves several parallel excitation paths, obfuscating attempts to extract a density of states from the absorption spectrum. We demonstrate that a simple variation of AC CPM can provide a more accurate means of determining the density of states than DC CPM.

## INTRODUCTION

Several groups have reported differences in the absorption spectrum of a-Si:H as measured by DC and AC CPM methods [1, 2, 3]. At photon energies $E_{ph} < 1.4$ eV, *i.e.* below the Urbach tail, the difference in absorption constant α, determined by these respective methods, can be as much as an order of magnitude in undoped material. The DC technique consistently returns a higher value. Sub-gap optical absorption measurements are often used to provide information on the density of states in the mobility gap as a diagnostic of the effects of fabrication parameters, annealing and light soaking processes. It is therefore vital to understand what may cause such discrepancies.

Hasegawa et al [2] report a 2- order of magnitude discrepancy in α between DC and AC methods, and point out that this results from an increase in response time for low values of $E_{ph}$. One possible explanation is briefly mentioned, involving (thermal) transitions from deep electronic states. Conte et al [3] noted the discrepancy but offered no explanation. Sládek and Thèye [1] note that the response time of the photocurrent increases as $E_{ph}$ is reduced so that an increasing photon flux is needed to maintain a constant photocurrent; hence this would return a lower value for α. The explanation given is in terms of the different occupancy of deep states under sub-gap excitation compared to super-gap excitation, leading to reduced recombination efficiency.

In a related experiment, Modulated Photoconductivity (MPC), such transitions have been included in dc and ac analyses, by several authors [4,5], but without completing the link to the CPM DC-AC discrepancy. The transitions in question involve optical excitation from valence band states into unoccupied defect states, with subsequent thermal emission to the conduction band. The latter transition produces a delay in the response, depending on the depth of the state. Such transitions are often discounted in CPM studies for a number of reasons [6]; the initial optical excitation does not produce a free electron; free holes produced have a low mobility. In our work, we include this possibility in our analysis, and allow for it in experiment by using low excitation rates such that quasi-Fermi level splitting is minimal, so that thermal emission is a probable process.

**THEORY**

Figure 1 shows schematically the transitions included in our small signal analysis. The rate equations for sub-gap excitation may be written, for ac conditions, for free electrons

$$\frac{dn(\omega)}{dt} = -\sum_i \frac{dn_{ti}(\omega)}{dt} - \omega_R n(\omega) \tag{1}$$

and for trapped electrons

$$\frac{dn_{ti}(\omega)}{dt} = -\omega_{ei} n_{ti}(\omega) + \omega_{ti} n(\omega) + G_{\omega,i}. \tag{2}$$

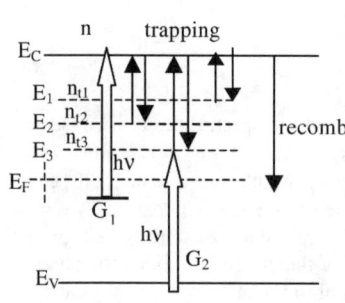
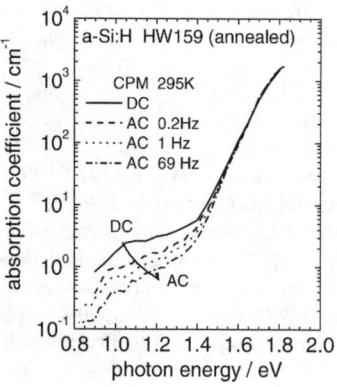

**Figure 1** Energy level scheme for the analysis of sub-gap excitation. Optical and thermal transitions shown for several levels. Typically, several hundred levels were employed in calculations.

**Figure 2** Measured DC and AC CPM for annealed HWCVD sample at 295K. The DC value for $\alpha$ exceeds the high frequency AC value at 1 eV by a factor of 6.

We have represented the DOS by a fine discrete array. The model allows for any given level to be involved in optical transitions as well as multi-trapping thermal transitions. The term $n(\omega)$ represents the excess free electron density as a function of frequency $\omega$, $n_{ti}(\omega)$ is the trapped electron density in level $i$, and $\omega_{ti}$, $\omega_{ei}$, $\omega_R$ are characteristic trapping and emission frequencies for trap '$i$' and a 'recombination frequency' respectively. The level-to-level sinusoidal excitation rates $G_{\omega i}$ are assumed to be proportional to the joint density of states for the occupied initial and empty final states '$i$', (one of which should be extended), separated by photon energy $E_{ph}$. The trap emission frequency $\omega_{ei} = v\exp(-E_n/kT)$, where $v$ is an attempt-to-escape frequency (value $10^{12}$ Hz assumed), $E_{ti}$ is the trap depth, $k$ is Boltzmann's constant and $T$ is the absolute temperature.

The ac solution to the above equation system is

$$n(\omega) = \left\{ \frac{\sum_k G_{\omega,k}/(1+j\omega/\omega_{ek})}{\left[j\omega + \sum_i \left(\omega_{ti} - \frac{\omega_{ti}}{1+j\omega/\omega_{ei}}\right) + \omega_R\right]} \right\}, \qquad (3)$$

and the measured AC photocurrent $I_{ph}(\omega) \propto n(\omega)$. Inspection of equation 3 reveals a neat insight. This result is the *same* as that for 'above-gap' excitation, i.e. the 'MPC' result for the *same* system, as reported earlier by Main [7] *except* for the appearance of multiple 'poles' in the generation rate, (see numerator of equation 3) corresponding to the process of emission which follows an electron transition to a trap

## EXPERIMENTAL DETAILS

PECVD films were prepared in an industrial reactor with: 5% hydrogen dilution, chamber pressure 500 mTorr, substrate temperature 200 °C. Hot-Wire (HWCVD) films were prepared in a laboratory system with: gas flow rate 8 sccm, chamber pressure 220 mTorr, tungsten wire temperature 1500 °C, substrate temperature 200 °C. Representative HW and PECVD films were measured after annealing at 180 °C for 3 hours, and again following exposure to simulated AM1 radiation for 10, 100 and 1000 minutes. Absolute CPM spectra were obtained following the procedures described by Vaněček *et al* [8] and calibrated with reference to optical transmission measurements *via* the Ritter-Weiser [9] formula. The photon flux used was typically less than $10^{13}$ cm$^{-2}$s$^{-1}$ so that the photocurrent was similar in magnitude to the dark current. The DOS was obtained by differentiation of the absorption curve, following Pierz *et al* [10]

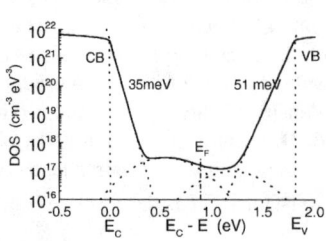 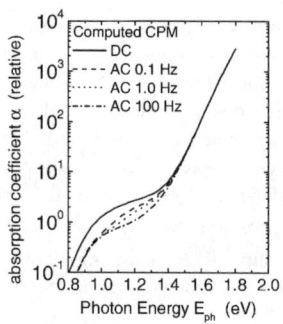

**Figure 3** Model DOS used to compute DC and AC CPM using equation 3.

**Figure 4** Computed DC and AC CPM for the model DOS of figure 3.

## RESULTS AND DISCUSSION

Figure. 2 shows the experimental plot of $\alpha$ vs $E_{ph}$ for a HWCVD sample in the annealed state, measured at DC and at AC chopping frequencies 0.2, 1.0 and 69 Hz. A factor of 6 difference is observed between the DC and 69 Hz measurement at $E_{ph}$ = 1eV. We also note that the AC curves converge at frequencies above about 5 Hz. Equation 3 was applied to the simple model DOS of figure 3, containing two band-tails, two broad Gaussian defect distributions and parabolic variation above the mobility edges to compute the DC and AC CPM plots of $\alpha$ vs $E_{ph}$ for a range of frequencies. Figure 4 shows the computed results. There is reasonably good agreement with the experimental data of figure 2.

The reason for the difference between the experimental $\alpha$ plots for DC and AC CPM is that the DC measurement includes transitions into unoccupied defect states which require subsequent thermal release to produce free carriers. This process is too slow to produce a measurable AC

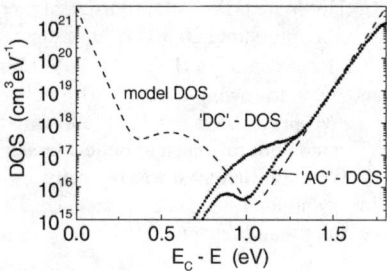 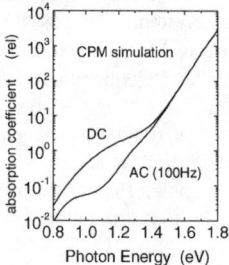

**Figure 5** Model DOS with high density of unoccupied defects above mid-gap, with computed DOS from DC and AC CPM spectra.

**Figure 6** DC and AC CPM spectra computed from the model DOS of figure 5.

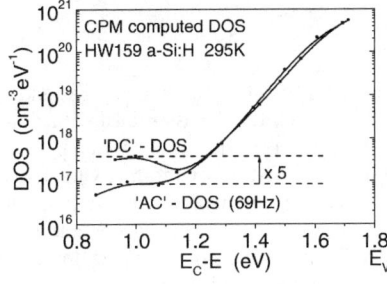
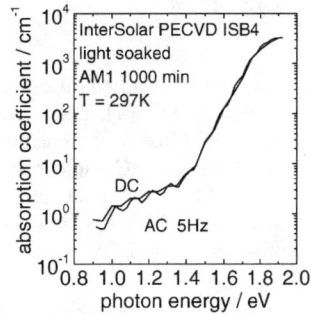

**Figure 7** DOS derived from DC and AC CPM measurements of Fig. 2, on HWCVD a-Si:H sample.

**Figure 8** Measured DC and AC CPM spectra for light-soaked PECVD sample at room temperature. Light soaking 1000 min at AM1.

response at frequencies above about 10 Hz in the case studied. since the associated response poles are at low frequency. A significant consequence of this is that the AC response should not take such transitions into account, and should arise only from transitions from occupied defects into the conduction band (or shallow tail states). This is usually the assumption made in deconvoluting CPM results to determine the DOS. We present here an illustrative example - the rather extreme model case depicted in figures 5 and 6, in which there is a high density of unoccupied defect states, and a low density of occupied defects. Fig. 6 shows the computed CPM spectra for DC and AC (100Hz) cases. While the DC simulation exhibits a defect 'shoulder' at photon energies below 1.4 eV, corresponding to generation of free electrons via *empty* defects (i.e. including a thermal emission), the AC simulation continues down the 'Urbach' slope to $E_{ph} = 1.1 eV$. The computed DOS for the DC and AC case is superimposed on the model DOS in Fig. 5, where it is clear that the DC case is incorrectly interpreted by the deconvolution process as a deep *occupied* set of defects. On the other hand, the AC CPM spectrum gives a DOS much closer to the actual DOS.

Using the experimental data of figure 2 to compute the DOS. we see in figure 7 a similar discrepancy between DC and AC-derived DOS for the HWCVD sample. The DC measurement returns an incorrect defect DOS which is a factor of 5 greater than that returned by the AC measurement. Measurements on different a-Si:H samples reveal varied behaviour. The HWCVD sample described above shows a similar discrepancy between DC and AC CPM after 1000 minutes AM1 light soaking, while a sample of undoped PECVD a-Si:H supplied by InterSolar UK, exhibited a similar discrepancy in the annealed state, but no such discrepancy in the light-soaked state, as shown in Fig. 8. It is possible that in this material, light soaking preferentially increases the density of the occupied defect states immediately above the valence band tail.

We have ignored throughout this work, the contribution made by holes to the measured photocurrent. It is possible that, even though the hole mobility is low, they may make a contribution to the low-frequency CPM response of the materials studied.

## CONCLUSIONS

We have reported marked differences in the absorption coefficient as measured by DC and AC CPM in undoped a-Si:H. The origin of this difference is proposed to be associated with slow phonon assisted free carrier creation, *via* unoccupied defects. This view is supported by our analysis of sub-gap AC photoconductivity which shows that CPM may be formally treated as super-gap MPC, but with the inclusion of distributed poles in the generation rate. A consequence of this is that the AC CPM measurement may actually allow a more accurate deconvolution to determine the defect DOS. Further work is required to include effects of defect occupancy and QFL shifts.

## ACKNOWLEDGMENTS

The authors acknowledge the support of EPSRC through research grant no. GR/M 16696, Dundee University and InterSolar UK for HWCVD and PECVD samples respectively. One author (AM) acknowledges part support from the University of Biskra, Algéria.

## REFERENCES

1. P. Sládek and M. L. Thèye, *Solid State Comms.* **89** 199, (1994).
2. S. Hasegawa, S. Nitta and S. Nonomura, *J. Non-Cryst. Solids* **198-200** 544 (1996).
3. G. Conte, F. Irrera, G. Nobile and F. Palma, *J. Non-Cryst. Solids* **164-166** 419 (1993).
4. H. Oheda, *Solid State Comms* **33** 203 (1980).
5. K. Abe, H. OkamotoK, Y. Nitta, Y. Tsutsumi, K. Hattori and Y. Hamakawa, *Philos. Mag* **B58** 171 (1988).
6. N. Wyrsch, F. Finger, T. J. McMahon and M. Vaněček, *J. Non-Cryst. Solids* **137-138** 347 (1991).
7. C. Main in MRS Symposium Proceedings,vol **467**, San Francisco, April 1997, edited by M. Hack, E.A. Schiff, S. Wagner, A. Matsuda and R. Schropp, (MRS, Pittsburgh 1997) Ch.143, p.167
8. M. Vaněček, J. Kočka, A. Poruba and A. Fejfar, *J. Appl. Phys.* **78,** 6203, (1995).
9. D. Ritter and K. Weiser, *Opt Commun.* 57 336 (1986).
10. K. Pierz, H. Mell and J. Terukov, *J. Non-Cryst. Solids* **97-98**, 547 (1985).

## Depth Profiling of Light-Induced Defects in Hydrogenated Amorphous Silicon by Transient Photocurrent Spectroscopy

Steve Reynolds, Charlie Main and Rudi Brüggemann[1]
School of Computing and Advanced Technologies, University of Abertay Dundee,
Bell Street, Dundee, U.K.
[1]Fachbereich Physik, Carl von Ossietzky Universität Oldenburg, D-26111 Oldenburg, Germany.

## ABSTRACT

The sensitivity of transient photocurrent measurements to the spatial location of native and metastable electronic defects in hydrogenated amorphous silicon films is demonstrated. The technique utilises red and green laser excitation to generate excess carriers in the bulk and at the surfaces of the film, respectively. In annealed films the defect density is found to be higher in the surface regions. Following white light soaking, the metastable defect density at the surface at which the light is incident is greater than that in the bulk, which in turn is greater than that at the exit surface. This is attributed to the white light absorption profile within the film. Green light soaking creates metastable defects at the incident surface, with the bulk of the film largely unaffected.

## INTRODUCTION

Transient photoconductivity (TPC) is a sensitive probe of the energetic distribution of localized states in amorphous semiconductors. [1]. The technique is based on the interpretation of the form of the decay in current $I(t)$ following a short flash of light, in terms of the multiple-trapping transport model. Recently, the present authors have shown that a degree of *spatial* as well as energetic resolution can be achieved using TPC [2,3]. This variant of the technique utilises the differing absorption depths of the light flash depending on the wavelength used. Green (510 nm) and red (640 nm) laser dyes were used giving, in the case of a-Si:H, typical absorption depths of 0.1 and 1.5 μm, respectively. Thus for a film thickness of order 1 μm, carrier generation by the red flash is fairly uniform throughout the film, but the green flash generates carriers close to the surface at which it is incident. As the diffusion length of carriers prior to undergoing deep trapping is of the order of 0.1 μm, the photocurrent decay will bear the hallmark of the local density of states (DOS) where they are generated.

Ghosh and Ganguly [4], using steady-state photoconductivity measurements, demonstrated a reduction in the $\eta\mu\tau$-product at short wavelengths in a-Si:H films following light soaking, in comparison with the annealed state. Although clearly visible following white light soaking, the effect was more pronounced following blue and violet illumination. Inspired by this observation, we have followed a similar approach, studying the decay in transient photocurrent following annealing and subsequent light-soaking with white light and green-filtered white light, using red and green laser pulses for carrier excitation, from both the silicon-air and the silicon-substrate sides of the film. This has enabled us to assess the correlation between defect creation and defect detection, and thus to place the use of spatially-resolved TPC on a firmer footing. No claims for uniqueness with regard to the depth profiling of electronic defects are made, however, since the constant photocurrent method [5], photothermal deflection spectroscopy and modulated photocurrent spectroscopy [6] have all been shown to be sensitive to such effects.

## EXPERIMENTAL

The sample studied was an a-Si:H film 1.7μm thick, prepared by InterSolar Ltd. in a commercial PECVD reactor and fitted with coplanar Al contacts. Light-soaking was carried out under simulated AM1 conditions either without filtering, or through a Spiers-Robertson BG18 filter with a sharp cut-off at 600 nm. The filter reduced the total power throughput by a factor of 10. Each light-soaking session was 750 minutes in duration, and annealing was carried out in air at 200 °C for 2 hrs. For the TPC experiment, 4 ns light pulses were obtained from a Laser Science VSL337 $N_2$ laser plus dye attachment. Wavelengths of 510 nm and 640 nm could be selected by insertion of the appropriate cuvette. The pulses were attenuated to give a total flux of $10^{11}$ photons $cm^{-2}$ at the sample, corresponding to a carrier density of order $10^{15}$ $cm^{-3}$ for red light. Each side of the sample could be optically probed by rotating the sample holder accordingly. A schematic of the coplanar sample configuration is shown in figure 1(a).

The TPC experimental system is shown in figure 1(b). Following appropriate pre-amplification using a Burr-Brown OPA637-based transimpedance amplifier, current transients were recorded on a Tektronix TDS3052 digital storage oscilloscope over the range 1 ns to 10s. Averaging of successive transients was used to reduce noise to an acceptable level. Data were transferred to a PC and processed into a single logarithmically-spaced file. All measurements were carried out at room temperature and with an applied electric field of 3800 V $cm^{-1}$.

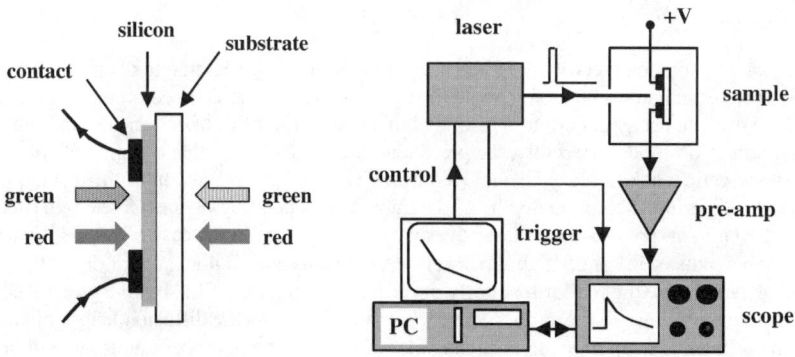

**Figure 1.** (a). Sample configuration – illumination from contact side or substrate side is possible; (b) TPC experimental system.

## RESULTS AND DISCUSSION

### Annealed film

Transient photocurrent decays obtained following annealing, taken with the laser directed at the contact side of the film are shown in figure 2(a). The form of the curves for both red and green laser pulses is quite similar, although it is evident that the current at approximately $10^{-6}$ s falls more rapidly in the case of green excitation, which can be attributed to carrier trapping in an increased density of deep defects [2]. This is borne out by the DOS plots shown in figure 2(b). The band tails are essentially identical, but there is a gradual departure at deeper energies that reaches a factor of 2 below 0.6 eV. It should be noted that essentially identical behaviour was observed when probing the film from the substrate side. As the relative permittivities of air and glass are approximately 1 and 5 respectively we believe that band-bending effects are not responsible and that the increase is associated with additional states within about 0.1 μm of the film surface, *i.e.* the absorption depth of the green light.

Integration of the two curves between 0.4 and 0.7 eV yields an estimate of the deep defect densities in each case, which amount to $0.7 \times 10^{16}$ and $1.1 \times 10^{16}$ cm$^{-3}$, a difference of $4 \times 10^{15}$ cm$^{-3}$. This figure should represent the apparent density of deep states, without need for rescaling; the reduction in effective film thickness is cancelled by the increase in initial carrier density as both laser pulses contain approximately the same number of photons. However, it is difficult to pursue this quantitative argument further as we have no reason to suppose the 'defect-rich' layer is 0.1 μm thick, it could be confined to a much thinner surface region.

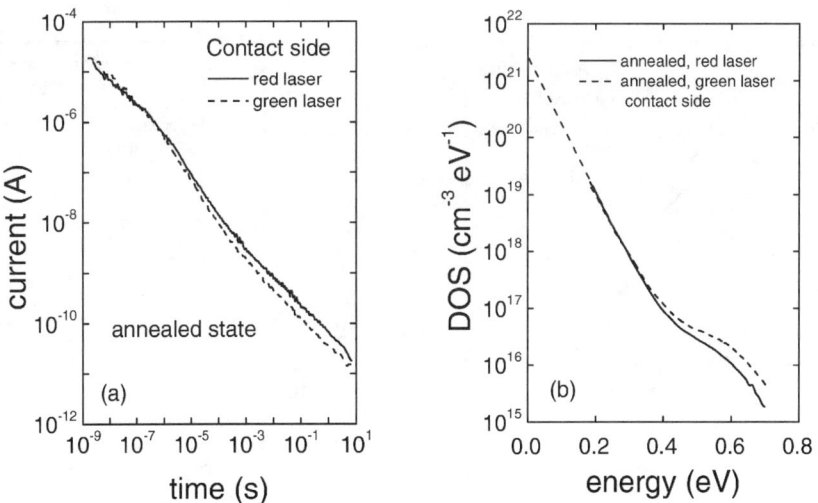

**Figure 2.** (a) Annealed film TPC decays, red and green laser from contact side; (b) corresponding DOS plots.

### White light soaking

Following exposure to white light, from the contact side, a profound change in the photocurrent decay is apparent, as shown in figure 3(a). The current falls much more rapidly than in the annealed case, beginning at about $10^{-8}$ s, due to greatly increased trapping in deep states associated with metastable defects. The decays measured with the laser incident from the substrate side are omitted for clarity, but the resulting DOS plots have been calculated as shown in figure 3(b). Integrating as before we obtain a density of deep defects of $6\times10^{16}$ cm$^{-3}$, an increase of about an order of magnitude over the bulk annealed value.

From figure 3 we see that, relative to the 'bulk' DOS obtained using the red laser (from either side), a small *increase* occurs when the green laser is incident from the *contact* side, and a rather larger *decrease* occurs when the green laser is incident from the *substrate* side. This can be understood as follows. The AM1 source contains a wide spectral content, and thus the carrier generation profile during light soaking in traversing the film will be quite complex. However, there will always be a higher generation rate at the incident surface, due to the enhanced absorption of the shorter wavelengths, and a lower generation rate at the exit surface, due to the filtering effect of the bulk of the film. Consequently we should anticipate a diminishing density of metastable defects, which is what our TPC results appear to indicate.

### Green light soaking

As a further test of the above hypothesis, a green filter was placed over the sample during light soaking, the effect being to remove wavelengths longer than 600 nm. The results can be

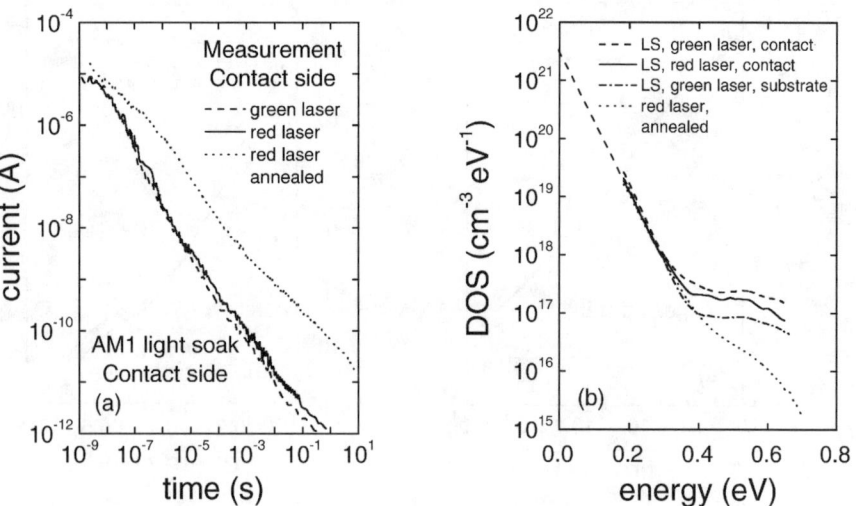

**Figure 2.** (a) White light soaked film TPC decays, red and green lasers from contact side, annealed decay shown for comparison; (b) corresponding DOS plots.

seen in figures 4(a) and 4(b). These show the effect of green light soaking from the substrate side. It is evident that the majority of metastable defects are created close to the incident surface, as revealed by the larger DOS calculated from the green laser decay. The decays obtained using the red laser from either side and the green laser from the contact side are not greatly different from those obtained with the film in the annealed state, confirming that few additional defects are created outside of the surface layer. It should be noted that green light soaking from the contact side produced a mirror-image of the results described above; a high density of defects recorded on the contact side, and little change compared with the annealed state elsewhere.

**Dark currents**

The sample dark current is a sensitive indicator of the presence of metastable defects. Typically, in an intrinsic sample, the dark current can decrease by two orders of magnitude on light soaking due to a combination of a reduction in carrier lifetime and a shift in the position of the Fermi level towards mid-gap. Here, the average annealed sample dark current was 180 pA, falling on white light soaking to approximately 2 pA. It is interesting to note that on light soaking through the green filter the dark current fell by only a factor of two, a further indication of the localised nature of defect creation.

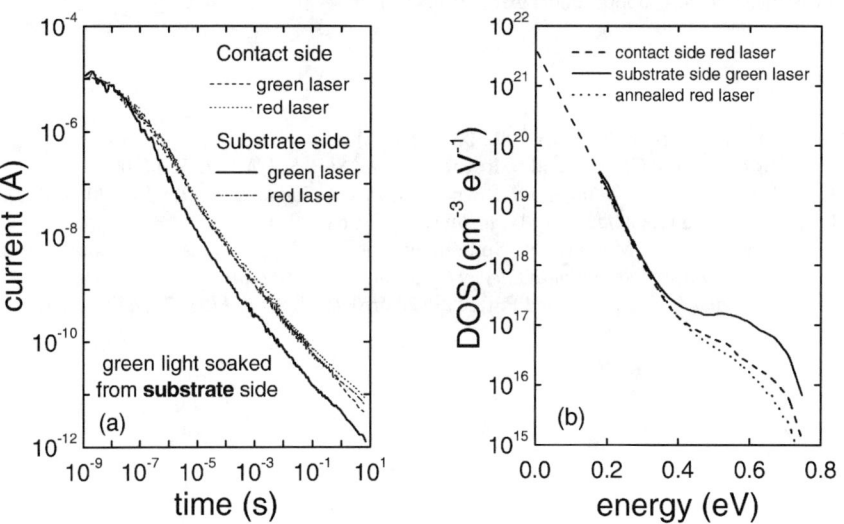

**Figure 4.** (a) Green light soaked film TPC decays, red and green lasers from contact side, annealed decay shown for comparison; (b) corresponding DOS plots.

## CONCLUSIONS

Transient photoconductivity measurements can be used to obtain energetic *and* spatial resolution of electronic defects in hydrogenated amorphous silicon films. The technique is sensitive to both native and light-induced metastable defects. The density at the air and substrate surfaces of annealed films is similar and is higher by at least a factor of two than in the bulk. White light-soaking of commercial PECVD films for 750 minutes under AM1 conditions increases the bulk defect density by one order of magnitude from the annealed condition, to $6 \times 10^{16}$ cm$^{-3}$. Significantly more defects are created at the incident surface of the film than in the bulk or at the exit surface. This can be attributed to the high absorption of shorter wavelengths at the incident surface and the resulting generation rate profile. Light soaking with green light creates metastable defects close to the incident surface and leaves the remainder of the film largely unaffected. We believe this approach is of use in the study of the kinetics of metastable defect creation, as well as in the investigation of the degradation of photovoltaic materials and devices.

## ACKNOWLEDGMENTS

The authors would like to thank InterSolar Ltd. for the provision of samples, and we are grateful for British Council support under ARC project 1185.

## REFERENCES

1. C. Main, *J.Non-Cryst. Solids* **299-302**, 525 (2002).
2. R. Brüggemann, C. Main and S. Reynolds *Phys. Stat. Sol. A* **191**, 530 (2002).
3. R. Brüggemann, C. Main and S. Reynolds *J. Phys.: Condens. Matter* **14**, 6909 (2002).
4. S. Ghosh and G. Ganguly, *J. Appl. Phys.* **68**, 5896 (1990).
5. P. Kounavis, D. Mataras, N. Spiliopoulos and D. Rapacoulias, *Proceedings of the 12$^{th}$ European Photovoltaic Solar Energy Conference, Amsterdam, 11-15 April 1994,* p144.
6. J.P. Kleider, C. Longeaud and P. Roca I Cabarrocas, *J. Appl. Phys.* **72**, 4727 (1992).

## A Study of Electronic Defects in Hydrogenated Amorphous Silicon Prepared by the Expanding Thermal Plasma Technique

Steve Reynolds, Charlie Main, Ivica Zrinscak, Zdravka Aneva[1] and Diana Nesheva[1],
School of Computing and Advanced Technologies, University of Abertay Dundee,
Bell Street, Dundee, U.K.
[1]G. Nadjakov Institute of Solid State Physics, Bulgarian Academy of Sciences,
Boul. Tzarigradsko chaussee 72, Sofia 1784, Bulgaria.

## ABSTRACT

The electronic properties of amorphous silicon films prepared by the expanding thermal plasma technique have been studied using steady-state and transient photoconductivity measurements. It is found that films deposited at a substrate temperature of 400°C have a conduction band tail slope of 29 meV, deep defect density of order $3\times10^{16}$ cm$^{-3}$, an Urbach tail slope of 65 meV, defect absorption of 5-10 cm$^{-1}$, and a mobility-lifetime product of $1.3\times10^{-7}$ cm$^2$ V$^{-1}$. A slight increase in defect density and reduction in mobility-lifetime product is observed on moderate light-soaking. The overall optoelectronic quality is somewhat poorer than commercial PECVD material, but there is scope for improvement as deposition conditions are further optimised.

## INTRODUCTION

The expanding thermal plasma (ETP) technique has been used to deposit hydrogenated amorphous silicon (a-Si:H) films of reasonable optoelectronic quality at growth rates as high as 10 nm/s [1]. This opens up the possibility of realizing substantially higher throughputs in the commercial production of photovoltaic cells. It has also been shown that the hole mobility, as measured by time-of-flight spectroscopy, is up to ten times higher in this material than in 'standard' plasma-enhanced or hot-wire CVD a-Si:H, but that the electron mobility is similar or a little lower [2,3]. Post-transit analysis of the time-of-flight photocurrent has been applied in order to obtain energy-resolved information on the mid-gap density of states (DOS). However, until very recently, these measurements have proved puzzling as they suggest an unfeasibly low DOS between mid-gap and the band tails. This has now been identified as an artefact associated with the photo-induced lowering of the contact potential between the Cr dot contact and the ETP material [4], which essentially gives rise to an error when subtracting the dark current.

In our work, we have used the related technique of transient photocurrent spectroscopy (TPC) [5] to study the DOS in ETP a-Si:H. Unlike post-transit measurements, which must be carried out on a sandwich structure with blocking contacts to ensure that only the primary photocurrent emitted from deep states is recorded, TPC measures the secondary photocurrent in a coplanar gap-cell where the metal contacts and sample geometry ensure an adequate supply of charge to maintain an ohmic response. This obviates the problems described above, and also permits a broader energy range to be studied (both defects and band-tails) since the sample capacitance is smaller, and the photocurrent decay at a given temperature is less rapid.

We have supplemented the TPC investigation with steady-state photoconductivity measurements, in which the electron quasi-Fermi level has been moved through the DOS by a combination of intensity and temperature changes, allowing the local slope of the DOS to be determined [6]. Both the above techniques give information on the majority-carrier kinetics and therefore reveal the localised-state density in the conduction-band 'half' of the bandgap. To obtain a complete picture we have used the 'absolute' constant-photocurrent method (ACPM) to probe the remaining states. Finally, we have made a preliminary investigation of the relationship between the DOS obtained using TPC and the thermally-stimulated current (TSC) observed as the temperature is ramped following the switch-off of illumination, by means of computer simulation.

## EXPERIMENTAL DETAILS

The ETP samples measured in this work were prepared at TU Eindhoven [1]. Samples were deposited on glass substrates held at 400°C at rates of 6-7 nm/s, to give a final film thickness of 1 µm as determined from optical interference measurements. The electrode geometry used in this work was a coplanar gap cell configuration with a 0.8 mm gap between Cr pads 2 mm square. The film was measured in one of two conditions; following annealing for 3 hours at 200°C, and following light-soaking under an AM1 source for 1 day.

For TPC measurements, an electrically screened Laser Science VSL-337 $N_2$ laser plus dye attachment (Rhodamin 101) was used to generate 4 ns pulses of 655 nm light, attenuated as necessary by means of neutral density filters. Following preamplification the photocurrent decay was recorded on a Tektronix TDS3052 storage oscilloscope and data transferred to a PC for analysis. To analyse the photocurrent decay, we used a Fourier transform technique [5] which allows the DOS to be extracted with a resolution of order $kT$.

Steady-state photoconductivity measurements were carried out using a red LED as a light source, driven through switchable series resistors and calibrated against a BPX65 pin photodiode. Fluxes in the range $10^{12}$ to $10^{17}$ cm$^{-2}$ s$^{-1}$ could be obtained. TSC signals were recorded following 50 mW cm$^{-2}$ white-light illumination at 77 K, with a sweep rate of 0.05 K s$^{-1}$.

The CPM system utilizes a dual monochromator plus appropriate order filters, illuminated by a 100 W tungsten-halogen bulb controlled by a programmable current source. The output was chopped at 5 Hz, enabling sample and detector currents to be measured by means of lock-in amplifiers. 'Absolute' CPM spectra were obtained by following the procedures described by Vaněček et al [7]. Calibration of the absorption scale was verified by comparison with overlapping optical transmission measurements, in the range 1.5 - 2.2 eV.

## RESULTS AND DISCUSSION

### Transient Photoconductivity

Transient photocurrent decays $I(t)$ measured on the light-soaked sample over a range of temperatures are shown in figure 1(a) (upper set of curves), and the DOS obtained from these data is displayed in figure 1(b). The overlap of the DOS sections obtained at different temperatures is quite good, which indicates that the choice of attempt-to-escape frequency $\nu_0$ of $10^{12}$ s$^{-1}$ is reasonable.

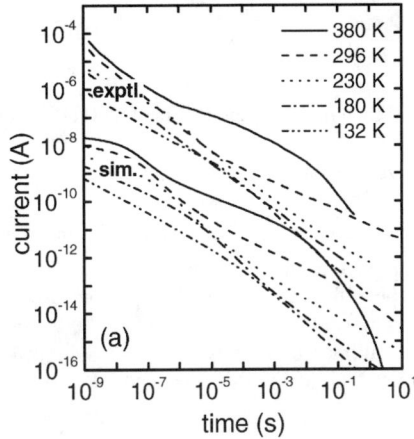
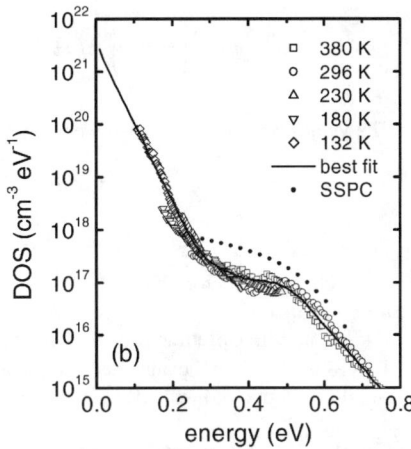

**Figure 1(a).** Experimental and simulated photocurrent decays. Simulated curves offset by a factor of 1000.

**Figure 1(b).** Density of states obtained from experimental photocurrent decays and steady-state photoconductivity (offset by factor of 3).

We have assumed a band-edge DOS of $4\times10^{21}$ cm$^{-3}$ eV$^{-1}$ and then extrapolated an exponential band-tail of 29 meV to this point. As a further check of this 'calibration', the sensitivity of the photocurrent decay to changes in laser pulse intensity was investigated. At 300 K, the $I(t)$ curves obtained at initial carrier densities of $10^{15}$ cm$^{-3}$ and $10^{16}$ cm$^{-3}$ scaled linearly, but a noticeable departure in shape occurs at $10^{17}$ cm$^{-3}$, suggesting a significant fraction of defect states are being filled between these two values. This observation supports the value of $3\times10^{16}$ cm$^{-3}$ obtained by integrating the DOS in figure 1(b) between 0.3 and 0.8 eV.

Also shown in figure 1(a) (lower set of curves) for comparison are the *simulated* $I(t)$ curves [8] obtained from the 'best-fit' DOS in figure 1(b) assuming a free carrier lifetime of $2\times10^{-6}$ s. The simulations match the shape of the experimental curves quite well, except at times $<10^{-7}$ s. It is likely that bandwidth limitations of the experimental set-up are responsible for this.

Results obtained from films deposited using a commercial PECVD reactor, normally used to manufacture photovoltaic plate, and from an optimised PECVD research reactor, are shown in figure 2 for comparison. The conduction band tail slope is similar for both ETP and commercial PECVD materials, but if the annealed states are compared the 'native' defect density is substantially higher in the ETP sample. The difference is not so large after light-soaking. The optimised research material has both a steeper tail and lower defect density.

## Steady-state photoconductivity

The photoconductivity *vs.* flux characteristics for the ETP sample in the light-soaked state as a function of temperature are shown in figure 3. The local gradient $\gamma$ of the curves on a log-log plot is related to the slope $E_0$ of a section of the DOS at $E = E_{Fn}$ by [6]:

$$E_0 = \left(\frac{1}{\gamma} - 1\right)^{-1} . kT \qquad (1)$$

The values of $E_0$ thus obtained were smoothed and then used to construct the DOS recursively:

$$DOS_{n+1} = DOS_n . exp(-(E_{n+1} - E_n)/E_0) \qquad (2)$$

The energy scale is given by $E = -kT \ln(\sigma_p/\sigma_0)$ where $\sigma_p$ is the photoconductivity and $\sigma_0$ is the conductivity prefactor (200 S cm$^{-1}$). The result is plotted in figure 1(b) adjacent to the TPC DOS for comparison, and can be seen to follow the shape of the defect 'shoulder' quite accurately.

The mobility-lifetime product was measured at 620 nm with a flux of $10^{15}$ cm$^{-2}$ s$^{-1}$ to be $1.3 \times 10^{-7}$ cm$^2$ V$^{-1}$ in the annealed sample at 293 K, falling only slightly on light-soaking, which should be compared with $10^{-6}$ cm$^2$ V$^{-1}$ or higher for good-quality PECVD material.

**Constant Photocurrent Method**

'Absolute' CPM measurements on the ETP sample are shown in figure 4. It can be seen that the Urbach tail slope is some 65 meV and the defect shoulder absorption coefficient is 5-10 cm$^{-1}$. Clearly the PECVD material is rather better, having an Urbach tail slope of 50 meV and reduced absorption in the 'defect shoulder' region. This is particularly noticeable for samples in the annealed state. Light-soaking makes little apparent difference to the ETP material whereas the defect shoulder absorption is increased by a factor of 5 for the PECVD sample. The ETP sample

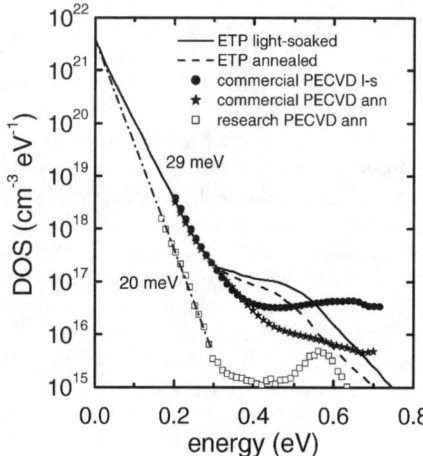
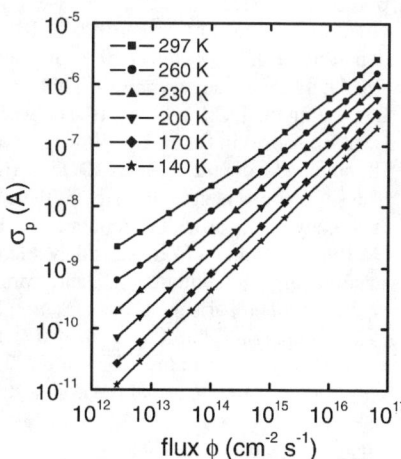

**Figure 2.** Comparison of TPC DOS plots for ETP and PECVD a-Si:H films.

**Figure 3.** Photoconductivity vs. flux for ETP sample as a function of temperature.

shows more pronounced optical interference, which may suggest a more inhomogeneous distribution of defects in this case [7]. This will be the subject of future work.

**Thermally-stimulated currents**

Measured and simulated TSC curves are shown in figure 5. The simulation solves the complete set of electron multiple-trapping rate equations for one-electron states, and although it does not calculate the contribution made by holes to transport, it has been shown [9] that this represents only a small fraction of the total current over the temperature range of interest. The input DOS is that obtained using TPC (figure 1(b)), with a free-carrier lifetime of $2\times10^{-6}$ s. It can be seen that the agreement between experiment and simulation is very poor, the experimental data suggesting the presence of a substantial density of deep defects (> 0.6 eV), which is not observed using TPC. We have no explanation for this, and we are currently seeking to verify the results from the simulation, using known analytic cases, and to check the measurement system.

## CONCLUSIONS

Hydrogenated amorphous silicon deposited by the expanding thermal plasma technique at a substrate temperature of 400°C has a conduction band tail slope of 29 meV, a deep defect density of order $3\times10^{16}$ cm$^{-3}$, an Urbach energy (valence band tail slope) of 65 meV, defect absorption coefficient of 5-10 cm$^{-1}$ and a mobility-lifetime product (from steady-state photoconductivity) of $1.3\times10^{-7}$ cm$^2$ V$^{-1}$. Only a small change in photoconductive properties, *e.g.* a factor of two

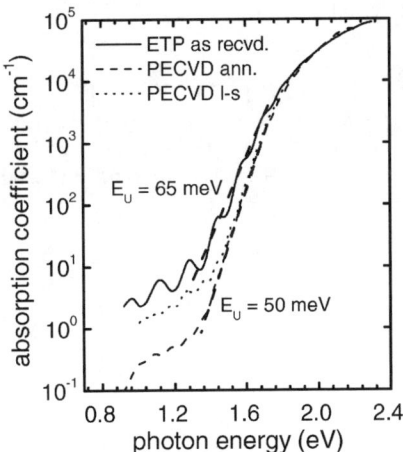
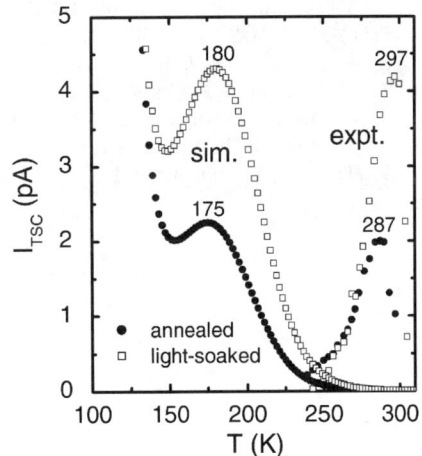

**Figure 4.** 'Absolute' CPM spectra for ETP and commercial PECVD samples. Lines are estimates of Urbach tail slopes.

**Figure 5.** Thermally stimulated currents – experiment and simulation.

reduction in $\mu\tau$-product, occurs after AM1 light-soaking for 1 day. When compared with PECVD material produced in a commercial reactor, the ETP material is somewhat inferior, especially with regard to valence-band tail slope and native defect density. Computer simulations of the thermally-stimulated current, based on the density of states deduced from transient photoconductivity, are in poor agreement with experiment, and this is the subject of current investigation. Future studies will seek to correlate process parameters such as substrate temperature with specific material properties.

## ACKNOWLEDGMENTS

The authors are indebted to the ETP group at TU Eindhoven for providing ETP samples, and would like to thank Erwin Kessels, Guy Adriaenssens and Monica Brinza for helpful and informative discussions. This work has been partially supported by a Royal Society grant.

## REFERENCES

1. W.M.M Kessels, R.J. Severens, A.H.M. Smets, B.A. Korevaar, G.J. Adriaenssens, D.C. Schram and M.C.M. van de Sanden, *J. Appl. Phys.* **89(4)**, 2404 (2001).
2. M Brinza, G.J. Adriaenssens, K. Iakoubovskii, A. Stesmans, W.M.M. Kessels, A.H.M. Smets and M.C.M. van de Sanden, *J. Non-Cryst. Solids* **299-302**, 420 (2002).
3. B.A. Korevaar, G.J. Adriaenssens, A.H.M. Smets, W.M.M. Kessels, H.-Z. Song, M.C.M. van de Sanden and D.C. Schram, *J. Non-Cryst. Solids* **266-269**, 380 (2000).
4. M. Brinza, G.J. Adriaenssens, W.M.M. Kessels, A.H.M. Smets and M.C.M. van de Sanden, MRS Spring meeting, April 21-25, 2003, poster A19.6.
5. C. Main, *MRS Symp. Proc.* **467**, 167 (1997).
6. R. Brüggemann, *J. Appl. Phys.* **92**, 2540 (2002).
7. M. Vaněček, J. Kocka, A. Poruba and A. Fejfar, *J. Appl. Phys.* **78**, 6203 (1995).
8. C. Main, J. Berkin and A. Merazga, in *New Physical Problems in Electronic Materials*, eds. M. Borissov, N. Kirov, J.M. Marshall and A. Vavrek (World Scientific, 1991), p55.
9. T. Smail, M. Aoucher and T Mohammed-Brahim, *J. Non-Cryst. Solids* **266**, 376 (2000).

# Thin Film Transistors
# and Imager Arrays

## Performance of Thin-Film Silicon MEMS Resonators in Vacuum

J. Gaspar[1,2], V. Chu[1], and J. P. Conde[1,2]
[1]INESC Microsistemas e Nanotecnologias, Rua Alves Redol 9, 1000-029 Lisbon, Portugal
[2]Dept. Materials Engineering, Instituto Superior Técnico (IST), Av. Rovisco Pais, 1049-001 Lisbon, Portugal

## ABSTRACT

This paper reports on the fabrication and characterization of microelectromechanical bridge resonators on glass substrates using thin-film technology and surface micromachining. All the processing steps are performed at temperatures below 110°C. The microbridges consist of either a single layer of heavily doped n-type amorphous silicon ($n^+$-a-Si:H) or bilayers of aluminum (Al) and intrinsic a-Si:H. The bridge is suspended over a gate electrode with a 1 µm air-gap. Applying a voltage between the bridge and an underlying Al gate electrode electrostatically actuates the microstructures. The resulting deflection is monitored optically. The resonance of the microbridges is measured in air and in vacuum. Resonance frequencies up to 70 MHz and quality factors up to 3000 are obtained at pressures below 1 Torr. The energy dissipation mechanisms of the resonators are discussed.

## INTRODUCTION

MEMS are a class of devices with electronic and/or mechanical capabilities. MEMS are usually fabricated using planar microelectronics fabrication techniques to produce three-dimensional structures. The most powerful aspect of MEMS is that it is possible to integrate on-chip sensing and actuation functions with control electronics [1]. Most current MEMS devices are fabricated using bulk micromachining of crystalline silicon (c-Si) substrates or using surface micromachining of poly-Si films, which requires processing temperatures as high as 550–900°C [2].

Thin-film MEMS exploits the advantages of thin-film technology such as low temperature processing (< 300°C), which allows the integration of MEMS with its control electronics, either as part of the backend processing of CMOS technology or with a thin film transistor (TFT) back plane. In addition, due to the low temperature processing, unconventional substrates such as plastics can be used. Thin-film MEMS devices such as thermal actuators [3], air-gap TFTs [4] and a-Si:H-based bolometers [5] have been developed.

Microresonators are of great technological importance, since they can be used as radio-frequency (RF) filters, sensitive mass detectors or pressure sensors [2,6], to name a few applications. This work reports on the fabrication and characterization of thin-film microresonators fabricated at temperatures below 110°C. The resonance response of the microstructures is analyzed in air and in vacuum and the elementary energy dissipation mechanisms are discussed. The experimental results demonstrate that thin-film resonators can achieve performances comparable to those of the best c-Si and poly-Si based microresonators.

## EXPERIMENT

Microbridges with an underlying gate electrode are fabricated using thin-film technology and surface micromachining on glass substrates at temperatures below 110 °C. The fabrication process is described in detail elsewhere [7]. The process begins with the sputter-deposition and patterning of a 100 nm-thick Al gate electrode. The sacrificial layer, a 1 µm-thick photosensitive polymer, is deposited by spin-coating and directly patterned by photolithography. The structural materials are bilayers of Al (100 nm) and intrinsic a-Si:H (300 nm) or single layer bridges of $n^+$-a-Si:H (400nm). The Al is deposited by magnetron sputtering and the a-Si:H by plasma enhanced chemical vapor deposition (PECVD). At the end of the fabrication process, the sacrificial layer is selectively etched leaving an air-gap between the bridge and the gate electrode. Fig. 1 shows a scanning electron microscope (SEM) micrograph of a microstructure. The length, $L$, of the microbridges varies from 4 to 100 µm and the width, $w$, is typically in the 3–15 µm range.

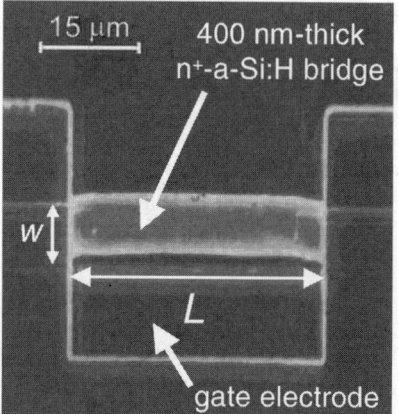

**Figure 1**. SEM micrograph of a 400 nm-thick $n^+$-a-Si:H bridge with $L = 35$ µm and $w = 10$ µm. The microbridge is suspended over an underlying Al gate counter electrode. The air-gap height (bridge-gate separation) is ~ 1 µm.

The structures are electrostatically actuated by applying a voltage between the gate electrode and the bridge. The resulting deflection is monitored optically [8]: a laser is focused on the top of the microstructure and the position of the reflected beam is measured by a photodetector. The photodetector output is proportional to the amplitude of the movement of the bridge. The frequency, $f$, of the actuation voltage is varied between 100 kHz and 100 MHz and the resonance frequency, $f_{res}$, of the microbridges is detected measuring the response of the photodetector with a spectrum analyzer. The resonance of microstructures with different geometries ($L$ and $w$) is characterized as a function of the pressure, $P$, from $10^{-6}$ to 760 Torr.

## RESULTS AND DISCUSSION

### Pressure dependence

Figure 2 shows the resonance peaks of a 10 µm-wide, 30 µm-long bilayer Al (100 nm)/a-Si:H (300 nm) bridge measured at $10^2$ Torr and $10^{-6}$ Torr. As the pressure is decreased from $10^2$ Torr to below $10^{-6}$ Torr, the amplitude of the resonance, $\delta_{res}$, increases more than 100 times, while the resonance frequency, $f_{res}$, decreases slightly, from 2.33 MHz to 2.30 MHz (~ 1.3%). In addition, as the pressure is decreased, the resonance peak narrows, resulting in an increase of the quality factor, $Q$, from 30 at atmospheric pressure to 700 at $10^{-6}$ Torr.

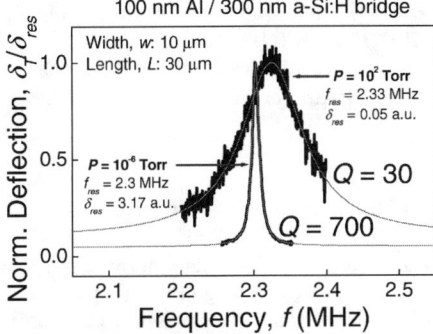
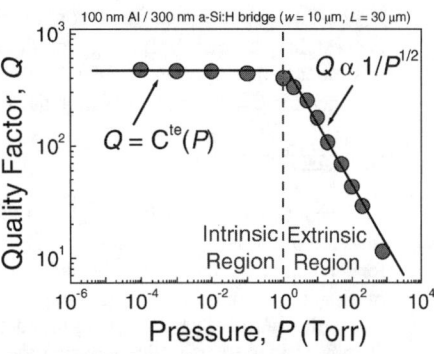

**Figure 2.** Normalized deflection, $\delta_f/\delta_{res}$, of a Al/a-Si:H bridge measured as a function of the driving force frequency, $f$, at $10^2$ Torr and $10^{-6}$ Torr. The resonance peak narrows as the pressure decreases.

**Figure 3.** Quality factor, $Q$, dependence on pressure, $P$. For $P < 1$ Torr, $Q$ is independent of $P$, while for $P > 1$ Torr, $Q \propto P^{-1/2}$, indicating air damping dissipation.

$Q$ is defined as $Q = f_{res}/\Delta f_{\text{-3dB}}$, where $\Delta f_{\text{-3dB}}$ is the bandwidth 3dB below the peak. $1/Q$ is proportional to the energy dissipated by the resonator [9] and increases with pressure, indicating air damping [10]. At high pressures, the air molecules exchange momentum with the resonator, leading to energy dissipation. As the pressure is lowered, the number of air molecules decreases and this extrinsic mechanism no longer limits the $Q$ of the resonator.

The quality factor, $Q$, is plotted as a function of the pressure, $P$, in Fig. 3. The extrinsic region, where air damping is the dominant dissipation mechanism, occurs at pressures above 1 Torr. Fluid mechanics calculations show that $Q$ is proportional to $P^{-1/2}$ in this pressure range [10]. For $P < 1$ Torr, the number of surrounding air molecules no longer limit the resonator dissipation and $Q$ becomes pressure independent. In this pressure range, $Q$ is limited by intrinsic mechanisms Such as thermoelastic dissipation, clamping losses, surface/interface losses, phonon-phonon and phonon-electron interactions [9,11].

## Length dependence of resonance in vacuum

An expression for the undamped resonance frequency, $f_{res}$, of the $n^{th}$ flexural mode can be found by solving the Euler-Bernoulli equation [9]. The resulting expression is given by

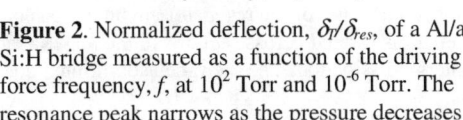

$$f_{res} = a_n^2 \frac{\pi}{4\sqrt{3}} \frac{t}{L^2} \sqrt{\frac{E}{\rho}}, \qquad (1)$$

where $a_n$ is a constant that depends on the structure boundary conditions (for clamped-clamped bridges, $a_1 \cong 1.5$, $a_2 \cong 2.5$, ..., according to the vibration mode number), $t$ and $L$ are the thickness and length of the bridge, respectively, and $E$ and $\rho$ are the Young's modulus and mass density of the structural material, respectively. The first flexural mode resonance frequency of 400 nm-thick n$^+$-a-Si:H microbridges is plotted as a function of the length in Fig. 4. The measurements are made at $10^{-6}$ Torr (undamped resonance) for widths of 10 μm and 7 μm. Fig. 4 shows that $f_{res}$ is proportional to $1/L^2$ and is independent of the width, $w$, in agreement with Eq. (1). The density,

$\rho$, of a-Si:H is ~ 2.3 g/cm$^3$ and from the fit of the experimental data to Eq. (1) (line in Fig. 4), it is possible to extract the Young's modulus of a-Si:H. $E_{\text{a-Si:H}}$ ~ 160 GPa is thus obtained, which is in similar to that of poly-Si [9] and c-Si [12]. Resonance frequencies as low as 100 kHz are obtained for $L$ = 70 µm and as high as 70 MHz for $L$ = 10 µm for Al/a-Si:H bilayer bridges.

**Intrinsic dissipation**

The intrinsic quality factor, $Q_i$, of the n$^+$-a-Si:H bridges, measured at 10$^{-6}$ Torr, is plotted as a function of the length in Fig. 5. A slight increase of $Q_i$ with $L$ is observed, and $Q_i$ reaches ~ 3000 for the longer structures. No dependence with width is observed. The major intrinsic dissipation mechanism in microresonators is expected to be thermoelastic damping [13]. Dissipation arises due to the thermoelastic coupling in materials, when a temperature change induces a volume change and vice-versa. Since, during vibration, some parts of the resonator are under tension and others under compression, a temperature gradient is formed across sections of the material and an irreversible heat flux occurs, leading to energy dissipation. The thermoelastic-limited quality factor, $Q_{ted}$, is given by

$$\frac{1}{Q_{ted}} = \frac{\alpha^2 TE}{\rho C_p} \frac{\left(2 f_{res} \rho C_p t^2 / \pi k\right)}{1 + \left(2 f_{res} \rho C_p t^2 / \pi k\right)^2}, \qquad (2)$$

where $T$, $\alpha$, $C_p$ and $k$ are the temperature, linear expansion coefficient, heat capacity and thermal conductivity of the material, respectively [13]. The calculated thermoelastic limit for a-Si:H resonators (with $\alpha$ ~ 2.6×10$^{-6}$ K$^{-1}$ [14], $C_p$ ~ 700 J kg$^{-1}$ K$^{-1}$ [15] and $k$ ~ 4.74 W m$^{-1}$ K$^{-1}$ [15]) is shown in Fig. 5 and experimental $Q$'s are expected to approach, but not exceed, this limit. Deviations of experimental data from Eq. (2) may be due to the uncertainty in the a-Si:H film properties. Fig. 5 also shows the range of quality factor values obtained for the high-quality resonators made of c-Si or poly-Si prepared at 700-1000°C [9,16]. The a-Si:H based microresonators fabricated at low temperatures (<110°C) described here have performances that are approaching those of single-crystal Si and annealed poly-Si microresonators.

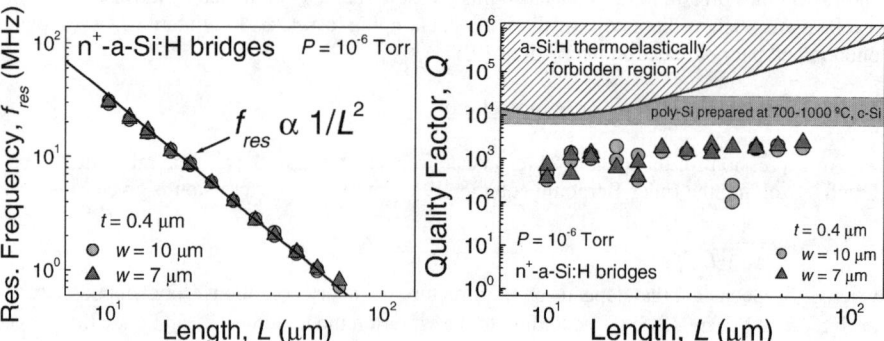

**Figure 4.** Resonance frequency, $f_{res}$, of n$^+$-a-Si:H bridges measured at 10$^{-6}$ Torr as a function of the length, $L$, for $w$ = 10 µm (circles) and $w$ = 7 µm (triangles). The line is a fit to the experimental data ($f_{res} \propto 1/L^2$).

**Figure 5.** Quality factor, $Q$, of n$^+$-a-Si:H bridges measured at 10$^{-6}$ Torr and plotted as a function of $L$. The thermoelastic limit for a-Si:H and the range of $Q$'s obtained for high-quality poly-Si and c-Si resonators are also shown.

## CONCLUSIONS

This work presents high performance microresonators fabricated at low temperatures using a-Si:H based microbridges and surface micromachining. The main advantages of this technology are its flexibility for integration with other technologies, namely CMOS, and the possibility of using substrates such as glass and plastic. Resonance frequencies up to 70 MHz and quality factors up to 3000 are observed with thin-film resonators on glass.

For applications such as RF filters or sensitive mass detectors, achieving a quality factor of $Q \sim 10000$ would make it possible to resolve 1 kHz in 10 MHz-bandwidth signals or to detect a $10^{-17}$ kg mass change in a $10^{-13}$ kg structure ($10^{-13}$ kg $\equiv$ typical mass of an average thin-film microbridge) with a measurable shift of the resonance frequency.

## ACKNOWLEDGMENTS

The authors gratefully acknowledge F. Silva, J. Bernardo, V. Soares and J. Faustino for help in sample/clean room processing. This work was supported by the Fundação para a Ciência e Tecnologia (FCT) through a Pluriannual Contract with INESC MN and through POCTI projects. J. Gaspar thanks FCT for his Ph.D grant.

## REFERENCES

1. Jack W. Judy, *Smart Mater. Struct.* **10**, pp. 1115-1134, 2001.
2. See for example, N. Malouf, *An introduction to Microelectromechanical Systems Engineering*. Artech House, Boston, MA, 2000.
3. J. Gaspar, V. Chu, N. Louro, R. Cabeça, and J. P. Conde, *J. Non-Crystalline Solids* **299-302**, pp. 1224-1228, 2002.
4. .M. Boucinha, V. Chu, and J. P. Conde, *Appl. Phys. Lett.* **73**, pp. 502-504, 1998.
5. A.J. Syllaios, T.R. Schimert, R.W. Gooch, W.L. McCardel, B.A. Ritchey, J.H. Tregilgas, *Mat. Res. Soc. Symp. Proc.* **609**, A14.4, 2000.
6. See for example, M. Elwenspoeck, and R. Wiegerink, *Mechanical Microsensors*. Springer, New York, 2001.
7. J. P. Conde, J. Gaspar, and V. Chu, *Thin Solid Films* **427**, pp. 181-186, 2003.
8. J. Gaspar, V. Chu, and J. P. Conde, *J. Appl. Phys.*, in press, June 2003.
9. See for example, A. N. Cleland, *Foundation of Nanomechanics – From Solid-State Theory to Device Applications*. Springer, New York, 2004.
10. William E. Newell, *Science* **161**, pp. 1320-1326, 1968.
11. J. Yang, T. Ono, and M. Esashi, *J. Microelectromech. Syst.* **11**, pp. 775-783, 2002.
12. L. M. Zhang, D. Uttamchandani, and B. Culshaw, *Sens. Act.* **29**, pp. 79-84, 1991.
13. R. Lifschitz, and M. L. Roukes, *Phys. Rev. B* **61**, pp. 5600-5609, 2000.
14. J. Fabian, and P. B. Allen, *Phys. Rev. Lett.* **79**, pp. 1885-1888, 1997.
15. S. Nakhmanson, and D. A. Drabold, *Phys. Rev. B* **61**, pp. 5376-5380, 2000.
16. K. Y. Yasumura, T. D. Stowe, E. M. Chow, T. Pfafman, T. W. Kenny, B. C. Stipe, and D. Rugar, *J. Microelectromech. Syst.* **9**, pp. 117-125.

## Area-dependent switching in thin film-silicon devices

Jian Hu, Warren Jackson[1], Scott Ward, Pauls Stradins, Howard M. Branz, and Qi Wang

National Renewable Energy Laboratory, 1617 Cole Boulevard, Golden, CO 80401, USA
[1]Hewlett-Packard Laboratories, 1501 Page Mill Rd., Palo Alto, CA 94304, USA

## ABSTRACT

We report on the area dependence of switching in both Cr/$p^+$a-Si:H/Ag(Al) and Cr/$p^+$μc-Si/Ag(Al) filament switches. The doped amorphous (a-Si:H) or microcrystalline (μc-Si) thin Si layers are made by hot-wire chemical vapor deposition. The device active region area (A) is varied over 5 orders of magnitude, from $10^{-7}$ to $10^{-2}$ cm$^2$, using photolithographically defined Ag and Al top contacts. Before switching, the resistance of 100-μm$^2$ devices is normally about 100 kΩ for μc-Si and 10 GΩ for a-Si:H. After switching with applied current ramps, the resistance decreases to a few hundred ohms in all a-Si devices and to a few thousands ohms in μc-Si devices. In both μc-Si and a-Si:H devices, the switching voltage ($V_{sw}$) decreases with increasing device area according to $V_{sw} \sim V_0 - \alpha \ln(A/A_0)$ with α=0.3V for a-Si:H and α=0.04V for μc-Si. For both materials, the switching current roughly obeys the power law $I_{sw} \propto A^\beta$ with β~1. A statistical model is proposed to explain the area scaling of the switching voltage and relate the parameters to the material properties.

## INTRODUCTION

Metal/hydrogenated amorphous silicon (a-Si:H)/metal thin film structures have been used widely in various applications such as analog memory switching devices [1-2] and thin film antifuses [3]. It has been shown that an electroforming process is necessary in order for these devices to function. During this forming, a part of the top metal contact diffuses or migrates into the a-Si:H matrix, resulting in the creation of a highly conducting filament and an irreversible decrease of device resistance from a high value of 1 MΩ~100 MΩ to about 1 kΩ or less [1-2,4]. We have reported previously [5] that in c-Si(p+)/a-Si(p)/Ag devices switching events occur when the metal/a-Si(p) contacts are blocking injection. We have recently extended our study into μc-Si thin film devices. In this paper, we shall report area-dependence of the switching characteristics in both metal/a-Si(p)/metal and metal/μc-Si(p)/metal thin-film structures.

## SAMPLE PREPARATION AND EXPERIMENTAL SETUP

Boron-doped a-Si:H and μc-Si layers are deposited by hot-wire chemical vapor deposition at 160 °C [6] on separate Corning 1737 glass substrates coated with 600 Å of Cr. The deposition rate for the a-Si:H layer is about 10 Å/s, and the μc-Si layer is about 3 Å/s. Typical thickness of the a-Si:H layer is about 100 nm, and the μc-Si layer is 60 nm. The conductivity of the a-Si layer is 2.98×10$^{-10}$ (Ω-cm)$^{-1}$, whereas the μc-Si layer is ~0.3 (Ω-cm)$^{-1}$. A 800-Å-thick Ag or Al top metal contact is deposited using electron-beam evaporation,. This top contact is patterned photolithographically to define many devices of various sizes. The device active area varies 5 orders of magnitude, from $10^{-7}$ to $10^{-2}$ cm$^2$. Switching is stimulated with a slow current-ramp from 0 to 50 mA (total ramping time ~0.5 sec). Before switching, the resistance of 100-μm$^2$ devices is normally about 100 kΩ for μc-Si and 10 GΩ for a-Si:H.

**RESULTS**

1. Area-dependent switching behaviour in a-Si:H devices.

Typical switching behaviour of amorphous silicon devices in response to a current ramp is shown in Fig.1. The current I increases with voltage V as I ~ $e^{AV}$ (A is a constant) until a critical current is reached, whereupon the voltage abruptly decreases. After the switching event, the current is roughly linear with voltage. During switching, the resistance decreases to a few hundred ohms in all devices. The current ($I_{sw}$) and voltage ($V_{sw}$) at which the switching occurs are measured for many devices of various areas. The median values for $I_{sw}$ and $V_{sw}$ as a function of area are shown in Fig. 2. The switching current $I_{sw}$ roughly scales as A, whereas the switching voltage $V_{sw}$~12-0.15ln(A/$10^{-7}$) ($V_{sw}$ in V, A in cm$^2$).

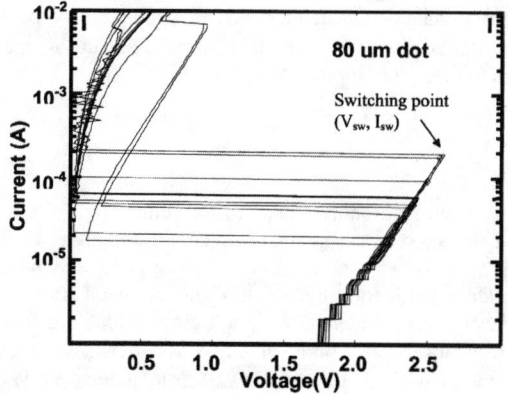

**Fig.1.** Current vs. voltage for a-Si switches for a number of devices.

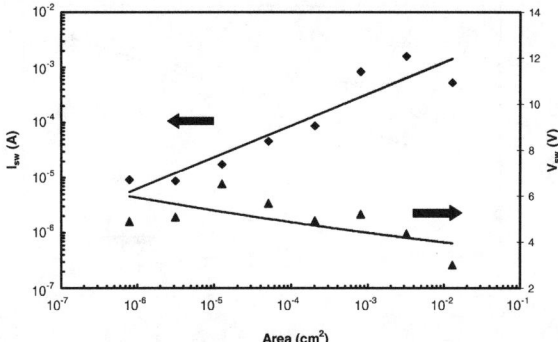

**Fig. 2.** Switching current $I_{sw}$ and $V_{sw}$ vs the top contact area for a-Si devices.

2. Area-dependent switching behaviour in μc-Si devices.

In μc-Si devices, the switching behaviour is notably different from that in a-Si devices. Fig.3 shows typical switching current-voltage characteristics for several different device sizes from 5x5 to 30x30 μm$^2$. At low biases (V ~ 0.3 V), the current is almost linearly proportional to voltage; at intermediate biases (V = 0.3 to about 2 V), the current changes with voltage superlinearly. When the voltage reaches a threshold value, the current abruptly increases, indicating the development of a low resistance current path. As the voltage increases further, additional current paths develop, reducing the voltage across the device. As current increases even further, the voltage becomes saturated and keeps almost constant at 2~2.5 V until the

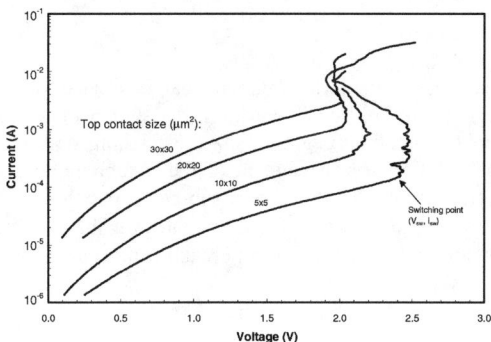

**Fig.3.** Typical switching current-voltage traces in μc-Si devices for 4 different sizes from 5x5 μm to 30x30 μm$^2$.

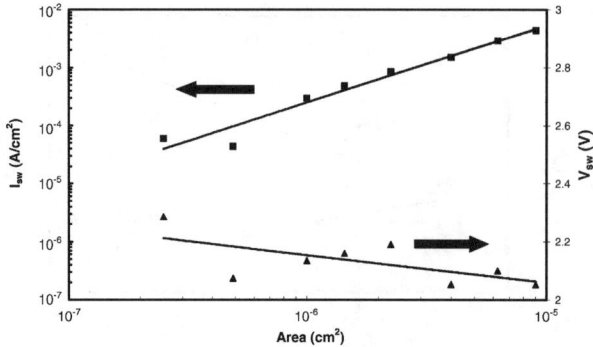

**Fig. 4.** Switching current $I_{sw}$ and $V_{sw}$ vs. the top contact area for μc-Si devices.

current reaches a critical point where voltage begins to drop. One such critical point is labelled in Fig. 3. Compared to a-Si:H devices, the reduction in voltage is limited, merely 0.2~0.5V, and more gradual. During this switching transition, the device resistance drops to a few kΩ, which is also considerably larger than the resistance in the switched a-Si:H devices (hundreds of ohms). The critical current (where the voltage first stops increasing) is found to increase linearly with the top contact area. The critical current density $J_{sw}$ remains roughly constant, $I_{sw} \propto A^{\beta}$ with β ~ 1, as shown in Fig.4. The switching voltage $V_{sw}$ increases only slightly as the area decreases, from 2 to 2.4 V.

## DISCUSSION

The area dependence of the switching event can be related to the statistical distribution of the subregion switching thresholds. We can divide a switch of overall area A into a parallel group of number $n=A/A_0$ small subregions of $A_0$. Switching occurs when the applied voltage exceeds the threshold of the weakest element of the n subregions as a result of field-induced breakdown. Assume that the probability that a subregion threshold is below V is given by the cumulative probability distribution $\Phi(V)$. Then, the probability that the minimum threshold voltage of of n subregions is below V (i.e., that breakdown occurs at V) is given by

$$\Phi_n(V) = 1 - (1 - \Phi(V))^n. \qquad (1)$$

The probability density that the minimum of n regions lies between V and V+dV is given by the derivative of Eq. (1) with respect to V. We would like to find the average minimum, <V> in terms of n because this average is the measured average switching voltage, $V_{sw}$. This probability distribution is highly peaked so that <V> is the same as the peak of the probability distribution.

The peak of the probability distribution minimum switching voltage for a device with n regions is approximately given by

$$n(1-\Phi(<V>))^{n-1} = 1/2. \qquad (2)$$

Solving for $\Phi(<V>) \ll 1$ approximately yields the result

$$\Phi(<V>) \cong 1/n. \qquad (3)$$

If the probability of the switching voltages P(V) is exponential with a voltage width $1/r$ up to a maximum of $V_0$, then,

$$\Phi(<V>) = \exp[-r(V_0 - <V>)] \approx 1/n \quad <V> < V_0. \qquad (4)$$

We have numerically verified this equation for exponential distributions. Solving for $<V>$ gives

$$<V> = V_{switch} = V_0 - \ln(n)/r = V_0 - \ln(A/A_0)/r. \qquad (5)$$

Thus, the average switching voltage decreases logarithmically as the area increases. The slope is the width of the distribution of switching threshold voltages. A Gaussian distribution yields a somewhat similar dependence but there is curvature to the $V_{sw}$ versus $\ln(A)$. Fitting the results of Figs.(2) and (4) to Eq.(5), indicates that the distribution of breakdown voltages has a width of 0.05 V and an estimated maximum of 2.3 V if the subregion area is 1 um$^2$. For a-Si switches, the distribution $1/r$ is much wider, being about 0.3 V and a maximum of 8-10 V at 1 um$^2$.

Switching voltage decreases with increasing area because the switching occurs first at the weakest element of the subregions when the applied voltage exceeds the threshold ($V_{sw}$). Naturally, the larger the device area, the more weak subregions that will be available for switching. Statistically, the switching voltage for a large device will be smaller than the voltage for small devices, which have a smaller number of subregions. As the voltage increases across the device, the leakage current increases through all the subregions, giving rise to the device subswitching characteristics. Obviously, with increasing area, contributions to the current from subregions also increase. As a result, current scales with the increase in the number of subregions, n, and eventually the overall switch area. In fact, we can define a device resistance R as $R = V_{sw}/I_{sw}$. Then, according to Eq.(5), the switching current $I_{sw}$ can be expressed as

$$I_{sw} = \frac{V_{sw}}{R} = \frac{V_0 - \alpha \ln(A/A_0)}{\rho d / A} = f(n) \bullet A, \qquad (6)$$

where $\rho$ and $d$ are the resistivity and thickness of device respectively. Here we use a simple expression for R: $R = \rho d/A$. Terms $V_0$, $\alpha$, $\rho$, and $d$ do not depend on A, whereas the factor f(n) relies mainly on the number of subregion, $n(=A/A_0)$, instead of A itself. From this simple analysis, we see that indeed $I_{sw} \sim A$. That is the current scaling law we observe.

With regard to the difference between a-Si and μc-Si devices, it is expected that the disorder of amorphous is greater than the disorder of microcrystalline, so the distribution should be wider. The width of the subregion distribution $1/r$ divided by its median value is roughly the same for a-Si and μc-Si. Hence, the percent variation of the two cases is the same. Finally, the fact that the final resistances are similar indicates that a filamentary path of similar conductance is formed in both a-Si and µc-Si. Because the device conductance is several orders of magnitude smaller for a-Si, the formation of one conductive channel greatly alters the device conductance and a large switching event is observed. In µc-Si, it takes many of such events to appreciably alter the device conductance. A high electric field may exist at grain boundaries, thus local heating or metal diffusion could occur at grain interfaces. Switching could relate to formation of multiple conduction channels (or filaments), resulting in the observed area dependence of switching current.

## CONCLUSION

We have observed that the switching voltage ($V_{sw}$) in both µc-Si and a-Si:H devices decreases with increasing device area according to $V_{sw} = V_0 - \alpha \ln(A/A_0)$, where A is the area of the top contact metal (Ag or Al), whereas the switching current ($I_{sw}$) in both roughly obeys the power law $I_{sw} \propto A^{\beta}$ with $\beta \sim 1$. A statistical model is proposed to explain the area scaling of $V_{sw}$ by assuming that each device consists of a many small elements that function as statistically independent devices. The total device characteristics represent an aggregation of the elements' statistically distributed switch characteristics, with the easiest-to-switch elements dominating the observed behavior. Switching current $I_{sw}$ comes from the contribution of all the subregions (or elements). Therefore, $I_{sw}$ increases with the increase in the number of these subregions (or total area). A good agreement between the model calculation and experimental data has been obtained.

## ACKNOWLEDGEMENT

The authors would like to thank Anna Duda for assistance with metallization.

## REFERENCES

[1] J. Hajto, A. E. Owen, A. J. Snell, P. G. LeComber, and M. J. Rose, *Amorphous and Crystalline Semiconductor Devices*, edited by J. Kanicki, (Artech House, 1992), p. 641. See also references cited therein.
[2] M. Jafar and D. Haneman, Phys. Rev. B **49**, 13611 (1994)
[3] K. Gordon and R. Wong, IEEE IEDM Tech. Dig., 27 (1993)
[4] A. J. Snell, P. G. LeComber, J. Hajto, M. J. Rose, A. E. Owen, and I. S. Osborne, J. Non-Cryst. Solids, **137&138**, 1257 (1991)
[5] J. Hu, H.M. Branz, R.S. Crandall, S.Ward, C.Perkins and Q.Wang, Res.Soc.Symp.Proc. vol.715, pp.763 (2002)
[6] A. H. Mahan, J. Carapella, B. P. Nelson, R. S. Crandall, and I. Balberg, J. Appl. Phys., **69**, 6728 (1991)

# PROPERTIES OF SILICON NITRIDE FILMS PREPARED BY COMBINATION OF CATALYTIC-NITRIDATION AND CATALYTIC-CVD

A. Izumi[*,**], A. Kikkawa[**], K. Higashimine[**] and H. Matsumura[**]
*Kyushu Institute of Technology
Fukuoka 804-8550, JAPAN, izumi@ele.kyutech.ac.jp
**JAIST (Japan Advanced Institute of Science and Technology)
Ishikawa 923-1292, JAPAN

## ABSTRACT

This paper reports about the interface of silicon nitride ($SiN_x$) formed on Si(100) prepared by combination of catalytic-nitridation and catalytic-vapor deposition method in a catalytic chemical vapor deposition system. It is found that flat interface of $SiN_x$/Si(100) is formed by inserting nitridation layer before growing the $SiN_x$ films.

## INTRODUCTION

Silicon nitride ($SiN_x$) films prepared at low temperatures below 400°C are widely applicable as gate dielectric films used in liquid-crystal displays (LCD) thin-film transistors (TFTs). Plasma-enhanced CVD (PECVD) method is one low temperature methods of deposition. However, PECVD has serious problems, such as plasma damage to the substrates. Moreover, large area deposition of $SiN_x$ films by the conventional radio frequency PECVD becomes difficult because the standing waves in radio frequency plasmas limit uniform power dissipation. Therefore, a novel method to obtain $SiN_x$ films in large area at low temperatures, without any help from plasma excitation, is required.

The catalytic chemical vapor deposition (catalytic-CVD) method [1], which is often called hot-wire CVD (HWCVD), is one method to answer above requirement. In this method, deposition gases are decomposed by the catalytic cracking reactions of a heated tungsten (W) catalyzer placed near substrates. $SiN_x$ films are formed at low temperature, as low as around 300°C, without any help from a plasma or photochemical excitation [2,3]. Low-temperature direct nitridation of crystalline Si below 300°C, named catalytic-nitridation can be performed using decomposed $NH_3$ species in a catalytic-CVD system [4]. It is well known that the quality of interface between gate dielectric films and semiconductors determines the electrical properties of TFTs. Inserting a nitridation layer before $SiN_x$ deposition may improve the quality of interface.

Recently, we have developed a novel $SiN_x$ deposition method by combination of catalytic-nitridation and catalytic-CVD method in a catalytic-CVD system [5,6]. It is found that inserting a catalytic-nitridation layer, injection-type hysteresis loop in capacitance versus voltage (C-V) curve and large threshold voltage shift ($V_{th}$) to the negative direction are drastically reduced.

In this work, the $SiN_x/Si(100)$ interface structure was investigated by high-resolution transmission electron microscope (HRTEM).

**EXPERIMENTAL**

A catalytic-CVD system is schematically illustrated in Fig. 1. A tungsten (W) wire is spread with an area of 100 mm x 100 mm at the distance about 60 mm from the substrates keeping it parallel to the substrate surface. The W wire is heated by supplying direct AC electric power and the temperatures are monitored by both an infrared thermometer and the electrical resistivity. CZ n-type (100) Si substrates with resistivity of 0.85-1.5 $\Omega$cm were used for nitridation and $SiN_x$ deposition. These Si substrates were degreased and cleaned by RCA method [7]. Then, they were dipped in 2% diluted HF for 2 min to H-terminate the surface. After the cleaning process, the pieces of wafer were immediately loaded into the catalytic-CVD chamber by way of the load lock chamber. Next, surface nitridation was performed, followed by growth of the 40 nm-thick $SiN_x$ film. The base vacuum pressure of Cat-CVD chamber was $6 \times 10^{-6}$ Pa. A gas mixture of $SiH_4$ (99.9999% pure) and $NH_3$ (99.999% pure) was used as the deposition gas, and the same $NH_3$ was used as the nitridation gas. The gas pressure during deposition or nitridation was fixed at 4 Pa. The deposition or nitridation conditions are summarized in Table I.

The films prepared by deposition and nitridation methods were characterized by C-V characteristics. Metal-insulator-semiconductor (MIS) diode structures for C-V measurements were completed by evaporating Al on $SiN_x$ through a metal mask. The electrode area was $3 \times 10^{-4}$ cm$^2$. C-V characteristics were measured at 1 MHz in dark at room temperature. The bias voltage was scanned from the depletion-inversion side to the accumulation side at a sweep rate of 0.1 V/s. In these structures, no post annealing treatment was performed.

The $SiN_x/Si(100)$ interface structure was observed by cross-sectional HRTEM. For the HRTEM observation, a HITACHI H-9000NAR TEM was used with an accelerating voltage of 300 kV.

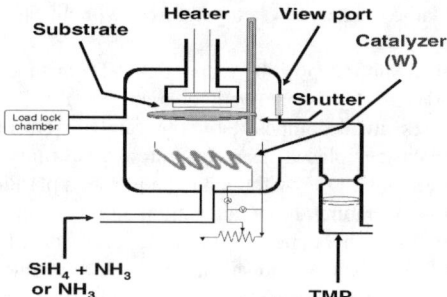

*Figure 1. Schematic diagram of a catalytic-CVD system.*

Table I. Conditions of nitridation and $SiN_x$ deposition

|  | Nitridation | $SiN_x$ deposition |
|---|---|---|
| Catalyzer temperature ($T_{cat}$) | 1000-1650°C | 1650°C |
| Catalyzer surface area | 25.9 cm$^2$ | 25.9 cm$^2$ |
| Susceptor temperature ($T_s$) | 250°C | 250°C |
| Gas pressure | 4 Pa | 4 Pa |
| SiH$_4$ flow rate | – | 2 sccm |
| NH$_3$ flow rate | 200 sccm | 200 sccm |

## RESULTS AND DISCUSSION

C-V characteristics

Figure 2 shows the dependence of $V_{th}$ and the hysteresis loop width on the catalyzer temperature [6]. The nitridation time was fixed at 1 min. The lowest value of the $V_{th}$ shift is obtained at the catalyzer temperature of 1350°C. We consider that conformal nitridation layer was not fabricated at the catalyzer temperature below 1350°C. On the other hand, in the case of catalyzer temperatures over 1350°C, the Si substrate was strongly etched by atomic hydrogen. As the result, defects were created near the interface and large negative $V_{th}$ shift was observed. However, the hysteresis loop width doses not increase. This can be explaining that atomic hydrogen terminated these defects.

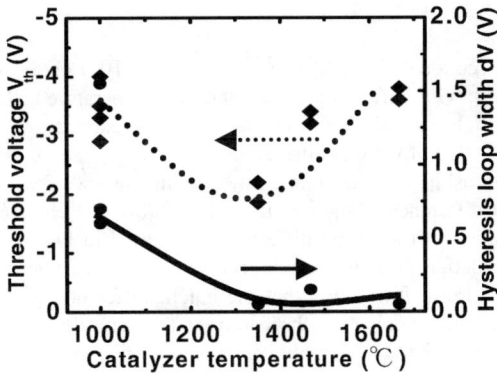

Figure 2. Catalyzer temperature dependence of threshould voltage $V_{th}$ and hysteresis loop width $dV$.

HRTEM observation

**(a) without nitridation**

**(b) with nitridation**

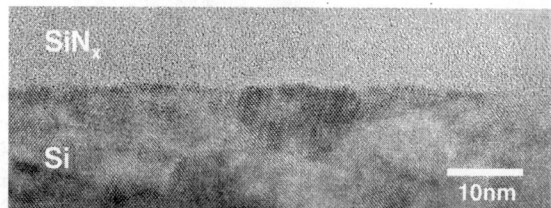

*Figure 3. Cross-sectional HRTEM images of $SiN_x/Si(100)$ obtained from (a): without nitridation and (b): with nitridation at the catalyzer temperature of $1350^\circ C$.*

Figure 3 shows cross-sectional HRTEM images of $SiN_x/Si(100)$ obtained from (a): without nitridation and (b): with nitridation at the catalyzer temperature of 1350°C. It is found that interface roughness of $SiN_x/Si(100)$ which is carried out nitridation treatment is more flat than that of without nitridation.

Figure 4 shows relationship between the catalyzer temperature of the nitridation treatment and the height difference of the interface of $SiN_x/Si(100)$ obtained from the TEM images. The definition of the height difference is inserted in Fig. 4. This figure shows that inserting nitridation layer with optimized nitridation treatment, the height difference became decreasing. It is found that the height difference of the interface becomes large in nitridation treatment at the catalyzer temperature of 1650°C. The Si(100) surface may be etched by atomic hydrogen which was generated by decomposition of $NH_3$. It is also found that the variation in a height difference is large in case of the samples without nitridation and with nitridation at the catalyzer temperature of 1650°C. To be compared with the result of C-V measurements as shown in Fig. 2, it becomes clear that there is strong correlation between the flatness of the $SiN_x/Si(100)$ interface and the C-V characteristics.

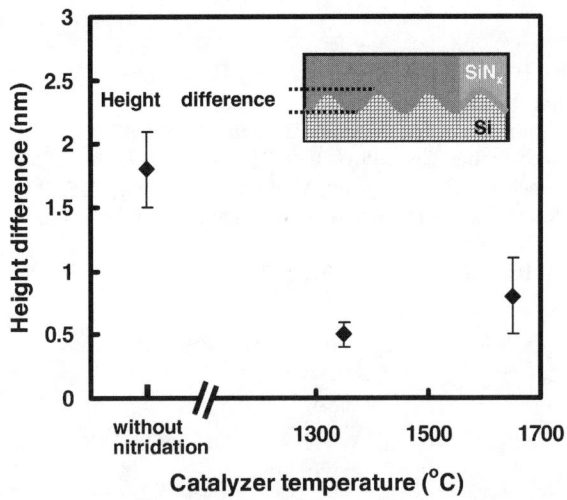

*Figure 4. Relationship between the catalyzer temperature the height difference of the interface of SiN$_x$/Si(100) obtained from the HRTEM images. Inserted figure shows definition of the height difference of HRTEM image.*

## CONCLUSIONS

The interface of SiN$_x$/Si(100) prepared by combination of catalytic-nitridation and catalytic-CVD method in a catalytic-CVD system was observed by HRTEM. It is found that flat interface of SiN$_x$/Si(100) is formed by inserting nitridation layer before growing the SiN$_x$ films.

## ACKNOWLEDGMENTS

This work is in part supported by the R&D Projects in Cooperation with Academic Institutions "Cat-CVD Fabrication Processes for Semiconductor Devices" entrusted from the New Energy and Industrial Technology Development Organization (NEDO) to the Ishikawa Sunrised Industries Creation Organization (ISICO) and carried out at Japan Advanced Institute of Science and Technology (JAIST). This work was also in part supported by Ozawa and Yoshikawa Memorial Electronics Research Foundation, the Foundation of Ando Laboratory and Grant-in-Aid for Scientific Research from the Ministry of Education, Science, Sports and Culture.

**REFERENCES**

[1] H. Matsumura and H. Tachibana, Appl. Phys. Lett **47**, 833 (1985).
[2] H. Matsumura, Jpn. J. Appl. Phys. **28**, 2157 (1989).
[3] S. Okada and H. Matsumura, Jpn. J. Appl. Phys. **36**, 7035 (1997).
[4] A. Izumi and H. Matsumura, Appl. Phys. Lett. **71**, 1371 (1997).
[5] A. Izumi, A. Kikkawa and H. Matsumura, Mat. Res. Soc. Symp. **715**, 491 (2002).
[6] A. Kikkawa, R. Morimoto, A. Izumi and H. Matsumura, Thin Solid Films (2003) *in press*.
[7] W. Kern and D. A. Poutinen, RCA Rev. **31**, 187 (1970).

# High-rate (> 1nm/s) and low-temperature (< 400 ºC) deposition of silicon nitride using an $N_2/SiH_4$ and $NH_3/SiH_4$ expanding thermal plasma

J. Hong, W.M.M. Kessels, and M.C.M. van de Sanden
Department of Applied Physics, Eindhoven University of Technology, P.O. Box 513, 5600 MB, Eindhoven, The Netherlands

## ABSTRACT

High-rate (> 1 nm/s) and low-temperature (50 – 400 ºC) deposition of silicon nitride (a-SiN$_x$:H) films has been investigated by the expanding thermal plasma (ETP) technique using SiH$_4$ as Si-containing and N$_2$ or NH$_3$ as N-containing precursor gases. The structural, optical and electrical properties of the a-SiN$_x$:H films have been studied by elastic recoil detection, spectroscopic ellipsometry, infrared spectroscopy, dark conductivity measurements and atomic force microscopy. The film properties of the ETP deposited a-SiN$_x$:H films in this low-temperature range are discussed in terms of deposition rate, atomic composition, UV-VIS optical and IR vibrational properties, conductivity, and surface topography of the films.

## INTRODUCTION

Deposition of amorphous silicon nitride (a-SiN$_x$:H) films has been one of the crucial technologies in the semiconductor industry. Owing to the wide range of material properties that can be deliberately tuned to the desired values, the a-SiN$_x$:H films have a vast number of functional applications: gate dielectrics in TFT, barrier layers to contact metals, dielectric layers for memory devices, masks for lithography, encapsulation layers, and optical coatings with tunable refractive indices, etc. [1]. Recently, a-SiN$_x$:H films have been applied as antireflection and passivation layers for Si-based photovoltaic devices. In our previous studies [2], high-rate deposited a-SiN$_x$:H films using the expanding thermal plasma (ETP) technique have revealed bulk passivation effects for multicrystalline Si solar cells. This a-SiN$_x$:H material is very promising for high-throughput solar cell production since the deposition rate can be higher than 1 nm/s. Another important issue is the deposition at very low temperatures (< 150°C), because a-SiN$_x$:H is a viable barrier against moisture and can possible be used in organic and polymer-based devices. In this paper, we address the two important technological aspects of a-SiN$_x$:H deposition: deposition at high-rate and at low-temperature.

## EXPERIMENTAL

The deposition of the a-SiN$_x$:H films is carried out using SiH$_4$ as Si-containing precursor gas, and N$_2$ or NH$_3$ as N-containing precursor gas. The ETP technique has extensively been treated in literature [2,3], hence it will only be described very briefly. A thermal plasma is created at subatmospheric pressure in pure Ar (when using NH$_3$) or in a mixture of Ar-N$_2$-H$_2$ (when using N$_2$) from a cascaded arc plasma source. This non-depositing plasma expands into a low-pressure (20 Pa) reactor due to the pressure gradient. In this reactor, either a NH$_3$/SiH$_4$ mixture (when using NH$_3$) or pure SiH$_4$ (when using N$_2$) is injected into the expansion region. The resulting reaction products deposit subsequently on a substrate that is positioned at a distance of 70 cm from the source.

The deposition conditions including flow rates of the processing gases have initially been chosen for a substrate temperature of 400 °C in order to produce a-SiN$_x$:H films with a visible refractive index around 2.0. The refractive index of 2.0 is a prerequisite for single layer antireflection coatings on Si substrates. Subsequently, deposition of a-SiN$_x$:H at lower temperatures has been investigated maintaining the other deposition conditions constant. Si(100) and Corning 7059 glass substrates (2.5×2.5 cm$^2$) have been used. The UV-VIS optical properties of the a-SiN$_x$:H films have been investigated by spectroscopic ellipsometry (J.A. Woollam, M-2000). IR vibrational properties of the a-SiN$_x$:H films have been analyzed by transmission infrared spectroscopy (Bruker Vector 22). The composition of the a-SiN$_x$:H films has been revealed by elastic recoil detection analysis using a 54 MeV $^{65}$Cu$^{8+}$ beam. DC dark conductivity measurements have been carried out with a polarization voltage of 50 V as a function of temperature (50–160°C) in vacuum. A Keithley 486 picoammeter is used together with Al contacts. The surface topography of the a-SiN$_x$:H films has been measured by atomic force microscopy (NT-MDT).

## RESULTS AND DISCUSSION

The influence of the substrate temperature $T_s$ on the deposition rate of the ETP a-SiN$_x$:H films deposited from N$_2$/SiH$_4$ and NH$_3$/SiH$_4$ is plotted in Fig. 1. The very high-rate (> 1 nm/s) of the deposition process is clearly demonstrated. It is observed that the deposition rate decreases as $T_s$ increases. Although in principle also the film density needs to be considered, we have determined the activation energy $E_a$ of the deposition rate from Arrhenius-type of plots for comparison purposes. The values determined are –0.053 and –0.062 eV for the ETP a-SiN$_x$:H films deposited from N$_2$/SiH$_4$ and NH$_3$/SiH$_4$, respectively. These values are clearly different from the positive values of 0.9 and 1.1 eV, which are reported for high temperature (>600 °C) thermo-CVD silicon nitrides [4]. A value of $E_a$ reported in the literature for remote PECVD a-SiN$_x$:H is –0.025 eV as obtained in the same temperature range (70–350 °C) [5]. The larger value of $E_a$ of the ETP deposited films may be attributed to the significantly higher deposition rate of the ETP technique. This might be related to the shorter time-scales available for the surface reactions but in addition, it is (partially) also closely linked to densification of the films with increasing substrate temperature.

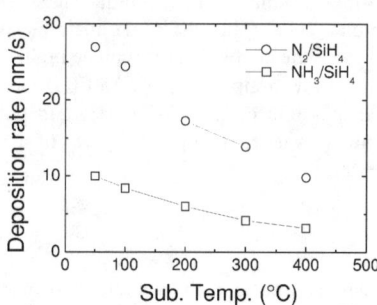

**Figure 1.** Deposition rate of ETP a-SiN$_x$:H films deposited from N$_2$/SiH$_4$ and NH$_3$/SiH$_4$ as a function of the substrate temperature T$_s$.

Figure 2 shows the atomic composition of the ETP a-SiN$_x$:H films deposited from N$_2$/SiH$_4$ and NH$_3$/SiH$_4$ as a function of $T_s$. It is observed that the films deposited from N$_2$/SiH$_4$ contain a considerable amount of oxygen atoms, which are due to the penetration of ambient moisture into the film when the films are exposed to air. This post-deposition oxidation process has been reported for PECVD a-SiN$_x$:H films in literature [6]. Taking into account that the oxygen atoms are not only detected on the film surface but also

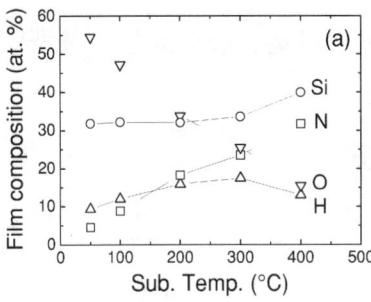 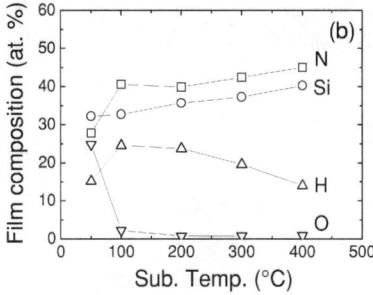

**Figure 2.** Composition of the a-SiN$_x$:H films deposited from (a) the N$_2$/SiH$_4$ and (b) the NH$_3$/SiH$_4$ reactant mixture as a function of the substrate temperature $T_s$.

throughout the bulk of the film, the presence of oxygen atoms reveals a porous microstructure of the ETP a-SiN$_x$:H films deposited from the N$_2$/SiH$_4$ mixture. Moreover, the quantity of oxygen atoms is found to decrease with $T_s$ indicating that the film structure becomes more and more porous when $T_s$ decreases. On the other hand, the films deposited from NH$_3$/SiH$_4$ contain almost no oxygen atoms (less than 1 at. %) for substrate temperatures higher than 200 ºC. A porous film structure is therefore only obtained at temperatures below ~100 ºC.

The densification of the a-SiN$_x$:H at higher temperatures is clearly displayed in Fig. 3, where the mass density of the films is calculated with and without taking the oxygen atoms into account. Note that the permeation of ambient oxygen atoms significantly influences the density of the films deposited from N$_2$/SiH$_4$ especially at low temperatures. On the other hand, for the films deposited from NH$_3$/SiH$_4$ the density difference due to oxygen atoms only appears below ~100 ºC as expected.

The optical properties of the films in the UV-VIS wavelength range are investigated by spectroscopic ellipsometry. The optical functions (refractive index $n$ and extinction coefficient $k$) of the a-SiN$_x$:H films are parameterized using Jellison and Modine's model [7]. Details of the ellipsometric analysis can be found elsewhere [8]. In Fig. 4, the optical functions of the a-SiN$_x$:H films are displayed. It is observed that the films deposited from the N$_2$/SiH$_4$ mixture are more absorbing (higher $k$ value) than those deposited from the NH$_3$/SiH$_4$ mixture. The temperature variation of the optical properties implies a structural change of the film properties which can be attributed to the densification of the films: Although the N/Si ratio of the films increases with $T_s$ (leading in principle to a lower refractive index $n$), the increase of $n$ with

**Figure 3**: Mass density of the ETP a-SiN$_x$:H films deposited from N$_2$/SiH$_4$ and NH$_3$/SiH$_4$ as a function of $T_s$. Mass densities are calculated including (filled symbols) and excluding (open symbols) the oxygen in the film.

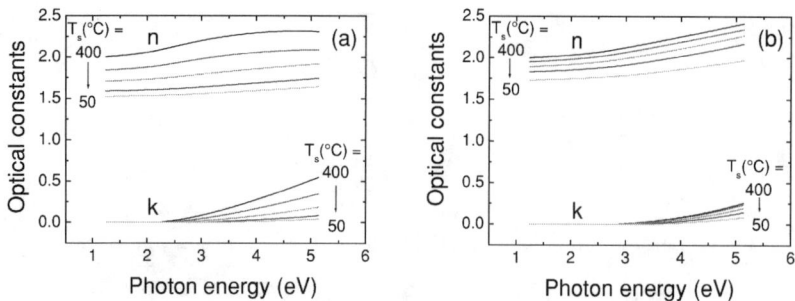

**Figure 4**: Refractive index $n$ and extinction coefficient $k$ of the a-SiN$_x$:H films deposited from (a) N$_2$/SiH$_4$ and from (b) NH$_3$/SiH$_4$ for different $T_s$.

substrate temperature is due to the densification of the films at higher temperatures. The simultaneous increase in $k$ with $T_s$ indicates that the films become also more absorbing and have a lower band gap at high temperatures. Furthermore, the changes in $n$ and $k$ with $T_s$ ranging 50–400 °C for the films deposited from N$_2$/SiH$_4$ are mainly related to the incorporation of oxygen atoms. As $T_s$ decreases from 400 to 50 °C, the film composition changes from a-SiN$_{0.79}$O$_{0.39}$H$_{0.32}$ to a-SiN$_{0.14}$O$_{1.72}$H$_{0.29}$. Especially, the film deposited at $T_s$=50 °C is an oxygen-rich silicon nitride film with a visible refractive index (~ 1.53) close to the value of SiO$_2$ (~1.46).

Figure 5 shows the infrared absorption spectra of the ETP a-SiN$_x$:H films deposited from N$_2$/SiH$_4$ and from NH$_3$/SiH$_4$ as a function of $T_s$. Basically, common IR absorption features are found for the two types of the films. The NH$_2$ modes at ~3450 cm$^{-1}$ are more clearly seen for the films deposited from NH$_3$/SiH$_4$ than those deposited from N$_2$/SiH$_4$, since very broad OH absorption bands are present in the region 3000–3700 cm$^{-1}$ for the films deposited from N$_2$/SiH$_4$. Large Si-N stretching bands are observed in the range of 700–1000 cm$^{-1}$. The intensity of this absorption mode is stronger for the films deposited from NH$_3$/SiH$_4$ than the ones deposited from N$_2$/SiH$_4$. This is in a good agreement with the higher atomic density of Si and N atoms for the films deposited from NH$_3$/SiH$_4$ (see Fig. 2). Furthermore, absorption bands at ~1070 cm$^{-1}$ due to the Si-O stretching modes are observed at lower temperatures. The observation of both Si-O bands and O-H bands is in good agreement with the ERD results for the a-SiN$_x$:H films deposited at low substrate temperatures.

The dissimilar microstructure of the two types of a-SiN$_x$:H films deposited from different N-containing precursor gases results in distinctive electrical properties. Due to the very high resistivity of the a-SiN$_x$:H films under examination, dark conductivity of only a few samples is measurable within the experimental limit of ~$10^{-15}$ ($\Omega$cm)$^{-1}$. The extrapolated room temperature conductivity of the a-SiN$_x$:H films deposited from N$_2$/SiH$_4$ at $T_s$=400 °C and 300 °C is ~$8\times10^{-14}$ and ~$3\times10^{-13}$ ($\Omega$cm)$^{-1}$, respectively. The a-SiN$_x$:H films deposited from NH$_3$/SiH$_4$ at $T_s$=400 °C is found to be less than the detection limit of $10^{-15}$ ($\Omega$cm)$^{-1}$. We observe a higher conductivity for the films deposited from N$_2$/SiH$_4$ than those deposited from NH$_3$/SiH$_4$ at a given $T_s$. This is mainly related to the fact that the films deposited from the N$_2$/SiH$_4$ mixture possess a higher Si content than those deposited from the NH$_3$/SiH$_4$ mixture. Plausible explanations can be found in literature [9] regarding the defects influencing the conduction paths in silicon nitride. The main defects in a-SiN$_x$:H are Si and the N dangling bonds. Since the N dangling bonds are highly

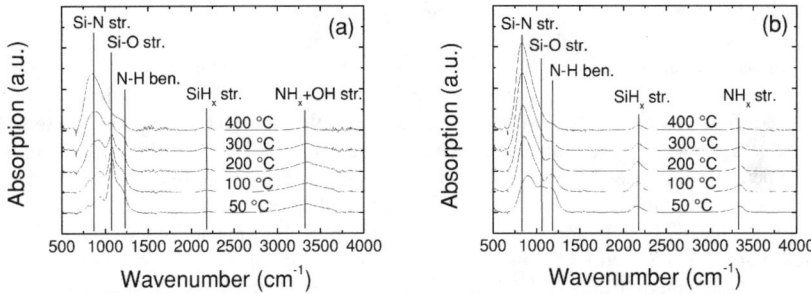

**Figure 5**: Infrared absorption spectra of the ETP a-SiN$_x$:H films deposited from (a) N$_2$/SiH$_4$ and from (b) NH$_3$/SiH$_4$ as a function of the substrate temperature.

localized, the Si dangling bonds are assumed to be mainly involved in the conduction mechanism. The Si dangling bonds can be carrier-trapping centers in the film network. The charge carriers can hop from one Si dangling bond to a nearby bond and the leakage current is higher in the Si-rich film than in the N-rich film. Furthermore, N incorporation into the a-SiN$_x$:H network reduces the formation of Si-related defects, *e.g.*, distorted Si-Si bonds and Si dangling bonds.

Finally, it is worth to mention something about the evolution of the surface roughness as a function of $T_s$. From AFM investigations, we observe that the surface roughness increases as $T_s$ decreases. This may be understood by a surface smoothening mechanism during growth for example by an enhanced surface mobility of adsorbed species at higher $T_s$. In addition, the surface roughness of the films deposited from the NH$_3$/SiH$_4$ mixture is significantly smaller than the one of the films deposited from the N$_2$/SiH$_4$ mixture at the same $T_s$, as shown in Fig. 6. This fact also indicates a structural difference between the a-SiN$_x$:H films originating from the two different N-containing gases.

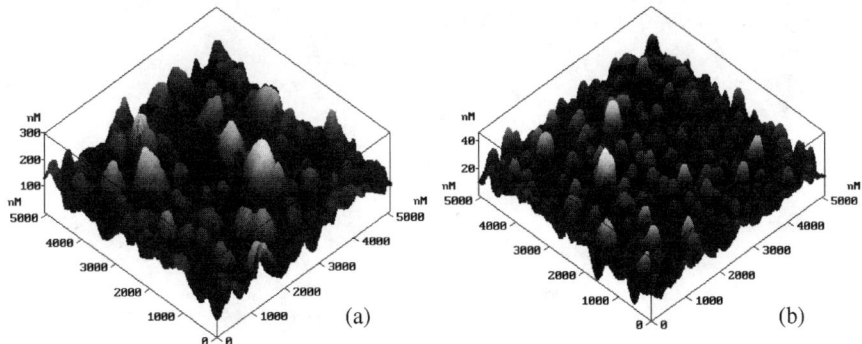

**Figure 6**: AFM images of the ETP a-SiN$_x$:H films deposited from (a) N$_2$/SiH$_4$ and from (b) NH$_3$/SiH$_4$ at T$_s$ = 100 °C. The rms roughness of the films is 37.6 and 5.2 nm, respectively and the thickness of both films is ~730 nm.

## CONCLUSIONS

This paper deals with the high-rate (> 1nm/s) and low-temperature (50 – 400ºC) growth of a-SiN$_x$:H films by the ETP technique. In particular, various properties of the a-SiN$_x$:H films deposited from different N-containing gases, i.e., N$_2$ and NH$_3$, are investigated. Several noticeable observations are found: i) Deposition rate decreases and mass density increases when the temperature increases. ii) The a-SiN$_x$:H films deposited from N$_2$/SiH$_4$ are more vulnerable to post-deposition oxidation than the films deposited from NH$_3$/SiH$_4$ due to the porous microstructure. iii) Conduction mechanisms in the a-SiN$_x$:H films are mainly determined by Si-related defects. iv) Structural stability can be maintained down to a temperature of ~150 ºC for the films deposited from NH$_3$/SiH$_4$. This indicates that these a-SiN$_x$:H (after further optimization) can possible be used in a variety of thin film applications requiring low substrate temperatures.

## REFERENCES

1. V.I. Belyi, L.L. Vasilyeva, A.S. Ginkvker, V.A. Gritsenko, S.M. Repinsky, S.P. Sinitsa, T.P. Smirnova, and F.L. Edelman, *Silicon nitride in electronics* (Elsevier, Amsterdam, 1988), p. 34.
2. J. Hong, W.M.M. Kessels, F.J.H. van Assche, H.C. Rieffe, W.J. Soppe, A.W. Weeber, and M.C.M. van de Sanden, *Prog. Photovolt: Res. Appl.* **11**, 125 (2003).
3. W.M.M. Kessels, J. Hong, F.J.H. van Assche, M.D. Moschner, T. Lauinger, W.J. Soppe, A.W. Weeber, D.C. Schram, and M.C.M. van de Sanden, *J. Vac. Sci. Technol. A* **20**, 1704 (2002).
4. P.Temple-Boyer, C. Rossi, E. Saint-Etienne, and E. Scheid, *J. Vac. Sci. Technol. A* **16**, 2003 (1998).
5. D. Landheer, N.G. Skinner, T.E. Jackman, D.A. Thompson, J.G. Simmons, D.V. Stevanovic, and D. Khatamian, *J. Vac. Sci. Technol. A* **9**, 2594 (1991).
6. W.S. Liao, C.H. Lin, and S.C. Lee, *Appl. Phys. Lett.* **65**, 2299 (1994).
7. G.E. Jellison, Jr. and F.A. Modine, *Appl. Phys. Lett.* **69**, 371 (1996), *ibid.* **69**, 2137 (1996).
8. S. Lee and J. Hong, *Jpn. J. Appl. Phys.* **39**, 241 (2000).
9. J. Robertson, *Phil. Mag. B* **69**, 307 (1994).

## Switch-On Transients And Static Characteristics Of Polymorphous And Amorphous Silicon Thin-Film Transistor

V. Tripathi , and Y. N. Mohapatra
Department of Physics, IIT Kanpur-208016,INDIA E-mail:vibha@iitk.ac.in
P. Roca i Cabarrocas
Laboratoire de Physique des Interfaces et des Couches Minces (UMR 7647 du CNRS),
Ecole Polytechnique, 91128 Palaiseau Cedex, France

## ABSTRACT

Hydrogenated polymorphous silicon (Pm-Si:H) being an admixture of amorphous and ordered phase silicon shows improved optical and electrical properties due to the presence of nanocrystallites. In order to compare the dynamic and steady state electrical properties in a-Si:H and pm-Si:H, bottom gate Thin Film Transistors (TFT) of these materials were fabricated with $SiO_2$ as the insulating layer. The active materials were deposited using plasma-enhanced chemical vapor deposition (PECVD) by varying pressure, temperature and hydrogen dilution. Transfer characteristics of TFTs made using pm-Si:H show lower leakage current, higher on-current and sharper volt per decade change as compared to similar TFTs made from a-Si:H. Density of states in pm-Si:H as calculated from field effect conductance using incremental method is observed to be an order of magnitude lower than in a-Si:H based devices. To compare dynamic characteristics, we studied the switch-on transient characteristics of polymorphous and amorphous silicon TFTs by pulsing the gate to different voltages in the temperature range of 150-300K. The switch-on transients are trap limited with overall better switching characteristics for pm-Si:H samples. An initial rising transient in case of pm-Si:H is activated with an effective energy of 0.3 eV. The origins of transients are interpreted in terms of trap limited carrier dynamics and charge redistribution within the distribution of localized states.

## INTRODUCTION

Polymorphous silicon, containing small nano-crystallites of silicon is a new generation variant of a-Si:H. It is deposited by PECVD under conditions close to powder formation in the chamber and nano-crystallites of silicon get embedded in the thin film. It is of interest to find out how the basic properties of pm-Si:H differ from a-Si:H. Meaudre et al[1]have shown bulk pm-Si:H to have much lower density of states than a-Si:H by using constant photocurrent method. Transfer characteristics of thin TFTs made by pm-Si:H exhibit superior mobility, and threshold voltage and better on/off voltage ratio [2] .

However, a comparison of dynamic characteristics of TFTs made from these types of materials is yet to be studied. Though there have been attempts in the past to provide detailed simulational description of switching characteristics of a-Si:H based TFTs, a convincing connection between density of states (DOS) in the material to the dynamic properties has not yet emerged. We use field effect conductance method to analyze static characteristics to compare DOS. For dynamic characteristics, we monitor the current transient when the channel is switched on by pulsing the gate voltage.

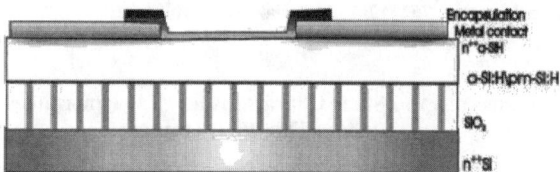

**Figure 1.** Cross-section of typical bottom gate TFT used. $n^{++}$ Si base is used as gate. $SiO_2$ and a-Si layers are both approximately 250nm thick. There is a 50nm $n^{++}$ a-Si:H layer for ohmic contact with metal. Exposed a-Si:H layer is encapsulated for protection

## SAMPLE PREPARATION AND EXPERIMENTS

Active layers for several polymorphous and amorphous Silicon TFTs were prepared in PECVD reactor [2,3]. Figure 1 shows the structure of the TFTs employed in this work. We used strongly hydrogen-diluted silane (2%) as the reactant gas source to deposit these films with an rf power density of 90 mW/cm$^2$ at a substrate temperature of 250 C. The total pressure was varied in the range from 600mT to 1400mT. The deposition rate was around 2 Å/s. The films were grown on a thermally oxidized n-type (100) silicon wafer to fabricate bottom gate TFT (Fig. 1). The thickness of both the $SiO_2$ gate dielectric and the active layer were ~250 nm. The top n-doped a-Si:H layer of ~50 nm was also deposited by PECVD while the source and drain contacts were obtained by thermal evaporation of a thin chromium layer in high vacuum. The TFTs were defined by photolithography. The width of the channel (W) was 137 μm and its length (L) 55 μm for an aspect ratio (W/L) of ~ 3[2].

The TFTs were scribed and bonded on TO5 header for convenient handling. Temperature

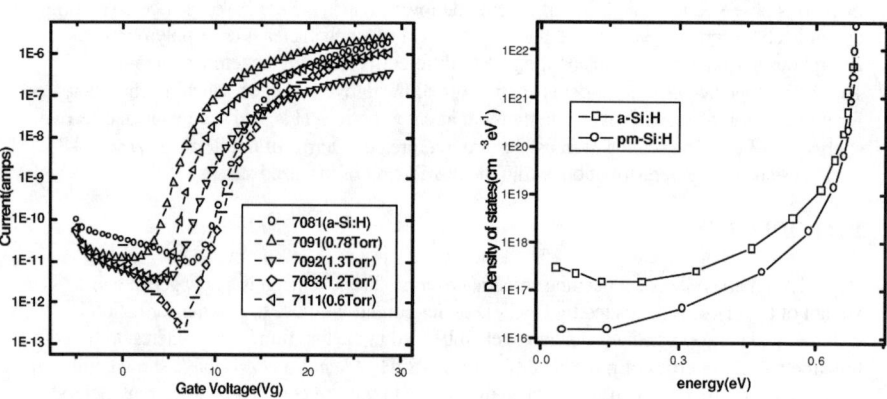

**Figure 2a.** Transfer characteristics of several a-Si:H and pm-Si:H devices prepared under identical experimental conditions. Amorphous silicon TFT(7081) shows higher off current and threshold voltage

**Figure 2b.** Density of States of a-Si:H and pm-Si:H TFTs obtained by field effect conductance measurements. a-Si:H has an order of magnitude higher DOS than pm-Si:H

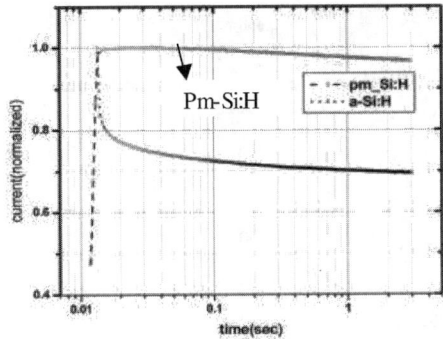

**Figure 3.** Comparison of switch on transient of pm-Si:H and a-Si:H at room temperature. Note the difference in shape after maximum has been achieved.

was varied from 150K to 400 K using a home made cryostat. We used combination of Keithley 236 Source Measure Unit (SMU) and Keithley 230 Programmable Voltage Source to measure the steady state characteristics. For dynamic characteristics, transients were recorded using Keithley 194 High Speed Voltmeter.

## RESULTS AND DISCUSSION

Figure 2a. shows transfer characteristics of TFTs with active layers as a-Si:H and several pm-Si:H grown under different pressures in PECVD. Note that the leakage current is at least an order of magnitude lower in pm-Si:H samples. Also, the rise of forward current per decade is better for pm-Si:H samples. Fig 2b. shows density of states obtained from field effect conductance method for a-Si:H and one of the pm-Si:H samples. The method of calculation is described by Suzuki *et al*[4]. The density of the states of other pm-Si:H samples is nearly the same. Undoubtedly, presence of nanocrystallites of ~3nm in a relaxed environment of a-Si:H causes reduction in DOS by an order of magnitude leading to an improvement in transfer characteristics.[1]

To compare dynamic characteristics, the switch-on transient characteristics of polymorphous and amorphous silicon TFTs are taken in the temperature range of 150-300K by pulsing the gate to different voltages while keeping the drain-source voltage constant. By doing this, we pulse the MOS interface from depletion to accumulation region while monitoring the current due to carrier accumulation. Care has been taken to minimize contribution from fast displacement current by suitably choosing ramp rate of pulsing without compromising response time within the time window of observation. Fig. 3 shows comparison between one each of typical device made using a-Si:H and pm-Si:H. The most noticeable differences between the two types of samples at room temperature is that the current decay on switching on the transistor is much less for pm-Si:H samples. This is in accordance with the general improvement of material quality and hence DOS for these samples.

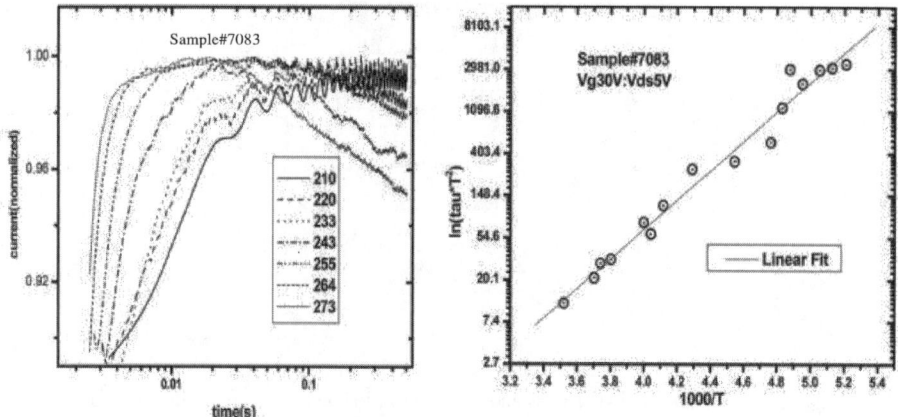

**Figure 4a..** Series of switch on transients of pm-Si:H transistor taken at temperatures varying from 190K to 300K. The gate was pulsed from 0 to 30V and source kept at 5V.
**Figure 4b.** Fig.4a was fitted to a single transient and the log of extracted time constant ($\tau$) multiplied with square of temperature was plotted with 1000/Temp. Slope of this plot gives us information about activation energy of trap levels to be ~0.3eV.

rising part of the current before peaking is approximately exponential with time. This time dependence is shown in Fig. 4a for temperature between 200-300 K. Note that this part of the transient is also thermally activated. The time constants obtained from fitting the rising transients are plotted in an Arrehenius plot (i.e., $\tau T^2$ vs 1000/T, where $\tau$ is the time constant and T is temperature)as shown in fig 4b. The activation energy obtained from the fitted line is 0.3 eV. Such dependence was not observed for a-Si:H samples.

Fig. 4(a) also shows that the normalized current decays slowly beyond the peak, and in fact roughly displaying ln(t) behavior. Fig. 5 shows normalized current decay characteristic of a TFT with a-Si:H as active layer. In the same range of temperature, the channel current decays faster showing highly non-exponential behavior. Clearly, both increase and decrease of current are related to trapping and emission at deep states.

There have been attempts in the past to understand switching behavior of a-Si:H TFTs through detailed simulation [4,5]. Initially the rise time was sought to be explained by invoking non-uniform accumulation of carriers at the interface and subsequent build up in the channel and in a two-dimensional fluid model of carriers [6]. It is by now well known that the current rise time is not limited by transit time between the drain and source, but by trapping time at the gap states in the channel [5].

A qualitative picture that emerges from work on a-Si:H TFTs is that changes in quasi-Fermi level in the channel slows down due to changes in occupancy at the distributed deep states. For addition of each carrier to the accumulation layer, many times more carriers need to be trapped at the deep states. As for the long-time decay of the current in the channel, it is attributed to the charge redistribution of carriers in the deep states through both capture and emission [7]. A carrier initially trapped at the shallower level emits the carriers to the band only to be recaptured at a deeper level eventually. Hence, the occupancy of the trapped charge distribution shifts to deeper energies. This process of charge redistribution is known to be highly non-exponential [7,8].

Our experimental results on a-Si:H are consistent with this qualitative description of current transients. In the context of pm-Si:H TFTs, the observation of thermally activated exponential rising current transient needs an additional explanation. We believe that though net density of states as derived from static characteristics are lower in pm-Si:H, the presence of nano-crystallites and their interface with a-Si:H medium introduces metastable states in the gap. The carrier capture at these traps is thermally activated as is normally found for many metastable states in semiconductors. The capture coefficient in such a case is activated due to an energy barrier due to phenomenon such as lattice relaxation or presence of grain boundaries. From photoconductivity measurements, Kwon *et al* [9] have proposed a metastable state at 0.7 eV in the gap, which they attribute to nanocrystallites acting as giant traps. Though in our experiments, we have not been able to measure emission activation energies, it is reasonable to assume that in the temperature range of our observation, the effective capture energy barrier to get captured into such a metastable state is ~0.30 eV. The quasi Fermi level in the channel gets temporarily pinned to these states slowing down appearance of the accumulation layer. As for the long-term decay of the channel current in the case of pm-Si:H, it can be attributed to decreased density of states responsible for slowing down the process of charge redistribution. Huang and Wu [5], have shown that for a-Si:H TFT, the mean trap filling level moves up in three stages first with log (t), then varies linearly with t and finally, with log (log t) to a steady state level. The comparison of long time decay of channel in the two cases simply shows that for the same window of observation in temperature and time, the carrier concentration changes as log t in the case of pm-Si:H, whereas it is faster than log t in the case of a-Si:H. Detailed simulation need be done to quantify these dependencies.

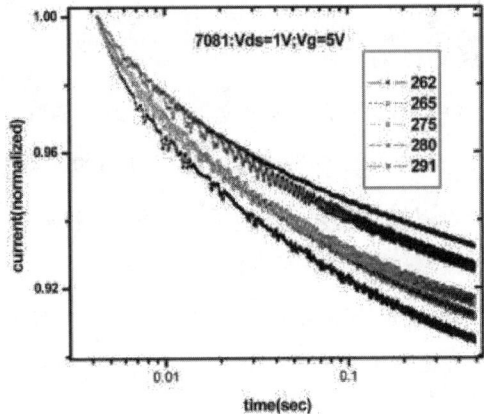

**Figure 5** Switch-on transients of a-Si:H TFTs. We observe a peak in current followed by a decay. Note that the change in current after maximum has been achieved is a small fraction of the total current

## CONCLUSION

We carry out comparison of static and dynamic characteristics of TFTs made with different active layers of PECVD grown a-Si:H and pm-Si:H containing nano-crystallites of mean size ~3nm. The static characteristics imply the DOS to be an order of magnitude less than that in a-Si:H. In both cases, the switch-on transients are trap limited. The initial rising transient appears to originate from metastable states associated with nanocrystals or their interfaces acting as giant traps with effective capture energy barrier of 0.3 eV. The origin of transients is interpreted in terms of trap limited carrier dynamics and charge redistribution within the distribution of localized states.

## ACKNOWLEDGEMENT

We thank Dr. J. Puigdollars, Electrical Engineering Department, Technical University of Catalonia for providing all the intermediate steps in fabrication of TFTs. Bonding and packaging of TFTs was done by Mr. Ajay Kumar from Solid State Physical Laboratory, Delhi. We also thank Department of Science and Technology, New Delhi for financial support under project SP/S2/M-14.

## REFERENCES

1. M.Meaudre, R.Meaudre, R.Butte, S.Vignoli, C.Longeaud, J.P.Kleider, and P.Roca i Cabarrocas, *J. Appl. Phys.* **86**, 946 (1999).

2. C.Voz, J.Puigdollers, A.Orpella, R.Alcubilla, A.Fontcuberta i Morral, V.Tripathi, and P.Roca i Cabarrocas, *J. Non-Cryst. Solids*, **299–302** 1345-1350, (2002).

3. P.Roca i Cabarrocas, J.B.Chévrier, J.Huc, A.Lloret, J.Y.Parey, and J.P.M.Schmitt, *J. Vac. Sci. and Technol. A* **9**, 2331 (1991).

4. T.Suzuki, Y.Osaka, and M.Hirose, *Jpn. J. Appl. Phys, part 2*, **21**, 159 (1982).

5. J.S.Huang and C.H.Wu, *J. Non-Cryst. Solids*, **76** (10), 5981 (1994).

6. C.Van Berkel, J.R.Hughes, and M.J.Powell, *J .Appl. Phys.*, **66** (9), 4488, (1989).

7. J.W.Farmer and Z.Su, *Phys. Rev Lett.*, **71** 18, (1993)

8. P.K.Giri, S.Dhar, V.N.Kulkarni, and Y.N.Mohapatra, *Phys. Rev. B* **57,** 57 (1998).

9. D. Kwon, C-C. Chen, J.D. Cohen, H.C Jin, E. Hollar, I. Robertson, and J. R.Abelson, *Phys. Rev. B* **60** . 4442, (1999).

## Dependences of Structural Parameters on the Characteristics of Poly-Si Thin-Film Transistors after Plasma Passivation

Cheng-Ming Yu[2], Tiao-Yuan Huang[2], Tan-Fu Lei[2], Horng-Chih Lin[1]
[1]National Nano Device Labs, 1001-1 Ta-Hsueh Road, Hsinchu, Taiwan, 30050, R.O.C.
[2]Institute of Electronics, National Chiao Tung University, 1001Ta-Hsueh Road, Hsinchu, Taiwan, ROC.

## ABSTRACT

The effects of $NH_3$ and $H_2$ plasma passivation on the characteristics of poly-Si thin-film transistors with source/drain extensions induced by a bottom sub-gate were studied. Our results show that significant improvements in device performance can be obtained by both passivation methods. Moreover, $NH_3$-plasma-treatment appears to be more effective in reducing the off-state leakage, subthreshold swing, compared to $H_2$ plasma passivation. $NH_3$ plasma treatment is also found to be more effective in reducing the anomalous subthrehold hump phenomenon observed in non-plasma-treated short-channel devices. Detailed analysis suggests that all these improvements can be explained by the more effective passivation of the traps distributed in both the front and back sides of the channel by $NH_3$ plasma treatment.

## INTRODUCTION

With their higher mobility, the use of polycrystalline silicon (poly-Si) thin-film transistors (TFT) allows the integration of the active switching element and the peripheral driver circuitry on the same substrate, which further improves system performance and reliability.[1,2] However, defects at the grain boundaries as well as inside the grains are known to cause device degradation.[3], resulting in poor device performance including low mobility and high off-state leakage current.[4,5] In order to obtain high-performance poly-Si TFTs, it is essential to reduce the trap density in the poly-Si channel. To this end, hydrogen plasma passivation is a well-known technique.[6,7] The atomic hydrogen can passivate defects in the poly-Si channel, thereby improves the device characteristics. In addition, nitrogen-containing plasma treatments in combination with hydrogen (e.g., $H_2/N_2$ mixture plasma[8], nitrogen implantation with $H_2$-plasma[9], and pre-oxidation $NH_3$ annealing with $H_2$-plasma[10], and $NH_3$ plasma[11]) have also been shown to further improve the device performance. The additional nitrogen passivation and/or the enhanced hydrogen passivation effects in the presence of nitrogen are presumably responsible for the observed improved characteristics. Concurrently, an effective method for reducing the off-state leakage current by employing an electrical drain junction induced by a sub-gate has been proposed.[12,13] It is shown that, with proper device structure design and operation conditions, the leakage current can be dramatically reduced without significantly compromising the drive current. Recently, we have proposed a novel TFT device with electrical drain induced by a bottom sub-gate.[13,14] In this work, we further explore the effects of plasma treatments in either $NH_3$ or $H_2$ ambient on the device characteristics.

## EXPERIMENTAL

Fig. 1 shows cross-sectional and top views of the test devices. The fork-shaped sub-gate has two split branches buried below the poly-Si active layer, and is used for electrically inducing the source and drain extensions. The device channel length, L, is thus set by the spacing between the two branches of the sub-gate. The length of the offset regions between implanted source/drain and the main gate is fixed at 1 μm in this study.

The key process flow is as follows. First, a 100-nm $n^+$-poly-Si layer was deposited on an oxidized silicon substrate by low-pressure chemical vapor deposition (LPCVD). The doped poly-Si film was then patterned to form the fork-shaped bottom sub-gate. Next, a 100-nm CVD nitride layer was deposited, followed by the deposition of a thick LPTEOS (550 nm) oxide layer. Chemical-mechanical polishing (CMP) was then applied to planarize the wafer surface and to expose the nitride layer on top of the sub-gate. Afterwards, a 50-nm CVD amorphous Si film serving as the active device layer was deposited at 550 ¢J, and subsequently transformed into polycrystalline phase by a solid-phase crystallization (SPC) treatment at 600 ¢J for 24 hours. A 20-nm CVD oxide layer was then deposited to form the gate insulator. An $n^+$ poly-Si film was deposited and patterned to form the top main-gate. Next, the "offset regions" were masked with photo resist by a lithographic step before $As^+$ with a dosage of $5 \times 10^{15}$ $cm^{-2}$ at 20 KeV and $BF_2^+$ with a dosage of $5 \times 10^{15}$ $cm^{-2}$ at 30 KeV were implanted to form the heavily-doped source/drain regions for n-channel and p-channel transistors, respectively.

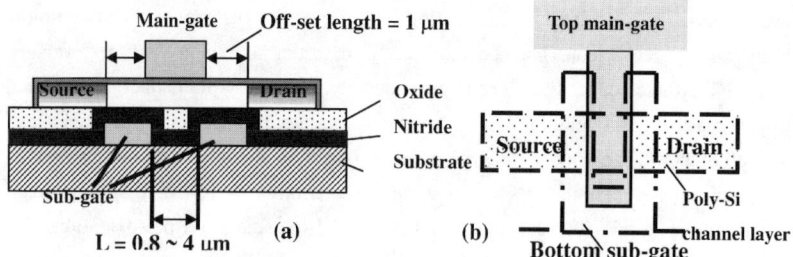

**Figure 1.** (a) Cross-sectional and (b) top views of the TFT device with bottom sub-gate sub-gate.

The implanted dopants were subsequently activated in $N_2$ ambient at 600°C for 12 hours. Finally, some devices were split to receive $NH_3$ or $H_2$ plasma treatment in a parallel-plate plasma reactor at 250°C for 3 hours before measurements.

## DISCUSSION

### Channel length effects

The effects of channel length on device performance for n- and p-channel devices are shown in Figs. 2 and 3, respectively. Two basic leakage components are identified. The first (i.e., Type-I) is the off-state leakage current that increases with increasing voltage difference between the gate and the drain (i.e., $V_{GD}$) as L < 2μm. While the other (i.e., Type-II) is the "hump" that appears in the subthreshold region as L < 1.5μm. The Type-I leakage has been well characterized in the literature, and could be ascribed to the trap-assisted thermionic emission[5] or field emission[15] conduction process.

Such leakage process is closely related to the trap density in the channel. Note that, due to the bottom sub-gate configuration in the device, the flow path for such leakage component is from the surface channel to the bottom extension drain junction, as is schematically shown in Fig. 4 (path 1). At a fixed channel length, the leakage for $H_2$-plasma-treated devices is higher than the $NH_3$-plasma-treated ones, as can be seen in Figs. 2 and 3. The extracted effective trap-state density ($N_t$) using the modified Levinson's method[10] is plotted as a function of channel length in Fig. 4.

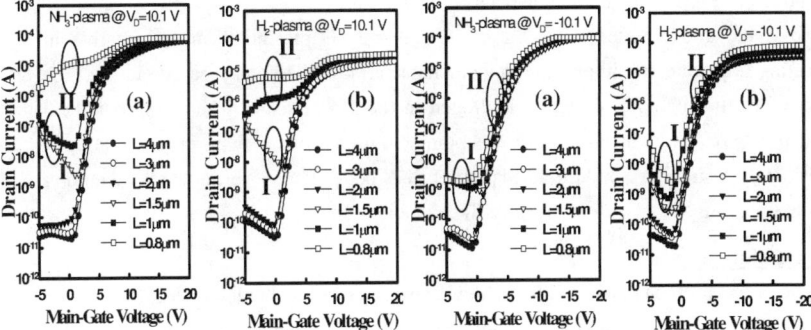

**Figure 2.** Subthreshold characteristics of n-channel TFT devices with various channel lengths at $V_D$=10.1V after (a) $NH_3$ plasma treatment, and (b) $H_2$ plasma treatment. Sub-gate bias is 40 V.

**Figure 3.** Subthreshold characteristics of p-channel TFT devices with various channel lengths at $V_D$=-10.1V after (a) $NH_3$ plasma treatment, and (b) $H_2$ plasma treatment. Sub-gate bias is -40 V.

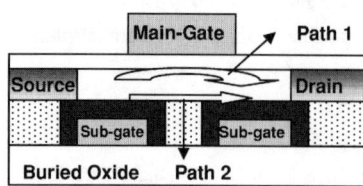

**Figure 4.** Illustration of leakage paths along the top (path 1) and bottom (path 2) interface of poly-Si active layer.

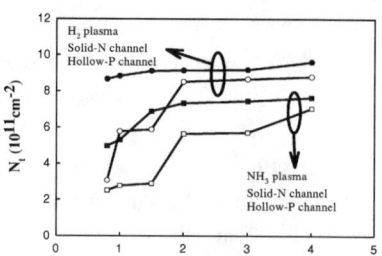

**Figure 5.** Trap-state density versus gate length for n- and p-channel TFT's with $H_2$-plasma and $NH_3$-plasma treatment.

It is observed that the $N_t$ values for $NH_3$-plasma-treated devices are lower than $H_2$-plasma-treated counterparts. Moreover, there is a reduction in $N_t$ for devices with a shorter channel. Devices with a shorter channel length may have more effective defect passivation in the middle of the channel simply because of a shorter diffusion path[16]. Nevertheless, the off-state leakage becomes significant as L < 2 μm because of the dramatic increase in field strength. The path of type-II leakage is believed to flow through the bottom surface of the channel layer, as illustrated in Fig. 4 (path 2). Some defects at or near the bottom interface generated during some processing steps (e.g., CMP) are believed to be responsible for such leakage.

From the above observations, it is clear that $NH_3$ plasma is more effective in passivating trap-state and improving device characteristics. This may be due to the high density of neutral or ionized atomic hydrogen in the $NH_3$ plasma. The nitrogen radical may also help passivate the defects. It was pointed out previously[8] that nitrogen pile-up at the $SiO_2$/poly-Si interface and the formation of Si-N bonds tend to terminate the dangling bonds at the grain boundaries in the poly-Si channel.

## Channel width effects

The effects of channel width on device subthreshold characteristics for n- and p-channel devices are shown in Figs. 6 and 7, respectively. It is interesting to note that the "hump" phenomena are not observed in devices with shorter widths. The improvement in device performance could be correlated with the diffusion path of the passivation species, as schematically shown in Fig. 8. To more clearly illustrate the situation, the top gate is deliberately not shown in the figure. Since the passivation species enter the device region from the edges of the channel, defect passivation takes place initially at the edges of the channel and then gradually extends to the central region away from the channel edges. For devices with a wider channel, the defects located at the central region may remain unpassivated after the plasma treatment. The unpassivated defects at the bottom central channel interface are believed to be responsible for the "hump" phenomenon observed in short-channel devices. When the width is narrow enough (e.g., ≤ 5 μm), the passivation process is complete (i.e., even the central channel region has become fully passivated) after the plasma treatment, so the leakage could be effectively suppressed.

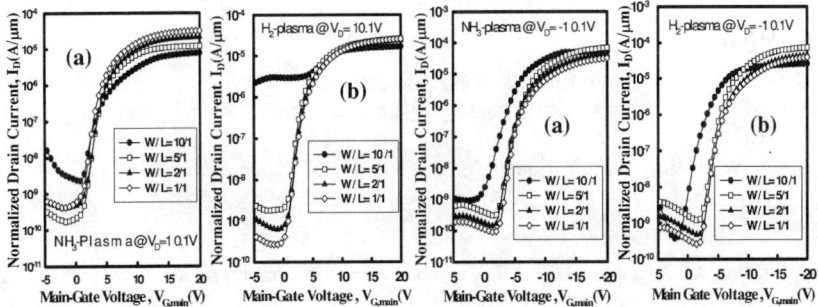

**Figure 6.** Subthreshold characteristics of n-channel TFT devices with various channel widths after (a) $NH_3$, and (b) $H_2$ plasma treatment. Sub-gate bias is 40 V.

**Figure 7.** Subthreshold characteristics of p-channel TFT devices with various channel widths after (a) $NH_3$, and (b) $H_2$ plasma treatment. Sub-gate bias is -40 V.

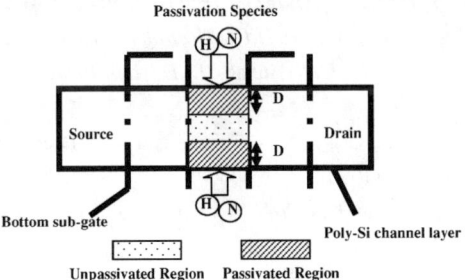

**Figure 8.** Illustration of the diffusion process for passivation species during plasma treatment.

## CONCLUSIONS

In this work, TFT devices with electrical source/drain extensions induced by a bottom sub-gate after treatment in $H_2$ or $NH_3$ plasma were characterized. $NH_3$ plasma was found to be more effective than $H_2$ plasma for passivating defects in poly-Si channel, thus resulting in lower off-state current, steeper subthreshold slope, and lower threshold voltage. The hump phenomenon observed in short-channel TFTs indicates the existence of additional leakage path formed on the bottom interface of the channel layer. Our results indicate that $NH_3$ plasma passivation is also more effective in suppressing such phenomenon. Moreover, the effects of channel width on the characteristics of both n- and p-type hydrogenated devices were also investigated. It is shown that the hump in I-V characteristics could be effectively suppressed for devices with narrow channel width.

**REFERENCES**

1. I. W. Wu, *Tech. Dig. Active Matrix Liquid Crystal Display,* 7 (1995).
2. T. Serikawa, S. Shirai, A. Okamoto, and S. Suyama, *IEEE Trans. Electron Devices,* **36**, 1929 (1989).
3. I. W. Wu, W. B. Jackson, T. Y. Huang, A. G. Lewis, and A. Ciang, *IEEE Electron Device Lett.,* **12**, 181 (1991).
4. W. G. Hawkins, *IEEE Trans. Electron Devices,* **33**, 477 (1986).
5. J. G. Fossum, A. O. Conde, H. Shichijo, and S. K. Banerjee, *IEEE Trans. Electron Devices,* **33**, 1518 (1986).
6. K. Baert, H. Murai, K. Kobayashi, H. Namizaki and M. Nunoshita, *Jpn. J. Appl. Phys.,* **32**, 2601 (1993).
7. A. Yin and S. J. Fonash, *IEEE Electron Device Lett.,* **15**, 502 (1994).
8. M. J. Tsai, F. S. Wang, K. L. Cheng, S. Y. Wang, M. S. Feng, and H. C. Chen, *Solid State Electronics,* **38**, 1233 (1995).
9. C. K. Yang, T. F. Lei, and C. L. Lee, *IEDM Tech Dig.,* 505 (1994).
10. C. K. Yang, T. F. Lei, and C. L. Lee, *IEEE Electron Device Lett.,* **15**, 389 (1994).
11. H. C. Cheng, F. S. Wang and C. Y. Huang, *IEEE Trans. Electron Devices,* **44**, 64 (1997).
12. T. Y. Huang, I. W. Wu, A. G. Lewis, A. Chiang, and R. H. Bruce, *IEEE Electron Device Lett.,* **11**, 244 (1990).
13. H. C. Lin, M. Yu, C. Y. Lin, K. L. Yeh, T. Y. Huang, and T. F Lei, *IEEE Electron Device Lett.,* **22**, 26 (2001).
14. M. Yu, H. C. Lin, G. H. Chen, T. Y. Huang, and T. F. Lei, *Jpn. J. appl. Phys., Part 1,* 5A, **41**, 1 (2002).
15. Sudhir K. Madan, and Dimitri A. Antoniadis, *IEEE Trans. Electron Devices,* **33**, 1518 (1986).
16. W. B. Jackson, N. M. Johnson, C. C. Tsai, I. W. Wu, A. Chiang, and D. Smith, *Appl. Phys. Lett.,* **61**, 1670 (1992).

Mat. Res. Soc. Symp. Proc. Vol. 762 © 2003 Materials Research Society     A18.10

## Above-threshold parameter extraction including contact resistance effects for $a$-Si:H TFTs on glass and plastic

Peyman Servati, Denis Striakhilev, and Arokia Nathan
Electrical and Computer Engineering Department, University of Waterloo
Waterloo, Ontario, Canada N2L 3G1

## ABSTRACT

This paper presents a fast and accurate method for extraction of the above-threshold physical parameters (such as threshold voltage, power parameter, effective mobility, and contact resistance) from measurement data in the linear and saturation regions of hydrogenated amorphous silicon ($a$-Si:H) thin-film transistors (TFTs) fabricated on glass and plastic substrates. The method of extraction is different from techniques that are currently used by virtue of the departure from the square law dependence of the current-voltage characteristics. In addition, a broader range of process-induced variation in material properties is expected, which is accentuated by the effects of different substrates, leading to wide-ranging device parameters. In particular, non-ideal parameters such as contact resistance may vary by orders of magnitude due to process variations, thus strongly influencing the extracted values of TFT parameters if its effect is not considered in the extraction method. In this paper, the effect of contact resistance and other non-ideal parameters is systematically identified and eliminated using TFTs with different channel length. The extracted values for TFTs on glass and plastic substrates clearly highlights the differences in material properties stemming from the different process conditions and substrate properties, and provide insight that is invaluable for subsequent device/process optimization.

## INTRODUCTION

In emerging application areas such as active matrix organic light-emitting diode (AMOLED) displays [1], the $a$-Si:H TFT also serves as an active analog element in addition to its conventional switching functionality. In these applications, compact TFT models and reliable methods for parameter extraction are critical for accurate design and simulation of circuits. Shur et al. [2], Leroux [3], Khakzar et al. [4], and other authors [5] have presented several models (e.g., AIM-SPICE) for the different operational regimes of $a$-Si:H TFTs. Although the models do predict the true terminal behaviour of TFTs, the physical interpretation of the extracted parameter values is somewhat debatable. Accurate knowledge of parameter values is vital for TFT optimization and to assess impact of process conditions. This signifies the importance of a physically-based model and associated extraction method, which may be more complex than techniques currently used [6] by virtue of departure from the square law dependence of the current-voltage characteristics. The associated power parameter ($\alpha$), if assumed to be 2, can lead to inaccuracy in extracted values. Cerdeira et al. [7] have included the effect of $\alpha$ in a unified extraction method for the above-threshold parameters using the AIM SPICE model. In this method, a mathematical technique is used to extract $\alpha$ and threshold voltage ($V_T$) simultaneously. However, the effects of non-idealities such as contact resistance are not rigorously treated. These non-idealities may vary by orders of magnitude due to process variations, particularly for different substrates, thus strongly influencing the extracted values, and hence should be systematically identified. For instance, the presence of a high contact resistance (> 0.5 kΩ-cm, an approximate per unit width value) stemming from the n$^+$ a-Si:H contact layers can significantly undermine the accuracy of extracted threshold voltage, mobility, and power parameter. Based on the compact model presented in ref. [5], this paper presents a fast and accurate method for extraction of the model parameters, including the effect of non-idealities such as contact resistance and channel length modulation. The extraction method is evaluated for different inverted-staggered a-Si:H TFTs fabricated on glass and plastic substrates and at different process temperatures. The fabrication process is described elsewhere [8] and is beyond the scope of this paper.

## CURRENT MODEL

In a-Si:H TFTs, drain-source current in the linear region ($V_{DS}$ = 0.1 V), $I_{lin}$, as a function of terminal voltages may be written as

$$I_{lin} = \mu_{eff} \zeta C_i^{\alpha-1} \frac{W}{L_{eff}} (V_{GS} - V_T - 0.5 V_{DS})^{\alpha-1} (V_{DS} - R_{DS} I_{lin}), \tag{1}$$

for symmetric source and drain contact resistances ($R_D = R_S = R_{DS}/2$) [5]. Here, $\mu_{eff}$ is the effective mobility, $C_i$ the gate capacitance, and $L_{eff}$ and $W$ the effective channel length and channel width, respectively. The power parameter $\alpha$ is given by $2V_{nt}/V_{th}$, where $V_{nt}$ is a measure for the slope of conduction band tail in a-Si:H, and $V_{th}$ the thermal voltage. In eqn. (1), $\zeta$ is a unit matching parameter and is a function of $\alpha$, i.e., $(4.1 \times 10^{-16} \alpha)^{1-\alpha/2}/(\alpha-1)$ [9]. Eqn. (1) forms the basis for extraction of the critical device parameters such as threshold voltage, power parameter, contact resistance, and effective mobility. Conversely, drain-source current in the saturation region, $I_{sat}$, may be written as

$$I_{sat} = \frac{\mu_{eff}}{\alpha} \zeta C_i^{\alpha-1} \frac{W}{L_{eff}} \gamma_{sat} \left( V_{GS} - \frac{R_{DS}}{2} I_{sat} - V_T \right)^{\alpha} x_{cm}. \tag{2}$$

Here, $x_{cm} = 1 + \lambda V_{DS}/L_{eff}$, where $\lambda$ is the channel length modulation parameter, and $\gamma_{sat}$ is the saturation current parameter.

## EXTRACTION METHOD

A comparison of the linear transfer characteristics for a-Si:H TFTs with different channel length can help identify the effects of contact resistance. Figure 1a shows the current for a long channel TFT ($L_{max}$ = 200 μm) in comparison to that for a small channel TFT ($L_{min}$ = 23 μm) showing the effect of contact resistance. The latter is scaled by the ratio of channel lengths, $k = L_{min}/L_{max}$ (=23/200), which is expected to yield the same current as the former given the same channel width for both TFTs. A more detailed study of the contact resistance effects is described elsewhere [9]. It is clear that the high contact resistance suppresses the current of the small channel TFT from its expected value. This can also be viewed in terms of the total measured resistance between drain and source terminals ($R = V_{DS}/I_{lin}$), which includes the channel resistance and the contact resistances. In small channel length TFTs, the contact resistance is the dominant part of the measured resistance. In contrast, for TFTs with channel length, e.g., $L$ > 200 μm, the channel resistance prevails. Figure 1b depicts the measured resistance for both TFTs.

The measured resistance for long channel TFT is now scaled by the channel lengths ratio $k$, which is the expected ratio for the resistances. As can be seen, the measured resistance for the short channel TFT is relatively higher and does not scale accordingly with channel length. Using eqn. (1), the difference between the two curves, which serves as a clear measure for the contact resistance, may be written as

$$R_{23} - kR_{200} = (1-k)R_{DS} + \frac{(1-k)\Delta L}{\mu_{eff} \zeta C_i^{\alpha-1} W(V_{GS} - V_T - 0.5V_{DS})^{\alpha-1}}. \tag{3}$$

where $\Delta L$ is the channel length enlargement factor that models the voltage dependent part of the contact resistance. From Figure 1b and eqn. (3), $R_{DS}$ and $\Delta L$ may be extracted for these TFTs, with $R_{DS}$ being the value that the dotted curve approaches at high voltages. Here, the dotted curve represents the difference in resistances, $R_{23} - kR_{200}$.

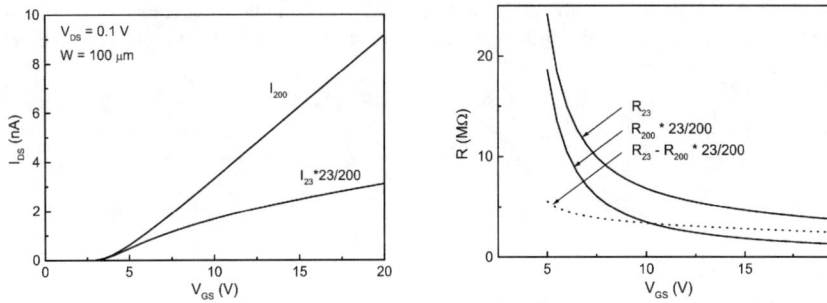

Figure 1 (a) Linear transfer characteristics ($V_{DS} = 0.1$ V) and (b) measured resistance for TFTs with channel length $L = 200$ μm and $L = 23$ μm.

Having extracted $R_{DS}$ and $\Delta L$, we now proceed to find $\alpha$ and $V_T$. From eqns. (1) and (2), we find

$$\frac{I_{lin}}{g_{mlin}} = \frac{V_{GS} - V_T - 0.5V_{DS}}{\alpha - 1} \frac{V_{DS}}{V_{DS} - R_{DS}I_{lin}} \quad \text{and} \quad \frac{I_{sat}}{g_{msat}} = \frac{V_{GS} - V_T + R_{DS}I_{sat}(\alpha-1)/2}{\alpha}, \tag{4}$$

where parameters $g_{mlin}$ and $g_{msat}$ are defined as

$$g_{mlin} \equiv \frac{\partial I_{lin}}{\partial V_{GS}} \quad \text{and} \quad g_{msat} \equiv \frac{\partial I_{sat}}{\partial V_{GS}}, \tag{5}$$

respectively. The right hand sides of eqn. (4) may be found from linear and saturation measurement data using definitions of eqn. (5). Assuming the contact resistance related terms are negligible (i.e., $R_{DS}I_{lin} \ll V_{DS}$ and $R_{DS}I_{sat} \ll V_{GS} - V_T$ for linear and saturation regions, respectively) in eqn. (4), we may write

$$\frac{I_{lin}}{g_{mlin}} = \frac{V_{GS} - V_T - 0.5V_{DS}}{\alpha - 1} \quad \text{and} \quad \frac{I_{sat}}{g_{msat}} = \frac{V_{GS} - V_T}{\alpha}. \tag{6}$$

For the low contact resistance case ($R_{DS}W < 0.5$ k$\Omega$-cm), according to (6), a plot of $I_{lin}/g_{mlin}$ or $I_{sat}/g_{msat}$ versus $V_{GS}$ must be a straight line whose slope and intercept on the abscissa yield $\alpha$ and $V_T$, respectively. This is similar to the result of the integral method [7]. However, for contact resistance values higher than 0.5 k$\Omega$-cm, which is a typical range for a-Si:H TFTs, the values extracted using eqn. (6) are inaccurate. Taking into account the contact resistance related terms, we can remove the assumptions used for derivation of eqn. (6). Here, based on eqn. (4), we modify the $I_{lin}/g_{mlin}$ and $I_{sat}/g_{msat}$ values calculated from measurement data to find $(I_{lin}/g_{mlin})_1$ or $(I_{sat}/g_{msat})_1$ using

$$\left(\frac{I_{lin}}{g_{mlin}}\right)_1 = \frac{I_{lin}}{g_{mlin}}\frac{V_{DS} - I_{lin}R_{DS}}{V_{DS}} \text{ and } \left(\frac{I_{sat}}{g_{msat}}\right)_1 = \frac{I_{sat}}{g_{msat}} - (1-1/\alpha)R_{DS}I_{sat}/2. \quad (7)$$

Now, plots of $(I_{lin}/g_{mlin})_1$ or $(I_{sat}/g_{msat})_1$ versus $V_{GS}$ contain minimal contact resistance induced errors. For saturation measurement results when $V_{DS} = 20$ V, channel length modulation effect can also cause error in the extracted parameters and hence should be considered. In this case, we modify the $I_{sat}/g_{msat}$ values for both contact resistance and channel length modulation effects using

$$\left(\frac{I_{sat}}{g_{msat}}\right)_2 = \left(\frac{I_{sat}}{g_{msat}} - (1-\kappa/\alpha)R_{DS}I_{sat}/2\right)/\kappa, \quad (8)$$

where

$$\kappa^{-1} = 1 - \frac{\lambda R_{DS}I_{sat}}{1+\lambda V_{DS}}. \quad (9)$$

Figure 2 illustrates the calculated values of $I_{lin}/g_{mlin}$ and $I_{sat}/g_{msat}$ and their modified versions as a function of gate-source voltage for the long channel TFT ($L = 200$ μm). As can be seen, the modified versions for both linear and saturation regions yield different slopes (i.e., different values of $\alpha$) showing the adverse effects of contact resistance on the extracted values. This is even more significant in the linear region. In summary, the value of $\alpha$ should be extracted from $(I_{lin}/g_{mlin})_1$ and then may be verified by the $(I_{sat}/g_{msat})_2$ results. It should be noted that $I_{sat}/g_{msat}$ is less sensitive to effects of contact resistance in comparison to $I_{lin}/g_{mlin}$.

With knowledge of the power parameter and threshold voltage, extraction of the effective mobility and $\gamma_{sat}$ is straightforward. The slope ($s$) of the best fit to the $(I_{lin}/V_{DS})^{1/(\alpha-1)}$ versus $V_{GS}$ curve for the long channel TFT may be used to find $\mu_{eff}$ as

$$\mu_{eff} = \frac{L_{eff}}{\zeta W}\left(\frac{s}{C_i}\right)^{\alpha-1}. \quad (10)$$

As for $\gamma_{sat}$, the slope ($S$) of the fit to the $I_{sat}^{1/\alpha}$ versus $V_{GS}$ curve for the long channel TFT may be used

$$\gamma_{sat} = \frac{\alpha S^\alpha}{x_{cm}s^{\alpha-1}}. \quad (11)$$

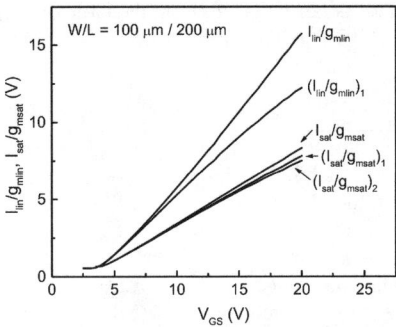

Figure 2  $I_{lin}/g_{mlin}$ and $I_{sat}/g_{msat}$ and their modified versions $(I_{lin}/g_{mlin})_1$, $(I_{sat}/g_{msat})_1$, and $(I_{sat}/g_{msat})_2$ showing the effect of contact resistance.

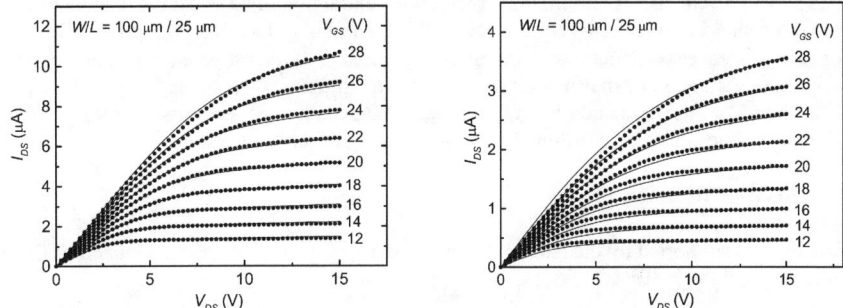

Figure 3  Comparison between simulation (solid lines) and measurement (dots) results for the output characteristics of TFTs fabricated on (a) glass and (b) plastic substrates.

## TFTS ON GLASS AND PLASTIC

Figure 3 illustrates a comparison of simulation results (solid lines), using model eqns. (1) and (2) along with the extracted parameter values (see Table 1) and measurement data for the output characteristics of TFTs ($L = 25$ µm) fabricated on glass and plastic substrates. The discrepancy is less than 5% thus clearly validating the model (and extraction procedure) for prediction of the true terminal characteristics of the TFT. Table 1 summarizes the extracted above-threshold parameters for TFTs fabricated on glass and plastic substrates. Clearly, the TFTs fabricated on plastic substrate suffer from higher contact resistance, higher threshold voltage, and lower effective mobility. This may be attributed to process issues related to the higher thermal expansion of the substrate resulting in misalignment of source and drain regions thus leading to high contact resistance. This is fairly evident from the large discrepancy in the values of $R_{DS}W$ (see Table 1) retrieved for TFTs on glass and plastic.

Table 1  Parameters extracted for TFTs fabricated on glass and plastic substrates.

| Model Parameter | TFTs on glass | TFTs on plastic | Unit |
|---|---|---|---|
| $\mu_{eff}$ – effective mobility | 0.94 | 0.78 | cm$^2$/Vs |
| $V_T$ – threshold voltage | 3.9 | 5.0 | V |
| $\alpha$ – power parameter | 2.085 | 2.096 | - |
| $\zeta$ – unit matching parameter | 4.03 | 4.82 | (C/cm$^2$)$^{\alpha-2}$ |
| $R_{DS}W$ – contact resistance | 0.23 | 9.0 | k$\Omega$-cm |
| $\Delta L$ – channel length enlargement | 5.4 | 5.5 | µm |
| $\gamma_{sat}$ – saturation parameter | 0.977 | 0.873 | - |
| $\lambda$ – channel length modulation | 0.61 | 0.75 | µm/V |

## CONCLUSIONS

This paper presents a fast and accurate method for extraction of physical parameters of the above-threshold characteristics of $a$-Si:H TFTs. In particular, the effect of non-idealities (*e.g.*, contact resistance) on parameter extraction is systematically investigated. Good agreement between modeling and experimental results is obtained with a discrepancy of less than 5% for TFTs fabricated on glass and plastic substrates. The extracted parameter values show the effect of process conditions and substrate properties on the TFT characteristics.

## ACKNOWLEDGEMENTS

This work is supported by the Natural Sciences and Engineering Research Council of Canada (NSERC) and the NSERC E.W.R. Steacie Fellowship.

## REFERENCES

1. P. Servati, S. Prakash, A. Nathan, and Ch. Py, J. Vac. Sci. Technol. A **20**, 1374 (2002).
2. M. S. Shur, H. C. Slade, M. D. Jacunski, A. A. Owusu, and T. Ytterdal, J. Electrochem. Soc. **144**, 2833 (1997).
3. T. Leroux, Solid-State Elec. **29**, 47 (1986).
4. K. Khakzar and E. H. Lueder, IEEE Trans. Electron Devices **39**, 1428 (1992).
5. P. Servati and A. Nathan, J. Vac. Sci. Technol. A **20**, 1038 (2002).
6. C.-Y. Chen and J. Kanicki, Solid-State Elect. **42**, 705 (1998).
7. A. Cerdeira, M. Estrada, R. García, A. Ortiz-Conde, and F. J. García Sánchez, Solid-State Elec. **45**, 1077 (2001).
8. D. Striakhilev, A. Sazonov, and A. Nathan, J. Vac. Sci. and Technol. A **20**, 1087 (2002).
9. P. Servati, D. Striakhilev, and A. Nathan, submitted to IEEE Trans. Electron Devices (2003).

## Mechanical Stress and Process Integration of Direct X-ray Detector and TFT in a-Si:H Technology

Czang-Ho Lee, Isaac Chan, and Arokia Nathan
Department of Electrical and Computer Engineering, University of Waterloo,
Waterloo, ON, N2L 3G1, Canada

## ABSTRACT

This paper presents an alternate strategy to reduce the mechanical stress issues pertinent to the process integration of molybdenum/hydrogenated amorphous silicon (Mo/a-Si:H) Schottky diodes and thin film transistors (TFTs), used as X-ray sensor pixels for medical imaging. The previous approach was to minimize the intrinsic stress in the Mo layer through appropriate process conditions and film thickness, but over narrow process latitude and with a compromise in X-ray sensitivity. Alternatively, the mechanical stress in Mo can be reduced by reducing and/or avoiding the extrinsic stress exerted on Mo by the underlying films through a different masking sequence in the fabrication. This modified process allows for a more flexible design of the Mo layer for enhanced X-ray sensitivity, while maintaining the mechanical integrity of the various layers. Also, the performance of the Schottky diode is improved, in terms of its forward current. The pixel shows good linearity in the X-ray response over the range of 40 ~ 100 kV$_p$.

## INTRODUCTION

Hydrogenated amorphous silicon (a-Si:H) flat panel technology for diagnostic X-ray imaging is a very promising technology for the creation of digital X-ray images, and can potentially replace conventional film imaging techniques [1-4]. We have previously developed a direct X-ray detection scheme based on a Mo/a-Si:H Schottky diode structure for low-energy X-rays [5]. Here, a Mo layer acts not only as a Schottky metal but also as a direct converter, which transforms X-rays into energetic electrons by virtue of the photoelectric effect. These electrons are injected into an a-Si:H layer. They are highly energetic and undergo various scattering events leading to generation of electron-hole pairs. Owing to the small diffusing length in the a-Si:H, the effective separation of the generated electron-hole pairs is enhanced by the electric field inside the reversed-biased a-Si:H depletion region, thereby producing an output signal [6]. The process of the Mo/a-Si:H Schottky diodes is also fully compatible with that of the a-Si:H thin film transistor (TFT) which is used as a switching element for charge readout from each pixel [7]. The operating principle and photograph of a fabricated pixel are shown in Figure 1.

One critical issue with the Mo/a-Si:H Schottky diode process is the mechanical stress in the Mo, both intrinsic and extrinsic, which constrains the process integration of the detector and TFT. Extensive characterization of the intrinsic stress in Mo has been reported [8, 9], in an attempt to reduce the stress through appropriate process conditions [8] and film thickness (~ 300 nm) [9]. However, this is achievable only over a narrow process window [8] and is further compounded by a compromise in X-ray sensitivity (15 % drop from the optimal Mo thickness value of 500 nm Mo) [9], which limits process flexibility for improving electrical performance. Therefore, an alternate process integration strategy is considered, to avoid the stress issues by taking into account the extrinsic stress acting on the Mo layer by the underlying films. This allows the realization of the desired Mo thickness for high X-ray sensitivity.

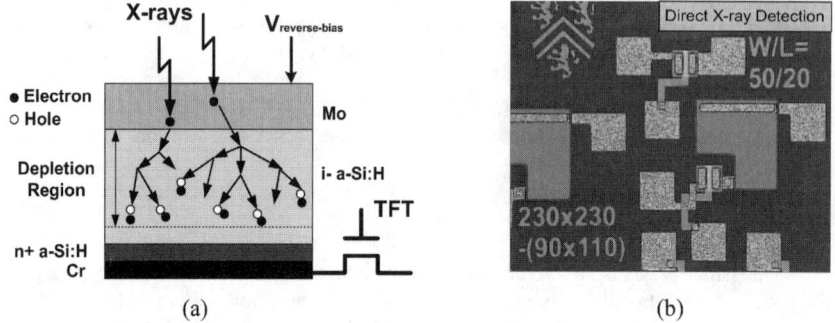

**Figure 1.** (a) Operating principle and (b) photograph of an X-ray detection pixel based on Mo/a-Si:H Schottky diode and a-Si:H TFT.

## PROCESS INTEGRATION AND MECHNICAL STRESS

All thin films were deposited on Corning 1737 glass substrates. Metal films such as Cr, Mo, and Al were deposited using a magnetron rf sputtering system. Silicon based thin films such as intrinsic a-Si:H, $n^+$ a-Si:H, highly doped hydrogenated microcrystalline silicon ($n^+\mu$c-Si:H), and hydrogenated amorphous silicon nitride (a-SiN$_x$:H) were deposited using an rf plasma enhanced chemical vapor deposition (PECVD) system. The fabrication was carried out using a fully wet etch process. The total process requires 7 mask steps and 9 lithographic steps. The Schottky diodes were fabricated prior to the TFTs. The fabrication details are reported elsewhere [7]. The deposition conditions of the various layers are shown in Table I.

**Table I.** Deposition conditions of various layers.

| Deposition system | Film | Temp. (°C) | Pressure (mTorr) | Power (W) | Gas flow (sccm) |
|---|---|---|---|---|---|
| RF sputtering | Cr, Mo, Al | RT | 5 | 400 | Ar (30) |
| PECVD | $n^+$ a-Si:H | 260 | 150 | 12 | 1%PH$_3$+SiH$_4$ (10) |
| | $n^+\mu$c-Si:H | 260 | 900 | 90 | 1%PH$_3$+SiH$_4$ (5) & H$_2$ (500) |
| | i-a-Si:H | 260 | 250 | 12 | SiH$_4$ (50) |
| | a-SiN$_x$:H | 260 | 400 | 110 | 50% NH$_3$+N$_2$ (160) & SiH$_4$ (4) |

In order to demonstrate the effect of extrinsic stress, the original and the modified processes are compared in terms of their impact on the structural integrity of the Schottky diode. In the original process (Fig. 2a), the 100 nm $n^+$ a-Si:H, 1 μm intrinsic a-Si:H, and 500 nm Mo are deposited on top of the patterned 100 nm Cr layer (Mask 1). Here, the extrinsic stress on the deposited Mo film by the Cr patterns is evident by the observed cracks and peel-off of Mo initiated from the corners (Fig. 3a). In the modified process (Fig. 2b), this stress can be avoided by changing the sequence of the first 3 masks, where there is no patterning of any underlying

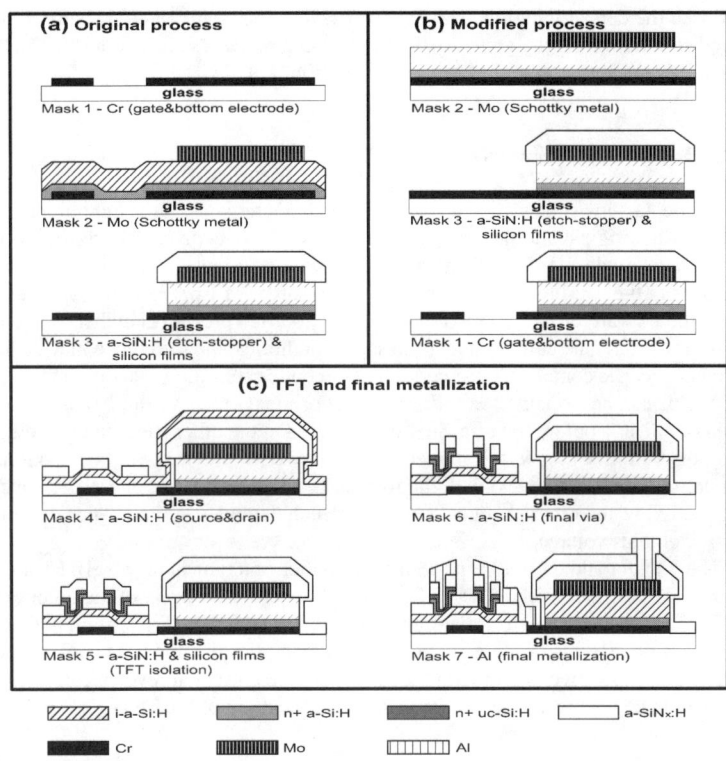

**Figure 2.** (a) Original and (b) modified Mo/a-Si:H Schottky diode, and (c) a-Si:H TFT and final metallization processes.

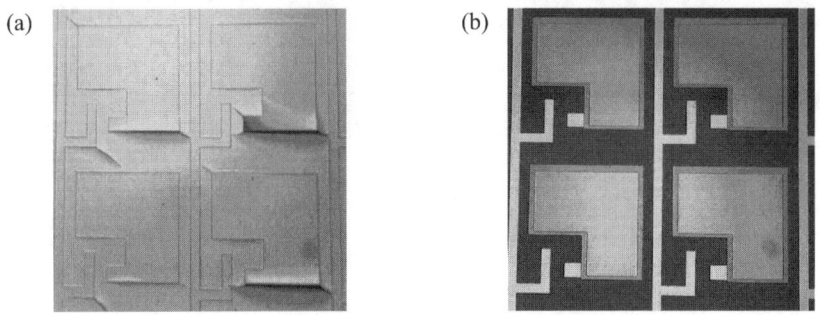

**Figure 3.** Photographs of X-ray pixels using (a) original and (b) modified fabrication processes.

films prior to the deposition of the Mo Schottky layer. The resulting diode patterns are formed with good structural integrity (Fig. 3b). The remaining process sequence for the TFTs and the final metallization is the same for both processes (Fig. 2c) [7].

**RESULTS AND DISCUSSION**

Figure 4a shows the static current-voltage characteristics of the Schottky diodes fabricated in the original and the modified processes. The data on each curve constitutes an average over 10 diode samples. There is no significant difference in the reverse leakage current of the diodes in the two processes, both being of the order of 1 pA at very low reverse bias. However, the forward current of the diodes in the modified process is higher. Since the Schottky barriers of the diodes are formed under the same conditions, the difference may be attributed to the different specific contact resistance, $R_{sc}$, of the bottom ohmic contact in the two processes (Fig. 4b). This can be explained as follows. In the original process (Fig. 2a), the Cr layer was patterned by Mask 1 before the $n^+$ a-Si:H deposition. On the other hand, in the modified process (Fig. 2b), the bottom Cr layer was immediately followed by the $n^+$ a-Si:H deposition. Because of the lithographic steps associated with the former, the patterned Cr layer may be covered with more native oxide and other chemical residues, which degrades $R_{sc}$ and hence decreases forward current at high bias voltage.

The transfer and current-voltage characteristics of a typical on-pixel TFT are illustrated in Figure 5a and 5b, respectively. The TFT shows a low leakage current of the order of $10^{-14}$ A and an ON/OFF current ratio of $10^8$ for a drain-source voltage of 5 V (Fig. 5a). In addition, no current crowding effect is observed (Fig. 5b), which implies that the $n^+$ μc-Si:H ohmic contact seems to be very effective, not only in blocking the hole current at reverse gate voltages, but also in reducing the contact resistance of the source and the drain of the TFTs.

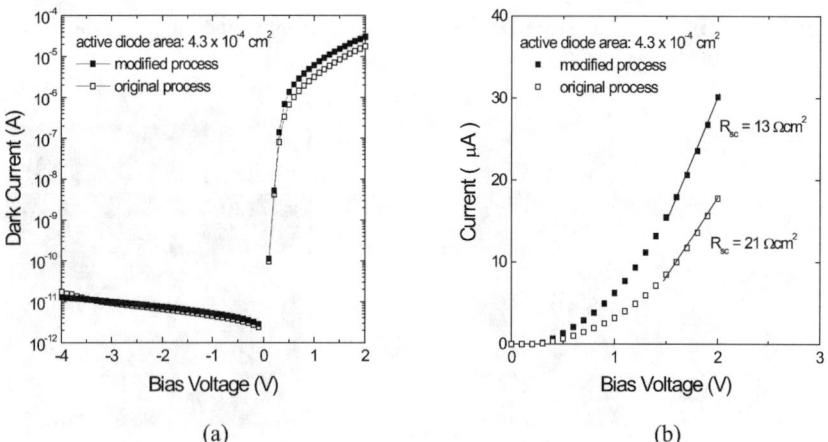

**Figure 4.** (a) Static current-voltage characteristics and (b) specific contact resistance of the Schottky diodes in the original and modified processes.

The X-ray response of the pixels fabricated in the modified process was measured in the form of a diode photocurrent over a range of X-ray source voltage (40 ~ 100 kV$_p$), while fixing the X-ray exposure time and the X-ray source current at 100 ms and 100 mA, respectively. The bias voltage of the Schottky diode was kept at -2 V to keep the diode metastability to a minimum [10]. For accurate readout, this photocurrent was amplified by a Keithley 427 Current Amplifier to a sufficiently large level and was then divided by the amplifier gain to obtain the actual current values. The TFTs were turned on at a gate bias of 5 V during the X-ray exposure. From the data (Fig. 6), the detection pixel showed good linearity with the X-ray source voltage.

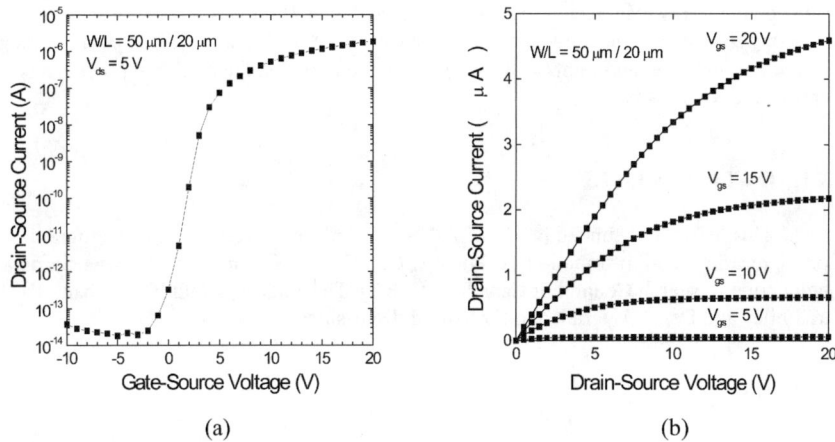

**Figure 5.** (a) Transfer and (b) current-voltage characteristics of a typical on-pixel TFT.

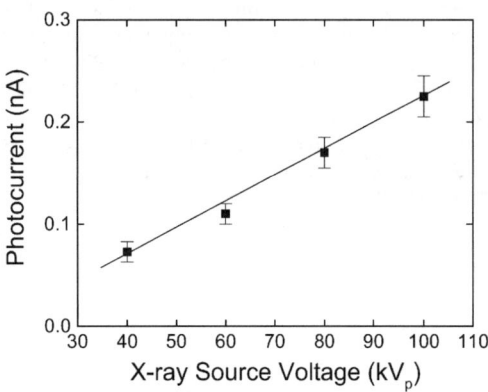

**Figure 6.** X-ray response for X-ray source voltage.

## CONCULSIONS

A modified process was designed to address the mechanical stress issues associated with the process integration of Mo/a-Si:H Schottky diodes and TFTs, for the realization of direct X-ray detection pixels. The patterned Cr layer in the original Schottky diode process exerts an extrinsic stress on the subsequent Mo Schottky layer, resulting in cracks or even peel-off at the corners of the device patterns. The modified process avoids this extrinsic stress by not patterning the underlying films prior to Mo deposition. This modification enables a Schottky diode process with good structural integrity and improved diode performance. The a-Si:H TFTs show a low leakage current (of the order of $10^{-14}$ A) and a high ON/OFF current ratio ($10^8$). Preliminary results on the X-ray response of the pixels, collected over 100 ms X-ray exposure time at 100 mA X-ray source current, show good linearity with the X-ray source voltage over the range of 40 ~ 100 kV$_p$. Further experiments will be carried out to optimize the modified process and to improve the X-ray sensitivity.

## ACKNOWLEDGMENTS

This research is funded by DALSA/NSERC Industrial Research Chair Program, Communications and Information Technology Ontario (CITO), and Natural Sciences and Engineering Research Council of Canada (NSERC). The authors would like to thank Dr. D. Striakhilev and Dr. Y. Vygranenko for helpful discussions.

## REFERENCES

[1] R. A. Street, S. Nelson, L. E. Antonuk, and V. Perez Mendez, Mater. Res. Soc. Symp. Proc. **192**, 441 (1990).
[2] R. A. Street and L. E. Antonuk, IEEE Circuits and Devices, 38 (1993).
[3] W. Zhao and J. A. Rowlands, Med. Phys. **22**, 1595 (1995).
[4] D. L. Lee, L. K. Cheung, and L. S. Jeromin, SPIE Proc. **2432**, 237 (1995).
[5] K. Aflatooni, A. Nathan, R. Hornsey, I. A. Cunningham, and S. G. Chamberlain, Technical Digest, IEEE, International Electron Devices Meeting, 197 (1997).
[6] A. Nathan, R. Hornsey, and K. Aflatooni, IEEE Trans. Electron Devices **47**, 2093 (2000).
[7] B. Park and A. Nathan, ECS Proc. **98-22**, 381 (1998).
[8] B. Park, K. S. Karim, and A. Nathan, J. Vac. Sci. Technol. **A 18**, 688 (2000).
[9] B. Park, R. V. R. Murthy, A. Sazonov, A. Nathan, and S. G. Chamberlain, Mater. Res. Soc. Symp. Proc. **507**, 237 (1998).
[10] K. Aflatooni, R. Hornsey, and A. Nathan, Mater. Res. Soc. Symp. Proc. **467**, 925 (1997).

## Stacked n-i-p-n-i-p heterojunctions for image recognition

M. Vieira, A. Fantoni, M. Fernandes, P. Louro, I. Rodrigues
Electronics Telecommunication and Computer Dept. ISEL, Rua Conselheiro Emídio Navarro, P
1949-014 Lisboa, Portugal Tel: +351 21 8317181, Fax: +351 21 8317114, E-mail: mv@isel.pt.

## ABSTRACT

This work aims to clarify possible improvements and physical limits of the Color Laser Scanned Photodiode image sensor when used as high sensitive non-pixel image reader. A new design based on a stacked n-i-p-n-i-p heterojunction is proposed and compared with the old single n-i-p sensing structure. Results show that a B-W image is acquired with an improved resolution. The readout frequency is optimized showing that scans speeds up to $10^4$ lines per second can be achieved without degradation in the resolution. A physical model is presented and supported by an electrical and a numerical simulation of the output characteristics of the sensor.

## INTRODUCTION

Amorphous silicon-carbon (a-SiC:H) is a material that exhibits excellent photosensitive properties. The possibility to modify the optical gap enables the detection from the UV to the IR part of the spectrum. This feature has been intensively used in the development of image sensors [1, 2]. In our group efforts have been devoted towards the development of a new kind of image sensor, the Color Laser Scanned Photodiode sensors (CLSP) [3, 4, 5]. The CLSP consists on one large cell detector and the image is scanned by sequentially detecting scene information at discrete XY coordinates. The advantages of this approach are quite obvious: the feasibility of large area deposition on different substrate materials (e.g. glass, polymer foil, etc.), the simplicity of the device and associated electronics, the high resolution, the uniformity of measurement along the sensor and the cost/simplicity of the detector. The design allows a continuous sensor without the need for pixel-level patterning, and so can take advantage of the amorphous silicon technology. It can also be integrated vertically, $i.\ e.$ on top of a read-out electronic, which facilitates low cost large area detection systems where the signal processing can be performed by an ASIC chip underneath.

This work aims to clarify possible improvements in the CLSP image sensor. The image capture device and the scanning reader are optimized and the effects of the sensor structure on the output characteristics is discussed.

## EXPERIMENTAL

### Sample preparation and characterization

Single and tandem a-SiC:H p-i-n diodes were fabricated in a three-chamber load-lock UHV-system by Plasma Enhanced Chemical Vapor Deposition at 13.56 MHz on ITO and Cr coated glass substrates, respectively. The back metal contacts define the active area of the sensor (4×4 $cm^2$).

The single structure is a p-i-n diode and the tandem one is composed by two stacked n-i-p diodes (see Figure 1). The p-layers of both single diode and of the second diode of the stacked structure were fabricated introducing methane during the deposition process, in order to decrease the conductivity of the material [4, 5].

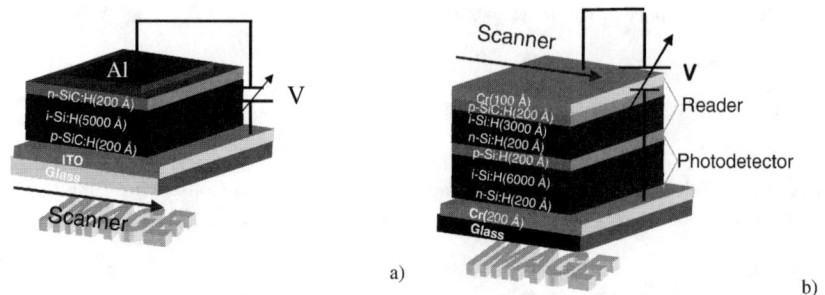

**Figure 1** Image reader and sensing element in: single (a) and double (b) CLSP image sensors.

Depending on the methane fluxes the n- and p-layers present conductivity and optical gaps between $1.9 \times 10^{-9}\ \Omega^{-1}\mathrm{cm}^{-1}$ - $8.2 \times 10^{-4}\ \Omega^{-1}\mathrm{cm}^{-1}$, 2.2 eV-1.8 eV, respectively. The i-layers have a dark conductivity of $7 \times 10^{-11}\ \Omega^{-1}\mathrm{cm}^{-1}$ and photosensitivity higher than $10^4$ under AM1.5 illumination (100 mW/ cm$^2$) [6].

All the structures were characterized by spectral response and current-voltage measurements, in dark and under different optical bias conditions ($0<\Phi_L<2$ mWcm$^{-2}$), as described elsewhere [6]. Open circuit voltages around 0.6 eV and 1.1 eV where found under AM1.5 illumination.

## Readout

The image readout process relies on the same principle for both single and stacked structures. An optical image is projected onto the active surface and is scanned by sequentially detecting scene information at discrete XY coordinates. In the single structure the image and the scanner are incident on the same side and in the double they are opposite, as shown in Figure 1. A 633 nm low power solid-state laser was used as scanner. The beam deflection is controlled by a two-axis deflection system capable of high speed scan (close to 1 kHz).

The current from the device is amplified by a current to voltage converter with selectable gain and converted to digital format by a signal acquisition card installed on a computer. Two additional photodiodes provide the signals synchronization for scanner position information, necessary for the image restoration process.

The data is stored as a matrix of photocurrent values which provide information about local illumination conditions on each XY position. Further processing algorithms like fixed pattern noise suppression are performed by software.

## RESULTS AND DISCUSSION

### Light to dark sensitivity in homo and hetero p-i-n structures

In Figure 2 the normalized spectral responses, $R_N$, with and without optical bias, $\Phi_L$, are compared for: a) a p-i-n homojunction based on a-Si:H material; b) a p-i-n heterojunction based on a-SiC:H,. Results show that in the homojunction the spectral sensitivity is light bias independent, while in the heterojunction it decreases as the light bias intensity increases. So, the heterojunction presents light to dark sensitivity while the homojunction is blind to a pattern of light projected onto its active surface. The light bias dependence presented by the heterostructures allows its use as image sensors [4].

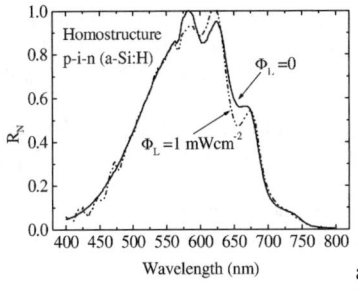 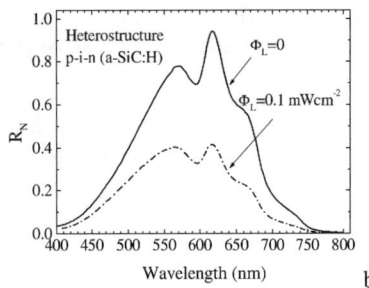

**Figure 2** Spectral response for a p-i-n homojunction based on a-Si:H (a) and for a heterojunction based on a-SiC:H (b).

## Single p-i-n sensing element operation

The operation and image representation are based on the analysis of the electrical field profile, induced across the capture device by a steady-state light pattern illumination [4]. Low local electrical fields are ascribed to illuminated regions and high electrical fields to dark zones. Figure 3 shows the signals obtained in a fast line scan using a single heterojunction under short circuit. In this experiment the central area of the sensor is illuminated while the outer region is kept in dark.

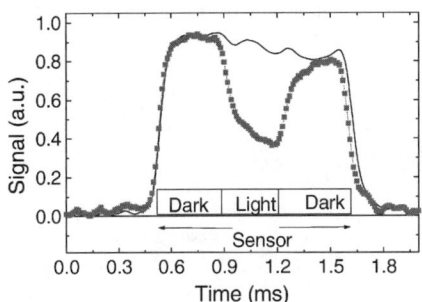

**Figure 3** Single line scans in dark (line) and crossing an illuminated region (squares).

When high resistive SiC:H doped layers are used the photogenerated carriers by the light pattern are confined to their generation regions [7]. So, the carriers generated by the scanner, in the dark regions, are separated by the electric field and collected (high signal), those generated inside the illuminated regions mostly recombine inside the bulk (low signal). So, by mapping the $ac$ component of the photocurrent during the scanning of the capture device, it is possible to reconstruct the projected light pattern.

In short circuit mode, the sensor can detect a black and white image with a spatial resolution less than 20 µm at a scan speed of 1000 lines per second.

## Double heterojunction operation

In the single p-i-n heterojunction the carriers generated by the light pattern and by the scanner are kept together and, depending on the resistivity of the doped layers, can diffuse away from the generated point and escape into the direction parallel to the junction, leading to image smearing. These lateral currents can be minimized if the resistivity of the doped layers is increased. However, high resistive doped layers leads to poor collection efficiencies and so, to a low image signal to noise ratio. The image signal, can be improved by adding to the n-i-p a-SiC:H reader a front photodetector (a n-i-p homojunction) as shown in Figure 1. In the resulting stacked structure the front i- layer has to be thick enough (>5000 Å) to absorb all the light incoming from the image and the rear thinner (<3000 Å) and based on a-SiC:H in order to enhance light transmission from the scanner.

The readout process for image acquisition is similar to the previous one with the difference that in this structures the image and scanner are incident on opposite sides of the sensor (Figure 1). This last approach simplifies the optical system since image and scanner have different optical paths. The front photodiode (the photodetector) confines the carriers inside the illuminated regions while the rear (the reader), driven by the optical scanner, gives information on their location (image shape), density (image intensity) and absorbed wavelength (image color). When the scan is performed with no image (dark) the photocurrent remains constant and only dependent on the scanner wavelength and applied voltage. Under these conditions there is no current generation in the photodetector. When the sensor is illuminated with some light pattern the photocurrent remains the same in dark zones and increases (proportionally to the light intensity) in the illuminated ones. In the last situation the charge generation occurs in both junctions at the same position so, the charges are generated across the whole sensor which justifies an increase in the image signal.

a) CLSP    b) D/CLSP

**Figure 4** Same image acquired using: (a) a single and (b) a double structure.

Figure 4 compares the same image acquired using a single and a double structure showing an improved resolution. The readout frequency was also optimised leading to a value of 10kHz, showing that scan speeds of 1000 lines per second can be achieved without a significant degradation of the resolution.

### Electrical and numerical simulation

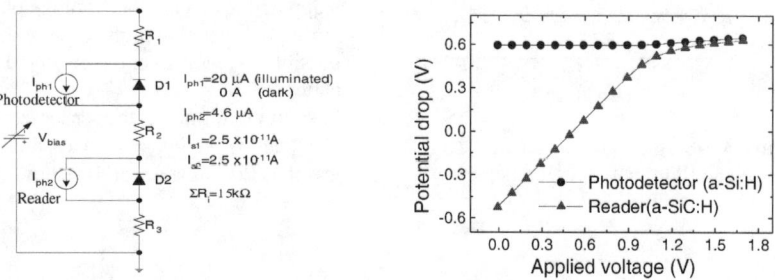

**Figure 5** a) Electrical model for the double CLSP. b) Simulated voltage drop across both diodes as a function of the applied voltage.

The proposed *dc* electrical model is presented in Figure 5a, and in Figure 5b the simulated voltage drop across both diodes, as a function of the applied voltage, is displayed.

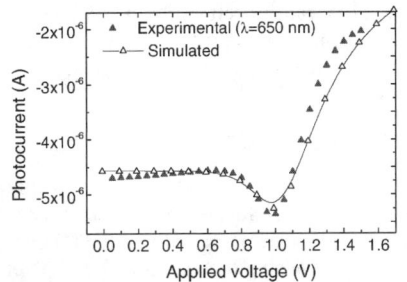

Assuming that the current across the photodetector and the reader is the same and that the applied bias is shared by both, we conclude that for voltages lower then the open circuit voltage (around 1.1 V) and under illumination, the homojunction is self forward biased and the heterojunction becomes reverse biased.

The carriers generated at the illuminated diode (load, ON state) are injected into the second one (photodiode, OFF state) where they recombine, are trapped or collected, depending on its reverse current. A good fit between both experimental and simulated data was achieved (see Figure 6).

**Figure 6** Experimental and simulated photocurrent as a function of the electrical bias.

To understand the transport mechanism under non-uniform illumination, in the double structure, we took into account the experimental characterization of the devices and some results from a computer simulation performed using the ASCA simulator[8]. Details about the program and the experimental input parameters are described elsewhere [9].

In Figure 7 the potential (a) and the electrical field (b) profiles across a stacked p-i-n-p-i-n structure [10] is shown, under different illumination conditions.

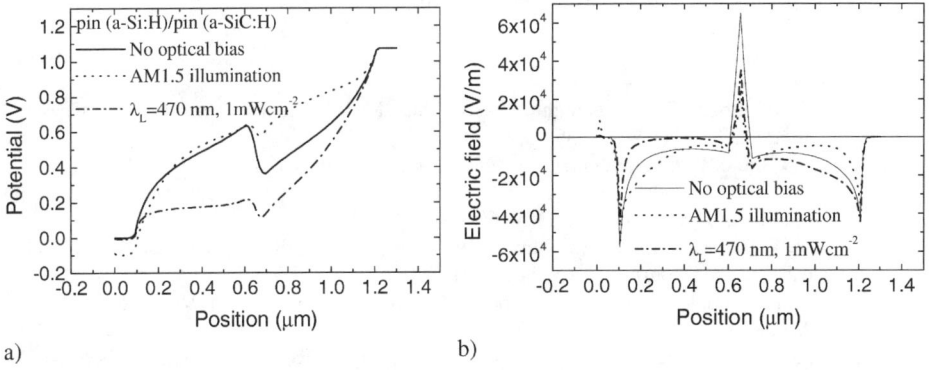

**Figure 7** Potential (a) and electrical field (b) profiles in dark and under AM1.5 and blue illumination inside a p-i-n (a-Si:H) / p-i-n (a-SiC:H) structure and under short circuit condition.

Results show that in dark the built-in potential is distributed unevenly across both diodes. It drops in a steep way inside the a-SiC: H p-i-n structure (reader) and slowly inside the homojunction (photodetector). At the internal p-n interface a reversal in the potential profile is observed, leading to carrier accumulation in this region.

Under illumination and depending on the light depth penetration the potential drop across the i-layer is reduced and even flattens at one or both illuminated diodes, which ensures the presence of a quasi neutral region inside the absorber layers. If the light is absorbed only in the front diode (blue light) the electrical field is always zero on the bulk of the photodetector and enhanced mainly at the reader whose i-layer becomes fully depleted. Under AM1.5 illumination the light is absorbed on both photodetector and reader changing the reader electrical field in an opposite way. In both cases an inversion layer is induced at the p-n internal junction leading to an increase of the electron population that delays the transition from the primary to the secondary photocurrent regime as it was observed in Figure 6.

## CONCLUSIONS AND FUTURE WORK

A new design based on a stacked n-i-p-n-i-p structure is proposed for the sensing element of the CLSP image sensor. Under short circuit condition a black and white image was acquired with improved resolution when compared with the single diode structure. Readout of 1000 lines per second was achieved allowing continuous and fast image sensing without the need of pixel patterning.
The stacked structure should be optimized in order to improve higher resolutions. Basic image processing algorithms should be applied for image enhancement and pattern recognition. Modules for brightness calibration and edge enhancement are still needed to improve the system. Further optimization of the optical scanning system includes dynamic characterization of the sensor (readout frequency, frame rate).

## ACKNOWLEDGEMENTS

We would like to thank IPE and Univ. Waterloo for the help during this study, as well as to M. Rakhlin and Yury Vygranenko for the film deposition. This work has been financially supported by. FLAD, IPL and POCTI/ESE/38689/2001 project.

## REFERENCES

[1] T. Neidlinger, R. Bruggemann, H. Brummak, M. B. Schubert *Journal of Non-Crystalline Solids* 227-230 (1998) 1335-1339.
[2] D. Caputo, G. Cesare *Journal of Non-Crystalline Solids* 198-200 (1996) 1334-1336.
[3] M. Vieira, M. Fernandes, J. Martins, P. Louro, A. Maçarico, R. Schwarz, and M. Schubert, Mat. Res. Soc. Symp. Proc., Vol. 609 (2000) A14.
[4] M. Vieira, M. Fernandes, J. Martins, P. Louro, A. Maçarico, R. Schwarz, and M. Schubert, IEEE Sensors Journal 1, No. 2 (2001)158-167.
[5] M. Fernandes, P. Louro, J. Martins, A. Maçarico, R. Schwarz, M. Vieira. Sensors and Actuators A. 92 (2001) pp.60-66.
[6] P. Louro, M. Vieira, Yu. Vygranenko, M. Fernandes, R. Schwarz, M. Schubert, Applied Surface Science, 184 (2001) 144-149.
[7] M. Vieira, M. Fernandes, A. Fantoni, P. Louro, Y. Vygranenko, R. Schwarz, M. Schubert. Applied Surface Science, 184 (2001) pp. 471-476.
[8] R. Martins, A. Fantoni, and M. Vieira, J. of Non-Cryst. Solids 164&166 (1993) 671.
[9] A Fantoni, M. Fernandes, P. Louro, R. Schwarz and M. Vieira, In Amorphous and Heterogeneous Silicon Thin Films-2001, Mat. Res. Soc. Symp. Proc., (S. Francisco, USA), Vol. 664 (2001) A25.11.
[10] A Fantoni, M. Fernandes, P. Louro, R. Schwarz and M. Vieira, In Amorphous and Heterogeneous Silicon Thin Films- 2001, Mat. Res. Soc. Symp. Proc., Vol. 664 (2001) A25.11.

## Development of Vertically Integrated Imaging and Particle Sensors

N. Wyrsch[1], C. Miazza[1], S. Dunand[1], A. Shah[1], N. Blanc[2], R. Kaufmann[2], L. Cavalier[2],
G. Anelli[3], M. Despeisse[3], P. Jarron[3], D. Moraes[3], A. G. Sirvent[3], G. Dissertori[4], G. Viertel[4],
[1] Institut de Microtechnique, CH-2000 Neuchâtel, Switzerland,
[2] CSEM SA, CH-8048 Zurich, Switzerland,
[3] CERN, CH-1211 Genève 23, Switzerland,
[4] ETH-Zurich, CH-8093 Zurich, Switzerland.

## ABSTRACT

Integrated imaging and particle sensors have been developed using thin-film on ASIC technology. For this purpose, hydrogenated amorphous silicon diodes, in various configurations, have been optimized for imaging and direct particle detection. These devices were first deposited on glass substrates and later on CMOS readout chips. With an optimization of the material properties and of the diode, a dark current of 1 pA/cm$^2$ could be achieved on p-i-n structures at reverse bias voltage of 1 V. CMOS imagers, incorporating these optimized diodes were then fabricated and characterized. Very thick diodes (with thicknesses up to 50 µm) were also optimized and deposited on glass and on CMOS readout chips. Particle detectors in TFA technology with 12 and 30 µm a-Si:H n-i-p diodes have been fabricated and characterized using light pulse illumination. Direct detection of single low-energy beta particles has been demonstrated.

## INTRODUCTION

Active pixel sensors (APS) in CMOS technology have recently gained a lot of attention. However, the fact that the pixel readout-electronics shares the die area with the photodiode array is a limiting factor for the imagers' light sensitivity. A vertical integration of the light sensor by the deposition of an amorphous silicon (a-Si:H) detecting layer on the CMOS readout chip can greatly improve the sensitivity [1]. This so-called thin-film on ASIC (TFA) or thin-film on CMOS (TFC) technology has therefore an interesting potential for high-sensitivity, low-level or high-dynamics imaging [2], for both small-area pixel imagers [3] as well as for large-area imagers for X-ray medical applications [4]. The high integration level of the detecting device and readout electronics also offer a good potential for reductions in system cost.

In particle physics, the increasing accelerator energies and fluences used in experiments call for radiation-hard particle detectors. a-Si:H has been proven to be a radiation-hard material [5,6,7] and is now regaining attention for applications in particle detection. In this context, the use of the TFA technology offers a much higher degree of detector integration as well as the additional possibility of constructing much larger detectors with full area coverage. Thus, a significant reduction of the system cost could be achieved. However, direct detection of particles requires very thick a-Si:H layers (thicknesses in the order of 20 to 50 µm,) to provide an adequate signal. At such high thickness values peeling of the layers due to mechanical stress and insufficient adhesion to the substrate often become a problem; also the deposition rates needs to be high to keep reasonable deposition times, without affecting the material quality [8,9].

For both fields of application (particle detection and imaging applications), a-Si:H detectors with very low dark currents are required. For this purpose, a-Si:H detectors have been developed and first deposited on glass substrates in various diode configurations (n-i-p, p-i-n and metal-i-p). Effects of top and back contacts and the impact of device geometry on dark current values have been studied. Similar devices have then been deposited on actual CMOS readout chips and characterized.

For particle detection, devices with thicknesses of up to 30 µm have been deposited on glass substrates and readout CMOS chips. A very high frequency plasma enhanced chemical vapor deposition (VHF PE-CVD) has enabled the deposition of thick layers at high rates with low defect densities, with low internal mechanical stress [10] and minimal powder formation [11].

## EXPERIMENTAL

All devices have been deposited by VHF PE-CVD at 70 MHz and 200°C using hydrogen dilution of silane. Devices were either deposited at a rate between 3 and 3.3 Å/s, or for thick devices (>10 µm) at a rate of 15.6 Å/s. Test devices were deposited on Cr- or Al-coated glass and the pixel areas were defined by a patterned ZnO or ITO (Indium Tin Oxide) top electrode. The patterning was done by a rubber stamping process followed by a wet etch of the transparent conductive oxide. A subsequent partial plasma etch of the a-Si:H layer was also carried out. In order to study test devices with a structure similar to that of TFA chips, "chip-like" test structures with small-size pixels (50 to 200 µm side length) and a common top electrode were fabricated by photolithography on glass substrates.

Devices for imaging application (in TFA technology) were deposited on CMOS readout chips designed by CSEM and fabricated in 0.5 µm technology of Alcatel-Mietec, while those for particle detection were deposited on CMOS readout chips designed by CERN and fabricated in 0.25 µm technology of IBM. A common top ITO electrode was used for both types of devices. Because the process was carried out on single chips, special procedures had to be developed for chip handling, uniform deposition and chip patterning.

Test structures have been characterised by measuring current vs. voltage (in the dark and under illumination), quantum efficiency and transient charge collection (time of flight). Regarding sensors for imaging applications, characterisation included determination of uniformity maps, of quantum efficiency, of dark current, of sensitivity and of linearity. Sensors for particle detection have been tested mainly for charge collection efficiency under very weak pulsed light illumination and with a beta source ($^{63}$Ni).

## RESULTS AND DISCUSSION

### Test structures

For high sensitivity sensors as well as for single particle detection, one aims at developing photodiodes with the lowest dark current $I_{dark}$. For this purpose, diodes (in various configurations) have first been optimised on glass substrates. Here best results were obtained with an a-Si:H i-layer deposited with an hydrogen dilution of R=[H$_2$]/[SiH$_4$]=3.5. In order to avoid the need for patterning the bottom doped layer in the TFA sensors (the layer that is first deposited on the CMOS chip and that may induce cross-talk effects between the pixels), low conductivity doped layers (either n or p, depending on the diode configuration) were developed. Alternatively, as a radical means of avoiding patterning and cross talk, metal-i-p structures were also studied. As shown in the upper part of Table 1, one µm thick devices with extremely low values of $I_{dark}$ were successfully fabricated.

Note that these low values were obtained without any introduction of carbon in the p-layer or at the p/i interface, in contrast to other work [12]. However, most devices, especially those without an n-layer exhibited a strong increase in the leakage current at high reverse bias voltages.

**Table 1.** Best dark current values obtained for three different i-layer materials (all a-Si:H) for various diode configurations and for 3 values of the reverse voltage. All diodes are 1 µm thick.

| i-layer material | Configuration | $I_{dark}$ at $-1$ V [Acm$^{-2}$] | $I_{dark}$ at $-3$ V [Acm$^{-2}$] | $I_{dark}$ at $-5$ V [Acm$^{-2}$] |
|---|---|---|---|---|
| Low rate 1 (3 Å/s) | n-i-p | 2.0·10$^{-12}$ | 3.6·10$^{-12}$ | |
| | p-i-n | 1.0·10$^{-12}$ | 3.0·10$^{-12}$ | 6.5·10$^{-12}$ |
| | metal-i-p | 5.5·10$^{-11}$ | 2.4·10$^{-10}$ | >10$^{-9}$ |
| Low rate 2 (3.3 Å/s) | p-i-n | 9.1·10$^{-12}$ | 2.2·10$^{-11}$ | 2.7·10$^{-11}$ |
| | metal-i-p | 4.4·10$^{-12}$ | 4.0·10$^{-11}$ | 6.1·10$^{-11}$ |
| High rate (15.6 Å/s) | n-i-p | 6.1·10$^{-12}$ | 1.6·10$^{-11}$ | |

Incorporation of a slightly different i-layer material (indicated as "low rate 2" in Table 1 and Fig. 1) resulted in much lower saturation current for p-i-n and metal-i-p diodes (see Fig. 1) and improved the collection efficiency, with only a very small increase of $I_{dark}$ at low reverse voltages (see Table 1).

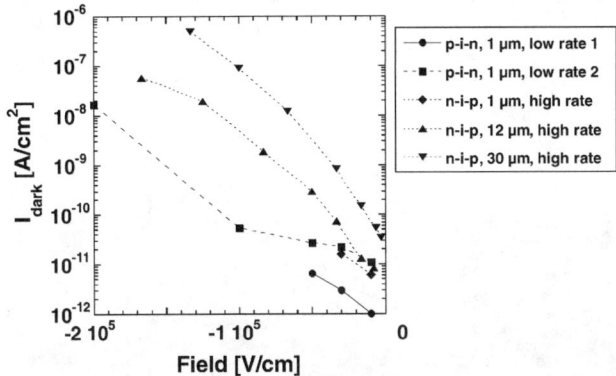

**Figure 1.** Dark current $I_{dark}$ as a function of reverse bias field and diode thickness for different i-layer materials, deposited at low (under 2 different deposition conditions) and at high rates.

In order to successfully fabricate very thick devices (needed for particle detection), another i-layer material was optimized at high deposition rate with a reduced hydrogen dilution of $R=[H_2]/[SiH_4]=0.35$, in a different VHF plasma reactor (mostly to avoid disturbance by the inevitable powder formation). Thereby, devices of various thicknesses (1 to 50 μm) were fabricated at a deposition rate up to 15.6 Å/s. This higher rate resulted in an increase of the dark current $I_{dark}$ (see Table 1) which is probably due to a slightly higher i-layer defect density. As we can observe in Fig. 1, $I_{dark}$ increases also with the device thickness, which seems to indicate that a significant contribution to $I_{dark}$ originates from defects in the bulk of the i-layer.

Most of the diodes have been deposited on a Cr back electrodes. Use of Al for the latter was not found to significantly affect the results; however, diodes with Al back contacts were found to be more prone to pinhole formation. As a top contact, both sputtered ITO and ZnO (deposited by sputtering or CVD) have been used without much effect on $I_{dark}$. Selection of one or the other material will mostly depend on the requirement for the spectral sensitivity of the sensor. For direct particle detection, a high-Z (atomic number) metallic contact is preferable as light shield, but also as a means to increase the sensitivity of the sensor [8].

<u>CMOS imagers</u>

After the optimization of n-i-p diodes on glass substrates, these structures have been deposited on a 64x64 pixel CMOS active-pixel readout chip (see inset of Fig. 2). The characterisation of this sensor for sensitivity were done at three different light wavelengths yielding the following sensitivities: 14 V/(μJ/cm$^2$) for red light (at 626 nm), 23 V/(μJ/cm$^2$) for green light (at 570 nm) and 16.8 V/(μJ/cm$^2$) for blue light (at 470 nm). However, besides this high sensitivity, dark current values were found to be much higher that those measured on test structures. As shown in Fig. 2 (left), $I_{dark}$ is more than 3 orders of magnitude larger than the best values obtained in test structures. In order to investigate the possible origin for this effect, "chip-like" test structures on glass "mimicking" the pixel configuration of a CMOS chip (with back contacts in wells through an an oxide layer and a common top electrode) were fabricated. On these test structures, we observed a dependency of $I_{dark}$ on the pixel size (as plotted in Fig. 2, right); $I_{dark}$ is found to decrease with an increasing size of the pixel, indicating that the pixel periphery is responsible for this effect. $I_{dark}$ was also found to depend on the depth of the wells in the oxide (i.e. the thickness of the oxide layer used for the insulation of the metallic connection between the pixel back contact pads and the bonding pads. The high value of $I_{dark}$ measured on TFA chips is therefore linked to peripheral pixel leakage through the n-layer, influenced by the thickness of the chip passivation layer. The use of planarized chips is expected to help reduce considerably $I_{dark}$. As an alternative option, metal-i-p structures will also be studied.

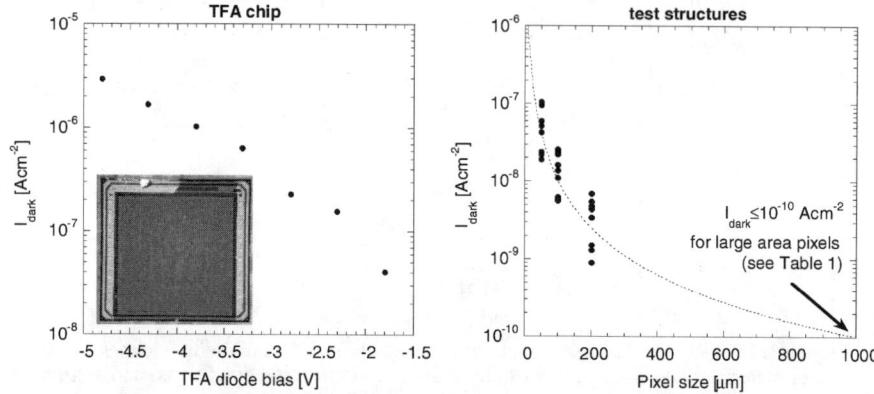

**Figure 2.** Dark current as a function of the photodiode reverse bias (left) measured on a pixel of the TFA sensor seen in the inset, and dark current at −1 V as a function of pixel size (right) for a "chip-like" test structure "mimicking" the configuration of a pixel chip. The dooted curve (left) corresponds to a model with a bulk contribution and a preipheral contribution to $I_{dark}$. In both case, the thickness of the photodiode i-layer is 1 μm. The inset shows a picture of a 64x64 active pixel sensor in TFA technology, with an a-Si:H n-i-p photodiode layer; the pixels have here a size of 20 μm with a pitch of 40 μm.

Particle detectors

For the direct detection of particles, 12 μm and 30 μm thick n-i-p diodes (first optimized on glass substrate) have been deposited on a CMOS chip with an array of very sensitive active feedback preamplifiers (AFP). The main technological problems were to achieve uniform deposition of thick a-Si:H photodiode layers on small (4x2 mm$^2$) single chips; the masking for the patterning was performed by hand (for the sake of simplicity), which is the reason for the rather poor esthetic appearance of the processed chip (see inset of Fig. 3). Nevertheless, very high sensitivity and very fast response was obtained, as shown in Fig. 3 for a 12 μm thick device.

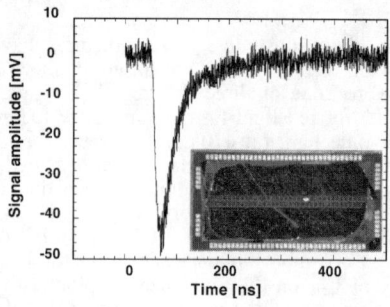

**Figure 3.** Signal amplitude measured at the pixel output of a 12 μm TFA AFP chip following a 2 ns light pulse (at 660 nm). The generated charge induced by the light pulse was here 1.5fC. A picture of the chip is displayed in the inset. The linear array consists of 32 pixels with a 100 μm pitch.

From the analysis of the current transient of the photodiode layer (Fig. 4), we can conclude that, even at 80 V, a full depletion of the a-Si:H n-i-p device is not yet attained. As the bias voltage is raised, the width of the depletion region as well as its electrical field increases, which results in an approximately constant electron collection time (given by the width of the peaks in Fig. 4, left). On the other hand, hole collection is less affected by the width of the depletion region and is facilitated by an increase of the bias voltage. Due to the much lower drift mobility of holes, their collection occurs over a much longer time and it is responsible for the slowly decaying tail seen in Fig. 4 (left) or better seen at longer times in Fig. 4 (right).

A preliminary study of the detection of single beta particles with a 12 µm TFA chip was also carried out. As demonstrated in Fig. 5, single beta particles emitted from the isotope $^{63}$Ni can be clearly detected down to an electron energy estimated to 15.6 keV. A clear relationship between the height of the peak and the particle energy is also observed. This experiment is to our knowledge the first example of single particle direct detection with a sensor fabricated with TFA technology.

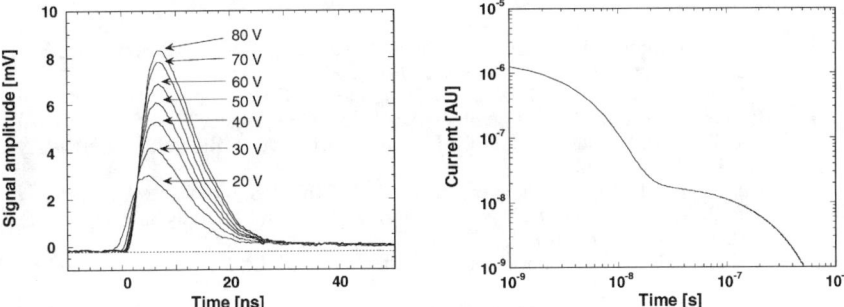

**Figure 4.** Current transient for a 2 ns light pulse, as measured for different reverse bias voltage values of the TFA photodiode layer (left); the signal waveforms include the active feedback amplifier response. The peaks correspond to the electron drift in the depletion region of the diode, while the slowly-decay tails at long times are attributed to hole collection towards the top electrode. On the right, the current transient for electrons and holes (obtained from the measured signal after correction for the response of the amplifier) is plotted for a bias voltage 60 V. The transient at short times is due to electron transport while at longer time it is dominated by holes.

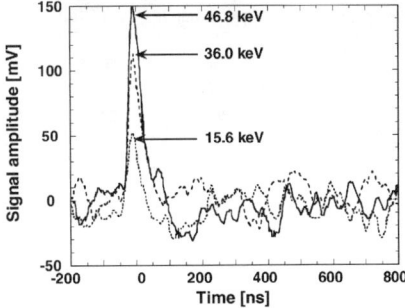

**Figure 5.** Signal measured at the pixel output of a 12 µm TFA AFP chip following absorption of single beta particles from the isotope $^{63}$Ni. The peak height amplitudes vary with the beta electron energy deposited in the depletion layer. Energy was estimated from the maximum energy of 50 keV.

## CONCLUSIONS

Several a-Si:H diodes with dark currents $I_{dark}$ as low as 1 pA/cm$^2$ have been deposited by VHF PE-CVD on glass substrates using hydrogen dilution of silane, but without an incorporation of a silicon carbide p-layer or of silicon carbide buffer layers at the p/i interface. Similar diodes deposited on a 64x64 pixel CMOS chips exhibited much higher values of $I_{dark}$ due to peripheral pixel leakage resulting from the non-planarity of the chips. Planarized chips, as well as the introduction of metal-i-p diode structures will be tested as a solution to this problem.

Since a-Si:H is known to be rather radiation resistant, this material is of interest for particle detection in high energy physics. In this context, TFA technology allows a high degree of integration with interesting cost and reliability benefits. First TFA detectors with an a-Si:H n-i-p diode thickness of 12 and 30 µm have been successfully deposited on CMOS readout chips. The detection of single beta particles with energies as low as 15 keV has been demonstrated.

## REFERENCES

[1] T. Lulé, S. Benthien, H. Keller, F. Mütze, P. Rieve, K. Seibel, M. Sommer, M. Böhm, IEEE Trans. on Electron Devices **47**, 2110 (2000).
[2] B. Schneider, P. Rieve, M. Böhm, in Handbook on Computer Vision an Applications, ed. B. Jähne, H. Haußecker, P. Geißler, (Academic Press, Boston, 1999)pp. 237-270.
[3] J. A. Theil, R. Snyder, D. Hula, K. Lindahl, H. Haddad, J. Roland, J. of Non-Cryst. Sol. **299-302**, 1234 (2002).
[4] R.A. Street, "Large Area Image Sensor Arrays" in Technology and Application of Amorphous Silicon, ed. R.A. Street, Springer Series in Materials Sciences **37** (Springer-Verlag, Berlin, 2000), p.147.
[5] S. Guha, J. Yang, A. Banerjee, T. Glatfelter, Proc. of the 2$^{nd}$ World Conf. and Exhibition on PV Solar Energy Conversion, Vienna, Austria, 3609 (1998).
[6] J. Kuendig, M. Goetz, J. Meier, P. Torres, L. Feitknecht, P. Pernet, X. Niquille, A. Shah, L. Gerlach, E. Fernandez, Proc. of the 16$^{th}$ EU PV Solar Energy Conf., Glasgow, UK, May, 986 (2000).
[7] L.E. Antonuk, J. Boudry, J. Yorkston, C. F Wild, M.J. Longo, R.A. Street, Nucl. Instr. And Meth. **A299**, 143 (1990)
[8] P. Chabloz, H. Keppner, V. Beartschi, A. Shah, D. Chatellard, J.-P. Egger, M. Denoréaz,E. Jeannet, J.-F. Germond, R. Vuilleumier, MRS Proc. **258**, 1057 (1992).
[9] W.S Hong, V.Petrova-Koch, J. Drewery, T. Jing, H.Lee, V.Perez-Mendez, MRS. Res. Symp. Proc. **377**, 773 (1995).
[10] P. Chabloz, H. Keppner, D. Fischer, D. Link, A. Shah, J. Non-Cryst. Sol., **198-200**, 1159 (1996).
[11] A. Shah, J.Dutta, N. Wyrsch, K. Prasad, H. Curtins, F. Finger, A. Howling, Ch. Hollenstein, MRS Proc. **258**, 15 (1992).
[12] S. Morrison, P. Servati, Y. Vygranenko, A. Nathan, and A. Madan, MRS. Res. Symp. Proc. **715**, 701 (2002).

# AN AMORPHOUS SILICON PHOTOCONDUCTOR FOR UV DETECTION

Matthias Hillebrand, Frank Blecher[1], Jürgen Sterzel[2], Markus Böhm
Institute for Microsystem Technologies (IMT), Universität Siegen, D-57068 Siegen, Germany
[1] now with LambdaLab, Kohlbettstraße 20, D-57072 Siegen, Germany
[2] now with Jena Optronik GmbH, Prüssingstraße 41, D-07745 Jena, Germany

## ABSTRACT

An amorphous silicon photoconductor to detect wavelengths between 180 nm to 550 nm without scintillator is presented. The photoconductor is based on a coplanar configuration of the electrodes, similar to measurement structures to determine material characteristics of amorphous layers, e.g. for the Constant Photocurrent Method (CPM). After passing through a thin transparent passivation layer, the incident radiation is directly absorbed in the intrinsic a-Si:H material. The carrier collecting electrical field is applied perpendicular to the incoming light. Test structures have been fabricated with 80 nm thick sputtered chromium contacts on top of a 60 nm carbonized hydrogenated i-layer and a $Si_xN_x$ passivation layer with a thickness of about 36 nm. The spacing between the Schottky contacts is varied between 3 µm and 100 µm. They are deposited on top or below the a-SiC:H layer. First experiments with this simple coplanar design show that with an increasing voltage a shift towards UV wavelengths can be observed. The new UV detector is applicable in the field of TFA image sensors (Thin Film on ASIC) and in the new Lab-on-a-Chip concept presently under development at the institute for microsystem technologies.

## INTRODUCTION

Conventional UV detectors are made of semiconductor materials such as SiC and GaN. The advantage is the solar blindness. UV-sensitive CCDs usually use scintillators, which results in a high responsivity for visible wavelengths. Especially for flame detection less cross responsivity for visible light is needed.

Amorphous silicon UV detectors are mostly based on ordinary pin-structures [1, 2, 3]. For this detector a TCO (Transparent Conductive Oxide) contact made of $ZnO_x$ or $SnO_x$ is applied as front electrode. Palma et al. [4] achieved solar blindness by using an aluminum grid as front contact, while the UV imager presented in [5] uses a transparent aluminum contact, supplemented with additional filters to shield the visible light.

All of the above radiation detectors use structures with a collecting electrical field parallel to the incoming light and therefore suffer from losses in the first doped layer. The concept presented here is based on a coplanar structure (COS), which has not yet been used for UV detection. These structures are more often used for the qualification of a-Si:H layers, for example for CPM measurements, resistance measurements, SSPG measurements and noise measurements [6, 7]. Other coplanar structures are used in two dimensional position sensitive detectors (PSD) [8] and in thin film transistors (TFT) [9]. The UV detector presented here may be used in low cost TFA imagers [10] and offers a wide range of applications for security and medical technology. Moreover this concept may be used in Lab-on-a-Chip concept applications [11].

## DEVICE STRUCTURES

Figure 1 shows the working principle of a coplanar radiation detector. The idea is to turn the normal pin diode by 90° such that the carrier collecting electrical field is applied perpendicular to the incoming light. An important parameter is the defect density at the surfaces, which cause recombination and thus a decrease of the photocurrent.

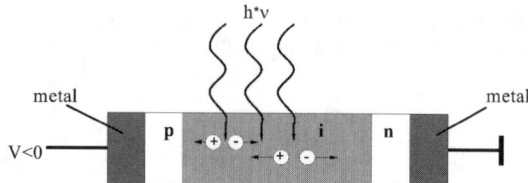

**Figure 1.** Coplanar radiation detector.

These surface effects may be suppressed using passivation layers in combination with more sophisticated device architectures. Figures 2 and 3 depict different cross sections for devices using double Schottky contacts or source gated devices. The two contacts of a detector pixel can be placed on top or below the photoactive layer or a mixed configuration may be employed. The arrangement of the two contacts may be varied according to technological requirements.

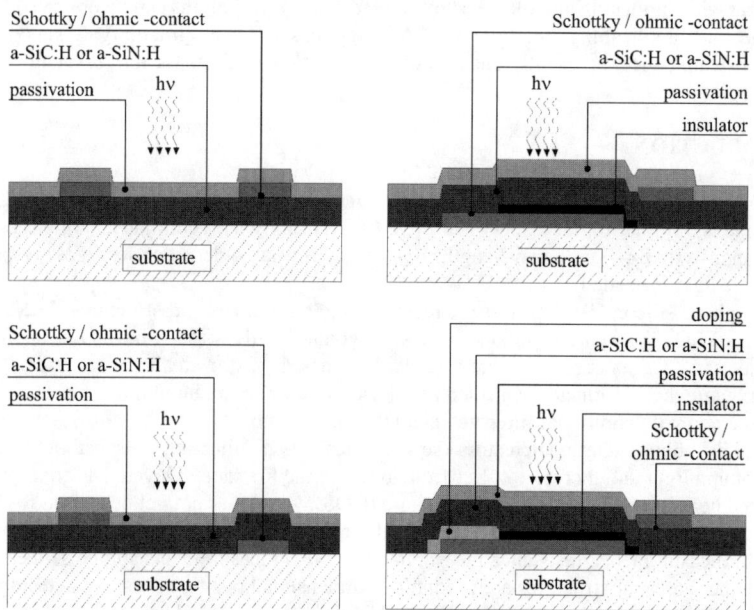

**Figure 2.** Double Schottky devices. **Figure 3.** Source gated devices.

The advantage of source gated devices is that the electrical field can be partitioned in a vertical and a horizontal field. This additional vertical field can separate the photo generated charge carriers such that recombination is reduced [12], providing a higher photocurrent.

## THEORETICAL

To calculate the dependency of the photocurrent on the surface an analytical simulation was performed. For the calculation some simplifications have been made: There is no optical reflection, no local lifetime profile and no vertical drift of the carriers due to the potential gradient in the vertical direction. Since amorphous silicon is used in the i-layer only electron conduction was assumed and the hole conduction was neglected. Figure 4 depicts the idealized configuration. In order to calculate the spectral photoconductivity of the COS a two dimensional boundary value problem was solved [13]. Figure 5 shows the calculated spectral photoconductivity.

The surface recombination velocities $s_b$ at the bottom and $s_f$ at the front are varied from 10 cm/s to 10,000 cm/s simultaneously. Furthermore, the figure indicates the penetration depth of the light for amorphous silicon with a band gap of 1.9 eV. The calculation is done with a drift length of 230 nm. The lifetime of the carriers has been assumed to be $2 \cdot 10^{-7}$ s. The spacing between the contacts is 5 µm and the thickness of the layer is 60 nm. A distinct decrease of the conductivity at the front surface towards short wavelengths is clearly detectable. A voltage dependency has not been considered in this calculation.

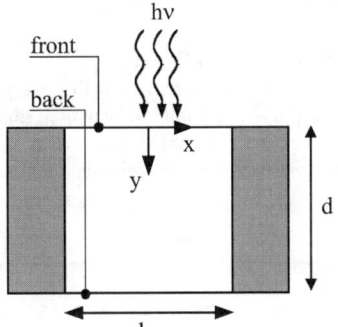

**Figure 4.** Geometry of the idealized device model.

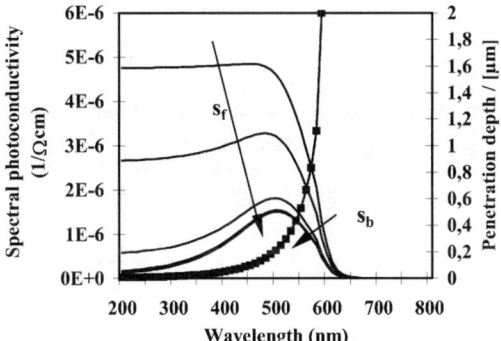

**Figure 5.** Calculation of the influence of the surface recombination velocities $s_b$ at the bottom and $s_f$ at the front on spectral photoconductivity. The doted line is the penetration depth of light.

## EXPERIMENTAL

The coplanar device, which was used for the model calculation, is shown in figure 6. The structure consists of a metal grid with Schottky contacts of chromium (84 nm), which is applied to a homogeneous 60 nm thick a-SiC:H i-layer (SiH$_4$:CH$_4$ / 20:40 sccm) without hydrogen dilution. The photosensitive area is $3.1 \cdot 10^{-3}$ mm² (length 1 mm, width 3.1 µm). On the amorphous silicon i-layer a passivation of a-$Si_{1-x}N_x$:H ($SiH_4$:$NH_3$ / 4:60 sccm) with a thickness of about 35 nm was deposited.

The I-V characteristics of the coplanar device for voltages $0..\pm 30$ V is shown in figure 7. The I-V characteristics were found to be symmetrical for positive und negative voltage. The photocurrent measurements were carried out with a XBO 150 W bulb as light source with and without a UG 11 filter to suppress the visible light.

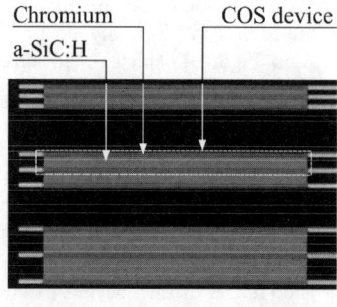

**Figure 6.** Photo of COS test device produced at the IMT.

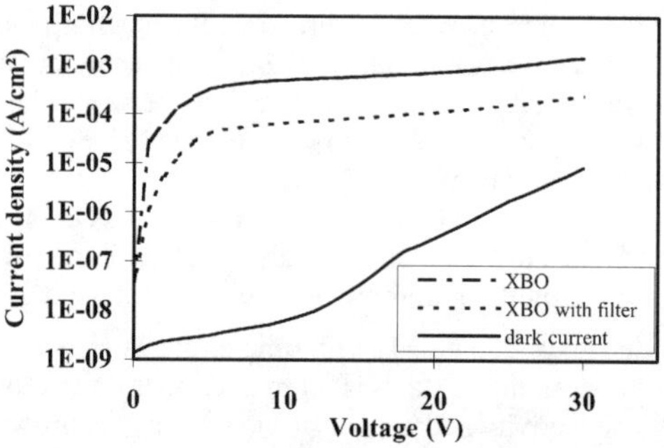

**Figure 7.** I-V characteristics of COS test device.

Figure 8 shows the spectral distribution of the XBO 150 W bulb with filter UG 11. The intensity of the radiation after passing the filter UG 11 is about 0.284 mW/cm². For this illumination density the calculated dynamic range of the COS is about 60 dB for wavelengths from 300 nm to 400 nm. The maximum of the spectral responsivity in the blue amounts to 1360 mAW$^{-1}$ at 410 nm at a voltage of 24 V. Figure 9 depicts the maximum of the responsivity as a function of the bias. A shift to smaller wavelengths is observed with increasing voltages.

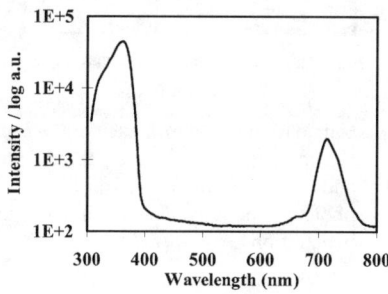

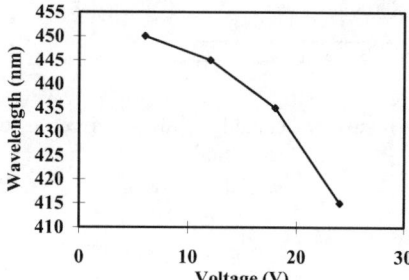

**Figure 8.** Spectral distribution of the XBO150W bulb with filter UG11.

**Figure 9.** Responsivity of test COS device for different voltages.

Dividing the current density by the voltage yields the spectral conductivity, which is shown in figure 10 normalized to its maximum. The graph shows clearly a shift of the responsivity towards short wavelengths with an increasing voltage. The increase of the voltage obviously effects the charge carrier mobility and the carrier lifetime. Before the generated carriers recombine, the carriers have to contribute to the current due to the electrical field. The recombination of the carriers occurs especially at the surface of the coplanar device.

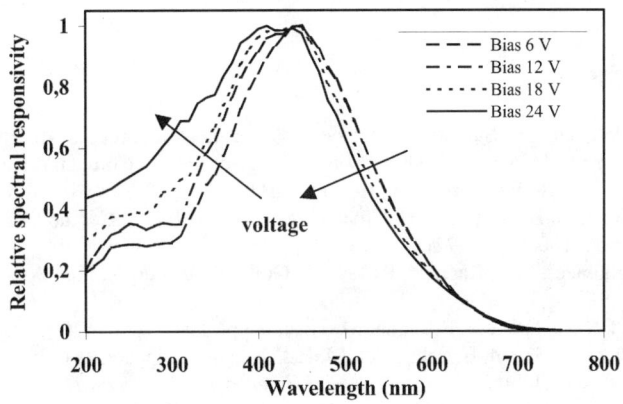

**Figure 10.** Relative spectral responsivity of COS test device.

The influence of the voltage increase for long wavelengths (450 nm to 750 nm) is not very high. For low voltages up to 12V the long wavelength slope of responsivity curve exhibits no voltage dependency. The device behaves like an ordinary photoconductor.

**Material dependency**

Due to the use of a-SiC:H for the absorber and a layer thickness of only the 60 nm the maximum spectral sensitivity is in the range of the visible light. It is expected that the use of a silicon nitride alloy will reduce the sensitivity in the visible spectrum. An advantage of a-$Si_{1-x}N_x$:H layers may be the possibility to vary the band gap in a wider range from less than 2 eV up to over 5 eV. An increase in band gap results in a deterioration of the electric characteristics of the material. However, a-$Si_{1-x}N_x$:H has a $\eta\mu\tau$-product, which is by an order of magnitude larger than that of a-SiC:H, although the dangling bond density is higher in a-$Si_{1-x}N_x$:H than in a-SiC:H [14].

## CONCLUSION

The presented coplanar device architecture for a detector for UV light offers some advantage compared to ordinary pin diodes, since it does not suffer from absorption losses due to TCO or metal front contacts. However, the surface properties of the absorbing layer are crucial to device performance and require optimization. A simple analytical model for the calculation of the spectral conductivity has been presented along with experimental results on a test device using a-

SiC:H for the absorbing layer The influence of the surface recombination velocities at the front and bottom of the absorber has been examined. It is anticipated that better solar blindness and better over all performance may be achieved by using a-Si$_{1-x}$N$_x$:H instead of a-SiC:H.

## ACKNOLEGEMENT

This work was supported by the Deutsche Forschungsgemeinschaft under contract DFG-Bo 772/4-1.

## REFERENCES

1. C.Y.W. Huang, C. Salupo, L.F. Szabo, G.P. Cesar, W. Javurek, in Amorphous Technology, edited by A. Madan, M.J. Thompson, P.C. Taylor, P.G. LeComber Y. Hamakawa, (Mater. Res. Soc. Proc. **118**, Pittsburgh, PA, 1988) pp. 411-416.
2. Y.K. Fang, S.B. Hwang, K.H. Chen, C.R. Lui, M.J. Tsai, L.C. Kuo, IEEE Trans. Elec. Dev., **39**(2), pp. 180-185 (1992).
3. P. Maddracci, M.L. Rastello, P. Rava, F. Guiliani, F. Giorgis, Thin Solid Films **337**, pp. 232-234 (1999).
4. G. de Cesare, F. Irrera, F. Palma, M. Tucci, Appl. Phys. Lett., **67**(3), pp. 234-236 (1995).
5. F. Mütze, K. Seibel, B. Schneider, M. Hillebrand, F. Blecher, T. Lulé, H. Keller, M. Wagner, P. Rieve, M. Böhm, in *Amorphous and Heterogeneous Silicon Thin Films: Fundamentals Device Technology*, edited by H.M. Branz, R.W. Collins, H. Okamoto, S. Guha, R. Schropp, (Mater. Res. Soc. Proc. **557**, Pittsburgh, PA, 1999) pp. 815-820.
6. A. Avila, R. Asomoza, Solid-State Electron. **44**, pp. 17-27 (2000).
7. A. Srivastava, S.C. Agrawal, J. Non-Crys. Solids **227-230**, pp. 259-262 (1998).
8. J.Y. Shung, K.Y.J. Hsu, Y.L. Jiang, H.J. Tsai, Thin Solid Films, **337**, pp. 226-231 (1999).
9. S.K. Kim, Y.J. Choi, K.S. Cho, J. Jang, IEEE Trans. Elec. Dev. **46**, pp. 1001-1006 (1999).
10. B. Schneider, P. Rieve, M. Böhm, in *Handbook of Computer Vision and Applications*, ed. by B. Jähne, H. Haußecker, P. Geißler (Academic Press, Boston, 1998) pp. 237-270.
11 H. Schäfer, S. Chemnitz, S. Schumacher, V. Koziy, A. Fischer, A.J. Meixner, D. Ehrhardt, M. Böhm, presented at the Proc. SPIE Microtechnologies for the New Millennium 2003 conference, Maspalomas, Spain (2003).
12. S.M. Gadelrab, S.G. Chamberlain, IEEE Trans. Elec. Dev., **44**(10), pp. 1789-1794 (1997).
13 M. Böhm, H.C. Scheer, H.G. Wagemann, Solar Cells, **13**, pp. 29-41 (1984)
14. T. Stapinski, G. Ambrosone, U. Coscia, F. Giorgis, C.F. Pirri, Physica B, **254**, pp. 99-106 (1998).

## Correlation Between the Tunnelling Oxide and I-V Curves of MIS Photodiodes

H. Águas, L. Pereira, A. Goullet[1], R. Silva, E. Fortunato, R. Martins
Departamento de Ciência dos Materiais, Faculdade de Ciências e Tecnologia, Universidade Nova de Lisboa and CEMOP, Campus da Caparica, 2829-516 Caparica, Portugal
[1]Laboratoire des plasmas et Couches Minces, Institut des Matériaux Jean Rouxel, 2 rue de la Houssinière, 44322 Nantes cedex 3 - France

## ABSTRACT

In this work we present results of a study performed on MIS diodes with the following structure: substrate (glass) / Cr (2000Å) / a-Si:H $n^+$ (400Å) / a-Si:H i (5500Å) / oxide (0-40Å) / Au (100Å) to determine the influence of the oxide passivation layer grown by different techniques on the electrical performance of MIS devices. The results achieved show that the diodes with oxides grown using hydrogen peroxide present higher rectification factor ($2\times10^6$) and signal to noise (S/N) ratio ($1\times10^7$ at -1V) than the diodes with oxides obtained by the evaporation of $SiO_2$, or by the chemical deposition of $SiO_2$ by plasma of HMDSO (hexamethyldisiloxane), but in the case of deposited oxides, the breakdown voltage is higher, 30V instead of 3-10 V for grown oxides. The ideal oxide thickness, determined by spectroscopic ellipsometry, is dependent on the method used to grow the oxide layer and is in the range between 6 and 20 Å. The reason for this variation is related to the degree of compactation of the oxide produced, which is not relevant for applications of the diodes in the range of ± 1V, but is relevant when high breakdown voltages are required.

## INTRODUCTION

Although the most common device structure for many of known applications is still the p-i-n structure [1], the MIS (metal-insulator-semiconductor) devices [2] present the advantage of being a simpler structure with high response times, able to support high breakdown voltages and with a higher yield than the p-i-n structure. Since a-Si:H can not stand temperatures higher than 300 °C without degradation, temperatures of 900 °C used in c-Si to grow a good quality oxide by thermal oxidation cannot be used in a-Si:H. We have observed by spectroscopic ellipsometry that when the a-Si:H is placed during 2h at 300 °C in a furnace with water vapour, its surface remains unchanged. This means that thermal oxidation must be excluded, since at low temperatures the diffusion of oxygen is very slow. To form a thin layer of oxide on the a-Si:H in a short period of time we have to consider the deposition of that oxide or to grow it chemically from the a-Si:H surface. In this work both approaches have been considered: deposition of a thin layer of oxide by PECVD, and thermal evaporation of $SiO_2$ assisted by Electron Gun; oxide chemically grown by reacting the a-Si:H surface with $H_2O_2$ (immersion and pulverisation), which is one of the most powerful oxidisers known. The surface of a-Si:H acts as a catalyser for two reactions of decomposition of the $H_2O_2$:

$$\{H_2O_2 \Leftrightarrow O_2 + 2H^+ + 2e^-\} + \{H_2O_2 + 2H^+ + 2e^- \Leftrightarrow 2H_2O\} = \{H_2O_2 \Leftrightarrow \tfrac{1}{2}O_2 + H_2O\} \quad (1)$$

resulting in the formation of $O_2$ adsorbed onto the a-Si:H surface that easily reacts to form $SiO_2$.
We have investigated the role of the oxide layer composition, structure and thickness on the electrical properties of the MIS photodiodes.

## EXPERIMENTAL DETAILS

The structure of the diodes consists in glass(substrate)/Cr(2000Å)/a-Si:H(n$^+$)(400Å)/a-Si:H(i) (5500Å)/oxide/Au(100Å). The Au dots form a 7mm$^2$ contact. The Cr layer was evaporated by the Electron Gun technique. The a-Si:H was deposited by PECVD [3], with the n and i layers deposited in different chambers in order to avoid cross contamination. The oxide was formed using four different techniques: deposition of SiO$_2$ by decomposition of hexamethyldisiloxane (HDMSO) in an helicon plasma PECVD reactor using a mixture of HDMSO and O$_2$ [4]; electron gun evaporation of SiO$_2$; chemical growth of an oxide layer on the a-Si:H through immersion of the sample in a H$_2$O$_2$ solution at 60 °C and by pulverization of H$_2$O$_2$ on the heated a-Si:H surface at 150°C with an H$_2$O$_2$ spray, using N$_2$ as carrier gas. In this last technique the H$_2$O$_2$ is vaporised as it reaches the hot a-Si:H surface, forming a very textured oxide on the a-Si:H surface by a technique that can be characterized a combination of thermal and chemical oxidation.

In all the cases, the oxide thickness was determined by spectroscopic ellipsometry using a JobinYvon UVISEL-DH10 ellipsometer. The acquisition of the ellipsometric angles $\Delta$ and $\Psi$ was performed from 1.5 to 5 eV with a step of 0.025 eV. The oxide thickness was simulated using the BEMA (Bruggemann Effective Medium Approximation) [5] with a combination of surface roughness (50% a-Si:H; 50% voids) layer and on the top, a thermal SiO$_2$ reference layer. The simulated curve was fitted to the experimental one, in order to minimise the error function $\chi^2$, which represents the minimum deviation between the experimental and the simulated curves. The fitting was first performed on the as deposited a-Si:H surface, to determine its roughness and then on the oxidised a-Si:H surface, adjusting the oxide thickness to achieve a minimum of $\chi^2$.

The diodes were electrically characterized by measuring the I-V characteristic between –1 and 1 V, under dark and AM1.5 illumination conditions. The breakdown voltage was determined by increasing the negative voltage applied to the diode until the breakdown occurs. The barrier height ($\phi_B$) and the diode quality factor ($\beta$) were determined by fitting the linear part of the I-V curve under direct polarisation to the equation [6]:

$$I \approx AR^*T^2 \exp(-q\phi_B/kT)\left[\exp\left(\frac{qV}{\beta kT}\right)\right] \tag{2}$$

Since the existence of nano size pinholes in the a-Si:H cannot be fully avoided, in order to improve the yield of the workable devices, we used a method able to eliminate small leakage currents on the diodes that consists in burning pinholes by applying a negative voltage (between 1.7 to 2.5 V) to the diode until the saturation current starts reducing. By doing so, a large current density passes through the nano size pinholes and burns them so that no current passes in the pinholes after their burning. This method allowed a reduction of about three orders of magnitude on the saturation current, especially in the Schottky diodes, where no oxide barrier is present.

## RESULTS

Fig. 1 shows the I-V curves of the diodes under AM1.5 illumination. The set of data achieved show that the ideal oxide thickness that leads to the increase of the open circuit voltage $V_{OC}$ without a decrease in the short circuit current density $J_{SC}$ depends on the oxidation method and subsequently on the nature of the oxide formed. Since the oxide deposited by HDMSO (Fig. 1a) was performed in a different laboratory in Nantes, we could not avoid the presence of a sub native oxide with about 5 Å bellow the deposited oxide. Apart from that, as this technique provides a very good coverage of the surface, we found that above a total thickness of 10 Å a strong decrease of the $J_{SC}$ and FF (fill factor) is observed. In the case where the oxide was formed by electron gun evaporation of $SiO_2$ (Fig.1b) the ideal oxide thickness was between 6 and 12 Å. In this case the native oxide was previously removed with an diluted HF solution. This technique allows also the formation of a dense and compact oxide, which results obtained agree with the ones obtained with the HDMSO plasma.

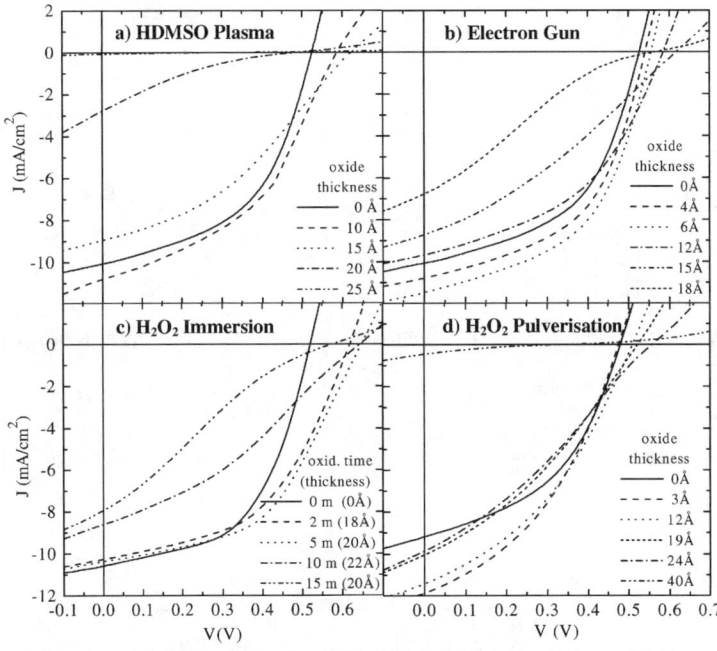

**Figure 1.** I-V curves of the diodes under AM1.5 illumination conditions with the oxide barrier formed by: a) HDMSO Plasma; b) Electron Gun; c) $H_2O_2$ Immersion; d) $H_2O_2$ Pulverisation.

When the oxide is formed by immersion of the sample in $H_2O_2$ (Fig. 1c), we noticed that the characteristics of the film produced are more dependent on the oxidation time than on the oxide thickness. In this case, short times lead to the formation of a porous oxide layer, whose porosity is reduced as the immersion time increases, until ≈ 15 min. Above this time the solution starts corroding the a-Si:H by action of the oxidation of the Cr under layer, through small pinholes in the a-Si:H film. For times longer than 20 min, we start observing the corrosion of the Cr bellow the a-Si:H, without causing any short circuit in the diodes. The $H_2O_2$ corrosion effect is also observed in Fig. 2c) where the saturation current increases after reaching a minimum for 5 min of oxidation. These results also show that for MIS diodes the density of the oxide is not an important factor. What is important is the passivation of the a-Si:H surface in which the chemical technique is very effective because the oxide is grown at expenses of the a-Si:H. This was the oxidation technique that allowed the achievement of the highest $V_{OC}$ without decreasing $J_{SC}$. Nevertheless, these results were achieved causing an increase of the series resistance of the diode, as observed for all the other oxidation techniques. The $H_2O_2$ pulverization technique (Fig.1d) is very effective in increasing $J_{SC}$. Curiously, the initial increment of $J_{SC}$ is not followed by an increase in $V_{OC}$, as observed in the other cases. What happens in this case is that the first layers of the oxide formed cause an increase of the surface roughness but not a passivation, promoting so light trapping at the diode surface, which leads to the enhancement of $J_{SC}$. The observed decrease in FF after the oxidation treatment can be due to the thermal gradient experienced by the sample during the oxidation process that may had cause some damage to the structure.

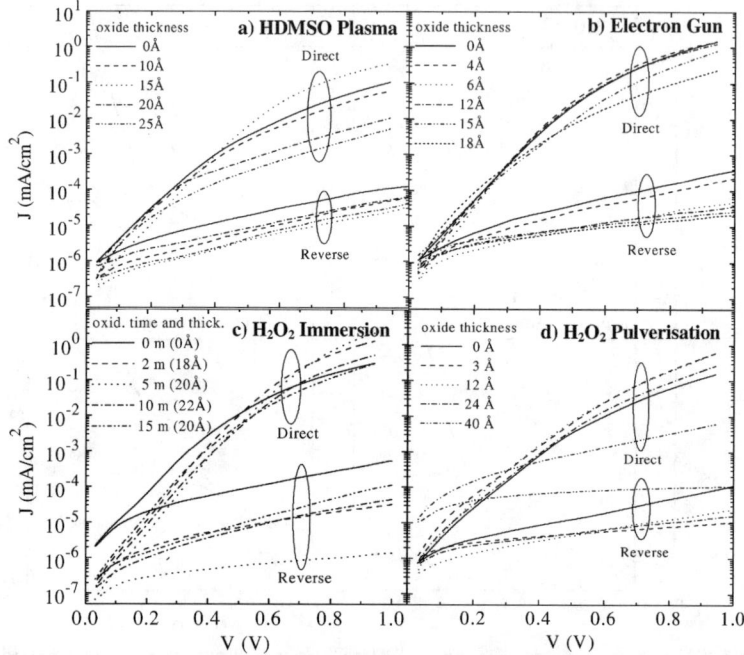

**Figure 2.** I-V curves of the diodes under dark conditions with the oxide barrier formed by: a) HDMSO Plasma; b) Electron Gun; c) $H_2O_2$ Immersion; d) $H_2O_2$ Pulverisation.

Fig. 2 shows the I-V curves of the diodes under dark conditions where the upper curves correspond to the direct polarisation and the lower curves to the reverse polarisation. Table I compares the values of the rectification factor (R.F.) i.e. the ratio between de direct and the reverse current at ± 1V respectively. We observe that the oxide barrier increases the R.F. by at least one order of magnitude, due to the decrease of the saturation current ($I_S$) and the increase of the direct current, caused by the surface passivation, which leads to the decrease of surface defects and shunt paths. Nevertheless, thicker oxides cause a decrease in the current, since carriers cannot tunnel through the oxide. The best results were achieved using the $H_2O_2$ immersion technique (Fig. 2c), but this technique is the most difficult to control, since an excess of oxidation time leads to the device degradation, due to the corrosion phenomenon.

On the other hand, the highest breakdown voltages (30V) were achieved using the oxide deposition techniques due to the effective shunt suppression caused by them.

Fig. 3 shows the difference between the reverse current of the diodes under AM1.5 illumination and dark conditions. Table I compares the signal to noise ratio (S/N) i.e. ratio between the illuminated and dark current of the diodes at 0 and -1V of polarisation.

**Table I.** Comparison of R.F. and S/N of diodes with different types of oxide barriers.

| Oxidation Method | R.F. (1.0V) Schottky | R.F. max. (1.0V) MIS | S/N (1.0V) Schottky | S/N max. (1.0V) MIS | S/N (0.0V) Schottky | S/N max. (0.0V) MIS |
|---|---|---|---|---|---|---|
| HDMSO Plasma | $1.1 \times 10^3$ | $1.5 \times 10^4$ | $9.2 \times 10^4$ | $3.8 \times 10^5$ | $1.0 \times 10^7$ | $6.3 \times 10^7$ |
| Electron Gun | $5.8 \times 10^3$ | $1.0 \times 10^4$ | $3.3 \times 10^4$ | $5.6 \times 10^5$ | $1.1 \times 10^7$ | $5.2 \times 10^7$ |
| $H_2O_2$ Immersion | $7.0 \times 10^2$ | $2.2 \times 10^6$ | $2.5 \times 10^4$ | $1.0 \times 10^7$ | $9.1 \times 10^6$ | $2.6 \times 10^8$ |
| $H_2O_2$ Pulver. | $1.8 \times 10^3$ | $1.0 \times 10^5$ | $1.3 \times 10^5$ | $1.5 \times 10^6$ | $1.8 \times 10^7$ | $4.2 \times 10^7$ |

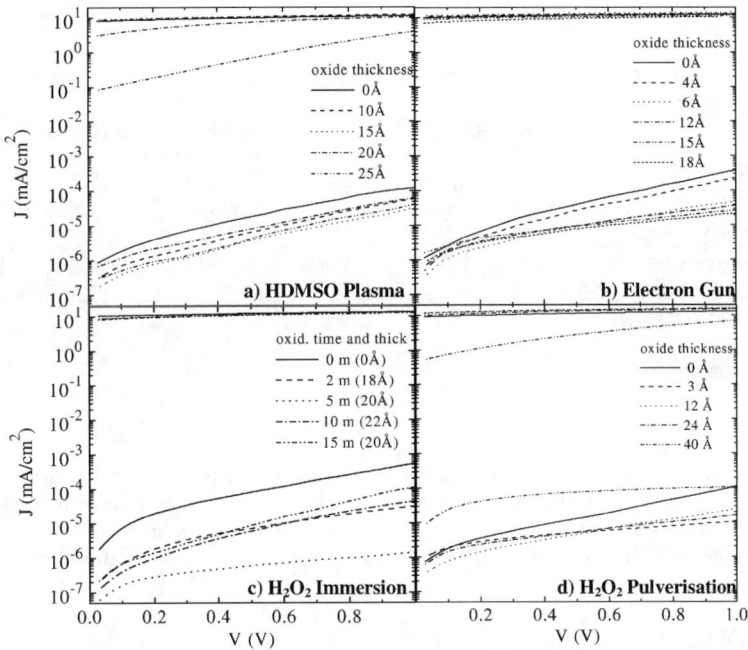

**Figure 3.** I-V curves of the diodes inversely polarised under dark and AM1.5 illumination conditions with the oxide barrier formed by: a) HDMSO Plasma; b) Electron Gun; c) $H_2O_2$ Immersion; d) $H_2O_2$ Pulverisation.

The data show that S/N is mostly sensitive to the dark current of the diodes. In average, the oxide layer barrier allows an increase of S/N between one to two orders of magnitude, being the highest increment achieved for the oxide grown by $H_2O_2$. Here, we also notice the lowest dependence of the reverse current on the bias, which is an indication of low defect interface.

Finally, table II shows the values of $\phi_B$ and $\beta$ of the diodes before (Schottky) and after the oxide formation process (MIS). We observe that the increase in light conversion efficiency and R.F. is only associated to a small increase of $\phi_B$. The reason for that is the high quality of the a-Si:H. Indeed, the Schottky diodes have already a high $\phi_B$ value, above the usually 0.7 eV found in literature [7]. On the other hand, $\phi_B = 1$ eV presented by the MIS devices is close to the highest values found in literature [7,8] for high quality MIS diodes. It is also seen in literature that high values of $\phi_B$ can be found in both Schottky and MIS diodes [8], being this basically related to the defects of the a-Si:H/Au or a-Si:H/oxide/Au interfaces. The data also show that the oxide has a more important role in determining the $\beta$ value, leading to a significant decrease to values close to the 1.5. That is, to a low role of the recombination/losses in the carriers transport mechanism. This behaviour is mostly due to the reduction of small shunt paths in the a-Si:H, promoted by the oxide formation. Nevertheless, comparing the different techniques used to grow the oxide layers, we do not observe significant changes of $\phi_B$ and $\beta$ values.

**Table II.** Comparison of the barrier height ($\phi_B$) and quality factor ($\beta$) of the diodes with the oxide barrier formed by the different methods.

| Oxidation Method | BEFORE OXIDATION | | AFTER OXIDATION (MIS) | | | |
|---|---|---|---|---|---|---|
| | $\phi_B$ (eV) Schottky | $\beta$ Schottky | $\phi_B$ (eV) (max eff) | $\phi_B$ (eV) (R.F. max) | $\beta$ (max eff.) | $\beta$ (R.F. max) |
| HDMSO Plasma | 0.98 | 1.96 | 1.01 | 1.02 | 1.64 | 1.68 |
| Electron Gun | 0.97 | 1.70 | 1.01 | 0.99 | 1.42 | 1.56 |
| $H_2O_2$ Immersion | 0.96 | 2.00 | 1.05 | 1.05 | 1.57 | 1.57 |
| $H_2O_2$ Pulver. | 0.98 | 1.97 | 1.00 | 0.98 | 1.42 | 1.58 |

## CONCLUSIONS

This works presents a comparative study between the performances of MIS diodes where the oxide layers were created by different techniques.

The diodes with the oxide layer chemically grown by $H_2O_2$ present better performances, due to a better surface passivation. On the other hand, the good surface coverage promoted by the oxide deposition techniques allows the achievement of higher breakdown voltages (30V). In all the cases the MIS diodes show an improvement in their properties in relation to their Schottky structures.

## ACKNOWLEDGEMENTS

The Portuguese authors would like to thank the financial support given by "Fundação para a Ciência e a Tecnologia" through pluriannual contract with CENIMAT, as well to the projects POCTI/CTM/35440/2000 and POCTI/ESE/38924/2001

## REFERENCES

[1] E. Fortunato and R. Martins, Solid State Phenomena **44-46,** 883 (1995).
[2] S. Arimoto, H. Yamamoto, H. Ohno, H. Hasegawa, Electronics Letters **19**, 628 (1983).
[3] H. Águas, E. Fortunato, V. Silva, L. Pereira, R. Martins, Thin Sol. Films **403-404,** 26 (2002).
[4] K. Aumaille, C. Vallée A. Granier, A. Goullet, F. Gaboriau, G. Turban, TSF **395,** 188 (2002).
[5] D.A.G. Bruggemann, Ann. Phys. (Leipzig) **24**, 636 (1935).
[6] S.M. Sze, *Physics of Semiconductor Devices,* Jonh Wiley, New York, cap. 5 (1981).
[7] E. Fortunato, A. Malik, A. Sêco, A. Maçarico, R. Martins, MRS. Symp. Proc. **467**, 949 (1997)
[8] K. Maeda, W. Chiyoda, I. Umezu, A. Kuroe, J. Appl. Phys. **75**, 3522 (1994).

## Numerical Simulation of the Influence of the Gap State of a-Si:H on the Characteristics of a-Si:H p-i-n/OLED Coupling Device

Chunya Wu[1], Yousu Chen[1], Juan Li[1], Guanghua Yang[1], Huidong Yang[1,2], Zhenhua Zhou[1], Ying Zhao[1], Zhiguo Meng[1], Xinhua Geng[1], Shaozheng Xiong[1] and Lizhu Zhang[3]
[1] Inst. of Photo-electronic Thin Film Devices and Technology,
Nankai University, Tianjin 300071, China
[2] Inst. Of Thin Film & Nano-material,
Wuyi Univ., Jiangmen, Guangdong 529020, P. R. China
[3] Tianjin Mechanical and Electronic Vocational Technical School, Tianjin, China
**Qi Wang**
National Renewable Energy Laboratory. 1617 Code Boulevard,
Golden, Colorado 80401-3393, USA

### ABSTRACT

The influence of the density of gap states and the band gap width of the intrinsic a-Si:H active layer on the characteristics of a-Si PIN/OLED coupling pair was analyzed by a-Si:H PIN/OLED CAD simulation model. The CAD simulation model was carried out based on a-Si PIN Hack & Shur model and OLED TCL transport model. At the same band gap width, for the intrinsic a-Si:H active layer with the higher density of gap states, the reverse current of a-Si PIN trended to be saturated at the higher reverse bias voltage. As a result, I-V curve of a-Si PIN/OLED around the turn point Vt became smoother with the increase of the density of gap states. At the same state density, the light induced current of a-Si PIN increased against the band gap width, assuming the input light had the same spectrum as AM1.5 solar light. Thus the luminance emitted from OLED increased with the decrease of the band gap width because OLED belongs to the light-emitting device controlled by current. The simulation results also showed that the influence of the state density intensified with the increase of the band gap of a-Si:H.

### INSTRODUCTION

Large area a-Si PIN/OLED image/sensor devices were reported in our previous paper [1]. The structure of one pixel is an a-Si PIN connecting in series but in opposite polarity with OLED. The a-Si PIN photodiode is used as a light sensor and converts the input light into a photocurrent. The photocurrent will be directly coupled to OLED which is biased in forward, and drive OLED to emit visible light. The luminance of the output light emitting from OLED is proportional to the intensity of the input light irradiating on a-Si PIN under the suitable voltage bias. So the matrix of such a-Si PIN/OLED can be used as optical spectrum converter, light-switch or space light modulator besides image sensor/display device. In this paper, the influence of the gap state density and the band gap of the intrinsic a-Si:H active layer on the characteristics of a-Si PIN/OLED coupling pair was analyzed by a-Si:H PIN/OLED CAD simulation model.

## SIMULATION MODEL

The CAD simulation model was carried out based on a-Si PIN Hack & Shur model and OLED TCL transport model. The opto-electronic characteristics of a-Si PIN were described with the Poisson's equation and the continuity equations for electrons and holes in one dimension space[2]. The standard difference methods were used to approximate the space derivatives in Poisson's equation and the continuity equations. Newton iteration was employed to solve the difference questions. Moreover, the vector modulus limitation was introduced to improve the solution astringency.

The input parameters of a-Si PIN CAD model, programmed in C Language, consist the structure parameters (such as the P, I, and N layer thickness), the density of state in the band gap, the opto-electronic characteristics and the work conditions (such as light intensity, light spectrum, working temperature and so forth). It can simulate J-V characteristics, as well as the space distributions of hole, electron, field, trapped carrier, and recombination rate in a-Si PIN.

Both tunneling dominated and bulk dominated mechanisms have been proposed to explain the carrier transport in OLED. Considerable debates continue over which model is appropriate, in particular among F-N tunneling and trap charge limited current (TCL) [3-5]. Our previous experiment results showed that the carrier transport mechanism of OLED I-V characteristics was dominated by the low field contact characteristics [6]. However it was found that the J-V curves of the state-of-the-art OLEDs or PLEDs fit well with TCL model in the progress of OLED research. Therefore TCL model was adopted to describe the carrier transport in OLED in this paper.

Based on J-V characteristics of a-Si PIN and OLED, the J-V of a-Si PIN/OLED coupled device is calculated with MathCAD 2000 by analyzing the equivalent circuit shown in reference 2.

## SIMULATION RESULT AND DISCUSSION

The density of gap states and the width of band gap are the major factors, which impact on the quantum conversion efficiency and reverse saturation current of a-Si PIN light sensor. The CAD simulation model mentioned above was used to analyze their influence on the J-V characteristics of a-Si PIN/OLED.

In the simulation model, the donor gap state $g_d(E)$ and the acceptor gap state $g_a(E)$ are supposed to be the exponential function of the electron energy, i.e.,

$$g_a(E) = \text{gmin}_a \times \exp((E - E_{mc})/E_a) \quad (1)$$

$$g_d(E) = \text{gmin}_d \times \exp(-(E - E_{mc})/E_d) \quad (2)$$

$$g\min_a = g\min_d = g\min/2 \quad (3)$$

$$\text{gmin}(N_n) = g\min(0) + K1 \times (N_n/K2)^{1/2} \quad (4)$$

where gmin(0) is the minimum state density in the deep energy state. $N_n$ is the doping concentration. K1 and K2 are two constants, K1 is about $10^{16}\text{cm}^{-3}\text{eV}^{-3/2}$ and K2 is about $10^{16}\text{cm}^{-3}\text{eV}^{-1}$ for a-Si:H thin film. $E_a$, $E_d$ and $E_{mc}$ have the different values for P, I and N type Si based thin films [7].

In this paper, the value of gmin(0) and $E_g$ of the intrinsic layer were varied to simulate the effect of the gap state, $E_{mc}$ of the intrinsic layer was kept at the half of $E_g$. As to the parameters describing the gap state distribution in P and N layer, only gmin(0) varied at the same scale as that of the intrinsic layer. All the other parameters were fixed in the simulation. The input light irradiating on a-Si PIN was assumed to have the same Global spectrum of AM 1.5, but its energy density was supposed to be 10mW/cm².

The J-V curves of a-Si PIN and the corresponding J-V curves of a-Si PIN/OLED were shown in Fig 1, Fig 2, Fig 3, Fig 4 and Fig 5 at Eg of 1.6ev, 1.65ev, 1.70ev, 1.75ev and 1.80ev with gmin(0) of 1e15cm⁻³, 5e15cm⁻³, 1e16cm⁻³, 5e16cm⁻³ and 1e17cm⁻³ respectively.

The simulation results showed that, with the variation of gmin(0) from 1e15cm⁻³ to 1e17cm⁻³ at the same band gap, the J-V curve of a-Si PIN trended to be saturated at the higher backward voltage bias, the filling factor and the open circuit voltage decrease with the increase of gmin(0). More and more light-induced carriers were trapped in the gap state, the space field distributed in a-Si:H PIN was reduced, and the quantum collection efficiency of the light-induced carriers was also reduced at the same backward bias voltage against the gap state density. As a result, with the increase of gmin(0), J-V curves of a-Si PIN/OLED around the turn point Vt became smoother, as

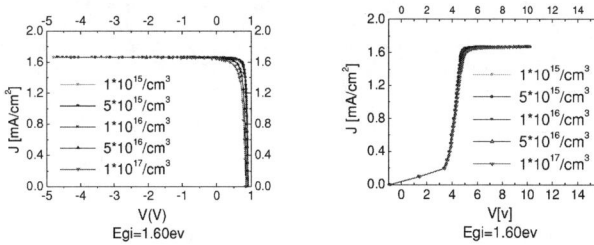

Fig.1 Simulated J-V curves of a-Si PIN (a) and a-Si PIN/OLED (b) with the different gap state densities shown in Fig 1-b through Fig 5-b. The current density became linear to the bias voltage at higher

Vt. As a image sensor/display device a-Si PIN/OLED had better work in the linear region. So the power consumption would be increased with the gap state density.

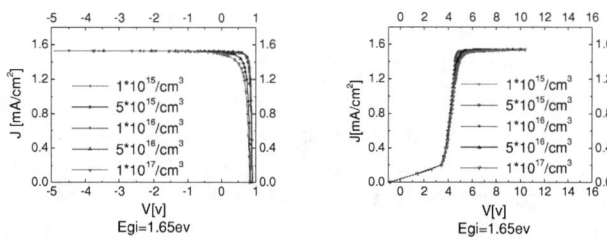

Fig.2 Simulated J-V curves of a-Si PIN (a) and a-Si PIN/OLED (b) with the different gap state densities at Egi=1.65eV

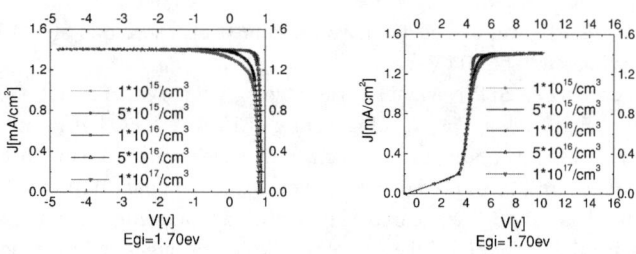

Fig.3 Simulated J-V curves of a-Si PIN (a) and a-Si PIN/OLED (b) with the different gap state densities at Egi=1.70eV

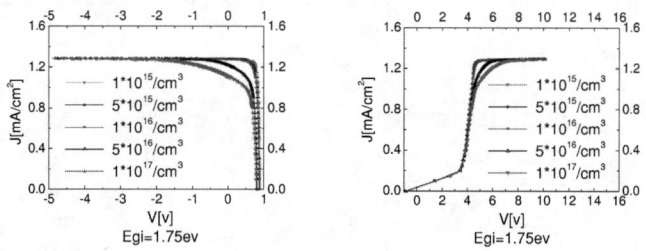

Fig.4 Simulated J-V curves of a-Si PIN (a) and a-Si PIN/OLED (b) with the different gap state densities at Egi=1.75eV

The simulation results also showed that the influence of the density of gap states intensified with the increase of Eg of a-Si:H (seeing Fig.1 through Fig.5). In the other words, the impact of the density of gap states on J-V curve of a-Si PIN/OLED in the saturated region at the narrower band gap was less than that at the wider band gap. Obviously this impact intensification is owned to the increase of the sum of the trap states.

At the same density of gap state, the light induced current of a-Si PIN increased with the decrease of the band gap width, while the input light has the same spectrum as AM1.5 solar light and the same energy density of 10mW/cm$^2$, as shown in Fig.6 and Fig.7. Besides the sum of the trap states was reduced, the band gap narrowing led to the absorption limit red-shifted. More photons of the light would be absorbed by a-Si PIN and more light induced carriers would be excited, which resulted in the increase of the reverse saturation current. Hence the luminance of OLED increased with the decrease of the band gap width since OLED the current controlled light-emitting device whose luminance is linearly proportional to the driving current density. This simulation results also implied that the sensitivity of a-Si PIN/OLED could be improved by utilizing the intrinsic Si-based thin film with the narrow band gap.

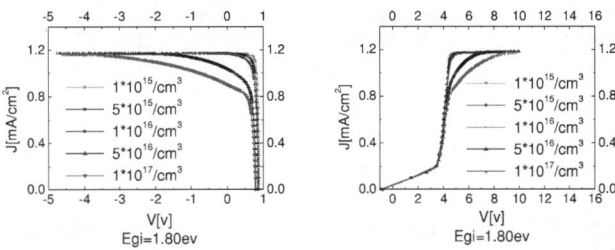

Fig.5 Simulated J-V curves of a-Si PIN (a) and a-Si PIN/OLED (b) with the different gap state densities at Egi=1.80eV

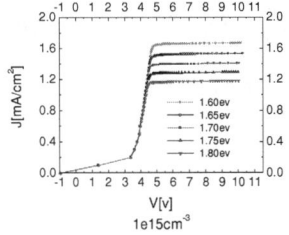

Fig.6 Simulated J-V curves of a-Si PIN/OLED at gmin(0)=1×10$^{15}$cm$^{-3}$

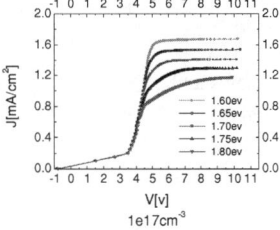

Fig.7 Simulated J-V curves of a-Si PIN/OLED at gmin(0)=1×10$^{17}$cm$^{-3}$

## CONCLUSION

The gap state density and the gap band width of the intrinsic a-Si:H active layer affect the opto-electronic characteristics of a-Si PIN, thereby affect the performance of a-Si PIN/OLED coupling device. At the same gap band width, for the intrinsic a-Si:H active layer with the higher gap state density, the reverse current of a-Si PIN trended to be saturated at the higher reverse bias voltage. J-V curve of a-Si PIN/OLED around the turn point Vt become smoother with the increase of the gap state density. At the same gap state density, the light induced current of a-Si PIN increased against the band gap width. Thus the luminance emitted from OLED would be enhanced with the decrease of the band gap width because OLED is the current controlled device. The influence of the gap state density intensified with the increase of the gap band width of a-Si:H. The opto-electronic characteristics of PIN/OLED image sensor/display device, such as the sensitivity, the quantum conversion efficiency and the power consumption, would be improved by utilizing the Si-based thin film with the low gap state density and the narrow gap band.

## ACKNOWLEDGEMENT

This paper was supported by china's NSFC (Proj. No. 60077011, 69907002), China's 863 plan (Proj. No. 2002AA303261), and Tianjin's Nature Science Funding (Proj. No. 023602011)

## REFERENCES

1. C. Wu, Y. Zhao, S. Xiong, E. Liu, W. Xie, L. Rong, and H. Cheng, "Design of a Novel a-Si PIN/OLED Image Sensor & Display Device", SID 99, pp.528-531, (1999).
2. Chunya Wu, Ying Zhao, et. al., "A Simulation Method for a-Si PIN/OLED coupling devices", Proceedings ASID 2002, The 7$^{th}$ ASID Symposium on Information Display, pp.217-221, 2002.
3. I. D. Parker, J. Appl. Phys., Vol.75, No.3, pp. 1656-1666 (1994).
4. A. J. Campbell, D. D. C. Bradley, and D. G. Lidzey, J. Appl. Phys., Vol.82, No.12, pp.6326-6342 (1997).
5. P. E. Burrows and S. R. Forrest, Appl. Phys. Lett., Vol.64, No.17, pp.2285-2287 (1994).
6. Shao-zhen Xiong, Ying Zhao, Chunya Wu, Yun Hao, Yousu Chen, Huidong Yang, zhenhua Zhou, and Gang Yu, Chinese Journal of Semiconductor, Vol.22, No.9, pp.1176-1181 (2001).
7. M. Hack and M.Shur, J. Appl. Phys., Vol.58, No.2, pp.997-1020 (1985).

## Low-temperature Growth of Poly-Si and SiGe Thin Films by Reactive Thermal CVD and Fabrication of High Mobility TFTs over 50 cm$^2$/Vs

Jun-ichi Hanna and Kousaku Shimizu
Imaging Science and Engineering Laboratory, Tokyo Institute of Technology,
Nagatsuta, Midori-ku, Yokohama, 226-8503, Japan

## ABSTRACT

We have established a new thermal CVD technique, *Reactive Thermal CVD*, for polycrystalline silicon (poly-Si) and silicon germanium (poly-SiGe) thin films aiming at thin film transistors (TFTs) applications, in which a low substrate temperature of 450°C enables us to use glass substrates. This technique achieved high crystallinity at very early stage of the film growth, resulting no amorphous incubation layer on the substrate surface. We fabricated bottom and top gate n-and p-channel TFTs with these of 200 nm thick films on SiO$_2$/Si wafers and glass substrates, respectively: the high field effect mobilities as high as 55 cm$^2$/Vs and 25 cm$^2$/Vs were achieved in the bottom-gate and top-gate TFTs, respectively. Here, we discuss the technical requirements in the low-temperature CVD technique for the large-area poly-Si thin films and how they can be achieved in the reactive thermal CVD.

## INTRODUCTION

In these 10 years the amorphous silicon TFTs have been well established as an active matrix for liquid crystal display (LCD) devices for computer outputs, and more recently for TV monitors. Due to the increasing demand for large-area and high definition LCDs, the needs for the switching elements faster than amorphous silicon TFTs, whose switching speed is limited by its mobility of 0.1~1 cm$^2$/Vs, are coming out. In addition, the monolysic fabrication of TFT array and the system circuits including scanning and signal drivers is launched in small-size quality displays for improving reliability and yields, and for the cost reduction. Furthermore, new needs for high mobility TFTs has emerged to realized the active matrix for organic light emitting displays, which request high current density for operation. Thus, there is an increasing demand for poly-Si TFTs whose mobility exceeds 100 cm$^2$/Vs.

Nowadays, the poly-Si thin films on the glass substrate for TFT applications are fabricated in two techniques, that is, excimer laser annealing and solid phase crystallization of a-Si thin films. As for the laser annealing technique, there remains a very serious problem of the cost-expensive production due to high installation and maintenance costs of the excimer laser systems, in addition to a low yield due to inhomogeneous crystallinity of the polycrystalline films caused by a laser scanning process. On the other hand, a high process temperature over 600°C and a long process time over tens hours are a big problem in the solid phase crystallization of amorphous silicon films. With the aid of metal catalysts such as Ni the crystallization process is improved very much, but the residual catalyst is harmful to off-current characteristics and a long-term reliability of the TFTs.

Furthermore, these on-going techniques need 3 steps to fabricate poly-Si films, which includes deposition of a-Si:H films by either plasma enhanced CVD or low pressure CVD (LP-CVD), dehydrogenation of the films by thermal treatment, and crystallization of the films by using either an excimer laser or a furnace.

In industrial fabrication point of view, the direct deposition of device-grade poly-Si thin films on the glass substrate is an ultimate goal for fabrication of poly-Si thin films in TFT applications.

Several CVD techniques have been proposed for the poly-Si thi films on the glass substrate so far, which include plasma-enhanced CVDs with silane and/or fluorosilane diluted with hydrogen [1, 2, 3, 4], Hot-wire CVD with a hydrogen-diluted silane [5], and reactive CVD featuring gas phase reaction of silane with fluorine [6]. However, no technique provides us with device-grade poly-Si thin films of a few sub-μm for TFT applications. In fact, the highest field effect mobility of 45 $cm^2$/Vs in n-channel top-gate TFTs was achieved, but 700nm thick poly-Si thin films is requested, which is prepared at $450^{\circ}$C by rf-plasma CVD with a mixture of silane and silicon tetrafluoride, and the mobility went down to 20 $cm^2$/Vs when the 200nm thick poly-Si films were employed [7, 8]. This is due to the poor crystallinity of the film deposited at the early stage, which is originated from the delayed nucleation on the substrate. Indeed, the film at the early stage of the deposition often accompanies an amorphous incubation layer. Thus, the resulting thick films exhibit inhomogenous crystallinity along with film deposition naturally, as shown in a transmittance electron microscope (TEM) image as shown in Fig. 1.

One can expect enhanced crystallinity of the film at the early stage by adopting any deposition condition promoting the crystallization intentionally. Indeed, this results in improved crystallinity of the film, but this causes the reduction of the grain size at the same time because of increased nucleation density at the early stage as shown in a schematic illustration of Fig. 2. Thus, we are in a dilemma whether to enhance the crystallinity or to give priority to grain size in the film. Only way we can solve this dilemma is to establish direct formation of nuclei on the substrate surface and grow them to the grains. This makes it possible to do well both in enhancement of grain size and crystallinity in the films, because the grain size is determined by an initial nucleation density on the substrate surface, as shown in the schematic illustration of Fig. 3.

Therefore, we have to establish such a CVD technique that enables us to grow the homogeneous crystallinity film from the bottom to the top-surface, which assures the high crystallinity in a very thin film less than 100nm suitable for TFT application, in addition to the large-area uniformity of the films. Furthermore, we can enhance its effectiveness, if we achieve a high deposition rate, the suppressed powder formation, and no deposition on the reactor walls at the same time.

**Fig. 1** Typical TEM image of low-temperature CVD poly-Si thin film on the glass substrate

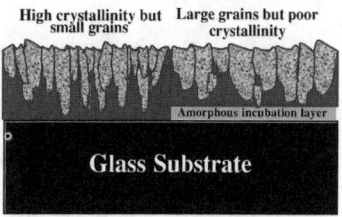

**Fig. 2** Ill Effect of promoting crystallization at the early stage of the film growth

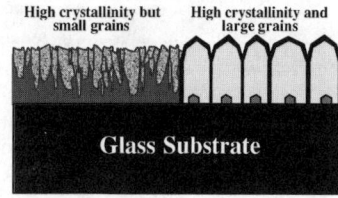

**Fig.3** Conventional and Ideal growth of poly-Si thin film for TFT application

## CVD TECHNIQUE FOR LOW-TEMPERATURE DEPOSITION FOR LARGE-AREA THIN FILMS

In order to establish such the CVD technique, we have to go back to the principle for low-temperature growth of the film in the CVD process and find a new way for that. Fig. 4 shows a conceptual diagram of chemical processes in CVD film growth: for the film growth, the source materials are decomposed into fragments, which are reactive species and film precursors; then, they are transported to the surface and chemisorbed on the surface; finally the chemisorbed film precursors relax into the network structure of growing film while the terminators are released. In these processes, the fragmentation of the source materials is the most energy consumption process, so that we

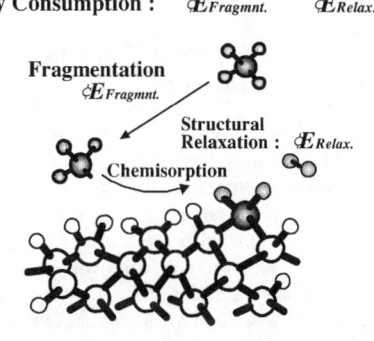

Fig.4  Conceptual diagram of the CVD process

can reduce the deposition temperature by activating the source materials to be decomposed with any kind of energy away from the substrate. This is a basic principle for reducing the growth temperature of films in the CVD process. In plasma enhanced CVD, which is the most popular low-temperature CVD technique, the source materials are decomposed by their elastic collision with a high-energy electron in the plasma. On the other hand, they are decomposed in contact with a high temperature filament in Hot-wire CVD and by light absorption in photo-CVDs. In fact, the considerable reduction of the film growth are achieved in these CVD techniques, but they have several disadvantage such as the ill effects of secondary reaction of the reactive species on the film quality, the powder formation, and the film deposition on the reactor wall, which are very serious especially in the large-area film deposition in terms of both a high yield and a high throughput. All these are attributed to the formation of the reactive species, i.e., the film precursors in the gas phase, which are inevitable because the generation of the reactive species is the principle for reducing the deposition temperature of the films in these CVD techniques.

One of the ways to escapee from the harmful effects of the secondary reactions associated with the reactive film precursors is to decompose the source materials in the limited region in the vicinity of the substrate surface. In this point of view, the thermal CVD is an ideal technique for the large-area film deposition, because the film precursors are generated in a very limited region close to the heated substrate surface. This technique leads to a large-area uniform film without film deposition on the wall and the powder formation as far as the substrate temperature is well maintained uniformly in a cold-wall type of reactor.

In the conventional approach to reduce the deposition temperature in the thermal CVDs, the major effort has been made to develop a new source material having a low pyrolitic temperature: the most successful example is the adoption of organometalic compounds in the metal-organic CVDs (MOCVDs). Poly-Si thin films are conventionally deposited at 600°C or higher by LP-CVDs, i.e., a kind of the thermal CVD, with silane as a source material. In fact, there have been several source materials for Si films with low pyrolitic temperatures lower than 500°C already: silane starts to decompose at around 500°C, disilane at around 400°C, and trisilane at around 350°C, but the problem is that no polycrystalline films deposit at the temperatures lower than 600°C, even if we adopt these source gases such as disilane and trisilane. This indicates that there must be something additional to promote the crystallization of the film deposited.

## REACTIVE THERMAL CVD FOR POLY-SiGe THIN FILMS [9-21]
### Materials preparation

We have proposed a new approach to overcome this problem, which we call *reactive thermal CVD*. In this technique, we feature a set of reactive gases as source materials, which are decomposed by thermally accelerated chemical reactions between the source materials selected, while the source materials are decomposed pyrolitically in the conventional thermal CVD. Therefore, the growth temperature of the films in this CVD is determined not by the pyrolitic temperature of the source materials but by a temperature where their chemical reactions become appreciable. This new technique has several advantages over the conventional low-temperature CVD processes including plasma CVDs and the Hot-wire CVD, e.g., less powder formation and no film deposition on the reactor wall, in addition to no need for external energy sources to activate the source materials and the easy technical basis backed by chemical engineering. Fig. 5 shows a schematic diagram of the reactor we have used. It is very simple and consists of a shower head type of nozzle, a substrate platform resistively heated, and a vacuum system, in addition to the load-lock chamber for substrate loading.

We selected disilane and germanium tetrafluoride as a set of source materials described above. This source gas system has unique features from chemical point of view: first of all, these gases can be a red-ox system and reactive, but the reaction rate is very low at ambient temperature and the appreciable film deposition takes place over 300°C, where germanium tetrafluoride is reduced into germanium with disilane and disilane is oxidized into silicon with germanium tetrafluoride: furthermore, disilane is decomposed thermally to afford amorphous silicon over 400°C, while germanium terafluoride hardly decomposes thermally even over 1000°C; germanium tetrafluoride exhibits etching activity for silicon and

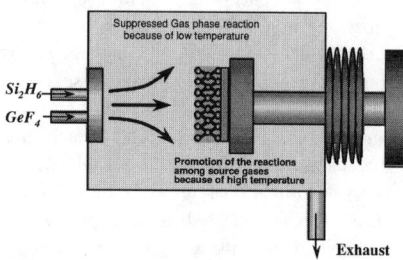

**Fig.5** Schematic diagram of a reactor for reactive thermal CVD

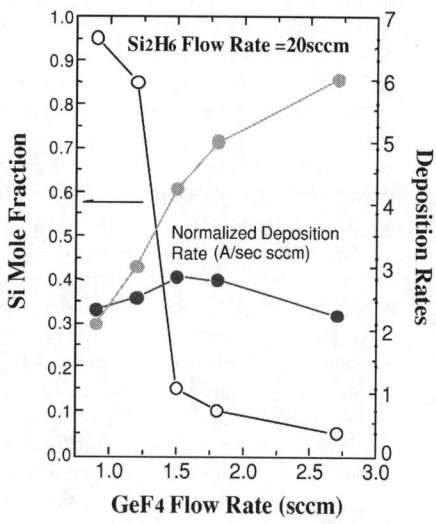

**Fig. 6** Example of film conten and froeth rate as a function of GeF$_4$ flow rte at 450°C

germanium. Indeed, all these properties are very much concerned with the growth of polycrystalline films in this CVD.

In this gas system, we can growth a wide range of films in terms of Si content of the films, from Si-rich film over 95 atm % of Si to Ge-rich films over 95atm % of Ge as shown in Fig.[6], as in the case of the conventional thermal CVD with disilane and germane ($GeH_4$). In order to deposit Si-rich film for TFT applications, a high ratio of disilane to germanium tertafluoride, e,g., over 20 is required typically.

In addition, the growth temperature over 400°C is a must requirement, while the Ge-rich films can be grown even at 320°C.

Thanks to a thermal CVD-based film growth, we can tune the growth condition with a great flexibility. A typical growth conduction fop Si-rich films is shown in Table 1. With an appropriate selection of gas flow rates including a flow rate of carrier gas, He, we can deposit the Si-rich polycrystalline films in a very wide range of the pressure: Fig. 7 shows two series of Raman spectra of the films prepared at different pressures. Although the film crystallinity is improved in general when the film is deposit at lower pressure, we can deposit a high crystallinity Si-rich film even at 5Torr where the substrate temperature is easy to control uniformly. Fig.[8] shows the TEM images of the Si-rich films prepared at 0.45 Torr and 5 Torr. The film thickness is about 200nm and 400 nm, respectively. It is clear that the polycrystalline film start to grow just from the interface of the substrate irrespective of the pressure, indicating that the nucleation takes place just on the substrate surface. In fact, we could observed the isolated nuclei on the substrate at very early stage of the film growth. The film textures arte quite different from those of the films prepared by conventional low-temperature CVDs including plasma enhanced CVDs and rather similar to those prepared over 600°C by LP-CVDs. [22] The growth rate is

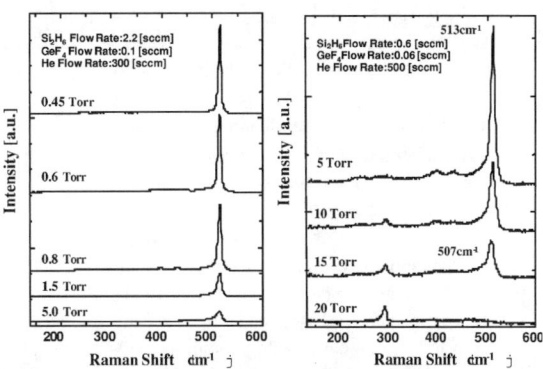

**Fig.7** Raman spectra of poly-SiGe thin films deposited at different gas flow rates as a function of pressure.

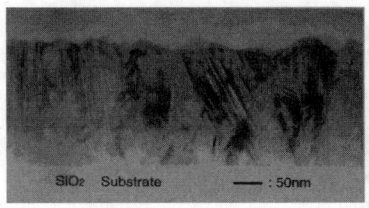

$Si_2H_6/GeF_4/He=2.2/0.1/500$ sccm
Press.= 0.45 Torr

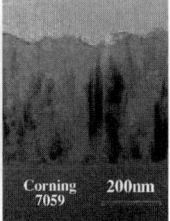

$Si_2H_6/GeF_4/He$
=3/0.03/3000 sccm
Press.= 5 Torr

**Fig.8** TEM images of typical poly-SiGe thin films prepared at low-and-high pressure conditions at 450°C.

**Table 1.** Typical growth conditions for Si-rich poly-Si films

|  | Low-pressure condition | High-Pressure condition |
|---|---|---|
| Pressure | 0.5 Torr | 5 Torr |
| $Si_2H_6/GeF_4$ | 2 sccm/0.2 sccm | 3 sccm/0.03 sccm |
| He | 500 sccm | 1000~4000 sccm |

## TFT fabrication

We have fabricated n- and p-channel bottom-and top gate TFTs with the Si-rich poly-SiGe tihn films of 200 nm prepared at 450°C. In the TFT fabrication, the post hydrogen treatment is a must process as in the case of on-going TFT fabrication using poly-Si thin films prepared by the laser annealing and solid state crystallization of amorphous silicon films, because the deposition temperature in this process is 450°C where the Si-H and Ge-H bonds are thermally dissociated. In fact, the as-deposited films have a high spin density attributed to Si and Ge dangling bonds over ~$10^{17}$ cm$^{-3}$, which is evaluated by electron spin resonance (ESR) measurements. Interestingly, we found that the spin density including both Si and Ge dangling bonds exhibits a linear relation as a function of the grain size of poly-Si films.

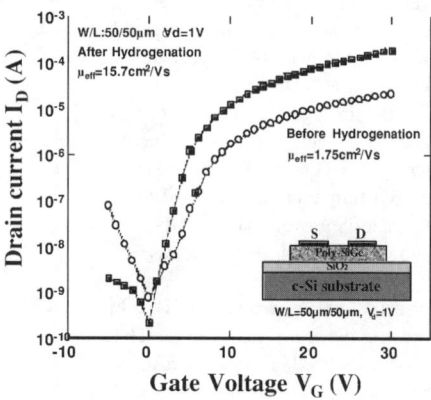

**Fig. 9** Effect of hydrogen treatment on n-channel bottom-gate TFT characteristics

This indicates that the dangling bonds are localized at the interface of the films including grain boundaries, which is a good sign for the TFT applications because it suggests that the defects states associated with the dangling bonds can be terminated by post hydrogen treatment effectively. In fact, we could confirmed this with a comparison of TFT characteristics before and after the hydrogen treatment as shown in Fig. 9. We adopted Hot-wire technique for the post hydrogen treatment for TFTs fabricated. The field effect mobility in n-channel bottom-gate TFTs, which is fabricated with 200 nm thick Si-rich poly-Si thin film s on a thermally oxidized SiO$_2$/Si wafer, is a very small and 1.75 cm$^2$/Vs before the hydrogen treatment. However, it is much improved up to 15.7 cm$^2$/Vs after the hydrogenation with the hot-wire technique for 1hr min at 250°C.

On the basis of the previous results on the post hydrogenation, we fabricated n- and p-type bottom gate TFTs with Si-rich poly-SiGe 200nm films. Fig. [10] shows the transfer characteristics for n-and p-type bottom-gate TFTs fabricated with the highest process temperature of 450°C for film deposition. The field effect mobility are 36 cm$^2$/Vs for n-channel TFT and 55cm$^2$/Vs for p-channel TFT after the hydrogen treatment. It should be noted that

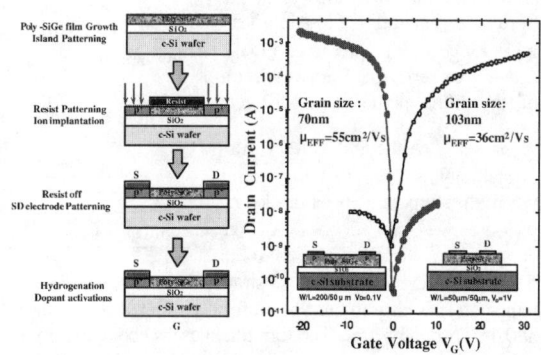

**Fig.10** Fabrication process of bottom-gate TFTs and transfer characteristics of n- and p-channel bottom-gate TFTs

such a high mobility over 30 cm$^2$/Vs can be achieved in th bottom-gate TFT in which the channel is formed at the very limited region of the film from the interface between substrate and the film deposited, even though the ideal gate insulator of a thermally oxide are used. This result demonstrates how high the film crystallinity at the early stage of the film deposition is in the present poly-SiGe film, which is very difficult to achieve in the conventional low-temperature CVDs. It looks that the p-channel TFT fabricated is superior to the n-channel one as far as we judge from the transfer characteristics as shown in the figure, but we think that it can be improved much because of inexperienced fabrication in the first trial.

We fabricated n-and p-channel top-gate TFTs with Si-rich 200nm thick poly-SiGe films on the corning 7059 glass according to the processes illustrated in Fig. 11. The maximum process temperature is 450°C for the film growth as well. The field effect mobility is 22cm$^2$/Vs and 25 cm$^2$/Vs for p-channel and n-channel TFTs after the hydrogenation, respectively. The transfer characteristics of n-and p- channel TFTs are very compatible and shows a good promise for the CMOS applications. It is very plausible that the mobility itself can be improved more because of a rather small grain of 70 nm in the present poly-SiGe film used for TFT fabrication.

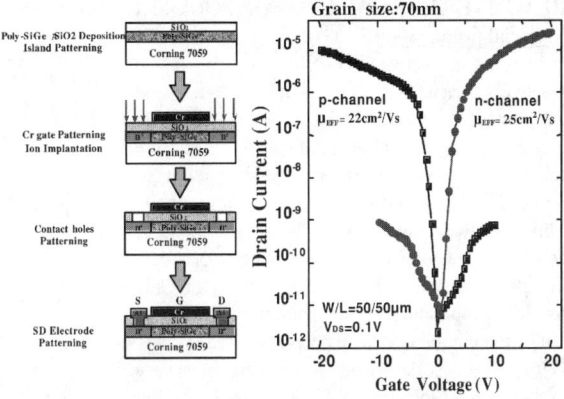

**Fig.11** Fabrication process of top-gate TFTs and transfer characteristics of n- and p-channel top-gate TFTs

**Table 2** TFT parameters in n-and p-channel TFTs

| poly-SiGe: Si > 0.95 :200nm Ts =450 | Bottom-gate | | Top-gate | |
|---|---|---|---|---|
| | n-channel W/L=50/50µm, V$_{ds}$=1V | p-channel W/L=200/50µm V$_{DS}$=0.1 | n-channel W/L=50/50µm, V$_{DS}$=0.1V | p-channel W/L=50/50µm, V$_{D}$=0.1V |
| Field Effect Mobility | 36 cm$^2$/Vs | 55 cm$^2$/Vs | 25 cm$^2$/Vs | 22 cm$^2$/Vs |
| Threshold Voltage | 1.8 V | -0.5 V | 1.9 V | -1.6 V |
| Sub threshold Swing | 1.4 V/dec | 0.55 V/dec | 0.45 V/dec | 0.56 V/dec |
| On-current | 4.9 ~10$^{-5}$ A | 3.78 ~10$^{-4}$ A | 1.5 ~10$^{-5}$ A | 6.5 ~10$^{-6}$ A |
| Off-current | 9.1 ~10$^{-11}$ A | 1.43 ~10$^{-10}$ A | 6.0 ~10$^{-12}$ A | 3.3 ~10$^{-12}$ A |
| On/Off ratios | 5.3 ~10$^5$ | 2.6 ~10$^6$ | 2.5 ~10$^6$ | 2.0 ~10$^6$ |

Table 2 is the summary of the performance of bottom and top-gate TFTs fabricated in the present study. These are fairly good results in terms of a threshold voltage, a sub-threshold swing, and mobility, compared with TFTs fabricated with poly-Si thin films deposited by the conventional low-temperature CVD techniques.

# REACTIVE THERMAL CVD FOR POLY-Si THIN FILMS [23-33]
## Material preparation

In the polycrystalline film growth in the reactive thermal CVD with disilane and germanium terafluoride, it is very clear that germanium tetrafluoride itself plays a decisive role for the low-temperature crystal growth of poly-SiGe, judging from a fact that amorphous films are always deposited in the absence of $GeF_4$ at 450°C, even if $GeH_4$ is used instead of $GeF_4$. We pay attention to the etching activity of $GeF_4$ for low-temperature crystallization of the films out of the unique chemical properties in $Si_2H_6$ and $GeF_4$ system as described previously. It is very probable that $GeF_4$ breaks Si-Si bonds during the film deposition because of its etching activity. Therefore, it is very likely that at the very early stage of the film growth, the resulting loose Si-network island is easy to reorganize into a nucleus on the substrate surface, while the crystal growth is well maintained for the loose Si-network, which promotes structural relaxation into a crystal-like structure via quasi-chemical equilibrium between Si-Si bond breaking and Si-Si bond formation mediated by $GeF_4$ as illustrated in the Fig. 12. We haven't confirmed how $GeF_4$ contributes to the crystal growth experimentally yet, but this idea reminds us of another possible promoter, fluorine, for the low-temperature crystal growth.

We proposed a new CVD technique for preparation of amorphous and μc-Si thin films with a gas hase reaction of silane with fluorine at reduced pressure about 15 year ago. Fig. 13 shows a beautiful chemi-luminescence from fluorine combustion flame, where silane is decomposed chemically through the gas phase reaction with fluorine. According to the present discussion, we have revived this red-ox gas system of silane and fluorine for low-temperature growth of poly-Si thin films. Indeed, we found that poly-Si films can be deposited in a different reaction mode from the gas phase reactions with silane and fluorine: in the gas phase reaction mode where silane and fluorine are used in one to

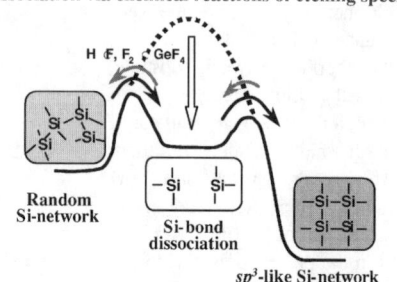

Fig.12 Conceptual diagram of low-temperature growth of polycrystalline film *via* chemical

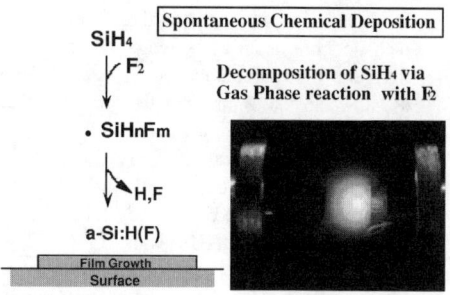

Fig.13 Chemical processes in Spontaneous chemical deposition of Si thin films with $SiH_4$ and $F_2$.

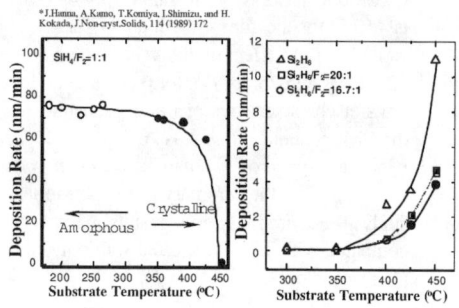

Fig. 14 Different deposition mode in Silane-$F_2$ systems

one molar ratio basically, the film deposition hardly depends on the temperature except for a high temperature range, where the etching of the films takes place as shown in Fig. 14(a) ; on the other hand, at a low gas ratio of disilanes to fluorine, e.g., 20 to 1, we see no film growth at the low temperature region below 400°C , but the film deposition takes place over 400°C as shown in Fig. 14 (b). This follows the film deposition in thermal CVD process with disilne basically, but interestingly poly-Si films grows over 425°C where amorphous films are always deposited with disilane alone. Fig. 15 shows a TEM image of the film deposited at 450°C with disilane in the presence of a small amount of fluorine, say 5 atm %, showing very uniform crystallinity from bottom to top-surface.

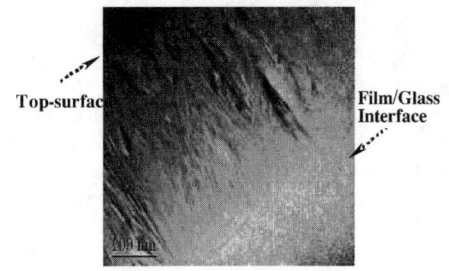

**Fig. 15** Typical TEM image of poly-Si thin films on glass substrates prepared with $Si_2H_6$ and $F_2$.

## TFT fabrication

We also fabricated n-channel bottom gate TFTs with a Si-wafer with thermal oxide as illustrated in Fig. 16 and we found that the TFTs after hydrogen treatment exhibit a very good transfer characteristics with field effect mobility as high as 54.5 $cm^2/Vs$. This result is probably attributed to the high crystallinity at the early stage of the film growth.

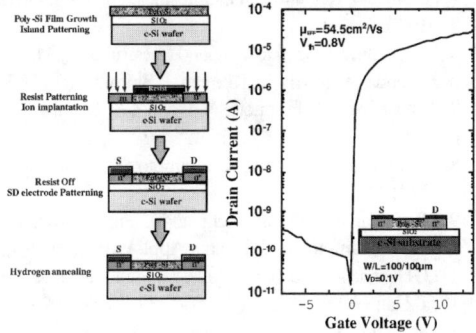

**Fig.16** Fabrication process of bottom-gate TFTs and transfer characteristics of n-channel bottom-gate TFTs

## CONCLUSIONS

In summary, we have established a new thermal CVD technique, i.e., *Reactive thermal CVD*, for low-temperature growth of poly-Si and SiGe thin films at 450°C with disilane and germanium terafluoride and with disilane and fluorine, respectively. The polycrystalline films exhibited high crystallinity even in very thin films less than 200nm and were successfully applied to bottom and top-gate n-and p-channel TFTs. The high field effect mobility of 36 and 54 $cm^2/Vs$ was achieved in bottom-gate n- and p-channel TFTs fabricated, and 25 and 22 $cm^2/Vs$ for top-gate n- and p-channel ones, respectively. This new CVD process is very promising for a direct growth of device-grade poly-Si thin films for TFT applications. Now we are establishing a good basis for challenging a high mobility TFTs over 100 $cm^2/Vs$ that no CVD technique has achieved yet.

## ACKNOWLEDGMENTS

We acknowledge Dr. Masaji Yamamoto and Dr. Masato Miyauchi for their early works to demonstrate the present CVD concept, Mr. Kunihiro Shiota for his establishment of the growth condition of Si-rich poly-SiGe films, Dr. JuanJin Zhang for his great contribution to TFTs fabrication, Mr. Jeong-Woo Lee and Mr. Toshiaki Kawazoe for their establishment of poly-Si films from silane and fluorine, and the students who gave serious contribution to make the present project successful since we started this project in 1992. We would like to express our sincere thanks for materials support, i.e., disilane to Mitsui Chemicals Inc. and germanium tetrafluoride and fluorine to Central Glass Inc.

## REFERENCES

1. A. Matsuda, J. Non-Cryst. Solids, **59&60**, 767 (1983).
2. A. Matsuda, S. Yamazaki, K. Nakagawa, K. Ohkushi, K. Tamnaka, S. Iizima, M. Matumura, and H. Yamamoto, Jpn. J. appl. Phys., **19**, L305 (1980).
3. C. C. Tsai, G. B. Anderson and R. Thonpson: J. Non-Cryst. Solids, **137&138**, 673 (1991).
4. N. Shibata, K. Fukuda, H. Ohtoshi, J. Hanna, S. Oda, and I. Shimizu, Jpn. J. Appl. Phys., **26**, L10 (1987).
5. H. Matsumura, Y. Hosoda, and S. Furukawa, Mat. Res. Soc. Proc., **283**, 623 (1993).
6. J. Hanna, A. Kamo, T. Komiya, I. Shimizu, and H. Kokado, J. Non-Cryst. Solids, **114**, 172 (1989).
7. T. Nagahara, K. Fujimoto, N. Kohno, Y. Kashiwagi, and H. Kakinoki, Jpn. J. Appl. Phys. **31**, 4555 (1992).
8. N. Kohno, T. Nagahara, K. Fujimoto, Y. Kashiwagi, and H. Kakinoki, Mat. Res. Soc. Proc., **283**, 629 (1993).
9. M. Yamamoto, M. Miyauchi, and J. Hanna: Appl. Phys. Lett., **63**, 2508 (1993).
10. M. Yamamoto and J. Hanna: Appl. Phys. Lett., **64**, 3467 (1994).
11. J. Hanna, T. Ohuchi and M. Yamamoto: Mat. Res. Soc. Proc., **358**, 877 (1995).
12. J. Hanna, Mat. Res. Soc. Proc., **377**, 105 (1995).
16. K. Shiota, D. Inoue, K. Minami and J. Hanna: J. Non-cryst. Solids, **227 & 230**, 1074 (1998).
15. K. Shiota, D. Inoue, K. Minami, M. Yamamoto, and J. Hanna: Mat. Res. Soc. Proc., **452**, 1001 (1997).
14. K. Shiota, D. Inoue, K. Minami, M. Yamamoto, and J. Hanna: Jpn. J. Appl. Phys., **36**, L989 (1997).
13. F. Yoshizawa, K. Shiota, D. Inoue, and J. Hanna, Mat. Res. Soc. Proc., **467**, 349 (1997).
17. J. Hanna, K. Shiota, and M. Yamamoto: Mat. Res. Soc. Proc., **507**, 945 (1998).
18. J. Hanna and K. Shimizu: J. Organomet. Chem. **611**, 531 (2000).
19. K. Shimizu, J.J. Zhang, J.W. Lee, and J. Hanna: Mat. Res. Soc. Symp. Proc. **638**, F14.13.1 (2001).
20. J. Zhang, K. Shimizu, and J. Hanna: J. Non-cryst. Solids, **299 & 302**, 163 (2002).
21. J. Zhang, K. Shimizu, and J. Hanna: Appl. Phys. Lett., **82**, 1745 (2003).
22. D. Meakin, J. Stoemenos, P. Migliorato and N.A. Economou: J. Appl. Phys. 61, 5031 (1987).
23. K. Shimizu, J.J. Zhang, J-W. Lee, and J. Hanna, Mat. Res. Soc. Proc., **686**, 107 (2002).
24. J. W. Lee, K. Shimizu, and J. Hanna: Mat. Res. Soc. Symp. Proc. **715**, A16.1.1 (2002).
25. J. Hanna, S. Oda, N. Shibata, A. Miyaichi, A. Tanabe, K. Fukuda, T. Ohtoshi, H. Nguyen, and I. Shimizu, Mat. Res. Soc. Proc., **70**, 11 (1986).

26. J. Hanna, A. Kamo, M. Azuma, N. Shibata, H. Shirai, and I. Shimizu, Mat. Res. Soc. Proc., **118**, 79 (1988).
27. J. Hanna, A. Kamo, T. Komiya, I. Shimizu, and H. Kokado, J. Non-cryst. Solids, **114**, 172 (1989).
28. J. Hanna, A. Kamo, T. Komiya, H.D. Nguyen, I. Shimizu, and H. Kokado, Mat. Res. Soc. Proc., **149**, 11 (1989).
29. T. Wadayama, H. Kayama, A. Hatta, W. Suetake, and J. Hanna, Jpn. J.Appl. Phys., **29**, 1884 (1990).
30. T. Komiya, A. Kamo, H. Kujirai, I. Shimizu, and J. Hanna, Mat. Res. Soc. Proc., **164**, 63 (1990).
31. C. Kawamura, I. Shimizu, and J. Hanna, J. Non-Cryst. Solids, **1137 & 138,** 697 (1991).
32. J. Hanna, New Functionality Materials, Ed. by T. Tsuruta, M. Doyama, and M. Seo, Vol. C Elsevier Science Publishers, 189 (1993).
33. K. Endo, M. Bunyo, I. Shimizu, and J. Hanna, Mat. Res. Soc. Proc., **283**, 641 (1993).

## Improvement of Gate Oxide Integrity in Low Temperature Poly Silicon TFT

Seok-Woo Lee, Dae Hyun Nam, Jin Mo Yoon, Hyun Sik Seo, Kyoung Moon Lim, and Chang-Dong Kim
LCD R&D Center, LG.Philips LCD, 533, Hogae-dong, Dongan-gu, Anyang-si, Kyongki-do, 431-080, Korea

## ABSTRACT

The electrical characteristics of $SiH_4$-based PECVD gate oxide have been investigated with respect to gate oxide integrity (GOI) and its reliability. It was found that the GOI of poly-Si TFT integrated on glass substrate strongly depended on the charge trapping and deep level interface states generation under Fowler-Nordheim stress (FNS). By applying elevated temperature post-anneal without vacuum break after the gate oxide deposition, highly reliable gate oxide was obtained. Under FNS, $I_D$-$V_G$ curve showed severe shift and degradation of subthreshold slope, which were reduced by adopting post-annealed gate oxide. Besides, the TFT with post-annealed gate oxide showed around 10 times higher charge to breakdown than that of as-deposited gate oxide. Charge to breakdown of MOS capacitors were also studied. By applying post-annealed gate oxide, charge to breakdown drastically improved, which could be explained by reduced charge trapping under FNS.

## INTRODUCTION

There have been surging demands on display appliances for mobile applications. To meet these demands, low power consumption is required and low temperature poly silicon (LTPS) thin film transistors (TFTs) with CMOS process are widely attracted for system-on-panel-integration. The improvement of gate oxide integrity (GOI) is one of key issues for improving LTPS TFT performances and reliabilities [1,2]. Nevertheless, not so many works were reported concerning GOI reliability for LTPS TFT application.

In this paper, the GOI reliability of PECVD $SiO_2$ films integrated in LTPS TFT fabricated on glass substrate was evaluated with respect to gate oxide reliability. Remarkable improvements of GOI and its device reliabilities could be obtained by applying post-anneal after gate oxide deposition.

## EXPERIMENTAL DETAILS

To evaluate the reliability of the gate oxides integrated in TFT, poly-Si TFTs ($W/L$ = 8/6 $\mu m$)

were fabricated using CMOS process. Poly-Si layer was made from 50 nm-thick PECVD amorphous silicon precursor deposited at 420 ℃ by XeCl excimer laser annealing with a wavelength of 308 nm. After defining the active area of poly-Si, 100 nm-thick gate oxides were formed at 420 ℃ using $SiH_4$-based plasma enhanced chemical vapor deposition (PECVD). After gate oxide deposition, post-annealing was performed at 470 ℃ for 10 minutes without breaking vacuum of cluster-typed chamber, and compared with as-deposited case, and then gate metal deposition was followed. No integration-related issue was found to integrate TFTs on Corning 1737 glass substrate although post-annealing was adopted.

To examine electrical characteristics of PECVD gate oxide, metal-oxide-semiconductor (MOS) capacitors were fabricated with the same thermal budget split conditions described above at gate oxide deposition step. The characteristics of MOS and TFT devices were measured with a HP 4145B semiconductor parameter analyzer.

## DISCUSSION

TFT device characteristics have been investigated with respect to GOI, and compared before and after FNS. Figure 1 shows (a) PMOS transfer curve variation by FNS and (b) threshold voltage variation ($V_{th\_FNS} - V_{th\_init}$) as a function of gate injected charge. Under FNS condition of (-) 2.1 $mC/cm^2$, a severe shift of transfer curve is observed in Figure 1(a) for TFTs employing as-deposited oxide as a gate insulator, which is drastically suppressed by applying of post-annealed gate oxide.

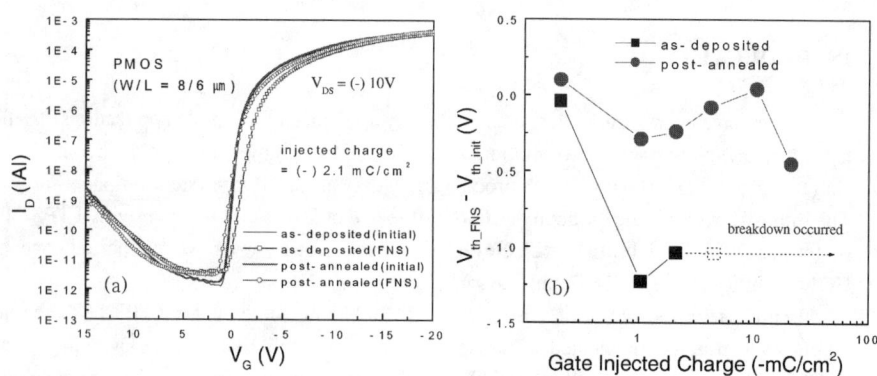

**Figure 1.** Comparison of (a) PMOS transfer curve variation and (b) $V_{th}$ variation : $V_{th\_FNS} - V_{th\_init}$ as a function of gate injected charge for as-deposited and post-annealed gate oxide, where the subscripts of $_{FNS}$ and $_{init}$ stand for after FNS and initial value, respectively.

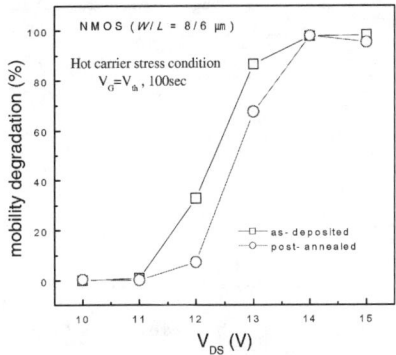

**Figure 2**. Hot carrier stress induced maximum field effect mobility degradation of NMOS TFT as a function of applied $V_{DS}$. $V_{DS}$ was biased variably with voltage step of 1V to a virgin NMOS for each step.

The shift of transfer curve continues with injected charge and eventually leads to gate oxide breakdown, which is understood by charge trapping in the gate oxide [3,4]. The threshold voltage variation induced by FNS is also much severer for as-deposited oxide as shown in Figure 1(b). Furthermore, gate oxide breakdowns occur over the injected charge of (-) 2.1 mC/cm$^2$ for as-deposited oxide. But, no breakdown is observed by applying post-annealed gate oxide so far as the injected charge of (-) 21 mC/cm$^2$, which is around 10 times larger than the charge to breakdown of as-deposited oxide integrated in TFT. Much degradation of sub-threshold slope ($S$) is also observed for as-deposited oxide. The initial $S$ is similar for both cases, but the degradation of $S$ ($\Delta S$) is 0.18 V/decade and 0.13 V/decade for as-deposited and post-annealed gate oxide, respectively. The shift of $I_D$-$V_G$ characteristics and $S$ degradation represent that the device degradation by FNS comes from charge trapping and deep level interface states generation [5-7].

Hot carrier stress (HCS) reliability of NMOS TFT was compared. Figure 2 shows maximum field effect mobility degradation as a function of drain voltage, $V_{DS}$. Larger degradation is observed for TFTs with as-deposited oxide comparing with post-annealed gate oxide. Besides, threshold voltage variation induced by HCS is smaller for TFTs with post-annealed gate oxide. These results represent the improvement of HCS reliability by post-annealing of gate oxide.

|  | Etch rate (nm/min) | thickness (nm) | Refractive index |
|---|---|---|---|
| as-deposited | 59.3 | 100.2 | 1.4703 |
| post-annealed | 59.1 | 101.1 | 1.4700 |

**Table 1**. Comparison of etch rate, thickness and refractive index between as-deposited and post-annealed oxide.

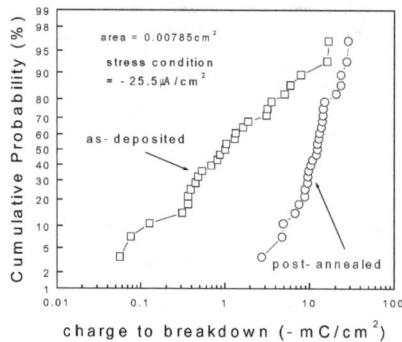

**Figure 3**. charge to breakdown ($Q_{bd}$) of as-deposited and post-annealed gate oxides under constant current FNS of (–) 25.5 µA/cm².

Measurements of etch rate, thickness and refractive index were performed on as-deposited and post-annealed gate oxides to examine the effect of thermal budget difference on their bulk properties. Table 1 compares the bulk properties between as-deposited oxide and post-annealed oxide. It is found that these bulk properties are almost same for both as-deposited and post-annealed oxide, indicating that the improved stability of TFTs is due to the improvement of GOI instead of bulk properties. Additionally, there was no process-related issue although post-anneal was applied.

For detailed study of PECVD $SiO_2$, metal-oxide-semiconductor (MOS) capacitors were fabricated on p-type Si wafer and electrical characteristics were compared between as-deposited and post-annealed gate oxide. Firstly, charge to breakdown ($Q_{bd}$) characteristics was investigated under constant current FNS of (-) 25.5 µA/cm². Figure 3 represents $Q_{bd}$ of as-deposited and post-annealed gate oxide. At 50 % failure, post-annealed oxide shows around 10 times larger $Q_{bd}$ than that of as-deposited oxide. To understand the difference of $Q_{bd}$, charge trapping characteristics were compared under constant current FNS. By monitoring gate voltage variation ($\Delta V_G$), charge trapping characteristics can be understood [8]. Figure 4 shows comparison of $\Delta V_G$ under constant current FNS between as-deposited and post-annealed gate oxide. As time elapse, the gate voltage variation due to charge trapping is severer for as-deposited oxide compared with post-annealed oxide. This also explains GOI dependence on charge trapping characteristics and the GOI of post-annealed gate oxide is drastically improved by reduced charge trapping under constant current FNS.

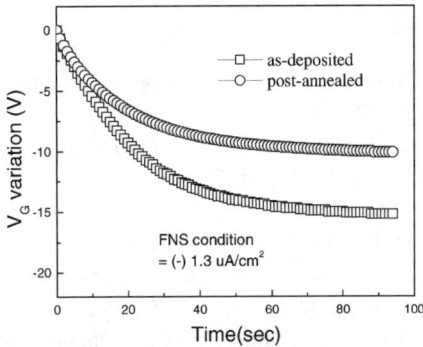

**Figure 4**. gate voltage variation as time elapse under constant current FNS. $\Delta V_G(t) \equiv V_G(t) - V_G(0)$, where $t$ and $V_G(0)$ represent stress elapsed time and gate voltage at time 0, respectively.

## CONCLUSIONS

Electrical characteristics of $SiH_4$-based PECVD $SiO_2$ were studied for TFT application, and gate oxide integrity and its reliability were compared between as-deposited and post-annealed gate oxide. By applying simple process of post-anneal after gate oxide deposition, highly reliable gate oxide was obtained, i.e., improved charge to breakdown, reduced gate voltage variation under constant current FNS, which could be explained by the reduction of charge trapping and deep level interface states generation compared with as-deposited case. Highly reliable CMOS TFT integrated on glass substrate was also obtained without process-related problems although post-annealing was adopted. These improvements of GOI reliability and TFT performances will give a solution for system on panel integration of flat panel displays.

## ACKNOWLEGMENTS

The authors would like to thank all of Research Group 2 members for helping us preparing test samples, especially Dr. Hae-Yeol Kim and Dr. Soo Young Yoon for their very useful discussions.

**REFERENCES**

1. Yasuhisa Oana, "Current and Future Technology of Low Temperature Poly-Si TFT-LCDs," *IDMC*, pp. 57-60, (2000).
2. N. Ibaraki, "Low-Temperature Poly-Si TFT Technology," *SID Digest*, pp. 172-175, (1999).
3. Masashi Goto, Masahiro Sakai, Munehiro Shibuya and Mikihiko Nishitani, "Characteristics of $SiO_2$ films prepared by TEOS based PECVD," *AMLCD*, pp. 143-146, (1999).
4. Akihiro Takami, Arichika Ishida, Junsei Tsutsumi, Tooru Nishibe and Nobuki Ibaraki, "Threshold voltage shift under the gate bias stress in low-temperature poly-silicon TFT with the thin gate oxide film," *AMLCD*, pp. 45-48, (2000).
5. F. V. Farmakis, J. Brini, G. Kamarinos, and C. A. Dimitriadis, "Anomalous Turn-On Voltage degradation During Hot-Carrier Stress in Polycrystalline Silicon Thin-Film Transistors," *IEEE Electron Device Letters*, Vol. 22, No. 2, pp. 74-76, (2001).
6. Mutsumi Kimura, Satoshi Inoue and Tatsuya Shimoda, "Dependence of Poly-Si TFT Characteristics on Oxide Interface Traps and Grain Boundary Traps," *AMLCD*, pp. 255-258, (2002).
7. Mutsumi Kimura, Satoshi Inoue and Tatsuya Shimoda, "Extraction of trap states in laser-crystallized polycrystalline-silicon thin-film transistors and analysis of degradation by self-heating," *Journal of Applied Physics*, Vol. 91, pp. 3855-3858, (2002).
8. Jao-Hsian Shiue, Joseph Ya-min Lee, and Tien-Sheng Chao, "A Study of Interface Trap Generation by Fowler-Nordheim and Substrate-Hot-Carrier stresses for 4-nm Thick Gate Oxides," *IEEE Trans. Electron Devices*, Vol. 46, No. 8, pp. 1705-1710, (1999).

# Threshold Voltage Performance of a-Si:H TFTs for Analog Applications

K.S. Karim, K. Sakariya, and A. Nathan
Department of Electrical and Computer Engineering,
University of Waterloo, Waterloo, ON N2L 3G1 Canada.

## ABSTRACT

Amorphous silicon (a-Si:H) thin-film transistors (TFT) used in emerging, non-switch applications such as analog amplifiers or active loads, often have a bias at the drain terminal in addition to the gate that can alter their threshold voltage ($V_T$) stability performance. At small gate voltages ($0 \leq V_{ST} \leq 15$ V) where the defect state creation instability mechanism is dominant, the presence of a bias at the TFT drain is found to decrease the shift in $V_T$ ($\Delta V_T$) compared to the $\Delta V_T$ in the absence of a drain bias. In this paper, a $\Delta V_T$ model accounting for TFT drain bias, is used to predict the performance of a-Si:H analog circuits in active pixel sensor (APS) medical x-ray imaging and active matrix, organic light emitting diode (AMOLED) display applications.

## INTRODUCTION

Unlike crystalline silicon transistors, a-Si:H TFTs exhibit bias induced metastability phenomena that can have adverse effects on circuit performance if the circuit is improperly designed or operated. The metastability concerns become particularly important when the TFT is used as an analog device. In contrast to traditional applications (e.g. in LCDs) where the TFT is used as a switch, analog applications require the device to withstand prolonged voltages on both drain and gate terminals. Defect state creation and charge trapping are the widely accepted mechanisms for $V_T$ instability in a-Si:H TFTs. Defect creation dominates at lower bias voltages while charge trapping in the gate insulator is dominant at higher biases. Low power requirements in emerging analog applications of a-Si:H TFT technology (e.g. APS x-ray imaging [1] and AMOLED display [2]) underscore the drive towards reducing circuit supply voltages to a level where defect state creation is the dominant $\Delta V_T$ mechanism. This research focuses on developing an understanding of circuit performance when dc bias voltages are applied simultaneously at the TFT drain and gate terminals in analog a-Si:H circuits.

## THEORY

Charge trapping occurs primarily in PECVD a-Si:N gate insulator TFTs where the high density of defects in the insulator can trap charge when the TFT gate undergoes bias stress [3][4]. In contrast, defect creation in the a-Si:H layer or at the a-Si:H/a-SiN interface due to a prolonged gate bias has some similarities to light induced defect creation [7] where the density of deep state defects increases. The point at which the charge trapping component of $\Delta V_T$ overtakes defect state creation has been shown to be a function of the gate nitride stoichiometry [4][5] and usually occurs at larger voltages for nitrogen rich gate dielectrics [6]. When a positive bias is applied to the gate of an a-Si:H TFT, electrons accumulate and form a channel at the a-SiN/a-Si:H interface where they predominantly reside in conduction band tail states [8]. These tail states have been identified as weak silicon-silicon bonds which, when occupied by electrons, can break to form silicon dangling-bonds (deep state defects) [9][10]. Deep state defect creation forms the basis of the defect pool model [11], where the rate of defect creation is a function of the barrier to defect formation, the number of electrons in the tail states and the density of the weak bond sites. Defect creation is generally characterized by a power law time dependence and is strongly affected by temperature[6]. In a uniform TFT channel, $\Delta V_T$ is expressed as [6][9],

$$\Delta V_T(t) = A(V_{ST} - V_{Ti}) t^\beta, \qquad (1)$$

where $A$ and $\beta$ are temperature dependent parameters, $V_{ST}$ is the gate bias stress voltage, $V_{Ti}$ is the $V_T$ of the TFT before bias stress is applied, and $t$ is the bias stress time duration.

## MEASUREMENTS AND DISCUSSION

For the initial experiments, $\Delta V_T$ was measured on in-house, bottom-gate, a-Si:H TFTs [12] [13] supplied with a gate bias where the drain and source terminals were grounded. The results for positive bias are presented in Fig. 1 and appear to corroborate the defect pool model, i.e. eqn. (1). Here, a linear relationship is observed between $\Delta V_T$ and $(V_{ST} - V_{Ti})$ while $\Delta V_T$ has a power dependence on bias stress time with $\beta \sim 0.3$.

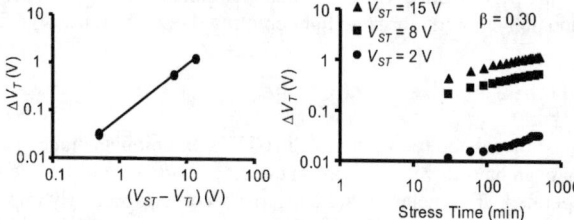

Fig 1. The relationship of $\Delta V_T$ to $(V_{ST} - V_{Ti})$ and stress time.

The $\Delta V_T$ results shown in Fig. 2 below further support the dominance of the defect creation mechanism in TFTs with nitrogen rich gate dielectrics [12] (as are the TFTs in this study) and relatively small stress voltages (15 V).

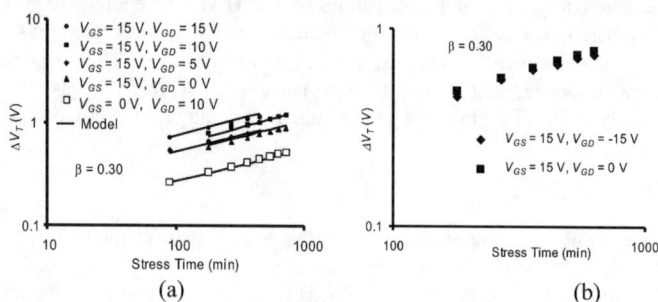

(a)            (b)

Fig. 2. Dependence of $\Delta V_T$ on $V_{GS}$ and $V_{GD}$. The value extracted for $\beta$ is 0.3. The solid lines represent the model shown in eqn. (2).

Unlike the linear mode of TFT operation where a uniform channel exists in the a-Si:H layer, the presence of an increasing bias on the TFT drain terminal increases the lateral electric field and decreases the carrier concentration in the channel near the drain. Around $V_{DS} = V_{GS} - V_T$, the TFT enters the saturation mode of operation beyond which the carrier concentration in the channel remains approximately constant. Fig 2(a) shows that the $\Delta V_T$ for a TFT entering saturation is smaller than that in the linear mode at the same gate bias voltage while driving the TFT deeper into saturation (i.e. $V_{DS} > V_{GS} - V_T$ or as $V_{GD} < V_T$) appeared to have little effect on $\Delta V_T$. This can be seen in Fig. 2(b) where $V_{GD}$ is changed from 0 to $-15$ V while $V_{GS} = 15$ V.

If the defect pool model is considered (where the $\Delta V_T$ is proportional to the number of carriers in the conduction band tail states), the decrease in TFT channel charge in saturation may help explain the smaller $\Delta V_T$ as compared to that in the linear region. Further, since once the TFT is saturated, there is no significant change in the concentration of channel charge as the TFT is driven further into saturation, the change in $\Delta V_T$ is negligible. Since $\Delta V_T$ appears to vary with the induced channel charge, simple MOS equations for channel charge were employed to develop a rudimentary $\Delta V_T$ model [13][15] for a TFT operating under both gate and drain bias voltages. Following the observed dependencies of $\Delta V_T$ on TFT bias, eqn. (1) is modified to include the effect of drain voltage by using the ratio of the channel charge at a given $V_{GD}$ and $V_{GS}$ bias to that of the channel charge in the TFT linear mode of operation,

$$\Delta V_T(t) = \left(\frac{Q_G}{Q_{G0}}\right) A(V_{GS} - V_{Ti}) t^\beta. \tag{2}$$

Here, $Q_{G0}$ is the channel charge when the TFT is in linear [15], $Q_G$ is the gate and drain bias dependent expression for channel charge [15], $V_{ST}$ has been replaced by $V_{GS}$, and again, $A$ and $\beta$ are temperature dependent parameters. Since $\Delta V_T$ is proportional to the normalized channel charge, the maximum (in linear mode) and minimum (in saturation mode) values for channel charge set the upper and lower limits of $\Delta V_T$ at a given $V_{GS}$. The model of eqn. (2) was found to be in good agreement with the measured data as shown in Fig. 2 (a).

## GAIN DEGRADATION OF AN AMPLIFIED PIXEL

As an example, the model developed in eqn. (2) is applied to the amplified pixel readout circuit shown in Fig 3. The amplified pixel was previously discussed in [1][14] and has the potential to reduce noise for digital fluoroscopic medical imaging. The pixel circuit of Fig. 3 (inside the dashed box) can be decoupled into two smaller circuits: the first is the RESET TFT switch and the second is the AMP/READ TFT output branch. The circuit of interest in the following discussion is the AMP/READ TFT branch since, in contrast to the RESET TFT switch, the AMP and READ TFTs suffer from both a gate and drain terminal bias.

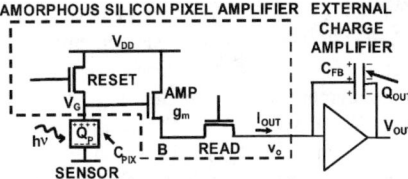

Fig 3. Amplified pixel readout circuit

The AMP and READ TFT form part of the readout circuit for the APS and connect directly to the column charge amplifier and thus, the gain is sensitive to the $\Delta V_T$ arising from these TFTs.

The gain of the amplified pixel was determined [14] to be a function of the circuit $g_m$ where the $g_m$ of the AMP TFT is given as $K_A(V_{GA} - V_T - V_{DS})$. Here, $K_A$ is the usual product of parameters $\mu_{EFF}C_GW/L$ for the AMP TFT, $V_{GA}$ is the AMP TFT gate voltage, and $V_{DS}$ is the voltage drop across the READ TFT switch [14]. $V_{DS}$ can be significant since a-Si:H TFT switch resistances usually have MΩ values and is obtained (see eqn. (3)) by equating the current through the AMP TFT in saturation and READ TFT in linear mode [15]. $V_{GRD}$ is the READ TFT gate voltage, $V_{TA}$ is the AMP TFT $V_T$, $V_{TRD}$ is the READ TFT $V_T$ and $K_{RD}$ is related to the READ TFT.

$$V_{DS} = \frac{\left[2K_A(V_{GA}-V_{TA})+K_{RD}(V_{GRD}-V_{TRD})-\left[K_{RD}\left(\begin{array}{c}-4K_AV_{GA}(V_{GRD}-V_{TRD}+V_{TA})+2K_{RD}V_{GRD}V_{TRD}\\+2K_A(V_{GA}^2-V_{TA}^2)-K_{RD}(V_{GRD}^2+V_{TRD}^2)\\+4K_AV_{TA}(V_{GRD}-V_{TRD})\end{array}\right)^{1/2}\right]\right]}{(K_{RD}+2K_A)}. \quad (3)$$

Two scenarios exist in the analysis of the AMP/READ TFT $\Delta V_T$, when the READ TFT is either OFF or ON. During the READ TFT OFF state, node B is charged up to $(V_{GA} - V_{TA})$ and thus, there is a minimal bias across the AMP TFT's gate and source terminals. Also, since the bias at the gate of the AMP TFT is close to that of the drain (i.e. $V_{GA} \sim V_{DD}$) after pixel reset, little $\Delta V_T$ occurs due to the $V_{GD}$ of the AMP TFT. Most of the $\Delta V_T$ occurs during the READ TFT ON state when a $(V_{GA} - V_{TA} - V_{DS})$ bias is at the AMP TFT gate where $V_{DS}$ is defined in eqn. (3). For the READ transistor, $V_{GRD}$ is a positive bias applied at the gate during the READ TFT ON state. During the ON state, the TFT operates in switch mode where the $V_{DS}$ across the switch is given by eqn (3). When readout is complete, $V_{GRD}$ is set to be a zero or small negative bias applied at the READ TFT gate to minimize leakage current in the OFF state. During the READ TFT OFF state, node B reaches a value of $(V_{GA} - V_{TA})$, which is then "frozen" at that node for the duration of the OFF state effectively increasing the reverse bias across the READ TFT gate-drain terminals. For a zero $V_{GRD}$ in the OFF state, the value across the gate and drain terminals of the READ TFT is $(-(V_{GA} - V_{TA}))$ while the value across the gate and source terminals (i.e. at the input of the charge amplifier) of the READ TFT stays at zero.

The $\Delta V_T$ model for positive gate and drain bias voltages given by eqn. (2) was supplemented with an empirical model [15] for negative drain and gate bias stress voltages in order to model the stability of complete readout circuit. In addition, any effects of TFT duty cycle [15][16] [17] are taken into account by modifying the $\Delta V_T$ expression of eqn. (1) or eqn. (2) as,

$$\Delta V_{T\_AC}(t) = (T_{ON}/T_{PERIOD}) \Delta V_T(t), \quad (4)$$

where $T_{ON}$ is the ON time of the pulse, and $T_{PERIOD}$ is the inverse of the pulse frequency. An empirical model [16] was used to simulate the effect of duty cycle on negative pulse bias stress.

In Fig. 4 (a), the shift in $g_m$ was calculated by substituting eqn. (2) and eqn. (3) into eqn. (4). A supply voltage ($V_{DD}$ = 8 V) is applied along with a READ TFT gate pulse of +12/-5 V at a duty cycle of 10% (clocked at 30 Hz). The $g_m$ measurements [15] shown in Fig. 4(a) were carried out on the pixel shown in Fig. 3 with the source of the READ TFT connected to ground and appear to be in good agreement with the model. Fig. 4(b) illustrates the effect of increasing duty cycle on $g_m$ for the AMP/READ TFT branch as a function of time for a $V_{DD}$ = 12 V and a $V_{READ}$ = +15/–5V, 30 Hz pulse at a duty cycle of 0.1 % for 10,000 hours. A large factor aiding the APS stability in medical imaging applications is the reduced TFT duty cycle. For example, in a 1000 x 1000 pixel real-time fluoroscopic imager, the TFTs are clocked 33 $\mu s$ every 33 $ms$ giving a duty cycle of 0.1%. Note that since the duty cycle is only 0.1%, the READ TFT is OFF 99.9% of the time. The relatively small $\Delta g_m$ values in Fig. 4(b) are primarily due to the duty cycle of 0.1% as well as due to the READ TFT switch, which acts as a compensation resistor. Here, as the AMP TFT $V_T$ increases, the $V_{DS}$ across the READ TFT decreases due to the smaller branch current. The reduction in $V_{DS}$ compensates the increase in $V_T$ and thus stabilizes the ($V_{GA}$ − $V_{DS}$ − $V_T$) drop and the branch current. The net effect is that $g_m$ shifts at a reduced rate.

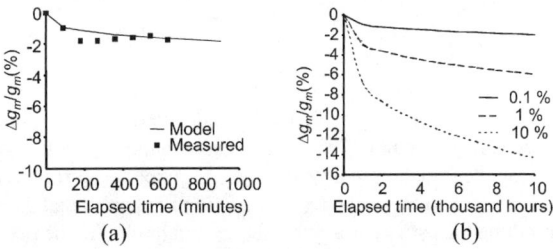

Fig. 4. (a) % change in $g_m$ for $V_{DD}$ = 8V, $V_{RD}$ = +12/-5, 30 Hz, 10% duty cycle, (b) % change in $g_m$ for $V_{DD}$ = 12V, and $V_{RD}$ = +15/-5 V at 30 Hz for varying duty cycle values.

## CURRENT RISE IN A CURRENT MIRROR BASED AMOLED PIXEL

Another example where the $V_{GD}$ dependence of $\Delta V_T$ affects circuit stability is in a current mirror based $\Delta V_T$ compensating AMOLED pixel circuit, shown in Fig. 5 (a).

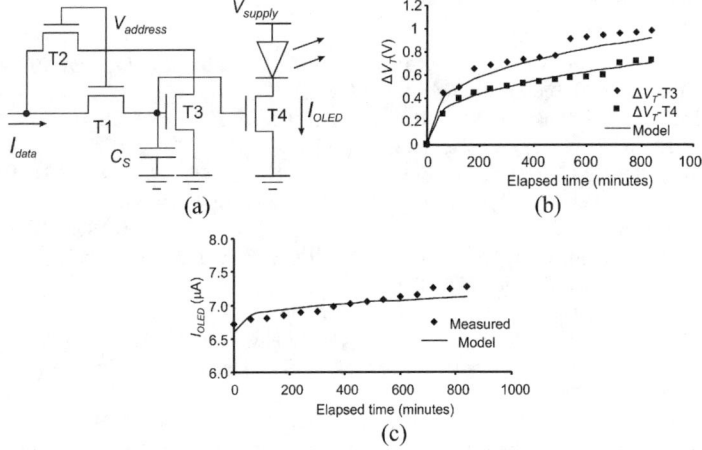

Fig. 5. (a) Current mirror based $\Delta V_T$ compensating AMOLED pixel circuit [18], (b) Differential $\Delta V_T$ in T3 and T4 leading to, (c) a rise in OLED current

Here, the programming current $I_{data}$ is used to set the OLED current ($I_{OLED}$) with any gain/attenuation depending on the ratio of transistors T4 to T3. When $V_{address}$ is ON, the capacitor $C_S$ at the gate of T3 is charged up by $I_{data}$ to allow all of $I_{data}$ to pass through T3. The final voltage on $C_S$ is not dependent on the $V_T$ of T3 and if the $V_T$ of T3 rises, $C_S$ is charged up to a larger voltage to keep the gate overdrive voltage ($V_{GS}$−$V_T$) constant [18]. The gate voltage on T3 mirrors $I_{data}$ to produce the required $I_{OLED}$ through T4. T4 is always biased in saturation (in the experiment, $V_{supply} = 20$V and the OLED is not present), whereas T3 is very close to saturation when $V_{address}$ is ON, but in linear mode when $V_{address}$ is OFF. Since T3 and T4 do not have the same $V_{GD}$ when the pixel is not being addressed, the $V_T$ of T3 (operating in linear mode) shifts at a higher rate than that of T4 (Fig. 5(b)), causing the circuit to over-compensate leading to a gradual rise in $I_{OLED}$ (Fig. 5(c)).

## CONCLUSIONS

$\Delta V_T$ experiments on bottom-gate a-Si:H TFTs indicate that defect state creation may be the dominant $\Delta V_T$ mechanism in our TFTs at lower bias voltages (< 15 V). An in-house $\Delta V_T$ model is used to model gain degradation in an APS imaging pixel and current rise in an AMOLED pixel circuit, and is found to be in good agreement with experimental data. The $\Delta V_T$ model for positive gate and drain bias stress presented here can help predict the performance of analog circuits for a variety of emerging a-Si:H TFT applications.

## ACKNOWLEDGEMENTS

This work is funded by the Natural Sciences and Engineering Research Council of Canada (NSERC), and Communications and Information Technology Ontario (CITO). The authors also thank Dr. Michael Hack (Universal Display Corporation) for stimulating discussions.

## REFERENCES

1. K.S. Karim, A. Nathan, J.A. Rowlands in *IEEE International Electron Devices Meeting* (IEDM 2002) Technical Digest, 215 (2002).
2. P. Servati, S. Prakash, A. Nathan, C. Py, *J. Vac. Sci. Tech. A*, **20**(4), 1374 (2002).
3. M.J. Powell, *App. Phys. Lett.*, **43**, 15 (1983).
4. M.J. Powell, C. Berkel, I.D. French, D.H. Nicholls, *App. Phys. Lett.*, **51**, 1242 (1987).
5. C. van Berkel, M.J. Powell, *App. Phys. Lett.*, **51**, 1094 (1987).
6. M.J. Powell, C. van Berkel, J.R. Hughes, *App. Phys. Lett.*, **54**, 1323 (1989).
7. M. Stutzmann, W.B. Jackson, C.C. Tsai, *Phys. Rev. B*, **32**, 23 (1985).
8. T. Leroux, *Sol. St. Elec.*, **29**, 47 (1986).
9. W.B. Jackson, M.D. Moyer, *Phys. Rev. B*, **36**, 6217 (1987).
10. M.J. Powell, S.C. Deane, W.I. Milne, *App. Phys. Lett.*, **60**(2), 207 (1991).
11. M.J. Powell, C. van Berkel, A.R. Franklin, S.C. Deane, W.I. Milne, *Phys. Rev. B*, **45**, 4160 (1992).
12. R.V.R. Murthy, P. Servati, A. Nathan, *J. Vac. Sci. Tech. B*, **18**(2), 685 (2000).
13. K.S. Karim, A. Nathan, M. Hack, W.I. Milne, "Drain bias dependence of threshold voltage stability of amorphous silicon thin-film transistors," *App. Phys. Lett.*, in review.
14. K.S. Karim, A. Nathan, *IEEE Elec. Dev. Lett.*, **22**(10), 469 (2001).
15. K.S. Karim, *Pixel Architectures for Digital Imaging Using Amorphous Silicon Technology*, Ph.D. Thesis, University of Waterloo, 2002.
16. C. Chiang, J. Kanicki, and K. Takechi, *Jpn. J. Appl. Phys.*, **37**(1), 9A, 4704 (1998).
17. C. Huang, T. Teng, J. Tsai and H. Cheng, *Jpn. J. Appl. Phys.*, **39**(1), 7A, 3867 (2000).
18. K. Sakariya, P. Servati, D. Striakhilev, and A. Nathan, Proc. EuroDisplay 2002: The 22nd International Display Research Conference, 609 (2002).

# Characteristics of Bottom Gate Thin Film Transistors with Silicon rich poly-$Si_{1-x}Ge_x$ and poly-Si fabricated by Reactive Thermal Chemical Vapor Deposition

Kousaku Shimizu, JianJun Zhang, Jeong-Woo Lee and Jun-ichi Hanna

Imaging Science and Engineering Laboratory, Tokyo Institute of Technology 4259 Nagatsuta Midoriku Yokohama 226-8503 Japan

## ABSTRACT

In the fabrication of thin film transistors (TFTs), little attention has been paid to the polycrystalline silicon thin films prepared at low temperatures where the glass substrates are adopted so far. Since the film quality is not sufficient to achieve high mobility, e.g., over 50 $cm^2/Vs$ in spite of high benefit in their industrial fabrication. We have fabricated bottom gate TFTs with poly-Si and poly- $Si_{1-x}Ge_x$ thin films deposited at 450°C by newly developed low-temperature LPCVD technique and characterized electrical characteristics of the TFTs: disilane and a small amount of either germanium tetrafluoride or fluorine were used as material gases and helium as carrier gas. Thermal annealing for dopant activation and atomic hydrogen treatment for defect passivation were carried out. We found that the defect elimination process is important for improving TFT performance significantly. Finally the mobility of p-channel and n-channel TFTs have attained 36.3-54.4 $cm^2/Vs$ and 57 $cm^2/Vs$, respectively.

## INTRODUCTION

Poly-crystalline Si (poly-Si) thin film transistors (TFTs) are being more important because of their applications for high resolution active-matrix liquid cyrstal displays (AM-LCDs). Especially poly-Si films prepared at low temperatures where glass substrates can be adopted are promising for extended application of the quality TFTs because of a good compatibility with conventional a-Si:H device fabrication processes. In the on-going processes for poly-Si TFTs on the glass substrate, poly-Si thin films are fabricated by either excimer laser crystallization of a-Si thin film or its solid-phase crystallization. However, these processes need deposition of a-Si thin films in prior to the crystallization. Therefore, direct deposition of device-grade poly-Si thin films on the glass substrate has a great benefit in industrial point of view. Several preparation techniques for low-temperature deposition of poly-Si have been intensively investigated, e.g. magnetron sputtering[1], plasma enhanced chemical vapor deposition (PECVD) [2], Hot-wire CVD (HW-CVD) [3], and photo-CVD [4]. These techniques, however, are still under investigation and remain some problems to be solved, i.e., poor crystallinity in very thin films adopted for TFT application, low uniformity of film thickness or low growth rates. We have proposed a new technique, *Reactive thermal CVD,* [5,6], in order to solve these problems, which is basically the thermal CVD but includes a new principle for reducing the growth temperature for crystalline films in the thermal CVD. In this CVD, a reactive source material such as $GeF_4$ and $F_2$ is decided to promote crystal growth at the low temperaturesare in addition to disilane. GeF4 reacts with disilane only when they are heated and F2 is is too highly diluted to react with disilane in the space though they react spontaneously when they are mixed. Therefore, the decomposition of the source materials is limited in the vicinity of substrate surface heated, where the activation of the source materials takes place. Thus, the film growth process is similar to that in the thermal CVD, so that the film growth hardly takes place on the reactor wall.

We report the fabrication of n- and p-channel bottom-gate TFTs with the polycrystalline films deposited by the present CVD and their characteristics, in addition to optimized conditions for thermal annealing for dopant activation and atomic hydrogen treatment for defect passivation. And the difference between poly-$Si_{1-x}Ge_x$ and poly-Si TFTs fabricated is discussed in terms of defects passivation and device performance.

**EXPERIMENT**

The Poly-Si and Poly- $Si_{1-x}Ge_x$ films were prepared with a typical cold-wall type of low-pressure CVD reactor and details are described elsewhere [7,8]. Disilane ($Si_2H_6$) and fluorine ($F_2$) were a set of the reactive source materials [7]. Germanium tetrafluoride ($GeF_4$) was used as an oxidant gas instead of $F_2$ for poly-$Si_{1-x}Ge_x$ film growth [8]. Interestingly, we found that the crystal growth takes place at 450 °C or less with compositional variety of Ge-rich to Si-rich and that those isolated nuclei are formed directly at the very early stage of the film growth on the $SiO_2$ substrate [6], when $Si_2H_6$ and $GeF_4$ were used for the source gases. The total pressure was kept at 5 Torr. The dilution gas, helium (He) was varied from 1000 to 4000 sccm for controlling the residence time, which corresponds to the apparent chamber residence times of 4-16s for our reactor volume ($V=4 \times 10^4$ $cm^3$). The reactive gas flow rates of pure $Si_2H_6$ and 10 vol.% diluted $GeF_4$ with He were fixed at 3.0sccm and 0.3 sccm for keeping the growth rate and Si content constant. The film thickness and the average grain size were measured by scanning electron microscopy (SEM). The Si content was determined by X-ray photoelectron spectroscopy (XPS). For observing crystal texture and comparing the difference of film growth process, transmission electron microscope (TEM) is used. Hydrogen content is evaluated by secondary ion mass spectroscopy (SIMS). Hydrogen effusion experiments are performed by thermal desorption
mass spectroscopy (TDS).

Inverted-staggered bottom-gate n- and p-channel TFTs were fabricated as shown in figure 1. In this series of experiments, the crystallinity and average grain size in 200 nm-thick poly-$Si_{1-x}Ge_x$ and poly-Si are more than 80% and 60-70 nm in diameter, respectively. The full width at
a half maximum in Raman spectroscopy at 520 wave number is 8-12 (/cm). The poly-$Si_{1-x}Ge_x$ or poly-Si films was deposited onto n-type (100) Si wafer with thermally oxidized 75 nm-thick $SiO_2$. After photo-resist patterning of source and drain regions, $2-5 \times 10^{15}$ ($ions/cm^2$) of phosphorus or boron ion was implanted at 30 keV, source/drain electrodes are patterned. To activate the dopant impurity, thermal annealing at 400 °C for 1 hour was performed. Finally hydrogenation were carried out to terminate dangling bonds by hot-wire technique at the substrate temperature of 150-200 °C and the filament temperature of 1100 °C under a $H_2$ flow rate of 200 sccm and 5 Torr for 1 h. The TFTs are fabricated with various channel length L (50-1000 μm) and channel width W (50-1000 μm). Figure 1 is an example of p-channel TFT, boron ion ($B^+$) is implanted. Phosphorus ion ($P^+$) in implanted and Chromium (Cr) is used for fabricating n-channel. Annealing can improve semiconductor-metal contact as well as activate dopant impurity.

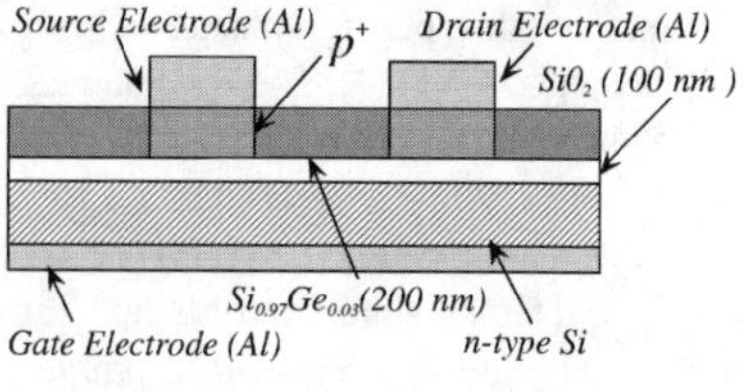

**RESULTS AND DISCUSSION**
*Deposition and hydrogenation*

Figure 1. Schematic cross sectional view of the poly-$Si_{1-x}Ge_x$ TFT.

One of the features of poly-$Si_{1-x}Ge_x$ and poly-Si films fabricated by the present CVD is that crystals grow from the early stage of deposition on a glass substrate and amorphous incubation layer is not observed in the film [9]. 10-110 nm-size of crystal grains are observed in the 200 nm-thick film, and the average grain size becomes larger as the gas residence time is smaller. Less than 20 seconds, 100 nm-size grains are fabricated and <110> orientation dominates. In our deposition system, however, deposition temperature is around 450 °C which is high enough for the most of the hydrogen to effuse out. Thus, the resulting defect density in 200 nm-thick film is $3.2-5.0 \times 10^{18}$ $cm^{-3}$. The hydrogen concentrations in poly-$Si_{0.97}Ge_{0.03}$ and poly-Si are about 80-90 ppm and 1000 ppm, respectively. Figure 2 shows the spin density dependence on grain size in around 1μm-thick films. The line in the figure 3 denotes -1 slope for the guide to the eye and it correlates with the data well. The -1 slope means that most of the defects in the film accumulates on the grain surface. This tendency is seen in figure 3 also. This figure shows the grain size dependence of Hall mobility and the line in the graph is slope 1, which means that the mobility strongly depends on grain size and corresponds to the total number of which the carrier crosses grain boundaries from one electrode to the other electrode. These results suggest that the defects should be terminated for device application.

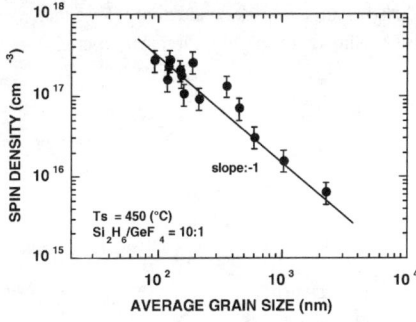

Figure 2. Spin density as a function of grain size of poly-$Si_{0.9}Ge_{0.1}$.

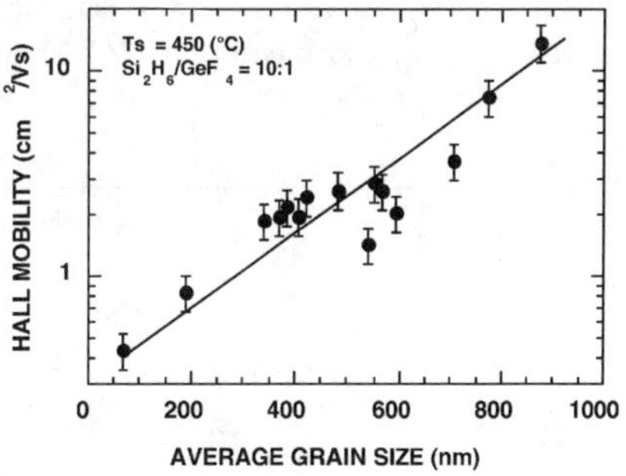

Figure 3. Hall mobility as a function of grain size of poly-$Si_{0.9}Ge_{0.1}$.

Hot wire technique is employed to decompose hydrogen molecule and to terminate dangling bond [3,9]. W investigated on/off current ratio in ni-diode structure as a function of i-layer thickness before and after hydrogenation in order to determine the effective termination depth for hydrogen. Figure 4 shows the on/off current ratio in ni-diode as a function of I-layer thickness. If hydrogen terminate effectively in the interface between n and i, the ratio would be improved more than as deposited. The ni-diode is fabricated by depositing poly-$Si_{0.97}Ge_{0.03}$ as an i-layer onto the n-type Si wafer, and 1mm-diameter of Al electrode is patterned. The ni-diodes with various i-layer film thickness is hydrogenated under the conditions that $H_2$ flow rate, gas pressure and substrate temperature are 100 sccm, 5 Torr and 150 °C, respectively. It is found that the effective hydrogenation depth is less than 300 nm and it suggests that i-layer thickness of TFT should be less than 300 nm despite the hydrogen diffusion length may be much longer. Figure 5(a) and (b) show the TDS spectra of (a) poly-Si and (b) poly-$Si_{0.97}Ge_{0.03}$, respectively. As shown in figure 4(a), TDS spectra have generally two peaks, low temperature (LT) at ~300 °C and high temperature (HT) ~600 °C, respectively, as seen in reference [10]. In the poly-Si fabricated by the present CVD, LT and HT peaks are observed at ~350 °C and ~640 °C and hydrogen concentration is around 0.1% before hydrogenation and become 0.65 % at maximum after hydrogenation.

In the case of poly-$Si_{0.97}Ge_{0.03}$, hydrogen concentration is less than 0.08% before hydrogenation and become 1.0 % at maximum. There are two kinds of effusion peaks at ~300 °C and ~380 °C, despite HT peak located near 600 °C is not observed. These two peaks in figure 5(a) may be originated from Ge-H and Si-H in the grain boundary or void according to Beyer's results [11]. The missing of HT peak in poly-SiGe indicates that there are few in-grain defects. Both the peaks increase with hydrogenation time by 30 minutes, the peaks, however, decrease when the hydrogenation treatment ics carried out over 40 min. The hydrogen peak areas increase at first and decrease when the the treatment is continued over 40 minutes, which correspond to the change in defect density and also change in dark conductivity. This change may be caused by the radial heat of 1100 °C heated tungsten wire or by highly chemical reactivity of atomic hydrogen.

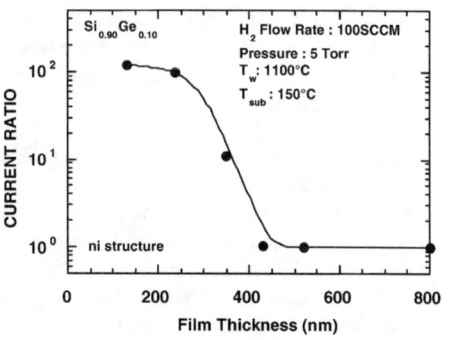

Figure 4. Effective hydrogenation depth.

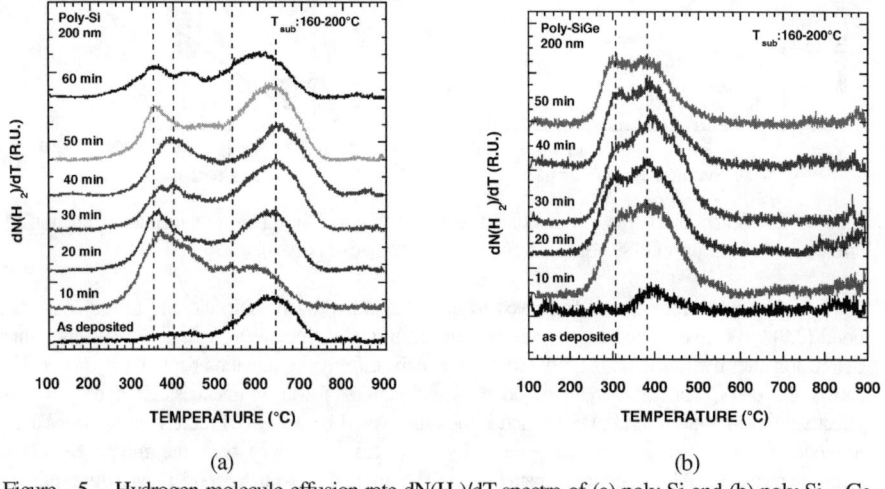

Figure 5. Hydrogen molecule effusion rate $dN(H_2)/dT$ spectra of (a) poly-Si and (b) poly-$Si_{0.97}Ge_{0.03}$ at the difference of hydrogenation time.

### *TFT characteristics of poly-$Si_{0.97}Ge_{0.03}$ and poly-Si TFT*

The transfer characteristic of TFTs as fabricated is very poor as shown in figure 6, which results from poor Ohmic contact between semiconductor and electrode metal as well as high defects in the active layer. After $H_2$ (or $N_2$) annealing for 1 hour, the characteristic is improved relatively. This may result from the improvement of Ohmic contact due to the activation of the dopant impurity, but many defects still exist in the active layer. By the hydrogenation, the characteristic is significantly improved as same as plasma hydrogenation [12]. It is found that the hydrogenation with hot wire technique is effective to terminate the dangling bonds, but prolonged hydrogenation, e.g., more than 40 minutes, makes the films degraded and defect density increased, which does not coincide with the case of plasma hydrogenation in the referennce [13,14].

Figure 7(a), (b) show the transfer characteristic before and after hydrogenation in (a) poly-Si (b) poly-$Si_{0.97}Ge_{0.03}$. In (a) poly-Si case, the field effect mobility and on-current are improved drastically, but threshold voltage, subthreshold swing and off current do not change significantly. This tendency of improvement indicates that dangling bonds in in-grain are dominant compared with those in inter-grain boundary, which is consistent with the results of TDS in figure 5(a). In poly-$Si_{0.97}Ge_{0.03}$ shown in figure 7(b), the characteristics concerned with field effect mobility, threshold voltage and subthreshold swing are significantly improved, which indicates that the defects in the inter-grain boundary and gate oxide interface are terminated well, deciding from

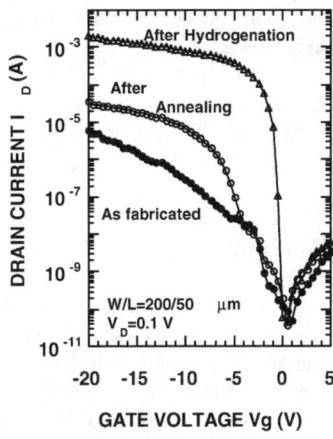

Figure 6. Transfer characteristics of p-channel TFT as shown in figure 1. The solid circles denotes as fabricated TFT's, open circles denote after annealing and open triangles denote after hydrogenation. (see text for detail)

numerical simulaion [15]. This would be found from the result that the off current near zero gate bias become leaky, which denotes the potential in inter-grain boundary becomes flat [16]. These results coincide with the results of hydrogen effusion that there are very few in-grain defects. For more exact

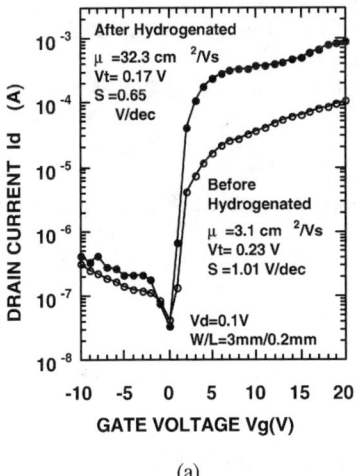

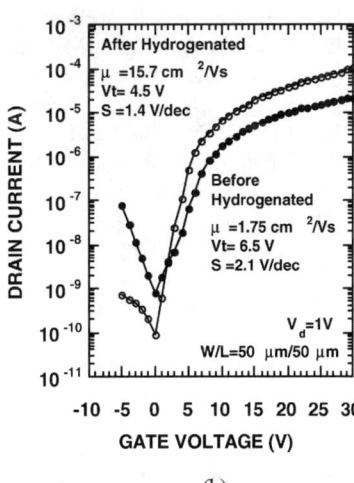

(a)                  (b)

Figure 7. Transfer characteristics of n-channel TFT of (a) poly-Si and (b) poly-$Si_{0.97}Ge_{0.03}$ at the difference of before hydrogenation (open circles) and after hydrogen (solid circles).

information about the defects and their passivation with hydrogenation and carrier conduction, further studies should be required.

In comparison with poly-Si and poly-$Si_{0.97}Ge_{0.03}$, it is found that defect distribution in the film between them is different, despite a total number of defects, the average grain size or the crystallinity is not so different. This difference can be originated from growth kinetics during growth process, such as impinging the nucleations or retarding the growth.

## SUMMARY

We have fabricated the inverted staggered TFT with poly-$Si_{0.97}Ge_{0.03}$ and poly-Si. The transfer characteristics of TFT are drastically improved by hydrogenation with hot wire technique. There is the optimum point in hydrogenation by hot wire with regard to film thickness and treating time. We have attained to fabricate bottom gate TFT with over 50 $cm^2$/Vs of field effect mobility.

The intensive and extended research are required to elucidate the phenomena of growth kinetics or to develop the TFT with much higher quality, such as more than 300 $cm^2$/Vs of field effect mobility in p- and n-channel TFT.

## ACKNOWLEDGMENT

The authors would like to thank Mitsui Chemicals, Inc. for supplying pure disilane gas and Central Glass Co., Ltd. for helium base 10%-tetrafluorinated germanium gas.

## REFERENCES

[1] Y. H. Yang and J. R. Abelson, Appl. Phys. Lett., 67, 3623 (1995).
[2] T. Hamasaki, H. Kurata, M. Hirose, and Y.Osaka, Appl. Phys. Lett., 37, 1084 (1980).
[3] H. Matsumura, Jpn. J. Appl. Phys., 30, L1522 (1991).
[4] S. Nishida, H. Tasaki, M. Konagai, and K. Takahashi, J. Appl. Phys., 58, 1427 (1985).
[5] J. Hanna, Proc. Mat. Res. Soc., 37, 105 (1995).
[6] J. Hanna, K. Shiota, and M. Yamamoto, Mat. Res. Soc. Proc., 507, 945 (1998).
[7] J. Lee, K. Shimizu and J. Hanna, Mat. Res. Soc. Symp. Proc., 715, 77 (2002).
[8] K. Shimizu, J. Zhang, J. Lee and J. Hanna, Mat. Res. Soc. Symp. Proc. 638, F14.13.1 (2001).
[9] K. Shimizu, J. Zhang, J. Lee and J. Hanna, Mat. Res. Soc. Symp. Proc., 636, 107 (2002).
[10] W. Beyer, "Tetrahedrally bonded Amorphous Semiconductors" ed. D. Adler and H. Fritsche (Plenum, New York, 1985), pp.129-145.
[11] W. Beyer, J. Non-Cryst. Sol. 198-200, 40 (1996)
[12] Kwon-Youg Choi and Min-Koo Han, J. Appl. Phys. 80, 1883 (1996).
[13] Kwon-Young Choi, Juhn-Suk Yoo, Min-Koo Han and Young-Sang Kim, Jpn.J. Appl. Phys 35, 915 (1996).
[14] I-Wei Wu, IEEE Electron Device Lett. 10, 123 (1989).
[15] Tsung-Kuan A. Chou and Jerzy Kanicki, Jpn. J. Appl. Phys 38, 2251 (1999).
[16] I-Wei Wu, IEEE Electron Device Lett. 12, 181 (1991).

## Optoelectronic detection of DNA molecules using an amorphous silicon photodetector

F. Fixe[1,2], D.M.F. Prazeres[2], V. Chu[1] and J.P. Conde[1,3]
[1]INESC Microsistemas e Nanotecnologias, Lisbon, Portugal
[2]Center for Biological & Chemical Engineering, Instituto Superior Técnico, Lisbon, Portugal
[3]Department of Materials Engineering, Instituto Superior Técnico, Lisbon, Portugal

## ABSTRACT

This work demonstrates the use of an amorphous silicon (a-Si:H) photodetector to measure the density of covalently-bound DNA molecules tagged with a fluorescent molecule. This device is based on the photoconductivity of a-Si:H in a coplanar electrode configuration. Excitation of a fluorescently-tagged biomolecule with near UV/blue light results in the emission of visible light. The emitted light is then converted into an electrical signal in the photodetector, thus allowing the detection of the presence of the tagged DNA molecules. The design, fabrication and characterization of this integrated a-Si:H-based bio-detector is described. The detection limit of the present device is of the order of 20 pmol/cm$^2$. A surface density of $\leq$ 30 pmol/cm$^2$ for DNA covalently-bound to an active silica layer was measured with the a-Si:H-based bio-detector.

## INTRODUCTION

DNA microarrays have the potential to revolutionize the acquisition and analysis of genetic information [1,2]. Microarray technology bridges fields such as materials science, microelectronics, biochemistry and physics. In these massively parallel microarrays, DNA hybridization can be tested with different capture single-strand DNA "probes" immobilized at specific sites in a matrix. Current DNA-chip data acquisition is based on the use of fluorescence microscope image capture of the emission from a fluorescent tag bound to DNA "target" molecules that hybridize with immobilized DNA "probe" molecules in the array [3-8]. Although these optical systems show high sensitivity, they require the use of complex image acquisition and processing systems. On-chip electronic data acquisition would improve both the speed and the reliability of DNA chip hybridization pattern analysis. This would be particularly important for applications such as clinical point of care diagnostics.

In this paper, a detection approach in which the presence of fluorescently-tagged DNA molecules is optoelectronically detected by an integrated thin-film a-Si:H-based photodetector is demonstrated.

## EXPERIMENTAL

### Device fabrication

The device (Figure 1) is fabricated on a glass substrate. The length of the aluminum parallel contacts ranges between 30 μm and 3 mm and their separation is between 5 μm and 500 μm. The Al metal lines are defined by photolithography in a clean room environment. The a-Si:H, silicon nitride (SiN$_x$), amorphous silicon-carbon (a-SiC:H) and silicon dioxide (SiO$_2$) thin-films are deposited by RF-PECVD. The SiO$_2$ film is functionalized by a silanization and cross-linking process (see below) and becomes an active layer upon which the probe DNA molecule can be immobilized.

## Spectral response measurements

One of the parallel Al electrodes is connected to a voltage source and the other to a lock-in amplifier. The chopped incident light (380 nm-800 nm) is normal to the a-Si:H detector and the spectral response is the ratio between the number of photogenerated electrons and the number of photons present in the incident light.

## Surface functionalization

The surface of the $SiO_2$ layer is first cleaned with cholic acid (12 hr at room temperature). The surfaces are then silanized [9] with 3-aminopropyltriethoxysilane (APTES) 2% (v/v) in acetone, for 2 hr at room temperature. This process results in a surface covered with reactive primary amines (-$NH_2$).

## DNA-labeling and purification

The single-strand DNA molecules are marked with PyMPO, SE (1-(3-(succinimidyloxycarbonyl)benzyl)-4-(5-(4-methoxyphenyl)oxazol-2-yl) pyridinium bromide) through the primary amine group present on the 3´-end. The labeling reaction is performed for at least 6 hrs and ethanol precipitation is used to remove excess fluorescent molecules [10].

## DNA-Calibration curve

The calibration of different surface densities of DNA molecules with the a-Si:H-based photodetector is performed by measuring the spectral response (SR) of the device after a known volume of DNA solution, with a specific concentration of DNA marked with PyMPO, is dried, forming a spot with a known area, on the top of the photodetector structure. After the measurement at that particular DNA surface density, the adsorbed DNA is removed by washing and the spectral response is measured again to confirm that the bio-detector returns to the initial state. This procedure is performed starting with the lowest concentration and is repeated for solutions with increasing concentration of DNA tagged with PyMPO. The calibration curve is the response of the a-Si:H-based biodetector at 400 nm ($\lambda_{exc}$ of PyMPO) plotted as a function of the DNA density (pmol/cm$^2$).

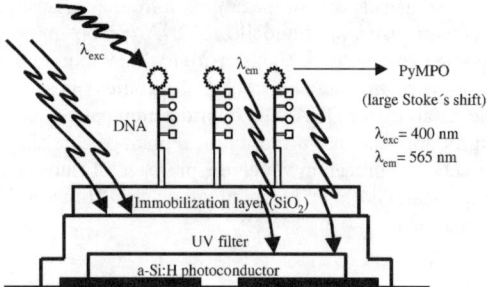

**Figure 1.** Schematic diagram of the integrated bio-detector with immobilized tagged DNA molecules: a-Si:H (photoconductor), UV filter ($SiN_x$ (insulator film), a-SiC:H (UV light filter)) and $SiO_2$ (functional layer for the DNA immobilization and hybridization).

## DNA immobilization

The $NH_2$ groups are further functionalized by covalently attaching the hetero-bifunctional cross-linker, sulpho succinimidyl 6-maleimidylhexanoate (sulfo-EMCS), for 2 hr at room temperature. The DNA probes are 17 base-pair oligonucleotides with a thiol group on one end (5´) and an $NH_2$ group on the other end (3´) (HS-5´- TTA ACT TTG TTA AAA AC - 3´- C7-$NH_2$). The probes are labeled with PyMPO at the $NH_2$-terminated end to enable optoelectronic detection of the immobilization. The labeled probes are covalently linked to the maleimide termination of the crosslinker monolayer on the chip *via* the thiol group [9].

## RESULTS AND DISCUSSION

### Incorporation of barrier and filter layers

Since the a-Si:H layer has a significant photosensitivity in the UV, to improve the sensitivity of the device an a-SiC:H ($E_g$=2.1 eV) UV filter was deposited above the a-Si:H photoconductor to lower its absorption of light at the PyMPO excitation energy (figure 2). A thin $SiN_x$ layer was used to reduce the spill-over of carriers photogenerated in the a-SiC:H into the a-Si:H.

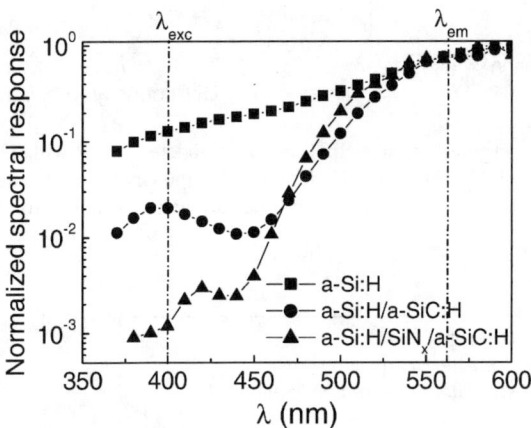

**Figure 2.** Improvement of the sensitivity of the device at 400 nm ($\lambda_{exc}$ of PyMPO), by adding a filter layer (a-SiC:H) and a barrier layer ($SiN_x$). The spectral response curves were normalized at the highest signal.

### Calibration of surface DNA concentration

Measurement of the spectral response curve of the detector in the presence of PyMPO shows an increase in the detector signal in the 400-450 nm range when PyMPO is present. This occurs because PyMPO, when excited at this wavelength, emits light with a spectrum centred in the visible ($\lambda_{em}$=565 nm), which is in turn absorbed by the a-Si:H photoconductor, resulting in a photocurrent. Because the detector has an integrated UV filter, the overall effect of the system in the presence of UV illumination is an apparent increase in the response at 400-450 nm when PyMPO or DNA tagged with PyMPO is present on the surface. The excitation light in this range is filtered, while the emitted light at 565 nm is transmitted and produces a photoresponse. Thus tagging DNA molecules with the dye PyMPO allows the use of an a-Si:H photodetector to optoelectronically detect the presence of DNA.

Different concentrations of DNA marked with PyMPO are adsorbed on the device surface and measured. After washing, the spectral response of the a-Si:H photodetector returns to the initial state (figure 3a, open circles). On figure 3b, the relative response of the device at 400 nm is plotted for different concentrations of adsorbed DNA. The DNA density on the surface is calculated taking into account the volume of the drop, the concentration of the DNA solution used, and the area of the drop. From figure 3b, the detection limit of the present device is estimated at approximately 20 pmol/cm$^2$.

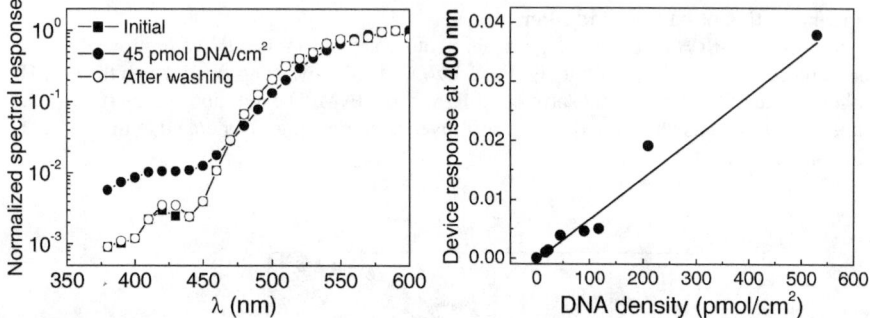

**Figure 3.** (a) Initial spectral response of the a-Si:H-based photodetector, SR with 45 pmol/cm$^2$ of DNA labeled with PyMPO, and SR after washing the DNA from the device. (b) Calibration curve of the device response after baseline subtraction as a function of the DNA surface density.

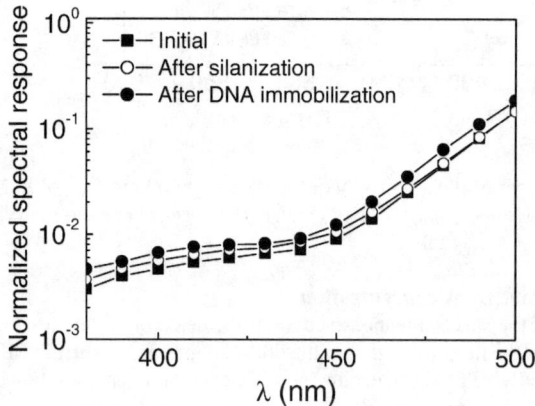

**Figure 4.** The normalized spectral response of the full detector system in the initial state (squares), after silanization (open circles) and after crosslinking and DNA immobilization (closed circles).

Detectors operating in the photoconductive mode can have gain if the recombination lifetime is larger than the transit time of the majority carrier through the device [11]. In our configuration, however, this is not the case as the recombination lifetime is of the order of

$10^{-7}$ s and the transit time was estimated to be approximately $10^{-4}$ s. Another factor which might have affected the calibration of the detector response is the dependence of $\sigma_{ph}$ on the photocarrier generation rate, G, in the a-Si:H, which is known to follow a power law dependence, $\sigma_{ph} \alpha\ G^{\gamma}$, where $\gamma$ is between 0.5-1. However, at the low illumination intensities used in this experiment, $\gamma$ is nearly 1.

**DNA immobilization**

After testing the sensitivity of the a-Si:H-based bio-detector to different surface concentrations of adsorbed PyMPO-tagged DNA molecules, the same device was used to quantify the surface density of DNA probes covalently immobilized on top of a functionalized $SiO_2$ film. Figure 4 shows that there is a small increase of the device response in the UV /blue when compared to the signal after silanization and cross-linking. This increase is close to the detection limit of the present device and can only be used to give an estimate of the upper limit of the immobilized DNA concentration. Inserting the response at 400 nm measured after DNA immobilization into the calibration curve (figure 3b), the immobilized DNA is estimated at $\leq 30$ pmol/cm$^2$. This surface density is in agreement with the values reported in the literature which give, for typical densities of DNA immobilized on a functionalized glass surface, values between 1-200 pmol/cm$^2$ [12,13].

To improve the sensitivity of the device, an interference filter with high reflection at 400 nm and high transmission at 565 nm is presently being developed and will be incorporated into a future device design.

## CONCLUSIONS

An a-Si:H-based photodetector was developed that can detect surface DNA concentrations down to 20 pmol/cm$^2$. Linear detection behaviour in the range of 20-500 pmol DNA/cm$^2$ was observed. Covalently-bound DNA immobilized on a functionalize $SiO_2$ surface layer was quantified ($\leq 30$ pmol/cm$^2$) although its surface density is close to the detection limit of the present detector. The detection system demonstrated in this paper could be the basis of an optoelectronic detector array that can enable rapid, reliable and inexpensive detection of nucleic acids in a wide variety of DNA microarray applications.

## ACKNOWLEDGEMENTS

The authors gratefully acknowledge F. Silva for help in some of the deposition steps and J. Faustino for wirebonding. This work was supported by the Fundação para a Ciência e Tecnologia (Plurianual and Programatico programs, POCTI projects and a Ph.D.grant to F. Fixe).

## REFERENCES

1. Schena, M., Shalon, D., Davis, R.W. and Brown, P.O., *Science* **270**, 467 (1995).
2. Duggan, D.J., Bittner, M., Chen, Y., Meltzer, P. and Trent, J., *Nature Genet.* **21**, 10 (1999).
3. Southern, E., Mir, K. and Shchepinov, M. *Nature Genet.*, **21** (Suppl.), 5 (1999).
4. Ramsay, G. DNA chips: state-of-the art, *Nature Biotechnol.*, **16**, 40 (1998).
5. Pease, A.C., Solas, D., Sullivan, E.J., Cronin, M.T., Homes, C.P. and Fodor, S., *PNAS*, **91**, 5022 (1994).
6. Marshall, A. & Hodgson, J. DNA chips: an array of possibilities, *Nature Biotechnol.*, **16**, 27 (1998).

7. Pividori, M.I., Merkoçi, A. and Alegret, S., *Biosens. Bioelectron.*, **15,** 291 (2000).
8. Wang, J., *Nucl. Acids Res.*, 28, 3011 (2000).
9. Fixe, F., Faber, A., Gonçalves, D., Prazeres, D.M.F., Cabeça, R., Chu, V., Ferreira, G. and Conde, J.P., *Mat. Res. Soc. Symp. Proc.* **723**, O2.3.1 (2002).
10. *in* http://www.molecularprobes.com
11. Rose, A., *Concepts in photoconductivity and allied problems,* John Willey & Sons, USA (1978).
12. Rogers, Y.H. Jiang-Baucom P. Huang Z. Bogdanov V. Anderson S. and Boyce-Jacino M.T. *Anal. Biochem.* **266** 23 (1999).
13. Beier, M. and Hoheisel, J., *Nucl. Acids Res.*, **27**, 1970 (1999).

# Fabrication of Novel TFT LCD Panels with High Aperture Ratio Using a-SiCO:H Films as a Passivation Layer

W.S. Hong[1], K.W. Jung[2], B.K. Hwang[3], G. Cerny[3], S.H. Yang[2], J.H. Choi[2], and K. Chung[2]
[1]Dept. of Electronics Engineering, Sejong University, Seoul, Korea
[2]R&D Team, AMLCD Division, Samsung Electronics, Yongin-city, Kyunggi-do, Korea
[3]Electronic Industry & Advanced Materials Business, Dow Corning, Midland, MI, U.S.A.

## ABSTRACT

Fabrication of a novel TFT-LCD panel, using amorphous silicon oxycarbide (a-SiCO:H) films as a passivation layer, was successfully demonstrated for the first time. The a-SiCO:H low-k films were deposited using a standard PECVD (plasma-enhanced chemical vapor deposition) reactor from a gas mixture of trimethylsilane[$Si(CH_3)_3H$] and $N_2O$. The resulting films have a dielectric constant between 2.7 and 3.5 and high optical transmittance in the range of visible light

The transfer characteristics of the TFT's having a-SiCO:H as a passivation layer were comparable with that of a conventional TFT with a PECVD-grown $SiN_x$ passivation layer. Stability of the resulting TFT was performed under prolonged bias conditions, and the source-drain current was fairly constant over the test period. The LCD panel with the a-SiCO:H passivation layer showed 30% higher brightness than that of the standard panel.

## INTRODUCTION

As the demand of the high definition TFT-LCD's (Thin Film Transistor Liquid Crystal Displays) increases in the market, the transmittance of the LCD panel must be enhanced to achieve both high brightness and resolution. The aperture ratio (or *fill factor*) of the TFT panel must be maximized by eliminating any inactive area and minimizing the misalignment with the color filter layer. As shown in Fig. 1(b), this high aperture ratio design requires that the pixel

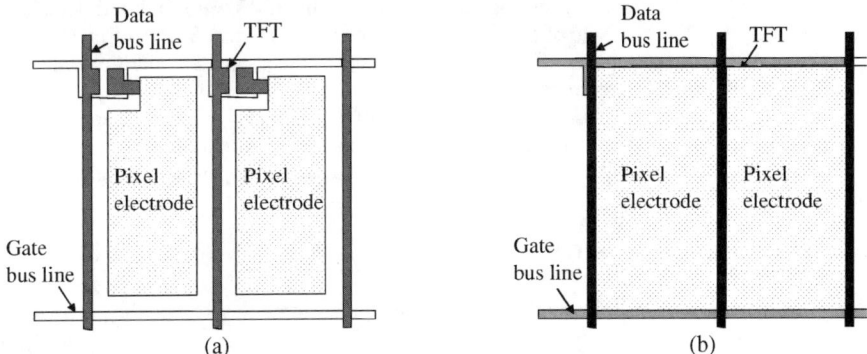

**Figure 1.** Plan view of the pixel structure of TFT-LCD panels: (a) conventional structure, (b) high aperture ratio

electrodes are overlapped with the gate and data signal lines. In such a structure, however, build-up of parasitic capacitance between the pixel electrodes and the signal lines may lead to a huge load capacitance and prevent the pixels from being charged to adequate signal levels. The passivation layer, located between the pixel electrode and the data electrode, must then provide a sufficiently low dielectric constant ($k$) and good insulation. However, the conventional $SiN_x$ (non-stoichiometric silicon nitride) passivation layer cannot be used in this structure because of its relatively high dielectric constant ($k$~7), slow deposition rate and high residual stress.

Efforts have been made to utilize organic insulators, such as acrylic, as a thick passivation layer[1,2]. However, the TFT's with organic passivation often show a high contact resistance between the pixel and the data electrodes, because the organic layer tend to be sputtered out and redeposit inside the via-holes during subsequent processes.

Therefore, an inorganic alternative is desired, which can be processed with the same equipments and process flow that the conventional $SiN_x$ uses. Silicon carbo-oxide(a-SiCO:H), so-called *low-k*, shows a remarkably low dielectric constant and high deposition rate, and has been drawing attention as a good candidate for inter-layer dielectrics[3]. The a-SiCO:H film can be deposited by the conventional PECVD technique, using $SiH(CH_3)_3$ (trimethylsilane, Z3MS™) and $N_2O$ (nitrous oxide) as source gas[4]. It has a much lower thermal expansion coefficient and better heat resistance and time stability than those of organic films. It also has much lower residual stress than $SiN_x$, and a layers thicker than 5000 Å can be stacked without peeling off.

In this paper, process variables and their effects on various properties of the a-Si:C:O film are studied. Also, characteristics of a TFT-LCD panel fabricated with the a-Si:C:O passivation layer is discussed.

## EXPERIMENTAL DETAILS

Layers of a-SiCO:H were deposited on Corning 1737 glass substrates using a standard PECVD (plasma enhanced chemical vapor deposition) reactor. Dow Corning Z3MS™ (trimethylsilane) and $N_2O$ were used as source gas. Argon was added to boost the deposition rate. Source gas composition was varied to investigate its influence on various properties of the a-SiCO:H film. Substrate temperature and chamber pressure were set to 250°C and 2 torr, respectively. Film thickness was measured by both the Alphastep profilometer and the ellipsometer, and the results of the two agreed very well.

The type and density of various bonding structure in the films were analyzed with Bio-Rad FTIR system. The backside of the substrate was scanned and the a-SiCO:H film stress was estimated from the curvature. Surface hardness was also measured by micro-indentation using a thermal $SiO_2$ film as a reference.

A 1.5 μm thick a-SiCO:H layer was applied to 17" XGA TFT panels. Display characteristics of a fully integrated LCD panel were measured with a BM5A photometer and a PR650 calorimeter, and were compared with those of a same size panel with organic passivation.

## RESULT AND DISCUSSION

The main objective of this study was to attain a pixel structure such that the pixel electrodes

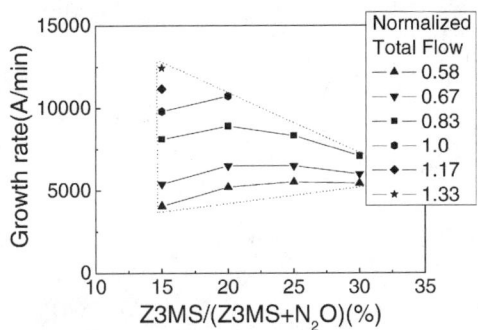

**Figure 2.** Variation of the growth rate with the Z3MS fraction in the total source gas, Z3MS/(Z3MS+N$_2$O). Each symbol represents corresponding level of total gas flow rate, normalized to that of a standard SiN$_x$ recipe.

and the signal lines are vertically separated as much as practical. Therefore, the first step was to obtain a fast-growing film with sufficiently good mechanical and electrical properties. It has been reported that the a-SiCO:H growth rate can be enhanced by increasing the plasma power[5]. However, in the TFT-LCD fabrication the large substrate size casts a limit on the capacity of the plasma power supply. For this reason, the gas composition was chosen to be the primary control variable in this study.

Figure 2 shows variation of the growth rate as a function of the Z3MS content in the source gas for different levels of the total gas flow participating in the reaction. We took the total flow rate used for standard SiN$_x$ deposition as a reference (dark hexagon) and normalized all other flow rates. As shown in the figure, the deposition rate shows a maximum with the increase in the fraction of Z3MS gas. As the total gas flow (Z3MS+N$_2$O) increases, the growth rate increases monotonically.

This result suggests that the supply of precursor radicals is the rate-limiting step of the growth kinetics. Existence of a maximum along the x-axis implies that the deficit of either of the species (Z3MS or N$_2$O) hinders the film growth. Increase in the total flow is believed to increase the growth rate by providing sufficient number of reactive radicals, but too many radicals available in the plasma may lead to particle generation before reaching to the substrate. Outside the gas composition range indicated by the dotted triangle on the graph, either the plasma could not be sustained or abnormal (powder-like) deposition occurred.

Variation of the dielectric constant ($k$) as a function of the Z3MS content is plotted in Figure 3 for different levels of total gas flow. The dielectric constant decreases monotonically with the Z3MS fraction and with the total gas flow, showing no minima or maxima. Also, the dielectric constant seems to converge to a value of ~2.7.

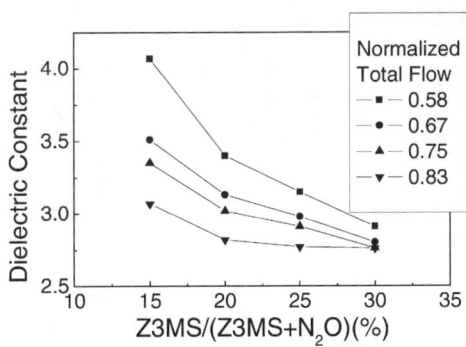

**Figure 3.** Variation of the dielectric constant with the Z3MS fraction in the total source gas, Z3MS/(Z3MS+N$_2$O). Each curve represents corresponding level of total gas flow rate, normalized to that of a standard SiN$_x$ recipe.

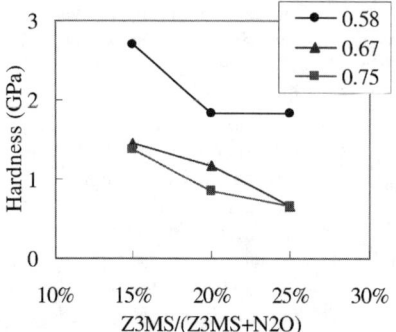

**Figure 4.** Indentation hardness as a function of the Z3MS content, plotted for different levels of the total gas flow.

The deposition rate and dielectric constant has a trade-off with the mechanical property of the film. The passivation layers in TFT's are subject to a great deal of shear stress during the fabrication and test processes. Therefore, a certain level of surface hardness must be ensured to protect the TFT's during subsequent process.

Figure 4 shows the indentation hardness as a function of the Z3MS content. The indentation hardness decreases as both the Z3MS content and the total gas flow increase.

It has been reported that the methyl group ($-CH_3$) contained in the Z3MS gas terminates the -Si-O- network, creating atomic-scale micro-voids around themselves[7]. The formation of these micro-voids takes on an effect of alloying with vacuum. Therefore, sufficient supply of the $-Si(CH_3)_n-$ radicals to the growing surface may help increase the deposition rate by enlarging the volume of the growing film and reduce the dielectric constant and hardness.

In order to verify this phenomenon, type and density of various chemical bonding inside the a-SiCO:H were analyzed by FTIR technique. The ratio of the densities of $Si-CH_3$ and $Si-O$ bonds, $[Si-CH_3]/[Si-O]$, were calculated from the FTIR peak area and used as a variable to estimate the content of micro-void.

As shown in Figure 5 (a), the deposition rate increases and tends to saturate as the

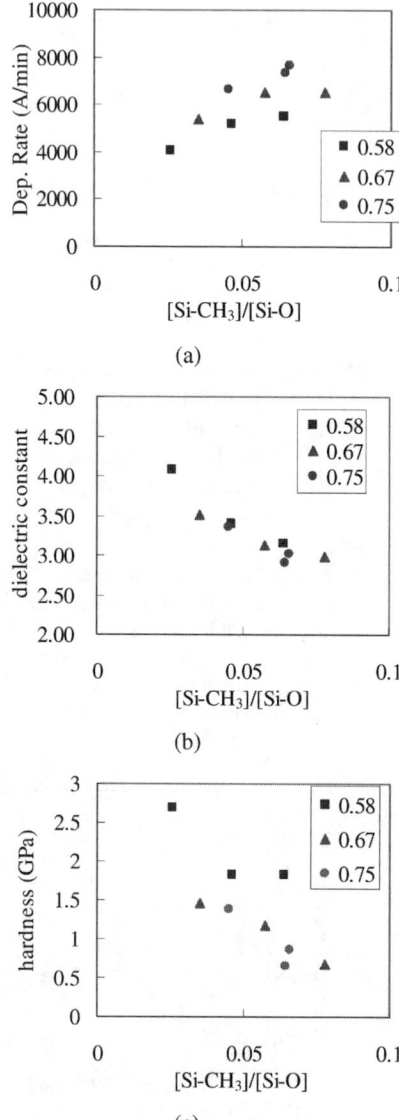

**Figure 5.** Effects of the Si-CH3 bonds in the film on various material characteristics of the a-SiCO:H (a) deposition rate, (b) dielectric constant, (c) indentation hardness

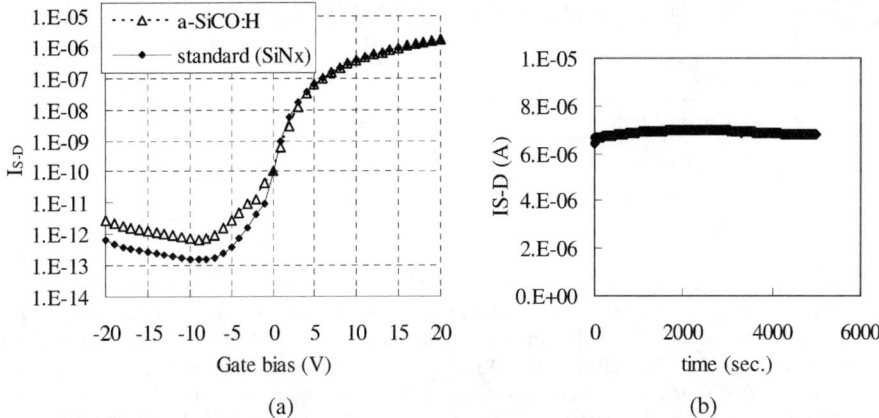

**Figure 6.** (a) Source-drain current ($I_{S-D}$) as a function of gate bias voltage, White triangle and dard square symbols represent the TFT's with 1.5μm-thick a-SiCO:H passivation and with 2000Å-thick $SiN_x$ passivation, respectively. (b) Time dependence of the TFT on-current for a-SiCO:H passivated sample. Gate bias of +20V was applied.

[Si-CH$_3$]/[Si-O] ratio increases. At the same [Si-CH$_3$]/[Si-O] ratio, the deposition rate increases with the total gas flow. Hence, the overall growth rate is rather dependent on the material transport over the growing surface than on expansion of the virtual volume by incorporating micro-voids.

On the other hand, as shown in Figure 5(b), the dielectric constant only depends on the [Si-CH$_3$]/[Si-O] ratio in the film and not on the total flow of the source gas, indicating that it only has to do with the micro-voids created by the -CH$_3$ group. The indentation hardness, as plotted in Fig. 5(c) shows a slightly random behavior with the [Si-CH$_3$]/[Si-O] ratio, but can be approximated as a single straight line, if a relatively high experimental error embedded in the measurement is considered. However, the graph implies that the growth kinetics may affect the hardness of the film, as the data points for the normalized total flow of 0.58 lie well above the others. If the growth kinetics is slowed down below a certain level by controlling the source gas supply, the hardness can be improved with the same content of the Si-CH$_3$ bonds in the film.

The minimum allowable indention hardness of the a-SiCO:H film was empirically determined to be ~1 GPa. From these results, the optimum gas composition was chosen to be the Z3MS/(Z3MS+N$_2$O) fraction of 25% and normalized total flow of 0.58.

Figure 6 (a) compares the transfer characteristics of the TFT's having a-SiCO:H passivation layer and that having standard $SiN_x$. Both TFT's show on-current values higher than 1 μm, but the one with a-SiCO:H passivation showed a higher off-current than the other. However, an on/off current ratio greater than $10^6$ was obtained from the device with a-SiCO:H, and this value was sufficiently high to drive LCD pixels. Stability of the a-SiCO:H passivated TFT was tested by applying 20Vdc to the gate electrode and 10Vdc between source and drain for prolonged time. As shown in Fig 6(b), the source-drain current was fairly constant over the test period.

Residual stress becomes significant in TFT-LCD manufacturing, because of the large

**Table I.** Display Characteristics of a 17" XGA (1024x768) panel

|  | a-SiCO:H | Organic Film |
|---|---|---|
| Max. Brightness (cd/m$^2$) | 450 | 480 |
| Contrast ratio | 360:1 | 390:1 |
| Panel Transmittance | 6.86% | 6.95% |
| Thickness | 1.5 µm | 3 µm |

substrate size. Only a small curvature caused by the overlaid film may lead to a severe deflection over a large area substrate and make it difficult to align during lithography. The residual stress of the a-SiCO:H film was estimated to be an order of magnitude lower than that of the standard SiN$_x$. Therefore, a-SiCO:H passivation layer can be grown ten times thicker than the standard SiN$_x$ without the concern of extensive bending or film failure.

In Table I, display characteristics are compared for two 17" XGA (1024 X 768 resolution) LCD panels made from the a-SiCO:H layer and from the organic film, respectively. The LCD panel with the a-SiCO:H passivation showed maximum brightness of ~450 cd/m$^2$ and contrast ratio of ~360:1, which were comparable to those of the same grade panel with organic passivation. This high aperture-ratio structure exhibited ~30% increase in brightness compared to a standard one.

## SUMMARY

A low-k inorganic insulator, a-SiCO:H, has been successfully processed and applied to high aperture-ratio TFT-LCD's by conventional PECVD technique. Deposition rates greater than 5000Å/min. and dielectric constant values as low as 2.7 could be obtained by controlling the source gas composition. A 17" XGA panel having a high aperture-ratio structure was fabricated using a 1.5µm thick a-SiCO:H layer. The panel exhibited display characteristics comparable to those of its organic counterpart.

## REFERENCES

1. J.H.Kim and H.S.Soh, *AMLCD '97*, pp.5 (1997)
2. M.Sakamoto, T.Ukita, A.Maeda and S.Ohi, *SID 96 Digest*, pp.681 (1996)
3. E.Korczynski, *Solid State Technology*, pp. 43, May 1999.
4. M.J.Loboda, J.A.Seifferly, C.M.Grove and R.F.Schneider, in *Environmental, Safety and Health issues in IC Production*, edited by R.Reif, M.Heyns, A.Bowling and A.Tonti, (Mater. Res. Soc. Proc. **447** Pittsburg, PA 1997) pp.145-151
5. G.Cerny, *Dow Corning Internal Report*, Nov. 2000.
6. J.Shi, M.A.Piano, T.Mountsier and S.Nag, *Semicon Korea Technical Symposium*, 2000, pp. 279 (2000)

## LEAKAGE CURRENT BEHAVIOR IN COMMON I-LAYER A-SI:H P-I-N PHOTODIODE ARRAYS

Jeremy A. Theil

Semiconductor Products Group, Agilent Technologies, 5301 Stevens Creek Blvd., MS 51L-GW, Santa Clara, CA, 95051, U.S.A.

### ABSTRACT

Hydrogenated amorphous silicon photodiode arrays form the basis of monolithic three-dimensional integrated circuit sensor technology. In these arrays, the intrinsic a-Si:H layer covers the entire area to maximize light collection. One technique by which the pixel diode is defined, is to pattern the bottom contact layer independently of the intrinsic layer. One of the most important characteristics of any diode array, however, is that the dark-state reverse bias leakage currents must be as low as possible to minimize diode noise. This study examines the leakage currents associated with the pixelated array. These structures are unique in that the edge of the diode is defined by the local electric field between diodes, rather than the physical surface of an a-Si:H film. The effect of the diode edge has been found to induce a field-dependent component to the reverse bias leakage current. For example, diodes with 5500Å i-layer, have a junction leakage of 14 to 20 $pA/cm^2$, at $5.0 \times 10^4$ V/cm, while the pixel edge-dependent current component can be as high as 30 $pA/cm^2$. In addition, it will be shown that the i-layer thickness and junction doping plays a key role in determining the behavior of the leakage currents.

## 1  INTRODUCTION

Over the last ten years, there has been increasing interest in the use of a-Si:H in photodiode arrays that are monolithically integrated onto integrated circuits [1-3]. Such integration allows a combination of 1) reduced imaging pixel area, 2) reduced sensor cost, 3) lower photodiode leakage, and 4) improved pixel sensitivity. As pixel-level complexity (hence area) grows, the advantages become more apparent. In order to maximize the photodiode light collection area of the array, the array is designed to have a continuous fully depleted intrinsic layer, and patterned electrodes on the integrated circuit side of the pixel to define the diode. One of the most important performance requirements of any imaging array is the dark-state reverse bias pixel leakage current, (a.k.a. dark current), as this sets the lower limit of dark spatial noise (shot noise). In most designs, the photodiode junction leakage current is a limiting factor, therefore, it is worthwhile to minimize this value. For a-Si:H photodiode arrays, leakage current sources include 1) junction leakage, 2) thermally generated bulk currents, 3) array edge injection currents, and 4) pixel edge leakage currents. Some work has been done on characterizing junction, bulk, and array edge injection currents for stacked diodes, however, little if any has been said about the pixel edge currents due to the difficulty of pixel fabrication [4-6]. It has been widely assumed though, that for the p-i-n diode stack, edge currents are dominated by surface states along the junction boundary. However, since that these diodes have continuous intrinsic layers, it is possible to decouple the effect of the of a-Si:H surface states from the edge of the array. This gives a unique opportunity to examine purely geometric effects of the pixel edge. The objective of this paper is to explore the diode edge effects on leakage current behavior as a function of electric field, i-layer thickness, and contact layer doping.

## 2  EXPERIMENTAL

Details of diode fabrication have been presented elsewhere, and are only briefly described here [1, 2, 7]. Diode array fabrication starts with a standard integrated circuit process flow. The photodiode array consists of n-type a-Si:H pixels overlapping a thin metallic pixel contacts with a common intrinsic and p-type a-Si:H layers that form an array of p-i-n junctions, (see Figure 1). Top contact to the array is made by using a transparent conductor layer that connects the top surface of the p-type a-Si:H layer to vias adjacent to the array, (monolithic top contact structure) [2]. Allowing the transparent conductor to be in contact with the edge of the intrinsic layer for the array simplifies sample construction, but puts the transparent conductor in direct contact with the intrinsic layer [1, 8]. This produces a contact junction that injects a dark current into the array, which overwhelms any surface state leakage formed by the physical array edge. However, all test structures are bounded by a ring diode that is held at the same bias as the measurement structure itself, so that it removes the injected current.

The test structures measured for this paper came from a general process development vehicle design for this process flow and consists of junction area ranging from 140 μm on a side up to 1940 μm on a side, and striped pixel-edge intensive structures with 940 μm sides ranging from $3.7 \times 10^3$ μm, to $1.2 \times 10^6$ μm. Two series of samples were made: 1) various i-layer thickness 3000, 4000, 5500, 7500 and 9000Å, 2) where the p-layer boron atomic concentration, was varied from $7.0 \times 10^{19}$ cm$^{-3}$ to $2.1 \times 10^{20}$ cm$^{-3}$, and the n-layer phosphorus concentration was varied from $2.0 \times 10^{20}$ cm$^{-3}$ down to $2.0 \times 10^{19}$ cm$^{-3}$, all measured by SIMS. In all samples the rest of the p-i-n junction stack contained a p-layer thickness of 200Å, an n-layer thickness of 500Å, and a 600Å transparent conductor layer on top of the p-layer. The a-Si:H layers are formed by very high rate PECVD deposition methods (> 20 Å/s), with the resultant films having an intrinsic defect density of $< 4 \times 10^{15}$ cm$^{-3}$ [9]. Unless otherwise noted, the dopant concentrations of the junction layers are: [B] is $7.0 \times 10^{19}$ cm$^{-3}$ for the p-layer, and [P] is $2.0 \times 10^{20}$ cm$^{-3}$, for the n-layer.

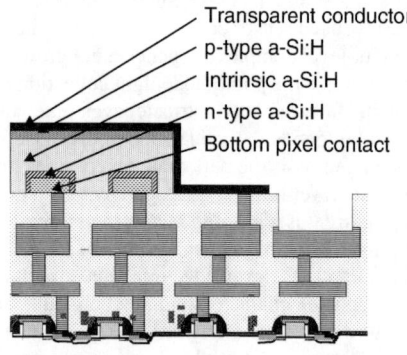

Figure 1: Schematic diagram of elevated a-Si:H photodiodes.

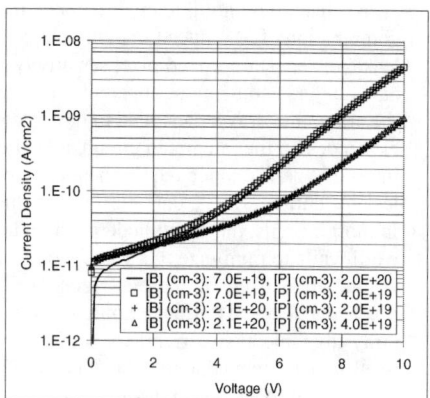

Figure 2: Reverse dark J-V plot of a-Si:H p-i-n junctions with various p- and n- doping concentrations. The i-layer thickness is 5500Å.

Two types of electrical measurements were made on the devices utilizing an Agilent 4156B. One was a reverse bias I-V measurement, in which the junction bias was varied from 0 to either 5 or 10V reverse bias in 0.1 V increments. At each measurement point, the measurement was made after a 65 second hold at bias and a long measurement integration time, thus allowing for 40fA sensitivity. The other was a low frequency transient decay measurement made under reverse bias conditions from 0.2 to 4V. In these experiments, the instrument was put into a sampling mode in which data was collected every 0.1s, and the current was monitored from the point at which the diode bias was switched from 0V to the set bias. In all cases, the guard-ring diode that surrounded the structure-under-test was driven with identical voltages but utilizing separate source measurement units, so that there was no current flow between the two devices. All measurements were made in complete darkness and a sample temperature of 21°C.

## 3  RESULTS

The effect of p and n layer doping levels on current density is shown in Figure 2. In this case, the solid curve has the same doping level as the thickness series experiment and each junction had an i-layer thickness of 5500Å. The J-V of Figure 3 shows two leakage current behaviors, with a transition point about that increases with i-layer thickness. Given the dopant concentrations in each junction, there is a correspondence of phosphorus doping and the low bias region in which the higher phosphorus concentration corresponds to lower junction leakage. Conversely, the boron doping concentration correlates to the higher bias region, in which higher boron concentration leads to lower junction leakage.

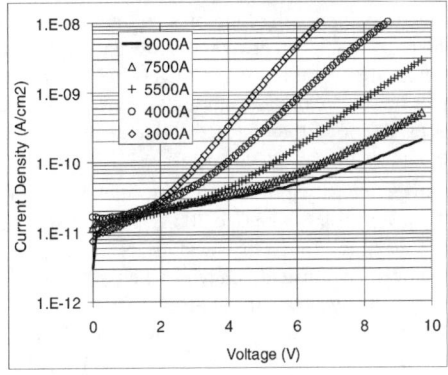 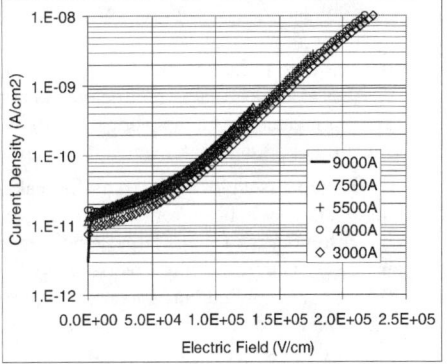

Figure 3: Reverse dark J-V plot of a-Si:H p-i-n junctions with various i-layer thickness.

Figure 4: Reverse dark J-E plot of a-Si:H p-i-n junctions with various i-layer thickness.

The reverse bias junction current density behavior of a 940 μm x 940 μm square diode as a function of i-layer thickness is shown in Figure 3. For each curve the junction shows qualitatively the same behavior, with a relatively voltage independent regime at low biases, then switching to more voltage dependent regime at higher voltages. The lower voltage regime is largely independent of thickness. For the 9000Å diode, the junction shows full depletion above 300 mV, even though the p-layer is only 200Å thick, and that point of full depletion decreases as the thickness decreases. At 1V reverse bias the current density is about 15 pA/cm$^2$, and at 2V it is about 20 pA/cm$^2$. Figure 4 shows the same data re-plotted for current density as a function of electric field. This plot demonstrates that the thickness dependence is largely related to the applied electric field, and shows that the critical electric field between the two current density regions is about $7 \times 10^4$ V/cm.

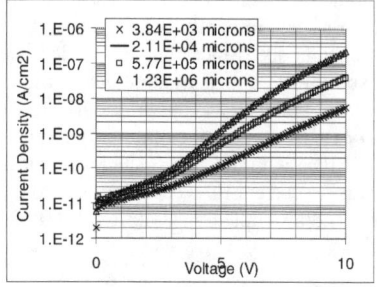

Figure 5: Effect of pixel edge length on J-V. The i-layer thickness is 5500Å.

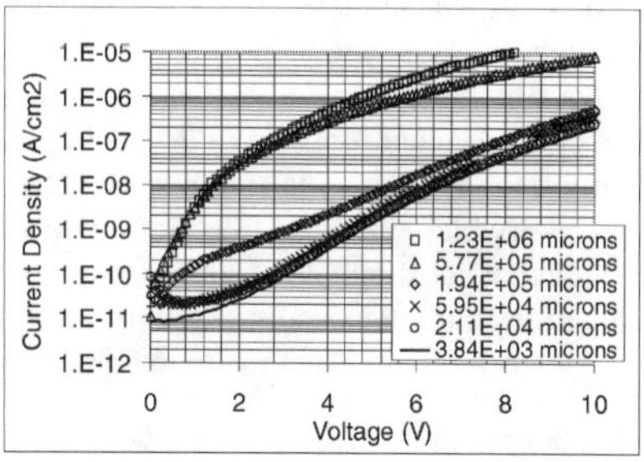

Figure 6: Effect of pixel edge length on J-V. The i-layer thickness is 3000Å.

While previous work by Schiff [4] and Street [5] have studied the effect of the diode stack edge length on reverse bias leakage current densities, Figure 5 shows the effect of the length of the pixel edge length, in which the edge terminates within the i-layer itself, on the leakage current density for a 5500Å thick i-layer. The leakage current density seen here is representative of the diode leakage seen in arrays that share a common i-layer. Figure 6 shows the effect of pixel edge length with a 3000Å i-layer. For short pixel edge lengths, the behavior shows a low bias voltage independent regime and an exponentially increasing high bias regime. As the pixel edge length increases, there is a decrease in the transition bias between the low and high bias behavior, until the low bias behavior is completely suppressed. For the longest pixel edge length devices an almost linear regime occurs at higher biases. In all cases, though the overall leakage current increases as a pixel edge length increases. At modest voltages, the change in leakage as a function of pixel length can result in a 1000x increase in leakage current.

Figure 7 is a plot of the transient leakage current density plot for a 5500Å thick i-layer p-i-n diode upon switching from 0V to 0.2, 1 and 4V reverse bias from 0 to 300 s. The plot shows that the current decay time is largely independent of the bias, though the steady state leakage increases as a function of reverse bias. It also shows the transient behavior when switching from 0 V to some applied bias appears that by 300s, the steady state current has not yet been reached. The 65s hold time was selected as the maximum clock time allowed by the instrument, and shows that its value is about 2x higher at 65s than at 300s for all values. Therefore, the behavior seen at 65s is directly proportional to that 300s. For the 1V reverse bias condition Figure 7 shows that the decay current is roughly 12 pA/cm$^2$ at 300s. The 0.2V bias shows 4 pA/cm$^2$, though this is for a non-fully depleted situation. In addition, there is no evidence of a long-term increase in the leakage current, which has been attributed to reverse bias injection across the junction contacts [10]. It has also been shown that the decay transient is independent of junction dopant concentration.

Figure 8 is a log-log plot of the leakage current density as a function of electric field, the same data as shown in Figure 4. Above $7 \times 10^4$ V/cm, all curves for all thickness i-layers follow the same exponential behavior. Below $7 \times 10^4$ V/cm though, for diodes, except the 3000Å one, show a slight but increasing field dependence as a function of field. The 4000Å thick diode shows no field dependence. The low-field dependence shown in Figure 8 is evidence that there are two distinct mechanisms driving the leakage current.

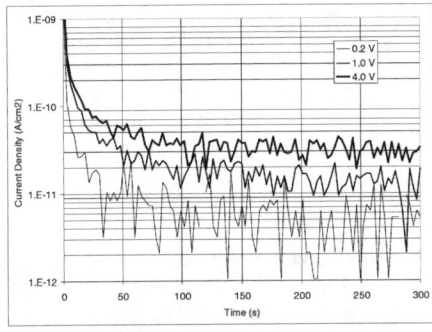

Figure 7: Current decay plot for 0.2, 1, and 4V reverse bias. The i-layer thickness is 5500Å.

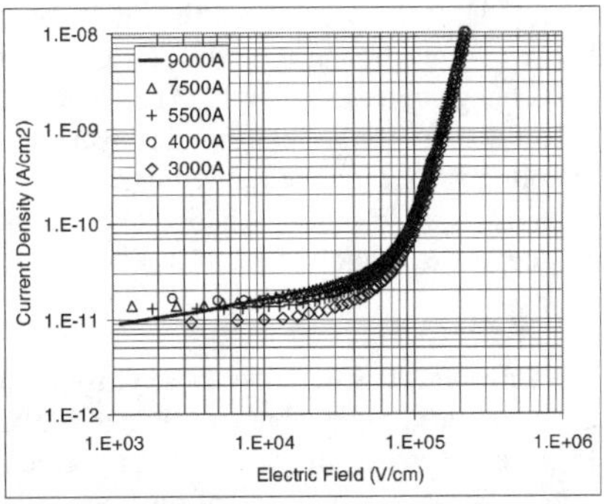

Figure 8: Reverse dark log J- log E plot of a-Si:H p-i-n junctions with various i-layer thickness.

## 4 DISCUSSION

The junctions presented here appear to push the quality given the leakage current densities and thickness of the p-layer. Work by Street has shown that previously, high-quality junctions with 500Å thick, doped layers and 1 micron i-layers fully deplete around 0.7 V, and have about 15 pA/cm$^2$ leakage currents at 2.0V [5]. The junctions created for this work were optimized for maximum quantum efficiency with a 200Å p-layer, however the diodes were fully depleted at 0.3 V, and a 20 pA/cm$^2$ after 65 s at 2.0V. Given the ~33% drop in leakage current density between 65 and 300s, the long-term steady-state value would be close to 14 pA/cm$^2$.

It is generally believed that the cause of time dependent behavior of a-Si:H p-i-n junctions is governed by leakage across the doped junctions, and thermal current generation [5, 10]. Thermal current generation alone cannot account for the field dependent behavior seen in the J-V plots, so most invoke a Poole-Frenkel effect. Ilie, however, has shown through calculation that the Poole-Frenkel effect does not account for the strong field dependence, and proposed instead an electron-lattice interaction mechanism [6]. This phenomenon is a tunneling phenomenon in which the tunneling rate is strongly dependent upon the carrier effective mass, and the electric field.

The effect of pixel edge length on J-V behavior also appears top be field-dependent. MEDICI finite element analysis of the pixel edge region shows a higher electric field than the region toward the pixel center, even when only 1 micron from adjacent pixels of an identical potential. To first order the effect is likely largely dependent upon the area ratio of p-type and n-type contact layers. Therefore, the increase in pixel edge length increases the applied electric field at the pixel periphery thus inducing higher field behavior than for area diodes. The form of the J-V plots is therefore a summation of low and high field regions within the pixel.

The behavior of the J-V curves as a function of junction dopant concentration is very interesting in that low bias behavior is governed by the n-layer doping, and the high bias behavior is governed by p-layer doping, and can be explained by the electron-lattice mechanism. Given the high doping density of the junctions the depletion depth into the junctions varies by about 10Å/V. Therefore for the p-layer, there still exists at least 100Å of non-depleted material at 10V. Therefore, it is unlikely that across junction leakage contributes. On the other hand, the defect pool model can be used to explain the doping dependence. The defect pool model suggests that the position of defect states within the gap near an interface is influenced by the local Fermi level [11]. If the depth from the interface of intrinsic material influenced by the contact layer Fermi level is greater than the carrier diffusion length, it can be expected that the effective mass will change correspondingly. Thus increased doping of the n-layer increases the valence band defect density of the interfacial i-layer thus increasing the e- effective mass. electron-lattice coupling and hence the tunneling rate will change. It is therefore likely that the low-field behavior seen in Figure 8 is due to the Poole-Frenkel effect, while high field behavior is caused by deep-state tunneling [6, 11, 12].

## 5 SUMMARY

This work has detailed the leakage current density behavior as a function of junction dopant concentration, pixel edge length, and i-layer thickness. I-layer thickness does not affect current leakage other than by modulating the electric field for a given potential. The high junction doping shows no evidence of the contact layer injection currents, even though the p-layer is only 200Å. This allows the steady state leakage current at 2.0 V to be as low as $14pA/cm^2$. Also, it has been demonstrated that the pixel edge does contribute to increased leakage currents even when adjacent pixels are held at constant voltage. It appears that the cause is field-enhanced emission. Since it appears that the interfacial a-Si:H region may well govern the leakage current, it will be necessary to take steps to control the evolution of these states during growth. Finally, it has been proposed that the low-bias behavior is governed by the Poole-Frenkel effect, while the high bias behavior is caused by electron-lattice coupling driven tunneling, can be modulated by deep-level interfacial states derived in response to the local Fermi level.

## REFERENCES

[1] J. A. Theil, R. Snyder, D. Hula, K. Lindahl, H. Haddad, and J. Roland, J. Non-Cryst. Sol., **299**, 1234 (2002).

[2] J. A. Theil, M. Cao, G. Kooi, G. W. Ray, W. Greene, J. Lin, A. J. Budrys, U. Yoon, S. Ma, and H. Stork, MRS Symp. Proc., **609**, A14.3.1 (2000).

[3] H. Fischer, J. Schulte, P. Rieve, and M. Böhm, Mat. Res. Soc. Symp. Proc., **336**, 867 (1994).

[4] E. Schiff, R. A. Street, and R. L. Weisfeld, J. Non-crystalline Solids, **198** (1996) 1155.

[5] R. A. Street, Appl. Phys. Lett., **57**(13) 1334 (1990).

[6] A. Ilie, and B. Equer, Phys. Rev. B, **57**(24), 15349 (1998).

[7] J. A. Theil, H. Haddad, R. Snyder, M. Zelman, D. Hula, and K. Lindahl, Proceedings of the SPIE, **4435**, 206 (2001).

[8] J. A. Theil, IEE Proc. Circuits, Devices, and Systems, **150**(4), accepted for publication (2003).

[9] J. Theil, D. Lefforge, G. Kooi, M. Cao, and G. Ray, J. Non-crystalline Solids, **266**, 569 (2000).

[10] R. A. Street, Hydrogenated Amorphous Silicon, Cambridge University Press: Cambridge, pp 370-372, (1991).

[11] S. C. Deane, F. J. Clough, W. I. Milne, and M. J. Powell, J. Appl. Phys., **73**(6), 2895 (1993).

[12] K. Winer, Phys. Rev. B, **41**(7), 12 150 (1990).

## Enhanced Blue Sensitivity in ITO/a-SiN$_x$:H/a-Si:H MIS Photodetectors

S. Tao, Y. Vygranenko, and A. Nathan
Electrical and Computer Engineering, University of Waterloo,
Waterloo, Ontario N2L 3G1, Canada

## ABSTRACT

We report an ITO/a-SiN$_x$:H/a-Si:H MIS photodetector with improved performance in terms of its dark current, stability, and spectral response in the blue region. The a-Si:H and a-SiN$_x$:H thin film layers were deposited by plasma-enhanced chemical vapor deposition (PECVD) on a glass substrate with patterned Mo back contact. The ITO was polycrystalline with a wide band gap (> 3.75 eV) and was deposited at room temperature by magnetron sputtering. SIMS (Secondary Ion Mass Spectrometer) measurements show that an ultra thin a-SiN$_x$:H film (a few nm) can effectively block the diffusion of oxygen from the ITO to the a-Si:H. In addition, the insulator layer provides a barrier for electrons, which serves to reduce the dark current. This is in contrast to the ITO/a-Si:H Schottky photodiode whose electrical and optical performance is impaired by the large defect density at the interface due to impurity diffusion from the ITO layer. At a reverse bias of 1 V, the dark current density of the MIS photodetector is as low as 4 nA/cm$^2$. Photoresponse measurements show a dramatically enhanced sensitivity in the UV/blue spectral region. A high quantum efficiency (~80%) is achieved at a wavelength of 440 nm, which can be attributed to reduction of both optical and recombination loses by virtue of the highly transparent polycrystalline ITO and the low defect density at the a-SiN$_x$:H/a-Si:H interface.

## INTRODUCTION

Hydrogenated amorphous silicon (a-Si:H) technology is widely used in large area imaging applications [1]. Here, the sensing element is either the p-i-n or Schottky photodiode [2]. The latter, with an ITO front contact, has advantages in terms of fabrication simplicity, high photosensitivity and fast response [3]. However, the diffusion of impurities (O, In, Sn) from the ITO to the a-Si:H undermines the integrity of the Schottky interface, which leads to high dark current [4][5]. It has been recently proposed that use of a thin insulator layer (e.g. a-SiN$_x$:H) between the ITO and the a-Si:H can suppress the electron injection from the ITO through the Schottky interface [6]. In this paper, an ITO/a-SiN$_x$:H/a-Si:H MIS photodetector is presented, which exhibits a low dark current, good stability, and enhanced short-wavelength photoresponse.

## EXPERIMENTS

A cross-sectional diagram of the ITO/a-SiN$_x$:H/a-Si:H MIS photodetector is illustrated in Figure 1. The n$^+$a-Si:H, intrinsic a-Si:H, and thin a-SiN$_x$:H films were deposited consecutively on a Mo coated 7059 corning glass substrate, followed by the sputter deposition of a 2000 Å-polycrystalline ITO film at room temperature to form the MIS structure. The ITO layer was patterned to define and isolate the MIS structure using reactive ion etching (RIE). All samples had the same thickness of n$^+$ and intrinsic a-Si:H layers, and were 1000 Å and 7500 Å, respectively. Different thicknesses of the a-SiN$_x$:H layers were deposited for comparison of device characteristics.

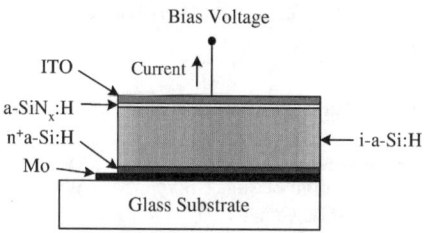

Figure 1. Schematic cross section of the ITO/a-SiN$_x$:H/a-Si:H MIS photodetector.

The quality and integrity of the a-SiN$_x$:H layer are critical since its thickness is only several atomic layers. At the initial stages of film deposition, nucleation happens at discrete spots on the surface, and then expands to the whole surface to form a continuous layer. The integrity of the a-SiN$_x$:H film has been studied by scanning transmission electron microscope (STEM). Figure 2 illustrates STEM pictures of the ITO/a-SiN$_x$:H/a-Si:H interface at different a-SiN$_x$:H deposition times. With no a-SiN$_x$:H, a clear interface between the ITO and the a-Si:H is observed as shown in Figure 2(a). In the first 10 seconds of deposition, there is no obvious a-SiN$_x$:H layer between the ITO and a-Si:H, as depicted in Figure 2(b). But roughness of the ITO/a-Si:H interface is observed, which indicates inhomogeneous growth of the a-SiN$_x$:H at the initial stages. After 20 seconds of deposition time, an a-SiN$_x$:H layer is observed. The thickness of the a-SiN$_x$:H layer is approximately 35 Å, as seen in Figure 2(c). SIMS measurements were performed to investigate the effect of the a-SiN$_x$:H layer on the diffusion of impurities. Here, both ITO/a-Si:H Schottky and ITO/a-SiN$_x$:H/a-Si:H MIS structures were considered with a 35 Å a-SiN$_x$:H layer for the latter. A decrease of the impurity concentrations near the interface is observed in the MIS structure (see Figure 3), which indicates the effectiveness of the thin a-SiN$_x$:H as a diffusion barrier.

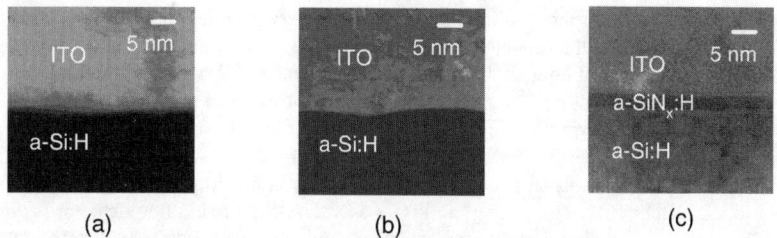

Figure 2 TEM pictures of ITO/a-SiN$_x$:H/a-Si:H cross-sections: (a) with no a-SiN$_x$:H, (b) after the first 10 seconds of a-SiN$_x$:H deposition, and (c) after 20 seconds of a-SiN$_x$:H deposition.

**RESULTS AND DISCUSSIONS**

Current-voltage measurements of the MIS photodetector were performed using a Keithley 236 source measurement unit. Figure 4 illustrates the dark and photo current densities as a function of bias. The dark current density is $4 \times 10^{-9}$ A/cm$^2$ and the photocurrent density $9.2 \times 10^{-6}$

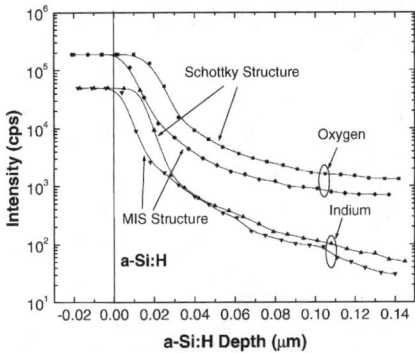

Figure 3. SIMS measurements of the oxygen and indium profiles in the a-Si:H layer near the interface, in the ITO/a-Si:H Schottky and ITO/a-SiN$_x$:H/a-Si:H MIS structures.

A/cm$^2$ at a reverse bias of 1.0 V. The ideality factor of the MIS photodetector is 2.6, calculated from the current under forward bias. This is high in comparison with the ideality factor of a Schottky diode (normally 1.1-1.2), and can be attributed to the blocking effect of the insulator layer on charge transport across the barrier. Figure 5 shows the dark current density of MIS structures of different a-SiN$_x$:H film thicknesses. It is observed that the dark current density in the MIS structure is approximately two orders less than that in the Schottky diode at a reverse bias of 1.0 V. However, there are no significant differences in dark current when the thickness of the insulator layer increases from 2.5 nm to 4.5 nm.

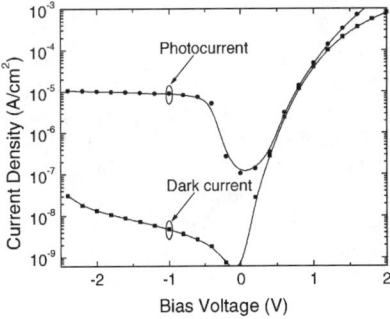

Figure 4. Current-voltage characteristics of the MIS photodetector. The size of the photodetector is 500 μm by 500 μm and the thickness of a-SiN$_x$:H layer is 35 Å. The light intensity used in the photocurrent measurement is 36 lux.

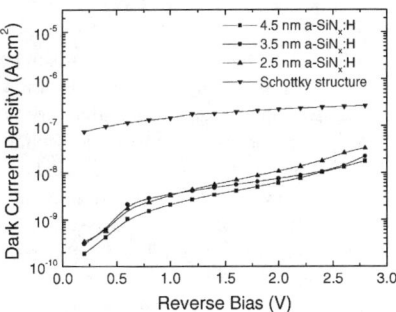

Figure 5. Current-voltage characteristics of the Schottky and MIS structures with different a-SiN$_x$:H film thicknesses.

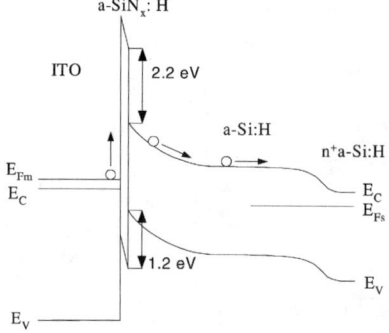
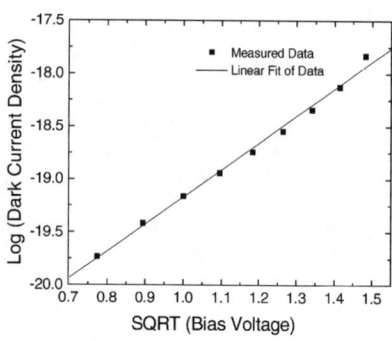

Figure 6. Schematic band diagram of the MIS structure under reverse bias.

Figure 7. Dark current density as a function of the square root of the applied voltage.

Figure 6 depicts the band diagram of the MIS structure under reverse bias. Carrier transport is composed of transport through the insulator layer and transport across the barrier in to the conduction band of the a-Si:H. Under reverse bias, the voltage drop across the insulator layer leads to high electric field inside the layer. For an a-SiN$_x$:H layer with thickness of 3.5 nm, this electric field can be as high as $3\times10^6$ V/cm under a reverse bias of 1 V. At such high fields, conduction in the insulator layer takes place that is characteristic of Poole-Frenkel or Richardson-Schottky emission processes [7]. The Poole-Frenkel emission is observed only when the conduction process is bulk-limited [8]. Because the insulator layer is ultra thin, the Richardson-Schottky emission can be assumed to be dominant. This is given by

$$J \sim A^*T^2 \exp\left(\frac{-q(\phi_B - \sqrt{qE/4\pi\varepsilon_i\varepsilon_0})}{kT}\right), \qquad (1)$$

where $A^*$ is the effective Richardson constant, $\phi_B$ the barrier height, $\varepsilon_i$ the insulator dynamic dielectric constant, $d$ the insulator thickness, and $E$ the electric field across the insulator layer. Figure 7 shows the exponential variation of dark current density with the square root of applied voltage, in accordance with the Richardson-Schottky emission model described in Equation (1).

Figure 8 shows the dark current as a function of time for different reverse bias voltages. The current decays initially due to removal of electrons from the conduction band and the transient release of the depletion charge from defect states [9]. Then, the current increases with time which can be attributed to the collapse of the depletion region and the associated increase of the electric field [10]. The reverse current eventually stabilizes at a certain value after an order of magnitude over a period of 60 seconds. High reverse bias leads to large dark current and long time to stabilize.

Figures 9 and 10 illustrate measurement results of spectral responsivity and quantum efficiency. The spectral response measurements were performed with a PC controlled setup using an Oriel's 77200 grating monochromator, Stanford Research System's SR540 light chopper and SR530 DSP lock-in amplifier. The light intensity was calibrated in the spectral range of 400-1100 nm using Newport's 1830C Optical Power Meter. Figure 9 compares the spectral responsivity of the MIS and Schottky photodiodes. The responsivity of the Schottky diode degrades in the short wavelength range (400 nm – 500 nm) due to the high recombination rate of

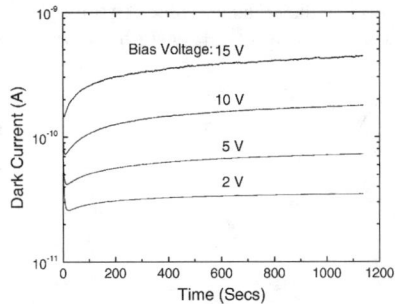

Figure 8. Time dependence of the dark current at different reverse biases.

photogenerated carriers near the a-Si:H/ITO interface. This can be attributed to impurity induced defect states near the interface. Compared to the Schottky photodiode, the MIS photodetector has an enhanced responsivity in the blue light region (see Figure 9). In MIS structures, the thin insulator a-SiN$_x$:H layer effectively suppresses the impurity diffusion from the ITO, consequently reducing the defect density near the interface [6]. In addition, a good quality a-SiN$_x$:H/a-Si:H interface is achieved through consecutive deposition of the two layers without breaking vacuum. Therefore, a low recombination rate is achieved relative to the Schottky diode, leading to enhanced responsivity. The enhanced responsivity can also be attributed to the polycrystalline ITO structure, which has a high transmittance in the blue region due to its wide band gap and no additional absorption loss due to the absence of a front doping layer. This layer limits the short wavelength responsivity of the p-i-n photodiode [11]. The responsivity also increases with increasing reverse bias by virtue of the depletion region extension in the a-Si:H layer, consequently increasing the charge collection efficiency. Figure 10 shows the quantum efficiency of the MIS photodetector as a function of wavelength. The quantum efficiency can be defined as

$$\eta = \frac{No.\,of\,collected\,electrons}{No.\,of\,incident\,photons} = 1.24 \frac{J_{ph}}{\lambda \xi}, \qquad (2)$$

where $J_{ph}$ is the photocurrent density, $\lambda$ the wavelength (in $\mu$m), and $\xi$ the light intensity (in $\mu$W/cm$^2$). At a wavelength of 440 nm, the quantum efficiency reaches the maximum value of 80%. The degradation of quantum efficiency at higher wavelengths is due to the decreasing absorption of photons in the intrinsic a-Si:H layer.

## CONCLUSION

This paper reported on the electrical and optical properties of ITO/a-SiN$_x$:H/a-Si:H MIS photodetectors. The insertion of a thin a-SiN$_x$:H layer (~35 Å) effectively suppressed impurity diffusion and reduced the electron injection from the ITO to the a-Si:H under reverse bias. Low reverse current density (4 nA/cm$^2$) and high stability in the dark current were achieved. The highly transparent and wide band gap ITO layer and low defect density at the a-SiN$_x$:H/a-Si:H interface lead to an enhanced quantum efficiency in the blue region ( maximum 80% at the wavelength of 440 nm). With the enhanced blue sensitivity, the MIS photodetector constitutes a promising alternative as a low cost optical sensor for short wavelength detection.

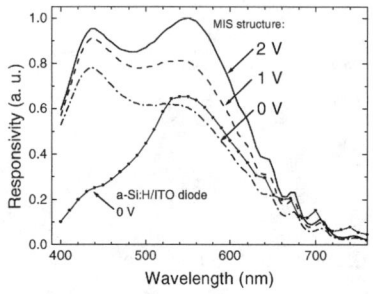

Figure 9. Spectral responsivity of the a-Si:H/ITO photodiode and MIS photodetector with a thickness of 3.5 nm a-SiN$_x$:H layer under different reverse biases.

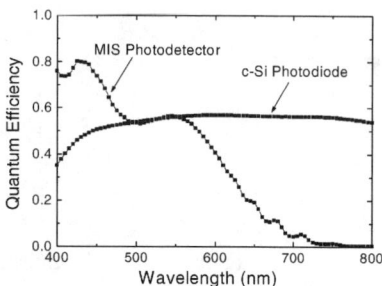

Figure 10. Quantum efficiency of the MIS photodetector in comparison with a commercial photodiode in the Newport's 1830C optical power meter.

## ACKNOWLEDGEMENTS

This work was funded by DALSA/NSERC Industrial Research Chair Program, Communications and Information Technology Ontario, and the Natural Sciences and Engineering Research Council of Canada.

## REFERENCES

[1]  R.A. Street, R. Weisfield, S. Nelson, and P. Nylen, *Mat. Res. Soc. Symp. Proc.*, vol. 258, pp. 1145-1150, 1992.
[2]  H. Kakinuma, M. Sakamoto, Y. Kasuya, and H. Sawai, *IEEE Transactions on Electron Devices*, vol. 37, pp. 128-133, 1990.
[3]  Y.K. Fang, S.B. Hwang, Y.W. Chen, and L.C. Kuo, *IEEE Electron Device Letter*, vol. 12, pp. 172-174, 1991.
[4]  Q. Ma, A. Nathan, and R.V.R. Murthy, *Mat. Res. Soc. Symp. Proc.*, vol. 558, pp. 231-236, 2000.
[5]  S. Ri, H. Fujioka, K. Takasaki, and K. Fujino, *Mat. Res. Soc. Symp. Proc.*, vol. 118, pp. 451-456, 1988.
[6]  S. Tao, Q. Ma, D.S. Striakhilev, and A. Nathan, *Mat. Res. Soc. Symp. Proc.*, vol. 609, pp. A12.2.1-6, 2001.
[7]  S. M. Sze, in Physics of Semiconductor Devices, John Wiley & Sons, New York.
[8]  J.G. Simmons, *Physical Review*, vol. 155, pp. 657-660, 1967.
[9]  R.A. Street, *Philosophical Magazine B*, vol. 63, no. 6, pp. 1343-1363, 1991.
[10] K. Aflatooni, R. Hornsey, and A. Nathan, *Mat. Res. Soc. Symp. Proc.*, vol. 467, pp. 925-930, 1997.
[11] S. Morrison, P. Servati, Y. Vygranenko, A. Nathan, and A. Madan, *Mat. Res. Soc. Symp. Proc.*, vol. 715, pp. A7.4.1-A7.4.6, 2002.

# Solar Cells

## Amorphous and Microcrystalline Silicon Based Solar Cells and Modules on Textured Zinc Oxide Coated Glass Substrates

Bernd Rech, Joachim Müller, Tobias Repmann, Oliver Kluth, Tobias Roschek, Jürgen Hüpkes, Helmut Stiebig, Wolfgang Appenzeller
Institute of Photovoltaics - IPV, Forschungszentrum Jülich GmbH, D-52425 Jülich, Germany

## ABSTRACT

This paper addresses scientific and technological efforts to develop highly efficient silicon thin film solar modules on glass substrates. We present a comprehensive study of μc-Si:H p-i-n single junction and a-Si:H/μc-Si:H stacked solar cells prepared by plasma-enhanced chemical vapour deposition (PECVD) at 13.56 MHz excitation frequency. In the first step cell development was performed in a small area PECVD reactor showing the relationship between deposition process and resulting solar cell performance. Subsequent up-scaling to a substrate area of 30×30 $cm^2$ confirmed the scalability to large area reactors. Moreover, we developed textured ZnO:Al films by sputtering and post deposition wet chemical etching as front contact TCO-material with excellent light scattering properties. A-Si:H/μc-Si:H tandem cells developed on this textured ZnO yielded stable efficiencies up to 11.2 % for a cell area of 1 $cm^2$. First solar modules were prepared in our recently installed process technology, which includes PECVD, sputtering, texture etching and laser scribing on substrate sizes up to 30x30 $cm^2$. Initial module efficiencies of 10.8 % and 10.1 % were achieved for aperture areas of 64 $cm^2$ and 676 $cm^2$, respectively.

## INTRODUCTION

Silicon thin film modules based on a-Si:H and μc-Si:H tandem cells are one of the most promising future thin film PV technologies. Necessary prerequisites for a cost-effective mass production of solar cells incorporating μc-Si:H films are the demonstration of high deposition rates and scalability to large areas. Moreover, effective light trapping is essential to obtain high cell efficiencies at small absorber layer thickness. The a-Si:H/μc-Si:H tandem cell concept was pioneered at the University of Neuchâtel [1] using the VHF-PECVD technique. Since then several research groups have demonstrated high efficiencies by applying different deposition techniques, light trapping concepts and substrates [1-5]. First commercial solar modules were demonstrated by Kaneka Co. [5]. However, these modules are only available on the Japanese market so far.

This contribution addresses scientific and technological efforts to develop highly efficient silicon thin film solar modules. The first part describes the development of textured zinc oxide films prepared by sputtering and post deposition etching as TCO material which provides excellent light trapping properties. The second part treats the development of microcrystalline silicon solar cells at high deposition rates using plasma enhanced chemical vapour deposition (PECVD) at 13.56 MHz plasma excitation frequency. Both techniques were developed on laboratory scale (substrate size: 10x10 $cm^2$) and are now implemented in solar modules with sizes of up to 30x30 $cm^2$. The recent status of our module development will be presented.

## EXPERIMENTAL DETAILS

All cells and modules presented in this study were prepared in the superstrate configuration starting from bare glass substrates. The substrate size for all films, TCO and module processes was either 10x10 $cm^2$ or 30x30 $cm^2$.

**Figure 1**. IPV in-line sputtering system for 30x30 cm$^2$ substrate size.

Textured ZnO coated glass substrates prepared in-house by magnetron sputtering and post deposition wet chemical etching were applied as front TCO in most cases. These films are smooth in the as-deposited state. A textured surface is realized by a simple (wet chemical) etching step in hydrochloric acid, which is either done manually in a chemical bath or by using an etching apparatus, especially designed for this purpose. The ZnO films were either prepared in a small area sputtering systems in static mode or in dynamic mode in our recently installed in-line sputtering system (Fig. 1), which is part of the 30x30 cm$^2$ process technology. This system is capable of rf, dc and mf sputtering and is described in more detail in [6].

The silicon layers of a-Si:H, µc-Si:H and a-Si:H/µc-Si:H solar cells were prepared either in a multi-chamber PECVD system for 10x10 cm$^2$ substrate size or in a large area (30x30 cm$^2$) PECVD reactor, which consists of two deposition chambers and 3 showerhead electrodes as sketched in Fig. 2. All silicon layers were deposited using 13.56 MHz plasma excitation frequency at substrate temperatures below 250°C. The back reflector of all cells and modules is a ZnO/metal double layer, also prepared by sputtering. All patterning steps for the module fabrication were performed using high speed laser scribing.

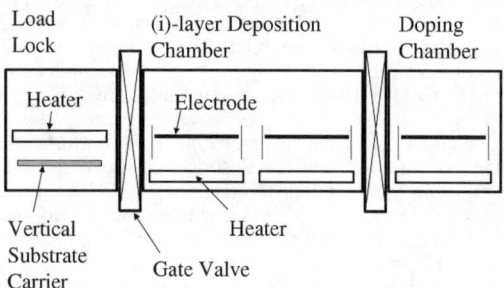

**Figure 2**. Schematic sketch (top view) of the PECVD system for 30x30 cm$^2$ substrate size.

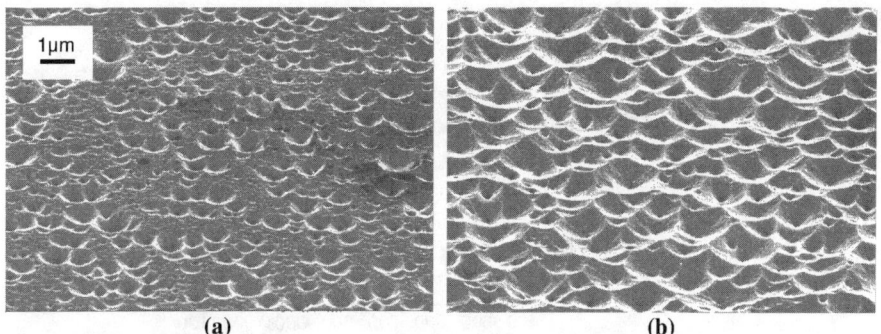

(a)                 (b)

**Figure 3**: SEM micrograph of an etched ZnO:Al film a) shortly dipped in diluted hydrochloric acid and b) after prolonged etching.

The solar cells and modules with areas up to 10x10 cm$^2$ were characterised by current-voltage (J-V) measurements under illumination (AM1.5, 100 mW/cm$^2$, 25 °C) using a class A double light source solar simulator (Wacom-WXS-140S-Super). For light soaking tests, cells and modules were exposed for several 100 hours to AM 1.5 illumination (100 mW/cm$^2$) at open-circuit condition at a constant temperature of 50 °C. This light soaking set-up provides a very homogenous solar spectrum and was also used to measure the illuminated J-V curves of our 30x30 cm$^2$ solar modules.

## TCO AND LIGHT TRAPPING

For silicon thin film solar cells in the p-i-n (superstrate) structure the transparent conductive oxide (TCO) films are required as a front contact material and have to combine low series resistance, high transparency and an adequate surface texture. While for large area amorphous silicon (a-Si:H) based solar modules tin oxide (SnO$_2$:F) coated glass substrates are commonly applied, doped ZnO has been established as a successful alternative on laboratory scale. The promising approach followed at the IPV is the use of magnetron sputtering together with a post-deposition chemical etching step [3,7,8]. Other examples are boron-doped ZnO prepared by low pressure chemical vapour deposition (LPCVD) [9] or intrinsic and aluminium-doped films deposited by expanding thermal plasma CVD [10].

### ZnO material properties and surface texture

The sputtering process leads to highly conductive and transparent but smooth ZnO films. A simple chemical etching step in diluted acid yields a textured surface which can be adjusted to give optimal light scattering over a wide wavelength range [3,7,8]. As an example Fig. 3 shows how the surface morphology of an rf-sputtered ZnO:Al film changes upon etching in diluted hydrochloric acid. Note that the deposition parameters for these films were adjusted to give an optimised surface texture after etching. The SEM micrograph in Fig. 3a) clearly illustrates that a short dip in hydrochloric acid leads to randomly distributed craters with diameters of up to 1 µm, while – on a microscopic scale - parts of the ZnO area are only slightly attacked and remain rather smooth. After prolonged etching the crater formation proceeds until a homogenously textured surface is obtained (see Fig. 3b). The root mean square roughness of the latter texture-etched ZnO:Al film is 150 nm, as determined by AFM. The sheet resistance $R_{sheet}$ after etching was only 7.2 Ω, which demonstrates the good electrical properties even after the etching step.

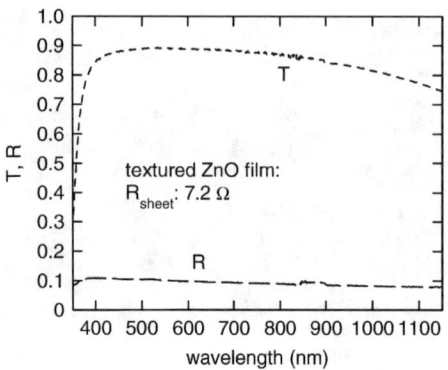

**Figure 4**. Transmission T and Reflection R of the texture-etched ZnO film shown in Fig. 3b). T and R were measured using $CH_2I_2$ as index-matching fluid.

The optical properties of the latter film were measured using a dual beam spectrometer. Note that the textured film was attached to another glass substrate with $CH_2I_2$ as index-matching fluid. This method avoids systematic measurement errors due to light scattering of the rough TCO (see e.g. [11] for a description of the method). The corresponding T and R curves, shown in Fig. 4, demonstrate the high optical transparency over a wide wavelength range. These texture etched ZnO films show high optical haze for a wide wavelength range, indicating good light scattering properties also for µc-Si:H based solar cells, which will be demonstrated in the next section.

**Solar cell results**

The textured ZnO films shown in Figs. 3 and 4 together with a smooth (as deposited) reference film were applied as TCO coated glass substrates for µc-Si:H p-i-n solar cells. The i-layer thickness was 1.1 µm. The external quantum efficiency (e.q.e.) curves (see Fig. 5) illustrate how the spectral response increases over the whole wavelength range by introducing an optimised surface texture. In the blue and green part of the spectrum the increased e.q.e. is due to a reduction of solar cell reflection, which is shown in Fig. 6. The textured ZnO surface provides an effective index grading layer between the ZnO and the silicon films. The increase in the red and infrared part of the spectrum can partly be attributed to this AR-effect, but the main gain in current must be attributed to light trapping. The spectral absorbance of a µc-Si:H film with 1 µm thickness was calculated from the absorption coefficient, as determined by photothermal deflection measurements [12]. By comparing the e.q.e.- and absorbance values at 900 nm one can estimate that the effective light path through the i-layer must be more than 10 µm for the optimised substrate (solid line). ZnO coated glass substrates similar to those presented in this section were applied for the development of µc-Si:H single junction and a-Si:H/µc-Si:H stacked solar cells which will be described later.

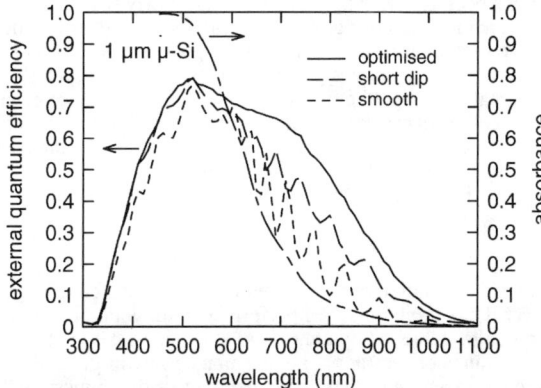

**Figure 5**: External quantum efficiency of μc-Si:H solar cells co-deposited on smooth and texture-etched ZnO coated glass substrates. Dash-dotted line: absorbance of 1μm μc-Si:H.

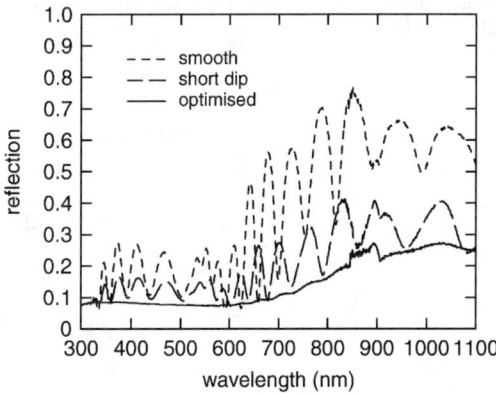

**Figure 6**: Reflection of μc-Si:H solar cells prepared on smooth and texture-etched ZnO coated glass substrates.

## Up-scaling issues

Further progress towards an application of textured ZnO for industrial a-Si:H solar module production requires large area high quality uniform ZnO films to be manufactured at high deposition rates and reasonable costs. This challenge of up-scaling small area ZnO films to substrate areas of the order of one square-meter was taken up in an R&D project [13], which focuses on a high rate reactive mid-frequency (mf) sputtering process for ZnO:Al [14]. Using the mf-technique, ZnO films can be manufactured at high deposition rates with good homogeneity and with excellent optical and electrical properties on a substrate area of 0.6 m$^2$. A further fine tuning of the surface texture is expected to lead to films with very good light scattering for silicon thin film solar cells, comparable to state-of-the-art small area ZnO substrates. More details are given in [13].

To further support this project and to show the viability of the ZnO sputtering technique for a-Si:H/μc-Si:H tandem cells and modules, sputtering processes developed on small areas in static mode are now being up-scaled using dynamic sputtering techniques on 30x30 cm$^2$ substrate areas. An example is given in an accompanying paper [6].

## MICROCRYSTALLINE SILICON BASED SOLAR CELLS

First efficient μc-Si:H p-i-n solar cells were developed at the University of Neuchâtel using the very high frequency (VHF) PECVD technique [1]. The VHF-technique is still widely used as a standard technique on laboratory level to achieve excellent μc-Si:H material quality and solar cell properties at reasonable high deposition rates [1-3]. However, these high frequencies (typically above 50 MHz) make an up-scaling to production scale (module sizes ~ 1 m$^2$) more difficult. For this reason we favour conventional RF (13.56 MHz) excitation frequency which is also compatible with existing deposition equipment. High rate deposition of μc-Si:H films in a high pressure regime at 13.56 MHz has been reported by Guo et al. [15]. Early results on µc-Si:H solar cells in this regime were achieved in our group revealing the key role of a "high pressure deposition regime" to achieve high quality solar cells at high growth rates [16,17].

In the first step, development of µc-Si:H solar cells was performed in a small area PECVD reactor. Later on, these cells were up-scaled to 30×30 cm$^2$ substrate size. The influence of the various i-layer deposition parameters on the solar cell characteristics was studied in p-i-n cells directly. In the following, we present some key results and discuss the way to obtain the best suited material for µc-Si:H solar cells using 13.56 MHz plasma excitation frequency. We will also present the status of our cell process in the 30x30 cm$^2$ reactor.

## Role of depostion parameters

Purpose of this work was to investigate and understand the influence of different deposition parameters applied for µc-Si:H i-layer deposition on device performance. In general, a transition between a-Si:H and µc-Si:H growth can be achieved by almost any deposition parameter. The most convenient way is to reduce the silane concentration, i.e. the silane to hydrogen gas flow ratio [SiH$_4$]/[H$_2$] in the plasma. For each variation of deposition parameters the highest µc-Si:H solar cell efficiencies were achieved in the µc-Si:H growth regime close to the transition to amorphous Si growth [3,17]. This is demonstrated by a series of solar cells where the transition between µc-Si:H and a-Si:H growth is either achieved by increasing [SiH$_4$]/[H$_2$] or by increasing the deposition pressure p$_{dep}$ (see Fig. 7). In both cases the increase in efficiency is due to an increase of V$_{OC}$ and FF with increasing [SiH$_4$]/[H$_2$] or p$_{dep}$ in the µc-Si:H growth regime, while J$_{SC}$ stays almost constant. The efficiency loss in the a-Si:H growth regime for the highest pressures is mainly caused by a sudden drop in J$_{SC}$.

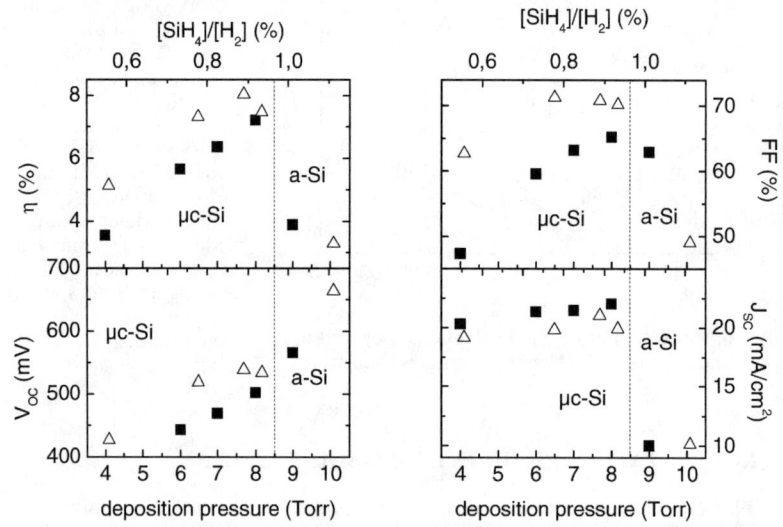

**Figure 7**. Efficiency η, fill factor FF, open-circuit voltage V$_{OC}$ and short-circuit current density J$_{SC}$ as a function of deposition pressure p$_{dep}$ (closed squares) and silane to hydrogen gas flow ratio [SiH$_4$]/[ H$_2$] (open triangles) applied during deposition of the i-layer.

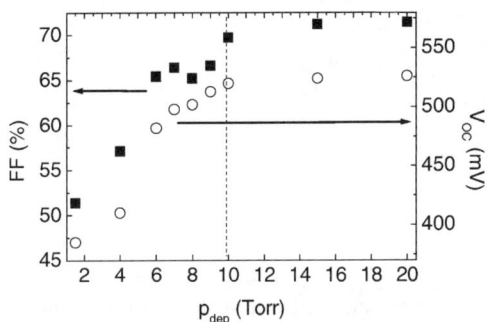

**Figure 8.** Fill factor FF and open-circuit voltage $V_{OC}$ of the best μc-Si:H solar cells achieved at various deposition pressures $p_{dep}$ (i-layer deposition rate 5±1 Å/s).

To separate the role of deposition pressure from the shift between the amorphous and microcrystalline growth regime, we prepared solar cells at different $p_{dep}$ and optimised the $[SiH_4]/[H_2]$ ratio to obtain optimal cell performance for each $p_{dep}$. Additionally, we adapted the plasma power to maintain similar growth rates (5±1 Å/s) over the whole $p_{dep}$ region. Fill factor FF and open-circuit voltage $V_{OC}$ served as a measure for the quality of the μc-Si:H i-layer material. Only the μc-Si:H solar cells exhibiting the highest FF and $V_{OC}$ for a given $p_{dep}$ are used for the following discussion. The results are plotted in Figure 8. For both FF and $V_{OC}$ there is an increase with rising deposition pressure from ~50 % FF and $V_{OC}$<400 mV at 1.5 Torr to ~70 % and 520 mV at 10 Torr, respectively. A further increase of the deposition pressure showed no further improved cell performance in this series. Only at deposition pressures above 10 Torr we achieved both high efficiencies of above 8 % and high growth rates of 5 – 6 Å/s. At $V_{OC}$-values above 500 mV these cells exhibit FFs of more than 70 % indicating good interface quality and electrical transport properties of the corresponding μc-Si:H i-layer material.

In further experiments we investigated the influence of the electrode distance and the effect of using pulsed instead of continuous wave (CW) plasma excitation in the small area PECVD reactor. The studies are described in [18,19] and here we just summarise the key results: Highly efficient solar cells were achieved at electrode distances between 5 and 20 mm. For small electrode distances (5 and 10 mm) solar cell quality is maintained at deposition rates up to 9 Å/s, while for 20 mm cell properties worsen for rates above 3 Å/s. Pulsed plasma deposition yielded good results for deposition rates up to ~5 Å/s, for higher rates a strong decrease of efficiency was observed. Although no advantage of a pulsed plasma for μc-Si:H growth was found in our work, pulsed plasma excitation can still be a method to reduce powder formation.

## Up-scaling of the PECVD process

Homogenous deposition of μc-Si:H in the high pressure and high power regime described above requires a careful design of electrode, gas inlet and pumping. First experiments on 30x30 cm² substrate size showed that the gas feeding plays an important role with regard to the homogeneity of thickness and structural material properties of the μc-Si:H films. As an example, Figure 9 shows a grey-scale photograph of a silicon film prepared in the μc-Si:H growth regime using a non-optimised showerhead electrode. An image of the holes in the electrode can be found directly on the film as different film thickness.

The showerhead electrode design developed for the 30x30 cm² reactor is based on three principles for the gas distribution unit [20]: symmetric design of the gas inlet holes with respect to each other, increasing number of gas inlets towards the plasma space and reduced speed of the gas molecules entering the plasma space by enlarging the feed-through area. Furthermore, the RF-feeding and the grounding of the carrier are centered with respect to the area of electrode and carrier, and the plasma shield is grounded symmetrically.

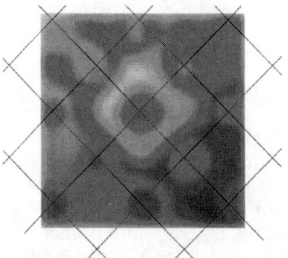 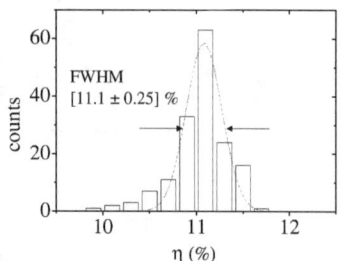

**Figure 9**. Photograph of μc-Si:H i-layer thickness inhomogeneity for non-optimised PECVD electrode design.

**Figure 10**. Distribution of initial efficiencies of 162 a-Si:H/μc-Si:H test cells of 1 cm² size on a 30x30 cm² substrate.

In the large area PECVD reactor we succeeded in preparing μc-Si:H i-layers with good homogeneity over the inner 27x27 cm² of the 30x30 cm² substrate. Only boundary effects caused by the substrate carrier limit the area of homogeneous deposition. The homogeneity in μc-Si:H i-layer thickness and structural properties was proven by both thickness measurements and cell depositions. One experiment is shown in Fig. 10: 162 a-Si:H/μc-Si:H tandem cells of 1 cm² size were evenly distributed over the inner 27x27 cm² of the 30x30 cm² substrate. For this experiment, the substrate was SnO$_2$ coated glass (Asahi U-type). These cells yielded an average initial efficiency of 11.1±0.25 %. FF, V$_{oc}$, and J$_{sc}$ showed also a very narrow distribution, which proves the homogeneity of the structural film properties.

### **Highly efficient solar cells**

In Fig. 11 we compare the light soaking behaviour of an a-Si:H/μc-Si:H tandem cell, an a-Si:H top cell and a μc-Si:H bottom cell. The a-Si:H top cell has a simple Ag back contact while for the two other cells ZnO/Ag back contacts were applied. The best stabilised cell efficiencies obtained in the 30x30 cm² PECVD reactor (on small area ZnO substrates) were 8.9 % and 11.2 % stabilised for a μc-Si:H pin cell and an a-Si:H/μc-Si:H tandem cell, respectively (see Fig. 12 for the J-V curves). The deposition rate for the μc-Si:H i-layers was 5 Å/s for both cells.

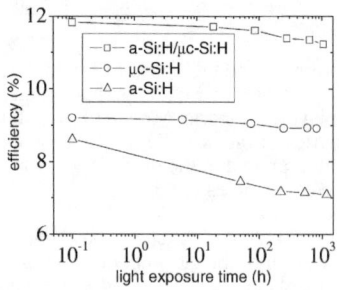 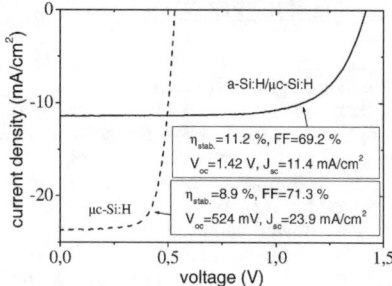

**Figure 11**. Efficiency of silicon thin film solar cells as function of light exposure time.

**Figure 12**. Illuminated J-V curve of a μc-Si:H and an a-Si:H/μc-Si:H solar cell measured after 900 h and more than 1000 h hours of light soaking, respectively.

**Requirements for the μc-Si:H i-layer deposition process**

Conventional 13.56 MHz PECVD is capable of producing device quality μc-Si:H films at high growth rates in a deposition regime of high plasma powers and deposition pressures. Comparing plasma properties and solar cell performance, it is evident that "soft" deposition parameters (i. e. low ion energy) are essential to obtain device quality μc-Si:H i-layers. Note that these "soft" conditions are also fulfilled for high efficiency μc-Si:H cells prepared by very high frequency (VHF)-deposition (e.g. [1-3]) or by Hot-Wire CVD. Using the latter technique Klein et al. recently achieved a μc-Si:H cell efficiency of 9.4 % [21]. In case of 13.56 MHz PECVD homogenous plasma properties can be achieved at very high pressures with high plasma densities. These conditions yield high growth rates ($\geq 5$ Å/s), but still provide sufficiently "soft" deposition for high material quality. If these boundary conditions are fulfilled, usually the silane concentration has to be adjusted to grow the μc-Si:H i-layer in a deposition regime close to a-Si:H growth. Hence it follows that the best μc-Si:H solar cells are generally prepared in a narrow deposition regime. Subsequent up-scaling to large areas requires carefully designed PECVD reactors with respect to power feeding (and substrate grounding), homogeneity of the gas supply and pumping.

## MODULE DEVELOPMENT

First a-Si:H/μc-Si:H tandem modules were realised on Asahi U-type $SnO_2$:F substrates in close co-operation with RWE SCHOTT Solar GmbH using their production technology for patterning and back reflector preparation, while the Si films were prepared at the IPV (see [22]). These 30x30 $cm^2$ modules showed initial aperture area efficiencies up to 9.7 %, demonstrating the compatibility of the a-Si:H solar module process with the a-Si:H/μc-Si:H tandem cell technology.

By the end of 2002 the 30x30 $cm^2$ process technology at the IPV had started operation and first a-Si:H/μc-Si:H modules could be realised on texture-etched ZnO coated glass substrates. All patterning steps were performed using high speed laser scribing. So far, the best aperture area initial efficiencies are 10.1 % and 10.8 % on substrate areas of 30x30 $cm^2$ and 10x10 $cm^2$, respectively (see Fig. 13 for the J–V curve of the 30x30 $cm^2$ module). Light soaking of the small area module has started and the efficiency of this mini-modules stabilises near 10 % after several hundred hours of light soaking (see Fig. 14). In parallel a 10x10 $cm^2$ a-Si:H/a-Si:H tandem module cut from an industrially produced module was included for comparison. The stabilised aperture efficiency is slightly above 6.5 %, which is the specified efficiency for this module. The comparison demonstrates the potential of the a-Si:H/μc-Si:H tandem cell technology on textured ZnO coated glass substrates for highly efficient large area thin film solar modules. However, the transfer of the technology to a cost-effective module production still requires strong efforts in research and technology development.

## CONCLUSIONS

ZnO:Al films prepared by magnetron sputtering and post deposition wet chemical etching were applied as front contact TCO-material with excellent light scattering properties for silicon thin film solar cells and modules. These substrates introduce an effective light trapping for μc-Si:H based solar cells and the textured surface reduces reflection losses at the ZnO/Si-interface. Moreover, low sheet resistances (< 10 Ω) can be achieved while high optical transparency for visible and near-infrared light is maintained.

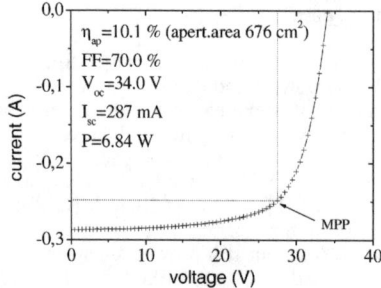
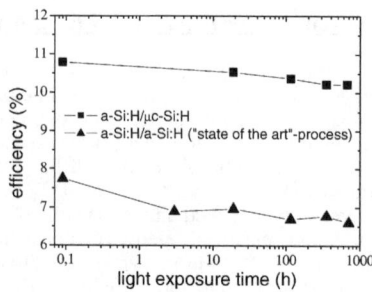

**Figure 13**. Illuminated I-V characteristics of an a-Si:H/μc-Si:H module on textured ZnO with a module size of 30x30 cm$^2$.

**Figure 14**. Aperture area module efficiency of two Si thin film 10x10 cm$^2$ modules as a function of light exposure time.

We developed μc-Si:H i-layers by plasma-enhanced chemical vapour deposition (PECVD) using 13.56 MHz excitation frequency, very high deposition pressures and high RF-powers. These conditions provide sufficiently "soft" deposition for μc-Si:H growth and yield high growth rates. Subsequent up-scaling to 30x30 cm$^2$ substrate area was performed in a carefully designed PECVD reactor.

Finally, highly efficient silicon thin film solar cells and modules were realised starting from bare glass substrates. A-Si:H/μc-Si:H tandem cells yielded stable efficiencies up to 11.2 % for a cell area of 1 cm$^2$. First solar modules were prepared in our recently installed process technology showing initial module efficiencies of 10.8 % and 10.1 % for aperture areas of 64 cm$^2$ and 676 cm$^2$, respectively.

## ACKNOWLEDGEMENTS

The authors thank W. Beyer, F. Birmans, J. Kirchhoff, S. Michel, W. Reetz, R. Schmitz, B. Sehrbrock, H. Siekmann, G. Schöpe, and C. Zahren for their contributions. The laser scribing processes for the solar modules were developed in close co-operation with RWE SCHOTT Solar GmbH, Phototronics. We gratefully acknowledge financial support by the BMWi (contract Nos. 0329854A and 0329885) and the EU (contract No. ENK6-CT-2000-00321).

## REFERENCES

1. J. Meier, E. Vallat-Sauvain, S. Dubail, U. Kroll, J. Dubail, S. Golay, L. Feitknecht, P. Torres, S. Faÿ, D. Fischer, A. Shah, *Sol. Energy Mater. Sol. Cells* **66**, 73 (2001), and references therein.
2. K. Saito, M. Sano, K. Matuda, T. Kondo, T. Nishimoto, K. Ogawa, I. Kajita, Proc. 2$^{nd}$ World Conference PVSEC, Vienna, Austria, 1998, p. 351.
3. O. Vetterl, F. Finger, R. Carius, P. Hapke, L. Houben, O. Kluth, A. Lambertz, A. Mück, B. Rech, H. Wagner, *Sol. Energy Mater. Sol. Cells* **62**, 97 (2000).
4. B. Rech, O. Kluth, T. Repmann, T. Roschek, J. Springer, J. Müller, F. Finger, H. Stiebig and H. Wagner, *Sol. Energy Mater. Sol. Cells* **74**, 439 (2002).

5. K. Yamamoto, M. Yoshimi, Y. Tawada, S. Fukuda, T. Sawada, T. Meguro, H. Takata, T. Suezaki, Y. Koi, K. Hayashi, T. Suzuki, M. Ichikawa, A. Nakajima, *Sol. Energy Mater. Sol. Cells* **74**, 449 (2002).
6. J. Hüpkes, B. Rech, O. Kluth, J. Müller, H. Siekmann, C. Agashe, H.P. Bochem, M. Wuttig, this conference.
7. A. Löffl, S. Wieder, B. Rech, O. Kluth, C. Beneking, H. Wagner, *Proc. 14$^h$ EC-PVSEC*, Barcelona (1997), p. 2089
8. O. Kluth, B. Rech, L. Houben, S. Wieder, G. Schöpe, C. Beneking, H. Wagner, A. Löffl, H.W. Schock, Thin Solid Films **351**, 247 (1999)
9. S. Faÿ, S. Dubail, U. Kroll, J. Meier, Y. Ziegler, A. Shah, *Proc. 16$^{th}$ EC-PVSEC*, Glasgow, U.K. (2000), p. 362.
10. R. Groenen, J. Löffler, P. M. Sommeling, J. L. Linden, E. A. G. Hamers, R. E. I. Schropp and M. C. M. van de Sanden, *Thin Solid Films* **392**, 226 ( 2001).
11. P. Lechner, R. Geyer, H. Schade, B. Rech and J. Müller, *Proc. 28$^{th}$ IEEE Photovoltaic Specialists Conference*, Anchorage, USA (2000) p. 861.
12. R. Carius, private communication.
13. J. Müller, G. Schöpe, O. Kluth, B. Rech, M. Ruske, J. Trube, B. Szyszka, X. Jiang and G. Bräuer, *Thin Solid Films* **392**, 327(2001).
14. B. Szyszka, *Thin Solid Films* **351**, 164 (1999).
15. L. Guo, M. Kondo, M. Fukawa, K. Saitoh, A. Matsuda, Jpn. J. Appl. Phys. 37, L1116 (1998).
16. B. Rech, T. Roschek, J. Müller, S. Wieder, H. Wagner, *Solar Energy Materials & Solar Cells* 66 (2001) 267-273
17. T. Roschek, T. Repmann, J. Müller, B. Rech, H. Wagner, *J. Vac. Sci. Technol.* **A 20** (2), 492 (2002)
18. T. Roschek, T. Repmann, O. Kluth, J. Müller, B. Rech, H. Wagner, Mat. Res. Soc. Symp. Proc. 715 (2002) A26.5
19. B. Rech, T. Roschek, T. Repmann, J. Müller, R. Schmitz and W. Appenzeller, *Thin Solid Films* **427**, 157 (2003).
20. T. Repmann, W. Appenzeller, T. Roschek, B. Rech and H. Wagner, *Proc. 28$^{th}$ IEEE Photovoltaic Specialists Conference*, Anchorage, USA (2000) p. 912.
21. S. Klein, F. Finger, R. Carius, T. Dylla, B. Rech, M. Grimm, L. Houben and M. Stutzmann, *Thin Solid Films* ( 2003), available online.
22. T. Repmann, W. Appenzeller, T. Roschek, B. Rech, O. Kluth, J. Müller, W. Psyk, R. Geyer, P. Lechner, *Proc. 7th European Photovoltaic Solar Energy Conf.*, edited by B. McNelis, W. Palz, H. A. Ossenbrink, P. Helm (WIP-Munich and ETA-Florence), Munich, 2001, p. 2836

# Bandtail Limits to Solar Conversion Efficiencies in Amorphous Silicon Solar Cells

K. Zhu, J. Yang,† W. Wang, E. A. Schiff,[*] J. Liang, and S. Guha†
Department of Physics, Syracuse University, Syracuse, NY 13244-1130
†United Solar Systems Corp., 1100 West Maple Rd., Troy, MI 48084

## ABSTRACT

We describe a model for a-Si:H based *pin* solar cells derived primarily from valence bandtail properties. We show how hole drift-mobility measurements and measurements of the temperature-dependence of the open-circuit voltage $V_{OC}$ can be used to estimate the parameters, and we present $V_{OC}(T)$ measurements. We compared the power density under solar illumination calculated with this model with published results for as-deposited a-Si:H solar cells. The agreement is within 4% for a range of thicknesses, suggesting that the power from as-deposited cells is close to the bandtail limit.

## INTRODUCTION

For most of the interval since its discovery thirty years ago, a fairly large proportion of basic research on hydrogenated amorphous silicon (a-Si:H) has been concerned with its "D-centers," or silicon dangling-bond defects. These defects are certainly electronically active, and they exhibit fascinating metastabilities (the Staebler-Wronski and related effects) that have eluded fundamental understanding for decades.

For solar cells, the fascination with defects obscures the possibility that a-Si:H solar cells may be fairly close to their "zero-defects" conversion efficiency. "Zero-defects" simply means the limit for solar cell parameters that would be achieved if the density of D-centers or other defects were zero. In Figure 1 we have illustrated some measurements on *pin* solar cells from United Solar Systems Corp. both in their as-deposited and light-soaked states [1]. We have also illustrated an idealized model calculation that uses parameters consistent with typical hole and electronic drift-mobility measurements, but that neglects defects altogether. Subsequently, we describe this model in more detail. The model calculation very accurately predicts the conversion efficiency of as-prepared cells. The agreement of the calculation and the measurements suggests that the power density of the as-prepared state can be largely understood without recourse to defects.

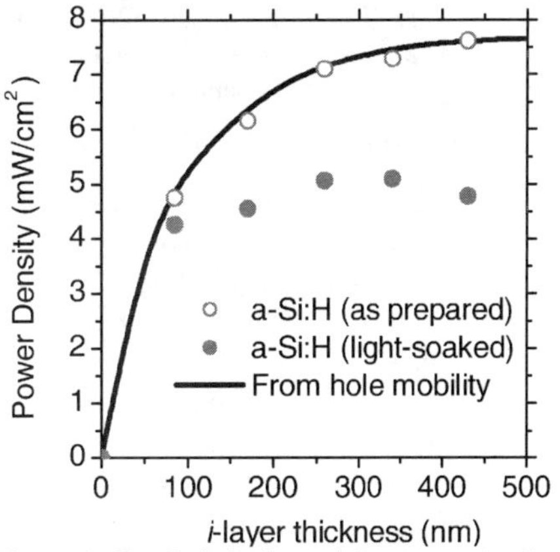

Indeed, because the electron drift-mobilities in a-Si:H are *much* larger than hole drift-mobilities, they are also largely irrelevant to the power-density, and this is why we used

Figure 1: Symbols indicated the power (under solar simulator illumination) for a-Si:H solar cells with varying absorber-layer thickness. The line is a model calculation described in the text.

---

[*] Corresponding author; easchiff@syr.edu .

the legend "from hole mobility" to label the model calculation in Figure 1. Since the very low drift-mobility of holes in a-Si:H is a consequence of the broad valence bandtail in a-Si:H, the "zero-defects" model is the "bandtail limit" to conversion efficiencies.

In this paper, we first discuss the relationship hole drift-mobility measurements to the valence bandtail parameters conventionally used for modeling of hole tr ansport in non-crystalline semiconductors. As we shall see, the valence bandtail parameters are not completely specified by the hole measurements. We then discuss the use of open-circuit voltage measurements to add additional about bandtails.

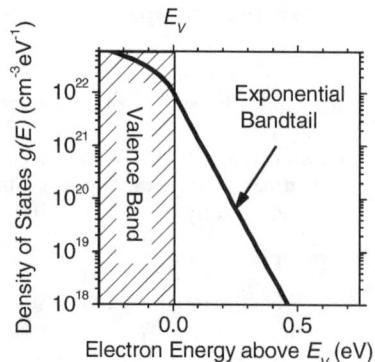

Figure 2: An exponential bandtail lies above the valence band in a-Si:H and other non-crystalline semiconductors; the bandtail leads to very low hole mobilities in a-Si:H.

## MODELING AND HOLE DRIFT MOBILITY MEASUREMENTS

*Valence Bandtail Parameters for Solar Cell Modeling*

Figure 2 illustrates the density of electronic states $g(E)$ near the edge of the valence band. Note the exponential bandtail that extends beyond the edge $E_V$ of the valence band. This figure is the basis for most electrical transport models for holes in amorphous semiconductors; four independent parameters are involved. Holes occupying valence band states are mobile, possessing a "microscopic" or "band" mobility $\mu_h^0$. Bandtail states (beyond $E_V$) act as traps that capture and immobilize holes moving in the valence band proper ($E < E_V$); the width of the bandtail $\Delta E_V$ is of course very important. Only two other parameters [2] are required to characterize hole transport:
1. The effective bandedge density-of-states $N_V$.
2. The capture coefficient $b_t$ that describes the rate of capture of a free hole to a particular bandtail trap; $b_t$ is usually assumed to be the same for all bandtail states.

*Parameterization of Hole Drift Mobility Measurements*

The "drift-mobility" of holes determined by measuring their time-of-flight across some specified distance is much lower than $\mu_h^0$ because of the trapping processes. In addition to $\mu_h^0$, the drift-mobility is determined by the width of the exponential bandtail $\Delta E_V$, and also by an "attempt-to-escape" frequency $\nu$. $\nu$ describes the rate $R$ at which a trapped carrier is thermally released; more specifically, $R = \nu \exp(-\delta E/kT)$, where $\delta E$ is the binding energy of the carrier to the trap. $\nu$ is equated by "detailed-balance" to the product $N_V b_t$.

In Table I, we summarize these three parameters as they have been reported for

Table I: Valence Bandtail Parameters from Hole Drift-Mobility Measurements

| Sample | $\Delta E_V$ (eV) | $\nu$ (s$^{-1}$) | $\mu_h^0$ (cm$^2$/Vs) | Ref. |
|---|---|---|---|---|
| PSU (1999) | 45 | 1.0×10$^{12}$ | 0.7 | 3 |
| ECD (1990) | 48 | 7.7×10$^{10}$ | 0.27 | 4 |

two a-Si:H materials. We shall take the ECD (1990) measurement as characteristic of earlier samples (cf. [4]). We shall take the PSU (1999) measurements as characteristic of "contemporary" materials [3]. The particular parameterizations are less significant than the fact that the PSU mobility (1999), and more generally the hole drift-mobilities in contemporary a-Si:H, have increased several times over values for earlier samples. There has not yet been a study of the best procedures for estimating the parameters from drift-mobility measurements, nor are there systematic studies of how the valence band parameters vary with deposition conditions.

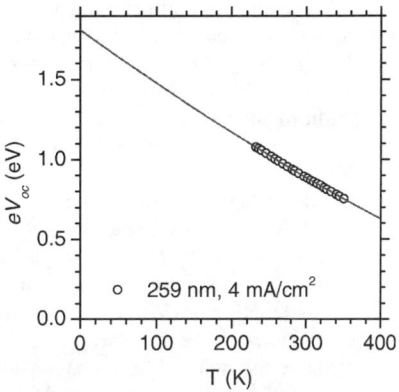

Figure 3: The symbols indicate the temperature-dependence of the open-circuit voltage (laser illumination at 685 nm, 4 mA/cm$^2$; sample thickness 259 nm). The solid line is the best quadratic fit.

## OPEN-CIRCUIT VOLTAGES AND THE BANDEDGE DENSITY-OF-STATES

Within the exponential bandtail model, it is interesting that the hole drift-mobility requires only 3 parameters for its description, whereas general hole transport processes require 4. The fundamental reason for this difference is that hole drift-mobilities are measured in a "linear response" regime, in which photocurrents depend linearly upon the intensity of illumination. The additional valence bandtail parameter is necessary to describe nonlinear effects, which certainly include operating solar cells. In addition, even the simplest bandtail-limited model also requires the bandgap $E_G$, the effective conduction band density-of-state $N_C$, and the recombination coefficient $b_R$ describing electron capture by a hole in the valence bandtail. The parameter $b_R$ has been estimated from high-intensity photoconductivity measurements; two independent measurements gave essentially the same value $b_R = 10^{-9}$ cm$^3$/s [5,6]. We neglect the conduction bandtail; electron drift-mobility measurements indicate the conduction bandtail does not affect electron transport near room temperature.

We now report on our effort to estimate the remaining three parameters, $N_V$, $N_C$, and $E_G$, by using temperature-dependent $V_{OC}$ measurements. This approach is based on the assumption that $V_{OC}$ can be modeled using only bandtail parameters, so that defects are negligible, and $p$ and $n$ layer interfaces are ideal. Given these assumptions, $V_{OC}$ may be analytically calculated for uniform photogeneration $G$ [7]:

$$eV_{OC} = E_G + \frac{kT}{2}\left\{\ln\left(\frac{G}{b_R N_C^2}\right) + 2\ln\left(\frac{G}{b_T N_V^2}\right)\right\} - \frac{(kT)^2}{2\Delta E_V}\ln\left[\frac{b_R}{b_T}\left(\frac{G}{b_T N_V^2}\right)\right] \qquad (1)$$

This expression is valid for $kT < \Delta E_V$. Note that the linear term in $T$ is determined primarily by $N_C$ and $N_V$; the exponential bandtail causes a curvature in the $V_{OC}$ vs. $T$ relation.

*Temperature-Dependent $V_{OC}$ Measurements*

We have measured $V_{OC}(T)$ for a series of three *pin* solar cells prepared at United Solar Systems Corp. with varying intrinsic layer thickness. The measurements were done with a diode laser ($\lambda$ = 685 nm) adjusted to maintain constant photocurrent density of 4 mA/cm$^2$ under reverse bias. Results for one cell are shown in Figure 3 along with a simple quadratic fitting.

The quadratic fitting parameters cannot be directly identified with the parameters in equation (1) because the bandgap itself is temperature-dependent. We adjusted the fitting parameters for the temperature-dependence of the bandgap published by Cody [8]. We measured the bandgap optically (using the peak of the electroabsorption spectrum) for one sample at 200 K and 300 K, and found that Cody's form was consistent with our measurements.

With this adjustment, we calculated the parameters in Table II from the quadratic fit. We set $b_T = N_V/\nu$ using the value of $\nu$ from the PSU (1999) sample in Table I. We assumed $N_C = N_V$; this assumption is arbitrary, but unavoidable at present. We have indicated some statistical errors in parenthesis. The thickest sample (599 nm) was not well described by the quadratic form, and we have not included fitting parameters. We do not know why the quadratic form failed in this case; one speculation is that the $p/i$ interface is degraded for the thicker sample.

Table II: Bandtail parameters estimated from $V_{OC}(T)$

| Sample | $N_C = N_V$ (cm$^{-3}$) | $\Delta E_V$ (meV) | $E_G^e$ (293K) |
|---|---|---|---|
| 259 nm | 4.2×10$^{20}$ | 49 (5) | 1.74 (0.01) |
| 445 nm | 4.5×10$^{20}$ | 56 (6) | 1.76 (0.01) |

The most interesting outcome of this fitting experiment is the value for $N_C$ and $N_V$, which is about 4×10$^{20}$ cm$^{-3}$. These values seem fairly compatible with estimates of the bandedge density-of-states $g(E_V) = 10^{22}$ cm$^{-3}$eV$^{-1}$ from electron photoemission experiments [9] (see endnote [2] for a formula connecting $g(E_V)$ and $N_V$).

One indicator of the systematic errors of this fitting procedure is the bandtail width that was estimated from the curvature of the $V_{OC}(T)$ relation. If the theoretical approach is correct, we expect these estimates to agree with those from photocarrier time-of-flight; in reality, they are somewhat larger. There are at least two possible sources for this modest systematic error in the analysis. First, the parameter $\nu$ is taken from hole drift-mobility measurements on different material than the solar cell measurement; we hope to rectify this deficiency in future work. Second, we have neglected both intrinsic-layer defects and interfaces in the theoretical expression for $V_{OC}$.

We do have evidence that defects are affecting $V_{OC}$ under the conditions of our temperature-dependence measurements. In Figure 4 we show the correlation of $V_{OC}$ with the midgap absorption coefficient $\alpha$ for the intrinsic layer. The different symbols represent successive states of light soaking. $\alpha$ was measured using the infrared photocurrent of the cell under reverse bias, and is an indication of the density of defects in the intrinsic layer of the cell.

The leftmost data in Figure 4 indicate the state of the sample following the $V_{OC}(T)$ measurement. In the low defect-density limit, we expect the line traced by the curve to be essentially horizontal (i.e. independent of the

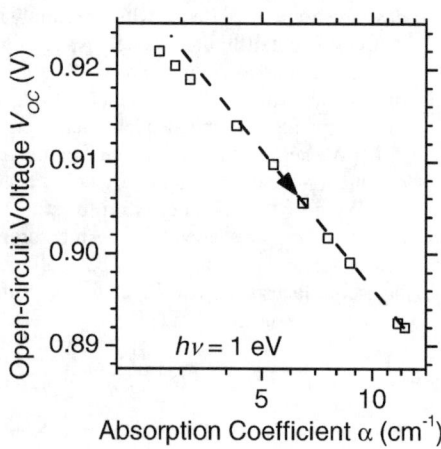

Figure 4: The decline in the open-circuit voltage ($\lambda = 685$ nm, 4 mA/cm$^2$) as light-soaking proceeds is plotted parametrically against the infrared absorption coefficient (1 eV photon energy). The intrinsic layer thickness was 445 nm.

defect density). It is evident that the measurements have not reached this limit.

## SOLAR CELL MODELING

The solid line in Figure 1 is a computer calculation for the power-density in a-Si:H solar cells with varying thicknesses based on the bandtail parameters introduced in this paper. The calculations were done using the AMPS 1D program (Pennsylvania State University®). In this section we summarize the main issues in this modeling.

We used the minimal set of intrinsic-layer parameters in Table III. These are based on the hole time-of-flight measurements (bandtail width, band mobility, and attempt-frequency; PSU [1999] sample) in Table I; since the AMPS 1D program limits bandtail widths to multiples of 10 meV, we needed to modify the fitting parameters, accepting a somewhat inferior fit. We took the bandgap and the

Table III: Summary of Bandtail Parameters

| Parameter | AMPS 1D Symbol | Value |
|---|---|---|
| Electrical Bandgap $E_G$ | EG | 1.74 eV |
| Conduction band density of states $N_C$ | NC | $4\times10^{20}$ cm$^{-3}$ |
| Electron band mobility $\mu_e$ | MUN | 2 cm$^2$/Vs |
| Valence band density of states $N_V$ | NV | $4\times10^{20}$ cm$^{-3}$ |
| Hole band mobility $\mu_h^0$ | MUP | 0.3 cm$^2$/Vs |
| Valence bandtail width $\Delta E_V$ | ED0 | 0.040 eV |
| Bandtail prefactor $g_V^0$ | GD0 | $6\times10^{21}$ cm$^{-3}$eV$^{-1}$ |
| Bandtail trapping cross-section $10^{-7} b_T$ | TSIG/PD | $1.3\times10^{-16}$ cm$^2$ |
| Bandtail recombination cross-section $10^{-7} b_R$ | TSIG/ND | $10^{-16}$ cm$^2$ |

effective band densities-of-states $N_V = N_C$ from the 259 nm sample in Table II, but did not use the bandtail width from this table; the time-of-flight measurements are plainly more appropriate. The bandtail-trapping coefficient $b_T$ was calculated using detailed balance ($b_T = \nu/N_V$). The recombination parameter $b_R$ was taken from high-intensity photoconductivity studies [5,6]. The modeling program actually uses cross-sections $\sigma = b/v_{th}$, where $v_{th}$ is (arbitrarily) set to $10^7$ cm/s. The exponential valence-bandtail prefactor $g(E_V)$ was calculated from $N_V$ and $\Delta E_V$ using the formula in endnote [2], which assumes that the bandedge $E_V$ lies within the exponential bandtail.

The p and n layer parameters will not be discussed here; we chose parameters that yielded "ideal" n and p layers that did not significantly affect the calculated results.

We assumed that the front surface and back surface reflectivity of the interfaces to the cell were zero; the optical absorption properties of the cell are "typical values" for a-Si:H prepared by RF plasma deposition, and were not specifically matched to the particular series of cells in Figure 1.

## DISCUSSION

The agreement between the calculated power density in Figure 1 and the experimental measurements on an as-deposited series of cells is striking. The quantitative agreement needs to be understood in the context of the generic optical properties (reflectivity and absorption) that were assumed by the model; it is probable that there are discrepancies of several percent in the absorbed photon flux for the model and the actual cells.

The power densities in Figure 1, for both the measured points and the calculated curve, tend to saturate for thicknesses greater than about 150 nm. Although we cannot discuss this in greater depth here, for the model this thickness is determined by the hole drift-mobility [7]. In essence, there is a space-charge region of slowly drifting hole photocarriers near the p/i interface, and nearly all of the electrical power generated by the cell is associated with photocarriers absorbed in this space-charge region. The agreement between the calculations and the as-deposited cells suggests that the as-deposited cells are close to the fundamental bandtail limit to conversion efficiencies, and thus that further improvements in the as-deposited cells will require improvements in valence bandtail properties.

The (intentional) limitation in these considerations is that they do not apply directly to the light-soaked state, which has a power density lower by about 30% than the as-deposited state. Still, it is odd that the light-soaked cells are as close to the bandtail limit as they are; while a 30% diminishment in cell efficiency is very harmful to device application, it is not a vast change in how the cell operates. One doesn't know that improving bandtail properties will lead to improvements in light-soaked properties as well, but the nearness of the light-soaked and bandtail-limited states for cells suggests that this will be an interesting direction for further device physics research.

## ACKNOWLEDGMENTS

We thank Joshua Pearce (Pennsylvania State University) for providing his optical absorption coefficient measurements, and Baojie Yan (United Solar Systems Corp.) for discussions and for sharing of an unpublished manuscript. This work has been supported through the Thin Film Photovoltaics Partnership of the National Renewable Energy Laboratory (NDJ-2-30630-19 & 24).

## REFERENCES

1. S. Guha, in *Technology and Applications of Amorphous Silicon*, edited by R. A. Street (Springer, Berlin, 1999), pp. 252-305,.
2. The bandtail density-of-states evaluated at the valence bandedge $g_V^0 \equiv g(E_V)$ is also required by some modeling programs such as AMPS 1D. The relationship to $N_V$ is [7]:

    $N_V = kTg_V^0 \left(1 - kT/\Delta E_V\right)^{-1}$. This relationship is valid only for $kT < \Delta E_V$.
3. S. Dinca, G. Ganguly, Z. Lu, E. A. Schiff, V. Vasilios, C. R. Wronski, Q. Yuan, in *Amorphous and Nanocrystalline Silicon-Based Films*, edited by J. R. Abelson, *et al.* (Materials Research Society, Symposium Proceedings Vol. 762, Pittsburgh, 2003), *in press*.
4. Q. Gu, Q. Wang, E. A. Schiff, Y.-M. Li, and C. T. Malone, *J. Appl. Phys.* **76**, 2314 (1994).
5. G. Juska, J. Kocka, M. Viliunas, and K. Arlauskas, *J. Non-Cryst. Solids* **164-166**, 579 (1993).
6. P. Stradins, H. Fritzsche, P. Tzanetakis, and N. Kopidakis, in *Amorphous Silicon Technology - 1996*, edited by M. Hack, *et al* (Materials Research Society Symposium Proceedings Vol. 420, Pittsburgh, 1996), p. 729.
7. E. A. Schiff, *Solar Energy Materials and Solar Cells, in press*.
8. G. D. Cody, T. Tiedje, B. Abeles, B. Brooks, and Y. Goldstein, *Phys. Rev. Lett.* **47**, 1480 (1982).
9. W. B. Jackson, S. M. Kelso, C. C. Tsai, J. W. Allen, and S.-J. Oh, *Phys. Rev. B* **31`**, 5187 (1985).

Mat. Res. Soc. Symp. Proc. Vol. 762 © 2003 Materials Research Society

A3.4

# CARRIER TRANSPORT AND RECOMBINATION IN A-SI:H P-I-N SOLAR CELLS IN DARK AND UNDER ILLUMINATION

J. Deng, J.M. Pearce, V. Vlahos, R.J. Koval[1], R.W. Collins, and C.R. Wronski
Center for Thin Film Devices, the Pennsylvania State University, University Park, PA 16802
[1]Currently at Intel Corp., 2200 Mission College Blvd., Santa Clara, CA 95054

## ABSTRACT

A study has been carried out on the forward bias dark current and the short circuit current - open circuit voltage characteristics of a-Si:H p-i-n solar cells over wide range of illumination intensities. Results are presented with superposition of these characteristics over extended current voltage regimes. This and the observed separation between these characteristics are consistent with the arguments presented based on first principle arguments. The conclusions drawn about the role of photo-generated carrier lifetimes, the densities of defects and the potential barriers in the i-layers adjacent to the n and p contacts are confirmed by numerical simulations. The key role of these potential barriers to the split in the characteristics offer new insight into both why the lack of superposition has been observed and the erroneous conclusions drawn about carrier transport for a-Si:H solar cells in the dark and under illumination.

## INTRODUCTION

A number of studies on a-Si:H solar cells have been reported in which it was inferred that it is not possible to relate the recombination in the dark with that under illumination. This conclusion was based on the absence of superposition between the dark forward bias current ($J_D$-V) and short circuit current versus open circuit voltages ($J_{sc}$-$V_{oc}$) characteristics found for a variety of cell structures [1,2]. However the presence of such superposition has also been reported [3-5]. To gain a better understanding of carrier transport and recombination mechanisms in a-Si:H solar cells, in the dark and under illumination, detailed studies were carried out on the $J_D$-V as well as $J_{sc}$-$V_{oc}$ characteristics obtained for a wide range of illumination intensities. In this paper first principle arguments are used to discuss the transport and recombination of the carriers photo-generated in the i-layers and compared to that for the cells in the dark. Results are then presented on $J_D$-V and $J_{sc}$-$V_{oc}$ characteristics on a-Si:H p-i-n solar cells which exhibit superposition over extended regions of currents and voltage. The results are found to be consistent with the drift transport of the photo-generated carriers and the diffusion of carriers from the high concentrations of electrons and holes, $n_0$ and $p_0$, at the n and p contacts. Possible reasons are presented for the absences of superposition that have been reported.

## CARRIER TRANSPORT AND RECOMBINATION IN THE DARK AND UNDER ILLUMINATION

From recent studies on $J_D$-V characteristics by Deng et al. [6], valuable insights were obtained about the defects in the i-layers; carrier recombination in the bulk and p/i interfaces; as well as the limitations imposed on the injection of carriers into the bulk. It was found that the forward bias currents determined by the bulk i-layers can be quantified with: the Shockley-Reed-

Hall (SRH) recombination model [7]; a uniform distribution of defects in the i-layers having a continuous distribution of states in the gap [8]; and the diffusion [9] of electrons and holes from their respective contacts. The $J_D$-V characteristics are given by the dependence of the recombination in these diffusive currents on the applied voltage V. This SRH recombination is determined by the concentration of the electrons and holes in the bulk whose densities change exponentially with potential drop across the i-layer, $\phi$ [6]. In a-Si:H solar cells with low defect densities and absence of externally applied bias, the built-in potential $V_{bi}$ distributes itself uniformly as $\phi$ across the bulk of the i-layer with small potential drops $\Delta V_n$ and $\Delta V_p$ at the n and p contacts due to the space charge associated with the high carrier densities $n_0$ and $p_0$, respectively. These potential drops constitute the barriers for the injection of electrons and holes from the n and p contacts into the bulk of the i-layer. Since $V_{bi}$-V=$\phi$+$\Delta V_n$+$\Delta V_p$ and both $\Delta V_n$ and $\Delta V_p$ are essentially independent of bias, $\phi$ decreases with voltage V as $V_{bi}$-V-$\Delta V_n$-$\Delta V_p$. For small values of $\Delta V_n$, $\Delta V_p$ relative to $V_{bi}$-V, $\phi$=$V_{bi}$-V, which leads to the exponential dependence of $J_D$ on V. At high current densities and forward bias, however, clear deviations from such a dependence are obtained. This occurs when $V_{bi}$-V=$\Delta V_n$+$\Delta V_p$ which corresponds to $\phi$=0. Subsequent increases in V invert the field across the i-layer thus introducing the field driven drift currents [9]. In addition, at high values of $J_D$ the potential barriers associated with $\Delta V_n$ and $\Delta V_p$ begin to limit the carrier injection into the i-layer from the contacts. The height of these barriers depends on $n_0$, $p_0$ and the defect density in the i-layer which is small in high quality cells. As a consequence, the effect is not significant until very high current levels. However, in low quality cells, where these barriers are high due to a large density of defects, their effects become significant at lower current levels.

The transport of the carriers in the case of photocurrents is distinctly different from those injected from the contacts since they are created throughout the bulk of the i-layer. They are subject to a drift field E and are swept out towards the contacts as a net drift current [10]. In i-layers with low defect densities, the space charge has only a small effect on the electric field distribution so that E=$\phi$/L where $\phi$ is the potential drop across the i-layer and L its thickness. The photo-current $J_L$ is determined by the densities of electrons and holes that reach the contacts before recombining, where the probability of undergoing recombination depends on their effective lifetime, $\tau$, and transit time $T_{tr}$. This current $J_L$ can be written as $J_L$=$J_G$-$J_{RG}$ where $J_G$ is the photocurrent created by the generation rate of carriers, G, due to the absorbed photons in the bulk. $J_{RG}$ is the recombination current in the bulk which is determined by $T_{tr}$ and $\tau$. $J_G$ depends only on the intensity of the illumination and is independent on the electric field E(V) while $J_{RG}$ depends on E since $T_{tr} \propto 1/E$. Under short circuit conditions the effective transit time $T_{tr} \propto L/V_{bi}$. In high quality a-Si:H cells $J_{RG}$ is significantly lower than $J_G$ where this is indicated by quantum efficiencies under short circuit conditions of virtually 100% for carriers generated in the bulk. For $J_{RG} \ll J_G$, $J_{sc}$=$J_G$ which makes $J_{sc}$ directly proportional to the illumination intensity as is observed in high quality cells.

Under a forward bias V across the i-layer, as $T_{tr}$ increases, so does $J_{RG}$. At the same time the balance between the drift and diffusive currents from the high densities of $n_0$, $p_0$ carriers is disturbed so that diffusion currents, $J_{diff}$, are introduced in the opposite direction to the photo-generated currents. $J_L$ then becomes:

$$J_L(V)=J_G-J_{RG}(V)-J_{diff}(V) \qquad (1)$$

Since the mechanism governing this $J_{diff}$ is exactly the same as that of the diffusion current $J_D$ in the dark, as discussed earlier, they increase exponentially with V. If the diffusive transport under illumination is the same as that in the dark $J_{diff}(V)=J_D(V)$. As both $J_{RG}$ and $J_{diff}$ increase with V, $J_L$ decreases until it becomes zero at $V=V_{oc}$. Under open circuit conditions the relation becomes

$$J_{sc} = J_G = J_{RG}(V_{oc}) + J_{diff}(V_{oc}) \qquad (2)$$

and if $J_{diff} \gg J_{RG}$,

$$J_{sc} = J_{diff}(V=V_{oc}) = J_D(V=V_{OC}) \qquad (3)$$

so that superposition is present between $J_{sc}$ versus $V_{oc}$ and $J_D$ versus V. This occurs even though the superposition principle is not valid in a-Si:H solar cells for *all* values of V due to the voltage dependent contributions of $J_{RG}$ to $J_L$.

Superposition between $J_{sc}$-$V_{oc}$ and $J_D$-V, however, will not also be present if at $V=V_{oc}$, $J_{RG}$ is comparable to or larger than $J_{diff}$. This can occur at high illumination levels when $V_{oc}$ approaches $V_{bi}$ so that the small values of E present in the bulk result in $T_{tr} > \tau$ which leads to large values of $J_{RG}$. At the same time the high forward bias currents in the $J_D$-V characteristics encounter limitations on carrier injection imposed by the potential barriers $\Delta V_n$ and $\Delta V_p$. Both of these effects contribute to a split between the two characteristics since the limitations of $J_{RG}$ on $V_{oc}$ move the $J_{sc}$-$V_{oc}$ to lower values of $V_{oc}$, and the limitations on carrier injection on the other hand move $J_D$-V to higher values of V. Increase in the density of defects in the i-layer, with the corresponding increases in $J_{RG}$ and $\Delta V_n$, $\Delta V_p$ at the n and p contacts, introduce a split between the $J_{sc}$-$V_{oc}$ and $J_D$-V characteristics at lower voltages. At sufficiently large defect densities this can thus limit such superposition to very low values of $V_{oc}$ and V.

## EXPERIMENTAL PROCEDURE

The a-Si:H p-i-n a-Si:H solar cells studied here were fabricated by RF plasma enhanced chemical vapor deposition at a substrate temperature of 200°C under conditions described previously [11]. The 4000 Å thick i-layers in the cell structures were deposited using $SiH_4$ diluted with $H_2$, with R=[$H_2$]/[$SiH_4$] of 0 and 10. In the case of the R=10 cells the first 200 Å were deposited with R=40. To minimize the contributions of shunts, cell structure areas of 0.02 $cm^2$ were used, which were defined by removing the top n µc-Si:H layers with reactive ion etching. The $J_D$-V characteristics measurements were carried out in 50 mV increments using a four-probe technique to eliminate any extraneous series resistance effects. Care was also taken to ensure that the equilibrium currents were measured at low biases. The $J_{sc}$-$V_{oc}$ measurements were obtained using red light illumination from an ELH lamp with a filter that passed $\lambda > 600$ nm. The intensity of illumination was changed using neutral density filters that allowed the range of $J_{sc}$ values from $10^{-9}$ to $10^{-2}$ A/$cm^2$ to be studied as well as the corresponding $V_{oc}$ from 0.2 V to 0.9 V.

## RESULTS AND DISCUSSION

Results are presented here on the superposition of $J_{sc}$-$V_{oc}$ and $J_D$-V characteristics obtained on high quality a-Si:H cells which illustrate the validity of the mechanisms just described.

Shown in Fig. 1 are the results for two p-i-n solar cells, one with R=0 intrinsic layer and the other with R=10 intrinsic bulk layer, which exhibit excellent superposition between $J_{sc}$–$V_{oc}$ and $J_D$–V characteristics. These characteristics, which are from the cells in the annealed state, have regions of superposition from 0.2 V to 0.7 V and 0.2 to 0.8 V, respectively. Because the R=0 i-layer has a smaller band gap (1.78 eV) than that of R=10 (1.86 eV), the intrinsic carrier concentrations in the R=0 cell are about a factor of ten higher. Since the defect densities in the two i-layers are very similar the dark forward bias currents are thus about ten times higher for the cell with R=0 i-layer. It can also be seen in Fig. 1 that the split between $J_{sc}$–$V_{oc}$ and $J_D$–V in the cell with R=0 i-layer occurs at a smaller voltage than that for the cell with R=10 i-layer. However these splits occur at essentially the same current level (~$10^{-4}$ A/cm$^2$) for both cells. This is an indication that the split between the $J_{sc}$–$V_{oc}$ and $J_D$–V characteristics is due to the dark currents becoming limited by the barriers adjacent to the n and p contacts. This effect is also reflected in the $J_{sc}$–$V_{oc}$ characteristic remaining in an exponential form while the $J_D$–V characteristics do not. It can be pointed out that such limitation on carrier injection can be identified and reliably characterized only by using the four-probe technique so that the effects of series resistance on $J_D$–V are eliminated.

The potential barriers that limit the carrier injection are determined by the space charge which is directly related to the densities of defects in the i-layer. Any increase in density of these defects leads to larger $\Delta V_n$ and $\Delta V_p$ which reduces the carrier injection. In order to confirm that the split in $J_{sc}$-$V_{oc}$ and $J_D$–V in Fig. 1 is due to such field distortions the densities of gap states in the i-layer of the R=10 of Fig. 1 were increased with light induced degradation. The results for the $J_{sc}$-$V_{oc}$ and $J_D$-V characteristics obtained for the annealed state and that after nine hours of degradation with 1 sun intensity of red light are shown in Fig. 2 over the voltage range of interest. It can be seen in Fig. 2 that after introducing the light induced defects, as expected, there is an increase in the bulk recombination currents in the dark and the separation between $J_{sc}$-$V_{oc}$ and $J_D$-V characteristics move to a lower voltage, from 0.8 to 0.6 V, and lower $J_{sc}$ from $10^{-4}$ to $10^{-6}$ A/cm$^2$. These higher densities of defects not only increase the field distortions and potential barriers at the n and p but also reduce the carrier lifetimes. Consequently the changes observed in Fig. 2 include both the effect although the increase in $J_{RG}$ in these cells is expected to be relative small.

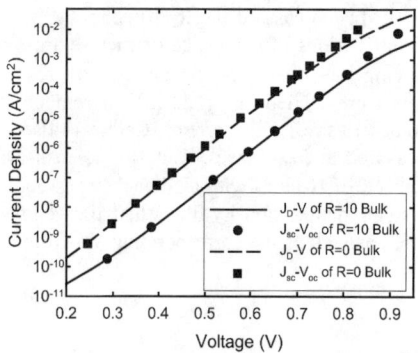

Fig. 1 $J_{sc}$–$V_{oc}$ and $J_D$–V characteristics of p-i-n cells with 4000 Å R=0, R=10 i-layers.

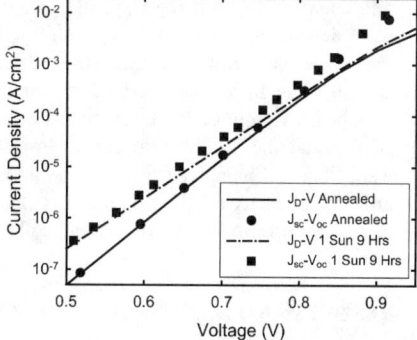

Fig. 2 $J_{sc}$–$V_{oc}$ and $J_D$–V characteristics of the p-i-n cell with R =10 i-layer of Fig. 1 before and after 1 sun red light degradation.

In order to verify the self-consistency of the mechanisms presented for the superposition of $J_{sc}$-$V_{oc}$ and $J_D$-V characteristics in the a-Si:H p-i-n solar cells numerical simulations were carried on these characteristics. The parameters used in the simulation were those by Jiao [12] except that the distribution near mid gap now consisted of both donor, $N_D$, and acceptor $N_A$, states. The electron capture cross-sections used are $10^{-16}$ cm$^2$ and $10^{-14}$ cm$^2$ for the $N_A$ and $N_D$ states respectively while their hole capture cross-sections are $10^{-14}$ cm$^2$ and $10^{-16}$ cm$^2$. The results obtained with $N_D$, $N_A$ of $5\times10^{15}$ cm$^{-3}$ and an intrinsic band gap of 1.86 eV, corresponding to the R=10 cell, are shown in

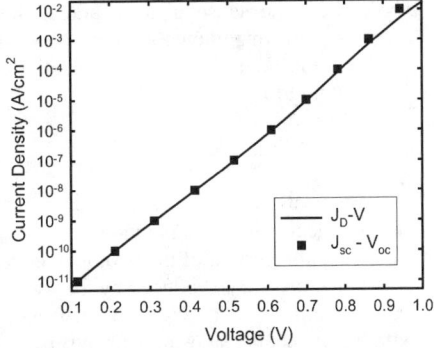

Fig. 3 Simulated $J_{sc}$-$V_{oc}$ and $J_D$-V characteristics of a p-i-n cell with a 4000 Å i-layer that has a band gap of 1.86 eV.

Fig. 3. It can be clearly seen in the figure that, consistent with the experimental results, there is excellent superposition between $J_{sc}$-$V_{oc}$ and $J_D$-V over the voltage and current regime such as in Fig. 1. Simulations were also carried out in which only the densities of the donor and acceptor states near mid-gap were changed. As the defect density of these states is increased, corresponding to a degradation of the cell, the split between $J_{sc}$-$V_{oc}$ and $J_D$-V moves systematically to smaller and smaller voltages until there is no superposition between the characteristics even at the lowest voltages. This is found to occur when the defect density exceeds $2\times10^{17}$ cm$^{-3}$, a forty-fold increase, which is shown in Fig. 4. To verify the key role played by the limitations imposed by the space-charge-induced barriers at the contacts in determining the absence of superposition, the space charge effects present in this case were reduced by decreasing the defect density from $2\times10^{17}$ cm$^{-3}$ back to $5\times10^{15}$ cm$^{-3}$, the level used for Fig. 3. However at the same time, by increasing the $N_D$, $N_A$ capture cross-sections by a factor of forty the carrier lifetimes were reduced by the same factor. The corresponding results are shown in Fig. 5 where it can be seen that now there is a regime of superposition between the $J_{sc}$-$V_{oc}$ and

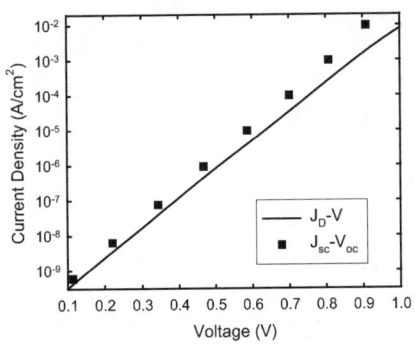

Fig. 4 Simulated $J_{sc}$-$V_{oc}$ and $J_D$-V characteristics of the p-i-n cell in Fig. 3 but with $N_D$, $N_A$ densities 40 times higher.

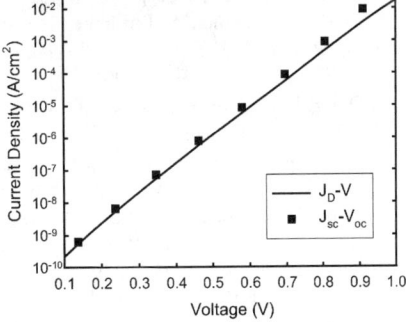

Fig. 5 Simulated $J_{sc}$-$V_{oc}$ and $J_D$-V characteristics of a p-i-n cell in Fig. 3 but with $N_D$, $N_A$ capture cross-sections 40 times higher.

$J_D$-V where the subsequent separation between them is much smaller than in Fig. 4. This illustrates the importance of the space-charge-induced barriers at the n, p contacts on the superposition and their effects relative to those of the photo-generated carrier lifetime contributing to $J_{RG}$.

CONCLUSIONS

Arguments based on first principles have been presented for the transport and recombination of photo-generated carriers in a-Si:H solar cells. It is shown that superposition is valid when the recombination current of the photo-generated carrier, $J_{RG}$, is significantly smaller than that of the diffusion current, $J_{diff}$, induced by the forward bias from the n, p contacts. This is confirmed with experimental results on the $J_{sc}$-$V_{oc}$ and $J_D$-V characteristics obtained on high quality a-Si:H cells with low defect densities in their intrinsic layers. The results are consistent with the mechanisms presented in terms of the contribution of the photo-generated carrier lifetime and in particular that of the space-charge-induced barriers in the i-layer adjacent to the n and p contacts. It is found that these potential barriers play a key role in the split between $J_{sc}$-$V_{oc}$ and $J_D$-V characteristics in these cells and this is subsequently verified by numerical simulations. This suggests that the presence of high potential barriers at the n and p contacts due to high densities of defects there is responsible for the observed absences of superposition between $J_{sc}$-$V_{oc}$ and $J_D$-V characteristics as well as diode quality factors greater than two.

ACKNOWLEDGEMENT

We acknowledge research support from the National Renewable Energy Laboratory under subcontract NDJ-2-30630-01.

REFERENCES

1. S. Hegedus, N. Salzman, and E. Fagen, J. Appl. Phys. **63**, 5126 (1988).
2. T. Brammer, F. Bermans, M. Krause, H. Stiebig, and H. Wagner, Mater. Res. Soc. Symp. Proc. **664**, A19.6 (2001).
3. C. R. Wronski, Jpn. J. Appl. Phys. **17**, 299 (1978).
4. H. Sakai, T. Yoshida, S. Fujikake, T. Hama, and Y. Ichikawa, J. Appl. Phys. **67**, 3494 (1990).
5. J. M. Pearce, R.J. Koval, A.S. Ferlauto, R.W. Collins, C. R. Wronski, J.Yang, and S. Guha, Appl. Phys. Lett. **77**, 3093 (2000).
6. J. Deng, J. M. Pearce, R. J. Koval, V. Vlahos, R.W. Collins, and C. R. Wronski, Appl. Phys. Lett. May 2003 (in press).
7. W. Shockley and W. T. Read, Phys. Rev. **87**, 835 (1952).
8. C. van Berkel, M. J. Powell, A. R. Franklin, and I. D. French, J. Appl. Phys. **73**, 5264 (1993).
9. K. Lips, Mat. Res. Soc. Symp. Proc. **377**, 455 (1995).
10. D. E. Carlson and C. R. Wronski, "Topics in Applied Physics", Vol. 36, M. H. Brodsky ed., Springer-Verlag, New York, 1979.
11. R. J. Koval, J. Koh, Z. Lu, L. Jiao, R. W. Collins, and C.R Wronski, Appl. Phys. Lett. **75**, 1553 (1999).
12. L. Jiao, "Charged Defect States in Hydrogenate Amorphous Silicon Materials for Solar Cells", PhD Thesis, The Pennsylvania State University, 1998.

## Hydrogenated Microcrystalline Silicon Single-Junction and Multi-Junction Solar Cells

Baojie Yan, Guozhen Yue, Jeffrey Yang, Arindam Banerjee, and Subhendu Guha
United Solar Systems Corp., Troy, MI 48084, U.S.A.

## ABSTRACT

This paper summarizes our recent studies of hydrogenated microcrystalline silicon (µc-Si:H) solar cells as a potential substitute for hydrogenated silicon germanium alloy (a-SiGe:H) bottom cells in multi-junction structures. Conventional radio frequency (RF) glow discharge is used to deposit hydrogenated amorphous silicon (a-Si:H) and µc-Si:H at low rates (~ 1 Å/s), searching for the highest efficiency. We have achieved an initial active-area efficiency of 13.0% and stable efficiency of 11.2% using an a-Si:H/µc-Si:H double-junction structure. Modified very high frequency (MVHF) glow discharge is used to deposit a-Si:H and µc-Si:H at high rates (~ 3-10 Å/s) for comparison with our a-Si:H/a-SiGe:H/a-SiGe:H triple-junction production technology. The deposition time for the µc-Si:H intrinsic ($i$) layer in the bottom cell should be less than 30 minutes in order to be acceptable for mass production. To date, an initial active-area efficiency of 12.3% has been achieved with the bottom cell deposited in 50 minutes. By increasing the deposition rate and reducing the bottom cell thickness, we have achieved an initial active-area efficiency of 11.4% with the bottom cell $i$ layer deposited in 30 minutes. The cell stabilized to 10.4% after prolonged light soaking. We will address issues related to µc-Si:H material, solar cell design, solar cell analysis, and stability.

## INTRODUCTION

Using the µc-Si:H cell as a potential substitute for the narrow bandgap a-SiGe:H bottom cell in a multi-junction structure has attracted significant attention in the last decade [1-10]. First, the material is reportedly stable against light soaking [1, 2]. Second, deposition of µc-Si:H films does not require the expensive GeH$_4$ source gas; less expensive SiH$_4$ gas is used. Third, µc-Si:H material has an excellent long wavelength response that makes it eminently suitable for the bottom cell in a multi-junction structure. On the other hand, the µc-Si:H solar cell has some drawbacks. Due to the indirect bandgap, the maximum absorption coefficient of µc-Si:H is lower than that of a-Si:H or a-SiGe:H. Therefore, a thick $i$ layer is necessary to obtain a high photocurrent. Thus, the deposition rate of µc-Si:H should be much higher than that of a-SiGe:H. Furthermore, µc-Si:H materials are usually made with high hydrogen dilution. It is known that the deposition rate decreases with increasing hydrogen dilution. Therefore, the challenge is to deposit good quality µc-Si:H $i$ layers at high deposition rates.

Many studies have been carried out on high rate deposition of µc-Si:H films. It is difficult to deposit good quality µc-Si:H films at a high deposition rate using conventional RF glow discharge. Meier *et al.* found that by using a VHF glow discharge technique, the deposition rates of a-Si:H and µc-Si:H can be increased significantly. They successfully incorporated µc-Si:H material into solar cells [1, 2] and achieved an initial efficiency of 13.1% using an a-Si:H/µc-Si:H double-junction structure [2]. Since then, many groups have used this technique to make µc-Si:H solar cells. A 14.5% initial efficiency on an a-Si:H/µc-Si:H double-junction solar cell and a 12.3% initial efficiency on an a-Si:H/µc-Si:H hybrid module have been reported [3, 4].

In our laboratory, we started the research of μc-Si:H solar cells in the summer of 2001. By optimizing the deposition conditions, we have successfully controlled the ambient degradation due to impurity diffusion [10]. We use conventional RF glow discharge to study the common issues related to μc-Si:H solar cell and search for the highest efficiency. We also use a modified very high frequency (MVHF) glow discharge to deposit μc-Si:H materials and solar cells at high rates to evaluate the possibility for future production. In this paper, we report our recent results on μc-Si:H solar cell design, deposition, device analysis, and stability.

## EXPERIMENTAL

A multi-chamber glow discharge system with conventional RF (13.56 MHz) excitation is used to deposit a-Si:H and μc-Si:H films and solar cells at a low rate of ~ 1 Å/s, and another system with both RF and VHF excitations for high-rate (~ 3-10 Å/s) depositions. The VHF frequency is 60-80 MHz. No gas purifier was used in either system. μc-Si:H single-junction and a-Si:H/μc-Si:H double-junction solar cells were deposited on specular stainless steel (ss), and Ag/ZnO or Al/ZnO back reflector coated ss substrates. Current-voltage (J-V) characteristics were measured under AM1.5 illumination at 25 °C and in the dark in the temperature range of 20 °C to 120 °C. Quantum efficiency measured in the range of 300 to 1000 nm is used to obtain short-circuit current density ($J_{sc}$). Light-soaking experiments were carried out under 100 mW/cm$^2$ white light at 50 °C.

## RESULTS AND DISCUSSION

### Thickness dependence

The advantages of μc-Si:H cell over a-SiGe:H cells are higher $J_{sc}$ and minimal light-induced degradation. However, some unsolved issues still limit further improvement of the cell efficiency. One drawback is the lower open-circuit voltage ($V_{oc}$), which is usually less than 0.5 V for cells with high $J_{sc}$ (e.g. > 24 mA/cm$^2$). In order to improve the cell efficiency further, we need to understand what are the limitations of $V_{oc}$ in μc-Si:H solar cells. For this purpose, we first study the thickness dependence of μc-Si:H in single-junction solar cells.

RF glow discharge was used to deposit μc-Si:H single-junction solar cells at a low rate (~1.0 Å). *nip* structures were deposited on Ag/ZnO coated stainless steel substrate with μc-Si:H *i* layer varied from 0.3 μm to 1.2 μm. Figure 1 shows the J-V characteristic parameters of the solar cells as a function of the *i* layer thickness. It is observed that $J_{sc}$

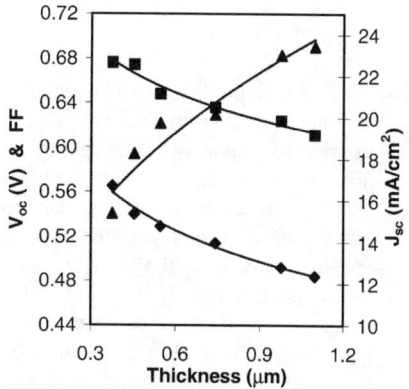

**Figure 1.** J-V characteristics of μc-Si:H single-junction solar cells versus the intrinsic layer thickness, where ■, ◆, and ▲ denote FF, $V_{oc}$, and $J_{sc}$, respectively.

increases, but FF decreases with the increase of the *i* layer thickness. This phenomenon is similar to that observed in a-Si:H solar cells. For a thicker cell, higher $J_{sc}$ results from more optical absorption, and lower FF from reduced collection. A significant difference from a-Si:H solar cells is that $V_{oc}$ decreases with the *i* layer thickness significantly. In a-Si:H solar cells, $V_{oc}$ also decreases with the *i* layer thickness, but the change is usually very small unless microcrystalline formation occurs [11]. For µc-Si:H cells shown in Fig. 1, $V_{oc}$ decreases with the *i* layer thickness dramatically. Although we do not fully understand what causes the thickness dependence of $V_{oc}$ at this moment, we have considered the following possible mechanisms. First, it is known that the amount of microcrystalline inclusion (grain size and crystal volume fraction) increases with film thickness. We have also found that $V_{oc}$ decreases with an increase of $H_2$ dilution in mixed-phase solar cells [12]. With increasing microcrystalline fraction, $V_{oc}$ varies from over 1.0 V in the fully a-Si:H phase to below 0.5 V in the substantially microcrystalline phase [12]. The mixed-phase solar cells have $V_{oc}$ between 0.5 V and 1 V, depending on the amount of crystal volume fraction. Based on the two observations, we conclude that a thicker µc-Si:H cell has a higher average crystal volume fraction, hence lower $V_{oc}$. Second, for µc-Si:H cells with an *i* layer thickness of over 1 µm, its $V_{oc}$ is normally below 0.5 V, which is much lower than that of a single crystalline silicon solar cell. Although we may consider micrograins to have a similar band gap as crystalline silicon, the distorted bonds in the grain boundary regions lead to increased band tail states and defects, thus causing a lower $V_{oc}$. In addition, impurity is another issue. We found a significant amount of oxygen impurity in some unoptimized µc-Si:H silicon films [10]; SIMS analysis showed a large tail diffusing into the film. Oxygen atoms can form n-type doping centers, which move the Fermi level toward the conduction band and lead to a high dark conductivity. In an *nip* structure, a higher dark conductivity causes higher dark current density and lowers $V_{oc}$. Since we found that the post-deposition impurity diffusion can be blocked by an a-Si:H layer [10], the oxygen problem is not a big issue for multi-junction solar cells, but the incorporation of oxygen during the deposition is still a concern.

## **Light trapping effect**

For µc-Si:H solar cells, due to the indirect optical transition in microcrystallites, the absorption coefficients are not high enough to generate high current density with a thickness such as 1 µm. Therefore, a high quality back reflector is very important for increasing the cell efficiency. Figure 2 shows quantum efficiency plots for three solar cells deposited on specular ss, and Al/ZnO and Ag/ZnO back reflector coated ss. It is noted that the solar cells deposited on the back reflectors have a significant response in the long wavelength region, especially the cell on the Ag/ZnO back reflector. However, the short wavelength response is lower, a phenomenon also observed in a-Si:H and a-SiGe:H solar cells. It is probably due to back diffusion of photo-generated electrons near

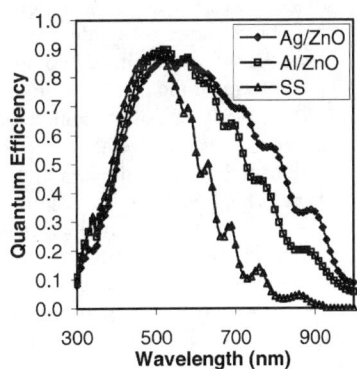

**Figure 2.** Quantum efficiency of µc-Si:H solar cells made on ss, Al/ZnO coated ss, and Ag/ZnO coated ss.

**Table I.** J-V characteristics of μc-Si:H solar cells deposited under the same condition but on various substrates.

| Substrate | Efficiency (%) | $J_{sc}$ (mA/cm$^2$) | $V_{oc}$ (V) | FF | FF (blue) | FF (red) |
|---|---|---|---|---|---|---|
| ss | 4.36 | 13.45 | 0.478 | 0.679 | 0.694 | 0.698 |
| Al/ZnO | 6.45 | 20.39 | 0.489 | 0.647 | 0.665 | 0.656 |
| Ag/ZnO | 7.25 | 22.83 | 0.489 | 0.649 | 0.658 | 0.640 |

the *p/i* interface for cells deposited on texture substrates. The low quantum efficiency at the short wavelength region corresponds to lower blue FF, as noted in Table I. In addition, the slightly higher density of photo-generated carriers results in the slightly higher $V_{oc}$ and lower FF for the cells deposited on the back reflectors than on specular ss. A higher density of photo-generated carriers produces a wider split of quasi-Fermi levels leading to a larger $V_{oc}$, and a higher recombination rate leads to a lower FF. Since μc-Si:H normally has a higher response in the long wavelength range, designing the anti-reflection coating so that the minimum reflection point is shifted to a longer wavelength would be beneficial for improving $J_{sc}$.

**Control of interfaces**

Superior material quality in all the layers of a device is essential, but high quality material does not necessarily ensure high efficiency solar cells. An optimized cell design is also very important [13]. For a μc-Si:H *nip* solar cell, we need to not only optimize the material quality for each layer, but also control the interface between layers. For example, four solar cells were made on Ag/ZnO back reflector with the same recipe except for different buffer layers at the *n/i* and *i/p* interfaces. The μc-Si:H *i* layer was deposited using RF at 1 Å/s with a thickness of ~0.9 μm. Figure 3 shows the light J-V characteristics of solar cells with different interface layers, where (a) is a cell with optimized *n/i* and *i/p* buffer layers, (b) incorporates an *n/i* buffer layer only, (c) incorporates an *i/p* buffer layer only, and (d) has no *n/i* or *i/p* buffer layer. The role of the *n/i* buffer layer is to reduce the thickness of the incubation layer, and the *i/p* buffer layer to reduce the shunt current. By comparing (a) and (b), we find that without the *i/p* buffer layer, the $V_{oc}$ is lower and the FF is also slightly lower. The reduced $V_{oc}$ is due to the higher shunt current shown in Fig. 4. It is very common to have a high shunt current for μc-Si:H solar cells if there is no specific treatment on the interface. Although the mechanism of a high shunt current is not very clear, controlling the *i/p* interface can significantly reduce the shunt current. By comparing (a) and (c) in Fig. 3, we find that for the cell with no *n/i* buffer layer, while $V_{oc}$ is slightly higher, $J_{sc}$ and FF are both lower than the cell with an *n/i* buffer layer. The higher $V_{oc}$ and lower $J_{sc}$ indicate a lower average microcrystalline volume fraction in the *i* layer for the sample with no *n/i* buffer layer. The poorer FF with no *n/i* buffer layer probably results from a poorer carrier transport due to an amorphous incubation layer. Normally we see a crossover between the dark and light J-V curves as shown in Figs. 3 (c) and (d). The crossover is due to an insufficient forward current as shown in Fig. 4, where the dark J-V levels off at high voltages for the samples with no *n/i* buffer layer. The cell with no buffer layer has the lowest efficiency with a large crossover between the dark and light J-V characteristics, and the dark J-V in Fig. 4 shows a high shunt current at low voltages and levels off at high voltages.

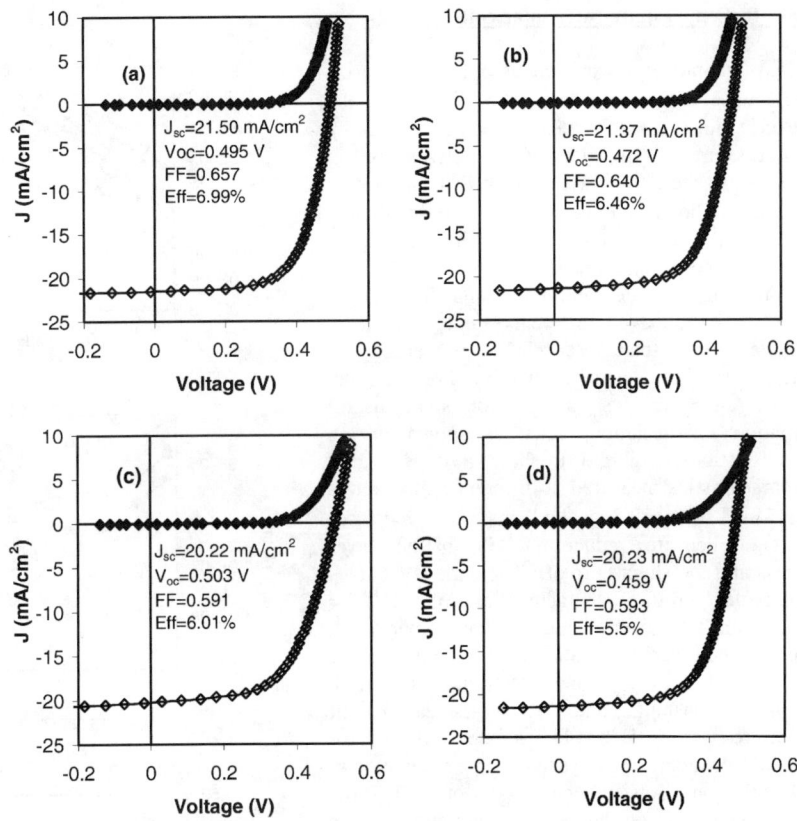

**Figure 3.** J-V characteristics of μc-Si:H single-junction solar cells deposited using RF glow discharge with different interface layers described in the text.

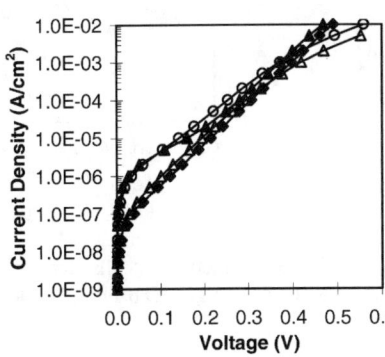

**Figure 4.** Dark J-V characteristics of μc-Si:H solar cells made with (◆) both $n/i$ and $i/p$ buffer layers, (▲) only $n/i$ buffer layer, (△) only $i/p$ buffer layer, and (○) no $n/i$ or $i/p$ buffer layer.

## Carrier transport in µc-Si:H solar cells

Many techniques have been used to characterize carrier transport properties in µc-Si:H solar cells. However, most measurements require specially designed samples with different thicknesses or on different substrates. It is well known that the growth of µc-Si:H depends on the substrate and thickness [14]. A thicker sample usually has a higher microcrystalline volume fraction than a thinner one. Therefore, one must be careful in using the material information obtained on different sample structures. We believe that the most reliable properties are obtained from the solar cells directly. For example, dark J-V characteristics can provide some useful information for designing cells, as shown in the previous sections. Figure 5 (a) shows a series of dark J-V characteristics measured at different temperatures for a µc-Si:H solar cell deposited on ss. The interfaces were controlled to reduce the incubation layer thickness and the shunt current. The dark J-V curves show perfect diode characteristics of $J(V) = J_0[\exp(qV/nkT) - 1]$ over a wide range of temperatures, where $J_0$ is the reverse saturation current density, n the diode quality factor, q the electronic charge, k Boltzmann constant, and T the measurement temperature. Figure 5 (b) plots $J_0$ versus 1000/T for two cells deposited using the same recipe but one on ss and the other on a Ag/ZnO coated ss. Both cells show a linear relationship on the semi-log plot and give the same activation energy of 0.65 eV. If the Fermi level is at the center of the band gap, then twice the activation energy value is the electronic band gap. Since most µc-Si:H materials show a weak n-type transport, we can only conclude that the electronic band gap in the µc-Si:H i layer is at least 1.3 eV. The $J_0$ versus 1000/T plot for the cell deposited on Ag/ZnO is a factor of 3 higher than the cell on ss, which could result from the large effective area due to the texture of the back reflector. Figure 5 (c) plots n versus T for the same two samples. The decrease of n with the increase of temperature indicates an increased diffusion length (or mobility) at higher temperatures. The larger n for the cell on Ag/ZnO is probably due to the larger effective path for carrier traversing across the i layer due to the texture of Ag/ZnO.

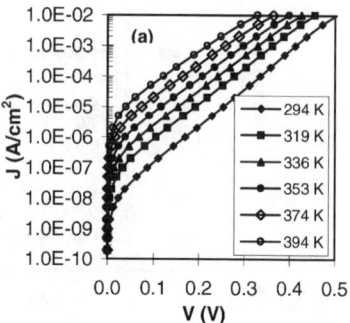

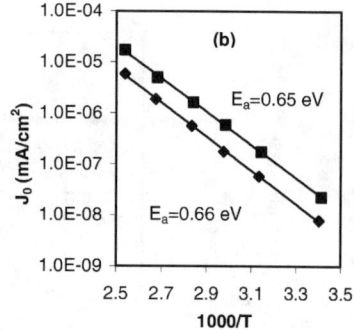

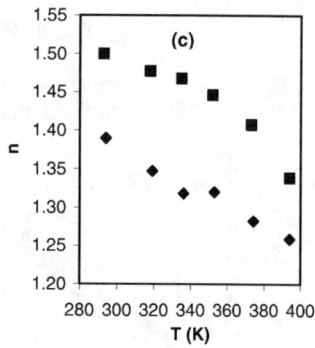

**Figure 5.** (a) dark J-V for a µc-Si:H cell on ss, and (b) and (c) show $J_0$ and n, respectively, for cells deposited on ss (♦) and Ag/ZnO back reflector (■).

## Optimization of RF low rate solar cells

One major effort is focused on the development of a high-efficiency μc-Si:H bottom cell using optimized material and cell structure. We have achieved an initial active-area efficiency of 7.4% ($J_{sc}$ = 23.0 mA/cm$^2$, $V_{oc}$ = 0.490 V, and FF = 0.653). The values of $J_{sc}$ and $V_{oc}$ are reasonably good and comparable with those reported in the literature [1-10]. However, the value of FF is lower. Two reasons for the lower FF have been identified. First, the sheet resistance of our ITO is high. Note that the quality of ITO has been optimized for a-Si:H alloy cells. Since μc-Si solar cell has a low $V_{oc}$ and high $J_{sc}$, it requires high conductivity ITO to obtain high FF. For a multi-junction solar cell structure in which the current density is relatively low, the role of the ITO resistance is not as critical. Since the ultimate goal is to produce multi-junction cells based on μc-Si:H, we did not change the deposition conditions for ITO. Second, for a single-junction μc-Si:H cell, the quality of the μc-Si:H *i* layer may degrade due to atmospheric impurity diffusions after the cell is exposed to the atmosphere following fabrication. Experimental results show that an a-Si:H alloy top cell deposited on top of the μc-Si:H cell can provide a cap to prevent post-deposition impurity diffusion [10]. Therefore, atmospheric contamination of the μc-Si:H cell is not an issue for the multi-junction structures. Consequently, we did not spend much time in optimizing the single-junction μc-Si solar cell. Instead, we use the a-Si:H/μc-Si:H structure to optimize the μc-Si:H cell.

As discussed above, there are two advantages of the a-Si:H/μc-Si:H double-junction structure for optimization of the μc-Si:H cell. First of all, the a-Si:H top cell can serve as a cap to prevent impurity diffusion, and no highly conductive ITO is needed. Figure 6 shows (a) the J-V characteristics and (b) quantum efficiency of the best a-Si:H/μc-Si:H double-junction cell with an initial active-area efficiency of 13.0%. The $J_{sc}$ is obtained from the quantum efficiency measurement of the current-limiting bottom cell, which is measured up to 1000 nm. At this wavelength, the cell has a response ~15%. The current would be higher if our measurement could go beyond 1000 nm. Figure 7 shows the J-V characteristics of the same double-junction structure under (a) blue light and (b) red light. Under the blue light, light absorption/current generation is predominantly in the top cell. Since the top and bottom cells are connected in

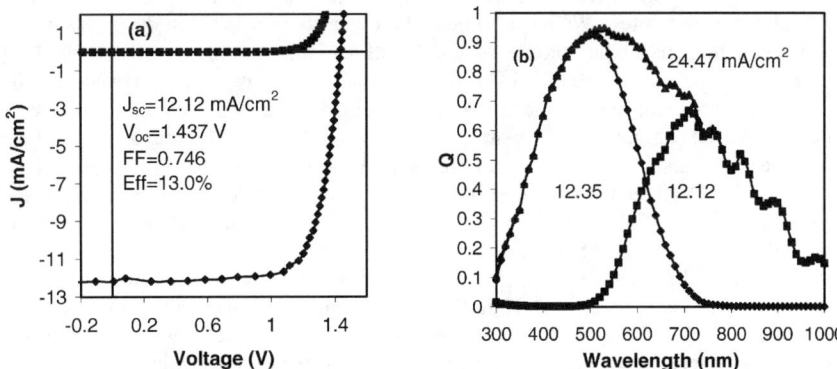

**Figure 6.** (a) J-V characteristics and (b) quantum efficiency of the best a-Si:H/μc-Si:H double-junction cell made using RF glow discharge at a low rate.

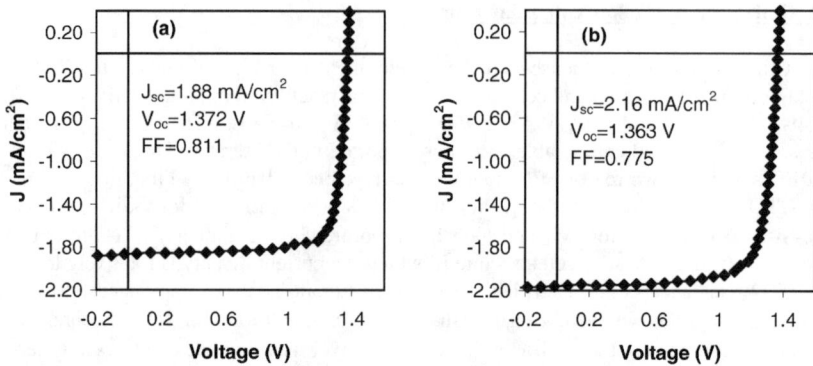

**Figure 7.** J-V characteristics of the best a-Si:H/μc-Si:H double-junction cell under an AM1.5 solar simulator with (a) a 585 nm cut-off filter and (b) a 610 nm cut-on filter.

series, the external current is limited by the bottom cell. In this case, the measured FF reflects the bottom cell FF under a reduced light intensity. From Fig. 7 (a), one can see that the bottom cell FF is 0.811, which is much higher than the single-junction μc-Si:H. This result suggests that the μc-Si:H bottom cell quality is indeed superior to that of the a-SiGe:H alloy bottom cell. Under the red light, on the other hand, the J-V characteristics reflect the top cell performance. As shown in Fig. 7 (b), the FF is 0.775, which is similar to the value of a single-junction a-Si:H top cell under a low light intensity.

## Optimization of MVHF high rate solar cells

The a-Si:H/μc-Si:H double-junction cell is a promising structure for high efficiency solar cells. However, high deposition rate is essential for transferring the research result to production. We use MVHF to deposit intrinsic μc-Si:H layer in *nip* single-junction and double-junction cells at high deposition rates. We have studied all the aspects mentioned in the previous sections and achieved a μc-Si:H single-junction solar cell with an initial active-area efficiency of 7.1% ($J_{sc}$ = 22.83 mA/cm$^2$, $V_{oc}$ = 0.504 V, FF = 0.616), where the *i* layer was deposited in 50 minutes. Using this recipe for making the bottom cell in an a-Si:H/μc-Si:H double-junction structure, we have achieved an initial active-area efficiency of 12.3%. Figure 8 shows (a) the J-V characteristics and (b) the quantum efficiency of the best a-Si:H/μc-Si:H double-junction cell made using MVHF at a high rate with the bottom cell *i* layer deposited in 50 minutes. This cell showed a FF of 0.822 under AM1.5 illumination with a 585 nm cut-off filter, and reflected a very high quality μc-Si:H bottom cell.

In order to be considered acceptable for production, we believe that the deposition time for the μc-Si:H *i* layer in an a-Si:H/μc-Si:H double-junction solar cell should be less than 30 minutes. Table II lists J-V characteristics for four solar cells with different bottom-cell deposition times. By reducing the time to 30 minutes, the efficiency dropped to 11.3% due to a lower current from the bottom cell. The strong bottom-limited current results in a better stable efficiency, which is discussed in the next section.

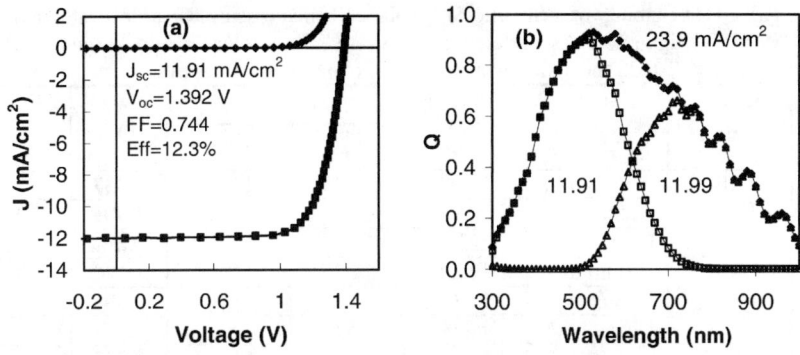

**Figure 8.** (a) J-V characteristics and (b) quantum efficiency of the best MVHF deposited a-Si:H/μc-Si:H double-junction solar cell at high deposition rates. The intrinsic μc-Si:H layer in the bottom cell was deposited in 50 minutes.

**Table II.** J-V characteristics of a-Si:H/μc-Si:H solar cells with different bottom cell deposition times. The underlined data indicate the limiting current.

| Sample No | Bottom cell deposition time (min.) | Efficiency (%) | Jsc (mA/cm$^2$) | | $V_{oc}$ (V) | FF |
|---|---|---|---|---|---|---|
| | | | top | bottom | | |
| 11569 | 60 | 12.04 | 12.09 | 12.07 | 1.359 | 0.734 |
| 11635 | 50 | 12.33 | 11.91 | 11.99 | 1.392 | 0.744 |
| 11797 | 35 | 11.34 | 11.77 | 10.73 | 1.385 | 0.763 |
| 11835 | 30 | 11.35 | 11.19 | 11.10 | 1.406 | 0.727 |

## Stability of a-Si:H/μc-Si:H double-junction solar cells

Optimized μc-Si:H solar cells have better stability against light soaking than a-Si:H or a-SiGe:H solar cells. However, in an a-Si:H/μc-Si:H structure, the a-Si:H top cells are usually over 300 nm thick to match the current from the μc-Si:H bottom cell. A thick a-Si:H solar cell usually shows a large degradation after light soaking. With the $i$ layer thickness of over 300 nm, the degradation is in the range of 20 - 40% for a single-junction a-Si:H solar cell; the amount of degradation depends on deposition conditions, especially the deposition rate. Therefore, the stability of a-Si:H/μc-Si:H is still questionable, and experimental data are needed to clarify the concern. From our experience on a-Si:H and a-SiGe:H multi-junction cell studies, we know that a certain mismatch with a-Si:H top cell limiting (a-Si:H top cell usually has better stability than a-SiGe:H middle and bottom cells) can improve the stable efficiency of multi-junction solar cells. In the a-Si:H/μc-Si:H case, the bottom cell has better stability than the top cell, thus a bottom cell limited a-Si:H/μc-Si:H double-junction cell should have a better stability than a top

**Table III.** Stability of a-Si:H/μc-Si:H solar cells made using RF at low rates and MVHF at high rates. As a reference, a low rate RF a-Si:H/a-Si:H double-junction cell is also listed. The underlined data indicate the limiting current.

| Sample No. | State | Efficiency (%) | $J_{sc}$ (mA/cm$^2$) Top | $J_{sc}$ (mA/cm$^2$) Bott. | $V_{oc}$ (V) | FF |
|---|---|---|---|---|---|---|
| RF 13986 | Initial | 12.99 | 12.35 | 12.12 | 1.437 | 0.746 |
|  | Stable | 10.88 | 11.57 | 11.88 | 1.405 | 0.669 |
|  | Degradation | 16.2 % | 6.3 % | 1.9 % | 2.2 % | 10.3 % |
| RF 14066 | Initial | 12.29 | 12.15 | 11.31 | 1.420 | 0.765 |
|  | Stable | **11.15** | 11.48 | 11.25 | 1.396 | 0.710 |
|  | Degradation | 9.3 % | 5.5 % | 0.5 % | 1.7 % | 7.2 % |
| MVHF 11570 | Initial | 11.76 | 12.09 | 12.07 | 1.359 | 0.717 |
|  | Stable | 9.30 | 11.37 | 12.11 | 1.323 | 0.618 |
|  | Degradation | 20.9% | 6.0% | -0.3% | 2.6% | 13.8% |
| MVHF 11568 | Initial | 11.62 | 12.15 | 10.96 | 1.386 | 0.765 |
|  | Stable | 10.24 | 11.47 | 10.89 | 1.351 | 0.696 |
|  | Degradation | 11.9% | 6.1% | 0.6% | 2.5% | 9.0% |
| MVHF 11834 | Initial | 11.30 | 12.09 | 10.77 | 1.404 | 0.747 |
|  | Stable | **10.42** | 11.70 | 10.67 | 1.374 | 0.711 |
|  | Degradation | 7.8% | 3.2% | 0.9% | 2.1% | 4.8% |
| MVHF 11835 | Initial | 11.35 | 11.19 | 11.10 | 1.406 | 0.727 |
|  | Stable | **10.42** | 10.85 | 10.96 | 1.378 | 0.697 |
|  | Degradation | 8.2% | 3.0% | 1.3% | 2.0% | 4.1% |
| a-Si/a-Si RF 7361 | Initial | 11.66 | 8.13 | 8.15 | 1.975 | 0.726 |
|  | Stable | **9.70** | 7.88 | 7.99 | 1.914 | 0.643 |
|  | Degradation | 16.8 % | 3.1 % | 2.0 % | 3.1 % | 11.4 % |

cell limited one. To confirm this, we have carried out light-soaking studies using a-Si:H/μc-Si:H double-junction cells with different current mismatches. This data is summarized in Table III. As a comparison, an a-Si:H/a-Si:H cell is also included in the table as a reference. The RF low rate cell (13986) has the best initial efficiency of 13.0%, which has a matched current density between the top and bottom cells in the initial state. This cell degraded by 16.2% and reached a stable efficiency of 10.9%. Another RF low rate cell (14066) has a large current mismatch with the bottom cell limiting the current. Even though the initial efficiency is lower than 13986, this cell only degraded by 9.3%, and reached a stable efficiency of 11.2%. The stable efficiency of this cell turns out to be better than the one with the highest initial efficiency. Similar behavior in terms of current mismatch and stability is found for MVHF high rate cells 11570 and 11568. The cell with a large current mismatch and bottom cell limiting (11834) degraded less. The cells with 40 minutes (11834) and 30 minutes (11835) bottom *i* layer deposition times showed a stable active-area efficiency of 10.4%.

An example of light-soaking kinetics for cells with different current mismatches is shown in Fig. 9 (a), where the bottom-cell limited (MVHF 11570) degraded by about 11.9%, but the current matched one (MVHF 11568) degraded by 21% due to a large degradation of $J_{sc}$ (6%). In

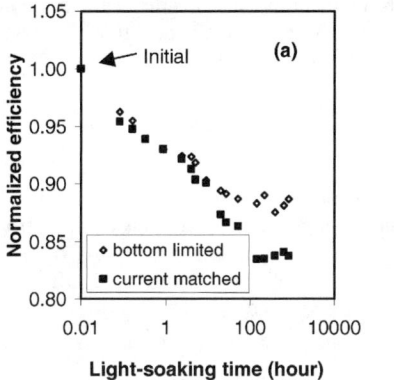 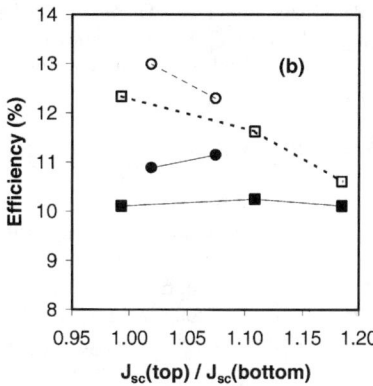

**Figure 9**. (a) Stability of a-Si:H/μc-Si:H double-junction solar cells, and (b) initial (open symbols) and stable (solid symbols) efficiencies as a function of current mismatch for cells made with RF (circles) and MVHF (squares).

addition, we found that the bottom-cell limited a-Si:H/μc-Si:H double-junction cell stabilized after about 20 hours, but the current matched one needed about 100 hours to reach stability. From this study, we conclude that a bottom-cell limited a-Si:H/μc-Si:H double-junction cell has better stability than a top-cell limited or current matched ones. Figure 9 (b) plots the initial and stable efficiency versus the current mismatch defined as $J_{sc}(top) / J_{sc}(bottom)$ for a-Si:H/μc-Si:H double-junction cells with a similar top cell thickness. It shows that reducing the bottom-cell current causes a lower initial efficiency, but the stable efficiency is not affected very much. Another advantage of bottom-cell limited structure is that the bottom cell can be made thin in a relatively short time, resulting in a better FF.

## SUMMARY

We have studied the μc-Si:H solar cell as a potential substitution for a-SiGe:H in multi-junction structures. Some key issues including the thickness dependence of the cell performance, light trapping, interface, and light-induced stability have been studied. While high quality material is an essential condition for achieving high efficiency solar cells, an optimized cell design is also very important. Controlling the interface between the doped layers and the *i* layer is especially critical for solar cell optimization. We find that increasing the *i* layer thickness can increase $J_{sc}$ but reduce $V_{oc}$ and FF. The optimized thickness of the μc-Si:H should be around 1 μm. A thick *i* layer also requires a thick top cell to match the current, which causes a lower stable efficiency. To date, we have achieved an initial active-area efficiency of 13.0% with an a-Si:H/μc-Si:H double-junction structure made using RF at a low rate of ~ 1 Å/s, and 12.3% using MVHF at a high rate. A current mismatch with the bottom cell limiting the current is desirable for a-Si:H/μc-Si:H double-junction solar cells. By optimizing the current mismatch, we have obtained a stable active-area efficiency of 11.2% using RF at a low rate, and 10.4%

using MVHF at a high rate. The *i* layer of the MVHF bottom cell was deposited in 30 minutes, a reasonably short time that we believe is acceptable for mass production. Triple-junction structures using a combination of amorphous and microcrystalline silicon based alloys are being investigated.

## ACKNOWLEDGMENTS

We thank S. R. Oshinsky for his continuous encouragement over the years, and K. Lord, K. Beernink, and H. Fritzsche for stimulating discussion. We appreciate outstanding efforts of the entire R & D group toward this program. This work was supported partially by NREL under the thin-film partnership program Subcontract No. ZDJ-2-30630-19, and by DOE under SBIR Subcontract No. DE-FG02-02ER8356371. The dedicated work of S. Sundquist for the manuscript preparation is acknowledged.

## REFERENCE

1. J. Meier, R. Flückiger, H. Keppner, and A. Shah, Appl. Phys. Lett. **65**, 860 (1994).
2. J. Meier, P. Torres, R. Platz, S. Dubail, U. Kroll, J. A. Anna Selvan, N. Pellaton Vaucher, Ch. Hof, D. Fischer, H. Keppner, A. Shah, K.-D. Ufert, P. Giannoulès, and J. Koehler, Mat. Res. Symp. Proc. **420**, 3 (1996).
3. K. Yamamoto, A. Nakajima, M. Yoshimi, T. Sawada, S. Fukuda, K. Hayashi, T. Suezaki, M. Ichikawa, Y. Koi, M. Goto, H. Takata, and Y. Tawada, Proc. of 29$^{th}$ IEEE Photovoltaic Specialists Conference, New Orleans, 1110 (2002).
4. K. Ogawa, K. Saito, M. Sano, A. Sakai, and K. Matsuda, Technical Digest of the International PVSEC-12, Jeju, Korea, 343 (2001).
5. Y. Nasuno, M. Kondo, and A. Matsuda, Appl. Phys. Lett. **78**, 2330 (2001).
6. Y. Nasuno, M. Kondo, and A. Matsuda, Technical Digest of the International PVSEC-12, Jeju, Korea, 791 (2001).
7. T. Roschek, T. Repmann, J. Müller, B. Rech, H. Wagner, Conf. Record of 28$^{th}$ IEEE Photovoltaic Specialists Conference, Anchorage, 150 (2000).
8. O. Vetterl, R. Carius, L. Houben, C. Scholten, M. Luysberg, A. Lambertz, F. Finger, and H. Wagner, Mat. Res. Symp. Proc. **609**, A15.2 (2000).
9. S. J. Jones, R. Crucet, X. Deng, D. L. Williamson, and M. Isu, Mat. Res. Symp. Proc. **609**, A4.5 (2000).
10. B. Yan, K. Lord, J. Yang, and S. Guha, Mat. Res. Symp. Proc. **715**, 625 (2002).
11. J. Yang, K. Lord, S. Guha, and S. R. Ovshinsky Mat. Res. Symp. Proc., **609**, A15.4 (2000).
12. K. Lord, B. Yan, J. Yang, and S. Guha, Appl. Phys. Lett. **79**, 3800 (2001).
13. A. Banerjee, X. Xu, J. Yang, and S. Guha, Mat. Res. Symp. Proc. **377**, 675 (1995).
14. S. Guha, J. Yang, D. L. Williamson, Y. Lubianiker, J. D. Cohen, A. H. Mahan, Appl. Phys. Lett. **74**, 1860 (1999).

# Electronic Properties of Microcrystalline Silicon investigated by Photoluminescence Spectroscopy on Films and Devices

R. Carius, T. Merdzhanova, F. Finger
Institut für Photovoltaik, Forschungszentrum Jülich GmbH
D-52428 Jülich

## ABSTRACT

Photoluminescence spectroscopy has been applied to investigate localized states in microcrystalline silicon (μc-Si:H) films and to address the problem of the changes of the electronic properties of this material upon changes of the hydrogen dilution during film growth. By a comparison of photoluminescence and Raman spectra on device grade sample series prepared at different silane concentration in hydrogen (SC) by PE-CVD and HW-CVD a correlation between the microstructure and the photoluminescence energy is found. It is proposed that the density of band tail states is reduced with increasing SC leading to the increase of the PL energy as well as to the increase of $V_{oc}$ of solar cells. The reason for the tails and their reduction is not clear but strain might play a crucial role and the amorphous hydrogenated phase might be effective for strain reduction.

## INTRODUCTION

Plasma enhanced chemical vapor deposition (PE-CVD) is at present the well established method for preparation of microcrystalline silicon (μc-Si:H) as this technique is highly successful for manufacturing device grade amorphous silicon films. Recently, a significant progress in the material and device quality prepared by hot wire (HW)-CVD has been achieved by lowering the substrate temperature to values typically used in PE-CVD processes [1,2]. Microcrystalline solar cells with this material as intrinsic layer show similar efficiencies as for PE-CVD material [1,2]. Moreover, very similar electronic and device properties are observed, such as the decrease of the dark conductivity and an increase of the open circuit voltage ($V_{oc}$) with decreasing crystalline volume fraction [1]. These effects are not understood and particularly the high $V_{oc}$ of more than 590 mV obtained for a HW-CVD cell at a still high fill factor and current density [2] is of current interest. We address this issues by investigation of the states at the band edges using photoluminescence spectroscopy (PL).

Already in the early stages of research on microcrystalline silicon PL measurements indicated similarities of microcrystalline and amorphous silicon, e.g. a broad featureless PL band at energies significantly below the band gap of crystalline silicon and a distribution of recombination lifetimes. The interpretation was based on transitions involving distributions of defect states [3]. In a later investigation a shift of the PL energy as a function of the plasma excitation frequency was observed but no shift of the band gap derived from absorption measurements was found [4]. Although transitions between tail states were discussed as a possible source for the PL band the similarities with well known defect bands (D1-D4) in crystalline silicon was taken as evidence to favor the defect model. More recent work revealed a systematic decrease of the PL energy as a function of the crystalline volume fraction [5, 6]. In Both cases evidence is given for radiative recombination of carriers trapped in band tail states as the cause of the PL band in the μc-Si:H films. A decrease of the PL energy has also been observed with increasing substrate temperature and correlated with an decreasing optical gap (defined at high absorption coefficient) and interpreted in terms of quantum size effects [7].

In the following we will present a study of the influence of SC (and thus the crystalline volume fraction) on the photoluminescence properties by comparing sample series of well characterized high quality microcrystalline silicon films prepared by PE-CVD and HW-CVD. By investigating photoluminescence and $V_{oc}$ in a series of high efficiency solar cells with an i-layer prepared by HW-CVD we will give evidence that the increase of $V_{oc}$ with increasing SC is very likely caused by a decrease of the density of band tail states.

## EXPERIMENTAL DETAILS

Undoped μc-Si:H sample series have been prepared by PE-CVD at very high frequencies (VHF) of 95 MHz at substrate temperatures ($T_s$) of 230°C (PE-CVD1) and 200 °C (PE-CVD2) or by HW-CVD at $T_s$=185 °C. Rough quartz or Corning glass served as substrate for photoluminescence measurements. Film thicknesses were about 0.5-0.7 μm or 2-3 μm for the different series. The solar cells were prepared by HW-CVD in a p-i-n sequence on textured ZnO substrates with a ZnO/Ag stack as back reflector (for more details see [2]). The silane concentration in hydrogen (SC) is used as the main parameter to vary the microstructure. Details of the deposition parameters and measurement procedures are described elsewhere [1,2]. The films have been characterized in detail by e.g. Raman and infrared (IR) spectroscopy, conductivity and X-ray diffraction (XRD) and the solar cells by I-V characteristics, Raman spectroscopy and XRD.

Photoluminescence (PL) spectra were measured at temperatures 10-300K using an FTIR spectrometer (Bruker FS66V) with a cooled germanium detector applying lock-in technique in a step scan mode to enhance the sensitivity when necessary. For excitation of the PL different laser wavelength of 488 nm (for the near surface region) and 642 nm and 633 nm (for homogenous carrier generation) have been used. The PL on the solar cells was excited and measured through the glass substrate. A rear silver contact of 1 mm diameter determined the contact area and the laser spot was adjusted to match this diameter. The maximum generation rate was chosen such that it provided a current density of about 100 mA/cm$^2$, i.e. about 5 times the current density of AM 1.5 illumination, and was reduced by neutral density filters to the indicated values. $V_{oc}$ and $j_{sc}$ were measured in the same configuration as the PL.

## RESULTS AND DISCUSSION

As the microstructure of the samples turns out to be the most important parameter when SC is varied at low $T_s$ Raman spectra will be shown together with PL spectra of the same sample series. This allows the comparison of the different samples because the relation between SC and the microstructure depends on many process parameters.

In figure 1 Raman spectra for the PE-CVD2 series, 2 %<SC< 8%, are shown. The spectra are normalized to the height of the narrow peak at 520 cm$^{-1}$, which is due to the LO-TO mode of the crystalline phase. An additional contribution from structural defects of the crystalline phase (stacking faults) leads to the hump at 492 cm$^{-1}$. The broad band located at about 480 cm$^{-1}$ is attributed to the amorphous phase. $I_c^{RS}$ defined as the ratio of the Raman intensity of the crystalline phase (fitted by two Gaussian lines at 520 cm$^{-1}$ and 500 cm$^{-1}$ to take care of the asymmetry) to the total scattering intensity (where the amorphous phase is described by a Gaussian line at 480 cm$^{-1}$) and serves as a measure for the crystalline volume fraction. With increasing SC, the

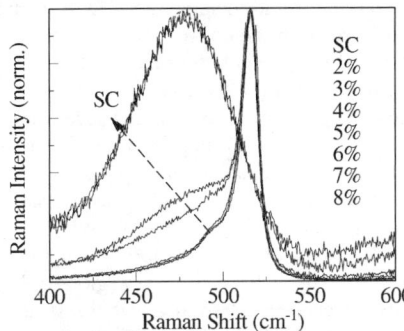
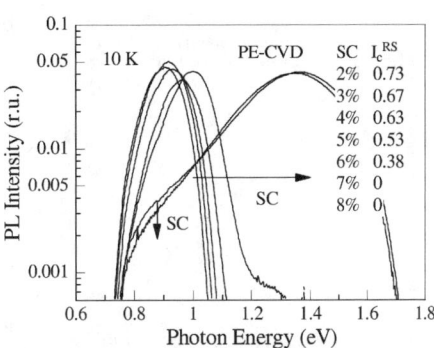

**Figure 1**. Raman spectra of μc-Si:H films for different silane concentration (see text for details)

**Figure 2**. Photoluminescence spectra for the same samples as in figure 1. Note the logaritmic intensity scale.

relative contribution of the broad band increases, i.e. the crystalline volume fraction decreases. For SC =2 - 4 % no contribution of the amorphous phase is observed (note that the tail of these Raman spectra results from strain and disorder at grain boundaries rather than from a well defined amorphous phase which leads to an overestimation of amorphous contribution in the fit). A significant increase of the amorphous contribution is found for the 5 % and 6 % sample and it dominates the Raman intensity for the 7 % and 8 % sample.

Photoluminescence spectra of the same series are shown in figure 2. The PL was excited with photons of 488 nm wavelength like the Raman spectra and therefore a similar sample volume is probed by the two experiments. A logarithmic scale has been chosen for the PL intensity to demonstrate two important effects: (i) the PL band at about 0.95 eV (μc-Si-band), characteristic for the microcrystalline phase, shifts continuously to higher energies with increasing SC by about 0.1 eV and (ii) the PL band at 1.35 eV, characteristic for the amorphous phase, is visible only in the 7% and 8% samples. The quantum efficiency of the PL is very similar for all films and slightly lower than high quality amorphous silicon, i.e. in the % range. The Raman and PL spectra of other PE-CVD and HW series are very similar to those in figures 1 and 2 [8].

In figure 3 the PL energy is shown as a function of the 'crystalline volume fraction', $I_c^{RS}$, for the three sample series. A clear trend is found, i.e. the PL energy decreases with increasing crystalline volume fraction. As the shift to lower energies is not accompanied by a decrease of the efficiency and the spin density also decreases with decreasing $I_c^{RS}$ [9] the shift can not be caused by enhanced non-radiative recombination. Therefore another effect must be the cause for this behavior. The increase of $V_{oc}$, the decrease of both the short circuit current density and the dark conductivity with increasing SC, i.e. decreasing $I_c^{RS}$ have been tentatively explained by an increase of the effective mobility gap towards amorphous growth conditions [10]. However, a careful analysis of the absorption spectra of many series do not give any evidence for such a shift in this energy range. Barriers between crystalline regions could cause a similar effect, but would affect the carrier mobility rather than the carrier density. We propose instead, that the shift of the μc-Si-PL-band is caused by a decrease of the density of band tail states and/or by the steepening of the band tails. Because we do not find a narrowing of the PL band but rather a constant width or a slight broadening, a steepening of the tail seems unlikely. The microscopic effect that leads to the tail states and the mechanism of their reduction is yet unknown. If strain is the major cause

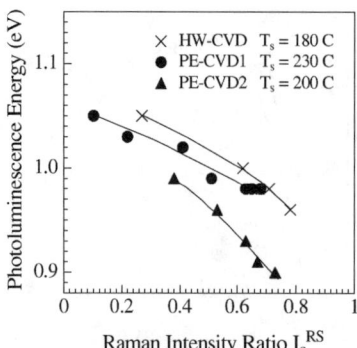

**Figure 3.** Photoluminescence energy as a function of the Raman intensity ratio for μc-Si:H prepared by PE- and HW-CVD

of the tail one could imagine that due to the increase of the hydrogen content strained bonds will be broken and saturated by hydrogen and the remaining lattice will relax more effectively.

However, a word of caution is needed as the 'picture' of the band tail is adopted from a material such as amorphous silicon which is homogeneous on a scale of several nm. For μc-Si:H this is unlikely and the material is considered to be inhomogeneous on the scale of tens of nm. Therefore it is not clear how clusters of defects, local strain or regions of high and low hydrogen concentration affect the states at and below the band edges. However, it is reasonable to analyze and interpret the experimental findings assuming localized band tail states at the present stage.

## PL on solar cells

The solar cells investigated in this study cover the structural composition range from fully crystalline (SC=3%) to more than 50% amorphous volume fraction in the i-layer [2]. The i-layer of the cells was prepared by HW-CVD and within this series the highest solar cell efficiency reported for microcrystalline HW cells are observed in addition to the highest $V_{oc}$ reported so far for microcrystalline solar cells. In order to avoid spurious PL signals from the glass, TCO and p-layer and to investigate only those carriers which are taking part in the photovoltaic process a modulation technique has been applied. The solar cell was continuously excited with the excitation light of full intensity and the applied voltage has been switched from $V_{oc}$ to 0V. Thereby the photocurrent changed from 0 mA (no carrier extraction) to $I_{sc}$ (full carrier extraction). The recombination rate of the carriers active in the photovoltaic process changed from full recombination to none. By applying this technique it was assured that only the recombination of the relevant carriers are probed. In figure 4 modulated PL spectra taken at 150 K for all samples of the series are shown on a logarithmic intensity scale as an example. Ignoring the fringe pattern in two of the samples, the shift of the spectra to higher energy with increasing SC is clearly visible. Such a shift is observed for all temperatures investigated (compare figure 6). The PL intensity and the width of the PL band does not change significantly indicating that the shift is not caused by an increasing defect density, consistent with the very similar $j_{sc}$ observed for all the samples. It should be noted here, that (i) the conventional PL spectra look almost identical and (ii) surprisingly a contribution of the amorphous volume to the PL

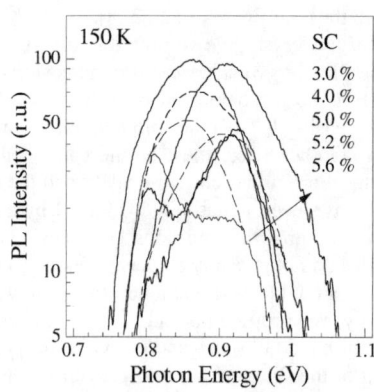

**Figure 4.** Modulated PL spectra of μc-Si:H solar cells

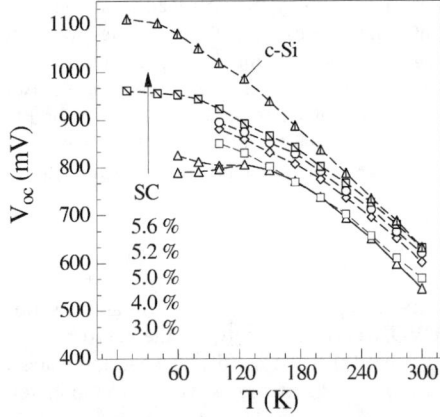

**Figure 5.** Open circuit voltage, $V_{oc}$, as a function of temperature for μc-Si:H solar cells and a c-Si solar cell as a reference

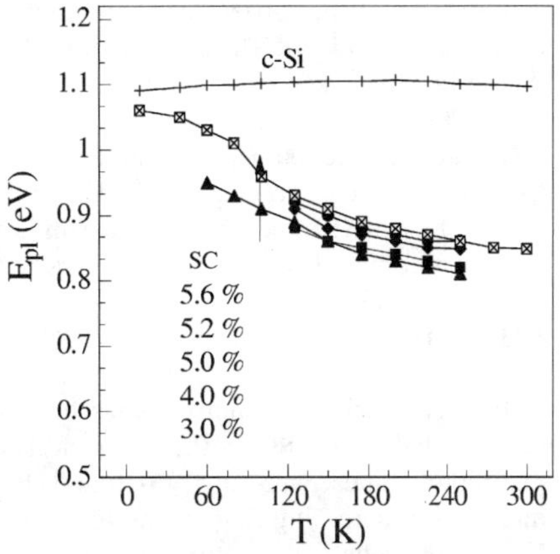

**Figure 6.** Energy of modulated photoluminescence spectra as function of temperature for the same samples as in figure 5

spectrum is absent in both types of measurements for the 5.6 % sample despite of the $I_c^{RS}$ of more than 50 % [2]. This points to an effective carrier transfer from the amorphous to the crystalline phase.

In figure 5 and 6 $V_{oc}$ and $E_{PL}$ as a function of temperature are summarized. $V_{oc}$ rises differently for different samples. The SC=3 % sample starts at the lowest value at 300 K and rises towards low temperature and bends over into saturation at about 150 K. With increasing SC the $V_{oc}$ is higher at all temperatures and the saturation occurs at successively lower temperatures. The sample with SC=5.6 % shows the highest $V_{oc}$ and bends over at the lowest temperature. For comparison a commercially available c-Si p-i-n diode (Hamamatsu) was investigated. The $V_{oc}$ of this diode is about 600 mV and increases to about 1100 mV at 10 K. It is obvious that with increasing SC the $V_{oc}$ (T) dependency of the microcrystalline diodes is getting closer to that of crystalline silicon. At the same time the crystalline volume fraction of the solar cells decreases. This can be interpreted by a reduction of the states below the gap of crystalline silicon as this effect allows the quasi-Fermi level to shift closer to the band edges and thereby giving rise to an increase of $V_{oc}$. The limitation in the rise of $V_{oc}$ in the microcrystalline solar cells found in figure 5 is accompanied by a strongly reduced carrier extraction, i.e. $j_{sc}$ decreases rapidly when the bend-over of $V_{oc}$ sets in. The same holds for the modulated PL intensity which is directly linked to the effectiveness of the carrier extraction.

In figure 6 $E_{PL}$ is shown as a function of temperature for comparison. $E_{PL}$ increases towards lower temperature for all samples and with increasing SC for a given temperature. A comparison of the figures 6 and 7 reveals that the sample with the lowest $V_{oc}$ exhibits the lowest $E_{PL}$ and the sample with the highest $V_{oc}$ exhibits the highest $E_{PL}$. Therefore, the trend of both quantities is similar but the shift of $E_{PL}$ as a function of temperature is much less than that of $V_{oc}$.

The increase of the PL peak position and the increase of $V_{oc}$ in the solar cells with increasing SC can be interpreted assuming a reduced density of the localized states which - for a constant carrier density - leads to a shift of the carrier distribution towards higher energies. An increasing carrier concentration due to an increasing carrier lifetime with decreasing temperature will also cause a shift of the carrier distribution towards higher energies and will lead to the observed increase of $E_{PL}$ and $V_{oc}$. The smaller shift of $E_{PL}$ as compared to $V_{oc}$ with increasing temperature is due to fact that radiative recombination of electrons an holes require occupied states which sets the lower limit of the recombination energy to the minimum energy difference of a reasonable amount of conduction band tail states and valence band tail states (if coulomb effects and Frank-Condon shifts can be neglected). The splitting of the quasi-Fermi levels of the electrons and holes which determines $V_{oc}$ is not similarly restricted but will tend to zero when the carrier density becomes close to the thermal distribution.

## CONCLUSION

By a comparison of photoluminescence and Raman spectra on high quality sample series prepared with different SC by PE-CVD and HW-CVD, a correlation between the microstructure and the photoluminescence energy is found. It is proposed that the density of band tail states is reduced with increasing SC leading to the increase of the PL energy as well as to the increase of $V_{oc}$ of solar cells.

The PL efficiency is very similar for all films so that a strong influence of non-radiative recombination can be ruled out and the data can be interpreted assuming a decrease of the localized band tail states as a result of the increasing silane concentration. The reason for the tails and their reduction is not clear but strain might play a crucial role and the amorphous hydrogenated phase might be effective for strain reduction.

## ACKNOWLEDGEMENTS

The authors gratefully acknowledge L. Baja Neto and S. Klein for providing excellent material, solar cells and experimental results and G. Schöpe for the textured TCO.

## REFERENCES

1. O.Vetterl, R. Carius, L. Houben, C. Scholten, M. Luysberg, A. Lambertz, F. Finger, H. Wagner, *Mat. Res. Soc. Symp. Proc.*. 609 (2000) A15.2.
2. S. Klein, F. Finger, R. Carius, B. Rech, L. Houben, M. Luysberg, M. Stutzmann, *Mat. Res. Soc. Symp. Proc.*. 715 (2002) A26.3
3. P. K. Bhat, G. Diprose, T. M. Searle, I. G. Austin, P. G. LeComber, W. E. Spear, Physica 117&118B, 917 (1983); P. K. Bhat, I. G. Austin, T. M. Searle, J. Non-Crystal. Solids, 59&60, 381 (1983)
4. R. Carius, Mat. Res. Soc. Meeting, San Francisco, (2000) unpublished
5. Gouzhen Yue, Daxing Han, L. E. McNeil, Qi Wang, J. Appl. Phys. 88, No. 8, 4904 (2000)
6. R. Carius, F. Finger, U. Backhausen, M. Luysberg, P. Hapke, M. Otte and H. Overhof, *Mat. Res. Soc. Symp. Proc.*. 467 (1997) 283.
7. A. Kaan Kalkan, S. J. Fonasch, Shang-Cong Cheng, Appl. Phys. Lett. 77, No. 1, 55 (2000)
8. R. Carius, T. Merdzhanova, F. Finger, S. Klein, O. Vetterl, Journal of Materials Science - Materials in Electronics, Kluwer Academic Publishers, 2003, in print
9. F. Finger, S. Klein, R. Carius, T. Dylla, O. Vetterl, A.L. Baia Neto, ibid.
10. T. Brammer, E. Bunte, H. Stiebig, F. Finger, H. Wagner, Proceedings of the 16th E-PVSEC, Glasgow, May 2000, edited by H. Scher, B. McNelis, W. Palz, H.A. Ossenbrink, P. Helm, (James&James, London, 2000) p. 545

## Localised States in Microcrystalline Silicon Photovoltaic Structures Studied by Post-Transit Time-of-Flight Spectroscopy

Steve Reynolds, Vladimir Smirnov, Charlie Main, Reinhard Carius[1] and Friedhelm Finger[1],
School of Computing and Advanced Technologies, University of Abertay Dundee,
Bell Street, Dundee U.K.
[1] Forschungszentrum Jülich, Institute for Photovoltaics, D-52425 Jülich, Germany.

## ABSTRACT

Post-transit time-of-flight spectroscopy has been used to study the density of states distribution in hot-wire CVD microcrystalline silicon pin solar cell structures. For an absorber layer Raman scattering intensity ratio $I_{CRS}$ of 0.4 or less, behaviour consistent with multiple-trapping carrier transport is observed and may be interpreted in terms of a conduction-band tail of some 18 meV slope plus a broad defect bump of order $10^{17}$ cm$^{-3}$ centered at 0.55 eV relative to the mobility edge. As $I_{CRS}$ is increased beyond 0.4, the temperature-dependence of the photocurrent transient becomes inconsistent with multiple-trapping and above 0.6 the decays are almost temperature-independent. By comparing data taken at 300 K, it may be inferred from the multiple-trapping model that localised states between 0.35 and 0.5 eV are associated with the presence of columns or clusters of nanocrystals and those deeper than 0.5 eV with the amorphous tissue. Results are compared with previous work on coplanar and sandwich structures.

## INTRODUCTION

Microcrystalline silicon (μc-Si:H) is an attractive choice as the carrier generation layer in single or tandem thin-film solar cells, where its increased optical absorption over amorphous silicon (a-Si:H) in the long-wavelength region of the solar spectrum offers potential improvements in conversion efficiency. It is also more resistant, but not impervious, to light-induced degradation. There have been several reports in the literature correlating optoelectronic properties, such as defect densities and photovoltaic conversion efficiencies, with deposition parameters, film structure and morphology [1,2]. For example, it has been demonstrated that the highest conversion efficiencies are obtained when the absorber layer composition lies within the transitional régime between microcrystalline and amorphous growth.

Previously, we have applied transient photocurrent spectroscopy (TPC) to the study of coplanar μc-Si:H films having varying degrees of crystallinity [3]. This work has now been extended to include pin structures, using post-transit time-of-flight (PT-TOF) measurements and analysis to probe transport properties relevant to an actual solar cell. This technique monitors the primary photocurrent decay in a reverse-biased device following the initial rapid transit of carriers, which may be related directly to the emission of carriers from deep traps, since following emission they will be extracted before deep re-trapping can occur and also cannot be replenished by the external circuit.

While PT-TOF has been widely and successfully used in the study of a-Si:H films, μc-Si:H presents additional complications owing to its mixed-phase composition of crystallites, voids and amorphous regions. Previous studies [4] have demonstrated the feasibility of the technique and the potential for application, however the extent to which the multiple-trapping (MT) transport model on which the PT-TOF analysis is based may legitimately be applied remains unclear. The aim of this work was to probe the density of localised states (DOS) in the upper portion of the band gap using PT-TOF over a range of crystallinities, as indicated by the Raman scattering intensity ratio ($I_{CRS}$) [5], and temperatures, looking for behavior consistent with MT, and to compare our findings with previous work.

## EXPERIMENTAL DETAILS

The pin samples studied were deposited at IPV Jülich in the following sequence: glass substrate/TCO/pin/TCO/1 mm Ag dot contact. The p- and n-layers were prepared using VHF PECVD and the i-layer by HWCVD, with a filament temperature of 1650 °C and substrate temperature below 220 °C to minimise defect density [1,6]. $I_{CRS}$ was varied between 0.32 and 0.60 by altering the gas concentration $r$ = [silane] : [silane + hydrogen] between 7 and 4%.

For PT-TOF measurements, an electrically screened Laser Science VSL-337 $N_2$ laser plus dye attachment was used to generate 4 ns pulses of 500 nm light, attenuated as required by means of neutral density filters, so that the photogenerated charge was less than the CV product. Following preamplification and dark current nulling, the photocurrent decay at 2 V dc reverse bias was recorded on a Tektronix TDS3052 storage oscilloscope and data transferred to a PC for analysis. Measurements between 100 K and 400 K were made in a screened cryostat.

## DATA ANALYSIS

The DOS $g(E)$ is related to the post-transit photocurrent $I(t)$ by the mathematically approximate expression [7]

$$g(E) = \frac{2g(0)I(t)t}{Q_0 t_0 \nu_0} \tag{1}$$

where $Q_0$ is the total photo-induced charge, $t_0$ the trap-free transit time, $\nu_0$ the attempt-to-escape frequency and $g(0)$ the DOS at the conduction band mobility edge. The energy scale relative to the mobility edge is related to the emission rate and is given by $E = kT \ln(\nu_0 t)$, where $k$ is Boltzmann's constant and $T$ the absolute temperature. Equation 1 is quite accurate in practice, provided the features of interest do not vary more steeply than $kT$. Other methods based on Fourier (and Laplace) transformations are available [8], which can utilize both pre- and post-transit data. Here, relatively minor differences were noticeable between the use of these methods and equation 1 although improved overlap between DOS sections as the temperature was varied was achieved by means of the Fourier technique. It is, however, more straightforward to interpret the $I(t)$ decay directly through equation 1; clearly, if the current is falling less steeply than a power law index of $-1$ on the log-log plot, then the DOS is rising, if it falls more steeply than this then the DOS is falling.

## RESULTS AND DISCUSSION

### **Low crystalline volume fraction**

PT-TOF decays from the sample with $I_{CRS} = 0.32$ is shown in figure 1(a), and the resulting DOS in figure 1(b). The data have been processed assuming $v_0 = 10^{12}$ s$^{-1}$, and it is evident there is a good degree of overlap between the DOS sections calculated at different temperatures. This indicates an appropriate choice has been made for $v_0$, at least for the defect distribution centred at 0.55 eV. At sufficiently low temperatures (< 200 K) it is possible to observe a section of the conduction band tail. Based on this evidence the tail slope is quite steep, some 18 meV. Calibration of the DOS can be achieved by extrapolating this section to an assumed band edge density $g(0)$ of $4 \times 10^{21}$ cm$^{-3}$ eV$^{-1}$. This yields a maximum value of $2 \times 10^{17}$ cm$^{-3}$ eV$^{-1}$, and an integrated density of about $5 \times 10^{16}$ cm$^{-3}$, for deep defects. It is also possible to obtain the DOS relative to $g(0)$ from equation 1; if appropriate values of $t_0 = 2$ ns (see below) and $Q_0 = 10^{-11}$ C are used the ratio of the peak defect DOS to $g(0)$ is approximately $10^{-4}$, in good agreement with the extrapolation.

Also shown in figure 1(b) for comparison is the DOS obtained for a (slightly thicker) a-Si:H pin device. It can be seen that the DOS is quite similar in shape, though the defect density is lower and the band tail slope slightly larger than for the µc-Si:H device. This suggests that at low film crystallinity the transport process prevailing in both devices is similar, and is consistent with multiple-trapping in tail and defect states.

To verify that the calculation of the DOS is consistent with MT theory, we have used the DOS obtained to *simulate* the $I(t)$ curves under post-transit conditions [9], as shown in figure 2. Best agreement with experiment is obtained by assuming a transit time of 2 ns. Bearing in mind that the laser pulse width is approximately 4 ns and that our signal recovery system is not optimised for use at short times, we cannot measure this directly. However, the implied drift mobility value of 2.5 cm$^2$ (Vs)$^{-1}$ is within the range reported by others for similar material [4].

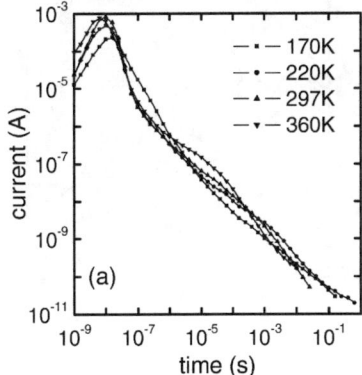 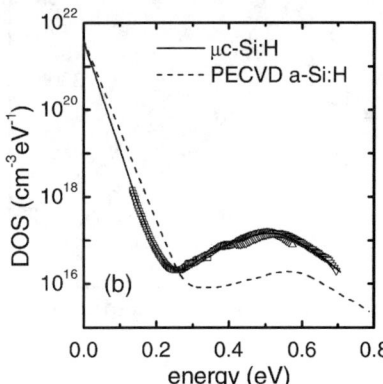

**Figure 1.** (a) TPC decays for µc-Si:H pin device, $I_{CRS} = 0.32$; (b) DOS, with that for a-Si:H device shown for comparison.

Although the agreement between experiment and simulation is not perfect, the general behaviour of the decays with temperature is well-reproduced and supports our interpretation.

**Highly crystalline films**

The effect of temperature on the $I(t)$ decays with increasing absorber layer crystallinity is illustrated in figure 3. There is a tendency for the curves to become temperature-independent for $I_{CRS}$ above 0.5, although an increase in gradient is apparent at times longer than $10^{-4}$ s. However, in more highly crystalline devices the reverse leakage current is of the order of $10^{-6}$ A, several orders higher than the photocurrent in this region, which should thus be viewed with caution.

In the context of single-carrier MT theory it is not possible to construct a realistic DOS that results in a temperature-independent $I(t)$ decay. However this does not necessarily mean that MT is not taking place, it could be that its manifestations within the post-transit current are obscured by other effects. We have previously shown [3] that the transient photocurrent decay in highly-crystalline *coplanar* films more closely resembles behavior observed in n-type a-Si:H. In that case, the photocurrent decay is controlled by the release of the deeply-trapped *holes* required for recombination to proceed. We have not yet carried out a full simulation of what effects this might have on the post-transit current. On the basis of thermopower measurements, the position of the Fermi level is believed to move toward the transport level by as much as 0.2 eV between 400 K and 200 K [10]. This could account for the temperature-independent break-point at $10^{-5}$ s in figure 3 because although the thermalisation energy sinks more slowly at lower temperatures it reaches the (raised) Fermi level at much the same time. The increased gradient at longer times might also be anticipated, reflecting the sharper Boltzmann tail. However, there is also evidence to suggest that transport below room temperature in highly crystalline films is by hopping [11,12], and the presence of a distribution of states at potential barriers between crystalline

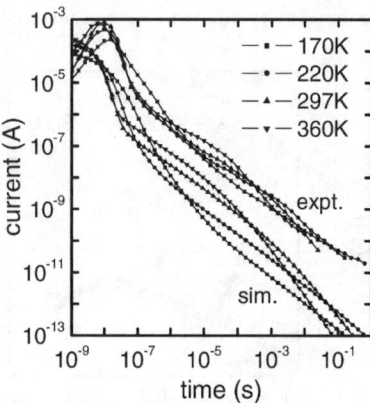

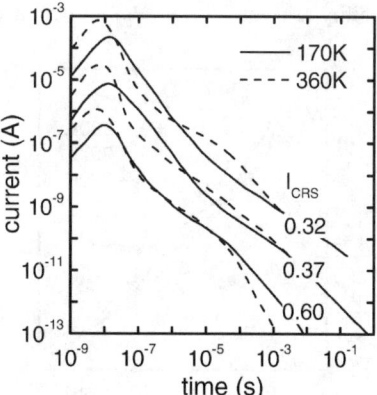

**Figure 2.** Comparison of experimental and simulated $I(t)$ using DOS of figure 1(b) and assuming transit time of 2 ns. Simulations are offset by a factor of 10.

**Figure 3.** Temperature–dependence of experimental $I(t)$ with increasing $I_{CRS}$. Successive groups of curves are offset by a factor of 30.

clusters or columns and the surrounding amorphous tissue has also been inferred from ESR [13]. At present our measurements and analysis cannot resolve these questions.

**Energetic distributions of defect states in a-Si:H and µc-Si:H**

Although the results presented in the previous section cannot be consistently interpreted in terms of MT theory, this transport mechanism may still prevail at room temperature (and above). If so, comparing room temperature $I(t)$ decays as the crystallinity is increased, as shown in figure 4, could give an insight into how the defect distributions evolve. As $I_{CRS}$ increases, the current increases below $10^{-4}$ s and decreases at longer times. Through equation 1, this may be interpreted as an increasing shallow DOS, and decreasing deep DOS, with increasing $I_{CRS}$. Linking this to structural changes, it is tempting to suggest that the shallow defects (0.35-0.5 eV) are associated with an increase in crystalline columns or granular clusters and their boundaries, and deeper defects (0.5-0.65 eV) with the amorphous tissue. This interpretation is supported to some extent by the comparison presented in figure 5 between this and previous work [3], and also by photoluminescence studies [14]. It can be seen that for both coplanar and pin amorphous silicon configurations, the defect distributions reach a maximum at about 0.6 eV, but for films containing microcrystalline silicon there is a shift to shallower energies. However, at present this interpretation must remain speculative.

## CONCLUSIONS

Post-transit time-of-flight measurements may be used successfully to study localised state distributions in microcrystalline silicon pin solar cell structures. For $I_{CRS} \leq 0.4$, the density of states is similar in form to that of amorphous silicon, although a somewhat higher defect density is indicated. As the crystallinity as measured by $I_{CRS}$ is increased, the temperature-dependence

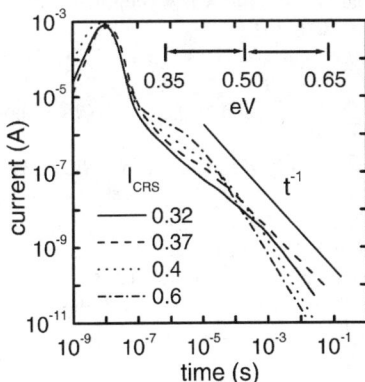

**Figure 4.** $I(t)$ decays obtained over a range of $I_{CRS}$ at 300 K. Energy scale assumes $v_0 = 10^{12}$ s$^{-1}$.

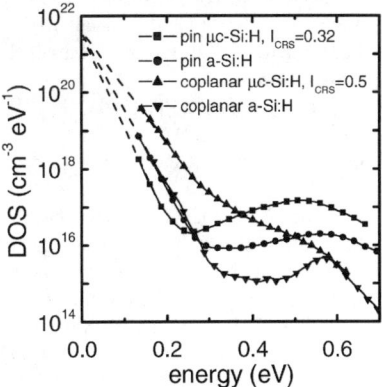

**Figure 5.** Comparison of DOS plots for µc-Si:H and a-Si:H in coplanar and pin configurations.

anticipated for multiple-trapping transport is no longer observed, and thus we conclude that, at least below room temperature, other transport mechanisms may play a significant role. If the photocurrent decays obtained at room temperature for pin samples prepared over a range of crystallinities are compared, there is a good correlation in behaviour, with a higher photocurrent observed at short times and a lower photocurrent at longer times as $I_{CRS}$ is increased. In terms of a multiple-trapping model this indicates that a higher density of shallow defects are involved in transport in the more crystalline material, and that in more amorphous material the defect density is lower and the distribution is deeper in energy. This general conclusion is supported by evidence from earlier transient photoconductivity studies of coplanar films of amorphous and microcrystalline silicon. Correlation with structural properties would suggest that the shallow defects are associated with columns, fibres or clusters of microcrystalline silicon, or the interface between these structures and the surrounding disordered material, and the deeper defects are the native dangling bonds that occur in amorphous silicon.

## ACKNOWLEDGMENTS

The authors thank research staff at IPV Jülich for preparation and characterisation of samples.

## REFERENCES

1. S. Klein, F. Finger, R. Carius, B. Rech, L. Houben, M. Luysberg and M. Stutzmann, *MRS Symp. Proc.* **715**, A21.2.1 (2002).
2. N. Wyrsch, L. Feitnecht, C. Droz, P. Torres, A. Shah, A. Poruba and M. Vaněček, *J. Non-Cryst. Solids* **266-269**, 1099 (2000).
3. S. Reynolds, V. Smirnov, C. Main, R. Carius and F. Finger, *MRS Symp. Proc.* **715**, A21.2.1 (2002).
4. N. Beck, P. Torres, J. Fric, Z. Remeš, A. Poruba, H. Stuchlíková, A. Fejfar, N. Wyrsch, M. Vaněček, J. Kočka and A. Shah, *MRS Symp. Proc.* **452**, 761 (1997).
5. L. Houben, M. Luysberg, P.Hapke, R. Carius, F. Finger and H. Wagner, *Philos. Mag. A* **77**, 1447 (1998).
6. F. Finger, S. Klein, T. Dylla, A.L. Baia Neto, O. Vetterl and R. Carius, *MRS Symp. Proc.* **715**, A16.3.1 (2002).
7. G.F. Seynhaeve, R.P. Barclay, G.J. Adriaenssens and J.M. Marshall, *Phys. Rev.* **B39**, 10196 (1989).
8. C. Main, *J. Non-Cryst. Solids* **299-302**, 525 (2002).
9. C. Main, J. Berkin and A. Merazga, in *New Physical Problems in Electronic Materials*, eds. M. Borissov, N. Kirov, J.M. Marshall and A. Vavrek (World Scientific, 1991), p55.
10. D. Ruff, H. Mell, L. Tóth, I. Sieber and W. Fuhs, *J. Non-Cryst. Solids* **227-230**, 1011 (1998).
11. J. Kočka, H. Stuchlíková, J. Stuchlík, B. Rezek, T. Mates, V. Švrček, P. Fojtík, I. Pelant and A. Fejfar, *J. Non-Cryst. Solids* **299-302**, 355 (2002).
12. A. Fejfar, N. Beck, H. Stuchlíková, N. Wyrsch, P. Torres, J. Meier, A. Shah and J. Kočka, *J. Non-Cryst. Solids* **227-230**, 1006 (1998).
13. F. Finger, J. Müller, C. Malten and H. Wagner, *Philos. Mag. B* **77**, 805 (1998).
14. G. Yue, J.D. Lorentzen, J. Lin, D. Han and Q. Wang, *Appl. Phys. Lett.* **75**, 492 (1999).

# Femtosecond far-infrared studies of carrier dynamics in hydrogenated amorphous silicon and silicon-germanium alloys

A. V. V. Nampoothiri,[1] B. P. Nelson,[2] and S. L. Dexheimer[1]

[1]Department of Physics, Washington State University, Pullman, WA
[2]National Renewable Energy Laboratory, Golden, CO

## ABSTRACT

We present femtosecond time-resolved studies of the photoexcited carrier response in the far-infrared spectral range in PECVD a-Si:H and a-SiGe:H thin films. The experiments are carried out using an optical pump / terahertz (THz) probe technique, in which a femtosecond pump pulse excites carriers in the sample and a time-delayed probe pulse measures the resulting change in the far-infrared optical properties as a function of time delay following the excitation. These measurements are sensitive to carrier processes at low energy, corresponding to a range of approximately 1 - 10 meV, a key energy scale in these materials. We find that the observed photoexcited carrier dynamics are consistent with trapping of carriers into band tail states on a picosecond time scale.

## INTRODUCTION

Despite considerable previous work on time-resolved optical measurements on amorphous silicon and related materials, important questions have remained regarding fundamental carrier processes in these materials, including the time scale and nature of the initial carrier localization processes. Most experiments on picosecond and femtosecond time scales have been carried out using a pump-probe technique, in which a short optical pump pulse excites carriers into the extended states, and a short, time-delayed probe pulse measures the resulting change in optical properties during the time evolution of the photoexcited carrier distribution. In a large body of previous work, the photoexcited carrier distribution has been probed in the near-infrared and visible spectral ranges, corresponding to relatively large transition energies. The response observed under these conditions is dominated by an induced absorption from the photoexcited carriers with a decay consistent with simple bimolecular recombination kinetics, though the recombination mechanism is still not well understood [1-3]. More recently, we have used femtosecond pump-probe measurements in the visible and near-infrared spectral ranges to resolve the initial carrier thermalization dynamics, in which carriers that are initially photoexcited into extended states in the conduction and valence bands relax in energy toward the band edge via phonon emission, and we have established that this process takes place on a time scale of ~ 150 fs [4]. In the work we report here, we have used recently developed methods for generating and detecting femtosecond pulses in the far-infrared, or terahertz (THz), spectral range to directly probe the photoexcited carrier distribution at energies of ~ 1 – 10 meV, a key energy scale for these systems, and one which has remained essentially unexplored in these materials. We find that measurements in the THz spectral range allow us to detect a new component of the carrier response, and that the observed dynamics reflect carrier trapping into the exponential distribution of band tail states.

## EXPERIMENT

Time-resolved optical pump / THz probe measurements were carried out using the apparatus shown schematically in Fig. 1. An amplified Ti:sapphire laser system provides optical pulses 35 fs in duration centered at 800 nm (1.55 eV) at a 1 kHz repetition rate. These pulses are split into three arms for the THz emitter, THz receiver, and optical pump beams. In the apparatus, THz pulses are generated in a <110> ZnTe emitter by optical rectification of the 35-fs near-infrared pulses, and are detected by free-space electro-optic sampling in a <110> ZnTe sensor [5]. The THz beam from the emitter is collected and collimated by a low f-number off-axis paraboloid, then focused onto the sample, and the transmitted beam is collected and focused onto the electro-optic sensor. The electric field of the THz pulse is detected via the Pockels effect by measuring the polarization rotation of a co-propagating optical sampling beam. The optical sampling beam for the receiver is variably delayed relative to the THz beam using a second computer-controlled delay stage to measure the electric field waveform of the THz pulse. Absorption from atmospheric water vapor in the THz spectral region has been alleviated by enclosing the apparatus and purging with dry nitrogen.

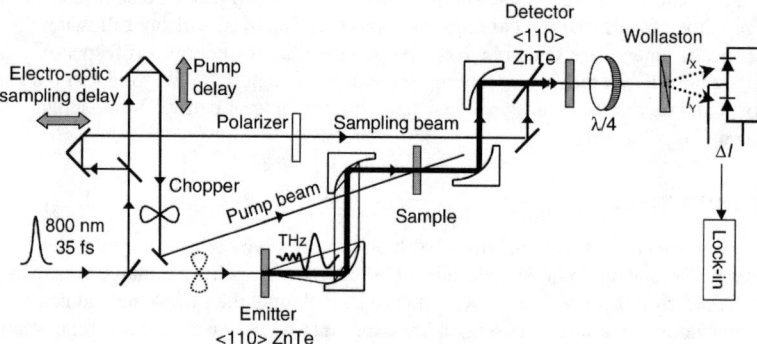

**Figure 1.** Time-resolved THz spectrometer, based on THz pulse generation via optical rectification and THz pulse detection via electro-optic sampling.

The electric field waveform of the THz pulse used as a probe is shown in Fig. 2. The waveform is a nearly single-cycle pulse with a central feature of ~ 400 fs width. The power spectrum of the pulse, determined by Fourier transformation of the waveform, is shown in Fig. 3. A frequency of 1 THz corresponds to a wavelength of 300 µm, and an energy of 4.1 meV or 33 cm$^{-1}$, and the pulse spectrum is seen to span the energy range of ~ 1- 10 meV.

For time-resolved optical pump / THz probe measurements, an optical pump beam is delayed relative to the THz beam using a computer-controlled delay stage, and is directed onto the sample at a small angle relative to the THz beam. Differential pump-probe measurements are performed by chopping the pump beam and detecting the change in the THz response in the presence and absence of the pump beam using lock-in amplification. The measurements presented here are "1-D pump scans" in which the differential signal is detected at the peak of the THz electric field waveform as a function of time delay between the pump and THz probe pulses. This type of measurement essentially reflects the change in absorption averaged over the frequency spectrum of the THz pulse, as long as there is not a large photoinduced index change or a strong frequency dispersion [6], which we have verified to be the case by more detailed 2-D frequency-dependent measurements, in which the full THz waveform is measured at constant pump-probe delay. To avoid contributions during the region of pump-probe pulse overlap in the 1-D scans, only the portion of the signal after ~ 1 ps delay is analyzed here.

Thin film samples of a-Si:H and a-$Si_xGe_{1-x}$:H (x = 0.5) of ~ 1 μm thickness were grown by PECVD. Sapphire substrates (1 mm thickness) were used instead of standard optical glass to allow transmission measurements in the THz spectral range.

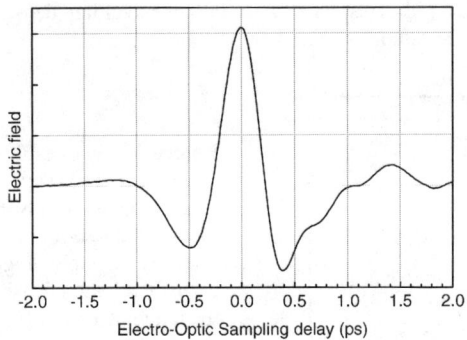

**Figure 2.** Electric field waveform of the THz probe pulse measured by electro-optic sampling.

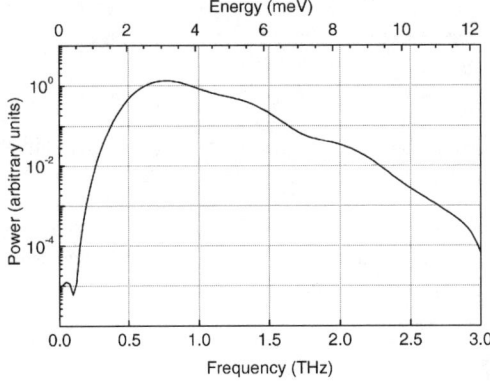

**Figure 3.** Fourier transform power spectrum of the THz pulse.

## RESULTS

Representative time-resolved optical pump / THz probe measurements on a-$Si_xGe_{1-x}$:H (x = 0.5) are shown in Fig. 4. In these measurements, carriers are generated by the 800 nm pump pulse at delay time t = 0. Since the 1.55 eV pump photon energy is well above the band gap, carriers are excited directly into the extended states. The individual traces shown in Fig. 4 are measurements made at a series of pump fluences ranging from 0.4 to 3.3 mJ/cm$^2$, corresponding to initial excitation densities in the range of $10^{18}$ to $10^{19}$ cm$^{-3}$. The observed time-resolved differential response is a net decrease in THz transmission following excitation, corresponding to an induced absorbance associated with the photoexcited carrier population. The induced absorbance signal decreases rapidly on a picosecond time scale. The dominant contribution to the response fits to a power law time dependence:

$$y(t) = a\,(t-t_0)^{-\alpha} \qquad (1)$$

The results of fits, shown as lines through the data traces presented in Fig. 4, yield values $\alpha \sim 0.6$ for the exponent. A slight variation in the response with initial excitation density reflects recombination processes.

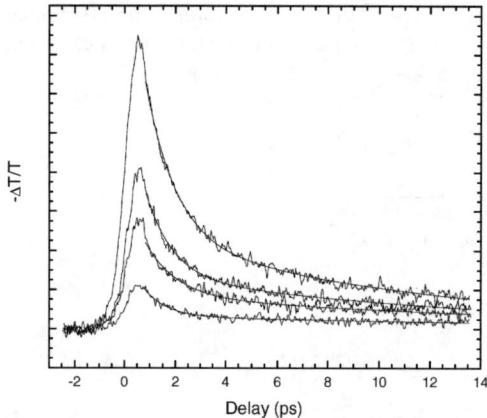

**Figure 4.** Time-resolved optical pump / THz probe measurements on a-$Si_xGe_{1-x}$:H (x = 0.5) at a series of pump fluences corresponding to initial carrier excitation densities in the range of $10^{18}$ to $10^{19}$ cm$^{-3}$. Solid lines represent fits to the power law model described in the text.

Time-resolved optical pump / THz probe measurements on a-Si:H give qualitatively similar behavior, showing a net induced absorbance that decays on a picosecond time scale. In a-Si:H, carriers are excited into the extended states by the 800 nm optical pump pulse via two-photon absorption [4], corresponding to an effective transition energy of 3.1 eV, well above the nominal 1.7 eV band gap of the material. The dynamics in a-Si:H are also well represented by the power law function of Eqn. (1), yielding values $\alpha \sim 0.35$ for the exponent.

## DISCUSSION

In the experiments presented above, carriers, which are initially photoexcited into extended electronic states, give rise to an induced absorbance in the far-infrared spectral range.

The time evolution of the induced absorbance measured at times greater than 1 ps following excitation is dominated by a simple power law decay. It is interesting to note that the time-resolved response of the photoexcited carrier distribution detected in the far-infrared differs markedly from the simple bimolecular decay kinetics detected at near-infrared and visible wavelengths, showing that the far-infrared spectral range is sensitive to a very different, and intriguing component of the carrier dynamics.

Since the far-infrared optical absorption associated with the photoinduced carriers is directly proportional to their conductivity, time-resolved THz spectroscopy measurements provide a non-contact method for measuring photoconductivity, as has been demonstrated in other semiconductor materials [6, 7]. In our measurements on a-Si:H and a-SiGe:H, we observe a power law time dependence for the far-infrared response, which is characteristic of conductivity processes involving trapping in an exponential distribution of band tail states [8]. This observed temporal response is consistent with the reduction of the effective mobility as the initial nonequilibrium free carrier distribution traps into the localized band tail states, establishing the time scale for this fundamental localization process. The physics of the carrier response is further reflected in the measured far-infrared spectral response of the photoexcited carrier distribution, which will be presented elsewhere [9].

## SUMMARY

We have carried out femtosecond optical pump / terahertz (THz) probe studies of the photoexcited carrier response in PECVD a-Si:H and a-SiGe:H thin films, which are sensitive to carrier processes at low energy, in a range of approximately 1 - 10 meV. We find that the observed photoexcited carrier dynamics on a picosecond time scale are dominated by a power law decay, consistent with the loss of mobility as the initially free carriers trap into an exponential distribution of band tail states.

## ACKNOWLEDGMENTS

This work was supported by subcontract AAD-9-18668-01 from the US Department of Energy National Renewable Energy Laboratory and by the National Science Foundation under grant DMR-9973615.

## REFERENCES

1. P. M. Fauchet, D. Hulin, R. Vanderhaghen, A. Mourchid and J. W. L. Nighan, *J. Non-Cryst. Solids* **141,** 76 (1992), and references therein.
2. A. Esser, H. Heesel, H. Kurz, C. Wang, G. N. Parsons and G. Lucovsky, *J. Appl. Phys.* **73,** 1235 (1993).
3. I. A. Shkrob and R. A. Crowell, *Phys. Rev. B* **57,** 12207 (1998).
4. S. L. Dexheimer, C. P. Zhang, J. Liu, J. E. Young and B. P. Nelson, *Materials Research Society Proc.* v. **715,** A2.1 (2002).
5. Zhang, X.-C., *J. Lumin.* **66-67,** 488 (1996).
6. M. C. Beard, G. M. Turner, and C. A. Schmuttenmaer, *Phys. Rev. B* **62,** 15764 (2000).
7. M. C. Beard, G. M. Turner, and C. A. Schmuttenmaer, *J. Appl. Phys.* **90,** 5915 (2000).
8. H. Scher and E. W. Montroll, *Phys. Rev. B* **12,** 2455 (1975).
9. A.V.V. Nampoothiri and S.L. Dexheimer, to be published.

## Micro-Raman Studies of Mixed-phase Hydrogenated Silicon Solar Cells

Jessica M. Owens[1], Daxing Han[1], Baojie Yan[2], Jeffrey Yang[2], Kenneth Lord[2], and Subhendu Guha[2]
[1]Department of Physics & Astronomy, University of North Carolina at Chapel Hill, Chapel Hill, NC 27599, U.S.A.
[2]United Solar Systems Corp., 1100 W. Maple Road, Troy, MI 48084, U.S.A.

## ABSTRACT

The open-circuit voltage ($V_{oc}$) of mixed-phase hydrogenated silicon solar cells has been found to increase after light soaking. In this study, we use micro-Raman to investigate the heterogeneous structure of solar cells in the amorphous-to-nanocrystalline transition region. For a cell with $V_{oc} = 0.981$ V, Raman spectra show a typical broad Gaussian lineshape around 480 cm$^{-1}$, a signature of typical amorphous material. A cell with $V_{oc} = 0.674$ V displays a sharp Lorentzian peak around 516 cm$^{-1}$, indicative of nanocrystallinity. A cell with $V_{oc} = 0.767$ V was systematically scanned for 20 different positions in 500 μm increments. Most spectra show a typical Gaussian lineshape around 480 cm$^{-1}$, several spectra reveal a hint of a nanocrystalline shoulder around 512 cm$^{-1}$, and one spectrum exhibits a distinct nanocrystalline peak. We conclude that the nanocrystallite distribution in the mixed-phase material is very non-uniform even within a mm dot. This result provides direct evidence supporting a recently proposed two-diode equivalent-circuit model to explain the light-induced effect.

## INTRODUCTION

We recently observed a significant light-induced enhancement in the open-circuit voltage ($V_{oc}$) of mixed-phase silicon solar cells in the amorphous-to-nanocrystalline transition region [1, 2]. Since the average grain size is in the order of a few nanometers in the mixed-phase material, we use the term nanocrystalline instead of microcrystalline silicon. The magnitude of the variation in $V_{oc}$ ($\Delta V_{oc}$) depends on the intrinsic ($i$) layer thickness, the deposition temperature, the initial $V_{oc}$, and the light-soaking intensity. Under intense light soaking, a $\Delta V_{oc}$ as large as 150 mV has been observed. Subsequent annealing of the solar cells substantially restored the original $V_{oc}$ values. *In-situ* photoluminescence (PL) spectral studies found that the PL intensity and peak-energy position associated with the amorphous component of the mixed-phase material increases upon light soaking. We proposed that a reduction of nanocrystalline volume fraction ($X_c$) or size is responsible for the observed $V_{oc}$ enhancement [1, 2]. However, recent studies by conventional Raman and X-ray diffraction spectroscopy found no observable change before and after light soaking, prompting us to carry out further investigations.

Raman spectroscopy is a sensitive tool that provides valuable structural information about hydrogenated amorphous silicon (a-Si:H) materials. In a-Si:H, all phonon modes of the transverse acoustic (TA), longitudinal acoustic (LA), longitudinal optical (LO), and transverse optical (TO) modes are Raman active. Thus, Raman spectra from a-Si:H at room temperature yield a reasonable spectral comparison with phonon density of state [3]. In crystalline silicon (c-Si), on the other hand, the Raman spectrum consists of a single TO mode with full-width-at-half-maximum (FWHM) of 4 cm$^{-1}$ at a peak of 520 cm$^{-1}$. All other modes are not Raman active because of the symmetry in the fcc lattice. In hydrogenated nanocrystalline silicon (nc-Si:H), a sharp peak or shoulder around 510-520 cm$^{-1}$ appears on the broad amorphous spectrum, indicating the inclusion of nanocrystallites. Decomposition of the Raman spectrum from nc-Si:H gives the average nanocrystalline volume fraction, and the shift from crystalline peak at 520 cm$^{-1}$ relates to the average grain size. However, for the mixed-phase material, the average crystalline volume fraction is low, and the distribution of grains might be non-uniform. Using conventional Raman to detect the existence of nano-grains is difficult. However, micro-Raman is a sensitive tool for detecting nano-grains and providing their distribution information. In this paper, we present the micro-Raman data on the mixed-phase solar cells, which correlates well the structural non-uniformity to the physical model of light-induced increase of $V_{oc}$ observed in the mixed-phase solar cells.

**EXPERIMENTAL**

Single-junction *n i p* solar cells were deposited onto 4 cm ×4 cm stainless steel (ss) substrates using a conventional *rf* glow discharge technique. Indium-tin-oxide (ITO) dots of area 0.25 cm$^2$ were deposited on the *p* layer as the top contact. The deposition parameters were adjusted to arrive at the amorphous-to-nanocrystalline transition region [2].

Raman measurements were made in backscattered geometry with a JY triple spectrograph fitted with a liquid-nitrogen cooled CCD detector. The spectra were collected under ambient conditions using the 514.5 nm line of an argon-ion laser. The penetration depth of the 514.5-nm light is ~ 60 nm for a-Si:H and is larger for nc-Si:H. The spectral resolution was approximately 0.5 cm$^{-1}$ in the frequency range of 100 to 1000 cm$^{-1}$. The frequency of the Raman lines were calibrated using the TO mode of a (111) c-Si wafer at 520 cm$^{-1}$. A low laser power was used to avoid the possibility of laser-induced crystallization. The incident beam was approximately three microns in diameter. To suppress the scattered light from the substrate, the light spot was focused on the *p* layer using a microscope. At first, four Raman scans were taken in the bottom quarter of the cell. The locations of the scans are shown in Fig. 1 (a). Since non-uniformity appears to be more pronounced for the cells with $V_{oc}$ ~ 0.75 V, twenty scans were made in steps of 500 μm in the same bottom quarter of the sample 13678-42, as indicated in Fig. 1 (b). The laser beam was moved to a new spot for each scan on the ITO dot. This allowed us to scan the sample under the same initial conditions for each scan and thereby avoiding any light-induced metastable effect of laser radiation at the previously irradiated spot.

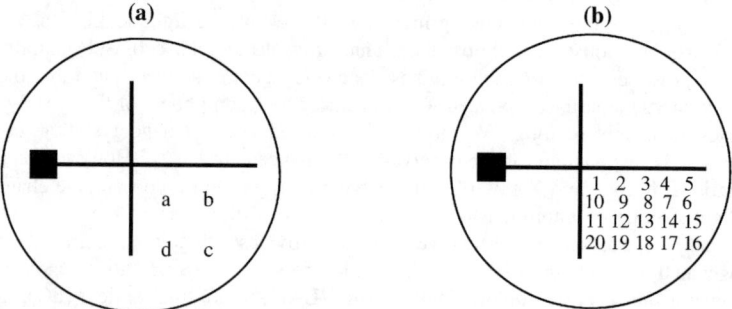

**Figure. 1.** Schematics of (a) 4 scans for all four samples, and (b) 20 scans for 13678-42. The diameter of the ITO dot is ~ 6 mm.

## RESULTS AND DISCUSSION

A cell's $V_{oc}$ can be conveniently used to determine if it is in the amorphous or mixed phase [1, 2]. Cells with high, low, and medium $V_{oc}$ corresponding to a-Si:H, nc-Si:H, and mixed-phase Si solar cells, were studied. Table I lists the $V_{oc}$ values and the Raman features for four scans on each cell. The four scans for sample 13687-32 with $V_{oc}$ = 0.981 V are the same and exhibit typical amorphous signature, as shown in Fig. 2. The spectrum can be decomposed into four broad Gaussian distributions: one centered at 480 cm$^{-1}$ with a FWHM of 67 cm$^{-1}$ representing the TO mode, one at ~ 140 cm$^{-1}$ with FWHM of 100 cm$^{-1}$ representing the TA mode, one around ~ 330 cm$^{-1}$ with FWHM of 117 cm$^{-1}$ representing the LA mode, and one around ~ 440 cm$^{-1}$ with FWHM of 100 cm$^{-1}$ representing the LO mode. The

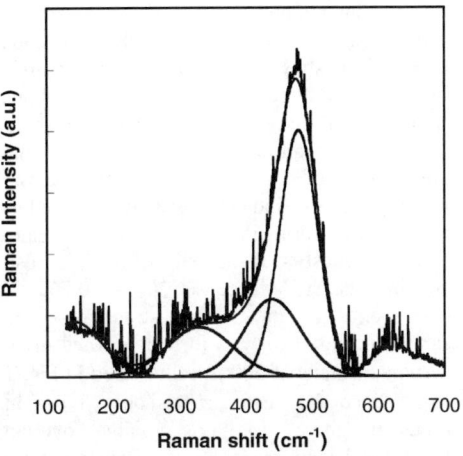

**Figure 2.** Raman shift of cell with $V_{oc}$ = 0.981 V.

four components in the Raman spectrum are typically observed for high quality a-Si:H material. Figure 3 shows the Raman TO mode for sample 13678-13 with a $V_{oc}$ = 0.674 V. All Raman spectra indicate a nc-Si:H material with a sharp c-Si TO mode around 516 cm$^{-1}$. We decompose the spectrum of the TO mode using [4]

$$X_c = (I_c + I_{gb})/[I_c + I_{gb} + y(L) I_a], \qquad (1)$$

where $I_c$, $I_{gb}$, and $I_a$ are integrated intensities of the peaks associated with c-Si, grain boundary [5] or intermediate order [6], and a-Si:H components, respectively. y(L) decreases from 1 to 0.4 when the grain size (L) grows from 3 to 30 nm. We obtained a nano-crystalline volume fraction of 35% - 56%.

**Table I.** $V_{oc}$ values and Raman lineshape for the four samples measured.

| Sample ID | $V_{oc}$ (V) | Raman lineshape for four scans |
|---|---|---|
| 13687-32 | 0.981 | All broad a-Si peak at ~ 480 cm$^{-1}$ |
| 13678-42 | 0.767 | All broad a-Si peak at ~ 480 cm$^{-1}$ |
| 13678-43 | 0.748 | One with c-Si peak at ~ 516 cm$^{-1}$ |
| 13678-13 | 0.674 | All sharp c-Si peak at ~ 516 cm$^{-1}$ |

When the mixed-phase cells with $V_{oc}$ ~ 0.767 V and ~ 0.748 V were scanned by Raman, interesting results are observed, as shown in Figs. 4 (a) and 4 (b). The four scans in Fig. 4 (a) show typical amorphous features. In Fig. 4 (b), three scans show the amorphous signature, but one scan clearly exhibits a sharp Lorentzian peak around 516 cm$^{-1}$. The fact that only one of the eight spectra shows nanocrystallinity indicates that the material is very non-uniform in the micrometer scale. One may question if the cell in Fig. 4 (a) also contains nc-Si:H, because $V_{oc}$ ~ 0.767 V corresponds to a mixed-phase structure. The sample in Fig. 4 (a) was further scanned in 20 systematic steps of 500 μm illustrated in Fig. 1 (b). Since the beam size is only 3 μm in diameter, there was no possible overlap between adjacent scans. The 20 Raman spectra shown in Fig. 5 exhibit typical Gaussian lineshapes centered around 480 cm$^{-1}$. Several spectra show a hint of a nanocrystalline shoulder around 512 cm$^{-1}$. Of the entire set of spectra, only one spectrum shows definite nanocrystallinity. This confirms the structural non-uniformity for both the mixed-phase cells

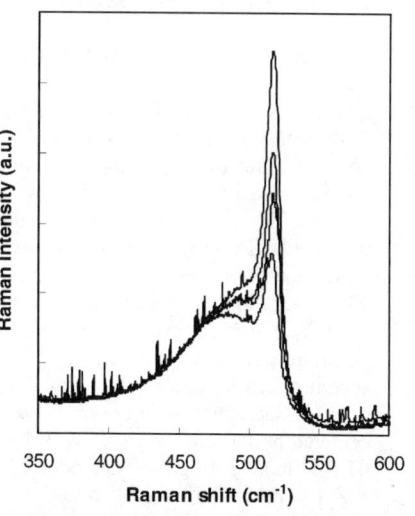

**Figure 3.** Raman shift of a cell with $V_{oc} = 0.674$ V.

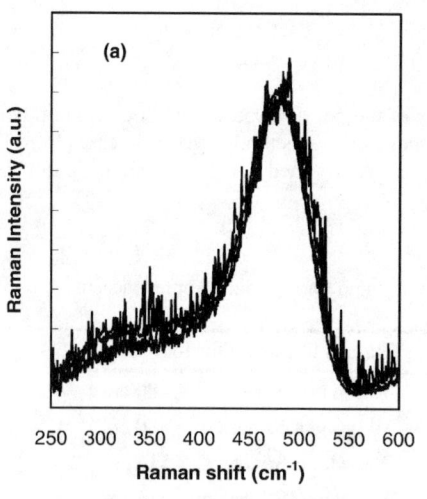

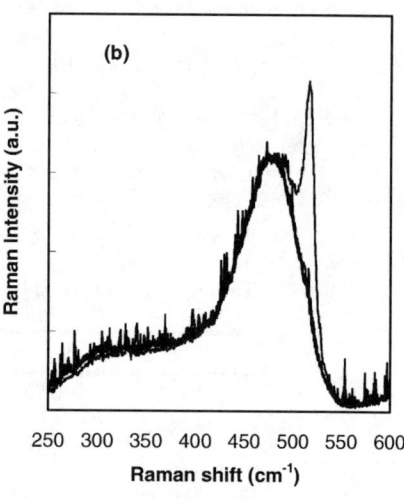

**Figure 4.** Raman shift of mixed-phase cells with (a) $V_{oc} = 0.767$ V and (b) $V_{oc} = 0.748$ V.

with $V_{oc} \sim 0.767$ V and $\sim 0.748$ V. Since only a small portion of the entire cell was scanned, the overall nanocrystalline volume fraction is difficult to determine. For the c-Si peak on 13678-43 and 13678-42, the frequency was at 516.7 cm$^{-1}$ and 514 cm$^{-1}$, respectively. When the c-Si grain size is as small as a few nm, the momentum conservation will be relaxed and Raman active modes will not be limited to the center of the Brillouin zone. Thus, the frequency could shift from 520 cm$^{-1}$ to $\sim 512$ cm$^{-1}$ with decreasing grain size from $\geq 10$ to $\sim 3$ nm [5, 7, 8]. Therefore, the significant shift of c-Si peak observed in the mixed-phase cells indicates that the grain size is small. Since the decomposition of the spectrum with a sharp c-Si peak in Fig. 5 gives a crystalline volume fraction of over 35% in that spot, we conclude that the small c-Si grains are not uniformly distributed in the a-Si:H matrix but are clustered in a micrometer-size spot.

In the two-diode model [9], a mixed-phase solar cell is equivalent to two diodes connected in parallel. One diode has the characteristics of an a-Si:H cell with a $V_{oc}$ around 1 V, and the other one has the characteristics of an nc-Si:H cell with a $V_{oc}$ below 0.5 V. It is assumed in the model that a certain area of the cell contains more nanocrystallites than the others. In this case, the lateral transport between the a-Si:H phase and the nc-Si:H phase becomes negligible. The non-uniform distribution of nanocrystallites obtained by micro-Raman measurements in this study provides a direct supporting evidence for the assumption. Although micro-Raman does not provide information on the average crystalline volume fraction in the mixed-phase solar cells due to the small area probed, the crystalline fraction is estimated to be very low. Since the nc-Si:H diode normally has high forward current, a small percentage of nc-Si:H could cause significant drops in $V_{oc}$. Good nc-Si:H solar cells have been reported to be stable against light soaking. However, the nc-Si:H grains in the mixed-phase cells are very small, and the effect of grain

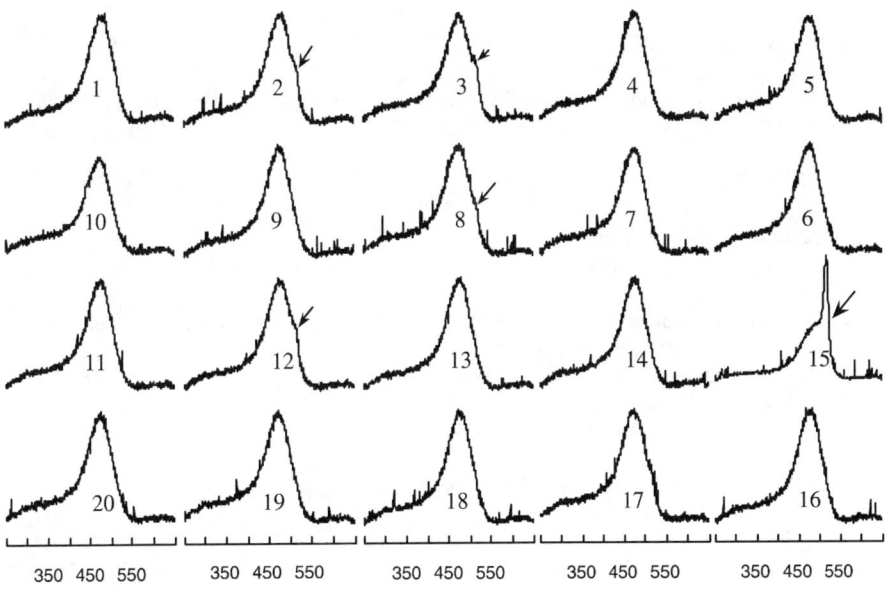

**Figure 5.** Raman shift of 20 scans on a mixed-phase cell 13678-42.

boundary could be very high. The light-induced changes in the grain boundary may effectively reduce the grain size [1, 2] and affect the carrier transport resulting in lower FF and forward current. In addition, if the transport path of nc-Si:H does not form throughout the entire thickness of the intrinsic layer, especially at the beginning of the deposition, the amorphous portion could act as a series resistor, which would limit the forward current at high forward voltages [10]. The degradation of the amorphous layer underneath nc-Si:H would cause low forward current in the nc-Si:H diode and reduce FF [10].

## SUMMARY

We use micro-Raman spectroscopy to study the degree and dimension of the structural uniformity of mixed-phase silicon solar cells. Low laser power was used to avoid possible thermal effects. For a cell with $V_{oc} = 0.981$ V, four scans on different locations of the cell show typical broad Gaussian lineshapes around 480 cm$^{-1}$, indicative of amorphous silicon structure. All scans of the cell with $V_{oc} = 0.674$ V show a sharp Lorentzian peak around 516 cm$^{-1}$, with an estimated 40% nanocrystalline volume fraction. For mixed-phase cells with $V_{oc} = 0.748$ and 0.767 V, one of twenty and one of four spectra, respectively, exhibited both a broad Gaussian lineshape around 480 cm$^{-1}$ and a sharp peak around 516 cm$^{-1}$. This suggests that a large fraction of a-Si:H component but small fraction of nc-Si:H grains within the mm dot of the mixed-phase cells. The spatial distribution of the nc-Si:H grains exist is clustered in a micrometer spot. The light-induced $V_{oc}$ enhancement has been observed in such mixed-phase cells. The structural non-uniformity observed in this micro-Raman study provides a direct supporting evidence for the two-diode model recently proposed for explaining the light-induced $V_{oc}$ enhancement in mixed-phase solar cells.

## ACKNOWLEDGEMENT

This work is supported by NREL Thin-film partnership program Subcontract No. ADJ-1-30630-09 at University of North Carolina and ZDJ-2-30630-19 at United Solar Systems Corp.

## REFERENCES

1. K. Lord, B. Yan, J. Yang, and S. Guha, *Appl. Phys. Lett.* **79** (2001) 3800.
2. J. Yang, K. Lord, B. Yan, A. Banerjee, S. Guha, D. Han, and K. Wang, *Mat. Res. Soc. Symp. Proc.* **715** (2002) 601.
3. J. S. Lannin, in *Semiconductors and Semimetals*, Vol. 21, part B, Chapter 6, edited by J. L. Pankove (Academic Press, Inc., London, 1984).
4. G. Yue, J. D. Lorentzen, J. Lin, Q. Wang, and D. Han, *Appl. Phys. Lett.* **75** (1999) 492.
5. S. Veprek, Z. Iqbal, and F. A. Sarott, *Philos. Mag.* B **45** (1982) 137.
6. D. V. Tsu, B. S. Chao, S. R. Ovshinsky, S. Guha, and J. Yang, *Appl. Phys. Lett.* **71** (2003) 1317.
7. S. Veprek, F. A. Sarott, and Z. Iqbal, *Phys. Rev.* B **36** (1987) 3344.
8. Y. He, C. Y. Yin, G. Cheng, L. Wang, X. Liu, and G.Y. Hu, *J. Appl. Phys*, **75** (1994) 797.
9. B. Yan, G. Yue, J. Yang, K. Lord, and S. Guha, submitted to *Appl. Phys. Lett.* (2003).
10. G. Yue, B. Yan, J. Yang, K. Lord, and S. Guha, *Mat. Res. Soc. Symp. Proc.* **762** (2003) in press.

# Hole Drift-Mobility Measurements in Contemporary Amorphous Silicon

S. Dinca, G. Ganguly,[1] Z. Lu,[2] E. A. Schiff, V. Vlahos,[2] C. R. Wronski,[2] Q. Yuan[*]
Department of Physics, Syracuse University, Syracuse, NY 13244-1130
[1] BP Solar, Inc., Toano, Virgina 23168
[2] Department of Electrical Engineering, Pennsylvania State University, University Park, Pennsylvania 18702

## ABSTRACT

We present hole drift-mobility measurements on hydrogenated amorphous silicon from several laboratories. These temperature-dependent measurements show significant variations of the hole mobility for the differing samples. Under standard conditions (displacement/field ratio of $2 \times 10^{-9}$ cm$^2$/V), hole mobilities reach values as large as 0.01 cm$^2$/Vs at room-temperature; these values are improved about tenfold over drift-mobilities of materials made a decade or so ago. The improvement is due partly to narrowing of the exponential bandtail of the valence band, but there is presently little other insight into how deposition procedures affect the hole drift-mobility.

## INTRODUCTION

The drift of electrons and holes in electric fields is central to most electronic devices. While the fundamental physics of drift is fairly well established for most crystalline semiconductors, for hydrogenated amorphous silicon (a-Si:H) and other disordered semiconductors our understanding remains provisional. Experimentally, electron and hole drift are generally measured using "time-of-flight" measurements of the transit time $t_T$ for a carrier across a specified displacement $L$ and at a specified electric field $E$; by definition, the drift-mobility is

$$\mu_D \equiv \frac{L/E}{t_T}.$$

For a-Si:H and related materials, the drift-mobilities of electrons and holes in a given material can generally be understood using a "bandtail multiple-trapping" model that invokes a band mobility $\mu^0$, the width $\Delta E$ of an exponential bandtail

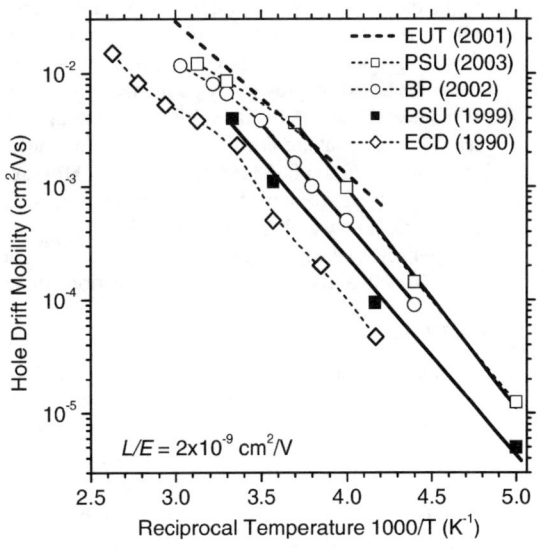

Figure 1: Temperature-dependent hole drift-mobilities for several a-Si:H materials; the mobilities correspond to a ratio $L/E = 2 \times 10^{-9}$ of the hole displacement $L$ and the electric field $E$. The solid and dashed lines are fits to simply activated behavior.

---

[*] Present address: dPix, Inc., Palo Alto, California.

of localized states extending into the bandgap from the bandedge, and an attempt-frequency $\nu$ describing the trapping dynamics of mobile carriers by the bandtail states. The drift-mobility is generally much smaller than the band mobility. This property is the consequence of "multiple-trapping" – the successive capture and release of mobile carriers by the bandtail states.

This multiple-trapping parameterization is essentially phenomenological: we have very little understanding of the fundamental physics of exponential bandtails. For this reason it is important to study how changes in underlying materials affect drift-mobilities and the multiple-trapping parameters. For example, for electrons, alloying with germanium or carbon diminishes the drift-mobility substantially; this alloying effect appears to be due primarily to an alloying-induced broadening of the conduction bandtail [1,2]. Holes are typically several hundred times less mobile than electrons in amorphous-silicon based materials, but alloying effects are also much smaller than for electrons [2]. To the best of our knowledge, there is no theoretical understanding of these drift-mobility observations – neither the asymmetry of electron and hole drift mobilities in "standard" a-Si:H, nor their quite different alloying effects.

The fact that alloying does not significantly affect hole drift mobilities suggested that there was little that might be done to improve them, but starting in the mid-1990's there have been several reports of significant increases in hole mobilities in "contemporary" materials. We summarize some of these measurements, including those being reported here, in Figure 1. The lowest curve, denoted ECD(1990), was reported in ref. 2; the curve was measured on a sample prepared at Energy Conversion Devices, Inc., and is quite close to measurements on samples made prior to 1990 in several laboratories. In 1995, Ganguly and Matsuda [3] published hole drift-mobilities on several samples of a-Si:H showing a *much* higher drift-mobility than the 1980's baseline. This material has not yet been reproduced by other laboratories, but the indication that significant improvement in hole drift-mobilities is possible in a-Si:H is consistent with more modest improvements reported subsequently. In Figure 1, we have shown drift-mobilities based on previously published measurements on "expanding thermal plasma" material [4] made at Eindhoven University of Technology (denoted EUT(2001)). We also show measurements that are newly reported here on materials prepared at Pennsylvania State University (denoted PSU(1999) and PSU(2003)) and at BP Solar, Inc. (denoted BP(2002)).

In this paper we next present some additional details on the samples. We then describe for one sample how we obtain drift-mobilities from transient photocurrent measurements. We also briefly discuss fitting of multiple-trapping fitting parameters to the transient measurements. We conclude with a discussion of future directions for hole drift-mobility research.

## SAMPLES

Several Schottky barrier diode samples were made in 1999 at Pennsylvania State University; substrates were $SnO_2$-coated glass. A 35 nm $n^+$ a-Si:H contact layer was first deposited onto the substrate. The undoped a-Si:H layer was then plasma-deposited (13.56 MHz) at a substrate temperature of 200 C and a hydrogen/silane dilution ratio $R$ of 10:1. A top, semi-transparent Schottky barrier was formed by thermal evaporation of Ni onto the intrinsic layer; the top surface was briefly etching with buffered hydrofluoric acid prior to the evaporation step. The measurements reported here were on a sample with an intrinsic layer thickness of 1.47 µm.

Two additional samples were made at Pennsylvania State University in 2003 using similar substrates. These samples were *pin* structures: 25 nm a-SiC:H *p*-layer, a-Si:H intrinsic layer, 35 nm nanocrystalline Si *n*-layer, semitransparent Cr top contact. The intrinsic layer of one sample was made using $R = 10$ dilution (thickness 0.53 µm); the second was made without hydrogen dilution (intrinsic layer thickness 0.69 μm). We have shown measurements for the undiluted sample, which had a somewhat larger hole drift-mobility than the diluted one.

The sample prepared at BP Solar, Inc. that was used in Figure 1 was prepared using DC plasma deposition; the intrinsic layer was prepared using a dilution ratio of 10, and was 0.91 μm thick. The sample had a *pin* structure (a-SiC:H *p*-layer), and was deposited onto $SnO_2$ coated glass. A semitransparent ZnO electrode was deposited onto the top *n*-layer. The sample was prepared under conditions similar to those used in the solar cell factory operated by BP Solar in Toano, Virginia. A second sample from BP Solar prepared in 1999 had quite similar drift properties; we don't report these here.

## HOLE DRIFT MEASUREMENTS AND BANDTAIL MULTIPLE TRAPPING

In this section we provide details of the measurements and analysis for the sample denoted PSU(1999) in Figure 1. Transient photocurrent measurements are shown for three temperatures in the upper panel of Figure 2. These measurements were done using a dye laser (3 ns pulsewidth, 500 nm). The sample was illuminated through the *n*-layer. The normalization $i(t)d^2/Q_0V$ involves the thickness of the *i*-layer $d$, the reverse bias voltage $V$ across the diode, and the total photocharge generated in the diode $Q_0$ (as estimated by integrating the transient photocurrent); the normalized photocurrent has the dimensions of a mobility ($cm^2$/Vs).

The photocurrent at 300 K shows an initial feature peaking at about 20 ns. This feature is due to the motion of electrons that are photogenerated in the top 10% (150 nm) of the sample. The electrons are swept to the top interface faster than can be resolved by the electronics, which had a response time of 60 ns for this sample. The longer-time photocurrent is due to the motion of holes. The "kink" in the hole-dominated section that occurs at a delay of about 1 μs corresponds to the time at which about half of the photogenerated holes

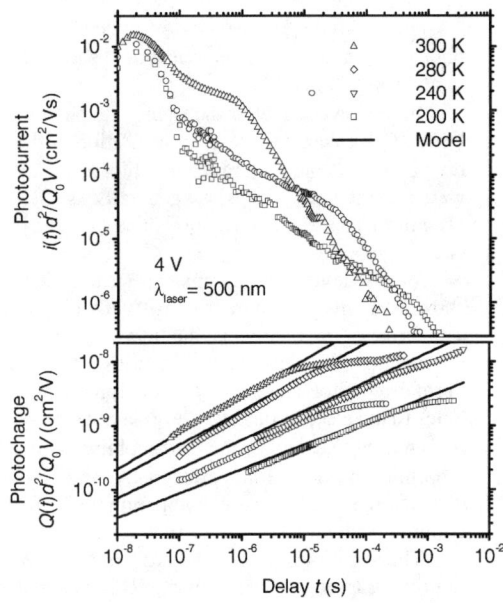

Figure 2: Normalized photocurrent ($i(t)$) and photocharge ($Q(t)$) transient measurements in one a-Si:H sample (PSU 1999) are shown at several temperatures as the open symbols. The photocurrent transients show an electron-transport feature peaking at about 20 ns that is due to the finite absorption-length of the laser; the longer-time behavior is the dispersive drift and sweepout of holes. We have subtracted the electron feature from the photocharge transients in the lower panel. The solid lines in the lower panel represent a fitting to the photocharge measurements using the bandtail multiple-trapping model.

have crossed the sample; this kink is often used to define a transit-time for calculating a hole drift-mobility.

These same features are also apparent in the transients at 240 K and 200 K. As the temperature falls, the "electron" feature becomes more prominent. This effect occurs because hole mobilities decline more with falling temperature than do electron mobilities. Following the electron feature, the hole photocurrent declines as a power-law ($t^{-0.44}$ at 240 K) through the kink (at

Table I: Valence Bandtail Parameters from Hole Drift-Mobility Measurements

| Sample | $\Delta E_V$ (eV) | $v$ (s$^{-1}$) | $\mu_h^0$ (cm$^2$/Vs) |
|---|---|---|---|
| PSU (1999) | 45 | $1.0 \times 10^{12}$ | 0.7 |
| ECD (1990) | 48 | $7.7 \times 10^{10}$ | 0.27 |

the transit time). A power-law decay of the photocurrent $t^{-(1-\alpha)}$ is the defining attribute for "dispersive" transport, where $\alpha$ is termed "the dispersion parameter;" in principle, the photocurrent following the transit time should fall as $t^{-(1+\alpha)}$. For the bandtail multiple-trapping model, $\alpha = kT/\Delta E_V$.

In the lower panel of the figure, the open symbols indicate the transient photocharge $Q(t)d^2/Q_0V$ calculated by time-integration of the photocurrent. The transients which saturate near the value $10^{-8}$ cm$^2$/V were recorded for 2 V bias. The transients saturating at $2.5 \times 10^{-9}$ cm$^2$/V were recorded at 8 V bias; we generally use high bias voltages to estimate the total photocharge $Q_0$. For this panel, we have also subtracted the early-time photocharge that is primarily due to electron motion.

We use photocharge transients such as these to calculate drift-mobilities; we have previously shown that this procedure is consistent with methods using direct transit-time measurements [1]. The virtue of the method is that, in a single measurement, one obtains drift-mobilities for a continuum of displacement-field ratios. Prior to the directly-measured transit-time, the normalized photocharge may be interpreted as the ratio $L(t)/E$ of the mean displacement $L(t)$ of holes to the electric field $E$. The delay $t$ corresponding to a specific displacement-field ratio $L/E$ is then used to calculate the drift-mobility from the definition $\mu_D = (L/E)/t$.[*] The hole drift-mobilities in Figure 1 all correspond to the particular value $L/E = 2 \times 10^{-9}$ cm$^2$/V; because of dispersion, it is essential in comparing the drift-mobilities for different materials to use a common value for $L/E$.

The solid lines in Figure 2 are a fit to the experimental measurements based on the bandtail multiple-trapping model. This model has been described elsewhere; the three parameters that are involved in the fitting are the width of the exponential bandtail of the valence band $\Delta E_V$, the band mobility of holes $\mu_h$, and the escape frequency $v$ describing hole trapping dynamics. The particular parameters chosen for Figure 2 are summarized in Table I, along with the parameters published for a sample (ECD 1990) prepared about ten years ago [2]. The equation to which these parameters apply is

$$L(t)/E = K(\mu_h/v)(vt)^{kT/\Delta E_V} , \qquad (1)$$

---

[*] Drift mobilities are often calculated using the conventional equation $\mu_D = d^2/Vt_T$, where $t_T$ is the transit time actually observed as a breakpoint or kink in a photocurrent transient. Such mobilities are twice as large as drift-mobilities calculated using the "$L/E$" procedure used here. We prefer the "$L/E$" definition. For dispersive transport, the photocharge at the breakpoint has reached only half its saturation value. This implies that the mean displacement of carriers is half the sample thickness at the time at which the breakpoint occurs, so $L/E = d^2/2V$, and $\mu_D = d^2/2Vt_T$.

where $K = \sin(\alpha\pi)/(\alpha\pi(1-\alpha))$ [5], and $\alpha \equiv kT/\Delta E_V$. While the fit in Figure 2 is imperfect, it is worth noting that it does account fairly well for measurements over a range of nearly $10^4$ in time and $10^2$ in photocharge. Hole drift is enhanced by increases in $\mu_h$ and by decreases in $\Delta E_V$; increases in $\nu$ diminish drift. We have not made a careful study of the errors in these fitting parameters.

## DISCUSSION

We first comment again on the measurements of Figure 1. The drift-mobilities at lower temperatures are simply activated, as we have illustrated with the heavy solid lines. For the higher temperatures, the drift-mobilities are generally lower than expected from this activated behavior; in this regime, the electrical response times of the samples (typically 50-100 ns) were approaching the times used to calculate the drift-mobilities. We haven't tried to deconvolute these response-time effects from the measured photocurrent transients, although it is possible to do so [6]. We think it plausible that the deviation from activated behavior at higher temperatures is attributable to these electrical response times; the alternative, which needs further exploration, is that simply activated behavior – and the bandtail multiple-trapping model – fails.

It is curious that the activation energies for the various samples, excepting the Eindhoven sample, are so similar. For the bandtail multiple-trapping model, this activation energy is $(\Delta E_V)\ln(L\nu/\mu_h E)$. If the improvement in hole drift-mobilities were due exclusively to narrowing of the valence bandtail width, and not to changes in the band mobility or attempt frequency, we would expect the samples with superior drift-mobilities to have noticeably smaller activation energies. This perspective, applied to the conduction bandtail, was actually quite successful in explaining the alloy-effect on electron drift-mobilities in a-SiGe:H [1]. However, for the present measurements on holes, it seems that the improvement in hole drift-mobilities must reflect changes in at least one other of the multiple-trapping parameters; the results in Table I suggest that $\nu$ varies substantially.

The drift-mobility measurements for the Eindhoven material indicate a much shallower activation energy than for the other samples. At first glance, this suggests a very different set of multiple-trapping parameters. In conjunction with the large dispersion parameter (0.7) reported for this material near room-temperature, one might conclude from the bandtail multiple-trapping expression $\alpha = kT/\Delta E_V$ that the samples had valence bandtails of width about 36 meV. This conclusion leaves unexplained the fact that the actual magnitudes of the drift-mobilities are fairly similar to the other samples over the range of measurement.

We draw two conclusions from the present work. First, contemporary materials generally have better hole drift mobilities than materials prepared ten to fifteen years ago. This improvement is not found only in "special" materials prepared under research conditions; the sample from BP Solar is typical of materials which that company used in its solar cell factory. Second, there is substantial, unexplained variation in hole drift mobilities. At present, we do not know which aspects of deposition cause this variation, nor do we know whether there are structural probes that would correlate well with the variations in hole properties. For example, we had anticipated that hydrogen-dilution of silane during plasma deposition might be essential to obtaining improved properties, but the highest mobility curve in Figure 1 corresponds to a sample made without dilution. This lack of insight into hole drift-mobilties, and presumably into the structure of the valence bandtail, is regrettable both scientifically and technically. In particular, we consider it possible that further improvement in hole drift-mobilities would further improve solar cells – but the program of hole drift-mobility measurements does not yet point in any particular direction for changing the deposition conditions in order to realize such improvements.

## ACKNOWLEDGMENTS

The authors thank Prasanna Rao (Syracuse University) for access to his unpublished drift-mobility measurements. This work has been supported through the Thin Film Photovoltaics Partnership of the National Renewable Energy Laboratory (NDJ-2-30630-24, NDJ-1-30630-1, and ZDJ-2-30630-10).

## REFERENCES

1. Qi Wang, H. Antoniadis, E. A. Schiff, and S. Guha, *Phys. Rev. B*, **47**, 9435 (1993).
2. Q. Gu, Q. Wang, E. A. Schiff, Y.-M. Li, and C. T. Malone, *J. Appl. Phys.* **76**, 2310 (1994).
3. G. Ganguly and A. Matsuda, *J. Non-Cryst. Solids* **198-200**, 1003 (1996); Ganguly's hole drift-mobility measurements are consistent with later measurements on the same materials by P. Rao at Syracuse University (private communication).
4. M. Brinza, G. J. Adriaennsens, K. Iakoubovskii, A. Stesmans, W. M. M. Kessels, A. H. M Smets, M. C. M. van de Sanden, *J. Non-Cryst. Solids* **299-302**, 420 (2002). In this paper, mobilities for several samples with varying thickness are reported for a common displacement-field ratio about $L/E = 9.5\times 10^{-9}$ cm$^2$/V. The fact that these mobilities are essentially the same for the three differing thicknesses is a very nice experimental confirmation that the displacement/field ratio determines drift-mobilities for a given type of material. These authors use a definition of the drift-mobility that yields values twice as large as the definition used in the present paper. In preparing Figure 1, we corrected for the lower value of $L/E = 2\times 10^{-9}$ cm$^2$/V by using the result for dispersive transport $\mu_D \propto (L/E)^{1-1/\alpha}$ [1]. The authors gave a room-temperature dispersion parameter of 0.7; for other temperatures, we assumed the dispersion parameter to be proportional to the absolute temperature, as expected for bandtail multiple-trapping.
5. The constant K is based on two sources. One is the expression connecting the transient photocurrent before carrier transit to the multiple trapping parameters (E. A. Schiff, *Phys. Rev. B* 24, 6189 (1981) – equation (6)); after integration, this expression yields:

$$i(t)d^2 \Big/ Q_0 V = \mu^0 \frac{N_V}{kTg_V^0} \frac{\sin(\alpha\pi)}{\alpha\pi}(\nu t)^\alpha ,$$

where $\alpha = kT/\Delta E_V$, $N_V$ is the effective density of states of the valence band, and $g_V^0$ is the density-of-states (cm$^{-3}$eV$^{-1}$) for the valence bandtail at the upper edge of the valence band. The second source is the relationship $N_V/g_V^0 = kT/(1-\alpha)$ that obtains from the assumption that the valence band edge lies within the exponential bandtail (E. A. Schiff, *Solar Energy Materials and Solar Cells, in press* (2003).
6. Q. Gu, E. A. Schiff, J.-B. Chevrier, and B. Equer, in *Amorphous Silicon Technology - 1993*, edited by E. A. Schiff, *et al* (Materials Research Society, Symposium Proceedings Vol. 297, Pittsburgh, 1993), pp. 425-430.

# RECOMBINATION IN n-i-p (SUBSTRATE) a-Si:H SOLAR CELLS WITH SILICON CARBIDE AND PROTOCRYSTALLINE p-LAYERS

V. Vlahos, J. Deng, J.M. Pearce, R.J. Koval[1], G.M. Ferreira, R.W. Collins, and C.R. Wronski
The Center for Thin Film Devices, Pennsylvania State University, University Park, PA 16802
[1]Intel Corp., 2200 Mission College Blvd., Santa Clara, CA 95054

## ABSTRACT

A study was carried out on hydrogenated amorphous silicon (a-Si:H) n-i-p (substrate) solar cell structures with p-a-SiC:H and highly diluted p-Si:H layers grown with different dilution ratios R=[$H_2$]/[$SiH_4$]. The contributions of the recombination at the p/i interfaces to the forward bias dark current characteristics were identified and quantified for the different cell structures. In both cell structures the role of the p/i interfaces was identified and it is found that the lowest p/i interface recombination is obtained with *protocrystalline* p-Si:H layers having no *microcrystalline* component. The results with p-Si:H layers are attributed not only to their properties but also to the subsurface modification of the intrinsic layer. Evidence is also presented that points to the beneficial effects of the high hydrogen dilution and power used in the deposition of these p-layers in creating the p/i interface regions. The limitations on 1 sun open circuit voltage ($V_{OC}$) imposed by the p/i recombination present in all the cell structures is consistent with the mechanisms proposed by Deng et al.[1]. The results presented here also point to why the 1 sun $V_{OC}$ in protocrystalline p-Si:H solar cells is higher than that in p-a-SiC:H cells.

## INTRODUCTION

The mechanisms responsible for the differences in the 1 sun open circuit voltage ($V_{OC}$) values reported for n-i-p cells having p-a-SiC:H and highly diluted p-Si:H layers [2,3] are as yet not well understood. It is well known that in p-a-SiC:H solar cells the magnitude of $V_{OC}$ depends on the properties of the p-layers as well as the nature of the p/i interface regions [4-6]. In their study on the role of p/i interfaces in a-SiC:H p-i-n cell structures, Pearce et al. [7] showed that in addition to obtaining systematic changes in 1 sun $V_{OC}$ it was possible to observe differences in the dark forward bias current ($J_D$-V) characteristics. This clearly indicated that $J_D$-V characteristics offer a possible method for characterizing different p/i interface regions and the carrier recombination responsible for limiting $V_{OC}$. Recently, in a detailed study of $J_D$-V characteristics on such cell structures Deng et al. [8] were able to clearly separate and quantify the contributions of both the bulk and p/i interfaces. For example, they clearly showed that for the intrinsic material used there are distinct differences in the diode quality factors between the currents due to bulk recombination and that at the p/i interfaces.

In the work reported here on n-i-p solar cells having identical bulk intrinsic layers with both p-a-SiC:H and highly diluted p-Si:H layers the contributions of p/i interface recombination to their $J_D$-V characteristics are clearly identified and quantified. Results are presented on a-SiC:H cell structures having identical p-layers where the recombination in the protocrystalline a-Si:H p/i interface regions are quantified and their limitation on 1 sun $V_{OC}$ identified. In the case of the cells with p-Si:H having different dilution ratios R=[$H_2$]/[$SiH_4$] it is clearly shown that the lowest p/i interface recombination is obtained with *protocrystalline* and not *microcrystalline* layers which lead to the highest 1 sun $V_{OC}$. The systematic changes of recombination obtained with the different p-a-Si:H layers are discussed in terms of their properties as well as those of the corresponding p/i interface region. Evidence is also presented for the importance of the subsurface modification in the intrinsic layer due to the high hydrogen dilution and power used in depositing the p-layers.

## EXPERIMENTAL DETAILS

The n-i-p structures in this study were deposited by plasma enhanced chemical vapor deposition (PECVD) at substrate temperatures of 200°C with bulk R=10 intrinsic layers under conditions previously described [9]. The p-a-SiC:H cell structures consisted of glass / Cr / n a-Si:H (350 Å) / i a-Si:H (4000 Å) / p-a-SiC:H (250 Å) / ITO, where the p-layer was deposited with a constant doping ratio of $D=[B(CH_3)_3]/[SiH_4]=0.005$ and with $Z=[CH_4]/\{[SiH_4]+[CH_4]\}=0.5$. A two-step i-layer process utilized in these structures to change the p/i interface region in a controlled way was obtained by terminating the last 200 Å of the i-layer with R values of 0 and 40. The p-Si:H cell structures consisted of glass / Cr / n μc-Si:H (350 Å) / i a-Si:H (4000 Å) / p-Si:H (~200Å) / ITO, where the p-layers were deposited with R varying from 100 to 200, and with a doping ratio of $D=[BF_3]/[H_2]=0.2$. The ITO top contact layers were sputter-deposited at 170°C. The structures tested had areas of 0.02 cm$^2$ for the p-a-SiC:H and 0.05cm$^2$ for the p-Si:H cells. The cells were characterized using $J_D$-V as well as light I-V measurements under 1 sun illumination conditions. Both of these measurements were performed at 25°C using a four-point probe technique in order to eliminate any extraneous series resistance effects which then allow the $J_D$-V characteristics to be accurately characterized at high forward bias.

## RESULTS AND DISCUSSION

### p-a-SiC:H solar cells:

In the n-i-p solar cell structures with different p/i interfaces, but identical p- and bulk i-layers, the respective contributions to carrier recombination could be identified and quantified from their $J_D$-V characteristics. The beneficial effects of incorporating a thin protocrystalline region prior to the deposition of the a-SiC:H p-layer is in agreement with the results reported for p-i-n structures [10]. Fig. 1 shows the forward $J_D$-V characteristics of two a-SiC:H n-i-p structures with different p/i interface regions, where the recombination at the p/i interface can be identified. The bandgaps ($α_{2000}$) of the R=10 and R=40 p/i interface regions incorporated in these cells are 1.86 and 1.95 eV respectively.

From the two regimes in the $J_D$-V characteristics in Fig. 1 the carrier recombination in forward bias currents that occurs in the bulk and at the p/i interface can be clearly separated as in the case of corresponding p-i-n cell structures [1,8]. At the lower voltages there is a diode quality factor of m=1.4 which then decreases to values closer to 1 at the higher voltages. The regimes where m=1.4 correspond to bulk recombination and as expected there is overlap in the currents over an extended voltage region. As the p/i interface is improved by introducing the R=40 p/i layer, the bulk recombination regime extends to a higher bias and the currents corresponding to the p/i interface recombination become lower. The key role played by this recombination in the p/i interface regions in determining the 1 sun $V_{OC}$ is reflected in their systematic decrease as R at the p/i interface is increased from 0 to 10 to 40 the corresponding 1 sun $V_{OC}$ increases from 0.86 to 0.90 to 0.92 V.

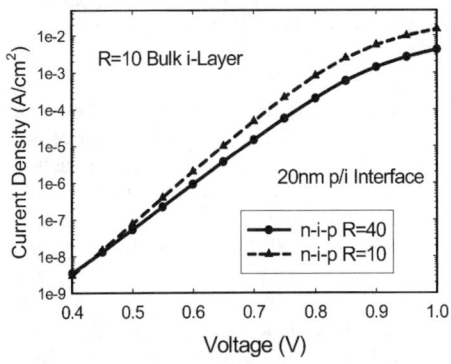

**Figure 1.** The $J_D$-V characteristics for p-a-SiC:H n-i-p cells with different p/i interface regions.

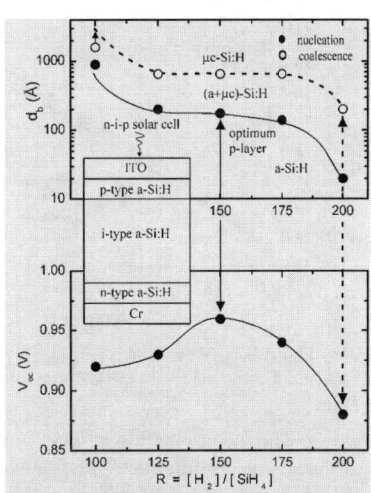

**Figure 2.** Extended phase diagram for D=0.2, p-Si:H as a function of R and thickness. Also shown is the 1 sun $V_{OC}$ for cells with corresponding ~200 Å p-layers.

*Highly-diluted p-Si:H n-i-p solar cells:*

The microstructure and its evolution of the p-Si:H layers in this study were characterized with real-time spectroscopic ellipsometry [11]. From the results a phase diagram was obtained for the evolution of the microstructure with thickness in the p-Si:H layers deposited with different R. The phase diagram for layers having doping ratio of D=0.2 deposited under conditions previously described [9] is shown in Fig. 2 where the regions of protocrystalline a-Si:H, mixed (a+μc)-Si:H, and purely microcrystalline Si:H phases can be clearly identified. Also shown in Fig. 2 are the 1 sun $V_{OC}$ values obtained from corresponding n-i-p solar cell structures fabricated with ~200 Å of these p-layers. The highest $V_{OC}$ is achieved by the cell having a R=150 p-layer which is in the protocrystalline a-Si:H phase and where its growth is terminated at or close to the transition into the mixed (a+μc)-Si:H phase [3]. In the case of R=100, the p-layer is amorphous throughout its thickness and the 200 Å thickness is nowhere close to the transition into a mixed phase. On the other hand for the R=200 cell, the p-layer is in the form of a mixed (a+μc)-Si:H phase right from the beginning and evolves into a purely microcrystalline phase at the thickness of ~200 Å. These results clearly point to the highest $V_{OC}$ being obtained with a protocrystalline a-Si:H p-layer which is deposited at a maximum R value that allows the desired thickness to be obtained without crossing the transition into the mixed (a+μc)-Si:H phase growth regime.

To obtain insight into the carrier recombination that may be limiting these 1 sun $V_{OC}$ values the $J_D$-V characteristics of the different cell structures were investigated. Shown in Fig. 3 are the forward bias $J_D$-V characteristics of the n-i-p solar cells with R=100, 150, and 200. In Fig. 3 the contribution of the p/i interface to recombination can be distinguished from those of the bulk. Just as in the case of the results in Fig. 1, the overlap in the characteristics at the lower voltages reflects the bulk recombination which is the same in all three cells. At the higher voltages the relative contribution of the recombination at the p/i interface regions in the three cells can be identified. This recombination can be seen to be the lowest for the R=150 cell,

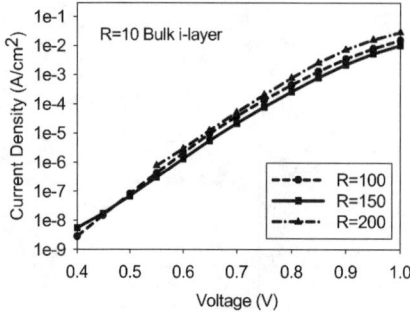

**Figure 3.** $J_D$-V characteristics of highly-diluted p-Si:H n-i-p cells with D=0.2 for different R.

highest for the R=200 cell and in between those for the R=100 cell. It is important to note here that the differences in 1 sun $V_{OC}$ in these cells can again be directly attributed to the changes in these recombination currents. In addressing the mechanisms limiting $V_{OC}$ in p-Si:H cells, unlike the case of the p-a-SiC:H cell structures, the effects of the changes in the p-layer have to be also considered. In addition to the effect of R on microstructural changes just discussed, there are also differences in bandgaps and the doping efficiencies of the p-layers, as well as subsurface reactions due to the very high hydrogen dilution used in the deposition of the p-layers.

The recombination at the p/i interface depends on parameters which include: the densities of defects at the interface itself as well as the p/i region, the bandgap in the p/i region and its alignment with that of the p-layer, and the built-in potential ($V_{bi}$) which depends on both bandgap alignment between the i- and p-layers as well as its doping efficiency. When R is increased from 100 to 150 in the two protocrystalline p-layers there is an increase in the gap as well as the doping efficiency which could increase the $V_{bi}$. An increase in $V_{bi}$ contributes to the reduction of recombination at the p/i interface since it decreases the concentration of electrons in the interface region, thus lowering the recombination there [8]. It should be noted here that the effect of moving the Fermi level as a consequence of the higher doping efficiency is counteracted by the increase in the p-a-Si:H bandgap. Another possible reason for the significant reduction in recombination seen in Fig. 3 is due to the differences in the p/i interfaces themselves in addition to those of the interface regions. The higher R, which improves the microstructure of the protocrystalline p-layer, can also result in a p/i interface with a lower density of defects that reduces the p/i interface recombination velocity.

Another consequence of the higher R is its effect in subsurface reactions which improve the i-layer adjacent to the p/i interface in an analogous way to that obtained with the protocrystalline a-Si:H buffer layers used in a-SiC:H cells. That this occurs is indicated by significant changes in the bandgap of ~200 Å in the intrinsic a-Si:H that are observed during the deposition of the protocrystalline p-layers. When R is increased from 150 to 200 and the p-layer is immediately formed in the mixed (a+μc)-Si:H phase not only does the nature of the p/i interface change but there is also a decrease in the gap of the p-layer. This decrease is fairly symmetrical for the conduction and valence band [12] so even though the doping efficiency remains high, the net band misalignment can in fact lead to a decrease of $V_{bi}$ and further enhance the p/i interface recombination, as indicated by the results in Fig. 3. The low recombination that is obtained at the p/i interface with the R=150 protocrystalline p-layer is significantly smaller than that for the R=40 buffer layer in the a-SiC:H n-i-p cells which was the lowest achieved for those cells. This is illustrated in Fig. 4 which shows the $J_D$-V characteristics between 0.6 and 1.0 V, where also as a consequence of this low p/i interface recombination in the protocrystalline p-Si:H cell the bulk recombination regime extends to higher voltages. It also explains why the 1 sun $V_{OC}$ increases from 0.92 to 0.96 V and suggests that in the latter case it could now actually be limited by bulk recombination in the i-layer. In comparing the recombination at the p/i

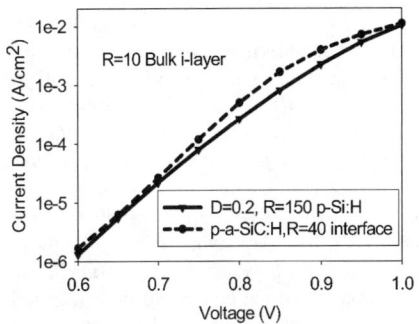

**Figure 4.** $J_D$-V characteristics of the best protocrystalline p-Si:H and p a-SiC:H cells.

interfaces in these two cell structures it is important to note the difference in the way they are formed. In the case of the a-SiC:H cell the top 200 Å of the i-layer is modified by increasing R from and subsequently depositing the p-layer. The deposition conditions under which the protocrystalline p-layer is grown are such that there are modifications in the subsurface region of the i-layer as discussed earlier. These are attributed to the very large concentrations of atomic hydrogen which is generated by the very high R and power levels utilized for the deposition of the p-layer.

A consequence of such concentration of atomic hydrogen is that the p/i interface region become far less sensitive to the presence of a barrier layer on the intrinsic a-Si:H than in the case of n-i-p a-SiC:H cell structures. This is illustrated in Fig. 5 where the light I-V characteristics are shown for the two types of cells with and without exposure of the i-layer to air for 24 hours before the p-layer deposition. It can be seen in Fig. 5a that the air exposure introduced to the p-a-SiC:H n-i-p cell has a drastic effect on its performance by significantly lowering the $V_{OC}$ and FF values as compared to an identical cell completed without an air-gap. Exposure of the sensitive p/i interface region to air leads to the formation of a thin native oxide whose effect is *not* eliminated by the subsequent deposition of a-SiC:H. On the other hand, virtually the same p/i interface regions are obtained in the protocrystalline cells with and without the exposure to air as is indicated by the identical light I-V characteristics, as shown in Fig. 5b.

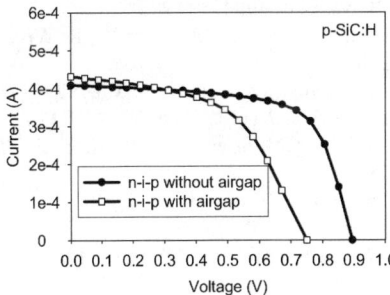
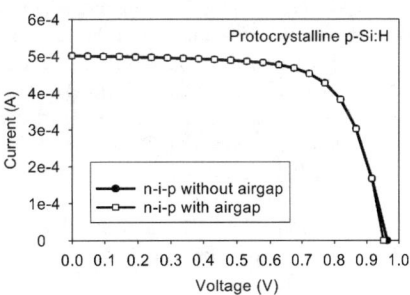

**Figure 5a.** 1 sun light I-V characteristics of p-SiC:H n-i-p solar cells with and without a vacuum break between the deposition of the i- and p-layers.

**Figure 5b.** 1 sun light I-V characteristics of protocrystalline-p n-i-p solar cells with and without a vacuum break between the deposition of the i- and p-layers.

## CONCLUSIONS

Carrier recombination in n-i-p (substrate) a-Si:H cell structures with a-SiC:H and highly diluted Si:H p-layers has been identified in their $J_D$-V characteristics and the recombination in their p/i interface regions has been quantified. The systematic study on the cells with highly hydrogen diluted p-Si:H clearly established that the lowest p/i interface recombination is obtained with *protocrystalline a-Si:H* and not layers containing any *microcrystalline* phase. Such p/i interface recombination is found to be significantly lower than that which could be achieved with the a-SiC:H structures. In all the cell structures the recombination in the p/i interface regions could be directly correlated to their 1 sun $V_{OC}$ values. This points to the reason why open circuit voltages obtained with protocrystalline p-Si:H layers are higher than those with p a-SiC:H. The results also substantiate the mechanisms proposed by Deng et al. [1] for the limitations imposed by carrier recombination in high quality a-Si:H cells. The significantly lower p/i interface recombination in the protocrystalline p-Si:H cells is attributed in large part to the subsurface modification of the intrinsic layer during the deposition of the p-contact. Evidence is presented that the high concentrations of atomic hydrogen present during the deposition of the protocrystalline p-Si:H layer are responsible for this beneficial effect.

## ACKNOWLEDGEMENTS

The authors gratefully acknowledge support for this research by the National Renewable Energy Laboratory under subcontract NDJ-2-30630-01.

## REFERENCES

1. J. Deng, J.M. Pearce, V. Vlahos, R.W. Collins and C.R. Wronski, *Mater. Res. Soc. Proc.* (2003, in press).
2. S. Guha, J. Yang, P. Nath, and M. Hack, Appl. Phys. Lett. 49, 218 (1986).
3. R. J. Koval, C. Chen, G.M. Ferreira, A.S. Ferlauto, J.M. Pearce, P.I. Rovira, C.R. Wronski and R.W. Collins, *Appl. Phys. Lett.* **81**, 1258 (2002).
4. H. Sakai, T. Yoshida, S. Fujikake, T. Hama, and Y. Ichikawa, *J. Appl. Phys.* **67**, 3494 (1990).
5. M. Isomura, T. Takahama, S. Tsuda, and S. Nakano, *Jpn. J. Appl. Phys.* **32**, 1902 (1993).
6. Y. Lee, A.S. Ferlauto, Z. Lu, J. Koh, H. Fujiwara, R.W. Collins and C.R. Wronski, *$2^{nd}$ World Conf. PVEC*, Vienna, Austria pp. 940-943 (1998).
7. J.M. Pearce, R.J. Koval, A.S. Ferlauto, R.W. Collins, C.R. Wronski, J. Yang and S. Guha, *Appl. Phys. Lett.* **77**, 3093 (2000).
8. J. Deng, J.M. Pearce, R.J. Koval, V. Vlahos, R.W. Collins and C.R. Wronski, *Appl. Phys. Lett.* (2003, in press).
9. R.J. Koval, C. Chen, G.M. Fereira, A.S. Ferlauto, J.M. Pearce, P.I. Rovira, C.R. Wronski and R.W. Collins, *Mater. Res. Soc. Proc.* **715**, A6.1 (2002).
10. J. Koh, Y. Lee, H. Fujiwara, C.R. Wronski and R.W. Collins, *Appl. Phys. Lett,* **73**, 1526 (1998).
11. J. Koh, A.S. Ferlauto, P.I. Rovira, C.R. Wronski and R.W. Collins, *Appl. Phys. Lett.* **75**, 2286 (1999).
12. J. Koval, A. S. Ferlauto, J. M. Pearce, R. W. Collins, and C. R. Wronski, *J. of Non-Cryst. Solids* **299-302**, 1136 (2002).

## Nanocrystalline silicon ( nc-Si ) from single ion beam sputtering

Z.B. Zhou *, G.M. Hadi, R.Q. Cui
Solar Energy Institute, Department of Physics, Shanghai Jiaotong University,
Shanghai 200030 China
Z.M. Ding, G. Li
Instrumental Analysis Center, Shanghai Jiaotong University,
Shanghai 200030 China

## ABSTRACT

Based on a small set of selected publications on the using of nanocrystalline silicon films (nc-Si) for solar cell from 1997 to 2001, this paper reviews the application of nc-Si films as intrinsic layers in p-i-n solar cells. The new structure of nc-Si films deposited at high chamber pressure and high hydrogen dilution have characters of nanocrystalline grains with dimension about several tens of nanometer embedded in matrix of amorphous tissue and a high volume fraction of crystallinity (60~80%). The new nc-Si material have optical gap of 1.89 eV. The efficiency of this single junction solar cell reaches 8.7%. This nc-Si layer can be used not only as an intrinsic layer and as a p-type layer. Also nanocrystalline layer may be used as a seed layer for the growth of polycrystalline Si films at a low temperature.

We used single ion beam sputtering methods to synthesize nanocrystalline silicon films successfully. The films were characterized with the technique of X-ray diffraction, Atomic Force Micrographs. We found that the films had a character of nc-amorphous double phase structure. Conductivity test at different temperatures presented the transportation of electrons dominated by different mechanism within different temperature ranges. Photoconductivity gains of the material were obtained in our recent investigation.

## INTRODUCTION

Much work has been done to study the structure and properties of the nanocrystalline silicon thin films (nc-Si). It has been reported that these films have many super opto and electrical properties [1]. Most of the films were deposited by plasma enhanced chemical vapor deposition (PECVD). There are also a few papers, which reported the nc-Si thin films deposited by the sputtering method and quite exciting results on the photovoltaic application of this material [2,3,4]. Actually, sputtering methods, especially RF, or DC sputtering, are quite favorable processing for production. For the final industrialization, we should try to understand the material thoroughly. For most situations, nanocrystalline silicon thin films we obtained from sputtering and PECVD can been described as a double phases structure. That means there are two kinds of microstructure in the films, one is crystalline nanometer size crystallite, another is amorphous tissue, which wraps around grains forms nano-crystal grain boundary. Electron transportation mechanism should be a coalescent result of the two. Considering the quantum refinement effect, and electron wave property in essence, we can reasonably predicate that the electron should show some unique character as they pass through this material with nanometer size double phase structure. Defects with high density should be planted into the films deposited with DC or RF sputtering process due to the direct contact between plasma and the surface of films [5]. To avoid processing induced defects, lowering the ion bombarding is an effective way. In this paper, we present our recent research result on the synthesis of the nanocrystalline silicon thin films by using single ion beam sputtering method. We also studied the transportation character of electrons passing through nanocrystalline silicon thin films, which were deposited by single ion beam sputtering method. We found that the film from ion beam sputtering showed a typical nanocrystalline structure mentioned above with the help of XRD and AFM probes. Their electronic transportation property exhibited a unique character at the temperature range from 500K to 150K. We attributed this result to the quantum wave property of electrons in the films. Non-decayed photoconductivity gain about 10 was tested through using a Source Meter of Keithley instruments.

## EXPERIMENTAL DETAILS

The deposition equipment for nanocrystalline silicon thin films is a set of single ion beam sputtering instrument with a Kaufman ion source. The synthesizing system also consists of a diffusion pump, water-cooled target holder and substrate holder. The background and operating pressure in the vacuum chamber are better than $1 \times 10^{-4}$ Pa and $3 \times 10^{-2}$ Pa, respectively. A high resistivity silicon wafer was used as sputtering target. Glass sheets coated with tin oxide thin films were used as substrates. High purity argon gas diluted with hydrogen was used for forming ion beam to bombard the target. The ratio of partial pressures of two gases Ar and $H_2$ in depositing chamber was 1 to 2. Argon gas was ionized and extracted by powerful electrical field produced by an accelerating grid electrode. An energetic ion beam was formed, which provided bombarding ions. Hydrogen gas mixed in argon was ionized at the same time to provided hydrogenating ion precursors. The energy and current intensity of the ions coming from the ion beam source were 400eV and 45mA/cm$^2$. The substrate holder, with an ion beam impact angle of about 85 degrees, between ion beam and normal line of substrate surface, consisted of an electronic controlled heater copper plate, which kept substrates at a temperature between ambient temperature and 300℃ during the deposition.

## TEST AND DISCUSSION

In order to demonstrate this nanocrystalline silicon thin film material with our novel depositing method, we used XRD, AFM probes and conductivity test under different temperature for our characterization motive. An AFM equipment of Digital Instrument Nanoscope was used to scan the surface of the sample. XRD data were obtained with a Shimadzu XD-3A instrument.

## Microstructure test

The average grain size and texture in the layer have been estimated by x-ray diffraction measurements. Figure 1 shows the XRD result for a typical sample. CuK$_\alpha$ radiation was used in the experiment. The XRD measurements shows that the films have double phase structure, amorphous and preferred orientation crystal phases. The crystal peak (111) has the highest intensity. Another smaller peak is (220). The average grain size can be obtained from the width of the diffraction peaks, being given by the Scherrer formula $\delta = \kappa\lambda/(w\cos\theta)$, where $\kappa$ is approximately 1, $\lambda$ is the wave length of x rays, w is the width of the band after correcting the instrumental contribution and $\theta$ is the Bragg angle of the diffraction planes. It is about 15 nm.

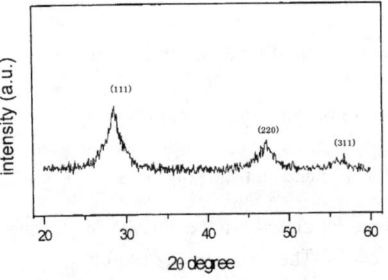

**Figure 1**. X-ray diffraction pattern of a typical nc-Si film deposited by ion beam sputtering method

## Electronic transportation property

Figure 2 shows the Arrhenius characteristics of the temperature-dependent conductivity for the typical sample. We found that the curve can been divided into two parts at the temperature of 490K, and two different excitation energy corresponding to two different structure phases. They are 0.28eV and 0.18 eV, respectively. Part one indicates a transportation region dominated by amorphous structure matrix which can be called phase one in the temperature range above 490K, and part two corresponds to another region dominated by the nanocrystalline grains embedded in amorphous matrix in the temperature range below 490K, as drawn in figure 1. The whole conductivity can be expressed as a combination of the partial conductivities from two parts in series. Both two phases have size of nanometer dimension. In this nanometer size system, quantum wave effect shows a distinctive contribution to electronic transportation process. We can recognize this fact from figure 2. At high temperature region above 490K, conductivity is dominated by nanometer size amorphous boundary, we can express it as expanding state conductivity.

$$\sigma = \sigma_0 \exp(-\frac{\Delta E}{k_B T}) \tag{1}$$

For phase two, nanocrystalline grains dominate conductivity, its curve of variations of log( α ) against the reciprocal environment temperature has a similar character compared with traditional amorphous silicon thin films. At the range from 490K to 200K, the transportation mechanism across the phase two can be described using heterojunction quantum tunneling (HQD) model. The conductivity can be expressed as [6,7]

$$\sigma_{HQD} = \sigma_0 \exp(-\Delta E / k_B T) \times erfc(e/\sqrt{8<q^2>}) \quad (2)$$

where $<q^2>$ is quantum fluctuation of charge distributing function in heterojunction. Because $erfc(e/\sqrt{8<q^2>})$ is a function of temperature, naturally the exciting energy of $\sigma_{HQD}$ varies with T. Below 200K, electron transportation mechanism in phase two is changed from HQD model directly to the variable-distance hopping conductivity between localized states near $E_f$. The conductivity can be expressed as

$$\sigma_{hop} = \frac{1}{6}\sigma_0 e^2 R^2 v_{ph} g(E_F) \exp(-2\alpha R) \times \exp(-\frac{W}{k_B T}) \quad (3)$$

where, e is the charge of electron, R is hopping distance, $\exp(-2\alpha R)$ is the probability of electron to transfer from one state to another. $\alpha$ is a quantity which is representative for the rate of fall-off the wave function at a site. $g(E_f)$ is the density of localized states near $E_f$ level. $Exp(-\frac{W}{k_B T})$ is the possibility of finding a phonon with excitation energy of W. Due to the emergence of nanoscale grains in the film, the contribution from transition between band tail states disappears. We called this as delocalization situation of CB and VB. We also tested the conductivity of the sample under a certain light source. Neglectable decay of the photoconductivity gain was measured, which was about ten for typical sample.

## Topography observation

Surface topography was determined by atomic force microscopy (AFM), and microstructure of the films was analyzed by combining AFM data with X-ray diffraction.

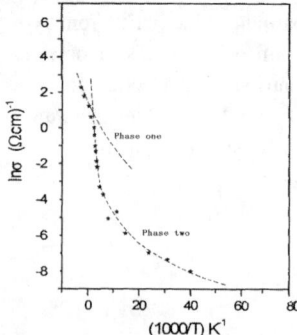

**Figure 2**. Arrhenius curves of nc-Si thin film

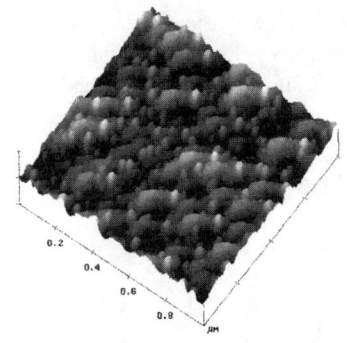

**Figure 3**. Atomic force micrographs of a typical sample from ion beam sputtering method

Figure 3 shows an AFM image of the surface topography of the typical sample about 500 nm thick. The grains about 20 nm on the average are observed from the image, which are embedded into the amorphous background. We can calculate the grain sizes with the XRD patterns using the Laue method and Scherrer formula. The average dimension is about 15 nm. We note that the Scherrer's results have slightly difference from the AFM results, because AFM image can only give us a shape and size of the grain, in which both crystalline and amorphous issues are included. For microstructure, we could only obtain reliable information through XRD diffraction peak analysis. The image of AFM also gives us a typical surface character of the film bombarded by oblique ion beam.

## CONCLUSION

Nano-crystalline silicon thin film is a promising material for the future development of highly efficient and low cost thin film solar cells. We developed a new deposition method—ion beam sputtering for synthesis of this thin film material from room temperature to 300℃. We investigated the microstructure, electronic transportation property and surface topography of the film. The results showed that the film was constructed with two phases, one is nanocrystalline grains, which are about 15nm in average, and the two is amorphous part, which forms a tissue structure surrounded the grains. Whole conductivity can been expressed as a combination of the partial conductivities from two parts in series, and the transportation mechanism alters from heterojunction quantum tunneling mechanism directly to the variable-distance hopping conductivity between localized states near $E_f$. Delocalization of band tail states occurs in this nano scale system.

## REFERENCES

1. Sukti Hazra, Swati Ray "Nanocrystalline silicon as intrinsic layer in thin film solar cells" Solid State Communications 109 (1999), 125~128
2. H. Motegi, Y. Ohshita, K. Yamaguchi, PVSEC-12 International Technical Digest, Korea, 2001, p443
3. K. Yamamto, T. Suzuki, Proc. of 14th E-PVSEC, WIP, Munchen, 1997, p1018

4. G. Grabosch, D. Borchert, Proc. of 14$^{th}$ E-PVSEC, WIP, Munchen, 1997, p1472
5. A. Achiq, R. Rizk, F. Gourbilleau, P. Voivenel "Effects of hydrogen partial pressure on the structure and properties of sputtered silicon layers" Thin Solid Films 348 (1999) 74~78
6. G.Y. Hu, Y.L. He, M. B. Yu, Phys. Rev., B59 (1999), p15352
7. G.Y. Hu, R.F. Connell, J. Appl. Phys. 78 (1995), p3845

## Correlation of Material Properties and Open-Circuit Voltage of Amorphous Silicon Based Solar Cells

Baojie Yan, Jeffrey Yang, Guozhen Yue, and Subhendu Guha
United Solar Systems Corp., Troy, MI 48084, U.S.A.

## ABSTRACT

Correlation of hydrogenated amorphous silicon (a-Si:H) alloy material properties and solar cell characteristics have been studied experimentally and by computer simulation. Simulation results show that all three solar cell parameters, short-circuit current density ($J_{sc}$), open-circuit voltage ($V_{oc}$), and fill factor (FF), decrease with increased defect density. For a given intrinsic layer thickness, a larger band gap ($E_g$) results in a higher $V_{oc}$ but a lower $J_{sc}$. However, FF does not depend on band gap. This allows us to distinguish the effect of change in band gap from that in defect density on the variation in $V_{oc}$. For solar cells with good interface characteristics, a linear relation FF = $\beta V_{oc} + \gamma$ is obtained by light soaking experiments and simulation with different defect densities. The slope $\beta$ is in the range from 2 to 3 $V^{-1}$ depending on cell properties and light soaking condition, and the intersect $\gamma$ depends mainly on the band gap. Comparing cells made with high $H_2$ dilution to no $H_2$ dilution, we find that a 58 mV enhancement in $V_{oc}$ with $H_2$ dilution is due to both widening of band gap and reduced defect density. Simulation results also show that a narrower valence band tail leads to a higher $V_{oc}$. We did not include this effect in the analysis due to lack of available data for correlation between $H_2$ dilution and band tail narrowing.

## INTRODUCTION

Hydrogen ($H_2$) dilution is now widely used to improve the performance of hydrogenated amorphous silicon (a-Si:H) based solar cells since the first report in 1981 by Guha *et al.* [1]. The best solar cells are made with a dilution ratio just before the onset of microcrystalline silicon formation [2]. The main improvements are on open-circuit voltage ($V_{oc}$) and fill factor (FF). It has been found that a-Si:H films made with high hydrogen dilution include chain-like structures and improved medium range order [3]. However, how these changes in material property affect the device performance is not well understood. For example, it is not clear as to how much of the enhancement of $V_{oc}$ is due to a wider band gap, narrower band tails, or lower defect density. In addition, the limiting factor of $V_{oc}$ has become an interesting topic lately [4,5]. Imperfect interfaces between the doped (*n* or *p*) layer and the intrinsic (*i*) layer would definitely affect $V_{oc}$, and are reflected on the insensitivity of $V_{oc}$ upon light soaking. However, for a device with reasonably good interface, the bulk properties of the *i* layer become the dominant factor. In this case, all the three characteristic parameters, short circuit density ($J_{sc}$), $V_{oc}$, and FF, of a solar cell degrade after light soaking. Schiff and his co-workers reported that the density of defect states does not affect the $V_{oc}$ very much [4], which is inconsistent with most of our experimental results [6].

In this study, we report hydrogen dilution effect on material property and solar cell performance. We focus on the enhancement of $V_{oc}$ and its correlation with band gap and defect density. The result shows that high hydrogen dilution indeed widens the band gap and reduces defect density; both effects contribute to the enhancement of $V_{oc}$.

## EXPERIMENTAL AND SIMULATION

Intrinsic a-Si:H and a-SiGe:H films and *nip* solar cells were deposited using a multi-chamber system with a 13.56 MHz excitation. The *i* layer was deposited with a gas mixture of $SiH_4$, $GeH_4$ and $H_2$. The deposition rate was kept at 1.2 Å/s for all the a-Si:H samples. For comparison, a few high rate a-SiGe:H alloy solar cells were also studied. Optical transmission and reflection were measured on samples deposited on Corning 7059 glass substrate with a film thickness about 5000 Å to determine the optical band gap. The *nip* solar cells were deposited on specular stainless steel substrate over an area of 4 cm × 4 cm. Sixteen 0.25 cm² indium tin oxide (ITO) dots were

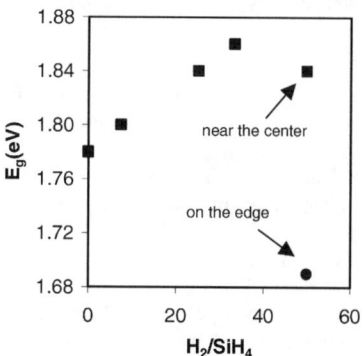

**Figure 1.** Optical band gap of a-Si:H as a function of $H_2$ dilution.

deposited on *p* layer as top transparent electrodes. Current-voltage (*J-V*) characteristics were measured under an AM1.5 solar simulator and in the dark at 25 °C. The solar cells were light soaked under a 100 mW/cm² white light at 25 °C and 50 °C for stability studies.

We simulate the solar cell performance by using AMPS-1D program [7] provided by Pennsylvania State University. The main parameters used in the simulation are similar to those used by Jiang *et al.* [4] except we included two types (donor-like and acceptor-like) of defects with Gaussian distribution in our study. We assume the same amount of donor-like and acceptor-like defects. The standard deviation of 0.15 eV for the Gaussian distribution is fixed for both the donor-like and acceptor-like defects. The capture cross-sections for electrons by donor-like defects and acceptor-like defects are $1.0\times10^{-14}$ cm² and $1.0\times10^{-15}$ cm², respectively, and $1.0\times10^{-15}$ cm² and $1.0\times10^{-14}$ cm² for corresponding holes. The band-tail characteristic energies of 0.03 eV for conduction band tail and 0.05 eV for valence band tail are fixed unless otherwise specified.

## RESULTS AND DISCUSSION

Figure 1 plots the optical band gap ($E_g$) of intrinsic a-Si:H films as a function of hydrogen dilution ratio. $E_g$ was deduced from $(\alpha h\nu)^{1/2}$ versus $h\nu$ plot, where $\alpha$ is the absorption coefficient of photons with energy $h\nu$. The optical band gap increases from 1.78 eV (no $H_2$ dilution) to 1.86 eV (high $H_2$ dilution). Further increasing $H_2$ dilution results in the decrease of optical band gap due to inclusion of microcrystallites. In this case, a large non-uniformity, as shown by the different values from the center to the edge on a 4 cm × 4 cm substrate, indicates a mixed phase of amorphous and microcrystalline in the material [8]. Although these results strongly confirm that $H_2$ dilution indeed widens the optical band gap when the material is still in the amorphous regime, we cannot precisely find out how much of the $V_{oc}$ enhancement is due to the widened band gap.

In order to search for a better understanding, we simulate the solar cell performance as a function of band gap (the same values for both optical band gap and mobility band gap are used), defect density, and band tail characteristics. Figure 2 compares the simulated *J-V* characteristics with measured results for an a-Si:H solar cell (a) under AM1.5 illumination and (b) in the dark. Since the simulation results fit both the light and dark $J$-$V$ characteristics very well, we believe that the parameters used in this simulation can represent the properties of the real solar cells. We have known for quite some time that all three characteristic parameters of a-Si:H solar cells degrade after light soaking, which means that defect generation degrades not only the FF but also the $V_{oc}$ and $J_{sc}$. Figure 3 gives an example of simulation results of solar cell characteristic parameters as a function of defect density for a band gap of 1.80 eV and a band gap of 1.77 eV. The results show that all three parameters decrease with increased defect density, and the decrease becomes steeper at high defect densities. The large decrease of solar cell performance with increased defect density at the high defect regime is explained below. In general, the photo-generated carriers recombine through both band tails and defects. In the low defect density regime, the recombination through defects may play a less important role, thus a relatively weak dependence of solar cell performance on the defect density. On the other hand, most recombination of photo-generated carriers is through defects in the high defect regime, which results in a strong decline of solar cell

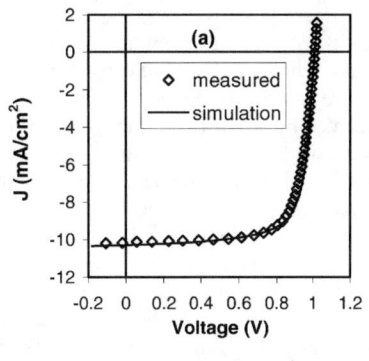

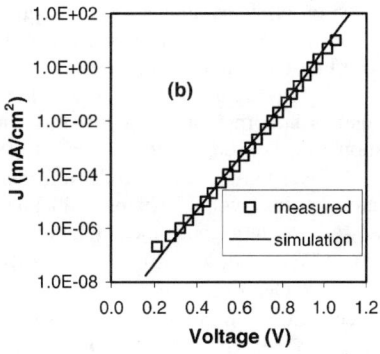

**Figure 2.** Comparison of measurement and simulation results of J-V characteristics under (a) AM1.5 illumination and (b) in the dark.

performance with the increase of defect density. In addition, an electric field free region appears when the defect density is very high, especially for thick cells. A higher defect density causes a wider field free region, which accelerates the decrease of cell performance with the increase of defect density. One important observation from Fig. 3 is that the curves of FF for the two solar cells with different band gaps coincide with each other, indicating that FF does not depend on the band gap in the given range. In this case, even though there is no simple correlation between FF and defect density, we still can use FF to infer the defect density in the intrinsic layer, at least for comparison between samples. One must keep in mind that we assume the same recombination cross-section for all samples, which may vary from sample to sample [9]. Figure 4 plots FF of three a-Si:H solar cells made with different $H_2$ dilution ratios and two simulation results as a function of $V_{oc}$. The experimental data were obtained by light soaking and simulation by variation of defect density. Both the experimental and simulation results show a linear relation of FF = $\beta V_{oc} + \gamma$, where $\beta$ and $\gamma$ are sample-depended constants. $\beta$ weakly depends on band gap with a value around 2.1 for all three cells, but $\gamma$ depends strongly on band gap. The high $H_2$ diluted solar cell has a $V_{oc}$ about 58 mV larger than that of the no diluted one at the initial state.

From these plots, we estimated that 28 mV of the 58 mV enhancement in $V_{oc}$ results from widened band gap as indicated by arrow (1), and 30 mV from defect density deduction as indicated by arrow (2) in the figure. From this result, we find that $H_2$ dilution reduces defect density and widens the band gap. Both effects lead to a similar enhancement of $V_{oc}$. Of course, we should point out that the amount of $V_{oc}$ enhancements due to a reduction of defect density and a widening of band gap might vary from sample to sample, because defect density and band gap depend not only on $H_2$ dilution but also on other parameters such as pressure and temperature used in the experiment.

The linear relationship between FF and $V_{oc}$ upon light soaking is very interesting. In order to understand the meaning of this linear relation, we measured a few series of a-Si:H and a-SiGe:H samples made under different $H_2$ dilution ratios with different thicknesses. Most cells show a good linear relationship as shown in Fig. 4, but with different slopes β and intersects γ. One exception is a 5000 Å a-Si:H thick cell with no dilution. Figure 5 shows FF versus $V_{oc}$ for three 5000 Å thick a-Si:H cells with high, medium, and no $H_2$ dilution. The cells with high and medium $H_2$ dilutions follow the linear relation, but the undiluted cell shows an irregular shape. This experiment was repeated for a few times and the same result was obtained. We have not understood the reason for the different behavior. Table I lists all values of β and γ for a-Si:H and a-SiGe:H solar cells. It seems that the thickness of the intrinsic layer does not affect β and γ values significantly (except for the 5000 Å thick undiluted one), but light-soaking temperature does. Light soaking at a lower temperature results in larger β values. Also, a-SiGe:H cells have larger β values than a-Si:H cells. The different β values may be related to different defect properties. Certain defects could affect FF more than $V_{oc}$. It has been reported that light-induced defects created at lower temperatures have lower annealing activation

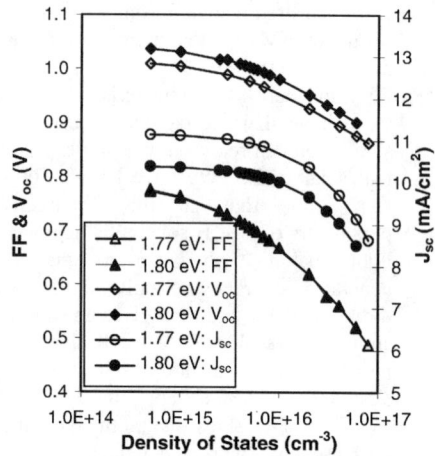

**Figure 3.** Simulated J-V characteristics of a-Si:H solar cells as a function of defect density.

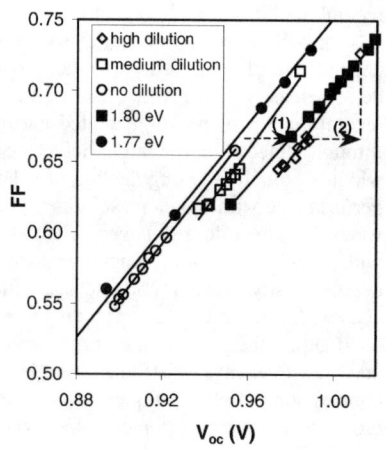

**Figure 4.** FF versus $V_{oc}$ for three a-Si:H solar cells with different $H_2$ dilution and for two simulation results with different band gaps.

**Table I:** list of the β and γ values for samples deposited with different H₂ dilution ratios and light soaked at different temperatures (T).

| Sample | Thickness (Å) | H₂ dilution | β ($V^{-1}$) | γ | T (°C) and comments |
|---|---|---|---|---|---|
| a-Si:H | 2200 | High | 2.2 | 1.5 | 50 |
| | | Medium | 2.1 | 1.4 | |
| | | No | 1.9 | 1.1 | |
| | 2200 | High | 2.8 | 2.1 | 25 |
| | | Medium | 2.6 | 1.8 | |
| | | No | 2.7 | 1.8 | |
| | 5000 | High | 2.7 | 2.0 | 25 |
| | | Medium | 2.4 | 1.7 | |
| | | No | 2.2 | 1.5 | |
| a-SiGe:H | 1600 | High | 2.4 | 1.1 | 50, $V_{oc}$ = 0.74 V |
| | 1800 | High | 2.6 | 1.3 | |
| | 2000 | High | 3.0 | 1.6 | 50, $V_{oc}$ = 0.77 V |
| | 3000 | High | 3.0 | 1.7 | |

energies than those created at higher temperatures. Therefore, we believe that a certain correlation between defect properties and the β value might exist, but we have not fully understood the physics behind the simple linear relation.

We should also point out that the width of the valence band tail could affect $V_{oc}$ significantly. Our simulation results, as shown in Fig. 6, show that a decrease of the valence band tail characteristic energy from 0.05 eV to 0.04 eV causes a 30 mV increase in $V_{oc}$. However, most measured optical results show that the width of valence band tail is around 50 mV and is not sensitive to deposition conditions such as H₂ dilution, unless the deposition rate is very high. In addition, light soaking does not change its value [10]. Furthermore, the width of the valence band tail is even not sensitive to the Si/Ge ratio for a-SiGe:H alloys [11], which may relate to

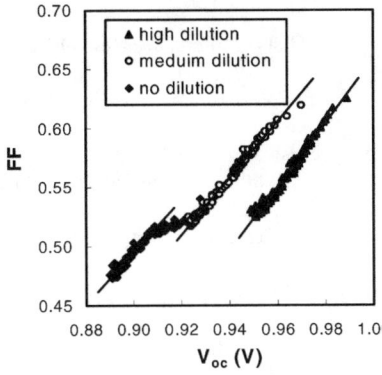

**Figure 5.** FF versus $V_{oc}$ for three 5000 Å a-Si:H cells deposited with different H₂ dilution ratios.

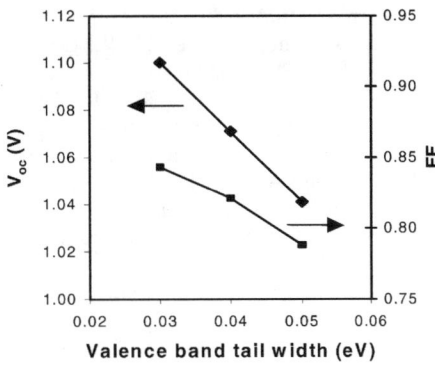

**Figure 6.** Simulated $V_{oc}$ and FF versus valence band-tail width. The simulation was made with no defect.

the constant hole drift mobility in a wide range of band gaps from a-SiGe:H to a-Si:H and a-SiC:H made in different laboratories [12]. Recently, Schiff and his coworkers found the hole drift mobility to be higher in $H_2$ diluted a-Si:H films than in undiluted ones, which may be an indication of valence band tail narrowing [13]. If we indeed take this into account, the assignments for widened band gap and reduced defect density by $H_2$ dilution as shown in Fig. 3 would be smaller.

## CONCLUSION

Hydrogen dilution widens the band gap and reduces the defect density in a-Si:H. Both effects improve solar cell performance, as reflected by high $V_{oc}$ and FF. Light-soaking experiments and simulation reveal a liner relation between FF and $V_{oc}$ for most of a-Si:H and a-SiGe:H solar cells. The slope and intersect of the linear relation depend on sample properties and light soaking conditions. We find that 28 mV of the 58 mV enhancement in $V_{oc}$ is due to widening of the band gap from $H_2$ dilution, and 30 mV from lowering the defect density. Narrowing of valence band tail could also improve $V_{oc}$ and FF significantly, but we have not included this effect in the analysis due to the lack of available data.

## ACKNOWLEDGMENTS

We gratefully acknowledge useful discussions with S. R. Ovshinsky, H. Fritzsche, E. Schiff, and K. Lord. This work is supported by NREL under subcontract No. ZDJ-2-30630-19.

## REFERENCES

1. S. Guha, K. L. Narasimhan, and S. M. Pieruszko, J. Appl. Phys. **52**, 859 (1981).
2. J. Yang, A. Banerjee, and S. Guha, Appl. Phys. Lett. **70**, 2975 (1997).
3. D. V. Tsu, B. S. Chao, S. R. Ovshinsky, S. Guha, and J. Yang, Appl. Phys. Lett. **71**, 1317 (1997).
4. L. Jiang, J. H. Lyou, S. Rane, E. A. Schiff, Q. Wang, and Q. Yuan, Mat. Res. Soc. Symp. Proc. **609**, A18.3 (2000).
5. E. A. Schiff, Proc. of 29th IEEE Photovoltaic Specialists Conference, New Orleans, USA, 1086 (2002).
6. X. Xu, J. Yang, and S. Guha, Proc. of 23rd IEEE Photovoltaic Specialists Conference, Louisville, USA, 971 (1993).
7. H. Zhu and S. J. Fonash, Mat. Res. Soc. Symp. Proc. **507**, 395 (1998).
8. J. Yang, A. Banerjee, K. Lord, and S. Guha, Proc. of 28th IEEE Photovoltaic Specialists Conference, Anchorage, USA, 742 (2000).
9. P. Stradins, S. Shimizu, M. Kondo, A. Matsuda, Mat. Res. Soc. Symp. Proc. **664**, A12.1 (2001).
10. J. Heath, S. B. Iyer, Y. Lubianiker, J. D. Cohen, and G. Ganguly, Mat. Res. Soc. Symp. Proc. **664**, A25.3 (2001).
11. S. Guha, J. S. Payson, S. C. Agarwal, and S. R. Ovshinsky, J. Non-Cryst. Solids **97 & 98**, 1455 (1987).
12. Q. Gu, Q. Wang, E. Schiff, Y. M. Li, and C. T. Malone, J. Appl. Phys. **76**, 2310 (1994).
13. E. Schiff, private communication.

## Simulations of Buffer Layers in a-Si:H Thin Film Solar Cells Deposited with an Expanding Thermal Plasma

A.M.H.N. Petit[1,2], M. Zeman[1], R.A.C.M.M. van Swaaij[1] and M.C.M. van de Sanden[2]
[1]Delft University of Technology, DIMES-ECTM,
P.O. Box 5053, 2600 GB Delft, The Netherlands.
[2]Eindhoven University of Technology, Department of Applied Physics,
P.O. Box 513, 5600 MB Eindhoven, The Netherlands.

## ABSTRACT

With an Expanding Thermal Plasma Chemical Vapor Deposition system (ETP-CVD), solar grade amorphous silicon (a-Si:H) can be deposited at high deposition rate (> 2 nm/s). We think that during the first stage of deposition, a material is grown with a higher defect density than the rest of the bulk creating a defect-rich layer (DRL). Therefore we analyzed, by the means of simulations, the influence of the position of the DRL on the performance of a p-i-n a-Si:H solar cell when moved from the p-i towards the i-n interface and as a function of its thickness. We investigate the effect of a buffer layer in between the p- and the i-layer on the external parameters of the solar cell. The presence of a buffer layer increases the electric field near the p-i interface, which leads to a higher collection of free charge carriers at the interface, although the electric field is then diminished deeper in the bulk. It appears that 10 nm thick buffer layer is sufficient to improve the performance. In case no buffer layer is applied, recombination losses at the p-i interface diminish the performance of the solar cell. We also observe that an increase of the DRL thickness results in a reduction of the solar-cell performance, which is more pronounced when the DRL is located in the region close to the p-i interface rather than close to the i-n interface.

## INTRODUCTION

In order to reduce the production cost of thin-film amorphous silicon solar cells, high growth rates for the deposition of the intrinsic layer are required. With expanding thermal plasma chemical vapour deposition (ETP CVD), deposition rates of up to 10 nm/s for solar grade a-Si:H have been achieved [1]. At high growth rates, however, higher deposition temperatures are required to obtain high-quality amorphous silicon. In turn, these high deposition temperatures lead to deterioration of the p-layer when these intrinsic layers are implemented in p-i-n solar cells [2]. It has been demonstrated experimentally [3] that implementation of a high-quality intrinsic layer (which we will refer to as the 'buffer' layer) in between the p- and high-rate ETP i-layer improves the performance of the solar cells, in particular due to an enhancement of the open-circuit voltage, $V_{oc}$, and the fill factor, $FF$. We think that this buffer layer protects the p-layer from thermal damage by reducing the diffusion of hydrogen out of the p-layer. However, a performance improvement was observed also for solar cells with a low temperature (250°C), high-rate (0.9 nm/s) intrinsic layer in which a buffer layer was implemented [3]. We think that with ETP CVD a thin defect-rich layer (DRL) is formed during the first stages of deposition, whereas the quality of the remainder of the intrinsic layer is good. Implementing the buffer layer in the cell moves the DRL away from the critical p-i interface, reducing the recombination and improving the *FF*.

In this article we present simulations, carried out with the Advanced Semiconductor Analysis (ASA) program [4], that aim to investigate the position dependence of the DRL on the solar-cell performance. By increasing the thickness of the buffer layer, the DRL is shifted through the intrinsic layer. We also investigate the influence of the DRL thickness on the cell performance. These simulations may shed light on the properties of the material deposited during the initial stages of the growth with ETP CVD. Other workers (e.g., see [5]) have simulated the effect of a buffer layer in the p-i region, with for example different band gap profiles, but in contrast to the work presented in this article, the aim was to optimize the cell performance.

The ASA program is an one dimensional (1-D) simulation program that calculates the internal electrical properties and external characteristics of multi-layer, heterojunctions a-Si:H solar cells by solving the system of semiconductor equations (Poisson's, continuity and transport equations) for the steady state. The main features of the ASA program include calculation of light generation profile, models describing a complete density of states distribution as function of energy, and calculation of the defect-state distribution in a layer according to the defect-pool model (DPM) [6]. The continuous change of all input parameters as a function of position in the device can be defined.

## SIMULATIONS AND RESULTS

### Solar cell structure

The structure of the cells simulated with the ASA program is presented in figure 1: a glass substrate covered with a transparent conductive oxide, an a-Si based p-, i- and n-layer and finally the aluminum back contact. The intrinsic layer is divided in three parts: the buffer layer with thickness varying between 0 and 390 nm, the initial grown DRL, which has a thickness of 30 nm, and finally the bulk of the ETP i-layer with a thickness varying between 410 and 20 nm. The total thickness of the intrinsic layer is kept constant at 440 nm.

### Input parameters

In table I, the input parameters for the ASA program are listed. Table I includes the band gap, $E_g$, the electrons and holes extended-state mobility, $\mu_e$ and $\mu_h$, the density of states at the mobility conduction band edge, $N_C^{mob}$ and at the mobility valence band edge, $N_V^{mob}$, and the characteristic energy $E_{C0}$ and $E_{V0}$ of the conduction and valence band tail, respectively. The effective densities of extended states for the conduction band, $N_C$ and for the valence band, $N_V$, were kept constant at $4 \times 10^{26}$ m$^{-3}$ and the electron affinity, $\chi_e$, was taken to be 4 eV. We assume that the Urbach energy of the DRL is higher than in the bulk of the intrinsic layer in order to account for the higher defect density.

The density-of-states distribution is calculated using the defect-pool model (DPM) from Powell and Deane [6]. The following DPM parameters have been used: the position of the peak of the defect pool, $E_P$, is taken at 1.25 eV, except for the p-layer in which case it is at 1.27 eV, the width of the defect pool, $\sigma$, can be found in table I. The correlation energy between the transition energy levels representing the dangling bonds is $E_{corr} = 0.2$ eV.

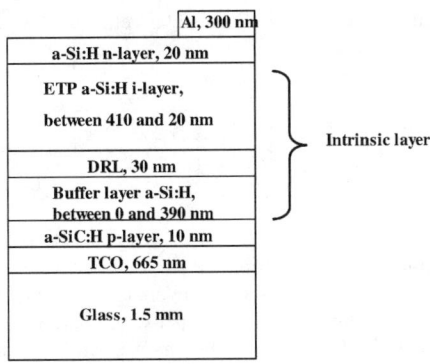

**Figure 1.** Schematic diagram of an a-Si:H solar cell as depicted in the ASA modeling.

### Effect of buffer layer

Two kinds of simulations were carried out. In the first series, the buffer-layer thickness was increased, moving the DRL away from the p-i interface. The DRL thickness was fixed at 30 nm.

The external parameters of the simulated solar cells can be found in table II. When a buffer layer is incorporated in the cell, the $V_{OC}$ first increases slightly, but then stays constant for a buffer-layer thickness in excess of 10 nm. The short-circuit current density, $J_{SC}$, decreases somewhat, though the $FF$ increases significantly. The efficiency, $\eta$, also increases. The same trends have been observed experimentally on ETP solar cells [3].

The band diagram of the two simulated cells, one with and one without buffer layer, is shown in figure 2. We can see that near the p-i interface, in the region denoted as A, the bending of the conduction and valence bands is less strong for the cell without buffer layer compared to the one with a 50 nm thick buffer layer. This strong bending implies that the internal electric field is stronger in that region when a buffer layer is applied (see figure 3) and therefore the charge carriers are swept out of this region faster than when there is no buffer layer. The chance of recombination is then less and the $FF$ increases. Moreover, when no buffer layer is incorporated in the cell, the defect density drastically increases at the p-i interface because of the DRL, creating a barrier for the holes to reach the p-layer for collection. The buffer layer permits to move the DRL away from the sensitive p-i area, thereby diminishing the recombination at the interface and improving the $FF$.

**Table I.** Parameters used in the ASA modeling.

| Layers | $E_g$ (eV) | $\mu_e / \mu_h$ ($m^2 V^{-1} s^{-1}$) | $N_V^{mob}$ ($m^{-3}.eV^{-1}$) | $N_C^{mob}$ ($m^{-3}.eV^{-1}$) | $E_{V0}$ (meV) | $E_{C0}$ (meV) | $\sigma$ (eV) |
|---|---|---|---|---|---|---|---|
| p-type | 1.97 | $1\times10^{-3} / 1\times10^{-4}$ | $1\times10^{28}$ | $1\times10^{28}$ | 80 | 70 | 0.179 |
| buffer | 1.75 | $2\times10^{-3} / 5\times10^{-4}$ | $5\times10^{27}$ | $7\times10^{27}$ | 50 | 32 | 0.160 |
| DRL | 1.75 | $1\times10^{-3} / 1\times10^{-4}$ | $5\times10^{27}$ | $7\times10^{27}$ | 80 | 53 | 0.165 |
| bulk ETP | 1.75 | $2\times10^{-3} / 5\times10^{-4}$ | $5\times10^{27}$ | $7\times10^{27}$ | 45 | 30 | 0.160 |
| n-type | 1.75 | $1\times10^{-3} / 1\times10^{-4}$ | $7\times10^{27}$ | $7\times10^{27}$ | 90 | 80 | 0.175 |

**Table II.** External parameters of the simulated solar cells with varying buffer layer thickness.

| Buffer layer (nm) | $V_{OC}$ (V) | $J_{SC}$ (mA/cm$^2$) | FF (-) | $\eta$ (%) |
|---|---|---|---|---|
| 0 | 0.795 | 11.9 | 0.627 | 5.94 |
| 10 | 0.801 | 11.7 | 0.658 | 6.18 |
| 50 | 0.801 | 11.4 | 0.688 | 6.27 |
| 75 | 0.802 | 11.4 | 0.693 | 6.32 |

Further in the cell, in the region denoted as B, we can see that the band bending is stronger for the cell without a buffer layer, resulting in a stronger electric field deeper in the cell. This means that, for the cell without buffer layer, more charge carriers are collected from deeper in the cell than for a cell with buffer layer at the expense of charge-carrier collection near the p-i interface. However, deeper in the cell fewer carriers are generated, but as these carriers experience a higher electric field, it is more likely that these carriers are collected, which implies that $J_{SC}$ is higher for the cell without buffer layer.

There are several competing effects that determine the performance of a cell: the strength of the electric field and the defect density, the generation and recombination profiles. When the electric field is high, the carriers have more chance to be separated and collected, but if the defect density is also high, the carriers have more chance to recombine. The collection probability depends on where the DRL is situated and how thick it is. In our case, it seems that a buffer layer of 75 nm thick permits to increase the $FF$, but the $J_{SC}$ is already slightly decreasing, so a thicker buffer layer would not increase the performances anymore.

**Effect of the defect-rich layer**

In the second series of simulations, we studied the effect of the variation of the thickness of the DRL, grown during the initial stages of the deposition with ETP. It appears that the thicker this defective layer, the lower the solar cell performance in terms of $V_{OC}$, $FF$ and $J_{SC}$ (see figure 4). A 10 nm thick buffer layer is sufficient to improve the $V_{OC}$. For thicker buffer layers, the $V_{OC}$

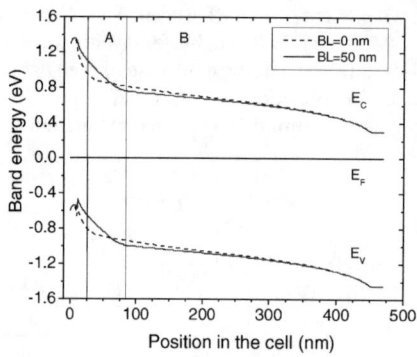

**Figure 2.** Band diagram of the simulated cells at thermal equilibrium with and without buffer layer. The DRL is 30 nm thick.

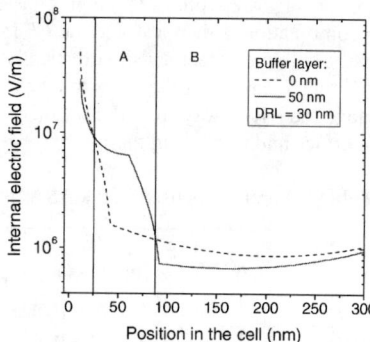

**Figure 3.** Internal electric field distribution for cells with and without buffer layer. The DRL is 30 nm thick.

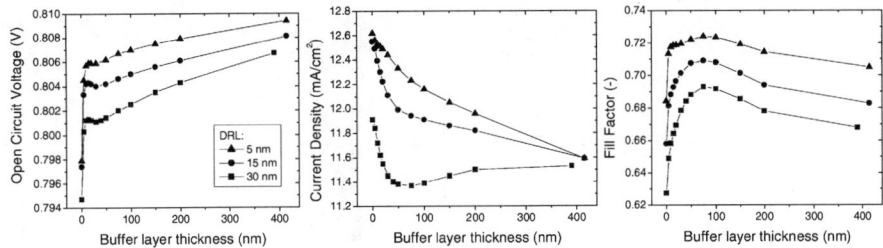

**Figure 4.** Simulations of the external parameters of a p-i-n cell as a function of the buffer layer thickness, this for three different thicknesses of DRL. The total thickness of the intrinsic layer is 440 nm.

increases only slightly. The $J_{SC}$ decreases to reach a minimum value, which does not depend on the DRL thickness. The $FF$ is largely affected by the DRL thickness and the same trend is observed for the three thicknesses. The $FF$ becomes maximal for a buffer layer of about 75 nm. With a buffer layer at the p-i interface, the DRL is situated further in the cell, so more carriers recombine in the middle of the cell, where the defect density is highest, which means that fewer carriers are collected and the current density decreases. Moreover, the $FF$ is first increased because more carriers are separated at the p-i interface. Above a certain buffer layer thickness (~75 nm), the DLR is too far from the p-i interface to have an influence on the recombination there but it starts to influence the i-n interface and the $FF$ diminishes.

If the DRL cannot be avoided, it is better to situate it within the first quarter of the intrinsic layer. The $FF$ is indeed optimum around 75 nm buffer thickness, $V_{OC}$ is more or less constant and $J_{SC}$ is decreasing.

To understand the effect of the thickness of the DRL on the cell performance, we compare two cells that have a buffer layer of 10 nm; the thickness of the DRL is respectively 5 nm and 30 nm. The band diagram of these cells at thermal equilibrium is presented in Figure 5. The DRL thickness has a little influence on the band diagram in the buffer layer, but in the ETP layer we can see that the thinner the DRL, the less strong the slope of the band. The DRL thickness does not influence the band diagram after 200 nm in the cell. Close to the p-i interface, the internal electric field is stronger for the cell with a thick DRL (see figure 6). This means that the carriers are swept away from this region faster than in the cell with a thin DRL. However, the $J_{SC}$ and the $FF$ are lower. We believe that this is due to the fact that more carriers are trapped or recombine in the DRL.

The presence of the DRL influences the current collection not only in this layer, but also in the rest of the cell. Deeper in the cell, the electric field is higher for cell with a thin DRL, which allows more carrier collection and thus a higher $J_{SC}$.

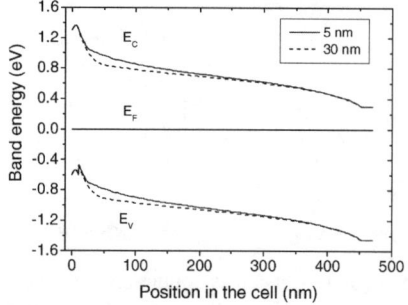 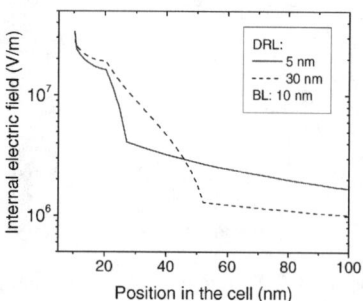

**Figure 5.** Band diagram of the simulated cells at thermal equilibrium with resp. a 5 and 30 nm thick DLR. The buffer layer is 10 nm thick.

**Figure 6.** Internal electric field distribution for cells with resp. a 5 and 30 nm thick DRL. The buffer layer is 10 nm thick.

## CONCLUSIONS

We have carried out two series of simulations. The first series shows that the incorporation of a buffer layer at the p-i interface increases the performance of the cell by augmenting the internal electric field at the interface and shifting the defect-rich layer (DRL) further in the bulk of the cell away from the sensitive p-i interface.

The second set of simulations shows that a thicker DRL suppresses the performance of the cell. The DRL has an opposite competing effect with the buffer layer. The simulations also show that if the DRL is situated close to the i-n interface its thickness has a strong influence on the *FF*, which is clearly lower for a thicker DRL. This means that a larger defective layer enhances the recombination at the i-n interface.

## ACKNOWLEDGEMENTS

This work has been partially funded through the Dutch 'Economy, Ecology and Technology' Program under contract EETK01019.

## REFERENCES

1. W.M.M. Kessels, R.J. Severens, A.H.M. Smets, B.A. Korevaar, G.J. Adriaenssens, D.C. Schram and M.C.M. van de Sanden, J. Appl. Phys. **89**, 2404 (2001).
2. B.A. Korevaar, C. Smit, A.H.M. Smets, R.A.C.M.M. van Swaaij, D.C. Schram and M.C.M. van de Sanden, Proc. of 28th IEEE-PVSC (2000), 916.
3. B.A. Korevaar, A.M.H.N. Petit, C. Smit, R.A.C.M.M. van Swaaij, and M.C.M. van de Sanden, Proc. of 29th IEEE-PVSC (2002), 1230.
4. M. Zeman, J.A. Willemen, L.L.A. Vosteen, G. Tao, and J.W. Metselaar, Solar Energy Materials and Solar Cells **46**, 81 (1997).
5. J. Zimmer, H. Stiebig and H. Wagner, J. Appl. Phys. **84**, 15 (1998).
6. M.J. Powell and S.C. Deane, Phys. Rev. B **48**, 10 815 (1993).

## Microcrystalline (Si,Ge):H Solar Cells

Jianhua Zhu and Vikram L. Dalal
Dept. of Electrical and Computer Engr. and Microelectronics Research Center
Iowa State University, Ames, Iowa 50011, USA

ABSTRACT

We report on the growth and properties of microcrystalline Si:H and (Si,Ge):H solar cells on stainless steel substrates. The solar cells were grown using a remote, low pressure ECR plasma system. In order to crystallize (Si,Ge), much higher hydrogen dilution (~40:1) had to be used compared to the case for mc-Si:H, where a dilution of 10:1 was adequate for crystallization. The solar cell structure was of the $p^+nn^+$ type, with light entering the $p^+$ layer. It was found that it was advantageous to use a thin a-Si:H buffer layer at the back of the cells in order to reduce shunt density and improve the performance of the cells. A graded gap buffer layer was used at the p+n interface so as to improve the open-circuit voltage and fill factor. The open circuit voltage and fill factor decreased as the Ge content increased. Quantum efficiency measurements indicated that the device was indeed microcrystalline and followed the absorption characteristics of crystalline (Si,Ge). As the Ge content increased, quantum efficiency in the infrared increased. X-ray measurements of films indicated grain sizes of ~ 10nm. EDAX measurements were used to measure the Ge content in the films and devices. Capacitance measurements at low frequencies (~100 Hz and 1 kHz) indicated that the base layer was indeed behaving as a crystalline material, with classical C(V) curves. The defect density varied between $1 \times 10^{16}$ to $2 \times 10^{17}/cm^3$, with higher defects indicated as the Ge concentration increased.

INTRODUCTION

Microcrystalline Si:H (mc-Si:H) and microcrystalline (Si,Ge):H (mc-(Si,Ge):H)alloys are potentially useful electronic materials. A number of groups have reported on the growth of mc Si:H materials and solar cells [1-5]. It has been shown that hydrogen is a critical element in controlling grain-boundary recombination, and that the material contains H primarily at the grain boundaries, with perhaps a thin a-Si:H tissue surrounding the grains. Excellent solar cells, with efficiencies approaching 10%, have been made in this material [4,5]. In contrast, much less work has been done to fabricate and understand the properties of mc-(Si,Ge):H materials or devices [6-8]. In this paper, we report on the growth and properties of mc-(Si,Ge):H solar cells across the entire bandgap region, from Si to Ge, so as to study the material for potential device applications.

GROWTH TECHNIQUE

The materials and devices were grown using a remote, hydrogen rich, low pressure ECR technique described earlier[6,7,9]. A mixture of silane, germane and hydrogen was used to grow the films and devices. Substrate temperature was kept in the range of 300 °C, the microwave power was 200 W, the pressure was 5 mTorr, and the typical gas flows were:
For Si: Hydrogen = 35 sccm , silane = 2.4 sccm
For (Si,Ge): Hydrogen = 50 sccm, silane = 2-0.6 sccm, and germane = 0.2-0.7 sccm

We found that it was always necessary to use a much higher hydrogen dilution in order to crystallize (Si,Ge), as opposed to the case for crystallizing Si, where as little as 10:1 hydrogen dilution was adequate.

The devices grown were of the substrate type geometry, deposited on stainless steel substrates, provided by USSC. The typical device geometry is shown in Fig. 1. It consists of first depositing a 100 nm thick amorphous $n^+$ Si layer on steel substrates, followed by a thin undoped a-Si:H layer. The purpose of the undoped a-Si:H layer was to reduce the shunt density in the device, and also provide a passivating layer between the n and $n^+$ interface. For Si cells, the amorphous layer was followed by a n-type microcrystalline Si:H base layer, followed by a graded gap amorphous layer to reduce recombination at the $p^+n$ interface. An amorphous $p^+$ layer and a top ITO contact completed the cell. For microcrystalline (Si,Ge):H solar cells, the amorphous Si layer at the back of the cell was followed by a graded bandgap microcrystalline (Si,Ge) layer, which was then followed by a constant gap n-type base layer. This base layer was followed by a buffer layer, a $p^+$ a-(Si,C):H layer and the ITO top contact. See Fig. 1.

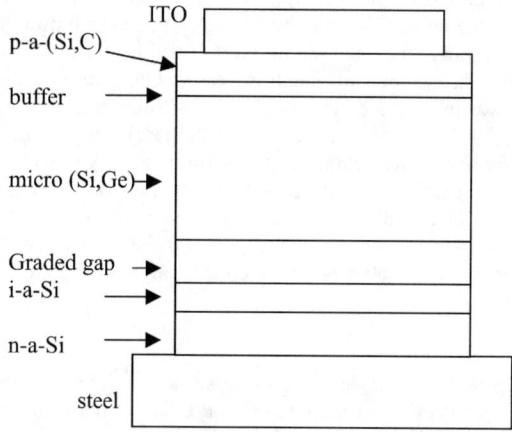

Figure 1 Device structure for microcrystalline (Si,Ge):H cell.

## FILM RESULTS

The Ge content in the films was determined using energy-dispersive x-ray spectroscopy (EDAX). In Fig. 2, we plot the Ge content in the film vs. Germane/silane ratio in the gas phase. From Fig. 2, we observe that a close to 1:1 Si:Ge decomposition from the gas to the solid phase appears to be achieved in the ECR discharge.

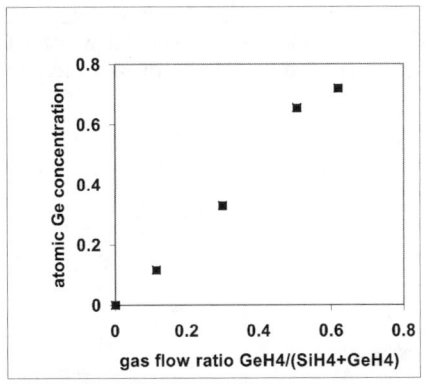

Figure 2 Ratio of Ge/Si in solid vs. Gas flow rate.

DEVICE RESULTS

The I(V) curve for a typical mc-Si:H device is shown in Fig. 3, and for a mc-(Si,Ge):H cell is shown in Fig. 4. The fill factor for the mc-Si:H cell is excellent. The thickness of the n layer was ~ 0.9 micrometer and the QE at 800 nm was 0.05. The current density for the mc-Si:H cell is 14 mA/cm$^2$ ( no back reflector) and the efficiency is 4.65%. For the microcrystalline (Si,Ge) cell, the thickness of the n layer was less, about 0.4 micrometer, and even then, the QE at 800 nm was higher (0.1). However, the current density was lower, about 10 mA/cm$^2$. The open circuit voltage decreased from 0.47 in mc-Si:H to 0.39 V for the mc-(Si,Ge):H cell.

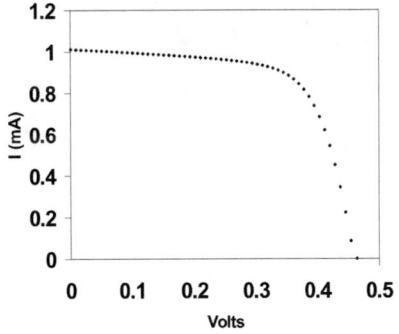

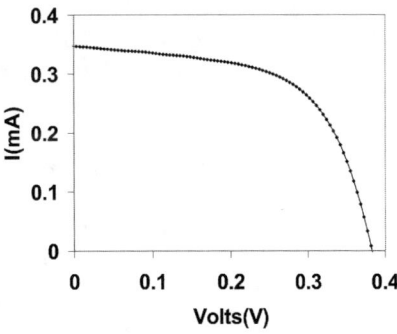

Figure 3 I(V) curve for mc- Si:H cell. FF-0.7 QE at 800 nm = 0.05.

Figure 4. I(V) curve for a mc- (Si,Ge):H cell FF =0.6, QE at 800 nm =0.1.

In Fig. 5, we plot the open circuit voltage for the cells as a function of germane/silane ratio, and in Fig.6, we plot the fill factor vs. germane/silane ratio. Clearly, both the open circuit voltage and the fill factor appear to decrease as germane concentration increases.

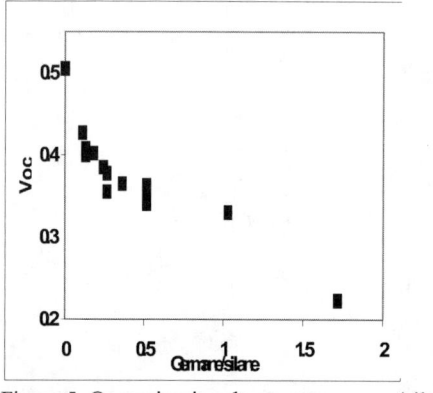

Figure 5 Open circuit voltage vs. germane/silane Ratio.

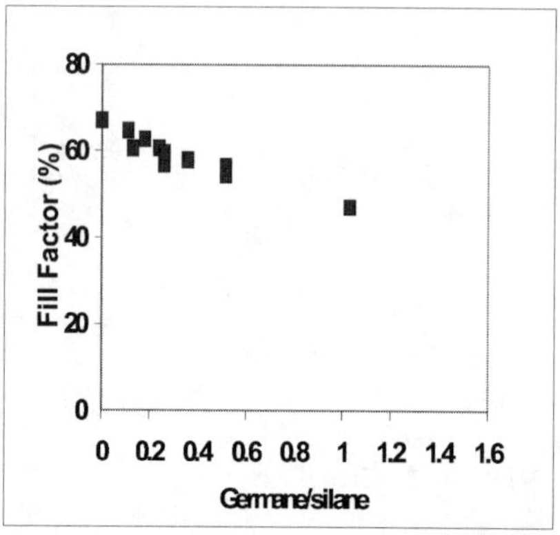

Figure 6. Fill factor vs. germane/silane ratio.

However, as expected from using a material with a smaller bandgap, quantum efficiency in the infrared, using QE at 800 nm as a guide, increases with Ge content, as shown in Fig. 7.

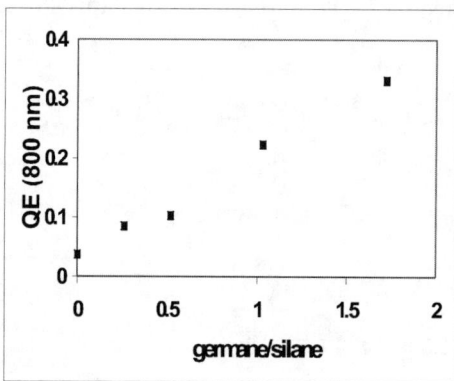

Figure 7 Quantum efficiency at 800 nm as a function of geramne/silane ratio.

QE vs. photon energy data are shown in Fig. 8 for various (Si,Ge) devices, and clearly, as Germane/silane ratio content increases, the edge of the QE curves shift to lower photon energies. All the QE curves show a non-exponential behavior, typical of crystalline Si and Ge, rather than the exponential behavior typical of a-Si:H and a-(Si,Ge):H. Thus, we confirm that the devices are indeed microcrystalline.

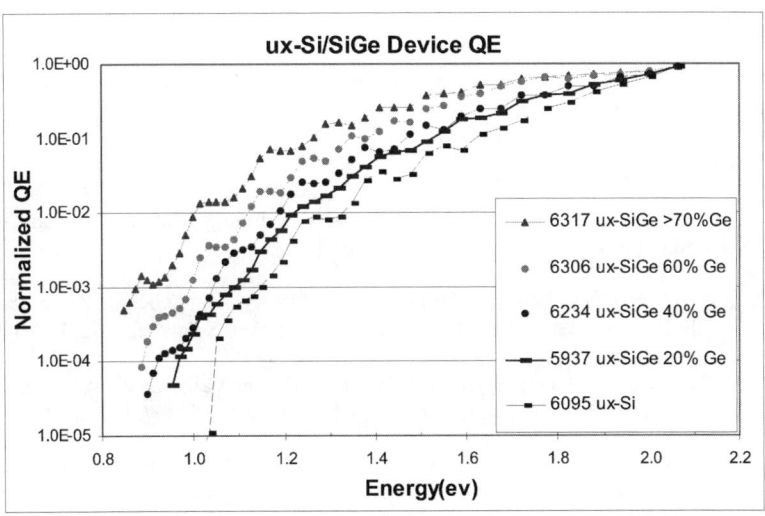

Figure 8 Quantum efficiency vs. photon energy curve for a series of devices, showing that as the germanium concentration increases, the edge of the QE curves shift to lower energies. All the curves have a non-exponential behavior with photon energy, implying crystalline-type absorption as opposed to amorphous.

We also measured C(V) curves in these devices as a function of voltage. We obtained classical $1/C^2$ vs. reverse voltage relationship, implying crystalline type behavior. The capacitance was measured as a function of frequency, with low frequency values being more reliable since low frequencies will pick up all the states in the gap. The measurements indicated relatively constant values for doping, with doping values depending on the material used as the base n layer. We obtained values of doping varying between a few $10^{16}/cm^3$ for mc-Si:H to 1-$2 \times 10^{17}/cm^3$ for mc-(Si,Ge):H. This higher defect density may be one reason why the fill factor and current density are poorer in mc-(Si,Ge):H cell.

DISCUSSION AND CONCLUSIONS

From the data presented, it is clear that while one can make (Si,Ge) to crystallize, the properties of the material, as indicated by the poorer fill factor, are inferior to those of mc Si:H.

The QE curves clearly indicate that the materials are crystalline; the absorption follows the crystalline type behavior ( rounded absorption curve on a log plot). However, the defect density is too high, and one needs to pay attention to the proper growth conditions which will lead to excellent passivation of nanocrystalline grains, by encompassing an amorphous tissue to surround the crystalline grain. One also needs to pay attention to the specific type of H bonding achieved in the material. Further experiments to improve the material are in progress.

ACKNOWLEDGEMENTS

This work was partially supported by NREL. We thank Keqin Han, Durga Panda and Puneet Sharma for their technical help.

REFERENCES

1. Y. Nasuno, M. Kondo and A. Matsuda, Solar Energy Mater. And Solar Cells, 74, 497-503(2002).

2. O. Vetterl, F. Finger, R. Carius, P. Hapke, L. Houben, O. Kluth, A. Lambertz, A. Mück, B. Rech and H. Wagner, Solar Energy Mater. And Solar Cells, 62, 97-108(2000).

3. J. Meier, S. Dubail, S. Golay, U. Kroll, S. Faÿ, E. Vallat-Sauvain, L. Feitknecht, J. Dubail and A. Shah, Solar Energy Mater. And Solar Cells, 74, 457-467(2002).

4. Kenji Yamamoto, Masashi Yoshimi, Yuko Tawada, Susumu Fukuda, Toru Sawada, Tomomi Meguro, Hiroki Takata, Takashi Suezaki, Yohei Koi, Katsuhiko Hayashi , Solar Energy Mater. And Solar Cells, 74, 449-455 (2002).

5. B. Rech, O. Kluth, T. Repmann, T. Roschek, J. Springer, J. Müller, F. Finger, H. Stiebig and H. Wagner , Solar Energy Mater. And Solar Cells, 74, 439-447(2002).

6. V. L. Dalal and K. Erickson, Proc. Of $28^{th}$. IEEE Photovolt. Spec. Conf.(2000) ,p.792-795.

7. K. Erickson and V. L. Dalal , Proc. of MRS, 507, 987(1998).

8. M. Isomura, K. Nakahata, M. Shima, S. Taira, K. Wakisaka, M. Tanaka and S. Kiyama, Solar Energy Mater. And Solar Cells, 74, 519-524 (2002).

## Investigation of the Causes and Variation of Leakage Currents in Amorphous Silicon P-I-N Diodes

Todd R. Johnson*, Gautam Ganguly, George S. Wood, and David E. Carlson
BP Solar, North American Technology Center, Toano, Virginia
*Princeton University, Princeton, New Jersey

## ABSTRACT

Excess leakage currents under reverse bias (known as shunting) and spontaneous reductions of this excess leakage under increased reverse bias (known as curing) were investigated in hydrogenated amorphous silicon (a-Si:H) based single junction p-i-n type diodes. An increase in the frequency of shunting was observed when the front contacts were switched from tin oxide to zinc oxide, most likely due to defects in the previously deposited zinc oxide coated glass was observed. Storage in the dark and light soaking up to 100 hours were both observed to independently increase the leakage current in previously leaking diodes. Models for the distribution of shunt-causing defects within a given cell area were considered. Comparing the measured frequency of shunting using cells of varying area (1 to 16 mm$^2$) to the models' predictions indicate a distribution of point defects separated by relatively large average distances that are slightly larger for tin oxide (5-6 mm) than for zinc oxide (4 mm).

## INTRODUCTION

The commercialization of a-Si:H based thin film solar cells has been going on for some time. [1,2] However, the consequent shunting in a-Si:H thin film solar cells has been an issue for manufacturers because the higher leakage currents in reverse bias results in a loss of power (efficiency). It has previously been observed that subjecting some cells to a higher voltage can "cure" these shunts and repair the cells to normal working order [3]. However, curing is not effective on all shunts and sometimes the shunts reappear or increase in leakage after some period of time [4]. In some cases, large shunts are caused by tin oxide particles [4]. The mechanisms of both shunting and curing have not been definitively explained and so systematic prevention of shunts in manufactured cells has not been possible. In this study, we looked to determine the distribution of the shunts themselves: whether they were grouped together in clusters, randomly distributed single point defects, or something else. Microscope observations were made of the cells at different stages of deposition and before and after curing to see if visible changes or shunt causing defects could be seen. We have also compared various front and back contacts to shed light on the causes and find possible cures for shunting.

## EXPERIMENT

a-Si:H based solar cells were fabricated by DC plasma enhanced chemical vapor deposition (PECVD) on transparent conducting oxide coated glass substrates. Some of the glass substrates were coated with smooth or textured tin oxide (~700 nm thick) by a commercial vendor while others were RF magnetron sputter coated with ~100nm of zinc oxide, in house. The back contact consisted of a zinc oxide layer of ~100 nm and /or ~200 nm of aluminum or silver. The deposition of silicon layers for these cells were carried out during that part of the cathode cleaning cycle where excessive shunts due to flaking-off of silicon from the cathode is known not to occur.

Cells of area 25 mm$^2$, which exhibited shunting when measured immediately after fabrication were monitored at intervals after sitting in the dark or after light soak. In order to model shunt distribution and location, statistical sampling was carried out on cells of 4 different areas viz. 16, 9, 4, and 1 mm$^2$. After fabrication, the cell performance was tested on a curve tracer and the percentage of cells exhibiting shunting (defined as a reverse bias leakage of > 50 µA at –2V) was observed. Curing was then attempted by increasing the reverse bias to 5V and the number of cells seen to cure was also recorded. The percentage of passing cells was normalized to their equivalent passing percentage if the cells were shrunk to 1 mm$^2$, which is equivalent to shunts per area. Comparisons of the shunting and curing rates of different sized cells were analyzed with respect to the expected rates with a given average shunt separation to find the average distance between shunts for the two different front contact materials (ZnO and SnO$_2$). The effects of using seamed or as-cut glass and textured or specular surfaces were also investigated.

## RESULTS AND DISCUSSION

### Time dependent shunting increases

It was seen that on average, under light soak, the leakage increased by approximately 100% until approximately 100 hours, at which point it leveled off (see Fig. 1a). However, we saw that there was an increase in leakage over time of ~ 50% even without light soak, which also leveled after over 100 hours (see figure 1b). Light soaking the cells after this period of darkness caused a second increase in leakage of ~ 50% (Fig. 1b). The end result is that the increase in leakage was the same (~ 100%) regardless of whether light soaking was at the beginning or end of the experiment so long as sufficient time (>100 hours) was given to allow the leakage to level out. From this data, we concluded that there are two independent causes of the increase in leakage after curing. Light soaking causes one, while the other is an ageing process that occurs over time regardless of the illumination conditions. They can occur simultaneously or independently, and the sum of their effects is the total change in leakage of the cell. Temperature dependence of these processes might yield more information.

### Effect of back contact fabrication

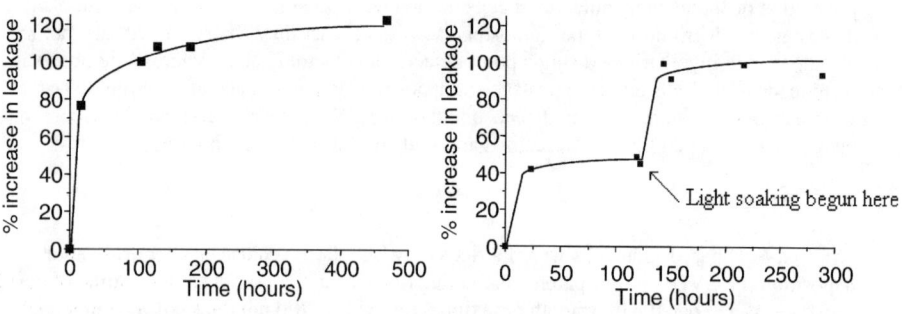

**Figure 1:** The change in leakage current with a) light soaking time and b) storage in the dark followed by light soaking. The lines are guides to the eye.

Eight different back contacts, using different combinations of silver, aluminum, and ZnO, were deposited on a-Si layers on specular glass to look for changes in passing and curing rates.

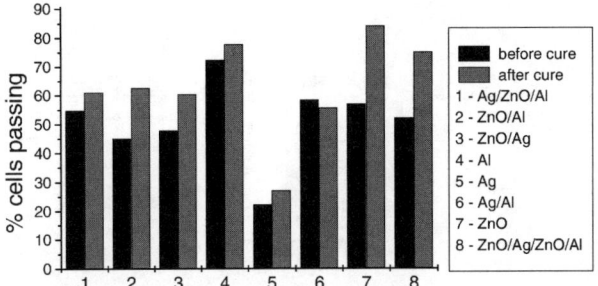

**Figure 2.** The percentage of cells that do not exhibit shunting, both before and after curing, for 8 different back contact materials and $SnO_2$ front contacts.

A variety of behaviors were observed. ZnO was seen to be a key material in curing properties. ZnO alone was the best contact for curing while the Al contact had the highest passing rate before curing (indicating fewest number of shunts present). Contacts with Al or Ag cured better when ZnO was present as a buffer than when it was absent (see Fig. 2). Combining Ag and Al together was the only sample that did not show any curing at all.

The percentage of cells passing after curing for cells of varying i-layer thickness was observed using four different mask sizes (16, 9, 4, and 1 $mm^2$) on both specular and textured $SnO_2$ surfaces (see Fig. 3). Both substrates showed a general increase in non-curable shunts with an increase in cell area. While the textured samples showed an increase in shunting with decreasing i-layer thickness from 2500Å, the specular samples did not show any thickness dependence down to 500Å (see Fig. 3). Since surface structural feature size decreases from ~1000-3000Å for the 'textured' to ~100Å on the 'smooth' $SnO_2$, the additional shunts on the 'textured' $SnO_2$ would appear to be due to the incomplete coverage of the 'peaks' with silicon.

Shunt distribution and size

The normalized values of the shunt frequency of the 4 different sizes of cells give a representation of the shunts per unit area. The results indicated dependence only on the front contact material ($SnO_2$ or ZnO) and not on the size of the cell (see Fig. 4). Since the size of the cell area does not affect shunt frequency, we infer that there is a shunt distribution where some average distance separates randomly dispersed shunt sources throughout the sample. Using the dimensions of the cells and sample, and the passing percentages on all four sizes, the average distance between the shunts was calculated to be ~5-6 mm for $SnO_2$ front contacts with ZnO/Al back contacts. Shunting in cells on ZnO front contacts (and ZnO/Al back contacts) is more frequent and the separation was calculated to about 4 mm.

We assume that the shunts are cured due to the burn-out of regions where the silicon layer is missing [5](i.e. the shunts consists of ZnO in contact with $SnO_2$ in regions where pinholes exist in the silicon layer). Using the resistances and current flows through the cells before and after curing, and using the thermal and electrical properties of ZnO, including heat capacity and melting point, we made a rough estimate of the diameter of the shunt path through the cell and found that the diameter of the shunt pathway through the i-layer was on the order of

0.6 mm. This is an order of magnitude estimate since the actual size of the shunts is likely to vary over a wide range.

Front contact materials

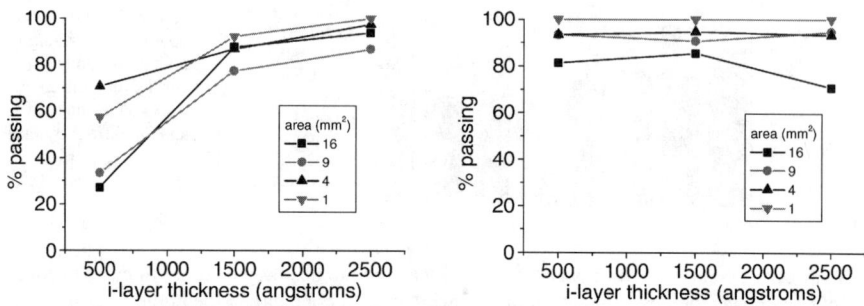

**Figure 3.** The percentage of samples passing after cure for a) textured $SnO_2$ and b) specular $SnO_2$

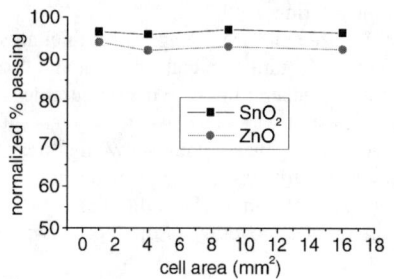

**Figure 4.** The normalized passing percentage for cells of different areas (equivalent to shunts per unit area) with $SnO_2$ and ZnO front contacts

Other than their inherent material properties, there is one obvious difference between the ZnO front contacts and $SnO_2$ front contacts we used. The $SnO_2$ is deposited on the hot glass in a commercial process as soon as the glass comes out of the float line furnace but the ZnO was deposited much later by us, at a lower temperature, in our own chambers. When we examine the $SnO_2$ and ZnO materials under the microscope, we see that the $SnO_2$ is very clean, with few to no defects or signs of debris. In comparison, the ZnO has a large number of noticeable defects, some of which are embedded debris in the ZnO layer (see figure 5a), (meaning they were present at the time of ZnO deposition) and others that are on top of the ZnO (meaning they fell on the sample after ZnO deposition). Those embedded in the ZnO can come loose at any time. If they come off before a-Si or back contact deposition, it will allow these layers to fill the hole and create a shunt. An example of one such hole is shown in figure 5b. The debris ranged in size but was of range consistent with the calculation above.

Observations of the samples after different steps of manufacture failed to find a specific step responsible for the sudden appearance of the debris that is believed to be causing these shunts. However, the rough-cut edges of the glass substrate are suspected of flaking off and falling onto the cell area before or during ZnO deposition. Past observations have also shown that glass with rough (un-seamed) edges can flake off at these edges when heated. We also observed that physical contact with the rough edges of the glass substrates knocks particles onto the surface. This means that each manufacturing step that requires physical contact with the substrate edges could contribute to the addition of glass debris onto the cell surface, which would increase the number of shunts.

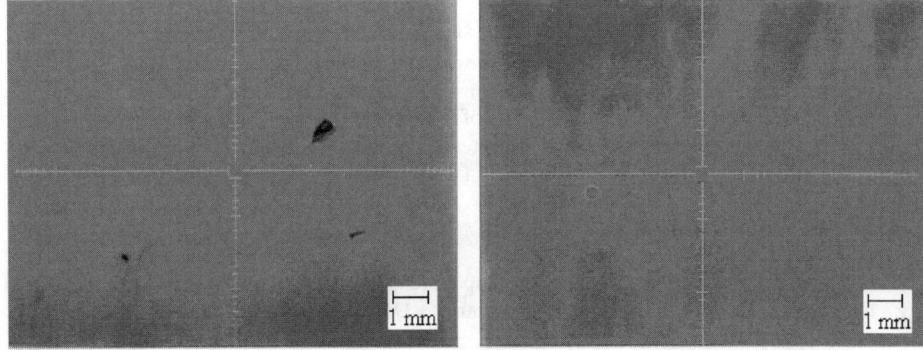

**Figure 5.** Microscopic images of the surface of ZnO layers with a) debris embedded in the ZnO layer and b) one of the large holes in the ZnO where the debris has fallen off.

In an attempt to confirm this hypothesis, a seamed (edges ground smooth after cutting) 12"x13" glass plate was run with ZnO and standard silicon layers before being cut into the standard 3"x3" sample size with rough edges. Throughout this process greater than normal care was taken to prevent surface contamination. These cells showed improved initial passing rates that were nearly equal to that of the commercially produced $SnO_2$ contacts. However, their curing rates were similar to previous ZnO front contact tests and did not improve to the $SnO_2$ level. This implies that the glass debris is increasing the number of shunts in the ZnO cells by opening holes for a-Si or the back contact material to form current paths, but that it is unrelated to the curing mechanism of the cells.

A second shunt-related issue is the curing of the cells that are initially shunted. About half of the cells with areas of 16 $mm^2$ and 25 $mm^2$ need to be cured. Of these, more than 80% usually cure to passing for $SnO_2$ but only about 30% cure for ZnO. Most of the ZnO cells show some decrease in leakage, but they often reach at best around 200 – 1000 µA of leakage at -2V.

While the first type of shunting is caused by holes in the front contact, the second type is caused by holes in the a-Si layers that allow a connection between the top and bottom contacts. It is possible that ZnO cures less frequently than $SnO_2$ because the $SnO_2$ / ZnO interface formed in this hole create a junction that is more conducive to heat generation, and therefore cures easier, but this was not confirmed. ZnO may be better than Al and Ag because of it's higher resistance, lower thermal conductivity, and low heat capacity, all of which help to generate heat and keep it contained locally. We surmise that the higher current applied to the shunt during curing causes the shunt to burn up or melt away, removing the low resistance pathway and allowing the cell to function properly. This was indicated by microscope observations of changes in surface appearance after curing, but was not unambiguous.

## CONCLUSIONS

The reverse bias leakage current in p-i-n type a-Si:H based solar cells was found to increase from a combination of storage time in the dark and under illumination, resulting in an average increase of 100%. Analysis of data on diodes of 4 different areas are consistent with shunts being distributed randomly throughout the sample surface with an average separation of ~5 mm for $SnO_2$ and ~4 mm for ZnO front contacts, respectively. Assuming thermal curing of shunts, their diameters are estimated to be on the order of about 0.6 mm. We identify two different aspects to the shunting problem. One issue is to prevent shunts that cannot be cured due mainly to debris on glass surface. This determines the percentage of the cells that pass before curing is attempted. Keeping the glass surface clean prior to ZnO deposition and using seamed glass substrates can prevent most of these shunts. The second type of shunting involving missing silicon regions is curable, and heat generation in the shunt path is suggested to be the mechanism of the cure.

## ACKNOWLEDGMENTS

We are grateful to S. Stone for sample preparation. This work was partially supported by the US DOE through NREL under sub-contract ZDJ-2-30630-10.

## REFERENCES

1. R.R. Arya and D.E. Carlson, Progress in Photovoltaics 10, 69 (2002).
2. D. E. Carlson, G. Ganguly, G. Lin, M. Gleaton, M. Bennett and R. R. Arya MRS Spring Meeting Symp. Proc. 664, pA11(2001)
3. T.J. McMahon and M.S. Bennett, Solar Energy Materials and Solar Cells, 41/42 465 (1996).
4. K.R. Lord II, M.R. Walters and J.R. Woodyard, Mat. Res. Soc. Symp. Vol 336 (1994)
5. J. Hu, H.M. Branz, R.S Crandall, S. Ward and Q. Wang, Mat. Res. Soc. Symp. Vol 715 (2002)

Deposition of device quality μc-Si films and solar cells at high rates by HWCVD in a W filament regime where W/Si formation is minimal

E. Iwaniczko, A.H. Mahan, B. Yan*, L.N. Gedvilas, D.L. Williamson**, and B.P. Nelson
NREL, 1617 Cole Blvd., Golden, CO 80401
* United Solar Systems Corporation, 1100 W. Maple Road, Troy, MI 48084
** Colo. School of Mines, Golden, CO 80401

## ABSTRACT

μc-Si has traditionally been deposited by Hot Wire CVD at a low filament temperature. At these temperatures, silicides rapidly form on the filament surface, leading in the case of a tungsten filament to both film reproducibility and filament lifetime issues. By depositing films consecutively using identical deposition parameters, these issues are chronicled for a filament temperature of ~ 1750°C. Upon increasing the filament temperature to ~1825-1850°C, these reproducibility and lifetime issues disappear and, by lowering both the substrate temperature and chamber pressure, device quality μc-Si is deposited at high deposition rates in a filament regime where tungsten silicide formation is minimal. Both single junction and tandem solar cells are fabricated using this material, confirming the validity of this approach.

## INTRODUCTION

Microcrystalline silicon (μc-Si) has recently emerged as a viable candidate as the active layer in both single junction cells, and as the low bandgap material in a tandem solar cell structure. The advantages of this material include both its low bandgap (~1.1 eV) and the virtual absence of a Staebler-Wronski Effect for films exhibiting a high microcrystalline volume fraction. However, due to the indirect nature of the bandgap, thick absorber layers (> 0.7 – 1.0μm) are traditionally used, and then film deposition rate ($R_d$) rapidly becomes an issue for all deposition technologies. In PECVD deposition of μc-Si, while a considerable success has been achieved by going to both high excitation frequencies and high chamber pressures, the best cell efficiencies are still deposited at deposition rates below 2-3Å/s [1]. This is also true in Hot Wire CVD (HWCVD), where the highest solar cell efficiency (9.4%), achieved using a HWCVD i-layer in a p-i-n single junction solar cell, was achieved with the i-layer deposited at an $R_d$ of ~ 1Å/s [2].

Common to all device quality HWCVD μc-Si i-layer depositions has been the use of a low filament temperature [2-4]. Two reasons for this have been advanced. The first is that the use of low filament temperatures limits the amount of radiative heating incident upon the growing film, and thus allows the use of a low substrate temperature ($T_{sub}$). This approach is best illustrated by the work of Klein et al. [2], where the substrate was heated entirely by filament radiation. Using a filament temperature of ~ 1650°C and a $T_{sub}$ of 195°C yielded the lowest i-layer ESR spin densities and the highest solar cell efficiencies; all higher filament temperatures and $T_{sub}$'s gave considerably higher spin densities and lower cell efficiencies. The possibility of increased H passivation of these defects at lower $T_{sub}$ has been advanced. The second deals with the possible generation of different radical species contributing to film growth. Although identification of such radicals as a function of deposition parameters is still in its infancy [5], it is generally believed that when lower filament temperatures are used, a smaller percentage of Si radicals (resulting from complete silane dissociation on the filament) are re-evaporated into the gas phase

[6]. As such radicals are believed detrimental to film growth, the correlation between silicide formation on a filament and an improvement of μc-Si i-layer properties may have some validity.

On the other hand, while silicide formation can be controlled somewhat by the use of Ta as a filament material [7], its 'production' continues when a W filament is used, and can lead, at least in a-Si:H, to a rapid deterioration with filament 'aging' in material properties and film deposition rate [8] as well as a dramatic shortening of filament lifetime. This paper documents these issues for the deposition of μc-Si using W filaments, and explores whether a deposition regime can be found where device quality μc-Si can be deposited at high deposition rates in the (virtual) absence of silicide formation.

## EXPERIMENTAL DETAILS

The load locked HWCVD chamber used in this study has been described elsewhere [9]. Two 0.5 mm diameter filaments, located 3.2 cm from the substrate holder, are used to decompose the $SiH_4$ and $H_2$ gas mixtures. Filament currents investigated (through each filament) were 12 amps and 13 amps, which correspond respectively to filament temperatures of ~1750°C and 1825-1850°C. After each i-layer we expose the filament to hydrogen to remove (control) silicide formation; during this treatment we raise the filament temperature to ~500°C above the processing temperature and run 160 sccm of $H_2$ for ~7 minutes at 25 mTorr. The substrates used in this study are a Ag/ZnO textured SS back reflector kindly provided by United Solar, as well as 1737 Corning glass and c-Si substrates. Details about the HWCVD dopant and 'edge' layers used in the NREL fabricated devices, as well as the device structure used, are provided elsewhere [9]. In this study, 100Å thick highly crystalline seed layers were routinely used before i-layer deposition to facilitate i-layer crystallinity [10]. Due to the thinness of the dopant layers in relation to the > 0.7μm i-layer thickness, both infrared (IR) spectroscopy and x-ray diffraction (XRD) measurements could be performed on actual device structures, from which i-layer film properties could be extracted. Such properties include the film H bonding (stretch mode position) and post deposition oxygen pickup as well as the XRD microcrystalline volume fraction (μ-volume fraction), the latter obtained by integrating the areas of the (111), (220), and (311) c-Si peaks and comparing these to the first and second order amorphous peak areas. No Raman measurements were performed on these samples due to the lack of routine measurement capabilities available at NREL. To examine the feasibility of using p-layers alternate to those previously optimized for a-Si:H device structures, a series of two half finished devices were shipped overnight express to United Solar; the first was fabricated into a single junction device and analyzed using their buffer layer/μc-$p^+$-layer structure, and when appropriate, a tandem structure was fabricated using the second half finished device.

## RESULTS AND DISCUSSION

Figure 1 shows (a) the film $R_d$ and (b) the crystalline volume faction (μ-volume fraction) of eleven i-layers (deposited in a device structure) deposited consecutively using identical deposition conditions (hydrogen dilution 14/1, $T_{sub}$ 340°C, chamber pressure 140 mT), as a function of filament usage. The filament current used in this case was 12 amps, corresponding to a filament temperature of ~ 1750°C, and has resulted in device efficiencies (at an $R_d$ > 5Å/s) exceeding 6% [4], with every layer deposited by HWCVD. As previously noted, at this filament current, silicide formation occurs rapidly with silane exposure [7-8]. In Fig. 1(b) two different methods of determining film crystallinity are presented. While the first (μ-volume fraction) is an absolute measurement, it was performed only on selected samples. On the other hand, as film crystallinity (for a film showing a dominant (220) crystallite orientation) evolves with increasing film thickness [11], a measure of the (220) peak height normalized to a standard (0.9 μm) film thickness and film crystallinity [12] can also serve as a relative measure of crystallinity with increasing filament usage. For the (high μ-volume fraction) films deposited beyond a filament usage of two hours, where the film $R_d$ stays relatively constant and viable solar cells have been fabricated, the grain sizes estimated from the Scherrer equation ((220) peak), parallel to the growth direction, are ~ 300Å.

As can be seen, film reproducibility and filament lifetime immediately become significant issues. Regarding the former, both $R_d$ and film crystallinity change with filament usage. From previous experience, the best devices using 12 amps filament current are fabricated with an 'aged' filament, and correspond in Fig. 1 to a filament usage in the range of 3-4 hr. While the (low) $R_d$ observed here is constant for a short while, we have also previously observed a change in crystallite orientation (from a dominant (220) orientation to one approaching a random (c-Si powder pattern) orientation) at this filament usage [13], so even in this constant $R_d$ region film deposition is not reproducible. Beyond the data presented in Fig. 1, the film $R_d$ drops so precipitously that the filament either breaks has to be replaced.

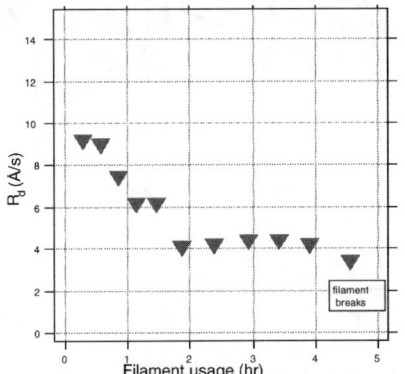

Fig. 1(a) Deposition rate versus filament lifetime for a filament operated at 12 amps.

Fig. 1(b) μ-volume fraction (solid symbols, left axis) and relative (220) Si-peak height (open symbols, right axis) versus filament lifetime for a filament operated at 12 amps.

Going to 13 amps filament current (corresponding to a filament temperature of ~ 1825-1850°C) has immediate benefits. Upon deposition of eight devices deposited consecutively, using the same $H_2$ dilution, $T_{sub}$ and chamber pressure conditions as for 12 amps (see above), i-layer film reproducibility becomes immediately evident. That is, the film $R_d$ is 17-18Å/s and does not drop, and from the first deposition the XRD μ-volume fraction is high (>75%) and also remains constant. In addition, filament lifetime also improves significantly, as more than 50 depositions have to date been performed using this filament, with minimal change in i-layer properties (film $R_d$, μ-volume fraction) when control depositions are repeated at selected filament usage intervals. However, device efficiencies utilizing these i-layers are very low (~0.3%). In particular, all solar cell parameters ($V_{oc}$ = 0.30 V, $J_{sc}$ = 2.1 mA/cm$^2$, FF = .31, yielding η ~ 0.2% for this cell) are representative of an i-layer quality that is not suitable for its incorporation into a state-of-the-art solar cell. An additional significant problem with these i-layers becomes evident from IR measurements. Fig. 2 shows an IR spectrum for such an i-layer, both measured immediately after growth, and then after a short exposure (1 day, 3-4 days) to atmospheric conditions. As can be seen, post deposition oxygen absorption, as evidenced by a broad peak occurring at ~ 1000 cm$^{-1}$, is clearly evident; such absorption, either during or after deposition, has previously been shown to be detrimental to device performance [14]. The sharp peak located at ~ 1100 cm$^{-1}$, seen in all samples, is due to a mismatch in oxygen contents between the reference c-Si substrate and the respective sample c-Si substrates, and is not due to the μc-Si layers. The grain sizes, once again estimated from the Scherrer equation ((220) peak), parallel to the growth direction, remain the same ((~ 300Å) as the filament current is increased.

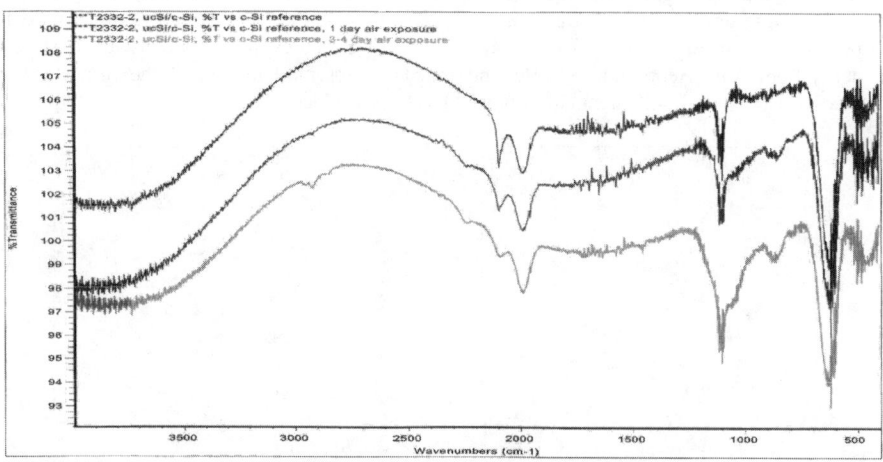

Fig. 2 FTIR spectra showing films experiencing post deposition oxidation with air exposure.

Following the results of Klein et al. [2], a significant improvement in device performance can be achieved at 13 amps by depositing i-layers at lower $T_{sub}$ and lower chamber pressures. While i-layer $R_d$ is sacrificed as the chamber pressure is reduced from 140 mT to 60 mT, significant oxygen pickup is no longer evident, as IR scans of devices exposed to atmospheric conditions over longer periods of time (> 3 months) now show no trace of the oxygen feature in the 1000 cm$^{-1}$ region. Furthermore (see Poster A6.9), the i-layer photoconductivity improves, from a value of 5-7 x 10$^{-7}$ S/cm$^2$ for $T_{sub}$ = 340°C and a chamber pressure of 140 mT, to 2-4 x 10$^{-5}$ S/cm$^2$ for $T_{sub}$ = 240°C and a chamber pressure of 60 mT. For the latter i-layer, $\sigma_L/\sigma_D$ > 100. Such conductivity values are now equivalent to those of Klein et al. [2], but for i-layers deposited at much different $R_d$ (1Å/s vs. ~8Å/s for the present films). The (220) grain size of the present samples has only decreased slightly, to ~ 230Å.

Devices performance utilizing these i-layers, which are deposited at an $R_d$ = 8Å/s, are shown in Fig. 3; both i-layer thicknesses are ~ 0.7 μm. Fig. 3(a) shows the efficiency of a device fabricated entirely at NREL; all layers are deposited by HWCVD, with the 'edge'/a-Si-p⁺-layer combination described previously. While the efficiency of the NREL device is only 4.1%, we comment that this i-layer is perhaps more suitable as the bottom layer in a tandem device structure than for a single junction device alone. In particular, the device $V_{oc}$ has previously been found to depend sensitively on the μ-volume fraction, and the value for the present device (~ 0.42 V) is consistent with that of an i-layer exhibiting a high μ-volume fraction [4]. Indeed, XRD measurements on this device show a μ-volume fraction > 75%.

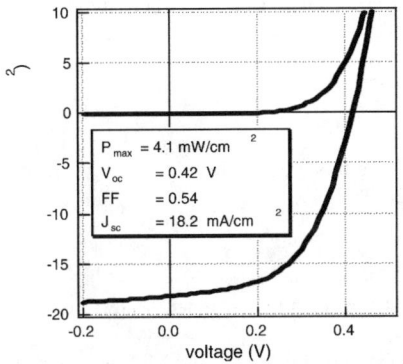 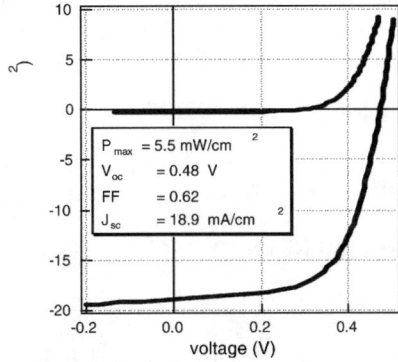

Fig. 3(a) JV characteristics for T2340 where all layers were grown by HWCVD at NREL.

Fig. 3(b) JV characteristics for T2358 where the n- and i-layers were grown at NREL and the buffer- and p-layer (and ITO) were grown at United Solar.

Using the same identical i-layer, an efficiency of 5.5% (Fig. 3(b)) has been achieved using a different top contact (the buffer layer/μc-Si-p⁺-layer structure optimized by United Solar). Improvements in performance can be attributed to enhancements in both device FF and device $V_{oc}$. It is difficult to determine which part of the United Solar top contact (buffer layer or μc-Si-p⁺-layer) is more beneficial to single junction device performance. However, using our 'n & k' apparatus as a sensitive measure of film crystallinity of the topmost layer in our device structure, we can say that, even though the NREL top contact 'recipe' has now been deposited onto a μc-Si i-layer instead of an a-Si:H i-layer, the top (p⁺) layer still seems to be amorphous. That is, the reflectance at ~ 270 nm remains broad and featureless, and is quite unlike when either a μc-Si i-layer by itself, or a μc-Si-p⁺-layer in a device, is measured. In the latter cases, a distinctive and relatively sharp reflectance feature appears at ~ 270 nm.

In addition, we also report tandem junction solar cell results, with United Solar depositing not only their buffer layer/μc-Si-p⁺-layer top contact on top of the NREL HWCVD μc-Si i-layer, but also an additional a-Si:H based top cell in a two tandem device structure. Preliminary results show an initial tandem efficiency of 9.6%. Individual solar cell parameters for this structure are $V_{oc}$ = 1.40 V, $J_{sc}$ = 9.8 mA/cm², and FF = 0.70. Further results will be reported at a later date.

## CONCLUSIONS

By depositing HWCVD µc-Si films consecutively at low filament temperature (~1750°C) using identical deposition conditions, a sharp drop in film $R_d$ as well as a change in film crystallinity are observed with filament usage. When using the same deposition parameters but a higher filament temperature (~1825-1850°C), film reproducibility issues now disappear, but these films pick up oxygen from ambient exposure, and the resultant solar cells are of very poor quality. However, by lowering both the substrate temperature and chamber pressure, device quality µc-Si is deposited at high deposition rates in a filament regime where tungsten silicide formation is minimal. Both single junction and tandem solar cells are fabricated using this material, confirming the validity of this approach.

## ACKNOWLEDGEMENTS

The authors thank J. Yang for his active interest in this work. This work is performed under DOE contracts DE-AC36-99-GO10337 and ZDJ-2-30630-19.

## REFERENCES

1. T. Roschek, T. Repmann, O. Kluth, J. Muller, B. Rech, and H. Wagner, Mater. Res. Soc. **715** (2002) A26.5.
2. S. Klein, F. Finger, R. Carius, B. Rech, L. Houben, M. Luysberg, and M. Stutzmann: Mater. Res. Soc. **715** (2002) A26.2.
3. R.E.I. Schropp, C.H.M. Van Der Werf, M.K. van Veen, P.A.T.T. van Veenendaal, R. Jimenez Zambrano, Z. Hartman, J. Loffler, and J.K. Rath: Mater. Res. Soc. **664** (2001) A15.6.
4. R.E.I. Schropp, Y. Xu, E. Iwaniczko, G.A. Zaharias, and A.H. Mahan: Mater. Res. Soc. **715**, (2002) A26.3.
5. See section 1 of the proceedings of the '1st International Conference on Cat-CVD (Hot-Wire CVD) Process', published in Thin Solid Films **395** (2001).
6. J. Doyle, R. Robertson, G.H. Lin, M.Z. He, and A.C. Gallagher, J. Appl. Phys. 64 (1988) 3215.
7. P.A.T.T. van Veenendaal, O.L.J. Gijzeman, J.K. Rath, and R.E.I. Schropp, Thin Solid Films **395** (2001) 194.
8. A.H. Mahan, A. Mason, B.P. Nelson, and A.C. Gallagher, Mater. Res. Soc. **609**, (2000) A6.6.
9. Q Wang, E. Iwaniczko, Y. Xu, B.P. Nelson, A.H. Mahan, R.S. Crandall, and H.M. Branz, Twenty-Eighth IEEE PV Specialists Conference (2000) 717-720.
10. G.A. Zaharias, A.H. Mahan, R.E.I. Schropp, Y. Xu, D.L. Williamson, M.M. Al-Jassim, M.J. Romero, and L.M. Gedvilas, Mater. Res. Soc. **715** (2002) A26.2.
11. J.K Rath, F.D. Tichelaar, H. Meiling, and R.E.I. Schropp, Mater. Res. Soc. **507** (1998) 879.
12. A. H. Mahan, private communication.
13. E. Iwaniczko, Y. Xu, R.E.I. Schropp, Q. Wang, and A.H. Mahan, NREL/CP-520-30476 (2002) 67.
14. J. Meier, P. Torres, R. Platz, S. Dubail, U. Kroll, J.A. Anna Selvan, N. Pellaton Vaucher, Ch. Hof, D. Fischer, H. Keppner, A. Shah, K.-D. Ufert, P. Giannoules, and J. Koehler, Mater. Res. Soc. **420** (1996) 3.

# ROOM TEMPERATURE RECOVERY OF LIGHT INDUCED CHANGES IN AMORPHOUS SILICON SOLAR CELLS

G. Ganguly, D.E. Carlson, M.S. Bennett, F. Willing, R.R. Arya and P. Stradins*
BP Solar, Toano, VA; National Renewable Energy Laboratory, Golden, CO

## ABSTRACT

We have observed the recovery of the performance of amorphous silicon (a-Si:H) based solar cell (especially the fill factor) at temperatures between 25°C and 170°C after ~600 hours of light soaking under 1 sun illumination at ~40°C. We find that there is some recovery of the fill factor of the cells even at temperatures below the temperature of light soaking. The recovery is significantly greater in cells of poor quality. There is also some evidence of enhancement of the recovery rate due to an external electric field. Above the light soaking temperature, a sort of thermally activated recovery of the fill factor is observed.

## INTRODUCTION

Light induced changes in amorphous silicon have been extensively studied since they were first observed in films and solar cells [1,2]. The changes are reversible through annealing at elevated temperatures ~90-200°C and the recovery is well documented as being thermally activated with a distribution of energies [3]. High light intensities, injection of carriers or electron irradiation also result is similar or more pronounced changes. The changes occurring under higher light intensities with the sample temperature increasing to 50-70°C have been observed to recover significantly at room temperature (25°C) [4] and the recovery is enhanced under electric fields [5]. Changes due to light soaking at temperatures significantly below room temperatures also result in some or complete recovery at room temperature [6]. Solar cell modules deployed in the field are exposed to intermittent periods of illumination and darkness as well as changes in temperature. Therefore, it becomes interesting to look for recovery effects at ambient temperatures and light soaking intensities that may be experienced under real operating conditions. Here we report on the observation of material quality dependent, field-enhanced but temperature-independent recovery of the fill factor of a-Si:H, a-SiGe:H as well as a-Si:H/a-SiGe:H solar cell devices at temperatures *below* the temperature of the devices during the prolonged 1 sun light soaking.

## EXPERIMENT

a-Si:H and a-SiGe:H single-junction and a-Si:H/a-SiGe:H tandem solar cells with the structure *glass/SnO$_2$/p/i/n/p/i/n /ZnO/Al* were fabricated in one of several double load-lock research reactors using DC plasma decomposition of silane and/or germane mixed with the appropriate dopant gases and/or diluted with hydrogen. The i-layer was deposited under a variety of deposition conditions. The plasma was generated over areas of $0.1m^2$ or $0.8m^2$ and the cells were fabricated using 8cm x 8cm pieces of commercial SnO$_2$ coated glass. The cells were light soaked using ~100mW/cm$^2$ illumination from sodium vapor lamps at a temperature of ~40°C. In some cases the effect of increasing the temperature to ~60°C was investigated and the degradation was reduced, but no significant differences in the recovery rate was observed. The samples were either left sitting under room light conditions or placed in an oven set to a fixed temperature between 35°C and 170°C. In order to estimate the effect of an electric field, cells were wired and a constant reverse bias of 4V was applied on one pair of 0.25cm$^2$ cells with a current of ~0.6mA being monitored. Another pair on the same superstrate was maintained under open circuit conditions. The temperature rise due to the reverse bias leakage current was ascertained to be negligible.

## RESULTS

The change of the average fill factor of six 0.25cm$^2$ a-Si:H/a-SiGe:H tandem solar cells has been plotted in Fig. 1 as function of logarithmic time after anneal at several temperatures. Similar data are shown for a-Si:H single junction solar cells in Fig. 2.

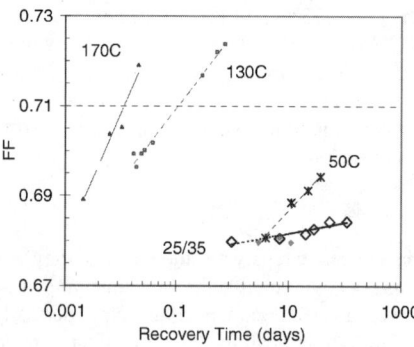

**Fig. 1**. The fill factor of a-Si/a-SiGe tandem solar cells as function of annealing time at the indicated temperature in air after ~600h light soaking.

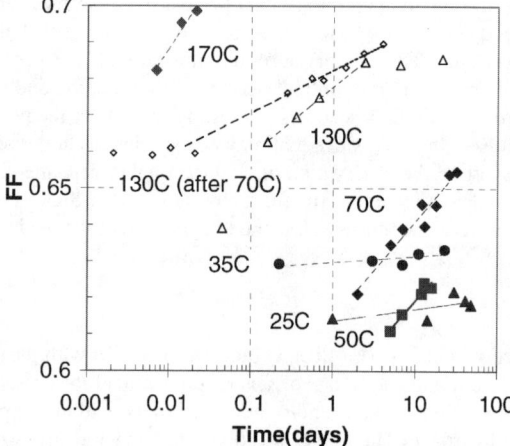

**Fig. 2.** The fill factor of a-Si solar cells as function of annealing time at the indicated temperature in air after ~600h light soaking. Also shown are two sets of data for annealing at 130°C after previous annealing at 70°C and 35°C.

The slopes of the lines of best fit are similar at 25 and 35°C. The slopes are similar but significantly steeper at temperatures between 50 and 170°C. It is significant that the light soaking temperature (40°C) separates these two groups.
In a single junction p-i-n device, the defect density is large enough after light soaking such that

the fill factor is limited not by the interfaces but by the defects in the bulk of the i-layer. We assume that the fill factor is determined by a set of defects with a distribution of thermally activated annealing energies (Fig. 3). The fill factor at annealing time, t, is assumed to be determined by the defects remaining. The demarcation energy, $E_d$ (t), between remaining and annealed states in this distribution, sweeps out the distribution as a function of time, t. If, for simplicity, we assume a constant density of defects with annealing energy (Fig. 3), the defects remaining after annealing time t, can be expressed as:

$$N_{rem} = N_{total} - A\,(E(t) - E_o). \tag{1}$$

The annealing rate is given as [3]

$$1/t = \nu_0 \exp(-E_d/kT), \tag{2}$$

so that,

$$N_{rem} = (N_{total} + AE_o) - A\,kT\,\ln(\nu_0 t). \tag{3}$$

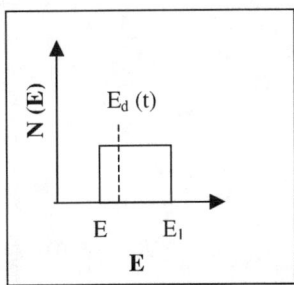

**Fig. 3** A schematic of a simplified, assumed energy distribution of defects (see text).

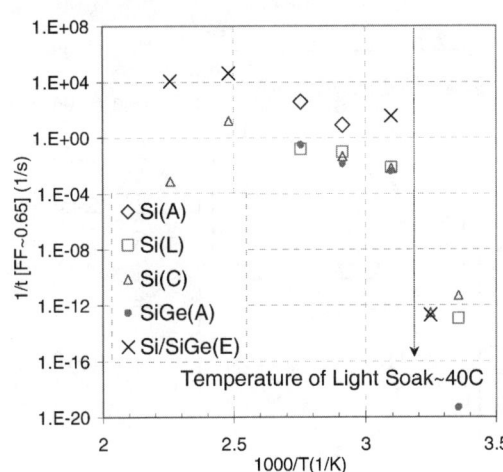

**Fig. 4** The reciprocal of the time required to anneal the fill factor to 0.65, derived from best fits including those shown in Figs. 1 and 2, plotted as function of the reciprocal temperature for solar cells with i-layers made using the indicated materials and on deposition systems designated by letters in parentheses.

This would explain the logarithmic time dependences observed for the fill factor in Figs. 1 and 2. In addition, Fig. 2 shows the effect of a pre-anneal at 70°C on the rate of anneal at 130°C. The slope is clearly shallower due to a previous, lower temperature anneal. This would imply that the annealing energy distribution is significantly modified. On the other hand, below the light soaking temperature, the effect of temperature on the annealing rate is not significant.

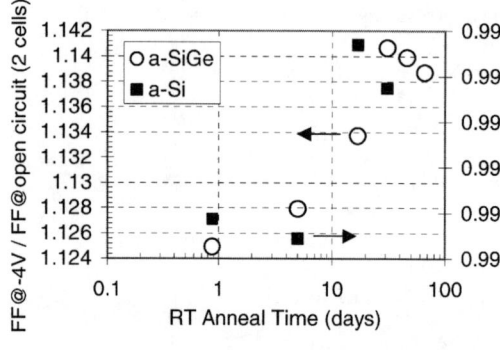

**FIG. 5** The ratio of the fill factor of a pair of diodes at –4V to that of a pair at open circuit for different durations at room temperature for solar cells fabricated with the indicated intrinsic layer materials.

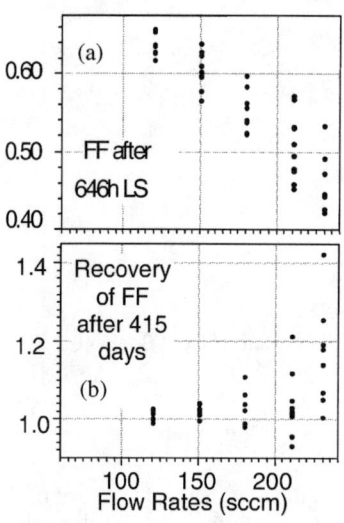

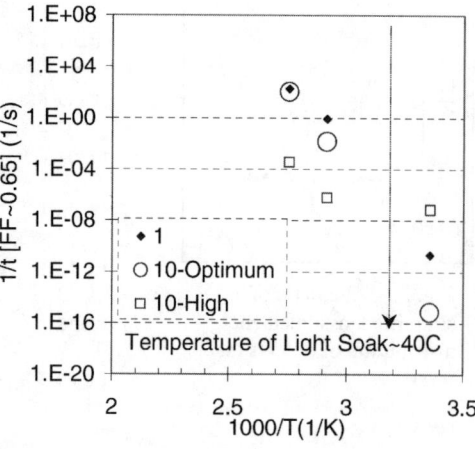

**Fig. 6** The (a) fill factor after light soak and (b) the ratio of FF after 415 days at RT to FF after light soak of a-Si:H solar cells deposited at ~10Å/s with the indicated silane flow rates.

**Fig. 7** The reciprocal of the time required to anneal the fill factor to 0.65 plotted as function of reciprocal temperature for a-Si:H solar cells with i-layers deposited at 1Å/s (1) and at ~10Å/s with a silane flow rate of 120 (10-optimum) and 230sccm (10-high).

This becomes obvious in Fig. 4 where the rate of annealing is plotted as function of the reciprocal temperature. Clearly, a variety of samples, drawn from several deposition systems, exhibit a sort of thermally activated annealing rate at temperatures above the light soaking temperature. The scatter in this plot derives from plots like those in Figs. 1 and 2 where each data point is an average of six diodes on a 8cm x 8cm superstrate containing 36 0.25cm$^2$ diodes. At temperatures below the light soak temperature, the rate of annealing drops significantly but does not depend systematically on the material.

We have also investigated the effect of applying an additional reverse bias on the annealing rate at room temperature. Figure 5 shows the effect of applying a reverse bias on the room temperature recovery of a-Si:H and a-SiGe:H single junction cells. Clearly, there is additional recovery due to the field out to about 20days and both materials exhibit similar behavior.

In order to investigate instances of enhanced rates of room temperature recovery of light induced degradation of the fill factor we compared a series of a-Si:H single junction cells made using growth rates of ~10Å/s. This series was made using optimum and excess flow of silane and its degradation behavior has been reported in detail elsewhere [7]. Briefly, the initial and stabilized fill factor and efficiency devices with i-layers fabricated at 10Å/s improve with increasing growth flow rate of a faxed ratio of silane and hydrogen [8]. Beyond an optimum flow rate, these quantities decrease with increasing flow rate. We have explained the improvement at lower flow rates as being due to the contribution of short lifetime radicals to growth due to depletion of silane [8]. Short lifetime radicals have high reactivity in the gas phase as well as on the growth surface. Therefore, they do not diffuse to find surface growth sites that result in the lower energy, ordered structures leading to fewer voids, smoother surfaces and reduced light induced degradation. When excess silane is available at increasing silane flow rates above the optimum flow rate, multiple silicon containing radicals are formed and contribute to growth. These heavier radicals do not diffuse as well as the lighter SiH$_3$ radicals and when they do bond, it is unlikely that all the silicon atoms will find suitable bonding locations. The situation becomes worse as the number of silicon atoms in the radical increases. Therefore, voids are formed and dihydride type bonding is observed along with increased light induced degradation.

In Fig. 6, we have plotted the fill factor after over 600h of light soaking and also the ratio of the fill factor after 415 days of sitting in a laboratory closet (at RT) to that after light soaking. We see a clear increase in the recovery when the i-layers are deposited with excess silane flow rates. The decrease in fill factor at excess flow rates has been attributed to formation of multiple silicon containing radicals contributing to growth [7]. The additional room temperature recovery observed for these samples suggests that the annealing energy distribution is significantly modified, especially in the low energy tail region. In Fig. 7 we have plotted the reciprocal of the time to anneal the cells to a fill factor of 0.65 for a-Si:H cells with i-layers deposited at 1Å/s and those deposited at 10Å/s with silane flow rates of 120 (optimum) and 230sccm (high). While the 1Å/s and 10Å/s sample deposited at optimum flow exhibit significant change in annealing rate below and above the light soaking temperature, the 10Å/s sample deposited using high silane flow rate exhibits a much smaller difference in rates. Clearly, when the material properties are significantly affected, the room temperature annealing of light induced defects at temperatures below the temperature of light soaking using ~1 sun illumination intensities becomes significant. Similar effects have been observed in the presence of high impurity concentrations and will be reported elsewhere.

## CONCLUSIONS

We have observed the annealing behavior of a-Si:H based solar cells at temperatures below and above the temperature during light soaking at intensities ~1sun. There is a small but significantly reduced rate of annealing below the light soaking temperature. The annealing rate is similar in a-Si:H and a-SiGe:H. The results can be understood in terms of the movement of the demarcation energy through a distribution of defects. In poor quality material, deposited for example at ~10Å/s under conditions of excess silane flow, the rate of annealing at temperatures below the light soaking temperature is enhanced significantly.

## ACKNOWLEDGMENTS

This work was partially funded by the DOE through NREL Subcontract ZDJ-2-30630-10. We would like to thank G. Wood, S. Stone, M. Dicolli, G. Mckinley, F. Jackson and R. Murphy for technical assistance with various parts of this work.

## REFERENCES

1. D.L. Staebler and C.R. Wronski, Appl. Phys. Lett. 31, 292 (1977)
2. D.L. Staebler, R.S. Crandall, and R. Williams, Appl. Phys. Lett. 39, 733 (1981)
3. M. Stutzmann, W. B. Jackson, and C.C. Tsai, Phys. Rev. B32, 23 (1985)
4. D. Fischer, N. Pellaton, H. Keppner, A. Shah and C.M. Fortmann, Mater. Res. Soc. Symp. Proc. Vol. 258, 893 (1992)
5. D.E. Carlson, and K. Rajan, Appl. Phys. Lett. 70, 2168 (1997)
6. P. Stradins and H. Fritzsche, Mater. Res. Soc. Proc. Vol. 297, 571 (1993)
7. G. Ganguly, R.S. Oswald and D.E. Carlson, NREL Final Technical Report, 2002 (unpublished)
8. G. Ganguly, D.E. Carlson and R.R. Arya, J. Non-Cryst. Solids. 299-302, 1123 (2002).

# Toward understanding the degradation without light soaking in hot-wire a-Si:H thin films and solar cells

Qi Wang[1], Keda Wang, and Daxing Han,
[1]National Center of Photovoltaic, National Renewable Energy Laboratory,
Golden, CO 80401-3305, U.S.A.
Department of Physics & Astronomy, Univ. of North Carolina at Chapel Hill,
Chapel Hill, NC 27599-3255, U.S.A.

## ABSTRACT

The non-reversible degradation without light soaking has been observed in both thin films and solar cells fabricated using the hot-wire CVD technique. For solar cells, a 9.8% initial efficiency became 9.2% when measured after a few weeks of being stored in the dark. For the intrinsic layers, the conductivity increased in the air but remained steady in vacuum up to 700 hours. Also, the conductivity increased and activation energy decreases from an initial ~0.95 eV to ~0.85 eV after several thermal cycles, even in a vacuum. We suggest that the degradation in both materials and solar cells come from the same origin: the upward shift of the Fermi-energy position from its initial value. In other words, the as-grown material is slightly p-type and gradually became slightly n-type after sitting in the air or after thermal annealing cycles in a vacuum. The shift in the Fermi-energy made the fill factor of the solar cell decrease and changes in the electronic properties of the i-layer. It is likely that adsorption of oxygen from water vapor is driving this degradation.

## INTRODUCTION

The advantages of the hot wire chemical vapor deposition (HWCVD) are the simplicity in gas decomposition, the absence of ion bombardment, and the high growth rate compared with the standard plasma enhanced (PE) CVD. High performance hydrogenated amorphous silicon (a-Si:H) based solar cells and thin-film transistors have been made using HWCVD [1, 2]. In addition, several new physical phenomena were also observed in the HW a-Si:H thin films [3-6]. For instance, the NMR studies showed a new hydrogen distribution in HW films compared to PECVD films [3], the internal fraction studies showed HW films have a much lower friction value than glass and are very close to the value of crystalline silicon [4]. In the high growth-rate films, the findings of the confinement effect on spin-spin interactions in nanogas systems [5] and nanovoid-related large red shift of photoluminescence peak energy have been reported [6].

The degradations in HW solar cells and films were different than in the PECVD ones. The features of light soaking effects in HW a-Si:H films were distinguished from those in conventional PECVD films. The photoconductivity (PC) decreased only slightly and the dark conductivity increased in the films deposited at a substrate temperature $\geq 320°C$ [7-9]. The increase of the density of the metastable dangling bonds ($\Delta D^o$) was found to be the same as that in PECVD films. On the other hand, it has been reported that in addition to the light-induced degradation the HW solar cells showed a few percent decrease in efficiency without exposure to the light [10]. Most degradation in fill factor has shown in the air but not in the vacuum or in liquid nitrogen environment. Those results suggest a possible post-oxidation in the devices. The degradation without light soaking is a more serious problem in the HWCVD samples than that in the PECVD samples. In this work, we studied in detail the variations in electronic properties of intrinsic HW films in terms of gas adsorption and thermal annealing in order to understand the phenomena of degradation in the dark.

## EXPERIMENTAL DETAILS

Intrinsic HW a-Si:H films with low initial dark conductivity and large $E_a \geq 0.85$ eV were selected as listed in Table I. One standard PECVD film, A7119, was used for comparison. Group 2 was a pair of films with and without a thin $SiN_x$ passivition layer for post-oxidation study. Group 3 was one sample that was cut into three pieces for gas adsorption and thermal annealing studies. Corning-1737-glass substrate was used for all the samples. The coplanar metal contacts were first deposited on the glass. The details of the HWCVD system are published elsewhere [1].

Table 1 Sample information

| Group | Sample ID | $T_s$ (°C) | Growth Rate (Å/s) |
|---|---|---|---|
| 1 | T773 | 360 | 10 |
|   | T779 | 360 | 10 |
|   | T837 | 400 | 12 |
| 2 | H1441 | 360 | 14 |
|   | H1442 | 360 | 14 |
| 3 | H1540a | 320 | 12 |
|   | H1540b | 320 | 12 |
|   | H1540c | 320 | 12 |
|   | A7119[a] | 200 | 1 |

[a] PECVD sample

The measurements were carried out in the following experiments: (a) Dark conductivity measurement at 300 K in environments of: vacuum ($10^{-4}$ Torr), dry-oxygen gas, and air. Thermal annealing cycles were followed after storage in the above environments. (b) Conductivity activation-energy measurement with sample temperature ranging from 300 to 420 K. Before taken data, the sample was heated to 420 K for 30 min to clean up the film surface adsorption. Then, the $E_a$ was deduced from the equation $\sigma_d = \sigma_o \exp(-E_a/kT)$. (c) Hot-probe measurement is a simple way to distinguish between n- and p-type semiconductors using a hot-probe such as a heated soldering iron, a cold-probe and a standard multimeter. Based on the thermal-electric effect of semiconductors, when applying the probes to n- or p-type materials one obtains a current in opposite directions. Finally, PC temperature dependence was studied for two typical HW and one PECVD film.

## RESULT AND DISCUSSION

For group 1 HW samples, the PC decreased only slightly and the dark conductivity increased upon light soaking [7-9]. We used hot probe to exam the conductive type of the intrinsic samples. We found that high $T_s$ HW samples appeared slightly p-type initially and changed to n-type after defect creation. Figure 1 shows the results of the hot-probe measurement for sample T837. One can see that the sample at the initial state was slightly p-type, and then changed to slightly n-type after the metastable defect creation. In corresponding to its $E_a$ value, the Fermi level moved up from 0.92 eV to 0.82 eV below the conduction band edge.

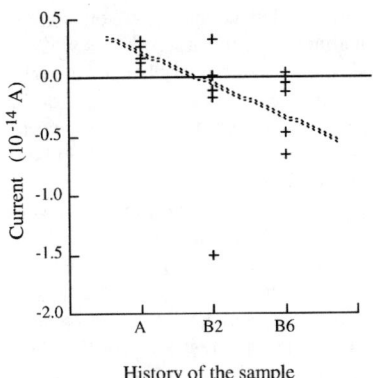

Figure 1. The sign of current for sample T837 measured by hot probe. The current was positive at the initial state where $E_a$ was as large as 0.92 eV. After the metastable defect creation at State B2 (light soaked using 200 mW/cm$^2$ white light for 2 hrs) and B6 (light soaked using 200 mW/cm$^2$ white light for 6 hrs), the current became negative, and the $E_a$ decreased to 0.82 eV.

This change is in an opposite direction from standard PECVD film (A7119) in that the $E_a$ increased from 0.73 eV to 0.77 eV after the metastable defect creation. The changes in the activation energy and the PC were partially recovered after annealing at 160 °C for an hour. The activation energy recovered to ~0.86 eV instead of 0.92 eV. So, there is some non-reversible degradation. In order to exam the effect of post-oxidation on the position of the Fermi level, we studied a pair of film with and without the SiN$_x$ passivition layer. We assumed that a thin layer of SiN$_x$ would prevent adsorption from air. The dark current as a function of temperature and annealing time are shown in Fig. 2.

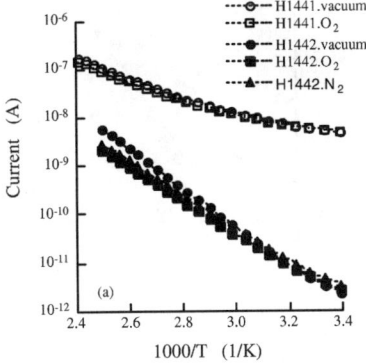
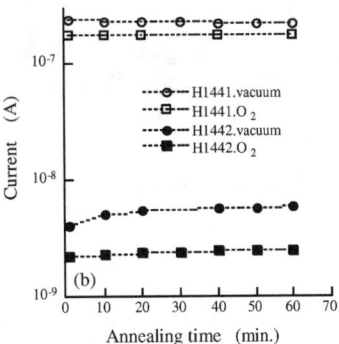

Figure 2. (a) Current-temperature dependence and (b) current as a function of annealing time at 420 K, for samples of H1441 and H1442 with- and without SiN$_x$ passivition layer. The samples were mounted in the vacuum and in pure oxygen or nitrogen, respectively.

Figure 2 shows the results of (a) conductivity temperature dependence in various gas environments, and (b) the conductivity as a function of annealing time at 420 K. As we expected, the H1442 without a $SiN_x$ passivation layer showed conductivity changes in both measurements in regard to the gases of $O_2$ and $N_2$; but, there were very little changes for the sample H1441 with $SiN_x$ layer. The 10 times difference in the value of the current between H1441 and H1442 can be attributed to the interface states between $SiN_x$ and a-Si:H. This interface effect dominates the conductivity measurement on H1441.

For solar cells, we have reported a 9.8% initial efficiency became 9.2% after a few weeks of being stored in the dark. The degradation mainly came from the fill factor (FF). In order to check whether the degradation took place in the intrinsic layer, we cut sample H1540 into three pieces of H1540a, H1540b and H1541c and studied the current changes in the vacuum (×), pure oxygen (O) and the air (Δ), simultaneously. Figures 3(a) and 3(b) show the changes of the dark current at 300 K and the corresponding activation energy, respectively. The crosses in Fig. 3(a) show that the conductivity in the vacuum of $10^{-4}$ torr did not change much up to a month. However, the conductivity increased about an order of magnitude after following thermal cycles heated up to 420 K. The change in conductivity corresponded to $E_a$ changes from 0.81 to 0.78 eV as shown in Fig. 3(b). The open circles show interesting changes of the conductivity in pure $O_2$: the initial dark current has the same value as in the vacuum but increased and then decreased. Whereas, the current value at 300 K increased about an order of magnitude after temperature increased to 420 K; again the changes in conductivity corresponded to $E_a$ changes as shown in Fig. 3(b). Finally, the open triangles show the conductivity measurement in the air: the dark current increased about 60 times more than that in the vacuum within one day and continuously increased in the air for 54 days. The current decreased an order of magnitude after $E_a$ measurements in the air. We

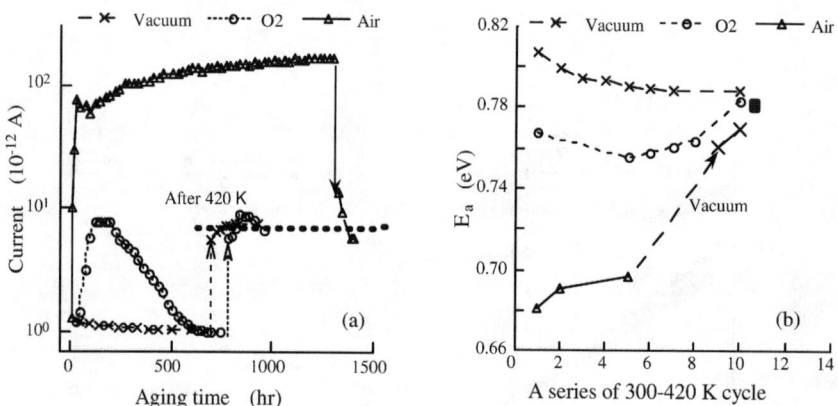

Figure 3 Degradation experiments: (a) the dark current at 300 K for samples of H1540a in vacuum (×), H1540b in pure $O_2$ (O), and H1540c in the air (Δ). The data following the arrows indicate the current after $E_a$ measurements. (b) Several measurements of $E_a$ obtained from those three samples after the current measurements at 300 K for about a month. The data following the arrow indicate the measurements were done in vacuum for H1540c.

believe that this is due to de-adsorption of humidity at elevated temperatures. As shown in Figs 3(a) and 3(b), the final values of both the current at 300 K and their activation energy $E_a$ tended to reach the same value. The above results suggest that $H_2O$ from the air has a stronger effect than the pure $O_2$. The source of the degradation in the dark can be related to trace oxygen incorporation at 300 K in the air or at 420 K in a $10^{-4}$ torr vacuum (we can not exclude the fact that the structural relaxation may also play a role) that results in an upward movement of the Fermi-level position. Consequently, the hole-mobility-lifetime product decreases and then the FF decreases.

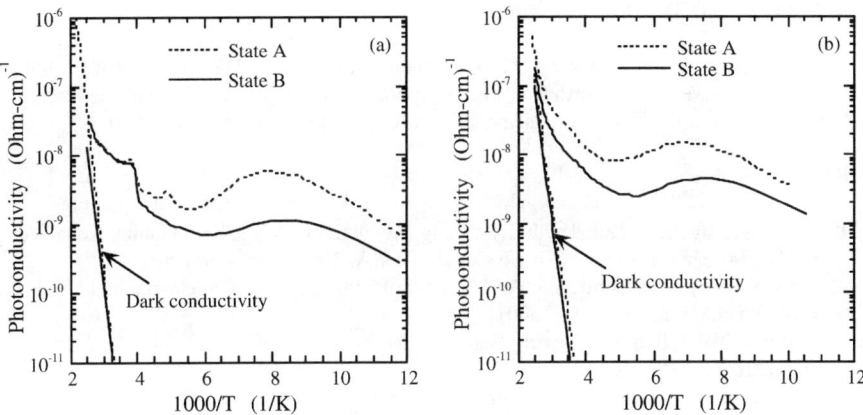

Figure 4 PC temperature dependence for (a) HW film T779 and (b) PE-CVD film A7119.

All of the above results showed that the changes in activation energy, in other words, the Fermi-level position were crucial to understanding the degradation in the dark. Since the gap states were crucial for the Fermi-level position, we compared the gap states between HW and PECVD films in Figure 4 using PC temperature dependence [12]. To show the opposite movement of the Fermi-level positions in these two films, the dark conductivity curves are also plotted in Figs 4(a) and 4(b). We found that there are two extra PC enhancement peaks at ~200 K and 250 K for HW film, and the ~200 K peak disappeared after light soaking. No matter what those states are, the disappearance of the deep traps is consistent with the upward movement of the Fermi level. One may also notice that the PC value does not decreases at ~ room temperature at State B for the HW film but it does for the PECVD film.

## CONCLUSIONS

For HW a-Si:H samples, we have observed the decreases in the FF of the solar cells and the dramatic changes in the dark conductivity for the i-layer regarding the various gases and thermal cycles. Three effects could be involved in the changes that are: surface adsorption, oxygen incorporation and thermal relaxation. We observed that the surface adsorption of $H_2O$ in the air resulted in a great rising of the dark conductivity as shown in Fig. 3(a). The value of activation energy was slightly decreased after temperature cycles in the vacuum.

When the activation energy decreased from ~0.9 eV to ~0.8 eV, the material changed from slightly p-type to slightly n-type, as shown in Fig. 1. This resulted in a decrease of the mobility-lifetime product of holes, and then a decrease of FF. We suggest that the degradation of both materials and solar cells may share the same origin. With prolonged storage time in the air at 300 K or being heated to 420 K in a $10^{-4}$ torr vacuum, oxygen incorporation resulted in the Fermi-level position moving upward. Essentially, the differences in both the microstructures [3,5,9] and the gap states (see Fig. 4) in HW- from PE-CVD samples could be in response to the degradation in the dark.

## ACKNOWLEDGMENTS

This work is supported by the U.S. DOE under Contract No. DE-AC36-99GO10337 and the work at University of North Carolina is supported by the National Renewable Energy Laboratory under the Thin Film Partnership, subcontract number ADJ-1-30630-09.

## REFERENCES

1. Qi Wang, Eugene Iwaniczko, Jeffrey Yang, Kenneth Lord, Subhendu Guha, Keda Wang, and Daxing Han, 29th PVSC IEEE Orlando, USA, 2002, to be published.
2. R.E.I. Schropp, B. Stannowski, A.M. Brockhoff, P.A.T.T.van Veenendaal and J.K. Rath, Mater.Phys.Mech.1, 73-82, (2000).
3. Y. Wu, J. Todd. Stephen, Daxing Han, J.M. Rutland, R. A. Crandall, and H. Mahan, Phys. Rev. Lett. **77** 2049-2052 (1996).
4. Jonathan Baugh, Alfred Kleinhammes, Daxing Han, Qi Wang & Yue Wu, Science, **294**, 1505-1507 (2001).
5. Liu, Xiao, B. E. White, Jr., R. O. Pohl, E. Iwaniczko, K. M. Jones, A. H. Mahan, B. P. Nelson, R. S. Crandall and S. Veprek, Phys. Rev. Lett. **78**, 4418 (1997).
6. Daxing Han, Guozhen Yue, Keda Wang, Jonathan Baugh, Yue Wu, Yueqin Xu, and Qi Wang, Appl.Phys. Lett. **80**, 40-42 (2002).
7. Daxing Han, Jonathan Baugh, Guozhen Yue, and Qi Wang, Phys. Rev. B **62**, 7169-7178 (2000).
8. Daxing Han, Guozhen Yue, H. Habuchi, Eugene Iwaniczko, Qi Wang, Thin Solid Films, **395**, 134-137 (Sept. 2001).
9. Daxing Han, Guozhen Yue, Qi Wang, and Tatsuo Shimizu, Thin Solid Films, **430**, 141-144 (2003).
10. P. Stradins et al, abstract in the 2$^{nd}$ International conference on CAT-CVD processing, (2002, CO) to be published in Thin Solid Films, (2003).
11. M. Hack and M. Shur, J. Appl. Phys. 58, 997 (1985).
12. Daxing Han and H. Fritzsche, J. Non-Cryst. Solids **59/60**, 397 (1983).

## Material Aspects of Reactively MF-Sputtered Zinc Oxide for TCO Application in Silicon Thin Film Solar Cells

Jürgen Hüpkes, Bernd Rech, Oliver Kluth, Joachim Müller, Hilde Siekmann, Chitra Agashe, Hans P. Bochem[1] and Matthias Wuttig[2]
Institute of Photovoltaics (IPV), Forschungszentrum Jülich GmbH, D-52425 Jülich,
[1]Institute of Thin Films and Interfaces (ISG), Forschungszentrum Jülich GmbH, D-52425 Jülich,
[2]Institute for Physics of New Materials - Department of Physics, RWTH, D-52056 Aachen

## ABSTRACT

Al-doped ZnO films were deposited on glass in an in-line system by reactive mid-frequency (MF) magnetron sputtering. The influence of substrate position on the film properties as well as the relation between static and dynamic deposition are studied. All films showed low resistivity (<4x10$^{-4}$ Ωcm) and excellent transparency (> 80 % in the visible region). The resistivity ρ for substrate positions above the sputter craters (race tracks) is up to a factor of two higher than on other positions where the smallest ρ is 1.9 x 10$^{-4}$ Ωcm. Major differences in statically deposited films as a function of the position on the substrate are found for the structural film properties as characterized by x-ray diffraction (XRD) and etching behaviour. The different surface textures obtained after etching are directly related to variations in the short-circuit current densities of amorphous silicon p-i-n solar cells prepared on these etched ZnO:Al films.

## INTRODUCTION

Silicon thin film solar cells in the p-i-n (superstrate) structure require a transparent conductive oxide (TCO) film which has to combine low series resistance, high transparency and an adequate surface texture [1]. Magnetron sputtered and texture-etched ZnO:Al films fulfill these requirements [2]. For industrial applications high rate large area sputtering techniques and in-line processing (dynamic mode) are required. In small area lab-type systems both substrates and targets are usually at a fixed position (static mode), whereas in industrial in-line processes moving substrates are passing fixed line targets, Therefore, the results obtained on small areas have to be transferred to these techniques. Szyszka et al. [3] have developed high quality ZnO:Al with high deposition rates by the reactive mid-frequency (MF) sputtering process. The success of this development led to the initiation of a joint R&D project with the objective of developing large area (0.6 m$^2$) textured ZnO:Al coated glass substrates for solar cell applications [4]. Recently, ZnO:Al films reactively sputtered in static and dynamic mode have been investigated with focus on the electrical and structural properties of the as-deposited films [5].

This paper addresses the material properties of reactively MF-sputtered ZnO:Al films for the application in thin film silicon solar cells. We investigate the relationship between growth process in dynamic and static mode, material properties and finally the surface texture obtained after etching. The experiments clearly reveal the decisive influence of the magnetron configuration on film structure and resulting surface texture after etching.

## EXPERIMENTAL DETAILS

ZnO:Al films were prepared on Corning glass (1737) in a vertical in-line sputtering system supplied by Von Ardenne Anlagentechnik GmbH (Dresden). Dual magnetron cathodes with a length of 750 mm were used. The ratio between oxygen and argon gas flow during the reactive sputtering process from metallic Zn/Al targets was controlled by plasma emission monitoring (PEM) [6]. For all films presented in this study the discharge power and pulse frequency were

4kW and 40kHz, respectively, leading to dynamic deposition rates as high as 44nm·m/min. In dynamic mode the substrate oscillates with a speed of 1m/min. For static deposition the holder is moved within about 10 seconds from the pre-sputtering position to the target position. Then it is fixed during deposition.

The electrical film properties were investigated by four point probe and room temperature Hall measurements. In case of static deposition (static print) the films were separated into stripes of 5mm width and 12.5mm length perpendicular to the direction of substrate motion for resistance measurements with high lateral resolution (2-point probe). The film structure was characterized by X-ray diffraction (XRD). From θ/2θ-scans the position and peak width of the (002)-peak were derived as a measure for stress and grain size, respectively. The films were etched in diluted hydrochloric acid (0.5% HCl) for 25 seconds and the resulting surface morphology was characterized by SEM. Finally, we applied the etched films as substrates for a-Si:H p-i-n solar cells. Solar cell characterization was performed with a Wacom sun-simulator under standard test conditions (AM1.5, 100 mW/cm$^2$, 25 °C).

## RESULTS

In this section, first the result of the as deposited, smooth films are presented and the properties obtained in dynamic and static mode are compared. After that the influence of the deposition mode and the position on the substrate on the etching behavior is described. Finally, we present the solar cell results obtained on these films and discuss the results.

### Smooth films

In the first step, the sputtering process was optimized in dynamic mode. All films show good electrical properties ($\rho<4.1\times10^{-4}$ Ωcm) and high optical transparency for visible light of more than 80 %. Fig. 1 shows the variation of specific resistance calculated from 2-point-probe measurements as a function of the position on the substrate. The positions on the substrate (in cm) are given with respect to the cathode system starting from the left target: in front of the left side of target (A), left racetrack (B), center of target between racetracks (C), right racetrack (D) and center of the whole cathode system between the two targets (E). While the specific resistance of the dynamic film (indicated by the dotted horizontal line) is about $3\times10^{-4}$Ωcm independent of the position on the substrate, for the static deposition pronounced maxima are found at positions B and D opposite to the racetracks. At positions A, C and E very low ρ-values around $2.3\times10^{-4}$Ωcm are found, which are comparable to the lowest resistivities reported in literature for sputtered ZnO:Al-films [7]. The electrical data are confirmed by Hall measurements for the maximum and minimum positions (triangles in Fig. 1, see also Table I). Table I also lists the film thickness, deposition rate and etch rate of five characteristic samples.

**Table I.** Electrical properties and etchrate of some ZnO:Al-films (ρ = specific resistance, n = carrier concentration, μ = carrier mobility). For film F the dynamic deposition rate is given.

| sample name | Position on the substrate (cm) | ρ ($10^{-4}$Ωcm) | N ($10^{20}$cm$^{-3}$) | μ (cm$^2$/Vs) | etchrate (nm/s) | thickness (nm) | dep. rate (nm/min) |
|---|---|---|---|---|---|---|---|
| A | 1.5 (left side of target) | 2.1 | 12.4 | 24.9 | 5.2 | 575 | 96 |
| B | 5 (left racetr.) | 3.8 | 7.7 | 21.2 | 3.0 | 840 | 140 |
| C | 8 (between racetr.) | 2.0 | 9.8 | 31.5 | 5.4 | 940 | 157 |
| D | 11 (right racetr.) | 4.1 | 6.8 | 22.5 | 2.8 | 1025 | 171 |
| E | 17 (between targets) | 1.9 | 11.1 | 29.9 | 6.6 | 849 | 142 |
| F | dynamic | 3.0 | 8.0 | 26.5 | 4.8 | 660 | 44 nm·m/min |

Films B and D, taken from positions above the racetrack, show reduced mobilities and at the same time lower carrier densities as compared to the films A, C and E. This indicates that both, film structure and film composition obtained opposite to and beside the racetrack area are different.

Detailed information about the structural properties are derived from XRD-analysis. XRD-measurements show that all films are well orientated with the c-axis perpendicular to the substrate surface. Position and peak width of the (002)-peak have been evaluated (Fig. 2). The angular position of this peak varies with the location of the measured spot on the substrate. Directly opposite to the racetrack the diffraction peak is shifted to lower angles by about 0.1° with respect to other locations. This shift probably results from compressive stress [8,9], caused by the impact of energetic oxygen [10] on the growing film. For the determination of absolute stress values it is necessary to know the relaxed lattice constant, which might be slightly different from their respective literature value of 34.4° [11], *e.g.* due to the Al-doping. The peak width is approximately 0.2°, which is the resolution limit of the diffractometer for the selected parameters. Hence, the corresponding grain size of 50 nm is a lower limit for the grain size perpendicular to the substrate surface for every position.

## Etched films

The etching behavior is directly correlated to the film structure. This is obvious from Fig. 3. Characteristic minima in the etching rate $R_E$ are obtained at substrate positions facing the racetracks during deposition, while $R_E$ at positions above the center of the targets and between the targets is about a factor of two higher. The asymmetry between the right and left side of the substrate can be explained by the fact that one side of the substrate has to move along the target (with plasma already on) before the final position for the static print is reached (see section 3). For dynamic deposition the etchrate is within the variations of the static print.

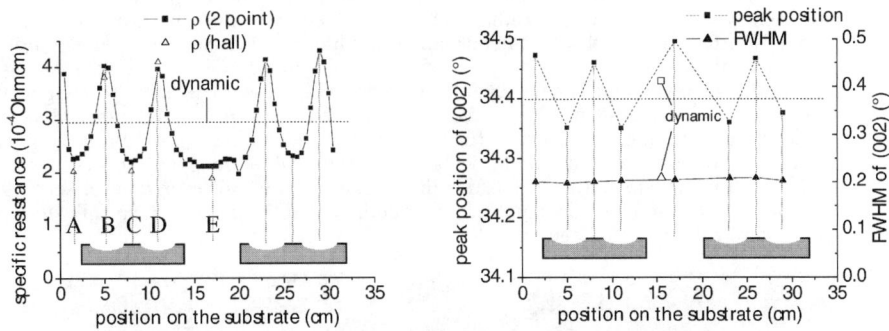

**Figure 1 (left).** Specific resistance as function of the position on the substrate (static print and dynamic mode). A, B, C, D and E represent five characteristic positions. Squares: 2-point-probe; triangles: Hall measurements; dotted horizontal line: Hall data of dynamic film.

**Figure 2 (right).** Peak position and FWHM of (002)-peak at different positions on the substrate for static and dynamic deposition. The reference value of the (002)-peak position for a zinc oxide powder sample is 34.4° [11], included as dotted horizontal line.

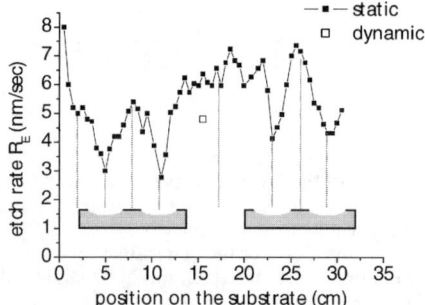

**Figure 3.** Etch rate of a static print as a function of the position on the substrate. For comparison, the etch rate of the dynamic film is added.

Three samples from these characteristic positions (B, C, E) were characterized by SEM. Figure 4 shows SEM micrographs of these etched surfaces compared to the surface of the dynamically deposited film F. The films grown near the target (B and C) develop a similar surface structure: Only a few large flat craters are randomly distributed over the area. Film E shows a more regular but sharp and deep surface structure. The dynamically deposited film shows also a crater structure. However, compared to B and C the surface roughness is higher and there are more smaller craters between the larger ones. It appears that the surface structure is somehow a mix of the structures that were observed on statically deposited etched films.

## Solar cells

A-Si:H p-i-n solar cells were prepared on these differently textured ZnO:Al-films. The results are compared with an rf-sputtered ZnO:Al film serving as a reference (see [12] for more details). Table II shows the photovoltaic parameters of these solar cells. The cell performance strongly depends on the related substrate. Although smaller differences in FF are also observed, the main difference between the samples is the short-circuit current density $J_{SC}$. In general, a rough surface structure of the substrate is essential, both to minimize reflection losses at the front side and to enhance the optical path length in the solar cell by scattering and subsequent trapping of the incident light (see e.g. [12]). Both effects enhance $J_{SC}$. On the other hand, $V_{OC}$, which is also influenced by the diode dark current, could be reduced by sharp surface structures due to an increased TCO/Si interface area leading to higher dark current values of the pin structure.

These effects can be observed in our samples. Films B and C show very similar performance what could be expected from the very similar appearance in the SEM micrographs. $V_{OC}$ is high, but $J_{SC}$ is quite low. Film E, on the other hand, yields a high $J_{SC}$-value due to good light scattering, but at the same time $V_{OC}$ is reduced due to the sharp surface structure and increased surface area of the film. The dynamic film F shows both, quite high $J_{SC}$ and $V_{OC}$. The good properties of different positions of the statically deposited films, good light scattering and also favorable structure, are partly combined in the dynamic film F. However, comparison with an low rate rf-sputtered reference sample shows that there is still quite a large potential to further optimize the mf-sputtered large area films.

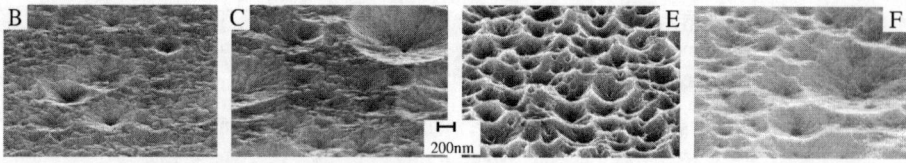

**Figure 4.** SEM-micrographs of ZnO:Al film surface after etching (magnification: 100k). Film F was prepared in dynamic mode; A, C, and E were taken from the static print. The positions correspond (from the left) to racetrack, center of one target and between the two targets (see also Fig. 1).

**Table II.** Illuminated I-V-parameters (efficiency η, fill factor FF, open circuit voltage $V_{OC}$ and short circuit current density $J_{SC}$) of a-Si:H pin solar cells deposited on different films (see also Table I and Fig. 1). An rf-sputtered film is added for comparison and as a reference.

| sample | η (%) | FF (%) | $V_{OC}$ (mV) | $J_{SC}$ (mA/cm$^2$) |
|---|---|---|---|---|
| B | 8.2 | 72.0 | 925 | 12.3 |
| C | 8.3 | 72.2 | 926 | 12.4 |
| E | 8.9 | 70.9 | 891 | 14.1 |
| F (dynamic) | 8.7 | 67.6 | 928 | 13.9 |
| reference | 9.9 | 68.4 | 923 | 15.6 |

## DISCUSSION

All films sputtered in dynamic and static mode in this work exhibit a columnar and dense film structure, which is typical for zone 2 of the Thornton model for sputtered metals [13]. The respective film structure and the etching behaviour correspond also very well to zone 2 of a modified Thornton model. The latter has been developed to describe the relationship between the growth and the resulting etching behaviour of RF sputtered ZnO:Al films on glass substrates as a function of substrate temperature and sputter pressure [14]. In the static prints, the variation in lateral position on the substrate leads to similar effects on etching behaviour like the variation of deposition pressure for the rf-sputtered films reported in [14].

The characterization of the static print clearly reveals the influence of the target configuration on the electrical and structural properties of the presented MF-sputtered films. The positions above the racetracks yield lower carrier mobility, lower carrier density and different etching behaviour with respect to etching rate and resulting surface texture. Moreover, the XRD measurements reveal that compressive stress is introduced into the film at positions above the racetrack. A similar peak-shift was observed by Sato et al. [10] for dc-sputtered ZnO:Al films. In literature the influence of target and cathode configuration on the film properties is widely discussed considering several phenomena. The variations are attributed to ion bombardment with energetic oxygen [5,15,16] and/or to the oxygen chemistry (e.g. activity of oxygen and amount of oxygen reaching the film surface) [17,18]. Another effect is simply geometric: the path length of sputtered particles on their way from the target to the substrate is different, which reduces or increases the probability of reactions in the plasma.

The properties of the dynamic film seem to be a mixture of the static print: All observed properties of the dynamic film are between the variation of the statically deposited film. This is even valid for the surface structure after etching. It shows similarity to both smooth flat craters (film B and C) and the high density of craters (film E). However, this means that an optimization of the dynamically sputtered films might require an examination of the static print on each position [19]. This is especially challenging for the design of optimised surface-textured ZnO films for thin film silicon solar cells.

## CONCLUSIONS

We prepared ZnO:Al films under the same deposition conditions in static and in dynamic mode, respectively, and studied the resulting film properties as a function of the position relative to the targets. We found that the lateral variations in film properties of the static print can be attributed to an image of the double cathode system. The racetrack positions showed lower mobility and lower carrier density resulting in a resistivity which is by a factor of two higher. In between the racetracks, we achieved minimum resistivity of $\rho=1.9\times10^{-4}$ $\Omega$cm close to the best published data and a high mobility of about $\mu\approx30$cm$^2$/Vs with carrier densities of $n\approx10^{21}$cm$^{-3}$.

XRD-measurements reveal different stress in the films depending on the substrate position. Upon etching in diluted hydrochloric acid, the films at positions above the racetracks turn out to be about a factor of two more resistant against the acid than at other positions. Films from positions near the target, facing the racetrack or the center of one target, remained flat after etching and only showed a few smooth craters randomly distributed over the surface. On the other hand, a rough surface with sharp steep structures develops on films taken from substrate positions between the two targets.

The properties of amorphous silicon p-i-n solar cells prepared on samples cut from the different positions of the static print were well correlated to the various surface structures of the etched substrates.

## ACKNOWLEDGEMENTS

The authors would like to thank O. Kappertz for fruitful discussion about XRD and growth behaviour of ZnO:Al films. We gratefully acknowledge financial support by the BMWi (contract Nos. 0329854A and 0329923A).

## REFERENCES

1. J. Müller, G. Schöpe, O. Kluth, B. Rech, M. Ruske, J. Trube, B. Szyszka, X. Jiang, and G. Bräuer, Thin Solid Films **392** 327-333 (2001)
2. Löffl, S. Wieder, B. Rech, O. Kluth, C. Beneking, and H. Wagner, Proceedings of the 14$^{th}$ European Solar Energy Conference, Barcelona p. 2089 (1997)
3. B. Szyszka, Thin Solid Films **351**, 164-169 (1999)
4. J. Müller, G. Schöpe, O. Kluth, B. Rech, V. Sittinger, B. Szyszka, R. Geyer, P. Lechner, H. Schade, M. Ruske, G. Dittmar, H. P. Bochem, Proc. 4th Int. Conf. on Coatings on Glass, pp. 505-512 (2002)
5. R.J. Hong, X. Jiang, B. Szyszka, V. Sittinger, S.H. Xu, W. Werner, and G. Heide, Jornal of Crystal Growth, In Press, Corrected Proof (2003)
6. J. Strümpfel, G. Beister, D. Schulze, M. Kammer, S. Rehn, 40$^{th}$ Annual Technical Conference of the Society of Vacuum Coaters, New Orleans USA, April 12-17, (1997)
7. K. Ellmer, J. Phys. D: Appl. Phys. **34**, 3097-3108 (2001)
8. K.N. Tu, and R. Rosenberg, academic press inc. ISBN 0-12-341827-5
9. O. Kappertz, R. Drese, and M. Wuttig, J. Vac. Sci. Technol. A **20** (6), (Nov/Dec 2002)
10. H. Sato, T. Minami, S. Takata, T. Mouri, and N. Ogawa, Thin Solid Films, **220** 327-332 (1992)
11. JCPDS card 80-0075
12. B. Rech, J. Müller, T. Repmann, O. Kluth, T. Roschek, J. Hüpkes, H. Stiebig, and W. Appenzeller, MRS Spring meeting, San Francisco 2003
13. J. A.Thornton, J. Vac. Sci Technol. **11**, 666 (1974)
14. O. Kluth, G. Schöpe, J. Hüpkes, C. Agashe, J. Müller, B. Rech, Proc. 4th Int. Conf. on Coatings on Glass, Braunschweig, Germany (2002) p. 299; accepted for publication in Thin Solid Films
15. K. Tominaga, K. Kuroda, and O. Toda, Jap. J. Appl. Phys. **27**, 1176 (1988)
16. T. Minami, H. Nanto, H. Sato, and S. Takata, Thin Slid Films, **164** 275-279 (1988)
17. Song D., P. Widenborg, W. Chin, and A. G. Aberle, Solar Energy Materials & Solar Cells **73** 1-20 (2002)
18. T. Minami, T. Miyata, T. Yamamoto and H. Toda, J. Vac. Sci. Technol. A **18** (4) (Jul/Aug 2000)
19. J. B. Webb, Thin Solid Films **136** 135-139 (1986)

# Growth Mechanisms, Hot Filament CVD and Microcrystalline Si:H Growth

# Combinatorial approach to thin-film silicon materials and devices

Qi Wang, Leandro R. Tessler[1], Helio Moutinho, Bobby To, John Perkins, Daxing Han[2], Dave Ginley, and Howard M. Branz

National Renewable Energy Laboratory (NREL), 1617 Cole Blvd., Golden, CO 80401 U.S.A.
[1]Instituto de Física "Gleb Wataghin," Unicamp, C. P. 6165, 13083-970 Campinas, SP, Brazil
[2]Department of Physics & Astronomy, University of North Carolina at Chapel Hill, Chapel Hill, NC, 27599 U.S.A.

## Abstract

We apply combinatorial approaches to thin-film Si materials and device research. Our hot-wire chemical vapor deposition chamber is fitted with substrate xyz translation, a motorized shutter, and interchangable shadow masks to implement various combinatorial methods. For example, we have explored, in detail, the transition region through which thin Si changes from amorphous to microcrystalline silicon. This transition is very sensitive to deposition parameters such as hydrogen-to-silane dilution of the source gas, chamber pressure, and substrate temperature. A material library, on just a few substrates, led to a three-dimensional map of the transition as it occurs in our deposition system. This map guides our scientific studies and enables us to use several distinct transition materials in our solar-cell optimization research. We also grew thickness-graded wedge samples spanning the amorphous-to-microcrystalline Si transition. These single stripes map the temporal change of the thin silicon phase onto a single spatial dimension. Therefore, the structural, optical, and electrical properties can easily be studied through the phase transition. We have examined the nature of the phase change on the wedges with Raman spectroscopy, atomic force microscopy, extended x-ray absorption fine structure (EXAFS), x-ray absorption near-edge spectroscopy (XANES), ultraviolet reflectivity, and other techniques. Combinatorial techniques also accelerate our device research. In solar cells, for example, the combinatorial approach has significantly accelerated the optimization process of p-, i-, n-, and buffer layers through wide exploration of the complex space of growth parameters and layer thicknesses. Again, only a few deposition runs are needed. It has also been useful to correlate the materials properties of single layers in a device to their performance in the device. We achieve this by depositing layers that extend beyond the device dimensions to permit independent characterization of the layers. Not only has the combinatorial approach greatly increased the rate of materials and device experimentation in our laboratory, it has also been a powerful tool leading to a better understanding of structure-property relationships in thin-film Si.

## Introduction

The combinatorial approach (high-throughput experimentation) has been applied with success in recent years to thin-film materials research [1-4]. However, the initial work on this approach was begun by Joseph Hanak almost 30 years ago [5]. Hanak made a continuously varied composition sample to study binary composition alloys. As now implemented, the method dramatically shortens the processing time compared to the conventional approach. Development of computer-automated, high-throughput characterizations makes possible full use of the combinatorial approach. Combinatorial sample deposition, measurements, and data analysis must all be optimized for successful research. Only recently has the combination of deposition tools, computing power, automation, and, more importantly, characterization techniques, become available to allow true combinatorial approaches. In our laboratory, research that previously required up to a year can now be completed in a few weeks.

In spite of more than 25 years of hydrogenated amorphous silicon (a-Si:H) research, the properties of thin-film materials and performance of devices for photovoltaic and related applications remain a significant technical challenge. For example, the properties of undoped thin-film Si that grows near the transition from amorphous to microcrysalline silicon [6, 7] not only depends on the atomic H content in the Si film, but also on many other deposition parameters such as substrate temperature, gas mixture, and gas pressure. Clear fundamental understanding of these better materials is still not established; however, the applications have advanced rapidly. The gradual nature of structural transition and the dependence on multiple deposition variables make the research more difficult. For example, most doped a-Si:H thin films have either B or P doped into a binary alloy (Si and H) or B doped into a tertiary alloy (Si, C, and H). The challenges are even greater because, in applications, the doped a-Si:H materials are normally very thin (tens of nm). In device applications, state-of-the-art solar cells also utilize materials that are grown very close to a structural phase transition to improve performance [7]. In addition to theses materials issues, the production of optimized devices is still laboratory dependent: certain procedures are more successful in one lab than others. Each laboratory must perform its own painstaking optimizations on its own deposition equipment. The properties of the materials and the device performance are still lacking strong causal correlations to guide this process to completion. Although there are some general guiding principles, device making is still more often art than science—device optimization has often required years of slow, empirical experimentation. Therefore, amorphous Si and related materials, and the devices made from them, are excellent candidates for using the combinatorial approach. In this paper, we will present our work on amorphous Si based materials and device research using our unique combinatorial hot-wire chemical vapor deposition (HWCVD) system. However, applications of this approach should not be limited to HWCVD. Combinatorial approaches to materials and device research can be used profitably in plasma-enhanced CVD, sputtering, and other deposition techniques.

## Combinatorial HWCVD system

In NREL's conventional HWCVD deposition system, a stainless-steel vacuum chamber with a base pressure of less than $10^{-6}$ Torr is used. A spiral 0.5-mm-diameter, 13-cm-long tungsten wire is heated to about 2000°C using an ac or dc current. The hot wire is normally placed 5 cm below a typically 2.5-cm x 2.5-cm or 6.35-cm x 6.35-cm heated Corning 1737 substrate. Process gases such as $SiH_4$ decompose on the wire surface. Atomic Si, H, and silane radicals react in the chamber, and reactive radicals form a film on a substrate. For a-Si:H deposition with $SiH_4$ flow rate of 20 sccm and a chamber pressure of 10 mTorr, a film of device-quality a-Si:H can be deposited at a deposition rate of about 10 Å/s [8].

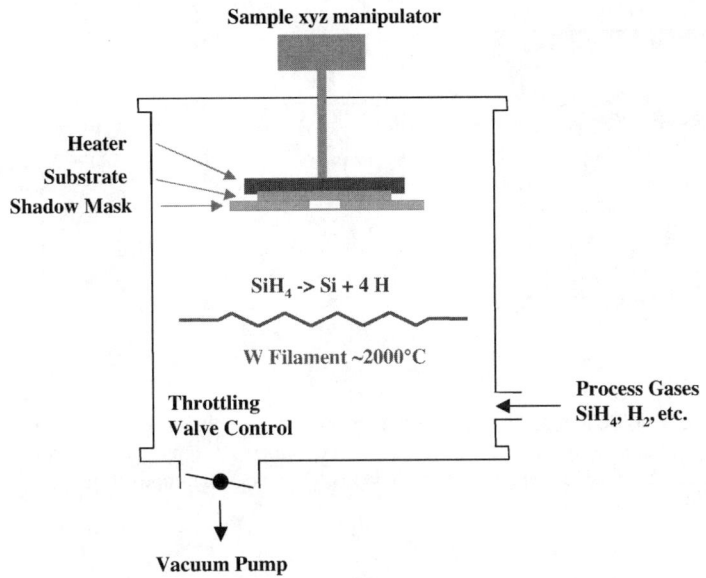

Figure 1. Schematic diagram of the NREL combinatorial HWCVD system.

Figure 1 is a schematic of our combinatorial HWCVD system [9] which is a modified version of NREL's conventional HWCVD system. There are two key modifications: The first is to add a physical mask or a set of masks to deposit the film only on the selected partial area of a substrate. The second is to add a sample manipulator (xyz) that allows a relative xy movement between the mask and the substrate so that the film can be deposited at any position on the substrate. The z-translations are used primarily to move the sample and then ensure a tight contact between the substrate and the mask. This prevents the deposition of materials "shadows" from reactant that would diffuse under loose-fitting masks. The vacuum chamber, filament, and gas inlet are identical to NREL's conventional system. The combinatorial deposition speeds up the sample deposition primarily because pump down cycles between samples are avoided. The system also adds considerable functionality, especially because we have developed a set of masks that are interchangeable in vacuum.

Figure 2 shows the simple example of making a library of discrete variable samples by combinatorial deposition. On the left is a shadow mask with a small open square in the middle. We move the substrate or the mask, using the xy manipulator, to position 1 on the substrate, as indicated in the center figure and deposit the first film. Then, we move the open square to the position 2 and make the second film. The mask isolates each sample from the others. This procedure was repeated for all 16 locations in this example. The optical image shows an actual combinatorial sample on 2.5-cm x 2.5-cm glass substrate. The various gray scales or different colors indicate the thickness or the optical index changes to the various deposition parameters. This approach greatly reduces the time spent on the cycle of loading, heating, and

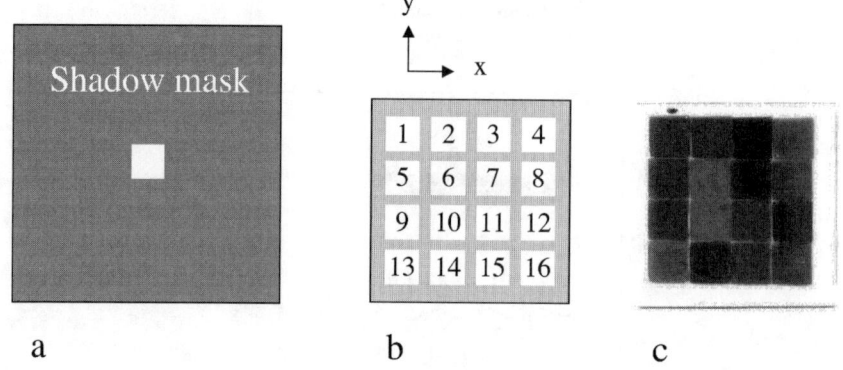

Figure 2. a) Schematic diagram of the mask used to deposit discrete square samples; b) schematic of the sample pattern obtained by stepping the sample beneath the mask between each of 16 depositions; and c) an optical image of the resulting film.

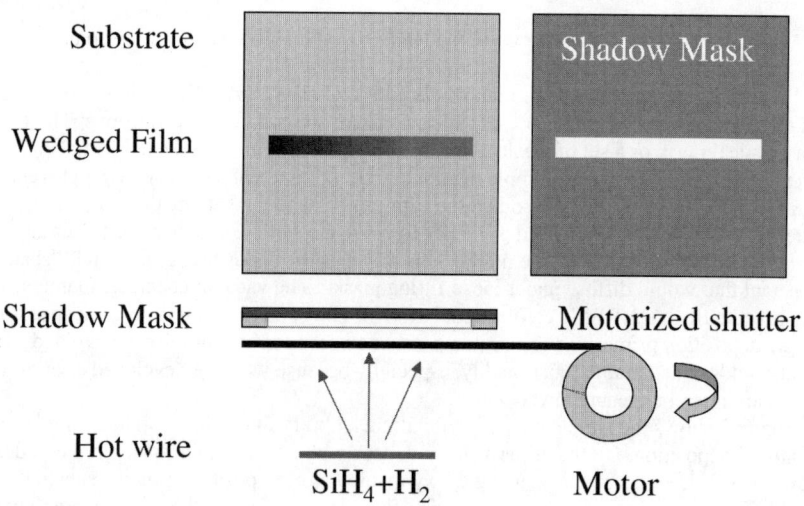

Figure 3. Schematic diagram of the process for making continuously wedged samples with continuously variable thickness. The shutter is placed close to the shadow mask. The motor speed and deposition rate determine the thickness grading.

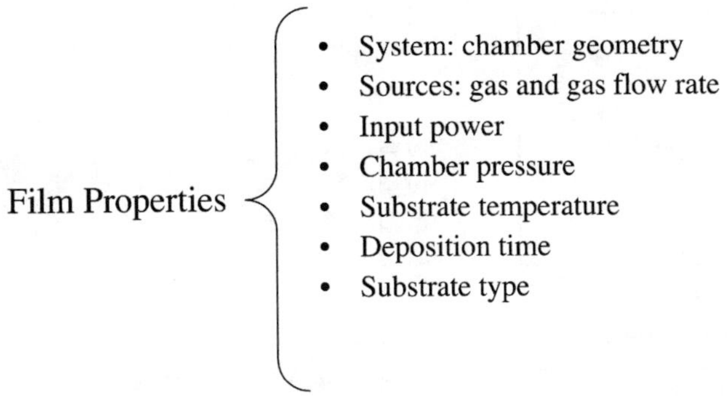

Film Properties
- System: chamber geometry
- Sources: gas and gas flow rate
- Input power
- Chamber pressure
- Substrate temperature
- Deposition time
- Substrate type

Figure 4. Some deposition factors directly or indirectly determine the thin film properties. This illustration shows the scope of deposition parameters that can be chosen in the combinatorial depositions.

cooling in between the samples in the conventional deposition. A factor-of-10 reduction of total deposition time can be achieved easily.

Figure 3 shows how we make samples with a continuously changing variable such as thickness. In principle, the substrate temperature (by means of the hot and cold ends) and the gas composition (by the means of timing the gas flow rate) can also be continuously changed. To make a thickness-graded stripe sample, a physical mask with a rectangular aperture of 5 x 50 mm and a motor-driven shutter are used. The thickness profile along the stripe is controlled by the motor speed and deposition rate. After the film of one stripe is deposited, the substrate can be moved to another location for deposition of another stripe. Libraries with as many as ten separate wedged stripes are made on a 6.35-cm x 6.35-cm substrate. This combination of discrete and continuous deposition provides a powerful tool to study the sensitive phase transition with regard to the various deposition parameters. We will show our work on the thickness-dependent phase transition from amorphous to microcrystalline silicon on one single substrate below.

In the hot-wire CVD process, many variables, also called "factors," can affect the amount of $SiH_4$ in the reaction region, the density of reactants at the film surface and the film formation reactions. These factors, taken together, determine the structure and properties of the films. For a given deposition system, chamber geometry, chamber pressure, source of gas, gas flow rate, direction of gas flow, gas mixture ratio, $T_{sub}$, substrate, deposition time, filament temperature, distance from filament to the substrate, number of filaments, and post treatments are all important factors. Figure 4 lists some of the important factors in any thin-film deposition that affect the properties of the films. From an engineering point of view, the combinatorial approach dramatically increases the throughput of experimentation and is an ideal tool to map the film properties as a function of many variables. The mapping can be done with a short time even if all the detailed mechanisms of growth are still unknown.

In one study, we selected three variables in the HWCVD process to create a small film library to study the phase transition from amorphous to microcrystalline silicon. Silane flow rate and hydrogen flow rate were obvious and convenient choices. We select the throttling valve position (TVP) as the third variable because it is an independent deposition factor with

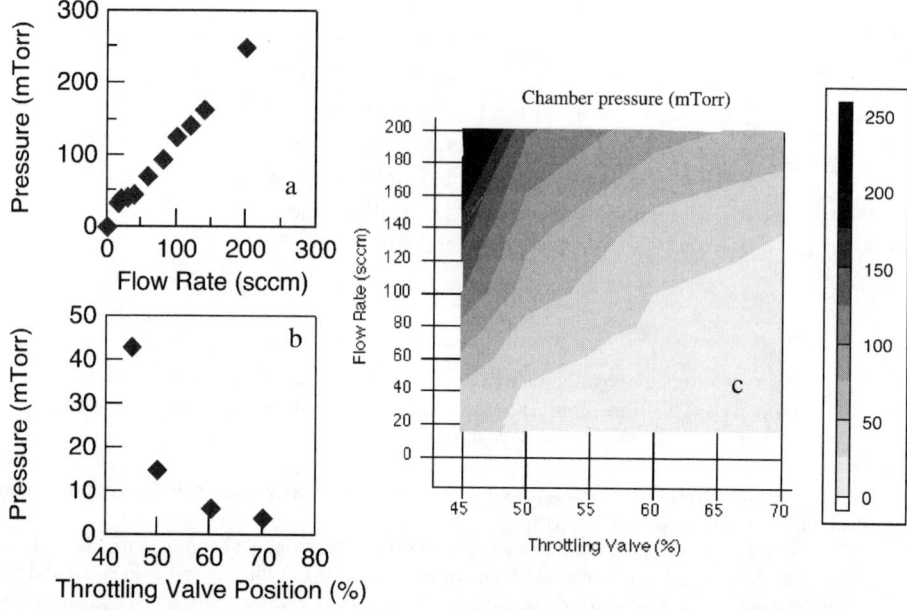

Figure 5. Chamber pressure dependence on the $H_2$ gas flow rate (a), throttling valve position (b), and in a 2-D contour plot (c). The gray scale on the right indicates the chamber pressure in mTorr.

which we partially control the chamber pressure. We use hydrogen as an example because of its high flow rate. It will apply to other gases, too. The chamber pressure of hydrogen in the chamber depends on the gas flow rate and the TVP. Figure 5 shows how pressure changes with gas flow rate at the fixed TVP of 45 % (Fig. 5a) and with pressure varying with TVP at a fixed gas flow rate of 15 sccm (Fig. 5b). Fig. 5c, shows a contour plot of chamber pressure as a function of a gas flow rate and TVP. To clarify the meaning of TVP, the valve is fully open at 100% and closed at 0%. Pressure has a linear dependence on the gas flow rate (for a fixed TVP of 45%) and rapidly increases with decreasing TVP below 50% (for a fixed gas flow rate of 15 sccm). Due to the nature of the pressure's dual dependence, one can achieve the same pressure by changing the gas flow rate or the TVP. However, the gas pressure alone is not enough to determine that the same material will always be produced. For example, for a-Si:H films with $SiH_4$ gas, different combinations of gas flow rate and TVP with the same pressure give different films. This is even more complex when using a mixture of gases such as $SiH_4$, $H_2$, and others. In this paper, we plot our results based on the factor, TVP, even though pressure has a more obvious physical meaning.

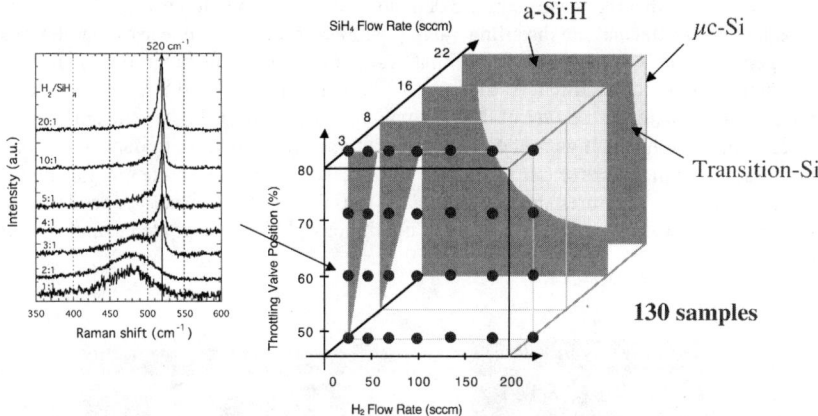

Figure 6. Left, Raman spectroscopy vs. H dilution at TVP of 60% and 3 sccm of silane flow rate. Right, amorphous-to-microcrystalline Si phase diagram as a function of $SiH_4$ flow rate, $H_2$ flow rate, and throttling valve positions. The dots represent some of the experimental points where the sample was made. The shaded area denotes a-Si, unshaded is area μc-Si, and the boundary in between is the transition region.

## Results and discussion

In this section, we examine several specific examples that demonstrate the power of the combinatorial approach in thin-film Si research. These are (1) a small-scale materials library, (2) a study of the thickness-dependent structure changes, and (3) the effect on a buffer layer at the solar cell p/i interface.

### 1. Amorphous-to-microcrystalline silicon library

Figure 6 presents the a-Si-to-μc-Si phase diagram as a function of three variables: silane flow rate, hydrogen flow rate, and throttling valve position (TVP, see above). The dots in the figure illustrate the location of the samples in the hydrogen-TVP plane for one of the four silane flow-rates we explored. In total, 130 samples were made sequentially on only 10 substrates (each substrate has less than 16 samples) for a total of 9 vacuum breaks. In each hydrogen-TPV plane, the shaded area represents materials in the amorphous phase and the unshaded area represents material with a microcrystalline phase. The four planes represent silane flow rates of 3, 8, 16, and 22 sccm, respectively. The boundary between the amorphous and microcrystalline phase is where the materials in transition are located. The structure of each sample was determined by the Raman spectroscopy and UV reflectance measurements (Model 1280, n&k Technology, Inc, Santa Clara, CA). Raman spectra at the left of the figure represent a series of H-dilution samples with silane flow rate of 3 sccm and TVP at 60%.

We found that the material phase depends nonlinearly on all the variables: hydrogen flow rate, silane flow rate, and the throttling valve position. From Fig. 6, one sees that the ratio of hydrogen to silane is not a unique factor that describes whether the material will grow amorphous or microcrystalline. At the lowest silane flow rates (3 and 8 sccm), it is easy to change to μc-Si with the mixture of only a small amount of $H_2$. At higher silane flow rates (16 and 22 sccm), it requires higher H dilutions and an open throttle valve to deposit μc-Si. This film library has guided us for our many HW solar cells applications. Note that this phase diagram is based entirely on measurement of thick (>200 nm) films. The thickness dependence of the phase transition was addressed in a study of continuously thickness-graded samples as described below.

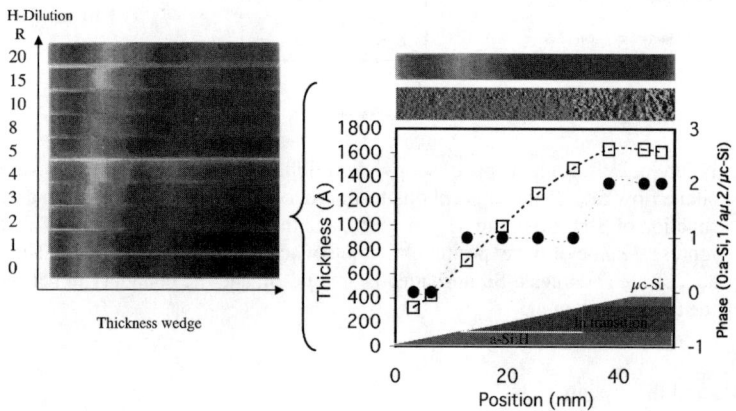

Figure 7. Thickness dependence of a-Si-to-μc-Si phase transition in samples continuously graded thickness. Left, an optical image of a combinatorial sample with a thickness grading in horizontal direction and H-dilution in the vertical direction, as indicated by the R for each stripe. Right, a summary plot on the stripe number 2 of R = 2. The optical image of the stripe is placed on the top, and a collections of six AFM pictures with roughly matched positions on the stripe is under that. The figure below shows the thickness variation and the phase changes along the stripe. The insert cartoon best describes the profile of the interesting stripe with a large portion of transition materials.

## 2. Thickness-dependent phase transition

The left picture in Fig. 7 is an optical-reflection image obtained by placing a ten-stripe continuously thickness-graded sample in a conventional color image scanner. This sample has discrete H-dilution stripes arrayed in the vertical direction, deposited with $H_2$-to-$SiH_4$ ratios (R) of 0, 1, 2, 3, 4, 5, 8, 10, 15, and 20. The $SiH_4$ flow rate was fixed at 8 sccm. Each stripe has a thickness grading ranging from about 200 to 1600 Å. This special sample varies by two deposition parameters: ten discrete H-dilutions and continuously varied thicknesses. From these parameters, we can construct an evolutionary-phase diagram as a function of H-dilution and thickness for HWCVD Si by measuring the Raman or UV reflectance of the sample [10, 11]. The substrate temperature is 200°C, and the chamber pressure ranges from 6 to 57 mTorr, depending on R. The total deposition time for the entire 10-stripe sample was just over 2 hours. The gray scale change along each stripe indicates the thickness variation. The gray scale or color shift among the stripes directly indicates both the decrease of deposition rate and an increase of μc-Si fraction in the film with increasing R.

By measuring UV reflectance and observing the peaks at 275 and 360 nm, we are able to quickly determine whether or not there is crystallinity within the top 100 Å of the film [3, 12]. We found that with R > 2, all the stripes are in μc-Si phase everywhere on the stripes. The stripe with R=0, as expected, is entirely the a-Si phase. The stripe with R=1 was amorphous along most of its length, showing the signal of μc-Si only at the thickest end. The stripe with R=2 has a long transition phase. For this reason, it is of the greatest interest, and we have simulated it accordingly to the need to develop high-throughput characterization techniques to measure this special stripe.

The right side graph in Fig. 7 is a summary plot of film properties of surface morphology, thickness, and corresponding phase along the stripe. The top picture is a scanner image of the stripe. Just below this image is a composite of six AFM images arranged to be in roughly the correct position on the stripe. This composite AFM image shows that the a-Si:H surface is rather smooth compared to the much rougher μc-Si. Interestingly, a large range in the middle of the stripe appears rather smooth, although the UV reflectivity and Raman measurements revealed that this material is already undergoing the phase transition at this thickness. The roughness is about 10 Å for a-Si:H, 15 Å for transition materials, and 46 Å in the μc-Si. The thickness of this stripe changes from 200 to 1600 Å across its 50-mm length, as shown in the main graph. The thickness increases quite linearly from 200 to 1600 Å along the positions from 0 to 40 mm. However, in the last 10 mm the thickness is quite constant because (1) the deposition rate decreases when μc-Si growth commences and (2) there is some deposition rate inhomogeneity in our HWCVD reactor. To plot phase results, we denote a-Si:H with a 0; transition or mixed phase Si as 1; and μc-Si as 2. For thicknesses less than 600 Å, the material is in a-Si:H phase. For thicknesses of more than 1500 Å, μc-Si starts to grow, and between 600 and 1500 Å, the material appears to be mixed phase (in transition). The wedge cartoon is a schematic illustration of the phase evolution from amorphous-to-microcrystalline silicon in this stripe. Our results are in qualitative in agreement with Collins' in-situ ellipsometry measurement for PECVD films [13].

Figure 8 shows the extended x-ray absorption fine structure (EXAFS) (Fig. 8a) and which x-ray absorption near-edge spectroscopy (XANES) (Fig. 8b) results on the stripe (Fig. 7) that has a large section of transition region from the amorphous-to-microcrystalline phase. The EXAFS spectra have been measured along the stripe from a-Si:H to μc-Si as indicated with sample coordinate x. From EXAFS spectra, one can perform numerical fitting to get the average interatomic distance ($r$) and the radial Debye-Waller factor ($\sigma^2_{incr}$). The detailed fitting procedure has been published elsewhere [14]. The results in Fig. 8a show the decrease of $\sigma^2_{incr}$ (which becomes negative) and the subtle decrease of $r$ as the crystalline fraction of the sample increases. The negative $\sigma^2_{incr}$ is a consequence of the imposed average coordination N = 4: the average coordination of both a-Si:H and μc-Si are expected to be below 4, and imposing a higher coordination is compensated by a decrease in $\sigma^2_{incr}$.

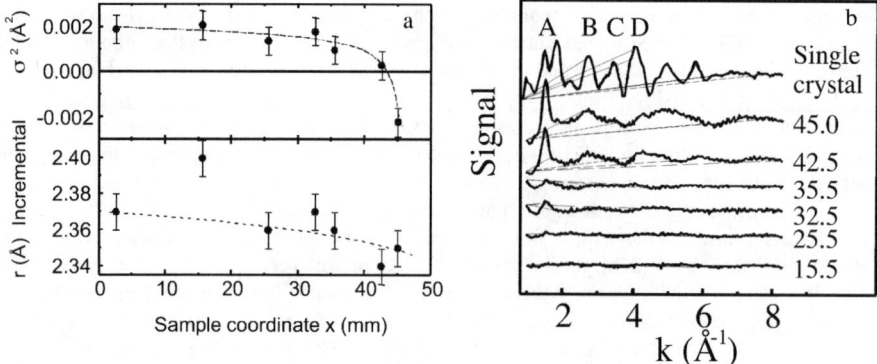

Figure 8. EXAFS study (Fig. 8a) and XANS study (Fig. 8b) on the strips sample in Fig. 7.a), incremental Debye-Waller factor (relative to single-crystal Si), and average first neighbor separation. Four fold coordination was imposed in the fitting procedure. The lines are guides for the eye. (b) x-ray absorption spectroscopy signal with the single scattering (EXAFS) contribution subtracted. The numbers in the figure represent the x coordinate, and the letters follow the peak assignment. The coordination is the same as in Fig.7.

Figure 8b show the XANS spectra at various positions on the stripe together with that of a c-Si reference [15]. In the single crystal, we can clearly identify the A, B, C, and D multiple scattering peaks that are attributed to double scattering paths. In the a-Si:H side of the sample, no peaks are found. In the $\mu$c-Si side, only peak A can be identified. In order to isolate the multiple scattering effects, we have subtracted all spectra from the one obtained at x = 2.5 mm, where only a-Si:H is present. In the a-Si part of the sample, indeed, there are no multiple scattering signals. In the $\mu$c-Si part, the dominating path is due to scattering by two adjacent atoms of the first coordination shell. This surprising result indicates that the bond-angle disorder of $\mu$c-Si is enough to suppress the multiple scattering signals from atoms beyond the first coordination shell. In amorphous silicon, bond-angle disorder suppresses even the contribution from the first shell. The main consequence of this result is that the $\mu$c-Si formed at the transition region and even at the end of the strip (highly $\mu$c-Si) is much more disordered than single crystal-silicon —to a point that the second-neighbor shell disorder is comparable to the first-neighbor shell disorder in a-Si:H.

## 3. Optimization of devices

In the thin-film devices such as solar cells, the device structure combines three or more layers. This structure is naturally suited for the combinatorial HWCVD approach because different masks can be used on each layer, thereby giving more information than in conventional devices. Certainly, all device-makers would like to know how the properties of each layer relate to device performance. By using the combinatorial approach, this kind of insight can be obtained.

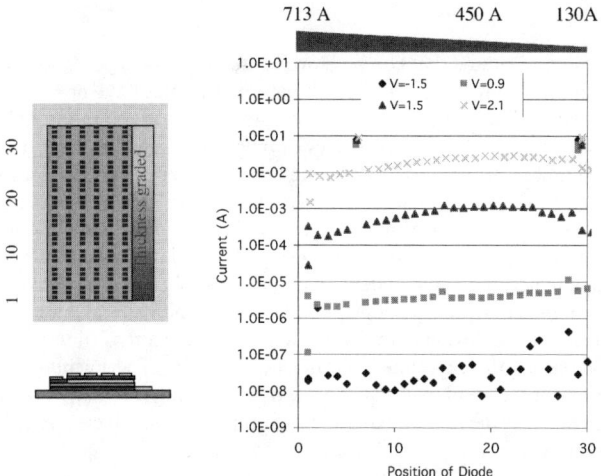

Figure 9. Effects of a buffer layer and its thickness at p/i interface on a diode in one combinatorial sample. Left, a schematic of device n-i-b-p structure and layer locations (top and side views) of the combinatorial sample. The top dot array is the top contact to evaluate the devices. The thickness-graded buffer layer is deposited at the p/i interface, but offset to have the buffer layer on the substrate. Right, the dark current as a function of the positions (the buffer layer thickness) at various voltage biases. The scale of the thickness is plotted on top of the figure.

Figure 9 shows a combinatorial device we made to study the effect of a single device layer— the thin buffer (b) layer between the p and i-layers — on diode performance. A schematic of the device structure is shown on the left. Using a single mask, the thickness graded buffer layer was offset from the n-i-p diode. In this way, part of the buffer layer was incorporated into the device and part was deposited directly on the substrate. The cross section of the device shows that n-i-p, n-i-b-p, and b can be measured independently. In the perpendicular direction, the thickness of the buffer layer was graded. The 50-mm-long wedge incorporates 30 top Pd contacts that we used to evaluate the device performance versus thickness. The thickness of the buffer layer was directly measured on the offset b-layer. The graph shows the logarithm of the dark current at various voltage biases, plotted as a function of diode position.
The diode position is directly related to the thickness of the buffer layer, as indicated schematically by the triangular wedge above the graph. The results show that the buffer layer has little effect on the reverse current but does affect the forward current. At V=1.5 V and with 450 Å of buffer layer the dark current, increases by about a factor of 10 when compared to either the zero- or 700-Å-thick buffer layers. With only one sample, the effect of buffer layer and its thickness on the device performance can be readily studied.

## Summary

We have demonstrated that combinatorial studies can be a powerful tool for materials and device research in thin-film silicon. The combinatorial samples have not only greatly speeded up the deposition process, but also stimulated development of many high-throughput characterizations to measure the samples. Developing fast characterization and data analysis tools for a large number of samples has become critical for optimizing processes and searching for new materials. With our combinatorial HWCVD tool, we efficiently created a material library of the transition from a-Si:H to μc-Si and were able to using this information to guide our device research. We were able to provide new XANES information about the transition material and hope to accelerate the understanding of ordering in the transition materials. For thin-film device applications, we demonstrated that the combinatorial approach can also make important contributions to device optimization and to correlation of films properties to device performance. Increasing the experimental efficiency, i.e., the experimental throughput, is a central motivation for the combinatorial approach. Based on our initial work, the combinatorial approach has speeded up the rate of experimentation by at least a factor of 10 in our HWCVD thin-film Si research.

## Acknowledgements

The author would like to thank Richard Crandall, Brent P. Nelson, and Yueqin Xu for discussions and help. This work is supported by the U.S. Department of Energy under subcontract No. DE-AC39-98-GO10337 and DDRD research funded at NREL.

## REFERENCES

1. X.-D. Xiang, X. Sun, G. Briceno, Y. Lou, K.-A Wang, H. Chang, W. G. Wallace-Freedman, S.-W. Chen, and P.G. Schultz, *Science*, **268**, p. 1738, 1995.
2. X-D. Xiang, *Materials Science & Engineering. B*, **56,** p. 247,1998.
3. Qi Wang, Guozhen Yue, Jing Li, and D. Han, *Solid State Commun.*, **113** p. 175, 2000.
4. Qi Wang, J. Perkins, H. Branz, J. Alleman, C. Duncan, and D. Ginley, *Applied Surface Science*, **189** p. 271, 2002.
5. J.J. Hanak, *J. Mat. Science*, **5** p. 964, 1970.
6. L. Yang and L.-F. Chen, *Mat. Res. Soc. Proc.*, **336**, p. 669. 1994.
7. J. Yang and S. Guha, *Mat. Res. Soc, Proc.,* **557**, p. 239, 1999.
8. A.H. Mahan, J. Carapella, B.P. Nelson, R.S. Crandall, and I. Balberg, *J. Appl. Phys.*, **69** p. 6728, 1991.
9. Qi Wang, *Thin Solid Films*, (2$^{nd}$ HWCVD Conference) to be published, 2003.
10. G.Z. Yue, J. D. Lorentzen, J. Lin, D.X Han, and Q. Wang, *Appl. Phys. Lett.*, **67** p. 3468, 1999.
11. G.Z. Yue, J. Lin, L. Wu, D.X. Han, and Q. Wang, *Mat. Res. Soc. Proc.*, **557**, p. 525, 1999.
12. D.L. Greenaway, G. Harbeke, Optical Properties of Semiconductors, Pergamon, New York, 1968.
13. R. W. Collins, J. Koh, A.S. Ferlauto, P.I. Rovira, Y. Lee, R.J. Koval, and C.R. Wronski. *Thin Solid Films*, 364, p.129. 2000.
14. Leandro R. Tessler, Q. Wang, and H.M. Branz, *Thin Solid Films*, (2$^{nd}$ HWCVD Conference) to be published, 2002.
15. J. J. Rehr, and R.C. Albers, *Rev. Mod. Phys.*, **72** p. 621, 2000.

# Calculations of SiH$_3$ diffusion and growth processes on a-Si:H surfaces

P Vigneron, P W Peacock, K Xiong and J Robertson*
Engineering Department, Cambridge University, Cambridge CB2 1PZ, UK
* jr@eng.cam.ac.uk

## Abstract

Surface diffusion of a growth species is needed to give the observed smooth surface of hydrogenated amorphous silicon (a-Si:H). But what diffuses, the weakly bound SiH$_3$ radical on the hydrogenated surface, or the bound SiH$_3$ at a growth site. Diffusion is complicated by the change in the surface termination of a-Si:H as temperature rises. We use total energy pseudopotential calculations on a variety of periodic Si:H surface configurations to show that it is the weakly bound SiH$_3$ that diffuses. We provide an overall energy scheme of the bound states and transport levels of SiH$_3$ on a-Si:H surfaces.

## Introduction

A key test of a growth mechanism of a-Si:H is to account for the surface roughness and its temperature dependence. A proposed model of growth is as follows [1-3]. The growth species is SiH$_3$ and the a-Si:H surface sites are essentially all terminated by hydrogen. The SiH$_3$ is assumed to adsorb weakly to the surface, diffuse over it, and abstract a hydrogen from a surface Si-H bond to create a surface dangling bond, Fig. 1(a). A second SiH$_3$ then arrives, adsorbs, diffuses over the surface to find the dangling bond and add to it, to form a 'bound SiH$_3$' and permanent Si-Si bond, to give growth. There are two important factors in this process to give a smooth surface. First, surface diffusion must occur, so that the SiH$_3$ can sample sites to react with (Langmuir-Hinselwood mechanism), rather than react immediately with the first site it hits (Eley-Rideal mechanism). Second, the H must be less stable at Si-H bonds in valley and kink sites, so that H is preferentially abstracted from these sites. This then gives valley filling [4].

In detail, roughness evolution during film growth is often described by a fractal process with a growth exponent β, which describes the scaling of the roughness W with the film thickness h,

$$W \sim h^\beta$$

For random deposition, so that an atom sticks where it hits, β=0.5. If it diffuses around and then sticks, β=0.25. If in addition the surface relaxes after sticking, then β=0. Applying this to growth of a-Si:H is complicated by the above two-step process, where two SiH$_3$ radicals are needed to allow one SiH$_3$ to stick to the surface. Clearly, abstraction by the first SiH$_3$ determines the roughness evolution and this process cannot be random, but must be preferential at the valley sites. Otherwise, the theory is the same.

Smets et al [5] recently measured the roughness exponent in detail for a-Si:H produced by an expanding thermal plasma (ETP) using spectroscopic ellipsometry. They found that β decreased with increasing deposition temperature from about β=0.5 at T=50C to β~0 at T=500C, as shown in Fig 2. They then modelled the growth process with a simple solid-on-solid model and found that an activation energy of 0.7 eV described this. They suggested that this energy was larger

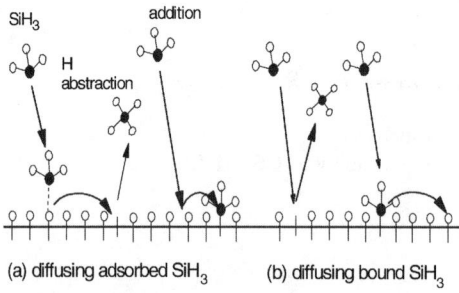

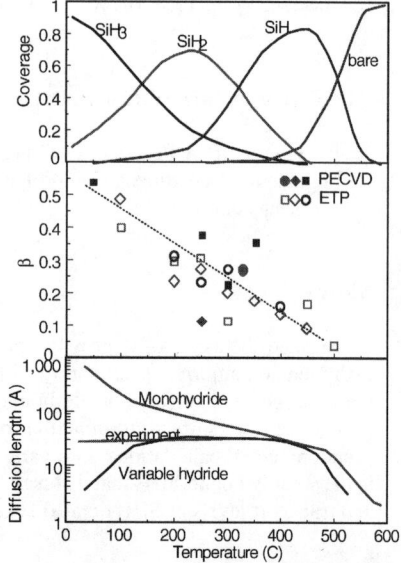

Fig. 1. Conventional 2-step growth model (a), compared to model (b) with diffusing bound SiH$_3$.

than normally associated with the diffusion of a weakly bound species SiH$_3$, compared to a value of 0.2 eV previously found in the simple growth model [3].

They proposed instead that either surface hydrogen diffuses, giving effectively a diffusion of the surface dangling bond, or that the fully bound SiH$_3$ diffuses [6], Fig. 1(b). To understand these processes in more detail, we carried out detailed calculations on various periodic surface configurations of Si:H. A key factor is that the surface H termination of a-Si:H

Fig. 2. Variation with temperature of surface hydride coverage [6], growth exponent for PECVD and expanding thermal plasma (ETP) [5], and SiH$_3$ diffusion length, experiment and theory [7].

varies with temperature [6](Fig 2a). Depending on growth conditions, roughly, surface SiH$_3$ groups dominate below 100C, SiH$_2$ groups dominate for 100-200C and SiH dominates for 200-500C, with bare surfaces above that.

## Method

The calculations are carried out on periodic super-cells of crystalline Si with surfaces that model the desired configurations. A Si(111):H surface is used to represent a surface with second neighbour Si-H groups [8]. We use a $2 \times \sqrt{3}$ super-cell with a slab of four (111) bilayers of Si, with its surfaces terminated by hydrogen, for a total of 32 Si atoms and 8 H atoms. The cell is 3.5 nm high, so the slabs are vertically separated by 2.23 nm of vacuum. The Si(110):H surface represents a surface with nearest neighbour Si-H groups, on which they form a zig-zag chain. This also has 32 Si atoms and 8 atoms. A Si(100):2H surface represents a surface of SiH$_2$ groups. A Si(111):H surface with a single SiH$_3$ group replacing a surface hydrogen represents a surface terminated by SiH$_3$ groups. A Si(111) surface is unstable if all surface Si atoms are SiH$_3$ groups [9], because of the steric hindrance of the hydrogens, but a single SiH$_3$ standing above the surface is possible.

The calculations are carried out using the CASTEP code [10], with atoms represented by Vanderbilt ultra-soft pseudopotentials, a plane wave basis with 350 eV cut-off energy and the spin-restricted local density approximation (LDA) and the generalised gradient correction (GGA) of the exchange-correlation potential. Total energies are evaluated using two special k points. In each calculation, atoms on the surface and in the top two Si bilayers can fully relax, and atoms in the lower two bilayers are fixed at their bulk positions. For transition state configurations, the symmetry was constrained and the height of the SiH$_3$ above the surface varied, to limit relaxations back to the ground state.

## Results

Fig. 3(a) shows the ground state of SiH$_3$ adsorbed onto a hydrogenated (111)Si:1H surface, with the formation of a direct Si-Si bond and the surface H displaced to the centre of a lateral Si-Si bond [8]. This configuration is –0.85 eV below that of a free SiH$_3$ in a vacuum. The SiH$_3$ can hop to the next surface SiH group, which could be either a first nearest neighbour (1nn) or second nearest neighbour (2nn). Fig 3(d) shows this same ground state on a (110)Si:1H surface, representing 1nn hops. The transition state for the 1nn hop is shown in Fig 3(e), and it lies at 0.25 eV above the ground state. This is a low barrier. A 2nn hop is more complex, as it could pass through various transition states such as those shown in Fig. 3(b) and (c). The energies of these above the ground state are plotted on the hexagonal surface cell in Fig 4, showing that a minimum energy of 0.75 eV is needed for migration. These hopping energies are summarised in Fig. 5.

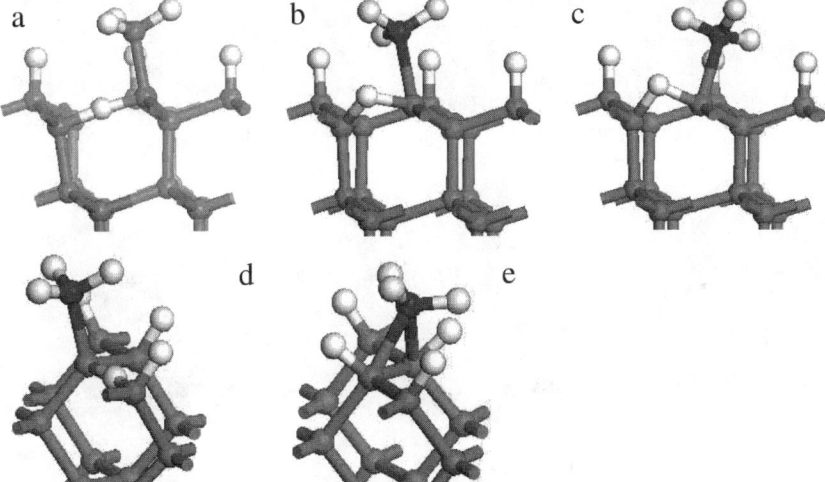

Fig. 3. (a) SiH$_3$ adsorbed on (111)Si:H surface. (b,c) Transition states of SiH$_3$ diffusion on (111) for 2nn hops. (d) Ground state of SiH$_3$ on (110)Si:H. Transition state of diffusion for 1nn hops.

At moderate temperatures, the a-Si:H is mainly covered with SiH$_2$ groups (Fig. 2). These are represented by the (100)Si:2H surface in Fig. 6(a). The close spacing of H's bonded to adjacent silicons creates the canted structure. Fig 6(b) shows the ground state of SiH$_3$ physisorbed on this surface. Its bonding is analogous to SiH$_3$ on Si:1H, but its binding energy is only –0.32 eV, much less than on Si:1H. This is also summarised in Fig 5. A second, less stable state is also found, with the Si-H-Si bond broken leaving a Si dangling bond, Fig 6(c). The SiH$_3$ can diffuse across this surface, with a barrier of about 0.3 eV. In other words, it nearly desorbs.

At the lowest temperatures, the a-Si:H is covered with SiH$_3$ groups, as represented by the surface in Fig. 6(d). A SiH$_3$ radical can also physisorb onto this group, creating a configuration analogous to Fig 3(a), with a H displaced to a bond centre. This configuration is at –0.28 eV, which is less stable than on Si:2H, fig 6(e). A slightly less stable configuration shown in Fig 6(f) lies at –0.25 eV. The physisorbed SiH$_3$ radical can hop to another site over a barrier of ~0.2 eV.

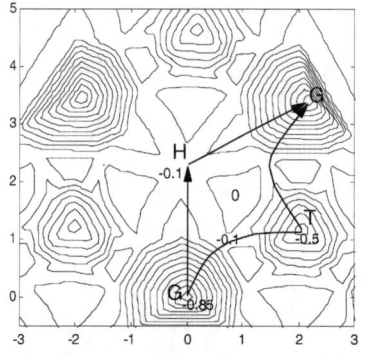

Fig. 4. Biding energies contours of SiH$_3$ on a (111)Si:H surface. 0.1 eV contours

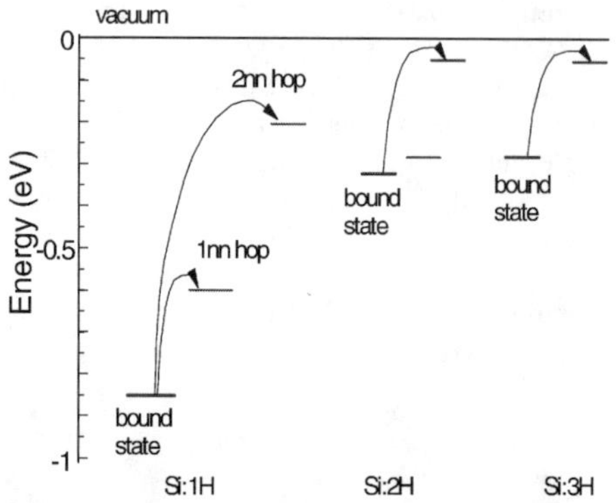

Fig. 5. Energy of bound and transition states of SiH$_3$ on various Si:H surfaces.

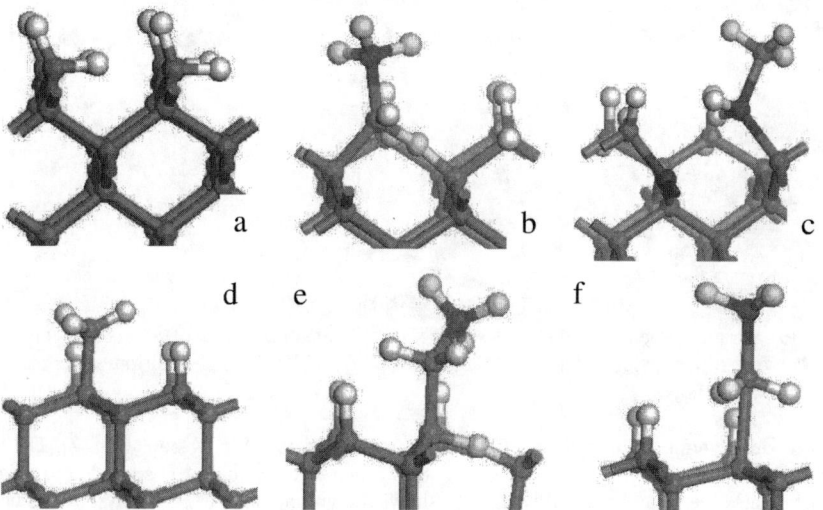

Fig. 6(a) The (100)Si:2H surface. (b) Ground state of SiH$_3$ adsorbed on (100)Si:2H. (c) Higher bound state of SiH$_3$ on Si:2H. (d) A Si:3H surface. (e) Bound state of SiH$_3$ radical on Si:3H surface group. (f) Weaker bound state of SiH$_3$ on Si:3H surface.

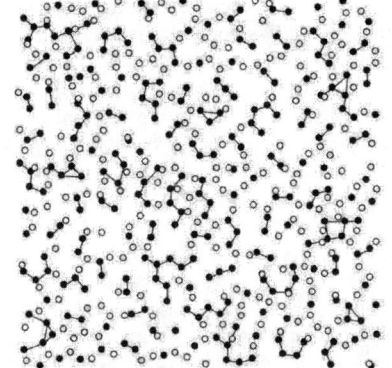

Fig. 7. Sites on a a-Si surface. Full circles are the top layer, open circles the next layer down. Nearest neighbour, top sites are joined by a line, showing that they do not form a percolation path.

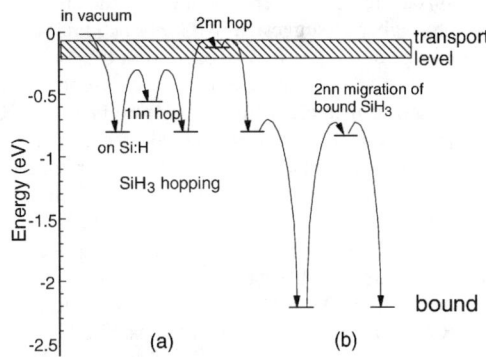

Fig. 8. Energies of (a) adsorbed and (b) bound states of SiH$_3$ compared.

Let us summarise these results. Fig. 5 emphasises that a SiH$_3$ has 1nn hops much more easily than 2nn hops. Can a SiH$_3$ cross a monohydride a-Si:H surface using only 1nn hops? Fig 7 shows that it cannot, because 1nn sites do not percolate over a a-Si surface. This surface was generated by slicing a 10,000 atom random network model of a-Si [11] so as to create only monovalent (ie monohydride) surface sites. Statistics on this large model show that 1nn surface sites do not percolate across it. Thus, the hopping of adsorbed SiH$_3$ on the surface is determined by 2nn hops, giving the larger barrier of 0.75 eV.

A second point is that on a fully monohydride surface, the SiH$_3$ radical is quite strongly bound, at –0.85 eV. But on dihydride or trihydride surfaces, it is much less strongly bound. Thus, SiH$_3$ will preferentially adsorb onto monohydride groups on a mixed surface.

Now consider how a bound SiH$_3$ group (after addition to a Si DB) can diffuse over the surface in Fig 1(b). We calculate that this has an energy barrier of 1.4 eV, Fig 8(b). This means the transition state is at roughly the same energy as the adsorbed SiH$_3$. This is as expected. The bound SiH$_3$ is a closed shell species with only normally coordinated atoms, and is therefore strongly bound, Fig 8. On the other hand, the adsorbed SiH$_3$ state and the transition state of the bound SiH$_3$ are all 'defect' configurations with *under- or over-coordinated* sites. They therefore have higher energies – like a deep tail state. Their energy of 1.4-1.5 eV above the bound state is typical of defect formation energies involving H in a-Si:H. These transition states of an adsorbed SiH$_3$ form a transport level just below the vacuum level as in Fig 8. This diagram makes it clear that the bound SiH$_3$ can never diffuse for an energy of less than about 1.4 eV.

**Discussion**

The above calculations show that SiH$_3$ radicals are bound on the a-Si:H surface and can diffuse over it [8,12]. This accounts for the smoothness of the a-Si:H surface. However, the actual activation energy of diffusion cannot be directly taken from an energy in Fig 4, because the surface hydride configuration varies with temperature. The adsorbed SiH$_3$ radicals avoid the surface regions covered with dihydride and trihydride groups, not because their diffusion barriers are higher as suggested by Gupta [7], but because the SiH$_3$ is less bound to them. An activation energy of order 0.7 eV needed to fit the roughness data [5] is reasonably consistent with Fig. 5.

The varying surface hydride configuration has a key effect [7]. It causes the SiH$_3$ diffusion length to increase with temperature below 250C as in Fig 1(c), unlike in the original model of [3]. The diffusion length equals the roughness correlation length, and experimentally this is found to increase slowly with temperature [13,14]. Above 300C, the diffusion length then decreases with increasing T, as in the simple model [3], due to evolution of surface H. Thus, the surface diffusion length has a maximum at 200-300C, as in the original Ganguly-Matsuda [15] model. Growth models need to be re-parameterised for this more complex situation.

Smets et al [5], believing that 0.7 eV was too large a diffusion energy, consider a different mechanism. An SiH$_3$ or atomic H abstracts a hydrogen from a surface Si-H group in an Eley-Rideal process. A second SiH$_3$ then adds to the resulting dangling bond to give a bound SiH$_3$, and they wish this species to diffuse across the surface. Our calculations do not support this mechanism. Once an SiH$_3$ has become bound, the activation energy is 1.4 eV or larger (Fig. 8), so it cannot diffuse as proposed. Other alternatives such as the diffusion of a surface H from Si-H to a surface dangling bond site, allowing the surface dangling bond to diffuse, also costs at least 1.5 eV. So this is forbidden. Note finally that the temperature dependence of β in Fig 1(b) is similar for PECVD and ETP conditions. ETP conditions have few atomic hydrogen species, so that atomic H mediated diffusion cannot be critical for the smoothening process.

In summary, total energy calculations suggest that the adsorbed SiH$_3$ radical is the only plausible diffusing species responsible for the smooth a-Si:H surfaces. However, the much weaker binding of SiH$_3$ to surface poly-hydride groups means that SiH$_3$ avoids such regions. This feature causes the increased roughness at low temperature and causes the diffusion length to have maximum at 200-300C.

### Acknowledgements
We are grateful to Prof N Mousseau and F el Mellouhi for the calculation of a neighbour map of the a-Si surface. P W Peacock acknowledges funding from EPSRC.

### References
1. A Gallagher, J App Phys **63** 2406 (1988)
2. A Matsuda, K Tanaka, J Non-Cryst Solids **97** 1367 (1987)
3. J Robertson, J App Phys **87** 2608 (2000)
4. D A Doughty, J R Doyle, G H Lin, A Gallagher, J App Phys **67** 6220 (1990)
5. A H M Smets, W M M Kessels, M C M van de Sanden, App Phys Lett **82** 865 (2003)
6. W M M Kessels, A H M Smets, D C Marra, E S Aydil, D C Schram, M C M van de Sanden, Thin Solid Films **383** 154 (2001)
7. A Gupta, K R Bray, G N Parsons, J Vac Sci Technol, to be published (2003)
8. R Dewarrat, J Robertson, App Phys Lett **82** 883 (2003)
9. J E Northrup, Phys Rev B **44** 1419 (1991)
10. M C Payne, et al, Rev Mod Phys **64** 1045 (1992)
11. G T Barkema, N Mousseau, Phys Rev B **62** 4985 (2000)
12. S Ramalingham, S Sriraman, E S Aydil, D Maroudas, App Phys Lett **78** 2685 (2001)
13. K R Bray, G N Parsons, Phys Rev B **65** 035311 (2001)
14. K Bray, A Gupta, G N Parsons, App Phys Lett **80** 2356 (2002)
15. G Ganguly, A Matsuda, Phys Rev B **47** 3661 (1993)

## The a-Si:H growth mechanism: Temperature study of the SiH$_3$ surface reactivity and the surface silicon hydride composition during film growth

W.M.M. Kessels, Y. Barrell, P.J. van den Oever, J.P.M. Hoefnagels, and M.C.M. van de Sanden
Dept. of Applied Physics, Eindhoven University of Technology, P.O. Box 513, 5600 MB Eindhoven, The Netherlands

## ABSTRACT

We report on two experimental studies carried out to reveal insight into the interaction of SiH$_3$ radicals with the a-Si:H surface as assumed essential in the a-Si:H growth mechanism. The surface reaction probability $\beta$ of SiH$_3$ on the a-Si:H has been investigated by spectroscopic means as a function of the substrate temperature (50 - 450 °C) using the time-resolved cavity ringdown technique. The silicon hydrides –SiH$_x$ on the a-Si:H surface during deposition have been studied by the combination of *in situ* attenuated total reflection infrared spectroscopy and argon ion-induced desorption of surface hydrogen. For SiH$_3$ dominated plasma conditions, it is found that the surface reactivity of SiH$_3$ is independent of the substrate temperature with $\beta$ = 0.30±0.03 whereas the silicon hydride composition on the a-Si:H surface changes drastically for increasing substrate temperature (from –SiH$_3$ to =SiH$_2$ to ≡SiH). The implications of these observations for the a-Si:H growth mechanism are addressed.

## INTRODUCTION

Although the efforts on constructing a growth model for a-Si:H have been fruitful and have contributed greatly to the understanding of the a-Si:H deposition process, there are several issues in the growth mechanism of a-Si:H that are still not completely unraveled [1]. The interaction of SiH$_3$ radicals with the a-Si:H surface under different surface conditions (as *e.g.*, determined by the substrate temperature) is one particular unresolved issue. Although recently several SiH$_3$ surface reactions have been proposed on the basis of *ab initio* calculations and simulations such as density-functional theory (DFT) calculations and molecular dynamics (MD) simulations [2,3,4,5,6], there is still insufficient experimental data. Therefore we have carried out dedicated experiments using the expanding thermal plasma (ETP) technique [7] which is well-suited for these kind of studies:
- the conditions in an Ar-H$_2$-SiH$_4$ plasma can be chosen such that a-Si:H film growth is approximately for 90% due to SiH$_3$ radicals as revealed from cavity ringdown spectroscopy and threshold ionization mass spectrometry [8,9];
- ion bombardment does not play a role because the low electron temperature in the plasma leads to a very low self-bias during deposition (<2 V) [10];
- surface reactions by atomic hydrogen H from the plasma are of very minor importance as there are clear indications that the H flux towards the a-Si:H is much lower than the SiH$_3$ flux [11].

In this paper, we will address the surface reactivity of SiH$_3$ and the silicon hydride composition of the a-Si:H surface as a function of the substrate temperature.

# SURFACE REACTION PROBABILITY OF SiH$_3$ ON THE a-Si:H SURFACE

The surface reactivity of SiH$_3$ has been determined by using time-resolved cavity ringdown spectroscopy as described in Ref. [12]. Briefly, the decrease in SiH$_3$ density at a distance of 5 mm from the substrate is monitored after a minor periodic modulation of the SiH$_3$ density by a rf power pulse. From the time-constant associated with the decrease, *i.e.*, the loss time of SiH$_3$, the surface reaction probability $\beta$ can be determined because it is shown that SiH$_3$ is not reactive in the gas phase [12]. Figure 1 shows the loss time of SiH$_3$ as a function of the reactor pressure for different substrate temperatures. The loss time increases with pressure because SiH$_3$ diffusion to the substrate is slower at higher pressures. At zero pressure however, the SiH$_3$ loss time is unaffected by diffusion ("free-fall limit") and reflects directly the surface reactivity of the SiH$_3$. The (extrapolated) loss time at zero pressure is therefore used to calculate the values of $\beta$ for the different substrate temperatures. In this calculation, necessary information about the diffusion geometry is derived from the slope of the lines in Fig. 1 and a (constant) gas temperature of 1500±200 K is used as derived from Doppler linewidth measurements of the (low-density) Si radicals present in the plasma [13].

Figure 2 shows the values of $\beta$ as obtained for the different substrate temperatures. The figure reveals that $\beta$ of SiH$_3$ is not significantly influenced by the substrate temperature. This is in good agreement with previous investigations which used a very indirect method (*i.e.*, not directly monitoring the SiH$_3$ radical itself) and for which is was not completely clear whether the $\beta$ values derived could (solely) be attributed to SiH$_3$ radicals [14]. Furthermore, the average value of $\beta = 0.30\pm0.03$ is in agreement with most of the values reported in the literature as listed in Table I [15]. This indicates that the $\beta$ value of SiH$_3$ is for the largest part a "radical property" although some influence of the conditions and type of plasma used on $\beta$ cannot be excluded. Moreover, we find a very good agreement with previous studies of $\beta$ of SiH$_3$ done in the same ETP setup using an indirect method under similar plasma conditions [15]. Furthermore, we have also found that the Si growth flux (*i.e.*, product of deposition rate and Si atomic density in the film) is independent of the substrate temperature in the range 50 - 450 °C. This implies that the sticking probability s ($\leq \beta$) is temperature independent [15].

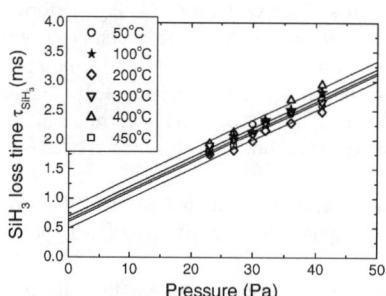
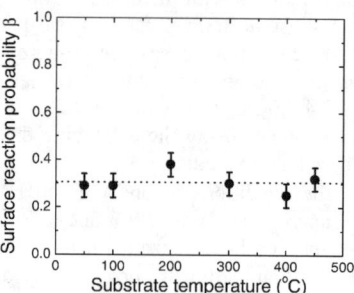

Figure 1. Measured loss times of SiH$_3$ as a function of the reactor pressure for different substrate temperatures.

Figure 2. Surface reaction probability $\beta$ of SiH$_3$ as a function of the substrate temperature.

Table I: Overview of the surface reaction probabilities $\beta$ reported for the SiH$_3$ radical as obtained under different experimental conditions and by several techniques (RT = room temperature).

| | $\beta$ (substrate temperature) | Experimental conditions | Technique applied | Ref. |
|---|---|---|---|---|
| indirect techniques | 0.10±0.01 (RT)<br>- 0.21±0.01 (350 °C) | Hg photo-CVD,<br>SiH$_4$ with Hg | grid | [16] |
| | 0.26±0.02 (240 °C) | rf triode, SiH$_4$ | grid | [17] |
| | 0.26±0.05 (RT – 480 °C) | rf triode, SiH$_4$ | grid and trench | [14] |
| | 0.29<br>0.33<br>0.37<br>(20 – 250 °C) | dc triode, SiH$_4$<br>dc anode, SiH$_4$<br>rf diode, SiH$_4$ | aperture-well assembly | [18] |
| | 0.18 (RT) | rf diode, H$_2$-SiH$_4$ (63%) | infrared laser absorption | [19,20] |
| | 0.28±0.05 (200 °C) | hollow cathode (CVD-like), SiH$_4$ | macroscopic trench | [21] |
| | 0.25±0.06 (250 °C - 325 °C)<br>0.33±0.05 (400 °C) | ETP plasma,<br>Ar-H$_2$-SiH$_4$ | aperture-well assembly | [15] |
| direct techniques | 0.05±0.01 (RT) | microwave,<br>He-Cl$_2$-SiH$_4$ (6%) | mass spectrometry | [22] |
| | 0.28±0.03 (300 °C) | rf diode (afterglow), SiH$_4$ | appearance potential mass spectrometry | [23] |
| | 0.15 (RT)<br><br>0.03 (RT) | rf diode,<br>He-SiH$_4$ (50%)<br>in afterglow | infrared laser absorption | [24] |
| | 0.30±0.03 (50 - 450 °C) | ETP plasma,<br>Ar-H$_2$-SiH$_4$ | cavity ringdown spectroscopy | present work |
| | 0.18 (RT) | - | molecular dynamics | [2] |

## COMPOSITION OF THE SILICON HYDRIDES ON THE a-Si:H SURFACE

The composition of the silicion hydrides $-$SiH$_x$ on the a-Si:H surface during deposition has been investigated by means of very sensitive *in situ* infrared absorption spectroscopy measurements using the attenuated total reflection (ATR) technique. Films of a-Si:H have been deposited on a GaAs ATR crystal for three substrate temperatures (100, 250, 400 °C) and surface specificity has been obtained by ion-induced desorption of the surface hydrides by means of exposing the film to a gentle Ar plasma for 10 s [25]. By comparing the infrared spectra before and after this desorption step, information is obtained about the surface hydrides removed from the a-Si:H surface and consequently about the surface hydrides initially present on the a-Si:H.

The results depicted in Fig. 3 show the relative composition of the $-$SiH$_x$ species on the a-Si:H surface for the three substrate temperatures. Conclusions about the hydrogen coverage of the surface cannot easily be deduced from the measurements [26] although the data indicate that the surface contains a considerable amount of hydrogen. The results are in good agreement with previous investigations for ICP plasma deposited a-Si:H films which revealed that the dominant silicon hydrides on the surface change from $-$SiH$_3$ to $=$SiH$_2$ to $\equiv$SiH for increasing substrate temperature [26]. A very striking result is however that this drastic change of the a-Si:H surface as a function of substrate temperature does not affect the surface reaction probability $\beta$ of SiH$_3$.

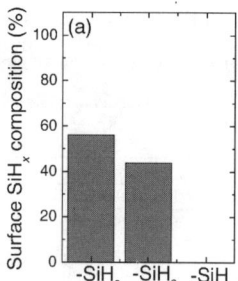 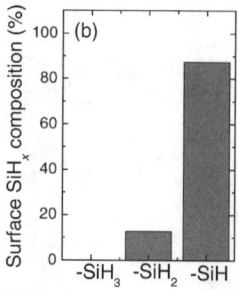 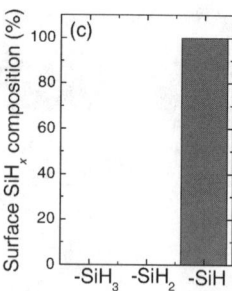

**Figure 3.** Relative surface hydride (–SiH$_x$) composition of a-Si:H as obtained by *in situ* attenuated total reflection infrared spectroscopy for substrate temperatures of (a) 100 °C, (b) 250 °C, and (c) 400 °C.

## DISCUSSION ON THE IMPLICATIONS FOR THE a-Si:H GROWTH MODEL

The experimental observations discussed above have direct consequences for the a-Si:H growth mechanism in terms of the SiH$_3$ surface reactions. We will discuss on these consequences by briefly considering some of the SiH$_3$ surface reactions reported in the literature.

First of all, Matsuda *et al.* have explained their observation that the surface reaction probability β of SiH$_3$ is independent of the substrate temperature by the so-called physisorbed state of SiH$_3$ [14]. Following Gallagher [27], they proposed that SiH$_3$ can adsorb (physisorb) everywhere on an almost fully hydrogenated surface by forming a three-center Si–H–Si bond on surface SiH$_x$ sites [Fig. 4(a)]. The SiH$_3$ in this physisorbed state subsequently diffuses over the surface until it finally sticks on a surface dangling bond [Fig. 4(a2)] or until it abstracts a H atom from the surface forming gaseous SiH$_4$ and a surface dangling bond [Fig. 4(a1)]. However, recent *ab initio* calculations have revealed that the three-center Si–H–Si bond is not stable [4] and therefore other surface reactions of SiH$_3$ with a-Si:H have been subject of investigation.

One reaction that has given particular consideration in the literature is the so-called "insertion" reaction of SiH$_3$ into strained Si–Si surface bonds [Fig. 4(b)]. This reaction, which does not require dangling bonds for the SiH$_3$ to stick at the surface, has been proposed on the basis of surface infrared studies [28] and has also been observed in DFT calculations and MD simulations [2,6]. However, it is expected that this insertion reaction is heavily substrate temperature dependent because it relies on the presence of strained Si–Si bonds on the a-Si:H surface. These strained bonds are more likely at high temperatures when the surface is composed of ≡SiH rather than at low substrate temperatures when the surface contains mainly –SiH$_3$ hydrides. Furthermore, the calculated activation energy of the insertion reaction is in the range of 0.7–0.9 eV [6] and is therefore also not compatible with the substrate temperature independent surface reaction probability β of SiH$_3$.

A reaction that would be more compatible with the temperature independent β of SiH$_3$ is H abstraction from the surface directly by SiH$_3$ from the gas phase [Fig. 4(c)]. This so-called Eley-Rideal type of reaction has been observed in MD simulations and DFT calculations and the calculated activation energy is relatively low (~0.09 eV) [3]. Direct H abstraction would lead to an (almost) temperature independent generation mechanism of dangling bonds on the a-Si:H surface and these dangling bonds can act as growth sites for other incoming SiH$_3$ radicals. The adsorption reaction of SiH$_3$ on top of these surface dangling bonds [Fig. 4(d)] is not expected to be temperature dependent. Therefore the combination of H abstraction and SiH$_3$ adsorption onto the dangling bond created can explain the temperature independent β of SiH$_3$ in the case that the H abstraction reaction is the rate-limiting step. In this reaction sequence, it is not required that the adsorption of the SiH$_3$ on the surface dangling bond takes place directly from the gas phase. Precursor-mediated sticking of SiH$_3$ is compatible with the observations (as long as it is not the rate-limiting step) and it might even be necessary in order to keep the dangling bond density on the a-Si:H surface sufficiently low [1]. Furthermore it has to be noted that the reaction sequence is also compatible with the changing $-SiH_x$ surface composition. Although, relatively more H atoms are present at the surface at lower temperatures, not all these H atoms will be available for H abstraction because also the incorporation rate of H into the a-Si:H is higher at low substrate temperatures (*i.e.*, the H content of a-Si:H increases when going to lower temperatures).

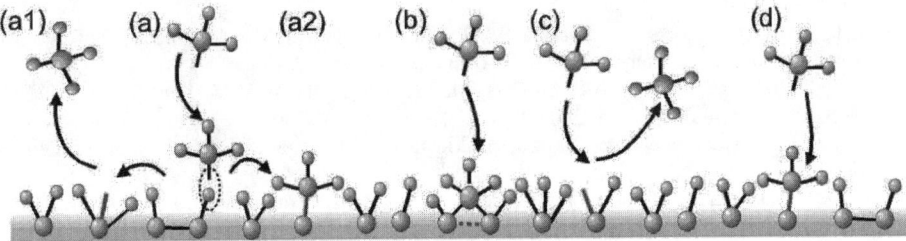

**Figure 4.** Proposed SiH$_3$ surface reactions: (a) physisorption with (a1) H abstraction and (a2) sticking; (b) insertion into strained Si–Si bonds; (c) direct H abstraction creating a dangling bond and SiH$_4$ molecule; (d) SiH$_3$ adsorption onto a dangling bond.

*Comment to the editor:* The authors realize that more views on a-Si:H growth exist (See, e.g., Ref. [5] and P. Vigneron, P.W. Peacock, and J. Robertson in these proceedings) and the different reaction mechanisms proposed in the literature can certainly not be excluded *a priori*. It remains however to be revealed how the different reactions mechanisms account for the experimental data presented in this paper. The intention of this paper is therefore to trigger discussion on the a-Si:H growth mechanism from different perspectives within the "boundary conditions" of the experimental results reported in the literature such that the a-Si:H growth mechanism will finally be resolved.

## ACKNOWLEDGMENTS

The authors acknowledge M.J.F. van de Sande, J.F.C. Jansen, A.B.M. Hüsken, and H.M.M. de Jong for their skilful technical assistance. This work was supported by the Netherlands Foundation for Fundamental Research on Matter. The research of W.K. has been made possible by a fellowship of the Royal Netherlands Academy of Arts and Sciences (KNAW).

# REFERENCES

[1] W.M.M. Kessels, A.H.M. Smets, D.C. Marra, E.S. Aydil, D.C. Schram, and M.C.M. van de Sanden, Thin Solid Films **383**, 154 (2001).
[2] S. Ramalingam, D. Maroudas, and E.S. Aydil, J. Appl. Phys. **86**, 2872 (1999).
[3] S. Ramalingam, D. Maroudas, E.S. Aydil, and S.P. Walch, Surf. Sci. **418**, L8 (1998).
[4] A. Gupta, H. Yang, and G.N. Parsons, Surf. Sci. **496**, 307 (2002).
[5] R. Dewarrat and J. Robertson, Appl. Phys. Lett. **82**, 883 (2003).
[6] S. Walch, S. Ramalingam, E.S. Aydil, and D. Maroudas, Chem. Phys. Lett. **329**, 304 (2000).
[7] W.M.M. Kessels, R.J. Severens, A.H.M. Smets, B.A. Korevaar, G.J. Adriaenssens, D.C. Schram, and M.C.M. van de Sanden, J. Appl. Phys. **89**, 2404 (2001).
[8] W.M.M. Kessels, M.G.H. Boogaarts, J.P.M. Hoefnagels, D.C. Schram, and M.C.M. van de Sanden, J. Vac. Sci. Technol. A **19**, 1027 (2001).
[9] W.M.M. Kessels, J.P.M. Hoefnagels, M.G.H. Boogaarts, D.C. Schram, and M.C.M. van de Sanden, J. Appl. Phys. **89**, 2065 (2001).
[10] W.M.M. Kessels, C.M. Leewis, M.C.M. van de Sanden, and D.C. Schram, J. Appl. Phys. **86**, 4029 (1999).
[11] W.M.M. Kessels, A. Leroux, M.G.H. Boogaarts, J.P.M. Hoefnagels, M.C.M. van de Sanden, and D.C. Schram, J. Vac. Sci. Technol. A **19**, 467 (2001).
[12] J.P.M. Hoefnagels, A.A.E. Stevens, M.G.H. Boogaarts, W.M.M. Kessels, and M.C.M. van de Sanden, Chem. Phys. Lett. **360**, 189 (2002).
[13] J.P.M. Hoefnagels, Y. Barrell, W.M.M. Kessels, and M.C.M. van de Sanden, to be published.
[14] A. Matsuda, K. Nomoto, Y. Takeuchi, A. Suzuki, A. Yuuki, and J. Perrin, Surf. Sci. **27**, 50 (1990).
[15] W.M.M. Kessels, M.C.M. van de Sanden, R.J. Severens, and D.C. Schram, J. Appl. Phys. **87**, 3313 (2000).
[16] J. Perrin and T. Broekhuizen, Appl. Phys. Lett. **50**, 433 (1987).
[17] J. Perrin, Y. Takeda, N. Hirano, Y. Takeuchi, and A. Matsuda, Surf. Sci. **210**, 114 (1989).
[18] D.A. Doughty, J.R. Doyle, G.H. Lin, and A. Gallagher, J. Appl. Phys. **67**, 6220 (1990).
[19] N. Itabashi, N. Nishikawa, M. Magane, S. Naito, T. Goto, A. Matsuda, C. Yamada, and E. Hirota, Jpn. J. Appl. Phys. **29**, L505 (1990).
[20] M. Shiratani, H. Kawasaki, T. Fukuzawa, Y. Watanabe, Y. Yamamoto, S. Suganuma, M. Hori, and T. Goto, J. Phys. D **31**, 776 (1998).
[21] A. Nurrudin, J.R. Doyle, and J.R. Abelson, J. Appl. Phys. **76**, 3123 (1994).
[22] J.M. Jasinski, J. Phys. Chem. **97**, 5037 (1993).
[23] J. Perrin, M. Shiratani, P. Kae-Nune, H. Videlot, J. Jolly and J. Guillon, J. Vac. Sci. Technol. A **16**, 278 (1998).
[24] M. Shiratani, H. Kawasaki, T. Fukuzawa, Y. Watanabe, Y. Yamamoto, S. Suganuma, M. Hori, and T. Goto, J. Phys. D **31**, 776 (1998).
[25] W.M.M. Kessels, D.C. Marra, M.C.M. van de Sanden, E.S. Aydil, J. Vac. Sci. Technol. A. **20**, 781 ( 2002).
[26] D.C. Marra, W.M.M. Kessels, M.C.M. van de Sanden, K. Kashefizadeh, E.S. Aydil, Surf. Sci. **530**, 1 (2003).
[27] A. Gallagher, Mater. Res. Soc. Symp. Proc. **70**, 3 (1986).
[28] A. Von Keudell and J.R. Abelson, Phys. Rev. B **59**, 5791 (1999).

## Effects of Excitation Frequency and $H_2$ Dilution on Cluster Generation in Silane High-Frequency Discharges

Masaharu Shiratani, Kazunori Koga, Atsushi Harikai, Takanori Ogata, and Yukio Watanabe
Department of Electronics, Kyushu University, Fukuoka 812-8581, Japan

## ABSTRACT

Reduction of cluster amount in silane discharges is the key to decreasing microstructure parameter $R_\alpha$ of a-Si:H films deposited with the discharges. The cluster amount is found to be reduced more than one order of magnitude using 60 MHz discharges instead of 28 MHz ones or using $H_2$ dilution of an $H_2$/$SiH_4$ ratio of 5. The cluster-suppressed plasma CVD using 60 MHz discharges realizes deposition of a-Si:H films of $R_\alpha \sim 0$ at a fairly high rate of 0.55 nm/s. Moreover, a downstream cluster collection method of high sensitivity has been developed for detecting a small amount of clusters formed under deposition conditions of $R_\alpha < 0.01$.

## INTRODUCTION

Particles below a few nanometers in size (clusters) generated in silane high frequency discharges have been pointed out to be closely related to the light-induced defects in the hydrogenated amorphous silicon (a-Si:H) solar cell [1-4]. Previously, the growth kinetics of clusters has been studied by using two in-situ methods for detecting them: the double pulse discharge (DPD) method [4,5] and the photon counting laser light scattering (PCLLS) method [3, 5-10]. Based on the results obtained in those studies, we have developed a cluster-suppressed plasma CVD method utilizing gas viscous and thermophoretic forces exerted on clusters and have demonstrated deposition of a-Si:H films of $R_\alpha \sim 0$ [11], while a deposition rate of 0.2 nm/s is low compared to a goal value of 2.0 nm/s. The VHF discharge and $H_2$ dilution have been conventionally employed to deposit films of device quality at a high rate. These facts motivate us to study effects of VHF discharge and $H_2$ dilution on cluster generation in the discharge. Moreover, by using the cluster suppression method together with the PCLLS method, we have found that a reduction in cluster amount in the discharge brings about significant decrease of microstructure parameter $R_\alpha$ [11], which is related to light-induced degradation of a-Si:H films [1, 2]. Since clusters under deposition conditions of $R_\alpha < 0.01$ are hardly detected by the PCLLS method, development of an alternative method of high sensitivity for detecting them is necessary for studying correlation between cluster amount and film quality in the regime of interest.

In this paper, we will report on experimental results regarding the effects of VHF discharge and $H_2$ dilution on cluster generation, which are obtained by using the DPD method, and describe a downstream cluster collection (DCC) method by which clusters can be detected even in a range of small cluster amount for $R_\alpha < 0.01$ is satisfied.

## EXPERIMENTAL METHODS

Experiments were carried out using two capacitively-coupled high frequency discharge reactors A and B: the reactor A is of the cluster-suppressed plasma CVD type and employed for depositing a-Si:H films and developing the DCC method; the reactor B is of conventional type and employed for studying effects of excitation frequency and $H_2$ dilution on cluster generation.

For the reactor A, the growth of clusters is suppressed by utilizing gas viscous and thermophoretic forces with reducing gas stagnation regions [12,13]. Figure 1 shows the schematic diagram of reactor A. For this reactor, a powered electrode, of stainless steel mesh, of 120 mm in diameter and a grounded plane electrode of 120 mm in diameter, were placed 20 mm apart in a stainless steel vessel of 315 mm in diameter and 250 mm in height. Two types of the powered electrode were used in the deposition experiments. The electrode 1 and 2 were designed without and with paying attention to the gas flow near the powered electrode, respectively. Gas of $SiH_4$ was supplied towards the center axis of the discharge column through 44 holes of 1 mm in diameter bored in a tube ring of 240 mm in diameter, which was placed at 10 mm above the grounded electrode as shown in Fig. 1. The flow rate and pressure were 5-30 sccm and 9.3 Pa, respectively. The excitation frequency and supplied power were 28-80 MHz and 2-15 W (0.018-0.13 $W/cm^2$), respectively. Gas was pumped out through the powered electrode by a molecular drag pump (pump A) and also through four ports on the side wall by another molecular drag pump (pump B). Gas flow through the powered electrode contributes to reducing cluster growth in the radical generation region around the plasma/sheath boundary. When the grounded electrode is heated up to about 250°C, the gas temperature gradient drives clusters above a few nm in size toward the powered electrode of room temperature. The pump B was employed to reduce accumulation of clusters due to their reflection at the wall and stagnation of gas flow. For the DCC method, the size and density of clusters, which were trapped on a cluster collecting mesh placed in the pumping port through the powered electrode as shown in Fig. 1, were measured using a TEM.

The reactor B was described in detail elsewhere [10,11]. Figure 2 shows the schematic diagram of reactor B. For this reactor, two stainless steel plane electrodes, of 85 mm in diameter, at room temperature were placed at a separation of 50 mm in a Pyrex glass vessel of 94.5 mm in inner diameter. Pure $SiH_4$ or $SiH_4$ diluted with $H_2$ was fed into the reactor at a flow rate of 5-100 sccm and a pressure of 13.3-266 Pa. The excitation frequency used was 13.56, 40 or 60 MHz,

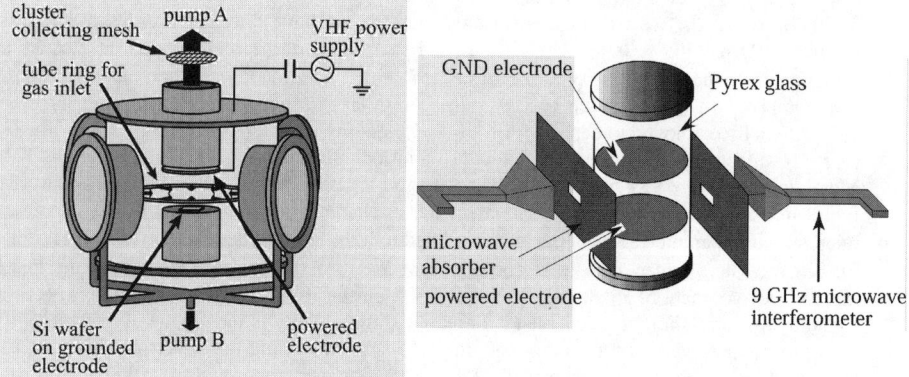

**Figure 1.** Schematic diagram of reactor A.      **Figure 2.** Schematic diagram of reactor B.

and the supplied power was 10W (0.18 W/cm$^2$) for 13.56 and 40 MHz, and 20W (0.35 W/cm$^2$) for 60 MHz. Langmuir probe and 9 GHz microwave interferometer were employed for measuring densities of plasma ions and electrons [9]. The DPD method was employed to detect clusters in the discharges [4-6]. For the DPD method, the size and density of clusters are determined by using both their density decay due to diffusion and that of plasma electrons due to attachment to them after turning off the discharge. In order to obtain information on radical generation rates, the emission intensity from Si was measured with an optical detection system composed of apertures, lens and a monochromator. For deposition experiments, the a-Si:H films of 1 μm in thickness were prepared at a substrate temperature of 250 °C. In these cases, Si (111) wafer with a high resistivity (1000-5000 Ωcm) was placed on the grounded electrode in the reactor A in order to measure the microstructure parameter $R_\alpha$ of a-Si:H film, which is defined as the ratio of absorption intensity at 2100 cm$^{-1}$ (Si-H$_2$ stretching mode) to that at 2000 cm$^{-1}$ (Si-H stretching mode) measured by Fourier-transform infrared (FTIR) spectroscopy. To obtain a high sensitivity for $R_\alpha$, the FTIR measurements were carried out in a vacuum more than 64 times for averaging, and also, baseline determination and curve fitting to deduce $R_\alpha$ were made carefully.

## RESULTS AND DISCUSSION

To study effects of excitation frequency $f$ on growth of clusters, time evolution of their size and density was measured for $f=$ 13.56, 40 and 60 MHz. The results are shown together with time evolution of electron density $n_e$, ion density $n_i$ and optical emission intensity of Si (288nm) $I_{Si}$ in Fig. 3. The maximum $n_e$ value increases from 1.7x10$^9$ cm$^{-3}$ for 13.56 MHz to 1.3x10$^{10}$ cm$^{-3}$ for 60 MHz. The $n_e$ value falls to a quasi-steady state value of about 1.5x10$^9$ cm$^{-3}$ for 40 and 60 MHz, whereas $n_i$'s for all $f$'s and $n_e$ for 13.56 MHz are nearly constant. Kinetic energy of electrons in a bulk regime of their energy distribution, corresponding approximately to the electron temperature $T_e$, becomes lower for the higher $f$. The electron attachment to clusters becomes notable for the higher $f$, since the attachment cross section increases considerably with decreasing $T_e$. Therefore, the $n_e$ begins to

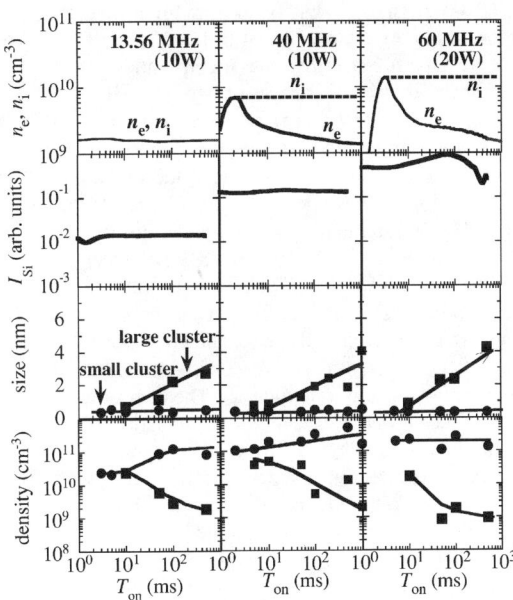

**Figure 3.** Time evolution of electron density $n_e$, ion density $n_i$, optical emission intensity of Si (288nm) $I_{Si}$ as well as size and density of clusters as parameter of $f$. Experimental conditions: SiH$_4$ (100%), 5 sccm, 13.3 Pa, and T= RT.

decrease at $T_{on}$= 2-3 ms after the discharge initiation, when clusters nucleate and start to grow for 40 and 60 MHz. The $I_{Si}$ becomes larger for the higher frequency, indicating that the radical generation rate or deposition increases with $f$. Now, we focus our attention to cluster growth. For three $f$ cases, the small clusters have an almost constant size of 0.5 nm and their density is about $1 \times 10^{11}$ cm$^{-3}$ in the steady state of $T_{on}$> 100 ms. The large clusters nucleate at $T_{on}$ ~ 5 ms for 40 and 60 MHz, and ~ 10 ms for 13.56 MHz. After nucleation, they grow at an almost same growth rate of 4 nm/s independent of $f$. Their steady state density is about $1 \times 10^9$ cm$^{-3}$. Effects of $f$ on the amount of clusters which were incorporated into the depositing a-Si:H films, were studied by evaluating a ratio of volume fraction of clusters observed in the discharge space to a deposition rate of the films on the substrate, since most clusters generated in gas phase are neutral and hence their amount is considered to be proportional to that of clusters incorporated into the films, if a sticking probability of clusters to the films is insensitive to the excitation frequency and cluster size. The results are shown in Fig. 4. The ratios for small and large clusters decrease by about two orders of magnitude when increasing $f$ from 13.56 MHz to 60 MHz. The ratio with respect to all clusters (total in Fig. 4) is close to that of large clusters. The results suggest that the incorporation of clusters is drastically suppressed for 60 MHz. Figure 5 shows dependence of $R_\alpha$ on deposition rate. Solid circles and squares indicate results for the powered electrode 1. The value of $R_\alpha$= 0.084 at a high deposition rate of 0.74 nm/s for 60 MHz still remains low compared to $R_\alpha$= 0.115 at a rate of 0.38 nm/s for 28 MHz. Open squares and triangles in Fig. 5 show results for the electrode 2. The $R_\alpha$ values are notably improved compared to those for the electrode 1. Especially, a-Si:H films of $R_\alpha$~ 0 is deposited at 0.55 nm/s for the electrode 2. These results show that the VHF discharge and reduction of gas stagnation region near the powered electrode are effective in reducing $R_\alpha$.

Effects of $H_2$ dilution on the amount of clusters which were incorporated into the depositing a-Si:H films, were studied by evaluating a ratio of volume fraction of clusters in gas phase to a deposition rate of the films. The results are shown in Fig. 6. The large clusters were hard to be detected for $P_{SiH4}$=20 and 50%. For the small clusters, the ratio decreases by about one fortieth when decreasing

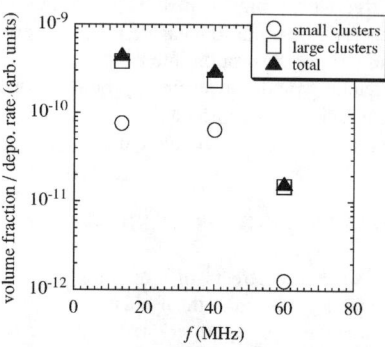

**Figure 4.** Frequency $f$ dependence of volume fraction of clusters divided by deposition rate. Experimental conditions: SiH$_4$ (100%), 5 sccm, 13.3 Pa, and T= RT.

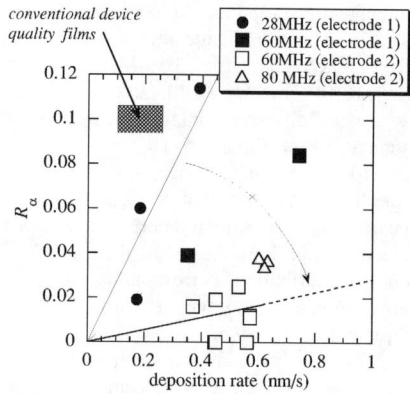

**Figure 5.** Dependence of $R_\alpha$ on deposition rate. Experimental conditions: SiH$_4$ (100%), 20- 60 sccm, 9.3 Pa, and T= 623 K.

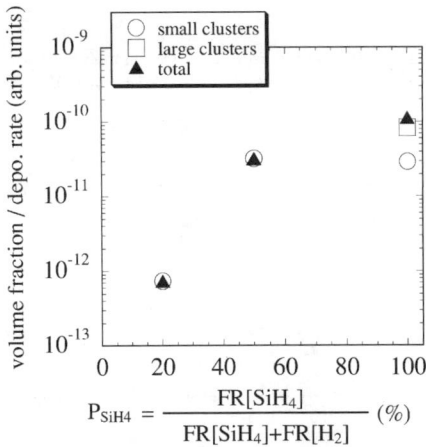

**Figure 6.** Dependence of volume fraction of clusters divided by deposition rate on $P_{SiH4}$.

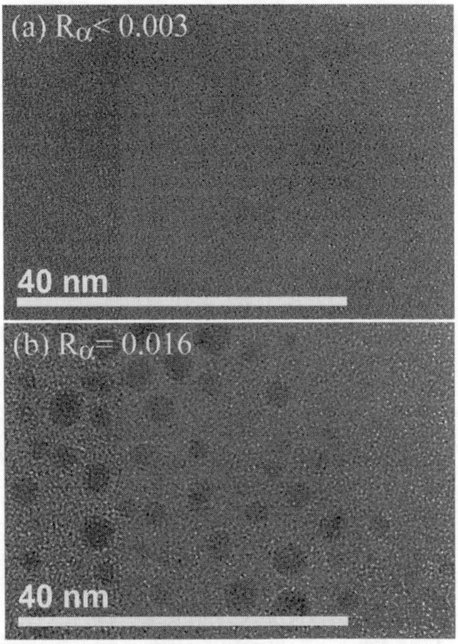

$P_{SiH4}$ from 100 % to 20 %. The results suggest that the incorporation of clusters is drastically suppressed by employing an appropriate $H_2$ dilution.

**Figure 7.** Typical TEM images of collected clusters for (a) $R_\alpha < 0.003$ and (b) $R_\alpha = 0.016$.

In the range of $R_\alpha < 0.01$ as established in our experiments, the clusters in the discharge space are hard to be detected using the PCLLS method. In order to detect them in such $R_\alpha$ range, we have developed the DCC method. The method has a high sensitivity for cluster detection, since most clusters passed through the powered electrode are accumulated on a mesh placed in the pumping duct during the whole discharging period. In other words, the method is considered to enhance its sensitivity by sacrificing its spatial and time resolutions. Figure 7 shows the typical TEM images of collected clusters for (a) $R_\alpha < 0.003$ and (b) $R_\alpha = 0.016$. While clusters of a few nm in size are observed for $R_\alpha = 0.016$, those are not for $R_\alpha < 0.03$. Table 1 shows the estimated size and density of clusters in the discharge region for $R_\alpha < 0.003$ and $R_\alpha = 0.016$. The discharge without observation of large clusters (> 1nm) results in deposition of a-Si:H films of $R_\alpha < 0.003$. Figure 8 shows dependence of $R_\alpha$ on amount of large clusters obtained in this study (solid squares) together with our previous one (open squares). All the results clearly show that the $R_\alpha$ decreases with the cluster amount. Furthermore, the results show that the DCC method has a lower detection limit of clusters than the PCLLS method.

| $R_\alpha$ | < 0.003 | 0.016 |
|---|---|---|
| size $d$ (nm) | < 1 | 3.8 |
| density $n$ (cm$^{-3}$) | <<1x10$^4$ @ 1nm | <1x10$^7$ |

**Table 1.** Cluster-size and -density in discharge region estimated for $R_\alpha < 0.003$ and $R_\alpha = 0.016$. Experimental conditions: SiH$_4$ 100%, 30 sccm, 9.3 Pa, 60 MHz, and T= 250°C.

## CONCLUSIONS

The cluster amount is found to be reduced more than one order of magnitude using 60 MHz discharges or using $H_2$ dilution of an $H_2/SiH_4$ ratio of 5. We have shown that the combination of cluster-suppression method and VHF discharge realizes deposition of a-Si:H films of $R_\alpha \sim 0$ at 0.55 nm/s. Moreover, we have developed the DCC method for detecting clusters formed even in the range of small cluster amount for which $R_\alpha$ is smaller than 0.01. Using the DCC method, we have found that the discharge without observing the large clusters (> 1nm) results in deposition of a-Si:H films of $R_\alpha \sim 0$

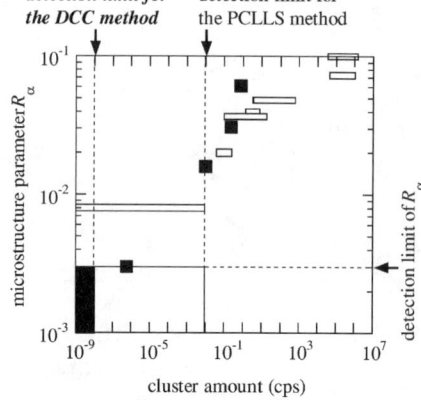

**Figure 8.** Dependence of $R_\alpha$ on cluster amount. Open and solid squares are data obtained by PCLLS and DCC method respectively.

## ACKNOWLEDGMENTS

This work was partly supported by a Grant-in-Aid for Scientific Research (A) from the Japan Society of the Promotion of Science (KAKENHI 14205047). We would like to acknowledge the assistance of Mr. T. Kinoshita who contributed greatly to the preparation of the experimental set-up.

## REFERENCES

1. M. Takai, T. Nishimoto, T. Takagi, M. Kondo, and A. Matsuda, J. Non-Cryst. Solids **266**, 90 (2000).
2. H. Miyahara, M. Takai, T. Nishimoto, M. Kondo, and A. Matsuda, Solar Energy Materials And Solar Cells **74**, 351, (2002).
3. M. Shiratani, S. Maeda, K. Koga, and Y. Watanabe, Jpn. J. Appl. Phys. **39**, 287 (2000).
4. K. Koga, Y. Matsuoka, K. Tanaka, M. Shiratani, and Y. Watanabe, Appl. Phys. Lett.**77**, 196 (2000).
5. K. Koga, M. Shisratani, and Y. Watanabe, Proc. Nano-technology Workshop, pp. 13-20, Dejun, Korea, Feb. 2002.
6. M. Shiratani and Y. Watanabe, Rev. of Laser Eng. **26**, 449 (1998).
7. M. Shiratani, T. Fukuzawa, and Y. Watanabe, Jpn. J. Appl. Phys. **38**, 4525 (1999).
8. Y. Watanabe, M. Shiratani, T. Fukuzawa, and K. Koga, J. Tech. Phys., **41**, 505 (2000).
9. M. Shiratani, T. Fukuzawa, K. Eto, and Y. Watanabe, Jpn. J. Appl. Phys. **31**, L1791 (1992).
10. Y. Watanabe, M. Shiratani, and K. Koga, Plasma Sources & Sci. Tech. **11**, A229 (2002).
11. Y. Watanabe, A. Harikai, K. Koga, and M. Shiratani, Pure Appl. Chem. **74**, 483 (2002).
12. K. Koga, M. Kai, M. Shiratani, Y. Watanabe, and N. Shikatani, Jpn. J. Appl. Phys. **41**, L168 (2002).
13. M. Shiratani, M. Kai, K. Koga, and Y. Watanabe, Thin Solid Films **427**, 1 (2003).

## Application of Deposition Phase Diagrams for the Optimization of a-Si:H-Based Materials and Solar Cells

R. W. Collins, A. S. Ferlauto, G. M. Ferreira, Joohyun Koh, Chi Chen,
R. J. Koval, J. M. Pearce, C. R. Wronski, M. M. Al-Jassim[1] and K. M. Jones[1]
Department of Physics, Department of Electrical Engineering, and Materials Research Institute, The Pennsylvania State University, University Park, Pennsylvania 16802, USA.
[1]National Renewable Energy Laboratory, Golden, Colorado 80401, USA.

## ABSTRACT

Real time spectroscopic ellipsometry (RTSE) has been applied to develop deposition phase diagrams that can guide the fabrication of hydrogenated silicon (Si:H) thin films at low temperatures (<300°C) for highest performance electronic devices such as solar cells. The simplest phase diagrams incorporate a single transition from the amorphous (a) growth regime to the mixed-phase amorphous + microcrystalline (a+μc) growth regime versus accumulated film thickness [the a→(a+μc) transition]. These phase diagrams have shown that optimization of amorphous silicon (a-Si:H) intrinsic (i) layers by rf plasma-enhanced chemical vapor deposition (PECVD) at low rates is achieved using so-called *protocrystalline* Si:H. This material is deposited with the maximum possible flow ratio of $H_2$ to $SiH_4$ while stopping short of the a→(a+μc) transition that inevitably occurs for film thicknesses greater than the desired value in the optimized device. The simplest phase diagrams can be extended to include in addition the thickness at which a roughening transition is detected in the amorphous film growth regime. It is proposed that optimization of a-Si:H in higher rate rf PECVD processes further requires the maximum possible thickness onset for this roughening transition.

## INTRODUCTION

Accumulated research results over the last several years have demonstrated new principles for the optimization of hydrogenated amorphous silicon (a-Si:H) intrinsic layers (i-layers) for highest performance and stability solar cells. In this previous research, it has been found that, in the rf plasma-enhanced chemical vapor deposition (PECVD) process, the optimum a-Si:H and related alloy films are obtained at the *maximum* possible value of the $H_2$-dilution ratio $R=[H_2]/[SiH_4]$ sustainable without entering the microcrystalline silicon (μc-Si:H) film growth regime [1-6]. More surprisingly, it has also been suggested recently that the optimum μc-Si:H i-layers for solar cells are deposited using the minimum possible R value while maintaining predominantly μc-Si:H, in other words without entering the a-Si:H film growth regime [7,8]. Owing to the critical importance of Si:H deposition near the boundaries between the a-Si:H and μc-Si:H growth regimes, the development and application of deposition phase diagrams is a research topic of great interest for the design of a-Si:H and μc-Si:H based solar cells [4,9,10].

The deposition phase diagram describes the accumulated thicknesses at which different microstructural and phase transitions are observed during the Si:H film growth process [10]. In such a diagram, the transition thicknesses are plotted as continuous functions of a key deposition parameter. In low temperature PECVD, the $H_2$-dilution gas flow ratio $R=[H_2]/[SiH_4]$ is used as the abscissa of the phase diagram since it exerts the greatest control over the phase of the film -- from a-Si:H at low R to μc-Si:H at high R. Over a wide range in R, however, Si microcrystallites have been observed to nucleate from within the growing a-Si:H phase after a critical phase-transition thickness that decreases with increasing R. As a result, the amorphous-to-(mixed-phase microcrystalline) [a→(a+μc)] transition boundary is not vertical, but instead exhibits a negative slope vs. R in the R-$d_b$ plane, where $d_b$ designates the bulk layer thickness. This in turn implies that the optimum preparation procedures for the a-Si:H and μc-Si:H i-layers depend on the desired thickness of the i-layers, and paves the way for multistep optimization [3].

The phase diagrams have led to the concept of *protocrystalline* Si:H deposition [3]. There are three important characteristics of this film growth regime [11].

(1) As its name implies, the protocrystalline growth regime is one in which a-Si:H is deposited initially, but given sufficient accumulated thickness, microcrystallites nucleate from the amorphous phase. Thus, the growing film will ultimately evolve first to mixed-phase (a+μc)-Si:H and finally to single-phase μc-Si:H. Once the a→(a+μc) transition is detected, however, the growing material is no longer considered protocrystalline.

(2) A second characteristic of the protocrystalline growth regime is the substrate dependence of the phase of the growing material. If the Si:H film grows in the protocrystalline regime on a freshly-deposited amorphous film substrate (such as R=0 a-Si:H), the same deposition conditions would lead to single-phase microcrystalline silicon growth on a freshly-deposited μc-Si:H substrate film. Thus, under protocrystalline growth conditions, local epitaxy is favored on a c-Si substrate; however, crystallite nucleation is suppressed on an amorphous substrate.

(3) A third characteristic of the protocrystalline growth regime is the observed enhanced degree of nuclei coalescence that yields the smoothest surfaces among a-Si:H films. Thus, the dielectric discontinuity between the ambient and bulk film is sharpest under protocrystalline growth conditions. Such a characteristic can best be detected, however, by using oxide-covered c-Si wafer substrates. (The presence of the oxide suppresses Si crystallite nucleation in the protocrystalline regime relative to clean c-Si or μc-Si:H so that the protocrystalline state can be observed.) By using c-Si substrates, the very smooth wafer surfaces prevent the development of substrate-induced roughness in the film that masks the nuclei coalescence behavior.

In addition to the unique evolutionary growth behavior exhibited under the protocrystalline Si:H deposition conditions, the protocrystalline material itself exhibits unique optoelectronic properties when deposited under optimized low rf plasma power conditions [3,4,12,13]. These properties are often difficult to measure because one must use the appropriate substrate and thickness to ensure that the film is protocrystalline throughout its thickness and has not crossed the a→(a+μc) transition during the growth process. First, the optical gaps of protocrystalline Si:H are larger than conventional materials due to the increase in gap with increasing $H_2$-dilution ratio R. In addition, the width of the broad Lorentzian-shaped peak in the imaginary part of the dielectric function $\varepsilon_2$ is narrower in protocrystalline Si:H than in conventional a-Si:H materials. This suggests that the relaxation time of the excited electron and hole in the bands is the longest. The Si bond-packing density is not the highest in protocrystalline Si:H, however, suggesting the presence of voids that are observed to increase in density with increasing R. Apparently these voids are not detrimental to the electronic properties. Perhaps the key feature of the protocrystalline Si:H is its relative stability to light induced degradation, as observed for films in their electron mobility-lifetime products and similarly for solar cells in their fill factors.

In this article, the application of real time spectroscopic ellipsometry (RTSE) is described for the development of phase diagrams that elucidate a-Si:H deposition processes and identify protocrystalline Si:H growth regimes. Illustrative phase diagrams are presented here that demonstrate a systematic approach for materials and device optimization under different deposition conditions.

## MICROSTRUCTURAL AND PHASE TRANSITIONS IN Si:H FILMS

### Overview

Figure 1 shows the overlying surface roughness layer thickness $d_s$ versus the bulk layer thickness $d_b$ for three Si:H depositions as deduced by RTSE using a rotating compensator multichannel ellipsometer. The depositions include (a) two i-layers, one prepared at R=0 and the other at R=10, that remain single-phase a-Si:H throughout growth, and (b) an i-layer prepared at R=20 that evolves through all three growth regimes: first single-phase a-Si:H, then mixed-phase (a+μc)-Si:H, and finally single-phase μc-Si:H. For these three depositions, the substrates were c-Si wafers with intact native oxides, held at a temperature of T=200 °C. The rf power was set at P=0.08 W/cm², the lowest level for a stable plasma. A low partial pressure of $SiH_4$ within the range of p($SiH_4$)=0.03-0.07 Torr was maintained versus R; thus, the total pressure increased from $p_{tot}$=0.07 Torr for R=0 to $p_{tot}$=0.9 Torr for R=40. With this combination of variables, the a-Si:H deposition rates were ~1.2 A/s at R=0 and 0.5 A/s at R=10 just before the transition to mixed-phase (a+µc)-Si:H deposition for a thick layer.

The smoothening behavior in the first ~100 Å of film growth for all three depositions in Fig. 1 is attributed to the coalescence of initial amorphous nuclei that form as clusters on the c-Si substrates. A detailed understanding of this process has been established previously [14,15]. The weak roughening onset at $d_b$~250 Å for the R=0 a-Si:H film in Fig. 1(a) corresponds to a surface morphological transition versus $d_b$, but without a change in the phase of the growing film. As a result, this transition is denoted as "a→a" to indicate that the film is a-Si:H on both sides of the transition. For this R=0 film, a simple two-layer [(uniform-bulk)/(surface-roughness)] optical model was applied to analyze the RTSE data set. For the R=20 film in Fig. 1(b), the roughening onset at $d_b$~300 Å corresponds instead to the a→(a+µc) transition, and the smoothening onset near $d_b$~1900 Å corresponds to the (a+µc)→µc transition. For this film, an optical model with a graded bulk layer was applied to analyze the RTSE data set, as will be described in greater detail later. For the R=10 deposition in Fig. 1(a), a stable surface is observed after initial nuclei coalescence (i.e., for $d_b$>100 Å), indicating that any roughening onset due to the a→a or the a→(a+µc) transition must occur for $d_b$>4000 Å, i.e., greater than the typical thicknesses of the i-layers used in a-Si:H solar cells. Thus, the RTSE data analysis for the R=10 deposition proceeded in the same way as that for the R=0 deposition.

The general features of the microstructural and phase evolution as deduced from the analysis of RTSE data such as those of Fig. 1 are enumerated (I-VI) with further details as follows.

(I) *Coalescence of initial amorphous nuclei.* Coalescence of nucleation-induced microstructure is observed as a smoothening effect during a-Si:H film growth in the first ~100 Å of bulk layer thickness. The magnitude of this smoothening effect can be characterized by $\Delta d_s$ = $d_s$(2.5 Å) − $d_s$(100 Å) [where $d_s$(x) is the value of $d_s$ when $d_b$=x]. The value of $\Delta d_s$ increases with increasing R in general; thus, the maximum $\Delta d_s$ values are observed for protocrystalline Si:H. This material exhibits the highest electronic performance for applications as i-layers in solar cells -- as long as protocrystalline growth is maintained throughout deposition to the desired thickness. An increase in $\Delta d_s$ has been proposed to reflect an increase in the surface diffusion length of the adsorbed radicals that form the film [15].

(II) *Coalescence of initial microcrystalline nuclei.* An even larger smoothening effect can be observed upon structural coalescence of clusters that nucleate directly on the substrate as microcrystalline silicon (µc-Si:H). This larger effect can be observed for depositions at higher R values than those depicted in Fig. 1. In this case, a much larger $d_s$ value is typically observed at the onset of bulk layer growth (~45-60 Å vs. 15-20 Å, for a-Si:H nucleation), due to a lower initial nucleation density compared to that of the a-Si:H films. The physical mechanisms that control amorphous and microcrystalline cluster coalescence are likely to be different (e.g., surface diffusion vs. competitive space filling, respectively).

(III) *Stable surface growth.* Under a narrowly-defined set of deposition conditions [specifically, in Fig. 1(a) for R=10, T=200°C, and minimum rf plasma power P=0.08 W/cm$^2$], the a-Si:H surface remains smooth and stable with < 1 Å change in the roughness layer thickness from the end of coalescence throughout thick film growth, e.g., from 100 to 4000 Å in the example of Fig. 1(a). When the stable surface regime is present, it is found to occur at an R value just prior to the a→(a+µc) transition for thick films. Under these conditions, highest performance and stability i-layer materials for solar cells are obtained.

(IV) *Amorphous roughening transition.* If one starts with a deposition exhibiting the stable-surface conditions, for example (R=10, T=200°C, P=0.08 W/cm$^2$) in Fig. 1(a), and gradually decreases R or increases P in successive depositions so that the material properties and stability degrade, then a roughening transition is detected that gradually shifts to lower $d_b$ in the successive depositions. The growing film is amorphous on both sides of the transition, and the associated surface microstructural changes are correlated with reductions in the performance and stability of such films as i-layers in solar cells. A shift in the a→a roughening transition to lower $d_b$ appears to reflect a reduction in the surface diffusion length of the adsorbed radicals.

(V) *Amorphous−to−(mixed-phase-microcrystalline) (a+µc) transition.* At moderate to high values of R, however, a different type of roughening transition is observed in which crystallites nucleate from the growing amorphous phase. Because the nucleation density is usually low and the crystallites grow preferentially, the crystalline protrusions generate a surface roughness layer that increases rapidly in thickness with $d_b$. Once the growing film crosses this transition and the crystallite volume fraction exceeds a critical value, the material becomes unsuitable as an i-layer component of an a-Si:H-based solar cell.

(VI) *(Mixed-phase)–to–(single-phase-microcrystalline) transition.* For thin films that have already undergone the a→(a+μc) transition, a second transition is possible that occurs at even greater bulk layer thickness. In this transition, the crystalline protrusions that extend above the surface have become large enough to make contact, leading to a crystallite coalescence process with continued film growth. This process is manifested in the data as a transition from surface roughening to smoothening during mixed-phase film growth. Once the crystallites have coalesced to cover the growing film surface completely, single-phase μc-Si:H growth proceeds with a resumption of surface roughening. For optimum μc-Si:H i-layers in solar cells, one generally seeks to deposit the film using the lowest R value possible while maintaining the film within the microcrystalline growth regime throughout the deposition [7,8].

The first phase diagrams developed to guide a-Si:H deposition vs. R included only feature (V), i.e., the amorphous–to–(mixed-phase-microcrystalline) (a+μc) transition [3,4]. More recently features (IV), (V), and (VI) in the above paragraphs have been included in so-called *extended* phase diagrams that provide deeper insights into the electronic quality of a-Si:H prepared near the a→(a+μc) transition, as well as provide the thickness ranges and conditions under which single-phase μc-Si:H films are obtained [9,10]. Owing to its importance in the optimization of the a-Si:H deposition process, additional discussion of the detailed behavior and physical mechanisms of the a→a transition (feature IV) and a→(a+μc) transition (feature V) will be presented next. Additional discussion of the a→(a+μc) transition appears in an accompanying article in these Proceedings [16].

## Amorphous roughening transition during film growth

In the first part of this section, it is demonstrated that the R=0 and R=10 a-Si:H films [Fig. 1(a)] can be described correctly in terms of a two-layer optical model [(uniform bulk)/(surface roughness)]. As a result, the roughening transition observed for the R=0 film describes a surface morphological transition without an evolution of the bulk properties of the growing material across the transition.

Figure 2(a) shows the unbiased estimator of the mean square deviation between the experimental and best fit ellipsometric spectra (1.5 - 4.7 eV) versus bulk layer thickness for the R=0 and R=10 depositions of Fig. 1(a). The unbiased estimator describes the quality of the best fit to the RTSE data, whereby the fit is based on the assumption of a two-layer optical model [(uniform bulk)/(surface roughness)] with a thickness-independent bulk layer dielectric function. This dielectric function is extracted by numerical inversion near $d_b$=200 Å – thus, the fit is nearly perfect for that thickness. Figure 2(a) shows that the fits remain good throughout the deposition process (for comparison see the R=20 deposition of the lower panel), with the result for the R=10 deposition exhibiting the best characteristics. It is concluded that the two-layer model is a close description of reality for the R=0 and R=10 films.

The results in Fig. 3 support the validity of our conclusions based on the statistical data in Fig. 2(a). Figure 3 shows dielectric functions for the R=0 film extracted at two different bulk layer thicknesses, $d_b$=200 Å (solid line) and $d_b$=1965 Å (broken line), before and after the a→a transition in the growth process. The results nearly superimpose on the scale of the figure and demonstrate explicitly that the properties of the a-Si:H film are nearly the same above and below the a→a roughening transition. The small differences in Fig. 3, and hence the gradual increase in the biased estimator in Fig. 2, may be attributable to a slightly higher H content in the film near the substrate interface, but this effect is not directly related to the roughening transition.

Next, it is important to emphasize the key role of the a→a roughening transition in the a-Si:H film deposition process and demonstrate typical results as a function of the deposition variable R. Figure 4 shows the surface roughness evolution as a function of bulk layer thickness for two series of a-Si:H films both deposited on c-Si substrates at T=200°C. The first series in

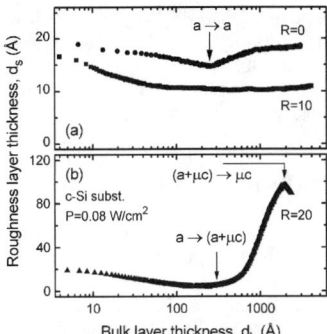

**Figure 1.** Surface roughness layer thickness ($d_s$) versus bulk layer thickness ($d_b$) from RTSE data collected during the deposition of (a) uniform a-Si:H with R=0 and R=10, and (b) structurally-graded (a→μc)-Si:H with R=20, all on c-Si substrates held at 200°C.

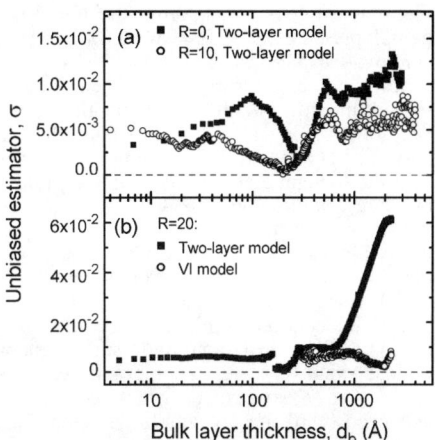

**Figure 2.** Unbiased estimator of the mean square deviation versus time obtained in the RTSE analyses of (a) the uniform R=0 and R=10 a-Si:H films of Fig. 1(a) applying a conventional two-layer optical model, and (b) the graded R=20 Si:H film of Fig. 1(b) applying a two-layer virtual interface model. Also shown in (b) are results obtained when the conventional two-layer model is applied to the R=20 film.

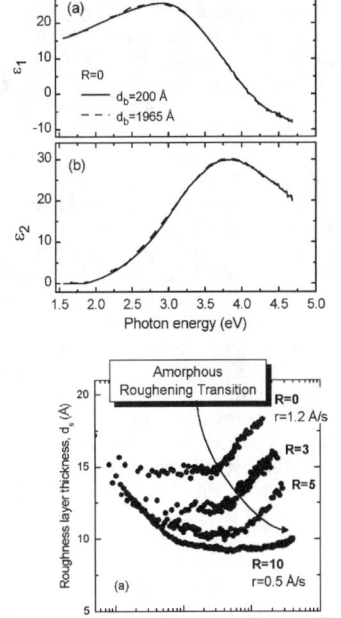

**Figure 3 (left).** (a) Real and (b) imaginary parts of the dielectric functions at 200°C for the R=0 a-Si:H film deposited on c-Si from Fig. 1(a). These results were obtained by RTSE, applying a conventional two-layer model and exact inversion at the thicknesses indicated.

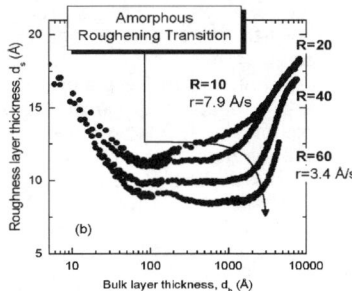

**Figure 4.** Surface roughness layer thickness ($d_s$) versus bulk layer thickness ($d_b$) for two series of a-Si:H films both deposited on c-Si substrates at T=200°C. In (a), the conditions of Fig. 1 are employed, i.e., minimum rf plasma power [P=0.08 W/cm$^2$] and low SiH$_4$ partial pressures [p(SiH$_4$)=0.03-0.07 Torr; $p_{tot}$<0.5 Torr]. In (b), elevated power [P=0.34 W/cm$^2$] and a high total pressure [$p_{tot}$=4 Torr] are employed.

Fig. 4(a) employs the conditions of Fig. 1 with minimum power [P=0.08 W/cm$^2$] and low SiH$_4$ partial pressure [p(SiH$_4$)=0.03-0.07 Torr; p$_{tot}$<0.5 Torr], and the second series in Fig. 4(b) employs elevated power [P=0.34 W/cm$^2$] and high total pressure (p$_{tot}$=4 Torr). For the latter series, the total pressure is controlled, rather than the partial pressure of SiH$_4$ as in the series of Figs. 1 and 4(a). With the combination of variables used for Fig. 4(b), the deposition rate is 3.4 Å/s for an i-layer at the maximum R value of 60 just before the onset of mixed-phase (a+µc)-Si:H deposition (assuming a desired 4000 Å thickness). This rate is a factor of ~7 higher than that obtained at the corresponding maximum R value (R=10) under the lowest power, low pressure conditions for the depositions of Fig. 4(a). In both parts of Fig. 4, an a→a roughening transition is clearly observed for each deposition, and this transition shifts to increasing bulk layer thickness with increasing R. In fact in Fig. 4(b) at R=60, just before the a→(a+µc) transition, the surface of the film is stable to a bulk layer thickness of d$_b$~3000 Å.

The shapes of the curves in Fig. 4, and the development of features (I), (III), and (IV) above can be understood from continuum models of film growth in the following general way (see, e.g., [17-20]). Low amplitude surface modulations A(λ,t) of spatial wavelength λ less than a critical value (λ$_0$) are unstable and decay as a function of time t (or thickness) according to A(λ,t) ~ A$_0$(λ)exp{ω(λ)t}, where ω(λ)<0 for λ<λ$_0$. Features with λ>λ$_0$, such that ω(λ)>0, are enhanced (until the low amplitude approximation breaks down). Typically, the highest rate of enhancement occurs for features with wavelength just above λ$_0$, and the development of features with this correlation length generates the a→a roughening transition. Under most deposition conditions, the correlation length associated with the initial nuclei is less than λ$_0$ and the smoothening of these surface features generates feature (I) described above.

In more specific models, the smoothening effects can be attributed to a *chemical* vapor deposition (CVD) effect, specifically surface diffusion which exhibits a (−a$_4$λ$^{-4}$) term in ω(λ), and the roughening effects can be attributed to a *physical* vapor deposition (PVD) effect, specifically shadowing due to the atomic size which exhibits a (+a$_2$λ$^{-2}$) term in ω(λ) [19,20]. Considering a simple model with only these two sources of roughness evolution, the diffusion length L is determined by the balance of the two terms (i.e., by ω=0 or L=λ$_0$). In more complicated models, the diffusion length does not enter in such a straightforward manner [17,18]. In any case, when the surface diffusion length increases in the a-Si:H deposition process, the onset of roughening shifts to longer times or greater thicknesses, and this correlates with observed improvements in the electronic properties of the films. Although advances in understanding a-Si:H film growth have been made over the years [21], uncertainties continue to exist concerning the diffusion mechanisms [22]. In spite of these uncertainties, it has become increasing clear from the RTSE studies that the longest precursor diffusion lengths are associated with the highest electronic quality and highest stability materials.

**Amorphous-to-(mixed-phase microcrystalline) roughening transition during film growth**

Once the film traverses the phase boundary from the a-Si:H growth regime into the mixed-phase (a+µc)-Si:H growth regime as a function of bulk layer thickness, the two-layer [(uniform bulk)/(surface roughness)] optical model used for analysis of the RTSE data for the R=0 and 10 depositions of Fig. 1(a) is no longer correct. This can be seen in Fig. 2(b) (closed squares) in which case the quality of the fit to the RTSE data for the R=20 film rapidly degrades for bulk layer thicknesses d$_b$>600 Å, i.e., in the mixed-phase (a+µc)-Si:H growth regime. In order to solve this problem and extract the volume fraction of µc-Si:H in the graded film throughout the mixed-phase growth regime, it is easiest to track the dielectric function of the top-most growing material throughout the film deposition. This is an ideal application for the virtual interface (VI) approach to RTSE data analysis [23,24]. In this analysis, one replaces the complicated history of the graded layer deposition by a *pseudo-substrate* whose optical properties are defined by a pseudo-dielectric function established assuming a single roughness layer on an opaque bulk material. With this model for the pseudo-substrate, the properties of the top-most growing material can be established with a two-layer model consisting of an outerlayer of fixed thickness (~30 Å in this case) and variable composition, and an overlying roughness layer of variable thickness. The interface between the pseudo-substrate and the outerlayer is the virtual interface, and it remains at a fixed depth from the surface (as long as the deposition rate and the roughness layer thickness remain constant) throughout the deposition. Figures 1(b), 5, and 6 show the final results of such an analysis as applied in the mixed-phase Si:H growth regime.

Figure 5 compares the (a) real and (b) imaginary parts of the dielectric functions ($\varepsilon_1$, $\varepsilon_2$) for the pure phases obtained in analyses of RTSE data for the R=20 deposition of Fig. 1(b) on oxide-covered c-Si. The dielectric function of the pure a-Si:H phase was determined from the RTSE data collected in the protocrystalline Si:H growth region $0<d_b<200$ Å, before the onset of the a→(a+μc) transition. The dielectric function of the pure μc-Si:H phase was obtained from data collected after crystallite contact during coalescence ($d_b>1900$ Å). In this latter case, the top ~200 Å of the film was characterized assuming a two-layer (outerlayer/roughness) model using a pseudo-substrate for the underlying graded layer. The correct surface roughness layer thickness in this model is determined by applying a global error minimization routine within which the entire VI analysis is embedded. The differences between the two dielectric functions in Fig. 5 are clear and provide support for the interpretation of the surface roughness evolution in Fig. 1(b) in terms of the a→(a+μc) and (a+μc)→μc phase transitions. The solid lines in Fig. 5 represent best fit Kramers-Kronig consistent analytical models for the dielectric functions [25].

Figure 6 shows the evolution of the volume fraction $f_{μc}$ of μc-Si:H in the R=20 film of Figs. 1(b) and 5 throughout the growth region starting from the a→(a+μc) transition and ending near the (a+μc)→μc transition. In the VI analysis used to deduce these results, a two-layer model similar to that of the previous paragraph is assumed. The details of this analysis are given elsewhere [16]. Apparently $d_s$ in Fig. 1(b) provides higher sensitivity to the presence of microcrystallites, since it leads to an a→(a+μc) transition of ~300 Å, whereas the μc-Si:H volume fraction in Fig. 6 first extends above zero for $d_b$~700 Å. The solid line in Fig. 6 is the result for $f_{μc}$ versus $d_b$ employing a model consisting of conical microcrystallites [16]. The conical microcrystallite geometry is evident in Fig. 7, which shows a cross-sectional transmission electron microscopy (XTEM) image from the same sample of Figs. 1(b), 5, and 6, but in a different probe region. One particular cone on the XTEM image is highlighted (lines), as are the positions of nuclei (dots). From the image the microcrystallite cone angle and nucleation density are estimated as θ=25° and $N_d=1.8 \times 10^{10}$ cm$^{-1}$. These estimates are in good agreement with those from RTSE (see solid line in Fig. 6), θ=19° and $N_d=1.1 \times 10^{10}$ cm$^{-1}$.

## DEVELOPMENT AND APPLICATION OF SI:H DEPOSITION PHASE DIAGRAMS

### Crystalline silicon wafers and amorphous silicon film substrates

The extended phase diagram is depicted in Fig. 8 for Si:H growth on oxide-covered c-Si wafer substrates held at 200°C. This diagram was deduced from results including those in Fig. 1, and so is appropriate for Si:H depositions at low rf power (P=0.08 W/cm$^2$) and a low, nearly constant, partial pressure of SiH$_4$ [p(SiH$_4$) = 0.03-0.07 Torr]. The diagram of Fig. 8 shows the bulk layer thicknesses at which the a→a, a→(a+μc), and (a+μc)→μc transitions occur as functions of R (lines). For R<10, the thicknesses corresponding to the a→a transition provide insights into the quality of the a-Si:H, irrespective of the substrate. As suggested by the discussion above, larger a→a transition thicknesses imply longer precursor surface diffusion lengths in the growth process and, thus, higher quality a-Si:H materials for the i-layers of solar cells. In fact, the highest electronic quality material is prepared in the narrow region near R=10 where the a-Si:H surface remains stable throughout the entire deposition to 4000 Å [see Fig. 1(a)]. (The short upward and downward pointing arrows in Fig. 8 indicate transitions that occur at thicknesses above and below the indicated values.)

The a-Si:H prepared at R=10 can be identified as protocrystalline Si:H owing to the fact that if the film were to continue accumulating, then the a→(a+μc) transition would eventually be traversed as suggested by ex situ measurements of films > 1 μm thick. Furthermore, if the R=10 deposition is performed on a single-phase μc-Si:H substrate (rather than on oxide-covered c-Si as in Fig. 8), then the μc-Si:H phase would continue to propagate [4]. Only by reducing R to 5 (outside the protocrystalline growth regime) can one effectively suppress the μc-Si:H phase.

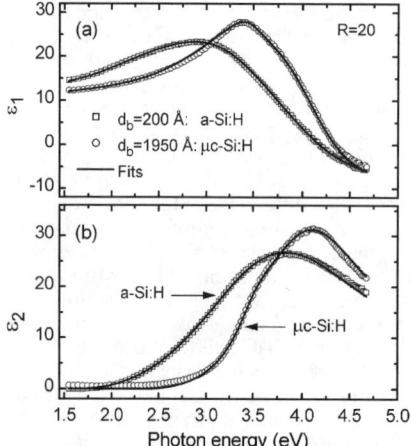

**Figure 5.** (a) Real and (b) imaginary parts of the dielectric functions at 200°C for the R=20 Si:H film deposited on c-Si from Fig. 1(b). The results for a thickness of 200 Å were obtained by RTSE assuming uniform film growth up to that point. The results for a thickness of 1950 Å were obtained applying a two-layer virtual interface model to describe the top ~200 Å of the film. The solid lines are fits using Kramers-Kronig consistent analytical models for the dielectric functions.

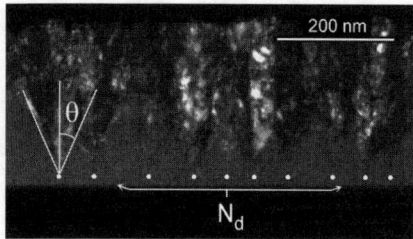

**Figure 7.** (a) Cross-sectional transmission electron microscopy image for the final R=20 Si:H deposition of Figs. 1(b) and 5-6. A representative conical microcrystalline structure is indicated, and the points near the substrate interface identify the nuclei from which the nucleation density is estimated.

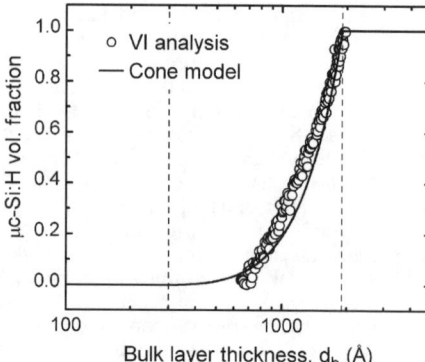

**Figure 6.** Depth profile in the volume fraction of the microcrystalline phase (points) throughout the mixed-phase (a+μc)-Si:H growth region for the R=20 Si:H deposition on c-Si of Figs. 1(b) and 5. Also shown as the solid line is the prediction of a conical microcrystallite growth model.

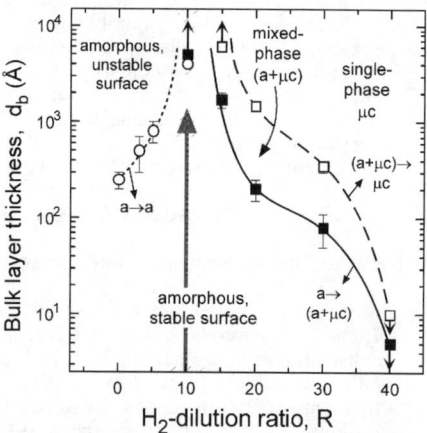

**Figure 8.** Extended phase diagram in the plane of R and $d_b$ for PECVD Si:H prepared at a low rf plasma power of P=0.08 W/cm$^2$ and a low partial pressure of p(SiH$_4$)=0.03-0.07 Torr. The substrates are oxide-covered c-Si wafers held at 200°C. The dotted, solid, and dashed lines identify the a→a, a→(a+μc), and (a+μc)→μc transitions, respectively. The (up, down) arrows connected to data points indicate that the transition occurs (above, below) the identified thicknesses.

From the discussion of the previous paragraph, it is evident that for R≥10 in Fig. 8, the phase evolution is very sensitive to the nature of the substrate, and in this deposition regime it is important to consider substrates that correspond more closely to the a-Si$_{1-x}$C$_x$:H p-layer often used for p-i-n cells (or the a-Si:H n-layer often used for n-i-p cells). Figure 9 shows the corresponding extended phase diagram for an R=0 a-Si:H substrate. This substrate is relevant when considering the optimization of i-layers for devices. However, the starting roughness on the a-Si:H substrate makes it difficult to detect the weak a→a roughening transition at low R as in Figs. 1(a) and 8. As a result, this transition is not shown in Fig. 9. Fortunately, insights in this region of low R can be obtained from the analyses of depositions on c-Si substrates, owing to the relative substrate independence of the a-Si:H growth process at the lower R values.

A comparison of Figs. 8 and 9 show clearly the substrate dependence of the phase evolution of the Si:H films. First, at the lowest R for which the a→(a+μc) is detected, R=15, this transition occurs at a lower thickness for the c-Si substrate (1500 Å for c-Si vs. 3000 Å for the R=0 a-Si:H substrate film). This suggests a higher microcrystallite nucleation density from the amorphous phase when deposition is performed on c-Si. This effect is preserved with increasing R. For R=40, microcrystallites nucleate immediately on the c-Si wafer substrate without an a-Si:H interlayer; however, the a→(a+μc), and (a+μc)→μc transitions occur near 200 and 750 Å for the R=0 a-Si:H film substrate, respectively. Overall the thickness and substrate dependences of the film properties, as well as the graded structures of films that enter the mixed-phase growth regime, lead to difficulties in characterizing the basic properties of the optimum solar cell materials. Under such conditions, identical layer thicknesses and similar substrates must be employed for both the materials and the device structures.

**Optimization principles for optimum solar cell fabrication**

In Fig. 9, the optimum one-step i-layers for solar cells are prepared in the protocrystalline regime at the maximum possible R value while avoiding the a→(a+μc) transition for the desired thickness. Thus, it is important to emphasize that the optimum conditions of i-layer deposition depend on the desired thickness. In fact, this is one of the most important insights provided by RTSE measurements and the resulting deposition phase diagrams [3,4] that had not been fully appreciated in previous studies. Returning to Fig. 9, the optimum H$_2$-dilution for an i-layer thickness of 4000 Å corresponds to R~10, the stable surface growth condition in Figs. 1(a) and 8. Figure 9 also demonstrates how optimum two-step i-layers 4000 Å thick can be designed for p-i-n cells on the basis of the phase diagram (see long vertical arrows) [3]. These steps include a 200 Å R=40 i-layer at the interface to the p-layer, followed by a 3800 Å thick bulk i-layer with R=10. This two-step i-layer provides the highest product of open-circuit voltage ($V_{oc}$) and fill factor (FF) in both the annealed and fully light-soaked states (see Fig. 10 for device studies) [3].

It is important to note, however, that the application of a single phase boundary as in Fig. 9 for guiding the two-step i-layer is an oversimplification. In fact, upon deposition of the R=40 i-layer, the phase boundary for the subsequent R=10 i-layer deposition is shifted to somewhat lower R compared with that for the R=0 substrate film. This effect arises owing to the substrate dependence of the phase boundary. The R=40 layer is more ordered than the R=0 layer, and this appears to promote a more ordered R=10 layer, and as a result, the a→(a+μc) transition occurs at lower $d_b$ for fixed R. The R=10 value for the second step i-layer is sufficiently low, however, to avoid being affected by this shift.

Figure 10 demonstrates more clearly the one-step and two-step optimization processes for a-Si:H p-i-n solar cells. This figure depicts $V_{oc}$ and FF, the latter in both annealed (circles) and degraded (100 hr AM1.5; squares) states, for Si:H p-i-n solar cells incorporating one-step (4000 Å) i-layers plotted as a function of i-layer R (open symbols) and for cells incorporating two-step (100-200 Å / 3900-3800 Å) interface/bulk i-layers plotted as a function of interface i-layer R (closed symbols). The vertical lines denote the R values corresponding to the a→(a+μc) transitions for the bulk i-layer with $d_b$=4000 Å (left) and for the interface i-layer with $d_b$=200 Å (right). In fact, $V_{oc}$ increases with increasing R for the near-interface material as long as the a→(a+μc) transition is not traversed. This effect is believed to be due to two mechanisms, namely the widening of the gap and the narrowing of the band tails with increasing R for the protocrystalline materials. This interpretation is supported by the Kramers-Kronig consistent analysis of the dielectric functions which reveals the former mechanism directly, and the latter indirectly through a narrowing of the Lorentz oscillator component of the protocrystalline Si:H dielectric function. Details of such analyses appear elsewhere [4,25]. The decreases in $V_{oc}$ and FF that occur for the one-step and two-step i-layers for R values above these transition lines are attributed to the development of microcrystallinity in the bulk and interface i-layers with increasing thickness. Thus, Fig. 10 clearly demonstrates that the optimum i-layers in one-step and two-step solar cells are deposited at the maximum possible R value that can be sustained without crossing the a→(a+μc) transition for the desired thickness.

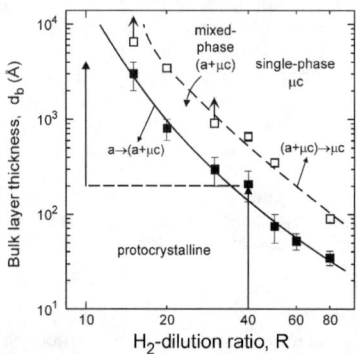 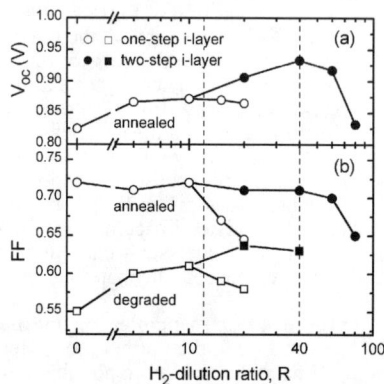

**Figure 9.** Extended phase diagram in the plane of R and $d_b$ for Si:H depositions with R ≥ 10 at a low rf plasma power of P=0.08 W/cm² and low partial pressure of p(SiH₄)=0.03-0.07 Torr. The substrates are R=0 a-Si:H films held at 200°C. The solid and dashed lines identify the a→(a+μc) and (a+μc)→μc transitions, respectively. The up arrows connected to data points indicate that the transition occurs above the identified thickness values.

**Figure 10.** (a) Open circuit voltage $V_{oc}$ and (b) fill-factor FF in annealed states (circles) and 100 hr AM1.5 degraded states (squares) for Si:H p-i-n solar cells incorporating one-step (4000 Å) i-layers versus the i-layer R (open symbols), and for cells incorporating two-step (100-200 Å)/(3900-3800 Å) interface/bulk i-layers versus the interface i-layer R (closed symbols). The vertical lines identify the a→(a+μc) transitions for 4000 Å and 200 Å thick i-layers.

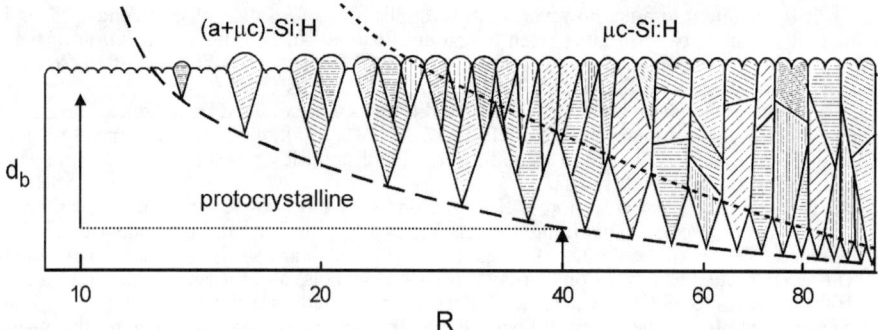

**Figure 11.** Schematic of the structure of Si:H films prepared as a function of R. The dashed and dotted lines identify the a→(a+μc), and (a+μc)→μc transitions, respectively. The arrows denote the two-step protocrystalline Si:H i-layer process found to optimize the performance of p-i-n solar cells as depicted in Fig. 9.

## SUMMARY

The surface microstructure and the phase of hydrogenated silicon (Si:H) thin films prepared by rf PECVD at low temperature evolve as a function of thickness, as is evident from continuous RTSE measurements and analysis. The evolutionary processes that have been reviewed here (and detailed in previous studies [3,4,9,10]) include (i) a nuclei coalescence effect in the initial stages of growth [surface smoothening], (ii) a surface morphological transition in the bulk growth regime without a change in phase [a→a, surface roughening], (iii) a nucleation transition in the bulk regime in which microcrystallites evolve from the growing amorphous phase [a→(a+μc); surface roughening], and (iv) a process of crystallite contact and coalescence in the bulk regime involving a transition from mixed-phase (a+μc)-Si:H to single-phase μc-Si:H [(a+μc)→μc; surface smoothening]. The thicknesses at which these transitions occur, as well as the magnitudes of the effects in some cases, correlate one with another, as well as with the optical, electrical, and device properties of the Si:H films.

Among the surface microstructural and phase transitions, the a→a roughening transitions, the a→(a+μc) roughening transitions, and the (a+μc)→μc smoothening transitions can be incorporated readily into deposition phase diagrams. In these diagrams, the Si:H bulk layer thicknesses at which these transitions occur can be plotted as continuous functions of a key deposition variable that is used to control the phase, typically the $H_2$-to-$SiH_4$ gas flow ratio R. The phase diagram depends not only on the other fixed deposition conditions, such as plasma power, substrate temperature, and total gas pressure, but also on the substrate since the latter exerts a strong influence over crystallite nucleation. Deposition phase diagrams are very convenient in the design of devices since they describe the regimes of layer thickness and deposition parameter space within which single-phase a-Si:H, (a+μc)-Si:H, and single-phase μc-Si:H are obtained. As a review of such phase diagrams, Fig. 11 shows the proposed schematic structure of ~5000 Å thick Si:H films on R=0 a-Si:H substrate films, given as a continuous function of R along with the thicknesses of the a→(a+μc) and (a+μc)→μc transition boundaries. In such structures, the cone angle for crystallite growth is relatively constant at 15-20° and the nucleation density increases rapidly with increasing R [16].

Correlations of the phase diagrams for intrinsic Si:H layers with the corresponding electronic properties and p-i-n device performance demonstrate that the optimum i-layers are obtained at the maximum possible R value for the desired thickness without crossing the a→(a+μc) boundary of the phase diagram into the mixed-phase growth regime. It should be emphasized that because the R value at this phase boundary depends on both the nature of the substrate and the i-layer thickness, these aspects of the materials or device structure must be specified in order to identify the optimum conditions. For i-layers deposited on amorphous film substrates (such as the p or n-layers of p-i-n or n-i-p solar cells), the optimum a-Si:H i-layer material has been described as *protocrystalline* Si:H. As its name implies, protocrystalline Si:H ultimately evolves into (a+μc)-Si:H if the film is allowed to grow beyond the desired thickness for which the deposition process was optimized. The unique characteristic of protocrystalline Si:H is its higher stability to light induced degradation as measured for both materials and devices. Finally, because the a→(a+μc) transition decreases in thickness with increasing R, two-step and even multistep i-layer processes can be designed on the basis of the phase diagram in order to optimize solar cells. Performance levels beyond those accessible in one-step processes have been demonstrated.

## ACKNOWLEDGMENTS

This research is supported by the National Renewable Energy Laboratory (under Subcontract Nos. NDJ-2-30630-01 and AAD-9-18668-09) and by the National Science Foundation (under Grant No. DMR-0137240).

# REFERENCES

1. Y. Lu, S. Kim, M. Gunes, Y. Lee, C. R. Wronski, and R. W. Collins, Mater. Res. Soc. Symp. Proc. **336** (1994) 595.
2. D. V. Tsu, B. S. Chao, S. R. Ovshinsky, S. Guha, and J. Yang, Appl. Phys. Lett. **71** (1997) 1317.
3. J. Koh, Y. Lee, H. Fujiwara, C. R. Wronski, and R. W. Collins, Appl. Phys. Lett. **73** (1998) 1526.
4. J. Koh, A. S. Ferlauto, P. I. Rovira, C. R. Wronski, and R. W. Collins, Appl. Phys. Lett. **75** (1999) 2286.
5. J. Yang, K. Lord, S. Guha, and S. R. Ovshinsky, Mater. Res. Soc. Symp. Proc. **609** (2001) A15.4.1.
6. D. V. Tsu, B. S. Chao, S. R. Ovshinsky, S. J. Jones, J. Yang, S. Guha, and R. Tsu, Phys. Rev. B **63** (2001) 125338-1.
7. O. Vetterl, F. Finger, R. Carius, P. Hapke, L. Houben, O. Kluth, A. Lambertz, A. Muck, B. Rech, and H. Wagner, Sol. Energy Mater. Sol. Cells **62** (2000) 97.
8. O. Vetterl, R. Carius, L. Houben, C. Scholten, M. Luysberg, A. Lambertz, F. Finger, and H. Wagner, Mater. Res. Soc. Symp. Proc. **609** (2000) A15.2.1.
9. A. S. Ferlauto, P. I. Rovira, R. J. Koval, C. R. Wronski, and R. W. Collins, Mater. Res. Soc. Symp. Proc. **609** (2000) A2.2.1.
10. A. S. Ferlauto, R. J. Koval, C. R. Wronski, and R. W. Collins, Appl. Phys. Lett. **80** (2002) 2666.
11. C. R. Wronski, J. M. Pearce, R. J. Koval, X. Niu, A. S. Ferlauto, J. Koh, and R. W. Collins, Mater. Res. Soc. Symp. Proc. **715** (2002) A13.4.1.
12. R. J. Koval, J. Koh, Z. Lu, L. Jiao, R. W. Collins, and C. R. Wronski, Appl. Phys. Lett. **75** (1999) 1553.
13. J. M. Pearce, R. J. Koval, A. S. Ferlauto, R. W. Collins, C. R. Wronski, J. Yang, and S. Guha, Appl. Phys. Lett. **77** (2000) 3093.
14. R. W. Collins in: H. Fritzsche, ed., Amorphous Silicon and Related Materials Vol. 1B (World Scientific, Singapore, 1988) 1003.
15. Y. Li, I. An, H. V. Nguyen, C. R. Wronski, and R. W. Collins, Phys. Rev. Lett. **68** (1992) 2814.
16. A. S. Ferlauto, G. M. Ferreira, R. J. Koval, J. M. Pearce, C. R. Wronski, and R. W. Collins, these Proceedings.
17. W. M. Tong and R. S. Williams, Annu. Rev. Phys. Chem. **45** (1994) 401.
18. B. J. Palmer and R. G. Gordon, Thin Solid Films **158** (1988) 313.
19. A. Mazor, D. J. Srolovitz, P. S. Hagan, and B. G. Bukiet, Phys. Rev. Lett. **60** (1988) 424.
20. R. W. Collins and B.-Y. Yang, J. Vac. Sci. Technol. B **7** (1989) 1155.
21. For a recent review, see: A. Matsuda, J. Vac. Sci. Technol. A **16** (1998) 365.
22. J. Robertson, J. Appl. Phys. **87** (2000) 2608.
23. S. Kim and R. W. Collins, Appl. Phys. Lett. **67** (1995) 3010.
24. H. Fujiwara, J. Koh, C. R. Wronski, R. W. Collins, and J. S. Burnham, Appl. Phys. Lett. **72** (1998) 2993.
25. A. S. Ferlauto, J. Koh, P. I. Rovira, C. R. Wronski, R. W. Collins, and G. Ganguly, J. Non-Cryst. Solids **266-269** (2000) 269.

## Influence of Filament and Substrate Temperatures on Structural and Optoelectronic Properties of Narrow Gap a-SiGe:H Alloys Deposited by Hot-Wire CVD

Yueqin Xu, Brent P. Nelson, D.L. Williamson*, Lynn M. Gedvilas, and Robert C. Reedy
National Renewable Energy Laboratory
1617 Cole Blvd., Golden CO 80401, USA
*Colorado School of Mines, Department of Physics
Golden, CO 80401, U.S.A.

## ABSTRACT

We have found that narrow-bandgap—1.25 < Tauc Gap < 1.50 eV—amorphous silicon germanium (a-SiGe:H) alloys grown by hot-wire chemical vapor deposition (hot-wire CVD) can be improved by lowering both substrate and filament temperatures. We systematically study films deposited using a one-tungsten filament, decreasing filament temperature ($T_f$) from our standard temperature of 2150° down to 1750°C, and fixing all other deposition parameters. By decreasing $T_f$ at the fixed substrate temperature ($T_s$) of 180°C, the Ge-H bonding increases, whereas the Si-$H_2$ bonding is eliminated. Films with higher Ge-H bonding and less Si-$H_2$ have improved photoconductivity. For the series of films deposited using the same germane gas fraction at 35%, the energy where the optical absorption is $1 \times 10^4$ (E04) drops from 1.54 to 1.41 eV with decreasing $T_f$. This is mainly due to the combination of an increasing Ge solid fraction (x) in the film, and an improved homogeneity and compactness due to significant reduction of microvoids, which was confirmed by small angle X-ray scattering (SAXS). We also studied a series of films grown by decreasing the $T_s$ from our previous standard temperature of 350°C down to 125°C, fixing all other deposition parameters including $T_f$ at 1800°C. By decreasing $T_s$, both the total hydrogen content ($C_H$) and the Ge-H bonding increased, but the Si-$H_2$ bonding is not measurable in the $T_s$ range of 180°-300°C. The E04 increases from 1.40 to 1.51 eV as $T_s$ decreased from 350° to 125°C, mainly due to the increased total hydrogen content ($C_H$). At the same time, the photo-to-dark conductivity ratio increases almost three orders of magnitude over this range of $T_s$.

## INTRODUCTION

We first reported at the 1998 MRS Spring Conference that we can deposit high-quality a-SiGe:H alloys with bandgaps over 1.5 eV at high deposition rates over 10 Å/s by hot-wire CVD[1]. We then fabricated a-Si:H/a-SiGe:H tandem solar cells using these mid-gap a-SiGe:H alloys and obtained a conversion efficiency over 11% [2]. In the earlier papers, we showed that substrate temperatures over 270°C produced high-quality mid-gap materials. However for a-SiGe:H alloys below 1.5 eV (which we call narrow gap), we observed that high $T_s$ and high $T_f$ were deleterious to the film quality. Therefore, recently we focused our research on this narrow-bandgap region, 1.25 eV< Tauc Gap ($E_{Tauc}$) < 1.50 eV, in different process regimes than used for our mid-gap materials [3, 4]. In this paper, we present the effect of $T_f$ and $T_s$ on the structural and optoelectronic properties of a-SiGe:H films grown by hot-wire CVD.

## EXPERIMENTAL DETAILS

We deposited two sets of films. The first set was deposited by reducing $T_f$ from 2150°C to 1750°C with all other deposition parameters were fixed. The second set was deposited by reducing $T_s$ from 350°C to 125°C with all other deposition parameters fixed. See Table 1 for a list of select deposition parameters, the film thicknesses and deposition rates (D.R.) for the films examined in this paper.

Table 1. Deposition Parameters of the Samples*.

| Sample (Set 1) | $T_f$ (°C) | $T_s$ (start) (°C) | Thick. (Å) | D. R. (Å/s) | Sample (Set 2) | $T_s$ (°C) | $T_f$ (start) (°C) | Thick. (Å) | D. R. (Å/s) |
|---|---|---|---|---|---|---|---|---|---|
| L902 | 2150 | 180 | 2976 | 9.92 | L908 | 350 | 1800 | 2919 | 3.04 |
| L904 | 2065 | 180 | 3434 | 8.18 | L894 | 300 | 1800 | 4087 | 3.45 |
| L905 | 1975 | 180 | 3315 | 6.50 | L895 | 250 | 1800 | 3669 | 3.08 |
| L907 | 1880 | 180 | 2997 | 4.16 | L896 | 200 | 1800 | 3622 | 2.92 |
| L911 | 1800 | 180 | 2128 | 2.03 | L897 | 150 | 1800 | 3501 | 2.84 |
| L913 | 1750 | 180 | 2085 | 0.98 | L898 | 125 | 1800 | 2856 | 2.14 |

* All samples are deposited using the same $GeH_4$ gas ratio of $GeH_4/(GeH_4+SiH_4) = 35\%$, the same $H_2$ dilution ratio of $H_2/(GeH_4+SiH_4) = 1$, and at the same deposition pressure of 15 mTorr.

The hot wire deposition chamber used in this study is a 10-cm-diameter × 30-cm-long stainless steel tube mounted inside of a standard high-vacuum 6-way cross [3]. The outside of this tube is wrapped with an encapsulated resistive heater in a pattern that provides an isothermal region in the center of the tube, the location of the substrate during deposition. The filament used in this study is one tungsten wire 0.38 mm in diameter and about 22 cm in straight length and 18 cm in coiled length. The spacing from the filament to substrate is 5 cm, and the $T_f$ is calculated from tables listing the filament diameter and current, and is also calibrated by a two-wavelength pyrometer under vacuum conditions.

We deposited a-SiGe:H films simultaneously on 1737F Corning glass and c-Si wafers. We evaporated coplanar (width to length = 0.05) Cr contacts on the films on the 1737F substrates for conductivity measurements. We also performed optical measurements using an n&k 1280 analyzer on the films grown on 1737F substrates to determine the thickness (Å), bandgap (E04), which is the photon energy where the optical absorption is $1 \times 10^4$. The Tauc bandgap is taken from the fitting of E vs. $(\alpha h\upsilon)^{1/2}$, in which $\alpha$ is calculated by the method of interference-free determination of optical absorption coefficient [5] on the raw transmission and reflectance data.

The FTIR absorption spectra were obtained from the films deposited on c-Si wafers by a Nicolet 510 system between 400 and 4000 $cm^{-1}$. The hydrogen content of these films was determined by calculating the integrated absorption of local vibration wagging modes of Si-H, Ge-H mono-hydrogen bonds at the peak positions of about 640 $cm^{-1}$ and 570 $cm^{-1}$, respectively [6, 7, 8].

For SAXS measurement, a duplicate set of films was deposited on high-purity aluminum-foil with conditions similar to those in Table 1. The total integrated SAXS intensity, $Q_T$, is a good measure of the overall film heterogeneity. The SAXS technique and analysis methods are described elsewhere [11]. The SIMS measurement was also taken on the films deposited on c-Si substrate by using a Cameca IMS-5F instrument to determinate the Ge solid-phase fraction (x) in the films [12].

## RESULTS AND DISCUSSION

Between wavenumbers 1700 and 2100 cm$^{-1}$ on each set of FTIR spectra, there are stretch modes of mono-hydrogen bonds of Si-H and Ge-H corresponding to the peaks at approximately 2000 and 1880 cm$^{-1}$, respectively, and di-hydrogen bonds of Si-H$_2$ and Ge-H$_2$ corresponding to the peaks at 2090 and 1980 cm$^{-1}$, respectively. Superpositions of Gaussians were used to fit these peaks. Figures 1-a and 1-b show these peak fittings for the two sets of films in their stretch band regions. We found no evidence that polyhydrides of germanium were incorporated in these films; otherwise there would be additional peaks at 830 and 760 cm$^{-1}$ corresponding to the bending modes of polyhydrides of germanium (GeH$_2$)$_n$ [9, 10], analogous to the two peaks at 890 and 845 cm$^{-1}$, which are a scissors mode corresponding to the bending modes of polyhydrides of silicon (SiH$_2$)$_n$. Table 2 lists the results of the various measurements made on these samples listed in Table 1.

### The effects of varying filament temperature

Figure 2 shows that the deposition rate (D. R.), E04, and hydrogen content (C$_H$) all monotonically decreased as T$_f$ decreased from 2150° to 1750°C. The C$_H$ in the films is relatively constant for this set (12-15 at.%) due to the fixed T$_s$, but a sharp decrease of D.R. can be seen from about 10 Å/s to 1 Å/s. This demonstrates that the D.R. is mainly dependent on the filament energy under the same deposition pressure. E04 also decreased from 1.54 to 1.41 eV with decrease of T$_f$. We observed that higher T$_f$, resulted in higher deposition rates but due to the fixed low T$_s$ at 180°C, the films are porous due to higher fraction of microvoids that usually coexist with polyhydrides of silicon- (SiH$_2$)$_n$. As can be seen, when T$_f$ is above 1900°C, there is a significant amount (~20% of C$_H$) of Si-H$_2$ in the film (Fig. 3), but below 1900°C, the films improved rapidly as demonstrated by the decrease in Si-H$_2$ and increase in photoconductivity (Fig. 3). The relative amount of Ge-H bonding (to total H-bonding) increases monotonically with decreasing T$_f$ (Fig. 3). This is an additional reflection of the improvement in film quality with decreasing T$_f$ as Ge-H bonding reflects favorable hydrogen passivation of Ge dangling bonds.

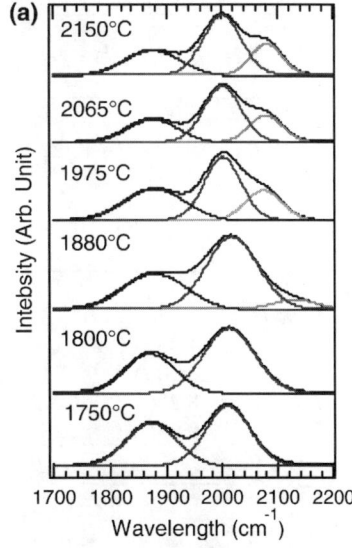

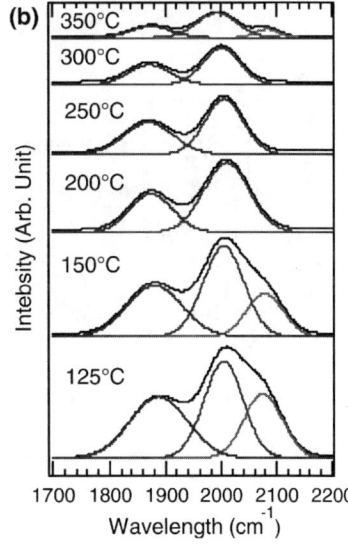

Fig. 1 Three-Gaussians fit to Si-H, Ge-H, and Si-H$_2$ bonding on the samples deposited under varying T$_f$ (Fig. 1-a) and T$_s$ (Fig. 1-b).

Table 2. Results of the Various Measurements Made on Samples Listed in Table 1.

| Sample (Set 1&2) | $\sigma_{photo}$ (cm$^{-1}\cdot\Omega^{-1}$) | Ratio $\sigma_{photo}/\sigma_{dark}$ | E04 (eV) | $E_{Tauc}$ (eV) | n (633nm$^{-1}$) | x (%) | $Q_T$ (10$^{24}$ eu/cm$^3$) |
|---|---|---|---|---|---|---|---|
| L902 | 2.39e-07 | 356 | 1.54 | 1.33 | 4.511 | 57.4 | 20.1 |
| L904 | 2.80e-07 | 800 | 1.54 | 1.33 | 4.504 | 57.4 | 18.3 |
| L905 | 3.56e-07 | 983 | 1.53 | 1.32 | 4.512 | 60.5 | 9.92 |
| L907 | 7.51e-07 | 736 | 1.50 | 1.30 | 4.685 | NA | 7.24 |
| L911 | 6.11e-06 | 382 | 1.45 | 1.26 | 4.783 | 66.3 | 2.61 |
| L913 | 5.11e-06 | 203 | 1.41 | 1.21 | 5.054 | 70.3 | 1.89 |
| L908 | 3.53e-07 | 1 | 1.40 | 1.21 | 4.833 | 59.7 | 9.91 |
| L894 | 8.93e-07 | 48 | 1.43 | 1.23 | 4.875 | 61.3 | 4.58 |
| L895 | 1.95e-06 | 179 | 1.45 | 1.26 | 4.875 | 63.4 | 2.18 |
| L896 | 2.10e-06 | 346 | 1.47 | 1.28 | 4.985 | 64.6 | 2.72 |
| L897 | 8.25e-07 | 292 | 1.49 | 1.29 | 4.691 | 64.5 | 7.11 |
| L898 | 9.65e-07 | 283 | 1.50 | 1.31 | 4.711 | 68.7 | 10.7 |

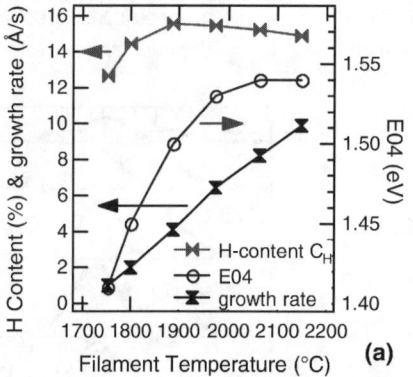
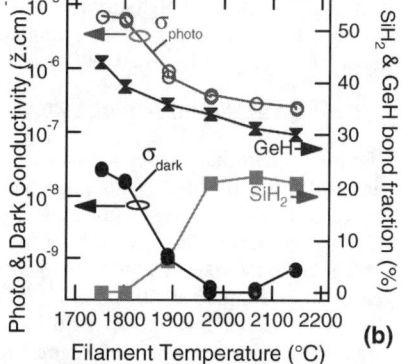

Fig. 2. Variation of deposition rate, E04, and hydrogen content as a function of $T_f$.

Fig. 3. The effects of $T_f$ on $\sigma_{photo}$, $\sigma_{dark}$, and Ge-H, Si-H$_2$ bonding fractions.

The total integrated SAXS intensity decreases monotonically with decreasing $T_f$, whereas the Ge solid-phase fraction x increases (Fig. 4). This inverse correlation is somewhat surprising, yet may be explained by the fact that the films with higher Ge are grown at lower deposition rates, and thus have improved microstructure.

The higher Ge at lower $T_f$ is due to the lower dissociation energy for GeH$_4$ relative to SiH$_4$. The decrease in E04 with decreasing $T_f$ (Fig. 2), is due primarily to the increase in Ge solid fraction (Fig. 4), but also to improved film homogeneity and compactness due to significant reduction of microvoids indicated by the SAXS measurements (Fig. 4).

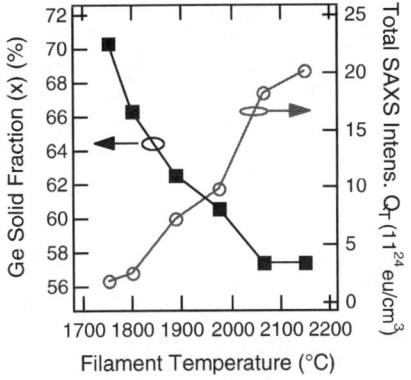
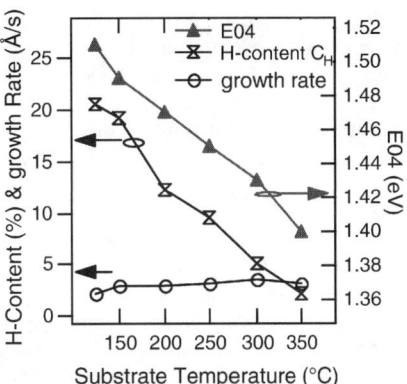

Fig. 4. Increase of Ge fraction x and decrease of total SAXS intensity $Q_T$ with decrease of $T_f$.

Fig. 5. Variation of deposition rate, E04, and hydrogen content as a function of $T_f$.

## The effects of varying substrate temperature

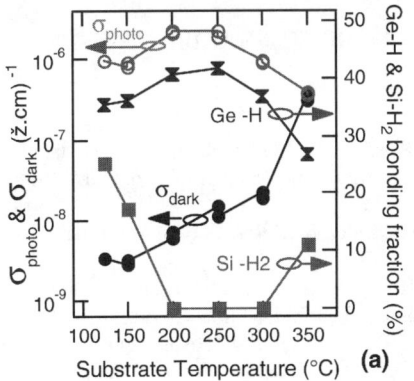
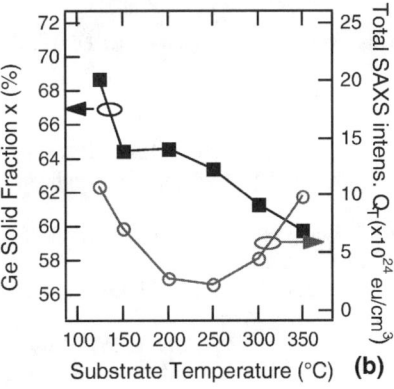

Fig. 6. The effects of $T_s$ on $\sigma_{photo}$, $\sigma_{dark}$, and Ge-H, Si-H$_2$ bonding fractions.

Fig. 7. Smaller increase of Ge fraction x and decrease of total SAXS intensity $Q_T$ at the $T_s$ of 200°-250°C.

In Fig 5, we show that both E04 and $C_H$ are dependent on $T_s$, whereas the growth rate is not strongly affected by Ts, remaining ~ 2-4 Å/s. When $T_s$ is between 200°- 300°C, Si-H$_2$ is below the detection limit of FTIR, which is accompanied by increased photoconductivity (Fig. 6).

The total hydrogen $C_H$ in the films increases monotonically from ~2% to ~20% as the $T_s$ is lowered from 350° to 125°C. This large change in $C_H$ has a strong influence on E04, which increases from 1.4 to 1.5 eV with the same decrease of $T_s$. The relative amount of Ge-H bonding is a maximum in the optimized temperature range of 200°- 300°C, which correlates nicely with higher photoconductivity (Fig. 6). The lowering of $T_s$ has a profound improvement (lowering) of the dark conductivity (Fig. 6).

The total integrated SAXS intensity is a minimum in the optimized temperature range of 200°- 300°C, whereas Ge increases monotonically with decreasing $T_f$ (Fig. 7). The improvement of optoelectronic properties in this optimized $T_s$ range correlates nicely with the SAXS intensity, even though there is a small increase of Ge fraction x (Fig. 7).

## CONCLUSIONS

Based on above evidence, lowering $T_f$ (<1850°C) and optimizing $T_s$ (200°-250°C) can improve the structural and optoelectronic properties of narrow-gap a-SiGe:H alloys deposited by HWCVD, in which a 0.38-mm tungsten wire was used, resulting in a D.R. of about 3Å/s. The improved material has about 65% Ge, an optical band gap around 1.3 eV, and an activation energy about 0.64 eV. This materials has a $C_H$ of about 10%, with over 40% of Ge-H bonding and about 60% of Si-H bonding but without detectable di-hydride bonding (Si-$H_2$, Ge-$H_2$). Therefore, this material has higher photoconductivity (~ $10^{-6}$) and photosensitivity (about 200-500). This material has much improved homogeneity as measured by SAXS, compared to all other such alloys with high Ge content.

## ACKNOWLEDGMENT

This work was performed under DOE contract number DE-AC36-99GO10337.

## REFERENCES

1. B.P. Nelson, Y. Xu, D.L. Williamson, B. von Roedern, A. Mason, S. Heck, A.H. Mahan, S.E. Schmitt, A.C. Gallagher, J. Webb, and R. Reedy, Mat. Res. Soc. Symp. Proc. 507 (1998) 447.
2. Q. Wang, E. Iwaniczco, J. Yang, K. Lord, and S. Guha, Mat. Res. Soc. Symp. Proc. 664 (2001) A7.5.
3. Y. Xu, B.P. Nelson, L.M. Gedvilas, and R.C. Reedy, Sept. 2002, 2nd Intern. Conf. on Cat-CVD (Hot-Wire CVD) Process, Denver, Colorado, Thin Solid Films (in press).
4. B. P. Nelson, Y. Xu, D. L. Williamson, D. Han, R. Braunstein, M. Boshta, and B. Alavi, Sept. 2002, 2nd Intern. Conf. on Cat-CVD Process, Denver, Colorado, Thin Solid Films (in press).
5. Yoshihilo Hishkawa, NoBoru Nakamura, Shinya Tsuda, Shoichi Nakano, Yasuo Kishi, and Yukinori Kuwano, Jpn. J. Appl. Phys. 30 (1991) 1008.
6. C.J. Fang, K.J. Gruntz, L. Ley, M. Cardona, F.J. Demond, G. Muller, S. Kalbitzer, J. Non-Cryst. Solids 35 & 36 (1980) 255.
7. M. Cardona, Phys. Stat. Sol. (b) 118 (1983) 463.
8. A.A. Langford, M.L. Fleet, B.P. Nelson, W.A. Lanford, N. Maley, Phys. Rev. B45 (1992) 13367.
9. Mohan Krishan Bhan, L.K. Malhotra, and Subhash C. Kashyap, J. Appl. Phys. 66 (1989) 2528.
10. G. Lucovsky, J. Non-Cryst. Solids 76 (1985) 173.
11. D.L. Williamson, Mat. Res. Soc. Symp. Proc. 377 (1995) 251.
12. R.C. Reedy, A.R. Mason, B.P. Nelson, Y. Xu, American Institute of Physics, NICH Report No. 27431 (1999) pp. 537-541.

## On the role of surface diffusion and its relation to the hydrogen incorporation during hydrogenated amorphous silicon growth

A.H.M. Smets, W.M.M. Kessels, and M.C.M. van de Sanden
Department of Applied Physics, Eindhoven University of Technology,
P.O. Box 513, 5600 MB Eindhoven, the Netherlands

### ABSTRACT

The incorporation of hydrogen in vacancies and at void surfaces during hydrogenated amorphous silicon growth from a remote expanding thermal plasma (ETP) is systematically studied by variation of the mass growth flux $\Gamma_{a\text{-Si:H}}$ and substrate temperature $T_{sub}$. An evident relation between the void incorporation and the growth parameters $\Gamma_{a\text{-Si:H}}$ and $T_{sub}$ has been observed. We speculate on a possible relation with the surface diffusion processes during deposition. An activation energy for surface diffusion during a-Si:H growth of 0.8-1.1 eV is obtained using this assertion, similar to the activation energy deduced from surface roughness evolution studies. For compact films hydrogen is predominantly present at vacancies, and a possible relation with the hydrogen removal mechanism during deposition is discussed.

### INTRODUCTION

In recent years the expanding thermal plasma (ETP) and its application in hydrogenated amorphous silicon (a-Si:H) growth has been extensively studied. These studies have been performed under typical ETP conditions in which $SiH_3$ radicals are responsible for >90 % of the a-Si:H growth [1] and ion bombardment is negligible. The growth can be considered as purely chemical in nature and this makes the ETP technique very suitable for fundamental studies of the surface processes during non-ion-assisted growth. Besides plasma studies, the research on a-Si:H growth by means of ETP has recently been focused on three topics: 1) radical interaction with the surface and its resulting surface composition [2], 2) impact of surface diffusion during growth [3] and 3) configuration types in which the hydrogen resides in the a-Si:H network [4]:

1) A study on the surface reaction probability $\beta$ of the dominant growth precursor $SiH_3$ revealed that $\beta=0.30\pm0.03$ and independent of the substrate temperature $T_{sub}$ [2]. Surface sensitive infrared studies showed that the surface hydride composition is temperature dependent, the dominant surface hydride crosses over from $-SiH_3$ to $=SiH_2$ down to $\equiv Si\text{-}H$ with increasing $T_{sub}$ [2], similar as observed for a-Si:H growth from an ICP source [5]. These results imply that the surface processes determine the surface composition during non-ion-assisted a-Si:H growth from $SiH_3$.
2) In Ref. [3] the temperature dependence of the scaling behavior of the surface roughness evolution during a-Si:H growth has been studied. The observed behavior revealed that the surface smoothening during ETP a-Si:H growth is ruled by surface diffusion of a yet unidentified surface species and the diffusion process is activated by an activation energy in the range of ~1.0 eV [3].
3) In Ref. [4] the dependence of the low (hydride) stretching mode (LSM, 1980-2020 $cm^{-1}$) and high stretching mode (HSM, 2070-2100 $cm^{-1}$) on the film mass density has been studied by means of infrared absorption spectroscopy. The results revealed that hydrogen is mainly present at vacancies or void surfaces [4]. For a hydrogen content $c_H>14$ at.% hydrogen is dominantly located on void surfaces (contributing only to the

HSM) [4]. The typical void size in ETP a-Si:H found are in the range of 4 nm [6] in line with the void sizes (1-4 nm) observed in HWCVD a-Si:H films [7]. For $c_H$<14 at.% hydrogen resides predominantly in divacancies. A divacancy is a lattice site at which 6 H atoms replace two missing Si atoms [8].

From micro-structural point of view, the essential issues in a-Si:H growth are: How is hydrogen incorporated into the matrix? How is the excess of hydrogen in the growing species SiH$_3$ at the surface removed during growth? Most hydrogen incorporation models are based upon random incorporation of Si-H bonds in the matrix [9-11], and therefore not in agreement with observation that H is located at vacancies or voids. To our opinion the question of how vacancies and voids are incorporated into the bulk is therefore of more relevance and if understood the hydrogen incorporation is probably understood as well. In this contribution we show that the void/vacancy incorporation depends on the mass growth flux $\Gamma_{a\text{-}Si:H}$ and substrate temperature $T_{sub}$. In line with the results on a-Si:H ETP growth studies summarized above, we will speculate that the surface diffusion process [3] is capable of explaining the hydrogen incorporation at void surfaces [4], whereas the vacancy incorporation is most probably related to the hydrogen removal mechanism on the surface [9,10].

## EXPERIMENTAL RESULTS AND DISCUSSION

In the study presented here, the ETP plasma has been used over a large parameter window: the mass growth flux is varied using 0.045 µgcm$^{-2}$s$^{-1}$ up to 2.7 µgcm$^{-2}$s$^{-1}$ (corresponding to 2 Å/s up to 120 Å/s) and the substrate temperature $T_{sub}$ has been varied from 100 °C up to 500 °C [12]. The easy accessibility of the ETP technique with respect to the mass growth flux-substrate temperature ($\Gamma_{a\text{-}Si:H}$-$T_{sub}$) parameter window is demonstrated in Fig. 1, in which the independence of the mass growth flux $\Gamma_{a\text{-}Si:H}$ on $T_{sub}$ is shown. The $c_H$ decreases with increasing $T_{sub}$ as presented in Fig. 2. At constant $T_{sub}$ the $c_H$ increases with $\Gamma_{a\text{-}Si:H}$. As shown in Ref. [4], at low $T_{sub}$ the $c_H$ is determined by hydrogen at void surfaces whereas at high $T_{sub}$ the $c_H$ is determined by hydrogen in vacancies.

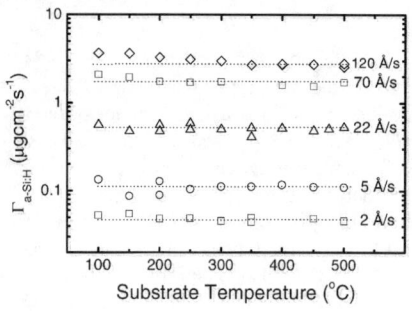
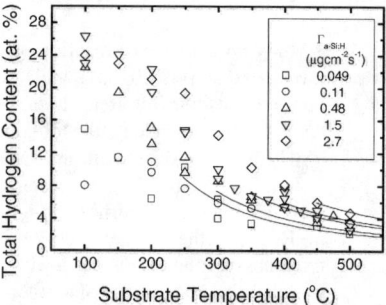

**Figure 1.** The mass growth flux $\Gamma_{a\text{-}Si:H}$ vs $T_{sub}$ for $R_d$ = 2, 5, 22, 70 and 120 Å/s.

**Figure 2.** The hydrogen content $c_H$ obtained from the 640 cm$^{-1}$ SiH$_x$ mode vs $T_{sub}$. The lines are fits using Eq. (2) as described in the text.

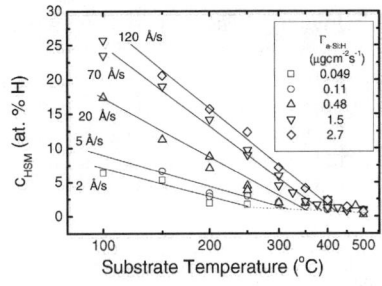
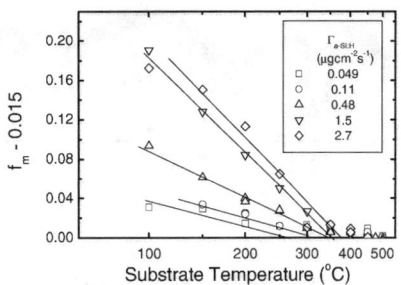

**Figure 3.** The $c_{HSM}$ vs $T_{sub}$

**Figure 4.** The mass deficiency $f_m$ vs $T_{sub}$

## INCORPORATION OF VOIDS

The void fraction has been determined from the hydrogen content contributing to the HSM $c_{HSM}$. The void fraction can also be deduced from the film mass deficiency defined as $f_m$ =(1-$\rho_{a-Si:H}$/$\rho_{a-Si}$) as discussed in Ref. [4]. Since, the $f_m$ includes the vacancy fraction as well, we define the void fraction as $f_m$ minus an offset: $f_m$-0.015 (for compact film growth the vacancy fraction is ~0.015). These quantities are shown in Figs. 3 and 4, respectively. At high temperatures almost no voids are incorporated in the material. In these compact films, H resides mainly at divacancies [4]. A critical temperature $T_c$, above which the growth can be considered as compact, can be defined. Figures 3 and 4 show that this critical temperature increases with increasing growth flux. The $\Gamma_{a-Si:H}$ and $T_c$ data of Figs. 3 and 4 are shown in an Arrhenius plot in Fig. 5. This figure implies that the void incorporation process is thermally activated for non-ion-assisted a-Si:H growth. In view of the observed importance of surface diffusion during a-Si:H growth [3], we will speculate on the surface diffusion to be the process, which controls the void incorporation. If present during growth, surface diffusion will result in smoothening of the surface. Furthermore, the fact that the surface roughness of a-Si:H films deposited by means of ETP decreases with increasing substrate temperature [13], implies that the smoothening mechanism is thermally activated [3]. Since the lateral correlation length of the ETP a-Si:H surface is temperature independent [12] the valleys on rougher surfaces become consequently steeper. Therefore the most plausible mechanism in which a void can be incorporated is the mechanism in which above a valley some overhangs grow. A void is created when these overhangs are able to confine a valley into the bulk (cf. Fig. 6). Consequently, if the diffusion is so fast that the valleys are filled, less or no voids can be

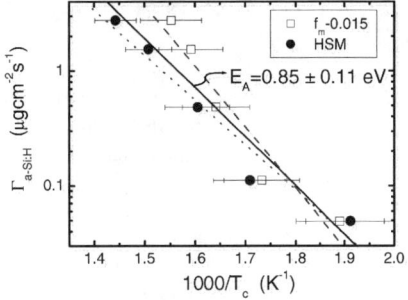

**Figure 5.** The mass growth flux $\Gamma_{a-Si:H}$ vs $1000/T_c$, with $T_c$ the critical temperature as defined in the text. The lines are fits using Eq. (2) on respectively HSM data (dotted), $f_m$-0.015 data (dashed) and both data (solid).

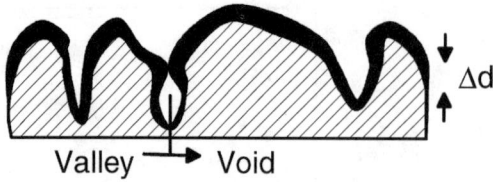

**Figure 6.** The proposed incorporation mechanism of a void. The shaded area indicates a film at which 3 valleys are present at the surface area zoomed in. One of the valleys is incorporated in to the film by the overhangs when growth continues (solid area).

incorporated ($T_{sub}>T_c$), whereas when the diffusion is slow ($T_{sub}<T_c$) the valleys cannot be filled and voids can be incorporated. This means that two timescales are in competition: one related to the surface diffusion $\tau_D$ and one related to the arrival rate of the growing species $\tau_R$. The characteristic time to grow one monolayer (with thickness $L_m$) is $\tau_R = L_m/R_d = L_m \times \rho_{a\text{-Si:H}}/\Gamma_{a\text{-Si:H}}$. The time scale for diffusion is $\tau_D = L_a^2/D$, where $D$ is the diffusion coefficient of the diffusing surface species, which is thermally activated by $D=D_0\exp(-E_{dif}/kT)$. $L_a$ is the lateral diffusion distance, which has been measured to be weakly dependent on $T_{sub}$ [12] for ETP growth and weakly dependent for PECVD growth [14]. When $\tau_D > \tau_R$ voids can be incorporated, whereas for $\tau_D < \tau_R$ a compact a-Si:H network is grown. At the critical temperature $T_c$: the two timescales are equal and the next relation can be deduced:

$$\Gamma_{a-Si:H} = \frac{D_0 L_m \rho_{a-Si:H}}{L_a^2} \exp\left(-\frac{E_{dif}}{kT_c}\right) \tag{1}$$

If we apply this equation to the data in Fig. 5, we obtain an activation energy for diffusion: $E_{dif} = 0.85 \pm 0.11$ eV, whereas if we apply Eq. (1) exclusive on the data based upon the HSM results we obtain $E_{dif}= 0.77\pm0.11$ eV and $E_{dif}=1.05\pm0.13$ eV for the data based upon the void fraction results. These values are consistent with the ~1.0 eV activation energy found from the temperature dependent scaling of the roughness evolution [3]. Note, that the HSM is a direct measure of voids, i.e. hydrogen in vacancies does not contribute to the HSM whereas within the definition of $f_m$ we have assumed a constant contribution of vacancies (~0.015). Therefore, the results based on the HSM data are more reliable than those based upon the $f_m$ data.

The fit in Fig. 5 also provides a prefactor, from which the absolute value for the lateral diffusion length $L_a$ can be obtained. If we assume that $D_0=\Lambda^2/\tau_{hop}$, with $\Lambda$ the lattice site distance (~$L_m$) and $\tau_{hop}$ the hopping time (typically $10^{-12}$ s), we find a $L_a$ ~ 150 nm. This value is in agreement with typical lateral correlation lengths found for ETP a-Si:H surfaces by means of AFM measurements [12]. The relative and absolute agreement between the $E_{dif}$ and $L_a$ values obtained from Fig. 5 and Refs. [3,12], supports the assertion that surface diffusion is the main cause for the void incorporation. The above suggested growth model shows similarities to the Street and Winer defect equilibrium model [15]. In contrast to our model, this latter model is not based upon surface diffusion but on bulk diffusion (of hydrogen) and these authors found an activation energy of 1.5 eV for bulk diffusion.

## INCORPORATION OF VACANCIES

Above $T_c$ compact films are grown and $\rho_{a\text{-Si:H}}$ depends slightly on $T_{sub}$. Nevertheless, the hydrogen content is not drastically, but only slightly, decreasing with increasing

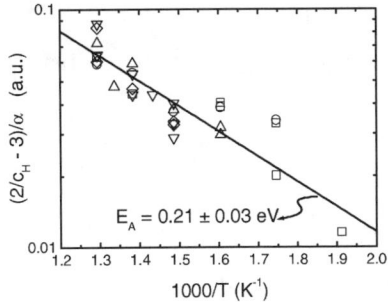

**Fig. 7** The quantity $(2/c_H-3)/\alpha$ for the $T_{sub}>T_c$ range is plotted vs $1000/T_{sub}$. The line is a fit following equation (2).

temperature (see Fig. 2). Since in this region hydrogen resides predominantly in divacancies ($c_H < 14$ at.%) [4], the vacancy incorporation mechanism rules the hydrogen incorporation. Since Si has the tendency to maximize its number of Si-Si bonds during its incorporation, a vacancy can be considered as a site at which this process did not occur. As SiH$_3$ is the dominant growth precursor for ETP growth, on average one H has to be removed at the surface to preserve the growth continuation, whereas the other two H atoms have to be removed for Si-Si bond maximization. Although many hydrogen removal mechanism have been suggested in literature [9-11], the net hydrogen removal reaction step, maximizing the Si-Si bonds in the surface region, should be: $2\text{Si-H}_{(s)} \rightarrow \text{Si-Si}_{(s)} + \text{H}_{2(g)}$. If we assume that this so-called cross-linking step is indeed mainly responsible for the hydrogen removal, we can apply the H incorporation formalism based upon cross-linking, as proposed by Kessels *et al.* [9], for $T_{sub}>T_c$. In this latter model the total hydrogen content can be described as a function of the activation energy $E_{cl}$ for the cross-link step and $T_{sub}$:

$$c_H = \frac{2}{3 + \alpha\exp\left(-\dfrac{E_{cl}}{kT_{sub}}\right)} \quad (2)$$

with $\alpha$ a temperature independent factor. Following Equation (2) the normalized quantity $(2/c_H-3)/\alpha$ for $T_{sub}>T_c$ range is plotted versus $1000/T_{sub}$. An empirical limitation in this $T_{sub}>T_c$ range is the fact that the relative scatter on the low $c_H$ values (<5 at.%) becomes larger. Nevertheless, a cross-link energy of $E_{cl} = 0.21\pm0.02$ eV could be deduced. This value is in line with the activation energy of ~ 0.21 eV found in Ref. [2,5] by means of *in situ* infrared surface sensitive absorption measurements on the surface hydride composition dependence on $T_{sub}$.

## CONCLUSIONS

In this contribution we speculate that in absence of sufficient surface diffusion, voids are incorporated into the a-Si:H material during SiH$_3$ dominated a-Si:H growth with negligible surface ion bombardment. From the observed dependence of the void fraction on $\Gamma_{\text{a-Si:H}}$ and $T_{sub}$ an activation energy for diffusion of 0.8-1.1 eV is deduced, and this energy is in line with results from Ref. [4]. Furthermore, we suggest that the hydrogen removal mechanism is related to the vacancy incorporation. If we assume that cross-linking is the ruling hydrogen removal mechanism, this process is activated with 0.21±0.02 eV for ETP a-Si:H growth. For the industrial application of high rate deposition of a-Si:H, a reduction of the critical temperature $T_c$ is desirable. Possible routes to achieve this goal could be ion assisted [16] or laser assisted a-Si:H growth. These additional energy supplies to the growth surface could enhance surface processes like diffusion and will be a topic for future experiments on the ETP growth.

## ACKNOWLEDGEMENTS

The technical assistance of Ries van de Sande, Jo Janssen, Bertus Hüsken and Herman de Jong is gratefully acknowledged. This work was sponsored by NOVEM, TDO and the EET Helianthos project.

## REFERENCES

[1] W.M.M. Kessels, A. Leroux, M.G.H. Boogaarts, J.P.M. Hoefnagels, M.C.M. van de Sanden, and D.C. Schram, J. Vac. S ci. Technol. A **19**, 467 (2001)
[2] W.M.M. Kessels, Y. Barrell, P.J. van den Oever, J.P.M. Hoefnagels, and M.C.M. van de Sanden, Mat. Res. Soc. Symp. Proc. Vol. **762**, A.9.3.1 (2003)
[3] A.H.M. Smets, W.M.M. Kessels, and M.C.M. van de Sanden, Appl. Phys. Lett. **82**, 865 (2003)
[4] A.H.M. Smets, W.M.M. Kessels, and M.C.M. van de Sanden, Appl. Phys. Lett. **82**, 1547 (2003)
[5] W.M.M. Kessels, A.H.M. Smets, D.C. Marra, E.S. Aydil, D.C. Schram, M.C.M. van de Sanden, Thin Solid Films **383**, 154 (2001)
[6] C. Smit, D.L Williamson, M.C.M. van de Sanden and R.A.C.M.M. van Swaaij, Mat. Res. Soc. Symp. Proc. **762**, A.15.3.1 (2003)
[7] A.H. Mahan, Y. Xu, D.L. Williamson, W. Beyer, J.D. Perkins, M. Vanacek, L.M. Gedvillas, and B.P. Nelson, J. Appl. Phys **90**, 5038 (2001)
[8] Z. Remes, M. Vanacek, A.H. Mahan and R.S. Crandall, Phys. Rev. B **56**, 12710 (1997)
[9] W.M.M. Kessels, R.J. Severens, M.C.M. van de Sanden, and D.C. Schram, J. Non-Cryst. Sol. **227-230**, 133 (1998)
[10] F.J. Kampas and R.W. Griffith, Appl. Phys. Lett. **39**, 407 (1981)
[11] A. Matsuda *et al.*, Surf. Sci. **227**, 50 (1990); J.R. Doyle *et al.* J. Appl. Phys. **68**, 4375 (1990); K. Maeda *et al.*, Mat. Res. Soc. Symp. Proc. Vol. **377**, 131 (1995); J. Robetson, J. Appl. Phys. **84**, 2426 (2000), A. Terakwa *et al.*, Phys. Rev. B **62**, 16808 (2000)
[12] A.H.M. Smets, PhD. Thesis, Eindhoven, The Netherlands (2002), http://alexandria.tue.nl/extra2/200211441.pdf
[13] A.H.M. Smets, D.C. Schram, and M.C.M. van de Sanden, J. Appl. Phys. **88**, 6388 (2000)
[14] K.R. Bray and G.N. Parsons, Phys. Rev. B **65**, 035311 (2001)
[15] R.A. Street and K. Winer, Phys. Rev. B **40**, 6236 (1989)
[16] V.L. Dalal, P Seberger, M. Ring and P. Sharma, in proceedings of 2nd conference of cat CVD (Hot-wire) process, Denver (2002)

# EFFECT OF TEMPERATURE AND TEMPERATURE UNIFORMITY ON PLASMA AND DEVICE STABILITY

G. Ganguly, M.S. Bennett, D.E. Carlson and R.R. Arya
BP Solar, Toano, Virginia

## ABSTRACT

We have investigated the changes in the cathode potential in a dc discharge of silane and hydrogen used to deposit the intrinsic layer of p-i-n type solar cells at deposition rates from 1 to 10Å/s with the superstrate temperature at 200°C and 250°C. Under plasma conditions that lead to higher deposition rates (5-10Å/s), fluctuations of the cathode potential which are suggestive of the formation and de-trapping of particulates in/from the plasma, are observed at 200°C but disappear at 250°C. Improvement of the temperature uniformity over the plasma region from 1.7°C/cm to 0.7°C/cm removes the fluctuations of the cathode potential even at 200°C, indicating that the particulates are formed predominantly at the plasma boundary. Consequently, the stability of solar cells with i-layers deposited at ~10Å/s in the center of the plasma region at the same superstrate temperature improved by 26% suggesting that multiple silicon containing molecules diffuse from the edge to the center of the plasma region.

## INTRODUCTION

The deposition rate of the amorphous silicon (a-Si:H) film continues to limit the throughput of photovoltaic module production plants [1]. It is well known that the stability of a-Si:H films and solar cells deposited at higher growth rates improve with increasing temperature [2-4]. Initially, it was assumed that increasing the temperature of the growth surface enhanced surface mobility of growth precursors, which resulted in an improved structure and hence more stable material [2,3]. However, it has been shown that the gas phase temperature can have a significant effect on the electronic properties of a-Si:H films, which was understood in terms of the thermal activation of the growth precursors [5,6]. Recently, the electron temperature in a silane plasma was observed to decrease with both increasing substrate temperature and hydrogen dilution of silane [7]. This reduction in electron temperature appears to reduce the incidence of multiple silicon containing molecules, which have been associated with light induced degradation [8]. Complementarily, these molecules are believed to evolve into particulates, when the number of silicon atoms increases to five or more [9], and their formation rate has been observed to decrease at higher gas temperatures [10]. We have found that the potential on the cathode in a dc silane gas discharge can be understood in terms of negative ion formation that alters the plasma impedance due to the lower mobility of ions relative to electrons [11]. Here, we report that the cathode potential variation suggests particulate formation under plasma conditions that lead to higher deposition rates. We show that increasing the superstrate temperature from 200 to 250°C suppresses this behavior. Further, we find that improvement of the temperature uniformity over the plasma region alleviates the instability even at 200°C, and yields significant improvements in the stability of solar cells.

## EXPERIMENT

Single-junction solar cells with the structure *glass/SnO$_2$/p/buffer/i/n/ZnO/Al* were fabricated in a double load-lock research reactor using DC plasma decomposition of silane mixed with the appropriate dopant gases and/or diluted with hydrogen. The i-layer was deposited at 1,3,5 or10Å/s. The other layers remained unchanged. The plasma was generated over an area of 34cm x 36cm while the cells were fabricated in the central 18cm x 18cm region using 8cm x 8cm pieces of commercial SnO$_2$ coated glass. The temperature distribution was measured over the 32cm x 30cm tin oxide coated glass superstrate as it sat in the deposition chamber in a flow of hydrogen at a pressure of 2Torr. The dc voltage applied to the cathode to maintain a set, constant current was monitored for the duration of the deposition processes. The effects of adding different gases to a silane discharge on the cathode potential have been discussed elsewhere [11].

## RESULTS

We have examined various growth conditions in an effort to improve the stable performance using higher growth rates for the i-layer. Time dependence of the cathode potential during the deposition of the i-layer under conditions yielding growth rates of 1,3,5 and 10Å/s are shown in Fig. 1 at superstrate temperatures of 200 and 250°C. Clearly, the cathode potential increases as the growth rate is increased through a combination of changes in the dc plasma current, the hydrogen dilution of silane and the gas pressure. What is remarkably different at and above a growth rate of 5Å/s is that the cathode potential decreases precipitously after the initial rise upon commencement of the discharge at a temperature of 200°C but not when the temperature is raised to 250°C. An increase of the cathode potential is consistent with the decrease in gas density with increase in temperature that is evident in the behavior of the cathode potential under the lower growth rate (1,3Å/s) conditions. However, under the deposition conditions that result in higher growth rates (5, 10Å/s), the change in cathode potential with increase in temperature from 200 to 250°C is much larger. At the lower

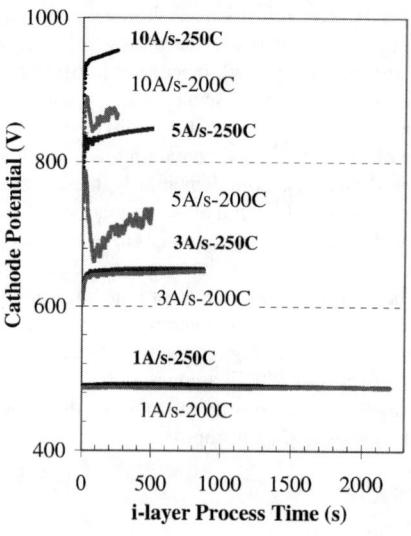

**Fig. 1.** The cathode potential in dc plasma discharges of silane and hydrogen that lead to the indicated growth rates at two different heater temperatures. The legends are arranged vertically in the same order as the curves they represent.

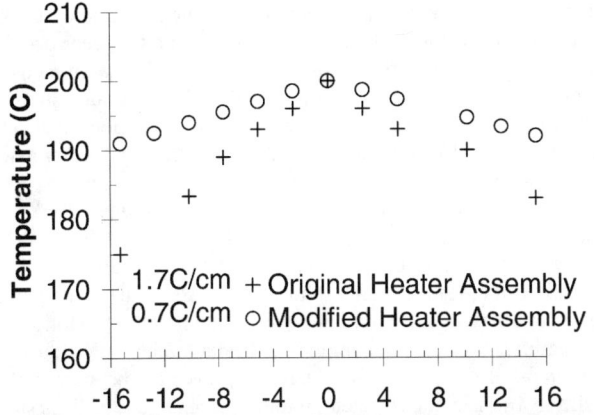

**Fig. 2.** The temperature measured across the glass superstrate using two different heater configurations.

temperature, the cathode potential continues to exhibit irregular fluctuations superposed on a gradual, increasing trend. We have associated the gradual increase with differences in pumping speeds for silane and hydrogen leading to a continuous increase in silane partial pressure at constant flow ratio of the gases [11]. The fluctuations suggest formation and de-trapping of particles in/from the plasma

The increase in temperature reduces the rate of exothermic reactions [7,12] that lead to the formation of multiple silicon containing molecules ($Si_nH_{2n+2-x}$) and hence the rate of particle formation. The rapid decrease in the cathode potential may thus reflect a reduction in the number of electrons attached to electronegative silane molecules, due to its dissociation followed by rapid growth into particulates.

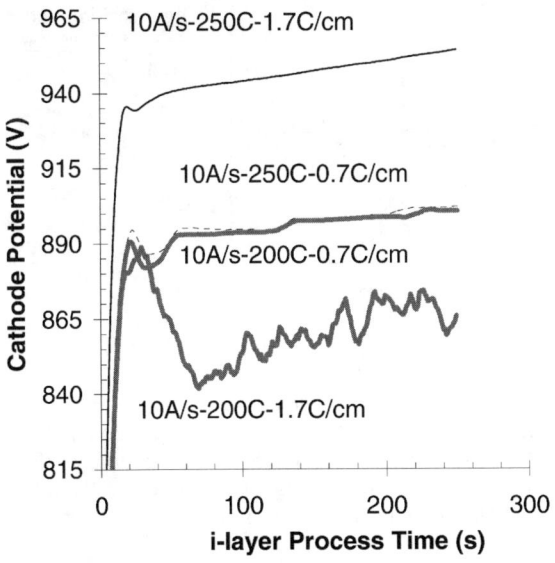

**Fig. 3.** The cathode potential in a dc plasma discharge under conditions that lead to a growth rates of 10Å/s at two different heater temperatures and two different heater configurations yielding the indicated temperature gradients across the superstrate. The legends are arranged vertically in the same order as the curves.

In Fig. 2 we show the variation of temperature measured on the tin oxide coated glass superstrate with the heater temperature set to 200°C. Clearly, the temperature decreases significantly towards the edges with a 1.7°C/cm drop on the cooler side. The asymmetry is ostensibly due to the heated load lock (entry) chamber on one side where the superstrates are preheated and a cooler load lock (exit) chamber on the other side where the deposited cells cool down. This 1.7°C/cm variation was significantly reduced when we altered the heater assembly to take into account the larger cooling rates at the edges and corners. The total heater power was kept the same with the distribution altered to supply more heat at the edges. The temperature profile with the improved heater assembly is also shown in Fig. 2. There is now a 0.7°C/cm drop that is much more symmetric. In Fig. 3 we have plotted the variation of the cathode potential under conditions that lead to a growth rate of ~10Å/s at heater temperatures of 200°C and 250°C using the original and improved superstrate temperature uniformity conditions. While there is a significant effect of changing the heater temperature from 200° to 250°C in the original condition (as also shown in Fig. 1), the improved temperature uniformity condition removes the sensitivity of the cathode potential to heater temperature. This suggests that the formation of multiple silicon containing molecules, and eventually particles, occurs predominantly at the edge of the plasma region where the temperature of the surfaces, that determine the gas molecule temperature, are the lowest. Additionally, the higher growth rate conditions deplete a significant fraction of the silane. At the edge of the plasma region, the lack of gas decomposition increases the partial pressure of silane. Hence the rate of the exothermic insertion reactions leading to multiple silicon containing molecules increases, due to both, lower temperatures and higher partial pressure of reactants.

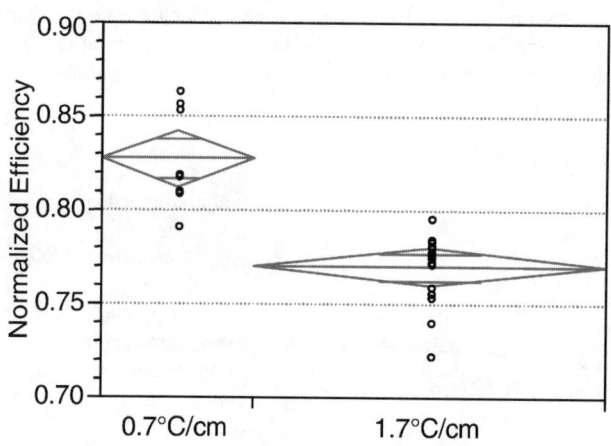

**Fig. 4.** The normalized efficiency loss of solar cells with 10Å/s i-layers after light soaking under 1sun at ~50°C for 600h

One may be tempted to think that if the multiple silicon containing molecules and the radicals formed from them are generated primarily at the plasma boundary region, in large area electrode systems, their effect on the material deposited in the central, hot region, would be minimal. In Fig. 4 we show the relative efficiency loss of single junction solar cells with i-layers deposited at growth rates of ~10Å/s using the original and improved temperature uniformity conditions with the same heater temperature. Each data point is an average of six 0.25cm$^2$ diodes on an 8cm x 8cm glass superstrate. The statistical comparison of the data was carried out using JMP® software, which generates the diamonds such that their vertical extents reflect the variation of the group and their horizontal extents reflect the number of data points in the group relative to the entire data set. Non-overlap of the diamonds vertically for two data groups indicate a 97% confidence level that the two groups are statistically, significantly different. Even though the temperature in the central region (7cmx7cm) differ <10°C between the two heating configurations, there is a significant reduction in the average relative degradation of the cells from 23% to 17%. This difference would imply that multiple silicon containing atoms that are formed at the plasma boundary, diffuse into the central region. Particulates are driven out from the hotter plasma region towards the cooler walls primarily by thermophoretic forces overcoming electrostatic confinement of the negatively charged particulates in the positive plasma potential. However, the motion of multiple silicon containing molecules would be determined by concentration gradients. Such molecules, diffusing into the central plasma zone, are dissociated into radicals that contribute to growth. Contribution of such radicals to growth has previously been associated with increased light induced degradation in a-Si:H [4,8].

## CONCLUSIONS

We have found that time dependent, random fluctuations in the cathode potential under higher growth rate conditions, that may be associated with formation and de-trapping of particulates in/from the plasma, are removed by elevation of temperature. Improvement of the temperature uniformity over the plasma region leads to removal of the fluctuations even at the lower temperature, suggesting that the reactions that lead to particulate formation occur predominantly at the plasma boundary, where the temperature is lower and the partial pressure of silane is higher. The multiple silicon containing molecules are conjectured to back diffuse into the central plasma region, where they are dissociated into radicals that contribute to growth of structures that lead to enhanced light induced degradation of solar cells.

## ACKNOWLEDGMENTS

This work was partially funded by the DOE through NREL Subcontracts ZAK-8-17619-02 and ZDJ-2-30630-10. We would like to thank G. Wood, S. Stone, M. Dicolli, G. Mckinley, F. Jackson and R. Murphy for technical assistance with various parts of this work.

## REFERENCES

1. D.E. Carlson, G. Ganguly, G. Lin, M. Gleaton, M. Bennett, and R.R. Arya, Mater. Res. Soc. Symp. Proc. **664** (2001) A11.4
2. G. Ganguly, and A. Matsuda, Phys. Rev. **B49** (1994) 10986
3. S. Okamoto, T. Takahama, M. Nishikuni, and S. Nakano, US Patent 5114498 (1992)
4. R. Hayashi, T. Takagi, G. Ganguly, M. Fukawa, M. Kondo and A. Matsuda, Proc. 2$^{nd}$. World Conference on PVSEC (European Commission, Ispra, Italy, 1998) p-929.
5. A. Matsuda, S. Yokohama, and K. Tanaka, Appl. Phys. Lett. **53** (1988) 1489
6. G. Ganguly, and A. Matsuda, Appl. Phys. Lett. **64** (1994) 3581
7. M. Takai, T. Nishimoto, M. Kondo, and A. Matsuda, Appl. Phys. Lett. **77** (2000) 2828
8. T. Takagi, R. Hayashi, G. Ganguly, M. Kondo, and A. Matsuda, Thin Solid Films **345**

(1999) 75
9. K. Koga, Y. Matsuoka, K. Tanaka, M. Shiratani, and Y. Watanabe, Appl. Phys. Lett. **77** (2000) 196
10. A. Bouchoule, A. Plain, L. Boulefendi, J. Ph. Blondeau, and C. Laure, J. Appl. Phys. **70** (1991) 1991
11. G. Ganguly, J. Newton, D.E. Carlson and R.R. Arya, J. Non-Cryst. Solids **299-302** (2002) 53; G. Ganguly, G. Wood, J.N. Newton, M. Bennett, D.E. Carlson and R.R. Arya Mater. Res. Soc. Symp. Proc. **715** (2002) 55.
12. A. A. Friedman, L. Boulefendi, T. Hbid, B.V. Potapkin, and A. Bouchoule, J. Appl. Phys, **79** (1996) 1303

## Hydrosilylation of Silicon Surfaces: Crystalline versus Amorphous

Andrea Lehner, Georg Steinhoff, Martin S. Brandt, Martin Eickhoff, Martin Stutzmann
Walter Schottky Institut, Technische Universität München, 85748 Garching, Germany

### ABSTRACT

Using thermally induced hydrosilylation, organic molecules were covalently bonded to H-terminated crystalline silicon (111) and hydrogenated amorphous silicon (a-Si:H) surfaces. The resulting chemical surface structure was analyzed by X-ray photoelectron spectroscopy (XPS) and compared to that of silicon surfaces covered by a native oxide or terminated with hydrogen. For both kinds of substrates, the presence of oxygen on the surface is found to hinder the hydrosilylation reaction. Stable H-termination as a starting point for a successful hydrosilylation can be obtained on a-Si:H surfaces with much less technological effort than on crystalline silicon surfaces. Photoconductivity measurements of the different a-Si:H surfaces at low intensity of illumination (monomolecular recombination regime) indicate that the hydrosilylated surface has less defects than the H-terminated surfaces or surfaces covered with native oxide. Spin-dependent photoconductivity measurements identify the dominant paramagnetic defect at the hydrosilylated a-Si:H surface to be the silicon dangling bond.

### INTRODUCTION

Stable, densely packed organic monolayers covalently bonded directly to silicon surfaces currently are of significant interest in the field of biosensor applications, e.g. in biochemistry and biophysics, as they in principle allow the detection and utilization of charge transport across the silicon/organic interface. Preparation of such surfaces can be performed by hydrosilylation of H-terminated silicon with alkenes or alkynes [1]. For hydrosilylated crystalline silicon (111) surfaces, a high chemical and thermal stability [2] as well as good electronic properties with interface defect densities as low as $2 \cdot 10^{11}$ cm$^{-2}$eV$^{-1}$ have been reported [3].

a-Si:H as an easily producible, large area electronic material is currently used for a variety of different applications, such as displays and solar cells. Organic surface modification of a-Si:H could therefore be an important issue for the fabrication of cheap biosensors based on silicon technology. Further, a-Si:H shows a smaller sensitivity to surface oxidation [4] and H-termination is easier to perform than on crystalline surfaces. The use of a-Si:H films for the fabrication of hydrosilylated silicon surfaces could have the additional benefit of requiring less stringent processing conditions during surface functionalization. In this study, we have modified a-Si:H surfaces by thermally induced hydrosilylation with 1-octadecene and compared their chemical surface properties to those of equally prepared silicon (111) surfaces by XPS investigations. In addition, the electronic properties of hydrosilylated a-Si:H surfaces were investigated by photoconductivity (PC) and spin-dependent photoconductivity (SDPC) measurements.

## SAMPLES AND EXPERIMENTAL DETAILS

(111)-oriented, n-type silicon with a resistivity $\rho$ = 1000...4000 $\Omega$cm and undoped a-Si:H layers with a thickness of 500 nm deposited on glass substrates by standard plasma-enhanced CVD were used in this work. All samples were cleaned ultrasonically with acetone and isopropanol, washed with deionized water and subsequently dried with $N_2$. For the H-termination, the native oxide was removed by etching the samples in HF-vapor (50%) for 90 seconds followed by a rinse in $H_2O$ and drying in a stream of $N_2$. Hydrosilylation was performed by exposing the samples for 90 minutes to liquid 1-octadecene at 150°C after exposure to HF-vapor. The fluid was bubbled with $N_2$ in order to suppress oxidation originating from dissolved oxygen [2]. Subsequently, the samples were washed five times in hexane, methanol and chloroform and finally dried in a stream of $N_2$.

XPS analysis was carried out in a UHV system at a background pressure of $5 \cdot 10^{-10}$ mbar using the Mg $K_\alpha$-radiation at 1253.6 eV for excitation and a hemispherical electron analyzer system (SPECTRA EA 10 N) to determine the kinetic energy of the photoelectrons. For each sample the O 1s, C 1s and Si 2p core level signals were recorded. The XPS spectra of the a-Si:H samples were corrected for an energy shift due to charging effects caused by the glass substrate by using the characteristic Si-Si photoelectron energy (99.6 eV) as a standard.

AC photoconductivity measurements were performed at 19 Hz under bias illumination using a lock-in amplifier to detect the photocurrent signal. Strongly absorbed light (wavelength $\lambda \cong 450$ nm) was used to enhance the surface sensitivity. To determine the photon flux, a silicon photodiode was used for a reference measurement with the same amount of light. Further details are given in [5]. For the SDPC measurements the samples were illuminated inside an ESR resonator with the same bias light. A microwave power of 2 W and a modulation amplitude of the magnetic field of 8 G were used.

## RESULTS AND DISCUSSION

### XPS Study

Figure 1 shows the XPS spectra of the hydrosilylated silicon (111) surface, compared to a surface covered with native oxide and a H-terminated surface. The surface covered with native oxide of about 1.0 nm thickness (determined from the O 1s peak area as described below) shows a large O 1s peak at 532.9 eV. On the hydrosilylated and H-terminated surfaces, a smaller O 1s peak at 532.2 eV is observed which can be attributed to remaining traces of oxide on these surfaces. Since the samples were handled under ambient air during the preparation, the formation of the native oxide on the surface apparently started instantaneously after the HF-vapor treatment. The large C 1s signal of the hydrosilylated sample at 285.0 eV clearly indicates the presence of organic molecules on its surface. The small C 1s peak detected at the oxidized and H-terminated surfaces can be assigned to spurious contamination by carbon. The Si 2p emission of the oxidized surface consists of two separate signals corresponding to the Si-Si (99.6 eV) and $SiO_2$ peak (103.8 eV). Neither the hydrosilylated nor the H-terminated surface show the presence of $SiO_2$, but both reveal a suboxide ($SiO_x$) peak at about 102.3 eV, indicating the formation of a nonstoichiometric native oxide [6]. Due to the formation of Si-C bonds, characteristic emission located at about 101 eV is expected on the hydrosilylated surface, but since this peak overlaps with the neighboring Si-Si and $SiO_x$ peaks, only a very small signature can be found.

The XPS spectra of the corresponding a-Si:H samples are shown in Figure 2. Again, a large O 1s peak is detected for the sample covered with native oxide of about 1.3 nm thickness. However, compared to the crystalline surface, the O 1s peaks of the hydrosilylated and H-terminated surfaces are much weaker, although these samples have been exposed to ambient atmosphere for the same period of time (approx. 5 min.) before being transferred into the XPS vacuum system. Thus, the surface of a-Si:H is much more resistant against oxidation than the crystalline silicon surface in the case of the simplified H-termination procedure employed here. This is confirmed by looking at the Si 2p peak. Neither a $SiO_2$ nor a $SiO_x$ contribution is observed for the hydrosilylated and H-terminated a-Si:H samples. The C 1s peak of the hydrosilylated a-Si:H surface seems to be even larger than for the corresponding crystalline sample, showing that the presence of a native oxide on the surface hinders the hydrosilylation reaction.

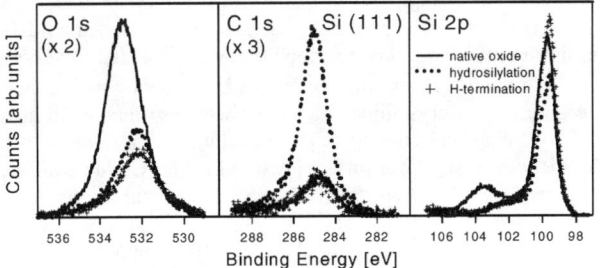

**Figure 1.** XPS spectra of Si (111) samples: (——) covered with native oxide, (+++) after H-termination and (····) after hydrosilylation. For each surface modification, the O 1s (multiplied by a factor of 2), the C 1s (multiplied by a factor of 3) and the Si 2p peaks are shown.

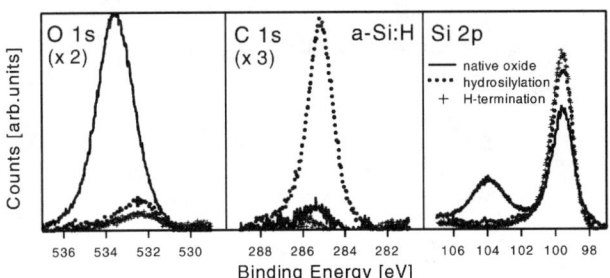

**Figure 2.** XPS spectra of a-Si:H samples: (——) covered with native oxide, (+++) after H-termination and (····) after hydrosilylation. For each surface modification, the O 1s (multiplied by a factor of 2), the C 1s (multiplied by a factor of 3) and the Si 2p peaks are shown. The O 1s signal of the hydrosilylated and H-terminated samples is noticeably smaller than that of the corresponding crystalline samples shown in Figure 1.

Although the a-Si:H surface shows a larger surface roughness of about 1.0 nm rms compared to 0.2 nm rms found for the crystalline silicon surface, the XPS signal intensities of both surfaces can be compared directly. Though a tilt of the surface results in an increased surface area, the XPS signal intensity stays constant since the effective overlayer thickness increases at the same time.

Having identified the surface oxide as a major obstacle for a complete hydrosilylation of crystalline and amorphous silicon, we have further studied the time dependence of the native oxide formation on H-terminated a-Si:H surfaces in more detail. For this purpose, the native oxide layer on a-Si:H samples was removed by HF-vapor and the surfaces were then exposed to air for various time intervals between five minutes and three months, after which XPS spectra were recorded. By determining the O 1s peak areas, the oxide thickness was calculated for each sample using the formalism described in [7]. Figure 3 shows the calculated oxide thicknesses of our samples as a function of the exposure time to air. For comparison, the results of Ponpon and Bourdon [4] for the thickness of native oxide grown on a-Si:H in air at room temperature obtained by ellipsometry are included in Figure 3. The oxidation rates obtained by both methods are in good agreement. The dashed line in Figure 3 shows that the oxide thickness $d$ increases as $d \propto t^{0.45}$, where $t$ is the exposure time to ambient atmosphere. This suggests that oxidation of a-Si:H is limited by diffusion through the oxide layer formed (with an expected time dependence of $d \propto t^{0.5}$ in the ideal case) [4]. While for H-termination of a-Si:H surfaces a simple treatment in HF-vapor is sufficient, H-termination of crystalline silicon surfaces resulting in atomically flat surfaces requires a series of oxidizing and etching steps, including a final etch with buffered $NH_4F$ [8]. The native oxide thicknesses on these surfaces determined by Gonda et al. with XPS measurements [9] are also shown in Figure 3 and are somewhat lower than those shown for a-Si:H surfaces up to exposure times of several hours. Obviously, for the a-Si:H surfaces the simple HF-vapor treatment results in a more or less equally stable H-termination than the multi-step procedure required on crystalline surfaces. On crystalline silicon surfaces, the simple HF-vapor treatment results in inferior H-termination as evident from the oxide thickness of 3 Å observed by XPS already after 5 min storage in ambient air also indicated in Figure 3.

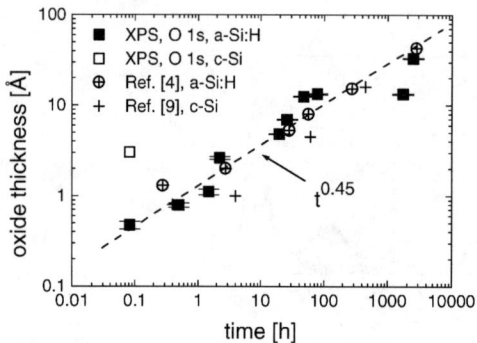

**Figure 3.** Thickness of the native oxide film on a-Si:H as a function of exposure time to ambient atmosphere obtained from the increase of the XPS O 1s peak area. For comparison, the results of Ponpon and Bourdon obtained by ellipsometry on a-Si:H are shown [4]. The corresponding data for native oxide growth on atomically smooth crystalline silicon surfaces are taken from [9].

**PC and SDPC Study**

Figure 4 shows the results obtained from the PC measurements as a function of the total photon flux for the three types of a-Si:H surfaces prepared: covered with native oxide, H-terminated and hydrosilylated with 1-octadecene. The hydrosilylated surface exhibits the highest photoconductivity $\sigma_{photo}$ over the whole illumination range, whereas the samples covered with native oxide and terminated with hydrogen are almost identical. At low levels of illumination, $\sigma_{photo}$ depends linearly on the photon flux or the density of carriers generated, indicating a dominating monomolecular recombination via defects (Shockley-Read-Hall recombination) [10-12]. Thus, $\sigma_{photo}$ is inversely proportional to the surface defect density. As expected, the recombination mechanism changes from monomolecular to bimolecular (band-to-band recombination) with increasing generation rate.

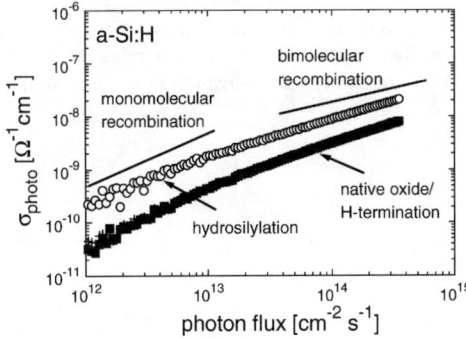

**Figure 4.** Photoconductivity of a-Si:H after different surface treatments as a function of the incident photon flux.

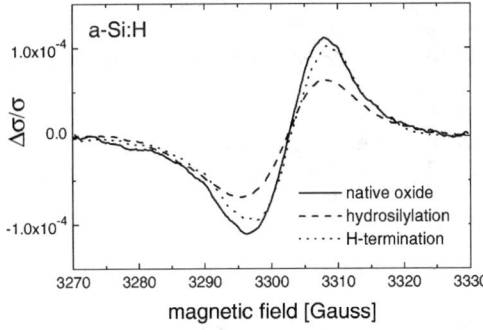

**Figure 5.** SDPC signal $\Delta\sigma/\sigma$ of a-Si:H after different surface treatments as a function of the magnetic field at a photon flux of $3.4 \times 10^{14}$ cm$^{-2}$s$^{-1}$ and a wavelength $\lambda = 450$ nm.

The microscopic nature of the dominant paramagnetic defect species responsible for recombination can be identified by SDPC. Figure 5 shows the SDPC signals $\Delta\sigma/\sigma$ of the different a-Si:H surfaces prepared as a function of the magnetic field. The characteristic resonance lines show a g-factor of 2.0051 and can be attributed to recombination at silicon dangling bonds (e-db line [13] ). As $\Delta\sigma/\sigma$ is proportional to the defect density, the hydrosilylated surface again shows the lowest defect density compared to the oxidized and H-terminated surfaces in agreement with Figure 4.

## CONCLUSIONS

We have shown by XPS measurements that monolayers of organic molecules can successfully be attached to H-terminated silicon (111) and a-Si:H surfaces by thermally induced hydrosilylation. The negative influence of an oxide layer on the surface on the hydrosilylation reaction was demonstrated. We confirmed by XPS measurements that stable H-termination of a-Si:H surfaces can be achieved with less technological effort than H-termination of crystalline surfaces. Thus, a-Si:H is a promising substrate for bio-functionalization procedures requiring less stringent process conditions.

## ACKNOWLEDGEMENTS

The authors thank S. Klein (Forschungszentrum Jülich) for providing the a-Si:H samples. This work was supported by the Deutsche Forschungsgemeinschaft (SFB 563).

## REFERENCES

1. M.R. Linford, P. Fenter, P. Eisenberger, C. Chidsey, J. Am. Chem. Soc. **117**, 3145 (1995).
2. A. Bansal, X. Li, S.I. Yi, W.H. Weinberg, N.S. Lewis, J. Phys. Chem. B **105**, 10266 (2001).
3. S. Kar, C. Miramond, D. Vuillaume, Appl. Phys. Lett. **78**, 1288 (2001).
4. J.P. Ponpon, B. Bourdon, Solid State Electronics **25**, 875 (1982).
5. A. Lehner, F. Kohl, S.A. Franzke, T. Graf, M.S. Brandt, M. Stutzmann, Appl.Phys. Lett. **82**, 565 (2003).
6. F.G. Bell, L. Ley, Phys. Rev. **37**, 8383 (1988).
7. A. Lehner, G. Steinhoff, M.S. Brandt, M. Eickhoff, M. Stutzmann, J. Appl. Phys., in print.
8. M. Hirose, T. Yasaka, M. Takakura, S. Miyazaki, Solid State Technology, 43 (Dec. 1991).
9. S. Gonda, M. Tanaka, T. Kurosawa, I. Kojima, Jpn. J. Appl. Phys. **37**, L 1418 (1998).
10. J.W.P. Hsu, C.C. Bahr, A. vom Felde, S.W. Downey, G.S. Higashi, M.J. Cardillo, J. Appl. Phys. **71**, 4983 (1992).
11. W. Shockley, W.T. Read, Phys. Rev. **87**, 835 (1952).
12. R.N. Hall, Phys. Rev. **87**, 387 (1952).
13. H. Dersch, L. Schweitzer, J. Stuke, Phys. Rev. B **28**, 4678 (1983).

## Present Status Of Hot Wire Chemical Vapor Deposition Technology

R.E.I. SCHROPP
Utrecht University, Debye Institute, SID - Physics of Devices, P.O. Box 80000,
3508 TA Utrecht, The Netherlands

## ABSTRACT

In the last few years, tremendous progress has been made in the field of Hot Wire Chemical Vapor Deposition (HWCVD): (1) It has been shown that there are no fundamental limitations in HWCVD with respect to substrate area. Using a periodic configuration of multiple short wires, good uniformity (± 7.5 %) has been demonstrated by Anelva over an area of 96 cm × 40 cm. (2) High quality microcrystalline Si can be produced. Solar cells in the n-i-p configuration are currently better than those made by PECVD. At Jülich, the efficiency of such cells is 9.4 %, and Utrecht has recently made the first HWCVD multibandgap *triple junction* solar cells, (3) HWCVD offers the potential of ultra high deposition rates. At NREL, a-Si:H has been deposited at rates in excess of 12 nm/s, and at Utrecht University μc-Si:H rates are in excess of 1 nm/s. (4) Thin film transistors (TFTs) with mobilities in excess of 1 $cm^2$/Vs with an a-Si:H channel have been shown to be stable and μc-Si:H TFTs have been made with mobilities in excess of 40 $cm^2$/Vs. (5) The efficient production of atomic H in HWCVD is beneficial in passivation processes, but it can also be applied in efficient etching processes. (6) Alloys of Si with various functions can be made, such as $SiN_x$ for antireflection and passivation coatings. Remarkably, all of the above results have been achieved without detailed knowledge about the primary reactions at the filament, the gas phase reactions, and the reactions with the growing film. The choice of filament material and its operation temperature have a large influence on the production of various reaction species and thus, on the structure of the resulting film. HWCVD is basically an ion-free deposition technique, which is an advantage for many kinds of thin films. HWCVD has also proven its feasibility in polymer deposition and nanotube formation.

## INTRODUCTION

Catalytic decomposition (Catalytic Chemical Vapor Deposition (Cat-CVD;[1])) of silane or silane/hydrogen mixtures at a resistively heated filament (therefore also called Hot Wire Chemical Vapor Deposition (HWCVD; [2])) takes place in a deposition regime that is fundamentally different from that where plasmas are involved. The last 10 years, technological and scientific developments are providing improved control of parameters (that are in the process of being defined), similar to the developments that Plasma Enhanced CVD has gone through in the 70's and 80's of the last century. HWCVD is becoming increasingly mature [3] and presently yields devices with state-of-the-art properties, even though our understanding of gas phase and growth reactions is far from complete. The high deposition rate of silicon-based thin films makes this method particularly interesting for application to, among others, low cost photovoltaic devices (solar cells) and thin film transistors (TFTs).

The high efficiency of $H_2$ dissociation at a tungsten wire has been utilized since the 1960's [4]. The first patent [5] and publication [6] on CVD appeared in 1979 ("thermal CVD") in the USA. About 6 years later, Matsumura and Tachibana [7] used this technique for the preparation of fluorinated amorphous silicon. The preparation of hydrogenated amorphous silicon (a-Si:H) was further investigated by Doyle et al. [8] and Matsumura [1,9,10] in the late 1980's, showing the high deposition rate as the prominent feature. Renewed interest in the deposition method came in 1991 due to Mahan et al. [2,11], who demonstrated for the first time the possibility to produce device-quality a-Si:H with a hydrogen concentration below 1 at.-%. Due to this development, many laboratories entered the field and presently, well over 30 laboratories have HWCVD deposition facilities. The first thin film silicon solar cells were made in 1993 at the University of Kaiserslautern and NREL, while the first TFTs were made in 1995 at Utrecht University and JAIST.

Within Europe, at least 10 groups are presently using the technology to create novel thin films and devices. The materials are also under study in other groups to compare their unique properties with those found in plasma-deposited material. The technology has yielded silicon thin films with amorphous, micro-, and polycrystalline properties [12,13]. Doped layers, both p-type and n-type, have been shown to be feasible. Alloys, such as a-SiGe:H and a-SiC:H have been demonstrated. Dielectric layers, such as silicon nitride with device quality properties are available. Even $SiO_2$ layers have been made with the HWCVD technique. Numerous kinds of applications exist, of which solar cells and TFTs are the most intensively studied, whereas applications in micro-electromechanical devices and in ULSI processes are lying ahead.

## HOT WIRE CVD PROCESS

In the HWCVD process the feedstock gases are very efficiently cracked into atomic radicals at the surface of a hot filament (usually tungsten or tantalum), which is held at a temperature higher than 1500 °C. The reactive species are subsequently transported to the substrates in a low pressure ambient (typically only 20 μbar for amorphous silicon). This enables a high deposition rate without gas-phase particle formation. Recently, it has been shown that very ultrahigh deposition rates can be achieved (more than 100 times faster than PECVD [14]).

The most frequently used filament materials are tungsten (W) and tantalum (Ta). Matsumura [10] reported on molybdenum (Mo), vanadium (V) and platinum (Pt) as filament materials. More recently, Duan et al. [15] and Veenendaal et al. [16] used rhenium (Re) as the filament material. Van Veenendaal showed that, except for the highest filament temperature $T_{fil}$ > 1950°C, polycrystalline silicon can be deposited with crystal orientation exclusively in the (220) direction. Finally, Morrison et al. [17] reported on the deposition of microcrystalline silicon using graphite (C) as the catalyzer. Brühne et al. [18] also reported on the use of graphite for the deposition of microcrystalline silicon with (220) orientation only. The deposited layers did however contain a considerable amount of carbon. Iridium (Ir) appears to be the most suitable filament for $SiO_2$ deposition [3].

Using tungsten filaments at sufficiently high filament temperature, silane is fully cracked into one Si and four H atoms. Only at temperatures below 1430 °C, $SiH_2$ and $SiH_3$ could be detected [19]. It is suggested that in this temperature regime, a Si/W alloy is formed on the filament [20]. From further experiments, we deduced that this alloy affects the decomposition of silane at the filament surface and virtually blocks the decomposition of $H_2$ [21]. Matsumura [22] also found that at filament temperatures above 1430 °C, the major species desorbed from the filament is the Si atom. The maximum production of Si atoms is observed at about 1530 °C. The latter results were obtained with W, Mo and Ta filaments. The Si atom is the only major species above $T_{fil} = 1430$ °C for all three filaments. The temperature dependence below $T_{fil} = 1430$ °C is large and different for these three filaments. Activation energies for Si atom desorption from the filament are found to be (251±63), (96±25) and (71±20) kJ/mol for Mo, Ta and W filaments, respectively [19]. The energy needed for Si atom production is much lower than 4 times the Si-H bond dissociation energy of 300 kJ/mol. Both the large differences for the different filaments and the small values of the activation energies indicate that the decomposition of $SiH_4$ on the hot filament is caused by catalytic reactions at the filament surface.

At low pressures (< 5 µbar), the Si and H atoms that come from the filament thermally diffuse to the substrate [23] with only few or no gas phase reactions. Duan et al. [24] reported on single photon ionization mass spectrometry measurements at 1.8 x $10^{-2}$ µbar at W filament temperatures of 1950 °C. The major silicon containing gas species detected is elemental Si itself, along with minor contributions of $SiH_3$ and $Si_2H_x$. However, these pressures are more than an order of magnitude lower than the pressures used during practical silicon deposition, e.g. the pressure used during deposition is in the order of 0.1 mbar and 0.02 mbar for microcrystalline and amorphous silicon, respectively.

At higher pressures (> 5 µbar), the silicon atom is highly reactive. It can abstract an H atom from silane, resulting in SiH and $SiH_3$, or it can insert into a Si-H bond [25]. Molenbroek described three possible insertion reactions, namely:

$$Si + SiH_4 \rightarrow SiH + SiH_3, \tag{1}$$

$$Si + SiH_4 \rightarrow HSiSiH_3 * \tag{2}$$

$$Si + SiH_4 \rightarrow 2SiH_2. \tag{3}$$

Because the first and third reactions are endothermic, they are unlikely to occur since the thermal energy of the radicals is only about 0.25 eV at most. $HSiSiH_3$ is formed through an exothermic reaction and will thus be the most probable species to be formed. The formation reaction of $HSiSiH_3$ has been studied by ab-initio molecular orbital calculations by Sakai et al. [26]. According to these calculations, $HSiSiH_3$ is unstable and will react with $SiH_4$ in the gas phase. There are three possible reactions, namely [19]:

$$HSiSiH_3 + SiH_4(+M) \rightarrow Si_3H_8(+M) \tag{4}$$

$$HSiSiH_3(+M) \rightarrow H_2SiSiH_2(+M) \tag{5}$$

$$HSiSiH_3 + SiH_4 \rightarrow SiH_2 + Si_2H_6 \tag{6}$$

In reactions (4) and (5), M stands for a third body (e.g. an atom or molecule). Up to now, no $Si_3H_8$ has been detected. Therefore, Inoue et al. suggest that reaction (4) can be neglected. However, Molenbroek [25] suggests that abstracting $H_2$ from the $Si_3H_8$ molecule forms $Si_3H_6$. The reaction product of reaction (5), $H_2SiSiH_2$, is a rather stable closed shell molecule and it has been expected to be an important precursor species for the film growth. In the experiments described by Inoue et al. [19], the most prominent species detected is $Si_2H_6$. They suggested that $SiH_2$, produced via reaction (6), further reacts with $SiH_4$, according to

$$SiH_2 + SiH_4(+M) \rightarrow Si_2H_6(+M). \tag{7}$$

The presence of atomic hydrogen in the reactor results in the following reaction:

$$H + SiH_4 \rightarrow SiH_3 + H_2. \tag{8}$$

The SiH$_3$ species does not react with SiH$_4$ and the only gas phase reaction of SiH$_3$ is self-recombination. Also Gallagher proposed a gas phase growth reaction, in which Si atoms react with silane [27]:

$$Si + SiH_4 \rightarrow Si_2H_4^* \rightarrow 2SiH_2, SiH + SiH_3, Si_2H_2 + H_2, Si_2H_4, \tag{9}$$

where Si$_2$H$_4$$^*$ is an unstable intermediate. Some of the reaction products will react with silane to produce more stable silanes, such as Si$_2$H$_6$ and Si$_3$H$_6$. Goodwin [28] pointed out, that the reaction yielding Si$_2$H$_2$ + H$_2$ is energetically favored, since it is an exothermic one by 110 kJ/mol. The relative contribution of Si$_2$H$_2$, being a closed-shell molecule, to film growth is however still very uncertain as the reaction probability is unknown.

In summary, the main gas phase reaction species are thus: SiH$_3$, Si$_2$H$_6$, Si$_3$H$_6$ and H$_2$SiSiH$_2$. It is expected that the detection of the actual gas phase reaction species will provide further insight.

## MAIN DIFFERENCES BETWEEN PECVD AND HWCVD

Whereas Plasma Enhanced CVD (PECVD) is presently the workhorse of the semiconductor industry, it has a number of limitations. First, the primary concern is the low deposition rate, in particular for μc-Si:H (indirect-gap) intrinsic layers. Attempts to achieve higher deposition rates have usually led to films with lower density (less compact) and higher void content. Although the application of the PECVD technique in the Very High Frequency (VHF) domain (VHF-CVD) and/or the high pressure domain leads to higher deposition rates, it has been difficult to scale up the VHF technique to very large area (> 1 m$^2$), due to the finite wavelength of the radio frequency excitation. Moreover, high power VHF generators are less standard than the 13.56 MHz generators, and therefore still more expensive, whereas the high-pressure regime at 13.56 MHz may suffer from non-uniformity upon scaling up.

The absence of *any* kind of plasma in HWCVD has three important advantages: (i) The first advantage is that HWCVD is inherently free of dust. Since there is no plasma, the most important source of microparticles that is present in conventional PECVD is eliminated. Plasmas are a source of dust since the positive potential in the bulk of the plasma tends to trap negatively charged particles that are likely to grow, leading to inferior films. To avoid dust in conventional production systems, e.g., in the Thin Film Transistor Liquid Crystal Display (TFT LCD) business or in the thin film a-Si/a-SiGe tandem solar cell business, depletion of silane must be avoided. This leads to very low source gas utilization rates in practice. In the case of TFT production, gas utilization is even as low as 1 %. HWCVD offers the possibility to almost completely deplete the source gases, and thus, gas utilization efficiencies of up to 80 % have been reported for a-Si:H deposition [29]. (ii) The second advantage is that no ions are involved and that damage due to energetic ions does not occur. This is particularly useful for passivation layers on electronic structures. (iii) The third advantage is that the substrate is not part of the deposition mechanism itself (similar to *remote* PECVD techniques), which permits the use of arbitrary substrates (insulating, conducting) as well as the use of transport mechanisms for moving the individual substrate panels or a continuous web through the deposition zones independently of the deposition mechanism.

In addition to the high deposition rate and high gas utilization, the elimination of expensive rf power supplies and matching boxes makes the Hot Wire CVD technique a low-cost production method for silicon thin films. For this technique to become more widely adopted, it is important to physically demonstrate the large area capabilities of the technique. With a special new design for a hot-wire assembly with a showerhead [29], Anelva has demonstrated a thickness uniformity of ± 7.5 % over 96 cm x 40 cm substrate area, and has thus overcome two main difficulties, the sagging of the filaments and the silicidation of the catalyst.

## RESULTS OBTAINED WORLD WIDE

The following is a summary of results reported at the 2$^{nd}$ International Conference on Cat-CVD (Hot Wire CVD), held in Denver, Colorado, September 2002. The papers are referred to by the presenting author and all papers referred to can be found in the Proceedings [3].

Progress has been made in the control of substrate heating and on filament condition. Filament lifetime enhancing treatments have been defined at JAIST and confirmed at NREL and Utrecht. An effective lifetime-enhancing treatment is the preheating of the filaments for several minutes at $T_{fil,treat} = T_{fil,process} + 500$ °C. Contamination from the filaments can virtually be eliminated by choosing the appropriate temperature regime, as discussed by Jean-Eric Bourée (Ecole Polytechnique, Palaiseau).

NREL showed further work on their ultra high deposition rate (> 100 Å/s) a-Si:H materials. These rates are achieved with 2 tungsten filaments located at 3.2 cm from the substrate. The saturated defect density of the a-Si:H is typically (2-4) × $10^{16}$ cm$^{-3}$ and *independent* of the deposition rate, up to 130 Å/s, though the void density increases by a factor ~ 100 [30]. Solar cells made on stainless steel with these films up to a deposition rate of 50 Å/s have similar stabilized efficiencies of 4.8 %.

A wide variety of filaments has been investigated for various purposes. Differences between tungsten and tantalum are under discussion [20], while a longer life for graphite filaments has been claimed. The application of Hot Wire CVD has been extended to silicon oxide deposition. For oxide deposition, iridium wires are reported to be most appropriate (Kazuya Saito, ULVAC).

The gas phase Si-chemistry has been extensively studied by in situ real time radical detection methods, which were described by Stacey Bent (Univ. Stanford). She presented evidence that the decomposition of SiH$_4$ is indeed catalytic. New insight into the catalytic action of the filaments was also presented by Mitsuo Koshi (Univ. Tokyo), who measured the temperature of the H atoms to be ~ 900 °C just after the total atomization of SiH$_4$. Alan Gallagher (Univ. of Colorado) complemented this work with a discussion of the reaction probability of atomic H at the surfaces and in the gas phase, and Vik Dalal (Iowa State Univ.) suggested that ion bombardment assisted HWCVD might improve the layers that are made at low temperature. The trade-off between low temperature, ion-assisted deposition and the benefits of a purely thermal, ion-free deposition regime still needs to be explored.

The high radical yield of HWCVD has been fully utilized in recent experiments on etching of crystalline silicon by H atoms, and on etching of Si$_3$N$_4$ and SiO$_2$ using NF$_3$. The etching rates can be very fast and HWCVD-dissociated NF$_3$ may thus replace plasma dissociated SF$_6$ for etching applications as well as for photoresist removal.

The truly unique properties of a-Si:H and μc-Si:H, deposited at high rates, are very attractive for application in solar cells and thin film transistors (TFTs). Bernd Schröder (Univ. Kaiserslautern) gave a comprehensive review of the status of HWCVD solar cells. The initial efficiencies for a-Si:H n-i-p and p-i-n cells with all layers deposited by HWCVD are currently equal and are 8.9 %. Particularly interesting is the performance of μc-Si:H cells made at IPV, Jülich. Stefan Klein presented a 9.4 %-efficient p-i-n cell, with a microcrystalline i-layer made by HWCVD, thus reaching a higher efficiency than that achievable for microcrystalline p-i-n's made by PECVD. The best material is made just over the microcrystalline/amorphous transition. Microcrystalline silicon solar cells with sufficient crystalline volume fraction in the i-layer do not show any light-induced degradation. The growth of microcrystalline silicon harbors many interesting aspects, including how a nucleation (or seed) layer affects crystal growth and orientation in the film bulk, and if there is a preferred orientation of the crystals for application in solar cells.

TFTs with Hot Wire deposited silicon and silicon nitride layers have been further improved. Bernd Stannowski (Utrecht University) presented bottom gate TFTs with a performance equal to conventional PECVD TFTs, but with an improved threshold voltage stability, and top gate TFTs with high as-deposited electron mobility. This aspect has been further exploited by Sony Corp., who reported [31] to have achieved a mobility as high as 40 cm$^2$/Vs by a back-etching technique.

The growth of passivating silicon layers in so-called HIT cells (C. Voz, University of Barcelona) and the deposition of silicon nitride layers on conventional multicrystalline silicon (J. Holt, Caltech, Pasadena) were shown to hold promise for breakthrough technology in the near future for this type of solar cells.

J. Roland Pitts (GVD Corp., NREL) described how HWCVD enables linear polymerization (in contrast to PECVD). Polymer coatings (PTFE) have been developed and will find applications in biohydrogen sensors. Conformal coating of polymers by HWCVD can also be used for catheters, implants, etc. Anne Dillon (NREL) described how HWCVD offers a new method for the fabrication of carbon nanotubes, both for clustered multi-wall carbon nanotubes (MWNT) and for isolated single-wall tubes (SWNT). The application where carbohydride dissociation and silane dissociation come together is in the field of microcrystalline SiC window layers, under development at Tokyo Institute of Technology and the Japan Advanced Institute of Science and Technology (JAIST).

## RECENT RESULTS OBTAINED AT UTRECHT UNIVERSITY

Our equipment consists of two HWCVD chambers that are connected to a multichamber ultrahigh vacuum (UHV) system that has an additional 3 PECVD chambers and another multichamber UHV system where the plasma- and HW-assemblies can easily be exchanged. This arrangement offers the opportunity to create different geometries and deposition parameters that are optimized for each type of intrinsic layer (amorphous/microcrystalline or polycrystalline). Tungsten filaments are used for poly-Si:H, tantalum filaments for a-Si:H and μc-Si:H. Polysilicon is obtained at substrate temperatures between 430 and 530 °C. Amorphous and microcrystalline silicon films are obtained between 220 °C and 430 °C. All intrinsic layers are incorporated in devices using plasma-deposited p-type and n-type layers, though presently also n-type μc-Si:H by HWCVD is available.

The gas flow in the HWCVD chambers is perpendicular to the length of the wires. The temperature of the wires is typically 1800 – 2000 °C, while for poly-Si:H deposition it is deliberately reduced to 1750 – 1850 °C. With the help of a heat transport model [32] the substrate temperature has been carefully calibrated. The substrate loading system is equipped with a shutter, which hermetically shields the substrate from any deposition during preheating.

## MATERIALS

At Utrecht University, the main focus in Hot-Wire CVD is on the deposition of a-Si:H [33], micro- and poly-Si:H [34] and $SiN_x$ [35]. Here, we discuss the latest results on micro- and poly-Si:H and $SiN_x$.

Microcrystalline silicon (μc-Si:H) and polycrystalline silicon (poly-Si:H) are attractive alternatives for a-SiGe:H as the low-bandgap absorbing component in tandem cells and triple cells, because (i) a band gap lower than 1.3 eV can be obtained without sacrificing electronic transport properties, (ii) the use of the quite expensive $GeH_4$ is avoided, and (iii) cells incorporating μc-Si:H do not show light-induced degradation.

We have recently identified the optimum filament conditions as well as the gas flow and gas mixtures that are needed for obtaining (220) oriented poly-Si:H. Highly oriented poly-Si has some unique properties, such as a very intrinsic nature (oxygen levels down to $3 \times 10^{18}$ cm$^{-3}$) and very good coalescence of the grains [36]. It has already been applied in thin poly-Si Thin Film Transistors [37] and (tandem) solar cells [38].

Van der Werf et al. [39] showed in the temperature range above 1500 °C that the filament temperature decreases upon exposure of the filament to hydrogen. This decrease is explained by the power consumption needed for the dissociation of hydrogen on the filament surface. At relatively low filament temperatures, the filament is covered by a silicon-rich silicide, as reported by van Veenendaal et al. [20,40]. For tungsten filaments, an increase in silicon content was shown in the near-surface region of a tungsten filament with time, while the silicon content on a tantalum filament shows saturation rather quickly (see Fig. 1).

Using the above combined results it is suggested [21] that the degree of coverage of the filament with Si primarily affects the catalytic dissociation of $H_2$ molecules and that the *reduced primary dissociation of $H_2$* in $SiH_4/H_2$ mixtures promotes the deposition of microcrystalline

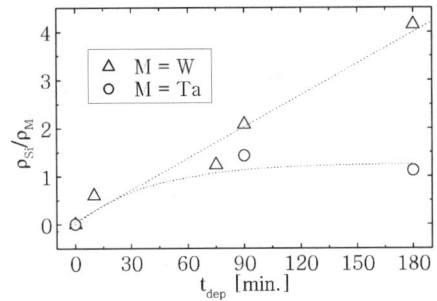

Fig. 1. Silicon content in the near-surface region of the filament ($\rho_{Si}/\rho_M$) as a function of deposition time ($t_{dep}$), for different filament materials (M). The lines are guides to the eye.

silicon with a preferred (220) orientation. Thin films with a dominant (220) orientation of the crystals are indeed obtained at tungsten filament temperatures at which silicide formation of the tungsten occurs and where we suspect that hardly any $H_2$ dissociation occurs. Indeed, for Ta filaments, (220) oriented material is reported [41] for filament temperatures *below* the threshold temperature for $H_2$ dissociation [39]. Furthermore, the use of graphite filaments is reported to *always* lead to (220) oriented material [18], while it is known from the field of hot filament deposition of diamond that graphite does not efficiently dissociate $H_2$ [42,43].

Further, the degree of depletion of the $SiH_4$ plays a role in the structure of the deposited layers. Silane depletion conditions generally prevent the annihilation of atomic hydrogen [44] through reaction (8). In Figures 2 and 3, using high $SiH_4$ flow, we observe a transition from random microcrystalline to amorphous structure with decreasing filament temperature. Under depletion conditions (lower graph of Fig. 2) we observe a transition from multi-oriented microcrystalline to purely (220) oriented microcrystalline nature. It is thus concluded that random crystallinity is enhanced by high atomic H production. For obtaining random crystallinity, $SiH_4$ depletion is less important. On the other hand, primarily (220) oriented material results if both the primary atomic H production is limited and the silane is depleted.

**Silicon nitride**

Amorphous silicon nitride (a-SiN$_x$:H) has been deposited by HWCVD from a mixture of silane and ammonia at substrate temperatures in the range of 300–500 °C. Layers deposited with an ammonia/silane gas-flow ratio of $R = 30$ are close to stoichiometry (N/Si = 1.33) with a hydrogen content around 10 at. %. These films have been implemented in Hot-Wire $a$-Si:H thin-film transistors. Deposition with $R > 30$ did not result in an increase of the N content, but led to more porous films. Infrared spectroscopy revealed that moisture penetrates these layers and that

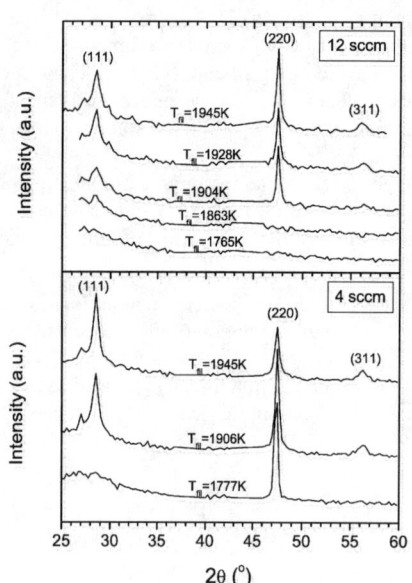

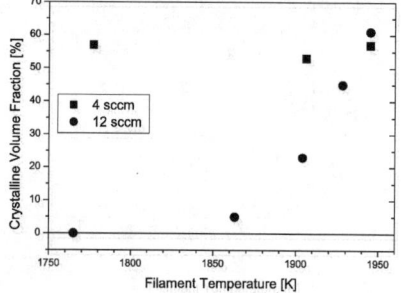

Fig. 2. XRD patterns for two series of samples deposited at a pressure of 0.1 mbar and a substrate temperature of ~500°C. In the top graph 12 sccm $SiH_4$ and 150 sccm $H_2$ were used, in the bottom graph 4 sccm $SiH_4$ and 150 sccm $H_2$.

Fig. 3. The crystalline volume fraction versus the filament temperature for the films deposited in these two series.

oxygen is incorporated in the network under air exposure. Cross-sectional transmission electron microscopy images showed that these layers contain spherical voids with diameters of several nanometers. In contrast, films deposited with a lower gas-flow ratio ($R < 30$) are inert and do not contain these voids. Both types of films show columnar growth. To better understand the deposition process, we used deuterated silane ($SiD_4$) as a source gas [45]. We found that no deuterium was incorporated in the films. This gives rise to the assumption that $SiH_4$ ($SiD_4$) is cracked effectively at the filaments. We infer that the ammonia species are scarcely dissociated at the filaments but rather in the gas phase by the atomic hydrogen (deuterium) originating from the dissociated silane. The abundance of atomic hydrogen in the gas phase is crucial for the breakup of ammonia and the incorporation of N in the film. We suggest that a similar mechanism is responsible for a-SiC:H deposition, where $CH_4$ is cracked by gas phase reactions, rather than dissociated at the filament.

## SOLAR CELLS

We present here, for the first time, triple junction multibandgap cells incorporating three intrinsic absorber layers that are all deposited by Hot Wire CVD. The bottom and middle cell have a microcrystalline silicon (µc-Si:H) absorber layer and the top cell has an amorphous silicon (a-Si:H) intrinsic layer. The cell is made in the configuration stainless steel/n-i-p/n-i-p/n-i-p/ITO and does not comprise a textured back reflector. We obtained a $V_{oc}$ of 1.725 V, $J_{sc}$ of 7.45 mA/cm$^2$, and a fill factor of 0.658. The open circuit voltage is appropriate for this structure, assuming a value of about 0.45 V for the two microcrystalline cells. The fill factor suggests that the structure is almost current-matched and the high $J_{sc}$ shows that there is true spectrum splitting. The triple junction cell, with an efficiency of 8.46 %, represents an improvement over our previously presented HWCVD µc-Si:H/a-Si:H tandem cells [38], which had an efficiency of 8.1 %. Further work is necessary, among others on back reflectors. Table 1 presents a collection of noteworthy cells that have been published thus far. It is by no means complete, but it serves to show various solar cell concepts that have great potential and have not been fully explored yet.

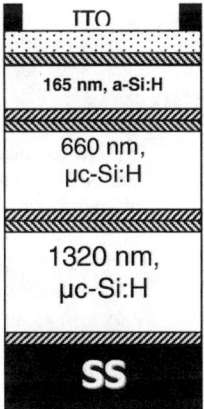

Fig. 4. Cross section of triple HWCVD cell

## THIN FILM TRANSISTORS (TFTs)

TFTs were made in bottom-gate and top-gate configurations [48]. Using the (220) oriented HWCVD poly-Si discussed above as the active material in TFTs, fabricated on Corning glass, an electron mobility of 4.7 cm$^2$/Vs was obtained in the top gate configuration [37]. This mobility was fully consistent with Hall effect measurements and time-resolved microwave conductivity (TRMC) measurements on equivalent individual layers [47]. A positive property of these TFTs is the low OFF current of 7.5 x 10$^{-11}$ A, which is not usually obtained for poly-Si TFTs and is due to the intrinsic nature of the poly-Si.

Table 1. A collection of noteworthy cells using HWCVD. The layers that are deposited by HWCVD are in *italics*.

| Laboratory | Structure of the cell | Remarks | η init. (%) | η stab. (%) |
|---|---|---|---|---|
| U. Kaiserslautern | glass/Asahi U/p-*i(a-Si)*-n/Ag | superstrate single | 10.2 | 7.0 |
| NREL/USSC | SS/Ag/ZnO/n-*i(a-Si)*-p/ITO/grid | substrate single | 9.9 | 7.8 |
| FZ Jülich | glass/text.TCO/p-*i(a-Si)*-n/p-*i(µc-Si)*-n/ZnO/Ag | superstrate micromorph tandem | 10.9 | |
| FZ Jülich | glass/text.TCO/p-*i(µc-Si)*-n/ZnO/Ag | superstrate microcryst. single | 9.4 | Δη ~ 10% |
| Utrecht Univ. | plain SS/n-*i(µc-Si)*-p/n-*i(µc-Si)*-p/n-*i(a-Si)*-p/ITO | substrate micro/micromorph triple junction | 8.5 | |
| Ecole Polytechnique | glass/text.TCO/n-*i(µc-Si)*-p/Ag | superstrate inverted microcryst. single | 6.1 | |

It should be mentioned that the thickness of the active layer was 1.5 µm. A rather thick layer was necessary because of the growth conditions employed, in which the (220) crystals extend conically and coalesce only after ~ 1 µm. There are two methods to reduce the thickness of the active layer. The first is to use a seed layer (or nucleation layer) (which has also been applied to solar cells; [48,49]). In this approach the incubation layer is avoided by creating a randomly oriented microcrystalline layer at high $H_2$ dilution conditions and a relatively low deposition rate (1 Å/s). Changing then to lower $H_2$ dilution conditions, the (220) seeds grow into larger crystals preferentially over other orientations. In another approach, as proposed by Sony Corporation [31], a previously deposited layer is back etched to a remaining thickness of about 100 nm before the channel region is created, and a mobility of up to 40 $cm^2$/Vs is obtained. This shows that the crystal formation is still proceeding at a large depth while the layer is being deposited. This suggests that atomic H, that is penetrating the film to very large depths [50], can structurally change the existing film and plays a role in the crystal formation below the growth zone. The diffusion depth of atomic H into amorphous silicon was estimated to be ~ 0.2 µm [51].

Nevertheless, using polycrystalline growth conditions even the first 10 nm already include a high density of nanocrystals, as is evident from the high electron mobility of 1.5 $cm^2$/Vs for *bottom gate* TFTs made under these growth conditions. Moreover, these TFTs are highly resistant against prolonged gate bias stress in that they show no threshold voltage instabilities. This opens the road to new applications, such as current-drive TFTs, e.g. for peripheral LCD row drivers or for OLEDs (Organic Light Emitting Diodes).

## CONCLUSION

With the successful demonstration of uniform large area deposition of silicon-based thin films, it has become likely that HWCVD finds its way to industrial implementation. Already now, a number of companies in Japan have set up (pilot) large area deposition systems. These systems offer the potential to drastically reduce the investment costs required to enter the 5th generation (125 cm wide) LCD fabrication technology. The first applications in products can be expected in passivation layers on silicon ULSI and GaAs devices, followed by HWCVD-produced TFTs and solar cells.

Yet, to open up the full potential of HWCVD it is important to improve our understanding of the catalytic decomposition reactions, the gas phase reactions, and the growth mechanism, as well as to overcome a number of fundamental technological hurdles such as substrate temperature control and control of the growth of alloyed films. With the open exchange of scientific results, HWCVD will have a bright future.

## ACKNOWLEDGEMENT

This research has benefited from the financial support from NOVEM and STW.

## REFERENCES

1. H. Matsumura, Jpn. J. Appl. Phys. **25** (1986) L949.
2. A.H. Mahan, J. Carapella, B.P. Nelson, R.S. Crandall, and I. Balberg, J. Appl. Phys. **69** (1991) 6728.
3. see: *Proceedings of the 2$^{nd}$ International Conference on Cat-CVD (Hot Wire CVD) Process*, Denver, USA, Sep. 10-13, 2002, Eds.: R.E.I. Schropp, M.B. Schubert, J.P. Conde, A.H. Mahan, and H. Matsumura, to be published in Thin Solid Films (2003).
4. J.N. Smith and W.L. Fite, J. Chem. Phys. **37** (1962) 898.
5. H.J. Wiesmann: US patent 4,237,150; Dec. 2, 1980.
6. H. Wiesmann, A.K. Ghosh, T. McMahon, and M. Strongin, J. Appl. Phys. **50** (1979) 3752.
7. H. Matsumura and H. Tachibana, Appl. Phys. Lett. **47** (1985) 833.
8. J. Doyle, R. Robertson, G.H. Lin, M.Z. He, and A. Gallagher, J. Appl. Phys. **64** (1988) 3215.
9. H. Matsumura, Mat. Res. Soc. Symp. Proc. **118** (1988) 43.
10. H. Matsumura, J. Appl. Phys. **65** (1989) 4396.
11. A.H. Mahan, B.P. Nelson, S. Salamon, and R.S. Crandall, J. Non-Cryst. Solids **137** & **138** (1991) 657.
12. R.E.I. Schropp and M. Zeman, *AMORPHOUS AND MICROCRYSTALLINE SILICON SOLAR CELLS: Modeling, Materials, and Device Technology,* (Kluwer Academic Publishers, Boston/Dordrecht/London, ISBN 0-7923-8317-6, 1998).
13. R.E.I. Schropp, K.F. Feenstra, E.C. Molenbroek, H. Meiling, and J.K. Rath, Phil. Mag B **76** (1977) 309.
14. B. Nelson, E. Iwaniczko, A.H. Mahan, Q. Wang, Y. Xu, R.S. Crandall, and H.M. Branz, Thin Solid Films **395** (2001) 292.
15. H.L. Duan, G.A. Zaharias, and S.F. Bent, Mat. Res. Soc. Symp. Proc. **664** (2001) A3.1.1.
16. P.A.T.T. van Veenendaal, C.M.H. van der Werf, J.K. Rath, and R.E.I. Schropp, J. Non-Cryst. Solids **299-302** (2002) 1184.
17. S. Morrison and A. Madan, Proc. of 17$^{th}$ European PVSEC (2001) 2951.
18. K. Brühne, M.B. Schubert, C. Köhler, and J.H. Werner, Thin Solid Films **395** (2001) 163.
19. K. Inoue, S. Tange, K. Tonokura, and M. Koshi, Thin Solid Films **395** (2001) 42.
20. P.A.T.T. van Veenendaal, O.L.J. Gijzeman, J.K. Rath, and R.E.I. Schropp, Thin Solid Films **395** (2001) 194.

21. C.H.M. van der Werf, A.J. Hardeman, P.A.T.T. van Veenendaal, M.K. van Veen, J.K. Rath, and R.E.I. Schropp, Thin Solid Films **427** (2003) 41.
22. H. Matsumura, Jpn. J. Appl. Phys. **37** (1998) 3175
23. N. Honda, A. Masuda, and H. Matsumura, J. Non-Cryst. Solids **266-269** (2000) 100.
24. H.L. Duan, G.A. Zaharias, and S.F. Bent, Thin Solid Films **395** (2001) 36.
25. E.C. Molenbroek, Ph.D.-thesis, University of Colorado (1995).
26. S. Sakai, J. Deisz, and M.S. Gordon, J. Phys. Chem. **93** (1989) 1888.
27. A. Gallagher, Thin Solid Films **395** (2001) 25.
28. D.G. Goodwin, Mat. Res. Soc. Symp. Proc. **557** (1999) 79.
29. K. Ishibashi, Thin Solid Films **395** (2001) 55.
30. A.H. Mahan, Y. Xu, D.L. Williamson, W. Beyer, J.D. Perkins, M. Vanecek, L.M. Gedvilas, B.P. Nelson, J. Appl. Phys. **90** (2001) 5038.
31. H. Matsumura, H. Umemoto, A. Izumi, and A. Masuda, to be published in [3].
32. K.F. Feenstra, R.E.I. Schropp and W.F. van der Weg, J. Appl. Phys. **85** (1999) 6843.
33. M.K. van Veen, Ph.D. thesis, Utrecht University, in print (2003).
34. P.A.T.T. van Veenendaal, Ph.D. thesis, Utrecht University (2002).
35. B. Stannowski, Ph.D. thesis, Utrecht University (2002).
36. J.K. Rath, H. Meiling and R.E.I. Schropp, Jpn. J. Appl. Phys. **36** (1997) 5436.
37. R.E.I. Schropp, B. Stannowski, J.K. Rath, C.H.M. van der Werf, Y. Chen, and S. Wagner, Mat. Res. Soc. Symp. Proc. **609** (2000) A31.3.
38. R.E.I. Schropp, C.H.M. van der Werf, M.K. van Veen, P.A.T.T. van Veenendaal, R. Jimenez Zambrano, Z. Hartman, J. Löffler, and J.K. Rath, Mat. Res. Soc. Symp. Proc. **664** (2001) A15.6.1.
39. C.H.M. van der Werf, P.A.T.T. van Veenendaal, M.K. van Veen, A.J. Hardeman, M.Y.S. Rusche, J.K. Rath, and R.E.I. Schropp, to be published in [3].
40. P.A.T.T. van Veenendaal, J.K. Rath, O.L.J. Gijzeman, and R.E.I. Schropp, Polycrystalline Semiconductors VI – Materials, Technologies, and Large Area Electonics, in: O. Bonnaud, T. Mohammed-Brahim, H.P. Strunk, J.H. Werner (Eds.), Scitech Publ., Solid State Phenomena **80-81** (2001) 53.
41. M. Fonrodana, D. Soler, J.M. Asensi, J. Bertomeu, and J. Andreu, J. Non-Cryst. Solids **229-302**, (2002) 14.
42. M. Sommer and F.W. Smith, J. Mater. Res. **511** (1990) 2433.
43. E. Zeiler, S. Schwarz, S.M. Rosiwal, and R.F. Singer, Materials Science & Engineering **A335** (2002) 236.
44. M. Kondo, M. Fukawa, L. Guo, and A. Matsuda, J. Non-Cryst. Solids **266-269** (2000) 84.
45. B. Stannowski, J.K. Rath, and R.E.I. Schropp, J. Appl. Phys. **93** (2003) 2618.
46. R.E.I. Schropp, B. Stannowski, and J.K. Rath, J. of Non-Cryst. Solids **299-302** (2002) 1304.
47. P.A.T.T. van Veenendaal, T.J. Savenije, J.K. Rath, and R.E.I. Schropp, Thin Solid Films **403 & 404** (2002) 175.
48. R.E.I. Schropp and J.K. Rath, IEEE Trans. Electron Dev. **46** (1999) 2069.
49. R.E.I. Schropp, Y. Xu, E. Iwaniczko, G.A. Zaharias, and A.H. Mahan, Mat. Res. Soc. Symp. Proc. **715** (2002) A26.3.4.
50. W. Beyer, Semiconductors and Semimetals **61** (1999) 165.
51. K.F. Feenstra, P. F. A. Alkemade, E. Algra, R.E.I. Schropp, and W. F. van der Weg, Prog. Photovolt. Res. Appl. **7** (1999) 341.

# PROPERTIES OF HIGH QUALITY p-TYPE MICRO-CRYSTALLINE-Si PREPARED BY Cat-CVD

Hideki Matsumura, Kouichi Katouno, Masaya Itoh and Atsushi Masuda
JAIST (Japan Advanced Institute of Science and Technology),
Tatsunokuchi, Ishikawa-ken 923-1292, JAPAN, E-Mail;h-matsu@jaist.ac.jp

## ABSTRACT

Properties of p-type µc-Si prepared by Cat-CVD (Catalytic Chemical Vapor Deposition), often called Hot-Wire CVD, are studied for possible application to window layer of a-Si solar cells. Electrical, structural and optical properties are investigated. It is concluded that Cat-CVD p-type µc-Si is a suitable material as a window layer for Cat-CVD a-Si solar cells.

## INTRODUCTION

It has been already widely known that Cat-CVD has a lot of advantages to prepare a-Si films. For instance, efficiency of gas use reaches to 60 to 80 % for $SiH_4$ source gas [1], and is much larger than that of the conventional plasma enhanced CVD (PECVD) by 5 to 10 times. The deposition rate is higher than that of PECVD, and cost of deposition apparatus is much lower than that of PECVD.

For these reasons, the number of reports for the fabrication of solar cells by this technique is increasing and progress has been seen in efficiency in recent years. For instance, it was reported that the efficiency of about 10 % is obtained by using Cat-CVD i-layer prepared with the deposition rates as high as 18 A/s [2]. Other recent promising data are summarized in the paper by Schroeder et al. [3]. However, since the history of research is not so old, the investigation on window materials of such solar cells is not sufficient at the present moment. Particularly, studies on µc-Si window layer have not been known for Cat-CVD solar cells, although extensive studies are carried out for PECVD case.

This paper is to reveal the properties of p-type µc-Si thin layer as a new window material for Cat-CVD a-Si solar cells. Electrical, structural and optical properties are studied to conclude that Cat-CVD p-type µc-Si is a suitable material as a window layer for Cat-CVD a-Si solar cells.

## EXPERIMENTAL SETUP

Table I. Deposition conditions of various Si films.

| $T_{cat}$ | 1550 °C |
|---|---|
| $T_s$ | 200 °C |
| $P_g$ | 0.33 – 48 Pa (mainly < 20 Pa) |
| FR(SiH$_4$) | 1 sccm fixed |
| FR(H$_2$) | 24 – 199 sccm |
| FR(B$_2$H$_6$) | 0 – 0.064 sccm |

The Cat-CVD apparatus is the same to that ever reported [4]. A tantalum (Ta) wire with a diameter of 0.7 mm and a length of 54 cm is used as a catalyzer. The distance between the catalyzer and the substrate is about 7 cm. Other deposition conditions for obtaining films are summarized in Table I, where $T_{cat}$, $T_s$, FR(G), and $P_g$ refer to the temperature of catalyzer, the temperature of substrate holder measured by thermocouple placed just beside substrates, the flow rate of the gas G and the gas pressure during deposition, respectively.

Electrical properties such as conductivity are measured for the films deposited on quartz substrates with the thickness of 0.5 to 1.0 μm. Structural properties are also measured by the Raman spectroscopy for the same kind of films. The Raman spectra are observed by using a He-Ne laser as excitation light. Optical properties are studied by measuring transmittance of the 20 nm thick films deposited on both quartz substrates and transparent electrodes of 10-nm thick ZnO-coated SnO$_2$ on glass substrates, which are supplied as Asahi-U.

## PROPERTIES OF INTRINSIC μc-Si

As same as the case of PECVD, it is known that the film is changed from a-Si to μc-Si, as a SiH$_4$ gas is diluted by H$_2$. Figure 1 shows the Raman spectra for films prepared with FR(SiH$_4$) of fixed 1 sccm and various FR(H$_2$)'s. By increasing H$_2$ dilution ratio, the crystalline peak at about 520 cm$^{-1}$ starts to grow, and it is found that at FR(H$_2$)=99 sccm the film becomes μc-Si. For FR(H$_2$) over 124 sccm, the crystallinity appears degraded again in the figure. For such FR(H$_2$)'s, $P_g$ begins to increase and the deposition conditions can not be kept at same. This is the reason of crystallinity degradation observed here.

Raman spectra are observed for the films whose thickness is about 1 μm. However, the thickness of window layer is as thin as 20 nm. It is a common sense that thicker films are even easier for poly- or μ-crystallization. Since the measurement of crystallinity for

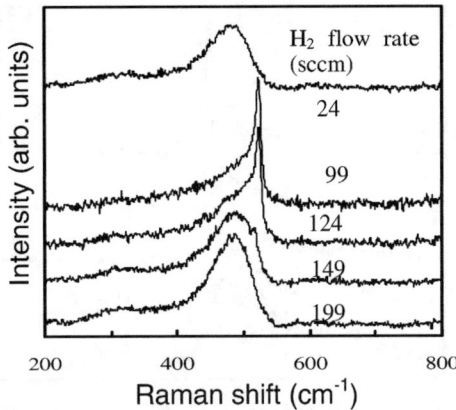

**Figure 1.** Raman spectra of Si films prepared with various $H_2$ flow rates.

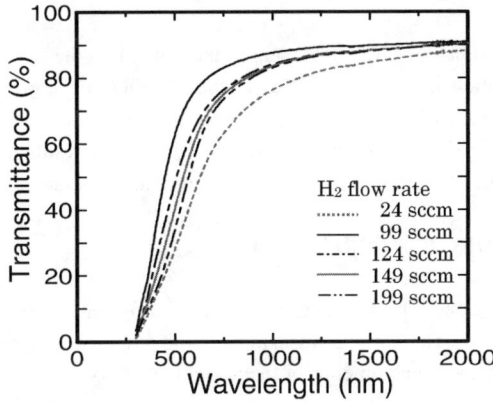

**Figure 2.** Transmittance of Si films prepared with various $H_2$ flow rates.

such thin films is not easy for Raman spectroscopy or other techniques such as X-ray diffraction spectroscopy. Thus, the crystallinity is checked by measuring the transmittance of thin films deposited on quartz substrates or on transparent electrodes of 10 nm-thick ZnO-coated Asahi-U substrates.

The results are summarized in Fig. 2. In the figure, the transmittance is shown as a function of the wavelength for the films prepared with various $H_2$ dilution ratios. When the film is crystallized, optical absorption mode is changed from direct to indirect and

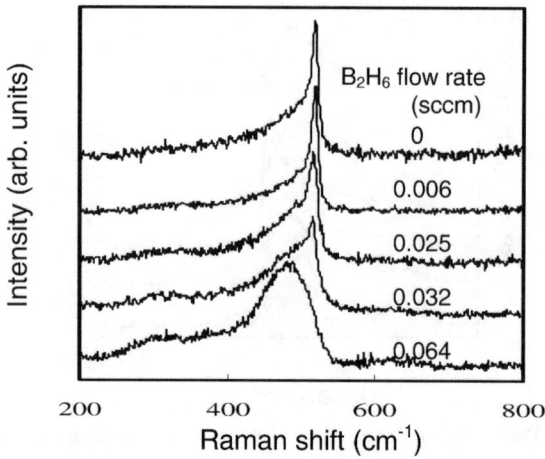

**Figure 3.** Raman spectra of boron-doped Si films.

the transmittance should be improved. It is clear in the figure that the film becomes more transparent as $H_2$ dilution increases, and that at the $H_2$ dilution ratio of 99:1 (that is, $FR(H_2)=99$ sccm for fixed $FR(SiH_4)=1$ sccm) it becomes most transparent. This means that the film is crystallized even for the thickness of 20 nm. The results for the films on ZnO-coated Asahi-U are same to this, and the crystallinity appears same.

## PROPERTIES OF BORON-DOPED P-TYPE µc-Si

For the films prepared with $H_2$ dilution ratio of 99:1, boron is mixed into films by $B_2H_6$ gas. Figure 3 shows Raman spectra again for the films of various $B_2H_6$ doping ratios. It is known that too much amount of incorporation of boron is likely to suppress the crystallinity. However, if $FR(B_2H_6)$ is kept below 0.032 sccm for $FR(H_2)=99$ sccm and $FR(SiH_4)=1$ sccm, the film is still in the phase of µc-Si. The conductivity of boron-doped films is plotted as a function of $FR(B_2H_6)$ in Fig. 4. It reaches to almost $10^0$ $Scm^{-1}$, large enough as a p-layer for $FR(B_2H_6)=0.025$ sccm.

It is known that boron-doping often degrades the transmittance of the films by narrowing band gap. Thus, finally, it is checked for boron-doped films.

The transmittance of such boron-doped films is also drawn in Fig. 5 as a function of wavelength for the films on TCO substrates. By doping of boron the transmittance is a little bit degraded, however, still it is rather transparent.

The solar cells are fabricated preliminarily by using this boron-doped p-type µc-Si

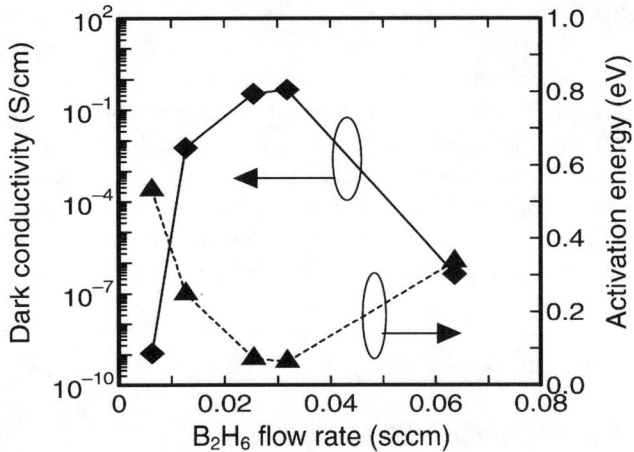

**Figure 4.** Conductivity and its activation energy of boron-doped Si films.

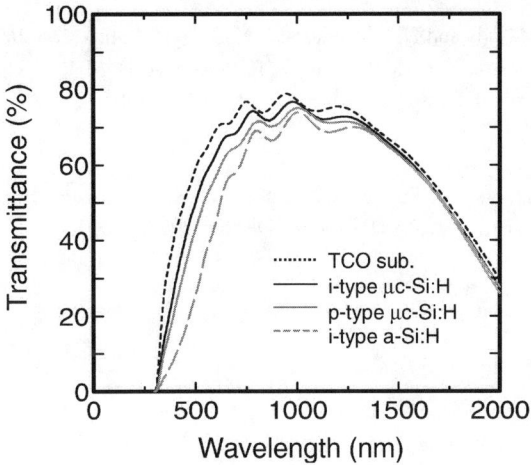

**Figure 5.** Transmittance of boron-doped Si films deposited on glass substrates.

window layer. The characteristics of such cells are not sufficient at the moment, and the efficiency is just over 5.3 %. However, the short circuit current density, which is a measure of quality of window layer, is likely to be improved to 16 mA/cm$^2$ or more for p-type µc-Si from 10 mA/cm$^2$ for p-type a-Si layer. Since this work is still going on, detailed information will be come out elsewhere.

## CONCLUSIONS

From the above results, the followings are concluded.
1) The conductivity of a boron-doped µc-Si p-layer, prepared by Cat-CVD, can be as high as $10^0$ Scm$^{-1}$. The value is sufficient as a p-layer of a-Si solar cell.
2) The transmittance of boron-doped µc-Si p-layer is a little bit degraded compared with intrinsic µc-Si layer. However, it is still higher than that of non-doped a-Si layer, whose transmittance is even more degraded by boron-doping.

Cat-CVD boron-doped µc-Si p-layer appears promising candidate as a window layer for a-Si solar cell.

## ACKNOWLEDGEMENT

This work is supported by NEDO, Japan.

## REFERENCES

1. N. Honda, A. Masuda and H. Matsumura, J. Non-Cryst. Solids, **266-269**, 100 (2000).
2. Q. Wang, E. Iwaniczko, Y. Xu, W. Gao, B. P. Nelson, A. H. Mahan, R. S. Crandall and H. M. Branz, Mater. Res. Soc. Symp. Proc., **609**, A4.3.1 (2000).
3. B. Schroeder: Ext. Abst. 2nd Int. Conf. Cat-CVD (Hot-Wire CVD) Process, Denver, 2002, p.7.
4. M. Itoh, K. Katohno, K. Sugita, A. Masuda and H. Matsumura: Ext. Abst. 2nd Int. Conf. Cat-CVD (Hot-Wire CVD) Process, Denver, 2002, p. 223.

## Investigations on the Real-Time Monitoring of the Crystallinity of Hydrogenated Microcrystalline Silicon Films

Christoph Ross, Friedhelm Finger and Reinhard Carius
Institute of Photovoltaics (IPV), Forschungszentrum Jülich,
D-52425 Jülich, GERMANY

### ABSTRACT

A method for monitoring the evolution of the crystallinity during the deposition of thin hydrogenated silicon films by using in situ spectroscopic ellipsometry is presented. The crystallinity of the topmost 10-20 nm of a film is derived from the analysis of the shape of ellipsometric spectra in the UV range. The values are closely related to parameters of the deposition process and in good agreement with Raman scattering results. Examples of different kinds of microcrystalline silicon films are shown. Improvements of the time resolution and/or accuracy are discussed. The method turns out to be well suited for process control.

### INTRODUCTION

Transitions between hydrogenated microcrystalline silicon (μc-Si:H) and amorphous silicon (a-Si:H) as a function of growth parameters have attracted a lot of interest during the past years [1-4]. For μc-Si:H, growth near the transition has resulted in optimized material for devices such as solar cells [5]. However, growth in this regime very critically depends on the substrate and on other process conditions [5,6]. Therefore, a real-time control of the crystallinity is desired.

We present an empirically derived method for extracting information about the crystallinity from ellipsometric data. As will be shown, this approach can be used as a real-time monitoring tool. We first report on as-deposited hydrogenated silicon (Si:H) with different compositions of crystalline and amorphous material. We then discuss the following examples of real-time measurement during the deposition: (1) growth of μc-Si:H on Si:H layers with different crystallinity, where the crystallinity evolution within the μc-Si:H layer is expected to be determined by the crystallinity of the underlying layer, (2) growth of device-grade intrinsic μc-Si:H as a part of a p-i-n solar cell, and (3) growth of a-Si:H/μc-Si:H multilayers.

### EXPERIMENT

Silicon films and solar cells are prepared in a multichamber system for RF and VHF plasma-enhanced chemical-vapor deposition by using hydrogen diluted $SiH_4$ [5]. SC = $[SiH_4]/([SiH_4] + [H_2])$ denotes the silane concentration in the source gas. The substrates for the different kinds of deposition are specified in the next section. Cr coating of glass substrates serves to reduce background signal in Raman spectroscopy measurements. ZnO:Al coated glass substrates are used for solar cells. In contrast to our standard μc-Si:H p-i-n cell preparation, parts of the ZnO:Al surface have not been texture-etched in order to reduce scattering and depolarization in ellipsometry measurements. Film thickness values have been determined after deposition.

One of the deposition chambers is equipped with a UV-visible ellipsometer of the rotating-polarizer type. The angle of incidence amounts to ≈70°. By using a monochromator with a scanning grating in connection with a single photodiode spectroscopic measurements are performed with a maximum photon energy of about 5 eV. Significant noise reduction can be achieved by accumulating raw data over several rotations of the polarizer.

Raman spectroscopy is used for an ex situ determination of the crystalline volume content. The photon energy of the incident laser beam is 2.54 eV. The crystalline content $I_C^{RS}$ is calculated from the integrated intensity values $I_{500}$ and $I_{520}$, which refer to contributions from crystalline phase at ≈500 cm$^{-1}$ and ≈520 cm$^{-1}$, respectively, and from $I_{480}$ referring to contributions from disordered phase such as a-Si:H:

$$I_C^{RS} = (I_{500} + I_{520}) / (I_{500} + I_{520} + I_{480}) \tag{1}$$

## RESULTS

### Crystallinity of as-deposited layers

The basic idea of our approach consists of (1) measuring optical data of the surface of a silicon film, (2) extracting information related to the crystallinity, and (3) performing this type of surface crystallinity measurement during the deposition of a film at sufficiently short time intervals, which should result in a real-time monitoring of the crystallinity profile. We first focus on the questions of the surface sensitivity and of the determination of the crystallinity, whereas the time resolution required for a real-time monitoring will be dealt with in more detail in the next subsection.

For spectroscopic analyses of Si:H films in the UV-visible range the minimum probe depth occurs in the UV. Here, the absorption coefficient is of the order of 10$^6$ cm$^{-1}$. This means that, if the monochromator is operated solely in the UV, only the topmost 10-20 nm of the film material contribute to the detected signal. Moreover, a significant part of the data acquisition time is saved compared to a full UV-visible scan.

Figure 1a shows two typical cos Δ spectra of Si:H films with different crystallinity. The ellipsometric angle Δ describes the sample-induced phase shift between the p and s components of the polarized light, where p and s correspond to parallel and perpendicular polarization with respect to the plane of incidence, respectively. It is our general observation that for amorphous films the shape of spectra can be well approximated by a straight line covering the shown UV range, whereas for films containing a significant fraction of crystalline phase a clear deviation from the linear shape occurs. The deviation can be described, for example, by the area A in figure 1a.

Similar spectra have been recorded for a series of films, where different SC values have been used

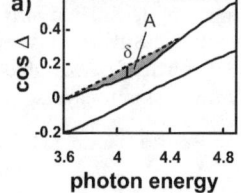
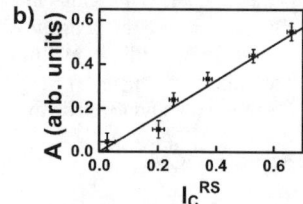

**Figure 1.** Determination of the crystallinity of ≈36 nm thick Si:H films grown on Cr coated glass: (a) UV cos Δ spectra. The top and bottom curves refer to μc-Si:H and a-Si:H, respectively. The curves are shifted vertically for clarity. For μc-Si:H, the deviation from a linear spectrum is represented by the area A or by the vertical line δ. (b) A versus Raman intensity $I_C^{RS}$ (as defined by equation 1) for the entire set of films prepared with different silane concentrations.

for film deposition. This results in a series of different crystalline volume contents as has been verified by Raman spectroscopy. Figure 1b contains the corresponding $I_C^{RS}$ values along with the size of the area A which has been introduced above. Both quantities appear to be closely related to each other by a linear relationship valid for the present set of samples.

## Time resolution

These observations show that the present approach is generally suited for gaining information related to the crystallinity of the surface region. This is an important prerequisite for the development of a real-time tool. However, in order to determine the A values shown in figure 1b we had to scan the photon energy in steps of 0.05 eV. Although this step width is useful for measuring optical functions in detail, a period as long as a few minutes is needed per A value. During this period, film material as thick as 10 nm or more is deposited at typical deposition rates of 0.1-0.2 nm/s. Thus, the present time resolution strongly limits the depth resolution of a real-time profile.

A straightforward way of replacing A by a quickly accessible quantity is shown in figure 1a. Here, $\delta$ denotes the difference between the dashed line and the cos $\Delta$ spectrum at the center of the spectral range covered by A. This quantity is expected to contain information on crystallinity in a similar way to A since both quantities describe the deviation from a linear shape. In contrast to A, for determining $\delta$ it is sufficient to record data at 3 points of the spectrum and to apply just a linear equation to these data. As a result, periods less than 1 minute per $\delta$ value can be obtained.

Another way of reducing the sampling time is decreasing the number of accumulations by which, for the present ellipsometer, data from consecutive rotations of the polarizer are collected and added up for noise reduction. However, lowering this number will result in a significant increase of noise. Figure 2 shows the noise performance of $\delta$ as a function of the rate at which $\delta$ is sampled. In principle, up to six $\delta$ values can be recorded per minute, which is done by omitting the data accumulation option. This is equivalent to a sampling rate of 0.1 Hz. Since in this case remarkable noise occurs, especially for low crystallinity, it might be reasonable to choose a lower sampling rate provided that the crystallinity profile is shallow enough to be sampled at a reduced speed.

## µc-Si:H growth on Si:H substrates with different crystallinity

The first real-time measurements using the approach described above have been performed on µc-Si:H films which are deposited onto different Si:H layers on glass. These substrates are part of a series of Si:H layers with different compositions of crystalline and disordered phase equivalent to the series shown in figure 1b.

In figure 3 the crystallinity of the substrate as quantified by means of A is indicated by an open symbol at thickness zero. The substrate crystallinity is highest for (a) and it decreases towards (d). The real-

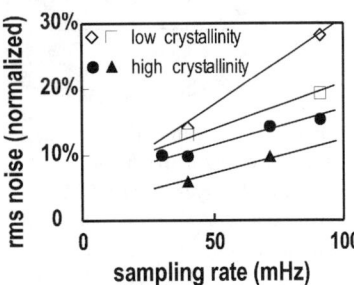

**Figure 2.** RMS noise of $\delta$ (see figure 1a) as a function of the rate at which $\delta$ values are sampled. Values on the vertical axis are normalized to the highest $\delta$ detected for µc-Si:H samples (see maximum crystallinity in figure 1b). Samples are µc-Si:H films on glass.

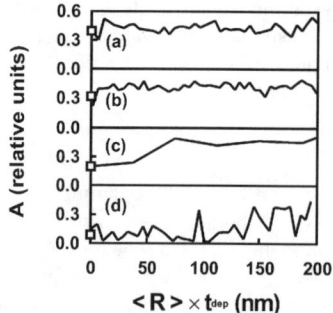

**Figure 3.** Real-time monitoring of the crystallinity of μc-Si:H films grown on glass substrates terminated by ≈36 nm thick Si:H layers with different crystallinity (see figure 1b). <R> denotes the average deposition rate as determined from the final thickness. The crystallinity A (see figure 1a) is determined continuously at times $t_{dep}$ of the deposition process.

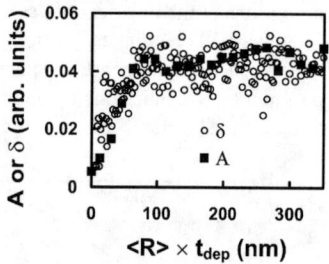

**Figure 4.** Real-time monitoring of A and δ (see figure 1a) during the deposition of i layers for μc-Si:H p-i-n solar cells on ZnO:Al coated glass without texture etching. A and δ refer to different depositions. The effective i layer thickness is given by the product of the average deposition rate <R> and the deposition time $t_{dep}$. Note that the noise of δ is consistent with that shown in figure 2 (≈10% rms deviation from the mean value at ≈60 mHz sampling rate).

time display reveals that A increases to approximately the same level in all four cases, but with different initial behavior. No significant initial increase can be observed for the two substrates with highest crystallinity (a,b), but a gradual increase is clearly detected for (c) and (d). In particular, if starting from a very low level as in case (d), it takes significantly longer time for A to reach the saturation value than in the case of a moderate crystallinity of the substrate, which is valid for (c). Further support for an enhanced time evolution of crystallinity by an enhanced crystallinity of the substrate arises from real-time ellipsometry measurements at a constant photon energy as described elsewhere [7].

## μc-Si:H i layer growth for solar cells

The deposition parameters used for the μc-Si:H films described in the last subsection are the same as for absorber layers of highly efficient p-i-n solar cells. The p layer of this type of cell has a comparatively low crystalline volume content. For the crystallinity profile of the i layer we therefore expect a gradual increase similar to that shown in figure 3c or d. Previous studies on the structure of the i layer have shown that a very high fraction of crystalline phase is present in the bulk [3,5]. In particular, this is seen from transmission electron microscopy (TEM) [3]. Therefore, a high crystallinity level is expected after the initial increase.

In figure 4 real-time crystallinity data from two solar cell depositions with identical process parameters are plotted. Here, A and δ have been recorded at sampling rates of ≈6 mHz and ≈60 mHz, respectively. Obviously, both quantities reproduce the expected crystallinity profile by starting from a low value and saturating at a high one. The high level has been found to remain for film thicknesses up to ≈1 μm. The very similar behavior of A and δ confirms our assumption that these two quantities contain similar information on crystallinity.

## Growth of a-Si:H/μc-Si:H multilayers

In order to further demonstrate the phase selectivity of the present approach we have performed Si:H depositions with varying silane concentration. Two SC values have been chosen in order to produce films with alternating crystallinity: SC=1.5% and SC=8%. For the present conditions it is known that the first value results in the growth of microcrystalline material even on a-Si:H substrates, whereas the latter value leads to amorphous growth even on predominantly crystalline surfaces. The average deposition rate amounts to 0.06 nm/s and 0.23 nm/s for SC=1.5% and SC=8%, respectively.

Figure 5 displays the SC program and the resulting evolution of $\delta$ as monitored during the deposition. Both are given as a function of the layer thickness, which has been calculated from the above-mentioned deposition rates. The response of $\delta$ to SC is clearly visible in a way that lowering SC initiates an increase of $\delta$ followed by a saturation and that raising SC results in the opposite process. By using Raman spectroscopy we have confirmed that at the end of a SC=8% period the film is completely terminated by a-Si:H, whereas a significant contribution from crystalline phase is detected at the end of the SC=1.5% intervals.

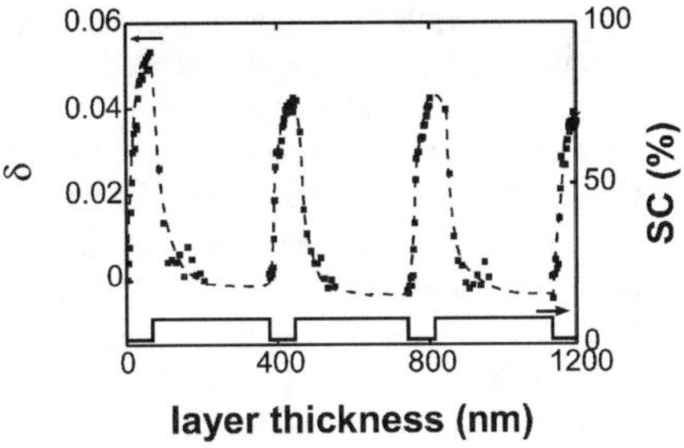

Since the behaviour of $\delta$ is in close correlation to the evolution of crystallinity as expected from SC and verified by Raman scattering, these results give further evidence that $\delta$ to a large extent contains information about the crystallinity of the film surface. More interestingly, in the present case we benefit from a comparatively large signal-to-noise ratio at a still reasonable sampling rate. In contrast to the $\delta$ measurement shown in figure 4 we have chosen a sampling rate as low as ≈25 mHz. Although $\delta$ values have been taken in steps of ≈2.5 nm and ≈9 nm for SC=1.5% and SC=8%, respectively, these intervals appear to be sufficient for monitoring in detail the evolution of crystallinity in these films.

**Figure 5.** Evolution of $\delta$ as monitored in real time along with SC for a film deposited with SC alternating between 1.5% and 8%. The layer thickness given on the horizontal axis is calculated from the deposition time and from the SC dependent deposition rates. The broken line is a guide to the eye. For simplicity, no data have been collected during the second half of each SC=8% interval. The substrate is a ≈190 nm thick a-Si:H film grown on glass. For the substrate, SC=100% has been used instead of SC=8%. This might explain the deviation of the first peak value of $\delta$, whereby the substrate roughness is considered to affect the $\delta$ value.

## DISCUSSION AND CONCLUSIONS

The question arises to what extent the quantities which have been introduced in the previous section are suited for describing the crystalline volume content. Here, in contrast to other spectroscopic measurements the fraction of crystalline phase has not been gained by fitting spectra from optical models to experimental data. Instead, we have empirically derived a crystallinity related quantity which might contain other contributions than merely the crystalline volume content, e.g. contributions from surface roughness.

First, we point out that the spectral feature, which is illustrated in figure 1a and quantified by A and $\delta$, occurs in a spectral range where the $E_2$ optical transition of crystalline silicon is located. It is well known that optical spectra of µc-Si:H films exhibit a more pronounced shape in this range than a-Si:H spectra [8]. Secondly, as evidenced by the present results, A and $\delta$ are in close correlation to (1) the crystalline volume content as determined by Raman spectroscopy, (2) parameters used in the deposition process such as SC, and (3) the microstructural evolution which is expected for the present solar cell i layer growth and which corresponds to observations from TEM [3]. Since different kinds of Si:H films with a broad range of crystallinity and thickness have been taken into account we can conclude from the results that the method is reliable for a qualitative and, to some extent, a quantitative determination of the crystallinity. This gives rise to the assumption that the crystallinity is the predominant information contained in A and δ. In particular, we assume that the crystallinity has a significantly larger effect on A and δ than the surface roughness, since reliable results have been achieved for a surface roughness as high as that of the 1.2 μm thick film (figure 5). Investigations on films with even higher surface roughness would provide more insight. Nevertheless, much attraction arises from the potential of an almost instantaneous evaluation of δ, which is an important requirement of a process control technique.

Although for the present experimental setup the time resolution is strongly limited by the scanning procedure of the monochromator grating, acceptable results have been obtained for deposition rates around 0.1-0.2 nm/s and in the case of gradual structural changes. In order to achieve an optimized accuracy at a reasonably high sampling rate, from the present results it is recommended to collect data at three spectral points (sufficient for the determination of δ) over an appropriate number of cycles of the rotating polarizer. As shown in the last example of the previous section, a sampling rate of ≈25 mHz may result in an optimum accuracy of the monitored profile for deposition rates around 0.1 nm/s. In this case the thickness by which the film grows during the sampling time amounts to ≈4 nm, which is significantly below the probe depth. The implementation of a multielement detector is under way and will speed up the measurement to achieve the same accuracy also at rates up to 0.5 nm/s.

In summary, we have presented an ellipsometry-based method for assessing the crystallinity of the topmost 10-20 nm of Si:H films. This method, by which the shape of UV optical spectra is analyzed, has been optimized for the present single-channel device. It turns out that the evolution of crystallinity in different kinds of μc-Si:H films can be monitored successfully. We conclude that this method is well suited for being implemented into a process control of μc-Si:H growth.

## ACKNOWLEDGEMENTS

The authours would like to thank J. Klomfaß, A. Lambertz, M. Hülsbeck, and S. Michel for their assistance. One of us (C.R.) gratefully acknowledges support by the Hermann von Helmholtz Association, Germany.

## REFERENCES

[1] J. Yang, K. Lord, S. Guha, S.R. Ovshinsky, Mater. Res. Soc. Symp. Proc. **609**, A 15.4 (2001).
[2] R.W. Collins, A.S. Ferlauto, G.M. Ferreira, C. Chen, J. Koh, R.J. Koval, Y. Lee, J.M. Pearce, C.R. Wronski, Solar Energy Materials and Solar Cells (2003), in press.
[3] L. Houben, M. Luysberg, P. Hapke, R. Carius, F. Finger, H. Wagner, Phil. Mag. A **77**, 1447 (1998).
[4] H. Fujiwara, M. Kondo, A. Matsuda, J. Appl. Phys. **91**, 4181 (2002).
[5] O. Vetterl, F. Finger, R. Carius, P. Hapke, L. Houben, O. Kluth, A. Lambertz, A. Mueck, B. Rech, H. Wagner, Solar Energy Materials and Solar Cells **62** (2000) 97.
[6] O. Vetterl, M. Hülsbeck, J. Wolff, R. Carius, F. Finger, Thin Solid Films **427**, 46 (2003).
[7] C. Ross, F. Finger, R. Carius, Proc. Int. Conf. "PV in Europe", p. 55 (2002).
[8] S. Kumar, B. Drévillon, C. Godet, J. Appl. Phys. **60** (1986) 1542.

# Towards microcrystalline silicon n-i-p solar cells with 10% conversion efficiency

L. Feitknecht, C. Droz, J. Bailat, X. Niquille, J. Guillet, A. Shah
Institut de Microtechnique, Université de Neuchâtel,
Breguet 2, CH-2000 Neuchâtel, Switzerland.

## ABSTRACT

High-performance microcrystalline and amorphous silicon solar cells are the key elements for a successful combination to form the "micromorph" tandem cell [1,2]. A microcrystalline silicon ($\mu$c-Si:H) solar cell in the n-i-p configuration was fabricated by the VHF PE-CVD deposition process. The cell has a conversion efficiency exceeding 9% ($V_{OC}$=520 mV, FF=73%, $J_{SC}$=24.2 mA/cm$^2$). This result was achieved by a successful combination of the following elements: first a fine-tuning of the silane concentration (SC) in hydrogen feedstock gas used for deposition of the intrinsic <i> absorber layer, second, the incorporation of an optimised back-reflecting substrate into the cell; and, third, the ideal combination of each of these key-components.

Compared to earlier results with n-i-p-type $\mu$c-Si:H solar cells, a substantial increase in $V_{OC}$ was now obtained, while maintaining reasonable $J_{SC}$-values. Earlier investigations on the role of the i-layer material had revealed a trade-off between cells with high $J_{SC}$ but low $V_{OC}$ or cells of low $J_{SC}$ and high $V_{OC}$. In the present contribution the authors now show the successful combination of a cell with an acceptable $V_{OC}$ and good $J_{SC}$ generation in the long-wavelength region (above 700 nm). This is mainly because of suitable light-diffusing back-reflectors which perform well with respect to both, optical and electrical aspects.

## INTRODUCTION

The fabrication of substrate-nip solar cells is substantially different from that of superstrate-pin solar cells, because of the change in the deposition sequence of the layers and its technological consequences (like initial n-type or p-type nucleation layer and optically transparent and doped window-layers). Whereas a relatively large number of papers have dealt with the design and fabrication of superstrate-pin cells, only a relatively small number of papers have looked at substrate-nip cells. The present work is a further contribution to latter topic and will deal with a specific problem in this context:

The interface of the back-TCO (transparent conductive oxide) to the n-type silicon film (we call it here the 'back-interface') takes on a very important role for the deposition of nip-type cells and this interface has an influence on the structure of the whole cell and its performance: Indeed, nucleation of silicon grains, electronic and optical properties are key issues which depend on the nature of the underlying back-TCO material [3,4].

Investigations on the interface of the back-TCO / n-layer of the n-i-p solar cell are not straight-forward since the interesting part is hidden by the solar cell and thus not directly accessible. Since most of the light is absorbed within the first 2 $\mu$m of the absorber, only the long wavelength part of the light spectrum reaches the back-interface under white-light illumination. A better cell characterisation is possible if the External quantum Efficiency (EQE) measurement is performed from both sides of the cell (double-sided illumination).

## EXPERIMENTAL

μc-Si:H silicon n-i-p solar cells were deposited in a single-chamber VHF-GD reactor at plasma excitation frequencies between 70 to 130 MHz. Typical deposition parameters for the intrinsic layer are: base pressure of the vacuum chamber $p_{base}$<4.1E-8 mbar, deposition pressure p = 0.1-0.9 mbar, applied plasma power P = 5-30 W, substrate temperature around 200 °C. A gas purifier was used to avoid incorporation of detrimental oxygen contamination [5]. These conditions lead to deposition rates in the region of 3 to 10 Å/sec.

Stainless steel and glass substrates were prepared for the back-reflectors which consist of a stack of sputtered silver (Ag) and zinc oxide (ZnO); silver is deposited at substrate temperatures around 400°C in order to increase the growth-induced surface-texture (RMS around 60nm). Current vs. voltage characterisation was performed under AM1.5 conditions at 100 mW/cm$^2$, using a two-source solar simulator. The short circuit current densities ($J_{SC}$) were calibrated by the scalar product of the External Quantum Efficiency (EQE) data with the AM1.5 sun spectrum within the range of 350 to 1000 nm. For a better characterisation, we swapped the solar cell in order to illuminate the cell not from the p-side (the normal configuration for thin-film silicon solar cells) but to illuminate the rear-side of the cell i.e. trough the glass substrate, the back-TCO and the n-layer. Under the n-side illumination, more information on the behaviour of the junction near the back-interface can be obtained.

## RESULTS AND DISCUSSION
### Interfaces

We deposited in the same deposition-run a nip-type μc-Si:H cell on substrates which were coated with four different back-TCOs: a) a raw stainless steel plate without any TCO material, b) sputtered ZnO, c) CVD-ZnO and d) CVD-ZnO covered with sputtered ZnO on top; see Figure 1.

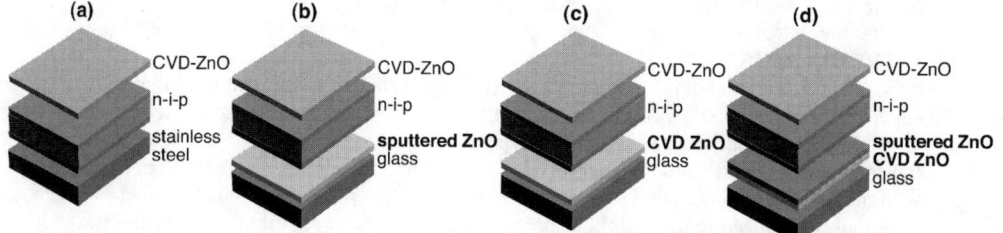

**Figure 1**: Investigations on the role of the back-interface material: in a single deposition-run, the substrates of (a) stainless steel, (b) glass/sputtered ZnO, (c) glass/CVD-ZnO and (d) glass/CVD-ZnO covered with sputtered ZnO were processed together.

In Table 1, the electronic cell-characteristics given there highlight the relation of the back-interface to the n-type film of the n-i-p cell. There is a close link between the type of TCO material used and the performance of the solar cell. The best conversion efficiency was achieved on the reference substrate without any ZnO, the next best cell is fabricated on the combined ZnO-back contact (a textured CVD ZnO covered with the non-intentionally textured sputtered ZnO). The worst conversion efficiency of this n-i-p cell is reported for the CVD-ZnO back-interface. The EQE curves of these respective cells are given in Figure 2.

**Table 1**: Electrical performance of the n-i-p solar cells deposited onto four different back-interface configurations:

| # | Back-interface | $V_{OC}$ [mV] | FF [%] | $V_{OC}$.FF [a.u.] | $J_{SC}$: n/p [mA/cm$^2$] | ratio [n/p] | η [%] |
|---|---|---|---|---|---|---|---|
| a) | Stainless steel | 450 | 71.6 | 322 | n.a. / 20.9 | n.a. | 6.7 |
| b) | Sputtered ZnO | 460 | 69.0 | 317 | 10.9 / 17.1 | 0.64 | 5.4 |
| c) | CVD-ZnO | 438 | 64.2 | 281 | 5.7 / 17.6 | 0.32 | 4.9 |
| d) | CVD + Sputtered ZnO | 454 | 68.3 | 310 | 11.9 / 18.7 | 0.64 | 5.8 |

**Double-sided quantum efficiency measurement**

For interpretation, the external quantum efficiency (EQE) data may be subdivided in three intervals, a first range up to 520 nm, the second from 600 nm to 800 nm and the third from 800 nm to 1000 nm. The EQE illuminated from the n-side is zero up to a wavelength of 520 nm on the back-contact with CVD-ZnO. On the two sputtered samples a considerably higher signal is measured and no difference between the two ZnO types is apparent, but the values obtained by the n-side EQE curve still remains below the values obtained for the p-side illuminated EQE curve of the cell. In the last interval only, the EQE curve of the cell with CVD-ZnO gains over the EQE curve of the cell with a sputtered ZnO back-contact: this is an effect of the good light scattering behaviour of the textured, rough CVD-ZnO.

In all curves of p-side illuminated EQE data, there is a slight dip in the region of 600 nm to 800 nm: this dip is a sign for the absence of a powerful light-trapping scheme, as one would obtain by e.g. optimised back-reflectors. The dip could also be avoided by increased cell absorber thickness (exceeding 2.5 µm).

**TEM investigations**

The Transversal Electron Microscope (TEM) allows for a close look on the nucleation of microcrystalline films on CVD and sputtered ZnO, see Figure. 3. Crystallographic observations help here to highlight one facet of the situation at the n-i interface.

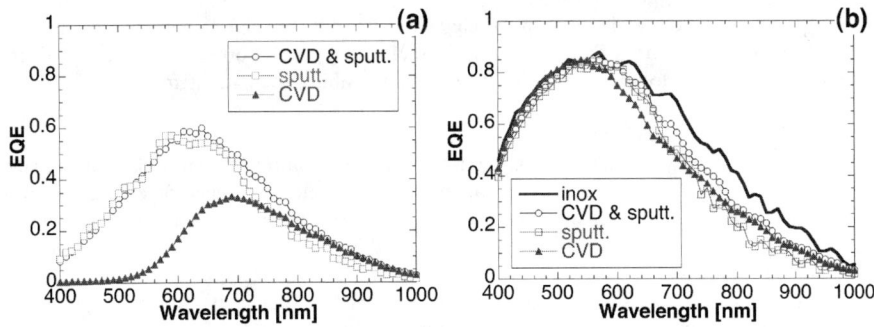

**Figure 2**: (a) Quantum efficiency measurement illuminated from the n-side and (b) illuminated from the p-side: stainless steel (inox), CVD-ZnO covered with sputtered ZnO, simply sputtered ZnO and CVD-ZnO on glass substrates were utilised respectively.

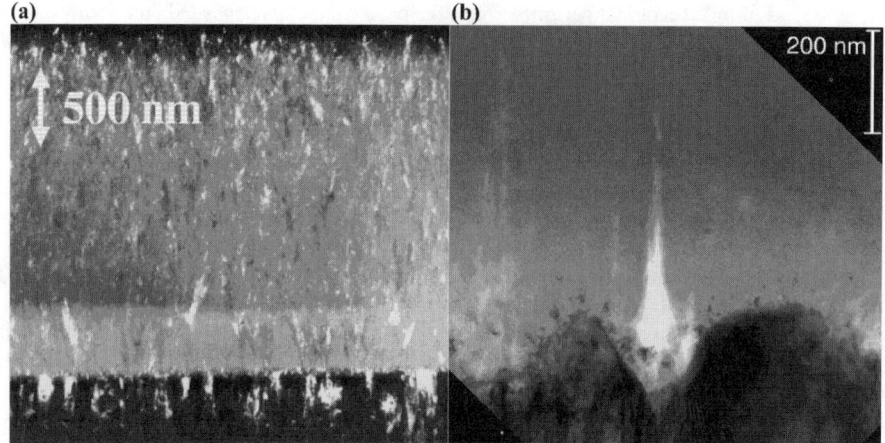

**Figure 3**: (a) TEM dark field micrograph of a microcrystalline n-i-p solar cell deposited on sputtered ZnO. ZnO is at the bottom of the picture, n-type μc-Si:H layer appears as a thin dotted layer. (b) TEM bright-field micrograph of the ZnO-n-i interface of a solar cell deposited on CVD ZnO. ZnO tetrahedrons are at the bottom of the picture. Note the disordered grainy contrast of the n-layer. A crack/void at the bottom of the ZnO valley appears bright; source: [4]

Measurements on another series of cells, namely on a "dilution series" of nip-type solar cells (i.e. n-i-p solar cells where the intrinsic <i> absorber layers were deposited at various silane concentrations) reveal a systematically denser crystalline structure of the n-layer on sputtered ZnO than on CVD ZnO. As a consequence of this different n-layer growth, in the initial stage of i-layer deposition, the beginning of the i-layer consists of a mixed-phase layer (amorphous and microcrystalline parts). This layer (also referred to as 'incubation layer') is reported to vary from 140 nm for cells on sputtered ZnO to 200 nm for cells on CVD-ZnO [4,6].

The back-TCO and the electrical characteristics are correlated: The observations of low EQE curves of cells on CVD-ZnO goes in parallel with a poorly formed n-film as visible in the TEM images. In the following section, we present possible explanations for this.

**Discussion on cell interfaces**

The difficulty to match a certain nip-type solar cell onto different back-contact materials is discussed in the precedent section. A cell might work well in one specific configuration (i.e. for a specific combination of substrate and cell) and the same cell may then show a considerable drop in performance if just the back-contact material is changed. The thin-film solar cell should not be considered as a combination of correctly optimised independent layers, but as a device of both layers and interfaces which have to be optimised on the device itself. Keeping in mind that not only the physical (e.g. crystallographic) but also chemical (out diffusion) and electronic (transport) aspects play important roles in this question, we draw attention to several possible explanations for the observed incompatibility between our present n-i-p cells and certain, specific back-contacts (e.g. CVD-ZnO, here):

1st If we consider that light of increasing wavelength is absorbed increasingly deeper inside the silicon film, a zone comprising the n-layer and the beginning of the i-layer can be made responsible for the non-existent carrier generation/collection in the spectral range of 400 to 520 nm , for the case of cells deposited on CVD-ZnO, as seen in Figure 2(a). This might be due to an optical problem (no photon absorption) or to an electronic problem (no electrical field, and thus no carrier separation). The junction would then lack an electrical field within something like 100 nm of the n-i interface. The most plausible explanation for poor n–film quality in these first 100 nm would be difference in surface roughness of these two ZnO types (rough CVD-ZnO and flat sputtered ZnO), since deposition onto a flat surface does not lead to the same layer microstructure, as growth on a textured surface [7]: All of the four investigated back-ZnO contacts have different surface roughness. starting from the flat sputtered ZnO (only 4 nm of rms roughness) up to the very rough CVD-ZnO (60 nm of rms roughness). Note that the surface roughness plays a decisive role in the nucleation of the silicon film. A poor nucleation of the silicon layer closest to the back-interface might cause a poor cell performance.

$2^{nd}$, cross-contamination due to the first stage of the i-layer growth might cause a poor junction within the region of the back-interface (i.e. the region within the n-type and the i-type μc-Si:H film). On the other hand, a phosphorus contamination due to the preceding deposition of the n-film can be excluded: The common problem of contamination in cells fabricated in a single-chamber deposition system would disturb the whole batch of cells. In our case here, cells of the same batch but deposited on CVD-ZnO show a poor EQE signal whereas cells deposited on sputtered ZnO back-contacts have a higher EQE signal, if we consider wavelengths below 520 nm. Since absorber layer contamination by a preceding n-film would harm the whole batch of cells, independently of the back-contact material, this type of contamination can indeed be excluded.

$3^{rd}$ The difference of the junction deposited onto the CVD-ZnO and deposited onto the sputtered aluminium doped ZnO layer might be due to an incompatibility of our n-type film with the CVD-ZnO combining physical (e.g. surface roughness) and chemical reasons.

$4^{th}$ Another explanation may be the ZnO surface states which may interact with the deposition plasma of the growing n-type film . Thereby, particles of zinc, oxygen, silicon and phosphorus might get mixed together in the plasma process and create an n-film of a less favourable quality in the case of the CVD-ZnO - something which apparently does not happen on the sputtered ZnO. The term of 'quality' refers here exclusively to crystallographic observations made by TEM, as no doping profile was analysed.

Up to now, not enough measured evidence for or against any of these hypothesis could be given (e.g. SIMS data were not satisfactorily clear). In our eyes, an optimisation of the n-i-p cell interface to the CVD-ZnO back-contact seems generally possible but was sofar beyond the reach of this study.

## CONCLUSIONS

The importance of 'interface-tailoring' for n-i-p solar cells has been made evident in this paper, by giving results on the investigation of four different back-contacts. One must always remember that a thin-film solar cell is not simply a stack of layers but a complex device resulting from an optimal fit of interacting layers. Special care is needed here at the "back-contact interface", i.e. at the interface substrate-silicon. The measurement of quantum efficiency illuminated from both sides contributed considerably to the detection of malfunctions of this "back-contact interface".

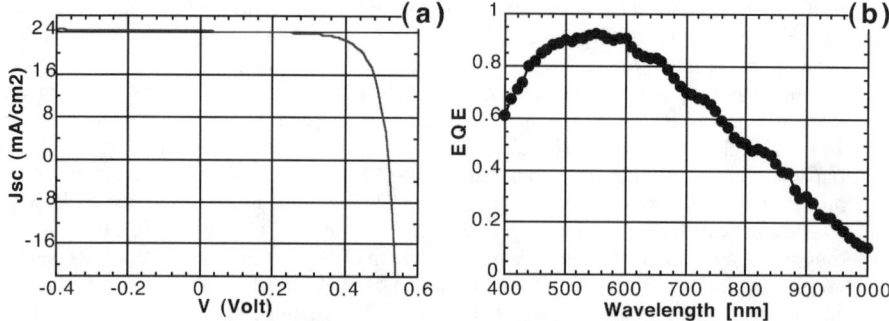

**Figure 4**: Our sofar best μc-Si:H solar cell in the n-i-p configuration deposited onto optimised light-scattering back-reflector of glass/silver/sputtered ZnO. AM1.5 conversion efficiency over 9%, $V_{OC}$=520 mV, FF=73%, $J_{SC}$ =24.2 mA/cm$^2$, cell surface=0.3 cm$^2$, thickness=2.5 μm.

For industrially relevant solar cell design, the introduction of highly scattering back-reflectors is necessary. Theses back-reflectors have to combine two constraints: a good optical performance as well as an optimally designed back-contact to the silicon layer.

The variation of silane concentration of the i-layer of the n-i-p cell is a rather old method (already known for the optimisation of amorphous solar cells [8]) but still applicable to solid μc–Si:H solar cell fabrication process.

The remarkably good μc-Si:H n-i-p solar cell (see figure 4) is the result of such an optimisation of the whole device at IMT Neuchâtel. A conversion efficiency exceeding 9% ($V_{OC}$=520 mV, FF=73%, $J_{SC}$ =24.2 mA/cm$^2$) shows exactly the combination of a good $J_{SC}$ with a high $V_{OC}$. The only drawback of this cell is the low deposition rate of 2.6 Å/sec which is mainly due to the rather low VHF-plasma frequency of 70 MHz. Nevertheless, this is a promising result of a n-i-p solar cell approaching the region of 10% conversion efficiency.

## ACKNOWLEDGEMENTS

This work was supported by the Swiss Federal Renewable Energy Program (grants 100 045 and 36 487) and the Swiss National Science Foundation (grant 66985.1).

## REFERENCES

1. L. Feitknecht, O. Kluth, Y. Ziegler, X. Niquille, P. Torres, J. Meier, N. Wyrsch, A. Shah. Sol. En. Mat. & Solar Cells, Vol. 66, 2001, pp. 397-403.
2. J. Meier, S. Dubail, S. Golay, U. Kroll, S. Faÿ, E. Vallat-Sauvain, L. Feitknecht, J. Dubail, A. Shah, Sol. En. Mat. & Solar Cells, Vol. 74, 2002, pp. 457-467.
3. H. Fujiwara, M. Kondo, A. Matsuda, J. Appl. Phys., 93 (5): 2400-2409 MAR 1 2003
4. J. Bailat, E. Vallat-Sauvain, L. Feitknecht, C. Droz, A. Shah, ICAMS 2001, Nice, France, 2001, J. Non-Crystalline Solids Vol. 299-302, pp. 1219-1223.
5. P. Torres et al., Appl. Phys. Lett. 69 (10) 1996 p. 1373
6. J. Bailat, E. Vallat-Sauvain, L. Feitknecht, C. Droz and A. Shah. J. Appl. Phys., Vol. 93, No. 9, 1 May 2003
7. Y. Nasuno, M. Kondo and A. Matsuda, Jpn. J. Appl. Phys. Vol. 40 (2001) L303-L305, Part 2, No. 4A, 1 April 2001
8. R. Platz, S. Wagner, C. Hof, A. Shah et al., J. Appl. Phys. 84 (7): 3949-3953 OCT 1 1998

## Hydrogenated Amorphous Silicon Thin Films with Nanocrystalline Silicon Inclusions

T. J. Belich[1], S. Thompson[2], C. R. Perrey[3], U. Kortshagen[2], C. B. Carter[3] and J. Kakalios[1]

[1]School of Physics and Astronomy, University of Minnesota, Minneapolis, MN 55455
[2]Department of Mechanical Engineering, University of Minnesota, Minneapolis, MN 55455
[3]Department of Chemical Engineering and Materials Science, University of Minnesota, Minneapolis, MN 55455

ABSTRACT

Thin films of hydrogenated amorphous silicon containing nanocrystalline silicon inclusions (a/nc-Si:H) have been synthesized in an RF capacitively coupled PECVD system using a mixture of hydrogen diluted silane and helium, under deposition conditions at the edge of powder formation within the plasma. High resolution TEM confirms the presence of nanocrystallites as small as 2 nm in these films. Measurements of the optical absorption spectrum using CPM and PDS indicates a broadening of the Urbach slope in the a/nc-Si:H, compared to a-Si:H films, but no appreciable increase in midgap absorption. Despite the deposition conditions for the a/nc-Si:H being very different from those associated with producing optimal quality a-Si:H, the dark conductivity and photoconductivity values, and the sensitivity to light-induced defect creation in the a/nc-Si:H films are comparable to those in a-Si:H.

INTRODUCTION

Light-induced defect creation, termed the Staebler-Wronski effect [1], leads to the degradation of the conversion efficiency of hydrogenated amorphous silicon (a-Si:H) based photovoltaic devices, and is a major barrier to a-Si:H fulfilling its technological potential [2,3]. Attempts to overcome the Staebler-Wronski effect have focused primarily on either the solar cell device architecture [2, 4], in order to minimize the deleterious effect of the light induced defects, or modifications to the deposition process, so as to "harden" the material against metastable defect formation [5-8]. Roca i Cabarrocas and co-workers have recently reported that a-Si:H films grown in a PECVD reactor with hydrogen diluted silane at high chamber pressures, at the edge of powder formation within the plasma, yield thin amorphous silicon films containing silicon nanocrystalline inclusions (a/nc-Si:H) with superior electronic transport properties [9-11]. In this paper we report experimental studies of the synthesis and characterization of undoped a/nc-Si:H.

SAMPLE PREPARATION

All samples examined here were synthesized at the University of Minnesota in a capacitively coupled RF (13.56 MHz) system using plasma-enhanced chemical vapor deposition (PECVD) of silane ($SiH_4$) diluted with helium and hydrogen ($H_2$). The upper electrode is both the RF electrode and a showerhead through which the deposition gases enter the reactor chamber. Substrates reside on the lower, grounded electrode. Both electrodes (each of area ~ 200 $cm^2$) are independently heated to enable a controlled temperature gradient during deposition. In this way any thermophoretic forces that may displace small clusters within the plasma away from the substrates can be minimized. The a/nc-Si:H films were deposited with both the RF electrode temperature and the substrate (grounded) electrode temperature of 250 °C. Pure a-Si:H films were also deposited in the same reactor without hydrogen dilution and at lower total gas pressures and RF power and with an unheated RF electrode (substrate electrode temperature = 250 °C). The various deposition conditions under which the films were synthesized are described in Table 1. Prior to deposition, a five minute hydrogen plasma was used in order to additionally clean the substrates. Films investigated with transmission electron microscopy (TEM) were 10 nm thick and deposited onto cleaved NaCl substrates. The films were then removed by dissolving the salt substrates in distilled water and floating the films onto a 3 mm copper TEM grid [12]. The samples were then allowed to air dry in desiccators prior to TEM study. Hydrogenated silicon thin films for bulk electronic characterization ranged in thickness from 0.3 to 0.8 µm, and were deposited onto Corning 7059 glass substrates. Coplanar Chromium electrodes ~ 50 nm thick, 10 mm long and either 2 mm or 4 mm apart were e-beam evaporated onto the films. These electrodes yielded linear current-voltage characteristics. All electronic characterization measurements, aside from CPM, were performed after a one hour anneal in the dark under vacuum at 170 °C, following which the sample was slowly cooled (cooling rate ~ 2-3 °C/min) to room temperature, to remove any effects of prior light exposure [1] or surface adsorbates [13].

Table 1: Deposition Conditions for a-Si:H and a/nc-Si:H films

| Pressure (mTorr) | Flow Rate $SiH_4$:He (5:95) (sccm) | Flow Rate $H_2$ (sccm) | Power (W) | Material |
|---|---|---|---|---|
| 100 | 19 | 0 | 5 | a-Si:H |
| 250 | 30 | 0 | 5 | a-Si:H[*] |
| 1450 | 40 | 100 | 20 | a/nc-Si:H |
| 1800 | 40 | 100 | 20 | a/nc-Si:H |

[*] While no nanocrystals were observed by TEM, this material behaves similar to the a/nc-Si:H films in many of its electronic properties.

## STRUCTURAL CHARACTERIZATION

Initial TEM studies to survey the crystallinity of the a/nc-Si:H films were performed using a Philips CM30 operating at 300 kV. Using displaced-aperture dark-field (DADF) imaging techniques, regions that satisfy the Bragg diffraction condition at the position of the objective aperture will appear bright [12]. Figure 1 is a DADF image of an a/nc-Si:H film deposited at a reactor chamber gas pressure of 1450 mTorr, that exhibits regions of crystallinity. One such region is denoted with an arrow in fig. 1. To ascertain the chemical identity of the film, electron energy loss spectroscopy (EELS) was performed in a Philips CM200 TEM equipped with a field emission gun (FEG). The resulting spectrum is shown in figure 2, clearly demonstrating only the presence of silicon.

High-resolution TEM studies were performed using a Philips CM200 FEG equipped with a spherical aberration corrector [14-18]. The TEM is used to record a focal series of images at a value of spherical aberration that decreases the contrast of the amorphous matrix [19]. Using this instrument, it is possible to identify nanocrystals in the a/nc-Si:H film. One such particle, with a nonuniform shape approximately 1 nm wide and 3 nm long is shown in fig. 3. Some of the silicon nanoparticles even exhibited planar defects such as a (111) twin boundary. Analysis of these images is aided through the use of the defocus series, which assists in the confirmation of crystalline material within the amorphous film.

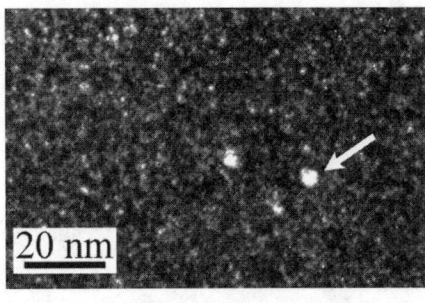

Fig. 1: TEM image illustrating the use of DADF to identify regions of crystallinity in an a/nc-Si:H film deposited at 1450 mTorr. One particle, ~5 nm in diameter, is indicated by an arrow.

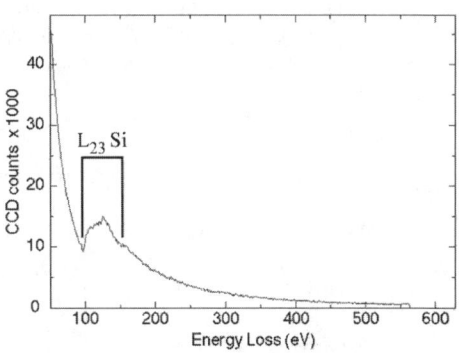

Fig. 2: Electron Energy Loss Spectrum from the nc-Si:H film confirming the presence of silicon.

## ELECTRONIC CHARACTERIZATION

The results above demonstrate that hydrogenated amorphous silicon deposited under heavy hydrogen dilution at high reactor chamber gas pressures does indeed contain nanocrystallite inclusions distributed throughout the thin film. The concentration and average diameter of these crystallites are found to depend on the details of the deposition process, as will be discussed elsewhere. We next address the bulk electronic transport properties of these a/nc-Si:H films, in particular the distribution of band-tail states and mid-gap defects, as reflected in measurements of the optical absorption spectra and their sensitivity to photodegradation.

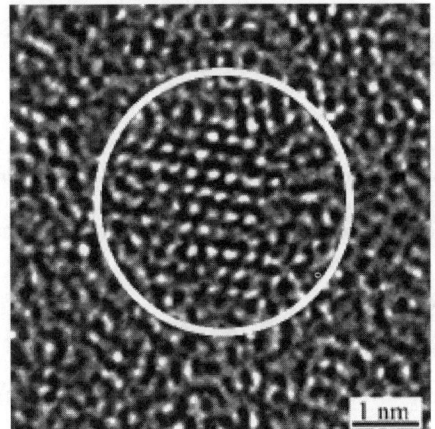

Fig. 3: HRTEM image of a silicon nanocrystal in an a/nc-Si:H film deposited at 1450 mTorr.

The optical absorption coefficient $\alpha$ was measured as a function of incident photon energy for the films described above using both Constant Photocurrent Method (CPM) at the University of Minnesota and Photothermal Deflection Spectroscopy (PDS) at the University of Utah. Prior to measurement by CPM, the films were dark annealed at 170 °C for one hour. The midgap absorption coefficient obtained from PDS and Urbach slope and optical gap as determined by a Tauc plot (both obtained from CPM) for all of the films investigated here are summarized in Table 2. Despite the fact that the a/nc-Si:H films were deposited under plasma conditions which are at the edge of powder formation, typically considered detrimental to the electronic quality of the films, the Urbach slope is only slightly broader for the a/nc-Si:H films compared to the a-Si:H reference sample. (The Urbach slope, $E_o = kT_o$, is obtained by fitting the sub-bandgap absorption coefficient to the expression $\alpha = \alpha_o \exp[-h\nu - E_1/kT_o]$, where both $E_1$ and $\alpha_o$ are independent of both thermal and structural disorder.) Similarly, the midgap density of states, as reflected in the absorption coefficient for $h\nu < 1.2$ eV, for the a/nc-Si:H film is comparable to that observed in the a-Si:H film. These $\alpha$ values corresponds to a defect density of $\sim 3\text{-}4 \times 10^{16}$ cm$^{-3}$, using the conversion factor of Amer and Jackson [20].

Table 2: Optical Absorption Characteristics for a-Si:H and a/nc-Si:H films

| Material | Pressure (mTorr) | $\alpha$ (1.2 eV) (cm$^{-1}$) | Urbach slope $E_o$ (meV) | Urbach slope $T_o$ (K) | Tauc Gap $E_{opt}$ (eV) |
|---|---|---|---|---|---|
| a-Si:H | 100 | 4* | 38 | 438 | 1.69 |
| a-Si:H** | 250 | 4 | 63 | 730 | 1.78 |
| a/nc-Si:H | 1450 | 3 | 54 | 626 | 1.77 |
| a/nc-Si:H | 1800 | 2.5 | 51 | 600 | 1.74 |

* $\alpha(1.2$ eV) for the 100 mTorr film is from CPM measurements
** Same comment as in table 1.

All of the films investigated here were synthesized in a reactor that has never been exposed to dopant gases. The dark conductivity for these materials is well described by an Arhennius expression $\sigma = \sigma_o \exp[-E_a/kT]$ where $E_a \sim 0.7\text{-}0.8$ eV and $\sigma_o \sim 2\text{-}5 \times 10^3$ $\Omega^{-1}$ cm$^{-1}$. All of the materials investigated here displayed a metastable decrease of the dark conductivity and photoconductivity following illumination with heat-filtered white light from a W-Ha lamp (intensity $\sim 75$ mW/cm$^2$). After one hour light exposure the dark conductance decreased by less than one order of magnitude for the a/nc-Si:H films, and the dark conductivity activation energy increased by $\sim 0.2$ eV, compared to a decrease of a factor of 40 for the a-Si:H film. The results of Arhennius measurements for these films are summarized in Table 3.

Table 3: Electrical Conductivity Characteristics for a-Si:H and a/nc-Si:H films

| Material | Pressure (mTorr) | $\sigma$ (320 K) State A ($\Omega^{-1}$ cm$^{-1}$) | $\sigma$ (320 K) State B ($\Omega^{-1}$ cm$^{-1}$) | $E_a$ (eV) State A | $E_a$ (eV) State B |
|---|---|---|---|---|---|
| a-Si:H | 100 | $6.4 \times 10^{-8}$ | $1.6 \times 10^{-9}$ | 0.74 | 0.94 |
| a-Si:H* | 250 | $1.5 \times 10^{-8}$ | $1.65 \times 10^{-9}$ | 0.70 | 0.92 |
| a/nc-Si:H | 1450 | $5.8 \times 10^{-9}$ | $5.1 \times 10^{-10}$ | 0.73 | 0.94 |
| a/nc-Si:H | 1800 | $1.5 \times 10^{-9}$ | $6.6 \times 10^{-10}$ | 0.81 | 0.99 |

* Same comment as in table 1.

Figure 4 shows a plot of the photoconductivity of the a-Si:H and a/nc-Si:H films investigated here measured at 320 K as a function of exposure time when illuminated with heat-filtered white light. The decay of the photocurrent during illumination is due to metastable defect formation [1] and is well described by a dispersive bimolecular time dependence [21] $\Delta\sigma_{ph} = \Delta\sigma_{pho} [1 + (t/\tau)^\gamma]^{-1}$ where $\gamma \sim 0.5$ and $\tau \sim 10 - 40$ sec for all of the films studied here. The photoconductivity values for the a/nc-Si:H are approximately equal to that for the a-Si:H film, while from table 3 the a-Si:H sample has a state A dark conductivity value roughly a factor of ten larger than the a/nc-Si:H materials. Consequently the a/nc-Si:H films actually have a higher photo-to-dark conductivity ratio than the a-Si:H reference film.

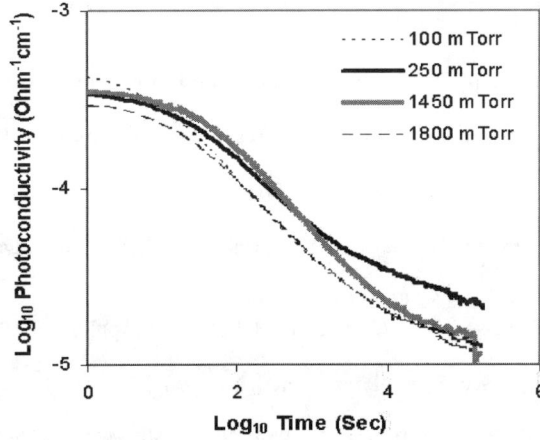

Fig. 4: Plot of the photoconductivity during illumination with heat-filtered white light as a function of exposure time for the a-Si:H and a/nc-Si:H films. The dotted line is the 100 mTorr a-Si:H film, the bold line is the 250 mTorr a/nc-Si:H film, the thick gray line is the 1450 mTorr a/nc-Si:H film, and the dashed line is the 1800 mTorr a/nc-Si:H film.

DISCUSSION

The results described here indicate that, in agreement with prior reports [9-11], deposition of amorphous silicon in a PECVD reactor with high $H_2$ dilution and relatively high reactor-chamber gas pressures does indeed produce silicon nanocrystallites embedded within a hydrogenated amorphous silicon matrix. The higher Urbach slope in a/nc-Si:H, compared to a-Si:H, indicates that these materials have broader band-tail states, and consequently a higher density of strained Si-Si bonds. This is consistent with the observation that a/nc-Si:H films exhibit an increased tendency to peel off the glass substrates when the film thickness exceeds 0.5 µm, while a-Si:H films typically do not peel off their substrates unless the film thickness is greater than 3 µm. Despite being synthesized under plasma conditions at the edge of macroscopic powder formation, which is typically considered far from the optimal conditions for depositing high electronic quality a-Si:H, the a/nc-Si:H films studied here have optical absorption coefficients, electronic conductivities and sensitivities to light-induced defect creation nearly identical to those observed in a-Si:H. However, while there is no appreciable degradation in the electronic quality of the a/nc-Si:H films, neither do they display significantly lower defect densities, higher conductances nor increased resistance to the Staebler-Wronski effect. Investigations are underway to determine the deposition conditions necessary to improve the electronic quality and decrease the photodegradation of the a/nc-Si:H films.

This research is supported by NSF under IGERT grant DGE-0114372. TJB and JK are primarily supported by NREL/AAD-9-18668-13 and the University of Minnesota. TJB, ST, UK and JK acknowledge partial support by the NSF MRSEC DMR-0212302. CRP and CBC acknowledge support from the 3M Heltzer Endowed Chair. They also thank Dr. M. Lentzen and Prof. Dr. K. Urban, IFF-IMF, Research Center, Julich for access to and assistance with the variable aberration HRTEM and Dr. U. Dahmen for access to the Philips CM200 at NCEM. The assistance of John Vinar and Prof. P. C. Taylor at the University of Utah for measurements of the PDS absorption spectra is gratefully acknowledged.

REFERENCES

1. D. L. Staebler and C.R. Wronski, Appl. Phys. Lett., **31**, 292 (1976); J. Appl. Phys. **51**, 3262 (1980).
2. See, for example, Mat. Res. Soc. Symp. Proc. (Materials Research Society, Pittsburgh, PA) **609,** (2000).
3. S. Wagner, D. E. Carlson and H. M. Branz, Electrochem. Soc. Symp. On Photovoltaics for the 21$^{st}$ Cent, (1999).
4. R. A. Street, *Hydrogenated Amorphous Silicon* (Cambridge University Press, Cambridge, 1991).
5. H. Matsumura, Japn. J. Appl. Phys. **30**, L1522 (1991).
6. A. H. Mahan, J. Carapella, B. P. Nelson, R. S. Crandall and I. Balberg, J. Appl. Phys. **69**, 6728 (1991).
7. L. Yang and L.-F. Chen, Mat. Res. Soc. Symp. Proc. (Materials Research Society, Pittsburgh, PA) **336**, 669 (1994).
8. Y. Lubianiker, J. D. Cohen, H.-C. Jin and J. R. Abelson, Phys. Rev. B **60**, 4434 (1999); D. Kwon, C.-C. Chen, J. D. Cohen, H.-C. Jin, E. Hollar, I. Robertson and J. R. Abelson, Phys. Rev. B **60**, 4442 (1999).
9. C. Longeaud, J.P. Kleider, P. Roca i Cabarrocas, S. Hamma, R. Meaudre, and M. Meaudre, J. Non Cryst. Solids, **227-230**, 96 (1998).
10. R. Butte, R. Meaudre, M. Meaudre, S. Vignoli, C. Longeaud, J.P. Kleider, and P. Roca i Cabarrocas, Philos.Mag. B, **79**, 1079 (1999).
11. P. Roca i Cabarrocas, A. Fontcuberta i Morral, and Y. Poissant, Thin Solid Films, **403-404**, 39 (2002).
12. D. B. Williams and C. B. Carter, *Transmission Electron Microscopy* (Plenum, New York) 1996.
13. M. Tanielian, Philos. Mag. B **45**, 435 (1982).
14. M. Haider, S. Uhlemann, E. Schwan, H. Rose, B. Kabius and K. Urban, Nature **392**, 768 (1998).
15. M. Lentzen, B. Jahnen, C. L. Jia, A. Thust, K. Tillmann and K. Urban, Ultramicrosc. **92**, 233 (2002).
16. K. Urban and M. Lentzen, Microsc. Microanal. **8** (sup. 2), 8 (2002).
17. M. Haider, H. Rose, S. Uhlemann, E. Schwan, B. Kabius and K. Urban, Ultramicrosc. **75**, 53 (1998).
18. C. L. Jia, M. Lentzen and K. Urban, Science **299**, 870 (2003).
19. C. R. Perrey, C. B. Carter and M. Lentzen, Microsc. Microanal. **9** (sup. 2) in press.
20. N. M. Amer and W. B. Jackson, in *Semimetals and Semiconductors*, edited by Jacques I. Pankove, (Academic Press, New York) vol. **21B**, 83 (1884).
21. D. Quicker and J. Kakalios, Mat. Res. Soc. Symp. Proc. (Materials Research Society, Pittsburgh, PA) **420**, 611 (1996).

## Microstructure and Optical Functions of Transparent Conductors and their Impact on Collection in Amorphous Silicon Solar Cells

G. M. Ferreira, Chi Chen, A. S. Ferlauto, P. I. Rovira, Ilsin An, C. R. Wronski, R. W. Collins, G. Ganguly,[1] Joong Hwan Kwak,[2] and Koeng Su Lim[2]
Materials Research Institute, Center for Thin Film Devices, and Department of Physics, The Pennsylvania State University, University Park, PA 16802;
[1]BP Solar, 3601 LaGrange Parkway, Toano VA 23168;
[2]Korea Advanced Institute of Science and Technology, 373-1 Guseong-dong, Yuseong-gu, Daejeon, 305-701, Korea.

## ABSTRACT

We have developed new procedures for determining the microstructure as well as the index of refraction and extinction coefficient spectra $\{n(E), k(E)\}$ for textured $SnO_2$ thin films on glass used as the top contact layers and superstrate for amorphous silicon (a-Si:H) p-i-n solar cells. These procedures combine (i) multichannel Mueller matrix spectroscopy using a dual rotating-compensator spectroscopic ellipsometer in reflection from the surface of the $SnO_2$, a measurement that is most sensitive to microstructure and $n(E)$, and (ii) transmission spectroscopy through a double-thick $SnO_2$ sandwich contacted with index-matching fluid, a measurement that is most sensitive to $k(E)$. An important optical loss in a-Si:H p-i-n solar cells is reflection from the $SnO_2$/p-layer interface. In this paper, we characterize this optical loss through modeling the solar cell optical quantum efficiency and demonstrate the extent to which microscopic roughness at this interface can serve as an anti-reflection layer for enhanced collection.

## INRODUCTION

An understanding of the limitations on the performance of amorphous silicon (a-Si:H) based solar cells due to optical collection can be achieved only through realistic optical simulation and fitting of experimental quantum efficiency spectra [1]. For multilayer solar cell structures that are ideal, i.e., structures consisting of uniform films with perfectly smooth, compositionally-abrupt interfaces, it would be straightforward to simulate and fit experimental spectra [2,3]. Interfaces are far from ideal, however; in reality, roughness is built into the solar cell structure intentionally in order to generate scattered light waves with increased optical path lengths [4]. It is important to emphasize that roughness exists on a wide range of in-plane scales, and multiple optical effects of intentional roughness (or so-called texture) may be generated [5]. For example, *microscopic* roughness has an in-plane scale $L$ much smaller than the central optical wavelength $\lambda_c$, i.e., $L < 0.1\lambda_c$, and can be modeled as a separate layer having optical properties determined by the Bruggeman effective medium theory. *Macroscopic* roughness has an in-plane scale on the order of the wavelength, i.e., $0.1\lambda_c < L < 10\lambda_c$, and can be modeled using a coherent superposition of scattered waves within the framework of diffraction theory. *Geometric* roughness has an in-plane scale much greater than the wavelength, i.e., $L > 10\lambda_c$, and greater than the lateral coherence of the light. As a result, the effects of the latter can be modeled within the framework of geometric optics and incoherent superposition of waves.

Roughness on different scales can be detected in different ways, in particular through its influence on polarization state changes upon reflection, and on the specularly reflected and scattered irradiances, as well as on the degree of polarization of the reflected beam [6]. In recent work, we have been focusing on the effects of microscopic roughness, especially its effect on the optical quantum efficiency of solar cells. In this article, we describe our improved measurements of the microstructure and optical properties of a standard transparent conducting $SnO_2$ film on a glass substrate by applying multichannel Mueller matrix spectroscopy and transmission and reflection spectroscopy. The results are in good agreement with direct structural measurements, and provide insights into the influence of microscopic surface roughness on optical collection in solar cells.

# EXPERIMENTAL DETAILS

## Optical measurements

In this study, the $SnO_2$ thin film was an Asahi U-type textured sample on glass, exhibiting roughness over the full range of in-plane scales. The free surface of this substrate/film was characterized in oblique (~70°) reflection using a dual rotating-compensator multichannel ellipsometer [7]. The output of this instrument is the 4x4 real Mueller matrix of the sample. We minimize the effects of macroscopic and geometric roughness on the deduced amplitude reflection coefficients $r_{ij}$ (i,j = p,s) by using a large $f/\#$ detection system and by filtering out the weak unpolarized component of the beam from the sample, thus analyzing only the specularly-reflected, purely-polarized component. This latter component is expected to be sensitive to microscopic scale roughness alone. In the analysis, we find that the p and s linear polarization states are the normal modes of the substrate/film, so that both cross-polarization coefficients $r_{ps} = |r_{ps}|\exp(i\delta_{ps})$ and $r_{sp} = |r_{sp}|\exp(i\delta_{sp})$ vanish. Thus, the two co-polarization angles $\psi \equiv \tan^{-1}(|r_{pp}|/|r_{ss}|)$ and $\Delta \equiv \delta_{pp} - \delta_{ss}$ can be used to characterize the $SnO_2$.

The primary disadvantage of even the most powerful ellipsometers is the poor sensitivity to the extinction coefficient $k$. As a result, normal-incidence transmittance $T$ is employed as a third parameter (i.e., in addition to $\psi$ and $\Delta$) to extract $k$ with higher sensitivity. Unfortunately, transmission and reflection (T&R) experiments are sensitive to macroscopic and geometric roughness due to the loss of irradiance via non-specular scattering. To eliminate such effects, a glass/$SnO_2$/$CH_2I_2$/$SnO_2$/glass sandwich was constructed for measurement by T&R, where the diiodomethane is used as an index-matching fluid to contact two samples of glass/$SnO_2$, cut from one substrate. A multichannel T&R spectrometer allowed rapid measurement of this structure.

## Data analysis

Microstructural (see Fig. 1) and optical models with five and eight free parameters, respectively, were employed to fit the spectra in ($\psi$, $\Delta$) and $T$ [8]. The free structural parameters for analysis of ($\psi$, $\Delta$) include the thicknesses of a microscopic interface layer, a bulk layer, and a microscopic surface roughness layer, ($d_i$, $d_b$, $d_s$). The interface and surface layers are also characterized by the volume fractions of $SnO_2$ in the layers ($f_i$, $f_s$). In contrast, the analysis of $T$ requires no additional structural parameters. In this case, the two $SnO_2$ films in the sandwich structure are considered as a single layer with a thickness given by $2d_{eff}$ where $d_{eff}$ is an effective thickness (or volume per unit area of material) given by $d_{eff} = f_i d_i + d_b + f_s d_s$. The analysis of the ($\psi$, $\Delta$) spectra proceeds using incoherent and coherent models for propagation within the glass and in the three-layer stack, respectively. The analysis of $T$ applies incoherent optics throughout. The resulting spectra for $SnO_2$ have been fitted using a Kramers-Kronig consistent expression for $\{n(E), k(E)\}$ that combines three components: (i) a Drude characteristic for intraband absorption with two parameters, (ii) a Tauc-Lorentz characteristic for interband absorption with four parameters [9], and (iii) a Sellmeier characteristic with two parameters for additional dispersion. These three components represent electronic oscillators with resonant frequencies (i) $\omega_0=0$; (ii) $\omega_0$ in the near ultraviolet; and (iii) $\omega_0$ well above the accessible range, respectively [10].

Figure 2 shows a flow chart of the overall analysis. First, trial values are selected for all parameters, and these are used in a least-squares regression analysis of the ($\psi$, $\Delta$) spectra. Once a best fit solution is obtained to the SE data, the same solution is applied as a trial in a second least-squares regression analysis fitting of the spectra in $T$. In this second fit, the only variable parameters are associated with the absorptive part of $SnO_2$ optical properties, namely, the six Drude and Tauc-Lorentz parameters. If this second analysis converges onto the same values, then a global fit is obtained and the analysis is complete. Otherwise, the improved best fit absorptive part is returned for a renewed fitting of the ($\psi$, $\Delta$) spectra, but now only the dispersive part of the optical properties, namely, the two Sellmeier parameters, along with the five structural parameters are varied. The entire coupled least-squares regression analysis is iterated until convergence onto a best fit solution is obtained. With this approach a Kramers-Kronig consistent optical expression is obtained in which the microstructural parameters and $n(E)$ are derived from ($\psi$, $\Delta$) and $k(E)$ is derived from $T$.

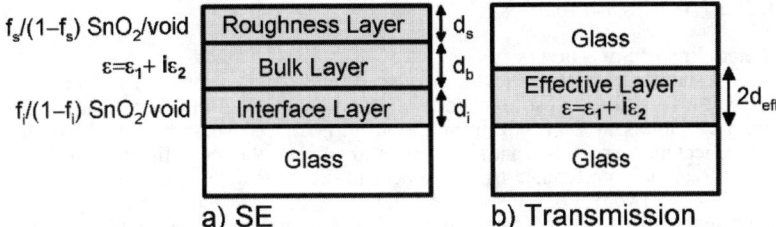

**Figure 1.** Structural models for SnO$_2$ films applied in the analysis of: (a) ellipsometric spectra ($\psi$, $\Delta$), measured in oblique reflection from the free surface of the film, and (b) T&R spectra measured at normal incidence for a sample consisting of a glass/SnO$_2$/CH$_2$I$_2$/SnO$_2$/glass sandwich. In (b), the SnO$_2$ in the sandwich structure is assumed to have a thickness equal to double the effective thickness deduced from the ellipsometric spectra.

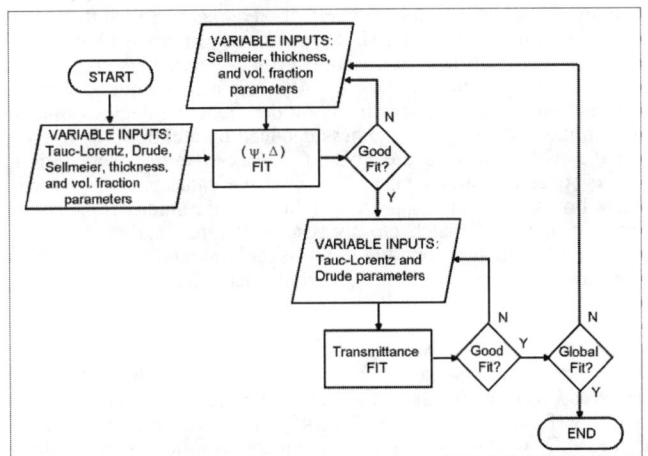

**Figure 2.** Flow chart describing the analysis of the spectra in the ellipsometric angles ($\psi$, $\Delta$) and in the transmittance $T$, consisting of two coupled least-squares regression analyses. In the fitting loop for ($\psi$, $\Delta$), the structural and Sellmeier optical parameters (that control $n$) are varied, and in the fitting loop for $T$, the Tauc-Lorentz and Drude parameters (that control $k$) are varied.

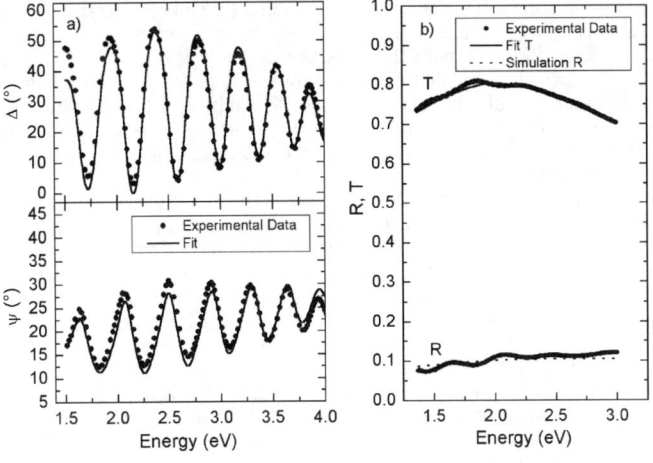

**Figure 3.** (a) Spectra in ($\psi$, $\Delta$) obtained by multichannel Mueller matrix ellipsometry at oblique incidence (~70°), and (b) data for transmittance and reflectance ($T$, $R$) obtained by multichannel spectroscopy at normal incidence. The solid line fits to ($\psi$, $\Delta$) and $T$ represent the outcome of the procedure of Fig. 2. The broken line overlapping $R$ is a simulation generated from the best fit structural and optical parameters.

# RESULTS AND DISCUSSION

## Optical modeling of SnO$_2$ layers

Figure 3 shows the best fits to the (a) ($\psi$, $\Delta$) and (b) $T$ spectra. The fringe pattern and average spectral variations in ($\psi$, $\Delta$) encode information on the film structure and $n(E)$, whereas the spectral variations in $T$ encode information on $k(E)$. Also shown in (b) is the normal-incidence reflectance spectrum and its simulation using the best fit structural and optical parameters. Residual interference fringes appear in the T&R data owing to multiple reflections within each of the two SnO$_2$ films of the sandwich; thus, the index matching is not perfect. Overall, however, there is good agreement in Fig. 3 between the fits/simulation and the data. The best fit structural parameters are given in Table I, and the best fit optical properties are provided in Fig. 4 for the range of 1.4 to 3.0 eV (over which the glass and CH$_2$I$_2$ are transmitting). Because the optical model is Kramers-Kronig consistent, it is seems possible to extrapolate the optical properties somewhat beyond the experimentally accessible range.

The results for the structure of the SnO$_2$ film are in very good agreement with those reported earlier on a different sample of Asahi U-type material [11], with the primary difference being a different bulk layer thickness $d_b$. In fact, one of the most critical parameters deduced in this analysis, the microscopic roughness layer thickness of $d_s$ = 455 Å is actually quite robust, being similar for different samples and not depending sensitively on the modeling details. In addition, this value is in agreement with the microscopic roughness deduced by atomic force microscopy (AFM). In the analysis of the AFM image of the SnO$_2$, the peak-to-valley excursions inside numerous 0.1x0.1 µm$^2$ image boxes are tabulated, and the average value is given by <$d_{pv}$(0.1 µm)> ~ 430 Å. In fact, this box size describes the upper limit of the microscopic roughness scale. Finally, the spectra in $n$ in Fig. 4 match closely results obtained earlier, however, the accuracy in the spectra in $k$ are much higher now due to (i) better suppression of scattering, (ii) improved separation of the absorption due to the SnO$_2$ and glass, and (iii) more advanced modeling.

## Optical modeling of p-i-n solar cells

The strongest reflection losses in the first pass through the a-Si:H p-i-n solar cell (i.e., before the retroreflector is encountered) will occur where the index of refraction discontinuity is the largest -- at the SnO$_2$/p-layer interface. The presence of the 455 Å microscopic roughness on the SnO$_2$ (which appears at the SnO$_2$/p-layer interface upon over-deposition of the p-layer) is expected to strongly influence the reflection at this interface. Table II shows the input structural parameters for two a-Si:H p-i-n solar cells, one with ideal, perfectly smooth interfaces, and the other with microscopically rough interfaces. The latter structure is constructed with three considerations in mind. (i) In accordance with the structure of Table I, 455 Å of roughness is incorporated at the SnO$_2$/p-layer interface. (ii) This roughness is propagated to all the other interfaces of the cell in accordance with previous real time SE studies [12]. (iii) The effective thicknesses or volumes of material per unit area for each of the layers is equivalent to the corresponding layer in the ideal structure. For example, the effective thickness of the bulk i-layer is taken to be 3000 Å for both structures.

Figure 5 shows the optical quantum efficiency predicted on the assumption that each photon absorbed in the i-layer is converted to a single electron-hole pair that is then collected. It is clear that the microscopic roughness at the SnO$_2$/p-layer interface leads to a significant reduction in the reflected irradiance over the range of 350 to 650 nm and thus, an enhancement in the optical quantum efficiency. In addition the interference oscillations in the quantum efficiency data are suppressed to levels comparable with those observed experimentally [1]. Figure 6 shows that the predicted gain in short circuit current collection by the solar cell due to the 455 Å thick microscopic roughness at the SnO$_2$/p-layer interface is ~1 mA/cm$^2$, a significant effect. Thus, the microscopic roughness component associated with textured SnO$_2$ can generate a considerable increase in current that is *unrelated* to scattering and optical path length enhancement. In fact, this is an *antireflection effect* due to the suppression of the dielectric discontinuity by microscopic roughness at the most strongly reflecting interface in the cell.

**Table I.** Best fit structural parameters obtained in an analysis of ($\psi, \Delta$) for Asahi U-type $SnO_2$.

| Interface layer | | Bulk layer | Roughness layer | |
|---|---|---|---|---|
| $d_i$ (Å) | $f_i$ | $d_b$ (Å) | $d_s$ (Å) | $f_s$ |
| 438 | 0.69 | 6777 | 455 | 0.57 |

**Table II.** Input parameters used in the optical modeling of two a-Si:H p-i-n solar cells with 3000 Å i-layers, one with an ideal structure, i.e., with perfect interfaces, and a second with a 455 Å thick microscopic roughness layer at the $SnO_2$/p-layer interface that propagates throughout the structure as established by real time SE measurements. Equivalent effective thicknesses $d_{eff} = \Sigma_j f_j d_j$ are used for the corresponding layers of the ideal and microscopically rough structures.

| Layer | Ideal - Thickness (Å) | Rough - Thickness (Å) | Composition |
|---|---|---|---|
| Glass | $1.2 \times 10^7$ | $1.2 \times 10^7$ | 1.0 glass |
| $SnO_2$ (interface) | - | 438 | 0.69/ 0.31 $SnO_2$/ void |
| $SnO_2$ (bulk) | 7338 | 6777 | 1.0 $SnO_2$ |
| $SnO_2$ / p-layer (interface) | - | 455 | 0.57/ 0.21/ 0.22 $SnO_2$/ p-layer/ i-layer |
| p-layer (bulk) | 100 | - | 1.0 p-layer |
| p-layer / i-layer (interface) | - | 12 | 0.38/ 0.62 p-layer/ i-layer |
| i-layer (bulk) | 3000 | 2808 | 1.0 i-layer |
| i-layer / n-layer (interface) | - | 125 | 0.68/ 0.32 i-layer/ n-layer |
| n-layer | 200 | 75 | 1.0 n-layer |
| n-layer / ZnO (interface) | - | 125 | 0.68/ 0.32 n-layer/ ZnO |
| ZnO (bulk) | 2000 | 1960 | 1.0 ZnO |
| Ag (bulk) | 2000 | 2000 | 1.0 Ag |

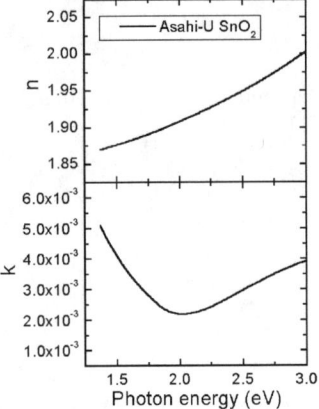

**Figure 4.** The optical properties of $SnO_2$ over the range from 1.4 to 3.0 eV, given in terms of the index of refraction and the extinction coefficient. These Kramers-Kronig consistent results were obtained in a best fit to ($\psi, \Delta$) and $T$ according to the procedure of Fig. 2.

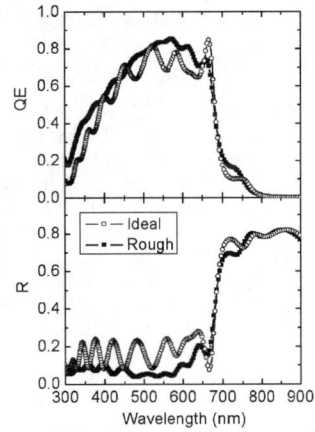

**Figure 5.** Simulated optical quantum efficiencies and spectroscopic reflectance for an ideal a-Si:H p-i-n solar cell structure, and for the corresponding cell structure with 455 Å of microscopic roughness at the $SnO_2$/p-layer interface that propagates to the succeeding interfaces of the cell, as determined by real time SE. The structural parameters for the two solar cells are given in Table II.

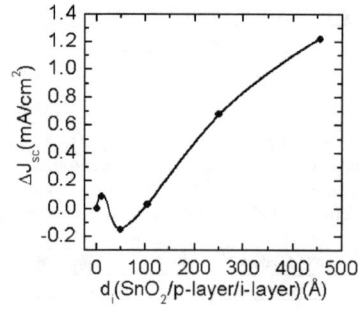

**Figure 6.** Enhancement of the short circuit current due to microscopic roughness layers of different thicknesses at the $SnO_2$/p-layer interface, as obtained in simulations such as the one in Fig. 5. For each solar cell, the roughness is assumed to propagate throughout the entire structure, as established by real time SE.

## SUMMARY

We have established improved procedures for determining the microstructure and optical functions of textured $SnO_2$ thin films incorporated into a-Si:H-based thin film solar cells. These procedures have combined multichannel Mueller matrix ellipsometry in oblique reflection from the top surface of the $SnO_2$ film and multichannel transmission spectroscopy at normal incidence through a double-thick $SnO_2$ sandwich contacted with index-matching fluid These techniques have been designed to extract accurate $\{n(E), k(E)\}$ spectra with excellent rejection of the scattered light that arises due to macroscopic and geometric scale roughness. A procedure has been developed that fits the ellipsometric and transmission spectra iteratively using a Kramers-Kronig consistent expression for $\{n(E), k(E)\}$. In this paper, we have compared the simulated optical quantum efficiency as well as the reflection losses for a-Si:H p-i-n solar cells having ideal, perfectly smooth interfaces and microscopically rough interfaces. From this study, we demonstrate the extent to which the microscopic roughness -- known to exist at the $SnO_2$/p-layer interface from the ellipsometric measurements -- can serve as an anti-reflection layer for enhanced collection. Such an effect must occur in conjunction with the enhancement in collection due to scattering by macroscopic roughness.

## REFERENCES

1. R. Schropp and M. Zeman, *Amorphous and Microcrystalline Solar Cells: Modeling, Materials, and Device Technology*, (Kluwer, Boston, 1998).
2. F. Leblanc, J. Perrin, and J. Schmidt, J. Appl. Phys. **75**, 1074 (1994).
3. L.A.A. Pettersson, L.S. Roman, and O. Inganäs, J. Appl. Phys. **86**, 487 (1999).
4. E. Yablonovitch and G.D. Cody, IEEE Trans. Electron. Devices **ED-29**, 300 (1982).
5. P. Beckmann and A. Spizzichino, *The Scattering of Electromagnetic Waves from Rough Surfaces*, (Pergamon, Oxford, 1963).
6. M.D. Williams and D.E. Aspnes, Phys. Rev. Lett. **41**, 1667 (1978).
7. J. Lee, J. Koh, and R.W. Collins, Rev. Sci. Instrum. **72**, 1742 (2001).
8. P.I. Rovira and R.W. Collins, J. Appl. Phys. **85**, 2015 (1999).
9. G. E. Jellison, Jr. and F. A. Modine, Appl. Phys. Lett. **69**, 371 (1996); **69**, 2137 (1996).
10. R. W. Collins and K. Vedam, in *Encyclopedia of Applied Physics*, Vol. 12, edited by G.L. Trigg (VCH, New York, 1995) p. 285.
11. G.M. Ferreira, A.S. Ferlauto, P.I. Rovira, C. Chen, H.V. Nguyen, C.R. Wronski, and R.W. Collins, Mater. Res. Soc. Symp. Proc. **664**, A24.6 (2001).
12. P.I. Rovira, A.S. Ferlauto, I. An, H. Fujiwara, J. Koh, R.J. Koval, C.R. Wronski, and R.W. Collins, Mater. Res. Soc. Symp. Proc. **557**, 719 (1999).

## Reaction Control in Amorphous Silicon Film Deposition by Hydrogen Chloride

Akihiro Takano[1], Takehito Wada[1], Shinji Fujikake[1], Takashi Yoshida[1], Tokio Ohto[1] and Eray S. Aydil[2]
[1]Fuji Electric Corporate Research and Development, Ltd., 2-2-1 Nagasaka, Yokosuka, Kanagawa 240-0194 Japan
[2]Department of Chemical Engineering, University of California Santa Barbara, Santa Barbara, CA 93106, U.S.A.

## ABSTRACT

HCl was added to $SiH_4$ containing plasmas to grow a-Si:H(Cl) films with dangling bonds terminated with Cl instead of H. Bulk and surface infrared spectra, film thickness and optical band gap were examined by *in situ* multiple total internal reflection Fourier transform infrared spectroscopy and *in situ* spectroscopic ellipsometry. $SiH_2Cl_2$ was also used as a conventional Cl source for reference a-Si:H(Cl) film deposition experiments. The introduction of HCl does not affect the deposition rate significantly, and the deposited a-Si:H(Cl) films contain over $10^{21} cm^{-3}$ Cl atoms. HCl addition to the gas phase changes the surface compositions of the growing films drastically from higher silicon hydride to chlorinated lower hydride. The surface reaction control eliminates unfavorable hydride bonding structures such as $SiH_2$ and/or SiH in voids in the deposited films. The a-Si:H(Cl) films deposited from mixtures of $SiH_4$ and HCl do not show significant optical band gap widening in spite of containing over $10^{21} cm^{-3}$ Cl atoms, a concentration that is comparable to that of hydrogen. In contrast, a conventional chlorine source of $SiH_2Cl_2$ increases the deposition rate significantly compared to HCl. The increase in the deposition rate results in monotonic decrease of the refractive index and the optical band gap widening.

## INTRODUCTION

Alternative dangling bond terminating atoms, such as D, F and Cl, have been extensively investigated aiming at the suppression of the photoinduced degradation in hydrogenated amorphous silicon (a-Si:H) solar cells [1-9]. These higher mass atoms are expected to have lower diffusion coefficients in the amorphous bulk network resulting in the suppression of the photoinduced degradation. Although $SiH_2Cl_2$ and $SiCl_4$ have been used as Cl sources for terminating dangling bonds by Cl, drastic improvements have not been reported. On the contrary, an excess introduction of $SiH_2Cl_2$ and $SiCl_4$ deteriorates the film properties and widens the optical band gap [3,10], which may be caused by excess SiCl bond introduction and formation of clustered SiCl and nanovoids.

Despite its simplicity and an applicability to conventional solar cell fabrication processes, HCl introduction as a chlorine source during film deposition had not been reported until we reported that chlorine containing hydrogenated amorphous silicon (a-Si:H(Cl)) film deposition from $SiH_4/HCl$ mixed gases without band gap widening is possible [11].

In this study, we investigate HCl introduction effects on amorphous silicon film deposition focusing on surface hydride and chloride composition, the resulting bonding features and optical properties in the films. HCl and conventional $SiH_2Cl_2$ were used as Cl source gases in a-Si:H(Cl) film deposition for comparison.

## EXPERIMENTAL

Films were deposited in an inductively coupled plasma enhanced chemical vapor deposition reactor with *in situ* process observation capabilities shown in Fig. 1. The growth surface and the bulk bonding configurations in the film were observed by multiple total internal reflection Fourier transform infrared (MTIR-FTIR) technique using GaAs substrates with a trapezoidal cross-sectional shape [11,12]. The film thickness and optical constants are derived from the spectroscopic ellipsometry data analysis using a three-layer model (GaAs substrate/ GaAsOx layer/ a-Si:H(Cl) bulk layer/ surface roughness layer). The bulk layer and the surface roughness layer were modeled by Tauc-Laurentz model and Bruggeman effective medium approximation with 50% (a-Si:H(Cl)) and 50% (void) in volume fraction [11,13,14]. Hydrogen and chlorine concentration in the films were determined by secondary ion mass spectrometry (SIMS) using $Cs^+$ ions.

Mixtures of $Ar(98\%)/SiH_4$ (2%), Ar and $Ar(75\%)/HCl$ (25%) were introduced in the reactor as source gases. Total flow rate ($Ar/SiH_4$ + Ar + Ar/HCl) and $Ar/SiH_4$ flow rate were kept constant at 40 and 20 standard cubic centimeters per minute (sccm), respectively. $HCl/SiH_4$ ratio was varied from 0 to 3.125. As reference experiments using a conventional chlorine source, mixtures of $Ar(98\%)/SiH_4$ (2%), Ar and $Ar(96\%)/SiH_2Cl_2$ (4%) were fed to the reactor in order to deposit a-Si:H(Cl) films. Total flow rate and $Ar/SiH_4$ flow rate were also kept constant at 40 and 20 sccm, respectively. $SiH_2Cl_2/SiH_4$ ratio was varied from 0 to 2. Deposition pressure was controlled at 6.65Pa (50 mTorr) by the position control of a throttle valve. The source gas was decomposed by the plasma which was sustained by applying 15 W rf power. The substrate temperature was 130°C.

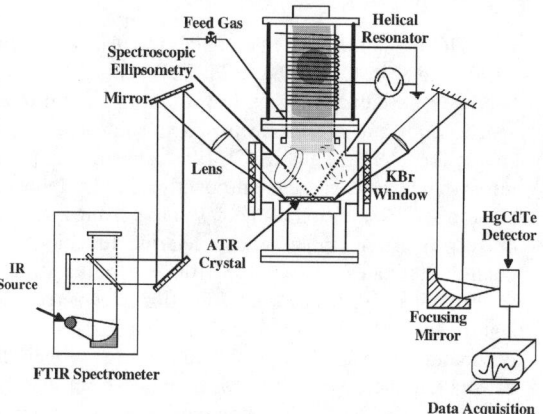

**Figure 1.** Schematic representation of the plasma-enhanced chemical vapor deposition reactor with in situ monitoring systems.

## RESULTS AND DISCUSSION

The effect of Cl atoms in the gas phase on the H and Cl bonding in the deposited films was investigated using MTIR-FTIR in multiple internal transmission mode [11,15]. Typical MTIR-FTIR spectra of a-Si:H(Cl) bulk films deposited using gas mixtures of $HCl/SiH_4=0$, 0.125 and 3.125 are shown in Fig. 2. Background spectra were collected just before starting the a-Si:H(Cl) film growth on an a-Si:H(Cl) layer deposited under the same conditions in order to avoid the initial stage growth affected by the substrate, cancel surface signals, and collect bulk spectra selectively. Higher hydride dominant bonding configurations ($SiH_3$ and $SiH_2$) were changed effectively into chlorinated mono hydride by HCl addition. a-Si:H(Cl) films deposited from $HCl/SiH_4>1$ mixed feed gas contain over $10^{21}$ $cm^{-3}$ Cl atoms [11].

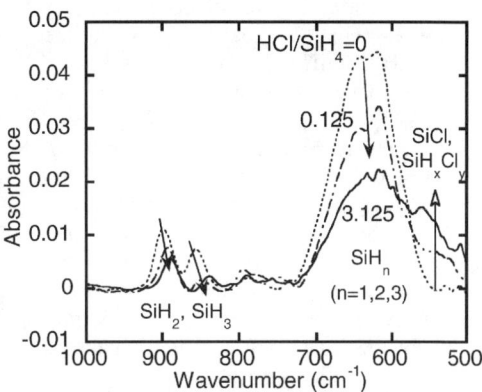

**Figure 2.** MTIR-FTIR sptectra of a-Si:H(Cl) films deposited using mixture gases of $HCl/SiH_4=0$, 0.125 and 3.125.

Surface bonding configuration was monitored by MTIR-FTIR in attenuated total reflection mode. The procedure for the surface infrared spectra collection is reported previously [12]. Briefly, reference spectra were collected after the film deposition. The film surface was exposed to a low energy Ar ions (15-25eV) from a low density Ar plasma to remove the surface bonding H and Cl. Surface infrared spectra were collected as difference spectra between before and after the removing of surface bonded species. Ar plasma was generated by applying 50 W rf power for 15 sec in this experiment. Ar plasma exposure conditions were optimized in the same manner reported in the previous publication [15]. The treatment time was set in order to maximize the surface signal intensities before the bulk components appeared in the infrared spectra. The observed surface infrared spectra are shown in Fig. 3. The surface bonded hydride is partially or fully chlorinated effectively by even small amount of HCl addition.

Deposition rate determined by spectroscopic ellipsometry was nearly independent of the $HCl/SiH_4$ ratio in the feed gas and is plotted in Fig. 4. As a reference, deposition rate in $SiH_2Cl_2/SiH_4$ system is also plotted in the same figure. Deposition rate in $HCl/SiH_4$ system shows a slight

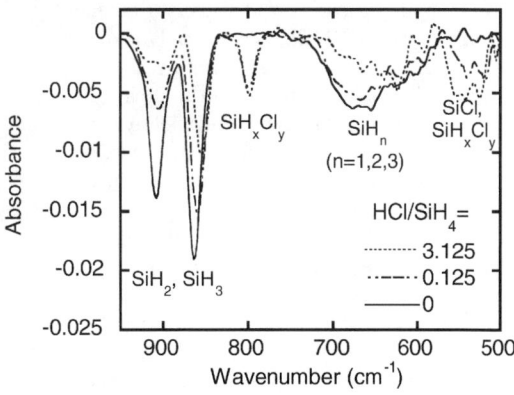

**Figure 3.** Surface IR spectra of a-Si:H(Cl) films deposited using mixture gases of $HCl/SiH_4=0$, 0.125 and 3.125.

decrease with the increase in Cl source gas concentration, whereas deposition rate in $SiH_2Cl_2/SiH_4$ system increases rapidly with the increase in the partial pressure of $SiH_2Cl_2$.

Figure 5 shows the dependence of refractive index at 500 nm on $HCl/SiH_4$ and $SiH_2Cl_2/SiH_4$ ratio in the feed gas. The refractive index in films deposited from $HCl/SiH_4$ increases with the small amount of HCl introduction, whereas the refractive index in films deposited from $SiH_2Cl_2/SiH_4$ shows a rapid monotonic decrease with increasing $SiH_2Cl_2$ partial pressure. The increase in the refractive index corresponds to the formation of higher density bulk network introduced by the bonding modification, such as selective elimination of unfavorable higher hydrides as shown in Fig. 2. In the $SiH_2Cl_2/SiH_4$ system, Cl containing precursors like $SiH_xCl_y$ find bonding sites on the growth surface after insufficient surface migration, which causes the rapid increase in deposition rate and the introduction of low density structure. Polymerized species formed in the $SiH_2Cl_2$ containing plasma could also contribute to film growth [16]. These polymerized species landing on the growing surface at a relatively low temperature used in this study do not migrate sufficiently and introduce the deposition rate increase resulting in the formation of low density network.

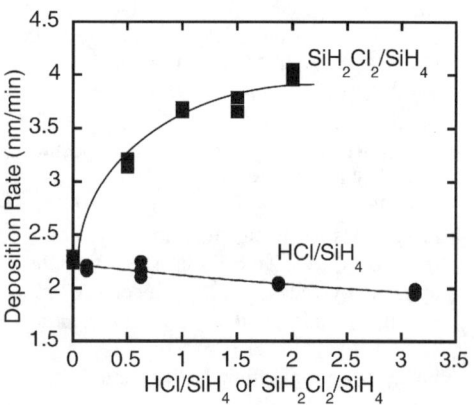

**Figure 4.** The dependence of deposition rate on $HCl/SiH_4$ or $SiH_2Cl_2/SiH_4$ in the feed gas.

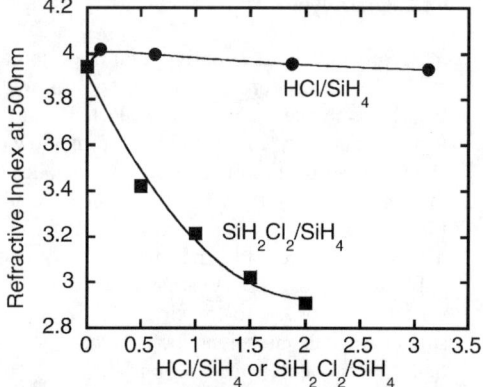

**Figure 5.** Refractive index at a wavelength of 500nm depending on $HCl/SiH_4$ or $SiH_2Cl_2/SiH_4$ ratio in the source gas.

In the spectroscopic ellipsometry data analysis, optical functions of the a-Si:H(Cl) layer are parameterized by Tauc-Lorentz model. As one of the five parameters, optical band gap can be determined by least-squares regression fitting. Figure 6 shows that the dependence of optical band gap on $HCl/SiH_4$ or $SiH_2Cl_2/SiH_4$ mixing ratio in the gas phase. In the conventional $SiH_2Cl_2/SiH_4$ deposition, a remarkable band gap widening was observed as reported previously [3,10], whereas the optical band gap is almost identical with the value of a-Si:H film without chlorine in spite of containing over $10^{21}$ $cm^{-3}$ chlorine atoms in films deposited using $HCl/SiH_4$.

In the a-Si:H(Cl) film deposition process using $SiH_2Cl_2/SiH_4$ and $SiCl_4/SiH_4$ as source gases, excess or clustered chlorine and resulting void incorporation may cause the abrupt optical band gap widening. In contrast, a-Si:H(Cl) films deposited from the mixture of HCl and SiH$_4$ did not show remarkable optical band gap widening in spite of containing over $10^{21}$cm$^{-3}$ chlorine atoms which is comparable to hydrogen concentration. We speculate that a combination of two mechanisms determine the optical band gap in a-Si:H(Cl) films. One possibility is that the optical band gap is altered due to the binding energy differences for species, such as Si-Cl (4eV), Si-H (3.4eV) and Si-Si (2.2-2.4eV) as already reported elsewhere [10,17-19]. Higher SiH and SiCl bond energies translate into wider gap. The binding energy difference affects the optical band gap widening by at most 0.02eV when over $10^{21}$ cm$^{-3}$ Cl atoms are introduced in the films by HCl as shown in Fig. 6. The other mechanism of the optical band gap widening could be from Cl local order. Excess localized clustered Cl may form a low density network structure including voids, which may affect on density of states (DOS) as in the case of hydrogen induced DOS change [20]. The films deposited from the conventional mixture of SiH$_2$Cl$_2$/SiH$_4$ or SiCl$_4$/SiH$_4$ may have bulk network structure with many Cl clusters and voids introduced by incomplete decomposition of chlorosilane and polymerized species formation in the gas phase. These Cl bonding configurations and voids can affect DOS and film density resulting in the change in the optical properties.

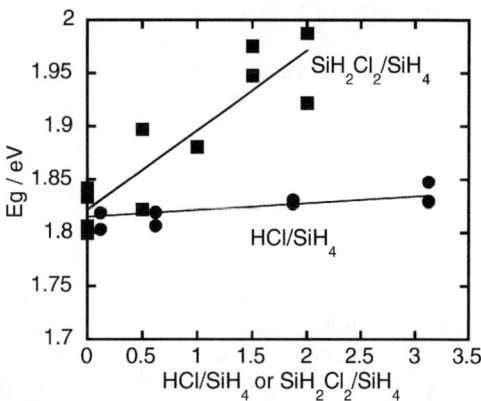

**Figure 6.** Optical band gap as a function of HCl/SiH$_4$ or SiH$_2$Cl$_2$/SiH$_4$ ratio in the feed gas.

The new a-Si:H(Cl) material from the HCl/SiH$_4$ mixed feed gas, in which Cl acts as a major dangling bond terminator together with H without optical band gap widening, is expected to be a promising material for amorphous silicon solar cells.

## CONCLUSIONS

We have introduced HCl into SiH$_4$ containing discharge to form a-Si:H(Cl) films in which chlorine and hydrogen are expected to terminate dangling bonds. Bulk and surface bonding structures and optical properties were monitored by *in situ* MTIR-FTIR and spectroscopic ellipsometry. The surface composition of the growing film changs drastically from a coverage dominated by higher silicon hydrides to chlorinated lower hydrides. The surface modification introduced by HCl eliminates unfavorable bulk SiH$_2$/SiH in voids selectively resulting in a denser Si network. Although the a-Si:H(Cl) films contain chlorine atoms whose concentration is comparable to that of hydrogen, these films do not show significant optical band gap widening. A-Si:H(Cl) films were also deposited using a conventional chlorine source of SiH$_2$Cl$_2$ as reference experiments. In the conventional SiH$_2$Cl$_2$/SiH$_4$ system, deposition rate increases rapidly, refractive index decreases monotonically and optical band gap widens with increasing in SiH$_2$Cl$_2$ partial pressure.

## ACKNOWLEDGMENTS

The authors would like to thank Dr. H. Fujiwara (National Institute of Advanced Industrial Science and Technology, Japan) for suggestions with spectroscopic ellipsometry data analysis. They are also grateful to S. Agarwal and A. R. Godfrey (University of California Santa Barbara) for technical help.

## REFERENCES

1. D. L. Staebler and C. R. Wronski, Appl. Phys. Lett. **31**, 292 (1977).
2. M. Nakata and S. Wagner, Appl. Phys. Lett. **65**, 1940 (1994).
3. I. S. Osborne, N. Hata, and A. Matsuda, Mater. Res. Soc. Symp. Proc. **377**, 113 (1995).
4. K. Dairiki, A. Yamada, and M. Konagai, Jpn. J. Appl. Phys. **38**, 4007 (1999).
5. M. Azuma, T. Yokoi, I. Shiiya, and I. Shimizu, J. Non-Cryst. Solids **164-166**, 47 (1993).
6. W. A. Nevin, H. Yamagishi, K. Asaoka, H. Nishio, and Y. Tawada, Appl. Phys. Lett. **59**, 3294 (1991).
7. T. Nishimoto, T. Takagi, M. Kondo, and A. Matsuda, Sol. Energy Mater. & Sol. Cells **66**, 179 (2001).
8. K. H. Lee, S. K. Kim, K. S. Lee, J. H. Choi, C. S. Kim, J. Jang, S. M. Pietruszko and M. Kostana, Sol. Energy Mater. & Sol. Cells **49**, 61 (1997).
9. N. Hata, I. S. Osborne, T. Ikeda, R. Durny, and A. Matsuda, J. Non-Cryst. Solids **198-200**, 415 (1996).
10. T. Oshima, K. Yamaguchi, A. Yamada, M. Konagai, and K. Takahashi, Mater. Res. Soc. Symp. Proc. **336**, 91 (1994).
11. A. Takano and E. S. Aydil, Jpn. J. Appl. Phys. **41**, L1357 (2002).
12. E. S. Aydil, D. Maroudas, D. C. Marra, W. M. M. Kessels, S. Agarwal, S. Ramalingam, S. Sriraman, M. C. M. Van de Sanden and A. Takano, Res. Soc. Symp. Proc. **664**, A1.1.1 (2001).
13. G. E. Jellison, Jr. and F. A. Mondine, Appl. Phys. Lett. **69**, 371 (1996).
14. H. Fujiwara, J. Koh, and R. W. Collins, Thin Solid Films **313-314**, 474 (1998).
15. W. M. M. Kessels, D. C. Marra, M. C. M. van de Sanden and E. S. Aydil, J. Vac. Sci. Technol. **A 20**, 781 (2002).
16. H. Shirai, C. Fukai, Y. Sakuma, and Y. Moriya, J. Non-Cryst. Solids **266-269**, 131 (2000).
17. S. Al-Dallal: J. Non-Cryst. Solids **59&60**, 361 (1983).
18. D. Adler: Semiconductors and Semimetals, ed. J. I. Pankove (Academic Press, Inc., Orlando, 1984) Chap. 14, p.295.
19. H. Feil, J. Dieleman, and B. J. Garrison: J. Appl. Phys. **74**, 1303 (1993).
20. D. C. Allan and J. D. Joannopoulos, in The Physics of Hydrogenated Amorphous Silicon II, edited by J. D. Joannopoulos and G. Lucovsky (Springer-Verlag, 1984), p.5-60.

## Material structure of microcrystalline silicon deposited with an Expanding Thermal Plasma

C. Smit[1,3], D.L. Williamson[2], M.C.M. van de Sanden[3], and R.A.C.M.M. van Swaaij[1]
[1]Delft University of Technology, DIMES-ECTM,
P.O. Box 5053, 2600 GB Delft, The Netherlands
[2]Colorado School of Mines, Department of Physics, Golden, CO 80401, USA
[3]Eindhoven University of Technology, Department of Applied Physics,
P. O. Box 513, 5600 MB Eindhoven, the Netherlands

## ABSTRACT

Expanding thermal plasma CVD (ETP CVD) has been used to deposit thin microcrystalline silicon films. In this study we varied the position at which the silane is injected in the expanding hydrogen plasma: relatively far from the substrate and close to the plasma source, giving a long interaction time of the plasma with the silane, and close to the substrate, resulting in a short interaction time. The material structure is studied extensively. The crystalline fractions as obtained from Raman spectroscopy as well as from X-ray diffraction (XRD) vary from 0 to 67%. The average particle sizes vary from 6 to 17 nm as estimated from the (111) XRD peak using the Scherrer formula. Small angle X-ray scattering (SAXS) and flotation density measurements indicate void volume fractions of about 4 to 6%. When the samples are tilted the SAXS signal is lower than for the untilted case, indicating elongated objects parallel to the growth direction in the films. We show that the material properties are influenced by the position of silane injection in the reactor, indicating a change in the plasma chemistry.

## INTRODUCTION

Thin silicon films receive much attention because they can be implemented in thin film solar cells, which can be produced at lower cost than conventional monocrystalline Si wafer based solar cells. Much research effort is directed towards amorphous as well as microcrystalline silicon films, because the optical band gaps of these materials (1.8 and 1.1 eV, respectively) allow an efficient use of the solar spectrum when combined in a tandem solar cell. Since large areas are required high throughput is needed to decreases production costs and therefore the deposition rate is an important item in thin silicon film research. For example, VHF (very high frequency) PECVD [1] and hot wire (HW) CVD [2] are techniques that aim for higher deposition rates. Furthermore, conventional RF PECVD deposition settings are optimized for high deposition rates [3].

With expanding thermal plasma chemical vapour deposition (ETP CVD) high deposition rates have been achieved in the past for the deposition of a-Si:H (around 0.7 nm/s successfully applied in solar cells [4]). This deposition technique makes use of a remote plasma source for the production of atomic hydrogen, which expands into a reaction chamber. Silane ($SiH_4$) is injected into the plasma expansion and the substrate is situated further downstream. The advantage is that deposition conditions like atomic hydrogen flux, $SiH_4$ flux, and pressure can easily be adjusted separately to optimize the material properties. We reported earlier on the deposition of microcrystalline silicon with ETP CVD [5]. In this paper we will show that we improved the material quality, especially the material density, and consequently the electrical behavior. The feed gas for the plasma source used to be a Ar-H$_2$ mixture, but here the expanding plasma consists of only hydrogen. The second improvement involved the position in the reaction chamber where SiH$_4$ is injected. Apparently the plasma chemistry in the reactor is different in case SiH$_4$ is injected close to the substrate than when injected further away. The composition of the mixture of radicals arriving at the substrate can influence the material properties, just as they are known to do in the deposition of amorphous silicon. The best films we can make at present are implemented in solar cells.

## EXPERIMENT

### Experimental set-up
The 'CASCADE' set-up is designed to prepare thin-film silicon solar cells in which the absorber layer is deposited with ETP CVD [6]. It consists of a load lock, a RF PECVD chamber for the deposition of doped layers, and a reaction chamber for the deposition of ETP CVD silicon films.

The cascaded-arc plasma source is positioned on top of the reaction chamber. It consists of a 4-mm diameter channel between three cathodes and an anode plate in which a plasma is created by a DC discharge current of 50 A at a pressure of about 0.1 bar. The plasma expands supersonically into the reaction chamber, which is at a pressure of about 0.25 mbar. The flow shocks when the expanding plasma collides with the background gas and flows towards the substrate with a speed that gradually decreases to about 100 to 200 m/s at substrate level. The precursor gas, SiH$_4$, is injected into the expanding plasma through a ring shaped (80 mm diameter) gas line, which is concentric with the reaction chamber, with holes (1 mm diameter) pointing to the axis of the reaction chamber. The atomic hydrogen in the plasma dissociates the SiH$_4$ molecules. The resulting radicals deposit on the substrate, which is positioned 410 cm below the plasma source exit.

### Film series
Two series of four films have been deposited using a hydrogen gas flow of 2000 sccm to feed the cascaded arc plasma source. The SiH$_4$ flow is varied to cover the transition from amorphous to microcrystalline material. The substrate temperature is set at 300°C. The layers are deposited simultaneously on Corning 1737 glass substrates and n-type crystalline Si wafers. The thickness of the films is approximately 700 nm. During the deposition of the first film series the SiH$_4$ injection ring was positioned close to the plasma source exit at 365 mm above the substrate, during the second the injection ring was positioned much closer to the substrate at 55 mm. Films of 5 μm thick have been deposited on high purity 10 μm thick Al foils using identical conditions as described above for the SAXS analyses.

### Film analysis
The material structure is analysed using various techniques. The crystalline fraction is obtained from Raman spectra using the subtraction method described in Ref. 7. Furthermore, the crystalline fraction is extracted from X-ray diffraction measurements following the method described in [8] and the crystallite size is determined using the Scherrer formula. Small angle X-ray scattering (SAXS) expose material inhomogeneities such as voids [9] and flotation density measurements reveal the mass density of the films [10]. Furthermore, the films are analysed refractive index at 2 eV. Dark- and photoconductivity (Oriel AM1.5 solar simulator)

measurements are carried out using co-planar electrode geometry. The films were additionally analysed with FTIR to obtain the hydrogen content. One of the films is analysed using HRTEM.

## RESULTS AND DISCUSSION

### Crystallinity

From the XRD analyses in Fig. 1 we can see the transition from microcrystalline to amorphous material around 15 and 22 sccm $SiH_4$ for the silane injection position of 365 mm and 55 mm above the substrate, respectively. The crystalline fractions can be determined from these patterns [8] and from Raman spectra (not shown) [7] and are listed in table I together with the other material properties. The crystalline fractions from XRD and Raman show a good agreement (within 10%). When comparing these results it should be realised that different substrates are used, which could lead to a different start of the growth. We assume that for the sample thickness used in these experiments this effect can be neglected. It can be noticed that the transition from μc-Si to a-Si occurs at higher silane flows when the silane is injected closer to the substrate. Apparently, when the $SiH_4$ is injected in the expanding hydrogen plasma close to the substrate there is a relatively short time for the $SiH_4$ to react with the atomic hydrogen before the substrate is reached. Therefore, less atomic hydrogen is consumed and, consequently, a higher atomic hydrogen flux reaches the substrate. A higher $SiH_4$ flow is then required to obtain amorphous material. The average crystallite size from the XRD analyses varies depending on which diffraction peak is used for the calculation. No clear trend is visible and the sizes range from approximately 5 to 20 nm. For a film deposited using the conditions of sample #3 a HRTEM analysis was carried out. From this image we estimated the average particle size to be 9.5 ± 4.5 nm.

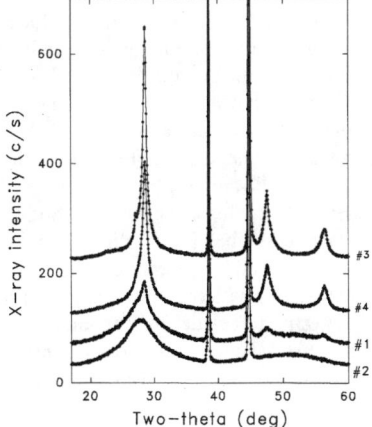
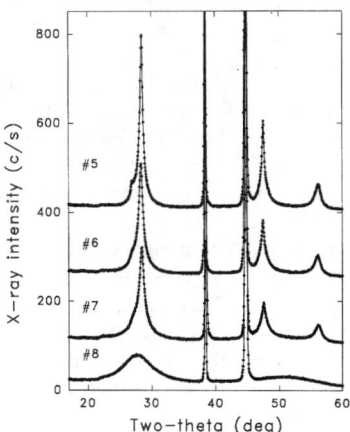

**Figure 1.** XRD analysis of 5-μm thick silicon films deposited using **(a)** (from top to bottom) 10, 14, 16, and 18 sccm $SiH_4$, respectively, injected into the reaction chamber 365 mm above the substrate and **(b)** using (from top to bottom) 14, 20, 22, and 28 sccm $SiH_4$, respectively, injected into the reaction chamber 55 mm above the substrate. The high peaks around 38° and 44° originate from the Al substrate, the sharp features around 28°, 48°, and 56° are signatures of c-Si oriented with the (111), (220), and (311) planes parallel to the film surface respectively. The broad peaks in the lowest patterns are typical for a-Si:H [8].

**Table I**. Experimental results on the material structure. Q is $SiH_4$ flow, $f_c$ and $f_v$ are the crystalline and void fractions, respectively, $\rho$ the mass density, 'flot.' indicates that this is determined by flotation-density measurements, $<D>$ is the average void size, n the refractive index, L the average crystallite size determined by applying the Scherrer formula to the (111), (220), and (311) peak, respectively. Because a second peak was needed to fit the 220 peaks correctly for the lower sample series, two crystallite sizes are presented.

| nr | Q ($SiH_4$) (sccm) | $f_c$ XRD | $f_c$ Raman | $\rho$ (g/cm$^3$) flot. | $f_v$ flot. | $f_v$ SAXS | $<D>$ (nm) | n at 2eV | $c_H$ (at%) | L (111) | L (220) | L (311) | rate (nm/s) |
|---|---|---|---|---|---|---|---|---|---|---|---|---|---|
| Injection @ 365 mm above substrate ||||||||||||||
| #3 | 10 | 0.58 | 0.57 | 2.19 | 0.060 | 0.084 | 7.6 | 2.97 | 3.1 | 17.2 | 6.4 | 7.0 | 0.91 |
| #4 | 14 | 0.50 | 0.46 | 2.20 | 0.056 | 0.071 | 8.1 | 3.10 | 5.9 | 10.1 | 5.5 | 7.0 | 0.99 |
| #1 | 16 | 0.12 | 0.11 | 2.17 | 0.069 | 0.052 | 8.5 | 3.22 | 8.5 | 6.6 | 6.9 | 7.2 | 1.09 |
| #2 | 18 | 0.00 | 0.00 | 2.17 | 0.069 | 0.063 | 8.0 | 3.38 | 7.8 | | | | 1.18 |
| Injection @ 55 mm above substrate ||||||||||||||
| #5 | 14 | 0.60 | 0.67 | 2.23 | 0.043 | 0.093 | 7.7 | 3.12 | 3.7 | 14.6 | 25/4.6 | 8.0 | 0.94 |
| #6 | 20 | 0.60 | 0.59 | 2.22 | 0.047 | 0.076 | 7.6 | 3.30 | 4.1 | 12.3 | 20/4.7 | 7.5 | 1.18 |
| #7 | 22 | 0.42 | 0.47 | 2.21 | 0.052 | 0.105 | 8.0 | 3.30 | 4.0 | 12.1 | 11/3.8 | 6.8 | 1.29 |
| #8 | 28 | 0.00 | 0.00 | 2.18 | 0.064 | 0.062 | 7.8 | 3.65 | 7.6 | | | | 1.52 |

## Film density and voids

The results of the flotation density measurements are shown in table I. The films that are deposited with the $SiH_4$ injected near the substrate generally have a slightly higher mass density, indicating a lower void fraction. The void fractions are calculated from the mass density deficits compared to c-Si (2.33 g/cm$^3$). The SAXS results, however, show no difference for the two injection positions, as shown in figure 2. An important single quantity extracted from the SAXS data is the integrated intensity [4], which is a good measure of the total inhomogeneity in each sample. The total scattering intensity is high compared to the device quality a-Si:H film (figure 2(b)). Assuming spherical scattering objects the size distribution can be fitted to the data. The average object sizes obtained from the non-tilted samples are shown as $<D>$ in table I. The significant drop in intensity for the tilted samples shows that oriented scattering objects that are elongated and aligned with the growth direction are found in all samples. This is confirmed by the TEM bright-field image of sample #3 in figure 3. Because of this anisotropic scattering a model based on ellipsoidal scattering objects is used to calculate the void fraction, assuming all scattering intensity is due to voids. Therefore, these values should be interpreted as an upper limit on the void fractions. This could explain the disagreement with the void fractions from the flotation density measurements. However, since the scattering intensities are the same for the amorphous and the microcrystalline samples we assume that the crystallites do not contribute significantly to the SAXS signal. Details on the interpretation of SAXS measurements are described in Ref. 9.

## Influence of precursor gas injection position

The change of the precursor gas injection position from 365 to 55 mm above the substrate leads to an increase in refractive index for films with comparable crystalline fractions (table I). This suggests that the material is denser as the injection position is lowered, which is corroborated by the flotation density measurements. Indeed a drop in void fraction of ~ 25% is observed for

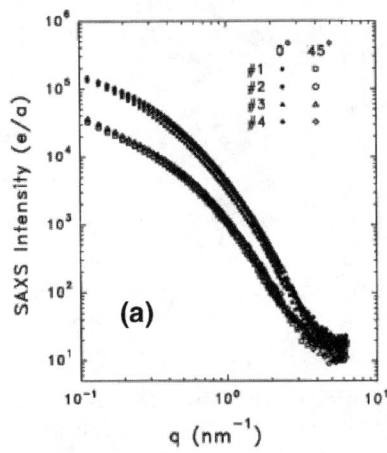 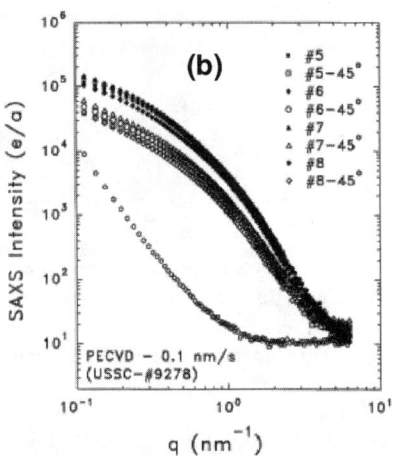

**Figure 2.** SAXS measurement of 5-μm thick silicon films deposited using (a) 10, 14, 16, and 18 sccm SiH$_4$ (nr. 3, 4, 1, and 2, respectively) injected into the reaction chamber at 365 mm above the substrate and (b) 14, 20, 22, and 28 sccm SiH$_4$ (nr. 5, 6, 7, and 8 respectively) injected at 55 mm above the substrate. Open symbols represent measurements with the sample tilted 45°. The momentum transfer q = (4π/λ)sinθ, where 2θ is the scattering angle and λ is the wavelength (0.154 nm). The unit for the scattering intensity is electrons/atom (e/a). For comparison in (b) the SAXS signal from a device quality a-Si:H sample from USSC is shown.

material deposited with the injection position near the substrate. However, the SAXS measurements indicate an equal void density in case all the scattering intensity is attributed to voids, as discussed above. Solar cell results also indicate that the change in injection position leads to a change in material quality. We prepared solar cells using the conditions of sample #3 and #6 (see table I) for the deposition of the intrinsic layer in pin cells. The results are shown in table II. The solar cell deposited with the SiH$_4$ injected close to the substrate performs much better, though there is still room for further improvement. Also, many cells are shunted when using a high injection position, which is not the case if the SiH$_4$ is injected near the substrate.

## CONCLUSIONS

The μc-Si films that are deposited using ETP CVD contain a void fraction of about 5% as determined using flotation density measurements. SAXS and TEM analyses show void shapes that are elongated in the growth direction. Optimisation of the position where the precursor gas

**Table II.** Solar cells consisting of Asahi U-type TCO coated glass, a 30-nm μc-Si p film, a 1000-nm μc-Si i film, a 30-nm μc-Si n film, and an Ag/Al back contact. The intrinsic film is deposited using the conditions of samples #3 and #6 for the high and low injection position, respectively.

| SiH$_4$ injection | $V_{oc}$ (V) | $J_{sc}$ (A/m$^2$) | FF | Efficiency |
|---|---|---|---|---|
| 365 mm | 0.22 | 10.6 | 0.33 | 0.10 |
| 55 mm | 0.38 | 75.0 | 0.44 | 1.24 |

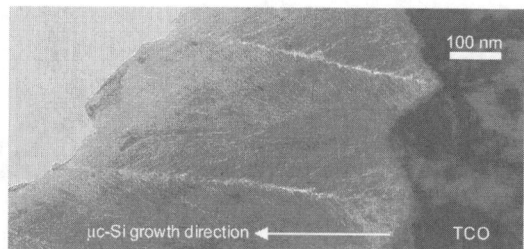

Figure 3. TEM bright field image of a solar cell containing an intrinsic µc-Si layer deposited at 300°C with the conditions of sample #3 (table I). The dark layer on the right is the TCO. The elongated, crack-like voids are typical, and a sample deposited at 250°C shows much wider voids. This suggests that the lighter contrast lines are not due to sample preparation.

$SiH_4$ is injected in the expanding hydrogen plasma improves the density of the films. Apparently the plasma chemistry changes when the injection position is varied; this needs further investigation. With this improvement the first working ETP CVD microcrystalline solar cells have been prepared with a conversion efficiency of 1.24%.

## ACKNOWLEDGEMENTS

The work at the Colorado School of Mines was supported by the National Renewable Energy Laboratory. Ries van de Sande and Jo Jansen (Eindhoven University of Technology) are acknowledged for the design and construction of the deposition system. Martijn Tijssen and Arjan Driessen (Delft University of Technology) are acknowledged for their skilful technical assistance. We thank Frans Tichelaar (Delft University of Technology) for the HRTEM analysis. This research was financially supported by Novem.

## REFERENCES

1. L. Feitknecht, O. Kluth, Y. Ziegler, X. Niquille, P. Torres, J. Meier, N. Wyrsch, and A. Shah, *Solar Energy Materials & Solar Cells* **66**, 397 (2001).
2. R.E.I. Schropp, *Thin Solid Films* **403-404,** 17 (2002).
3. B. Rech, T. Roschek, J. Müller, S. Wieder, and H. Wagner, *Solar Energy Materials & Solar Cells* **66**, 267 (2001).
4. B.A. Korevaar, A.M.H.N. Petit, C. Smit, R.A.C.M.M. van Swaaij, and M.C.M. van de Sanden, Proc. 29th IEEE Photovoltaic Specialists Conference, 2002.
5. E.A.G. Hamers, A.H.M. Smets, C. Smit, J.P.M. Hoefnagels, W.M.M. Kessels and M.C.M. van de Sanden, *Mat. Res. Soc. Proc.* **664**, A4.2.1 (2001).
6. B.A. Korevaar, C. Smit, R.A.C.M.M. van Swaaij, A.H.M. Smets, W.M.M. Kessels, J.W. Metselaar, D.C. Schram, and M.C.M. van de Sanden, Proc. of 16th European Photovoltaic Solar Energy Conference, 2000, B119.
7. C. Smit, R.A.C.M.M. van Swaaij, H. Donker, A.M.H.N. Petit, W.M.M. Kessels, and M.C.M. van de Sanden, accepted for publication in *J. Appl. Phys.*
8. D.L. Williamson, *Mat. Res. Soc. Symp. Proc.* **557**, 251 (1999).
9. D.L. Williamson, *Mat. Res. Soc. Symp. Proc.* **377**, 251 (1995).
10. D.L. Williamson, *Solar Energy Materials and Solar Cells* (in press 2003).

# Microcrystalline Silicon Thin Film Growth by Electron Cyclotron Resonance Chemical Vapour Deposition at 80°C for Plastic Application

Ian Y.Y. Bu, A.J. Flewitt, J Robertson and W.I. Milne

Cambridge University Engineering Department, Trumpington Street, Cambridge, XB2 1PZ, UK

## ABSTRACT

Microcrystalline silicon deposited at low temperatures (<150°C) is a candidate material for use as the channel layer in thin film transistors deposited on plastic substrates. This would enable driver electronics to be integrated onto cheap flexible AMLCD panel.

In this study microcrystalline silicon was deposited by Electron Cyclotron Resonance Plasma Enhanced chemical Vapour Deposition (ECR-PECVD) at a temperature of 80°C, compatible with most plastic such as PET and PEN. A source gas mixture of $SiH_4$ and $H_2$ was employed. The structural and optical properties of samples deposited under a range of deposition conditions were measured.

## INTRODUCTION

Currently thin film transistors (TFTs) fabricated for use in active matrix liquid crystal displays (AMLCDs) hydrogenated amorphous silicon (a-Si:H) channel owing to the large area capability of this material. Future displays will however demand a material with a higher field effect mobility than a-Si:H, and while polycrystalline silicon (poly-Si) can meet these requirements it does so at the expense of large area or low temperature substrate compatibility [1]. Recently TFTs on lightweight and robust plastic substrates are in high demand for small and medium size displays for mobile electronic applications such as mobile phones and personal digital assistants. The fabrication of TFTs on plastic demands a very low maximum process temperature < 150°C. [2-4]

Low temperature microcrystalline silicon (μc-Si:H) is a biphasic material consisting of crystalline regions in an amorphous matrix and maybe deposited over a large area, while providing an improvement in field effect mobility. μc-Si:H is conventionally deposited by rf-Plasma Enhanced Chemical Vapour Deposition (rf-PECVD) using $SiH_4$ and $H_2$ gas mixture. Recently, the use of Electron Cyclotron Resonace (ECR) PECVD for the formation of μc-Si:H thin films has attracted much attention since a high rate deposition is possible at low substrate temperatures.[5]

The ECR plasma operates at low pressure (1-10 mTorr). Therefore, neutral species, such as $SiH_3$ radicals, are able to reach the surface of the substrate without colliding with the other species. Consequently, the formation of undesirable Si configuration is suppressed, which results in better films, [5]-[6] Silicon growth at 80°C has been examined here. Structural and optical properties of the films have been studied with respect to pressure and microwave power. Raman spectroscopy, SEM and TEM was used to monitor the structure of the film. The aim of this study was to demonstrate the possibility of deposition of high quality µc-Si:H at 80°C.

## EXPERIMENTAL DETAILS

µc-Si:H films were deposited by ECR-PECVD using the reactor shown schematically in AsTEX PlasmaQuest series III

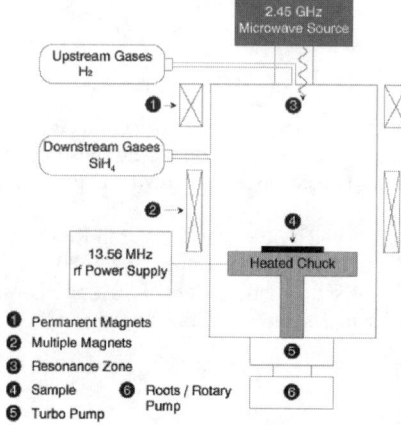

**Figure 1.** A Schematic diagram of the ECR-PECVD reactor

The plasma was generated by a microwave power supply operating at 2.45GHz. The typical experiment condition were as follows: a microwave power of 1200 W, a total pressure, adjusted by valve, of 1 mTorr and a gas flow rate $SiH_4/H_2 = 5/200$ sccm, the substrate temperature was kept at 80°C.

Samples were deposited at low doped p-type, double side polished Si (111) and corning 7059 glass substrates. The substrate were cleaned by ultrasound in consecutive baths of acetone, isopropanol and deionised water. In the case of the silicon substrate, this was followed by a dip in hydrofluoric acid to remove native oxide.

The crystallinity of the µc-Si:H films was measured by Raman spectroscopy at room temperature using a Renisaw RM 1000.

Fourier Transform Infra-red (FTIR) were made using an ATI matteson FTIR spectrometer over a scanning range of 400-4000 cm$^{-1}$.

Scanning electron microscopy (SEM) studies were carried out with a HITACHI S-5000 in-lens cold field emission SEM

## RESULTS AND DISCUSSION

The planar SEM picture, figure 2a) and 2b) shows a typical µc-Si:H sample produced by ECR-PEVCD consisting of crystallites in an amorphous matrix. From figure 2a) the grain size is observed in the range of 10-50 nm

**Figure 2 a)** Top-view SEM of a typical µc-Si:H deposited by ECR-PECVD

**Figure 2 b)** Cross-view SEM of a typical µc-Si:H deposited by ECR-PECVD

Figure 2 b) shows the cross-sectional SEM image of µc-Si:H film with thickness of approximately 300nm, deposited at a rate of 10nm/min. Such a deposition rate is comparable with that of a-Si:H deposited by rf-PECVD.

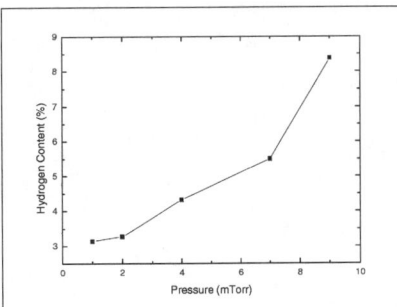 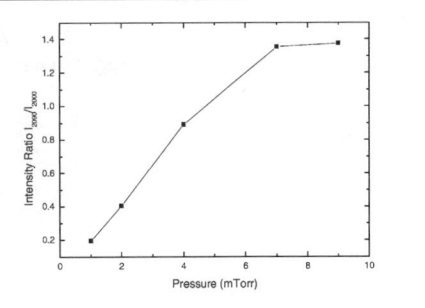

**Figure 3 a)** Hydrogen content as a variation with pressure estimated from FTIR

**Figure 3 b)** The bonding intensity of the peaks $I_{2000}$ & $I_{2090}$ estimated from FTIR

The hydrogen content was estimated from FTIR peak after the thickness was measured with profilometer. The hydrogen content is shown in figure 3 a) as a function of pressure.

The intensities $I_{2000}$ and $I_{2090}$ of SiH and $SiH_2$ are shown on figure 3 b). A low value of 2090 $cm^{-1}$ to 2000 $cm^{-1}$ ratio is often taken as an indication of high quality in the H bonding in a-Si:H. The films becomes more disordered with increasing pressure, may due to higher deposition rate, it is more likely to incorporate cluster voids in the film and disorder increases. The higher hydrogen content with disorder would also suggest void incorporation.

Raman spectroscopy was used to analyse the crystalline volume fraction of the µc-Si:H sample deposited on corning glass. There are usually two Raman peaks appear in the hydrogenated silicon in the range of 400-600$cm^{-1}$, the amorphous silicon peak at 480$cm^{-1}$ and the crystalline silicon peak at 520$cm^{-1}$. If the integrated intensity of the amorphous peak is $I_A$, and that of crystalline peak $I_C$ then the crystalline volume fraction $X_C$ maybe estimated by:

$$X_c = \frac{I_C}{I_C + I_A} \quad (1)$$

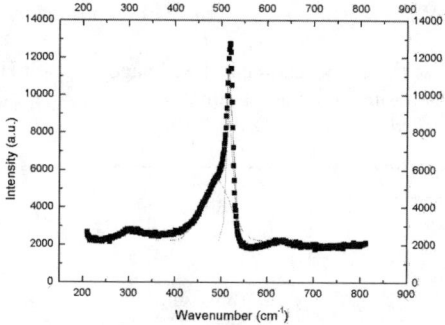

**Figure 4** Raman spectra of a typical film deposited at 80°C

From the samples evaluated during the optimisation of µc-Si:H, figure 5, only samples deposited at low pressure and microwave power of 1200 W exhibit crystallinity. Figure 4 shows the Raman scattering spectra of the µc-Si:H with crystalline volume fraction of 77 %. This correlates with the theory, ECR generate neutral radicals such as $SiH_3$ or H atoms are able to reach on the surface of the substrate without colliding so many times with the other speices. As the result, the formation of higher radicals, which deteriorate the film quality is suppressed. [5]

## CONCLUSION

The low temperature μc-Si:H film deposition by ECR-CVD with hydrogen dilution is very successful. The largest grain of μc-Si:H film occur at low pressure and high microwave power. Raman spectra of the observed sample shows good crystalline volume fraction at 77 %, due to the clean long lived and dense plasma produced by ECR-PECVD at low pressure.

## ACKNOWLEDGEMENT

The author would like to thank C. Casiraghi for her help in Raman spectra measurements. The author would also like to thank professor Oda and Reki Nakamura (Tokyo institute of technology) for the SEM measurements. This work was supported by EPSRC and Philips research laboratories

## REFERENCES

1. Yoon, S.Y., Park, S.J., Kim, K.H., and Jang, J. Thin Solid Films **383**, 34-38 (2001).

2. Choi, J.B., Yun, D.C., Park, Y.I., and Kim, J.H. Journal of Non-Crystalline Solids **266**, 1315-1319 (2000).

3. Gleskova, H., Wagner, S., Gasparik, V., and Kovac, P. Journal of the Electrochemical Society **148**, G370-G374 (2001).

4. Gosain, D.P., Noguchi, T., and Usui, S. Japanese Journal of Applied Physics Part 2-Letters **39**, L179-L181 (2000).

5. Murata, K., Kikukawa, D., Hori, M., Goto, T., and Ito, M. Journal of Vacuum Science &Technology a-Vacuum Surfaces and Films **20**, 953-956 (2002).

6. Birkholz, M., Conrad, E., and Fuhs, W. Japanese Journal of Applied Physics Part 1-Regular Papers Short Notes & Review Papers **40**, 4176-4180 (2001).

## Evolution of Crystallinity in Mixed-Phase (a+µc)-Si:H as Determined by Real Time Spectroscopic Ellipsometry

A. S. Ferlauto, G. M. Ferreira, R. J. Koval, J. M. Pearce, C. R. Wronski, R. W. Collins,
M. M. Al-Jassim[1], and K. M. Jones[1]
Department of Physics, Materials Research Institute, and Center for Thin Film Devices,
The Pennsylvania State University, University Park, PA 16802
[1]National Renewable Energy Laboratory, 1617 Cole Boulevard, Golden, CO 80401

ABSTRACT

The ability to characterize the phase of the intrinsic (i) layers incorporated into amorphous silicon [a-Si:H] and microcrystalline silicon [µc-Si:H] thin film solar cells is critically important for cell optimization. In our research, a new method has been developed to extract the thickness evolution of the µc-Si:H volume fraction in mixed phase amorphous + microcrystalline silicon [(a+µc)-Si:H] i-layers. This method is based on real time spectroscopic ellipsometry measurements performed during plasma-enhanced chemical vapor deposition of the films. In the analysis, the thickness at which crystallites first nucleate from the a-Si:H phase can be estimated, as well as the nucleation density and microcrystallite cone angle. The results correlate well with structural and solar cell measurements.

## INTRODUCTION

In previous research, extended phase diagrams have been developed that describe the accumulated thicknesses at which different microstructural and phase transitions are observed during plasma-enhanced chemical vapor deposition (PECVD) of hydrogenated silicon (Si:H) thin films [1-3]. In such diagrams, the transition thicknesses are plotted as continuous functions of the $H_2$-dilution gas flow ratio $R=[H_2]/[SiH_4]$. This is the key PECVD parameter that controls the phase of the film -- from amorphous silicon (a-Si:H) at low R to microcrystalline silicon (µc-Si:H) at high R. Thus, the phase diagram can be used to identify the regimes of film thickness and R within which so-called *protocrystalline* Si:H is obtained.

Protocrystalline Si:H is a higher stability form of a-Si:H optimized versus R for electronic devices [1]. It is deposited using the maximum possible R value that can be sustained without crossing the amorphous-to-(mixed-phase microcrystalline) [a→(a+µc)] transition boundary of the phase diagram for the desired bulk layer thickness $d_b$. Because this transition boundary exhibits a negative slope as a function of R in the R-$d_b$ plane, microcrystalline nuclei develop within the protocrystalline Si:H films after a critical phase-transition thickness. As a result, the growing film evolves with accumulated thickness from protocrystalline Si:H to mixed-phase Si:H [(a+µc)-Si:H], and eventually to single-phase µc-Si:H. Owing to the importance of Si:H deposition near the a→(a+µc) and (a+µc)→µc boundaries of the phase diagram for the optimization of the intrinsic (i) layers in a-Si:H and µc-Si:H solar cells [1-7], techniques are needed to determine the evolution of the crystalline Si:H content as a continuous function of the accumulated thickness. In this article, the application of real time spectroscopic ellipsometry (RTSE) is described for this purpose.

## EXPERIMENTAL DETAILS

The Si:H films in this investigation were prepared by rf PECVD for use as the i-layers of a-Si:H p-i-n solar cells. The deposition parameter of greatest interest is the $H_2$-dilution ratio R which was varied from R=0 to 40. The Si:H films studied by RTSE were deposited in a uhv-compatible, single-chamber system; the solar cells were fabricated in a load-locked multichamber system. The phase diagrams for multichamber Si:H deposition, as deduced from ex situ SE and device studies, compare well with those for single-chamber deposition, as deduced in real time SE studies [8]. Specular $SnO_2$-coated glass served as the substrate for p-i-n cells, and the p-layers were prepared without $H_2$-dilution from $[SiH_4]+[CH_4]+[B(CH_3)_3]$.

RTSE was performed using a rotating-compensator multichannel ellipsometer with a spectral range from 1.5 to 4.5 eV and a minimum acquisition time for full spectra in ($\psi$, $\Delta$) of ~50 ms [9]. Crystalline Si (c-Si) or R=0 a-Si:H substrates were used for RTSE studies of the evolution of microcrystallinity. The smooth surfaces of the c-Si substrates provide the highest sensitivity to the microstructural and phase transitions, whereas the R=0 a-Si:H substrates serve to simulate i-layer deposition on p-layers for p-i-n cells.

## RESULTS AND DISCUSSION: RTSE AND MICROSTRUCTURAL STUDIES

The conventional two-layer [(uniform bulk)/(surface roughness)] optical model used in RTSE data analysis is no longer correct once the film traverses the phase boundary from the single-phase amorphous (protocrystalline) Si:H growth regime into the mixed-phase (a+μc)-Si:H regime. In order to extract the volume fraction of μc-Si:H in the graded mixed-phase regime, it is easiest to track the dielectric function of the top-most growing material throughout the film deposition. This is an ideal application for the virtual interface (VI) approach to RTSE data analysis [10]. Figures 1-3 depict the final results of such an analysis.

Figure 1 compares the (a) real and (b) imaginary parts of the dielectric functions ($\varepsilon_1$, $\varepsilon_2$) for the pure phases obtained in analyses of RTSE data acquired during the deposition of an R=20 Si:H film on a c-Si substrate. This film evolves from single-phase a-Si:H to mixed-phase (a+μc)-Si:H and then to single-phase μc-Si:H throughout growth to 2300 Å in thickness. The dielectric function of the pure a-Si:H phase in Fig. 1 was determined by applying the conventional two-layer (uniform bulk)/(roughness) model to analyze the RTSE data collected in the protocrystalline Si:H growth regime $0<d_b<200$ Å, before the onset of the a→(a+μc) transition. The dielectric function of the pure μc-Si:H phase in Fig. 1 was obtained from data collected after crystallite contact, i.e., during coalescence ($d_b>1900$ Å). In this latter case, only the top ~200 Å of the film was analyzed, also by applying a two-layer growth model (bulk)/(roughness). In this case, however, the complicated history of the underlying graded layer is buried within a fictitious uniform substrate [or pseudo-substrate] located beneath a fictitious interface [or virtual interface]. The correct surface roughness layer thicknesses -- required in order to extract the ($\varepsilon_1$, $\varepsilon_2$) spectra of Fig. 1 -- were determined by applying a global error minimization routine within which the entire VI analysis was embedded. The differences between the two dielectric functions in Fig. 1 support an interpretation in terms of phase evolution during film growth. The solid lines in Fig. 1 represent best fit analytical models for the two dielectric functions [11].

Figure 2 shows the evolution of the volume fraction of μc-Si:H in the R=20 film of Fig. 1 throughout the growth regime starting from the a→(a+μc) transition and ending near the (a+μc)→μc transition. In the analysis that led to these results, a two-layer VI model similar to that described in the previous paragraph was applied. This model consists of a pseudo-substrate, incorporating the history of graded layer deposition, a ~30 Å thick outerlayer, modeled as a

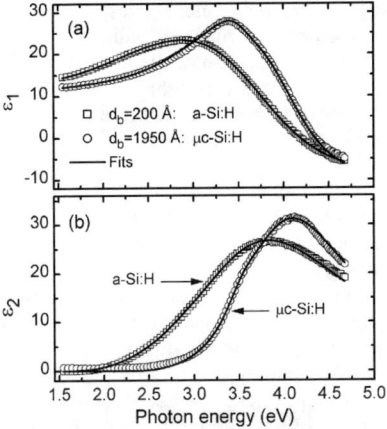

**Figure 1.** (a) Real and (b) imaginary parts of the dielectric functions of the pure phases for an R=20 Si:H deposition on c-Si. The solid lines are fits using analytical models for the dielectric functions [11].

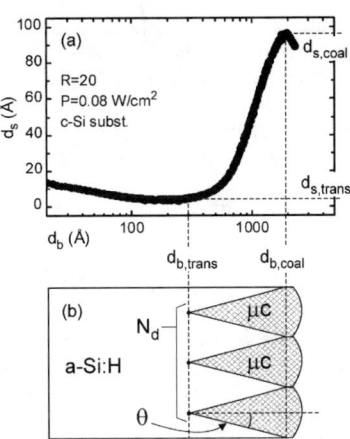

**Figure 3.** (a) Surface roughness thickness versus bulk layer thickness for the R=20 Si:H deposition of Figs. 1 and 2; (b) schematic of the cone growth model used to estimate the microcrystallite nuclei density and cone angle.

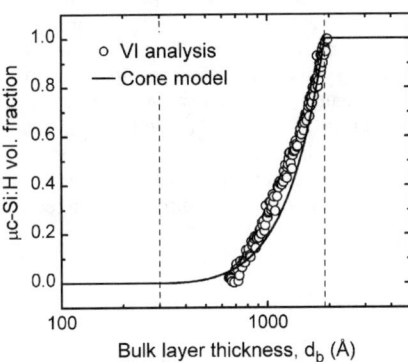

**Figure 2.** Depth profile in the volume fraction of the microcrystalline phase throughout the mixed-phase (a+μc)-Si:H growth regime for the R=20 Si:H deposition on c-Si from Fig. 1 (points). In order to extract the volume fraction of μc-Si:H, the dielectric function of the top-most growing material was tracked throughout the film deposition. Also shown as the solid line is the prediction of the microcrystallite cone growth model depicted in Figure 3.

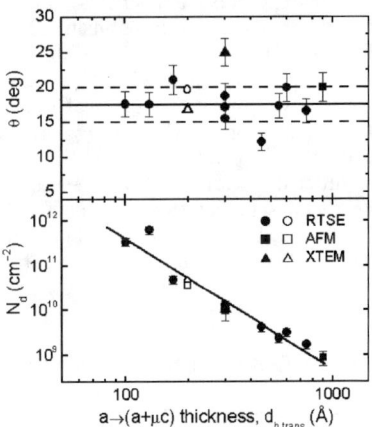

**Figure 4.** Cone angle θ (top panel) and nucleation density $N_d$ (bottom panel) versus the a→(a+μc) transition thickness $d_{b,trans}$ for Si:H films deposited under different conditions on both c-Si wafers and R=0 a-Si:H. The values for θ and $N_d$ are deduced from RTSE (circles), XTEM (triangles) and AFM (squares). Data are included not only from this study (closed symbols), but also from the study of Fujiwara et al. (open symbols) [12].

mixture of a-Si:H and μc-Si:H of variable volume fractions $1-f_{\mu c}$ and $f_{\mu c}$, respectively, and a surface roughness layer, modeled as a mixture of the outerlayer material and void with fixed volume fractions of $1-f_{sv}$ and $f_{sv}$, respectively. The dielectric function of the pseudo-substrate is determined as its pseudo-dielectric function (corrected for surface roughness, but not for the graded structure). The dielectric functions of the two layers are determined from their component volume fractions and dielectric functions via the Bruggeman effective medium theory.

Evidently, the evolution of the surface roughness layer thickness $d_s$ shown in Fig. 3(a) for the deposition of Figs. 1 and 2 provides higher sensitivity to the presence of microcrystallinity, since it reveals an a→(a+μc) transition at ~300 Å, whereas the μc-Si:H volume fraction in Fig. 2 first extends above zero for $d_b$~700 Å. The solid line in Fig. 2 is the result for $f_{\mu c}$ versus $d_b$ established using a cone model of microcrystallite evolution. Figure 3(b) identifies how this microcrystallite cone model is constructed. In this model, it is assumed that all microcrystalline nuclei originate at the a→(a+μc) transition layer thickness [~300 Å, from Fig. 3(a)]. The area density of such nuclei is assumed to be $N_d$, and the nuclei are assumed to grow preferentially at the expense of the surrounding a-Si:H phase with a constant, thickness-independent cone angle, θ. The cones are assumed to be spherically capped whereby the cap radius r evolves with bulk layer thickness according to $r = d_b - d_{b,trans}$. Applying this geometry, $N_d$ and θ can be deduced from the values of $\Delta d_b = d_{b,coal} - d_{b,trans}$ and $\Delta d_s = d_{s,coal} - d_{s,trans}$. Here, $d_{b,trans}$ and $d_{s,trans}$ are the bulk and surface roughness layer thicknesses at the a→(a+μc) transition and $d_{b,coal}$ and $d_{s,coal}$ are the corresponding values at the (a+μc)→μc transition. For example, for the deposition of Figs. 1-3, values of $\theta=19°$ and $N_d=1.1 \times 10^{10}$ cm$^{-2}$ are determined.

Figure 4 presents results for the cone angle θ and the nucleation density $N_d$ plotted as a function of the a→(a+μc) transition thickness for a series of Si:H films prepared on both c-Si and R=0 a-Si:H substrates under different conditions of H$_2$-dilution, plasma power, and substrate temperature. Results deduced solely from RTSE using the approach of Figs. 1-3 are compared with those from cross-sectional transmission electron microscopy (XTEM) and from atomic force microscopy (AFM). In Fig. 4, the solid and open symbols represent results from this work and from Fujiwara et al. [12], respectively. Evidently the nucleation density decreases significantly with increasing a→(a+μc) transition thickness, yet the crystallite cone angle is nearly constant between 15° and 20°. Figure 4 reveals consistency between the indirect (but real time) optical measurements and the direct (but ex situ) structural measurements, providing strong support for the generality of the cone growth model for microcrystallinity as depicted in Fig. 3(b).

## RESULTS AND DISCUSSION: DEVICE COMPARISONS

Although Figs. 1-3 were obtained for a Si:H film deposited on c-Si, similar analyses can be performed for films deposited on R=0 a-Si:H (see Fig. 4). Then comparisons with Si:H solar cell performance are possible as long as the doped layers in the cells are amorphous and prepared near R=0. To demonstrate this capability, Fig. 5 shows RTSE results for the evolution of $f_{\mu c}$ near the a→(a+μc) transition for three two-step (R=40 interface)/(R=20 bulk) i-layers on R=0 a-Si:H substrate films. These structures are distinguished by the thickness $d_i$ of the R=40 interface i-layer, selected as $d_i$=400, 150, and 0 Å, whereas the total i-layer thickness is fixed at 4000 Å. Figure 6 shows results for the dark current-voltage (J-V) characteristics of the corresponding Si:H p-i-n solar cells of this series, i.e., cells having two-step (R=40 interface)/(R=20 bulk) i-layers. For the cell with the highest currents in Fig. 6 (inverted triangles, with $d_i$=400 Å), the associated two-step i-layer structure measured by RTSE exhibits the a→(a+μc) transition after

200 Å of R=40 deposition and the (a+μc)→μc transition after 950 Å of total i-layer deposition (see Fig. 5). For comparison purposes, the cell with the lowest currents (squares) was fabricated using an optimum two-step (R=40 interface)/(R=10 bulk) a-Si:H i-layer (hereafter designated the "optimum cell").

Computer simulations of the J-V characteristics for these cells were also carried out, and the results are shown in Fig. 6 as the solid lines. The optimum cell could be considered in terms of a uniform a-Si:H i-layer with a mobility gap of 1.86 eV. Results for the other three cells on the other hand were modeled assuming an a-Si:H i-layer at the p/i interface with the same mobility gap as the i-layer of the optimum cell and a bulk μc-Si:H i-layer with a mobility gap of 1.15 eV. This simple model for the solar cells reproduces the overall features of the J-V curves. Table I includes a comparison of the RTSE and J-V simulation results for the three Si:H p-i-n solar cells with two-step (R=40)/(R=20) 4000 Å i-layers of Figs. 5 and 6. The table compares the thickness of the a-Si:H i-layer component nearest the p/i interface, as deduced from the device modeling, with the i-layer thickness required to reach a microcrystalline volume fraction of $f_{\mu c}$ = 0.2, as deduced from RTSE. This value was chosen because it gave the best overall agreement with the device modeling results (which exhibit quite large error bars due to correlation in the simulations). This good agreement suggests that once $f_{\mu c}$ approaches ~0.2, transport is dominated by the microcrystalline phase.

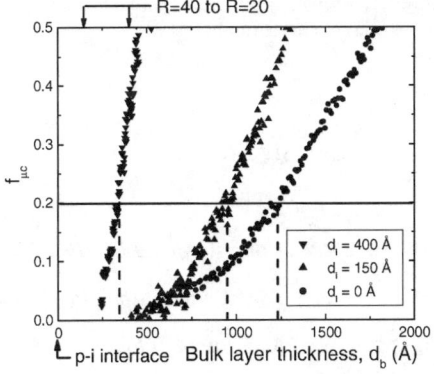

**Figure 5.** Depth profile of the microcrystalline Si:H volume fraction $f_{\mu c}$ in the i-layer for 4000 Å thick two-step R=40/20 i-layer structures on R=0 a-Si:H substrates. The thicknesses $d_i$ of the R=40 p/i interface layer are $d_i$=400, 150, and 0 Å. These results were deduced from a two-layer virtual interface model consisting of a pseudo-substrate, a 30 Å thick outerlayer (from which $f_{\mu c}$ is extracted), and a surface roughness layer.

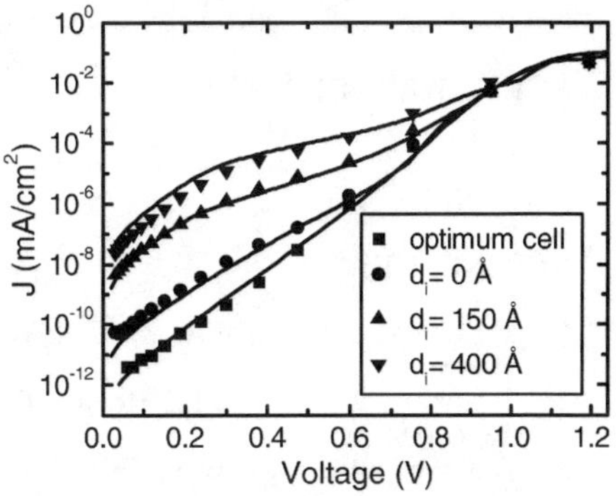

**Figure 6.** Experimental (symbols) and simulated (lines) dark J-V characteristics for four Si:H p-i-n solar cells. The squares denote a standard cell that incorporates an optimized [R=40 (200 Å)]/[R=10 (3800 Å)] a-Si:H i-layer, and the three other cells incorporate [R=40 ($d_i$)]/[R=20 (4000–$d_i$) Å] Si:H i-layers where $d_i$ = 0 Å (circles), $d_i$=150 Å (triangles), and $d_i$= 400 Å (inverted triangles).

**Table I.** Correlation of RTSE and device modeling results for p-i-n cells with 4000 Å two-step R=40/20 i-layers in which the R=40 interface thickness is varied.

| First step: R=40 thickness (Å) | RTSE: $d_b$ when $f_{\mu c}=0.2$ (Å) | Device Modeling: Heterojunction position from p/i (Å) |
|---|---|---|
| 0 | 1240 ± 30 | 1300 ± 300 |
| 150 | 950 ± 30 | 700 ± 200 |
| 400 | 330 ± 20 | 400 ± 100 |

## SUMMARY

Real time spectroscopic ellipsometry (RTSE) studies of plasma-enhanced chemical vapor deposition of Si:H films that undergo an amorphous-to-(mixed-phase amorphous + microcrystalline) transition were performed. Novel approaches for RTSE data analysis were conceived and developed that provide (i) the nucleation density and cone angle characterizing the evolution of the Si microcrystallites in the mixed-phase amorphous + microcrystalline (a+μc)-Si:H regime; and (ii) the volume fraction of the μc-Si:H phase as a function of the accumulated thickness, as well as (iii) the optical properties of the films in the amorphous, mixed-phase (a+μc)-Si:H, and single-phase μc-Si:H growth regimes. The resulting information correlates well with that from ex situ structural and device performance studies.

## ACKNOWLEDGMENTS

The authors acknowledge support of this research by NSF under Grant No. DMR-0137240 and by NREL under Subcontracts AAD-9-18-668-09 and NDJ-2-30630-01.

## REFERENCES

1. J. Koh, Y. Lee, H. Fujiwara, C.R. Wronski, and R.W. Collins, *Appl. Phys. Lett.* **73**, 1526 (1998).
2. J. Koh, A.S. Ferlauto, P.I. Rovira, C.R. Wronski, and R.W. Collins, *Appl. Phys. Lett.* **75**, 2286 (1999).
3. A.S. Ferlauto, R.J. Koval, C.R. Wronski, and R.W. Collins, *Appl. Phys. Lett.* **80**, 2666 (2002).
4. D.V. Tsu, B.S. Chao, S.R. Ovshinsky, S. Guha, and J. Yang, *Appl. Phys. Lett.* **71**, 1317 (1997).
5. J. Yang, K. Lord, S. Guha, and S.R. Ovshinsky, *Mater. Res. Soc. Symp. Proc.* **609**, A15.4 (2001).
6. C. Koch, M. Ito, V. Svrcek, M.B. Schubert, and J.H. Werner, *Mater. Res. Soc. Symp. Proc.* **609**, A15.6 (2001).
7. O. Vetterl, F. Finger, R. Carius, P. Hapke, L. Houben, O. Kluth, A. Lambertz, A. Muck, B. Rech, and H. Wagner, *Sol. Energy Mater. Sol. Cells* **62**, 97 (2000).
8. R.J. Koval, J.M. Pearce, A.S. Ferlauto, R.W. Collins, and C.R. Wronski, *Mater. Res. Soc. Symp. Proc.* **664**, A16.4 (2001).
9. R.W. Collins, J. Koh, H. Fujiwara, P.I. Rovira, A.S. Ferlauto, J.A. Zapien, C.R. Wronski, and R. Messier, *Appl. Surf. Sci.* **154**, 217 (2000).
10. H. Fujiwara, J. Koh, C.R. Wronski, and R.W. Collins, *Appl. Phys. Lett.* **70**, 2150 (1997).
11. A.S. Ferlauto, J. Koh, P.I. Rovira, C.R. Wronski, R.W. Collins, and G. Ganguly, *J. Non-Cryst. Solids* **266-269**, 269 (2000).
12. H. Fujiwara, Y. Toyoshima, M. Kondo, and A. Matsuda, *Phys. Rev. B* **60**, 13598 (1999).

# Structural Characterization of Microcrystalline Silicon Solar Cells Fabricated by Conventional RF-PECVD

Liwei Li, Yuan-Min Li[1], J. A. Anna Selvan[1], Alan E. Delahoy[1], and Roland A. Levy
Department of Physics, New Jersey Institute of Technology, Newark, NJ 07102
[1]Energy Photovoltaics, Inc., 276 Bakers Basin Road, Lawrenceville, NJ 08648

## ABSTRACT

Direct structural characterization of single junction *p-i-n* type µc-Si:H solar cells prepared in a single chamber, batch process type RF-PECVD system has been carried out using Raman scattering, XRD, and AFM. The overall degree of microcrystallinity of µc-Si:H *i*-layers is presented in terms of the ratio of peak intensities (*Ic/Ia*) of Raman shift at around 520 cm$^{-1}$ and 480 cm$^{-1}$, respectively. Strong correlations among device performance, *i*-layer structural properties, and uniformity have been established using information provided by such direct characterization. Our data support the notion that stable, high quality µc-Si *i*-layers are grown near the 'edge' of microcrystalline-to-amorphous phase transition. Solar cells made from such optimal areas exhibit moderate microcrystallinity (moderate *Ic/Ia* values). Preferential orientation corresponding to Si (220) planes was observed on those optimal solar cells, which also exhibit less-regular surface morphologies and lower surface roughness compared to that observed on solar cells with mixed-phase or highly crystalline Si:H *i*-layers.

## INTRODUCTION

Over the past few years, hydrogenated microcrystalline silicon (µc-Si:H) based materials and devices, especially µc-Si:H solar cells, have attracted extensive attention due to their good optical and transport properties, demonstrated high stability under light soaking, and successful application as the narrower-bandgap component in tandem photovoltaic devices [1-2]. Structural characterization has shown that µc-Si:H is a highly complex material which can take on a variety of microstructures and exhibit very different qualities depending on the exact deposition conditions [3-4].

Generally, stand-alone films (rather than actual devices) deposited on special substrates (e.g., Corning 7059 glass, etc.) are used to characterize properties of µc-Si:H and amorphous silicon (α-Si:H). However, unlike α-Si:H, which can be grown readily on various substrates, formation of µc-Si:H and performance of µc-Si:H based solar cells are highly affected by reactor geometries, processing parameters, as well as the nature and surface morphology of substrates [5-7]. Effect of accumulated bulk layer thickness on the nucleation and growth of µc-Si:H has also been recognized in recent studies [8-9]. Properties obtained from stand-alone films may not be necessarily translated into the µc-Si:H intrinsic layers (*i*-layers) within solar cells due to the aforementioned limitations. Therefore, this study has focused on the direct structural characterization of actual µc-Si:H *p-i-n* type single junction solar cells (rather than stand-alone films). Our study has not only revealed the actual properties of µc-Si:H *i*-layers incorporated in multilayer device structures, but also established correlations among device performance, *i*-layer structural properties, and their spatial non-uniformities. Direct structural characterization can provide valuable insight into device fabrication processes.

## EXPERIMENTAL DETAILS

Single junction $p$-$i$-$n$ type solar cells with μc-Si:H $i$-layers were deposited on commercial grade $SnO_2$/sodalime-glass superstrates at low temperatures (near 200 °C) by glow discharge of hydrogen diluted silane in a single chamber, non-loadlocked conventional RF-PECVD system which is capable of simultaneously coating 4 plates equal in size of 15"×12". For simplicity, all solar cells in this research have Al back contacts without any TCO (ZnO) as a rear reflector.

Device fabrication and performance testing, including light and dark I-V characteristics, quantum efficiency (QE), and accelerated light soaking (simulating 47 suns), were carried out at Energy Photovoltaics, Inc. (EPV). Structural characterization using Raman scattering, X-ray diffraction (XRD), and atomic force microscopy (AFM), were conducted directly on actual solar cells at New Jersey Institute of Technology. Strong red-light spectral response (QE at 800 nm or longer) and Raman shift at about 520 $cm^{-1}$ were taken as signatures of μc-Si:H. Raman scattering was performed with 830 nm diode laser excitation such that deep penetration can be achieved to probe the overall microcrystallinity of the Si:H absorber layer (response is not confined to the surface layer). In view of the unavoidable contributions from substrates, $i$-layer texture, doped layers, as well as large variations of optical absorption with $i$-layer properties, we prefer presenting the overall microcrystallinity in terms of the ratio of peak intensities ($Ic/Ia$) of Raman shift corresponding to μc-Si:H ($Ic$) and α-Si:H ($Ia$) rather than deduction of crystalline volume fraction. The latter method is more commonly adopted by other groups. A Rigaku D/MAX-B XRD system and a Nanoscope IIIa AFM system were used to study film microstructures and surface morphologies of the Si:H (including μc-Si:H) solar cells.

## RESULTS AND DISCUSSION

### Raman scattering

A wide variety of Raman scattering spectra have been observed depending on solar cell microstructures. Since both doped $n$- and $p$-layers are very thin compared to $i$-layers, Raman spectra of μc-Si:H solar cells are mainly determined by $i$-layer (absorber) microstructures. Typical Raman spectra of solar cells with different $i$-layer microstructures are shown in figure 1.

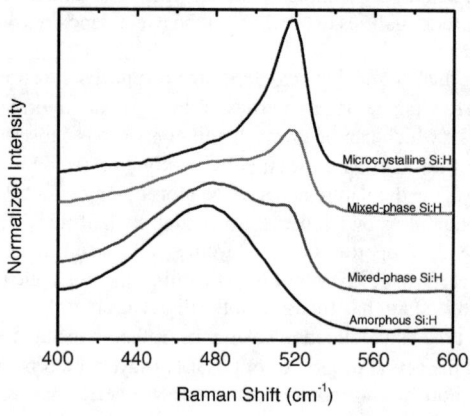

**Figure 1**. Raman spectra of solar cells with various $i$-layers

Clear correlations have been demonstrated between solar cell performance and $i$-layer microcrystallinity *(Ic/Ia)*. Generally, solar cells with mixed-phase $i$-layers exhibit higher open circuit voltage ($V_{oc}$), lower fill factor, and poorer red-light spectral response compared to solar cells with μc-Si:H $i$-layers. Such solar cells also show the worst stability under light soaking. As an example, figure 2 illustrates the correlations between device performance parameters ($V_{oc}$ and red-light response) and microcrystallinity *(Ic/Ia)* of μc-Si:H solar cells. As shown in figure 2 (a), moderate $V_{oc}$ (~500 mV) was obtained from solar cells with μc-Si:H $i$-layers showing microcrystallinity over a range with moderate *Ic/Ia* values (about 1.8~1.9). Correspondingly, highest red-light response (QE at 800 nm) was also observed over the same *Ic/Ia* range. Under deposition conditions resulting in reasonable carrier collection (good fill factor and short circuit current density), high performance and stable μc-Si:H solar cells were obtained over this 'optimal' *Ic/Ia* range. However, it is also evident from figure 2 (b) that large variations in red-light response occur when microcrystallinity is near or higher than this optimal *Ic/Ia* range. Low red-light response was also observed in solar cells with higher microcrystalline $i$-layers (poor carrier collection). Thus, the range of optimal $i$-layer microcrystallinity is quite narrow.

The uniformities of $i$-layer microstructure and device performance over the entire plates were also investigated by Raman scattering. As an example, figure 3 illustrates large variations of microcrystallinity and PV device parameters over a plate with very high non-uniformity. Such non-uniformity was caused by plasma parameters including gas flow pattern in the reactor. (Strong spatial variations of device properties can be highly instructive for research purposes.) As shown in figure 3, inferior device performance was observed in solar cells made from both mixed-phase (including the strong phase transition region near the top of the plate) and highly microcrystalline areas. Such data establish general correlations between microcrystallinity and device performance, and they support the notion that the best (and stable) μc-Si:H solar cells are obtained with $i$-layers deposited near the microcrystalline-to-amorphous (mixed-phase) transition 'edge'. Such 'edge' materials correspond to the previously mentioned 'optimal' films with moderate *Ic/Ia* ratio.

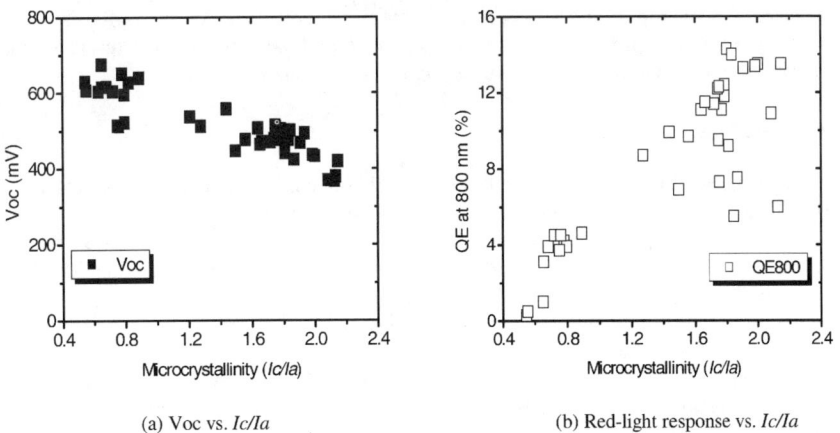

(a) Voc vs. *Ic/Ia*              (b) Red-light response vs. *Ic/Ia*

**Figure 2.** Correlations between Si:H solar cell parameters and microcrystallinity

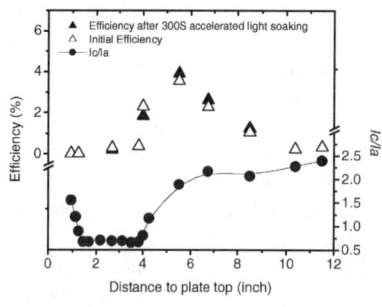
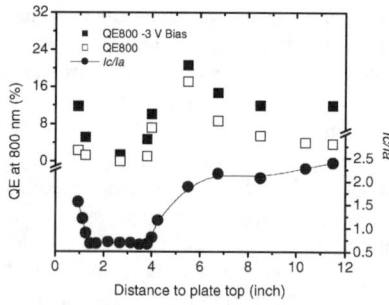

(a) Efficiency and $Ic/Ia$ vs. device position    (b) Red-light response and $Ic/Ia$ vs. device position

**Figure 3**. Variations of structural and solar cell parameters with sample position on a substrate

The reasons for the inferior performance of solar cells with more-microcrystalline $i$-layers have not been identified. As shown in figure 3 (b), at $Ic/Ia$ values higher than the optimal range, red-light response under zero bias is lower than that in the optimal microcrystallinity range, which is consistent with figure 2 (b). However, under negative electrical bias, the red-light response in higher crystalline areas can be improved by a greater percentage than that in the area with optimal $Ic/Ia$ values. Therefore, the possible explanation for the inferior performance of solar cells with higher crystalline $i$-layers relates to the suppression of photo-carrier collection, which is probably a consequence of enhanced carrier trapping and recombination at unidentified sites (by unclear mechanism). Such possibilities may include more effective contamination in the 'grains' (as opposed to 'grain boundaries', if such concepts are at all appropriate for really nano-crystalline Si films), insufficient hydrogen passivation of defects (along 'grain boundaries'), or higher density of $p/i$ interface defects resulting from energetic, highly etching hydrogen-rich plasma.

The advantages of direct characterization of real devices over characterization of stand-alone films are evident from figure 3. Direct structural measurements in device configuration are more informative and less prone to misinterpretation when multilayer device structures, precise deposition sequences, and various processing parameters are involved.

## X-ray diffraction

Typical XRD spectra of solar cells with μc-Si:H and mixed-phase Si:H $i$-layers are shown in figure 4. XRD peaks at 2θ around 28.5° and 47.4° were taken as signatures of Si (111) and Si (220) planes, respectively. For the mixed-phase Si:H sample, no detectable XRD Si peaks were observed even though the corresponding Raman spectrum showed a slight shoulder at about 520 cm$^{-1}$. The average grain sizes of μc-Si:H $i$-layers can be roughly calculated using the Scherrer formula. The calculated grain sizes for the samples shown in figure 4 with $Ic/Ia$ values of 1.88 and 2.39 are about 90 and 130 Å, respectively.

Crystallite grains and grain boundaries are generally considered to strongly affect carrier transport. It is not clear what role is played by the preferential crystalline orientation of the film. As shown in figure 4, Si (220) planes orientation is favored by the 'edge' sample with optimal microcrystallinity. While it has been shown that preferential growth strongly depends on plasma conditions, and the competition between selective etching and growth has been proposed as the growth kinetics of favorable crystal directions [10-12], detailed and precise mechanism remains elusive. More evidences are needed to firmly establish the correlation between preferential crystalline orientation and device performance of μc-Si solar cells.

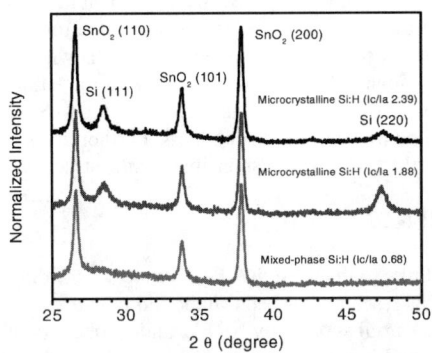

**Figure 4**. XRD spectra of solar cells with μc-Si:H and mixed-phase Si:H *i*-layers

**Atomic force microscopy**

Different surface morphologies have been observed by AFM. Figure 5 depicts typical AFM surface morphologies of solar cells with μc-Si:H and mixed-phase Si:H *i*-layers. Less-regular surface morphology and lower surface roughness were seen in solar cells with μc-Si:H *i*-layers compared to that observed in solar cells with mixed-phase Si:H *i*-layers. An increase of surface roughness (root means square, RMS) has been observed in phase transition regions. Compared to areas with optimal *Ic/Ia* values which exhibit the lowest surface roughness, higher crystalline areas show slightly higher surface roughness.

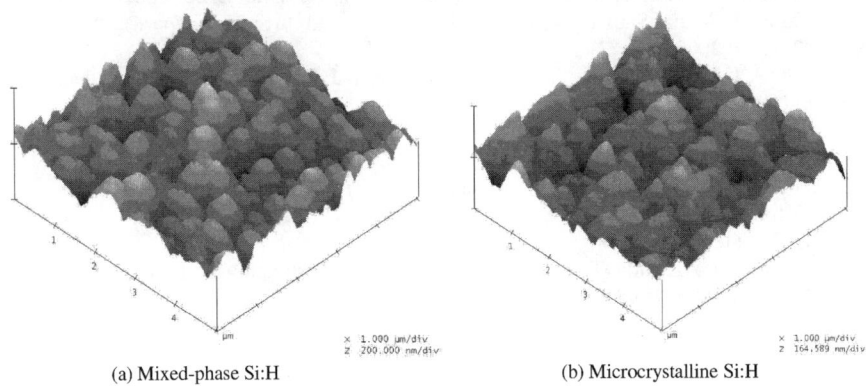

(a) Mixed-phase Si:H　　　　　　　　　　(b) Microcrystalline Si:H

**Figure 5**. AFM morphologies of solar cells with μc-Si:H and mixed-phase Si:H *i*-layers

## CONCLUSIONS

Strong correlations among device performance, $i$-layer structural properties, and the related uniformity behavior of Si:H solar cells have been established by direct structural characterization using Raman scattering, XRD, and AFM. It has been demonstrated that stable, high performance μc-Si:H solar cells were prepared with μc-Si $i$-layers grown near the microcrystalline-to-amorphous phase transition (or near the 'edge') where an optimal microcrystallinity range with moderate $I_c/I_a$ values was observed. Compared to solar cells with mixed-phase or higher microcrystalline Si:H $i$-layers, solar cells made from areas with moderate, 'optimal' microcrystallinity were found to exhibit preferential orientation corresponding to Si (220) planes, less-regular surface morphologies, and lower surface roughness. Further investigations are necessary to determine the film growth mechanism responsible for the structural and device properties.

## ACKNOWLEDGEMENTS

The authors would like to thank Andrei Foustotchenko for device measurements and other technical assistance. This research has been supported in part by the U.S. DOE under grant No. DE-FG02-00ER45806. Partial support by NREL, under subcontract No. ZDJ-2-30630-28, for the work performed at Energy Photovoltaics, Inc. (EPV) is also gratefully acknowledged.

## REFERENCES

1. A. Shah, J. Meier, E. Vallat-Sauvain, C. Droz, U. Kroll, N. Wyrsch, J. Guillet, and U. Graf, *Thin Solid Films* **403-404**, 179 (2002).
2. O. Vetterl, F. Finger, R. Carius, P. Hapke, L. Houben, O. Kluth, A. Lambertz, A. Muck, B. Rech, and H. Wagner, *Sol. Energy Mater. & Sol. Cells* **62**, 97 (2000).
3. C. Ross, J. Herion, and H. Wagner, *J. Non-Cryst. Solids* **266-269**, 69 (2000).
4. E. Vallat-Sauvain, U. Kroll, J. Meier, A. Shah, and J. Pohl, *J. Appl. Phys.* **87**, 3137 (2000).
5. K. Mori, T. Yasuda, M. Nishizawa, S. Yamasaki, and K. Tanaka, *Jpn. J. Appl. Phys.* **39**, 6647 (2000).
6. Y. Nasuno, M. Kondo, and A. Matsuda, *Jpn. J. Appl. Phys.* **40**, L303 (2001).
7. J. Bailat, E. Vallat-Sauvain, L. Feitknecht, C. Droz, and A. Shah, *J. Non-Cryst. Solids* **299-302**, 1219 (2002).
8. S. Guha, J. Yang, D. L. Williamson, Y. Lubianiker, J. D. Cohen, and A. H. Mahan, *Appl. Phys. Lett.* **74**, 1860 (1999).
9. A. S. Ferlauto, R. J. Koval, C. R. Wronski, and R. W. Collins, *Appl. Phys. Lett.* **80**, 2666 (2002).
10. A. Matsuda, K. Kumagai, and K. Tanaka, *Jpn. J. Appl. Phys.* **22**, L34 (1982).
11. S. Veprek, Z. Iqbal, O. Kuhne, P. Capezzuto, F. A. Sarott, and J. K. Gimzewski, *J. Phys.* C **16**, 6241 (1983).
12. K. Nakahata, A. Miida, T. Kamiya, Y. Maeda, C. M. Fortmann, and I. Shimizu, *Jpn. J. Appl. Phys.* **37**, L1026 (1998).

A6.14

## Hot-Wire Chemical Vapor Deposition for Epitaxial Silicon Growth
## On Large-Grained Polycrystalline Silicon Templates

M. S. Mason, C.M. Chen and H.A. Atwater
Thomas J. Watson Laboratory of Applied Physics
California Institute of Technology
Pasadena, CA 91107, U.S.A.

## ABSTRACT

We investigate low-temperature epitaxial growth of thin silicon films on Si [100] substrates and polycrystalline template layers formed by selective nucleation and solid phase epitaxy (SNSPE). We have grown 300 nm thick epitaxial layers at 300°C on silicon [100] substrates using a high $H_2$: $SiH_4$ ratio of 70:1. Transmission electron microscopy confirms that the films are epitaxial with a periodic array of stacking faults and are highly twinned after approximately 240 nm of growth. Evidence is also presented for epitaxial growth on polycrystalline SNSPE templates under the same growth conditions.

## INTRODUCTION

Hot-wire chemical vapor deposition has been shown to be a promising method for fast, low-temperature (<600°C) epitaxy [1, 2]. Previously, we showed that direct deposition by HWCVD on $SiO_2$ produced small grains (~40-80 nm), even with the addition of $H_2$ to a diluted mixture of 1% silane in He [3]. SNSPE layers formed by the use of nickel nanoparticles as nucleation sites for the solid-phase crystallization of phosphorus-doped amorphous silicon on $SiO_2$ display grain sizes on the order of 100 µm, and have been successfully used as seed layers for epitaxial growth by molecular beam epitaxy at 600°C [4,5]. We will discuss the microstructural properties of epitaxial films grown on silicon substrates and SNSPE templates.

## EXPERIMENT

Silicon films of 300 nm thickness were grown on silicon (100) and $SiO_2$ substrates and SNSPE templates by HWCVD using 70 mTorr $H_2$ at 20 sccm and 100 mTorr diluted $SiH_4$ in He at 20 sccm. A tungsten filament of 0.5 mm diameter was heated to 1850°C and placed 5 cm from the substrates. These conditions were chosen to produce amorphous silicon films on $SiO_2$, similar to those investigated by Seitz et al. [2]. Growth temperatures of 300°C – 450°C were chosen to investigate the effect of different levels of hydrogen surface passivation on the resulting epitaxial films. Substrates were UV ozone cleaned for 10 minutes and HF-dipped, then heated to 200°C in vacuum to desorb hydrocarbons. Ultrahigh purity gas mixtures were used and the base pressure of the growth chamber was below $10^{-6}$ Torr.

## RESULTS

Cross-sectional transmission electron microscopy of films grown on silicon (100) substrates at 300°C confirms the presence of epitaxial growth, as shown in Figure 1. The rough film-substrate interface is believed to have been caused by etching of the surface during growth by atomic hydrogen produced by the wire [3]. The roughened appearance of the silicon substrate in cross-section may be due to the presence of hydrogen platelet defects arising from the diffusion of hydrogen into the film during growth, although the exact structure of the defects has yet to be determined. Epitaxy continues to a thickness of approximately 240 nm, after which the film becomes highly twinned. The epitaxial films exhibit a periodic array of stacking faults which gives rise to the higher-order spots seen in the diffraction pattern in Figure 2.

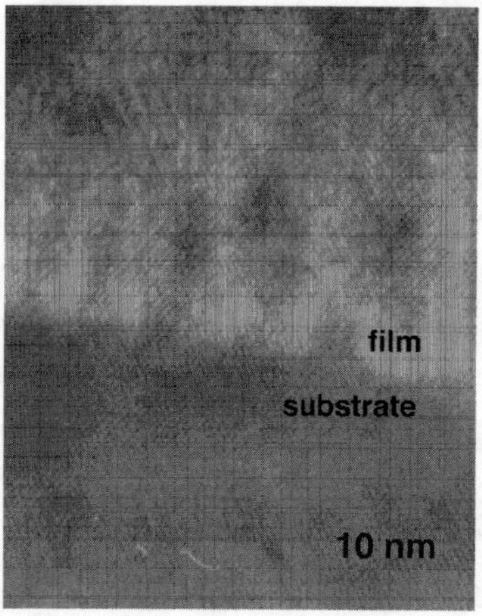

**Figure 1.** Cross-sectional TEM of HWCVD-grown Si on Si (100) at 300°C. The epitaxial films display a periodic array of stacking faults.

The results reported here for hydrogen-diluted epitaxial growth on Si (100) are broadly consistent with work reported elsewhere. Theisen et al. had previously observed epitaxial growth at temperatures between 195-325°C in which stacking fault defects were observed [1], while Seitz and Schröder observed no stacking faults or surface roughening in their epitaxial films grown between 280-360°C [2]. Both experiments were done using approximately 10 mTorr of pure $SiH_4$ and no additional hydrogen. Although Theisen et al. postulate that low-temperature epitaxy by HWCVD is possible because the growth species is believed to be $SiH_3$, we believe that for our dilute silane conditions the dominant growth species are silicon atoms [3]. It is possible that low-

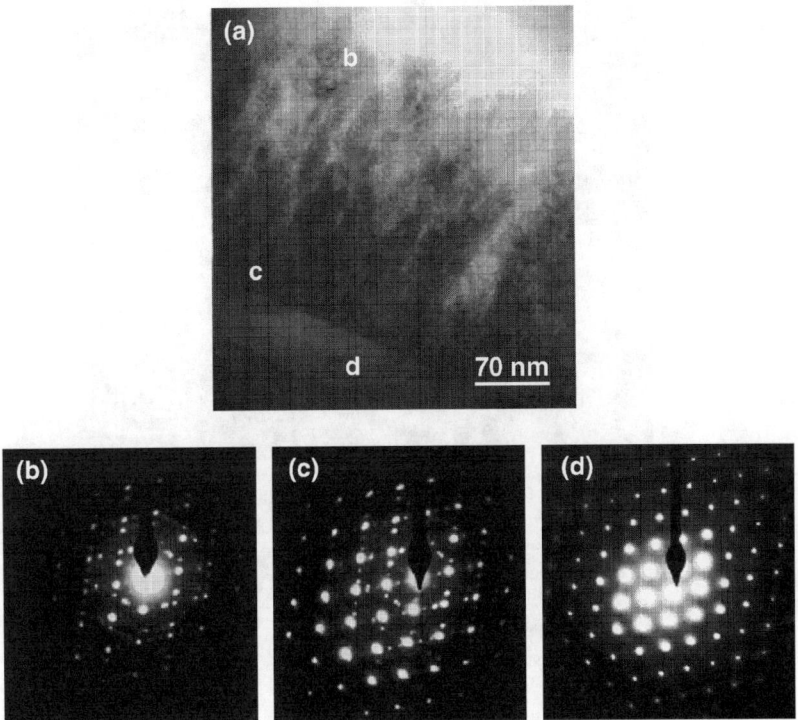

**Figure 2.** (a) Cross-sectional TEM of HWCVD-grown Si on Si (100) at 300°C. The films become highly twinned after a thickness of approximately 240 nm. Labels b, c, and d refer to areas from which selected-area diffraction patterns were obtained. (b) Selected area diffraction from HWCVD film and amorphous glue layer. (c) Selected area diffraction from HWCVD film and Si (100) substrate. (d) Si (100) substrate. Higher order spots in (b) and (c) are due to the periodic array of stacking faults in the epitaxial film and twinning in the uppermost regions of the film.

temperature epitaxial growth under these high hydrogen dilution conditions may be enabled by the preferential etching of amorphous regions by atomic hydrogen from the wire [6].

Silicon films 300 nm thick on SNSPE templates were investigated through plan-view transmission electron microscopy. The results, seen in Figures 3 and 4, are consistent with low-temperature epitaxy

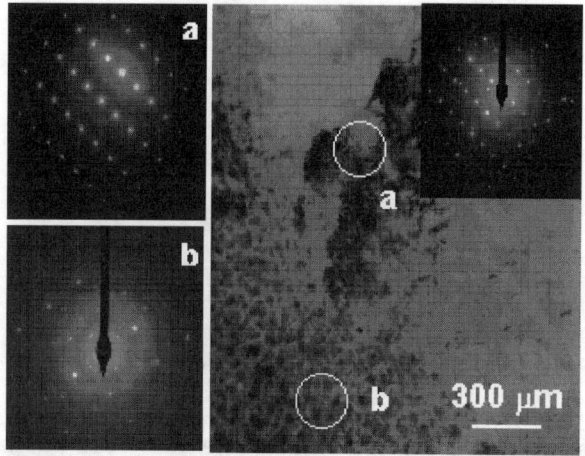

**Figure 3.** Plan-view TEM of HWCVD epitaxial film (T=300°C) on SNSPE template. (a) Selected area diffraction pattern from underlying SNSPE template. (b) Selected area diffraction pattern from HWCVD film on SNSPE template. (c) Bright-field image indicating selected area diffraction regions. Inset: diffraction from entire area.

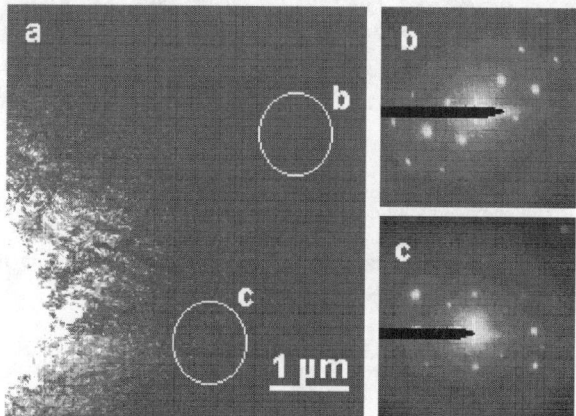

**Figure 4.** (a) Bright-field image of HWCVD film (T=300°C) on SNSPE template showing selected area diffraction regions. (b) Selected area diffraction from HWCVD film on (100)-oriented grain. (c) Selected area diffraction from HWCVD film on a grain of different orientation.

on the scale of the 100 μm grains of the SNSPE templates. Epitaxial breakdown is observed in the diffraction pattern of the HWCVD film, but some of the underlying higher-order diffraction spots are visible, making it likely that the underlying film has a morphology similar to that observed in the HWCVD films on Si (100). The effect of the orientation of the underlying grain structure of the SNSPE template on the morphology of the HWCVD film can be seen in Figure 4.

Cross-sectional analysis of these films reveals some areas of epitaxial growth as well as some areas of columnar growth. Before HWCVD growth, the SNSPE templates were cleaned in a solution of 3:7 $HNO_3$: H2O, which has been shown by Auger spectroscopy to remove elemental nickel from the template surface [5]. The lack of epitaxy in some areas is thus more likely to have been caused by the presence of ordinary surface contaminants, such as carbon and oxygen. These may be able to be removed by cleaning the surface for a short time with atomic hydrogen from the wire before growth. The SNSPE template layers have been successfully used as seed layers for epitaxy by MBE at 600°C [4]. It may be that a growth temperature of 300°C is not sufficiently high for epitaxial growth on the template layers; however, we anticipate that by increasing the growth temperature in the range from 300-600C, epitaxial growth may be achieved with HWCVD with a microstructure consistent with that achieved by MBE.

## CONCLUSIONS

Epitaxial films characterized by a periodic array of stacking faults have been grown at high hydrogen dilution by HWCVD on Si (100) substrates. The limiting thickness for epitaxy is shown to be approximately 240 nm at a growth temperature of 300ºC, after which the films are highly twinned. Evidence has been presented for epitaxial growth on SNSPE templates with 100 μm grains, although surface contamination may have resulted in columnar growth in some areas. Epitaxial growth on these templates could lead to the development of large-grained thin-film polycrystalline photovoltaic devices. Future work includes efforts toward increased growth rates (currently only 0.15 Å/s) and evaluation of the minority carrier lifetimes of the SNSPE templates and resulting epitaxial films to determine their suitability for photovoltaic applications.

**ACKNOWLEDGEMENTS**

This work was supported by the National Renewable Energy Laboratory. Expert technical assistance from Carol Garland and Jason Holt is gratefully acknowledged.

**REFERENCES**

1. J. Thiesen, E. Iwaniczko, K.M. Jones, A.H. Mahan, and R. Crandall, Appl. Phys. Lett. 75, 1999.
2. H. Seitz and B. Schröder, Sol. State Comm. 116, 2000.
3. J.K.Holt, M. Swiatek, D.G. Goodwin, R.P. Muller, W.A. Goddard III, and H.A. Atwater, Thin Solid Films 395, 2001.
4. C.M. Chen, Ph.D. thesis, California Institute of Technology, 2001.
5. R.A. Puglisi, H. Tanabe, C.M. Chen and H.A. Atwater, Mat. Sci. Eng. B 73, 2000
6. H. N. Wanka and M.B. Schubert, Mat. Res. Soc. Symp. Proc. 467, 1997.

# Film Growth Related to Devices

## Helium versus hydrogen dilution of silane in the deposition of polymorphous silicon films: Effects on the structure and the transport properties

O. Saadane[1], S. Lebib[2], A.V. Kharchenko[2], V. Suendo[2], C. Longeaud[1] and P. Roca i Cabarrocas[2]

[1]Laboratoire de Génie Electrique de Paris, (UMR8507 CNRS), Ecole supérieure d'Electricité, Université Paris VI et XI, Plateau de Moulon, 91192 Gif sur Yvette, France
[2]Laboratoire de Physique des Interfaces et des Couches Minces (UMR7647 CNRS), Ecole Polytechnique, 91128 Palaiseau Cedex, France

## ABSTRACT

We compare the deposition rate, hydrogen incorporation and optoelectronic properties of hydrogenated polymorphous silicon films produced either by the decomposition of silane-hydrogen or of silane-helium mixtures. Our results clearly show that He dilution allows to drastically reduce the RF power needed to achieve the same deposition rate as in the case of $H_2$ dilution. Infrared spectroscopy and hydrogen effusion experiments show clear differences in the hydrogen bonding and content in both series of films. Interestingly, both He and hydrogen dilution result in films with improved transport properties, in particular the hole diffusion length, with respect to standard amorphous silicon. These results indicate that He dilution is a good alternative to $H_2$ dilution to prepare intrinsic layers for solar cells.

## INTRODUCTION

In the last few years we have focused on a nanostructured material deposited with high hydrogen dilution of silane under conditions close to the powder formation. This material, made of a small fraction of nano-crystallites embedded in an amorphous matrix and named after that polymorphous silicon (pm-Si:H), presents enhanced transport properties and, under some deposition conditions, an excellent stability of the hole diffusion length upon light-soaking (LS) [1]. However, despite these properties promising for solar cell application, the use of hydrogen present some drawbacks when these layers are used in PIN solar cells. First, the high RF power and ion bombardment during deposition result in a reduction of the tin oxide substrate and the formation of a damaged P-I interface [2]. Second, the use of hydrogen dilution leads to films with a high hydrogen content and a wider band gap than standard a-Si:H. To overcome these drawback, we have studied the deposition of pm-Si:H films from the dissociation of silane-helium mixtures. In this paper, we present optical, structural and electronic properties of pm-Si:H deposited from either $H_2$-silane or He-silane mixtures.

## SAMPLES AND EXPERIMENTS

The standard RF glow discharge (13.56 MHz) was used to produce the silicon thin films from a mixture of silane and hydrogen or a mixture of silane and helium. The plasma was confined between two electrodes (12 cm in diameter, 2.5 cm apart) surrounded by a metal cylinder. The samples obtained under $H_2$ dilution were prepared with a gas flow of 10 sccm of $SiH_4$ and 90 sccm of $H_2$, a pressure P ranging from 0.6 Torr to 4 Torr, a temperature of 200 °C and a RF power of 22 W. The samples obtained under He dilution were prepared with a gas flow of 10 sccm of $SiH_4$ and 90 sccm of He, a pressure P ranging from 0.3 Torr to 4 Torr, a temperature of 250 °C and a lower RF power (4 W) than the one used with $H_2$ dilution.

The optical gap of the films has been deduced from spectroscopic ellipsometry measurements by fitting the measured real and imaginary part of the dielectric function with the Tauc-Lorentz dispersion law [3]. Hydrogen bonding and content were investigated by means of Fourier Transform Infra-Red (FTIR) spectroscopy and hydrogen effusion. Transport properties have been studied by means of dark and photo-conductivity measurements, the Modulated PhotoCurrent (MPC) technique [4], the Constant Photocurrent Method (CPM) [5] and the Steady State Photocarrier Grating (SSPG) [6] experiment. Transport properties were studied in the as-deposited (AD), light-soaked to saturation and light-soaked+annealed states of the layers.

## RESULTS

### Deposition

The deposition rate ($r_d$) as a function of the gas pressure for both series of samples is presented in Fig. 1. For the $H_2$ diluted films $r_d$ increases slowly with pressure in the range 0.6 < P < 1.6 Torr [7]. For P > 1.6 Torr $r_d$ sharply increases due to the $\alpha/\gamma$ transition [8], and reaches a maximum for P > 2 Torr. He diluted films behave roughly the same way except that the whole curve is shifted towards lower pressure. The $\alpha/\gamma$ transition occurs around 1 Torr and saturation is reached around 1.5 Torr. It is worth to note that despite of the much lower RF power for the He series, we can achieve a similar deposition rate. Small RF power results also in a drastic reduction of the ion energy $E_i$. Indeed, $E_i$ can be estimated from the RF voltage amplitude and the self-bias values. For a gas pressure of 1 Torr, $E_i$ reaches 160 eV for the $SiH_4+H_2$ gas mixture and 60 eV for the $SiH_4+He$ gas mixture. Clearly films prepared under He dilution will suffer less ion bombardment damage, which should lead to a less stressed material and should reduced the damage at the interface between two successive layers.

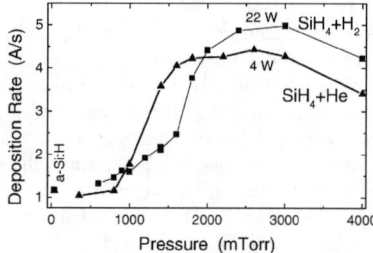
Fig. 1 : Deposition rate as a function of gas pressure for both series. Lines are provided as a guide to the eye.

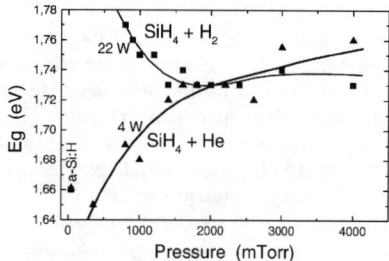
Fig. 2 : Optical band gap $E_g$ as a function of gas pressure for both series. Lines are provided as a guide to the eye

Fig. 2 presents the optical gap as a function of the gas pressure for both series of samples. For pressure lower than 1.8 Torr the He diluted films exhibit a smaller band gap than $H_2$ diluted layers. Above 1.8 Torr, the band gap are almost identical and above 3 Torr the He diluted materials exhibit a larger band gap than the $H_2$ diluted films. Taking account of these results one may expect that solar cells incorporating an intrinsic layer made of He diluted material deposited at a pressure P<1.8 Torr should present a large spectral absorption and consequently a better conversion efficiency.

The lower ion energy for the same deposition rate along with the smaller optical band gap pushed us to concentrate on layers deposited at the same pressure of 1.4 Torr.

**Structural properties**

Figure 3 shows the hydrogen effusion spectra of pm-Si:H samples deposited at 1.4 Torr. On this figure are also illustrated the hydrogen effusion spectra of standard a-Si:H deposited at 200 °C and 250 °C. These spectra usually exhibit a broad band with a maximum value at a temperature around 550 °C, related to a dense a-Si:H matrix, followed by a shoulder at higher temperature characteristic of the crystallization of the film. The features observed for the two pm-Si:H films reveal a very different behavior compared to standard a-Si:H. The main features are:

i) The hydrogen evolution for hydrogen and helium dilution samples starts at lower temperature than for standard a-Si:H.

ii) The presence of a sharp peak at about 420°C, the intensity of which depends on the diluting gas ($H_2$ or He) and is more pronounced for samples deposited with hydrogen dilution.

iii) For the hydrogen diluted sample we have performed a phenomenological decomposition of the spectrum. From this deconvolution (triangles) we found many other features at 450, 500, 575, 600 and 630 °C. On the contrary, the sharp peak for the He diluted sample is followed by only one large and intense band located at about 540 °C.

These results suggest a different hydrogen environment in the pm-Si:H films compared to standard a-Si:H.

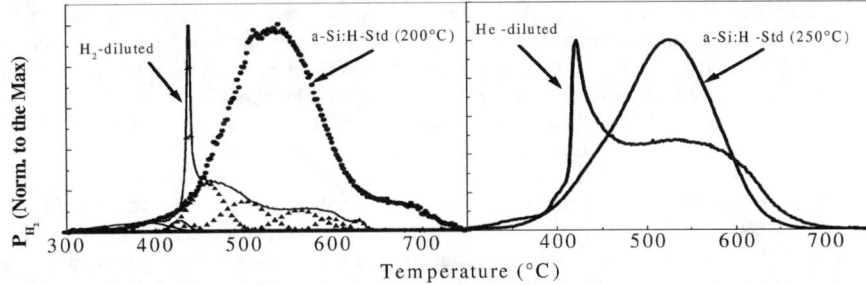

Figure 3 : Hydrogen effusion spectra for two samples deposited at 1.4 Torr. These spectra are compared to that of standard a-Si:H deposited at the same temperature.

This different hydrogen environment is also detected by the IR measurements. Table I shows the results of the analysis of the IR absorption spectra of the pm-Si:H films for both samples. The hydrogen contents of standard a-Si:H films deposited at 200 °C and 250 °C are also given as a reference.

For the $H_2$ diluted sample we find that the hydrogen content is 20 %, far above the ~11 % obtained for a standard a-Si:H. This increase is accompanied by a shift of the wagging band from ~ 638 $cm^{-1}$ for standard a-Si:H to lower frequencies (~ 626 $cm^{-1}$). This shift has been related by some authors to the presence of nanocrystallites in the amorphous network [9]. The stretching region has been simulated by three bands at about 2000, 2040 and 2090 $cm^{-1}$. The band at about 2040 $cm^{-1}$ has been recently related to the sharp peak observed in the hydrogen evolution spectra and associated to the presence of hydrogen bonded at the surface of clusters and crystallites [10].

For the He diluted sample, the hydrogen concentration deduced from the wagging is approximately the same as that of standard a-Si:H (~ 12 %), though the maximum of this band appears at a lower wavenumber than in a-Si:H. Moreover, the stretching region can be decomposed in two bands at about 2000 and 2088 $cm^{-1}$ as in the case a-Si:H. However, while for the standard film the total hydrogen content of the stretching band equals that of the wagging band, the sum of the hydrogen bonded at the 2000 and 2088 cm$^{-1}$ bands is much higher than that of the wagging band.

**Table I :** Results of the IR spectroscopy performed on standard a-Si:H samples and on pm-Si:H samples of the H$_2$ and He series

| Films | $C_{H630}$ (%) | $\omega_1$ (cm$^{-1}$) | $\omega_2$ (cm$^{-1}$) | $C_{H2000}$ (%) | $\omega_3$ (cm$^{-1}$) | $C_{H2090}$(%) | $\omega_4$ (cm$^{-1}$) |
|---|---|---|---|---|---|---|---|
| *Standard Films* | | | | | | | |
| **200 °C** | 11 | 638 | 1996 | 9.9 | 2090 | 1 | - |
| **250 °C** | 11.5 | 635 | 1996 | 9.8 | 2093 | 0.9 | - |
| **H$_2$ Dil.** | 19.39 | 627 | 2000 | 11 | 2089 | 5.3 | 2042 |
| **He Dil.** | 12 | 630 | 1997 | 15 | 2084 | 14 | - |

**Transport properties**

The values of the dark conductivities and of the activation energies we obtained were characteristic of undoped samples ($E_a \approx 0.95$ eV). In Table II are listed, for all the states of the films (AD, light-soaked and annealed), the room temperature values of $L_d$, and the electron mobility-lifetime product $\mu_e\tau_e$, deduced from the SSPG and photoconductivity experiments respectively. Note that, in the AD state, samples obtained by either H$_2$ or He dilutions give equivalent $L_d$ values. We observe for the sample of the He series almost no influence of LS and subsequent annealing, this sample exhibiting a rather good stability of the hole transport properties. On the opposite, $L_d$ of the H$_2$ diluted sample is strongly affected by the light-soaking process. Finally, for both samples, annealing results in a slight improvement of $L_d$.

For both samples, LS results in a strong decrease of $\mu_e\tau_e$, that reaches almost the same value of the order of $10^{-7}$ cm$^2$ V$^{-1}$, independently of the deposition conditions. Annealing at 460 K restores the $\mu_e\tau_e$ product to its initial value for both samples. From Table I one can see that the evolution of $\mu_e\tau_e$ is directly linked to the evolution of N(0.5).

**Table II.:** Dark conductivity ($\sigma_{dark}$) at 303 K, MPC-DOS measured at 0.5 eV below the conduction band edge N(0.5), $L_d$ and $\mu_e\tau_e$ in the as-deposited (AD), light-soaked (LS) and annealed (Ann) states of the H$_2$ and He samples.

| Samples | State | $\sigma_{dark}$ (303 K) (S. cm$^{-1}$) | N(0.5) (cm$^{-3}$eV$^{-1}$) | $L_d$ (nm) | $\mu_e\tau_e$ (cm$^2$ V$^{-1}$) |
|---|---|---|---|---|---|
| | AD | 0.2 x 10$^{-11}$ | 9.3 x 10$^{16}$ | 175 | 0.2 x 10$^{-6}$ |
| **H$_2$ Dil.** | LS | - | 1.5 x 10$^{17}$ | 135 | 0.8 x 10$^{-7}$ |
| | Ann | 8.9 x 10$^{-11}$ | 5.2 x 10$^{16}$ | 190 | 0.2 x 10$^{-6}$ |
| | AD | 4.0 x 10$^{-11}$ | 1.2 x 10$^{16}$ | 170 | 1.2 x 10$^{-6}$ |
| **He Dil.** | LS | 7.1 x 10$^{-11}$ | 1.2 x 10$^{17}$ | 160 | 1.2 x 10$^{-7}$ |
| | Ann | 5.8 x 10$^{-9}$ | 1.2 x 10$^{16}$ | 180 | 1.5 x 10$^{-6}$ |

Density of states (DOS) obtained from the MPC measurements in the AD, light-soaked, and annealed states for samples of the H$_2$ and He dilution series are shown in Figures 4.a and 4.b respectively. Absolute values of the DOS have been obtained assuming that the mobility of electrons in the extended states and the attempt-to-escape frequency are equal to 10 cm$^2$/V/s and $10^{12}$ s$^{-1}$, respectively. LS results in a strong increase of the deep defect density for both samples. One can also notice an enlargement of the conduction band tail (CBT). For both samples, annealing at 460 K during 24 h restored the deep defect density, but the CBT remains broadened.

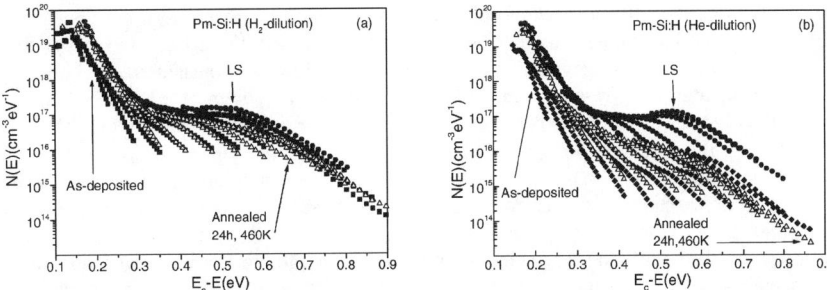

**Figure 4** : DOS deduced from MPC measurements on samples deposited with $H_2$ (a) and He (b) dilution in their AD (diamonds), light-soaked (dots) and annealed (open triangles) states.

The defect density below the Fermi level has been estimated by the CPM technique. Though both samples exhibited the same Urbach energy of the order of $56 \pm 5$ meV, it was clear that $H_2$ diluted sample exhibited a larger band gap as expected from the results of Fig. 2 and from the high hydrogen content. For both samples a strong increase of the absorption in the low energy part of the spectra, related to an increase of the deep defect density below the Fermi level, resulted from light-soaking. Within experimental error, no changes in the absorption exponential region were detected during the LS/annealing cycle.

## DISCUSSION

The deposition study shows that the $\alpha/\gamma$ transition occurs at a lower pressure for the He diluted samples than for the $H_2$ diluted ones. For this reason, at the same pressure of 1.4 Torr the deposition rate of the latter films is half that of the He diluted layers. Technologically, He layers present therefore a definite advantage. Moreover, since the RF power of this series is five times lower than that used for the $H_2$ series the ion bombardment is largely reduced when He is used as a diluting gas. From figure 2 it can also be seen that He layers present a slightly lower band gap than the $H_2$ layers. This behaviour may be a consequence of the lower hydrogen content of the He diluted film.

As far as hydrogen incorporation is concerned our experiments indicate a different hydrogen environment in pm-Si:H with respect to a-Si:H. In particular, the effusion of hydrogen starts at temperatures in the 300-400 °C range with a large effusion peak at 420 °C. The presence of this peak is explained as follow : hydrogen evolution at low temperature causes the formation of bubbles in which the pressure induced by molecular hydrogen builds up and eventually results in the explosion of the film, evidenced by the sharp peak at about 420 °C. Taking account of the peculiar structure of the material (nano-crystallites in an amorphous matrix) we believe that, prior to the sharp effusion peak, diffusing hydrogen may be stored at the interface between crystallites and amorphous matrix. It must be underlined that the H-effusion spectra after this effusion peak are very different in each dilution. Whereas, one could defined different structures in the spectrum of the $H_2$ diluted films, the spectrum of the He films presents only a single and large structure that we have attributed to the fact that the sample is more compact and dense than the $H_2$ diluted one.

IR measurements performed on the $H_2$ diluted films showed a new absorption band in the stretching region peaked at around 2040 cm$^{-1}$. This band is another evidence that hydrogen incorporation is peculiar to pm-Si:H and we have attributed this band to hydrogen bonded in clusters at the interface crystallites/amorphous matrix. Though this band does not appears in He diluted layer it does not means that there is no crystallites in these films but rather that these interfaces are less hydrogenated and thus not detectable from IR measurements.

The most interesting results come from the transport measurements. If one considers both materials in the AD states, the hole diffusion lengths are quite high and almost the same but the electron $\mu_e\tau_e$ product is higher for the He diluted in agreement with the lower MPC DOS. The influence of light-soaking on the diffusion length is much more pronounced for the $H_2$ diluted sample than for the He diluted one that presents a rather stable value upon LS/annealing cycle. However, it has to be noted some irreversible modification of the DOS upon this process, in particular a broadening of the CBT though the annealing was performed at a temperature lower than the deposition temperature. This behaviour was encountered in most of the pm-Si:H samples and was attributed to the peculiar hydrogen environment in this material [11]

## CONCLUSION

Optical, structural and transport properties of pm-Si:H films prepared from $H_2$-silane and He-silane mixtures have been investigated. He diluted material presents some clear advantages compared to $H_2$ diluted samples. First, high deposition rates can be reached with a low RF power minimizing the ion bombardment during deposition. Second, He diluted samples have a low hydrogen content resulting in a lower band gap. Third, He diluted material presents the same transport properties as $H_2$ diluted material in the as-deposited state (even slightly better) along with a better stability of the minority carriers diffusion length upon LS/annealing cycle. These properties are in favour of the use of He diluted pm-Si:H as intrinsic layer of solar devices.

## REFERENCES

[1] C. Longeaud, J. P. Kleider, M. Gauthier, R. Brüggemann, Y. Poissant, and P. Roca i Cabarrocas, Mat. Res. Soc. Symp. Proc. **557**, 501 (1999).
[2] Y. Poissant, P. Chatterjee and P. Roca i Cabarrocas, J. Non-Cryst. Solids **299-302**, 1173 (2002).
[3] G.E. Jellison, Jr and F.A. Modine, Appl. Phys. Lett. **69**, 371-373 (1996).
[4] J. P. Kleider and C. Longeaud, Solid-State Phenomena **44-46**, 597 (1995).
[5] M. Vanecek, J. Kocka, J. Stuchlik, Z. Kozisek, O. Stika and A. Triska, Solar Energy materials **8**, 411 (1982).
[6] D. Ritter, E. Zeldov and K.Weilson, Appl. Phys. Lett. **49**, 791 (1986).
[7] A.V. Kharchenko, V. Suendo, P. Roca i Cabarrocas, Thin Solid Films, In Press (2003).
[8] J. Perrin, P. Roca I Cabarrocas, B. Allain, and J.-M. Friedt, Jap. J.Appl.Phys. **27**, 2041 (1998).
[9] A. H. Mahan, J. Yang, S. Guha, and D. L. Williamson, Mat. Res. Soc. Proc. **557**, 269 (1999).
[10] S. Vignoli, A. Foncuberta i Morral, R. Butté, R. Meaudre, M. Meaudre, J. Non-Cryst. Solids **299-302**, 220 (2002).
[11] O. Saadane, S. Lebib, A.V. Kharchenko, C. Longeaud, and P. Roca i Cabarrocas, accepted for publication in J. Appl. Phys.

## Characterization of Nanocrystalline Silicon Film grown by LEPECVD for Photovoltaic Applications

M. Bollani, S.Binetti, M. Acciarri, L. Fumagalli, A. Arcari, S.Pizzini, H. von Känel [1]
INFM and Dept. of Material Science, Università di Milano-Bicocca, via Cozzi 53, I-20125 Milano, Italy
[1] INFM and L-NESS, Dept. of Physics, Politecnico di Milano, Via Anzani 52, I-22100 Como, Italy

### ABSTRACT

This work deals with the structural properties of nanocrystalline (nc) silicon films for solar cell applications, grown using a new PECVD process based on an arc discharge plasma characterized by low ion energies, called LEPECVD (Low energy PECVD). This process permits to increase the intensity of the plasma discharge in the growth region and thus to achieve higher growth rates while avoiding ion-induced surface damage of films. The structural properties of the LEPECVD grown films were studied as a function of the deposition parameters (substrate temperature, growth rate, hydrogen dilution) by Raman Spectroscopy, SEM, and HRTEM analysis. The results of this work allowed us to identify the process requirements suitable for the growth of nc-grains in an amorphous matrix.

### INTRODUCTION

It is known that nanocrystalline silicon (nc-Si) is a biphasic material consisting of a dispersion of silicon nanocrystals in a matrix of amorphous silicon (a-Si)[1].

Considering its potential application in photovoltaics, nc-Si represents an interesting alternative to amorphous silicon (a-Si) thanks to its better transport, optical and electronic properties. In fact, the carrier mobility is larger than that of a-Si silicon and the same holds for the carrier lifetime and the optical absorption coefficient. Moreover, both the optical and electronic properties can be modulated by a careful choice of the deposition conditions. Another advantage of nc-Si in comparison with crystalline silicon is the possibility of forecasting a continuous process for the production of solar cell modules, like in the case of a-Si, with substantial cost reduction, using glass or even plastic foils for the substrate. In spite of these potential merits, nc-Si was not considered in the past [2] a serious alternative to traditional materials, in view of the deposition rates achieved with the PECVD (Plasma Enhanced Chemical Vapor deposition) technique [3-4] or other variants of the CVD (Chemical Vapor deposition) techniques. Using silane as the silicon precursor, these rates were far too low (around 0.1 nm/s) for useful industrial developments.

In this work, we have grown thin films of nanocrystalline (nc) silicon using a new PECVD process based on an arc discharge plasma characterized by low ion energies, called LEPECVD (Low energy plasma enhanced chemical vapor deposition) [5]. Thanks the low discharge voltages, the ion energies in the plasma turn out to be sufficiently low, on the order of 10-15 eV, to avoid any ion-induced damage of the growing film, even when the substrate is fully exposed to the dense plasma. The main application of LEPECVD has so far been the epitaxy of SiGe heterostructures on Si, where impressive mobilities and transistor performance have been achieved [5], justifying the attempt of extending its use to nanocrystalline silicon. The deposition rate is not, however, the only condition to be satisfied, since the optical and electrical properties of nc-Si strongly depend on the nanocrystals size, on the ratio between the crystalline and amorphous phase volumes, on the residual hydrogen concentration in the solid mixture, on the width of the interfacial region between crystalline and amorphous phase, and on the density of (residual) dangling bonds. In turn, nanocrystal size, the ratio between crystalline and amorphous phase volumes, residual hydrogen concentration and growth rate depend on the deposition parameters, like substrate temperature and silane concentration and possibly on the doping gases as diborane or phosphine present in the inlet mixture.

We will show that the use of LEPECVD does allow the growth of good quality nanocrystalline films at relatively high growth rates and at substrate temperatures compatible with the use of low cost substrates. Careful design of the experimental deposition conditions and accurate determination of the factors, which determine the electrical and optical properties of nc-Si, are crucial for the development of a solar cell based on nanocrystalline silicon films. This preliminary work was therefore carried out with the aim of setting-up the deposition protocol.

## EXPERIMENTAL DETAILS

For the experiments p-type <100> oriented oxidized Si wafers were used (oxide thickness 1.6µm), in order to simulate the growth on an amorphous substrate. Before the growth, the samples were cleaned in a diluted HF (DHF) solution for 30 s. After loading into the LEPECVD system they were outgassed at 350°C for 10 min. Subsequently, a 2-min hydrogen plasma clean was applied at the same temperature using a gas flow of 5 sccm $H_2$ and an Ar flow of 50 sccm. The films were then deposited in the temperature range 450°C - 230°C, using different $SiH_4$ and $H_2$ flows leading to different dilution ratios $d = \Phi(SiH_4) / (\Phi (SiH_4) + \Phi (H_2))$. Silane flows ranged between 5 and 1 sccm. Growth rates were obtained by measuring the weights before and after the growth and the thickness calculated was confirmed also by SEM analysis on cross-sectioned samples. Table 1 shows the main features and growth parameters of the samples considered.

**Table I.** Summary of the growth parameters

| Sample | Thickness (µm) | Growth rate (nm/sec) | dilution ratio | Substrate temperature (°C) |
|---|---|---|---|---|
| 6344 | 1.6 | 1.20 | 25 % | 450 |
| 6577 | 1.5 | 1.08 | 4.2 % | 400 |
| 6581 | 1.4 | 1.12 | 4.2 % | 375 |
| 6578 | 1.6 | 1.03 | 4.2 % | 350 |
| 6579 | 4.2 | 0.96 | 4.2 % | 300 |
| 6580 | 1.8 | 1.19 | 4.2 % | 280 |
| 6731 | 2.1 | 1.00 | 4.2 % | 250 |
| 6732 | 1.5 | 0.99 | 4.2 % | 230 |
| 6733 | 1.4 | 0.53 | 1.96 % | 280 |

The growth rate is shown to be almost independent of the substrate temperature and roughly proportional to the silane flux, but decreases with increasing hydrogen content of the mixture. This indicates that, despite the efficient enhancement of the growth kinetics by means of the plasma, the growth can still be influenced by the presence of high concentrations of hydrogen.

The µRaman spectra were measured at room temperature using a Dilor Joben Yvon Spex spectrometer equipped with a CCD Spectrum one detector using the 633 nm line of an He-Ne laser. The energy resolution was 1 cm$^{-1}$ and the laser spot size 1x1 µm$^2$. Scanning electron microscopy (SEM) analyses were performed using a Cambridge Stereoscan mod. 250MK2.

Finally high-resolution micrographs of cross-section were recorded with a Philips CM 30 FEG transmission electron microscopy (TEM) in order to evaluate the grain structure.

## EXPERIMENTAL RESULTS AND DISCUSSION

The films appear homogeneous, compact and free of macroscopic defects. This can be observed in the SEM micrograph of a cross sectional view of one of the samples (fig.1). The columnar structure of the film is also evident from this micrograph.

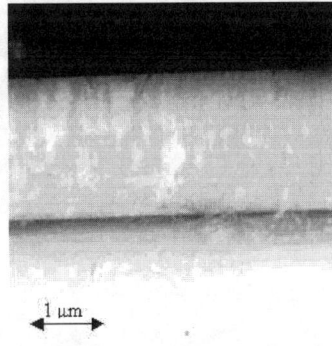

**Figure 1.** SEM cross-section image of sample #6344.

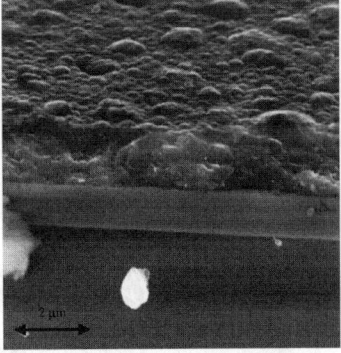

**Figure 2.** SEM image of sample #6344

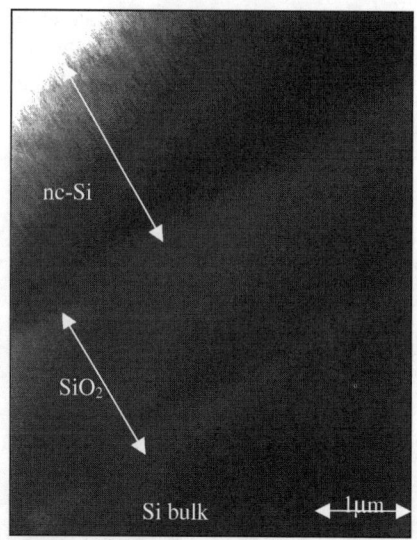

**Figure 3.** Cross-section TEM image of nc-Si /SiO$_2$/Si region #6580.

In the top view SEM micrograph of the same sample #6344 (Fig.2) it can be observed also that the surface the film has a natural roughness which would probably favor light trapping.

Fig. 3 shows the local morphology of sample #6580 grown at 280°C, investigated by transmission electron microscopy. It is characterized by elongated (columnar) "grains".

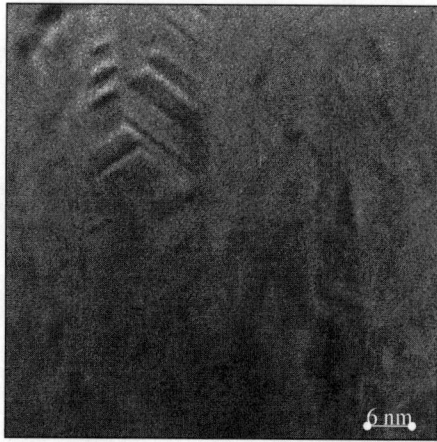

**Figure 4.** cross-section HRTEM image of a particular region of sample #6580. Some stacking faults and some crystalline regions, with different crystallographic orientation are evident.

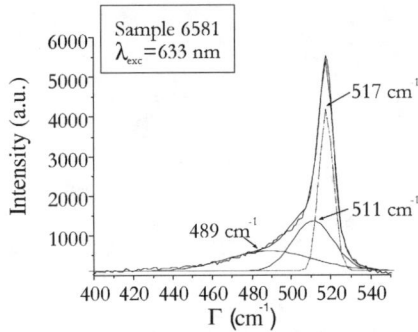

**Figure 5.** Representative Raman spectra of the films grown

Actually, the high resolution TEM image of the same sample (Fig 4) shows a random distribution of crystalline regions, inside the amorphous phase. The nanocrystals are randomly oriented and have different shapes and sizes. Fig.5 displays a Raman spectrum typical of the samples of the series. The gaussian deconvolution of the spectrum shows the crystalline silicon peak at 517 cm$^{-1}$ and two bands at 511 and 489 cm$^{-1}$ which could be assigned to the interaction between phonons and dangling bonds and to TO scattering involving the amorphous phase of the films respectively [7].

The Raman spectra taken at different positions on the samples were reproducible, indicating good homogeneity of the films. The crystalline volume fraction has been calculated [6] according to the following equation:

$$X_C = \frac{I_C + I_{GB}}{I_C + I_{GB} + \beta I_A} \qquad (1)$$

where $\beta$ is a scattering correction factor (equal to 1 [7]) and $I_C$, $I_A$ and $I_{GB}$ are the areas of the Raman emissions related to the crystalline, amorphous and interface phases, respectively.

As the measurements were done with a laser wavelength of 633 nm, the films deposited on thermal oxidized silicon were not thick sufficiently to prevent the laser beam from penetrating into the silicon substrate. To remove any possible doubt, measurements were also carried out using a 514.5 nm Ar$^+$ laser with a shorter penetration depth. As the Raman spectra obtained in this way do not differ significantly from the previous ones, this could suggest a higher absorption coefficient of the nc-crystalline silicon films compared to crystalline silicon, in agreement with some literature data [8]. From the Raman spectra the crystalline volume fraction ($X_c$) could be extracted using eq.1. It is reported in Fig.6 as a function of the growth temperature.

One can observe that $X_c$ increases almost linearly from 58 % to 74 % as the substrate temperature increases. At 280 °C (see #6733 and #6580 samples), the degree of crystallinity increases with decreasing dilution.

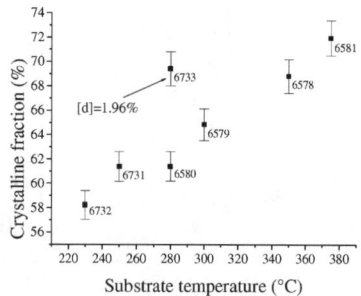

**Figure 6:** Crystalline volume fraction (%) versus substrate growth temperature

## CONCLUSIONS

The results of this work show, as expected, that the degree of crystallinity decreases with decreasing substrate temperature. Further work is needed to optimize the deposition protocol for temperatures below 230°C. It should be remarked, however, that the degree of crystallinity is surprisingly high even at low deposition temperature and high deposition rates. Future work will focus on the study of samples deposited on zinc oxide covered glass substrates which will be used as back contact for pin solar cells and on the influence of dopants on photoelectrical film properties. It is therefore possible to conclude that the LEPECVD technique allows the high deposition rates and homogeneity required for both optoelectronic and photovoltaic applications and, possibly, a better flexibility than other techniques.

## ACKNOWLEDGEMENTS
The authors wish to thank Dr. F. Meinardi for technical assistance in Raman spectroscopy.

## REFERENCES

1. J. Bailat, E. Vallat-Sauvain, L. Feitknecht, C. Droz, M. Goerlitzer and A. Shah, J. Appl. Phys. **93**, 5727 (2003).
2. A. Shah, E. Vallat-Sauvain, P. Torres, J. Meier, U. Kroll, C. Hof, C. Droz, M. Goerlitzer, N. Wyrsch, M. Venecek, Materials Science and Engineering, B **69-70,** 219 (2000).
3. H.R. Khan, H. Frey, F. Banhart, Nuclear Instruments and Methods in Physics Research B **112**, 84 (1996).
4. M. Kondo, M. Fukawa, L. Guo, A. Matsuda, Journal of Non-Crystalline Solids **266-269**, 84 (2000).
5. M. Kummer, C. Rosenbland, A. Dommann, T. Hackbarth, G. Höck, M. Zeuner, E. Müller, and H. Vo Känel, Materials Science and Engineering. B **89**, 288 (2002).
6. G. Yue, J.D. Lorentzen, J. Lin, D.Han , Q.Wang Appl. Phys. Lett. **75**, 492 (1999).
7. E.Bustarret, M.A. Hachica, M. Brunel, Appl. Phys. Lett. **52** (20) 1675, (1988).
8. A.V. Shah, R. Platz, H. Keppner, Solar Energy Mat. And Solar Cells **38,** 501 (1995).

# Influence of Substrate Temperature and Hydrogen Dilution Ratio on the Properties of Nanocrystalline Silicon Thin Films Grown by Hot-Wire Chemical Vapor Deposition

H.R. Moutinho, C.-S. Jiang, B. Nelson, Y. Xu, J. Perkins, B. To, K.M. Jones, M.J. Romero, and M.M. Al-Jassim
National Renewable Energy Laboratory
1617 Cole Blvd.
Golden CO 80401, USA

## ABSTRACT

We have studied the influence of substrate temperature and hydrogen dilution ratio on the properties of silicon thin films deposited on single-crystal silicon and glass substrates. We varied the initial substrate temperature from 200° to 400°C and the dilution ratio from 10 to 100. We also studied the effectiveness of the use of a seed layer to increase the crystallinity of the films. The films were analyzed by atomic force microscopy, X-ray diffraction, Raman spectroscopy, and transmission and scanning electron microscopy. We found that as the dilution ratio is increased, the films go from amorphous, to a mixture of amorphous and crystalline, to nanocrystalline. The effect of substrate temperature is to increase the amount of crystallinity in the film for a given dilution ratio. We found that the use of a seed layer has limited effects and is important only for low values of dilution ratio and substrate temperature, when the films have large amounts of the amorphous phase.

## INTRODUCTION

Nanocrystalline silicon (nc-Si) has been receiving special attention lately because it is cheaper to produce than crystalline silicon, does not seem to present the degradation problems of amorphous silicon ($\alpha$-Si) [1], and can be doped p- and n-type [2,3]. Furthermore, because of its bandgap, it can be used in tandem solar cells with $\alpha$-Si [4]. Among the deposition methods, hot-wire chemical vapor deposition (HWCVD) [5] has the advantage of higher deposition rates when compared to other conventional techniques, such as plasma-enhanced CVD.

In general, nc-Si is highly anisotropic, and it is deposited as a mixture of amorphous and crystalline phases. Furthermore, depending on the deposition conditions, the crystalline phase varies from nanocrystalline to large columnar grains [6]. Extensive work is still necessary before this material can be produced with controlled properties and is able to produce solar cells that can compete with more traditional ones. In the present work, we investigate the effects of different substrate temperatures and hydrogen dilution ratio on the structural properties of the films. We also investigate if the use of a very thin seed layer [7], deposited with a high value of dilution ratio, and different substrates (Si and glass) affect the growth process.

## EXPERIMENTAL DETAILS

The films were grown by HWCVD, using a double filament, with a current of 13A passing through each filament, resulting in a temperature around 1850°C. Films were grown at three ranges of substrate temperatures, $T_{sub}$: 200°-320°C, 300°-383°C, and 400°- 435°C. The first temperature is the one at the time when the shutter was open, and the last temperature is the one at the end of deposition. This increase in temperature is caused by the proximity between substrates and filaments. As expected, this effect is more pronounced at lower substrate temperatures. The dilution ratio (R) between hydrogen ($H_2$) and silane ($SiH_4$) was controlled by varying the flux of $SiH_4$. The flux of $H_2$ was kept around 250 sccm, while the flux of $SiH_4$ was varied from 25 to 2.5 sccm, for R varying between 10 and 100, respectively. The deposition pressure was 150 mTorr. The films deposited on seed layers were grown with similar parameters, and the seed layers ($\cong$ 12 nm thick) were deposited with R equal to 100 and the above substrate temperatures. At the end of the deposition of the seed layer, the value of R was adjusted for the subsequent growth of the film. The films were deposited on 1737 Corning glass and (100)-oriented crystalline Si substrates. The films were characterized by atomic force microscopy, in tapping mode, with a Digital Instruments DI 3100 scanning probe microscope; X-ray diffraction, with a Scintag X1 diffractometer; Raman spectroscopy, with a single-grating Spex 270M spectrometer; and transmission and scanning electron microscopy (TEM and SEM), using a Philips CM30 TEM and a JEOL 6320F field-emission SEM, respectively.

## RESULTS AND DISCUSSION

There is a considerable thickness variation of the films, that is, the film structure changes as the film grows, such that there can be a different type of film at the top surface from that at the substrate. For this reason, it is important to consider film thickness when comparing the properties of two films. Unless mentioned in the text, when we compare films, we make sure that the thickness is the same or that it is not playing a role in that analysis.

### Deposition rate

Deposition rates were calculated from thickness values measured by SEM. As expected, the deposition rates decreased with an increase in the value of R. An interesting observation is that there was no major influence of the substrate temperature on the deposition rate, as shown in Fig. 1. Because temperature influences the surface diffusion of adsorbed species, in general it is a critical parameter on the nucleation and growth of thin films. The results shown in Fig. 1 indicate that, in the temperature range used in this work, the availability of the source material is the main factor in the deposition rate of nc-Si films.

### X-ray diffraction

All the films in this study had the cubic structure (JCPDS 27-1402). We found that, regardless of the growth temperature, the films had the same general behavior as the dilution ratio was varied. Films with very low R were amorphous. As R increased, a (220)-oriented phase would appear. As R continued to increase, the (220) orientation would decrease, and, for higher values of R, the film would become practically randomly oriented. This behavior, shown in Fig. 2, had been observed in a previous work for films grown at 400°C [8]. The value of R for which the film was amorphous was a function of the temperature (Fig. 2). Although films deposited with R equal to 14 and $T_{sub}$ equal to 200° and 300°C did not show any diffraction

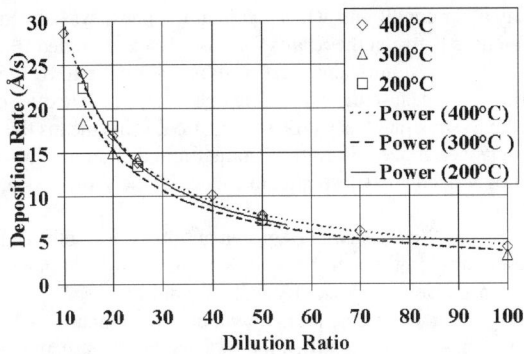

Figure 1. Deposition rate versus dilution ratio for Si films. The data were fitted to power functions, as denoted in the legends.

peak, an increase in $T_{sub}$ to 400°C would produce films with a sharp (220) peak. A possible explanation is that at 400°C the adsorbed atoms have a higher surface mobility, and more atoms will be able to reach crystalline regions. Nevertheless, because at these low values of R, the nucleation of amorphous material still dominates the deposition process, the film will still be highly amorphous. For the 400°C films, a value of R equal to 10 would result in XRD patterns without any diffraction peak. For all the temperatures, films deposited with R equal to 20 already present a decrease in the (220) texture, and films deposited with R equal to 50 are already randomly oriented.

The effect of a seed layer was observed only in some situations when the film naturally would be amorphous (low R). For instance, for R equal to 14 and $T_{sub}$ equal to 300°C, the seeded film had a small (220) peak, whereas the thicker unseeded film seemed to be completely

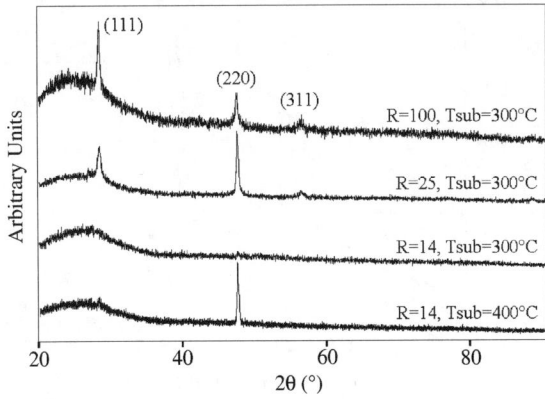

Figure 2. X-ray diffraction patterns for Si films grown at different conditions.

amorphous. For any other conditions, when a crystalline phase was already present, there were no significant differences between the results for unseeded and seeded films.

We also did not find any significant effect of substrate on the XRD measurements. This result is not so surprising, because the crystalline phase has (220) texture or is randomly oriented, whereas the Si substrate has a (100) orientation. This means that, if there is epitaxial growth, it is not extensive. Indeed, in spite of some limited epitaxial growth observed in TEM analysis, we have not noticed any substantial influence of the substrates on the properties of the films.

We observed a peak at 26.9° in samples grown at 400°C, for both seeded and unseeded films. Because it was present only in films grown with R larger than 14, it must be associated with the randomly oriented phase. Nevertheless, this peak was not observed for films deposited at 200° or 300°C, and its origin is at the moment unknown, due to the nature of XRD analysis, which in general requires few peaks for the identification of a crystalline structure.

**Raman spectroscopy**

We observed that the crystallinity in the films improves for higher substrate temperatures and dilution ratios. Nevertheless, as for XRD, an increase in R seems to be more effective than an increase in $T_{sub}$. Also, an increase in R affects the film more strongly at the lowest $T_{sub}$. As shown in Fig. 3, samples go from amorphous to highly crystalline as R and $T_{sub}$ increases. It is important to mention that the samples deposited at 200°C, in Fig. 3, were about 25% thinner than the ones deposited at 400°C. This probably results in the latter samples seeming to be relatively more crystalline than they would if all samples had the same thickness. Comparison of the results using red and green lasers indicates that, in general, the crystallinity increases with sample thickness. Nevertheless, because the measurements were done on the films deposited on

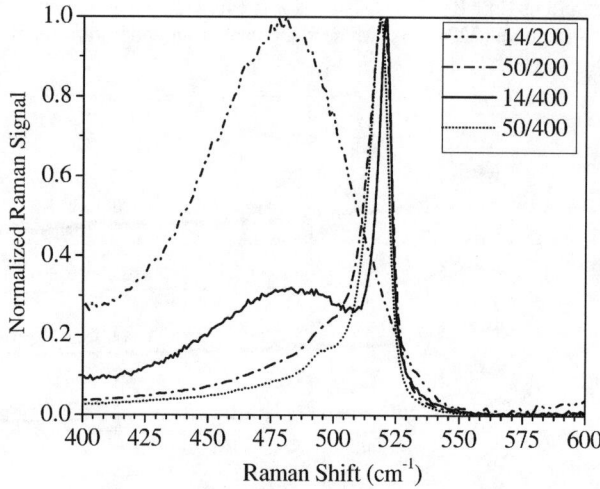

Figure 3. Raman spectra for Si films deposited at different conditions. In the legend, the number on the left is the dilution ratio and the one on the right is the substrate temperature.

Si, and because many of the films deposited at 200° and 300°C were not thick enough to prevent the red laser from reaching the substrate, we cannot confirm that this behavior occurs for all films analyzed in this work.

We have only analyzed seeded films deposited at 400°C. As in the XRD measurements, we only found evidence of the effectiveness of the use of a seed layer for films with large amounts of the amorphous phase, in which the seed layer increases the amount of crystalline phase in the film. No significant differences were found in Raman spectra of seeded and unseeded films for R equal or larger than 20.

## Atomic force microscopy and transmission electron microscopy

The AFM images of the $\alpha$-Si phase show structures that resemble grains. For this reason, it is very difficult to unmistakably assign grain-like structures to crystalline material in AFM images without the aid of other kinds of analysis.

From the AFM and TEM data, we observed that the structure and morphology of the samples vary with changes in R, independently of the substrate temperature. For low values of R, the film is completely amorphous. As R increases, some (220)-oriented grains start to grow at the interface with the substrate. These grains grow as columns while the film is being deposited, also growing laterally. They appear as elongated grains in AFM images. In this way, the amount of the crystalline phase increases as the film grows. For larger values of R, the density of these grains increases, increasing the amount of the crystalline phase. At this point, there is also the appearance of a randomly oriented phase, observed in the XRD analysis. For further increase in R, the film nucleates preferably as nc-Si, and the number of columnar grains decreases, until they almost disappear for large values of R. As the substrate temperature increases, the amount of crystalline material in the film increases, for the same values of R. This is more evident for lower values of R (up to 25), where the number of elongated grains increases with temperature, as observed in Fig. 4. For large values of R, when the density of elongated grains decreases, it is difficult to distinguish between the amorphous and the nanocrystalline phase in AFM images. It is important to notice that, for intermediate conditions of R and $T_{sub}$, three distinct phases are present in the films: $\alpha$-Si, nc-Si, and the columnar grains, which are large crystals, extending over the whole thickness of the film.

In general, the morphology of the films does not change for seeded and unseeded films. As observed before, the seeded layer makes a difference, increasing the amount of the crystallinity, only for samples with large amounts of the amorphous phase. Finally, we did not notice any major effect of the substrate on the morphology of the films.

## CONCLUSIONS

The dilution ratio is the major parameter for controlling the structural properties of silicon thin films. As the value of this parameter increases, the film goes from completely amorphous, to a mixture of amorphous and crystalline, to nanocrystalline. We have also seen a temperature dependence, where increasing the substrate temperature increases the crystallinity of the films. The use of a seed layer is only effective in improving the crystallinity of films grown with a high concentration of the amorphous phase.

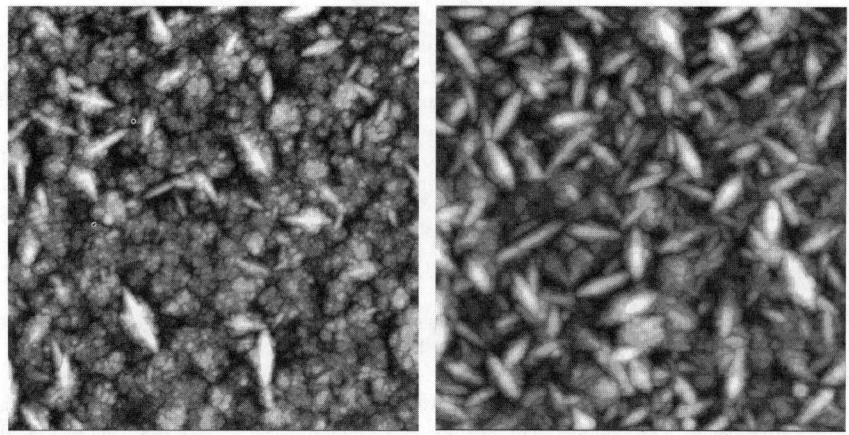

Figure 4. AFM images of Si film grown with R equal to 25 and substrate temperature equal to 200°C (left) and 400°C (right). The images have a 2µm x 2µm scale.

**ACKNOWLEDGMENTS**

This work was supported by the U.S. Department of Energy under Contract number DE-AC36-99GO10337.

**REFERENCES**

1. J.P. Kleider, C. Longeaud, R. Bruggemann, and F. Houze, *Thin Solid Films* **383**, 57 (2001).
2. J. Puigdollers, J. Cifre, M.C. Polo, J.M. Asensi, J. Tertomeu, J. Andreu, and A. Lloret, *Appl. Surf. Sci.* **86**, 600 (1995).
3. S.C. Saha, J.K. Rath, S.T. Kshirsagar, and S. Ray, *J. Phys. D: Appl. Phys.* **30**, 2686 (1997).
4. Y. Hamakawa and H. Takakura, *Proc. Twenty-Eighth IEEE Photov. Spec. Conf.*, 766 (Anchorage, 2000).
5. M. Konagai, T. Tsushima, Y. Ide, K. Asakusa, T. Jujisaki, M.K. Kim, Y. Wakita, and A. Yamada, *Proc. Twenty-Eighth IEEE Photov. Spec. Conf.*, 788 (Anchorage, 2000).
6. H.R. Moutinho, C.-S. Jiang, J. Perkins, Y. Xu, B.P. Nelson, K.M. Jones, M.J. Romero, and M.M. Al-Jassim, *Thin Solid Films* (2003) (in press).
7. J.-H. Zhou, K. Ikuta, T. Yasuda, T. Umeda, S. Yamasaki, and K. Tanaka, *Appl. Phys. Lett.* **71**, 1534 (1997).
8. H.R. Moutinho, M.J. Romero, C.-S. Jiang, Y. Xu, B.P. Nelson, K.M. Jones, A.H. Mahan, and M.M. Al-Jassim, *Proc. Twenty-Ninth IEEE Photov. Spec. Conf.* (New Orleans, 2002) (in press).

# Electrical Properties of Phosphorus-Doped and Boron-Doped Nanocrystalline Germanium Thin-Films for p-i-n Devices

William B. Jordan and Sigurd Wagner
Department of Electrical Engineering, Princeton University, Princeton, NJ 08544

## ABSTRACT

Nanocrystalline germanium thin-films deposited on glass by plasma-enhanced chemical vapor deposition from germane and hydrogen were doped with phosphorus and boron. We report some electrical transport and structural properties of Ge films as a function of dopant species and doping levels. The dark conductivities of the phosphorus- and boron-doped films are approximately three to four orders of magnitude higher than the intrinsic nanocrystalline germanium. In the solid phase, phosphorus comprised about 1 atomic percent in the Ge bulk over the range of source gas ratios used, and the conductivity remained fairly constant, indicating saturated conditions. Boron comprised about 10 atomic percent in Ge at the highest dark conductivity, while increased doping turned the films amorphous. To test the doped layers for device applications, an all-nanocrystalline germanium p-i-n diode was constructed and showed rectification when measured in the dark at room temperature.

## INTRODUCTION

Nanocrystalline germanium (nc-Ge:H) has recently attracted interest for possible use in p-i-n devices such as low gap solar cells, long-wavelength photodetectors, thermophotovoltaics, and thin-film diodes due to its high optical absorption coefficient and conductivity, and its low optical gap. Similar to µc-Si:H, nc-Ge:H will likely not exhibit the light-induced degradation of amorphous material. A working prototype nc-Ge:H p-i-n photodetector has been demonstrated [1], but which used µc-Si:H and a-Si:H for the p-type and n-type doped layers, respectively. Consequently, the intrinsic nc-Ge:H layer was graded from Ge to Si at each of the contact layers to avoid an abrupt interface.

There have been no studies to date reporting either n-type or p-type doping of nc-Ge:H. It has been shown [2] that at high hydrogen dilutions Ge films as thin as 10 nm can be substantially nanocrystalline, indicating that very thin layers of doped nc-Ge:H could be achieved. Studies on growing doped c-Ge thin films by molecular beam epitaxy have shown that phosphorus and boron are excellent n-type and p-type dopants, respectively, in Ge.

Previous studies [3-5] have shown that the crystalline structure of intrinsic nc-Ge:H is a strongly dependent upon the combination of deposition temperature and hydrogen dilution. A sequence of four phases *a-nc-a-nc* is found as deposition temperature is increased from room temperature to 350°C, and the temperatures at which the transitions occur are dependent upon the amount of hydrogen dilution. Dark conductivity of the intrinsic nc-Ge:H films deposited between 150-200°C was approximately $1\times10^{-4}$ S/cm [5], while the dark conductivities of the n- and p-doped films reported here are approximately three to four orders of magnitude higher. In addition, at deposition temperatures above 310°C, intrinsically-grown nc-Ge:H films have been found to be very strongly p-type by both hot probe and Hall effect measurements, with an electrical conductivity of approximately 10 S/cm, a thermal activation energy of about 0.040 eV, and a hole concentration of approximately $10^{18}$ cm$^{-3}$ [5].

## EXPERIMENT

Thin films of germanium were deposited on Corning 1737 glass substrates by RF (13.56 MHz) excited PECVD using $GeH_4$ and $H_2$ source gases, a chamber pressure of 800 mTorr, RF power of 43 mW/cm$^2$ of cathode area, and a $H_2/GeH_4$ flow ratio of 420 (210 sccm $H_2$, 0.5 sccm $GeH_4$). The best device quality material is achieved at high hydrogen dilutions [1, 5]. The dopant source gases were $PH_3$ and $B_2H_6$, and their flows were varied to the extent possible. The ratio of $B_2H_6/GeH_4$ was varied between zero (intrinsic reference) and 0.10, while the $PH_3/GeH_4$ ratio was varied between zero and 0.30. The doped films were all approximately 500 Å thick.

Crystallinity was measured by UV reflectance [6,7]. The electrical conductivity in the dark $\sigma_d$ and its thermal activation energy $E_{act}$ were measured between coplanar Al contacts under vacuum at 105°C to 35°C and extrapolated to room temperature. The thickness of the films was determined by plasma etching to the glass substrate and measuring the step height with a surface stylus.

For measurement of solid phase B, P, O, and H concentrations in the Ge bulk as a function of the most promising temperatures and gas flow ratios, a special stack of Ge films was grown on one glass substrate. Starting at the glass substrate, the first layer was deposited at 190°C and with 0.03 sccm of $B_2H_6$, the second layer at 150°C and 0.04 sccm $PH_3$, the third layer at 150°C and 0.03 sccm $B_2H_6$, and the fourth layer at 150°C and 0.1 sccm $PH_3$. A 150°C a-Si:H layer was added for encapsulation against surface contamination. Atomic concentrations were determined by secondary ion mass spectroscopy (SIMS) using a Ge standard implanted with H, B, and P [8].

Hall effect measurements were attempted to determine the carrier concentrations. However, even though the conductivities of the materials were high, the large resistance of the thin layers to be used in device applications prevented reliable results. Since the crystal structure [2] and conductivity [5] of nc-Ge change with thickness, doped layers much thicker than those for device applications were not tested.

To test the doped material, an all nc-Ge p-i-n diode was constructed and its current-voltage characteristics measured in the dark. A bottom contact of Al and Cr was thermally evaporated on a Corning 1737 glass substrate, and an all nc-Ge p-i-n stack was grown at the above-mentioned deposition conditions. The p-layer was deposited at 190°C and 0.03 sccm $B_2H_6$, the i-layer was deposited at 150°C, and the n-layer was deposited at 150°C with 0.1 sccm $PH_3$. The top contact consisted of 1.5 mm circles of Al evaporated through a shadow mask. Using the top contact as an etch mask, the surrounding material was plasma-etched only partway through the i-layer to avoid edge current dominating the measurement.

## RESULTS AND DISCUSSION

Figure 1 shows the UV reflectance of the doped films. The presence of a peak at approximately 275 nm indicates the $E_2$ transition of crystalline Ge. The measurements indicate all films are nanocrystalline except for the two boron-doped films with a B/Ge gaseous atomic ratio of 0.2 (0.05 sccm $B_2H_6$ flow), which are amorphous. Growth rates for the intrinsic and doped films ranged between 0.20 and 0.40 Å/sec. Figure 2 shows the dark conductivity, $\sigma_d$, as a function of deposition temperature and the dopant atom/Ge atom ratio in the PECVD gas flows. The phosphorus-doped, or n-type, material shows a fairly constant conductivity over the range tested, with the lower temperature material exhibiting a slightly higher conductivity. This indicates that the conductivity is not a strong function of source gas ratio at this high ratio of

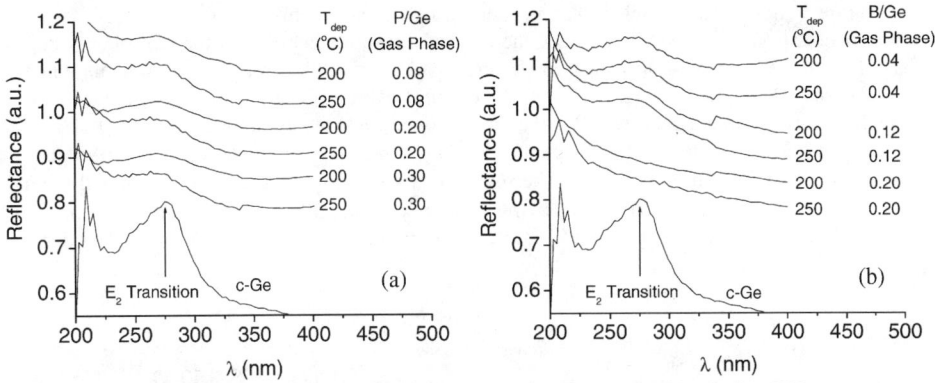

**Figure 1.** UV reflectance of nc-Ge:H films doped with phosphorus (a) and boron (b). UV reflectance of crystalline germanium is shown for comparison. The deposition temperature $T_{dep}$ and the ratio of dopant to Ge atoms in the PECVD gas phase are indicated for each film.

$PH_3/GeH_4$, a strong indicator that the doped material has reached a saturated condition. The boron-doped, or p-type, material shows a maximum conductivity at a dopant/Ge atom ratio of 0.12 in the source gases, and, in contrast to the n-type material, the higher deposition-temperature p-type material exhibits a higher conductivity. At an atomic ratio of 0.04 in the source gases, the material exhibited a conductivity similar to the intrinsic material, while at a ratio of 0.20, the material becomes amorphous. Due to the constraints of the PECVD machine, source gas ratios of less than 0.02 could not be tested while holding the hydrogen dilution at 420.

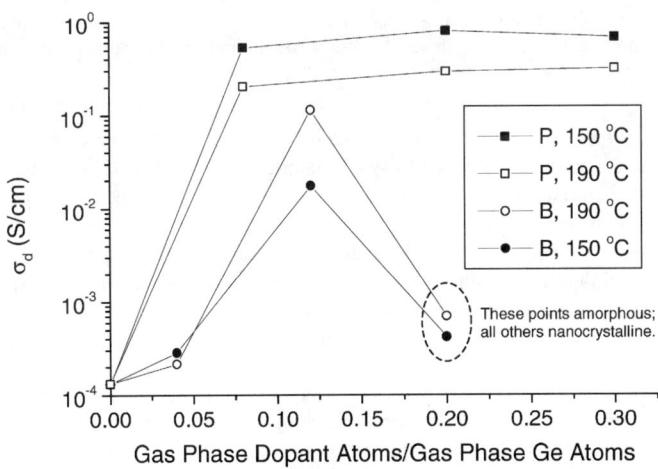

**Figure 2.** Room temperature dark conductivity, $\sigma_d$, as a function of the ratio of dopant atoms to germanium atoms in the gas flows within the PECVD chamber. The four series are indicated as doped with either phosphorus or boron, and at two different deposition temperatures.

Figure 3 shows the thermal activation energies of the same films as in Figures 1 & 2. As expected from the conductivity data, the thermal activation energies of the phosphorus-doped films are very constant over the dopant range, and are much lower than that of the undoped material. At a ratio of 0.04, the boron-doped films exhibit properties similar to the intrinsic nc-Ge material, while at a ratio of 0.2, the boron-doped films are amorphous and show a thermal activation energy for that of amorphous Ge. In contrast to the P-doped films, the B-doped film grown at 190°C exhibited a higher thermal activation energy than the film grown at the same ratio, but at 150°C. A Fermi level swing of approximately 500 meV between the n-doped and p-doped films deposited between 150-200°C is inferred from the electrical conductivity data.

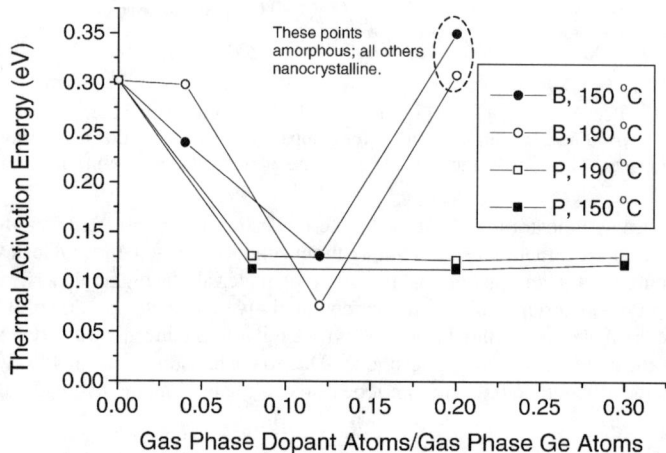

**Figure 3.** Thermal activation energy as a function of the ratio of dopant atoms to germanium atoms in the gas flows within the PECVD chamber. The four series are indicated as doped with either phosphorus or boron, and at two different deposition temperatures.

The results of the conductivity measurements indicate that the most promising films for the doped contact layers would be P-doped (n-type) material grown at 150°C and a ratio of 0.2, and B-doped (p-type) material grown at 190°C and a ratio of 0.12. A previous study [5] showed that an optimal intrinsic material is obtained at 150°C.

Figure 4 shows the results of the SIMS analysis and Table 1 lists the growth conditions of each layer shown in Figure 4. Using the value of $4.42 \times 10^{22}$ atoms/cm$^3$ for c-Ge, the concentrations of B are each about 15 atomic percent in the Ge bulk. The concentrations of P and H are each about one to two atomic percent in the Ge bulk over the temperature and doping ranges. In contrast, the intrinsic films were previously reported [5] to contain approximately 0.7% H in the Ge bulk for films grown at 150°C, and 0.5% H at 190°C. The data in Table 1 show that the H content decreases with increasing temperature, and also that the incorporation of B in to Ge decreases with increasing temperature. As expected, incorporation of P into Ge increased with increasing source gas ratios. The oxygen concentration decreased by over an order of magnitude in the phosphorus-doped layers.

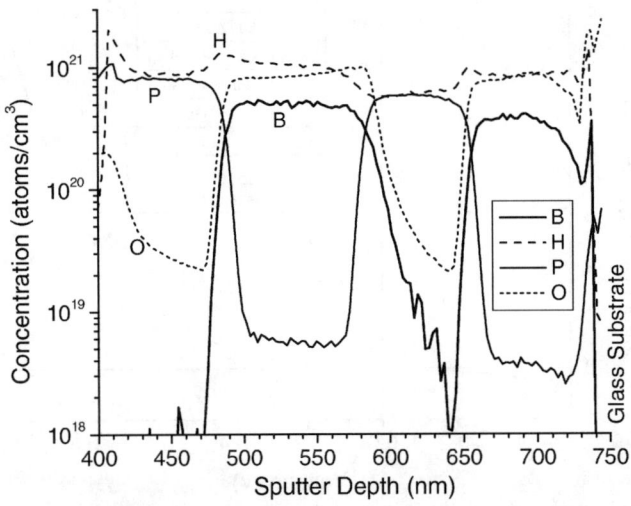

**Figure 4.** SIMS profile of doped stack, showing the B, P, H, and O concentrations in the Ge bulk. Table 1 lists the growth conditions for each layer. A 400 nm a-Si cap was grown over the Ge layers to protect against surface contamination.

**Table 1.** Growth conditions for doped stack measured by SIMS. The flows of $GeH_4$ and $H_2$ were 0.5 sccm and 210 sccm, respectively, for all layers.

| Layer | $T_{dep}$ (°C) | $B_2H_6$ flow (sccm) | $PH_3$ Flow (sccm) | Dopant/Ge atoms in gas flow |
|---|---|---|---|---|
| 4 | 150 | | 0.10 | 0.20 |
| 3 | 150 | 0.03 | | 0.12 |
| 2 | 150 | | 0.04 | 0.08 |
| 1 | 190 | 0.03 | | 0.12 |
| Glass Substrate | - | - | - | - |

(growth direction: bottom to top)

Figure 5 shows the current-voltage curve of a preliminary all nc-Ge p-i-n diode measured in the dark at room temperature. A potential barrier is clearly present, providing further evidence that the contact layers were doped n-type and p-type.

## CONCLUSIONS

The high conductivities and low thermal activation energies of the nc-Ge:H films doped with phosphorus and boron, in conjunction with observed rectification of an all nc-Ge p-i-n device, indicate that nc-Ge:H can be doped p-type with boron and n-type with phosphorus.

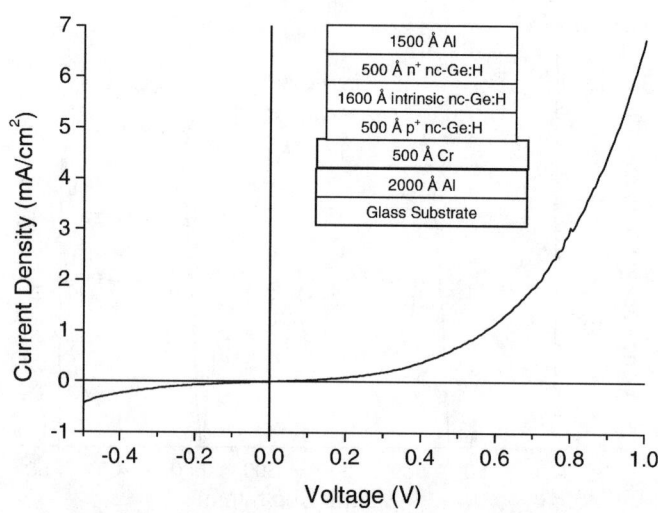

**Figure 5.** J-V curve of all nc-Ge p-i-n diode measured in the dark at room temperature. The device structure is shown as an inset.

## ACKNOWLEDGEMENTS

The authors gratefully acknowledge support of this research by NSF and EPA. The authors would also like to thank Jamie Bond and Darren Yamaguchi for their assistance with the dark conductivity measurements.

## REFERENCES

1. M. Krause, H Stiebig, R. Carius, H. Wagner, Mat. Res. Soc. Symp. Proc. **664**, A26.5 (2001).
2. W.B Jordan, E.D. Carlson, T.R. Johnson, and S. Wagner, Mat. Res. Soc. Spring Meeting, April, 2003; San Francisco, CA; paper A6.5.
3. P.R. Poulsen, M. Wang, J. Xu, W. Li, K. Chen, G. Wang, and D. Feng, J. Appl. Phys. **84**, 3386 (1998).
4. J. Jiang, K. Chen, D. Feng, and D. Sun, Thin Solid Films **230**, 7 (1993).
**149**, 93 (1989).
5. W.B Jordan and S. Wagner, Mat. Res. Soc. Symp. Proc. **715,** 509 (2002)
6. P.Y. Yu, M. Cardona, Fundamentals of Semiconductors, Springer-Verlag, 1996. pp. 375-392.
7. E.D. Palik, ed., Handbook of Optical Constants of Solids. Academic Press, 1985. pp. 472-473.
8. Evans East, East Windsor, New Jersey.

## P- and N-Type Microcrystalline SiC Fabricated by rf Plasma CVD with Ethane Gas

T. Toyama, Y. Nakano, T. Kosuge, A. Asano, H. Okamoto
Department of Systems Innovation, Graduate School of Engineering Science, Osaka University
Toyonaka, Osaka 560-8531, Japan

## ABSTRACT

We have been investigated p- and n-type microcrystalline $Si_{1-x}C_x$ (μc-SiC) films fabricated by a conventional rf (13.56 MHz) plasma CVD method with a use of a new carbon source of $C_2H_6$ gas at a low substrate temperature on a glass substrate. The Si crystallites incorporated in μc-SiC films retain with a carbon content up to 9 at.%. Both of p- and n-type μc-SiC films show relative high dark conductivities of on the order of $10^{-3}$ S/cm with optical energy gaps, $E_{04}$, of ~2 eV. In infrared spectra, any pronounced features due to C-$H_n$ vibration mode are not found in $C_2H_6$-based μc-SiC films, which is different from the case of $CH_4$-based μc-SiC films.

## INTRODUCTION

Highly conductive p- and n-type microcrystalline silicon-carbon alloys (μc-SiC) consisting of Si crystallites embedded in amorphous $Si_{1-x}C_x$ (a-SiC) are the key materials for optoelectronic devices due to their wide optical energy gap, and have been already applied to solar cells [1,2] as well as thin-film light-emitting diodes [3,4] in R&D level. In these experiments, the μc-SiC films were prepared by electron cyclotron resonance (ECR) plasma chemical vapor deposition (CVD) with a mixture of $SiH_4$ and $CH_4$ gas sources highly diluted by $H_2$ gas [1–4]. Quite high dark conductivity of >$10^{-1}$ S/cm and wide optical gap of >2.1 eV have been simultaneously achieved with a low substrate temperature of ~250°C. However, μc-SiC deposited by ECR plasma CVD has some disadvantages, e.g., with respect to deposition on transparent conductive oxide (TCO) layers, excess hydrogen radicals as well as significantly slow deposition rate of ≤~0.01 nm/s damage the TCO layer. On the other hand, some trials have been performed for fabrication of μc-SiC films by conventional rf (13.56 MHz) plasma CVD with a mixture of $SiH_4$ and $CH_4$ gases [5–8]. The highly conductive p- and n-type μc-SiC films with optical energy gaps up to ~2.1 eV have been achieved by the rf plasma CVD with a high $H_2$ dilution, however, the crystalline volume fraction of deposited SiC films made by rf plasma CVD with $CH_4$ gas as a carbon source markedly decreases with an increase in carbon content of the SiC film even if the carbon content is slightly over 3 at.% [8]. Quite recently, $C_2H_6$ has been proposed as a new carbon gas source for fabricating μc-SiC with hot-wire (catalytic) CVD [9].

Here we will demonstrate the results on p- and n-type μc-SiC fabricated by rf plasma CVD with the new carbon source of $C_2H_6$ gas. The Si crystallites are still incorporated in the deposited μc-SiC with a carbon content of 8 at.% showing high conductivities of on the order of $10^{-3}$ Scm$^{-1}$ as well as an optical energy gap, $E_{04}$, of ~2.0 eV. Additionally, the deposition rate of the μc-SiC remains ~0.1 nm/s.

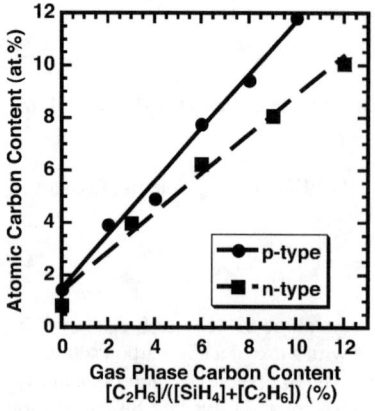
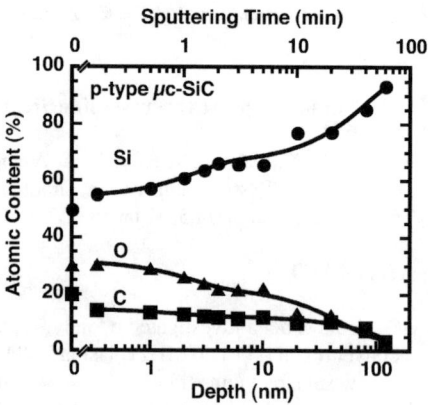

**Figure 1.** Atomic carbon contents of p- (●) and n-type (■) μc-SiC films plotted against gas phase carbon content.

**Figure 2.** Depth profile of atomic content in p-type μc-SiC film (Si; ●, C; ■, O; ▲). Solid lines are guide for the eye.

## EXPERIMENTAL DETAILS

The μc-SiC film was deposited on a Corning 7059 glass substrate or a FZ silicon substrate by a cross-field rf plasma CVD method at a deposition temperature of 225°C [10]. The applied rf power of 220 W was roughly 5–6 times larger than that of 35–45 W for an a-SiC film of this plasma CVD apparatus. A mixture of $SiH_4$ (10% diluted in $H_2$), $C_2H_6$ (1% diluted in $H_2$), $H_2$ and $B_2H_6$ (0.05% diluted in $H_2$) or $PH_3$ (0.05% diluted in $H_2$) gases was used as a source gas. The doping gas concentrations, $[B_2H_6]/([SiH_4] + [C_2H_6])$ and $[PH_3]/([SiH_4] + [C_2H_6])$ were set to be 1%. Hydrogen dilution, $([SiH_4] + [C_2H_6])/[H_2]$, was fixed to be 0.635%. Gas phase carbon concentration, $[C_2H_6]/([SiH_4] + [C_2H_6])$, was varied from 0 to 12%. Total gas pressure was kept at 133 Pa (1.0 Torr).

X-ray photoemission spectrometry (XPS) was performed for elemental analysis with Kratos AXIS-HSi. Raman scattering spectrum was measured at room temperature with the use of a microprobe system including a 514.5-nm $Ar^+$ laser (JEOL JRS-SYSTEM1000). Optical absorption spectra were derived from transmittance spectra measured with a spectrometer (Shimadzu UV-3100PC). Fourier transform infrared (FTIR) measurements were also carried out by means of Shimadzu FTIR-8100.

## DISCUSSION

Elemental analysis has been done employing XPS measurements. Figure 1 shows atomic carbon contents of deposited μc-SiC films derived from the integrated intensities of the $C_{1s}$, $Si_{2p}$, and $O_{1s}$ peaks in XPS spectra, C/(Si +C + O), because the oxygen contamination cannot be excluded in this experimental series as shown in figure 2. Figure 2 shows a typical depth profile

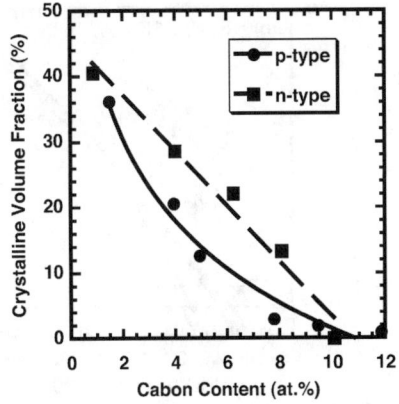
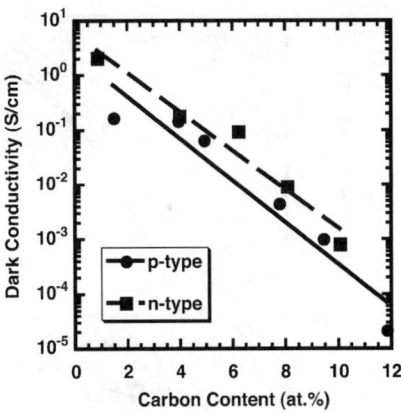

**Figure 3.** Crystalline volume fractions deduced from Raman scattering spectra plotted against carbon contents in μc-SiC (p-type; ●, n-type; ●).

**Figure 4.** Dark conductivities as a function of carbon content of both p- (●) and n-type(■) μc-SiC.

of atomic component of the p-type μc-SiC film with the carbon content of 4 at.% used in figure 1. At the surface before the $Ar^+$-ion sputtering, carbon content is extremely high ~20% due to adsorption of contamination carbon. The $C_{1s}$ peak shows a chemical shift to the location corresponding to the C-C bonding component. In figure 1, even the μc-SiC film deposited without the ethane gas flow, the carbon contents is ~1% because of the influence of the contamination. Additionally, although oxygen is not intentionally introduced, its content is high maybe due to contamination oxygen at deposition and/or post-deposition oxidation [7,11]. After the $Ar^+$-ion sputtering, the oxygen content monotonously decreases. Finally the oxygen concentration becomes within a few atomic percents, and at this point in depth, the $C_{1s}$ peak consists of C-Si bond component. Therefore, we define the carbon content in figure 1 as C/(C + Si + O) after surface sputtering until the $C_{1s}$ peak shifts to the peak corresponding to the C-Si bond component.

As shown in figure 1, the atomic carbon content almost linearly increases with increasing the gas phase carbon content. This tendency is also found in deposition of μc-SiC from a $SiH_4$ + $CH_4$ mixture gas [8]. However, the gas phase concentration of $C_2H_6$ against that of $SiH_4$ is much lower than that of $CH_4$ [7,8], showing the efficient gas decomposition of $C_2H_6$, which agrees with the binding energy of C-C bond being lower than that of C-H bond [9]. As the results, the deposition rate is as high as ~0.1 nm/s.

Figure 3 shows crystalline volume fraction deduced from the Raman scattering spectra at the Si-Si TO phonon mode of ~520 $cm^{-1}$ plotted against the carbon content in μc-SiC. The crystalline volume fraction is defined in accordance with semi-quantitative definition, i.e., $I_c/(I_c + I_a)$, where $I_c$ denotes the integrated intensity of the crystalline part, and $I_c$ that of the amorphous part [12]. With increasing the carbon content, the crystalline volume fraction decreases for both p- and n-type μc-SiC, however, crystalline part remains even in μc-SiC with the carbon content up to 9

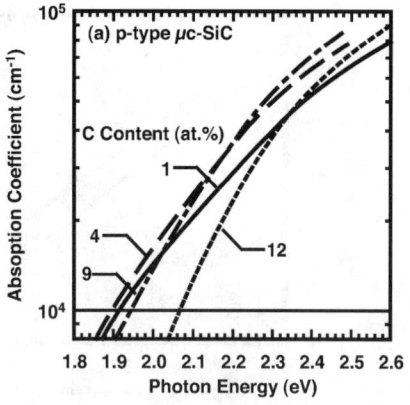 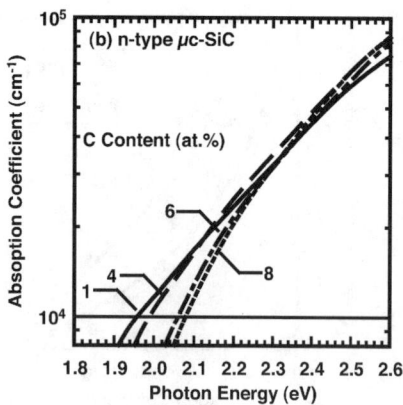

**Figure 5.** Optical absorption spectra at room temperature of p- (a) and n-type (b) μc-SiC as a function of carbon content.

at.%. The carbon content in μc-SiC deposited by rf plasma CVD employing a $SiH_4 + CH_4$ mixture gas in literature is 5 at.% [7] or below 3 at.% [9], so that the higher carbon content of μc-SiC fabricated from $SiH_4 + C_2H_6$ would arise from the low decomposition energy of C-C bond. On the other hand, at the same carbon content, the crystalline volume faction in n-type μc-SiC is higher than that in p-type μc-SiC, which is in agreement with the tendencies of the crystalline volume fraction against doping gases found in μc-SiC with $CH_4$ gas source [7], and also found in μc-Si [13]. In these samples, no pronounced signal corresponding to Si-C or C-C was found in either the Raman scattering spectra or the Raman scattering spectra of μc-SiC made from $CH_4$ using rf plasma CVD [7].

Figure 4 shows dark conductivities taken at room temperature as a function of the carbon content for both p- and n-type μc-SiC. With increasing carbon content, i.e. with decreasing crystalline volume fraction, the dark conductivities tend to decrease for both p- and n-type μc-SiC films. Furthermore, as mentioned above, differences of decomposition of carbon source gas also influence on the degree of the crystallinity. As a result, the higher crystalline fraction results in higher dark conductivities on the order of $10^{-3}$ S/cm for μc-SiC films with carbon content up to 9 at.%.

In figure 5, optical absorption spectra of p- and n-type μc-SiC are plotted as a function of the carbon content. In p-type μc-SiC, the optical absorption edge slightly shifts to red at the carbon content of 4 at.%, and the spectral tail becomes steeply. At carbon contents over 4 at.%, the absorption edge tend to shift to blue. In n-type μc-SiC, the optical absorption edge at ~$10^4$ cm$^{-1}$ monotonously shifts to blue with increasing carbon content.

Figure 6 shows the optical energy gaps, $E_{04}$, being the photon energy at the optical absorption coefficient of $10^4$ cm$^{-1}$ as a function of the carbon content. In spite of the larger carbon content compared to μc-SiC made from a $SiH_4 + CH_4$ mixture, $E_{04}$ of μc-SiC fabricated from a $SiH_4 + C_2H_6$ mixture does not indicate any significant increase. Difference in hydrogen termination is one of the possible reasons for the relative low optical gaps. Figure 7 shows FTIR spectra of p-type μc-SiC with different carbon contents. Strong features are observed at ~640 cm$^{-1}$ and

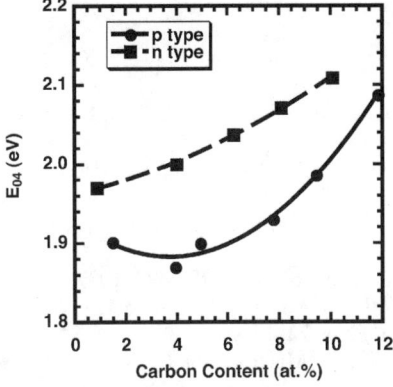
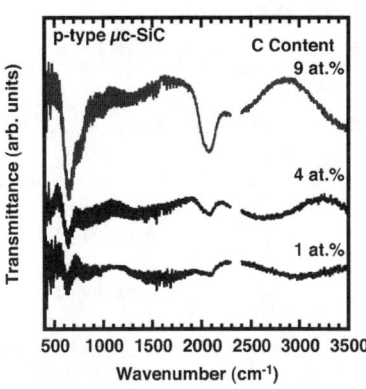

**Figure 6.** Optical energy gaps, $E_{04}$, as a function of carbon content (p-type; ●, n-type; ■).

**Figure 7.** FTIR spectra of p-type μc-SiC as a function of carbon content.

~2100 cm$^{-1}$, and ascribed to Si-H rocking/wagging mode and Si-H$_2$ stretching mode, respectively [7,14]. Additionally, weak shoulders are found at ~780 cm$^{-1}$ which can be attributed to Si-CH$_3$ rocking/wagging mode [15] and/or Si-C stretching mode [16]. Meanwhile, at 2800–3000 cm$^{-1}$ band corresponding to C-H$_n$ stretching mode, no detectable feature is observed. Furthermore, the 1250 cm$^{-1}$ spectral peak arising from Si-CH$_3$ symmetric bending mode is also not observed. No detectable feature related to methyl groups is largely different from the case of CH$_4$-based μc-SiC [7], CH$_4$-based a-SiC [14], and even from the case of C$_2$H$_6$-based μc-SiC fabricated by hot-wire CVD [9]. Thus the feature found at ~780 cm$^{-1}$ is assigned to be Si-C stretching mode. In figure 7, as the carbon content increases, or as the crystalline volume fraction decreases, the magnitude of the Si-H related peaks increases, and this tendency is also observed in FTIR spectra of n-type μc-SiC. Consequently, the increase in the optical gap is likely to be due to an increase in hydrogen incorporation as well as volume fraction of amorphous part rather than an increase in carbon content.

## CONCLUSIONS

Highly conductive p- and n-type microcrystalline $Si_{1-x}C_x$ (μc-SiC) films were fabricated by a conventional rf (13.56 MHz) plasma CVD method using a new carbon source of $C_2H_6$ at a low substrate temperature on a glass substrate. The Si crystallites incorporated in the films retain in μc-SiC with carbon content up to 10%. Both p- and n-type μc-SiC films show high dark conductivities of on the order of $10^{-3}$ S/cm with optical energy gaps, $E_{04}$, of ~2 eV with a relatively high deposition rate of ~0.1 nm/s. Being different from μc-SiC fabricated from CH$_4$ gas, methyl related vibrational absorption is not observed in the FTIR spectra, which may contribute to the higher dark conductivity as well as the lower optical energy gap.

**REFERENCE**

[1] Y. Hattori, D. Kruangam, T. Toyama, H. Okamoto, and Y. Hamakawa, J. Non-Cryst. Solids **97&98**, 1079 (1987).
[2] M.-K. Han, Y. Matsumoto, G. Hirata, H. Okamoto, and Y. Hamakawa, J. Non-Cryst. Solids **115**, 195 (1989).
[3] D. Kruangam, T. Toyama, Y. Hattori, M. Deguchi, H. Okamoto, and Y. Hamakawa, J. Non-Cryst. Solids **97&98**, 293 (1987).
[4] T. Futagi, M. Katsuno, N. Ohtani, Y. Ohta, H. Mimura, and K. Kawamura, Appl. Phys. Lett, **58**, 2948 (1991).
[5] K. Hanaki, T. Hattori, and Y. Hamakawa, Proc. of 3rd Intern. PVSEC (1987) p.49.
[6] C. Wang, G. Lucovsky, and R.J. Nemanich, J. Non-Cryst. Solids, **137&138**, 741 (1991).
[7] F. Demichelis, C.F. Pirri, E. Tresso, J. Appl. Phys. **72**, 1327 (1992).
[8] T. Wada, M. Kondo, and A. Matsuda., Sol. Energy Mater. & Sol. Cells, **74**, 533 (2002).
[9] T. Itoh, T. Fujiwara, Y. Katoh, K. Fukunaga, and S. Nonomura, J. Non-Cryst. Solids, **299–302**, 880 (2002).
[10] T. Toyama, T. Matsui, Y. Kurokawa, H. Okamoto, and Y. Hamakawa, Appl. Phys. Lett. **69**, 1261 (1996).
[11] S. Veprek, Z. Iqbal, K.O. Kuhne, P. Capezzuto, F.A. Sarrot, and J.K. Gimzewski, J Phys. C **16**, 6241 (1983).
[12] T. Matsui, M. Tsukiji, H. Saika, T. Toyama, and H. Okamoto, Jpn. J. Appl. Phys. **41**, 20 (2001).
[13] A. Matsuda, S. Yamasaki, K. Nakagawa, H. Okushi, K. Tanaka, S. Izima, M. Matsumura, and H. Yamamoto, Jpn. J. Appl. Phys. **19**, 1305 (1980).
[14] Y. Tawada, K. Tsuge, M. Kondo, H. Okamoto, and Y. Hamakawa, J. Appl. Phys. **53**, 5273 (1982).
[15] H. Wieder, M. Cardona, and C.R. Guarnieri, Phys. Stat. Sol. (b) **92**, 99 (1979).
[16] Y. Katayama and T. Shimada, Jpn. J. Appl. Phys. suppl. **19-2**, 115 (1980).

# Polymorphous Silicon Films Produced in Large Area Reactors by PECVD at 27.12 MHz and 13.56 MHz

H. Águas, L. Raniero, L. Pereira, E. Fortunato, P. Roca i Cabarrocas[1], R. Martins
Departamento de Ciência dos Materiais, Faculdade de Ciências e Tecnologia, Universidade Nova de Lisboa and CEMOP, Campus da Caparica, 2829-516 Caparica, Portugal
[2]Laboratoire de Physique des Interfaces et des Couches Minces, Ecole Polytechnique, 91128 Palaiseau Cedex, France

## ABSTRACT

This work refers to a study performed on polymorphous silicon (pm-Si:H) at excitation frequencies of 13.56 and 27.12 MHz in a large area PECVD reactor. The plasma was characterised by impedance probe measurements, aiming to identify the plasma conditions that lead to produce pm-Si:H films. The films produced were characterised by spectroscopic ellipsometry, infrared and Raman spectroscopy and hydrogen exodiffusion experiments, which are techniques that permit the structural characterisation of the pm-Si films and to study the possible differences between the films deposited at 13.56 and 27.12 MHz. Conductivity measurements were also performed to determine the transport properties of the films produced. The set of data obtained show that the 27.12 MHz pm-Si:H can be grown at higher rates with less hydrogen dilution and power density, being the resulting films denser, chemically more stable and with improved performances than the pm-Si:H films grown at 13.56 MHz.

## INTRODUCTION

Polymorphous silicon (pm-Si:H) [1,2] is a material with very similar properties to amorphous silicon (a-Si:H) but presents a structural difference that consists of having about 2% of small crystallites (2 to 5 nm) imbedded in an amorphous tissue that are identified by the presence of a sharp peak at about 693 K (420°C) in hydrogen exodiffusion experiments [3]. The peculiar structure of pm-Si allows it to have a low defect density ($\sim 5 \times 10^{14}$ cm$^{-3}$) and higher resistance to light soaking than a-Si:H films. It also presents better transport properties than a-Si:H, which is highly suitable for the production of optoelectronic devices such as solar cells [3].

pm-Si:H can be deposited in a PECVD diode system similar to that used to deposit a-Si:H. The key factor to deposit pm-Si:H is to keep the plasma close to the so-called α-γ' transition regime conditions [4] in order to take profit of the small particles 2- 5 nm that are formed in the bulk of the plasma that can be incorporated in the growing film if the residence time of the nanometer particles is bellow the threshold (limit) time for their coagulation [5]. To keep the plasma close to the α-γ' transition regime, high dilutions of $SiH_4$ in hydrogen ($H_2$), about 3%, are used. Monitoring of the plasma impedance can give information about the α-γ' transition, since it is known that a change in the plasma impedance occurs and it becomes more resistive [6].

In spite of all the studies performed so far in pm-Si:H, no significant attention has been put on the role of the excitation frequency in the deposition of pm-Si:H or on what happens when the films are deposited in large area reactors. Until now the main studies have been performed in small laboratory reactors [3] working at 13.56 MHz. The results presented in this paper show that the increase of the discharge frequency results in structural changes to pm-Si:H that improves the transport properties, besides allowing the increase of the deposition rate.

## EXPERIMENTAL DETAILS

The pm-Si:H films were deposited in a diode type PECVD reactor. This reactor uses a central r.f. electrode that consists of three short-circuited aluminum plates with holes, to allow the gas to flow between the plates. This configuration, called SGE (Short-circuited Grid Electrode), was previously studied [7] and has the property of enhancing the electrode effective surface area, leading to a self-dc bias of nearly 0V. The substrate and the r.f. electrode areas are both equal to 25 cm × 40 cm. The process parameters used were: 160 Pa ≤ pressure ≤ 240 Pa; 100 mW/cm$^2$ ≤ r.f. power ≤ 150 mW/cm$^2$, substrate temperature of 473 K; 20 sccm ≤ silane flow ≤ 45 sccm and hydrogen (H$_2$) flow of 340 sccm. The plasma impedance was monitored with an Advanced Energy Z-Scan probe, inserted on the connection cable between the matching box and the reactor. This probe allows the measurement of the plasma impedance, which gives information related to the particle formation and coagulation in the plasma [8].

The pm-Si:H were characterized by SE (spectroscopic ellipsometry) using a Jobin Yvon UVISEL DH10 ellipsometer. The pm-Si:H film was modeled with the Tauc-Lorentz [9] model, where five parameters are required to describe the real $<\varepsilon_1>$ and imaginary $<\varepsilon_2>$ parts of the dielectric function of the material: $E_0$, $C$ and $\varepsilon_1(\infty)$ are the resonance energy, the broadening term of the Lorentz oscillator and the real part of the dielectric function at high energies, respectively. $E_g$ is the optical gap and the constant $P$ is the Cauchy principal part of the integral. $A$ is a fifth fitting parameter that is related to the density of the material. The hydrogen content (%H) and microstructure factor ($R_f$), i.e. the ratio of SiH$_2$ to the total (SiH + SiH$_2$) bonds, were determined by IR-FTIR (ATI Mattson Genesis).

The hydrogen exodiffusion was performed in a vacuum furnace pumped down to 1.33×10$^{-6}$ Pa (10$^{-6}$ Torr) by a turbomolecular pump. The heating rate of the sample was fixed at 10 K/min. Raman spectra were recorded for 300 s in order to have a good signal to noise ratio with a micro-Raman spectrometer DILOR XY. The excitation was done by a He-Ne laser at 632 nm. The laser beam was focused on the surface of the sample with a magnification of 50× and the power kept below the threshold of crystallization, by monitoring the spectra of an amorphous silicon film used as reference [10]. Coplanar conductivity measurements as a function of temperature, under dark and AM1.5 illumination conditions were also performed.

## RESULTS

During the first seconds of the discharge the plasma impedance can change very fast, and this change carries information on the nanoparticle formation and coagulation into powders. Fig. 1 shows the resistance ($R$), and reactance (j$X$) dependence on time for a pm-Si plasma condition of a) 13.56 MHz discharge and b) 27.12 MHz discharge. Comparing both data we notice that the changes in $R$ and j$X$ during the first seconds of the 13.56 MHz discharge are smaller than the ones observed at 27.12 MHz. These changes are related to the particle formation and coagulation in the plasma [8]. Initially, for the 27.12 MHz discharge, it is observed a slightly decrease in R. This behavior is attributed to the growth of the first clusters with sizes of about 2 nm, which causes a momentary increase of the electron density, but, when the clusters reach a size of about 5 nm, they start coagulating into larger particles also called powders that attach electrons. This leads to an overall decrease of the electron density and so, $R$ increases, as it is observed during the 2 to 9 s of the 27.12 MHz discharge. When $R$ increases, due to the powder formation, the effective current density coupled to the plasma decreases slightly inverting the powder formation process.

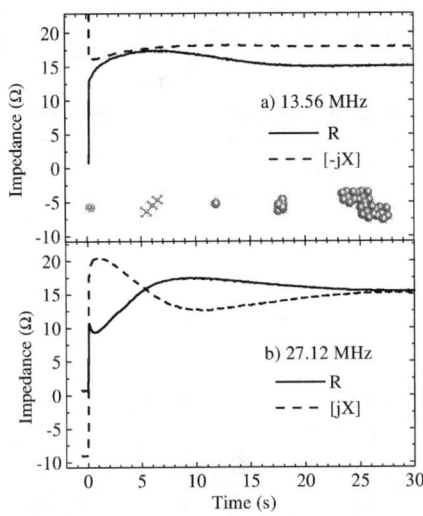

**Figure 1.** Variation of the plasma impedance (R and jX) during the first seconds of a discharge at a) 13.56 MHz and b) 27.12 MHz that lead to pm-Si:H deposition.

This causes $R$ to slightly decrease into an equilibrium situation where it finally stabilizes. We noticed that for the 27.12 MHz the stabilization of the plasma impedance occurs for longer times than with the 13.56 MHz discharge and also that the variations in the impedance are larger. Thus, the particles formed have enhanced residence times and so, they can contribute in much larger quantities to the growth process of pm-Si:H films before coagulating into powders. This allowed us to increase the $SiH_4$ fraction in the plasma (without going to amorphous silicon deposition conditions), enhancing so the number of clusters formed in the plasma, which lead to the increase of the deposition rate. We also verify that in our deposition system the plasma impedance changes from capacitive (negative $jX$, at 13.56 MHz, to inductive at 27.12 MHz (positive $jX$). This variation was also observed by other authors [11] and can be related to the ratio of the excitation frequency to the collision frequency, which leads to the increase of $jX$ with the excitation frequency.

Fig. 2 shows the differences between the H exodiffusion spectra of pm-Si:H deposited at 13.56 MHz and 27.12 MHz and of a-Si:H films. The a-Si:H is characterized by a broad band at about 550°C (823 K) associated with the breaking of isolated tightly Si-H bonds, and a second narrow peak at around 400°C (673 K) associated to weak H bonds [12]. The $\mu c$-Si:H is characterized by the appearance of a sharp peak at about 450°C (723 K) plus a shoulder at about 570°C (843 K). The position of the sharp peak decreases as the porosity of the $\mu c$-Si:H increases. The shift of the peaks towards high temperatures is usually related to more chemically stable structure. To contrast with the simplicity of these spectra, pm-Si:H presents a highly complex spectrum with 6 or 7 different H effusion modes, being the main characteristic the sharp peak located around 420°C (693 K). This peak is usually called the fingerprint of pm-Si:H since it has been associated with the effusion of the H bonded to the surface of clusters and crystallites [3].

Fig. 3 shows the IR spectra for the same samples. For the pm-Si:H produced at 13.56 MHz we observe that the spectra is deconvoluted by three Gaussians, respectively centered at 1997, 2030 and 2093 $cm^{-1}$, exhibiting average half-width respectively of about 85, 67.5 and 66.5 $cm^{-1}$. Here, the band centered at around 2030 $cm^{-1}$ is currently associated to the H bonded to the clusters and crystallites presented in the film [3]. On the other hand, the IR spectra of the pm-Si:H produced at 27.12 MHz can be properly fitted using only 2 Gaussians, centered at 2004.5 $cm^{-1}$ and 2104.5 $cm^{-1}$. Apart from that, the half-width of each Gaussian is respectively around 100 $cm^{-1}$ and 55 $cm^{-1}$. That is, for 27.12 MHz pm-Si:H films the deconvolution of the spectra do not show the band centered at 2030 $cm^{-1}$, but the other peaks are slightly shift towards high energies, meaning that the hydrogen content of the films should be enhanced as observed by other workers using plasma beam deposition techniques, where a high atomic hydrogen flow is present [13]. Apart from that, the half-width of the first band is slightly widened while the half-width of the third band is slightly decreased. This could mean that the type and nature of clusters formed and how H is bonded to them or to the nanocrystals are different, suggesting the existence of stronger tight bonds for the 27.12 MHz than for the 13.56 MHz pm-Si:H. This is also confirmed by the exodifusion data where a shift of the initial peak towards high temperatures is observed, for the 27.12 MHz pm-Si:H. These results clearly suggest a different structure for the 27.12 MHz pm-Si:H, that in spite of having been deposited at high rates and having a higher H content, it is constituted by a highly dense, relaxed and ordered structure with a less intense peak at 2090-2090 cm$^{-1}$ than the 13.56 MHz pm-Si:H.

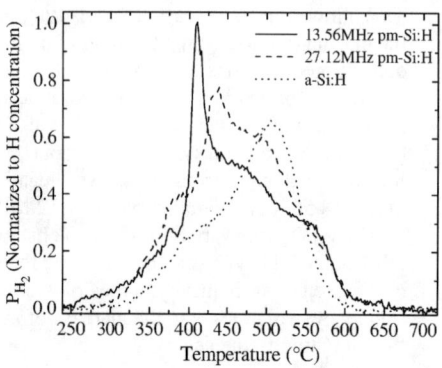

**Figure 2.** Comparison between the hydrogen exodiffusion spectra of pm-Si:H deposited at 13.56 and 27.12 MHz and of a-Si:H.

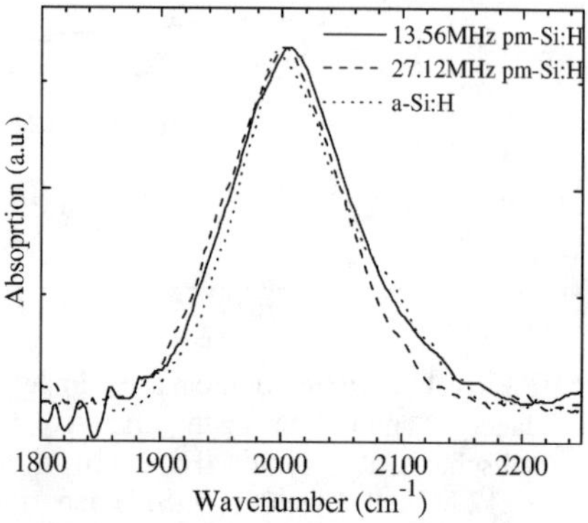

**Figure 3.** Comparison between the infrared stretching modes of pm-Si:H deposited at 13.56 and 27.12 MHz and of a-Si:H.

Fig. 4 compares the center of the Raman shift and the half-width of the TO mode of an a-Si:H a 13.56 MHz and a 27.12 MHz pm-Si:H film. The amorphous state of silicon is characterized by a broad band at 480 cm$^{-1}$, corresponding to the transverse optic (TO) mode [4]. In pm-Si:H we found that the TO mode is shifted to higher wavenumbers (483-484 cm$^{-1}$), indicating the presence of small microcrystallites detected by transmission electron microscopy on pm-Si films [14]. Besides that, the half-width of the bands of the pm-Si:H films produced at 27.12 MHz are smaller than what is usual obtained in standard a-Si:H and about 1.5 cm$^{-1}$ less than what is observed in the 13.56 MHz pm-Si:H films. This indicates that the 27.12 MHz pm-Si:H has a more ordered structure since the decrease of the half-width is correlated with a decrease of the bond angle distortion [15], explained by the fact that we are depositing the film in conditions close to µc-Si formation, favored by the use of high frequency discharge. This also could be an indication that the number of the small crystallites in the film increases, leading to a higher ordered structure. These results agree with the spectroscopic ellipsometry data shown in Fig.5. There, we show the dielectric function of an a-Si:H film, at 13.56 MHz and at 27 MHz pm-Si films, extracted from modeling the experimental data with the Tauc-Lorentz function. The results of the fitted parameters are shown in Table I. The amplitude of $<\varepsilon_2>$ is related to the density of the film and proportional to the constant $A$, centered in $E_0$, while the constant $C$ is related to the spread in the bond angle. Thus, low values of $C$ are related to an improved order of the film, while high values of the amplitude of $<\varepsilon_2>$ are related to denser films. The analyses of these data show that the maximum of $\langle\varepsilon_2\rangle$ is shifted to high energy in the pm-Si:H films, when compared to standard a-Si:H films. However, the $\langle\varepsilon_2\rangle$ maximum is higher for the 27 MHz pm-Si:H, which means that the 27.12 MHz pm-Si:H have a denser and more ordered structure (small $C$) than the 13.56 MHz pm-Si:H, in spite having a higher H content and $E_{op}$. Apart from that, the shift of the $\langle\varepsilon_2\rangle$ curves towards high energies can be ascribed to the presence of a small crystalline fraction, (about 6%) in the pm-Si:H, since the $\mu$c-Si:H has the maximum of the $\langle\varepsilon_2\rangle$ function at 4.2 eV, or to a decrease in weak Si-Si by substitution with stronger Si-H bonds [16].

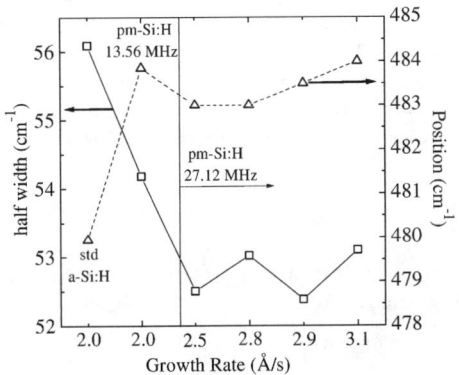

**Figure 4.** Comparison of the center of the Raman TO mode and its half-width between a standard a-Si:H, a 13.56 MHz pm and four 27.12 MHz pm-Si:H films growth at different rates.

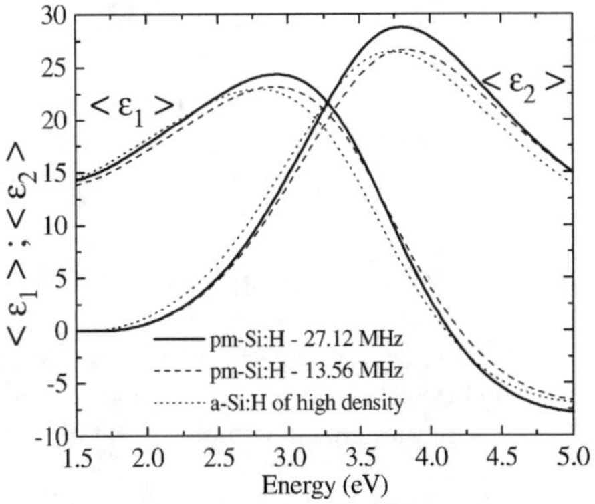

**Figure 5.** Comparison of the real $\langle\varepsilon_1\rangle$ and imaginary $\langle\varepsilon_2\rangle$ parts of the dielectric function for a 27.12 MHz a 13.56 MHz pm-Si:H and a standard high-density a-Si:H.

**Table I.** $E_{op}$, spectroscopic ellipsometry fitting parameters (Tauc-Lorentz), hydrogen content (%H) and $R_f$ of an a-Si:H film, a 13.56 MHz pm-Si:H and a 27.12 MHz pm-Si:H films.

| Sample | r (Å/s) | $E_{op}$ (eV) | Spectroscopic Ellipsometry | | | | | Infrared | |
|---|---|---|---|---|---|---|---|---|---|
| | | | $E_g$ | $\varepsilon_{2\,max}$ | A | $E_0$ | C | %H | $R_f$ |
| a-Si:H | 2.0 | 1.660 | 1.660 | 26.44 | 216.0 | 3.619 | 2.290 | 12 | 0.050 |
| 13.56MHz pm-Si:H | 1.8 | 1.819 | 1.724 | 26.61 | 210.5 | 3.717 | 2.290 | 19 | 0.105 |
| 27.12MHz pm-Si:H | 3.1 | 1.782 | 1.724 | 28.07 | 216.5 | 3.652 | 2.076 | 22 | 0.015 |

Another indication of the high density of the 27.12 MHz pm-Si:H films is given by the micro structure factor $R_f$ that exhibits a value close to zero. This behavior can only be explained by a denser amorphous network were the weak Si-Si bonds are replaced by strong Si-H bonds, causing the decrease of voids, where typically the SiH$_2$ bonds occur.

Concerning the dark and photoconductivity measurements performed, the data in Table II show that the activation energy ($\Delta E$) is higher in the 27.12 MHz pm-Si:H than in 13.56 MHz. The photosensitivity data show that although both films exhibit values in the same range, the photoconductivity is higher in the 27.12 MHz pm-Si:H. The same happens concerning the pre-exponential factor $\sigma_0$ that for the 27.12 MHz pm-Si:H reaches values of about $10^8$, four orders of magnitude higher than the typical values.

**Table II.** Conductivity measurements of the 13.56 MHz pm-Si:H and of the 27.12 MHz pm-Si:H

| Sample | $\sigma_0$ $(\Omega cm)^{-1}$ | $\Delta E$ (eV) | $\sigma_d$ $(\Omega cm)^{-1}$ | $\sigma_{ph}$ $(\Omega cm)^{-1}$ | $S$ $(\sigma_{ph}/\sigma_d)$ |
|---|---|---|---|---|---|
| 13.56MHz pm-Si:H | $6.0 \times 10^4$ | 0.83 | $6.7 \times 10^{-10}$ | $3.1 \times 10^{-4}$ | $4.7 \times 10^5$ |
| 27.12MHz pm-Si:H | $1.1 \times 10^8$ | 0.96 | $3.2 \times 10^{-9}$ | $2.4 \times 10^{-3}$ | $7.5 \times 10^5$ |

## CONCLUSIONS

The results presented suggest that the 27.12 MHz pm-Si:H has a more stable, relaxed and ordered structure with high H effusion temperatures and lower $SiH_2$ content than the 13.56 MHz pm-Si:H films in spite of having a higher H content. This is confirmed by the electronic and optical properties of the films where values of $\sigma_0$ of $10^8$ $(\Omega cm)^{-1}$ and $\Delta E$ of 0.96 eV were obtained for 27.12 MHz pm-Si:H, suggesting a very low density of gap states for these films. In spite of this, further study is still required to gain a better understanding on the reason of the differences pointed out.

## ACKNOWLEDGEMENTS

This work was performed in the frame of the European project - H-Alpha Solar (ERK6-CT-1999-00004) and under the financial support given by "Fundação para a Ciência e a Tecnologia" through pluriannual contract with CENIMAT.

## REFERENCES

[1] P. Roca i Cabarrocas, P. Gay, A. Hadjadj, J. Vac. Sci. Technol. **A 14**, 655 (1996).
[2] J. Costa, P. Roura, P. Roca i Cabarrocas, G. Viera, E. Bertran, Mat. Res. Soc. Symp. **507**, 499 (1998).
[3] P. Roca i Cabarrocas, A. Fontcuberta i Morral, Y. Poissant, Thin Solid Films **403-404**, 39 (2002).
[4] P. Roca i Cabarrocas, A. Fontcuberta i Morral, S. Lebib, and Y. Poissant, Pure Appl. Chem. **74**, 359 (2002).
[5] A.A. Fridman, L. Boufendi, T. Hibid, B.V. Potapkin, A. Bouchoule, J. Appl. Phys. **79**, 13031 (1996).
[6] B. M. Jelenkovic, A. Gallagher, J. Appl. Phys. **82**, 1546 (1997).
[7] H. Águas, V. Silva, I. Ferreira, E. Fortunato and R. Martins, Phil. Mag. B. **80**, 475 (2000).
[8] L. Boufendi, J. Gaudin, S. Huet, G. Viera, M. Dudemaine, Appl. Phys. Lett. **79**, 4301 (2001).
[9] G. E. Jellison, M. F. Modine, Appl. Phys. Rev. Letts. **69**, 415 (1996).
[10] J.S. Lannin: *Raman Scattering for Amorphous Si, Ge and their alloys* (chapter 6), Semiconductors and Semimetals, 21, Part B, Academic Press, Inc. (1984).
[11] M. Heintze, Solid State Phenom. **44-46**, 181 (1995).
[12] W.B. Jackson, A.J. Franz, H.-C. Jin, J.R. Abelson, J.L. Gland, J. of non-Cryst. Solids, **227-230**, 143 (1998).
[13] R. Severens, G. Brussaard, M. Van de Sanden, D. Schram, App. Phys. Lett. **67**, 491 (1995).
[14] D.V. Tsu, B.S. Chao, S.R. Ovshinsky, S. Guha, J. Yang, Appl. Phys. Lett. **71**, 1317 (1997).
[15] D. Beeman, R. Tsu, M.F. Thorpe, Phys. Rev. **B 32**, 874 (1985).
[16] A. Fontcuberta i Morral: *Croissance, propriétés structurales et optiques du silicium polymorphe*, PhD Thesis, Ecole Polytechnique (2001), Paris.

## Surface Roughness Study of Low-temperature PECVD $a$-Si:H.

George T. Dalakos[1], Joel L. Plawsky[2], and Peter D. Persans[3]
[1]General Electric Global Research Center, Niskayuna, NY
[2]Department of Chemical Engineering, Rensselaer Polytechnic Institute, Troy, NY
[3]Department of Physics, Rensselaer Polytechnic Institute, Troy, NY

## ABSTRACT

Surface topography of $a$-Si:H thin films, deposited at 75°C by Plasma-enhanced Chemical Vapor Deposition (PECVD) has been examined using helium/silane feedstock mixtures under different substrate bias conditions. Notable differences in the surface roughness evolution are shown for films deposited in "cathodic" versus "anodic" mode – where the substrate is placed on the powered and grounded electrode respectively. Smooth and apparently featureless surfaces result from deposition on RF powered surfaces, upon which a self-bias induces high-energy ion bombardment. Rougher surfaces result from films deposited on electrically grounded surfaces. These anodic films show that after a transition period, surface roughness grows linearly with processing time, exhibiting mounded type growth as evidenced by 2-D power spectral density functions of surface height measurements. Linear growth in roughness has been predicted for shadow growth models assuming film precursor sticking coefficients of one and random angle approach of film precursor species. Growth of this nature has not been reported before in $a$-Si:H studies, which usually assume directional deposition conditions and sticking coefficients less than unity – occurring even at low processing temperatures.

## INTRODUCTION

Thin films with relatively smooth surfaces are highly desirable and even critical in some high-performance optical thin film applications. For example, applications in thin film waveguides and laser mirrors, which inherently involve many reflections or require very high reflectance. Surface roughness leads to diffuse scattering and power loss, which gets worse as the refractive index contrast at the optical interface increases. When silicon is considered, whose high refractive index is advantageous in optical applications, scattering problems become more important versus other materials [1].

One successful method for fabricating thin films with smooth surfaces is chemical vapor deposition, albeit processed at relatively high temperatures. These high temperatures provide smooth interfaces from very good film precursor diffusion on the surface. Highly mobile film precursors are able to move about the surface and able to fill in voids suppressing overhangs and columnar morphologies, both of which lead to surface roughness.

Although high substrate temperatures are favorable processing conditions for forming smooth film interfaces, they are incompatible with temperature-sensitive materials such as some optical plastics. Despite this, smooth surfaces with a featureless microstructure are still obtainable under certain low-temperature deposition conditions. Indeed, by providing kinetic energy to the surface, ion bombardment can be used to make up for the lack in thermal energy to achieve smooth interfaces. In fact, many studies have shown the beneficial results of ion-assisted growth in PECVD a-Si:H thin films for a wide range of film properties including defect density, monohydride bond formation, film density, and deposition rate [2].

A common method for depositing amorphous silicon, which can offer high ion bombardment of the surface, is PECVD – especially when the substrate is placed upon the powered electrode. This configuration is typically not used in semiconductor device manufacturing in order to avoid electrical device damage [3]. For optical thin films however, this only presents a problem at very high energies where sputtering and ion implantation is appreciable.

We have previously studied the affect of ion bombardment with an argon/helium-diluted silane process to obtain very smooth surfaces [4]. Argon is commonly used in ion bombardment studies and was felt to be critical in the observed smooth surfaces. With argon dilution, we reported a square root time dependence of the RMS surface roughness at low ion bombardment conditions while observing featureless surface structure at high ion bombardment. This study seeks to answer on whether it would be possible under helium dilution only, to obtain mirror-like surfaces under high ion bombardment and have a square root-time dependence of the roughness under low ion bombardment.

**EXPERIMENTAL**

A capacitively coupled 13.56MHz Plasma-Therm 790 deposition/etch system was used to deposit amorphous silicon thin films on glass substrates. One aspect of this parallel plate system design allows either the top or bottom electrode surface to be powered while electrically grounding the other via a pneumatic air switch. We refer to the powered bottom electrode configuration as running in "cathodic" mode, with the other being the "anodic" mode. A blocking capacitor located between the matching network and the RF electrode creates a large negative self-bias voltage of 280V on the electrode surface [5]. Substrates were placed on the bottom electrode surface. It should be noted that even in anodic mode, the substrate is also being bombarded with ions, albeit at much lower energies (estimated at a fraction of an eV). A 2% silane in helium gas mixture was introduced through the top showerhead electrode plate. The system pressure and RF input power was set at 200mTorr and 200W (0.33 W/cm$^2$) respectively. The substrate temperature was set at a constant 75°C.

Anodic and cathodic film samples under the above processing conditions were prepared over a range from 15 to 180 minutes and deposited on Corning 1737F aluminosilicate glass substrates.

Tapping-mode Atomic Force Microscopy (AFM) analyses were performed using a Digital Instruments Dimension 3100 Microscope with Nanoscope IIIa Controller and Tap300 etched Si tips from Nanodevices. A scan size of 2μm x 2μm was used for all the samples. Power spectra could be calculated from the raw AFM height data.

Spectroscopic ellipsometry was also used to investigate the surface roughness of the films. A Tauc-Lorentz (TL) single oscillator model was used to model the fully dense bulk film. Effective media approximations using the TL model and air were used to model an intermediate layer between the substrate and bulk (we have previously observed an interfacial layer through electron microscopy in some of our samples [4]), voids in the bulk, and a top surface roughness layer.

# RESULTS

Both AFM and ellipsometric analyses showed the RMS roughness to increase as a function of processing time for the anodic samples. The surface roughness of the cathodic samples was independent of time. In fact, no true height data could be resolved from the noise in the cathodic samples. Deposition rates were constant with time, calculated to be approximately 120 and 140 Å/min for the anodic and cathodic samples respectively.

The RMS roughness data from both SE and AFM analysis are shown below in figure 1.

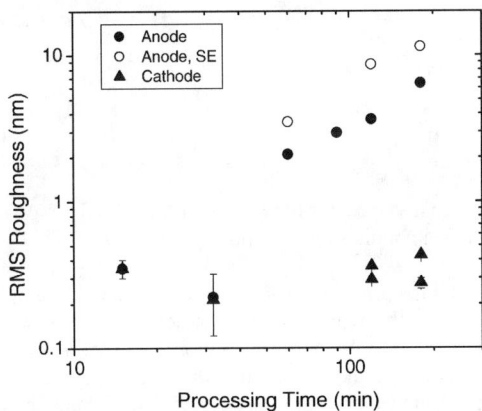

**Figure 1.** RMS roughness measured by AFM and Spectroscopic ellipsometry (SE) of the anodic and cathodic films as a function of deposition time.

The anodic-grown films show a systematic increase in roughness after about 30 minutes of processing. Before this transition, the films are very smooth and the roughness cannot be differentiated from the noise in the AFM analyses. SE analysis of the same data (below 30 minutes of deposition time) minimizes the fitting error best without including a top roughness layer in the model. While the roughness grows at the same rate, which is of immediate interest, the magnitude of roughness from the SE analysis is larger than the AFM measurements. This discrepancy of the RMS roughness obtained from SE analysis and AFM measurements has been observed before [6,7].

The 2-D isotropic power spectral density (PSD) function plotted against the wavenumber, $K$ ($=2\pi/\lambda$) of the anodic sample set yields a distinct peak as clearly shown below in figure 2, which becomes more prominent with time after 30 minutes. Furthermore, the PSD peak position and width clearly appear to change with time.

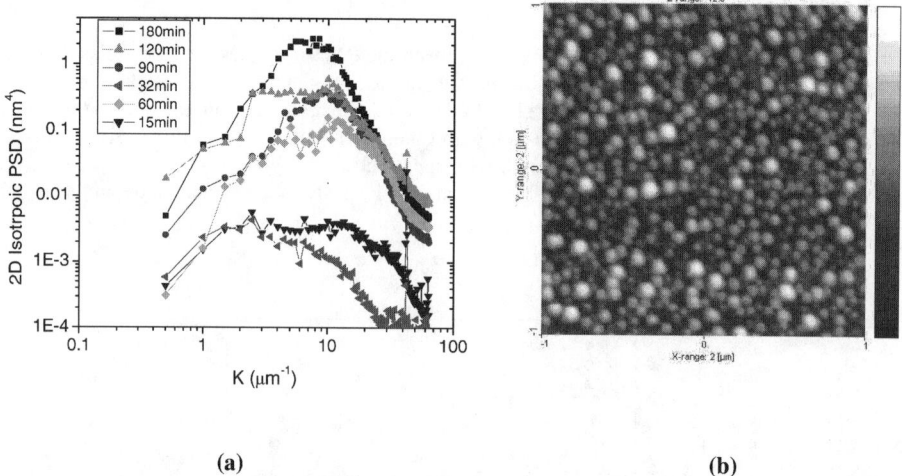

**Figure 2.** (a) 2-D power density spectra of the anodic films as a function of processing time. (b) Raw height surface map of an anodic sample deposited for 180min.

A distinct peak in the 2D isotropic power spectrum of the anodic samples is an indication of mounded growth -- an unusual observation for plasma-deposited amorphous silicon growth. Studies of $a$-Si:H (and $a$:Si) have previously reported only self-affine film growth, which uses three parameters to uniquely define the surface growth front [8-13] and should not exhibit a peak in the 2-D PSD function. Mounded growth is associated with a characteristic feature size, requiring only two time-dependent parameters to uniquely define the evolving surface [14].

The two mounded growth parameters are the average mound separation, $\zeta$, which can be obtained from the PSD peak position, $k_0$, ($\zeta = 2\pi/k_0$) and the RMS roughness, $w$. These parameters scale with time in the following manner:

$$\zeta \propto t^\gamma \quad \text{and} \quad w \propto t^\beta$$

where $\gamma$ and $\beta$ are the lateral coarsening exponent and surface roughness exponent respectively. The average mound separation, measured after 30minutes, increases from 0.51μm to 0.83μm with $\gamma = 0.475 \pm 0.15$. The surface roughness exponent, calculated from the AFM data set (excluding the first 30 minutes of deposition) is $\beta = 1.13 \pm 0.07$. The SE data also shows linear growth of the roughness, despite the magnitude difference.

In addition to the scaling behavior of the mounded growth parameters, the system correlation length, which is inversely proportional to the PSD peak width increases with deposition time. Increases in the system correlation length reflect a more ordered surface topology. Our samples exhibit more uniform distribution of mound sizes as the deposition time increases.

A relatively high value of beta that we report for longer deposition times for the anodic sample set is also predicted the zeroeth-order growth deposition model (ZOG) – a deposition model which predicts $\beta = 1$ to occur [15]. Zeroeth order growth results from assuming a sticking coefficient of one (i.e., no surface diffusion or re-emission) and a random incidence (non-directional) film precursor flux. A frequently reported upper limit of $\beta = 0.5$ from other $a$-Si:H studies cite the random deposition model to explain low-temperature conditions[8-13]. The random deposition model also assumes a sticking coefficient of one but differentiates from the ZOG model by assuming a normal incidence flux.

Fluxes in CVD processes are inherently non-directional (versus line-of-sight PVD processes) where an isotropic cosine distribution is a good representation of the incoming particles [16]. Assuming this to be the case, mechanisms must exist to smooth out the surface in order to achieve a $\beta < 1$ reported in other CVD processes [4,8-11].

The other assumption of no diffusion or re-emission may not be as justifiable in our system. Even at low temperatures, film precursor species (notably, $SiH_3$) should exhibit some diffusion on the hydrogen-covered surface due to the weekly bonded nature on the hydrogenated surface [17].

The following discussion applies to the anodic samples only. All cathodic samples appeared mirror-smooth – independent of processing time. Apparently, under these conditions there is adequate energy to prevent any surface structure to evolve.

It is believed that cathodic samples processed under lower ion energy conditions would exhibit surface roughness intermediate between this anodic and cathodic sample set. This experimental regime would be useful to examine more closely to better understand surface smoothing mechanisms.

## CONCLUSIONS

The following work has shown unusual growth behavior of $a$-Si:H prepared by PECVD. We have quantitatively shown a significant reduction in the surface roughness and apparent zero roughness evolution through the cathodic sample set – usually observed at much higher temperatures. The anodic films follow a mounded-type growth, linearly increasing in RMS roughness as a function of time and corresponding to $\beta = 1$ (at longer deposition times). This magnitude of the surface roughness exponent has not been reported before for $a$-Si:H deposition. In addition to the unusually high roughness exponent, mounded-type growth is clearly shown in the 2-D isotropic PSD functions of the anodic films. Details of the mounded-type growth were extracted from the time-dependent PSD function such as system correlation and mound separation lengths. There is an existing zeroeth-order growth model which predicts a $\beta = 1$, reflected in the anodic sample set, though we cannot rule out diffusion occurring on the $a$-Si:H surface. However, the assumption of non-directional film precursor flux in the ZOG model is believed to better represent CVD-type processes. It is apropos to suggest that similar $a$-Si:H CVD processes are better represented by the assumption of isotropic incidence flux of the film precursor species and smoothening mechanisms, if applicable. This is in contrast to reported self-affine surfaces with an upper limit of a random deposition model which places an upper limit of $\beta = 0.5$. We are currently investigating the mechanisms behind having both mounded-type growth and $\beta = 1$ as well as the contribution of ion energy to the smoothening of these films.

## ACKNOWLEGEMENTS

We wish to acknowledge J. Drotar, A.H.M. Smets and M.C.M. van de Sanden who provided helpful insight into surface roughness processes and J. Teetsov (GE GRC) for conducting AFM scans.

## REFERENCES

1. A.M. Agarwal et al., J. Appl. Phys., **80** (11), 6120, (1996).
2. W.G.J.H.M. van Sark, in *Handbook of Thin Film Materials*, edited by H.S. Nalwa *Volume 1: Deposition and Processing of Thin Films*, (Academic Press, 2002), pp.1-79.
3. M. Kondo, M. Fukawa, L. Guo, and A. Matsuda, J. Non-Cryst. Solids **266-269**, 84 (2000).
4. G.T. Dalakos, J.L. Plawsky, and P.D. Persans, Mat Res. Soc. Symp. Proc., **715**, A.19.4. (2002).
5. K.Kohler, J.W. Coburn, D.E. Horn, E. Kay, and J.H. Keller, J. Appl. Phys. **57** (1) 59 (1985).
6. J. Koh, Y. Lu, C.R. Wronski, Y. Kuang, R.W. Collins, T.T. Tsong and Y.E. Strausser, Appl. Phys. Lett. **69** (9), 1297 (1996).
7. J. Aue and J.Th. M. De Hosson, Appl. Phys. Let., **71** (10), 1347 (1997).
8. D.M. Tanenbaum et al., Phys., Rev., B **56**, 4243 (1997).
9. A.J. Flewitt et al., J. Appl. Phys., 85, 8032 (1999).
10. M. Kondo et al., J. Non-Cryst. Solids **227-230**, 890 (1998).
11. K. Ikuta et al., Mat. Res. Soc.Symp. Proc. **420**, 413 (1996).
12. Karabacak, Y.-P. Zhao, G.-C. Wang and T.-M. Lu, Phys. Rev. B, **64**, 085323-1, (2001).
13. A.H.M. Smets, D.C. Schram and M.C.M. van de Sanden, Mat. Res. Soc. Symp. Proc. **609**, A.7.6.1, (2000).
14. Y.-P. Zhao, G.-C. Wang, and T.-M. Lu, *Characterization of Amorphous and Crystalline Rough Surfaces: Principles and Applications* (Academic Press, San Diego, 2000).
15. J.H. Yao and H. Guo, Phys. Rev. E **47**, 1007 (1993).
16. T.S. Cale and V. Mahadev, in *Modeling of Film Deposition for Microelectronic Applications*, Thin Films, Vol 22, ed. S. Rossnagel and A. Ulman (Academic Press, San Diego, 1996), p. 192.
17. J. Perrin, M. Shiratani, P.Kae-Nune, H. Videlot, J. Jolly, and J. Guillon, J. Vac. Sci. Tech. A, **16**(1), 278 (1998).
18. R. Ditchfield and E.G. Seebauer, Phys. Rev. B **63**, 125317 (2001).

# Hollow electrode enhanced RF glow plasma generation and its application to the fast deposition of microcrystalline silicon films

Toshihiro Tabuchi, Masayuki Takashiri, Yasumasa Toyoshima, Hiroyuki Mizukami
Research Division, Komatsu Ltd.
1200 Manda, Hiratsuka, Kanagawa 254-8567, Japan

## ABSTRACT

A hollow electrode enhanced RF glow plasma excitation technique has been newly developed. In this technique, the reactor is divided into a capacitively-coupled RF glow discharge space and a processing space by the counter electrode, which includes a hollow structure and is placed between a RF electrode and the substrate. After introducing hydrogen gas into the chamber and applying RF power to the electrode, high intensity plasma emission is observed near and inside the hollow structure attached to the counter electrode. By using hollow RF electrode excitation in addition to the hollow counter electrode technique, it is found that plasma emission is further enhanced. The application of these discharge types for semiconductor processing is studied in the case of plasma enhanced chemical vapor deposition (PECVD) of hydrogenated microcrystalline silicon thin films. High crystallinity, photo-sensitivity and a maximum deposition rate of 4.9nm/s can all be achieved at a plasma excitation frequency of 13.56MHz and a temperature of 300°C. Properties of these plasmas are investigated by observing the plasma emission pattern and optical emission spectrum analysis. It is found that, using additional hollow RF electrode discharge, faster processing of device grade hydrogenated microcrystalline silicon thin films can be achieved under lower RF power compared to hollow counter electrode technique alone.

## INTRODUCTION

A RF excited hollow cathode discharge is a promising plasma technique not only for fast film deposition but also for etching [1,2]. Previous studies [3-5], have reported that a hollow electrode discharge, which is induced in a hollow structure prepared at a counter electrode (so-called anode), can generate high intensity plasma. It has also been shown that microcrystalline silicon (μc-Si) thin films with good photosensitivity can be deposited at 4.9nm/s using this technique named Hollow Electrode Enhanced Plasma Transportation (HEEPT) [5]. In order to improve processing speed and reduce plasma damage to films and/or substrates, a new HEEPT technique has been developed. In this article, the new HEEPT technique and a corresponding silicon thin film deposition process are presented. A plasma source is based on a hollow counter electrode enhanced RF glow discharge accompanied by a hollow RF electrode discharge. Both high plasma intensity and rapid deposition of well crystallized silicon thin films are easily achieved at lower RF power simply adjusting the diameter of holes for gas distribution arranged at the RF electrode. Using $SiH_4$ and $H_2$ gases, it is found that μc-Si thin films with good photosensitivity can be deposited at 6.0nm/s.

## APPARATUS

A schematic diagram of the HEEPT reactor is shown in figure 1. The reactor is made of stainless steel, and its inner diameter is 210mm. Both the RF electrode and counter electrode are made of aluminum. The RF electrode has a showerhead structure for gas distribution and also has a cave structure of 5mm in height. In this work, two types of aluminum showerhead are prepared. One has holes of 1mm in diameter for gas distribution ($d_C$=1mm), and the other has 5mm holes in diameter ($d_C$=5mm). The diameter of the showerheads is 180mm and the thickness is 2mm. The counter electrode has a diameter of 208mm, thickness of 7mm and it surrounds the RF electrode. The substrate holder has a diameter of 160mm and is set up under the counter electrode. Both the counter electrode and the substrate holder are grounded together with reactor wall. Substrates, which are placed on the substrate holder, can be heated during experiments. An orifice is prepared at the center of the counter electrode, and a straight aluminum tube (nozzle) is attached to it. The total length of the nozzle including the thickness of the counter electrode is 19mm. The internal diameter of the nozzle ($d_A$) is 13mm. The distance between the RF electrode and counter electrode is 13mm, and the distance from the bottom end of the nozzle to substrate holder is 15mm.

Processing gases first flow into the cave structure of the RF electrode. The cave structure is surrounded by an aluminum electrode, and the showerhead for gas distribution is one of its parts. Under some conditions, the hollow RF electrode discharge can be induced inside this cave structure (sub-discharge chamber). Next, through the showerhead, gases flow into the RF discharge space (discharge chamber) surrounded by the RF electrode and grounded counter electrode. Finally, through the nozzle, gases are carried to the plasma processing space (deposition chamber). When RF power (f = 13.56MHz) is applied to the RF electrode, a plasma is generated at the discharge chamber. This plasma is enhanced in the inside of the nozzle and reaches to the substrate, which is located in the deposition chamber. The plasma can be further enhanced inside of the sub-discharge chamber only by using the showerhead with 5mm holes. In order to observe the plasma emission shapes in both the discharge chamber and sub-discharge chamber, a small rectangular quartz window was prepared in a small portion of a grounded guard ring and the side of the RF electrode, respectively.

These experiments were performed for both $d_C$=1mm and $d_C$=5mm. In the reactor shown in figure 1, $d_C$ can be easily converted to the other by changing the showerhead electrode.

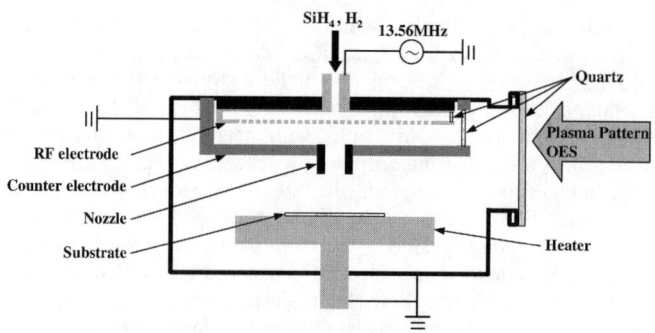

**Figure 1**. Schematic diagram of the HEEPT reactor.

## EXPERIMENTAL DETAILS

First, the plasma emission pattern and its optical emission spectroscopy (OES) were investigated with a $H_2$ flow rate of 100sccm at a pressure of 0.4Torr. The plasma emissions passed through the quartz window to an optical fiber for OES. The optical axis of this fiber is tuned to become parallel to the substrate and also is adjusted to the position 10mm above from the substrate. The OES signal is effectively a line integral of the emission across the chamber. OES from the plasma was measured for $H_\alpha$ (656.3nm).

Next, silicon thin films were deposited under the conditions of Table I. Corning 7059 glass, thickness 1.1mm, was used as substrates, which were heated up to 300°C during deposition. The deposition rate, conductivity and crystallinity of silicon films were evaluated. The deposition rate was calculated from the thickness of the region just below the nozzle. Conductivity was measured by a two-terminal method. In order to make contact with the silicon films, aluminum electrodes were deposited in parallel at intervals of 1mm using an electron beam evaporation technique. Photo-conductivity was measured under the conditions of AM1.5 at 100mW/cm$^2$. Crystallinity was investigated with Raman spectroscopy. The Raman crystallinity, Ic/Ia, was obtained from the intensity ratio for crystalline peak at 520cm$^{-1}$ (Ic) and amorphous peak at 480cm$^{-1}$ (Ia).

Table I. Growth conditions.

| RF frequency (MHz) | 13.56 |
| --- | --- |
| Input RF Power (W) | 20-100 |
| Gas pressure (Torr) | 0.4 |
| SiH$_4$ flow rate (sccm) | 7 |
| H$_2$ flow rate (sccm) | 100 |
| Sub. Temperature (°C) | 300 |

## RESULT AND DISCUSSION

### Observation of plasma patterns

Figure 2 shows the plasma emission pattern both in the discharge chamber and sub-discharge chamber. When $d_C$ is 1mm, a mushroom-shaped intense discharge is generated near and inside the hollow structure (nozzle) in the counter electrode. However, no bright spot and no irregular discharge can be seen either in the discharge chamber, especially just below the showerhead electrode, or inside the sub-discharge chamber. It is also confirmed that the same plasma is generated when a flat aluminum plate with no hole is used as the RF electrode. For this experiment, Hydrogen is supplied through the side of the reactor. Therefore, it is thought that only a hollow counter electrode enhancement is induced (Single Hollow Electrode: SHE).

On the other hand, when $d_C$ is 5mm, many bright funnel-shaped irregular discharges are clearly generated in the discharge chamber near and inside some of the holes in the showerhead. A similar strong inhomogeneous plasma is also observed inside the sub-discharge chamber. Every irregular spot converges at the holes in the showerhead. It is, therefore, obvious that the hollow RF electrode discharge is induced at the holes in the RF electrode in the case of $d_C$=5mm. As intense plasma is also seen near and inside the nozzle in the counter electrode, hollow counter electrode enhancement is also induced (Dual Hollow Electrode: DHE). The authors speculate that igniting a hollow counter electrode discharge inside the nozzle may facilitate an avalanche ignition of the hollow RF electrode in this set up. In both SHE and DHE mode, another high intensity plasma can be seen in the deposition chamber, especially near the exit of the nozzle, apparently from nozzle to chamber wall.

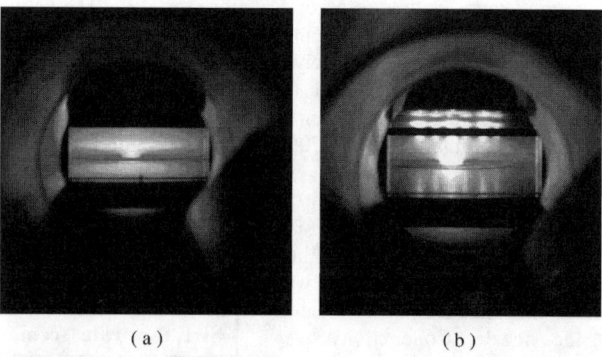

**Figure 2**. $H_2$ plasma emission in the discharge chamber and sub-discharge chamber for (a) $d_C = 1$mm, (b) $d_C = 5$mm, at 0.4Torr and 30W.

**OES study of the plasma**

Figure 3 shows the RF power dependence of $H_\alpha$ intensity obtained from SHE and DHE modes. It can easily be seen that the $H_\alpha$ intensity of DHE is at least three times stronger than that of SHE. The behavior of $H_\alpha$ intensity is quite different in the two modes, especially over 75W. In DHE mode, the intensity of $H_\alpha$ increases linearly with RF power, while in SHE mode it shows a tendency to saturate over 75W. It is also found that, in DHE mode, a light emitting strength and its extent in the sub-discharge chamber are both gradually increasing with the RF power. Furthermore, additional ignitions at holes in the showerhead and/or an increase of plasma extent in the sub-discharge

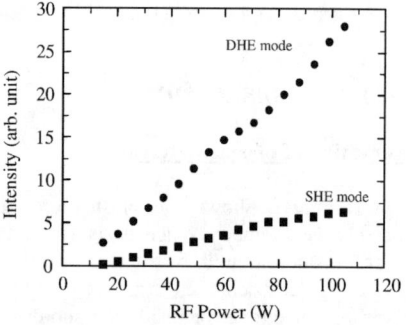

**Figure 3**. Optical emission intensities of $H_\alpha$ as a function of the input RF power. SHE mode and DHE mode correspond to $d_C = 1$mm and $d_C = 5$mm, respectively.

chamber can be observed with the increase of the RF. The authors expect that the $H_\alpha$ intensity in DHE mode would continue to increase linearly if we could provide higher RF power to the equipment. We therefore speculate that it will be possible to control both plasma intensity and its dynamic range not only by process conditions but also by showerhead design, e.g. number of holes, arrangement and/or diameter of holes, thickness, cavity height etc.

## Deposition of silicon thin films

The silicon thin films are deposited using gas mixtures of $SiH_4$ and $H_2$. Figure 4 shows the deposition rate, the conductivity, and the Raman crystallinity, Ic/Ia, as a function of input RF power. In both SHE and DHE mode, the deposition rate increases rapidly up to 35W and saturates until 50W. Within this RF power region, DHE mode is about 50% faster than SHE mode. For DHE at 35W, the deposition rate is 6.0nm/s with a Raman crystallinity, Ic/Ia, of 3.6. This is also the maximum deposition rate of this work. For SHE at the same RF power, the deposition rate is 4.0nm/s, but the film remains amorphous. In SHE mode, more than 50W are needed to make films crystallize. As the maximum deposition rate for SHE mode, 4.9nm/s with Ic/Ia=3.3 is obtained. The deposition rate decreases over 50W in DHE mode, while it is almost saturated in SHE mode. Correspondingly, both the conductivity and the crystallinity decrease in DHE mode over 50W as shown in figure 4.

The authors speculate that the faster deposition speed in DHE mode below 50W is due to the higher plasma intensity, which indicates a greater concentration of those radicals that contribute to silicon deposition, such as $SiH_3$. Above 50W, the $H_\alpha$ intensity in DHE mode increases, as shown in figure 3. It is well known that hydrogen atoms can etch silicon. It is, therefore, thought this is the reason why the deposition speed, the conductivity and the crystallinity decrease over 50W in DHE mode.

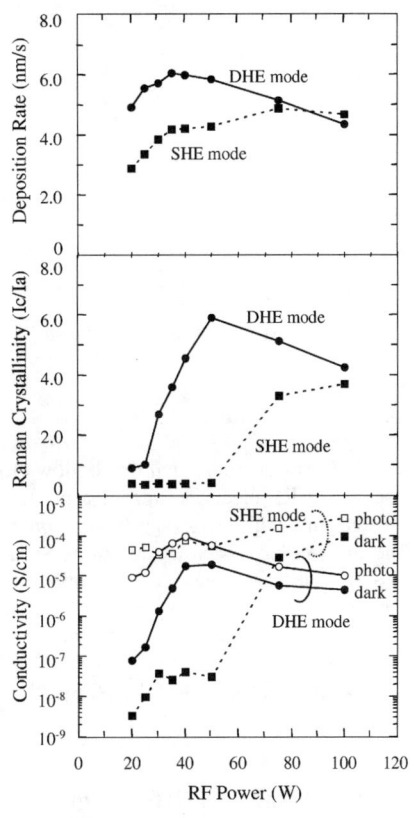

**Figure 4.** The deposition rate (top), Raman Crystallinit (center), and conductivity (bottom) as a function of the input RF power.

Figure 5 shows the Raman spectroscopy of μc-Si film of 1.75μm thickness deposited by our apparatus in DHE mode at 50W. These results indicate that it is possible for us to deposit photosensitive device grade μc-Si films at a high speed by utilizing both hollow RF electrode and hollow counter electrode discharge enhanced plasma.

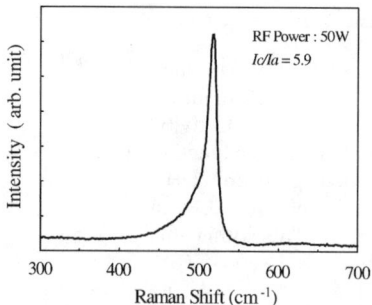

**Figure 5.** Raman profile of a 1.75μm thick μc-Si film deposited at 5.8nm/s on glass

## CONCLUSIONS

It was demonstrated that hollow electrode enhanced high intensity plasma can be induced both in a RF electrode and in a nozzle attached to the counter electrode, which divides the reactor into a discharge space and a processing space. Highly crystallized and photo-conductive μc-Si films have been fabricated by $SiH_4$ and $H_2$ at a high deposition rate of 6.0nm/s, using this hollow electrode plasma at 13.56MHz, 35W, and 0.4Torr. Comparative studies of the $H_\alpha$ intensity between two types of hollow electrode enhanced plasma were carried out. The results suggest that the plasma intensity can be increased if we can provide stronger RF power and can be controlled through the design of the RF electrode. However, excess hydrogen atoms may be rather harmful for the fast deposition of device grade μc-Si.

## ACKNOWLEDGMENT

The authors wish to thank K. Ishida for his technical assistance, S. Sano and K. Ogaki for their continuous encouragement.

## REFERENCES

1. C. M. Horwitz, S. Boronkay, M. Gross and K. Davies, J. Vac. Sci. Technol. **A6**, 1837 (1988).
2. L. Bardos, Surface and Coatings Technology **86-87**, 648 (1996).
3. T. Tabuchi, K. Ishida, H. Mizukami, M. Takashiri, Japan Patent P2000-83838 (2000).
4. T. Tabuchi, K. Ishida, H. Mizukami, M. Takashiri, U.S. Patent 09/730,813 (2000).
5. T. Tabuchi, M. Takashiri, M. Mizukami, Surface and Coatings Technology, (in press).

# The Reliability of Measurements on Electron Energy Distribution Function in Silane rf Glow Discharges

Kuixun Lin, Xuanying Lin, Linfei Chi, Chuying Yu, Yunpeng Yu, and Shi Liu
Department of Physics, Shantou University, Shantou 515063, China

## ABSTRACT

Electron energy distribution function (EEDF) is directly proportional to the second derivative of the probe $I$-$V$ characteristics. Because of an amplifying effect of unavoidable noises in the experimental probe $I$-$V$ curves on the derivation process, the experimental $I$-$V$ curves should be smoothed before performing the numerical derivation. This article investigates the effect of adjustable factors used in the smoothing process on the deduced second derivative of the $I$-$V$ curves, and an optimum group of the adjustable factors is selected to make the rms deviation of the smoothed $I$-$V$ curves from the measured ones less than 1 %. A simple differentiation circuit is designed and used to measure the EEDF parameter straightforwardly. It is the first time, so far as we know, to measure the EEDF parameters simultaneously by means of both numerical and circuit derivative methods under the same discharge conditions and on the same discharge equipment. The deviation between two groups of mean electron energy ($E$) and electron density ($n_e$) obtained by the above different methods is within about 7 %. This apparently improves the reliability of the measurements on the EEDF parameters.

## INTRODUCTION

Electron energy distribution function (EEDF) in silane rf glow discharges is a key parameter for studying the spatial reaction process in a plasma and the surface reaction process on the substrate. For these reasons, there have been extensive studies of EEDF in capacitively coupled rf glow discharge plasma [1-5].

Langmuir probe is the most commonly used diagnostic tool in the measurement on the EEDF of the plasma. The EEDF ($f(\varepsilon)$) is directly proportional to the second derivative ($d^2I/dV_p^2$) of the probe $I$-$V$ characteristic [6]:

$$[f(\varepsilon)]_{(\varepsilon=-eV)} = -\frac{4}{A_p e^2}\sqrt{\frac{-m_e V}{2e}}\frac{d^2 I}{dV_p^2} \tag{1}$$

Where $I$ and $V_p$ are the collection current and bias voltage of the probe, respectively. $A_p$ is the probe surface area exposed to the plasma. $V$ is the probe potential relative to the plasma potential. $m_e$ and $\varepsilon$ are the mass and energy of the electron, respectively. The mean electron energy $E$ and electron density $n_e$ can be evaluated from the normalized EEDF.

A tuned and heating Langmuir probe has been used to restrain both interference and contaminate effects successfully to provide an accurate measurement of the probe $I$-$V$ characteristic [7,8]. Even so, there are unavoidable statistic noises in the measured $I$-$V$ characteristic, and an amplifying effect of these noises on the derivative of the $I$-$V$ curves [5]. In order to improve the accuracy and reliability of the measurement on EEDF parameters, a key problem is to restrain this amplifying effect. This is the main subject of this article.

The second derivative of the probe $I$-$V$ curve can be either obtained by numerical treatments or straightforwardly measured by a differentiation circuit. Until now, the reported EEDF parameters of the plasma in a certain discharge equipment have been obtained by only one of the above methods, either numerical or circuit-derivative method. Therefore, these measured results cannot be compared with each other. In this paper, both the above differentiation methods are for the first time, so far as we know, used to measure the EEDF parameters of the plasma on the same discharge equipment under the same discharge situation simultaneously. Both the measured results are compared and agreed with each other well; therefore, the reliability of the measurements on EEDF parameters in silane rf glow discharges is apparently improved.

## EXPERIMENTAL DETAILS

The probe diagnostic system in the capacitively rf glow discharge equipment is shown in elsewhere [7]. The probe is made of a tungsten filament with a diameter of 0.45mm. The probe tip of a 5 mm length is exposed to the plasma, other part of the probe is electrically shielded by a thin glass sleeve. A parallel unit of L, C is connected to the probe and tuned at 13.56MHz for restraining the rf interference on the probe and the probe is positioned with its tip at the geometric center in the bottom- and up-electrode gap space and heated by thermal radiation from the heater located in the up-electrode for restraining the contamination effect of the probe.

The probe $I$-$V$ characteristics are measured and displayed automatically using a computer controlling system.

## MEASUREMENTS OF EEDF

### Numerical Differentiation Method

Numerical differentiation of the probe $I$-$V$ characteristic is an important method for measurement on the EEDF parameters of the plasma. Various numerical differentiation methods have been carried out, and a comparison between them has been made [3,4,5,9].

In order to restrain the amplifying effect of the statistic unavoidable noises on the derivative of the $I$-$V$ characteristics, the experimental $I$-$V$ curves should be smoothed before carrying out the derivative. However, the second derivatives of the smoothed $I$-$V$ curves may change with the selected adjustable factors in the smoothing process leading to an uncertainty of the deduced EEDF parameters. In this section, a smoothing method developed by Hayden [10] is used to describe this effect for searching an optimum group of the adjustable factors in the smoothing process and improve the reliability of measurement on the EEDF parameters of the plasmas.

If we measure a function $y(x)$, what we really measure is not such a function but its convolution product with the instrument function of the experimental system:

$$h(x) = y^* g = \int_{-\infty}^{\infty} y(t) g(t-x) dt \qquad (2)$$

Where $h(x)$ is the measured function and $g(x)$ is the instrument function.

In most cases, the instrument function $g$ can be approximated by a Gaussian distribution function with a standard deviation of $\sigma$. The function $g_n$ is:

$$g_n(x) = \sum_{k=1}^{n} (\frac{n}{k})(-1)^{k+1} \frac{1}{\sqrt{2\pi k}\sigma} e^{-\frac{x^2}{2\sigma^2 k^2}} \qquad (3)$$

The $h_n(x)$-$x$ is the smoothed probe $I$-$V$ characteristic of the measured one and corresponding to the $n$th approximation of the $y(x)$-$x$.

Tow group second derivative curves of the smoothed $I$-$V$ characteristics are shown in Figure 1 and 2, corresponding to $n=1,3$ at fixed $\sigma = 2$ and $\sigma = 2,4,6$ at fixed $n=2$ respectively. These curves are closely relative to the selected adjustable factors ($\sigma$ and $n$) used in the smoothing process of the measured I-V characteristics.

In Figure 1, the curve of second derivative, corresponding to $n=3$, $\sigma = 2$, appears severe oscillation because the probe $I$-$V$ characteristic is distorted in the smoothing process.

A group of the plasma parameters $E$ and $n_e$ evaluated from the data of the second derivatives curves shown in Figure 2, are listed in table I. These evaluated plasma parameters are quite different depending on the selected adjustable factors $\sigma$ and $n$ in the smoothing process. The rms deviations between the measured and its smoothed $I$-$V$ characteristics are 0.7 %, 1.2 % and 2% corresponding to a group of the adjustable factors $\sigma = 2,4,6$ at fixed $n=2$, respectively. An optimum group of the adjustable factors is selected as $\sigma = 2$ and $n=2$ depending on a minimum rms deviation (less than 1%), and the corresponding evaluated plasma parameters are $E$ of 4.2 eV, $n_e$ of $7.5 \times 10^{14} m^{-3}$, respectively.

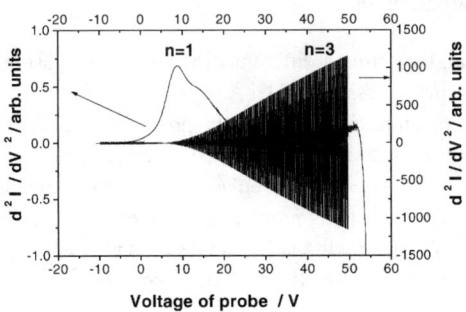

**Figure 1.** Plots of the second derivatives of the smoothed probe $I$-$V$ curves, $n = 1, 3$ at $\sigma = 2$.

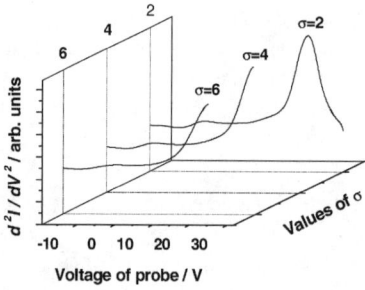

**Figure 2.** Plots of second derivatives of the smoothed $I$-$V$ curves, $\sigma = 2, 4, 6$, at fixed $n = 2$.

**Table I.** Change of mean electron energy $E$ and electron density $n_e$ with $\sigma$ at fixed $n = 2$.

| $\sigma$ | 2 | 4 | 6 |
|---|---|---|---|
| $E$/ eV | 4.2 | 6.1 | 8.4 |
| $n_e/10^{14}\,\mathrm{m^{-3}}$ | 7.5 | 12.8 | 13.5 |

## Circuit Differentiation Method

V. A Goyak et al. have used a differentiation circuit to measure the second derivative of the probe $I$-$V$ characteristic successfully [11]. A rapid scanning voltage (20 V/ms) was biased on the probe to avoid the contamination effect of the probe and reduce the phase retardation effect of the circuit. We design a more simple differentiation circuit consisted of two operation amplifiers, because both of the rf interference and contaminate effects on the probe have been restrained by means of using a tuned and heating probe diagnostic circuit. The designed differentiation circuit is shown in Figure 3, and joined to the tune unit of L and C of the probe. The output voltage ($V_o$) of the differentiation circuit is directly proportional to $d^2I/dV_p^2$.

$$V_o = K R_3^2 C_3^2 R_y \, d^2I/dV_p^2 \qquad (4)$$

Where $K$ is the ratio of the voltage increment to the time increment biased on the probe.

The sampling resistance $R_y$ of 2 k$\Omega$ provides an enough big input signal for the differentiation circuit, but does not affect the measurement accuracy of the circuit, because $R_y$ is much less than the resistance of probe sheath in the plasma. The series circuit of $R_3$ (2 k$\Omega$) and $C_3$ (4700μF) and the parallel

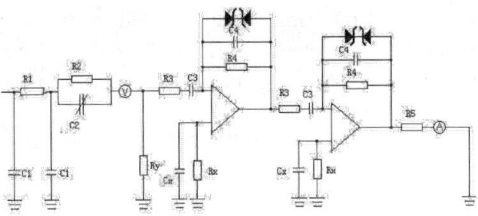

**Figure 3.** Diagram of the differentiation circuit

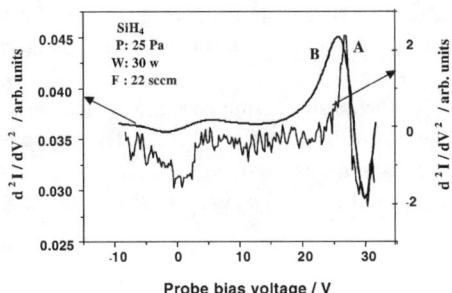

**Figure 4.** Plot of the second derivatives of the probe *I-V* characteristics
Plot A and B are measured by the circuit and numerical differentiation method, respectively.

circuit of $R_4$ (10 kΩ) and $C_4$ (100PF) are the input and feedback units of the operation amplifiers, respectively. The amplifying effect of the noises on the measured EEDF functions is very weak, because $R_3 << 1/(\omega C_3)$, $R_4 << 1/(\omega C_4)$ in normal operation situation. Along with increase in frequency (ω), the input and feedback units consisted of such component parameters will reduce the amplification of the closed differentiation circuit, therefore, restrain the effect of high frequency noises, and improve the stability of the circuit operation.

Second derivative of the *I-V* probe characteristic was straightforwardly measured by using the differentiation circuit, and a plot of the second derivative is shown in Figure 4 and labeled A.

In order to make a comparison, another plot of second derivative of an *I-V* probe characteristic smoothed using the optimum group of the adjustable factors is shown in Figure 4 also and labeled B. Both of plot A and B are measured on the same discharge equipment and under the same discharge conditions: operation pressure P=25 Pa, rf power W=30w, silane flow rate of F=22 sccm.

The parameters of mean electron energy (*E*) and electron density ($n_e$) evaluated from the

EEDF straightforwardly measured by using the differentiation circuit are 4.4 eV, $8.1\times10^{14}/m^3$, which are little higher than the parameters of 4.2 eV, $7.5\times10^{14}/m^3$ evaluated from the EEDF obtained by using the numerical differentiation method. The deviation between the two groups of parameters is less than 7%.

## CONCLUSIONS

It is possible to effectively measure the EEDF parameters of the plasmas by means of performing the second derivative of the probe $I$-$V$ characteristics. The key problem is to smooth the measured probe $I$-$V$ characteristic with a less than 1 % of rms deviation between the measured and its smoothed characteristics before performing the numerical derivative.

It is also possible to straightforwardly measure the EEDF parameters of the plasmas by means of a differentiation circuit. The key problem is to select an operation amplifier input and feedback units consisted of an optimum group of component parameters leading to restrain the effect of noises on the measurements.

Two group parameters of the mean electron energy and electron density evaluated from the EEDF functions are well agreed with each other. These EEDF functions were measured by using the above two different methods, respectively, in the same discharge equipment and under the same discharge conditions; therefore, this apparently improves the reliability of the measurement on EEDF parameters.

## ACKNOWLEDGES

The authors would like to thank the support of the State Key Development Program for Basic Research of China (Grant No G2000028208).

## REFERENCES

1. M. Surendra, D. B. Graves and I. J. Morey, *Appl. Phys. Lett.* **56,** 1022 (1990).
2. Z. D. Wang, J. L. Allan and H. C. Ronald, *IEEE Trans. Plasma Sci.* **26,** 59 (1998).
3. L. F. Chi, K. X. Lin, X. Y. Lin, *Vacuum,* **6,** 23 (1999).
4. R. H. Yao, L. F. Chi, X. Y. Lin, W. Z. Shi, K. X. Lin, *Acta Phys. Sin.* **49,** 922 (2000).
5. L. F. Chi, K. X. Lin, R. H. Yao, X. Y. Lin, *Acta Phys. Sin.* **50,** 1313 (2001).
6. M. J. Druyvesteyn, *Z. Phys.* **64,** 781 (1930).
7. Q. X. Lin, X. Y. Lin, Y. P. Yu, H. Wang and J. Y. Chen, *J. Appl. Phys.* **74 (8),** 4889 (1993).
8. Q. X. Lin, X. Y. Lin, Y. P. Yu, H. Wang, S. M. Hu, *J. Funct. materials,* **27(5),** 392(1996).
9. J. I. F. Palop, J. C. V. Ballestros and M. A. Hemandz, *Rev. Sci. Instrum.* **64(9),**4626 (1995).
10. H. C. Hayden, *Comput. Phys.* **1,** 74 (1987).
11. V. A. Godyak, 1990 *in Plasma Surface Interactions and Processing of Materials.* p95.

# High Rate Deposition of Stable Hydrogenated Amorphous Silicon in Transition from Amorphous to Microcrystalline Silicon

Guofu Hou[1], Xinhua Geng[1], Xiaodan Zhang[1], Ying Zhao[1], Junming Xue[1], Huizhi Ren[1], Jian Sun[1], Dekun Zhang[1] and Yueqin Xu[2]
[1]Institute of Photoelectronics, Nankai University, Tianjin, 300071, P.R.China
[2]National Renewable Energy Laboratory, 1617 Cole Blvd, Golden CO80401, U.S.A

## ABSTRACT

High rate deposition of high quality and stable hydrogenated amorphous silicon (a-Si:H) films were performed near the threshold of amorphous to microcrystalline phase transition using a very high frequency plasma enhanced chemical vapor deposition (VHF-PECVD) method. The effect of hydrogen dilution on optic-electronic and structural properties of these films was investigated by Fourier-transform infrared (FTIR) spectroscopy, Raman scattering and constant photocurrent method (CPM). Experiment showed that although the phase transition was much influenced by hydrogen dilution, it also strongly depended on substrate temperature, working pressure and plasma power. With optimized condition high quality and high stable a-Si:H films, which exhibit $\sigma_{ph}/\sigma_d$ of $4.4 \times 10^6$ and deposition rate of 28.8Å/s, have been obtained.

## INTRODUCTION

To reduce production cost and then to enlarge its applications, especially in solar cells, high rate deposition of a-Si:H films has attracted much attention recently[1-6]. Now many techniques have been successfully introduced to realize high rate deposition, such as radio-frequency PECVD with high pressure and high plasma power[1], microwave CVD[2], remote expanding thermal plasma CVD[3], hot wire CVD[4,5] and very high frequency (VHF) PECVD[6,7,8]. Since its compatibility with the well-established RF-PECVD, VHF-PECVD has been regarded as the most promising option for high rate deposition of a-Si:H films and solar cells[6,7,8].

Generally, in silane plasma $SiH_3$ is considered as the main film precursor, while $SiH_2$ is responsible for light-induced degradation of a-Si:H. When the frequency increases from RF to VHF, the high frequency helps to reduce electron temperature and increase electron density[8]. The lower electron temperature and higher electron density benefit for the production of $SiH_3$ and prevent the formation of $SiH_2$ and their offspring through successive reactions with silane----Higher order silane related chemical species (HSRS). Thus the light-induced degradation is much prohibited and the stability is much improved in high rate deposited a-Si:H. Furthermore, the DC self-bias voltage automatically applied to the cathode will also decrease as excitation frequency increases. This will reduce ion bombardment energy and increase ion flux toward the growing film surface and this will also be beneficial for the stability.

Nowadays hydrogen dilution has attracted special attention for the fabrication of high stable amorphous silicon films and solar cells[9,10,11]. The dilution level of silane in hydrogen is one of the key parameters for the deposition of silicon films. With the decrease of silane concentration monotonously, the phase transition from a-Si:H to μc-Si:H occurs. Device quality high stable films, deposited near the phase at a certain dilution level while still remained amorphous, not only have high photosensitivity but also have much more ordered structure. A high stable a-Si:H solar cell with initial conversion efficiency of 14.6% and stable conversion efficiency of 13% has been obtained[12].

In this paper high quality and stable hydrogenated amorphous silicon (a-Si:H) films have been prepared at high deposition rate by very high frequency plasma enhanced chemical vapor deposition (VHF-PECVD) method. The effect of hydrogen dilution on opticelectronic and structural properties of these films was also investigated by Fourier-transform infrared (FTIR) spectroscopy, Raman scattering and constant photocurrent method (CPM). A particular discussion about the experiment results was carried out.

## EXPERIMENTAL DETAILS

Thin silicon films used in this study were prepared by capacitively-coupled VHF-PECVD technique with the excitation frequency of 60MHz on 7059 glass substrate and crystalline silicon for Raman scattering and IR measurement, respectively. The background pressure of the deposition system was pumped down below $5\times10^{-4}$Pa. The samples were prepared with plasma power of 15~25Watt, working pressure of 60~180Pa and substrate temperature of 200~250°C. Silane concentration ($SC=SiH_4/SiH_4+H_2$) in the feed gas varied from 2% to 15% diluted by hydrogen.

The room-temperature photo- and dark conductivity and activation energy of dark conductivity were measured with a coplanar contact setup under and dark conditions. The light-induced degradation of conductivity was also measured under AM1.5 illumination. The thickness of these films was calculated from optical reflection spectrum. Fourier-transform infrared (FTIR) spectroscopy, Raman scattering and constant photocurrent method (CPM) were utilized to study the microstructure of thin films. All the IR absorption spectra were calculated from IR transmission spectra according to theory mentioned by Brodsky, et al [13].

## RESULTS AND DISCUSSION

### Raman spectra analysis

Raman scattering spectra of films, deposited at power of 25Watt, pressure of 180Pa and substrate temperature of 250°C, are shown in Fig.1. When the silane concentrations are 15% and 8%, the Raman spectra mainly showed amorphous phase characterized by scattering broadband with peaks at 484.17$cm^{-1}$ and 483.12 $cm^{-1}$, respectively. At silane concentrations of 7% and 6%, besides amorphous broadband at 480$cm^{-1}$ obvious narrow crystalline band peaked at 520$cm^{-1}$ also appeared, which indicated a transition from a-Si:H to μc-Si:H. Increasing hydrogen dilution further, amorphous broadband disappeared by and by while the width of crystalline band figure, too. It is obvious that hydrogen dilution helps to crystallinity of films. Even at SC=2%, μc-Si:H films have been obtained at rate of 5.4Å/s, which is much larger than that deposited by conventional RF-PECVD.

### Infrared spectra analysis

In order to investigate the effect of hydrogen dilution on optic-electronic properties and hydrogen bond configurations, another series of films were prepared on c-Si wafer with substrate

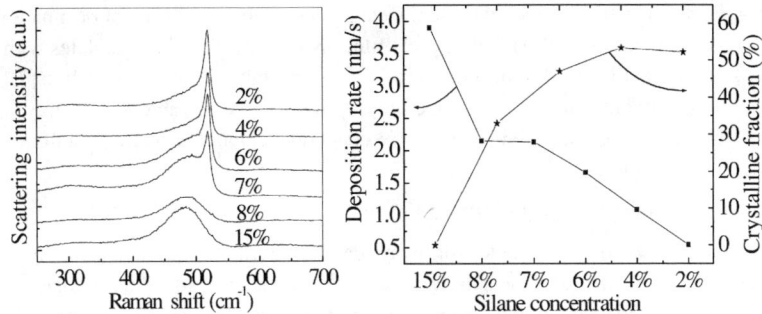

**Figure 1.** Raman scattering spectra of films at different silane concentration

**Figure 2.** Deposition rate and crystalline fraction vs. silane concentration.

temperature of 200°C, pressure of 60Pa and plasma power of 15Watt. Samples parameters are showed in Table 1 and IR absorption spectra are plotted in Fig.3.

It shows that typical vibration modes at 630/640cm$^{-1}$, 840/890cm$^{-1}$ and 2000/2100cm$^{-1}$, corresponding to wagging modes, bending modes and stretching modes of Si-H bonds respectively, can be found[7,15]. In addition, increasing hydrogen dilution resulted in increasing intensity of bands at ~1050cm$^{-1}$. This bands should pertain to the infrared stretching peak associated with the Si–O groups since the oxygen doping[6,10].

From Fig.3 we can find that with increasing hydrogen dilution, the stretching modes at ~2000cm$^{-1}$ decreased and it turned to be dominated by ~2100cm$^{-1}$ gradually. According to theory

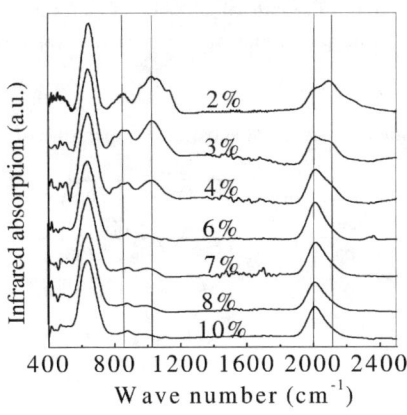

**Figure 3.** IR absorption spectra of samples at different hydrogen dilution

of Cardona[17], Shah[7], Sheng[18] and Deng[19] etc, there are different origination of ~2100cm$^{-1}$ modes in a-Si:H and μc-Si:H films.While ~2100cm$^{-1}$ peak relates with $(SiH_2)_n$ groups in a-Si:H films[17], it originates from microcrystallinity in μc-Si:H films[7,17,18,19]. Combining with our Raman and conductivity results showed in Table 1 we firmly suggest that films with silane concentration less than 6% contain an amount of microcrystalline particles and have diphasic structures.

### Electrical properties and stability

Table I shows some parameters of hydrogen-diluted films. With increasing hydrogen dilution, the deposition rate and photosensitivity all decreased. But even for films at silane concentration of 2% and 3%, they are still photosensitive. And this is very important for solar cells.

Sub-bandgap absorption spectra for some of these films were also performed by constant photocurrent method (CPM). The characteristic energy, $E_u$, of Urbach valence bandtail can be obtained from the exponential part of the absorption spectra. Since the Urbach bandtail originates from bonding and thermal disorder in the materials, $E_u$ may be used to study the width of bandtail and the disorder in the a-Si:H network structures[20]. From Table 1 it is apparent that for sample 12124 $E_u$ equals 47.6meV, which is smaller than those of sample 12121 and 12122. So This indicated that a high hydrogen dilution helps to form more order silicon network.

Fig.4 shows the light-induced change of conductivity for sample12121. It can be seen that only a slight degradation and a fast process to reach stable state for this material occurs. After more than about $10^4$ minutes of light soaking, the photosensitivity remained around $10^5$.

To get films with higher quality, we adjusted the deposition parameters carefully. Under a optimized condition, including substrate temperature of 200°C, working pressure of 90Pa, plasma power of 20W and silane concentration of 15%, high quality a-Si:H films have been got with $\sigma_{ph}/\sigma_d$ of $4.4\times10^6$ and deposition rate of 28.8Å/s. In fact, this kind of high photosensitive materials can always be obtained at various plasma power, as mentioned in our previous work[10].

**Table I** Properties of samples deposited at substrate temperature of 200°C and pressure of 60Pa.

| Sample NO. | Silane Concentration | Deposition Rate (Å/s) | Thickness (μm) | Photosensitivity | $E_u$ (meV) | $E_a$ (eV) |
|---|---|---|---|---|---|---|
| 12121 | 10% | 16 | 1.246 | 1.6E+5 | 48.3 | 0.7 |
| 12122 | 8% | 10.5 | 1.259 | 8.8E+4 | 49.6 | 0.67 |
| 12123 | 7% | 7.4 | 1.242 | 4.86E+4 | ----- | 0.68 |
| 12124 | 6% | 4.7 | 1.256 | 2.4E+4 | 47.6 | 0.68 |
| 01021 | 4% | 2.8 | 1.321 | 8.1E+3 | ----- | 0.57 |
| 01022 | 3% | 2.4 | 1.587 | 4E+3 | ----- | 0.54 |
| 01131 | 2% | 1.6 | 1.025 | 7.8E+2 | ----- | ---- |

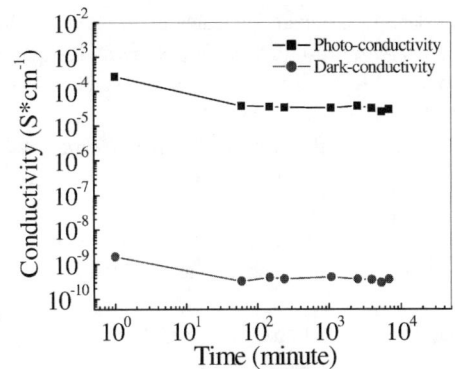

**Figure 4.** Light-induced degradation of conductivity for high hydrogen diluted a-Si:H film

## CONCLUSION

Hydrogen dilution has been adopted to deposit high quality a-Si:H films and to study its effect on optic-electronic and structural properties of materials. With increasing hydrogen dilution, Raman spectra shows that TO peaks of silicon films shifted from 480cm$^{-1}$ to ~520cm$^{-1}$, and FTIR spectra shows that the stretching modes at ~2000cm$^{-1}$ decreased and it turned to be dominated by ~2100cm$^{-1}$ gradually. These all indicated a phase transition from a-Si to μc-Si occurs. The above results also demonstrate that besides hydrogen dilution, substrate temperature, working pressure and plasma power also play an important role in the phase transition. While the first series samples show phase transition around silane concentration of 7%, the phase transition occurs at silane concentration of 3% for the second series samples. Under optimized the deposition parameters, high quality a-Si:H films have been got with $\sigma_{ph}/\sigma_d$ of $4.4\times10^6$ and deposition rate of 28.8Å/s.

## ACKNOWLEDGMENTS

The authors are grateful to Guanglin Kong from Institute of Semiconductors, Chinese Academy of Sciences for the CPM measurements. The work in this paper was supported by the National Key Basic Research Project of P.R.China (No.G2000028202, G2000028203) and the Key Project of Education Bureau (No.02167).

## REFERENCES

[1] B.Rech, T.Roschek, J.Muller, S.Wieder, H.Wagner, *Solar Energy Materials & Solar Cells* **66**, 267 (2001).
[2] S.Guha, X.Xu, J.Yang, A.Banerjee, *Appl. Phys. Lett.* **66** (5), 595-597 (1995).
[3] B.A.Korevaar, *et al*, *J. Non-Cryst. Solids*, **266-269**, 380 (2000).
[4] H.Matsumura, Japanese J. of Appl. Phys. **37**, 3175 (1998).

[5] B.P.Nelson, E.Iwaniczko, A.H.Mahan, Qi Wang, et al, *Extended Abstract of the 1st International Conference on Cat-CVD process,* Kanazawa, Japan, *2000,* 291.
[6] M.Kondo, T.Nishimoto, M.Katai, S.Suzuki, Y.Nasuno and A. Matsuda, *Technical Digest of the International PVSEC-12*, Jeju, Korea, *2001*, 41.
[7] U.Kroll, J.Meier, P.Torres, J.Pohl and A.Shah, *J. Non-Cryst. Solids*, **227-230**, 68 (1998).
[8] M. Takai, T. Nishimoto, T. Takagi, M. Kondo, A. Matsuda, *J. Non-Cryst. Solids,* **266-269**, 90 (2000).
[9] Qi Wang, Guozhen Yue, Jing Li, Daxing Han, *Solid state communication* **113**, 175 (2000).
[10] Gguofu Hou, Yaohua Mai, Xinhua Geng, Xiaodan Zhang, et al, *Physica.Status.Solidi B*, (unpublished )
[11] Swati Ray, Chandan Das, S.Mukhopadhyay, S.Csaha, *Solar Energy Materials & Solar Cells* **66**, 393 (2002).
[12] J.Yang, A.Banerjee and S.Guha, *Appl. Phys. Lett.* **70** (22), 2975-2977(1997).
[13] M.H.Brodsky, M.Cardona and J.J.Cuomo, *Phys. Rev.* **B16**, 3556 (1977).
[14] C.Droz, E.Vallat-Sauvain, J.Bailat, L.Feitknecht, A.Shah, *European PVSECE-17*, Munich, Oct.22~26, *2001*, 1.
[15] S.Klein, F.Finger, R.Cariu, H.Wagner, *Thin Solid Films* **395**, 305 (2001).
[16] R.Martins, H.Aguas, I.Ferreira, E.Fortunato and L.Guimaraes, *Solar Energy* **69** (1–6), 257-262 (2000).
[17] M.Cardona, *Physica.Status.Solidi*, **B118**, 463 (1983).
[18] Shuran Sheng, Xianbo Liao, Guangli Kong and Hexiang Han, *Appl. Phys. Lett.* **73** (3), 336-338 (1998).
[19] X.M.Deng, *Phys. Rev.* **B43**, 4820 (1991).
[20] Shuran Sheng, Xianbo Liao and Guangli Kong, *Appl. Phys. Lett.* **78** (17), 2509-2511 (2001)

# PROPERTIES OF NANOCRYSTALLINE GERMANIUM-CARBON FILMS AND DEVICES

X.J. Niu*, Vikram L. Dalal* and Max Noack+
Iowa State University
*Dept. of Electrical and Computer Engr., + Microelectronics Research center
Ames, Iowa 50011

## ABSTRACT

Nanocrystalline Germanium-Carbon alloys, denoted by nc- (Ge,C):H, are a potentially useful new electronic material whose bandgap can be varied by changing the Ge:C ratios. We have shown previously that nanocrystalline (Ge,C):H films can be grown using remote ECR plasma deposition. In this paper, we report on the crystal structure, electron mobility and some device related properties of these materials. The materials were grown using mixtures of either Germane and methane, with significant hydrogen dilution, or from mixtures of ethylene and germane, also with significant hydrogen dilution. X-ray diffraction measurements indicated a predominantly <111> crystal structure. The grain size was in the range of 10 nm. Raman measurements clearly show the 300 cm-1 Ge peak in the films. Electron Hall mobilities were measured in these films and were found to be in the range of 2.5-3 cm2/V-sec. Proof-of-concept p+nn+ junction devices were fabricated and showed distinct photovoltaic properties. The open circuit voltage was found to be a strong function of the itnerfaces between n+ and n layers, and between p+ and n layers. The use of an amorphous n+/n interface at the back of the improved the device performance significantly. Capacitance measurements indicated that the device behaved according to standard p/n junction theory, with the n- doping in the base layer being in the $10^{17}/cm^3$ range. Thus, the base layer was not intrinsic, but rather n type, as may be expected for a crystalline as opposed to amorphous material. Quantum efficiency data indicated that as C was added to the material, the edge of the QE curve shifted to higher photon energies, indicating larger bandgaps.

## INTRODUCTION

Nanocrystalline (Ge,C):H materials are of some interest for photovoltaic, image sensing and graded bandgap electronic devices. We have shown previously that the optical properties of the material behave are similar to the optical properties of crystalline Ge, with addition of C leading to a shift of the bandgap to higher energies[1,2]. It appears that both the L and the Γ valleys in the conduction band are moving simultaneously upward as more C is added to the material. In this paper, we study the fabrication of devices in this material system and report on a novel interface design used to improve device performance. We also study some fundamental properties such as doping density and electron mobilities. We also report on quantum efficiency measurements to study the influence of adding C on the movement of the absorption edge.

## MATERIAL GROWTH AND PROPERTIES

The materials and devices were grown using a remote, low pressure ECR technique previously described [3]. Basically, a remote, hydrogen rich discharge is generated in an ECR reactor at low pressures ( ~ 5 mT) and the material is grown using mixtures of germane (prediluted in hydrogen for safety reasons), methane and hydrogen. We have also sued ethylene as a source of carbon, but the results in this paper are for films and devices prepared using methane. The substrate temperatures were in the range of 300 C, and pressure in the reactor was 5 mTorr. Higher pressures ( 10 mTorr and above) led to the growth of amorphous films. Microwave power was 200 W. The typical flow rates were:Hydrogen = 55 sccm, germane =0.4 sccm and methane=0-4 sccm. It was always necessary to use very high hydrogen dilutions to achieve nanocrystallinity in Ge, in contrast to the case for Si, where much lower hydrogen dilution values(10:1) could be used. The ratio of methane to germane was kept high because methane is much more difficult to decompose than germane.

In Fig. 1, we show the Raman spectrum for a nc-(Ge,C):H film whose bandgap was ~ 1.0 eV. The classical 300 cm-1 Ge peak is clearly seen in the figure. The crystalline C peak is not visible because of fluorescence in that region of the spectrum. In Fig. 2, we show the x-ray diffraction peak for another (Ge,C):H sample prepared using similar conditions. The <111> peak at 27 degrees is clearly visible. We estimate the grain size, using Scherrer's formula, to be ~ 7.5 nm. The sharp peaks at 38 and 44 degrees are diffraction from the steel substrate on which the film was grown.

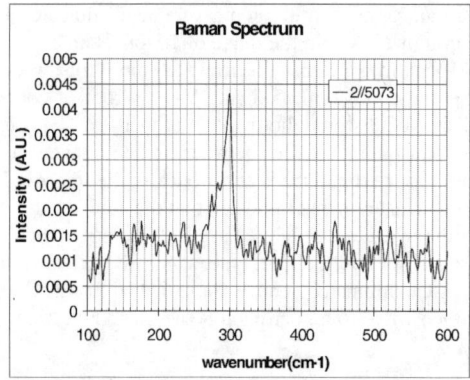

Fig. 1 Raman spectrum of nc-(Ge,C):H film

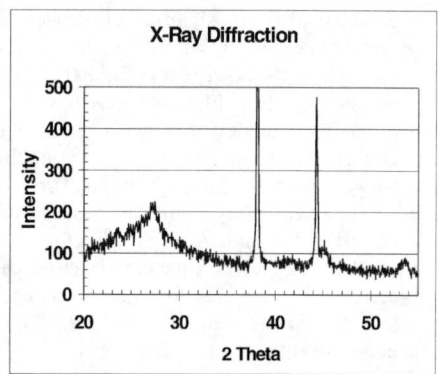

Fig. 2 x-ray diffraction spectrum of (Ge,C) Film deposited on steel

We measured the Hall mobility in these materials using the double-electrometer set up described in the literature from Keithley Instruments[4]. The low mobilities in these materials make the use of standard Hall measurement apparatus impossible because of the small signals, and one has to resort to carefully shielded configurations to measure the small signals. The electron Hall mobility was found to be 2.6 cm$^2$/V-sec. The material was n type with a carrier concentration of $1 \times 10^{17}$/cm$^3$.

## DEVICE FABRICATION AND PROPERTIES

p/n junction devices on stainless steel substrates were fabricated in these materials. The device structure is shown in Fig. 3. It consists of a n+ amorphous Si layer, followed by a thin undoped a-Si:H layer, which in turn is followed by a graded gap nanocrsytalline (Si,Ge) buffer layer. This is followed by the n-type nc-(Ge,C):H base layer. This layer is followed by a buffer layer consisting of a nanocrystalline (Si,Ge) layer with increasing bandgap and a graded gap a-(Si,C):H layer. The p layer is an amorphous (Si,C):H layer. An ITO contact completes the device. The purpose of the thin a-Si:H layer next to the n+ layer is to provide a compact layer to seal up any shorts. The graded gap buffer layers at the front and the back of the base layer provide for bandgap matching and also serve to provide a field to drive the holes away from the back (n+/n) interface and electrons away from the front (p+/n) interface, thereby reducing recombination at these interfaces. We found that the use of back amorphous buffer layers was absolutely critical for achieving devices with reasonable fill factors. Without them, the short density was too high and the fill factors and voltages deteriorated significantly.

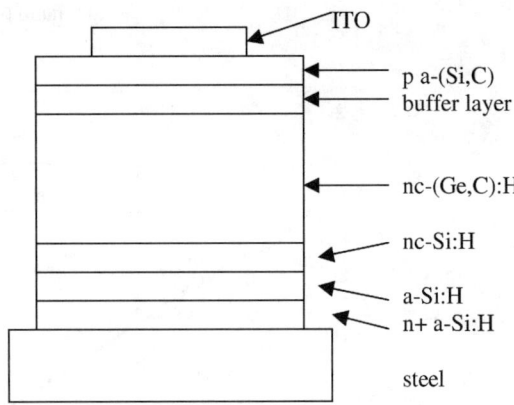

Fig. 3 Schematic diagram of solar cell

In Fig. 4, we show the I(V) curve for a nc-Ge:H cell, and the corresponding quantum efficiency (QE), plotted on a log scale, is shown in Fig. 5. The QE data clearly shows that the material follows the absorption of crystalline Ge. In Fig. 5, we also show the QE curve for a nc-(Ge,C):H cell. Upon comparing the two QE curves in Fig. 5, we see that the addition of C has clearly moved the edge of the QE curve to higher energies, implying a higher bandgap by about 0.2 eV.

The capacitance-voltage curve for the nanocrystalline Ge:H device is shown in Fig. 6. It shows a built-in voltage of 0.62 V and a donor density of $2.6 \times 10^{17}/cm3$.

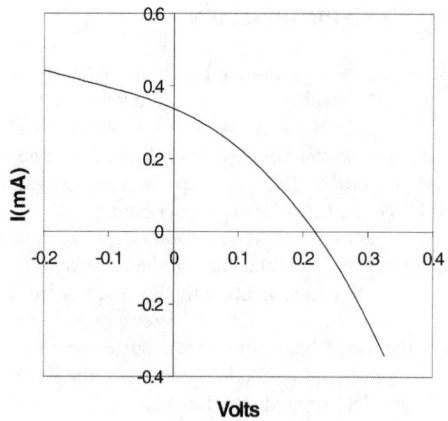

Fig. 4  I(V) curve for nano Ge:H cell

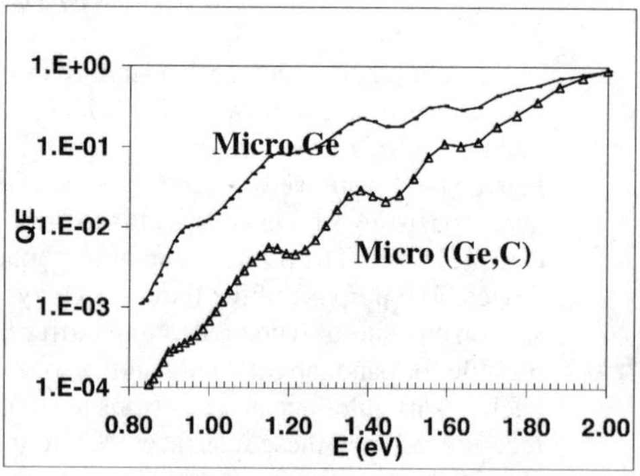

Fig. 5 Quantum efficiency curves for nano Ge and nano (Ge,C) cells.

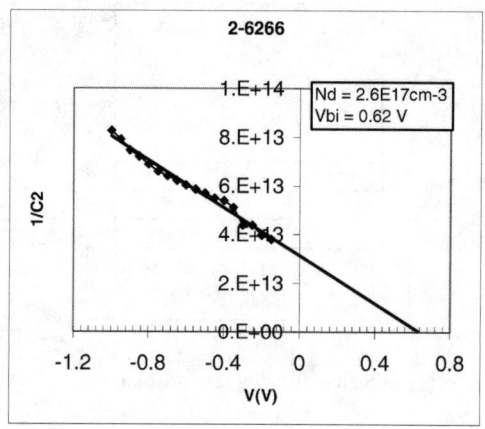

Fig. 6 Capacitance-voltage data taken at 1 kHz for nc-Ge:H cell.

## CONCLUSIONS

In conclusion, we have fabricated the first ever p/n junction photovoltaic devices in both nc-Ge:H and in nc-(Ge,C):H. The materials and devices were grown using remote ECR plasma discharge techniques with a high hydrogen dilution. The quantum efficiency data clearly show increasing bandgap as C is added to Ge. To improve device performance, we had to use a thin graded gap amorphous buffer layer at the n+/n interface. Both capacitance and Hall measurement data show a doping density in the $1\text{-}2\times10^{17}/cm^3$ range in the n layers. For best devices, this value will have to be reduced. Electron Hall mobility is ~ 2.6 $cm^2$/V-sec.

## ACKNOWLEDGEMENTS

This work was partially supported by NREL. We thank J. H. Zhu, Keqin Han, Durga Panda and Puneet Sharma for their technical help.

## REFERENCES

1. J. Herrold and V. L. Dalal, J.Non-Cryst. Solids., 270, 255(2000)

2. J. Herrold and V. L. Dalal, Proc. of MRS 507, 561(1999)

3. K. Erickson and V. L. Dalal, Proc. of MRS, 507, 987(1998)

4. Keithley Instruments, Catalog 2002,p.200.

# Structural Evolution of Nanocrystalline Germanium Thin Films with Film Thickness and Substrate Temperature

William B. Jordan, Eric D. Carlson, Todd R. Johnson, Sigurd Wagner,
Department of Electrical Engineering, Princeton University, Princeton, NJ 08544

## ABSTRACT

The structure of germanium thin films prepared on glass by plasma enhanced chemical vapor deposition was characterized by Raman spectroscopy, atomic force microscopy (AFM) and field emission scanning electron microscopy (SEM). Crystallinity, surface roughness, and grain size were measured as functions of film thickness and deposition temperature. Grain nucleation was apparent for films as thin as 10 nm. Over the thickness range studied, grain size increased with film thickness, whereas average surface roughness started to increase with film thickness, but then remained fairly constant at approximately 1 nm for a film thickness greater than 25 nm.

## INTRODUCTION

Following from past work with microcrystalline silicon (µc-Si:H) and the desire to find new materials for possible use in devices such as low gap solar cells and long-wavelength photodetectors, nanocrystalline germanium (nc-Ge:H) has recently attracted fresh interest. This stems from nc-Ge:H having a high optical absorption coefficient and conductivity, and a low optical gap. A working prototype nc-Ge:H p-i-n photodetector has already been demonstrated [1]. Similar to µc-Si:H, nc-Ge:H, may not exhibit the light-induced degradation of amorphous silicon. Also, the high optical absorption coefficient and electrical conductivity will be beneficial to certain applications.

As with µc-Si:H, there is a need to understand nc-Ge:H nucleation and microstructural film evolution to control film structure for device applications [2,3]. This is especially important for thin n-type and p-type doped layers of nanocrystalline material for p-i-n applications. Crystallized films have better electrical and structural properties such as higher mobility [4,5]. When deposited by PECVD from mixtures of silane with much hydrogen, silicon films typically begin growing with an amorphous ($a$) structure and then gradually evolve nanocrystalline ($nc$) as the film grows several 100-nm thick [6,7,8]. Nanocrystallization has been observed to begin directly on the substrate [9], and even to proceed backwards into the initially-deposited bottom amorphous layer [10]. In distinction to the silicon materials, $a$-Ge:H [11,12] and $nc$-Ge:H [4,13-21] remain poorly understood.

We characterized the surface structure of nanocrystalline germanium (nc-Ge:H) grown by plasma-enhanced chemical vapor deposition (PECVD) from germane ($GeH_4$) and hydrogen source gases. Atomic force microscopy (AFM) and field emission scanning electron microscopy (SEM) were used to image films with varying thickness and substrate temperature. It was previously shown using Raman spectroscopy and UV reflectance that germanium films exhibit a succession of $a$-$nc$-$a$-$nc$ phase transitions over a temperature range of 30°C through 310°C [22]. Our AFM and SEM images verify the $nc$ phases, and also show a small fraction of crystallites in the amorphous films. Nucleation directly on the substrate is shown.

## EXPERIMENT

Thin films of germanium were deposited on Corning 1737 glass substrates by RF (13.56 MHz) excited PECVD using $GeH_4$ and $H_2$ source gases, a chamber pressure of 800 mTorr, RF power of 43 mW/cm$^2$ of cathode area, and a $H_2/GeH_4$ flow ratio of 420. For the temperature series of films, the deposition temperature ($T_{dep}$) of successive films was varied between 30°C and 310°C. The thickness series was grown at 150°C with films between 10 nm and 160 nm in thickness. Crystallinity was measured by Raman spectroscopy (excitation $\lambda = 514$ nm) and UV reflectance [23,24]. The thickness of the films was determined by plasma etching to the glass substrate and measuring the step height with a surface stylus. Surface morphology was investigated by scanning electron microscopy (SEM) using a Philips XL30 FEG-SEM and by atomic force microscopy (AFM) using a Digital Instruments Nanoscope IIIa in tapping mode [25]. Grain size was determined from AFM and SEM micrographs using the mean lineal intercept method [26]. Average surface roughness, $R_a$, was calculated using the AFM software over a scanning area of 1 µm$^2$.

## RESULTS AND DISCUSSION

### Thickness Series

From the Raman and UV reflectance measurements, all films in the thickness series were found to be nanocrystalline [2]. Figure 1(a) shows grain diameter as a function of film thickness. With both measurement techniques, we found that grain diameter increases with film thickness. The SEM and AFM measurements of grain size are in remarkable agreement. However, the AFM measurements are larger for most samples. This is attributed to the finite size of the AFM tip. AFM micrographs can be interpreted as a convolution of true surface topography and the tip shape. It has been shown that when the surface features are less than two times the diameter of the tip, significant distortion of the image can occur [28-30]. Measured surface roughness is also less. In this study, the AFM tips had a diameter of approximately 20 nm. The grains ranged from 20 to 40 nm in diameter, 1 to 2 times the diameter of the tip.

Figure 1(b) shows average surface roughness from AFM measurements as a function of film thickness. Surface roughness increases with film thickness for films under 50 nm in

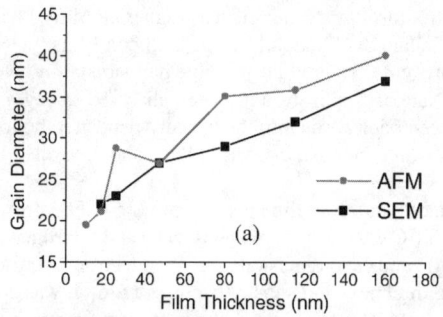
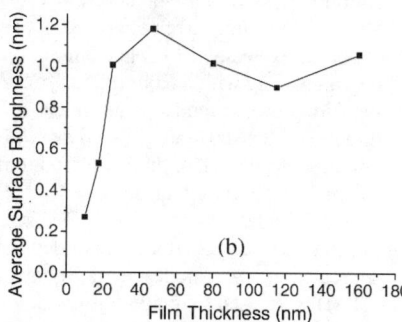

**Figure 1** (a) shows the increase in grain diameter with film thickness from both AFM and SEM microscopy. (b) shows the average surface roughness determined by AFM.

thickness. Above 50 nm, surface roughness remains relatively constant. This may be associated with the process of crystallite coalescence [3]. Studies using transmission electron microscopy (TEM) cross sections found lateral grain sizes to be 10 to 100 nm for film thickness of 300 nm to 1000 nm [5,27].

Figure 2 presents AFM images of the thickness series films for thicknesses of 17 nm, 80 nm, and 160 nm, respectively. Magnification of the 17 nm thick sample image shows that most grains share boundaries and most of the surface is covered with nc-Ge grains. Investigation of a 10 nm thick sample (not shown) also exhibited the beginning of grain nucleation over most of the sample, indicating that grains nucleate on the substrate. Figure 3 shows similar results with SEM images of the same samples. By a thickness of 25 nm, the surface is fully nucleated.

## Temperature series

Germanium thin film samples were grown over a temperature range of 30 to 310°C. Raman spectroscopy and UV reflectance measurements of crystallinity show that the films grow amorphous at $T_{dep} = 30°C$, nanocrystalline between 80°C and 190°C, amorphous again at 240°C and 280°C, and nanocrystalline again at 310°C [2]. The film thickness and growth rates are listed in Table I. Film growth rates ranged from 0.13 Å/sec to 0.26 Å/sec, and generally decreased with increasing temperature.

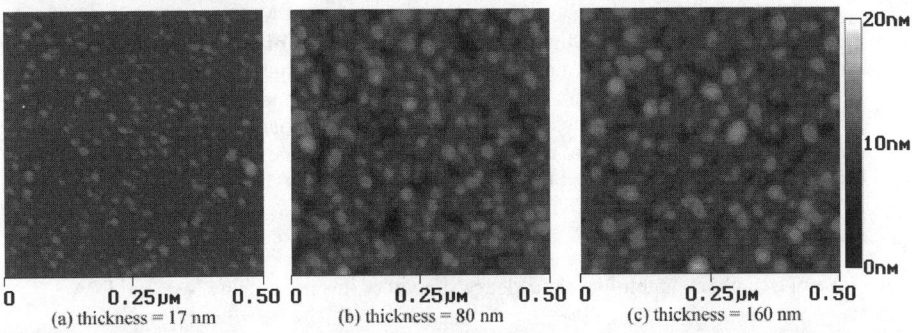

**Figure 2.** AFM images of nc-Ge:H films with increasing thickness

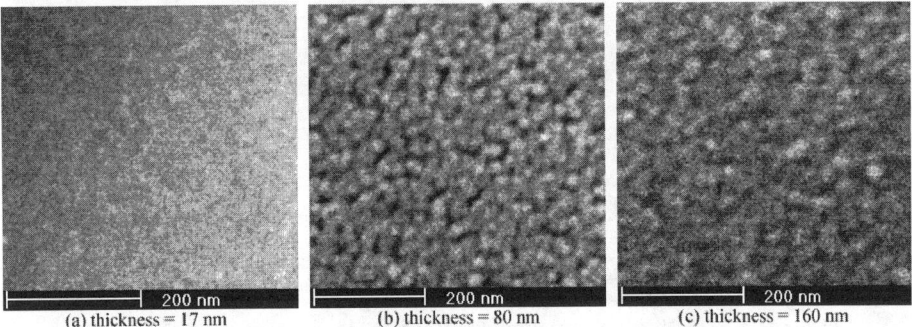

**Figure 3.** SEM images of nc-Ge:H films with increasing thickness

**Table I.** Ge film substrate temperature during growth, and associated thickness and growth rate

| $T_{dep}$ (°C) | 30 | 80 | 120 | 150 | 190 | 240 | 280 | 310 |
|---|---|---|---|---|---|---|---|---|
| Thickness (Å) | 920 | 1020 | 850 | 800 | 650 | 600 | 720 | 510 |
| Growth Rate (Å/sec) | 0.24 | 0.26 | 0.24 | 0.22 | 0.18 | 0.17 | 0.20 | 0.13 |

The relation between grain size and deposition temperature for the nanocrystalline samples is shown in Figure 4(a). The variation in film thickness limits an interpretation of the trend between 80°C and 190°C. However, a sharp increase in grain size is seen at $T_{dep}$ = 310°C in Figure 4(a). This corresponds with a sharp increase in dark conductivity as seen in a previous study of these samples [2]. This 310°C material is also strongly p-type, with a carrier concentration around $10^{18}$ cm$^{-3}$. We have not yet determined a specific cause of this result. Figure 4(b) shows surface roughness as a function of temperature. Again, there is a large increase in roughness at $T_{dep}$ = 310°C.

Although the samples with $T_{dep}$ = 30°C, 240°C, and 280°C were measured as amorphous by Raman spectroscopy and UV reflectance, these films also had significant roughness and granular features.

## CONCLUSIONS

We have shown via surface characterization studies by AFM and SEM that the nucleation of PECVD deposited nc-Ge:H thin films begins within 10 nm of the substrate surface for high hydrogen dilutions. Grain size increases with film thickness up to at least 160 nm thick. Surface roughness initially increases with thickness but then remains relatively constant, indicating grain coalescence. This study indicates that the SEM and AFM produce reliable results in close agreement and that thin layers of nc-Ge can be grown for device applications, opening the way for use in low gap solar cells and long-wavelength photodetectors.

## ACKNOWLEDGEMENTS

The authors gratefully acknowledge support of this research by NSF and EPA.

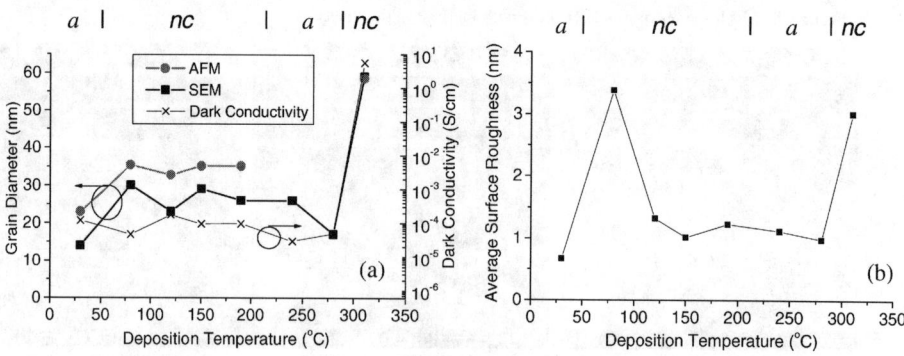

**Figure 4.** (a) nc-Ge:H grain diameter from both AFM and SEM microscopy with dark conductivity and (b) average roughness as functions of substrate temperature. The change of phase is shown above.

# REFERENCES

1. M. Krause, H Stiebig, R. Carius, H. Wagner, Mat. Res. Soc. Symp. Proc. **664**, A26.5 (2001).
2. A.S. Ferlauto, P.I. Rovira, R.J. Koval, C.R. Wronski, R.W. Collins, Mat. Res. Soc. Symp. Proc. **609**, A2.2.1 (2000).
3. H. Fujiwara, Y. Toyoshima, M. Kondo, A. Matsuda, Mat. Res. Soc. Symp. Proc. **609**, A2.1.1 (2000).
4. J. Jiang, K. Chen, D. Feng, and D. Sun, Thin Solid Films **230**, 7 (1993).
5. S. Wagner, H. Gleskova, I-Chun Cheng, M. Wu, Thin Solid Films, in press.
6. Joohyun Koh, A. S. Ferlauto, P. I. Rovira, C.R. Wronski and R.W. Collins, Appl. Phys. Lett. **75**, 2286 (1999).
7. E. Vallat-Sauvain, U. Kroll, J. Meier, A. Shah, J. Pohl, J. Appl. Phys. **87**, 3137 (2000).
8. J. Kočka, H. Stuchlíková, J. Stuchlík, B. Rezek, V. Švrček, P. Fojtík, I. Pelant and A. Fejfar, in *Polycrystalline Semiconductors VI*, Solid State Phenomena **80-81**, 213 (2000), edited by O. Bonnaud, T. Mohammed-Brahim, H. P. Strunk, and J. H. Werner (Scitec Publications, Switzerland 2001).; J. Kočka, A. Fejfar, V. Vorlicek, H. Stuchlíková, and J. Stuchlík, Mater. Res. Soc. Symp. Proc. **557**, 483 (1999).
9. S. Hazra, I. Sakata, M. Yamanaka and E. Suzuki, Appl. Phys. Lett. **80**, 1159 (2002)
10. P. Roca i Cabarrocas, N. Layadi, T. Heitz, B. Drevillon and I. Solomon, Appl. Phys. Lett. **66**, 3609 (1995).
11. F.H. Karg, H. Böhm and K. Pierz, J. Non-Cryst. Solids **114**, 477 (1989).
12. W. Paul, S.J. Jones, F.C. Marques, D. Pang, W.A. Turner, A.E. Wetsel, P. Wickboldt, and J. Chen, Mat. Res. Soc. Symp. Proc. **219**, 211 (1991).
13. A.D. Stewart, D.I. Jones, G. Willeke, Phil. Mag. B **48**, 333 (1983).
14. J. Gonzalez-Hernandez, G.H. Azarbayejani, R. Tsu, F.H. Pollak, Appl. Phys. Lett. **47**, 1350 (1985).
15. B. Drevillon, C. Godet, A.M. Antoine, Mat. Res. Soc. Symp. Proc. **75,** 341 (1986).
16. J.R. Woodyard, J. Gonzalez-Hernandez, R.T. Young, J. Piontkowski, Mat. Res. Soc. Symp. Proc. **70,** 65 (1986).
17. C. Godet, B. Drevillon, C. Senemaud, J. Non-Cryst. Solids **97,** 431 (1987).
18. B. Drevillon, C. Godet, J. Appl. Phys. **64**, 145 (1988).
19. C. Godet, P. Roca i Cabarrocas, S.C. Gujrathi, P.A. Burret, J. Vac. Sci. Technol. A **10,** 3517 (1992).
20. T. Aoki, Y. Nishikawa, K. Fukasawa, W.Q. Sheng, M. Hirose, J. Non-Cryst. Solids **164-166**, 91 (1993).
21. P.R. Poulsen, M. Wang, J. Xu, W. Li, K. Chen, G. Wang, and D. Feng, J. Appl. Phys. **84,** 3386 (1998).
22. W.B. Jordan and S. Wagner, Mat. Res. Soc. Symp. Proc. **715**, A18.2.1 (2002).
23. P.Y. Yu, M. Cardona, Fundamentals of Semiconductors, Springer-Verlag, 1996. pp. 375-392.
24. E.D. Palik,ed., Handbook of Optical Constants of Solids. Academic Press, 1985.pp. 472-473.
25. J. Xu, K. Chen, D. Feng, S. Miyazaki, M. Hirose, Thin Solid Films **335** 130 (1998).
26. R. T. DeHoff, *Applied Metallography*, edited by G. F. Vander Voort (Van Nostrand Reinhold Company, New York, 1986), pp. 89-99.
27. J. González-Hernández, Rev. Mex. de Fisica **35** 648 (1989).
28. K.L. Westra and D.J. Thomson, J. Vac. Sci. Technol. B **13**(2), 344 (1995).
29. K.L. Westra and D.J. Thomson, J. Vac. Sci. Technol. B **12**(6), 3176 (1994).
30. D.L. Sedin and K.L. Rowlen, App. Surf. Sci. **182**, 40 (2001).

# PROCESS PARAMETERS FOR POLY-SILICON DEPOSITION AT A HIGH GROWTH RATE (1-7 nm/s) BY HOT-WIRE CHEMICAL VAPOUR DEPOSITION

J.K. Rath, A.J. Hardeman, C.H.M. van der Werf, P.A.T.T. van Veenendaal, M.Y.S. Rusche and R.E.I. Schropp
Utrecht University, SID - Physics of Devices, P.O. Box: 80000, 3508 TA Utrecht, The Netherlands. J.K.Rath@phys.uu.nl

## ABSTRACT

High silane to hydrogen flow ratios and optimum wire temperatures are the key process parameters to achieve high growth rate poly-silicon films by hot wire chemical vapour deposition (HWCVD) using a four-wire hot-wire assembly. Four tungsten wires, 4 cm apart from each other, were used as catalytic filaments. Growth rates higher than 7 nm/s have been achieved at a substrate temperature of ~510 °C. The increase in deposition rate was accompanied by deterioration of two physical properties i.e., decrease in photoresponse and increase in oxygen incorporation in the film, which is attributed to high porosity in the material that is commonly observed in these high growth rate materials. The process conditions to incorporate a high hydrogen content into the material for passivation of defects and donor states have been identified as high hydrogen dilution and lower wire temperature. With these procedures, poly-Si films deposited at 1.3 nm/s showed a high ambipolar diffusion length of 132 nm. Incorporating such poly-Si films as the i-layer in an n-i-p solar cell on a stainless steel substrate, *without back reflector*, showed an efficiency of 4.4 % and a high open circuit voltage of 0.58 V, which is attributed to effective passivation of defects and dopants by incorporated hydrogen.

## INTRODUCTION

Since the success of the "micromorph" cell concept and its rapid industrial adaptation [1], the demand for a high deposition rate for the microcrystalline/polycrystalline cell, which constitutes the bottom cell of a micromorph cell, has become acute. A modification of the plasma CVD concepts, such as deposition under a high pressure depletion condition [2], has shown the possibility of achieving this target using the existing well known deposition process that can be easily transferred to industrial environment. This success of large area module efficiencies has helped to advance this proposition [3]. The main philosophy behind this concept is to have lower ion energy on the growing surface, which is brought about by cooling the plasma in a high-pressure condition. In other words, it would still be beneficial to employ a process that has extremely cooled ions [1], such as expanding thermal plasma (ETP), microwave CVD (MWCVD), and electron cyclotron resonance (ECR). Out of this, only MWCVD has been successful in achieving high efficiency cells at high growth rate [4]. Another possibility is to use ion-free deposition, namely photo-CVD and hot-wire chemical vapour deposition (HWCVD). We have already reported that compact poly-silicon films can be made at a high deposition rate of 0.5 nm/s, which is already comparatively high, and solar cells implementing this truly polycrystalline material (with no amorphous phase) have shown 4.4 % efficiency [5]. This paper probes into a high growth rate regime (> 1nm/sec), facilitated by a modified filament assembly. This would allow for the deposition of the i-layer of a poly-Si cell in less than 25 minutes.

## EXPERIMENTAL

Poly-Si films were made by HWCVD in a multichamber system called PASTA [5]. The hot-wire deposition chamber consists of a modified hot-wire assembly designed for fully utilizing the parameter space for enhanced deposition rate. The modification is achieved by increasing the catalytic area for decomposition by using four tungsten wires instead of two wires that we earlier used for poly-Si materials (which we earlier called *Poly2* and *Poly1* [6]). The structural properties of the materials were characterised by Fourier transform infrared (FTIR) spectroscopy (hydrogen content, oxygen incorporation), Raman spectroscopy (crystalline volume fraction) and X-ray diffraction (XRD) (grain orientation, size). The electronic properties were obtained from dark- and photoconductivity and steady state photocarrier grating (SSPG) technique (ambipolar diffusion length). The crystalline fraction is obtained from the Raman spectra using the simple ratio of intensities i.e., $V_f = \{I_c/(I_c+I_a)\}$ where $I_c$ and $I_a$ are the intensities of the crystalline-like peaks (at 520 and 510 cm$^{-1}$) and amorphous-like peak (at 480 cm$^{-1}$), respectively, of Si-Si TO stretching vibrations. In this case, the value of the crystalline volume fraction is considerably underestimated. The thickness was measured by a surface profiler (Dektak) or optically by reflection/transmission (R/T) measurement. The optical absorption and the band gap were determined from R/T and photothermal deflection spectroscopy (PDS). The indirect band gap is obtained from the $(\alpha)^{1/2}$ vs h$\nu$ plot, where $\alpha$ is the optical absorption coefficient and h$\nu$ is the energy.

Solar cells were made on stainless steel (SS) substrates in the same configuration as our earlier HWCVD poly-Si cell reports [5]. The cell structure is substrate type; SS/n-type μc-Si/profiled poly-Si/buffer a-Si:H/p-type μc-Si/ITO/Au (gridlines). Current-Voltage (I-V) measurements were made under AM1.5 100mW/cm$^2$ white light from a dual beam solar simulator (WACOM, Japan)

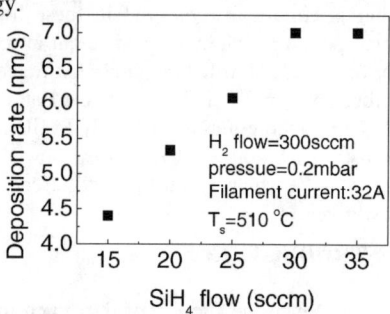

Figure 1: Dependence of deposition rate with SiH$_4$ gas flow

## RESULTS

One of the major effects of increasing the number of filaments is the increase of the catalytic area and the effective use of the gas in the deposition process. As the depositions in the microcrystalline regime of HWCVD are under depletion conditions, the deposition rate is simply related to the supply gas. Fig 1 shows the typical case of supply-limited growth. The depletion condition is maintained by employing a sufficiently high filament temperature (filament current per wire: 16 A). It is clearly observed that with a high SiH$_4$ gas flow rate of 30-35 sccm, a deposition rate as high as 7 nm/s can be achieved for a material with a

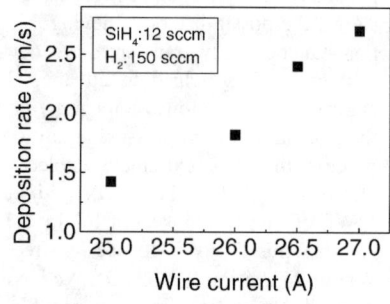

Figure 2 : Dependence of deposition rate on filament current though 4 filaments in parallel

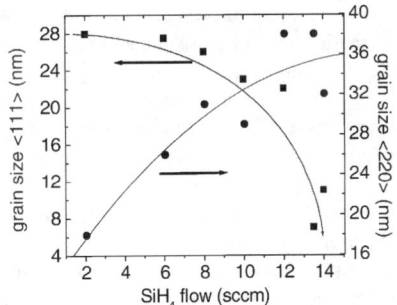

Figure 3 : Grain size variation with SiH$_4$ gas flow rate

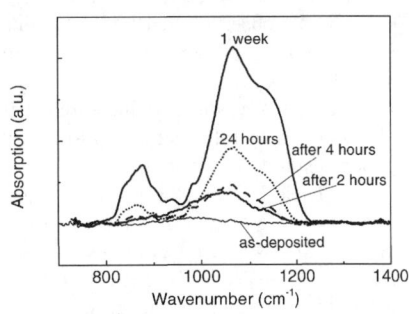

Figure 4: Increase of post-deposition oxygen incorporation at different times after deposition

crystalline volume fraction, V$_f$, of ~50 %. However, these high deposition rate materials have so far not shown the required device quality for solar cells. Nevertheless, it should be noted that for poly-Si at a deposition rate of 7 nm/s, the photosensitivity of 55 is worth considering future development.

We used a lower gas flow to obtain a deposition rate that is sufficient to achieve our goal of > 1nm/s, without compromising the electronic quality. Here again we found that gas flow rates and filament temperatures are the main parameters for deposition rate (Figure 2), and almost a linear relation between deposition rate and gas supply and filament current is observed. It is clear that the deposition rates are lower than in the case of Fig.1, however they are still as high as 3 nm/s. There are major differences in these types of materials compared to the poly-silicon films we have been reporting previously using the two-filament assembly [6]. For the "two wire" assembly case, our best poly-silicon (that we called *Poly2*) had an exclusive (220) oriented growth and grains as large as > 100 nm in diameter and the thickness of the film in length. In case of the "4-wire" assembly in the current study, the grains are small and essentially randomly oriented or slightly (220) preferentially oriented. However this may not necessarily be bad for solar cells, because microcrystalline silicon with randomly oriented grains have shown good efficiencies [1]. The second difference is that the grains are rather small (less than 40 nm, see Fig. 3). This would obviously manifest in a large amount of grain boundary surface area and defects. The third difference is that Si-H stretching vibration modes are predominantly at 2100 cm$^{-1}$. All these aspects point to a structure that is porous or non-compact. This is in contrast to the highly compact structure that we have been making by the "two wire" assembly [6] at lower deposition rate. In fact, structurally the current set of poly-silicon films resembles very well the poly-silicon deposited at high hydrogen dilution (called *Poly1* [6]) for the "two wire" assembly case.

Table 1: Physical properties of poly-silicon film made at optimum hydrogen dilution (SiH$_4$/H$_2$=6/150 sccm, Wire current: 27A

| Physical properties | |
|---|---|
| E$_{gap}$ (eV) by PDS | 1.01 |
| V$_f$ (%) by Raman | 60 |
| σ$_d$ (Ω$^{-1}$ cm$^{-1}$) | 8.23x10$^{-8}$ |
| σ$_{ph}$ (Ω$^{-1}$ cm$^{-1}$) | 1.96x10$^{-5}$ |
| Photoresponse (σ$_{ph}$/σ$_d$) | 239 |
| Dep. rate (nm/s) | 1.4 |
| Physical thickness (nm) | 2100 |
| R* | 1.000 |
| E$_a$ (eV) | 0.53 |
| Grain size <111> (nm) | 27 |
| <220> (nm) | 26 |
| Ambipolar diffusion length (nm) | 53 |
| [H] (%) by FTIR | 3.1 |

In order to justify these materials for the use in solar cells, the photosensitivity needs to be high enough. We observed that the crystalline volume fraction and photoresponse are substantially varied with the amount of hydrogen dilution. The dilution series was made by changing the $SiH_4$ flow from 2 to 14 sccm while keeping the $H_2$ flow constant at 150 sccm. While crystalline volume fraction steadily increased with increased dilution (decreasing silane flow), the photosensitivity follows that up to an optimum dilution ($SiH_4/H_2$ = 6/150 sccm), above which photosensitivity is lower again.

Table I shows the physical characteristics of the best photosensitive material made at $SiH_4/H_2$ = 6/150 sccm. It is clear that the electronic properties are as good as the *Poly2* sample made with "two wire" assembly [6], and in fact the photosensitivity is better than the *Poly2* case (photosensitivity of 140). Contrary to the expectation, these materials do not deliver the expected solar cell efficiencies. The reason lies in the structural difference. Whereas the columnar grains with exclusive (220) oriented growth in the *Poly2* material allowed the electronic transport of photogenerated carriers in the solar cell to bypass the grain boundary defects, for the case of poly-silicon made at high deposition rate by this "4 wire" assembly, the randomly oriented grains and defect states, and possible donor states, at the grain boundary deteriorate the transport properties in the cell structure. Moreover, the presence of the Si-H mode at 2100 cm$^{-1}$ as well as the high post deposition oxygen in-diffusion (Fig. 4) confirm the porosity and the large surface area of grain boundaries. To compensate the structural limitation of these materials, it is necessary to passivate these grain boundary defects and the possible donor states. This can be achieved by incorporating a larger amount of hydrogen into the film. It is known from literature that hydrogen not only passivates the grain boundary defects but also passivates the impurity induced donor states [1]. Normally, for the randomly oriented or small grain poly-silicon case, this is mostly achieved by a lower substrate temperature deposition [1]. However, our goal is to make poly-silicon films at a high growth rate, which we believe, is more easily achieved at high substrate temperature. Thus, other process conditions have to be invoked to achieve this objective. We found two such processes; hydrogen dilution and lower filament temperature, however at the cost of a slightly decreased deposition rate.

Figure 5 and 6 show the hydrogen content in the poly-silicon film made at different hydrogen dilutions and filament temperatures, respectively. It is clear that both parameters have substantial effect in manipulating the hydrogen content in the film. Our approach was to decrease the filament temperature after optimising the dilution in order to obtain the highest hydrogen content in the film. It was observed that the crystalline volume fraction decreased only marginally while decreasing the wire current from 27 A to 25 A. What was clearly visible however is that the deposition rate steadily decreased, concomitant with the increased hydrogen content in the film, when the filament temperature was decreased. Hence, we obtained a very high quality film (photosensitivity of 760) made at filament current of 25 A, but the deposition

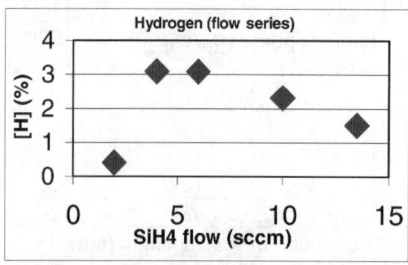

Figure 5 : Dependence of hydrogen content in the film on silane flow rate. $H_2$ : 150 sccm

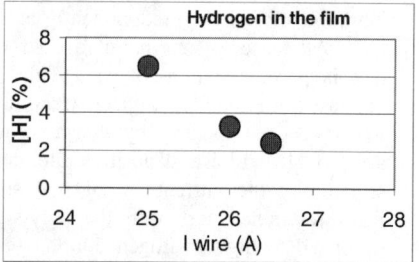

Figure 6 : Dependence of hydrogen content on wire current (temperature)

rate of 0.4 nm/s did not satisfy our goal. Thus a compromise between the deposition rate and the device quality had to be made. This was achieved for films made at 26 A filament current. The deposition rate of 1.3 nm/s is considerably high. The physical properties are crystalline volume fraction ($V_f$)= 53 % and diffusion length ($L_D$) = 132 nm. The XRD diffraction of this material is shown in Figure 7. These materials are again only slightly (220) oriented and contains small grains (25 nm). However, the hydrogen passivation allows these materials to show good solar cell behaviour. Fig. 8 shows the solar cell structure. The doped layers are made by plasma CVD. The structure is on plain stainless steel and does not

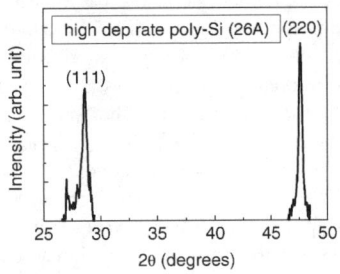

**Figure 7 : X-ray diffraction pattern of a poly-Si film made at $SiH_4/H_2$=6/150 sccm and filament current of 26A**

have any back reflector. Fig. 9 shows the I-V characteristics of the poly-silicon cell. An efficiency of 4.37% has been achieved. This shows that we have been able to achieve the same efficiency as our previously reported poly-silicon cell with "two wire" assembly [5], while increasing the deposition rate by more than 2 times. One important observation to be made is that the open circuit voltage and FF are both improved. The limitation is the low short circuit current density. This is not only due to lack of back reflector, but (especially comparing with the *Poly2* cell) also due to lack of low surface texture and optical enhancement that we had observed in the *Poly2* sample [5].

As mentioned above, one of the highlights of this cell is the high open circuit voltage of 0.58 V. Figure 10 shows the comparison of the $V_{oc}$ of this sample in comparison to the poly-Si cells reported in the literature [1]. The high $V_{oc}$ of the pn junction needs large grain size. However, at the other end of grain sizes smaller than 100 nm, the high $V_{oc}$ has been possible in pin or nip structure. It is also to be noted than orientation is not a limiting factor for high $V_{oc}$ as both (220) oriented as well as random oriented grains have shown high $V_{oc}$. In our earlier cells we observed that decreasing the deposition temperature (see HT and LT in fig. 10) has an effect on the grain environment that increases the $V_{oc}$ (though other factors like stability of the n-layer is also an important aspect). However, for the current cell, the deposition temperature is high (>500 °C). Other possibility is the presence of amorphous regions near the p/i interface, because literature shows that presence of such amorphous silicon buffer can improve $V_{oc}$. The

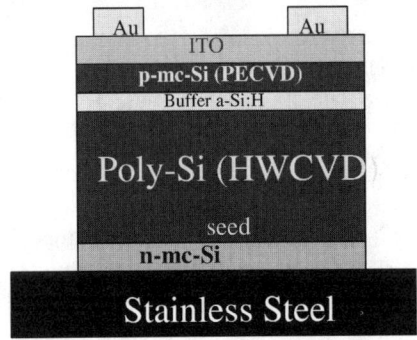

**Figure 8: Schematic of the cell structure**

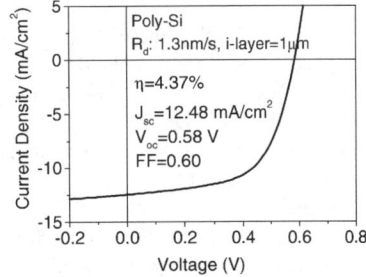

**Figure 9: I-V characteristics of the cell with polysilicon i-layer shown in fig. 7**

amorphous region near p/i interface is possible if the growth of the material in this region is amorphous due to decay of filaments in due course (towards the end) of deposition. To ascertain this, we measured Raman spectrum of the cells from the p side (top side). We observed a strong crystalline peak at 520 cm$^{-1}$ and no appreciable amorphous nature. Considering the fact that penetration depth of the 514.5 nm Ar ion laser (used in Raman spectroscopy) is ~200 nm, we can confirm that there is no significant amorphous region in the front region. We can conclude that the combination of small grains and effective passivation (hydrogen) is responsible for smaller band bending at the grain boundary as well as reduced trapping state density (notably the donor states) [1]. This changed grain environment is responsible for achieving high Voc.

## CONCLUSION

An efficiency of 4.37 % has been achieved for poly-Si cell deposited by HWCVD at a growth rate 1.3 nm/s. These cells have tremendous potential to show high efficiency when a back reflector is used due to the fact that the present cells are mainly limited by lower short current density due to low surface texture and lack of surface/internal optical enhancement. The high open circuit voltage of 0.58 V is attributed to the combination of small grain size and effective grain boundary passivation. High hydrogen dilution and low filament temperature have been identified as process parameters for effective hydrogen incorporation into the film.

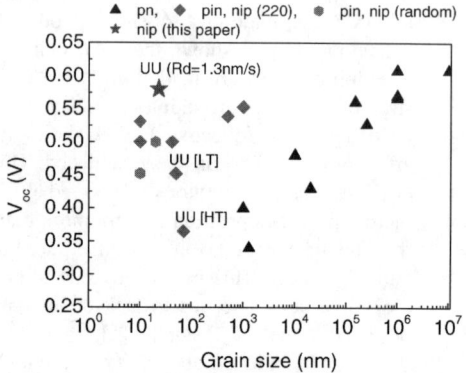

**Figure 10 : Open circuit voltage of the cell in the present study compared to others in literature [1]**

## ACKNOWLEDGEMENT

This research was partially funded by NOVEM (Netherlands Organisation for Energy and Environment) and EET. We thank Karine van der Werf for deposition of all the films.

## REFERENCES

[1] J.K.Rath, Solar Energy Material and Solar Cell, **76**, 431 (2003).
[2] L.Guo, M.Kondo, M.Fukawa, K.Saitoh, A.Matsuda, Jpn. J. Appl. Phys., **37**, L1116 (1998).
[3] B.Rech, O.Kluth, T.Repmann, T.Roschek, J.Springer, J.Muller, F.Finger, H.Stiebig, H.Wagner, Tech. Digest. Int'l PVSEC-12, Jeju, Korea, (2001) p339.
[4] S.R.Jones, R.Crucet, R.Capangpangan, M.Izu, A.Banerjee, Mat. Res. Soc. Symp. Proc., **664**, A4.5.1 (2001).
[5] J.K.Rath, F.D.Tichelaar, R.E.I.Schropp, Solid State Phenomena, **67-68**, 465 (1999).
[6] J.K.Rath, NATO SCIENCE SERIES VOLUME: Photovoltaic and Photoactive Materials - Properties, Technology and Applications, p171-182 (2002), KLUWER ACADEMIC PUBLISHERS, The Netherlands.

# High Temperature n- and p-type Doped Microcrystalline Silicon Layers Grown by VHF PECVD Layer-by-Layer Deposition

A. Gordijn, J.K. Rath*, and R.E.I. Schropp
Utrecht University, Debye Institute, SID-Physics of Devices,
PO box 80.000, NL-3508 TA Utrecht, The Netherlands

* corresponding author: j.k.rath@phys.uu.nl

## ABSTRACT

Due to the high temperatures used for high deposition rate microcrystalline ($\mu$c-Si:H) and polycrystalline silicon, there is a need for compact and temperature-stable doped layers. In this study we report on films grown by the layer-by-layer method (LbL) using VHF PECVD. Growth of an amorphous silicon layer is alternated by a hydrogen plasma treatment. In LbL, the surface reactions are separated time-wise from the nucleation in the bulk. We observed that it is possible to incorporate dopant atoms in the layer, without disturbing the nucleation. Even at high substrate temperatures (up to 400 °C) doped layers can be made microcrystalline. At these temperatures, in the continuous wave case, crystallinity is hindered, which is generally attributed to the out-diffusion of hydrogen from the surface and the presence of impurities (dopants).

We observe that the parameter window for the treatment time for p-layers is smaller compared to n-layers. Moreover we observe that for high temperatures, the nucleation of p-layers is more adversely affected than for n-layers. Thin, doped layers have been structurally, optically and electrically characterized. The best n-layer made at 400 °C, with a thickness of only 31 nm, had an activation energy of 0.056 eV and a dark conductivity of 2.7 S/cm, while the best p-layer made at 350 °C, with a thickness of 29 nm, had an activation energy of 0.11 V and a dark conductivity of 0.1 S/cm. The suitability of these high temperature n-layers has been demonstrated in an n-i-p microcrystalline silicon solar cell with an unoptimized $\mu$c-Si:H i-layer deposited at 250 °C and without buffer. The $V_{oc}$ of the cell is 0.48 V and the fill factor is 70 %.

## INTRODUCTION

Microcrystalline silicon deposited by Plasma Enhanced CVD (PECVD) or Hot Wire CVD (HWCVD) is a promising material for solar cell applications [1]. High initial efficiencies up to 14.1 % for a micromorph tandem cell [2] have been achieved. Nowadays most effort is in the development of high growth rate material. The optimum deposition temperature for microcrystalline silicon is still under discussion. A high deposition rate i-layer may require a high substrate temperature. Matsuda [3] claims that hydrogen coverage of the growth surface is necessary for crystallization. The crystallinity is limited at high temperatures (above 350 °C) by the reduced hydrogen coverage due to the hydrogen out-diffusion. Vepřek, however, observes that if the material is pure, crystallization can take place up to very high temperatures [4].

The deposition temperature of the i-layer has consequences for the doped layers in a device. High growth rate and high temperature i-layers require temperature-stable doped layers in devices, to prevent damage of the doped layer itself [5] and the underlying layers such as the $SnO_2$ transparent conductive oxide layer due to hydrogen diffusion through the doped layer [6]. It is known that the hydrogen diffusion coefficient of a layer is related to the deposition temperature [7]. This mechanism implies the need of doped layers deposited at a high substrate temperature. Here, we present the first results in the development of microcrystalline silicon p-type and n-type layers made at relatively high temperatures in order to achieve temperature-stable doped layers. We use a layer-by-layer process (LbL) [8, 9] to deposit the doped layers. In such a process, amorphous silicon growth by a silane/hydrogen plasma is alternated by hydrogen plasma treatments. The crystal nucleation is induced by the hydrogen treatments. In this way, film growth reactions are separated in time from nucleation reactions due to which the nucleation process is expected to be less dependent on temperature and the presence of dopant atoms. The LbL method has already been proven to be a successful method to study the role of hydrogen in the deposition of microcrystalline silicon (intrinsic) layers and to deposit doped layers at substrate temperatures around 250 °C [10]. In this paper, we show that with the LbL method, microcrystalline silicon doped layers can be made also at high temperatures (up to 400 °C), whereas this is difficult in standard continuous wave (CW) depositions.

**EXPERIMENTAL**

The layers are deposited in the ultra high vacuum multi-chamber system (called ASTER [11]) by means of very high frequency (VHF) PECVD at 50 MHz. The LbL process is implemented as follows: a computer automated system lets in the process gasses $SiH_4$, $H_2$, and a dopant gas for the growth cycle. The plasma is operated for typically 10 seconds, and subsequently the pump valve is opened in order to pump away all the gasses. The same procedure is repeated for the hydrogen treatment cycle, where the typical treatment time is 100 seconds. The process is repeated until the desired layer thickness is reached. As dopant gasses, we use trimethylboron (TMB) for p-type doping and phosphine ($PH_3$) for n-type doping. The substrate temperature is varied from 250 to 400 °C. Solar cells with the n-i-p structure have been made on plain stainless steel substrates. The i-layer is made with a high hydrogen diluted silane plasma by VHF PECVD. A standard 200 °C CW μc-Si:H p-layer is used. The front contact consists of indium tin oxide (ITO) with a gold grid.

The thin doped layers are characterized by UV-VIS spectroscopy (to determine the thickness and optical properties), temperature dependence of the dark conductivity, and Raman spectroscopy. The crystalline ratio $R_c$ is used as a measure for the crystallinity:

$$R_c = \frac{I_c(510) + I_c(520)}{I_a + I_c(510) + I_c(520)}$$

were $I_c(510)$ and $I_c(520)$ are the integrated intensities of the 510 $cm^{-1}$ and 520 $cm^{-1}$ contribution of the TO peak of the Raman spectrum, whereas $I_a$ refers to the amorphous contribution at 480 $cm^{-1}$. The mobility and the charge carrier concentration are determined with Hall mobility measurements. All characterizations are performed on thin layers (similar thickness as used in the solar cell) because microcrystalline silicon is known to be very inhomogeneous in the direction perpendicular to the surface, especially in the initial growth phase. The solar cell test structures are analyzed by current voltage (IV) measurements.

## RESULT AND DISCUSSION

Figure 1 shows the optical absorption spectra of optimized microcrystalline silicon p-layers deposited under different conditions. All p-layers are between 20 and 30 nm thick. The LbL p-layer shows the lowest optical absorption. This is attributed to the fact that the incubation phase is small and crystalline fraction high. The LbL method shows that it is suitable for tuning the structural and electrical properties of the microcrystalline silicon material, especially for very thin layers.

The hydrogen treatment time appears to be an important parameter to tune the material. Figure 2 shows three series for different TMB dopant gas flows for LbL p-layers at the standard substrate temperature of 250 °C in which the hydrogen treatment time is varied and the number of LbL cycles is kept constant at 15. The figure shows that with increasing treatment time, the material thickness per cycle decreases while the crystalline fraction is increasing. It shows that the crystallinity of the material is hardly affected by the addition of dopant gas to the plasma. This is an important difference to the CW case, where the crystallinity is hindered by the addition of dopant atoms. We attribute this difference between the LbL and the CW case to the difference in nucleation mechanism: in the CW case both growth and nucleation take place at the same time at the growth surface, due to which the nucleation is influenced by the plasma process. In the LbL case, on the other hand, the nucleation takes place in the bulk, induced by the hydrogen treatment. Optimal structural and electrical properties were achieved when the layer thickness of the sub-layers after treatment (the effective sub layer thickness) is around 1.4 nm. When thinner effective sub layers are tried, no film growth occurs.

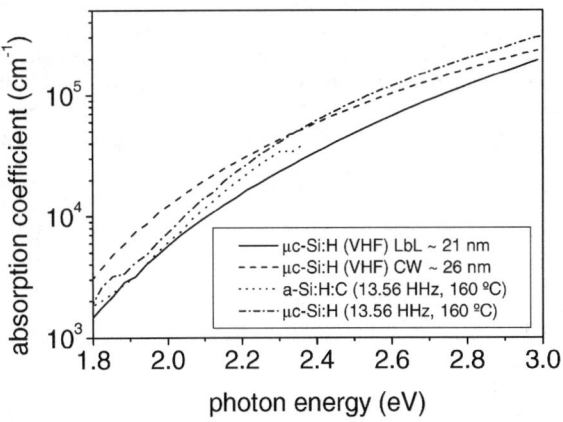

**Figure 1**. Optical absorption spectra of optimized p-layers deposited under different conditions.

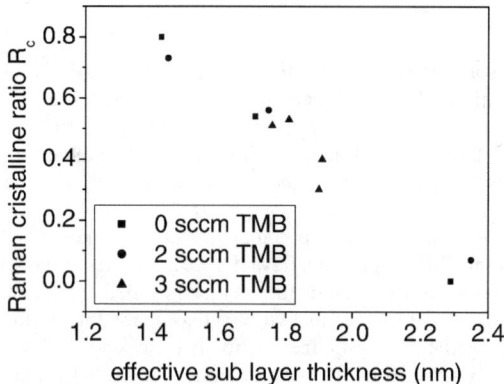

**Figure 2.** Relation between the thickness per cycle and the crystalline fraction for LbL p-layers at 250 °C under the variation of the hydrogen treatment time.

## High temperature doped layers

The substrate temperatures were increased from 250 °C to 400 °C. We re-optimized the doped layers at intermediate temperatures by adapting the hydrogen treatment times to the slower etching rate at elevated temperatures. The electrical properties of the high temperature thin layers made with LbL show the behavior of thin good doped layers (see Table 1). In the CW case, it is difficult to deposit a microcrystalline thin doped layer because the crystallinity is hindered at these temperatures, which has been attributed to the out-diffusion of hydrogen [3] or the presence of impurities [4].

In the optimization process of the layers at higher temperatures, we do see some minor effects of the addition of the dopant atoms. For n-layers the material properties are less critically dependent on the duration of the hydrogen treatments whereas for p-layers this process window is relatively small. Moreover, at high temperatures, longer hydrogen treatment times are necessary for p-layers than for n-layers. This is because for p-type material the crystallization is more adversely influenced by increasing temperatures. Moreover, in p-type material higher crystalline fractions are needed for good electrical properties. This is related to the observation that in n-type material percolation takes place with a lower crystalline material content compared to p-type material (see Figure 3).

**Table 1.** Electrical properties of LbL high temperature microcrystalline silicon doped layers.

| type | p | n |
|---|---|---|
| $T_{substrate}$ (°C) | 350 | 400 |
| Thickn (nm) | 29 | 31 |
| Raman $R_c$ | 0.26 | 0.12 |
| $E_a$ (eV) | 0.11 | 0.056 |
| $\mu$ (cm$^2$/Vs) | 0.13 | 0.35 |
| $N_{carrier}$ (cm$^{-3}$) | $1.2 \times 10^{19}$ | $1.0 \times 10^{20}$ |

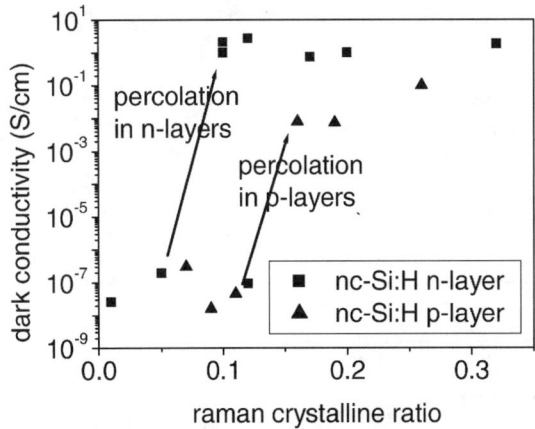

**Figure 3.** Dark conductivity versus the Raman crystalline ratio for p- and n-layers. For n-layers we see that percolation through the interconnected crystals occurs at lower crystalline fraction than for p-layers

A microcrystalline silicon n-i-p solar cell has been deposited to demonstrate the suitability (the doping efficiency and the resistance) of the high temperature LbL µc-Si:H n-layer. The structure is as follows: flat stainless steel (SS) / LbL µc-Si:H n-layer, 400 °C (50 nm; 0.1 Å/s) / VHF µc-Si:H i-layer, 250 °C (1000 nm; 1.0 Å/s) / VHF CW µc-Si:H p-layer, 200 °C (20 nm; 0.4 Å/s) / ITO / Au. No back reflector is used to enhance the optical absorption. The i-layer is made at a very high hydrogen dilution and the material has a high crystalline volume fraction (unlike the type of material made at the transition to amorphous). The $V_{oc}$ of the cell is 0.48 V (which is a high value considering no buffer is used at the p-i interface) and the fill factor is 70 % (see Figure 4). These values show the good doping efficiency of the high temperature doped layer deposited by LbL. The efficiency of 5.6 % is low due to the low current. It has the potential to be 40 % higher when the light trapping is enhanced by means of a back reflector, which means an efficiency of 7.8 % can be achieved.

## CONCLUSION

We have shown that with a layer-by-layer method in VHF PECVD it is possible to grow thin doped layers with an incubation phase that is thin enough for obtaining good electrical properties in a broad temperature range from 250 to 400 °C. The quality of the high temperature n-layer is demonstrated in a n-i-p solar cell with a good open circuit voltage and fill factor. The advantages of the layer-by-layer method compared to continuous wave growth are the thin incubation phase, the possibility to add dopants without limiting the crystalline fraction and the possibility to make doped microcrystalline silicon at high temperatures. The layer-by-layer experiments at high temperatures show that hydrogen coverage of the growth surface is not a prerequisite for nucleation.

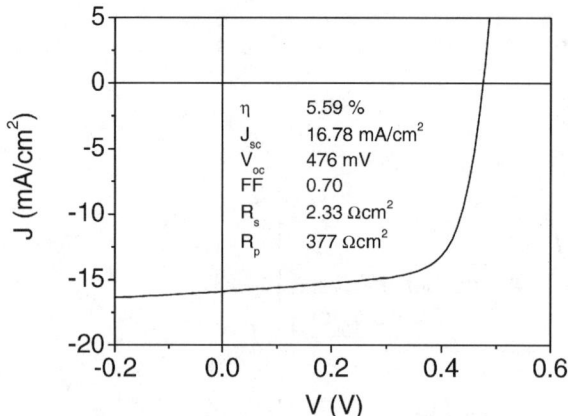

**Figure 4.** IV characteristics of a microcrystalline silicon n-i-p solar cell showing the quality of the electrical properties of the doped μc-Si:H n-layer.

## ACKNOWLEDGEMENTS

The authors would like to thank Jeroen Francke for the deposition of the doped layers in the ASTER system, Jan Winkeler at Akzo Nobel Chemicals for the assistance with the Hall mobility measurements, and the Netherlands Agency for Energy and the Environment (NOVEM) for the partial support of this research.

## REFERENCES

1. J.K. Rath, Solar Energy Materials and Solar Cells **76**, 431-387 (2003).
2. K. Yamamoto, M. Yoshimi, and A. Nakajima, Solar Energy Materials and Solar Cells **74**, 449 (2002).
3. A. Matsuda, J. Non-Cryst. Solids **59&60**, 767 (1983).
4. S. Vepřek, F.-A. Sarott, and M. Rückschloß, J. Non-Cryst. Solids **137-138**, 733 (1991).
5. J.K. Rath, F.D. Tichelaar and R.E.I. Schropp, Solid State Phenomena **67-68**, 465 (1999).
6. K.F. Feenstra, J.K. Rath, and R.E.I. Schropp, Proceedings of the 2nd World Conference on Photovoltaic Solar Energy Conversion, p. 956 (1998).
7. W. Beyer, J. Non-Cryst. Solids **198-200**, 40 (1996).
8. A. Asano, Appl. Phys. Lett. **56**, 533 (1990).
9. H. Shirai, J. Hanna, and I. Shimizu, Jpn. J. Appl. Phys. **30**, L881 (1991).
10. S. Hamma, P. Roca i Cabarrocas, Solar Energy Materials and Solar Cells **69**, 217 (2001).
11. Wilfried G.J.H.M. van Sark, Thin Films and Nanostructures, **30** (2002).

Deposition of device quality µc-Si films and solar cells at high rates by HWCVD
in a W filament regime where W/Si formation is minimal

E. Iwaniczko, A.H. Mahan, B. Yan*, L.N. Gedvilas, D.L. Williamson**, and B.P. Nelson
NREL, 1617 Cole Blvd., Golden, CO 80401
* United Solar Systems Corporation, 1100 W. Maple Road, Troy, MI 48084
** Colo. School of Mines, Golden, CO 80401

## ABSTRACT

µc-Si has traditionally been deposited by Hot Wire CVD at a low filament temperature. At these temperatures, silicides rapidly form on the filament surface, leading in the case of a tungsten filament to both film reproducibility and filament lifetime issues. By depositing films consecutively using identical deposition parameters, these issues are chronicled for a filament temperature of ~ 1750°C. Upon increasing the filament temperature to ~1825-1850°C, these reproducibility and lifetime issues disappear and, by lowering both the substrate temperature and chamber pressure, device quality µc-Si is deposited at high deposition rates in a filament regime where tungsten silicide formation is minimal. Both single junction and tandem solar cells are fabricated using this material, confirming the validity of this approach.

## INTRODUCTION

Microcrystalline silicon (µc-Si) has recently emerged as a viable candidate as the active layer in both single junction cells, and as the low bandgap material in a tandem solar cell structure. The advantages of this material include both its low bandgap (~1.1 eV) and the virtual absence of a Staebler-Wronski Effect for films exhibiting a high microcrystalline volume fraction. However, due to the indirect nature of the bandgap, thick absorber layers (> 0.7 – 1.0µm) are traditionally used, and then film deposition rate ($R_d$) rapidly becomes an issue for all deposition technologies. In PECVD deposition of µc-Si, while a considerable success has been achieved by going to both high excitation frequencies and high chamber pressures, the best cell efficiencies are still deposited at deposition rates below 2-3Å/s [1]. This is also true in Hot Wire CVD (HWCVD), where the highest solar cell efficiency (9.4%), achieved using a HWCVD i-layer in a p-i-n single junction solar cell, was achieved with the i-layer deposited at an $R_d$ of ~ 1Å/s [2].

Common to all device quality HWCVD µc-Si i-layer depositions has been the use of a low filament temperature [2-4]. Two reasons for this have been advanced. The first is that the use of low filament temperatures limits the amount of radiative heating incident upon the growing film, and thus allows the use of a low substrate temperature ($T_{sub}$). This approach is best illustrated by the work of Klein et al. [2], where the substrate was heated entirely by filament radiation. Using a filament temperature of ~ 1650°C and a $T_{sub}$ of 195°C yielded the lowest i-layer ESR spin densities and the highest solar cell efficiencies; all higher filament temperatures and $T_{sub}$'s gave considerably higher spin densities and lower cell efficiencies. The possibility of increased H passivation of these defects at lower $T_{sub}$ has been advanced. The second deals with the possible generation of different radical species contributing to film growth. Although identification of such radicals as a function of deposition parameters is still in its infancy [5], it is generally believed that when lower filament temperatures are used, a smaller percentage of Si radicals (resulting from complete silane dissociation on the filament) are re-evaporated into the gas phase

[6]. As such radicals are believed detrimental to film growth, the correlation between silicide formation on a filament and an improvement of μc-Si i-layer properties may have some validity.

On the other hand, while silicide formation can be controlled somewhat by the use of Ta as a filament material [7], its 'production' continues when a W filament is used, and can lead, at least in a-Si:H, to a rapid deterioration with filament 'aging' in material properties and film deposition rate [8] as well as a dramatic shortening of filament lifetime. This paper documents these issues for the deposition of μc-Si using W filaments, and explores whether a deposition regime can be found where device quality μc-Si can be deposited at high deposition rates in the (virtual) absence of silicide formation.

## EXPERIMENTAL DETAILS

The load locked HWCVD chamber used in this study has been described elsewhere [9]. Two 0.5 mm diameter filaments, located 3.2 cm from the substrate holder, are used to decompose the $SiH_4$ and $H_2$ gas mixtures. Filament currents investigated (through each filament) were 12 amps and 13 amps, which correspond respectively to filament temperatures of ~1750°C and 1825-1850°C. After each i-layer we expose the filament to hydrogen to remove (control) silicide formation; during this treatment we raise the filament temperature to ~500°C above the processing temperature and run 160 sccm of $H_2$ for ~7 minutes at 25 mTorr. The substrates used in this study are a Ag/ZnO textured SS back reflector kindly provided by United Solar, as well as 1737 Corning glass and c-Si substrates. Details about the HWCVD dopant and 'edge' layers used in the NREL fabricated devices, as well as the device structure used, are provided elsewhere [9]. In this study, 100Å thick highly crystalline seed layers were routinely used before i-layer deposition to facilitate i-layer crystallinity [10]. Due to the thinness of the dopant layers in relation to the > 0.7μm i-layer thickness, both infrared (IR) spectroscopy and x-ray diffraction (XRD) measurements could be performed on actual device structures, from which i-layer film properties could be extracted. Such properties include the film H bonding (stretch mode position) and post deposition oxygen pickup as well as the XRD microcrystalline volume fraction (μ-volume fraction), the latter obtained by integrating the areas of the (111), (220), and (311) c-Si peaks and comparing these to the first and second order amorphous peak areas. No Raman measurements were performed on these samples due to the lack of routine measurement capabilities available at NREL. To examine the feasibility of using p-layers alternate to those previously optimized for a-Si:H device structures, a series of two half finished devices were shipped overnight express to United Solar; the first was fabricated into a single junction device and analyzed using their buffer layer/μc-p$^+$-layer structure, and when appropriate, a tandem structure was fabricated using the second half finished device.

## RESULTS AND DISCUSSION

Figure 1 shows (a) the film $R_d$ and (b) the crystalline volume faction (μ-volume fraction) of eleven i-layers (deposited in a device structure) deposited consecutively using identical deposition conditions (hydrogen dilution 14/1, $T_{sub}$ 340°C, chamber pressure 140 mT), as a function of filament usage. The filament current used in this case was 12 amps, corresponding to a filament temperature of ~ 1750°C, and has resulted in device efficiencies (at an $R_d$ > 5Å/s) exceeding 6% [4], with every layer deposited by HWCVD. As previously noted, at this filament current, silicide formation occurs rapidly with silane exposure [7-8]. In Fig. 1(b) two different methods of determining film crystallinity are presented. While the first (μ-volume fraction) is an absolute measurement, it was performed only on selected samples. On the other hand, as film crystallinity (for a film showing a dominant (220) crystallite orientation) evolves with increasing film thickness [11], a measure of the (220) peak height normalized to a standard (0.9 μm) film thickness and film crystallinity [12] can also serve as a relative measure of crystallinity with increasing filament usage. For the (high μ-volume fraction) films deposited beyond a filament usage of two hours, where the film $R_d$ stays relatively constant and viable solar cells have been fabricated, the grain sizes estimated from the Scherrer equation ((220) peak), parallel to the growth direction, are ~ 300Å.

As can be seen, film reproducibility and filament lifetime immediately become significant issues. Regarding the former, both $R_d$ and film crystallinity change with filament usage. From previous experience, the best devices using 12 amps filament current are fabricated with an 'aged' filament, and correspond in Fig. 1 to a filament usage in the range of 3-4 hr. While the (low) $R_d$ observed here is constant for a short while, we have also previously observed a change in crystallite orientation (from a dominant (220) orientation to one approaching a random (c-Si powder pattern) orientation) at this filament usage [13], so even in this constant $R_d$ region film deposition is not reproducible. Beyond the data presented in Fig. 1, the film $R_d$ drops so precipitously that the filament either breaks has to be replaced.

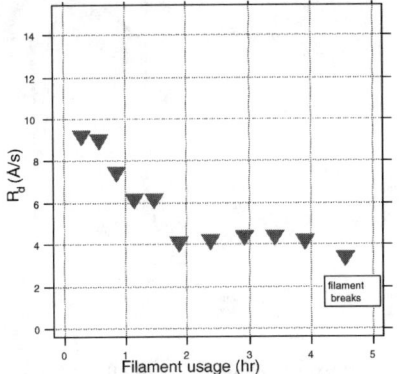

Fig. 1(a) Deposition rate versus filament lifetime for a filament operated at 12 amps.

Fig. 1(b) μ-volume fraction (solid symbols, left axis) and relative (220) Si-peak height (open symbols, right axis) versus filament lifetime for a filament operated at 12 amps.

Going to 13 amps filament current (corresponding to a filament temperature of ~ 1825-1850°C) has immediate benefits. Upon deposition of eight devices deposited consecutively, using the same $H_2$ dilution, $T_{sub}$ and chamber pressure conditions as for 12 amps (see above), i-layer film reproducibility becomes immediately evident. That is, the film $R_d$ is 17-18Å/s and does not drop, and from the first deposition the XRD μ-volume fraction is high (>75%) and also remains constant. In addition, filament lifetime also improves significantly, as more than 50 depositions have to date been performed using this filament, with minimal change in i-layer properties (film $R_d$, μ-volume fraction) when control depositions are repeated at selected filament usage intervals. However, device efficiencies utilizing these i-layers are very low (~0.3%). In particular, all solar cell parameters ($V_{oc}$ = 0.30 V, $J_{sc}$ = 2.1 mA/cm$^2$, FF = .31, yielding $\eta$ ~ 0.2% for this cell) are representative of an i-layer quality that is not suitable for its incorporation into a state-of-the-art solar cell. An additional significant problem with these i-layers becomes evident from IR measurements. Fig. 2 shows an IR spectrum for such an i-layer, both measured immediately after growth, and then after a short exposure (1 day, 3-4 days) to atmospheric conditions. As can be seen, post deposition oxygen absorption, as evidenced by a broad peak occurring at ~ 1000 cm$^{-1}$, is clearly evident; such absorption, either during or after deposition, has previously been shown to be detrimental to device performance [14]. The sharp peak located at ~ 1100 cm$^{-1}$, seen in all samples, is due to a mismatch in oxygen contents between the reference c-Si substrate and the respective sample c-Si substrates, and is not due to the μc-Si layers. The grain sizes, once again estimated from the Scherrer equation ((220) peak), parallel to the growth direction, remain the same ((~ 300Å) as the filament current is increased.

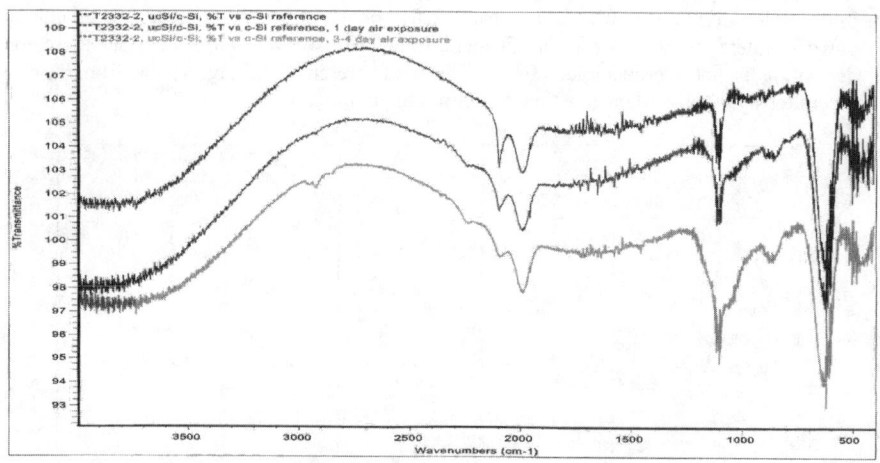

Fig. 2 FTIR spectra showing films experiencing post deposition oxidation with air exposure.

Following the results of Klein et al. [2], a significant improvement in device performance can be achieved at 13 amps by depositing i-layers at lower $T_{sub}$ and lower chamber pressures. While i-layer $R_d$ is sacrificed as the chamber pressure is reduced from 140 mT to 60 mT, significant oxygen pickup is no longer evident, as IR scans of devices exposed to atmospheric conditions over longer periods of time (> 3 months) now show no trace of the oxygen feature in the 1000 cm$^{-1}$ region. Furthermore (see Poster A6.9), the i-layer photoconductivity improves, from a value of 5-7 x 10$^{-7}$ S/cm$^2$ for $T_{sub}$ = 340°C and a chamber pressure of 140 mT, to 2-4 x 10$^{-5}$ S/cm$^2$ for $T_{sub}$ = 240°C and a chamber pressure of 60 mT. For the latter i-layer, $\sigma_L/\sigma_D$ > 100. Such conductivity values are now equivalent to those of Klein et al. [2], but for i-layers deposited at much different $R_d$ (1Å/s vs. ~8Å/s for the present films). The (220) grain size of the present samples has only decreased slightly, to ~ 230Å.

Devices performance utilizing these i-layers, which are deposited at an $R_d$ = 8Å/s, are shown in Fig. 3; both i-layer thicknesses are ~ 0.7 μm. Fig. 3(a) shows the efficiency of a device fabricated entirely at NREL; all layers are deposited by HWCVD, with the 'edge'/a-Si-p$^+$-layer combination described previously. While the efficiency of the NREL device is only 4.1%, we comment that this i-layer is perhaps more suitable as the bottom layer in a tandem device structure than for a single junction device alone. In particular, the device $V_{oc}$ has previously been found to depend sensitively on the μ-volume fraction, and the value for the present device (~ 0.42 V) is consistent with that of an i-layer exhibiting a high μ-volume fraction [4]. Indeed, XRD measurements on this device show a μ-volume fraction > 75%.

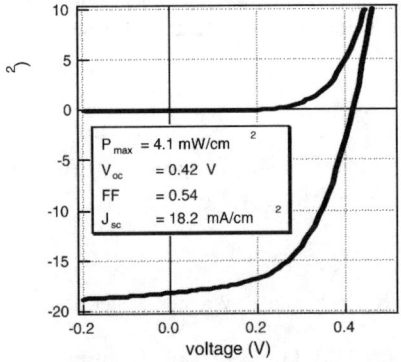 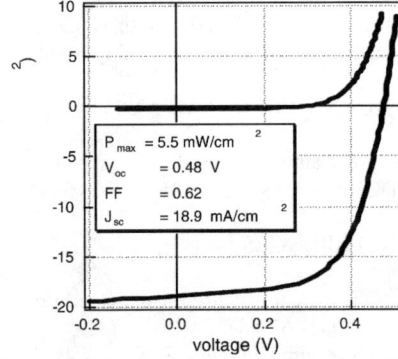

Fig. 3(a) JV characteristics for T2340 where all layers were grown by HWCVD at NREL.

Fig. 3(b) JV characteristics for T2358 where the n- and i-layers were grown at NREL and the buffer- and p-layer (and ITO) were grown at United Solar.

Using the same identical i-layer, an efficiency of 5.5% (Fig. 3(b)) has been achieved using a different top contact (the buffer layer/μc-Si-p$^+$-layer structure optimized by United Solar). Improvements in performance can be attributed to enhancements in both device FF and device $V_{oc}$. It is difficult to determine which part of the United Solar top contact (buffer layer or μc-Si-p$^+$-layer) is more beneficial to single junction device performance. However, using our 'n & k' apparatus as a sensitive measure of film crystallinity of the topmost layer in our device structure, we can say that, even though the NREL top contact 'recipe' has now been deposited onto a μc-Si i-layer instead of an a-Si:H i-layer, the top (p$^+$) layer still seems to be amorphous. That is, the reflectance at ~ 270 nm remains broad and featureless, and is quite unlike when either a μc-Si i-layer by itself, or a μc-Si-p$^+$-layer in a device, is measured. In the latter cases, a distinctive and relatively sharp reflectance feature appears at ~ 270 nm.

In addition, we also report tandem junction solar cell results, with United Solar depositing not only their buffer layer/μc-Si-p$^+$-layer top contact on top of the NREL HWCVD μc-Si i-layer, but also an additional a-Si:H based top cell in a two tandem device structure. Preliminary results show an initial tandem efficiency of 9.6%. Individual solar cell parameters for this structure are $V_{oc}$ = 1.40 V, $J_{sc}$ = 9.8 mA/cm$^2$, and FF = 0.70. Further results will be reported at a later date.

## CONCLUSIONS

By depositing HWCVD μc-Si films consecutively at low filament temperature (~1750°C) using identical deposition conditions, a sharp drop in film $R_d$ as well as a change in film crystallinity are observed with filament usage. When using the same deposition parameters but a higher filament temperature (~1825-1850°C), film reproducibility issues now disappear, but these films pick up oxygen from ambient exposure, and the resultant solar cells are of very poor quality. However, by lowering both the substrate temperature and chamber pressure, device quality μc-Si is deposited at high deposition rates in a filament regime where tungsten silicide formation is minimal. Both single junction and tandem solar cells are fabricated using this material, confirming the validity of this approach.

## ACKNOWLEDGEMENTS

The authors thank J. Yang for his active interest in this work. This work is performed under DOE contracts DE-AC36-99-GO10337 and ZDJ-2-30630-19.

## REFERENCES

1. T. Roschek, T. Repmann, O. Kluth, J. Muller, B. Rech, and H. Wagner, Mater. Res. Soc. **715** (2002) A26.5.
2. S. Klein, F. Finger, R. Carius, B. Rech, L. Houben, M. Luysberg, and M. Stutzmann: Mater. Res. Soc. **715** (2002) A26.2.
3. R.E.I. Schropp, C.H.M. Van Der Werf, M.K. van Veen, P.A.T.T. van Veenendaal, R. Jimenez Zambrano, Z. Hartman, J. Loffler, and J.K. Rath: Mater. Res. Soc. **664** (2001) A15.6.
4. R.E.I. Schropp, Y. Xu, E. Iwaniczko, G.A. Zaharias, and A.H. Mahan: Mater. Res. Soc. **715**, (2002) A26.3.
5. See section 1 of the proceedings of the '1st International Conference on Cat-CVD (Hot-Wire CVD) Process', published in Thin Solid Films **395** (2001).
6. J. Doyle, R. Robertson, G.H. Lin, M.Z. He, and A.C. Gallagher, J. Appl. Phys. 64 (1988) 3215.
7. P.A.T.T. van Veenendaal, O.L.J. Gijzeman, J.K. Rath, and R.E.I. Schropp, Thin Solid Films **395** (2001) 194.
8. A.H. Mahan, A. Mason, B.P. Nelson, and A.C. Gallagher, Mater. Res. Soc. **609**, (2000) A6.6.
9. Q Wang, E. Iwaniczko, Y. Xu, B.P. Nelson, A.H. Mahan, R.S. Crandall, and H.M. Branz, Twenty-Eighth IEEE PV Specialists Conference (2000) 717-720.
10. G.A. Zaharias, A.H. Mahan, R.E.I. Schropp, Y. Xu, D.L. Williamson, M.M. Al-Jassim, M.J. Romero, and L.M. Gedvilas, Mater. Res. Soc. **715** (2002) A26.2.
11. J.K Rath, F.D. Tichelaar, H. Meiling, and R.E.I. Schropp, Mater. Res. Soc. **507** (1998) 879.
12. A. H. Mahan, private communication.
13. E. Iwaniczko, Y. Xu, R.E.I. Schropp, Q. Wang, and A.H. Mahan, NREL/CP-520-30476 (2002) 67.
14. J. Meier, P. Torres, R. Platz, S. Dubail, U. Kroll, J.A. Anna Selvan, N. Pellaton Vaucher, Ch. Hof, D. Fischer, H. Keppner, A. Shah, K.-D. Ufert, P. Giannoules, and J. Koehler, Mater. Res. Soc. **420** (1996) 3.

## Influence of Hydrogen Dilution on Properties of Silicon Films Prepared by D.C. Saddle-Field Glow-Discharge: Observation of Microcrystallinity

T. Allen[1], I. Milostnaya, D. Yeghikyan, K. Leong, F. Gaspari, N.P. Kherani, T. Kosteski, S. Zukotynski
Department of Electrical and Computer Engineering, University of Toronto, Toronto, Ontario, CANADA, M5S 1A4
[1] Department of Physics, Geology and Astronomy, University of Tennessee at Chattanooga, Chattanooga, TN 37403 U.S.A.

## ABSTRACT

In the D.C. saddle field glow discharge deposition the transition from amorphous to microcrystalline silicon thin films occurs when the silane concentration in the gas phase drops below 10%. We report here the results of Raman spectroscopy, SEM, TEM, and HRTEM studies of the film morphology. We estimate the average crystallite size to be in the range of 5 to 7 nm and the crystalline volume fraction of 25 to 35%.

## INTRODUCTION

Amorphous Si is widely used for large area photovoltaic and microelectronic applications [1]. The use of microcrystalline Si is expected to improve stability against light-induced degradation and provide more efficient doping over that offered by amorphous silicon. Recently, we reported on the growth of mixed phase amorphous-microcrystalline silicon using the D.C. saddle field glow discharge deposition method [2]. The films were grown using hydrogen dilution of silane during the deposition. We were able to identify the growth conditions and the types of substrates that promote microcrystallinity. In this work we present the structural properties of saddle field glow discharge deposited microcrystalline Si films as a function of hydrogen dilution. The films were studied using Raman spectroscopy, SEM, TEM and high-resolution TEM.

## EXPERIMENTAL

The deposition process is described in [2]. The samples studied in this paper were grown on pyrex substrates at a temperature of 250°C, a chamber pressure of 200 mTorr, and an anode current of $I_{an}$= 30 mA. The source gas was a mixture of silane and hydrogen. The silane concentration in percent, C=[SiH$_4$]/([SiH$_4$]+[H$_2$]), ranged from 2.4% to 100%. For most films the flow rate was 10 sccm, but one deposition with a flow rate of 5 sccm was also performed.

The thickness of the films was measured with a profilometer (Tencor) and by SEM (Hitachi model S4500). The film thickness ranged from 0.2 to 0.5 microns and was uniform within ±10%. SEM was also used to study the film morphology.

Raman spectra were collected at room temperature in the near back-scattering geometry using the 514.32 nm line of an Ar$^+$ laser at a power output of 140 mW. To ensure that the sample surface was not thermally damaged, the laser beam was attenuated using an absorbing filter. A visual check of the surface was carried out before and after each measurement using an optical microscope. The crystalline Si Raman signal at 520 cm$^{-1}$ was used to reference the spectra.

In order to prepare samples for TEM measurements, films were lifted from glass substrates using hydrofluoric acid, and then mounted on TEM copper grids. TEM images as well as electron diffraction patterns were taken using a Hitachi H-800 TEM. High-resolution TEM measurements were done using a JEOL 2010F.

## RESULTS AND DISCUSSION

Hydrogen dilution of silane plays a critical role in the formation of the film structure during its growth. Fig. 1a shows the evolution of Raman spectra as a function of silane concentration, C. Fig. 1b shows the position of the Raman peak as a function of C.

The experimental Raman spectra were deconvolved using a minimal number of Gaussian functions to provide a fit consistent with the entire data set. For samples prepared using pure silane (C=100%) we were able to fit the experimental data using the LO band at ~400 cm$^{-1}$ and the primary TO band at 471 cm$^{-1}$. This indicates that the structure is purely amorphous. The sample with C=16.7% shows peaks at 501 and 483 cm$^{-1}$ and a residual LO shoulder. An attempt to fit all other samples with two Gaussians was not successful; so three Gaussians were used to fit the principal TO band (see Fig. 2.). For samples with C<10% the first Gaussian corresponds to the Raman active $\Gamma_{25'}$ mode of crystalline silicon (511-514 cm$^{-1}$) [3]. The second peak is observed at 465-485 cm$^{-1}$. It has been associated with grain boundaries [3], but more recently it has been attributed to pure amorphous material [4, 5]. The third peak occurs at 497-506 cm$^{-1}$. We refer to this as the transition phase. Veprek et al. [4] attribute this to grain boundaries.

When C increases from 2.4% to 9.1%, the position of the microcrystalline peak shifts from 514 cm$^{-1}$ to 511 cm$^{-1}$ (Fig. 1b). We use the position of the Raman peak to estimate the crystallite size, as suggested in [3], and find that for C=2.4% the crystallites are about 5 nm and that they decrease to about 3 nm for C=9.1%. Veprek et al. [6] suggest that a crystallite size of 3 nm represents the thermodynamic limit of stability for the silicon diamond lattice. When C increases beyond 9.1%, we no longer observe the crystalline peak. For C=16.7% the Raman signal is a mixture of the transition phase (peak at about 500 cm$^{-1}$) and the amorphous phase. When C increases further, the Raman signal becomes amorphous.

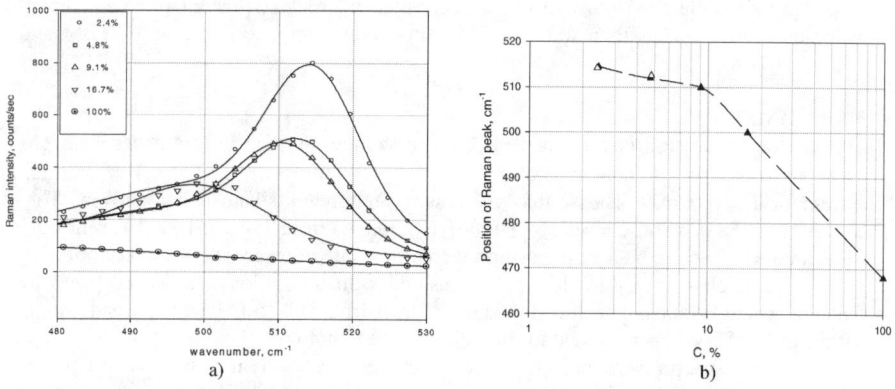

Figure 1: a) Raman spectra for films prepared at different dilution C. Solid lines represent fitted data using Gaussian functions. b) Position of the Raman band vs. hydrogen dilution of the process gas. The line is a guide for an eye. Open symbols refer to repeated depositions.

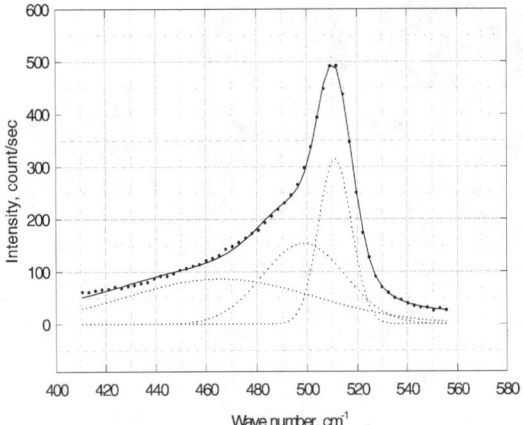

Figure 2. Deconvolution of the Raman spectrum for the film prepared at C=9.1%. The Gaussians represent crystalline (peak at 511 cm$^{-1}$), transitional (peak at 498 cm$^{-1}$) and amorphous (peak at 465 cm$^{-1}$) structural environments.

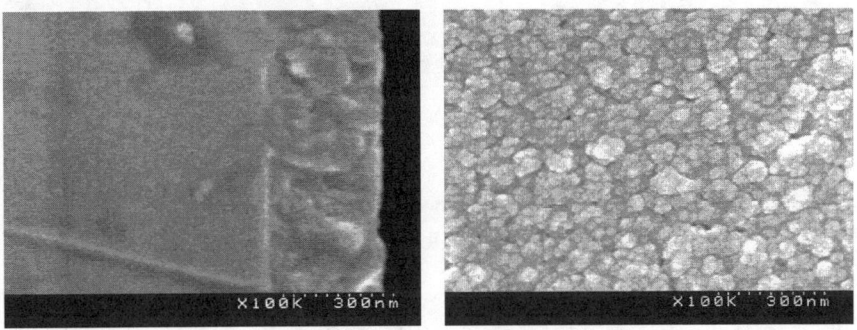

Figure 3. SEM images of the cross sectional (left) and planar (right) views of the silicon film prepared at C=9.1%.

Fig. 3 presents SEM micrographs of cross-sectional and planar views of the sample with C=9.1%. All of the films grown at high dilution show a granular surface morphology. These granular features were investigated using TEM. Fig. 4a gives the dark field image for the same sample as in Fig.3. Fig. 4b gives the corresponding electron diffraction pattern. Ring diameters and orientation yield a lattice constant of 0.54(5) nm; this is in good aggrement with 0.54(3) nm for crystalline silicon. Dark field TEM images were analysed for crystallite size distributions. Fig. 5 shows grain size distributions for three samples. The mode of the distributions is 5 nm and the average grain size is about 7 nm. $X_i$ is the volume fraction of the appropriate phase, as explained below.

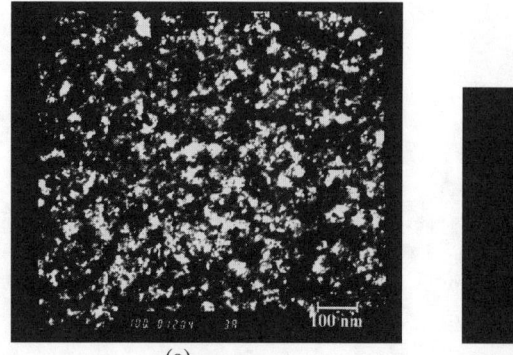
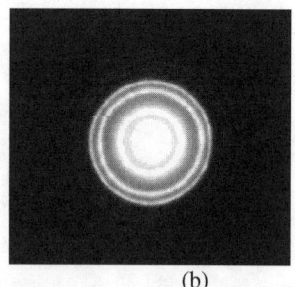

(a) (b)

Figure 4: (a) Dark field TEM image and (b) electron diffraction pattern for the film shown in Fig. 3.

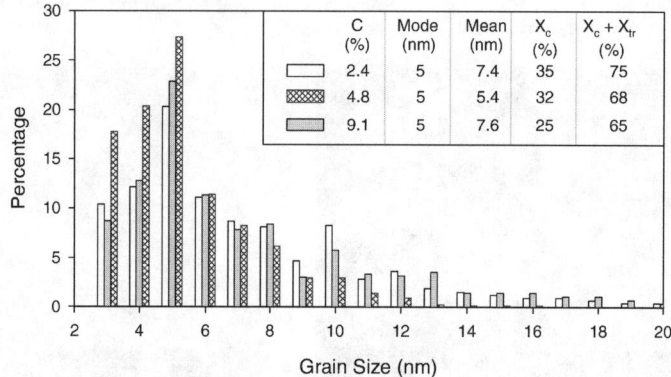

Figure 5: Grain size distributions from dark field TEMimages.

Fig. 6 shows a high resolution TEM image of silicon crystal planes. The crystallite dimensions were observed to be in the range of 5 to 7 nm in the examined sample. The Raman and dark field TEM results are in agreement with the high resolution TEM studies.

It appears that for samples with C less than 10%, the three Gaussians obtained from Raman spectra represent three distinct structural environments in the material. Fig.7 shows the volume fraction of each phase estimated from the integrated intensity of the corresponding Gaussian:

$$X_i = I_i/(I_c+I_t+I_a), \qquad (1)$$

where $I_i$ is the intensity of the crystalline, transitional or amorphous phase, respectively. Samples grown with hight dilution contain 25-35% of crystalline phase, according to this calculation.

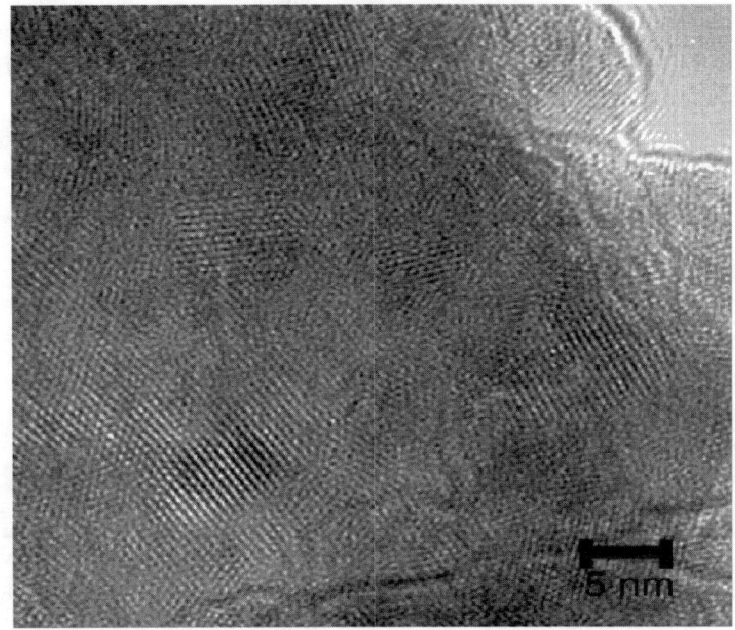

Figure 6: High Resolution TEM image of crystalline planes of Si for film prepared at C=9.1%.

There are several papers describing the calculation of the volume fraction of the crystalline phase in a two-phase system from integrated intensities, which make corrections for integrated backscattering cross section over the measured frequency range [7, 8]. Yue et al. [9] modified the two-phase calculation [8] and applied it to a three phase system, using the correction paramenter y=0.1+exp(-L/250), where L is the crystalllite size in Å. Our calculations using the same approach and the estimated size of the crystallites L agree within 3% with the data shown in Fig.7. The increase in the crystalline volume fraction with hydrogen dilution is also observed in IR studies of the same samples reported elsewhere [10].

Fig. 8 gives preliminary results of a study on the influence of the gas flow rate on the structure of the material. Reducing the flow rate from 10 sccm to 5 sccm leads to almost complete elimination of the transition phase. Further studies are needed to better understand the growth dynamics and its influence on film properties.

## CONCLUSIONS

We have demonstrated that dual phase hydrogenated amorphous-microcrystalline silicon films can be prepared with hydrogen dilution in a DC saddle-field glow-discharge. The presence of silicon crystallites, estimated average size of 5 to 7 nm, has been shown using TEM and HRTEM. Microcrystalline volume fraction, in the range of 25 to 35%, has been estimated from Raman measurements.

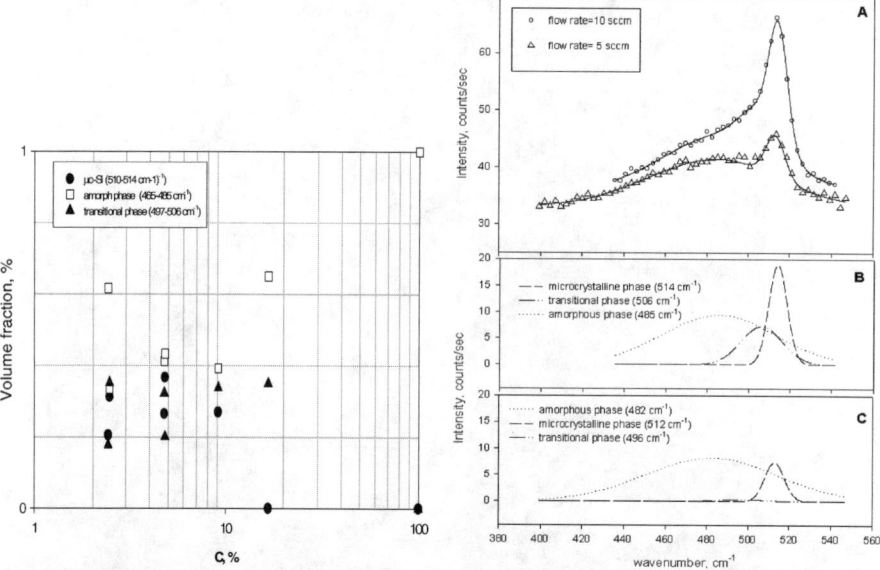

Figure 7: Volume fractions of the structural phases calculated according to (1).

Figure 8: Influence of the gas flow rate on the structural properties: B) 10 sccm; C) 5 sccm.

## ACKNOWLEDGEMENTS

The authors would like to thank Mr. Fred Neub and Dr. Fred Pearson for their help with the TEM and HRTEM measurements, respectively. This work was supported by ARISE Technologies Corporation, Materials and Manufacturing Ontario and the Natural Sciences and Engineering Research Council of Canada. T.A. would like to acknowledge support from the National Science Foundation (grant # DMR0074682), the Research Corporation Cottrell College Science Award (grant # CC4536), and support from the UC Foundation, University of Tennessee at Chattanooga. A NATO post-doctoral fellowship, administered by the Natural Sciences and Engineering Research Council of Canada, supported I.M.'s participation in the project.

## REFERENCES

1. R. A. Street (Ed.), Technology and Application of Amorphous Silicon, Springer-Verlag, New York (2000).
2. T. Allen, F. Gaspari, N.P. Kherani, T. Kosteski, K. Leong, I. Milostnaya, D. Yeghikyan, S. Zukotynski, Mat. Res. Soc. Symp. Proc., **744** (2003) M5.22.
3. Z. Iqbal, S. Veprek, J. Phys. C., **15** (1982) 377.
4. S. Veprek, F.-A. Sarrot, Z. Iqbal. Phys. Rev. B, **36** (1987) 3344.
5. G. Yue, J.D. Lorentzen, J. Lin et al, Appl. Phys. Lett, **75** (1999) 492.
6. S. Veprek, Z. Iqbal, H. R. Oswald, F.-A. Sarrot, et al., Solid State Commun., **39** (1981) 509.
7. A. T. Voutsas, M. K. Hatalis, J, Boyce, A. Chiang. J. Appl. Phys, **78** (1995) 6999.
8. E. Bustaret, M. A. Hachicha, and M. Brunel. Appl. Phys. Lett., **52** (1988) 1675.
9. G. Yue, J. D. Lorentzen, J. Lin, D. Han, Q. Wang. Appl. Phys. Lett., **75** (1999) 492.
10. I. Milostnaya, T. Allen, F. Gaspari, N.P. Kherani, D. Yeghikyan, W.L. Roes, T. Kosteski and S. Zukotynski, Mat. Res. Soc. Symp. Proc., **762** (2003) A6.15.

## Reactive Pulsed Laser Deposition
## of Microcrystalline Ge-based Thin Films

Matthew R. Wills[1], Ruth Shinar[2], and Alan P. Constant[1]
[1]Department of Materials Science and Engineering
[2]Microelectronics Research Center
Iowa State University, Ames Iowa 50011

## ABSTRACT

Pulsed laser deposition (PLD) was used to grow microcrystalline thin films of germanium (Ge) and Ge-carbon (Ge,C) alloys on fused quartz and silicon substrates at substrate temperatures 25 °C ≤ $T_s$ ≤ 325 °C. The films were analyzed structurally with x-ray diffraction (XRD), optically, electrically with four-point probe measurements, and chemically with x-ray photoelectron spectroscopy (XPS). XRD results displayed a varying degree of crystallinity, with the most crystalline films obtained at $T_s$ > 150 °C. The resistivity of the Ge films decreased with increasing temperature, displaying a significant decrease for the films deposited at $T_s$ ≥ 230 °C. The growth conditions for Ge films served as a starting point for low-temperature deposition of Ge,C alloys with up to 5% C. The effects of $T_s$ and carbon concentration on film properties are discussed.

## INTRODUCTION

Germanium (Ge) is an attractive material for electronic and optoelectronic applications due to its high electron and hole mobilities, 3900 and 1900 cm$^2$/Vs, respectively, as well as its higher absorption coefficient in comparison to silicon (Si). Low temperature growth of amorphous to epitaxial Ge-based thin films has received great attention for photovoltaic/optoelectronic applications such as band gap engineering and development of high-speed devices [1-3].
However, the widespread application of Ge in microelectronics is limited due to the lattice mismatch of Ge (5.6575 _) and Si (5.4307_), and the small bandgap of Ge (0.67 eV), making it more susceptible to thermal noise and less suitable for photovoltaic applications. Fabrication of high quality Ge-based films, such as Si-Ge with carbon, may offer solutions to these issues. By varying the concentration of substitutional carbon the Si-Ge-C bandgap can be extended by 21-26 meV per at. %C [4]. Growth methods of Ge and Ge-based films include chemical vapor deposition (CVD)-based techniques, molecular beam epitaxy (MBE), and pulsed laser ablation [5-7].
Pulsed laser deposition (PLD) has emerged as in important tool for deposition of high quality films, varying from amorphous to crystalline. Advantages of PLD include the ability to transfer a complex stoichiometry from a target to a growing film, as demonstrated for superconducting oxides [8]. Growth occurs under extreme non-equilibrium conditions with minimal surface reorganization of the atoms impinging on the substrate. Though low surface mobility can adversely affect the quality of the growing film, it may allow fabrication of crystalline films of thermodynamically unstable compounds, such as germanium carbide (Ge,C) with the possibility of a direct transition band structure for C compositions of 4-11 % [9] and compensate inherent strain due to

lattice mismatch with an Si substrate. Ge,C films have been fabricated with CVD-based and MBE processes with substitutional carbon concentrations of up to 2.5 at. %C [10,11].

As a starting point for investigating PLD of Ge,C films we evaluated fabrication conditions of Ge films grown on fused quartz substrates through measurements of their structural and electronic properties. As detailed in the following sections, using PLD, it was possible to fabricate 400 nm thick, microcrystalline films of Ge at a growth rate of ~1.4 Å/s and temperatures as low as 25-150 °C. Information from these experiments was used for fabrication of microcrystalline Ge,C films using reactive-PLD. Initial results for PLD grown Ge,C films indicated microcrystalline films containing up to 5% carbon.

## EXPERIMENTAL

The deposition system used in this investigation was a turbo-pumped commercial Neocera PLD system. Laser ablation was carried out using a Lamda Physik COMPex 201 KrF excimer laser (248 nm), focused onto a rotating (100) crystalline Ge target at an incidence angle of 45°, with a substrate-target distance of 75 mm. The laser was pulsed at 10 Hz, with a pulse width of less than 50 ns, and an energy density of ~.8 J/cm$^2$. Base pressures prior to deposition were $4 \pm 3 \leftrightarrow 10^{-7}$ Torr. Prior to film deposition, the target was ablated for 2-3 min to remove contaminations. Ge films were deposited on fused quartz substrates at temperatures, $T_s$, of 25, 150, 235, 320, and 325 °C with deposition times of 30 and 60 minutes.

Ge,C films were grown under similar conditions in the presence of methane (research grade, 99.995%) at pressures, $P_{CH4}$, of 0, 1, 5, and 10 mTorr. The films were deposited onto single-crystal (100) Si, with ~770 _ silicon dioxide, at $T_s$ of 25 °C, and a deposition time of 30 minutes.

Film thicknesses of Ge samples were measured using a stylus profilometer. Thicknesses varied from 245 nm (30 min deposition) to 420 nm (60 min deposition), yielding a deposition rate of ~1.2 to 1.4 _/s or 0.12 to 0.14 _/pulse at 10 Hz.

The films structure was analyzed using x-ray diffraction (XRD) with a Siemens D500 x-ray diffractometer employing monochromatic Cu-K$_\alpha$ radiation and sample rotation. The Ge films were analyzed with a single scan from 2_ (_ – angle of incidence) of 15° to 85° at a step rate of 0.02°/sec and Ge,C films from 25° to 55° at a step rate of 0.02°/sec. 2_ offset of the Ge,C films was negated using the (220) peak of the Si substrate. The Ge films were analyzed electrically, using a four-point probe at 0.10, 0.25, 0.50, 0.75, and 1.0 µA, and optically, using a Perkin Elmer UV-VIS spectrophotometer. Bandgap estimations were made using the $E_{04}$ approximation. A depth profile through a Ge film grown at 25 °C was obtained using a PHI model 6300 secondary ion mass spectrometer (SIMS).

The Ge,C films were analyzed chemically with x-ray photoelectron spectroscopy (XPS) to determine the percentage of C alloyed in the film. Measurements on Ar$^+$ cleaned samples were made with PHI 5500 spectrometer employing Al-K_ radiation. Subtitutional C was verified by analyzing the energetic bands of the 1s core level (C1s). C1s bands display a peak at 285 eV for single carbon, 284 eV for graphite, and 283 eV for carbides [12].

## RESULTS AND DISCUSSION

### Ge films

Figure 1 shows the XRD spectra for the Ge films deposited at various substrate temperatures, $T_s$.

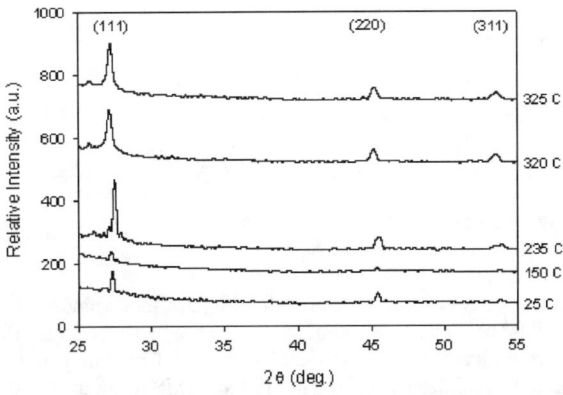

**Figure 1.** XRD scans of PLD deposited Ge films at $T_s$= 25, 150, 235, 320, and 325 °C representing a cubic structure with no preferred orientation during film growth.

Peak positions and relative peak intensities indicate a cubic Ge structure with no preferred orientation for all $T_s$. As seen in the figure, a varying degree of film crystallinity is evident, with a large increase for films grown at $T_s$ > 150 °C. At $T_s$ > 235 °C, the degree of crystallinity of the films further increased with the appearance of the (400) peak
Resistivity measurements (see Figure 2) were in agreement with the XRD results, displaying a decrease in resistivity with deposition temperature. The resistivity for a Ge film grown at 25 °C was ~95 _-cm and decreased strongly to ~0.5 _-cm, for a film grown at $T_s$ = 235 °C. The resistivity continued to decrease for the film grown at $T_s$ = 325 °C to a value of ~0.05 _-cm. For comparison, the measured resistivity of a p-type (Gallium) crystalline Ge wafer was 0.001 _-cm. The significant drop in resistivity with increasing $T_s$ can be attributed to an increase in crystallinity of the films, evident from the XRD results (see Figure 1) The resistivity data, along with the XRD results, indicate that Ge films with some degree of crystallinity can be grown at temperatures as low as 25-150 °C, with the most crystalline films grown at $T_s$ ~235 –325 °C.

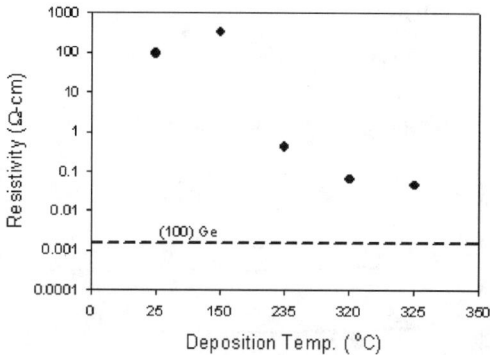

**Figure 2.** Resistivity of PLD Ge films as a function of deposition temperature with the resistivity of a (100) Ge wafer noted by the dashed line.

Optical absorption measurements agreed well with published values [13], with an absorption edge at ~0.80 eV and a 3-4 decade increase in the absorption coefficient. The difference in observed optical bandgap compared with that of crystalline Ge can be attributed to the microcrystalline structure of the films. SIMS depth profiling of the Ge film grown at 25 °C indicated a uniform Ge film, practically free of carbon; i.e., the carbon level did not exceed the background level at the ~$10^{-9}$ Torr base pressure of the SIMS system.

### **Ge,C films**

Figure 3 illustrates the XRD structure of the Ge,C films as a function of methane press positions from those found for the Ge films (see Figure 1). However, the relative peak intensities are in contrast to those found for the Ge films, suggesting the incorporation of C results in preferential orientation during film growth. The peak near 2_~32 is can be attributed to the Si substrate.

Carbon content of the Ge,C films after $Ar^+$ cleaning was quantified with XPS. For methane pressures up to 10 mTorr, percentage of C alloyed varied linearly with methane pressure, $P_{CH4}$ (see Figure 4). A 0 mTorr sample showed less than 0.2 % C indicating low C background. It is worth noting that recent Raman spectroscopy (not shown) further verifies Ge-C bonding, with a Ge,C vibrational mode at ~560 $cm^{-1}$.ure, $P_{CH4}$. The spectra show a single cubic crystalline phase with no change in peak .

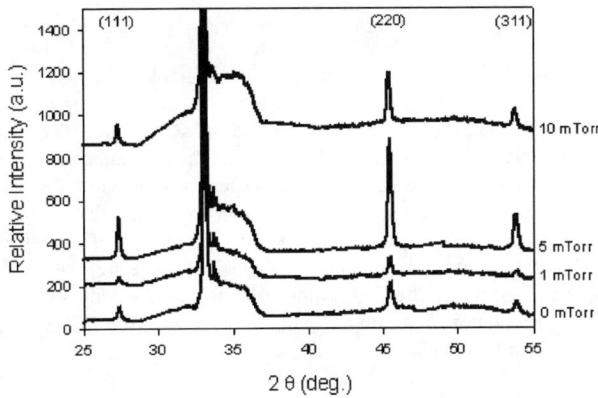

**Figure 3.** XRD scans of PLD deposited Ge,C films at $P_{CH4}$= 0, 1, 5, and 10 mTorr indicating a cubic structure with a preferred orientation during film growth.

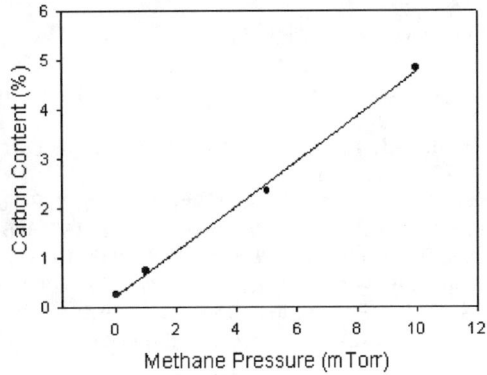

**Figure 4.** Percentage C alloyed in PLD Ge,C films as a function of methane pressure.

## CONCLUSIONS

The results from the structural and electrical measurements indicate that ~245-420 nm Ge thin films with a varying degree of crystallinity can be grown using PLD at low substrate temperatures of 25-325 °C with deposition rates of 1.2-1.4 Å/s. Increasing the substrate temperature improved the film crystallinity, with strongly enhanced electrical properties for films grown at or above ~230 °C. XRD spectra of the Ge films indicate a cubic crystalline phase with no preferred orientation. The growth conditions for the Ge thin films served as a starting point for low-temperature PLD growth of Ge,C films. Preliminary results indicate growth of Ge-C phase at room temperature, with carbon contents up to 5%. XRD spectra of the Ge,C films indicate a single cubic crystalline phase with a preferred orientation. Ongoing TEM and high resolution EELS studies are expected to directly determine the phase structure of the Ge,C films.

## REFERENCES

1. P. Sutter, U. Kafader, and H. Kanel, *Sol. Energy Mater. Sol. Cells* **31**, 541 (1994).
2. M. Lill and B. Schroder, *Appl. Phys. Lett.* **74**, 1284 (1999).
3. S. Bozzo, J.-L. Lazzari, C. Coudreau, A, Ronda, F. Arnaud d'Avitaya, J. Derrien, S. Mesters, B. Hollander, P. Gergaud, and O. Thomas, *J. Cryst. Growth* **216**, 171 (2000).
4. A.S. Amour, C. Liu, J. Sturn, Y. Lacroix, M. Thewalt, *Appl. Phys. Lett.* **67**, 3915 (1995).
5. D. J. Eaglesham, *J. Appl. Phys.* **77**, 3597 (1995).
6. C. Mukherjee, H. Seitz, and B. Schroder, *Appl. Phys. Lett.* **77**, 3457 (2001).
7. Y. Zhang, Z. Iqbal, S. Vijayalakshmi, S. Qadri, and H. Grebel, *Solid State Comm.* **115**, 657 (2000).
8. B. Ma, M. Li, R.E. Koritala, B.L. Fisher, S.E. Dorris, V.A. Maroni, D.J. Miller, U. Balachandran, *Physica C.* **377**, 501 (2002).
9. J. Kolodzey, P.R. Beger, B.A. Orner, D. Hits, F. Chen, A. Khan, X. Shao, M.M. Waite, S. Ismat Shah, C.P. Swann, K.M. Unruh, *J. Crystal Growth* **157**, 386 (1995).
10. M. Todd, J. Kouvetakis, D.J. Smith, *App. Phys. Lett.* **68**, 2047 (1996).
11. B.-K. Yang, M. Krisnamurthy, W.H. Weber, *J. Appl. Phys.* **82**, 3287 (1997).
12. J. Moulder, *Handbook of XPS*, (Perkin Elmer, New York, 1992), p. 95.
13. J.T. Herrold, V.K. Dalal, *J. Non-Crys. Sol.* **270**, 255 (2000).

# Electrical and Optical Properties of Amorphous and Microcrystalline Hydrogenated Silicon Films Deposited Using Saddle Field Glow Discharge

I. Milostnaya, T. Allen[1], F. Gaspari, N.P. Kherani, D. Yeghikyan, W.L. Roes[1], T. Kosteski and S. Zukotynski
Department of Electrical and Computer Engineering, University of Toronto, Toronto, ON M5S 1A4 CANADA
[1]Department of Physics, Geology and Astronomy, University of Tennessee at Chattanooga, Chattanooga, TN 37403 U.S.A.

## ABSTRACT

Amorphous (a-Si:H) and microcrystalline (μc-Si:H) hydrogenated silicon films were obtained using DC saddle field glow discharge. The structure of the films was determined by Raman spectroscopy, SEM and TEM. The optoelectronic characteristics of both a-Si:H and μc-Si:H were investigated using FTIR, UV/VIS spectroscopy, dark electrical conductivity ($\sigma_d$) and photoconductivity ($\sigma_{ph}$) measurements. Boron and phosphorous doping of a-Si:H and μc-Si:H films was also investigated. The results show that both a-Si:H and μc-Si:H undoped films are highly resistive ($\sigma_d=10^{-8}$-$10^{-10}$ $\Omega^{-1}cm^{-1}$). The doping efficiency of μc-Si:H films is much higher than a-Si:H films. The Tauc gap for a-Si:H was in the range 1.8-1.9 eV and for μc-Si:H films it was in the range 1.9-2.5 eV. The photoconductivity measurements of undoped films indicate a higher photosensitivity of a-Si:H films ($\sigma_{ph}/\sigma_d=10^4$) than that of μc-Si:H films ($\sigma_{ph}/\sigma_d=10$-$100$).

## INTRODUCTION

Hydrogenated amorphous silicon (a-Si:H) is used in a variety of optoelectronic applications, such as solar cells, thin film transistors, and color sensors [1]. However, many devices made from a-Si:H are observed to degrade with time which is commonly associated with hydrogen related defect states [1]. It has been observed that, by increasing the hydrogen dilution in the precursor gas used in the plasma, one can obtain microcrystalline silicon (μc-Si:H), which is a composition of crystalline grains embedded in an amorphous silicon matrix [2]. This material is more stable than a-Si:H and has the same advantages of low-temperature deposition.

The commonly used method for growth of a-Si:H and μc-Si:H is RF plasma-enhanced chemical vapour deposition (PECVD) [1, 2]. DC plasmas are also often used [3, 4, 5].

We have obtained both a-Si:H and μc-Si:H hydrogenated silicon films using DC saddle field glow discharge [6, 7]. The structural nature of the films was studied by Raman spectroscopy, SEM and TEM [7, 8]. IR and UV/VIS spectroscopy, dark electrical conductivity and photoconductivity measurements were used to obtain information on the bond structure, optical energy gap, electrical conductivity, doping efficiency and photosensitivity.

## EXPERIMENTAL

The DC saddle field glow discharge technique is described elsewhere [6]. In the saddle-field configuration a semitransparent central anode is positioned between two semitransparent cathodes; a substrate holder is positioned behind each cathode. This configuration allows for independent control of plasma parameters and substrate environment.

The gas source was a silane-hydrogen mixture. The main deposition parameters influencing microcrystalline growth are silane concentration $C=[SiH_4]/([SiH_4]+[H_2])$, gas pressure p, and the substrate holder bias [7]. Doping was achieved by using phosphine or diborane in the gas mixture.

The amorphous silicon samples were prepared using pure silane at p=0.1-0.2Torr, flow rate FR=5-30sccm, and anode current $I_{an}$=10-30 mA. Microcrystalline samples were prepared using silane-hydrogen mixtures. Microcrystallinity was observed for C<10%, p=0.2Torr, FR=10 sccm, $I_{an}$=30 mA. The structure of the films depends strongly on the bias of the substrate [7]. Therefore, to achieve similar self-bias conditions for all substrates we placed the conductive substrates on an insulating layer during deposition.

The films were deposited at a substrate temperature of 250°C on crystalline silicon (c-Si) wafers, pyrex glass, corning 7059 glass and fused silica. The films were 30 to 800 nm thick. The growth rates varied between 0.1 Å/s and 0.7 Å/s.

Structural characterization of the films was performed by Raman spectroscopy, SEM, TEM and HRTEM. The results of this analysis is presented elsewhere [7, 8]. Infrared analysis of the films grown on crystalline silicon substrates was carried out using a Perkin-Elmer FTIR spectrometer. Measurements of transmittance and reflectance were carried out on films deposited on pyrex and fused silica using a Perkin-Elmer UV/VIS spectrometer. The optical absorption obtained from these data was then plotted vs. the photon energy and the optical gap $E_g$ was obtained using the Tauc extrapolation method. Dark conductivity, $\sigma_d$, and photoconductivity, $\sigma_{ph}$, measurements were performed using the Van der Pauw's method. Photoconductivity measurements were done using quartz halogen lamp illumination at 100mW/cm$^2$.

## RESULTS AND DISCUSSION

Experimental data for undoped a-Si:H and μc-Si:H films are summarized in Table I. The data for doped films are presented in Table II.

Figure 1 shows Tauc plots for an amorphous film and a microcrystalline film. The optical gap is obtained by extrapolating the linear portion of the plot as shown in the figure. Tables I and II list the optical gaps measured for all the samples. Most amorphous samples exhibit an $E_g$ in the range of 1.8-1.9eV, while the μc-films exhibit $E_g$ in the range of 1.9-2.5eV. Figure 2 shows the dependence of the optical gap of the μc-films on the silane concentration. It can be seen that the optical gap decreases with decreasing H-dilution.

Figure 3 shows the infrared stretching mode spectra for four films with different silane concentrations.

Kroll *et al.* [9] have investigated the transition from an amorphous to a microcrystalline structure due to hydrogen dilution and they observed that the area under the infrared stretching mode decreases with increasing $H_2$-dilution. They linked this with a trend towards microcrystallinity using Raman and TEM analyses. Furthermore, they noted a shift of the stretching mode towards higher frequencies. We observe a similar trend in our films.

The microstructure factor $R=I_{2090}/(I_{2000}+I_{2090})$, where $I_x$ represents the integrated area of the deconvolved peak centered at x [10], is commonly used to characterize the quality of a-Si:H. However, it can also be used as an indicator of the level of microcrystallinity, keeping in mind that the peak at ~2090 cm$^{-1}$ represents monohydride and/or dihydride bonds on (100) and (111) surfaces in silicon crystallites [10]. The relation between R, C and $E_g$ is tabulated next to Figure 3.

## Table I  Undoped Films

| Film # | Structure | C % | Thickness (on Pyrex) μm | Growth Rate Å/s | $E_g$ eV | $\sigma_d$ $\Omega^{-1}cm^{-1}$ | $\sigma_{photo}$ $\Omega^{-1}cm^{-1}$ | $\sigma_{ph}/\sigma_d$ |
|---|---|---|---|---|---|---|---|---|
| 39 | a-Si:H | 100 | 0.18 | 0.25 | 1.83 | | | $1\times10^{+4}$ |
| 42 | a-Si:H | 100 | 0.20 | 0.28 | 1.90 | $4.11\times10^{-9}$ | $3.4\times10^{-5}$ | $1.00\times10^{+4}$ |
| 45 | a-Si:H | 100 | 0.80 | 0.93 | 1.8 | $2.11\times10^{-9}$ | $6.5\times10^{-6}$ | $3\times10^{+3}$ |
| 49 | a-Si:H | 100 | 0.50 | 0.25 | 1.80 | $5.2\times10^{-10}$ | | $1.00\times10^{+4}$ |
| 59 | a-Si:H | 100 | 0.23 | 0.64 | 1.82 | $2.3\times10^{-10}$ | | $1.00\times10^{+4}$ |
| 9 | μc-Si:H | 2.4 | 0.15 | 0.09 | 2.35 | | | 113 |
| 10 | μc-Si:H | 4.8 | 0.2 | 0.12 | 2.14 | | | |
| 11 | μc-Si:H | 9.1 | 0.32 | 0.19 | 1.99 | | | |
| 12 | μc-Si:H | 16.7 | 0.5 | 0.39 | 1.93 | | | |
| 46 | μc-Si:H | 2.4 | 0.04 | 0.06 | 2.50 | $8.6\times10^{-10}$ | | 10 |
| 48 | μc-Si:H | 2.4 | 0.03 | 0.05 | 2.53 | $8.5\times10^{-9}$ | | 10 |
| 50 | μc-Si:H | 4.8 | 0.07 | 0.13 | 2.25 | $7.7\times10^{-8}$ | $3.7\times10^{-5}$ | 480 |
| 52 | μc-Si:H | 4.8 | 0.12 | 0.18 | 2.40 | $1.6\times10^{-9}$ | $1.6\times10^{-8}$ | 100 |

## Table II  Doped Films

| Film # | Doping and Structure | C' % | $C_d$ % | $C_d'$ % | Thickness (on Pyrex) μm | Growth Rate Å/s | $E_g$ eV | $\sigma_d$ $\Omega^{-1}cm^{-1}$ |
|---|---|---|---|---|---|---|---|---|
| 53 | n-μc-Si:H | 4.64 | 2.50 | $1.19\times10^{-1}$ | 0.06 | 0.15 | 2.30 | 0.91 |
| 54 | n-μc-Si:H | 4.70 | 1.22 | $5.8\times10^{-2}$ | 0.08 | 0.16 | 2.20 | 1.08 |
| 55 | n-μc-Si:H | 4.74 | 0.56 | $2.6\times10^{-2}$ | 0.08 | 0.20 | 2.20 | 0.21 |
| 57 | n-μc-Si:H | 4.75 | 0.28 | $1.3\times10^{-2}$ | 0.06 | 0.15 | 2.30 | 0.24 |
| 58 | n-μc-Si:H | 4.76 | 0.14 | $6.6\times10^{-3}$ | 0.04 | 0.13 | 2.30 | 0.01 |
| 60 | p-μc-Si:H | 4.64 | 2.50 | $1.19\times10^{-1}$ | 0.06 | 0.28 | 2.10±.05 | $1.90\times10^{-4}$ |
| 61 | p-μc-Si:H | 4.70 | 1.22 | $5.8\times10^{-2}$ | 0.10 | 0.21 | 2.10±.05 | $3.80\times10^{-4}$ |
| 64 | p-μc-Si:H | 4.75 | 0.31 | $1.5\times10^{-2}$ | 0.28 | 0.17 | 2.10±.05 | $3.18\times10^{-3}$ |
| 65 | p-μc-Si:H | 4.76 | 0.16 | $7.4\times10^{-3}$ | 0.12 | 0.07 | 2.10±.05 | $3.84\times10^{-3}$ |
| 66 | p-μc-Si:H | 4.75 | 0.08 | $3.7\times10^{-3}$ | 0.16 | 0.10 | 2.10±.05 | $2.29\times10^{-3}$ |
| 1a | p-a-Si:H | 98 | 2.04 | 2.0 | 0.50 | 0.70 | 1.74 | $6.4\times10^{-6}$ |
| 2a | p-a-Si:H | 99 | 1.01 | 1.0 | 0.45 | 0.63 | 1.80 | $3.8\times10^{-6}$ |

C'=[SiH$_4$]/([SiH$_4$]+[H$_2$]+[Dopant Gas]), C$_d$=[Dopant Gas]/[SiH$_4$],
C$_d$' =[DopantGas]/([SiH$_4$]+[H$_2$]+[Dopant Gas]).

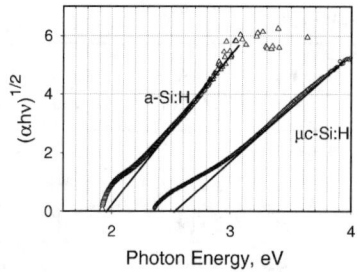

**Figure 1.** Tauc plots for an amorphous (sample 45), and a microcrystalline (sample 50) film.

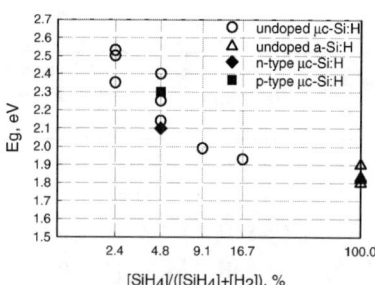

**Figure 2.** Tauc gap as a function of silane concentration for µc-Si:H films. The optical gap of a-Si:H films (C=100%) and two doped µc-Si:H samples are also shown for comparison.

The conductivities are listed in Tables I and II. Both a-Si:H and µc-Si:H undoped films show low conductivities at room temperature ($\sigma_d = 10^{-8}\text{-}10^{-10}\ \Omega^{-1}\text{cm}^{-1}$). Figures 4a and 4b show the change in conductivity with doping fraction for n-type (4a) and p-type (4b) microcrystalline films. The conductivity of two p-type amorphous films is also shown in Figure 4b. We observe efficient doping of microcrystalline films in both cases. Phosphorous doping appears to result in a maximum $\sigma_d$ of about $1\Omega^{-1}\text{cm}^{-1}$ at $C_d'=0.06\%$; boron doping has a peak value of $\sigma_d$ of about $4\times10^{-3}\ \Omega^{-1}\text{cm}^{-1}$ at $C_d' =0.007\%$. Amorphous films show a much lower doping efficiency.

Figure 5 shows the photosensitivity of amorphous and microcrystalline samples as a function of the optical gap. It can be seen that the amorphous films show a much higher sensitivity (at least 2 orders of magnitude).

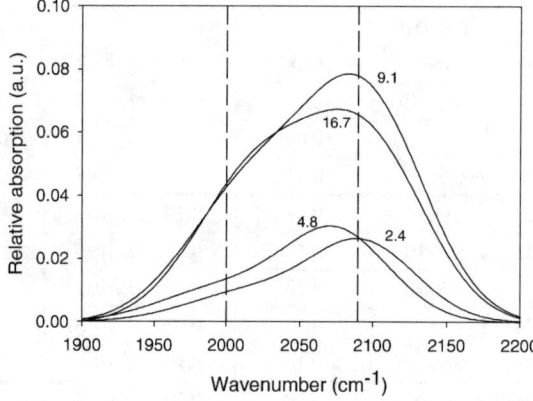

| C (%) | R | $E_g$ (eV) |
|---|---|---|
| 16.7 | 0.54 | 1.85 |
| 9.1 | 0.61 | 1.96 |
| 4.8 | 0.69 | 2.15 |
| 2.4 | 0.71 | 2.35 |

**Figure 3.** Infrared spectra for 4 samples grown with different silane concentration (C). The value of C is indicated next to the appropriate spectrum. The dashed lines indicate the two characteristic peaks for a-Si:H (2000 and 2090 cm$^{-1}$). The variation of $E_g$ and the microstructure factor R with C is tabulated beside the figure.

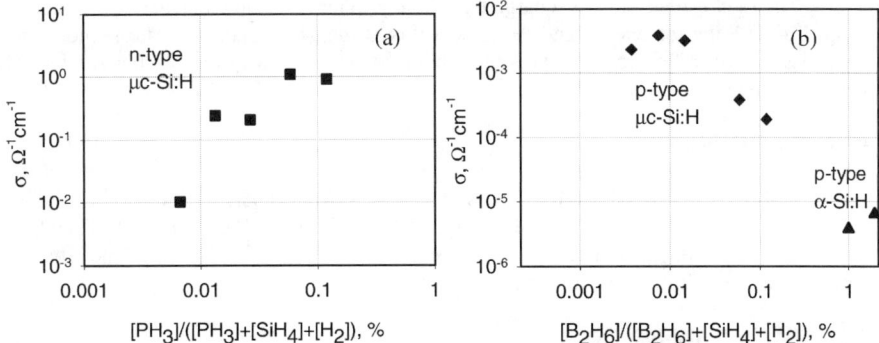

**Figure 4.** Conductivity vs. dopant concentration for phosphorous doped (a), and boron doped (b) microcrystalline films. Two boron doped amorphous films are also shown for comparison (triangles).

The optical gap of the amorphous films examined in this work (1.8-1.9 eV) is consistent with the literature. However, we observe relatively large values for the optical gap in μc-Si:H films (1.9-2.5 eV), with the larger gaps corresponding to a higher H-dilution and higher crystalline volume fraction. As shown in Figure 3, while the IR signal for our films decreases with higher H dilution, we observe an increase in the microcrystalline volume fraction [8]. This is confirmed by Raman and high-resolution TEM measurements [8]. Large optical gaps are also observed in our doped samples. The dopant concentration does not seem to have an effect on the gap, except for a generally lower value when compared with those of the intrinsic material. These gap values appear higher than those reported by Han *et al.* [10] and Alpuim *et al.* [11] but they are comparable with those reported by Platz *et al.* [4] and Nishida *et al.* [12].

Alpuim *et al.* [11] observed Tauc gaps of 1.6 - 2.1eV in μc-Si:H deposited by hot wire CVD and rf-CVD. Nishida *et al.* [10] use photochemical vapour deposition and report optical gaps of 2.0eV and 2.3eV in n-type and p-type films, respectively. They suggest that microcrystallization of a-Si is due to the presence of a large concentration of hydrogen radicals generated by the UV photons. They further suggest that photochemical vapour deposition eliminates ion damage associated with most glow discharge techniques. Enhanced incorporation of hydrogen into the film at the grain boundary and in the amorphous phase results in an

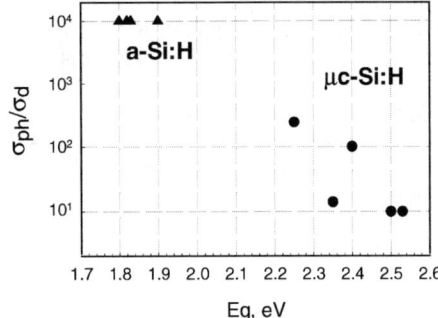

**Figure 5.** Photosensitivity vs. optical gap for amorphous and microcrystalline films.

The saddle-field glow discharge technique provides a remote plasma. It also permits the control of both the ion energy and the radical flux at the growth surface. Furthermore, the very high dissociation efficiency increases the amount of H radicals in the plasma similar to those obtained with photo-CVD.

## CONCLUSIONS

µc-Si:H films obtained by DC saddle field glow discharge exhibit a relatively wide optical band gap of 1.9 to 2.5eV, indeed higher than a-Si:H. Both undoped a-Si:H and µc-Si:H films have high dark resistivity. µc-Si:H films show much higher doping efficiency than a-Si:H. However, a-Si:H films have a higher photosensitivity.

## ACKNOWLEDGEMENTS

This work was supported by ARISE Technologies Corporation, Materials and Manufacturing Ontario and Natural Sciences and Engineering Research Council of Canada. T.A. would like to acknowledge support from the National Science Foundation (grant # DMR0074682), the Research Corporation Cottrell College Science Award (grant # CC4536), and support from the UC Foundation, University of Tennessee-Chattanooga. I.M.'s participation in the project was supported by a NATO Fellowship via NSERC.

## REFERENCES

1. R.A. Street (Ed.), "Technology and Application of Amorphous Silicon", Springer Verlag, New York, (2000).
2. A.H. Mahan, in "Amorphous Silicon and its Alloys", T. Searle (Ed.) , INSPEC (1998) 39.
3. L. Yang, M. Bennett, L. Chen, K. Janse, J. Kessler, Y. Li, J. Newtown, K. Rajan, F. Willing, R. Arya, and D. Carlson, Mat. Res. Soc. Symp. Proc. **420** (1996), 839.
4. R. Platz, S. Wagner, C. Hof, et al., J.Appl.Phys. **84** (1998) 3949.
5. J.A Anna Selvan, D. Grützmacher, E. Müller, M. Rebien, M. Kummer, H. von Känel, and J. Gobrecht, Mat. Res. Soc. Symp. Proc. **609** (2000).
6. S. Zukotynski, F. Gaspari, D. Manage, V. Pletnev, E. Sagnes, Mat. Res. Soc. Symp. Proc. **595** (2000) 239.
7. T. Allen, F. Gaspari, N.P. Kherani, T. Kosteski, K. Leong, I. Milostnaya, D. Yeghikyan, S. Zukotynski, Mat. Res. Soc. Proc. **744** (2002) M.5.21.
8. T. Allen, I. Milostnaya, D. Yeghikyan, K. Leong, F. Gaspari, N.P. Kherani, T. Kosteski, S. Zukotynski, Mat. Res. Soc. Symp. **762** (2003) A6.10.
9. U. Kroll, J. Meyer, A. Shah, S. Mikhailov, and J. Weber, J. Appl. Phys. **80** (1996) 4971.
10. D. Han, G. Yue, J.D. Lorentzen, J. Lin, H. Habuchi, and Q. Wang, J. Appl. Phys. **87** (2000) 1882.
11. P. Alpuim , V. Chu, and J.P. Conde, J. Appl. Phys. **86** (1999) 3812.
12. S. Nishida, H. Tasaki, M. Konagai, and K. Takahashi, J. Appl. Phys. **58** (1985) 1427.
13. A. Matsuda, M. Matsumura, S. Yamasaki, H. Yamamoto, T. Imura, H. Okushi, S. Itzima, and K. Tanaka, Jap. J. Appl. Phys. **20** (1981) L183.

# Crystallized Films

# Stress Effects on Nanocrystal Formation by Ni-Induced Crystallization of Amorphous Si

Yaocheng Liu, Michael D. Deal, Mahmooda Sultana[*], and James D. Plummer
Center for Integrated Systems, Stanford University, CA
[*]Department of Chemical Engineering, University of Southern California, CA

## ABSTRACT

Metal-induced crystallization (MIC) of amorphous Si is gaining increased interest because of its potential use for low-temperature fabrication of integrated circuits. In this work, the MIC technique was used to make Si nanocrystals and the effects of stress on the crystallization were studied. Amorphous Si films were deposited onto the Si substrate with thermal oxides on top by low-pressure chemical vapor deposition (LPCVD) and then patterned into nanoscale pillars by electron beam lithography and reactive ion etching. A conformal low-temperature oxide (LTO) layer was deposited to cover the pillars, followed by an anisotropic etch back to form a spacer, leaving only the top surface of the pillars exposed to the 5 nm Ni sputtering deposition afterwards. An HF dip was used to partially remove the LTO spacers on the pillars, leading to different LTO thicknesses on different samples. These samples were then annealed to crystallize the amorphous Si pillars, forming Si nanocrystals. Transmission electron microscope (TEM) observations after anneal found a clear dependence of the crystallization rate on the pillar size as well as the LTO thickness. The crystallization rate was lower for pillars with thicker LTO spacers, while for the same LTO thickness the crystallization rate was lower for pillars with narrower width. A model based on the stress in the pillars is proposed to explain this dependence. This model suggests some methods to control the nickel-induced crystallization process and achieve higher quality Si nanocrystals.

## INTRODUCTION

Metal-induced crystallization (MIC) has become an attractive technology due to its applications in three-dimensional integrated circuits and heterogeneous integration [1-4]. It has been reported to be possible to make high quality polycrystalline Si with low-temperature process by metal-induced crystallization [5, 6]. Among different metals, Ni is of particular interest because $NiSi_2$ has a fluorite structure with a very small lattice mismatch with Si. Large Si grains with low defect density can be achieved through $NiSi_2$-mediated solid-phase epitaxy [7, 8].

The main focus of this work is to obtain single-crystalline Si structures of sub-µm to nm scales and study the related mechanism. It was found in our experiments that stress plays an important role in metal-induced crystallization and can be used as a tool for process control to improve crystal quality.

## EXPERIMENTAL DETAILS

A thin oxide layer was first grown on (100) Si wafers by thermal oxidation. Then amorphous Si (a-Si) was deposited onto the wafers by low-pressure chemical vapor deposition

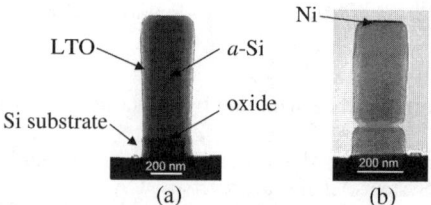

**Figure 1.** Pillar structures (a) before Ni deposition and (b) after Ni deposition.

(LPCVD). Electron beam lithography was used to pattern the wafers and reactive ion etch (RIE) was used to transform the patterns into pillar structures of different dimensions. The wafers were then covered with a conformal layer of low-temperature oxide (LTO), which was deposited with LPCVD at 300°C. The LTO deposition was followed by an anisotropic plasma etch-back to form spacers covering the sidewalls of the pillars. After that, the wafers were loaded to a sputtering chamber and a 50Å Ni layer was deposited. The pillar structures before and after Ni deposition are shown in Figure 1. The pillars were annealed at 450°C for 15h in a furnace for crystallization. The annealed pillars were studied with transmission electron microscope (TEM). Since the LTO spacers are covering the sidewalls of the a-Si pillars during Ni deposition, the a-Si pillars are seeded by Ni only from the top. The MIC growth front is expected to propagate downwards from the top of the pillars.

**RESULTS AND DISCUSSION**

Under the chosen annealing conditions, the crystallization was relatively slow, so different degrees of crystallization were observed, as shown in Figure 2. From this figure, a trend can be

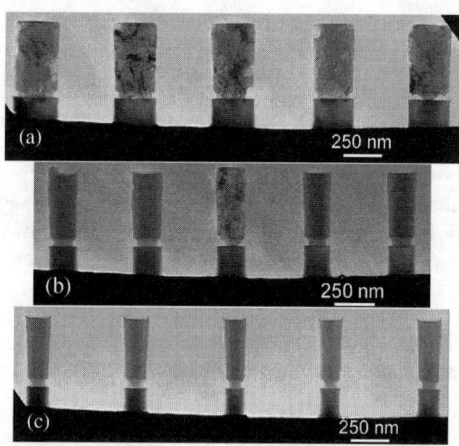

**Figure 2.** Pillars of different crystallization statuses after 450°C 15h anneal. (a) Completely crystallized pillars. (b) One out of five pillars crystallized. (c) Not crystallized pillars.

seen that crystallization becomes slower as the pillar width decreases. In order to get a more quantitative view of the crystallization behaviors of the pillars, we did statistical TEM study on pillars with different widths and different LTO spacer thicknesses. Figure 3 shows the statistical results.

Two basic observations can be made from Figure 3. First, for pillars with same width but different LTO thicknesses, crystallization becomes slower as the LTO thickness increases. Second, for pillars with different widths but same LTO thickness, crystallization becomes slower as pillar width decreases.

The changes in dimensions of pillars and LTO thickness correspond to changes in stress states in the Si pillars. Curvature measurements were carried out at room temperature on non-patterned wafers after a-Si deposition and LTO deposition, respectively. The results showed that the as deposited LTO film provided some compressive stress to the a-Si underneath. Besides that, since there is a thermal mismatch between Si and oxide (thermal expansion coefficients: $\alpha_{a\text{-}Si}=3.0\times10^{-6}/°C$ [9], $\alpha_{ox}=0.5\times10^{-6}/°C$), there exists some additional compressive stress in the a-Si pillars when they were heated up to 450°C. When the LTO thickness increases or the pillar width decreases, the compressive stress in a-Si increases. So the conclusion from the statistical results is that the compressive stress slows down metal-induced crystallization.

This conclusion is somehow counter-intuitive since crystallization of amorphous Si corresponds to some volume shrinkage. To understand the above results, we have to consider the mechanism of MIC. In thermodynamics, the driving force always exists for a-Si to transform into crystalline phase, even without any metal in it. However, crystallization of pure a-Si cannot occur at low temperatures such as 450°C as the atomic motion of Si is very slow. When some metal, e.g. Ni, is added into the a-Si, the atomic rearrangement is speeded up. In the case of Ni-induced crystallization, $NiSi_2$ moves into a-Si region, leaving behind a trail of crystalline Si. The crystallization enhancement by metal is actually an enhancement in atomic mobility. The atomic motion corresponds to some activation volume. The relationship can be expressed by

$$\frac{\partial(\ln D)}{\partial P} \propto -\frac{V_a}{kT}$$

where D is the diffusivity, P is the pressure and $V_a$ is the activation volume. When a

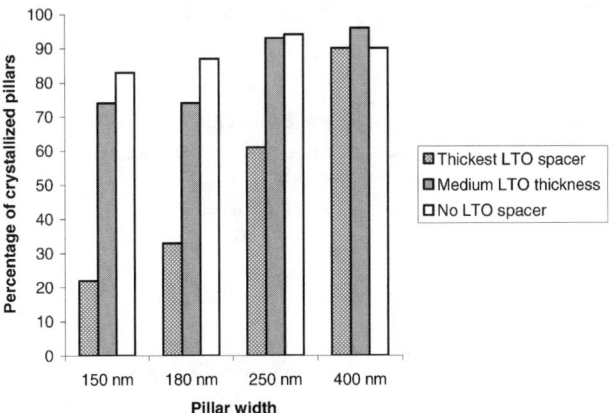

**Figure 3.** Statistical results of TEM study on pillars after 450°C 15h anneal.

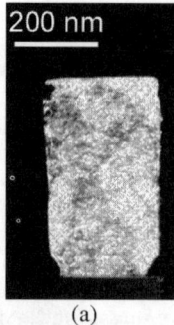 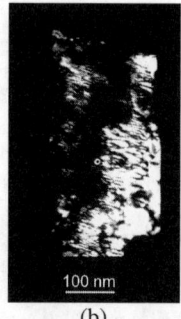

**Figure 4.** Dark field TEM images of (a) a single-crystalline Si pillar after two-step MIC and (b) a pillar with more than one grain after one-step MIC.

compressive stress, i.e., a pressure, is applied to the system, the diffusivity decreases. In other words, compressive stress slows down the atomic motion and this retards metal-induced crystallization.

Realizing that compressive stress retards metal-induced crystallization, we can use desired stress distributions to control the crystallization process and improve the possibility of getting single-crystalline structures. As shown in Figure1 (a), the LTO spacer covers the sidewalls of the a-Si pillars up to regions close to the top. As a result, during the crystallization anneal, most parts of the pillars are loaded with compressive stress except the top surface region, which is virtually stress free. Therefore, a two-step MIC technique can be applied to these pillars structures. During the first step, the crystallization of a-Si is suppressed by the compressive stress. But in the top surface region, the reaction to form $NiSi_2$ and the grain growth of $NiSi_2$ are not affected. After some time, the $NiSi_2$ grows into a single-grain layer without crystallizing the a-Si. Then as further anneal is carried out, single-crystalline Si can be obtained through silicide-mediated solid-phase epitaxy with the single-grain $NiSi_2$ as the template. Figure 4 shows a single-crystalline Si pillar after the as described two-step MIC process and a structure with more than one grain after direct one-step MIC without compressive stress controlling the process.

## CONCLUSIONS

A pillar structure is used to study the stress effects on metal-induced crystallization for sub-µm to nm structures. It is found that compressive stress retards metal-induced crystallization by reducing the atomic mobility. A two-step MIC technique has been developed to make single-crystalline Si structures using the stress effects to control the process.

## REFERENCES

1. S. Bae, and S. J. Fonash, *Appl. Phys. Lett.* **76**, 595 (2000).
2. K. Makihira, and T. Asano, *Appl. Phys. Lett.* **76**, 3774 (2000).

3. Z. Meng, M. Wang, and M. Wong, *IEEE Trans. Electron Devices* **47**, 404 (2000).
4. H. Huh, and J. H. Shin, *Appl. Phys. Lett.* **79**, 3956 (2001).
5. S. Y. Yoon, S. J. Park, K. H. Kim, J. Jang, and C. O. Kim, *J. Appl. Phys.* **87**, 609 (2000).
6. V. W. C. Chan, and P. C. H. Chan, *IEEE Electron Device Lett.* **22**, 80 (2001).
7. C. Hayzelden, and J. L. Batstone, *J. Appl. Phys.* **73**, 8279 (1993).
8. J.L. Batstone, C. Hayzelden, *Inst. Phys. Conf. Ser.* **n134: section 4**, 165-172 (1993).
9. A. Witvrouw and F. Spaepen, *J. Appl. Phys.* **74** 7154 (1993).

# FIELD-ASSISTED GERMANIUM INDUCED CRYSTALLIZATION OF AMORPHOUS SILICON

J. Derakhshandeh, S. Mohajerzadeh, N. Golshani, E. Asl Soleimani and *M.D. Robertson
Thin Film Laboratory, Department of Electrical and Computer Engineering, University of Tehran, Tehran, Iran Tel/Fax: +98 (21) 801 1235, *Department of Physics, University of Acadia, Canada
Email: smohajer@sun1.vlsi.uwaterloo.ca,

## ABSTRACT

A field-assisted germanium-induced crystallization of amorphous silicon on glass is reported at temperatures below 500°C. Silicon films with a thickness of 0.1um are covered with 500Å of germanium as the seed of crystallization. Applying an electric field enhances the growth from both cathode and anode sides. XRD, SEM and TEM analyses have been used to study the crystallinity of the samples which have been treated under different annealing conditions, all confirming the polycrystalline nature of the annealed silicon films. The value of the applied voltage plays a crucial role in the crystalline quality of Si layers. While samples treated without an external voltage are not polycrystalline, an electric voltage of 10 V applied for a 1cm separation between anode and cathode, seems suitable for achieving good poly-crystalline Si layers. The size of grains varies between 0.1 and 0.2µm, as observed using SEM.

## INTRODUCTION

Polycrystalline silicon has been of significant interest to many researchers for its prospective applications in realizing thin film transistors. Despite the usefulness of such material for large area electronics, its use has been limited to special glasses, mainly due to a high processing temperature needed for crystallization. Several techniques have been introduced to lower the annealing temperature during crystal growth, the most important of which are excimer laser annealing (ELA) and solid phase crystallization (SPC). While ELA is a low temperature process, it requires expensive facilities [1]. On the other hand SPC is energy consuming and requires temperatures above 550°C [2].

Metal induced crystallization (MIC) has been successful in reducing the crystallization temperature to values around 400°C [3,4]. Many researchers have attempted various types of MIC to lower the treatment temperature and realize thin film transistors on glass [5]. Metal contamination, however, is a main concern in all of these approaches. Metal induced lateral crystallization is a modified alternative of MIC to reduce the unwanted incorporation of metallic atoms in the growing layer [6-9]. It has been suggested that lateral diffusion of metallic atoms causes the crystallization of amorphous silicon. Also the annealing temperature is slightly above that of MIC. The use of germanium as a clean seed of crystallization has been attempted by Subramanian et.al. [10]. A lateral growth of the order of 0.5µm has been achieved at temperatures above 500°C and small geometry TFT's have been realized. Insignificant lateral growth and a high annealing temperature seem to be the main obstacles in using this approach for many applications [11].

In this paper we use a bi-layer of silicon and germanium to achieve silicon crystallization at temperatures as low as 480°C. The top Ge layer is the seed of crystallization for the underlying silicon film. The presence of an external voltage between anode and cathode has a dramatic effect on the crystallinity of the silicon film. Various samples have been realized for this study and the electrical conductivity of the annealed samples has been examined to study the crystallization of the silicon films. SEM, XRD and TEM were used to further examine the physical characteristics of the films. All analyzing tools seem to corroborate the results of our conductivity measurements. In the following sections, we first describe the experimental setup used for this investigation and then the results of an electrical and structural study of the silicon films are presented.

## EXPERIMENTAL SETUP

The substrates used for this study are 125µm-thick ordinary glasses cleaned using RCA#1 solution for 15min., followed by rinsing in DI water and blow-drying by air. After cleaning, samples are loaded in the vacuum chamber and heated up to 400°C for a baking period. Once a base pressure of $1 \times 10^{-6}$ torr is achieved, the deposition of silicon film begins. After depositing 0.1µm of amorphous silicon at 400°C and without breaking the vacuum, a germanium film with a thickness of about 500Å is evaporated onto the silicon film. The time interval between two consecutive depositions must be minimal for an intimate contact of Ge and Si layers. Finally a 500Å thick $SiO_2$ layer protects the surface of the Ge film during post thermal treatments.

Annealing of the samples has been performed on a hot plate in air atmosphere and in the presence of an external voltage applied between two metallic contacts with 1cm separation. Fig.1 schematically displays the sample structure and the post thermal treatment.

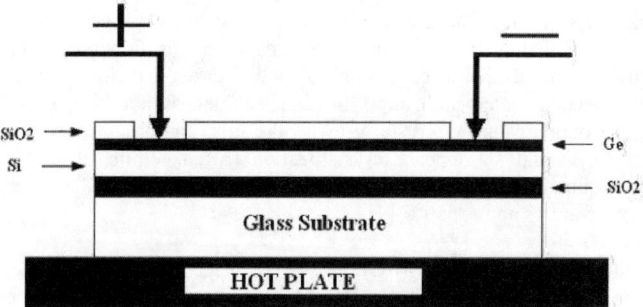

**Figure 1**: The schematic setup used for sample treatment in this study.

A thermocouple is placed beside the sample to monitor its temperature during post treatment. All samples are annealed for a period of 24 hours. Samples have been subjected to various voltages ranging from a few volts to a few hundred volts. Also annealing at temperatures between 350°C and 600°C has been carried out. While annealing at excessive temperatures causes considerable damage to the substrate, low temperature processes do not give considerable enhancement in the film conductivity. The electrical resistance of the sample is also monitored during the annealing period. Since Ge shows a higher conductance than the underlying silicon film, this *in-situ* observation contains the information about the germanium film and not much about the underlying silicon.

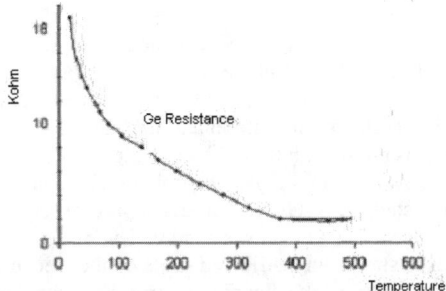

**Figure 2**: Resistance of the sample versus annealing temperature which is mainly due to germanium overlayer.

To measure the electrical resistance of the silicon film, one has to etch away the protective $SiO_2$ and the top Ge films, respectively. The resistance of the Si film has been measured to be about $2 \times 10^9 \Omega$ for the sample treated with an external voltage of 10 volts whereas for the sample treated with a higher voltage of 100V, this value increases to $10^{10} \Omega$ indicating a poorer conductivity for the latter case.

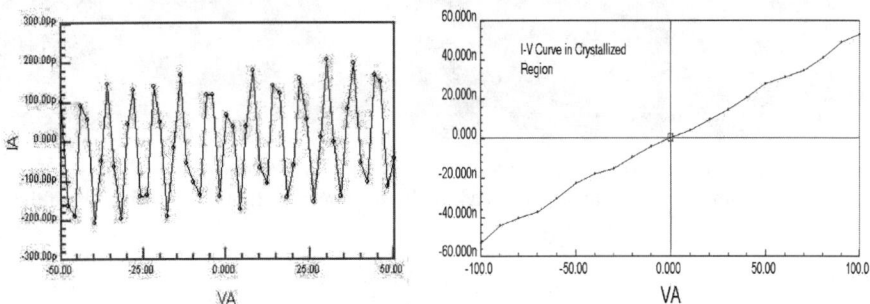

**Figure 3**: I-V curve in crystallized and non-crystallized region. The non-crystallized sample shows insignificant current due to a high resistance.

For the sample treated with no external voltage, the resistance is three orders of magnitude higher than for the previous samples and cannot be measured with our Keithley parameter analyzer. The electrical resistance of various samples have been measured and plotted in Fig3. The improved conductivity in the sample annealed with an applied voltage of 10V is believed to be due to improvement in the carrier mobility of the silicon film and reflects the crystallization of the silicon layer.

## PHYSICAL STUDY

The surface morphology of the sample post treated with an applied voltage of 10V has been investigated using SEM analysis and the results have been collected in Fig.4. The sample prepared using 100V has also been examined and much less features have been observed on its surface and hence its results are not presented here. For SEM specimen preparation, first the protective oxide layer is removed followed by etching away the germanium film. The sample is then soaked for three seconds in a crystallographic etching solution to reveal (111) and (110) orientations. For this study a Sirtl solution, containing an equal mixture of $CrO_3$ and HF in water, has been employed. Fig.4 (a) shows a large view of the sample close to cathode side. The crystallization seems to start from distributed seeds on the silicon surface and grow laterally in a circular fashion. In image (b) of this figure one can see a closer look at the structure indicating a granular structure with a grain size of the order of 0.2μm.

**Figure 4**: Surface morphology of the sample treated with 10V during annealing, (a) lateral growth of Si from distributed seeds in a circular fashion. (b) Higher magnification of the image showing a grain size of the order of 0.2μm.

The mechanism behind this type of growth is not yet understood and more investigation is needed before a satisfactory explanation can be reached. However, we speculate that a combination of lateral mechanical stress due to the over-lying germanium film and the electric field due to the applied voltage is responsible for this unique observation.

We have also investigated the crystallinity of the samples using XRD analysis and the results are provided in Fig.5. For this study, we have also removed the top oxide and Ge layers to avoid peaks from the crystalline germanium film. In this figure one can see the spectrum of the sample prepared using a voltage of 10V during post treatment. The (220) and (311) peaks of Si are evident in this figure, however the (111) peak is not discernible. The small thickness of the Si layer and distribution of grains can be responsible for this observation. This figure also corroborates the better crystallinity observed for the sample with a lower voltage of 10V. The sample treated without any external voltage has also been examined and no discernible peak is observed in its spectrum. We believe that by using thicker silicon films and longer annealing periods, we can obtain more significant XRD results.

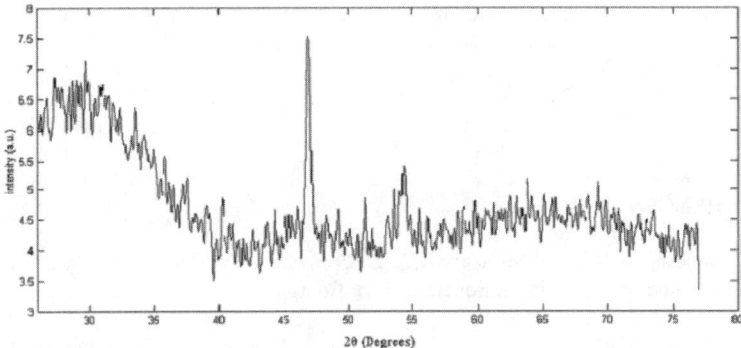

**Figure 5:** The XRD spectra for the sample treated with an applied voltage of 10V. The <220> and <311> peaks are discernible in the spectrum of sample

Fig.6 provides the results of electron diffraction patterns of the sample treated with a voltage of 10V. For TEM specimen preparation we have cleaved the samples to observe their cross-sectional view. The sample treated with 10V during annealing shows sharp and well-identified rings due to a polysilicon structure.

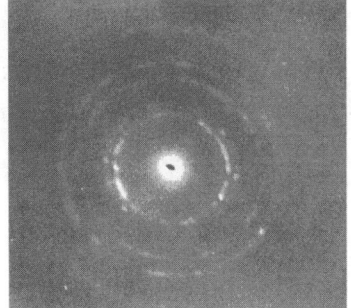

**Figure 6:** The electron diffraction pattern of the sample prepared using an external voltage of 10 V during post thermal treatment.

**SUMMARY AND CONCLUSIONS**

We report a germanium induced crystallization technique to grow polysilicon films on glass at temperatures as low as 480°C. Germanium acts as the seed of crystallization and no metal contamination is expected. Unlike field aided metal induced lateral crystallization, where the growth starts from the negative electrode (cathode) and progresses towards the other side, the growth initiates from randomly distributed seeds and propagates in a circular fashion. The density of such circular seeds is more significant in the area close to the cathode side than the anode one. We believe that the energy imparted by the electrons enhances the chance of crystallization at reduced temperatures. Also the germanium seed on top imparts mechanical compressive stress on the growing silicon layer. This stress may assist the formation of polysilicon grains, hence enhancing the overall crystallization phenomenon. The results of current-voltage characteristics as well as SEM, XRD and TEM analyses corroborate the crystallization of silicon.
Authors wish to thank Mr. Karvaneh for XRD and Mr. Kiani for SEM analyses. This work has been supported by a grant from Ministry of Industry and partial support from Faculty of Engineering, University of Tehran.

**REFERENCES**

[1] T. Sameshima, S. Usui, M. Sekiya, *IEEE Electron Device Lett.*, no.7, 276-278 (1986).
[2] K. Zellama, P. Germain, S. Squelard, J. C. Bourgoin, P. A. Thomas, *J. Appl. Phys.* 50, 6995 (1979).
[3] L. Rezaee, S. Mohajerzadeh, A. khakifirooz, S. Haji, E. Asl Soleimani, MRS Symp. Proc. 664 (2001).
[4] S. Y. Yoon, S.J. Park, K. H. Kin, J. Jang, *Thin Solid Films*, vol. 383, pp. 34 (2001).
[5] S. W. Lee, S. K. Joo, *IEEE Electron Device Lett.* 17, (1996).
[6] J.B. Lee, C.J. Lee, and D.K. Choi, *Jpn. J. Appl. Phys.*, vol. 40, 6177, (2001).
[7] T. K. Kim, G.B. Kim, B.I. Lee ands.K. Joo, *IEEE Electron Device Lett.*, vol21, No. 7, (2000).
[8] T. H. Ihn, B.I. Lee, S.K. Joo, and B.C. Jeon, *Jpn. J. Appl. Phys.*, vol. 36, p. 5029 (1997).
[9] A. Khakifirooz, S. Mohajerzadeh and S. Haji, *J. Vac. Sci. & Technol.* A, vol. 19, no. 5 (2001).
[10] V. Subremanian and K. C. Saraswatt, *IEEE Trans. Electron Devices* 45, 1934 (1998).
[11] V. Subremanian, M. Toita, N. R. Ibrahim, S. J. Souri, K. C. Saraswat, *IEEE Electron Device Lett.*, 20, 341 (1999).

# Prediction of the Interface Response Functions for Amorphous and Crystalline Phases of Silicon and Germanium

Erik J. Albenze, Laura A. Matejik[1], Nick F. Fynan, and Paulette Clancy
School of Chemical and Biomolecular Engineering, Cornell University, Ithaca, NY 14853, U.S.A.
[1]Department of Mechanical Engineering, Worcester Polytechnic Institute, Worcester, MA 01609, U.S.A.

## ABSTRACT

Interface response functions that govern the solidification kinetics of amorphous and crystalline phases of Si and Ge have been determined for reparameterized versions of the Stillinger-Weber (SW) potential. The strength of the three-body term in the SW potential and the energy scaling parameter were modified to obtain agreement with the experimental melting temperatures of both the amorphous and crystalline phases. These modified models were used to produce predictions of the interface response function for both Si and Ge that adequately fit the few known experimental data.

## INTRODUCTION

Silicon and germanium form both a crystalline and amorphous solid phase. The melting points of the crystalline phases are 1685 K for Si and 1210 K for Ge. Values of the melting points of the amorphous phases range from 1420 to 1500 [1, 2] for silicon, and 960 to 975 K for germanium [3]. In common with many other substances, these group IV semiconductors exhibit a phenomenon known as *explosive crystallization*. It is generally accepted that this self-sustaining process is mediated by a thin liquid layer converting the amorphous phase to the crystalline phase. Driving this reaction are both the melting temperature and enthalpy differences between the amorphous and crystalline phases. As shown in Figure 1, once an initial liquid layer has been created (*e.g.*, by the action of a laser), crystallization at the crystal/liquid interface will release heat which is then thermally conducted through the liquid to the liquid/amorphous boundary. Melting of the amorphous phase then withdraws heat from the system, encouraging further crystallization, and thus the process auto-propagates. This chain reaction process self-selects the fastest direction in which to propagate (the [100] direction). The overall speed of explosive crystallization is dictated by the kinetics of liquid phase crystallization which is expressed in the so-called *interface response function* which links the velocity of melting and solidification to the thermodynamic driving force of interfacial overheating and undercooling, respectively. The velocity of the moving solid/liquid interface will be defined to be positive for solidification and negative for melting. Our interest here is to determine the interface response functions for germanium and silicon as a precursor to using atomic-scale simulation to study explosive crystallization.

Past studies of the interface response

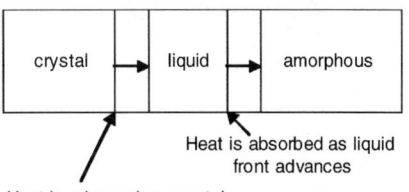

Figure 1: Schematic of Explosive Crystallization.

function (IRF) for Si include fitting the limited experimental data that are available (chiefly, the amorphization velocity, maximum velocities for crystallization and amorphization, and the melting points) to a Wilson-Frenkel equation [4]; or combining the Wilson-Frenkel form with heat-flow calculations and time-resolved experimental data [4]. Atomic-scale simulations of the IRF have been carried out for the Stillinger-Weber (SW) model of the crystalline phase [5,6] and the Environment-Dependent Interatomic Potential (EDIP) model for the amorphous phase [7]. Similar predictions for the behavior of germanium are unknown. We have recently shown that none of the commonly used interatomic potential models for Si (SW, EDIP, MEAM) are capable of reproducing experimental maximum regrowth velocities and simultaneously predicting accurate melting points for the crystal and amorphous phases [8]. Thus it is unlikely that these potential models would be useful for a study of explosive crystallization since the melting point difference is a key driving force for the process.

Nevertheless, atomic-scale simulation is an attractive option for the determination of the kinetics of solidification, since experimental measurements of the velocity-undercooling relationship are difficult, due to the high temperatures involved and the fast motion of the interface (up to 15 m/s). In atomic-scale simulations, the speed of the moving interface is determined from the location of the solid/liquid interface as a function of time. This process assumes we can differentiate between atoms as either solid-like or liquid-like (using an appropriate order parameter).

## POTENTIAL MODEL

One aim of this study was to develop a potential which can adequately represent all three condensed phases of silicon and germanium; a severe challenge for any potential, especially an empirical model. Our focus on an empirical model is dictated by the future need to consider the relatively large system sizes (several thousand atoms) involved in a representation of explosive crystallization. The Stillinger-Weber [9] potential is known to predict the crystalline IRF accurately in comparison to experimental data for the equilibrium melting point and maximum growth velocity. Unfortunately, the prediction of the melting point of the amorphous phase is much too low [8]. Results for other commonly used empirical models of Si, the Environment-Dependent Interatomic Potential [10-12] and the Modified Embedded Atom Method model [13] also give poor predictions of the interface response functions [8]. No determination of the interface response function has yet been made for germanium. Given the results obtained for Si, we chose to use the Stillinger-Weber model for Ge with parameters taken from [14, 15].

It is not surprising that the Stillinger-Weber potential proved to be the most successful of the three models investigated in ref. 8 in modeling melting and solidification processes. The SW model was specifically designed to reproduce the *c-/l*- melting transformation of Si, and thus gives a good description of both the liquid phase and the energetics of the crystal for silicon. It has a simple form consisting of a two- and three- body term. For Si, the strength of the three-body term was increased by 15% and the melting points were corrected by adjusting $\varepsilon$. This value was found to give the best compromise between fitting the melting point of the amorphous phase without denigrating the reproduction of other two phases [8].

# DETERMINATION OF THE INTERFACE RESPONSE FUNCTION

Molecular Dynamics simulations of the melting and solidification process were performed using a constant pressure-constant temperature ensemble with periodic boundary conditions in all three Cartesian directions using a leap-frog algorithm proposed by Brown and Clark and a Nosé-Hoover thermostat. The simulations consisted of

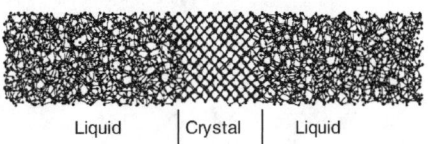

**Figure 2**: A typical sandwich box for the crystal IRF

'sandwich' configurations of solid and liquid phases, shown in Figure 2 for the solidification and melting of the crystalline phase. The samples were set up so that the growth (or melting) would be in the [100] direction since this is the direction which would be self-selected during explosive crystallization. An analogous 'sandwich' was constructed for the amorphous phase. The systems ranged in size from 1536-3072 atoms, with aspect ratios ranging from 3.0-10.0, depending on the sample size deemed necessary to capture adequately the motion of the interface. The samples were evolved at a pre-set undercooling to encourage solidification (or overheating, for melting) and the motion of the solid-liquid interface was recorded by determining its location using appropriate order parameters. Here we used the average bond angles to neighboring atom, deviations in the average bond angle, the coordination number and an angular order parameter, described below in more detail.

The location of the interface was determined by dividing each sample into slices of between 1.6 and 2.0 atomic layers thick (depending on sample size and aspect ratio) and assigning each slice to be solid, liquid, or interfacial, determined by inspection of the order parameters. Once the location of the interface was known at any given time, the velocity of the planar interface by determining the distance moved as a function of time.

The angular order parameter [16], mentioned above, designates an atom as being 'solid-like' or 'liquid-like'. Solid-like atoms must have four nearest neighbors; the nearest neighbor cutoff being chosen as the first minimum in the radial distribution function and a value for A < 0.4 determined by the following equation:

$$A = \sum_i \left(\cos\theta_i + \tfrac{1}{3}\right)^2 \tag{1}$$

where the sum is taken over the triplet sets of the four nearest neighbors and $\theta_i$ is the angle made by the triplet set. If these criteria are not met, the atom is designated as liquid-like. Within each slice, the fraction of atoms considered to be solid was compared to characteristic values for crystal (for which the fraction of solid is characteristically >99%) for liquid (0-45% solid-like), and for amorphous (60-90% solid-like) in order to determine the nature of the slice. Interface determination for the liquid/crystal samples was unambiguous, since there was essentially a step change in the percentage of solid with the distance along the sample, clearly indicating the location of the interface. In contrast, determination of the location of the interface in the amorphous/liquid samples was more difficult. The liquid and amorphous regions were often separated by ambiguously ordered slices (falling in the undefined 45-60% solid-fraction range). For consistency, the position of the a-/l- interface was determined to be the point where the slices first became unambiguously amorphous.

# RESULTS FOR ORIGINAL SW MODELS AND REPARAMETERIZATION APPROACH

Once the interface response functions were determined using the original parameter sets (*i.e.*, those of Stillinger and Weber for Si [9] and Yu and Clancy [14,15] for Ge), the equilibrium melting point for either the crystalline or amorphous phase was identified as the point at which the velocity-temperature curve crossed the *x*-axis. The melting points were determined to be 1650 K and 1000 K for the crystalline and amorphous phases for silicon, respectively, and 1400 K and 925 K for germanium, respectively. The corresponding IRF plots are shown in Figure 3. It should be noted that only the melting line for the IRF of the amorphous phase could be determined; the potentials are apparently not strong enough to encourage interfacial growth of the amorphous material into a liquid phase. In any case, there are no experimental measurements of velocity-temperature data for the amorphous phase. For silicon, the melting point of the crystal is correctly identified (as expected) but the amorphous melting point is in significant error (31% too low). For Ge, the opposite trend was observed; the melting point of the crystal is incorrect (16% too high) but the amorphous melting point is relatively accurate.

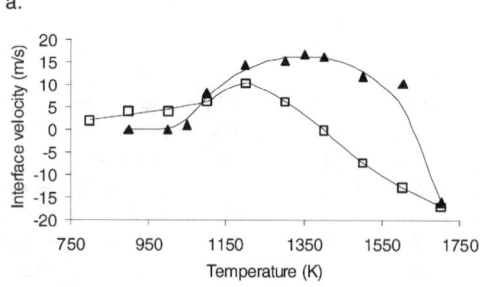

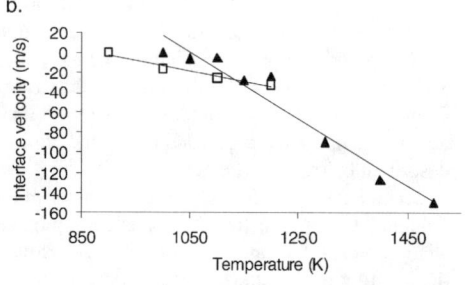

**Figure 3**: Interface response functions as determined for the original SW potential for silicon (filled triangles) and Yu and Clancy's set of parameters for germanium (open squares) for the crystalline phases in (a) and the amorphous phases in (b). Trendlines are added to guide the eye.

If we are to simulate explosive crystallization, it will be important to reproduce the melting point difference with better accuracy. Thus the SW potentials for Si and Ge will need to be reparameterized. Since both amorphous and crystalline phases possess tetrahedral structure (locally), increasing the three-body term will increase the melting points while decreasing the relative difference. The energy parameter can be used to scale the melting temperatures to appropriate values. However, the effect of the increased three-body term on the structural properties will need to be examined to ensure that the virtual samples are still in good agreement with the structural properties of the real phases. The reparameterization procedure is an iterative one, requiring increasing the three-body term, adjusting the energy scaling parameter to correct the crystal melting point, checking the

**Table I**: Refitted parameters for Ge and Si for the SW potential

| Parameter | Silicon | Germanium |
|---|---|---|
| λ | 24.15 | 23.625 |
| γ | 1.20 | 1.20 |
| A | 7.049556277 | 7.049556277 |
| B | 0.602224558 | 0.602224558 |
| a | 1.80 | 1.80 |
| p | 4 | 4 |
| q | 0 | 0 |
| ε | 35 kcal/mol | 25.79 kcal/mol |
| σ | 0.20951 nm | 0.21789 nm |

amorphous melting point and verifying the structural properties of solids and liquid. The process is repeated until a suitable set of parameters is obtained. New sets of parameters for Si and Ge are presented in Table I.

## RESULTS USING REFITTED SW MODELS

Using the prescription described above, the interface response functions for the new parameterization sets for SW models of Si and Ge were determined, as shown in Figure 4. The melting temperatures for the new parameterizations as well as experiment and the original SW potential for Si and Ge are shown in Table II. Better representations of the melting points were obtained for the refitted SW models. Sufficient reproduction of structural property data (omitted for brevity) was present to warrant the adoption of the new parameters. The melting points of both solid phases are now accurately reproduced for Ge (5% too low for the amorphous phase and 3% too low for the crystalline phase), but the melting point of amorphous silicon is still in significant error (9% too low). For silicon, any further increase of the three-body potential term results in only moderate improvement in the melting point accompanied by significant loss in accuracy of the liquid phase structural properties [8]. Accordingly, we limited ourselves to a 15% increase of the three-body potential for Si. Radial distribution functions for the refitted potentials were close to experimental data (results not shown).

**Table II**: Melting Temperatures

| Silicon | $T_{m,cryst}$ | $T_{m,amorph}$ |
|---|---|---|
| Stillinger and Weber | 1650 | 1000 |
| This paper | 1700 | 1320 |
| Experiment[2,3] | 1680 | 1450 |

| Germanium | $T_{m,cryst}$ | $T_{m,amorph}$ |
|---|---|---|
| Yu and Clancy | 1400 | 925 |
| This paper | 1175 | 915 |
| Experiment[1] | 1210 | 965 |

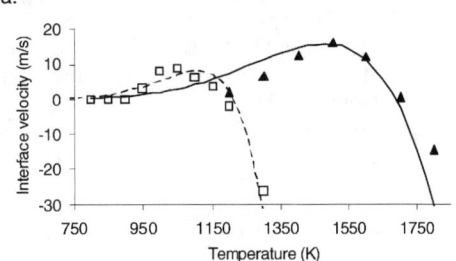

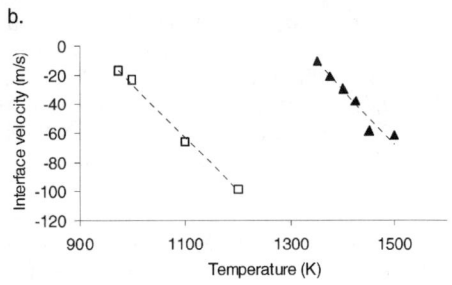

**Figure 4**: Interface response functions for the new parameter sets: a) crystalline germanium (open squares) and silicon (filled triangles) and Wilson-Frenkel equation fits for germanium (dashed line) and silicon (solid line). b) amorphous germanium (open squares) and silicon (filled triangles) with trendlines to guide the eye.

Figure 4 also demonstrates the 'prediction' of a Wilson-Frenkel equation fitted to key experimental data (amorphization velocity, maximum explosive crystallization and amorphization velocity, and equilibrium melting points) taken from reference 4 for silicon. There is not enough experimental data available for germanium to determine a similar set of Wilson-Frenkel parameters for the amorphous phase curve.

## CONCLUSIONS

We have presented a new parameterization of the SW potential which constitutes an improvement over the original model in representing all three condensed phases of silicon and germanium simultaneously. This parameterization was used to determine the interface response functions for both the crystal and amorphous phases of silicon. The results from atomic simulation agree qualitatively with a Wilson-Frenkel equation form fitted to experimental data.

## ACKNOWLEDGEMENTS

The authors thank the National Science Foundation for a KDI award (9980100). LAM and NFF thank the Cornell Center for Materials Research and the Cornell Learning Initiatives for Future Engineers program for funding undergraduate research internships.

## REFERENCES

1. M. O. Thompson, G. J. Galvin, J. W. Mayer, P. S. Peercy, J. M. Poate, D. C. Jacobson, A. G. Cullis, and N. G. Chew, Phys. Rev. Lett. **52**, 2360 (1984).
2. E. P. Donovan, F. Spaepen, D. Turnbull, J. M. Poate, and D. C. Jacobson, Appl. Phys. Lett. **42**, 698 (1983).
3. F. Spaepen and D. Turnbull, in *Laser-Solid Interactions and Laser Processing-1978* eds. S.D. Ferris, H. J. Leamy, and J. M. Poate (*American Institute of Physics Conf. Proc.* No. **50**, New York, 1979) p. 73; B.G. Bagley and H.S. Chen, ibid, p. 97.
4. P. A. Polk, A. Polman, and W. C. Sinke, Phys. Rev. B **47**, 5 (1993).
5. M. D. Kluge and J. R. Ray, Phys. Rev. B **39**, 1738 (1989).
6. Q. Yu, M. O. Thompson, and P. Clancy, Phys. Rev. B **53**, 8386 (1996).
7. L. Brambilla, L. Colombo, V. Rosato, and F. Cleri, Appl. Phys. Lett. **77**, 2337 (2000).
8. E. J. Albenze and P. Clancy, submitted to Phys. Rev. B, 2003.
9. F. H. Stillinger and T. A. Weber, Phys. Rev. B **31**, 5262 (1985).
10. M. Z. Bazant and E. Kaxiras, Phys. Rev. Lett. **77**, 4370 (1996).
11. M. Z. Bazant, E. Kaxiras, J. F. Justo, Phys. Rev. B **56**, 8542 (1997).
12. J. F. Justo, M. Z. Bazant, E. Kaxiras, V. V. Bulatov, and S. Yip, Phys. Rev. B **58**, 2539 (1998).
13. T. J. Lenosky, B. Sadigh, E. Alonso, V. V. Bulatov, T. Diaz de la Rubia, J. Kim, A. F. Voter, and J. Kress, Modeling Simul. Mater. Sci. Eng. **8**, 825 (2000).
14. Q. Yu and P. Clancy, J. Cryst. Growth **149**, 45 (1995).
15. Q. Yu and P. Clancy, Modeling Simul. Mater. Sci. Eng. **2**, 829 (1994).
16. M. J. Uttormark, in *Melting Kinetics of Small Crystalline Clusters in the Liquid by Molecular Dynamics*, Ph. D. Thesis, Cornell University, 1992.

## Advanced Lateral Crystal Growth of a-Si Thin Films by Double-Pulsed Irradiation of All Solid-State Lasers

Toshio Kudo, Koji Seike, Kazunori Yamazaki, Hirohito Komori, Sachi Yawaka[+], Shiro Hamada[+], and Cheng-Guo Jin[*]
Research & Development Center and [+]Laser System Division, Sumitomo Heavy Industries Ltd., 19 Natsushima-cho, Yokosuka-shi, Kanagawa-ken 237-8555, JAPAN
[*]ACT Center, TIC Corporation,
2-20-29 Takanawa, Minato-ku, Tokyo 108-0074, JAPAN

## ABSTRACT

A compact annealing machine with all solid-state green lasers has been developed, which has the advantage of widely adjustable solidification rate through the delay time control of two long pulses (pulse width ~100ns). Advanced lateral crystal growth (ALCG) process has been proved by the double-pulsed all solid-state laser annealing. The laser beam has a line shape 0.1mm wide and 17mm long, and the beam profile on the short axis is quasi-Gaussian (FWHM 0.1mm). Scanning the line beam along the short axis at the 86% overlapping ratio, the lateral crystal growth area of width 14μm, parallel to the long axis, is sequentially formed at the pitch of 14μm towards the scanning direction. The advanced lateral growth mechanism is easily explained as follows: (1) At the first irradiation, twin seed lines of width 4μm, parallel to the long axis, generates at a boundary between a near-complete melting region and a complete melting region. (2) At the second irradiation of scanning step 14μm, the front seed line in the scanning direction grows symmetrically toward both sides. (3) At the third irradiation of scanning step 2x14μm, the seeds laterally grow until stopped by the growing of seeds on both sides. Finally the ALCG process by the scanning line-beam technique like the current ELA enables us to produce the laterally grown Si thin-films sequentially arranging the belt-shaped texture at the pitch of 14μm. The quality of the laterally grown Si films is quite well except for the projections generated by the bump of lateral growing seeds.

## INTRODUCTION

In the roadmap of low-temperature poly-Si (LTPS) technologies [1], the high performance of thin-film transistors (TFTs) is to be demanded with the transition of TFT generations. High-quality LTPS films naturally are required for the high performance of TFTs. The current excimer laser annealing (ELA) method for crystallization, which is one of the key technologies in the LTPS-TFT process, is not able to apply in the next generation because of small Si grains of 0.5μm at maximum. In order to obtain high-quality poly-Si films, several crystallization methods by an excimer laser [2,3] and also by a reliable solid-state laser [4-6] were proposed, and are founded on controllable grain size and grain boundary position.

We propose a challenging crystallization method, appending the function of solidification rate control to the scanning line-beam technique like the current ELA. Our crystallization method, named advanced lateral crystal growth (ALCG), is based on double-pulsed laser irradiation controlling a delay time between two pulses of long duration, emitted from LD-pumped pulsed solid-state lasers. The all solid-state green lasers are expected to be of higher power in the near future. Here we describe a basic concept of the ALCG method and discuss the quality of laterally grown Si thin-films.

## CONCEPT OF ALCG

Advanced lateral crystal growth (ALCG) is the Si thin-film process of growing grains laterally by the scanning line-beam technique such as the present ELA and double-pulsed laser irradiation that is enabled to control the solidification rate. Controllable timing between long-duration pulse shots in the double-pulsed laser irradiation contributes to the generation of larger seeds and also to the more lateral growing of the seeds. The exact controlling of delay time and pulse energy in the double-pulsed laser irradiation holds the key to the ALCG process that has no need for a high performance stage, a high precision mask, and a complex optical system.

The ALCG process has two steps for lateral crystallization: to generate seeds at a steep slope of quasi-Gaussian profile on a short axis of a line beam, and to grow the seeds laterally toward both sides by the scanning of a line beam with overlapping until the growing seeds bump against each other. The temperature gradient due to the steep slope of the quasi-Gaussian beam shape produces continuously/ a partial-melting region, a near-complete melting (NCM) region, and a complete melting (CM) region with increasing temperature. On account of the multiplication effect of the temperature gradient and the double-pulsed laser irradiation, large seeds are generated at a boundary between the NCM region and the CM region. The first pulsed-laser shot forms twin-lines of seeds along a long axis of the line beam. It is expected from the simulation results shown in figure 1 that a double-pulse (DP) shot with a delay time is more effective for the generation of wide seeds than a single-pulse (SP) shot, because of its generating wider seeds more than two times. The second pulsed-laser shot at the next location scanned grows, preferentially and laterally, the front seeds in the scanning direction. The local temperature gradient due to a difference of optical absorption between the seeds and the fine texture becomes the driving force of lateral crystal growth. The frequent laser shots grow the seeds sequentially in the same manner until stopped by the growing of seeds on both sides. In the final stage the whole area scanned by the line-beam overlapped irradiation is periodically covered with the laterally grown, belt-shaped texture.

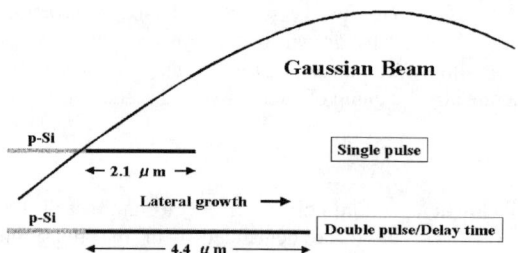

Fig.1. Simulation results of typical seeds generated by a single-pulse laser shot and a double-pulse laser shot.

## EXPERIMENTAL DETAILS

Our double-pulsed laser annealing system has been reported in detail elsewhere [7]. Before the development of the laser annealing system, the suitability of laser sources for the double-pulsed laser irradiation was confirmed with the simulation [8,9]: the influence of pulse-energy and delay-time fluctuations in grain growth and the potentiality for a single-pulse laser shot with long-duration. We roughly estimated the performance level that the two laser sources should need. The all solid-state green lasers adopted have the following performance: (i) long pulse duration of about 100ns; (ii) pulse energy stability less than 0.3 %; (iii) delay-time fluctuation less than 3ns. The high performance such as the all solid-state green lasers satisfies the level required as laser sources in the double-pulsed laser irradiation.

Amorphous Si samples of film thickness 50nm, having an underlayer of $SiO_2$ on a glass substrate, were dehydrogenated at first, and crystallized by the double-pulsed laser annealing system. The double-pulsed lasers have a line shape 0.1mm wide and 17mm long and the quasi-Gaussian profile with the full width at half maximum (0.1mm). The crystal of texture of seeds and laterally grown Si films was observed with an optical microscope and a field emission scanning electron microscope (FE-SEM). The quality, surface roughness and the degree of crystallization also was analyzed with an atomic force microscope (AFM) and Raman spectroscopy (RS) with a beam spot of diameter about 1μm. The FWHM of a Raman peak represents the ratio standardized by the Si(100) wafer.

**Seed formation**

Seeds for lateral crystal growth were formed by means of a single-pulse (SP) laser shot and a double-pulse (DP) laser shot. Figure 2 observed with the optical scope shows typical seed patterns formed by both SP and DP laser shots. Twin seed lines at a steep slope of quasi-Gaussian profile on a short axis of the line beam are formed parallel to the long axis. The width of seed lines is about 2μm for a SP laser shot and about 4μm for a DP laser shot. The values of line width come up to the expectation of the simulation as shown in Fig. 1.

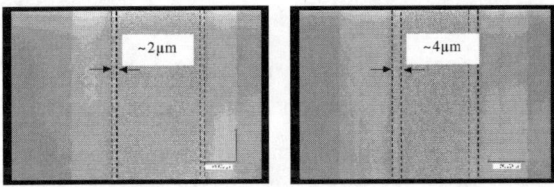

Fig.2. Optical scope images of typical seeds generated by a single-pulse laser shot and a double-pulse laser shot.

SEM images shown in figure 3 reveal which way the seeds are formed at a boundary between the NCM region and the CM region. The seeds are formed from the right side, the NCM region to the left side, the CM region. The seed formation against the temperature gradient starts from Si grains less than 1μm at the left end of the NCM region, and is blocked by Si grains generated in the CM region.

Figure 4 shows AFM images of a DP laser shot. The 2-D and 3-D images of AFM play an auxiliary role of the SEM images in Fig. 3. The left side of a seed line is the CM region made up of fine Si grains, and the right side is the NCM region of medium-size grains. The grain size in the NCM region becomes greatest at the left end. The seeds are formed from the greatest grains in the NCM region against the temperature gradient. The CM region has the mean surface roughness (Ra) of 5.3nm and the NCM region the Ra value of 5.2nm. From a practical point of view, however, the maximum surface roughness slightly increases with the increase of grain size. The mean surface roughness (Ra=2.9nm) in the formation area of seeds is very smooth compared with the CM region and the NCM region.

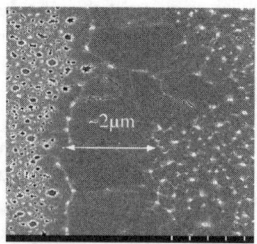 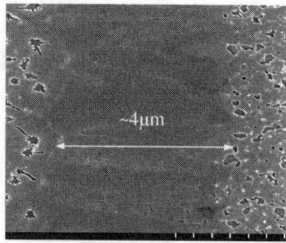

Fig.3. SEM images of typical seeds generated by a single-pulse and a double-pulse laser shot.

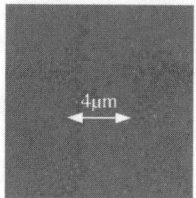 

Fig.4. 2-D and 3-D AFM images ($15 \times 15 \mu m^2$) for a double-pulse laser shot under irradiation condition: E1/E2 (800mJ/cm$^2$, 700mJ/cm$^2$), delay time (850ns), overlap ratio (-50%).

As shown in figure 5, the RS spectrum analyses of the seeds, formed by a DP laser shot, indicate as follows: (a) The seeds hold a low level of defects, deduced from the estimation of FWHM, near to the standard of the single Si crystal. (b) The crystal of texture of seeds looks homogeneous as the result of the small fluctuation of Raman peak shift and FWHM.

### Advanced lateral crystal growth

Figure 6 shows the SEM images of Si films, grown laterally by the ALCG method. The whole scanning area, overlapped at the ratio 86% by the line beam irradiation of double-pulsed lasers, is covered with the belt-shaped texture grown laterally at a pitch of 14μm. The pitch is estimated from the line-beam width 0.1mm and the overlapping ratio 86%. When the lateral-growing seeds bump against each other, the lines of pitch 14μm are formed as the traces. There is a starting line of lateral growth in the center of the belt-shaped texture 14μm wide. The lateral grain growth is proved by the SEM images enlarged at the both sides of the starting line.

AFM images in figure 7 show clearly a starting line of lateral growth and the projections generated by the bump of lateral growing seeds. On both sides of the starting line the lateral Si grain growth is symmetric and the twin seams of lateral growth exist on the way. The existence of seams in the lateral growing process gives evidence that the belt-shaped texture 14μm wide is formed by two-time double-pulsed laser shots after the generation of the seeds. The transition from the seeds to the symmetrical lateral growth, however, is still disputable. The projections are arranged with separation on the lines of pitch 14μm. The square area of 20μm x 20μm has the mean surface roughness of 3.6nm. From a practical point of view, the maximum surface roughness at the seams is 1.3 times as high as that in the most smooth lateral-growth area, the value Rmax at the starting line is 2.2 times, and the value Rmax at the projections is 5.1 times.

As shown in Fig.5, the RS spectrum analyses of Si thin-films, formed by the ALCG process, describe as follows: (a) The fluctuation of FWHM and Raman peak shift in the Si thin-films consisting of the smooth lateral-growth area and the projections is rather large than the fluctuation in the seeds. (b) The defect level of the smooth lateral-growth area is near to the single crystal, while the defect level at the projections generated by the bump of lateral growing seeds is higher than the defect levels of seeds and the smooth lateral-growth area.

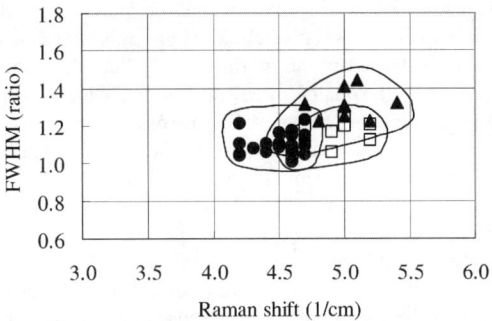

Fig.5. RS spectrum analyses of textures in Si thin-films: seeds (□), smooth lateral-growth area (●), projections (▲).

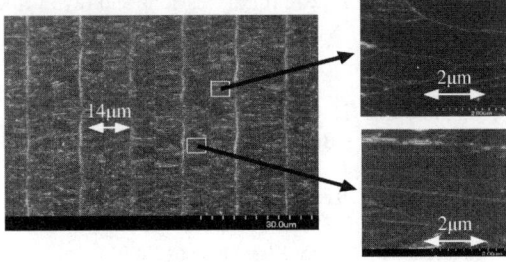

Fig.6. A SEM image and magnified images (x 9) of Si thin-films laterally grown by double-pulse irradiation: E1/E2 (800mJ/cm$^2$, 700mJ/cm$^2$), delay time (850ns), overlap ratio (86%).

## SUMMARY

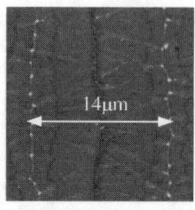 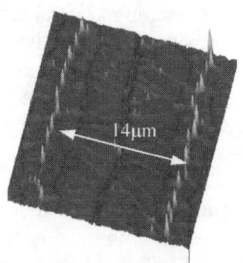

Fig.7. 2-D and 3-D AFM images (20x20μm$^2$) of Si thin-films laterally grown by double-pulse irradiation: E1/E2 (800mJ/cm$^2$, 700mJ/cm$^2$), delay time (850ns), overlap ratio (86%).

The double-pulsed laser irradiation is very useful for growing Si grains laterally and largely, especially when all solid-state green lasers used as the reliable laser sources have the excellent pulse-energy stability and pulse-timing jitter, and laser pulses emitted have the long duration above 100ns. The high potentiality is given to a double-pulsed laser shot; high-quality Si grains above length 4μm are formed by the optimum irradiation. The excessive overlapping irradiation, however, becomes the quality of Si grains worse with increasing the ratio.

The advanced lateral crystal growth (ALCG) process by the scanning line-beam technique (line-beam size: 0.1mm x 17mm, overlap ratio: 86%) like the current ELA, enables us to produce the laterally grown Si thin-films sequentially arranging the belt-shaped texture at the pitch of 14μm. The quality of the laterally grown Si thin-films, evaluated by surface roughness and the degree of crystallization, is quite well except for the projections generated by the bump of lateral growing seeds. The maximum surface roughness of the projections is about 5 times as large as that of the standard smooth surface. It is essential for the ALCG process to improve the surface roughness due to the bump of lateral growing seeds, because the surface roughness leads to gate-insulator leakage current [10].

## ACKNOWLEDGMENTS
We would like to thank Dr. T. Akashi for discussions and review of the manuscript.

## REFERENCES

1. S. Uchikoga, MRS Bulletin **27**, 881 (2002).
2. J. S. Im and R. S. Sposili, MRS Bulletin **21**, 39 (1996).
3. M. Matsumura, Proc. Eurodisplay'99 (1999) p.351.
4. A. Hara and N. Sasaki, IEDM. Tech. Digest **943**, 301(1999).
5. Y. Helen et al, Thin Solid Films **383**, 143 (2001).
6. M. Hatano, T. Shiba and M. Ohkura, SID. Digest (2002) p.158.
7. K. Yamazaki et al, AM-LCD2002 Digest (2002) p.149.
8. T. Kudo, D. Ichishima and C-G. Jin, AM-LCD2000 Digest (2000) p.125.
9. T. Kudo (private communication)
10. T. Fujimura et al, AM-LCD2001 Digest (2001) p.175.

## Influence of Laser Annealing on Hydrogen Bonding in Disordered Silicon Thin Films

N. H. Nickel and K. Brendel
Hahn-Meitner-Institut Berlin
Kekuléstr. 5, D-12489 Berlin, Germany.

## ABSTRACT

The influence of laser dehydrogenation and crystallization of hydrogenated amorphous silicon ($a$-Si:H) on H bonding is investigated. Depending on the deposition temperature the amorphous starting material contains a H concentration of up to 44 at.%. Laser crystallization lowers the H content significantly. Fully crystallized poly-Si contains H concentrations of up to 17 at.%. This reservoir of hydrogen can be used to passivate additional grain boundary defects by annealing the specimens at low temperatures in vacuum. Information on hydrogen bonding is obtained from hydrogen effusion measurements.

## INTRODUCTION

Polycrystalline silicon (poly-Si) produced by laser crystallization of amorphous silicon is of great interest for device applications such as thin-film transistors and solar cells. Commonly, amorphous silicon with a low hydrogen concentration is used for laser crystallization because the presence of large amounts of H causes difficulties during the crystallization process. In order to obtain large grained material a laser energy density exceeding the melt-through threshold must be applied. This, however, releases large amounts of hydrogen in a very short time and the $a$-Si film ablates. A low temperature furnace anneal for several hours can prevent the destruction of the $a$-Si film during laser crystallization [1]. However, a low temperature furnace anneal is not practical for the fabrication of hybrid devices where amorphous and polycrystalline silicon based devices are placed next to each other on the same substrate. Furthermore, lowering the H content using a conventional furnace anneal is a very time-consuming processing step.

While $a$-Si with a low H content can be crystallized with a single laser pulse a major drawback of the resulting poly-Si is the fact that grain-boundary defects have to be passivated with hydrogen to obtain device-grade material. On the other hand, hydrogenated amorphous silicon typically contains 8 – 15 at.% hydrogen. To avoid explosive out-diffusion of H during laser crystallization and thus, destruction of the silicon film a dehydrogenation and crystallization procedure is employed [2,3]. This technique takes advantage of the fact that the amount of H evolving from the sample increases with increasing laser fluence. The resulting poly-Si samples contain a residual hydrogen concentration of about 5 at.% [3].

In this paper, we investigate the influence of the hydrogen content in the amorphous starting material on hydrogen bonding and defect passivation in laser crystallized poly-Si. After laser dehydrogenation and crystallization the specimens contain a residual H concentration of $8\times10^{21}$ cm$^{-3}$ to $1.5\times10^{22}$ cm$^{-3}$. From H effusion measurements the H density of states distribution is derived that exhibits up to four peaks. Interestingly, the deposition temperature at which the amorphous starting material is deposited affects the H density-of-states distribution.

## EXPERIMENTAL DETAILS

Hydrogenated amorphous silicon films were deposited on quartz substrates by plasma enhanced chemical vapor decomposition of silane. To vary the hydrogen content the $a$-Si:H films were grown at substrate temperatures between 100 and 335 °C. The samples had a thickness of 0.13 to 0.2 µm. Laser crystallization was performed at room temperature using a XeCl excimer laser ($\lambda$ = 308 nm) by employing a step-by-step crystallization scheme. Details are described elsewhere [2,3]. For all scans a shot density of 100 per unit area was used. Information on hydrogen bonding of the amorphous starting material and of completely crystallized poly-Si was obtained from hydrogen effusion experiments. In these experiments the samples were heated with a heating rate of 20 K/min in ultra high vacuum while the $H_2$ flux was measured with a mass spectrometer. Information on the grain boundary defects was obtained from electron paramagnetic resonance (EPR) measurements.

## RESULTS AND DISCUSSION

The large amount of residual hydrogen in fully crystallized poly-Si [2] can be used to passivate additional silicon-bond defects. The as-crystallized poly-Si films reveal a spin density of $N_S \approx 3\text{-}6\times10^{18}$ cm$^{-3}$; the deposition temperature of the a-Si:H does not influence $N_S$ significantly. A subsequent low-temperature vacuum anneal results in a further decrease of $N_S$. Fig. 1 shows the concentration of additionally passivated silicon dangling bonds, $\Delta N_S$, as a function of the annealing temperature $T_A$ for poly-Si samples where the amorphous starting

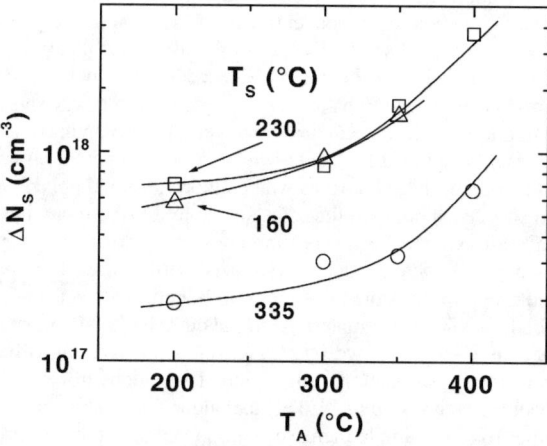

Fig. 1. Total change of the spin density, $\Delta N_S$, vs. post-annealing temperature $T_A$. Annealing was performed in vacuum. The amorphous silicon layers were deposited at the indicated substrate temperatures, $T_S$, and subsequently crystallized using a XeCl excimer laser.

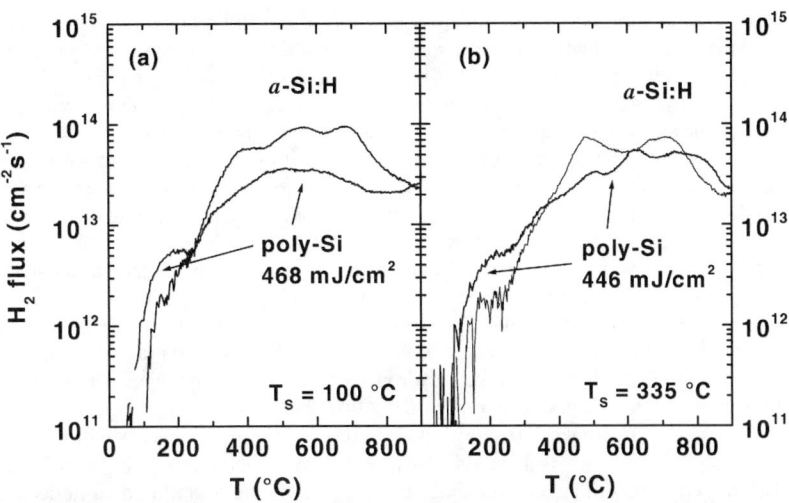

Fig. 2. Hydrogen effusion spectra before ($a$-Si:H) and after laser crystallization (poly-Si). (a) and (b) represent samples where the starting amorphous silicon layer was deposited at 100 and 335 °C, respectively. The final laser fluences used to crystallize the specimens are indicated in the figure.

material was grown at the indicated temperature $T_S$. All samples were annealed until the spin density reached a minimum value. As the annealing temperature increases, the number of passivated grain boundary defects increases. The temperature dependence of $\Delta N_S$ is independent of the substrate temperature at which the amorphous silicon layers were deposited. However, $T_S$ has an effect on the concentration of passivated dangling bonds at a given annealing temperature, $T_A$. $\Delta N_S$ increases by about a factor of 6 for poly-Si where the amorphous starting material is deposited at $T_S \leq 230$ °C (squares and triangles in Fig. 1) compared to poly-Si with $T_S = 335$ °C (circles in Fig. 1). This can be related to a reduction of the total hydrogen concentration in the amorphous starting material and/or a significant change in hydrogen bonding. Commonly, it is believed that an increase of the deposition temperature of hydrogenated amorphous silicon results in a change in H bonding. Since the data plotted in Fig. 1 where taken on fully crystallized poly-Si one might expect that laser crystallization preserves these changes.

Insight into hydrogen diffusion and hydrogen bonding was obtained from H effusion measurements. In Fig. 2 the molecular hydrogen flux is plotted as a function of annealing temperature for hydrogenated amorphous silicon and laser crystallized poly-Si samples. Comparing the spectra measured on the amorphous starting material shows that the sample deposited at $T_S = 100$ °C contains a vast amount of H that migrates out of the film at temperatures below 400 °C. The corresponding effusion spectrum reveals three peaks located at approximately 390, 558, and 683 °C. On the other hand, the $a$-Si:H film deposited at 335 °C shows only two peaks in the molecular hydrogen flux that are located near 470 and 720 °C. Laser crystallization of the amorphous samples alters the effusion spectra significantly for low and high deposition temperature of the $a$-Si:H film. At temperatures exceeding 300 °C the $H_2$ flux decreases significantly. Moreover, independent of $T_S$ of the amorphous starting material, laser annealing redistributes hydrogen in the specimens such that a low temperature peak emerges in the effusion spectra in the temperature range around 200 °C.

From the hydrogen effusion spectra, the total amount of hydrogen can be obtained by integrating the spectra, taking into account the heating rate and the sample thickness. Completely crystallized poly-Si contains a hydrogen concentration of up to 17 at.%. A fraction of the residual hydrogen content can be activated in a low temperature anneal to passivate additional silicon dangling-bonds at grain boundaries (Fig. 1). The residual H concentration in poly-Si depends on sample thickness and deposition temperature of the initial amorphous silicon layer. Knowledge of the sample thickness is important since step-by-step crystallization produces a depth dependent hydrogen distribution. Near the surface the hydrogen concentration is about a factor of 5 - 6 lower than at the interface to the substrate [4]. Initially, the crystallization procedure produces a stratified structure with a poly-Si layer close to the surface and amorphous silicon at the interface to the substrate. Hydrogen diffusion measurements performed on such samples show that the H flux decreases by more than one order of magnitude at temperatures below 500 °C. The hydrogen flux recovers when the top of poly-Si layer is removed by chemical wet-etching [5]. The H flux is a given by by $F = D_{eff} \times S / l$ where $D_{eff}$ is the effective diffusion coefficient, $S$ the solubility, and $l$ the thickness. Essentially the solubility does not change during the laser dehydrogenation procedure until a considerable amount of H diffuses out of the sample. Hence, the decrease of $F$ must be due to a decrease of $D_{eff}$. This translates into an increase of the diffusion activation energy that is given by $E_A = -kT \ln(D_{eff}/D_0)$, where $k$ is the Boltzmann constant, $T$ the temperature, and $D_0$ the microscopic diffusion prefactor. Changes in $D_{eff}$ and hence, $E_A$ can only happen when the H density-of-states distribution is modified.

From the hydrogen diffusion data the hydrogen density-of-states, $N_H$, can be determined by measuring the hydrogen concentration as a function of the hydrogen chemical-potential, $\mu_H$. The hydrogen density-of-states is derived using the relation [6]

$$N_H(\mu_H) = \frac{\partial C_H(\mu_H)}{\partial \mu_H} , \qquad (1)$$

where $C_H$ is the hydrogen concentration. The position of the hydrogen chemical-potential is estimated from the $H_2$ flux

$$F = F_0 \exp\left(-\frac{E^* - \mu_H}{kT}\right) , \qquad (2)$$

with the hydrogen transport sites located at $E^*$. $k$ is Boltzmann's constant, and $T$ is the temperature of the effusion experiment. The prefactor is given by $F_0 \approx 2a\nu N_{surf}/d$ [6]. With a mean free path of a $\approx 3\times10^{-8}$ nm, an attempt frequency of $\nu \approx 10^{13}$ s$^{-1}$, sample thickness $d$ and a density of surface states of $N_{surf} \approx 10^{15}$ cm$^{-2}$ the prefactor can be estimated to $F_0 \approx (1/d) \, 6\times10^{20}$ cm$^{-2}$s$^{-1}$.

This analysis was applied to the hydrogen diffusion data shown in Fig. 2 and the derived hydrogen density-of-states distributions are summarized in Fig. 3. The hydrogenated amorphous silicon samples exhibit three peaks in the hydrogen density-of-states distribution for substrate

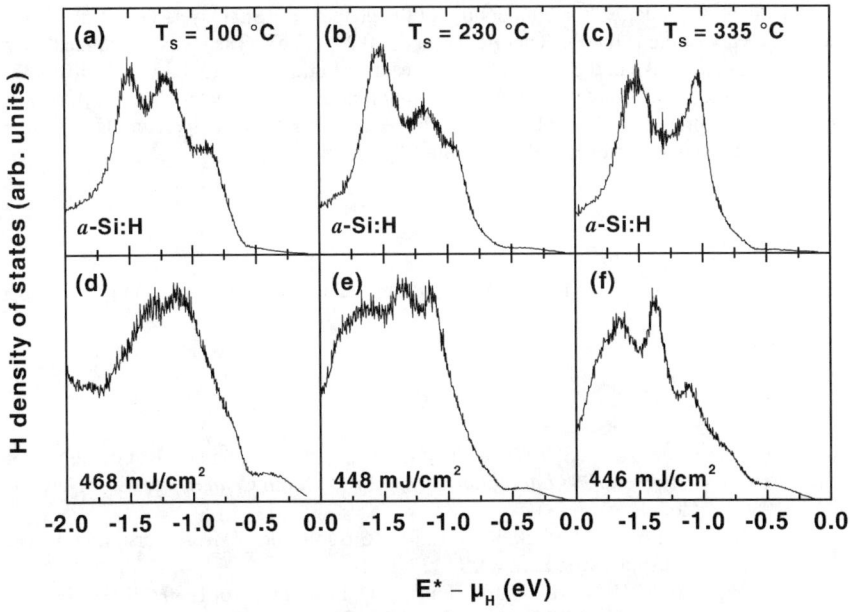

Fig. 3. Hydrogen density-of-states distribution as a function of the H chemical potential, $E^*-\mu_H$, for amorphous (a, b, c) and laser crystallized polycrystalline silicon (d, e, f). $T_S$ denotes the deposition temperature of the $a$-Si:H layers.

temperatures $T_S \leq 230$ °C. On the other hand, samples deposited at 335 °C reveal only two peaks at about $E^* - \mu_H = -1.0$ and $-1.5$ eV; the intermediate peak disappeared.

Laser crystallization of the amorphous silicon layers results in a shift of the hydrogen density-of-states distribution to larger binding energies by approximately 0.2 – 0.3 eV. This increase in H binding energy is independent of the initial deposition temperature. The sample prepared from low temperature amorphous silicon reveals 2 peaks in the H density-of-states that are located at $-1.1$ eV and $-1.35$ eV. These peaks are also observed in poly-Si that was prepared from $a$-Si:H deposited at higher temperatures. With increasing deposition temperature the peaks become more pronounced and a third peak emerges at about $-1.65$ eV. Although an assignment of the peaks to specific H complexes is difficult we believe that a vast amount of H is accommodated in platelet-like structures due to the observed large binding energies.

## Summary

In summary, we have shown that a fraction of the residual hydrogen content of poly-Si can be activated to passivate additional Si dangling bonds. Although Raman backscattering is not sensitive to H local vibrational modes in laser crystallized poly-Si information on H bonding and the H density-of-states distribution was obtained from hydrogen effusion measurements.

Independent of the deposition temperature of the precursor $a$-Si:H layers laser annealing results in an increase of the H binding energies by about 0.2 – 0.3 eV. The laser crystallized poly-Si samples reveals peaks in the H density-of-states distribution at −1.1, -1.35, and −1.65 eV. An increase of the deposition temperature of the amorphous starting material shows that these peaks become more pronounced. In addition with increasing deposition temperature H is preferentially accommodated in complexes with larger binding energies.

**Acknowledgment**

The authors are indebted to W. B. Jackson for valuable discussions and to H. Schöppe for technical support.

**References**

[1] M. Yuki, K. Masumo, S. Takafuji, T. Asakawa, N. Imajyo, and M. Kumigita, in *Proceedings of the 1988 International Display Research Conference*, IEEE New York, 1988, p. 220.
[2] P. Mei, J. B. Boyce, M. Hack, R. A. Lujan, R. I. Johnson, G. B. Anderson, D. K. Fork, and S. E. Ready, Appl. Phys. Lett. **64,** 1132 (1994).
[3] P. Lengsfeld, N. H. Nickel, and W. Fuhs, App. Phys. Lett. **76,** 1680 (2000).
[4] P. Lengsfeld and N. H. Nickel, in *Laser Crystallization of Silicon*, edited by N. H. Nickel (Academic Press, San Diego, to appear 2003).
[5] H. Heise and N. H. Nickel, J. Non-Cryst. Sol. **299-302,** 226 (2002).
[6] W. B. Jackson, A. J. Franz, H.-C. Jin, J. R. Abelson, and J. L. Gland, J. Non-Cryst. Sol. **227-230,** 143 (1998).

# METAL CONTAINING LINK FORMED IN AMORPHOUS SILICON METAL-TO-METAL ANTIFUSE

Frank Hawley, Farid Issaq, Jeewika Ranaweera, Roy Lambertson, John McCollum
Actel Corp., 955 E. Arques Ave., Sunnyvale, Ca. 94086, USA

## ABSTRACT

A Metal-to-Metal (M2M) antifuse is formed using hydrogenated amorphous silicon as a dielectric material between two refractory metal electrodes. The M2M antifuse is used as a programmable device in an FPGA, where it is placed between two interconnect metal layers of a Logic CMOS process. The M2M device may then be programmed to interconnect logic circuits. The resulting programmed link is an alloy of amorphous silicon and barrier metal that forms a low resistance path between the logic circuits.

## INTRODUCTION

A Metal-to-Metal (M2M) antifuse is formed using hydrogenated amorphous silicon plus a thin dielectric film as the antifuse material between two metal electrodes [1]. The metal electrodes can be typical barrier metals used in the standard CMOS processing, such as Titanium Nitride (TiN) or Tungsten (W). The M2M antifuse is used as a programmable device in an FPGA (Field Programmable Gate Array), where it is placed between two interconnecting metal layers of a Logic CMOS process. See Fig 1a and Fig 1b, It is then programmed by the users to connect their required logic circuits [2,3,4,5].

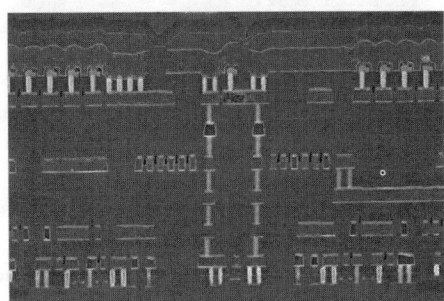

Figure 1a 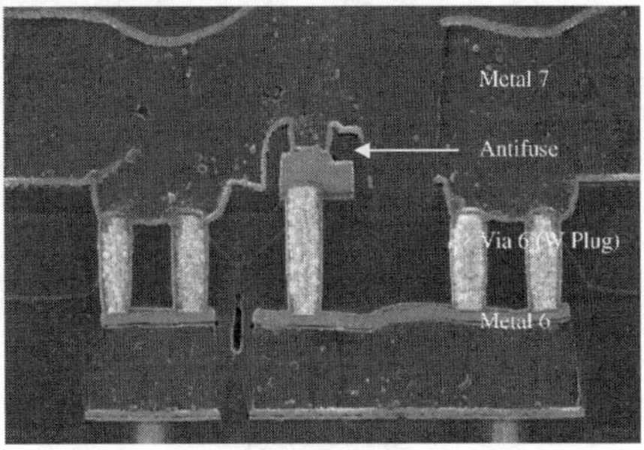            Figure 1b

**Figure 1a:** SEM of an FPGA product, using a 7-layer metal, 0.15um CMOS process, with the M2M Antifuse between Metal 6 and Metal 7.
**Figure 1b:** Higher magnification of the Metal 6, the M2M Antifuse and Metal 7.

The Unprogrammed M2M antifuse has a very low leakage of <0.1pA at Vcc, the FPGA operating voltage bias (see Fig 2), and has a very small capacitance of <1fF per antifuse cell. The low leakage of the M2M antifuse enables very low standby power for the FPGA. The low capacitance of the M2M antifuse results in a very small load for the FPGA circuit, enabling maximum performance. Referring to Fig 2, the IV curve is shown for an unprogrammed M2M antifuse. As the voltage is ramped across the antifuse the leakage increases to the point of breakdown. The breakdown voltage is approximately 1MV/cm. This is the voltage where the antifuse becomes programmed.

The Programmed M2M antifuse has a low resistance value with an average of between 20 to 250 ohms depending on the Ipp (applied programming current). See Fig 3. The large drop in resistance is due to the formation of a metal link between the electrodes.

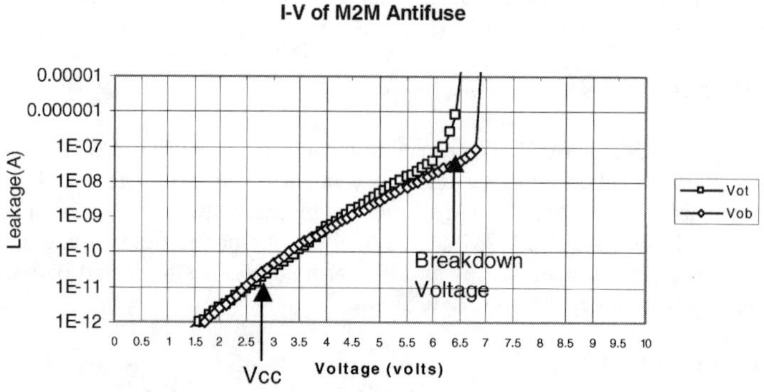

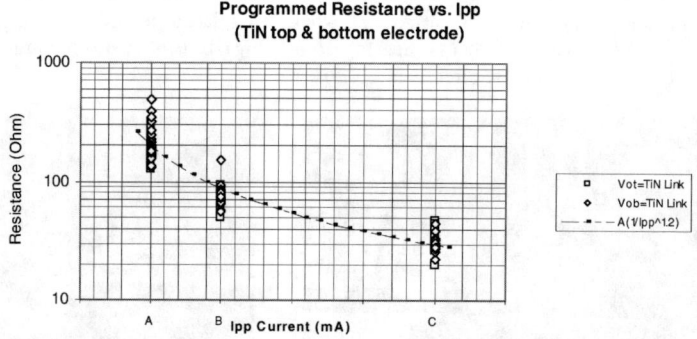

**Figure 2:** IV Curve of M2M Antifuse, leakage is dependent on voltage. The breakdown occurs at approximately 1MV/cm.

**Figure 3:** Plot of the Programmed Resistance vs. Ipp for TiN top and bottom electrode. Higher Ipp has a lower resistance link

The hydrogenated amorphous silicon is deposited at 350 degrees C in a PECVD system compatible with the aluminum metalization process. The amorphous silicon source is Silane ($SiH_4$) with Argon dilution.

## PHYSICAL STUDY

The Programmed M2M antifuse resistance and link material have been studied. Several CMOS Logic technology generations are included in the evaluation, ranging from 0.5-micron down to 0.15-micron feature size. Results are dependent on the electrode and the structure and not the specific technology.

Two different M2M antifuse structures were studied. Case one is with a TiN top and W bottom electrode and case two is with a TiN top and bottom electrode. To confirm the source of the link metal material the direction of the programming was reversed for both case structures. First, the resistance was compared between the Vot (positive voltage on the top electrode) and the Vob (positive voltage on the bottom electrode). Then, TEM analysis and EDS analysis were done to characterize the programmed link.

## RESULTS

Referring to Fig 4, the case one structure, TiN top and W bottom electrode resistance is plotted. Programming with Vot produces higher resistance than with Vob. This is in contrast when compared to the second case structure with TiN as top and bottom electrode (see Fig 3), which shows that Vot and Vob resistance are approximately equal.

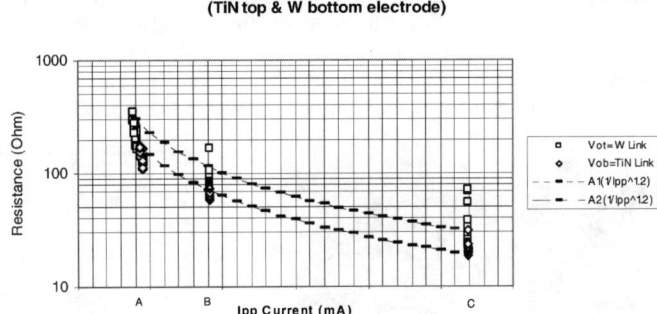

**Figure 4:** The M2M Programmed Resistance vs. Ipp for TiN top and W bottom electrode. Vot resistance is higher than Vob resistance.

In all cases the programmed link resistance is dependent on the Ipp and is generally described by Equation (1) where $R_{link}$ is the programmed link resistance (ohms), $I_{pp}$ is the programming current (amperes) and n=1.2.

$$R_{link} \alpha \, I_{pp}^{-n} \qquad (1)$$

TEM and SEM analysis of the Vot and Vob link confirms that the source material for the link is from the cathode electrode and the metal is pulled into the link by the electron wind, as previously described [6]. Referring to Fig 5, the TEM for the case two, TiN top and bottom structure, with Vot programming shows the link source is the bottom metal, which is the cathode. The metal source of the link is considered to be the electrode that shows the large melt depth merged with the link, from which the metal is pulled into the amorphous silicon and the link is formed. For the case one, TiN top and W bottom structure, the Vot link is W metal based and the Vob link is TiN metal. Thus, the difference in resistance for the TiN top and W bottom structure can then be explained due to the fact that the link material is different for each. The resistance for case two, TiN top and bottom structure, both Vot and Vob, are approximately equal. This is due to the fact that the link is TiN for both programming directions.

The EDS (Energy Dispersive Spectrometer) analysis on the TEM cross-sections finds that the link contains a large amount of Ti metal mixed with the amorphous silicon antifuse material. The Fig 5 EDS spectrum has two major peaks recorded, the first is Si and the second is Ti. The relative peak height indicates the respective concentration of the elements detected. In this case the Ti:Si ratio is 1.8. The location of this measurement is the center of the link and is shown in the Fig 5 TEM as the white cross. Fig 6 shows a representation of the case two, TiN top and bottom structure, programmed link, top to bottom and left to right with the link metal concentration shown as a Ti:Si ratio. Using the link Center data as reference the results clearly show that for a higher Ipp the metal Ti:Si ratio in the link is higher. The Ti:Si ratio varies from a ratio of 0.7 for Ipp=A up to a ratio of 1.8 for an Ipp=C, where Ipp of A<B<C, all in the mA range. The left, right, top and bottom are not measured in the same exact relative locations for each link. The variations of the Ti:Si ratio for these locations is due to the fact that these measured locations are near the border of the link. The left and right locations are near the edge of the link and the surrounding amorphous silicon material. The top and bottom are at the point the electrode merges into the link. The emphasis from this data is: The left and right show lower Ti:Si ratios for Ipp = B and C indicating that the metal is concentrated at the center of the link and the width of the link is limited in the lateral direction. The top and bottom data indicates that the electrode has a much higher metal Ti:Si ratio than the link center.

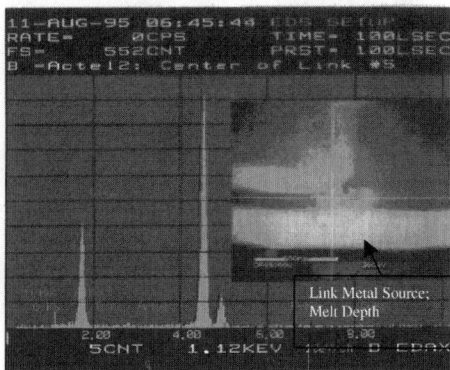

**Figure 5**

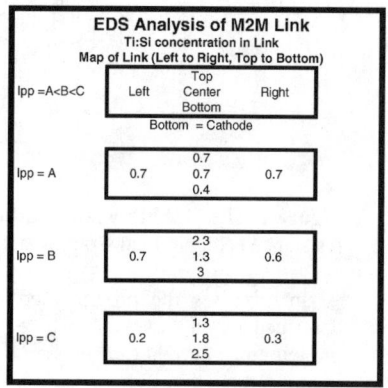

**Figure 6**

**Figure 5:** EDS and TEM of the case two M2M Link programmed Vot; Source metal for link is Cathode. The white cross in the TEM is the location of the EDS spectrum results for the link center.

**Figure 6:** Map of case two antifuse link (top to bottom and left to right). Ti to Si ratio is dependent on Ipp.

The difference in link resistance relative to Ipp can be explained in that the higher Ipp has a higher metal Ti:Si ratio in the link and thus a lower resistance, this can be seen in the Fig 7 plot. From TEM analysis it was found that the link melt depth into the cathode was dependent on the Ipp, the higher the Ipp the deeper the link was melted into the cathode. In Fig 8 the link metal Ti:Si ratio is correlated to the link melt depth into the cathode. The deeper the links melt depth the higher the metal Ti:Si ratio. The explanation for link melt depth correlation to metal Ti:Si ratio is that the source of the Ti metal pulled into the link comes from the cathode and in fact the deeper depth is the source of the higher Ti concentration in the link. Further, as a higher Ipp is applied the link becomes hotter and melts deeper into the cathode.

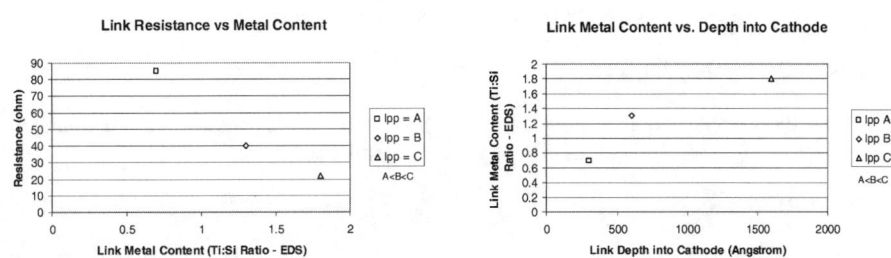

**Figure 7**

**Figure 8**

**Figure 7:** Link Resistance vs. Metal Content (Ti:Si ratio). Higher Ti:Si ratio has lower resistance.

**Figure 8:** Link Metal Content vs. Depth into Cathode. Higher Metal Content (T:Si ratio) has a deeper melt into cathode.

When the link resistance is plotted relative to the link melt depth there is a nice correlation showing that a deeper link has a lower resistance, shown in Fig 9. Finally, using the link depth, amorphous silicon thickness, and typical width from TEM data the link volume is plotted and found to have a reasonable correlation to the Ipp, where higher Ipp has a larger link volume, see Fig 10.

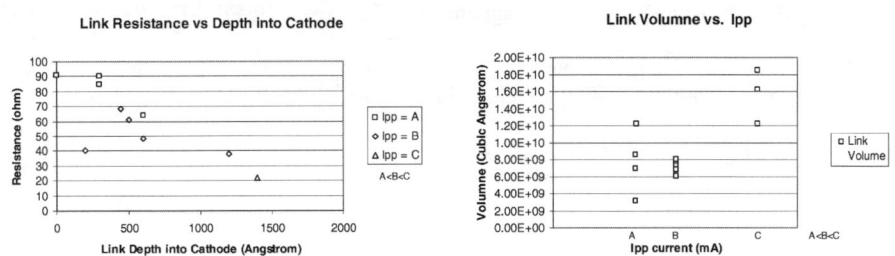

**Figure 9**

**Figure 10**

**Figure 9:** Link Resistance vs. Depth into Cathode. A deeper link depth has a lower resistance.
**Figure 10:** Link Volume vs. Ipp. The higher Ipp has a larger link volume.

## CONCLUSIONS

The programmed M2M antifuse link contains a metal silicon alloy. The metal source for the link is the cathode electrode. The link is a mixture of metal and amorphous silicon with the metal concentration (Ti:Si ratio) increasing as the Ipp increases. The link resistance is dependent on Ipp. The higher the Ipp the lower the measured resistance is due to the higher metal content in the link. The higher metal content in the link comes from a deeper melt depth into the cathode due to higher Ipp where higher Ipp generates more heat and more melting. The link resistance is also dependent on the electrode material with a TiN electrode having a slightly lower resistance than a W electrode. The metal concentration is highest at the link center and is lower further away from the center, horizontally. A higher Ipp creates a larger volume link.

## ACKNOWLEDGEMENTS

The authors would like to thank Chih-Ching Shih for his assistance with programming antifuse devices and for the TEM and EDS analysis and helpful discussions.

## REFERENCES

[1] C. Shih, R. Lambertson, F. Hawley, F. Issaq, J. McCollum, E. Hamdy, H. Sakurai, H. Yuasa, H. Honda, T. Yamaoka, T. Wada, C. Hu, Characterization and Modeling of a Highly Reliable Metal-to-Metal Antifuse for High Performance and High-Density Field-Programmable Gate Arrays, IEEE IRPS Proceedings, pg 25-33 (1997)
[2] E.Hamdy, J. McCollum, S. Chen, S. Chiang, S. Eltoukhy, J. Chang, T. Speers, A. Mohsen, Dielectric Based Antifuse for Logic and Memory IC's, IEEE IEDM Technical Digest, pg 786-789 (1988)
[3] C. Hu, (invited) Interconnect Devices for Field Programmable Gate Arrays, IEEE IEDM Technical Digest, pg 591-594 (1992)
[4] S.Chiang, R. Forouhi, W.Chen, F. Hawley, J. McCollum, E. Hamdy, C. Hu, Antifuse Structure Comparison for Field Programmable Gate Arrays, IEEE IEDM Technical Digest, pg 611-614 (1992)
[5] M. Takagi, I. Yoshii, N. Ikeda, H. Yasuda, K. Hama, A Highly Reliable Metal-to-Metal Antifuse for High-Speed Field Programmable Gate Arrays, IEEE IEDM Technical Digest, pg 31-34 (1993)
[6] K. Gordon, J. Wong, Conducting Filament of the Programmed Metal Electrode Amorphous Silicon Antifuse, IEEE IEDM Technical Digest, pg 27-30 (1993)

# Direct Electrical Characterization of Metal Induced Lateral Crystallization Regions by Spreading Resistance Probe Measurements

Alexandre M. Myasnikov[1], Vincent M.C. Poon[2], Vincent T.C. Leung[2], Mansun Chan[2], and Lawrence C.F. Cheng[2]
[1]Institute of Semiconductors Physics, Novosibirsk, Russia
[2]Department of Electrical and Electronic Engineering, Hong Kong University of Science and Technology, Hong Kong

*Abstract*-- Material characterization of metal induced lateral crystallization (MILC) process of amorphous silicon (a-Si) has been performed by using the spreading resistance probe (SRP) measurements. It was found that carrier mobility in boron ion implanted layer, formed in MILC region, is up to 65 % in comparison with mobility in boron ion implanted layer, formed in single crystalline silicon. It was also observed in this work that prolongation of MILC process from 1 hour to 2 hours had induced the increasing of mobility from 24 $cm^2/Vs$ to 34 $cm^2/Vs$.

## 1. INTRODUCTION

Films formed by metal induced lateral crystallization (MILC) of a-Si have attracted considerable interest as material for TFT, SOI and other applications [1-4]. However, almost all electrical characterization of these films was carried out by the creation of devices in films and the attempts to analyze the characteristics of these devices [1, 3-5]. But, it is common knowledge that the characteristics of the devices are excellent with the availability of standard technology. When we have no reliable technology, then using the formation of the device is barrier for process control, because the device process flow is very long, and the parameters of devices are integral amounts depending on many processes.

At the moment MILC films consist of separated recrystallized regions [2, 3, 5], and standard methods of electrical characterization are unacceptable for the study of areas of the size of about some tens micrometers. Even the characterization of continuous MILC films from a set of metal seeds will be troublesome due to varying properties dependent on the distance from the seeds.

As it appears to us that the solution for the electrical characterization of MILC films lies in the use of spreading resistance probe (SRP) measurements. During the past 30 years the SRP technique has become a well-established method for the measurement of electrical characteristics of different layers [6]. SRP method has gained wider credibility, since an error of less than 10% (standard deviation) is routinely achievable for typical results irrespective of where, when, or by whom the measurement is performed.

## 2. EXPERIMENTS

A layer of 3000 Å a-Si was deposited on a silicon substrate oxidized up to 7000 Å. On a-Si layer a low-temperature oxide (LTO) was deposited at 450 °C with thickness about 3000 Å and in this layer windows were formed as contacts at nickel deposition. 50 A thickness nickel island seeds were used with different patterns. The temperature of recrystallization was kept at 625 °C for 1 and 2 hours. The sizes of the MILC regions were 14 micrometers for 1 hour and 28 micrometers for 2 hours. After recrystallization and removing the LTO and nickel layer, a silicon film was ion implanted at a dose $10^{15}$ cm$^{-2}$ and an energy 40 keV with B$^+$ ions. Activation of ion implanted boron was carried out at 1100 °C for 10 s by rapid thermal annealing (RTA).

Using the same processes we prepared the standard sample, which was a single crystalline wafer of n-type with carrier concentration about $10^{14}$ cm$^{-3}$. This standard sample was used for calibration at SRP measurement and for recalculation of the SRP data to mobility.

Then the MILC regions were measured with a step of about 2.5 μm by standard spreading resistance profiler by Solid State Measurements, Inc. at the distance between probes 100 μm. The profile lengths of SRP measurement on the surface were changed depending on the MILC size and carried out perpendicular to the nickel/MILC regions.

3. RESULTS AND DISCUSSION

The SRP measurements for two different sizes of MILC regions are shown in fig.1 and 2 in alignment with the microscope photos with the curves. In these figures we also show the mobility distributions. Table 1 contains the summary of the SRP data, which were measured and calculated for different measurements and for different regions.

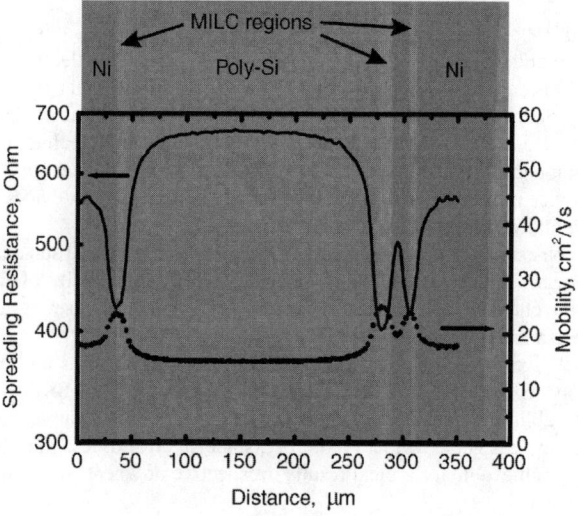

Fig.1. Alignment of top view photo and the spreading resistance (line) and mobility (dots) curves in dependence on distance. The sample was crystallized at 625 °C for 1 hour and additionally annealed at 1100 °C for 10 s.

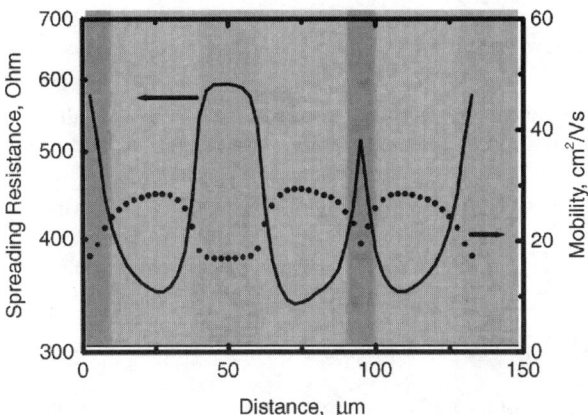

Fig.2. Alignment of top view photo and the spreading resistance (line) and mobility (dots) curves in dependence on distance. The sample was crystallized at 625 °C for 2 hours and additionally annealed at 1100 °C for 10 s.

In order to establish the application of SRP measurements for MILC characterization, some explanations are needed.

In the first, a uniform ion implantation of boron was carried out in the whole surface of wafer with MILC, the non-recrystallized regions, and also in the regions which were initially under nickel islands in the process of crystallization ("nickel" regions). Consequently we think that any non-uniformities on surface are related to the non-uniformities of different regions and to the behaviors in these regions.

In the second, RTA at 1100 °C for 10 s resulted in a full activation of the implanted boron without redistribution. It means that the carrier concentration in the doped layer is equal to the amount of implanted boron atoms and has the depth distribution of as-implanted layer. Thus, for a standard sample of single crystalline silicon we should have the defined distribution of carrier with defined mobility. As is known from [7], for our dose and energy the carrier concentration has a Gaussian distribution with an average concentration about $3 \times 10^{19}$ cm$^{-3}$ and a mobility about $\mu_{scs} = 51.3$ cm$^2$/Vs, which is almost constant along the depth.

At this high temperature annealing the regions of a-Si, which were non-recrystallized at 625 °C temperature, were turned to polycrystalline silicon (poly-Si) regions with measurable resistance, however, the visible change of MILC regions size was not found.

The third assumption is related to edge effect at SRP measurements.

For semi-infinite uniformly doped samples with resistivity $\rho$ the spreading resistance R of a non-penetrating circular (radius a) Ohmic probe contact is known from theory [8] and given by

$R = CF \times \rho/2a$

Table 1.
Summary of the data of SRP measurements.

| Sample/Region On Surface | Heat Treatment | R, Ohm | $\mu$, cm$^2$/Vs | $\mu/\mu_{Scs}$ |
|---|---|---|---|---|
| Single Crystalline Silicon | 1100 °C, 10 s | 194 - 196 | 51.3 | 1 |
| A-Si/MILC | 625 °C, 1hour + 1100 °C, 10 s | 400 - 424 | 23.6 - 25 | 0.46 - 0.49 |
| A-Si/Poly-Si | 625 °C, 1hour + 1100 °C, 10 s | 661 - 670 | 14.9 - 15.1 | 0.29 |
| A-Si/"Nickel" | 625 °C, 1hour + 1100 °C, 10 s | 459 - 565 | 17.6 - 21.8 | 0.34 - 0.42 |
| A-Si/MILC | 625 °C, 2hours + 1100 °C, 10 s | 298 - 363 | 27.5 - 33.6 | 0.54 - 0.65 |
| A-Si/Poly-Si | 625 °C, 2hours + 1100 °C, 10 s | 469 - 593 | 16.9 - 21.3 | 0.32 - 0.42 |
| A-Si/"Nickel" | 625 °C, 2hours + 1100 °C, 10 s | 347 - 515 | 19.4 - 28.8 | 0.38 - 0.56 |

In our case, the electrical properties of the probe – semiconductor contact depend on many different physical factors, which were included in a correction factor (CF). The correction factor CF [9], which corrects the measured resistance data for the current spreading effect, equals 1 for a uniformly doped semi-infinite layer.

When the sample has contact with the isolator, the CF increases if the probe is moved towards the boundary, and reaches 2 for the limiting case when the boundary is positioned exactly under the probe center. If the isolating boundary is replaced by a perfectly conducting one, the correction factor CF becomes smaller than 1 and decreases towards zero when the probe touches the conducting boundary.

For our measurements, when we have boundaries between MILC and poly-Si regions and between MILC and "nickel" regions, we have an intermediate case of boundary between two different conductors with comparable resistances. In this case CF for MILC regions is varied in the range from 1.06 to 1.07 [9].

Figure 1 shows the dependence of the spreading resistance with distance for the sample of a-Si, which was crystallized at 625 °C for 1 hour. It is clearly seen that for motion of probes across of MILC region the spreading resistance drops from 565 Ohm to 424 Ohm. Removing the probes to the poly-Si region the spreading resistance firstly raises to 670 Ohm, which is almost constant along all poly-Si region, then decreases to 402 Ohm in MILC region. In going from MILC region to the "nickel" region the value of the spreading resistance increases to 505 Ohm, and then drops again in MILC region to 421 Ohm. Finally, the motion to "nickel" region results in the increasing to 567 Ohm.

For the recalculation of spreading resistance data in mobility we have used the spreading resistance, which was obtained on the standard sample that was prepared together with the sample with MILC regions. For the standard single crystalline silicon the spreading resistance was equal 195 Ohm with an accuracy of about 1%. As well as the mobility in standard single crystalline silicon sample was of about 51.3 cm$^2$/Vs [7] and it is keeping in mind that the carrier concentration is independent on the regions, then by using equation

$$\mu = 1/e\rho n,$$

where $\mu$ - mobility, e - charge of electron, $\rho$ - resistivity, and n - carrier concentration, we can relate the spreading resistance on standard sample with mobility and use this relation, including correction factor CF of edge effect, to recalculate the values of spreading resistance to mobility of carriers in MILC, poly-Si and "nickel" regions.

Thus, a resistance curve in fig.1 was recalculated with mobility data depicted at the same plot.

In fig.2 the similar curves of spreading resistance and mobility are shown for the sample recrystallized for 2 hours at 625 °C. Also we have the higher spreading resistance at "nickel" and poly-Si regions and the lower at MILC regions.

It needs to emphasize that on the sample recrystallized for 2 hours the spreading resistances are lower than on the sample recrystallized for 1 hours as for MILC regions as for poly-Si regions, i.e. the mobility of carriers at 2-hour recrystallization is higher than the one at 1-hour recrystallization. As is seen in Table 1, which was summarized the data from different SRP measurements for different regions and at different conditions, the 2 hour crystallization results in the increasing of mobility in MILC regions up to 40 % or from 24 $cm^2/Vs$ to 34 $cm^2/Vs$ in comparison with 1 hour recrystallization and in this case mobility in MILC regions may be 65 % of mobility in doped layer on single crystalline silicon.

## 4. CONCLUSION

Thus, we have demonstrated the application of spreading resistance probe measurements for characterization of recrystallization process of a-Si. It was shown that carrier mobility in boron ion implanted layer, formed in MILC region, is up to 65 % in comparison with mobility in boron ion implanted layer, formed in single crystalline silicon. Also it was shown that for different conditions mobility can change from 24 $cm^2/Vs$ to 34 $cm^2/Vs$ depending on time of recrystallization.

## 5. ACKNOWLEDGEMENTS

The authors thank W.Y.Chan and C.W.Lung for sample preparation and measurements.

## 6. REFERENCE

1. Z.Meng, M.Wang, M.Wong, "High-performance low-temperature metal-induced unilaterally crystallized polycrystalline silicon thin film transistors for system-on-panel applications", IEEE Transactions on Electron Devices, vol.47, no.2, p.404-409, 2000.
2. S.Y.Yoon, S.J.Park, K.H.Kim, J.Jang, C.O.Kim, "Structural and electrical properties of polycrystalline silicon produced by low-temperature Ni silicide mediated crystallization of the amorphous phase", Journal of Applied Physics, vol.87, no.1, pp.609-611, 2000.
3. T.-K.Kim, G.-B.Kim, B.-I.Lee, S.-K.Joo, "The effect of Electrical Stress and temperature on the properties of polycrystalline silicon thin-film transistors fabricated by metal induced lateral crystallization", IEEE Electron Device Letters, vol.21, no.7, pp.347-349, 2000.

4. H.M.Wang, M.Chan, S.Jagar, M.C.Poon, M.Qin, Y.Y.Wang, P.K.Ko, "Super Thin-Film Transistor with SOI CMOS Performance Formed by a Novel Grain Enhancement Method", IEEE Transactions on Electron Device, vol.47, no.8, pp.1580-1586, 2000.
5. M.Wong, Z.Jin, G.A.Bhat, P.C.Wong, H.S.Kwok, "Characterization of the MIC/MILC interface and its effects on the performance of MILC thin-film transistors", IEEE Transactions on Electron Devices, vol.47, no.5, pp.1061-1067, 2000.
6. T.Clarysse, P.De Wolf, H.Bender, W.Vandervorst, "Recent insight into the physical modeling of the spreading resistance point contact", Journal of Vacuum Science and Technology, vol.B14, no.1, pp.358-368, 1996.
7. H.Ryssel, I.Ruge, "Ion implantation", Wiley, 1986.
8. T.Clarysse, W.Vandervorst, "Need to incorporate the real micro-contact distribution in spreading resistance correction schemes", Journal of Vacuum Science and Technology, vol.B18, no.1, pp.393-400, 2000.
9. P.De Wolf, T.Clarysse, W.Vandervorst, "Qualification of nanospreading resistance profiling data", Journal of Vacuum Science and Technology, vol.B16, no.1, pp.320-326, 1998.

## Effect of SiO$_2$ Capping Layer on a Laser Crystallization of a-Si Thin Film

Myung-Koo Kang, Hyun Jae Kim, Sook Young Kang, Su-Kyung Lee, Chi-Woo Kim and Kyuha Chung
LTPS Group, Flat Panel Display R&D Team, Samsung Electronics Co.,
Yongin-City, Gyeonggi-Do, Korea 449-711

## ABSTRACT

Effect of SiO$_2$ capping layer(C/L) on recrystallization of amorphous Si (a-Si) film was investigated. When a thick C/L over 500 Å was deposited on an a-Si film before crystallization, fine p-Si grains less than 100nm were obtained at full range of energy density window. However, when a thin C/L below 200 Å was used, Si-melt spouted out through C/L at over critical energy density. When Si-melt started spouting, abrupt change of grain size also occurred. These large grains could be explained by a non-uniformity of heat flow caused by Si-melt spouting. With this polycrystalline Si (p-Si) material having appreciable grain size protected by C/L, fabrication of low-cost Low Temperature Poly Si (LTPS) without additional cleaning of p-Si surface could be successfully developed.

## INTRODUCTION

Polycrystalline-silicon (p-Si) thin-film transistors (TFTs) are used in a variety of applications, including large-area electronics [1] and vertically stackable components for three-dimensional integration [2]. P-Si is typically fabricated from amorphous Si (a-Si) thin-film deposited on an inexpensive glass, such as Corning 1737, which has a quoted working range below 600°C. Various recrystallization technologies of a-Si film have been developed to meet this temperature limitation of glass. Among various recrystallization technologies, the excimer laser method has been commonly adopted in commercial fields. In this method, a pulsed laser is used for irradiating on an a-Si precursor film. It is absorbed by a film surface and induces instant melting and solidification. The short pulse duration of the laser beam prevents sustained heating of the underlying substrate, and makes the process compatible with glass substrate [3]. The crystallinity is superior to that of other crystallization techniques such as SPC (Solid Phase Crystallization) because the transformation from a-Si to p-Si is occurred via melting and re-solidification process.

In the mean time, the property of p-Si TFTs is mainly dependent on grain boundaries within channel and a surface morphology. To minimize the number of grain boundaries within the channel area, new technologies such as SLS [4] (Sequential Lateral Solidification) or MILC

(Metal Induced Lateral Crystallization), which can make large lateral grains, have been developed. Grain boundary passivation technique [5] has been also developed to decrease a boundary trap density. However, it is more difficult to improve the surface structure because a surface morphology after laser crystallization is hard to control [6].

For making this rough surface smooth and LTPS process simpler, usage of $SiO_2$ capping layer (C/L) has been tried [7]. $SiO_2$ film transmits laser beam and can act as a gate insulator. If stable $SiO_2$ C/L exists before and after crystallization, cleaning step of p-Si surface could be skipped. However, there have been not many studies on this topic because grain size was too small to be used. Viatella [8] et al explained that a small grain size was due to a heat absorption by C/L. For a C/L process to be used successfully in production, appreciable grain size should be obtained.

In this paper, effect of C/L with various thicknesses on the recrystallization of a-Si was studied. When optimum thickness of C/L is used, Si-melt spouts out through $SiO_2$, and large grains could be obtained at relatively lower energy density for a less number of shots. With this architecture, successful development of low cost LTPS fabrication process without an additional cleaning step of p-Si surface could be developed.

**EXPERIMENTAL DETAILS**

$SiO_2$ blocking layer (B/L) deposited glass was used as a substrate. The role of B/L is to protect the impurities from a glass substrate during following processes. 500Å-thick a-Si precursor was deposited on the B/L. Dehydrogenation was carried out to decrease hydrogen content inside a film. Upon a-Si precursor, various thickness of $SiO_2$ C/L was deposited. Sequential deposition from B/L to C/L and dehydrogenation could be possible without a vacuum break in the system.

XeCl excimer laser with a wavelength of 308nm was irradiated on this film. After irradiation, an extensive microstructural analysis was carried out using SEM and TEM. For a composition analysis, EDX analysis was also performed using HRTEM. With p-Si capped by C/L after crystallization, TFTs were fabricated following typical LTPS process sequence; active patterning, gate oxide deposition, gate patterning, inter insulating layer deposition, contact and data metal patterning. Before gate oxide deposition, an additional cleaning step of p-Si surface was skipped.

## DISCUSSION

Fig. 1 shows microstructures of p-Si of small grains with various C/L thicknesses. As is shown in figure 1(a), relatively thick C/L (1000 Å) suppressed grain growth. The grain size is less than 100nm at over the whole laser energy density window of our system (up to 450mJ/cm2). This grain size is too small to be used for a high performance application. Viatella [8] et al explained that a C/L could act as a heat sink and suppress a grain growth. From their work and our results, grains with appreciable size cannot be expected with thick C/L. When thin C/L was used and laser energy density was relatively low, the microstructure was also composed of very fine grains as is shown in figure 1(b).

However abrupt change of grain size occurred when a laser energy density is over critical value (in our experimental condition, it was 300mJ/cm2 for thin C/L deposited sample, as is shown in figure 2(a). There are two noticeable features in this microstructure. One is that large grain could be obtained at relatively less energy density for small number of shots. These kinds of large grains (~1μm) shown in figure 2(a) can be obtained at over 400mJ/cm2 for 20 shots when C/L isn't used. But it could be obtained only at 330mJ/cm2 for 7 shots when C/L was used.

Another noticeable feature of Fig 2(a) is that there are some protrusions on the surface that is indicated by arrows. These protrusions can be seen clearly by SEM observation of an un-etched sample that is tilted as is shown in Fig 2(b). As can be seen, many protrusions are formed on the surface of which shape is similar to droplet. From the point that an appearance of large grains and protrusion formation started at the same energy density, it can be easily speculated that the formation of protrusion induced grain growth.

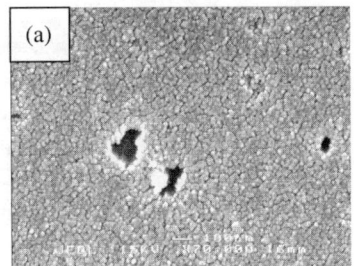
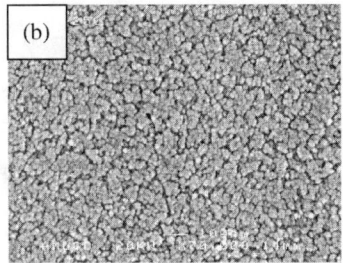

Figure 1. Fine p-Si microstructure with different C/L thickness and laser energy density; (a) C/L thickness is 1000 recrystallized at 380mJ/cm2 and (b) C/L thickness is 150 recrystallized at 250mJ/cm2.

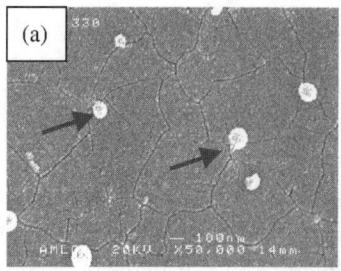 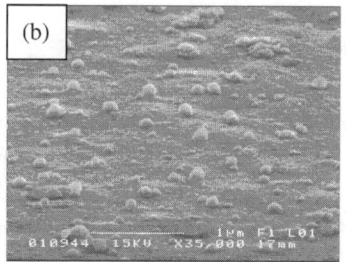

Figure 2. Microstructures of 150Å C/L deposited a-Si crystallized at 330mJ/cm2; (a) top view of Secco etched sample and (b) tilted view of non-etched sample.

For the composition of protrusions, composition analysis was performed using HRTEM. Figure 3(a) is a low-magnification TEM microstructure showing that a droplet-like protrusion on $SiO_2$ C/L. Even though C/L thickness was relatively thin (150Å), it is observed to be stable on a p-Si surface after crystallization. Figure 3(b) is a high magnification TEM microstructure indicating crystalline structure of protrusion. As can be seen, the protrusion was composed of many nano-size Si crystalline phases; arrows indicate some of their directions. The EDS composition analysis also showed that a main composition is Si.

From the TEM results, it can be known that Si-melts spouted out from Si layer during crystallization. These protrusions can induce non-homogeneous heat flow in the surface region, which can affect solidification process. Large Si grains can be easily formed in an environment where there is a thermal gradient. For a thin C/L deposited samples, this kind of thermal gradient could be possible because of non-homogeneous heat flow by Si-melt spouting.

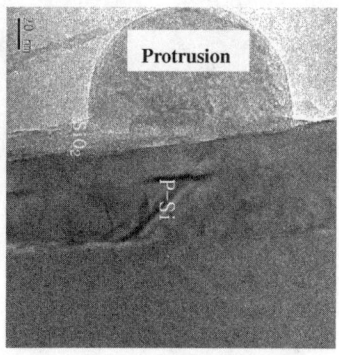 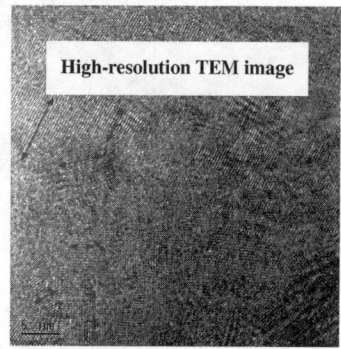

Figure 3. HRTEM microstructure of protrusion; (a) low magnification and (b) high magnification.

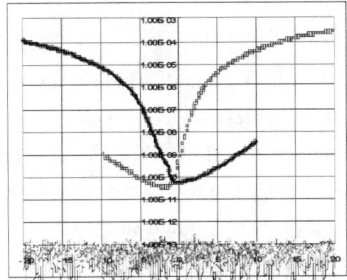 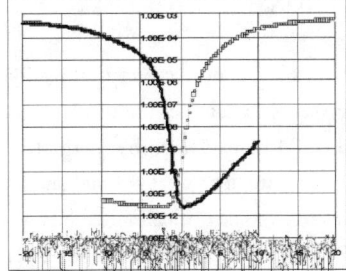

Figure 4. TFT transfer characteristics fabricated without additional cleaning of p-Si surface after crystallization for; (a) typical process and (b) C/L process proposed by this work.

Table 1. TFT performance summary of figure 4

|  | Mobility (cm2/Vs) | | Vth (V) | |
| --- | --- | --- | --- | --- |
|  | n | p | n | P |
| Typical process w/o cleaning | 28.1 | 18.7 | 5.3 | -9.3 |
| C/L process w/o cleaning | 156 | 53.5 | 0.47 | -4.1 |

With this large-grained p-Si capped by $SiO_2$, fabrication of low cost LTPS process was successfully developed. P-Si surface tends to be contaminated easily in a typical LTPS process, where the vacuum break is inevitable between crystallization and gate insulator deposition process. Because of that, an additional cleaning step is essential in a typical LTPS process. However if our proposed architecture is used, the interface between p-Si and gate insulator is protected until a gate insulator is deposited. This means that an additional cleaning step before gate insulator deposition can be skipped.

Figure 4 shows TFT characteristics without an additional cleaning before gate insulator deposition for typical LTPS process and C/L process proposed by this study. As can be seen in figure 4(a), a TFT characteristic is degraded significantly for the typical LTPS process without an additional cleaning because of the interface contamination. On the other hand, no degradation can be found in our architecture as is shown in figure 4 (b). The skip of an additional cleaning can make it possible to decrease a fabrication cost and non-uniformity caused by wet cleaning contamination. A specific TFT performance for both cases is summarized in table 1.

## CONCLUSIONS

The effect of C/L on the recrystallization of a-Si film was studied. For thick C/L (>500Å), grain size was so small over the whole energy range. This result is correspondent with Viatella's work and could be explained by thermal flow from l-Si to C/L. If C/L thickness becomes thin (<200 Å), Si protrusions were formed and an abrupt change of grain size occurred when the laser energy was over some critical value. Below that energy density, grain size was quite small. This abrupt change of grain size could be explained by non-homogeneous heat flow caused by Si protrusion formation. With this large grained p-Si protected by $SiO_2$ C/L, low cost LTPS fabrication process without an additional cleaning step could be successfully developed.

## REFERENCES

1. T. J. King, M. G. Hack and I. W. Wu, J. Appl. Phys., 75, 908 (1994)
2. B. Faughnan and A. C. Ipri, IEEE Trans. Electron Devices, 36, 101 (1989)
3. James S. Im and Robert S. Sposili, Mater. Res. Bull., 21, 39 (1996)
4. James S. Im, Robert S. Sposili and M. A. Crowder, Appl. Phys. Lett., 70, 3434 (1997)
5. Yeh-Jiun Tung, James Boyce, Jackson Ho, Xuejue Huang and Tsu-Jae King, IEEE Electron Device Letters, 20, 387 (1999)
6. Mutsumi Kimura, Tsukasa Eguchi, Satoshi Inoue and Tatsuya Shimoda, Jpn. J. Appl. Phys., 39, L775 (2000)
7. C. K. Chen, M. W. Geis, M. C. Finn and B-Y. Tsaur, Appl. Phys. Lett., 48, 1300 (1986)
8. J. Viatella, R. K. Singh, Mater. Sci. and Eng., B47, 78 (1997)

## Laser interference structuring of a-GeN for the production of optical diffraction gratings

M. Mulato[1], A. R. Zanatta[2], D. Toet[3,*], and I. E. Chambouleyron[4]
[1] Departamento de Física e Matemática, Faculdade de Filosofia Ciências e Letras de Ribeirão Preto, Universidade de São Paulo, Av. Bandeirantes 3900, Ribeirão Preto, SP, Brazil
[2] Instituto de Física de São Carlos, Universidade de São Paulo, São Carlos, SP, Brazil
[3] FlexICs Inc.,165 Topaz Street, Milpitas, CA, 95035, USA
[4] Instituto de Física Gleb Wataghin, Universidade Estadual de Campinas - Unicamp, Campinas, SP, Brazil
* Present address: Photon Dynamics, Inc., 17 Great Oaks Blvd., San Jose CA 95117, USA

## ABSTRACT

In this work, we study the pulsed laser crystallization of hydrogen-free amorphous germanium-nitrogen alloys (a-GeN). We discuss the role of nitrogen during phase transitions and the possible application of the resulting structure as an optical diffraction grating. The crystallized region results of pure microcrystalline germanium ($\mu$c-Ge). An indication that Ge-N bonds have broken and nitrogen outdiffused of the film is obtained from infrared spectroscopy and confirmed by Raman spectra. A pattern of alternating a-GeN and $\mu$c-Ge lines with a period of about 4 $\mu$m acts as an optical diffraction grating due to the difference in optical properties between the two materials, and the three dimensional surface profile, caused by $N_2$ effusion, that is formed on the sample.

## INTRODUCTION

In the last decade a growing interest in laser processing technologies was observed at the semiconductor industry [1-19]. Amorphous semiconductors have been crystallized by treatments with short laser pulses. This process has attracted a lot of attention since it enables the fabrication of high performance devices base on polycrystalline materials that are produced on low temperature substrates, e.g. for flat panel display applications [9-11]. The same process could be applied in the near future for the development of new generation X-ray medical imaging systems. Experimental and theoretical studies have shown that laser crystallization involves ultra-fast melting and solidification processes occurring far from thermal equilibrium [12-17]. In addition to that, when more than one laser beam is used for the crystallization process, new interesting phenomena take place also. Bringing two laser beams to interference on the surface of an amorphous film results in a sinusoidal modulation of the light intensity. This modulation of the light intensity leads as a consequence to a modulation of the temperature of the sample. The last one thus controls the selective crystallization process. As a result, a pattern consisting of alternating amorphous and polycrystalline lines (dots are obtained when 3 beams are used) is obtained. In summary, the technique seems to be very promising for controlled grain growth, and reduced lithographic processes in industrial applications, among others.

This technique was first demonstrated on hydrogen-free amorphous silicon (a-Si) in 1994 [18], and was soon applied to other materials as well. For instance, hydrogen free amorphous germanium (a-Ge) films were investigated three years latter, in 1997 [12,15]. As it is well known, amorphous silicon and germanium films typically contain more than 10 at. % hydrogen when grown by Plasma Enhanced Chemical Vapor Deposition (PECVD). This is undesirable for laser crystallization, since the rapid heating caused by the absorption of the laser radiation results in explosive effusion of the hydrogen. This leads to disruption of the film surface, causing roughness, and, in the case of a-Ge:H, the formation of a free standing film [19]. To overcome this problem, either another deposition technique such as sputtering has to be used, or a controlled hydrogen effusion by oven or low-fluency laser anneal is necessary. Besides the main amorphous networks and its alloys with hydrogen, few other alloys have been studied in recent years.

It is the objective of the present authors to report some data about the influence of nitrogen on the laser crystallization of amorphous nitrogenated alloys. In this specific contribution we focus mainly on germanium alloys with nitrogen. In other words, we study the pulsed laser crystallization of hydrogen-free amorphous germanium-nitrogen alloys (a-GeN). We discuss the role of nitrogen during phase transitions and the possible application of the resulting structure as an optical diffraction grating, when two laser beams are used for the interference structuring of the sample. Results related to silicon-nitrogen alloys will be further presented in another contribution.

## EXPERIMENTAL DETAILS

The samples of the present work were produced by the rf sputtering technique, using a germanium target in an Ar+$N_2$ atmosphere [20]. The nitrogen partial pressure during deposition was 7 x$10^{-3}$ mbar and the total pressure 15 x$10^{-3}$ mbar. The DC bias was 640V. Corning 7059 glass and c-Si bar substrates were used, and held at 230 $^0$C during deposition. Typical samples are 0.5 µm thick. The nitrogen concentration in the alloy is about 30% [20].

The spectral transmittance [21] of an a-GeN sample (solid line) is compared to that of the glass substrate and that of a typical a-Ge:H sample in Figure 1. The higher optical gap of a-GeN is apparent from the figure [20]. It is worth to point out that other physical properties of a-GeN, as well as those of a-GeN:H alloys can be found in the literature [20,22-25].

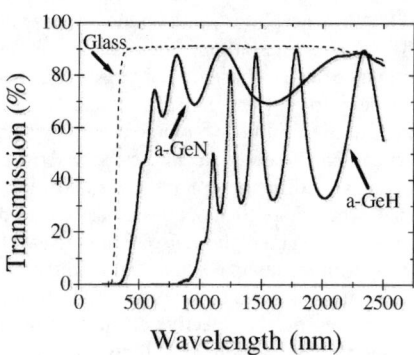

**Figure 1** - *Optical transmission spectra of glass, a-GeN and a-Ge:H samples.*

For the laser treatment of the samples we used a single shot of either one or two simultaneous 10ns laser pulses (150mJ/cm$^2$) of the third harmonic (355nm) of a Nd-YAG laser. For the production of the gratings, the laser output was split into two beams, which were then brought to interference at the surface of the sample. Time resolved experiments were not performed for monitoring the laser processes, which were performed at room temperature in air.

After the processing the samples were investigated using Infrared Spectroscopy, Raman Scattering, and Atomic Force Microscopy (AFM). In addition to that the fabricated amorphous and microcrystalline gratings were used as optical diffraction gratings using a low power HeNe laser as discussed in the next section.

## DISCUSSION

Figure 2 shows the relative intensity of the absorption band corresponding to the asymmetric Ge-N stretching vibration mode of an as-deposited a-GeN sample (solid line). The dashed line in the same figure indicates the strength of the same absorption band after laser irradiation. It is apparent from Fig. 2 that, after crystallization, the total number of Ge-N bonds has decreased in the sample. The area ratio indicates a loss of around 23%. This result suggests that nitrogen has either effused from the sample or is trapped inside the germanium matrix as $N_2$ molecules. As discussed next, nitrogen effusion seems to be the most plausible effect, in a way similar to the explosive effusion of hydrogen in Si:H (Ge:H) alloys upon laser crystallization.

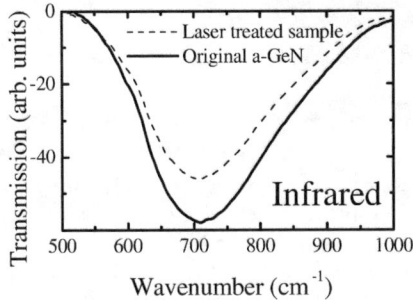

***Figure 2*** - *Infrared Ge-N stretching absorption band of an a-GeN film.*
*solid line: as-deposited sample; dashed line: laser treated sample.*

Figure 3 shows the results of Raman backscattering experiments performed using the 488 nm wavelength of an Argon laser. The figure displays the spectra of a control c-Ge (111 oriented) sample, an as-deposited a-GeN film, and a laser crystallized a-GeN sample, respectively as solid line, circles and squares. The difference between the control c-Ge and the as-deposited a-GeN sample, which does not show any peak, can be clearly seen in Figure 3. The c-Ge spectrum shows a peak corresponding to the TO-phonon, at 300 cm$^{-1}$. A similar peak also appears in the laser crystallized samples. However, the crystallized samples present a broader signal and a Raman shift smaller than c-Ge. This indicates that the laser treated sample consist of a distribution of small crystallites rather than a mono-crystalline Ge film.

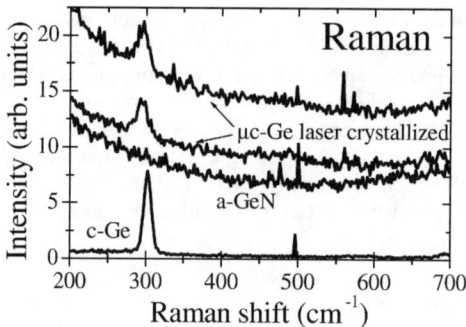

**Figure 3** - *Raman spectra of a reference c-Ge (111) ( bottom curve); as-deposited a-GeN (second from bottom to top); and laser crystallized a-GeN samples (two top curves).*

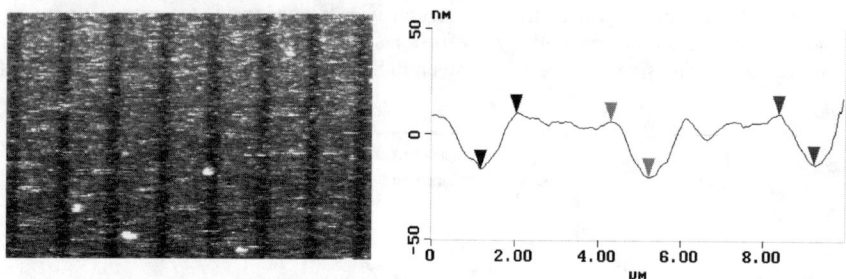

**Figure 4** - *Atomic force microscopy (AFM) measurements of the line pattern of an a-GeN laser irradiated film using two interfering beams. Left: surface profile for a length of 30 µm; right: vertical profile corresponding to a horizontal scanning along the top image. The hills correspond to a-GeN and the valleys to µc-Ge. Note the height variation of about 25 nm between the valley and the top of the hills. The line period is about 4µm.*

The left part of Figure 4 shows a two dimensional surface profile, measured by AFM, of an a-GeN film exposed to 2 interfering beams. The left figure corresponds to a surface image of the sample with a horizontal scale of 30 µm. The right part of Figure 4 corresponds to a horizontal scanning along the surface, thus leading to a vertical profile that helps analyzing the three dimensional nature of the sample. The darker lines in the left part of Figure 4 correspond to the laser crystallized part of the sample, resulting in a µc-Ge strip, while the clear (and broader) lines are unaffected by the laser and remain amorphous GeN. The white circular dots appearing in the left part of Figure 4 correspond to defects of the original sample and are not related to the laser treatment.

The profile shown in the right part of Figure 4, which corresponds to a horizontal scan across the lines, indicates that the crystallized portion of the film is lower than the amorphous part by about 25 nm. This can be explained by a partial ablation of the Ge film caused by explosive laser-induced nitrogen effusion, in analogy to what occurs in the case of hydrogen in a-Si:H. Note that no formation of a free standing membrane was observed in the present case, contrary to what was found in hydrogenated Ge samples. We believe that the laser irradiated lines are not crystallized through the whole thickness of the sample. Instead, only a thin surface layer of the sample might be converted to µc-Ge, whereas the bottommost part of the same lines still remains a-GeN. More investigations are needed to confirm this hypothesis.

The particular structure of the sample shown in Figure 4, that has a line period of around 4µm with a line width of about 1 µm, can be used as an optical diffraction grating because it combines two effects: the different surface optical properties of the two materials (µc-Ge and a-GeN) and a three dimensional profile of the lines.

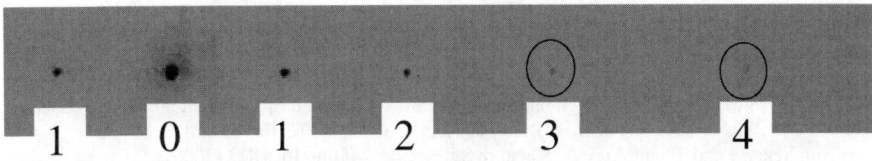

*Figure 5* - *Digital photograph of the diffraction pattern of a He-Ne laser beam ($\lambda$=633nm) with a circular spot. The snap-shot of the diffracted beams was taken with the grating at a distance of 20.5cm from the sample. The numbers identify the diffraction maxima (m). The distances from m = 0 are 3.4, 7.0, 11.3, 17.3 and 27.8 cm for m=1, 2, 3, 4 and 5, respectively. A grating period of 3.92 ±0.03µm is thus obtained.*

In order to check this possibility, we used a He-Ne laser beam (633nm) with a circular profile, and measured the diffracted pattern at a distance of 20.5 cm from the sample. The resulting diffraction pattern was photographed and is shown in Figure 5. The distance between the maxima was used to calculate the diffracted angle $\theta$, and used in the equation $d \sin\theta = m \lambda$, where d is the grating period, $\lambda$ is the wavelength of the incident light, and m is the number of the diffraction maximum, to obtain d. From the experiment, the resulting grating period was d = 3.92 ± 0.03 µm, in agreement with the AFM measurements.

## CONCLUSIONS

In summary, the present letter reports on the consequences of nitrogen effusion in laser processed a-GeN alloys, which behaves in a way similar to that of hydrogen in laser irradiated Si and Ge hydrogenated alloys. As a consequence of nitrogen effusion, the crystallized surface of a-GeN results in a pure µc-Ge region of a reduced thickness. This thickness difference and the different optical properties of the crystallized and the amorphous regions was found to behave as an efficient optical diffraction grating, with a line period of around 4 µm under the present experimental conditions. Further experiments based on the analysis of the integrated intensity of the interference maxima are needed to distinguish which is the main physical effect.

## ACKNOWLEDGEMENTS

The authors are indebted to Dr. Paulo Santos for introducing then to the laser processing technique almost seven years ago, also to their colleagues for fruitful discussions, and to E. de Paula for experimental help. This work has been supported by the Brazilian agencies FAPESP and CNPq.

## REFERENCES

[1] C. W. Lin, C. H. Tseng, T. K. Chang, C. W. Lin, W. T. Wang, H. C. Cheng, IEEE Electron Device Letters **23**, 133 (2002).
[2] Y. Hatanaka, M. Niraula, A. Nakamura and T. Aoki, Applied Surface Science **175**, 462 (2001).
[3] D. Toet, P. M. Smith, T. W. Sigmon, M. O. Thompson, Applied Physics Letters **77**, 307 (2000).
[4] J. P. Lu, P. Mei, R. T. Fulks, J. Rahn, J. Ho, Y. Wang, J. B. Boyce and R. A. Street, Journal of Vaccum Science and Technology A- Vacuum Surfaces and Films **18**, 1823 (2000).
[5] R. S. Sposili and J. S. Im, Applied Physics A- Materials Science and Processing **67**, 273 (1998).
[6] J. S. Im, R. S. Sposili and M. A. Crowder, Applied Physics Letters **70**, 3434 (1997).
[7] R. S. Sposili and J. S. Im, Applied Physics Letters **69**, 2864 (1996).
[8] T. Sameshima, Applied Surface Science **96-8**, 352 (1996).
[9] D. P. Gosain, A. Machida, S. Usui and M. Arai, Polycrystalline Semiconductors IV Materials, Technologies and Large Area Electronics **80-81**, 169 (2001).
[10] D. P. Gosain, T. Noguchi and S. Usui, Japanese Journal of Applied Physics Part 2 - Letters **39**, L179 (2000).
[11] P. G. Carey, P. M. Smith, S.D. Theiss and P. Wickboldt, Journal of Vacuum Science and Technology A - Vacuum Surfaces and Films **17**, 1946 (1999).
[12] M. Mulato, D. Toet, G. Aichmayr, P.V. Santos, and I. Chambouleyron, Appl. Phys. Lett. **70**, 3570 (1997).
[13] G. Aichmayr, D. Toet, M. Mulato, P.V. Santos, A. Spangenberg, S. Christiansen, M. Albrecht, and H.P. Strunk, phys. stat. sol. (a) **166**, 659 (1998).
[14] G. Aichmayr, D. Toet, M. Mulato, P.V. Santos, A. Spangenberg, and R. B. Bergman, J. of Non-Cryst. Solids **227-230**, 921 (1998).
[15] M. Mulato, D. Toet, G. Aichmayr, A. Spangenberg, P.V. Santos, and I. Chambouleyron, J. of Non-Cryst. Solids **227-230**, 930 (1998).
[16] G. Aichmayr, D. Toet, M. Mulato, P.V. Santos, A. Spangenberg, S. Christiansen and M. Albrecht, J. Appl. Phys. **85**, 4010 (1999).
[17] V. V. Gupta, H. J. Song and J. S. Im, Applied Physics Letters **71**, 99 (1997).
[18] M. Heintze, P.V. Santos, C.E. Nebel and M. Stutzmann, Appl Phys. Lett. **64**, 3148 (1994).
[19] M. Mulato, D. Toet, G. Aichmayr, P.V. Santos and I. Chambouleyron, J. Appl. Phys. **82**, 5159 (1997).
[20] A. R. Zanatta and I. Chambouleyron, Phys. Rev. B **48**, 4560 (1993).
[21] R. Swanepoel, J. Phys. E: Sci Instrum **16**, 1214 (1983).
[22] F.C. Marques, I. Chambouleyron and F. Evangelisti, J. of Non-Cryst. Solids **114**, 561 (1989).
[23] C. Guanchua and Z. Fangqing, This Solid Films **185**, 231 (1990).
[24] F.C.Marques, R.G. Lacerda, M.M. Lima Jr. and J. Vilcarromero, phys. stat. sol(b) **192**, 549 (1995).
[25] A. R. Zanatta, I. Chambouleyron and P.V. Santos, J. Appl. Phys. **79**, 1 (1996).

# Observation and Annealing of Incomplete Recrystallized Junction Defects due to the Excimer Laser Beam Diffraction at the Gate Edge in Poly-Si TFT

Woo-Jin Nam, Kee-Chan Park, Sang-Hoon Jung, Soo-Jeong Park and Min-Koo Han
School of Electrical Engineering, Seoul National University, Seoul, 151-742, KOREA.

## ABSTRACT

Incomplete recrystallized junction defects of self-aligned, excimer laser annealed polycrystalline silicon (poly-Si) thin film transistor (TFT) was investigated by high-resolution transmission electron microscopy (HR-TEM). TEM observation and simulation result verify that the laser irradiation intensity decreased remarkably at the junction due to diffraction of laser beam at gate electrode edge. We proposed oblique-incidence excimer laser annealing method and successfully eliminated the residual junction defects.

## INTRODUCTION

Polycrystalline silicon thin film transistors (Poly-Si TFT) recrystallized by excimer laser have attracted considerable attention for various device application. Recently crystalline defects in the poly-Si channel layer at source/drain junction have been reported to degrade the TFT characteristics such as stability and on-current [1]. It is well known that the source/drain region of TFT is amorphized during the ion implantation for source/drain doping and then recrystallized by excimer laser annealing. We have observed that considerable lattice disorder of amorphized silicon film at the gate edge, which results in junction defects, still exists even after laser beam annealing as shown in Fig. 1 [2]. It is noted that the junction defects maybe originated from residual amorphized silicon generated due to insufficient laser activation energy at the source/drain junctions of the poly-Si film.

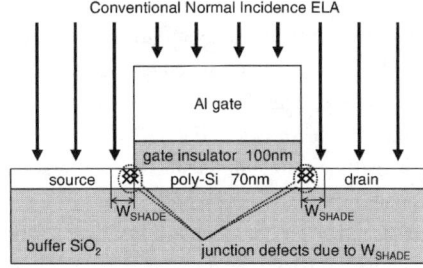 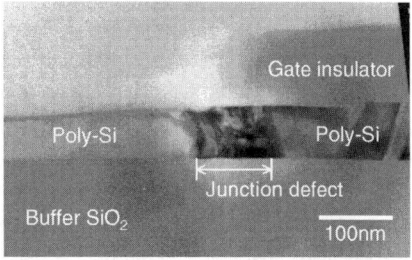

Figure 1. (a) The schematic of incomplete recrystallized junction defects formation, (b) TEM image of junction defects

The purpose of our work is to observe the junction defects near the source/drain junction are resulted from insufficient activation energy due to the laser beam diffraction at the gate edge. We also propose a new method of eliminating the junction defects by simple excimer laser annealing. Thickness of Al gate has been varied from 1000 Å to 10000 Å in order to evaluate laser beam diffraction effect on the junction defects. The laser beam is rather diffracted when the gate thickness is higher, which agrees with our simulated result. By high-resolution TEM investigation we observed that distribution range of the junction defect is wider as the gate thickness increases. We propose a simple method of oblique incident laser beam annealing which successfully eliminates the residual junction defects. The laser beam is irradiated with oblique incident angle and undesirable laser beam diffraction effect at the gate edge is fairly reduced.

## EXPERIMENTS

### Sample Preparation

In order to investigate the effect of laser beam diffraction on the self-aligned ELA poly-Si TFTs, TFTs with various gate thicknesses were fabricated as follows. 70nm thick a-Si films were deposited on oxidized silicon wafer substrates by PECVD at 350°C and dehydrogenated in a furnace at 450°C. Ten shots of XeCl excimer laser were irradiated with the energy density of 320mJ/cm$^2$ to crystallize the a-Si films. 100nm thick gate oxide layers were deposited using TEOS source by PECVD. 100nm, 300nm, 700nm and 1μm thick aluminum layers were sputter-deposited as gate electrodes. The gate electrodes and the underlying gate oxide layers were patterned by RIE. Arsenic ions (As$^+$) were implanted self-aligned with the gate pattern at 30keV to a dose of $5\times10^{15}$ cm$^{-2}$ for source/drain doping. The source/drain poly-Si regions were amorphized by ion implantation process. Six shots of XeCl excimer laser were irradiated again at 300mJ/cm$^2$ to anneal the implantation damage. Finally, the source/drain junctions of the fabricated poly-Si TFTs were observed by a cross-sectional HR-TEM. The Al gate electrodes were removed for easy preparation of the TEM samples.

### Laser Beam Diffraction at the Gate Edge

When a coherent light like laser beam passes an edge, it is diffracted near the edge [3]. Thus a diffraction pattern may appear on the upper surface of the silicon layer during the source/drain ELA of the poly-Si TFT due to the diffraction of the laser beam irradiated on the gate edge structure shown in Fig. 2(a). In order to evaluate the laser intensity on the source/drain silicon layer, the gate structure of the TFT was simply modeled as shown in Fig. 2 (b). The laser intensity on the incident plane of the upper surface of the source/drain silicon layer was calculated based on the Fresnel diffraction of the laser beam at an edge [4].

The normalized laser intensity irradiated on the upper surface of the silicon layer has been evaluated [3] and plotted in Fig. 3. The normalized laser intensity on the source/drain region fluctuated due to the diffraction and decreases noticeably near the source/drain junctions. The width of diffraction shade ($W_{SHADE}$) is defined as the length from the junction to half-intensity point as shown in Fig. 3. Calculated $W_{SHADE}$ values for XeCl excimer laser beam ($\lambda$=308nm) are listed in Table 1 for various $T_{GATE}$ from 200 nm to 1100nm. Typical values of $T_{GATE}$ are between 400nm and 800nm, and the corresponding $W_{SHADE}$ are about 87nm and 123nm respectively.

## RESULTS AND DISSCUSSION

### Incomplete Recrystallized Junction Defects at the Source/Drain Junction

Fig. 4 shows cross-sectional TEM images near the source/drain junctions of the fabricated poly-Si TFTs. The source/drain silicon layer amorphized by high dose ion implantation was completely annealed through melting and resolidification by the second ELA. However the black and white speckles around the junction region indicates that lattice orientation is not continuous and a considerable amount of crystalline defects are present, which manifests that the amorphized silicon layer did not completely melt near the junction due to insufficient thermal energy and thus the implantation damage was not fully annealed. The residual implantation damage is observed even underneath the gate oxide edge where the direct implantation of the dopant ions are blocked by the gate, due to the lateral straggle of the implanted $As^+$.

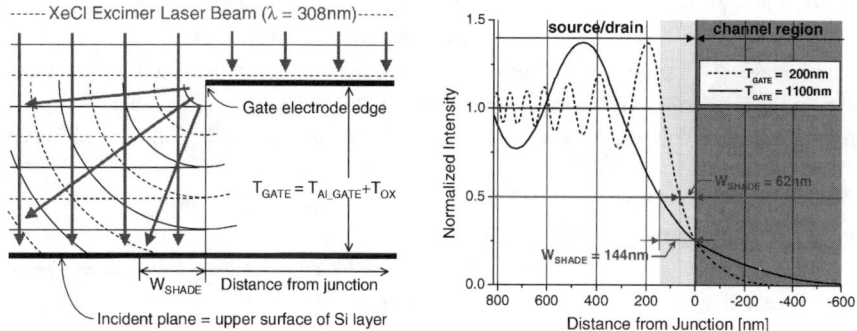

Figure 2. (a) The laser intensity on the upper surface of silicon layer was evaluated for the simple model based on the Fresnel diffraction. (b) Normalized laser intensity on the upper surface of silicon layer.

Table 1. Calculated $W_{SHADE}$ for gate thickness ($T_{GATE}$) of 200nm, 400nm, 800nm and 1100nm.

| $T_{GATE}$ (nm) | 200 | 400 | 800 | 1100 |
|---|---|---|---|---|
| $W_{SHADE}$ (nm) | 62 | 87 | 123 | 144 |

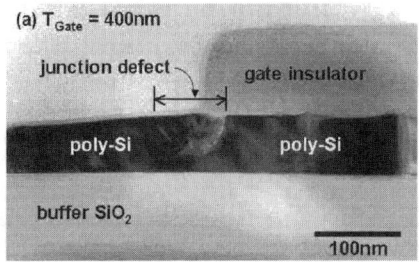 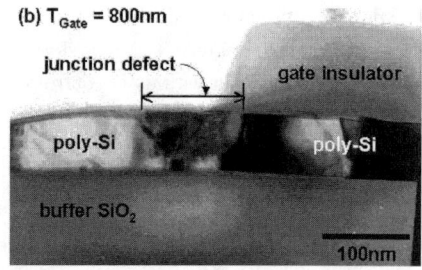

Figure 4. Cross-sectional TEM images of residual ion implantation dagmage at source/drain junctions of self-aligned ELA poly-Si TFTs entitled "junction defects".

For very thin gate, the residual damage is not so clearly observed. However the width of residual damage region increased as the thickness of the gate ($T_{GATE}$) increased. For 400nm thick gate, the width of residual damage region indicated in Fig. 4(a) was 85nm and it was 118nm in Fig. 4(b) for 800nm thick gate. The dependence of the residual damage on $T_{GATE}$ may be attributed to the heat dissipation to Al gate electrode and the decrease of the laser intensity near the junction due to the laser beam diffraction at the gate electrode edge of the TFT.

The increase of the width of the residual implantation damage with increasing $T_{GATE}$ shown in Fig. 4 may be ascribed to the increase of the width of the diffraction shade ($W_{SHADE}$). The values of $W_{SHADE}$ listed in Table 1 correspond well with the widths of the residual implantation damage region shown in Fig. 4. Thus it can be concluded that laser beam diffraction at the gate electrode edges is the main cause of the incomplete annealing of the ion doping damage and the resultant residual crystalline defects at the source/drain junctions of the self-aligned ELA poly-Si TFTs.

**Proposed Oblique Incident ELA method**

In order to avoid the effect of the diffraction shade and to irradiate full laser energy at the source/drain junction of the poly-Si TFT, a new post-implantation ELA method entitled "Oblique-Incidence ELA (OI-ELA)" is proposed. The schematic diagram of the proposed OI-ELA method is shown in Fig. 5(a). By irradiating the laser beam obliquely, the diffraction shade is formed in the channel region and no longer covers the junction region. Therefore the laser intensity does not decrease near the junction and the laser beam irradiated on the junction has sufficient intensity to completely anneal the ion doping damage.

The larger the incidence angle of the laser beam during the OI-ELA, the deeper inside the channel region forms the diffraction shade and the ion doping damage can be annealed more effectively. However, as the incidence angle increases, the energy density of the irradiated laser per unit area decreases and the ratio of the reflected laser energy to the incident energy also increases [3]. In addition, the edge of the gate is damaged by the OI-ELA for too large angle of incidence, because the molten channel poly-Si protrudes through the gate oxide and the aluminum gate is evaporated as the temperature of the underlying silicon layer rises too high. In this experiment, the angle of incidence was fixed to 55° that was the largest value not to damage the Al gate electrode.

Fig. 5(b) shows the cross-sectional TEM image near the source/drain junctions of the poly-Si TFT fabricated by the OI-ELA. The Al gate electrodes were removed for easy preparation of the TEM samples. The black and white speckles around the junction region shown in Fig. 4 are not observed, which indicates that lattice orientation is continuous and the amorphized silicon layer has been completely melted near the junction owing to the sufficient laser energy and thus the implantation damage has been completely annealed. The implantation damage underneath the gate oxide edge due to the lateral straggle of the implanted $As^+$ has been eliminated as well.

Fig. 6 is the enlarged high-resolution TEM images of the source/drain junction annealed by the conventional and the proposed ELA method. In the proposed annealing method, the lattice image is clear and continuous throughout the whole rectangular region, which manifests that the crystallographic orientation is invariant and the crystalline defect density is very low. In conclusion, the ion doping damage near the source/drain junctions of the self-aligned ELA poly-Si TFT can completely annealed by the proposed OI-ELA method.

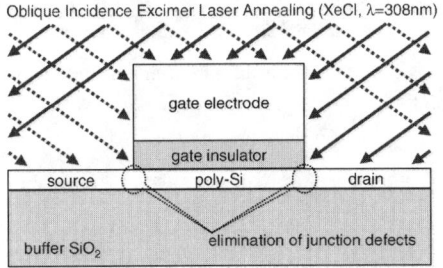 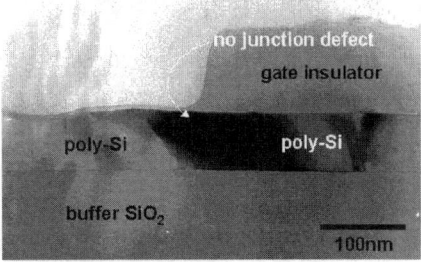

Figure 5. (a) Schematic diagram of the proposed Oblique-Incidence ELA (OI-ELA) method, (b) Cross-sectional TEM image of source/drain junction of OI-ELA poly-Si TFTs.

## CONCLUSTIONS

We have investigated the incomplete recrystallized junction defects of self-aligned excimer laser annealed poly-Si TFTs by High-resolution TEM observation. The junction defects are originated from residual ion implant damages and the width of defects region increases as the gate thickness increases. TEM observation and simulation result manifest that the laser beam diffraction at the gate electrode edges is the main cause of the incomplete annealing of the ion doping damages. We proposed simple annealing method "Oblique-Incidence ELA" in order to elminate the junction defects. In the proposed ELA method, the laser beam is irradiated obliquely on the substrate so that the laser intensity does not decrease near the junction and the ion doping damages are completely annealed.

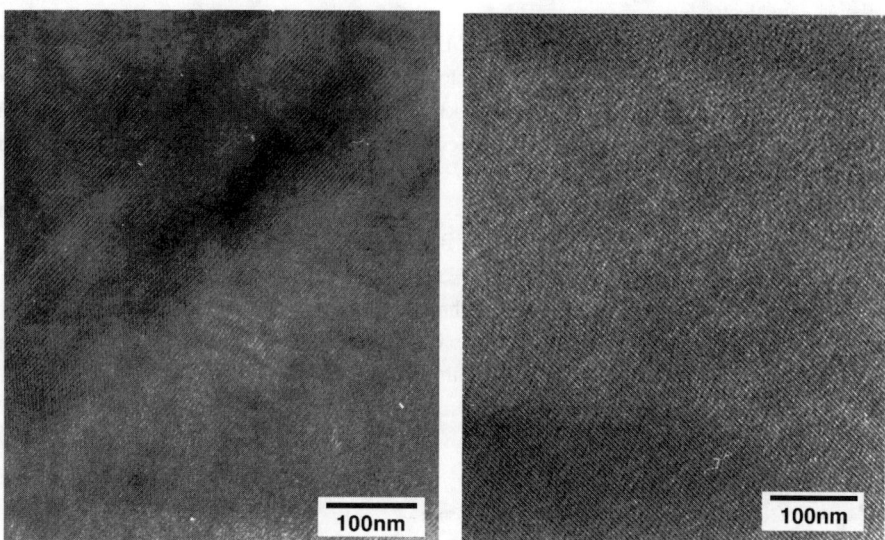

Figure 6. Enlarged high-resolution TEM images of the source/drain junction annealed by (a) the conventional and (b) the proposed ELA method.

**REFERENCES**
1. A. Mimura, J.I. Ohwada, Y. Hosokawa, T. Suzuki H. Kawakami and K. Miyata, IEEE Trans. Elec. Dev. **35**, 418 (1988)
2. K.C. Park, J.S. Kim, W.J. Nam and M.K. Han, Mater. Res. Soc. Spring Meeting (2002)
3. K.D. Moller, *OPTICS*, (University Science Books, Mill Valley, CA, 1988) p. 172.
4. Y. Tahkubo et al, Japan Display 89, pp. 584-587 (1989)

# 2-DIMENSIONAL CONTROLLED LARGE LATERAL GRAIN GROWTH ON THE FLOATING AMORPHOUS SILICON FILM BY EXCIMER LASER RECRYSTALLIZATION

In-Hyuk Song, Su-Hyuk Kang, Woo-Jin Nam and Min-Koo Han
School of Electrical Engineering, Seoul National University, Seoul 151-742, Korea
Phone: +82-2-880-7992, Fax: +82-2-883-0827, E-mail: ihsong@emlab.snu.ac.kr

## ABSTRACT

We have successfully obtained large lateral grains with well-controlled grain boundary. The proposed excimer laser annealing (ELA) method produces 2-dimensionally controlled grain growth because the temperature gradient is induced in two directions. Along the channel direction, the floating active structure produces large thermal gradient due to very low thermal conductivity of the air-gap. Along the perpendicular direction to the channel, the surface tension effect also produces thermal gradient. The proposed ELA method can control the grain boundary perpendicular and parallel to current path with only one laser irradiation.

## INTRODUCTION

A polycrystalline silicon (poly-Si) thin film transistor (TFT) employing excimer laser annealing (ELA) is a promising device for high-resolution flat panel display such as active matrix liquid crystal display (AMLCD) and active matrix organic light emitting diode (AMOLED). XeCl ELA of amorphous silicon (a-Si) results in poly-Si film with relatively low defect density. However, the grain size of poly-Si is not large enough (typically less than 1μm) for high performance poly-Si TFTs. The field-effect mobility and driving current of TFTs are critically dependent on the grain size and the potential barrier of grain boundary. It is noted that high quality poly-Si films with controlled grain size and grain boundary location are required in order to improve the performance. Various efforts such as sequential lateral solidification (SLS), optical phase shift mask and pre-patterned laser mask have been reported to control grain size as well as grain boundary. However, difficulties in attaining controllability are anticipated due to the process complexity in these methods.[1-3]

We have already reported ELA method with floating active structure which produces the large lateral grain and controls the location of grain boundary.[4] However, grain boundaries parallel to the current flow may disturb the carrier drift motion during operation of poly-Si TFT.

The purpose of our work is to report a new ELA method which can produce large lateral grains and control the grain boundary parallel to current path. A new ELA process also decreases the number of grain boundaries parallel to current flow by employing pre-patterned multi-channel structure on a selectively floating amorphous silicon layer.

**EXPERIMENTS AND RESULTS**

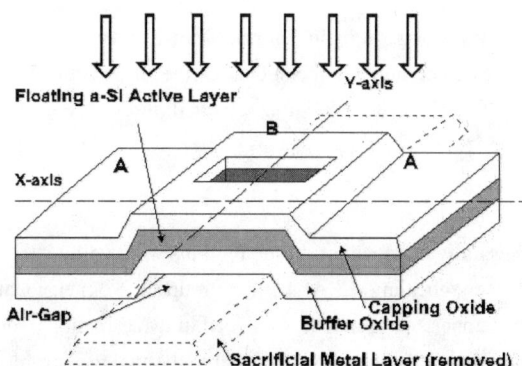

**Fig. 1**. Schematic diagram of the proposed ELA method

The proposed ELA process is illustrated in Fig. 1. A 50nm-thick metal sacrificial layer was deposited on oxidized Si substrate by sputter. After patterning of rectangular metal layer, a 100nm-thick buffer oxide was deposited. A 80nm-thick hydrogenated amorphous silicon (a-Si:H) film was deposited by plasma-enhanced chemical vapor deposition (PECVD) at 280°C and dehydrogenated at 450°C for 3 hours in order to prevent the rapid evolution of hydrogen atoms during the laser irradiation. The a-Si layer was covered with a thin capping oxide in order to protect the floating Si region from evaporation during the laser irradiation. The capping oxide, the a-Si layer and the buffer oxide were sequentially patterned across the sacrificial metal pattern in order to form an active island. The sacrificial metal layer under the a-Si active layer was completely removed by metal wet etchant. As a result, the a-Si layer (region B in figure 1), which has been located on the sacrificial metal pattern, is selectively floating. This bridge structure was irradiated with 30nsec-pulse XeCl excimer laser ($\lambda$=308nm) in a vacuum chamber. After the capping oxide layer was stripped, the samples were secco-etched. The grain structure of the laser-annealed poly-Si film was evaluated by SEM (scanning electron microscope) image.

It is well known that lateral temperature gradient in a melted Si film would induce the lateral grain growth [5]. In our proposed ELA method employing floating active structure and multi channel structure, temperature gradient is induced in two directions (x-axis, y-axis). Along the x-axis, an air-gap, of which thermal conductivity is very low, blocks the heat conduction to the substrate so that the lateral temperature gradient is successfully induced. In the Si region adherent to the substrate (region A), the heat absorbed in the molten Si film is conducted through the underlying substrate so that the melted Si cools off rapidly and begins solidification. However, the vertical cooling rate in the selectively floating Si region (region B) is remarkably reduced due to low thermal conductivity of the air-gap. The heat stored in the floating Si region would conduct laterally through the neighboring Si region adherent to the substrate, so that the significant lateral temperature gradient is occurred between the region A and B. The solidified Si grains at the border between the region A and B act as nucleation seeds and the grains grow in the lateral direction to the center of the floating Si region. Along the y-axis, temperature gradient also occurs because of surface tension effect [6] so that the number of parallel grain boundary is significantly reduced due to pre-pattern effect. The schematic diagram of grain structure after ELA process is shown in Fig. 2(a) and SEM image clearly indicated well-controlled parallel grain boundary as shown in Fig. 2(b). The length of floating active is 16μm and the width of each channel is 2μm.

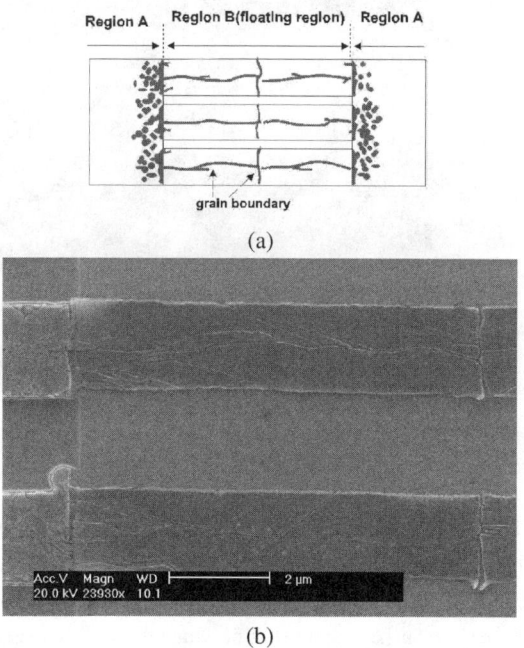

**Fig. 2.** (a) Schematic diagram of the grain structure after ELA process, (b) SEM image of grain structure (each channel width = 2μm, length of floating region = 16μm)

However, the width of each channel must be optimized in order to reduce the grain boundary parallel to current path. As shown in Fig. 3, the reduction of grain boundary parallel to current path is not observed. It is because that temperature gradient induced by pre-patterned a-Si thin film (along the x-axis) is smaller than that by floating active structure (along the y-axis). In our experiment, the temperature gradient induced by the effect of pre-patterning can reduce the number of grain boundaries parallel to current path when the width of each channel is less than 2μm.

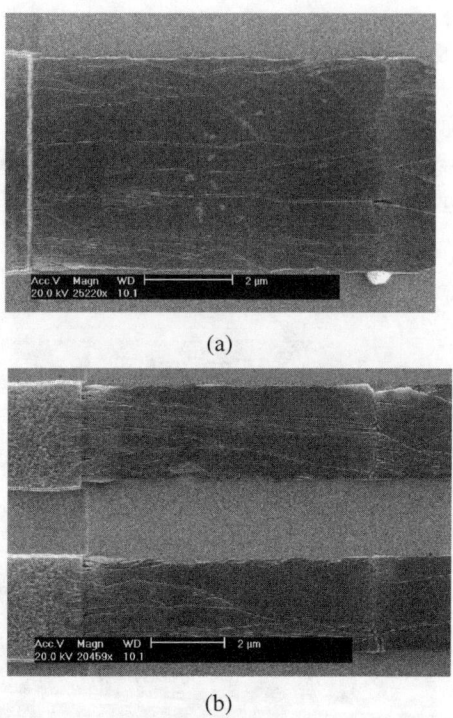

**Fig. 3**. SEM image of (a) channel width = 5 μm (b) channel width = 3 μm

Fig. 4 shows the form of active island along the y-axis in case of wide channel and narrow channel during the ELA process. Deformation of active island after laser irradiation occurs by surface tension effect. Surface tension is a property of liquids arising from unbalanced molecular cohesive forces at near the surface, as a result of which the surface tends to contract and has properties resembling those of a stretched elastic membrane. In laser irradiation on a pre-patterned active island, the surface tension effect has been studied by G. K. Giust.[6]

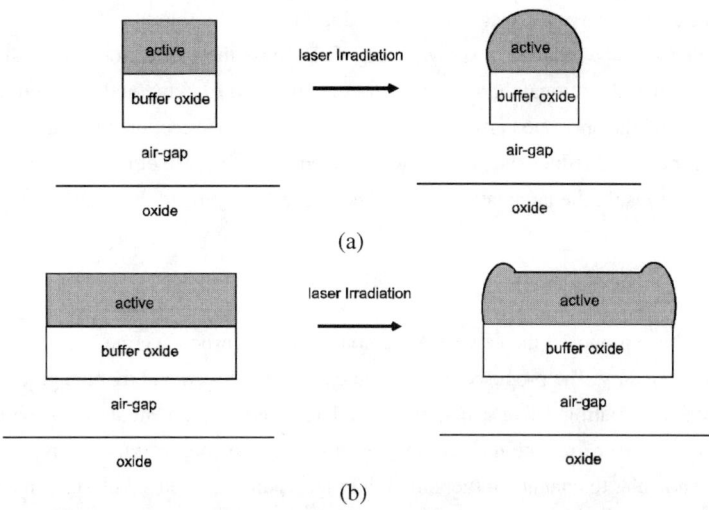

**Fig.4**. Deformation of the active island during the ELA process in case of (a) narrow width and (b) wide width

Fig. 4(a) represents the deformation in the narrow active pattern during the ELA process. The center of the active pattern is thicker than edge due to the surface tension effect. Therefore, the center of the active pattern may store the heat induced by laser irradiation for the longer time than edge. Solidification direction is from edge of the pattern to center of it. In the case of wide active pattern, surface tension effects cause the active island to shrink and pull up at the edge after laser irradiation as shown Fig. 4(b).

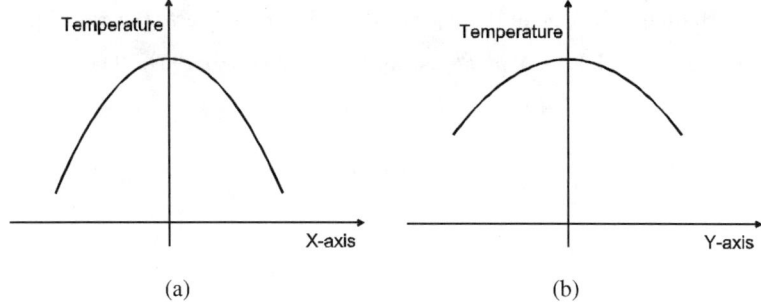

**Fig. 5**. Temperature gradient along x-axis and y –axis

This effect is adapted to the proposed ELA method. As shown in Fig.2 and Fig.3, the parallel grain boundary can be reduced in case of the width of the channel is less than 2μm due to the surface tension effect as represented in Fig. 4(a). In this case, 2-directional thermal gradients can be induced as shown in Fig. 5. Because the x-directional thermal gradient is very lager than y-directional thermal gradient, the proposed ELA method can produce the large lateral grains and the only one perpendicular grain boundary. Y-directional thermal gradient is small, but can control and reduce the parallel grain boundary in the case of narrow channel width (less than 2μm). In result, the large lateral grains and the well-controlled grain boundaries can be obtained.

## CONCLUSION

We have proposed the new ELA method which can produce large lateral grains and control the grain boundary. The proposed ELA method induce 2-directional thermal gradient. The selectively floating active structure induce large thermal gradient along the parallel to channel during the laser irradiation due to the very low thermal conductivity of air-gap. Along the perpendicular to channel, rather small thermal gradient is induced by surface tension effect on a pre-patterned active island. As a result, 2-dimensional grain growth occurs with large lateral grain boundary and well-controlled grain boundary.

## REFERENCES

[1] M. A. Crowder, P. G. Carey, P. M. Smith, R.S. Sposili, H. S. Cho and J.S. Im, *IEEE Electron Device Letter*, vol. 19, p. 306, 1998.

[2] C. H. Oh and M. Matsumura, *IEEE Electron Device Letter*, vol. 22, p. 20, 2001.

[3] J. H. Jeon, M. C. Lee, K. C. Park, S. H. Jung and M. K. Han, *IEDM Tech. Digest*, p. 213, 2000.

[4] C. H. Kim, I. H. Song, S. H. Jung and M. K. Han, *IEDM Tech. Digest*, p751, 2001

[5] J.S. Im, and H.J. Kim, *Appl. Phys. Lett.*, vol. 63, no. 14, p. 1969, 1993.

[6] G. K. Giust and T. W. Sigmon, *J. Appl. Phys.*, vol. 81, No. 3, p. 1204, 1997

# IMPROVED ELECTRICAL PROPERTIES IN NANOCRYSTALLINE SI FORMED BY METAL INDUCED GROWTH

Chunhai Ji and Wayne A. Anderson
University at Buffalo, The State University of New York, Dept of Electrical Engineering, Buffalo, NY.

## ABSTRACT

In recent work, the electrical properties of metal-induced-grown (MIG) Si thin films were studied by using current-voltage (I-V) data from a metal/Si Schottky contact. It was found that controlling the doping level of the films and annealing in forming gas (15% $H_2$ and 85% $N_2$) can improve the quality of the nc-Si films. From SIMS analysis on the nc-Si film deposited from a highly doped target, the nc-Si can duplicate the doping level of the sputtering target. Study of p-type doped nc-Si films showed that the fabrication of Schottky diodes on nc-Si films made from an extremely high-doped target ($\sim 10^{20}$ cm$^{-3}$) or low-doped target ($\sim 10^{15}$ cm$^{-3}$) was not successful. For highly doped p-type films, tunneling causes Ohmic conduction instead of rectifying conduction. For the nc-Si film deposited from a low-doped p-type target, the film shows conversion to n-type characteristics when measured by a hot probe. This might be due to defects or oxygen in the film. N-type films at the middle doping level ($\sim 10^{17}$ cm$^{-3}$) gave good Schottky diodes after annealing the film in forming gas at 700°C. The Schottky diodes fabricated by high work-function metal (Au) gave the rectifying ratio of $\sim 10^3$. Several techniques, e.g. slow/fast two-step sputtering at low working pressure and surface polishing, were used to improve the photo response of Schottky photodiodes. The open-circuit voltage ($V_{oc}$) of 0.164V and short-circuit current density ($J_{sc}$) of 2.5 mA/cm$^2$ were achieved under 100mW/cm$^2$ illumination.

## INTRODUCTION

Poly-Si compared to a-Si:H, has a higher stability and higher carrier mobility [1]. Fabrication cost and process temperature are two important concerns in the fabrication of poly-Si thin films and solar cells. Metal induced growth (MIG) is a promising method which could reduce the cost and decrease the thermal budget. In MIG, a metal layer is used as a seed layer. Poly-Si or nc-Si films epitaxally grow from the metal silicide, which reacted from metal and first-coming Si atoms. This kind of epitaxial growth needs less thermal energy so that a lower temperature is needed. The MIG growth mechanism was studied intensively [2,3]. It showed that the MIG process can produce various Si microstructures, from poly-Si thin film to nanowires [4]. So far, two metals were suitable for the MIG process, i.e. Ni and Co, in which Co shows the advantages of less diffusion and producing uniform Si grains [5]. However, there has been a problem when making Schottky devices for film characterization and investigation of photovoltaic properties, i.e. a linear I-V curve is found instead of a rectifying curve. Recent results found that the combined effects of high doping and grain boundary leakage may be two main problems. Based on the above analysis, a good Schottky diode was fabricated by controlling film doping level and annealing the film in forming gas. It shows that the film has a good value of carrier lifetime. With the modification of film structures by using some techniques, the Schottky photodiode from MIG poly-Si film had a $V_{oc}$ of 0.164V and $J_{sc}$ of 2.5 mA/cm$^2$. The research is still in progress to improve the photovoltaic properties and fabricate the MIG Si on different substrates.

## EXPERIMENTAL

Similar to the previous studies [4,5], the metal induced grown Si thin film was on a single-crystal Si (c-Si) wafer covered with 200nm-thick PECVD SiO$_2$. The polished c-Si wafer provides a smooth surface, while SiO$_2$ provides a barrier against metal seed-layer diffusion into the c-Si. The Ni and Co metal films were thermally evaporated on the SiO$_2$/Si-wafer substrates. Silicon was then deposited by DC magnetron sputtering from a boron doped p-type Si target or a phosphorus doped n-type Si target in order to achieve differently doped Si thin films. During the sputtering process, the samples were heated to the temperature range from 525°C to 625°C. Annealing in forming gas at 700°C allows the Si thin film to have better electrical properties than the films as grown. The Schottky diodes were fabricated on both the p-type and n-type Si thin films. With a specially designed thin film layout, shown in Fig.1, the current-voltage characteristics were measured in a vertical direction. The I-V results from these Schottky diodes were used to analyze the carrier transport in the metal/Si junction. The different Si targets with different doping density were studied. The range of the resistivity of doping level ranged from 0.001Ω-cm to 0.1Ω-cm. The SIMS analysis was done in order to determine the dopant concentrations and impurities in the Si thin films. The Au/n-Si Schottky photodiodes, with ~100 Å thick metal, were fabricated in order to study the photo response of the Si thin film devices. Some new techniques were used to improve the photo response. Sputtering with low-rate (~70Å/min) at the beginning and high rate (~300 Å/min) later give a thicker Si film with large grain size. Sputtering at low pressure (0.5mTorr) not only improved the grain size but also reduced the impurities in the films. Surface polishing smoothed the film surface and removed the surface states. The improved I-V is shown in the next section. Lifetime measurement together with SIMS was carried out at the National Renewable Energy Lab (NREL).

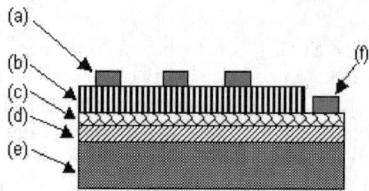

Fig. 1 Schematic representation of thin film layout and Al Schottky contact, (a) top Al Schottky contact on (b) p-type MIG nc-Si thin film, (c) metal silicide layer formed in sputtering, (d) SiO$_2$ layer as the barrier against metal diffusion into (e) c-Si wafer, (f) Ohmic contact under the nc-Si.

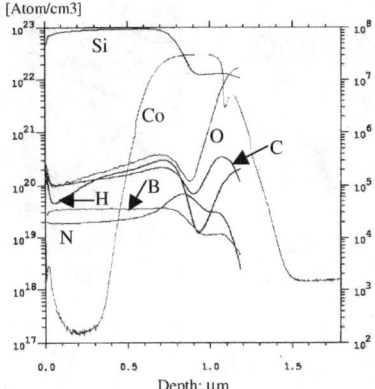

Fig. 2 SIMS analysis results on Si thin film sputtered from a 0.001 Ω-cm p-type target.

## RESULTS AND DISCUSSION:

### Carrier lifetime measurement

Three kinds of n-type Si thin-films were measured for carrier lifetime: Co seed layer as grown, Co seed layer annealed and Ni seed layer as grown. The result was 0.458μs for the Co seed layer as grown film, 1.33μs for the Co seed layer annealed film and 0.309μs for the Ni seed layer as grown film. The low value for the Ni seed layer sample is possibly due to the high diffusion of Ni atoms in the Si thin films [5]. The longer lifetime for the annealed film could explain why the electrical properties of the annealed film are improved.

### SIMS analysis of metal induced thin films

The dopant concentrations in the grown Si thin films were tested by SIMS to determine the difference in the doping level between the sputtering target and Si thin film. One of Si thin films which was sputtered from a B-doped 0.001 Ω-cm target was examined. The SIMS result (Fig. 2) shows that the doping concentrations from the Si thin film was about $2 \times 10^{19}$ cm$^{-3}$. By checking the graph of doping vs. resistivity, the target doping level should be about $10^{19} \sim 10^{20}$ cm$^{-3}$ [6]. It can be concluded that the MIG process allows the Si thin film to duplicate the doping level of the Si target. This result can be used to determine the effects of dopant concentration in Si thin films on Schottky I-V behavior. Furthermore, from the SIMS analysis, it can also be seen that the O and C content in the Si thin film is at a relatively low level.

### SEM ANALYSIS

SEM micrographs in Fig. 3 show surface and cross-section of poly-Si thin films, which were made by the two-step sputtering process. The large grain and thick-film structures is due to the high-rate sputtering. The twin structure observed in the grains shows the traces of grain growth. Poly-Si grown on a metal substrate had similar microstructures to the one grown on a Si wafer substrate.

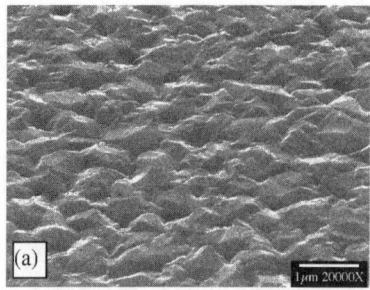 

Fig.3 SEM micrographs of poly-Si thin film: (a) Poly-Si grown on a tungsten substrate, (b) Cross-section view to show the poly-Si grown on an oxide-coated wafer substrate

## Doping level effects on the Schottky diodes

Si targets with different doping levels were used for growing the metal induced Si thin films. Schottky diodes were fabricated. Al was used for p-type Si and Au was used for n-type Si to examine the photovoltaic potential of the thin films. At the beginning of the study, a highly doped p-type target (0.001 ohm-cm) was used. It was found that by using this type of target, it is difficult to get a rectifying I-V curve from Al Schottky diodes. As shown in Fig. 4(a), a linear I-V was shown from this type of Si thin film. According to the SIMS analysis discussed in the last section, it is possibly due to the very high dopant concentration causing tunneling effects in the Schottky devices. This is similar to the so-called "degenerate semiconductor", which is also caused by very high doping. Based on the above analysis, a lower doped Si target was used next. For a 0.02 ohm-cm n-type Si target, the rectifying I-V was achieved for an Au Schottky diode as shown in Fig. 4(b). The rectifying I-V results were achieved after annealing at 700 °C for 2 hours in forming gas (15% $H_2$). The barrier height and ideality factor were 0.65eV and 3, respectively. When changing the target to an even lower doped n-type, the similar rectifying result was achieved. However, the lower doped p-type target, e.g. 0.1~1 Ω-cm, didn't show a rectifying I-V. By testing the Si thin film type by the hot-probe method, it was realized that the Si thin film was n-type instead of p-type. Based on the hot-probe results, an Au diode was fabricated on the Si thin film which was deposited from a 0.1~1 Ω-cm p-type Si target. A rectifying I-V was given by this Au diode, which proves that the Si thin film from a very low-doped p-type Si target is n-type instead of p-type. The abnormal conversion of Si thin film doping type from p-type to n-type has been reported [7]. The impurities and oxygen present in the Si thin film probably cause this phenomenon.

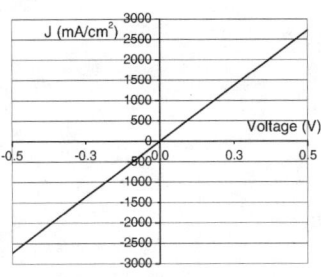
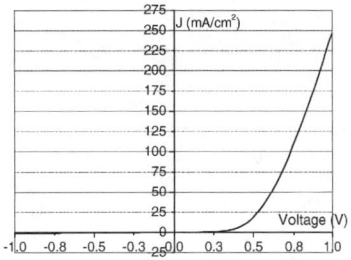

(a) Al/p-Si contact (0.001 Ω-cm target)    (b) Au/n-Si diode (0.02 Ω-cm target)

Fig. 4 Doping concentration causes the differences in I-V of Schottky diodes.

### Effects of annealing on Schottky I-V

The I-V characteristics of Schottky diodes were compared for as-deposited and annealed samples in Fig. 5. The annealing process was set to 700 °C for 2 hours in forming gas (15% $H_2$). Except for the very high-doped samples (p-type 0.001 Ω-cm), the annealing improved Schottky I-V data very much. There are two possible reasons that the annealing process improves the Schottky I-V characteristics. One reason is the passivation effect of hydrogen. Another reason is the diffusion of dopants inside the film from grains to grain boundaries. These dopants will passivate the defects in the grain boundaries to reduce leakage current along the grain boundaries.

### Recent progress in photovoltaic properties of Au/n-Si Schottky photodiode

In the previous studies [5], a preliminary result showed that for a Si film with 0.5μm thickness, the open-circuit voltage ($V_{oc}$) and short-circuit current density ($J_{sc}$) were 0.1 V and 0.3 mA/cm$^2$, respectively, under 100mW/cm$^2$ illumination. Such a thin film will absorb only about 23% of the incident light. The most recent study shows much better photovoltaic properties by further manipulating the sputtering process and surface polishing. Sputtering in a low working pressure (0.5mTorr) increased the sputtering rate and enhanced the growth of Si grains with fewer impurities in the films. A two-step sputtering, which has a low rate at the beginning and a high rate later, gave a thick Si film (5μm) with larger grain size, which is good for photo-response of the Schottky photodiodes. On the other hand, Schottky device quality relies on the surface condition of the Si film. A preliminary polishing procedure is shown to improve $V_{oc}$ and fill factor, as shown in Fig.6. The improvement of $J_{sc}$ is due to the increase of film thickness and the enlargement of the grain size. The $V_{oc}$ is partially affected by the leakage current which still exists because the surface condition of the film is still not perfect for fabricating the devices.

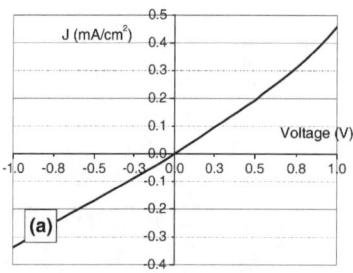 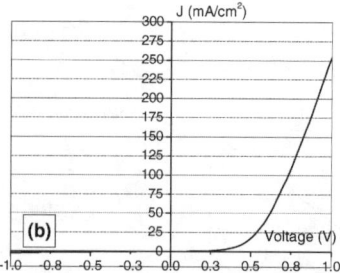

Fig. 5 Effect of annealing the film on the Schottky diode quality, (a) As deposited, results show a linear I-V curve, (b) After annealing, the film gave an improved I-V

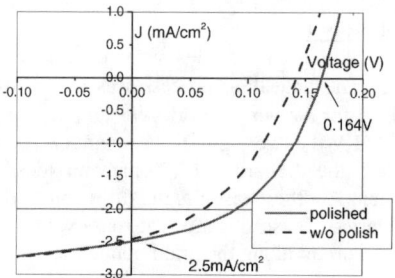

Fig. 6 Improved photo I-V of an Au/n-Si Schottky photodiode which shows $V_{oc}$ and $J_{sc}$ value of 0.143 V and 2.46 mA/cm$^2$, respectively.

## CONCLUSIONS

The latest thin Si films gave much improved solar cell performance. The photo-current can be further increased by using a thicker Si film, a P/N junction rather than Schottky to reduce photon loss and adding an antireflection coating. Passivation of defects will reduce leakage current to increase $V_{oc}$. With the new procedure, the films are microcrystalline rather than nanocrystalline. We have already shown that this process will work on a metal substrate to give a potentially low-cost, thin film device.

## REFERENCES:

1. J. Meier, R. Fluckiger, H. Keppner, A. Shah, *Appl. Phys. Lett.* **65**, 7, 860 (1994)
2. E. A. Guliants, C. Ji, W.A. Anderson, J. Elect. Mat'l, **31**, 5, 466 (2002)
3. E. A. Guliants, C. Ji, W.A. Anderson, J. Appl. Phys, **91**, 9, 6077 (2002)
4. C. Ji and W.A. Anderson, 29th IEEE PVSC, New Orleans, LA (2002)
5. C. Ji and W.A. Anderson, MRS fall meeting, Boston, USA, (2002)
6. L. Shon-Roy, A. Wiesnoski and R. Zorich, Advanced Semiconductor Fabrication Handbook, 1-16 (1998)
7. H.J. Stein, S.K. Hahn, J. Appl. Phys. 75, 3477 (1994)

# A Simple Lateral Grain Growth of Poly-Si by Single Excimer Laser Crystallization of Amorphous Silicon Film Deposited on Polygon Shaped Trench

**Sang-Hoon Jung, Su-Hyuk Kang, Hee-Sun Shin and Min-Koo Han**
School of Electrical Engineering, Seoul National University,
San 56-1 Shinlim-dong, Kwanak-gu, Seoul 151-742, Korea E-mail : jsh@emlab.snu.ac.kr

## ABSTRACT

A simple lateral grain growth of polysilicon employing single excimer laser irradiation is proposed. In order to increase the size of silicon grain and to control the location of the large lateral grain, the oxide trench is employed under the amorphous silicon film in the proposed method. The proposed oxide trench, which is shaped like a triangle or a polygon with an acute angle, induces temperature gradient on the molten silicon film during the solidification. It was verified by SEM that about 2 µm-long silicon grains are successfully achieved near the oxide trench edge and the locations of lateral grains are controlled by the angular points of the diagram.

## INTRODUCTION

Polycrystalline silicon thin film transistors (poly-Si TFTs) employing excimer laser annealing (ELA) of amorphous silicon (a-Si) film are promising devices in a various flat panel displays [1,2]. The characteristics of poly-Si TFT are dependent on the grain size of poly-Si and the grain boundaries in the channel [3]. Various efforts, such as sequential lateral solidification (SLS) or µ-czochralski (grain-filter) process, have been reported to increase the grain size and to control the location of grain boundaries [4,5,6]. However, SLS requires a rather sophisticated beam scan process and grain-filter process requires extremely narrow hole (about a diameter of around 100 nm) with rather deep trench (about 1 µm).

The purpose of our work is to report a simple lateral grain growth method employing single excimer laser irradiation. Oxide trench, which is shaped like a triangle or a polygon with an acute angle, is employed in our work. After the oxide trench formation, a-Si film was deposited and crystallized by single shot ELA. The thickness difference of a-Si film for lateral grain growth is induced by oxide trench and the location of grain is determined by the location of the angular point of the polygon. The proposed lateral grain growth employs simple process with shallow trench, which does not require sophisticated laser beam scan or an accurate process control of sub-micron holes.

## EXPERIMENTS AND RESULTS

The proposed lateral grain growth method is shown in Figure 1. Photolithography was performed on the buffer $SiO_2$ layer, then 300nm-deep $SiO_2$ trench (oxide trench) was formed by dry etching process. 200 nm-thick buried $SiO_2$ layer was deposited by TEOS PECVD at a temperature of 390 °C. Then, in order to achieve good step coverage of active film, 200 nm thick a-Si layer was deposited by LPCVD at 550 °C. Finally, a-Si active layer was crystallized by single XeCl excimer laser annealing ($\lambda$=308nm). At this time, excimer laser energy density was 700 mJ/cm$^2$ and the temperature of the substrate was 350 °C.

It is noted that this laser energy density is higher than the complete melting condition of the active a-Si film deposited. Because the proposed method is performed in the complete melting condition, the margin of the process temperature is wider than that of conventional ELA method. At the edge of the trench, effective thickness of a-Si film is larger than that of other region due to the slope of the edge and the laser energy absorbed is also effectively lowered. These non-uniform conditions induce the lateral growth of silicon grain as shown Figure 1 (f).

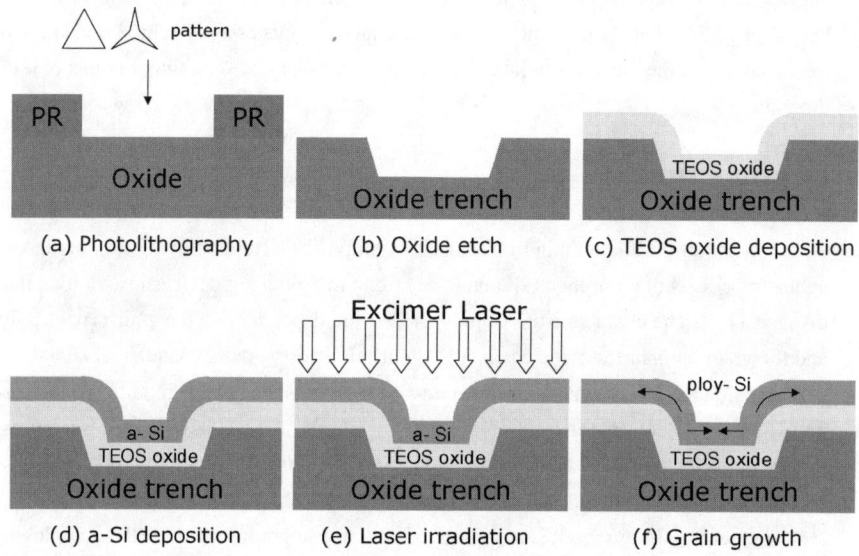

**Figure 1. The process sequence of the proposed ELA method.**

After the laser irradiation, Secco-etched poly-Si film was observed by using SEM. In our experiment, a regular triangle pattern, of which a side is 6 μm, and a polygon with an acute angle were used for the oxide trench. However, real patterns were expressed like Table 1 due to the rounding effect in the photoresist pattern. The angular point of the triangle is rounded because of the optical limit during the UV exposure process.

**Table 1. Geometrical mask patterns used in the experiment and real patterns after the photolithography**

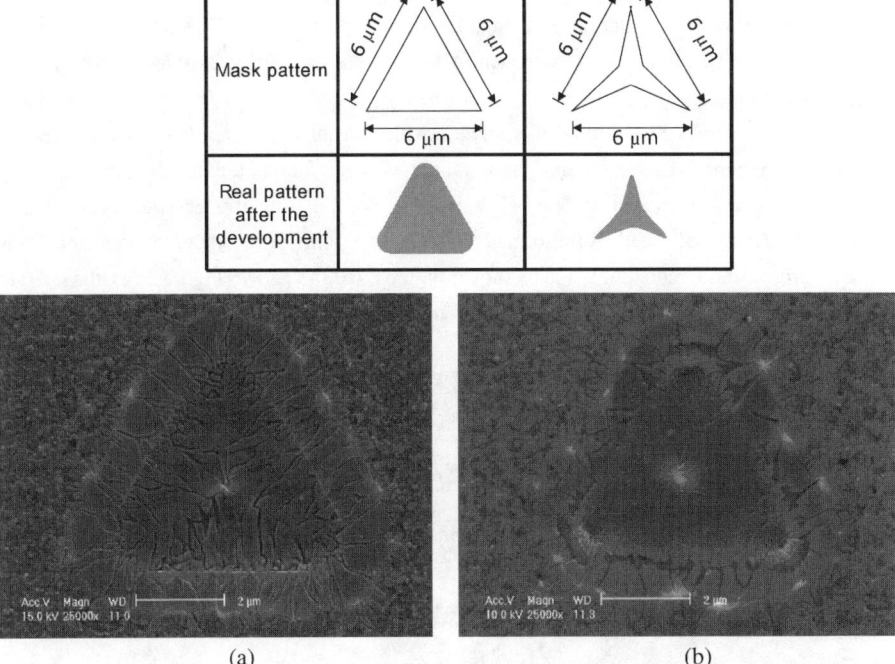

**Figure 2. SEM images of a regular triangle sample after the irradiation of excimer laser (a) without buried TEOS oxide, (b) with 200 nm-thick buried TEOS oxide deposited before the a-Si deposition.**

Figure 2 shows SEM images of a triangular trench sample after the irradiation of excimer laser. Figure 2 (a) is the result of laser irradiation when the buried TEOS oxide was not deposited and Figure 2 (b) is when 200 nm-thick buried TEOS oxide was deposited before the deposition of a-Si layer. In the case of Figure 2 (a), narrow and long poly-Si grains are observed. There is no difference at the angular point because of rounding effect. The depth and surface conditions of oxide trench are similar, so that nucleation was occurred on all the sides of the triangular trench simultaneously. As a result, many narrow silicon grains were grown from the each starting points.

In the case of Figure 2 (b), small grains are observed near the angular points of the triangle, however, wide and long silicon grains were observed except for the angular points. The buried TEOS oxide, which is deposited before a-Si layer deposition, makes the angular points of the diagram acuter. It means that the effective laser energy density at the angular points is lower than that of the other region due to the optical limit during the laser annealing. Thus, the nucleation occurs earlier at the angular points than at the sides of the diagram. This is the difference between Figure 2 (a) and (b). Consequently, lateral grains of Figure 2 (b) are larger than those of Figure 2 (a).

Figure 3 shows SEM image of a polygon with an acute angle and its schematic view. Although Figure 2 (b) and Figure 3 have undergone same fabrication process, the diagram of Figure 3 has acuter shape than that of Figure 2 (b). So, above-mentioned effect occurs more severely. As a result, seeds were formed at the angular points of the diagram and lateral grains were grown to the radial direction as shown Figure 3 (b). As shown Figure 3 (a), the lateral grains from each angular point impinged on the center of the diagram.

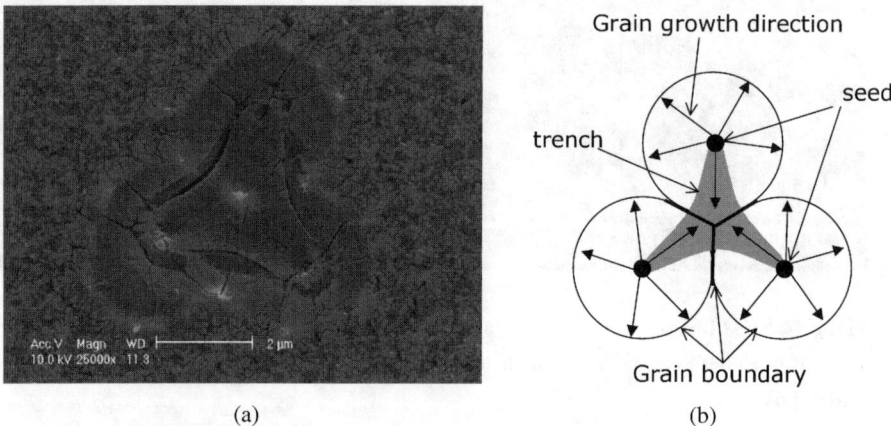

**Figure 3. SEM image of a polygon with an acute angle when 200 nm-thick buried TEOS oxide deposited before the a-Si deposition (a) and its schematic diagram (b)**

Our experimental results show that about 2 μm-long silicon grains are easily achieved near the oxide trench edge and that the locations of lateral grains are controlled by the angular points of the diagram. For the fabrication of high performance poly-Si TFT, the proposed method would be applicable as shown Figure 4. Figure 4 (a) and (c) are schematic diagrams for the long channel poly-Si TFTs and Figure 4 (b) and (d) are for the short channel poly-Si TFTs. Especially, circular grains of Figure 3 are applicable to honeycomb poly-Si array as shown Figure 4 (c).

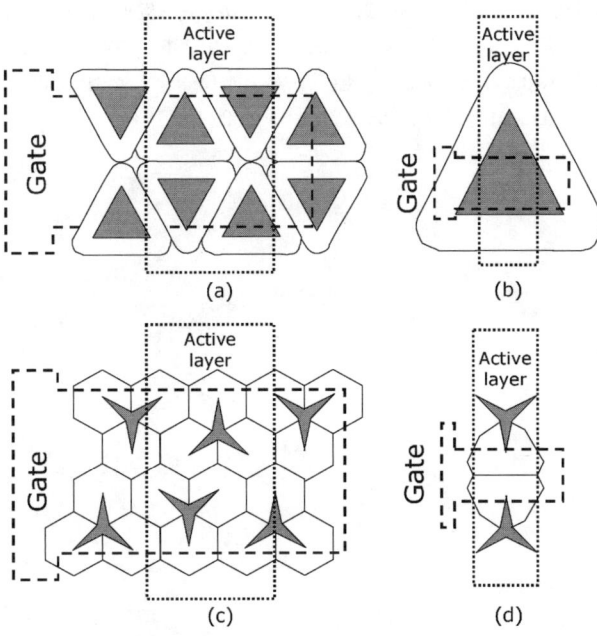

**Figure 4. Schematic diagrams applicable as an active layer for the poly-Si TFT**

**CONCLUSION**

A simple lateral grain growth of polysilicon employing single excimer laser irradiation is proposed. The proposed oxide trench, which is shaped like a triangle or a polygon with an acute angle, induces temperature gradient on the molten silicon film during the solidification. Consequently, lateral grain growth occurred and the location of the large lateral grain is controlled by the location of the oxide trench. It was verified by SEM that about 2 μm-long silicon grains are successfully achieved near the oxide trench edge and the location of the lateral grain is controlled by the angular points of the diagram.

As the proposed lateral grain growth method employs simple photolithography and comparative shallow trench, it would be competitive one for fabrication methods of high performance poly-Si TFTs

**REFERENCE**
[1] T. Sameshima, S. Usui, and M. Sekiya, *IEEE Electron Device Lett.*, 7, 276, 1986
[2] H. C. Cheng, L. J. Cheng et al. *AMLCD Tech. Dig.,* 281, 2000
[3] N. Kodama, K. Ikeda et. al, *Extended Abstracts of SSDM*, 431 ,1993
[4] R. S. Sposili and J. S. Im, *Appl. Phys. Lett.*, 69, 2864, 1996
[5] M. A. Crowder, J. S. Im et al, *IEEE Electron Device Lett.*, 19, 306, 1998
[6] R. Ishihara et. al, *AMLCD Tech. Dig.* 53, 2002

# Formation of large, orientation-controlled, nearly single crystalline Si thin films on SiO$_2$ using contact printing of rolled and annealed nickel tapes

Hwang Huh and Jung H. Shin
Department of Physics, Korea Advanced Institute of Science and Technology (KAIST), 373-1 Kusung-dong, Yusung-gu, Daejon, Korea

## ABSTRACT

Amorphous silicon ($a$-Si) films prepared on oxidized silicon wafer were crystallized to a highly textured form using contact printing of rolled and annealed nickel tapes. Crystallization was achieved by first annealing the $a$-Si film in contact with patterned Ni tape at 600 °C for 20 min in a flowing forming gas (90 % N$_2$, 10 % H$_2$) environment, then removing the Ni tape and further annealing the $a$-Si film in vacuum for 2 hrs at 600 °C. An array of crystalline regions with diameters of up to 20 μm could be formed. Electron microscopy indicates that the regions are essentially single-crystalline except for the presence of twins and/or type A-B formations, and that all regions have the same orientation in all 3 directions even when separated by more than hundreds of microns. High resolution TEM analysis shows that formation of such orientation-controlled, nearly single crystalline regions is due to formation of nearly single crystalline NiSi$_2$ under the point of contact, which then acts as the template for silicide-induced lateral crystallization. Furthermore, the orientation relationship between Si grains and Ni tape is observed to be Si (110) $\parallel$ Ni (001)

## INTRODUCTION

Polycrystalline silicon films (poly-Si) have been extensively studied because of its high carrier mobility and current carrying capacity in applications of active matrix liquid crystal display (AMLCD) and solar cell. For such cases, poly-Si are required to be fabricated below ~ 600 °C. For achieving low-temperature crystallization, there are many different methods- solid phase crystallization [1], excimer laser annealing [2], hydrogen dilution [3], etc. Among many different methods, silicide-mediated crystallization (SMC) of amorphous silicon has attracted much attention due to its advantages of low cost and large grain size. It is a kind of crystallization of metal-doped amorphous silicon. Crystallization of Ni-doped amorphous silicon has the advantage of low structural defect concentration. And so, nobilities as high as ~ 300 cm$^2$/Vs have been reported from n-channel thin film transistors fabricated using poly-Si prepared by Ni-SMC [4].

The ability to control both the location and the orientation of the crystal grains is rarely discussed in crystallization of a-Si. Since a poly-Si with controlled grain orientation is tantamount to a single crystalline Si film, developing such orientation-controlled poly-Si films will not only improve the performances of poly-Si based devices, but also expand their uses to high-performance devices that so far have been limited to single crystalline Si wafers. Controlling the location of crystal grains is relatively easy. Orientation-controlled crystallization is difficult because the characteristic of nucleation in solid phase is stochastic. We have shown that orientation-controlled crystallization of amorphous silicon deposited on a cold-rolled and annealed Ni tape due to nearly single-crystallinity of Ni tape [5]. However, such Ni tapes are not compatible with fabrication process for commercial TFT-based devices. We report the result of using Ni tapes as templates for location- and orientation-controlled crystallization of a-Si deposited on oxidized Si wafers.

## SAMPLE PREPARATION AND EXPERIMENTAL CONDITIONS

Commercial nickel tapes were cold-rolled at room temperature to 300 μm thickness and subsequently annealed at 1000 °C. This preparation process is known to produce nearly single crystalline Ni tapes [6]. This Ni tape usually can be used as substrates for preparation of superconducting films ($YB_2Cu_3O_{7-x}$) [7]. By X-ray poly figure analysis, the Ni tapes were confirmed to be nearly single crystalline with the (001) axis tilted by ~21 ° off the surface normal (not shown). Raised mesas were defined using photolithography and wet chemical etching. 100 nm thick $a$-Si:H films were deposited on oxidized silicon wafers using inductively coupled plasma of $SiH_4$. The base pressure, deposition pressure, deposition temperature, and the plasma power were $1\times 10^{-6}$ Torr, $8 \times 10^{-3}$ Torr, 130 °C (nominal), and 600 W, respectively. The annealing process was divided into two steps. First, nucleation was initiated by pressing the Ni tapes with mesas onto the a-Si:H film and annealing them at 600 °C for 20 min in a flowing forming gas (90 % $N_2$ and 10 % $H_2$) environment. Next, the Ni tapes were removed, and the a-Si films were further annealed at 600 °C for 2 h in vacuum ($1\times 10^{-4}$ Torr). After annealing, the crystallized films were analyzed using scanning electron microscopy (SEM) and transmission electron microscopy (TEM). Samples for SEM analysis were prepared by dipping the annealed films in a solution of $CrO_3$, deionized water, and HF (15.15 g : 200 mL : 2 mL) at room temperature for 7 min. This $CrO_3$ solution is well known to preferentially etch $a$-Si[8]. A sample for planar TEM analysis was prepared by backside etch and ion milling.

## RESULTS

Figure 1 (a) shows the low-resolution SEM image of the annealed film. Rectangular, crystallized areas can readily be identified, indicating that crystallization occurred only near the points of contact between $a$-Si and Ni tape. Figure 1 (b) shows one of the crystallized areas in detail. The crystallized area is larger than the area of the mesas. We can identify two distinct regions: the center region where the Ni tape contacted the *a*-Si film directly, and the edge region which was crystallized laterally from the center region. Figure 1 (c) shows the center region in detail. Figure 1 (d) shows the crystallization front of the edge region in detail. We find that the front is not smooth. Instead, it consists of many overlapping needles that are growing into the *a*-Si matrix. Because of needlelike morphology, the area near the crystallization front is not solid but contains voids, indicating that the crystallization was incomplete there. Away from the crystallization front, however, the crystallized area is solid with very few voids.

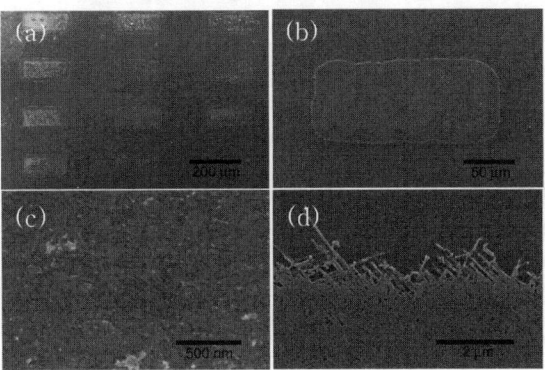

Figure 1 SEM images of the annealed and subsequently wet-etched sample (a) low-resolution image; (b) one crystallized area; (c) the center region of a crystallized area; and (d) the edge region of a crystallized area

This needlelike morphology is the typical observation in $NiSi_2$-mediated crystallized of *a*-Si [9]. $NiSi_2$ precipitates form in *a*-Si matrix at temperatures as low as ~325 °C, and acts as nucleation sites for Si grains. After nucleation, $NiSi_2$ migrate through the *a*-Si matrix in the <111> directions of Si grains, leaving crystalline needles behind. Hayzelden *et al.* have argued that the film can develop a <110> texture because only those grains that have the <110> axis normal to the surface have four <111> axes parallel to the film and therefore can grow unimpeded by the surface. Full crystallization of the film is achieved with overlapping and lateral growth of the needles.

The source of Ni is the Ni tape that was pressed onto the film during the initial step of the anneal. Thus, the whole crystallization process may be described by (1) formation of Ni silicides at the point of contact; (2) lateral growth of c-Si needles, mediated by migration of $NiSi_2$ precipitates; and (3) complete lateral crystallization by overlapping and lateral growth of needles. The schematic of crystallization process is presented in Fig. 2.

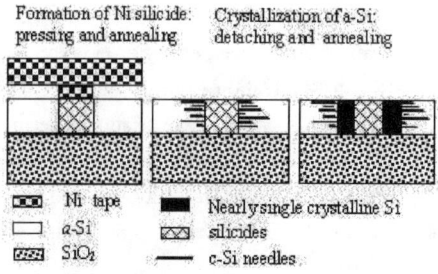

Figure 2. Schematic of the whole crystallization process

As full crystallization occurs via growth and overlap of such needles, we can expect the final crystal grains to be nearly single crystalline. Such a possibility is investigated using TEM. Figure 3 shows a bright-field TEM image of a crystallized region. Inset (a) shows a selected area diffraction pattern. We obtained a <110> diffraction pattern when the sample was tilted by ~ 21° off the surface normal. This is in agreement with our previous work that showed the <110> axis of the nucleated Si grains to be parallel to the <001> axis of the Ni tapes, and confirms that orientations of Si grains were controlled by the Ni tape [5]. More important, as shown in inset (b), we obtained the same <110> diffraction pattern even when no aperture was used, thus allowing the electron beam to illuminate the entire 20 × 20 μm area. This indicates that the entire area shown in Fig. 3 is nearly single crystalline, and confirms that using rolled, annealed Ni tapes to control the orientation of the crystal Si nuclei, large (~ 400 μm$^2$), highly textured crystalline grains can be obtained on SiO$_2$.

Furthermore, such orientation-control is observed no only within a crystallized region, but also between regions crystallized by different mesas on the Ni tape. This shown in Fig. 4, which shows the migrating directions of needles randomly selected from 12 different crystallized regions. We find that the migrating directions are not random, as is expected for lateral crystallization of randomly nucleated, <110> oriented grains [9]. Instead, 26 of 27 needles fall in five distinct groups, labeled α–η. More important, the angular relationships between the groups follow a strict pattern. Groups α and β are ~ 110° apart, which corresponds to the angles between the <111> directions. Furthermore, groups γ and δ are oriented such that they are again ~ 110° apart from α and β, respectively. Such orientation relationships may be ascribed to the occurrence of twinning. It is also possible that they are due to the occurrence of type A and B SMC, in which the NiSi$_2$ particle is rotated 180° about the Si <111> axis [10], as γ and δ may be described as 180° rotation of β about α direction and α about β direction, respectively. The direction of group η is somewhat ambiguous, because it is oriented in such a way that it is ~

110 ° apart from both δ and γ. It is also possible that it represents a 180 ° rotation of α about γ direction.

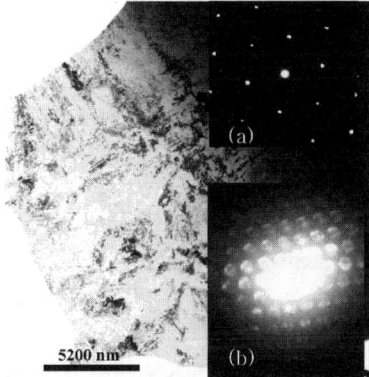

Figure 3. Bright-field TEM image of crystallized region. (a) Inset shows a selected area diffraction pattern. (b) Inset shows the electron diffraction pattern obtained without diffraction aperture. The electron illuminated area is nearly 400 μm$^2$ to obtain this pattern.

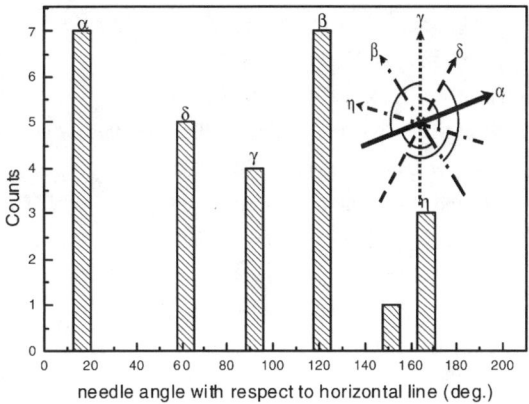

Figure 4 The distribution of migrating directions in 12 crystallized regions. Nearly all needles can be divided into five groups, labeled α −η. The inset shows the angular relationships of migrating directions. Curved lines indicate that the angle between directions is nearly 110 °

Thus, Fig. 4 indicates that including twins and possible occurrence of type A-B SMC, nearly all crystalline Si needles belong to the same <111> family of directions even when they were crystallized by different mesas on the Ni tape, and therefore were nearly 1 mm apart. Because the full crystallization of $a$-Si film proceeds by overlapping and lateral growth of such needles, Fig. 4, together with Fig. 3, indicates that all regions crystallized via contact with the Ni tape have the same orientation in all three directions. Thus, the entire crystallized film Si very highly textured.

## CONCLUSION

We have demonstrated location- and orientation-controlled crystallization of $a$-Si film on a glassy substrate using a contact-printing of cold-rolled and annealed Ni tape to provide a nearly single crystalline template. Nearly single crystalline regions with sizes in the range of 20 $\times$ 20 µm and identical orientation in all three directions were obtained after an anneal of 2 h and 20 min at 600 °C through formation and migration of $NiSi_2$ in epitaxial relationship with the Ni tape used to initiate nucleation.

## REFERENCES

1. R. B. Iverson and R. Reif, *J. Appl. Phys.* **62**, 1675 (1987).
2. J. S. Im and H. J. Kim, *Appl. Phys. Lett.* **64**, 2303 (1994).
3. J. Hanna and I. Shimizu, *MRS Bull.* **21,** 35 (1996).
4. K. Makihira, M. Yoshii, and T. Asano, *Device Research Conference 2001*, 189 (2001).
5. H. Huh and J. H. Shin, *Appl. Phys. Lett.* **79,** 3956 (2001).
6. D. Detert, P. Dorsch, and H. Migze, *Z. Metallkde.* **54**, 263 (1963).
7. S. H. Oh, J. Yoo, K. H. Lee, and D. J. Youm, *Physica C.* **308**, 91 (1999).
8. J. H. Ahn and B. T. Ahn, *J. electrochem. Soc.* **148**, H115 (2001).
9. C. Hayzelden, J. L. Batstone, and R. C. Cammarata, *Appl. Phys. Lett.* **60**, 225 (1992).
10. J. M. Gibson, J. L. Batstone, R. T. Tung, and F. C. Unterwald, *Phys. Rev. Lett.* **60**, 1158 (1988)

# Lifetime measurements of stain etched and passivated porous silicon

Ricardo Guerrero-Lemus, Fathi A. Ben-Hander[1], Cristoffer Ballif[2], Ali Kenanoglu[2], Dietmar Borchert[2], Cecilio Hernández-Rodríguez, Tomás Rodríguez[3] and José M. Martínez-Duart[1]
Dept. Física Básica, Universidad de La Laguna, Av. Astrofísico Francisco Sánchez s/n, 38204 La Laguna, S/C Tenerife, SPAIN.
[1]Dept. Física Aplicada C-XII, Universidad Autónoma de Madrid, 28049 Madrid, SPAIN.
[2]Laboratory- and Servicecentre, Institut für Solare Energiesysteme, Fraunhofer Institut, Haydnstr. 19, 45884 Gelsenkirchen, GERMANY.
[3]Dept. Tecnología Electrónica, Universidad Politécnica de Madrid, Av. Complutense s/n, Ciudad Universitaria, 28040 Madrid, SPAIN.

## ABSTRACT

In this work we present the first experimental study of photocarrier lifetimes in p-type and n-type Si substrates in which stain etched porous silicon (PS) has been formed on the surface. The lifetime values have been obtained before and after the surface passivation of the samples. The surface pasivation has been produced by two different techniques: (i) hydrogen passivation by immersion of the samples in a HF solution; and (ii) deposition of $SiN_x$ in a plasma enhanced chemical vapour deposition system. The results show a degradation of the photocarrier lifetime when the porous layers are not adequately passivated. This lifetime degradation is mainly associated to a large concentration of rapid recombination centres located at the Si/PS interface. We have also detected a weak influence of the PS outermost dangling bonds to the photocarrier lifetimes.

## INTRODUCTION

The discovery at the beginning of the 90's that porous silicon (PS) has photoluminescent [1] and electroluminescent [2] properties has sparked hopes that fabrication of optoelectronic circuits integrated on a single substrate may become feasible in the near future. Most of the research on PS, however, has been focused on electrochemically etched PS, even though it has been known that stain etched PS shows very similar photoluminescent and electroluminescent properties [3,4]. Furthermore, in order to obtain the stain etched PS samples it is not needed to realize a metallic contact (electrode) on the backside of the Si substrates. Recent works show that the formation process is reproducible when the stain etched samples are obtained in $HF/HNO_3$ solutions in which the nitric acid is present at very low concentrations [5].

Potential advantages of porous Si in solar cell structures have been previously pointed out by several authors [6-8]. In fact, recent works about Si antireflection coatings for solar cells show that the reflectance of the PS samples is under 5% in the 300 – 1100 nm spectral range for etching times lower than 30 seconds [9]. This result is very promising compared to the etching times (10- 20 minutes) and reflectance values (~ 11%) needed to form inverted pyramids in standard anisotropic etching of crystalline Si [10]. The characteristics of the formation process and the reflectance results obtained favour the stain etched PS for applications as antireflection coating in Si-based solar cell industrial processes [5,9]. A clear tendency towards thinner active cell structures [6] and simplified processing schemes with low thermal budget techniques is observable within contemporary crystalline Si PV research. However, improved light trapping and a lower surface recombination velocity [6] are crucial for crystalline Si-based cell structures composed by thin active regions.

In order to apply the stain etched process for Si-based solar cells, two main tasks must be previously solved: (i) realization of low resistance electric contacts on top of the PS layer [11]; and (ii) passivation of the PS layers to obtain low values of surface recombination velocity of photocarriers [12].

To characterise the quality of the surface passivation, one experimental technique is based in the measurement of the lifetimes of the excess charge carriers photoinduced by light pulses. However, very few research works in Si substrates with PS surfaces have been published: photovoltage [13], contactless laser-induced transient-grating measurements [14] and contactless Microwave transient Photoconductivity Decay (MWPCD) [15-16]. All of these experimental results have been obtained on electrochemical etched PS.

With respect to the MWPCD technique, the first experimental work referred in the literature showed a rise of the Si substrate lifetime when PS was formed on the Si surface, and this effect was attributed to a generation of charged traps at the Si/PS interface which avoid the charge carriers to reach the interface [15]. However, the results exposed by other authors [16] are in contradiction with the previous ones, showing a decrease of the lifetime values when PS was formed in the Si surface. The lifetime decrease is more severe when the PS layers are aged [16]. In this work, lifetime experimental results for PS stain etched formed in crystalline Si surfaces is presented for the first time. In order to study the influence of the Si/PS interface and the state of the PS surface, the surface has been hydrogen and $SiN_x$ passivated. The lifetime values have been obtained by measuring the microwave photoconduction decay when the samples are illuminated by short light pulses.

**EXPERIMENTAL DETAILS**

The stain etching process was applied to p-type Si wafers Cz <100>, 0.5-1.0 Ω·cm and 1-10 Ω·cm, and n-type Si wafers Cz <100> 10-20 Ω·cm. The wafers were immersed in a HF (48% wt.)/HNO$_3$ (65% wt.) 1:0.005 solution during different etching times. After the etching process, the samples were rinsed in deionised water and dried in a N$_2$ flux.

The nitridation process was carried out after de porous layers were formed on top of the Si substrates. The silicon nitride layers were deposited in a Plasma Enhanced Chemical Vapour Deposition (PECVD) system (Centrotherm prototype) at 13.56 MHz, with plane-parallel graphite electrodes (45 × 45 cm$^2$). The precursor gases were SiH$_4$/NH$_3$ ½ + 50% H$_2$. The thickness of the silicon nitride layers was around 100 nm and the refractive index 2.4. The porous layers and the silicon nitride layers were formed on both sides of de Si wafers.

The reflectance measurements were obtained using a Varian Spectrometer CARY SE (UV-Vis-NIR). The equipment used for the MWPCD measurements was an AMECON JANUS –

2000 μ-PCD system, where the effective lifetime range is 500 ns – 10 ms, and the spatial resolution is 10x10 μm$^2$.

## RESULTS

Table I shows the lifetime values for the Si substrates before and after the silicon nitride passivation process. The results show a large increase of the lifetime (1 – 2 orders of magnitude) when the substrates are passivated with the silicon nitride layer. These results evidence the importance of the passivation processes to increase the probability of photocarrier collection and, thus, to increase the radiation conversion efficiency for devices based in Si substrates.

**Table I.** lifetime values for the Si substrates before and after SiN$_x$ passivation.

| Lifetime (μs) | Before SiN$_x$ passivation | After SiN$_x$ passivation |
|---|---|---|
| Si-n, 10 – 20 Ω·cm | 24.4 | 584 |
| Si-p, 1 – 10 Ω·cm | 8.00 | 681 |

Figure 1a shows the values of the total reflectance spectra for a Si substrate and the same substrate with a PS stain etched layer on top. These spectra show the antireflection behaviour of the porous layer and, consequently, the potential application for Si-based solar cells. Figure 1b shows the reflectance spectra for very thin stain etched PS layers, i.e. immersed during few seconds. The significant decrease in the values of reflectance of the stain etched layers makes this process very competitive compared with other chemical procedures used in the photovoltaic industry.

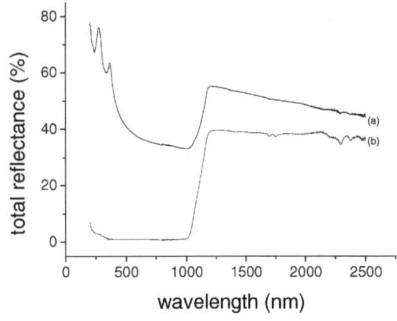
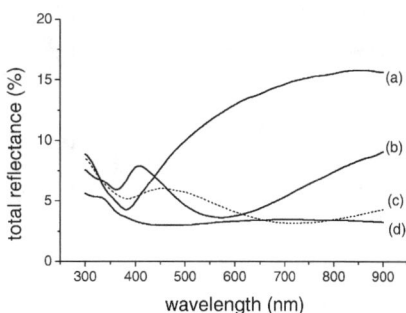

**Figure 1a.** total reflectance spectra of (a) the Si substrate, and (b) PS stain etched 3 minutes.

**Figure 1b.** total reflectance spectra of PS stain etched (a) 1 sec.; (b) 5 sec.; (c) 9 sec.; and (d) 45 sec.

Table II shows the lifetime values for p-type and n-type Si substrates with PS formed on their surface. It can be observed (compare table I and Table II) that the lifetime values for non-passivated samples remains similar before and after PS is formed. When silicon nitride is deposited upon the porous layers, the lifetime values decrease appreciably. This result indicates that the silicon nitride layers do not act as passivation layers on the samples with PS in the surface. Also, the lifetime results show that an increase of the porous layer thickness (by increasing the etching time) do not influence significantly the lifetime values.

**Table II.** lifetime values for the Si substrates with PS formed in the surface, before and after $SiN_x$ passivation.

| Lifetime (µs) | Before $SiN_x$ passivation | After $SiN_x$ passivation |
|---|---|---|
| Si-n, stain etched 30" | 21.4 | 21.4 |
| Si-p, stain etched 30" | 10.7 | 7.93 |
| Si-n, stain etched 90" | 24.1 | 18.1 |
| Si-p, stain etched 90" | 10.3 | 9.78 |

Figure 2 shows the lifetime values for Si substrates (p-type 0.5 – 1.0 Ω·cm) and for the same substrates with PS in the surface. The lifetime values have been measured before and just after a short immersion (3 minutes) in a HF solution. The lifetime measurements have also been obtained for the same samples just immersed in HF and exposed to ambient conditions for several minutes. It can be observed that the hydrogen passivation increases the lifetime values experimentally measured with respect to the samples without hydrogen passivation. Also, it can be observed that the passivation effect in the lifetime values is more pronounced for Si substrates without PS on the surface. In addition, the lifetime values progressively decrease with exposition time for samples exposed to ambient conditions, which is a consequence of the metastable passivation behaviour of the hydrogen bonds.

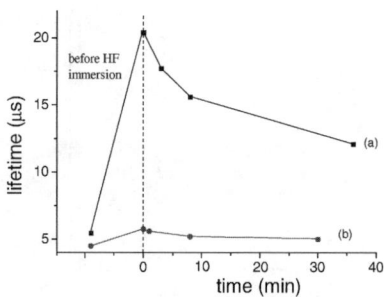

**Figure 2.** lifetime measurements for different aging times in ambient conditions for (a) p-Si substrate 0.5 – 1.0 Ω·cm, and (b) the same substrate after PS formed on top. The measurements were obtained before and after the samples were immersed during 3 minutes in a HF solution.

As we have observed above, the increase of the porous layer thickness does not produce a valuable variation of the lifetime values. This result can be interpreted as a consequence of the prevalence of the Si/PS interface as the region where the rapid recombination centres more strongly influences the lifetime values. This conclusion is also reinforced by the fact that the hydrogen passivation produces a higher lifetime increase for crystalline Si surfaces than for PS surfaces. Also, the great effect that the nitridation process has on the lifetime values for the Si substrates opposite to the negligible effect in the case of porous surfaces, is an argument which favours the preponderant role of the Si/PS interface in the lifetime values obtained.

To reduce the deleterious effect of the Si/PS interface and the PS layers to the lifetime values of the photocarriers, different alternatives can be proposed: (i) the anodic oxidation of the PS layers is a low-cost alternative, but it requires to make electric contact to Si; (ii) the thermal oxidation of the PS layers increases the energy budget; (iii) the chemical oxidation can turn out to be suitable for industrial processes because it does not need electrical contact, but it needs further development; (iv) removing a photocarrier type from the Si/PS interface by means of the control of the doping profile close to this region could be an alternative which is usually employed in standard procedures; and (v) an alternative chemical passivation process, similar to the hydrogen passivation, but with a greater stability in contact with the atmosphere could be also implemented [17].

## CONCLUSIONS

In this work we present, for the first time, results of lifetime for PS stain etched layers. The results indicate that the Si/PS interface is the major responsible for the low lifetime values obtained. Thus, it is necessary to find experimental procedures that would improve the lifetime values obtained for the stain etched samples and bring them to values comparable to the ones obtained in the case of direct $SiN_x$ passivation on the bare Si wafers. The improvement of the lifetime values for Stain etched porous silicon is an essential task to apply the PS stain etched layers as antireflection coatings on Si-based solar cells.

## ACKNOWLEDGEMENTS

The authors are grateful to the Technology Division of the Instituto Astrofísico de Canarias (IAC) for the support in the reflectance measurements. This work has been financed by the Canarias Government (PI2000/074) and the Ministerio de Ciencia y Tecnología (MAT2002-00044).

## REFERENCES

1.- L.T. Canham, Appl. Phys. Lett. **57**, 1046 (1990).
2.- N. Koshida and H. Koyama, Appl. Phys. Lett. **60**, 314 (1992).
3.- R.W. Fathauer T. George, A. Ksendzov y R.P. Vasquez, Appl. Phys. Lett. **60**, 995 (1992).

4.- J. Sarathy, S. Shih, K. Jung, C. Tsai, K.-H. Li, D.-L. Kwong, J.C. Campbell, S.-L. Yau and A.J. Bard, Appl. Phys. Lett. **60**, 1532 (1992).
5.- R. Guerrero-Lemus, C. Hernández-Rodríguez, F. Ben-Hander and J.M. Martínez-Duart, Solar Energy Mat. And Solar Cells **72**, 495 (2002).
6.- A. Rohatgi, E. R. Weber and L. C. Kimerling, J. Electron. Mat., **22**, 65 (1993).
7.- Y. S. Tsuo, Y. Xiao, M. J. Heben, X. Wu, F. J. Pern and S. K. Deb, Proc. 23rd IEEE-PVSC, 287 (1993).
8.- S. M. Vernon, N. M. Kalkhoran, H. P. Maruska and W. D. Halverson, Proc. 1st IEEE-WCPEC, 1583 (1994).
9.- R. Guerrero-Lemus, F. Ben-Hander, L. Vázquez, C. Hernández-Rodríguez and J.M. Martínez-Duart, Phys. Status Solidi A (in press) (2003).
10.- M.A. Green, *"Solar Cells: Operating Principles, Technology and System Applications"*, Ed. Prentice Hall, Inc. (1981). ISBN 0-13-822270-3.
11.- R.J. Martín-Palma, J. Pérez-Rigueiro, R. Guerrero-Lemus, J.D. Moreno and J.M. Martínez-Duart, J. Appl. Phys., **85**, 583 (1999).
12.- R. Guerrero-Lemus, F.A. Ben-Hander, C. Hernández-Rodríguez and J.M. Martínez-Duart, Mat. Sci. Engineer. (in press) (2003).
13.- S.T. Wu, Y.H. Wang and Q.H. Shen, Appl. Surf. Sci. **158**, 268 (2000).
14.- G. Kopitkovas, I. Mikulskas, K. Grigoras,I. Simkienÿ and R.Tomasiunas, Appl. Phys. A **73**, 495 (2001).
15.- V. Subramanian, J. Sobhanadri, A. Gupta, and V.K. Jain, 1996 Asia-Pacific Microwave Conference Proceedings. R.S. Gupta, Univ. Delhi South Campus, New Delhi, India vol.3, 1139 (1996).
16.- L. Stalmans, J. Poortmans, H. Bender, M. Caymax, K. Said, E. Vazsonyi, J. Nijs and R. Mertens, Progress in Photovoltaics **6**, 233 (1998).
17.- W.J. Royea, A. Juang, and N.S. Lewis, Appl. Phys. Lett. **77**, 1988 (2000).

## Anomalous behaviour of stain etched porous silicon photoluminescence

Ricardo Guerrero-Lemus, Fathi A. Ben-Hander[1], Cecilio Hernández-Rodríguez and José M. Martínez-Duart[1]
Dept. Física Básica, Universidad de La Laguna, Av. Astrofísico Francisco Sánchez s/n, 38204 La Laguna, S/C Tenerife, SPAIN.
[1]Dept. Física Aplicada C-XII, Universidad Autónoma de Madrid, 28049 Madrid, SPAIN.

## ABSTRACT

In this work we present a comparative study of porous silicon (PS) photoluminescence for samples stain etched and electrochemically etched. The etching parameters for both types of samples have been adjusted to obtain similar porous structures. The photoluminescence spectra have been obtained varying the excitation energy between 2.48 and 3.54 eV. The variation of the excitation energy produces differences in the evolution of the emission energy maximum between both types of PS. This behaviour is attributable to the differences in oxidation level in the porous structure. Also it has been established a higher concentration of luminescent centers for stain etched PS.

## INTRODUCTION

Porous silicon (PS) has been studied from the middle of the 50's [1]. However, the discovery at the beginning of the 90's that PS shows notorious photoluminescent [2] and electroluminescent [3] properties has sparked hopes that fabrication of optoelectronic circuits integrated on a single substrate may become feasible in the near future.

Electrochemically etched PS (EEPS) has been the procedure mainly used to produce Si luminescent nanostructures due to its low cost and because it allows a suitable control of the formation process [2]. Nevertheless, the electrochemical process presents the difficulty of realizing good electrical contacts to its surface.

At the beginning of the 90's the formation of stain etched PS (SEPS) showing luminescence was achieved by the chemical attack in $HF/HNO_3$ solutions [4-5]. Initially, the stain etching technique was not very successful, due to difficulties in the control of the process [6]. However, later studies demonstrated that in low concentration $HNO_3$ solutions, the SEPS samples showed excellent and reproducible luminescent properties [7]. Also, the achievement of reflectance values below 4 % (300 - 1100 nm) confirmed the potential advantages of SEPS as antireflection coating for silicon-based solar cells [7]. Due to its photoluminescent properties [4], SEPS can also convert high energy solar photons into photons of lower energy, increasing the photoconversion efficiency of the solar cells [8]. Nevertheless, few works have been published on the photoluminescent properties of SEPS [9-10].

In this work it is shown a comparative study of the SEPS and EEPS luminescence properties. Previously, the etching parameters have been adjusted to assure that the porous layers present similar structural characteristics for both kinds of samples. Then, the differences in the luminescence mechanisms are discussed for both kinds of samples.

## EXPERIMENTAL DETAILS

The silicon wafers used in this work were p-type <100> 0.1 - 0.5 $\Omega$·cm, and the etching solutions were composed by HF (48 % wt.) and $HNO_3$ (65 % wt.). To obtain SEPS, the etching solutions were selected in the range in which the polishing process is not present simultaneously with the formation of the pores: 0.3 % - 0.8 % of $HNO_3$ in HF [7]. For EEPS the current densities used were 15, 50 and 75 mA/cm$^2$. All the porous layers were formed in the dark. After the formation process, all the porous samples were rinsed in deionised water two minutes and subsequently dried in a $N_2$ flux.

The thicknesses of the porous samples were measured by cross section Scanning Electron Microscopy (SEM) in a microscope Philips model XL-30. The samples were submitted to anodic oxidation at a current density of 15 mA/cm$^2$. The surface area of the porous samples can be directly evaluated considering the charge transferred in the oxidation process until a certain voltage (2 V in our case) is reached.

The photoluminescence was measured for the samples just dried at excitation energies between 2.48 and 3.54 eV. The lower excitation energy was limited by the detection of luminescence signal from the stain etched samples. The photoluminescent measurements were obtained with a luminecent AMINCO Bowman series 2 computer controlled spectrometer using a xenon lamp of 150 W as excitation source.

## RESULTS

The cross section SEM images show that the SEPS thickness is approximately 300 nm for all the samples (figure 1). To obtain a similar thickness in the EEPS, the etching times for samples anodized at current densities of 15, 50 and 75 mA/cm$^2$ were 15, 4.37 and 2.31 seconds, respectively.

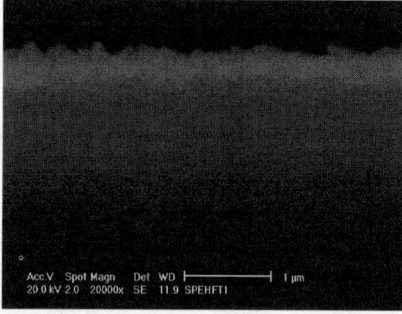

**Figure 1.** SEM image of PS stain etched during 1 minute in a 1:0.005 HF/$HNO_3$ solution.

Table I shows the charge transferred during the anodic oxidation processes for the different samples. The samples 2 and 6 are the most similar in thickness and surface area; therefore, the photoluminescent results exposed below are obtained from these two groups of samples.

**Table 1.** Measurement of the charge transferred to the porous layers by anodic oxidation for samples formed by electrochemical etching (samples 1 - 3) and stain etching (samples 4 - 6).

| sample | Conditions | Etching time | Charge (mC) |
|---|---|---|---|
| 1 | 15mA/cm$^2$ | 15.00 s | 108.06 |
| 2 | 50mA/cm$^2$ | 4.37s | 70.05 |
| 3 | 75mA/cm$^2$ | 2.31s | 46.95 |
| 4 | 1:0.003 HF/HNO$_3$ | 60.00 s | 128.55 |
| 5 | 1:0.005 HF/HNO$_3$ | 60.00 s | 79.02 |
| 6 | 1:0.008 HF/HNO$_3$ | 60.00 s | 72.35 |

Since photoluminescence was not observed in sample 2, the etching time was set to 3 minutes in order to increase the number of luminescent centers. For our samples it has been experimentally determined and it is considered that increasing the etching time for EEPS only the intensity of the photoluminescence is varied, without any appreciable shift of the photoluminescence emission maximum [11]. In this case, variations of the depth structure with etching time can only be considered for longer etching times and related to changes in the HF concentration inside the pores [12]. Since sample 6 shows luminescence this result indicates that the concentration of luminescent centers is significantly higher for SEPS, with respect to EEPS.

Figure 2 shows the evolution of the energy of the luminescence maxima for the PS stain etched (1 minute) and PS electrochemically etched (3 minutes) samples. For the sample electrochemically etched, the emission maximum increases when the excitation energy increases. Based on the quantum confinement model of nanostructured silicon [2], this behaviour is attributed to the excitation of increasingly small nanocrystals when the excitation photon energy increases. According to this model, the smaller nanocrystals emit photons of higher energies. It can be observed that the increase of the energy of the emission maxima is slower than the excitation energy increase. This behaviour has been attributed to the influence of excitonic and electronic transitions (formed by the presence of oxygen and hydrogen in the porous structure) in the quantum confinement luminescent processes [13].

On the contrary, for the SEPS samples the energy of the emission maxima weakly decreases when the excitation energy increases (figure 2). This result appears to be in contradiction not only with the quantum confinement model [2] but also with other models which associate the luminescence to surface compositional effects [14-16]. We have also observed that the samples formed by stain etching are hydrophobic and show a higher oxidation level than those electrochemically etched [9].

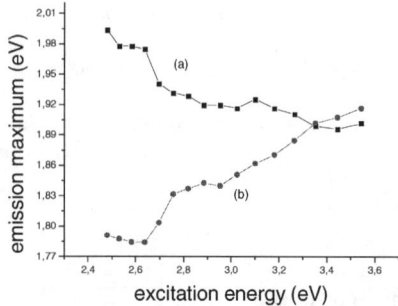

**Figure 2.** Evolution of the emission maximum for PS (a) stain etched in a 1:0.008 HF/HNO$_3$ solution 1 minute; and (b) electrochemically etched at 50 mA/cm$^2$ 3 minutes.

Based on the mostly elaborated luminescence model exposed above [13], it is known that the oxygen presence in the EEPS samples diminishes the shift of the emission energy maximum when the excitation energy varies. Therefore, the higher oxidation level of the SEPS might result in a decrease of the emission maximum energy when the excitation energy increases. As it can be observed in figure 3, this behaviour for the SEPS samples is also produced varying the concentration of the nitric acid in the etching solution and the cleaning process.

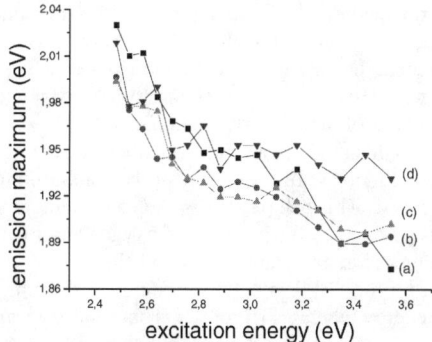

**Figure 3.** Evolution of the luminescence maximum with respect to the excitation energy, for SEPS etched 1 minute in a 1:0.003 HF/HNO$_3$ solution and cleaned in (a) deionised water or (b) in ethanol; or etched 1 minute in a 1:0.008 HF/HNO$_3$ solution and cleaned in (c) deionised water or (d) in ethanol.

Figure 4 shows the evolution of the luminescent intensity with respect to the excitation energy. As it is expected, the luminescent intensity increases when the excitation energy increases, as the excited nanocrystals are increasingly small. Thus, the concentration of luminescent nanocrystals increases when the excitation energy increases.

Figure 5 shows the FWHM decrease of the luminescent spectra when the excitation energy increases for both types of PS. For the EEPS this evolution is explained in terms of a larger proportion of smaller nanocrystals excited. This can be interpreted as a decrease of the probability of emission process assisted by excitonic and electronic states, which would make the spectral dispersion of the emitted photons to diminish. For SEPS the FWHM behaviour is similar. Nevertheless, in this case, the FWHM values are significantly larger, which can be due to a higher oxidation of the porous surface, and thus, making the concentration of the nanocrystal excitonic and electronic states involved in the photoluminescence increase.

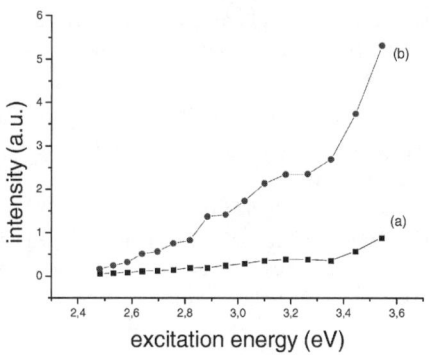

**Figure 4.** Evolution of the luminescent intensity with respect to the excitation energy, for PS (a) stain etched 1 minute in a 1:0.008 HF/HNO$_3$ solution; and (b) electrochemically etched 3 minutes at 50 mA/cm$^2$.

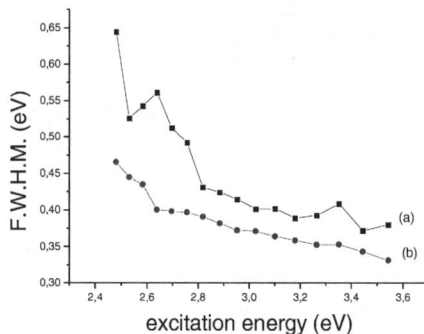

**figure 5.** Evolution of the FWHM with respect to the excitation energy for PS (a) stain etched 1 minute in a 1:0.008 HF/HNO$_3$ solution; and (b) electrochemically etched 3 minutes at 50 mA/cm$^2$.

## CONCLUSIONS

Photoluminescent porous silicon samples with similar inner surface areas have been formed by stain etching and electrochemical etching techniques. The SEPS shows a decrease of the emission energy maximum when the excitation energy increases, which means an anomalous behaviour with respect to the better known EEPS. This difference in behaviour is observed varying the stain etching concentrations and after subjecting the samples to various cleaning processes after formation. The variation in comportment is associated to the larger surface oxidation and hydrophobic character of the SEPS, in contrast with the electrochemically etched samples . Also it has been verified a larger concentration of luminescent nanocrystals for SEPS. Thus, a manifest difference in the photoluminescent characteristics between both types of PS is deduced, though in both cases the luminescent mechanisms are based on the quantum confinement model.

## ACKNOWLEDGEMENTS

The authors are grateful to the Technology Division of the Instituto Astrofísico de Canarias (IAC) for the support in the reflectance measurements. This work has been financed by the Canarias Government (PI2001/074) and the Ministerio de Ciencia y Tecnología (MAT2002-00044).

## REFERENCES

1.- A. Uhlir, Bell Syst. Tech. J. **35**, 333 (1956).
2.- L.T. Canham, Appl. Phys. Lett. **57**, 1047 (1990).
3.- N. Koshida and H. Koyama, Appl. Phys. Lett. **60**, 314 (1992).
4.- R.W. Fathauer T. George, A. Ksendzov y R.P. Vasquez, Appl. Phys. Lett. **60**, 995 (1992).
5.- J. Sarathy, S. Shih, K. Jung, C. Tsai, K.-H. Li, D.-L. Kwong, J.C. Campbell, S.-L. Yau and A.J. Bard, Appl. Phys. Lett. **60**, 1532 (1992).
6.- T. George, M.S. Anderson, W.T. Pike, T.L. Lin, R.W. Fathauer, H. Jung and D.L. Kwong, Appl. Phys. Lett. **60**, 2359 (1992).
7.- R. Guerrero-Lemus, C. Hernández-Rodríguez, F. Ben-Hander and J.M. Martínez-Duart, Solar Energy Mat. and Solar Cells **72**, 495 (2002).
8.- V.Y. Yerokhov and I.I. Melnik, Renewable and Sustainable Energy Rev. **3**, 291 (1999).
9.- R. Guerrero-Lemus, F. Ben-Hander, J.L.G. Fierro, C. Hernández-Rodríguez and J.M. Martínez-Duart, Phys. Status Solidi A (in press) (2003).
10.- R. Guerrero-Lemus, F. Ben-Hander, L. Vázquez, C. Hernández-Rodríguez and J.M. Martínez-Duart, Phys. Status Solidi A (in press) (2003).
11.- M.B. Robinson, A.C. Dillon and S.M. George, Appl. Phys. Lett. **62**, 1493 (1993).

12.- M. Thonissen, S. Billat, M. Kruger, H. Luth, M.G. Berger, U. Frotscher and U. Rossow, J. Appl. Phys. **80**, 2990 (1996).
13.- M.V. Volkin, J. Jorne, P. Fauchet, G. Allan and C. Delerue, Phys. Rev. Lett. **82**, 197 (1999).
14.- H.D. Fuchs, M. Stutzmann, M.S. Brandt, M. Rosenbauer, J. Weber, A. Breitschwerdt, P. Deak and M. Cardona, Phys. Rev. **B48**, 8172 (1993).
15.- S. Tong, X. Liu and X. Bao, Appl. Phys. Lett. **66**, 469 (1995).
16.- S. Banerjee, Phys. Rev. **B51**, 11180 (1995).

## Scattering Rings in Birefringent Porous Silicon

Claudio J. Oton, Zeno Gaburro, Mher Ghulinyan, Nicola Daldosso, Lucio Pancheri[1], Paolo Bettotti, Luca Dal Negro and Lorenzo Pavesi
INFM and Department of Physics, University of Trento, Via Sommarive 14, 38050 Povo (TN), Italy
[1] Department of Information and Telecommunication Technology, University of Trento, Via Sommarive 14, 38050 Povo (TN), Italy

## ABSTRACT

We report the observation of strongly anisotropic scattering of laser light at oblique incidence on (100)-oriented porous silicon layers. We performed angle-resolved light scattering measurements and three concentric rings were observed. Modeling porous silicon by means of nanometric columnar air pores and an effective anisotropic uniaxial dielectric constant explains the observed phenomenon, and besides, the observation of the angle aperture of these rings allows a direct measurement of relative birefringence. We finally study the changes of optical anisotropy after different modifications of the structure.

## INTRODUCTION

Porous silicon (PS) is a nanostructured material that has been very actively investigated during the last decade [1,2]. Its refractive index can be continuously tuned over a wide range by varying the porosity, *i.e.* the volumetric fraction of air in the layer. Therefore controlled variations of porosity with depth allow fabrication of inexpensive silicon-based high quality dielectric structures with many potential applications, such as distributed Bragg reflectors (DBRs) [3], microcavities [4] and waveguides [5].

The small size of the pores allows to apply an effective medium approximation in which the medium behaves as optically homogeneous. However, special care must be taken when the structure of the inhomogeneities is anisotropic. This fact leads to an anisotropic dielectric constant. In fact, it is well known that PS is optically anisotropic for this reason [6, 7]. The sign and the orientation of anisotropy depends on the crystallographic orientation of the silicon wafer. (100)-oriented porous silicon shows positive uniaxial birefringence with the optic axis perpendicular to the surface, thus only oblique incidence allows to observe polarization-dependent propagation of light.

Light can be scattered either on the air/PS/silicon interfaces, or on the walls of the pores present in PS. In heavily doped PS, the latter mechanism prevails because of the larger size of the pores and better flatness of the air/PS/Si interfaces. In this work we report the observation of light scattering rings in PS which allows direct measure of birefringence [8]. We also study the birefringence changes induced by filling the pores with a liquid, by thermal annealing, and by chemical etching of a PS layer.

## EXPERIMENT

The samples were fabricated starting from (100)-oriented heavily doped p-type silicon wafers (0.01 Ω cm resistivity). The electrolyte was prepared by mixing 31% volumetric fraction of aqueous HF (48 wt. %) with ethanol. We forced a constant electrochemical current with a Keithley 2400. A current density of 50 mA/cm$^2$ was applied for 22 minutes to make PS layers of about 30 μm. The samples were rinsed in ethanol and pentane successively, and finally dried in air.

We have investigated the geometry of the microstructure with transmission electron microscopy (TEM). The micrographs shown in fig. 1 correspond to small pieces extracted by scraping the sample with a metallic blade. Left panel shows a top-view of the sample, and right panel shows a cross section. The diffraction patterns can also be observed in the insets. They correspond respectively to [001] and [011] crystalline orientation, ensuring a good alignment of the sample in each case.

We illuminated another 30μm-thick layer obliquely (incidence angle $\theta_0 \sim 45°$) with a He-Ne collimated laser beam (λ=633 nm) and we placed a white screen some centimeters far from the sample. For unpolarized input light, three rings were observed, but for polarized light, the pattern changed between the transverse-electric (TE) and the transverse-magnetic polarization (TM). For each polarization we observed two rings, as shown in fig. 2. Both polarizations gave a scattering ring tangent to the incident and reflected beams (zero-order ring). However, for incoming TM-polarized light, a smaller concentric ring (internal ring) was also observed, whereas for incoming TE-polarized light a larger concentric ring (external ring) was observed instead. In free-standing samples, the scattering rings could be observed in transmission with the same radii and polarization [8].

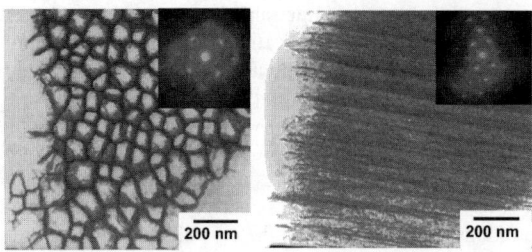

**Figure 1.** Top view (left panel) and cross section (right panel) TEM micrographs of small pieces scraped away from the porous silicon sample. The insets show the selected area diffraction pattern. Left panel: [001] zone axis, right panel: [011] zone axis.

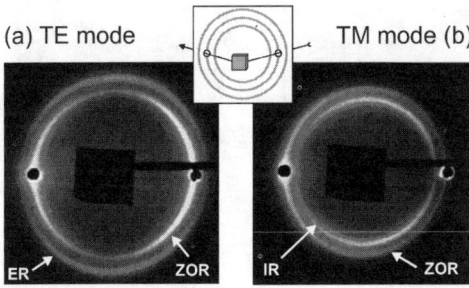

**Figure 2.** Photographs of reflected scattering rings for both TE (panel a) and TM (panel b) polarization. The external ring (ER), zero-order ring (ZOR) and internal ring (IR) are pointed with arrows. The inset sketches the setup.

In order to make a quantitative analysis of the pattern we made angle-dependent light scattering measurements. We collected the light with a fiber bunch mounted on a rotating table that varied the angle along a plane which was perpendicular to the incidence plane and to the surface, and we detected the light intensity with a silicon photodiode. Figure 3 shows the obtained data (0° corresponds to normal direction).

We measured the scattered pattern after different modifications of the structure. First, we filled the pores with pentane adding a drop and covering the sample with a glass plate to avoid fast evaporation. We show in fig. 4a the pattern of a dry and a wet sample. We annealed another sample at 1100°C for 30 s in air in order to partially oxidize the internal surface of PS, obtaining the pattern shown in fig. 4b. Finally we made a less porous sample (applying 30 mA/cm$^2$ for 22 minutes) and we immersed it in the HF electrolyte for different times in order to increase gradually the porosity by means of chemical etching of PS, and we measured the data shown in fig. 4c.

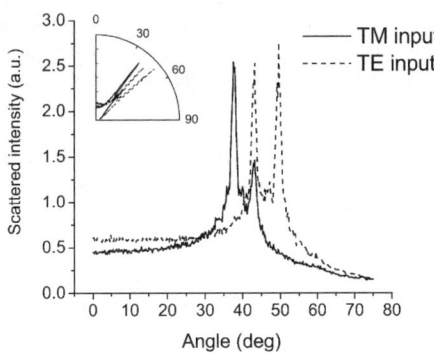

**Figure 3.** Quantitative measurement of scattered light varying the out-of-plane angle. Both polarizations TM (solid) and TE (dashed) are shown. The inset shows the same plot in polar coordinates.

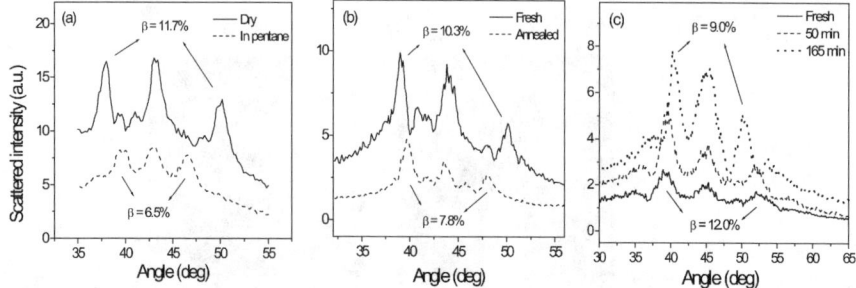

**Figure 4.** Scattering ring pattern of different samples after various treatments (unpolarized illumination). The arrows point the external and internal rings, which are used for calculation of relative birefringence (β). (a) Dry sample (solid line) and wet sample with pentane (dashed line). (b) Fresh (solid line) and thermally annealed sample (dashed line). (c) Fresh sample (solid line), same sample after 50 min chemical etching (dashed line) and after 165 min (dotted line).

## RESULTS AND DISCUSSION

As one can see in the TEM micrographs shown in fig. 1, this type of PS samples can be modeled as air parallel mesopores with polygonal section embedded in a silicon matrix. Despite the small size of the pores with respect to the wavelength of light, one could expect a small amount of light to scatter on the walls of the pores. As these are vertical, the scattered light conserves the latitude angle (angle with respect to the pore axis) for any pore section shape. This scattering generates the zero-order ring, since it forms the same angle as the incident light. It can also be observed in reflection because the whole pattern can be reflected at the bottom interface.

The secondary rings are due to PS optical anisotropy. When polarized light (TE or TM) enters the medium, it propagates with its correspondent refractive index ($n_{ord}$ or $n_H$). When a scattering event occurs, the polarization can be partially lost. The light that conserves the polarization generates the zero-order ring, and the crossed-polarized light undergoes a refraction event because of the change of refractive index. This explains an external ring for light that scatters from TE to TM polarization, and an internal ring for light that scatters from TM to TE [8]. Both cases are shown schematically in fig. 5.

To calculate analytically the angles, one has to solve three refraction events taking into account that the extraordinary mode propagates with a refractive index that depends on the angle. This problem is analyzed elsewhere [8] and it is found that the angles of the internal and external rings are given by the following equations:

$$\sin(\theta_{ext}) = (\beta + 1)\sin(\theta_0) \quad (1)$$
$$\sin(\theta_{int}) = (\beta + 1)^{-1}\sin(\theta_0) \quad (2)$$

where β is the relative birefringence, defined as:

$$\beta = \frac{n_{ext} - n_{ord}}{n_{ord}} \quad (3)$$

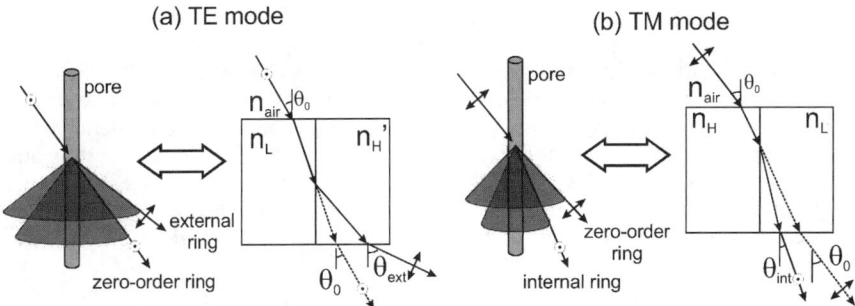

**Figure 5.** Mechanism of formation of the internal and external rings for incident TE (panel a) and TM polarization (panel b). The diagram on the left shows the effect and the schematic drawing on the right shows all the path of the light that comes from outside with a fixed angle $\theta_0$ and gets refracted 3 times.

where $n_{ext}$ and $n_{ord}$ are the extraordinary and ordinary refractive indices.

In fig. 2, the internal, zero-order and external rings are clearly visible as maxima in the pattern, and they formed respectively 37.8°, 43.2° and 49.6°. The incident angle was equal to the zero-order ring angle. With these angles one obtains a β value of (11.5 ±1)%. Critical angle measurements were also performed coupling the sample with a prism of higher refractive index than the sample. This allowed an independent and more accurate estimation of the relative birefringence, and we obtained 12.4 %, which agrees with the measurements of the scattering rings.

We show in fig. 4 how the angles of the rings change after different treatments. Firstly, fig. 4a reports the scattering pattern when the pores are filled with a liquid (pentane, refractive index ≈1.36). The rings get closer, therefore the birefringence decreases. Analyzing the ring angle, we obtained a decrease of β from 11.7% to 6.5%. When the pores are filled with pentane the index contrast of the dielectric mixture diminishes, thus we expect to observe a birefringence reduction. Figure 4b shows also a birefringence diminution after thermal annealing. We observed a decrease from 10.3% to 7.8% because a layer of silicon oxide grows on the walls of the pores, thus reducing the index contrast. Finally we can see in fig. 5c the decrease of anisotropy after chemical etching. The fresh sample was 12.0% birefringent, and as the etch time increases, the anisotropy decreases gradually until 9.0% after 165 min of chemical etching. The main effect of chemical etching of PS is an enlargements of the pores. This increases the porosity of the sample, but as we deal already with high porosities, even higher porosities unbalance more the mix between silicon and air hence the optical anisotropy decreases [9]. We believe that the increase in intensity of the rings is due to a decrease of the absorption coefficient since higher porosity leads to less absorption in the visible range [10].

## CONCLUSIONS

Scattering of light in (100)-oriented $p^+$-type porous silicon layers was investigated and bright scattering rings were observed. The rings are formed when light scatters on the vertical walls of the pores. Optical anisotropy of porous silicon is the origin of internal and external rings, due to a refraction event of the light that changes polarization. This fact allows to make straightforward measurements of optical anisotropy by only measuring the ring aperture. We performed relative birefringence measurements of porous silicon when pores were filled with pentane, after thermal annealing, and after chemical etching for different times, and in all cases a decrease was observed.

## ACKNOWLEDGEMENTS

We acknowledge INFM section E project PAIS-RANDS 2001 for financial support. We thank L. Zampedri for the m-line measurements and S. Gialanella for the TEM measurements.

## REFERENCES

1. L. T. Canham, *Appl. Phys. Lett.* **57** (10), 1046 (1990).
2. A. G. Cullis, L. T. Canham and P. D. J. Calcott, *J. Appl. Phys.* **82** (3) 909 (1997).
3. M. G. Berger, R. Arens-Fischer, M. Kruger, S. Billat, H. Luth, S. Hilbrich, W. Theiβ and P. Grosse, *Thin Solid Films* **297**, 137 (1997).
4. L. Pavesi, Riv. *Nuovo Cimento* **20**, (10) 1 (1997).
5. P. Ferrand, R. Romestain, and J. C. Vial, *Phys. Rev. B* **63**, 115106 (2001).
6. F. Ferrieu, A. Halimaoui and D. Bensahel, *Solid State Commun.* **84** (3) 293 (1992).
7. I. Mihalcescu, G. Lerondel and R. Romestain, *Thin Solid Films* **297**, 245 (1997).
8. C. J. Oton, Z. Gaburro, M. Ghulinyan, L. Pancheri, P. Bettotti, L. Dal Negro, and L. Pavesi, *Appl. Phys. Lett.* **81** (26) 4919 (2002).
9. F. Genereux, S. W. Leonard, H. M. Van Driel, A. Birner and U. Gösele, *Phys. Rev. B* **63**, 161101 (2001).
10. I. Sagnes, A. Halimaoui, G. Vincent, and P. A. Badoz, *Appl. Phys. Lett.* **62**, 1155 (1993).

# Light and Thermally Induced Metastabilities in Nanocrystalline Silicon

N. P. Mandal and S. C. Agarwal
Department of Physics, Indian Institute of Technology, Kanpur, 208016, India.

## ABSTRACT

Light soaking for short durations and thermal quenching produce metastable states having a higher dark current, higher photo current, a larger photoluminescence and a smaller electron spin resonance signal in porous silicon (PS). Long exposures, however, have the opposite effect. All metastabilities can be removed by annealing at $150^0 C$ (1 h), but not by exposure to infrared light. Micro-Raman spectroscopy shows the presence of a-Si:H in the PS sample. However, a closer look shows that our results can not be explained in terms of a-Si:H alone. Our experiments suggest that structural changes involving the movement of hydrogen present on the surface of PS or on PS/a-Si:H interface may be responsible for these effects.

## INTRODUCTION

Porous silicon (PS) consists of nanometer size crystallites of silicon and shows photoluminescence (PL) in the visible at room temperature. It has attracted quite a lot of attention because of the possibility of it's wide range of applications in optoelectronic devices, bio-sensors [1] and gas sensors [2-5]. PS is sensitive to various treatments, e.g., exposure to light or ambient gases. These give rise to metastabilities, which persist for a long time, even at room temperature. For example, short exposures to light, produce a state whose dark current (DC) is higher than the annealed state (persistent photoconductivity, PPC) [6-9] and persists for several days at room temperature. Some workers have found that PL does not change when the light soaking is done in vacuum or in nitrogen atmosphere [10,11]. On the other hand, Matsumoto et al. [12] reported that for long exposures (in vacuum) the PL decreases and dangling bond density (ESR) increases for their sample.

In this paper we report, for the first time, an increase in PL intensity in our PS after light exposures of short durations, when the sample is in vacuum. This is accompanied by an increase in DC, photocurrent (PC) and a decrease in ESR. Further, for long exposures, the changes are in the opposite direction and are similar to the well known Staebler-Wronski effect [13] in hydrogenated amorphous silicon (a-Si:H). Rapid cooling from high temperatures to room temperature also gives an increase in DC, PC and PL and a decrease in ESR signal. Although the increase in DC after fast cooling and short exposures in the case of PS is similar to PPC that has been observed [14-16] in doped a-Si:H, no decrease in ESR has so far been reported for a-Si:H. Since, micro-Raman measurements show the presence of a-Si:H phase in our PS [17] one might argue that the metastabilities in PS may be caused by a-Si:H. A closer look, however, suggests that this may not be so. For example, we find that the combined effect of $NH_3$ vapor exposure and LS on PS yield a state, which depends upon the order in which they are performed. This is contrary to the observation in a-Si:H [18]. Our inability to quench LS state by exposure to IR suggests that the metastable changes caused by the light exposures and thermal quenching might be related to the structural changes in PS or PS/a-Si:H interface.

## EXPERIMENTAL

Free standing PS layers (~2 x 5mm) were prepared by the electrochemical anodization of boron doped (100) crystalline silicon wafers having resistivity ~1 $\Omega$-cm [19]. Atomic force microscopy (AFM) shows that our PS consists of crystallites of sizes 2-5 nm and bigger [19]. Raman spectroscopy reveals the presence of a-Si:H (broad peak ~ 480cm$^{-1}$) in our PS [17]. PS samples were mounted on Corning 7059 glass substrates by putting silver paint at the two ends of the sample, which also served as electrodes for current measurements. For the PL and ESR measurements, PS flakes were sealed in 10$^{-1}$ torr Helium atmosphere in a quartz tube. The samples were annealed at 150$^0$C for 1 hour in vacuum, for removing the effect of previous light exposures, if any, and slowly cooled (rate$\approx$ 0.5 K/min) to room temperature (state A). ESR was measured using a X–band (Varian E-3) spectrometer at microwave power of 5 mW. For PL, the sample was excited by a He-Ne laser (632 nm, 15 mW) and the emitted light was analyzed using a grating spectrometer and a CCD detector. A bias of 40 volts was applied for the DC and PC measurements. The samples showed a symmetric, sub ohmic I-V behavior upto this voltage. Light soaking (LS) was done in vacuum, using heat filtered white light from a 250W tungsten halogen lamp, kept at a distance of about 15 cm, from the sample.

For thermal quenching experiments, the samples were annealed in vacuum at 150$^0$C (1 hr). Then the heater was switched off and chilled water was circulated through the copper block, upon which the sample had been mounted using silver paint, without breaking the vacuum.

## RESULTS AND DISCUSSION

Fig. 1 shows the ESR signal of our PS in the annealed state (state A). This signal is

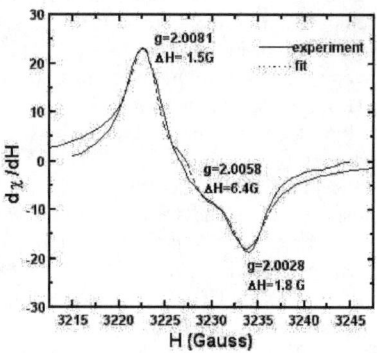

**Figure 1.** ESR signal in state A. Full line is the experiment and dotted line is the fit to three lines with g values and width shown.

asymmetric and is similar to that reported by others [20]. It can be fitted (Fig.1, dotted line) to three signals; an isotropic signal with g ≈ 2.0058 ±0.0006 (Lorentzian, width~ 6.4 ± 0.5G) corresponding to the dangling bonds in a-Si and two anisotropic signals having $g_\perp$ ≈ 2.0081 ±0.0006 (Gaussian, width ~ 1.5 ± 0.5G) and $g_\parallel$ ≈ 2.0028 ±0.0006 (Gaussian, width ~ 1.8 ± 0.5G). The last two be attributed to the dangling bonds at the surface of PS [21].

The PL shows a single peak at ≈ 800 nm with a full width at half maximum (FWHM) ≈ 130 nm. We see (Fig. 2) that DC, PC and PL increase for short light exposures but decrease for longer exposures with a maximum at about 2700s exposure time. The new dark current persists for several hours even at room temperature. Exposure to infrared light (< 1.2 eV) is unable to restore state A. Annealing at $150^0 C$ (1 hr), however, brings the sample to state A. A similar effect on DC for short as well as long exposures has been observed in doping modulated [22] and compensated [23] a-Si:H. We see that the largest increase in DC is about a factor of 10 for an exposure of about 2700s. Further, for the 2700s exposure, PL increases by about a factor of 3, and ESR decreases by ≈ 60% compared to the initial state A. On the other hand, we see that for longer exposures ESR increases with a decrease in DC, PC and PL. Interestingly, ESR shows a minimum at about the same exposure duration (2700s) at which DC, PC and PL show a maximum. This indicates that the dangling bonds probably act as the nonradiative recombination centers. The effect of long exposures is similar to the Staebler-Wronski effect in a-Si:H [13]. Furthermore, our long exposures results are in agreement with Matsumoto et al. [12] who also found that for long exposures the dangling bond density (ESR) increases and PL decreases. Raman spectroscopy on our PS shows the presence of a-Si:H along with the nanocrystals of silicon. Furthermore, heating to high temperature results in the evolution of hydrogen and Si-H species from PS [17]. Therefore, porous silicon can be visualized as a mixture of nanocrystalline silicon with a-Si:H tissues surrounded by hydrogen [17]. The light absorption occurs in the nanocrystalline silicon core and the photocarriers diffuse to the surface, where these recombine. The dangling bonds act as nonradiative recombination centers. Thus, if the density of dangling

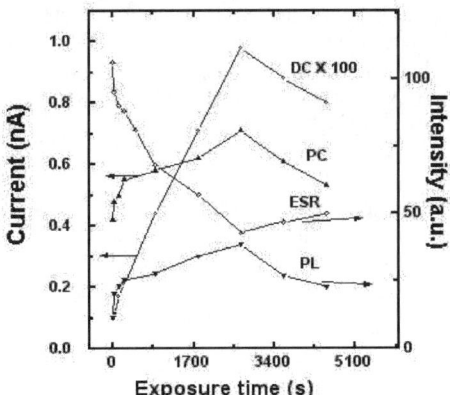

**Figure 2.** Effect of light exposure on DC, PC, PL and ESR as a function of exposure time. The slope of all curves changes sign at the exposure time of 2700s.

bonds (ESR) decreases, PL should increase and vice versa. Indeed, we find (Fig.2) that ESR decreases for short exposures, with an increase in PC and PL. This is possible, as the hydrogen present on the surface may provide the flexibility needed to rearrange the bond structure. The rigid four-fold coordinated silicon cannot give such flexibility at room temperature [24]. The breaking of hydrogen from the Si-H bond at room temperature is not possible, as it requires a large ($\approx$ 3 eV) energy [24]. However, the energy needed for the switching of Si-H bond is likely to be much smaller [25]. For short exposures, a hydrogen atom may move and bridge two nearby Si dangling bonds forming a Si-H-Si, which may be strained, but give a decrease in number of dangling bonds (process-I). For long exposures (>3000s), ESR increases; accompanied by a decrease in PC and PL and therefore the usual Staebler-Wronski [13, 25] type mechanism might be operative. In this case, hydrogen moves from one Si-H into a weak Si-Si bond creating one Si-H-Si metastable bond and one dangling bond (process-II) [25]. In general both these processes are likely to be operative during LS. Our results can be explained, if the process-I is weaker but faster than process-II. Thus, initially process-I dominates which decreases ESR. As LS is done for longer periods, process-II takes over and we see a net increase in dangling bonds (ESR).

Let us now look at the thermal quenching results. Fast cooling from $150^0$ C to room temperature (400 K/min) brings PS to another metastable state (C), which has higher DC, PC and PL than the annealed state A (Table 1). This is also accompanied by a decrease in the ESR signal. The state A can be recovered by annealing at $150^0$ C (1h) and slow cooling, but not by IR exposure. In state C, obtained after thermal quenching, the decrease in the number of dangling bonds is responsible for the increase in PC and PL. An increase in DC by thermal quenching has also been reported in P-doped a-Si:H [15,16]. Hydrogen movement in PS may be responsible for this behavior, as has been suggested in the case of a-Si:H [14].

Although the effect of LS and thermal quenching in PS are qualitatively similar to those reported for a-Si:H, a closer look shows differences between the two. For example, for short exposures the reduction in ESR found in PS, has not been reported so far in a-Si:H. Therefore, it appears that the presence of nanocrystalline Si also plays a role. In order to see, if the instabilities are caused mainly by the presence of a-Si:H in PS, we exposed our sample to cycles of LS and ammonia vapor a' la Tanielian [18]. Fig. 3 shows (cycle I) that the annealed state A (2.2 pA) goes to state LS after light soaking (30 min) with a higher DC (3.1 pA). A subsequent exposure to $NH_3$ results in the final state F, which has DC value 7.0 pA. Annealing brings the state F back to the state A. Now, if we reverse the order of LS and $NH_3$ exposure, we obtain the

Table 1
DC, PC, PL and ESR measured at room temperature of a PS in the annealed state A, and after fast cooling (state C)

|  | State A (Annealed) | State C (Quenched) |
|---|---|---|
| Dark Current (DC) | $1.0 \times 10^{-12}$ A | $5.0 \times 10^{-11}$ A |
| Photo Current (PC) | $4.6 \times 10^{-10}$ A | $5.5 \times 10^{-10}$ A |
| Photoluminescence Intensity (PL) | 3.3 ± 0.2 a.u. | 8.8±0.2 a.u. |
| Dangling bonds (ESR) | 20.0±0.2 a.u. | 16.6±0.2 a.u. |

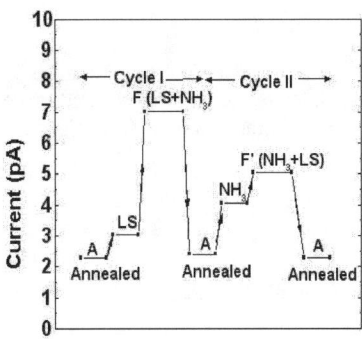

**Figure 3**: Effect of LS followed by ammonia exposure (Cycle I) and effect of ammonia followed by LS (Cycle II) on DC of a free standing PS.

state F' (cycle II, Fig.3) whose DC (4.8 pA) is smaller than that of state F be about a factor of 1.5. Annealing restores the sample to state A, as before. This observation is at variance with the results of Tanielian [18] who found that in a-Si:H, F and F' are the same and the order of LS and ammonia exposure is unimportant. This is expected if light exposure creates bulk defect states in a-Si:H [13], whereas ammonia vapor affects the surface only. Contrary to a-Si:H, we find that in PS the final state depends on the order in which LS and ammonia exposure are performed. Thus, our results can not be understood in terms of the presence of a-Si:H in our sample. Further, in contrary with a-Si:H, it may imply that LS in PS affects the surface.

## CONCLUSIONS

An increase in DC, PC and PL is observed when PS is exposed to light for a short duration (< 2700 s). This is accompanied by a decrease in ESR signal. For exposures longer than 3000 s we get the opposite effect. Fast cooling of PS to room temperature from $150^0$ C gives a state with higher DC, PC and PL, and is also accompanied by a decrease in the ESR signal. Further, all metastable states can be annealed out at $150^0$ C (1h), but are stable against exposure to sub band gap (IR) light. This indicates that the observed changes cannot be explained by trapping of charges alone. Though these results are qualitatively similar to those observed in a-Si:H, a more careful look reveals the differences. For example, a decrease in ESR signal after LS or thermal quenching as we observed in PS, has not been reported in a-Si:H. Similarly, we find that in PS the metastable state obtained by LS and ammonia exposures depends on the order in which they are performed. This is contrary to the results in a-Si:H [18]. This may mean that LS affects the surface in PS, in contrast to the a-Si:H where it is believed to be a bulk effect. This is not surprising as PS is an open system with large surface area, and there is evidence [26] that LS gives rise to fatigue and photochemical changes at the surface of PS. We suggest that a local structural rearrangement of bonds involving movements of hydrogen at PS or PS/a-Si:H interface is responsible for the LS and thermal quenching effect in PS.

## ACKNOWLEDGEMENTS

We are grateful to Dr. W. Fuhs for providing the crystalline Si wafers used in this study. One of us (SCA) would like to thank Professor H. Fritzsche and Professor F. Koch for fruitful discussions. This work is supported by a grant from the Council of Scientific and Industrial Research, New Delhi, India.

## REFERENCES

1. For a review, see for example, O. Bisi, S. Ossicini and L. Pavesi, Surf. Sci. Reports, **38**, 1 (2000).
2. L. Seals, J. L. Gole, L. A. Tse and P. J. Hesketh, J. Appl. Phys. **91**, 2519 (2002).
3. V. Mulloni and L. Pavesi, Appl. Phys. Lett. **76**, 2523 (2000).
4. P. Allcock and P. A. Snow, J. Appl. Phys. **90**, 5052 (2001).
5. W. E. Carlos, S. M. Porks, J. L. Gole and L. Seals, Phys Rev. B **62**, 1878 (2000).
6. W. H. Lee, H. Lee and C. Lee, J. Non-Crys. Solids, **164-166**, 965 (1993).
7. M. N. Islam, S. C. Agarwal, Y. N. Mohapatra and S. Kumar, Proc. Solid State Phys. (DAE) Symp. **37C**, 208 (1994).
8. T. Frello, E. Veje and O. Leistiko, J. Appl. Phys. **79**, 1027 (1996).
9. N. P. Mandal and S. C. Agarwal (12$^{th}$ International School on Condensed Matter Physics, 2002, Varna, Bulgaria and to be published).
10. Z. Y. Xu, M. Gal and M. Gross, Appl. Phys. Lett. **60**, 1375 (1992).
11. M. A. Tishler, R. T. Collins, J. H. Stathis and J. C. Tsang, Appl. Phys. Lett. **60**, 639 (1992).
12. T. Matsumoto, M. Kondo, S. V. Nair and Y. Masumoto, J. Non-Cryst Solids, **227-230**, 320 (1998).
13. D. L. Staebler and C. Wronski, Appl. Phys. Lett. **31**, 292 (1977).
14. R. A. Street, J. Kakalios, C. C. Tsai and T. M. Hayes, Phys. Rev. B **35** 1316 (1987).
15. Y. H. Song, B. S. Yoo, C. Lee and J. Jang, J. Non-Cryst. Solids, **114,** 666 (1989)
16. P. Agarwal and S. C. Agarwal, J. Appl. Phys. **81**, 3214 (1997).
17. S. C. Agarwal, in : B. K. Rao, S. M. Bose, M. P. Das and S. N. Sahu (Editors), Science and Technology of Nanostructred Materials, Nova, N.Y, 2001, p.101.
18. M. Tanielian, Phil. Mag. B **45,** 435 (1982).
19. N. P. Mandal, S. Dey and S. C. Agarwal, MRS Proceeding, Vol. **737**, Fall 2002.
20. H. J. von Bardeleben, M. Chamarro, A. Grosman, V. Morazzani, C. Ortega, J. Siekja and S. Rigo, J. Lumin. **57**, 39 (1993).
21. J.C. Mao, Y.Q. Jia, J.S. Fu, E.Wu, B.R. Zhang, L.Z. Zhang and G.G. Qin, Appl. Phys. Lett. **62**, 1408, (1993).
22. J. Kakalios and H. Fritzsche, Phys. Rev. Lett. **53**, 1602 (1984).
23. S.C. Agarwal and S. Guha, Phys. Rev. B **32**, 8469 (1985).
24. J. Kakalios, R. A. Street and W. B. Jackson, Phys. Rev. Lett. **59**, 1037 (1987).
25. M. Stutzmann, W. B. Jackson and C. C. Tsai, Phys. Rev. B **32**, 23 (1985).
26. Al. L. Efros, M. Rosen, B. Averbkouch, D. Kovalev, M. Ben-Chorin and F. Koch, Phys. Rev. B **56,** 3875 (1997).

## Physicochemical Characterization of Porous Silicon Surfaces Etched in Salt Solutions of Varying Compositions and pH

Mariem Rosario-Canales[1], Ana R. Guadalupe[1], Luis F. Fonseca[2], and Oscar Resto[2]
[1] University of Puerto Rico, Río Piedras Campus, Department of Chemistry, P.O. Box 23346, San Juan, Puerto Rico, 00931-3346
[2] University of Puerto Rico, Río Piedras Campus, Department of Physics, P.O. Box, 23343, San Juan, Puerto Rico, 00931-3343

## ABSTRACT

We prepared porous silicon (PSi) structures by standard electrochemical processes using aqueous sodium fluoride (NaF) solutions. We report the dependence of the porous structure on the variation of pH and salt concentration of the etching solution, and the applied current density. The PSi structures were characterized by Scanning Electron Microscopy (SEM) and Secondary Ion Mass Spectroscopy (SIMS) to determine the pore size and distribution and the surface chemical composition. Results obtained from SEM show that the PSi grown has two different structures depending on the current density. Low current densities produce a uniform, high-density arrangement of pores while high current densities yield a sponge-like structural network. SIMS results indicate that the porous framework is covered with a silicon oxide layer.

## INTRODUCTION

Several types of electrolytes are commonly used in electrochemical etching baths for silicon. The principal class consists of aqueous electrolytes including hydrofluoric acid and ammonium fluoride solutions [1, 2]. HF solutions typically contain ethanol to overcome the problem of hydrogen evolution during anodization thereby improving the layer uniformity [3]. Organic electrolytes consist of combinations of HF with organic solvents such as acetonitrile or dimethylformamide [1, 2]. Alkaline etchants are another major group and comprise aqueous solutions of inorganic compounds like LiOH, NaOH, KOH, or $NH_4OH$ as well as aqueous solutions of organic compounds, such as ethylenediamine, hydrazine, and tetramethyl ammonium hydroxide [4, 5].

The etching process of c-Si in HF solutions results in a hydride, and thus hydrophobic surface, populated with Si-$H_x$ groups (x = 1, 2, 3) [6]. However, as a result of the reaction with ambient air, PSi oxidation takes place leading to the formation of Si-O-Si, O-Si-H, and $O_3$-Si-H groups [7]. A chemical dissolution process leads to a surface covered with Si-OH and Si-H groups in alkaline solutions [8]. The pH also plays an important role, especially for $NH_4F$ electrolytes because the solution pH influences whether atomically flat, smooth or structured surfaces are obtained [9, 10].

In our case we are interested on the use of Si nanostructured surfaces for applications in the mass production of cost-effective biological sensors and devices. Based on our experience with both HF and NaF, we found that surfaces prepared with the latter showed a microwell structure and also exhibited mechanical stability for extended periods of time in aqueous media. We studied the preparation and morphology of such surfaces when grown under different anodization conditions. To our present knowledge we are not aware of systematic studies with NaF.

## EXPERIMENT

In this work, p-type (Boron doped), c-Si wafers with a <100> orientation, a resistivity of 20-30 $\Omega \cdot$cm, and an aluminum ohmic contact on the back side were used as substrates. The wafers with an anodization exposed area of 0.13 $cm^2$, were purchased from Virginia Semiconductors, Inc. They were first cleaned for 10 min in a 20% HF solution to remove any surface oxide. Aqueous NaF solutions at different concentrations and pH values were used as etchants to perform the electrochemical etching in a single-tank Teflon cell using a platinum wire as the counter electrode. The range of NaF concentrations used was: 0.76, 0.50, 0.30, and 0.10 M with pH values of 7.71, 7.34, 7.01, and 6.85, respectively. Ethanol was not added to the etching solutions because bubbles release is insignificant in these solutions and do not interfere with the current flow.

When working at a concentration of 0.76 M, the anodizations were performed at current densities of 25, 50, 100, and 150 $mA/cm^2$ while etching times were varied between 600 to 2400 s to determine the effects of current density and etching time on the properties of the porous structure. At the lower NaF concentrations of 0.50, 0.30, and 0.10 M, etching was performed at current densities of 25 and 50 $mA/cm^2$ and anodization times of 1800 s. After electrochemical etching, the anodized samples were left overnight in nanopure water followed by drying in vacuum for two days before performing the corresponding analyses. To assess the effect of lower solution pH on the PSi structure, HF 48-50 wt% was added to the 0.50 M NaF solution while keeping the total fluorine concentration constant.

Sample characterization was done by SEM and SIMS to determine the type of porous structure, pore size and distribution, the porous layer thickness, and the surface chemical composition. SEM analyses were done in a JEOL JSM 35CF instrument while SIMS was done with a PHI560 Quadrupole SIMS 600 instrument.

## RESULTS

### Scanning electron microscopy

Figure 1 shows SEM micrographs corresponding to upper surface views of PSi obtained after etching in 0.76 M NaF solution at 100 $mA/cm^2$ for 2400 s (a) and 50 $mA/cm^2$ for 1800 s (b). At 100 $mA/cm^2$ a sponge-like structural network was obtained while at 50 $mA/cm^2$ a high-density arrangement of defined wells is observed. Although not shown, the morphology at 150 $mA/cm^2$ was the same as that obtained at 100 $mA/cm^2$. SEM cross-section views of the samples showed a porous layer thickness between 5.0 to 8.3 microns. When working with the lower NaF concentrations at a current density of 50 $mA/cm^2$, an array of wells was also obtained with the 0.50 M solution as with the 0.76 M solution. At 0.30 M, the wells appear deformed into more elongated structures while at 0.10 M, well formation was not observed. These results are demonstrated in figure 2 that shows upper view SEM micrographs of surfaces after etching in each of the four NaF solutions. All samples grown so far are observed to be non-luminescent.

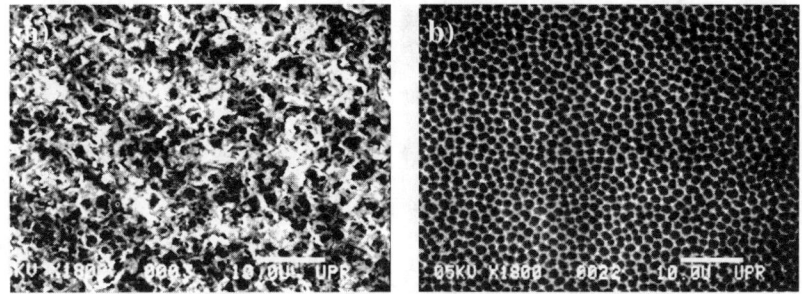

**Figure 1.** SEM micrographs showing upper surface views of samples etched in 0.76 M NaF solution at 100 mA/cm$^2$ for 2400 s (a) and 50 mA/cm$^2$ for 1800 s (b).

Figure 3 shows upper surface SEM micrographs of samples etched in 0.50 M NaF solution after adjusting the pH in each case using HF while maintaining total fluorine concentration constant. Electrochemical etching was done at 50 mA/cm$^2$ for 1800 s. The purpose of adding HF was to determine if its presence induced any morphological changes or produced thicker porous layers than those obtained with plain NaF solutions. At pH 6.27 (micrograph a), a well-defined array of wells was not observed. Lowering the pH to 5.56 results in a clear well structure comparable to that observed at pH 7.34 and this is seen in micrograph b. However, a higher density of wells is observed at this lower pH. This result could be attributed to the presence of HF in the etching solution. The silicon oxide layer that forms during the etching process is continuously removed when HF is present leaving more sites available for the electrochemical process to occur at the surface. Micrograph c corresponds to the sample etched in 0.50 M NaF solution at pH 5.24. At this particular pH, a polishing effect occurred as the porous structure was etched away due to the higher amount of HF in the etching solution. The porous layer thickness was not affected by the presence of HF.

Figure 4 shows histograms depicting the well size distributions for samples prepared in 0.76 M NaF solution (a) and in 0.50 M NaF solution at pH 7.34 and 5.56, (b) and (c) respectively. Average well size for the sample etched in 0.76 M NaF solution is 1.8 microns, while at 0.50 M average size is slightly smaller, 1.7 microns. Lowering the pH of the 0.50 M solution to 5.56 using HF 48-50 wt % resulted in a decrease in the average well size to 1.2 microns. The well shape was mostly rectangular and square.

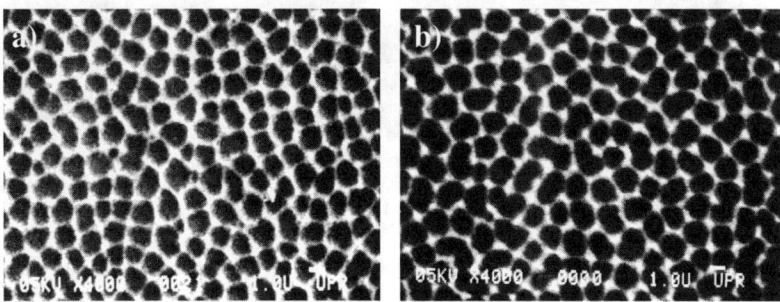

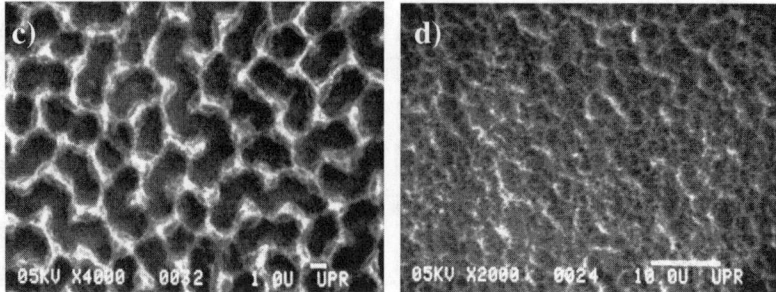

**Figure 2.** SEM micrographs showing upper views of samples etched at 50 mA/cm$^2$ for 1800 s in NaF solutions at different concentrations: 0.76 M (a), 0.50 M (b), 0.30 M (c), and 0.10 M (d).

## Secondary ion mass spectroscopy

Samples for SIMS analyses were sputtered with argon ions at 5 keV and at a current density of 1x10$^{-5}$ A/cm$^2$. The SIMS survey of the sample shown in micrograph b of figure 2 is presented in figure 5. Results show that the PSi structure is oxidized as evidenced by the presence of Si-O and Si-O-Si species. Thus, this leads to non-luminescent samples as mentioned before. Similar surveys were obtained for the samples etched in the other NaF solutions. The spectra for the 0.50 M NaF sample was compared to the SIMS survey obtained for a 20-30 $\Omega$·cm sample etched in 25% HF solution (1:1:2, HF:H$_2$O:ethanol, respectively) for 2400 s at 100 mA/cm$^2$ shown in figure 6. The abundance of SiO species in the sample prepared in the NaF solution is about one order of magnitude greater than that for the PSi grown in the 25% HF system. In both samples fluorine appears bonded to the silicon matrix forming SiF species. However, there is an increase of abundance of F and SiF species for the sample etched in the 0.50 M NaF solution in comparison to that prepared in 25% HF. On the contrary, the abundance of Si species is greater for the 25% HF sample.

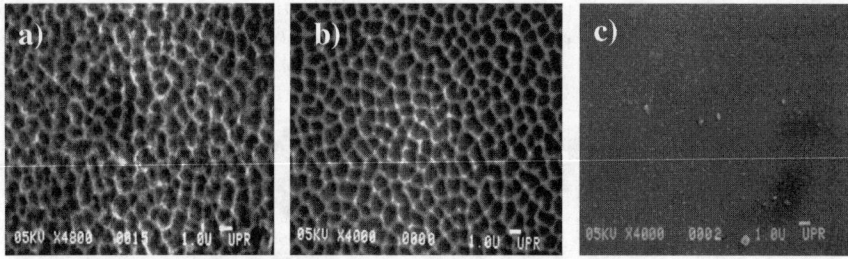

**Figure 3.** SEM micrographs showing upper surface views of samples etched at 50 mA/cm$^2$ for 1800 s in 0.50 M NaF solutions at different pH: 6.27 (a), 5.56 (b), and 5.24 (c).

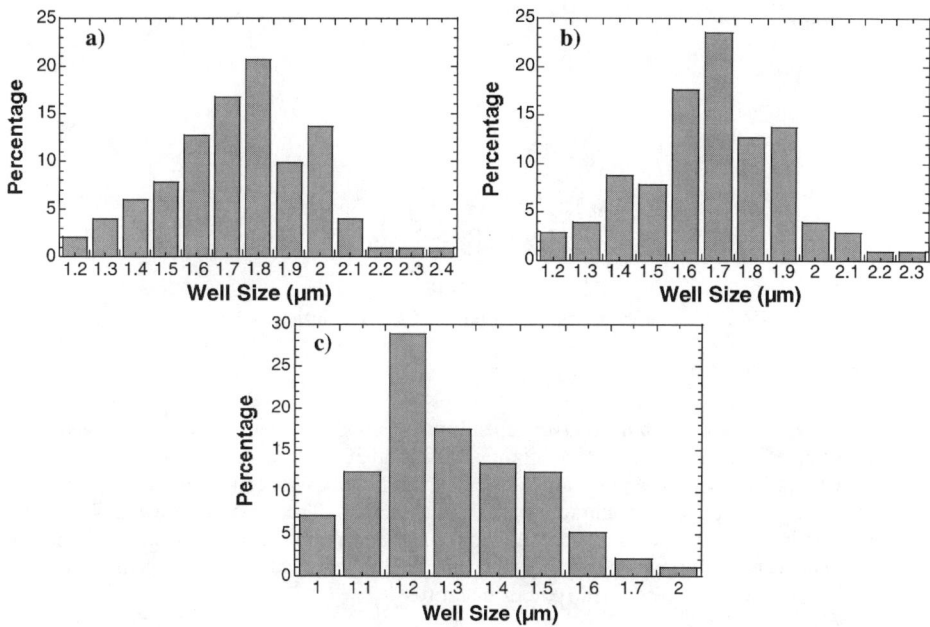

**Figure 4.** Well size distributions for samples etched at 50 mA/cm$^2$ for 1800 s in 0.76 M NaF solution (a) and 0.50 M NaF solution at pH 7.34 (b) and pH 5.56 (c).

Previously we had mentioned that the sample prepared in the 0.50 M NaF solution at pH 5.24 had undergone a polishing effect. It is reasonable to attribute such outcome due to the fact that the etching solution for this sample contained the highest amount of HF. Thus, the anodization solution would have the capacity of completely dissolving the silicon oxide framework, etching away the porous structure.

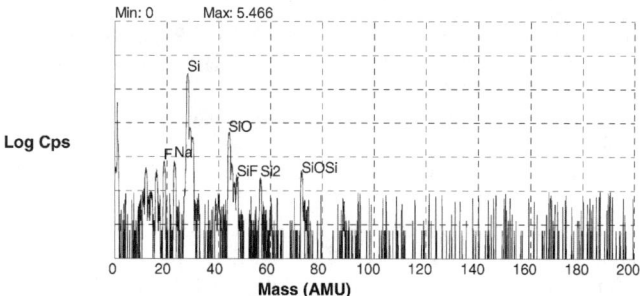

**Figure 5.** SIMS survey of sample presented in micrograph b of figure 2. Sample was etched in 0.50 M NaF solution for 1800 s at 50 mA/cm$^2$.

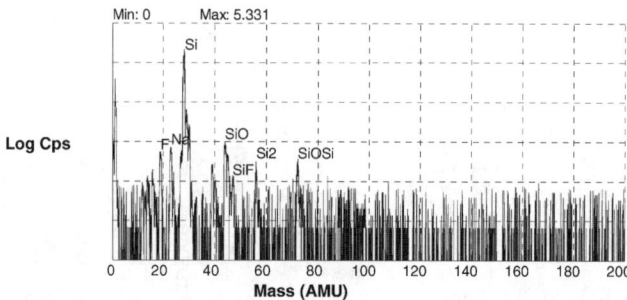

**Figure 6.** SIMS survey for sample prepared in 25% HF solution at 100 mA/cm$^2$ for 2400 s.

## CONCLUSIONS

Morphology of the porous layer depends on the applied current density as was demonstrated for samples etched in 0.76 M NaF solution at 100 and 50 mA/cm$^2$ and also on the NaF solution concentration. Etching at concentrations below 0.30 M does not follow morphology trend observed at the higher concentrations of 0.76 and 0.50 M. The samples grown in NaF solutions showed high porous densities, structural uniformity and can be kept in aqueous environments for extended periods of time without any mechanical damage. In addition, they do not need special drying procedure to preserve the porous integrity.

## ACKNOWLEDGEMENTS

Mariem Rosario-Canales would like to thank NASA PRSGC grant number NGTS-40091 for her graduate fellowship. This work was supported by DOE grant number DE-FG02-01ER45868, the US Army Research Office under grant 41002-MS-DPS, and NASA NCC5-518.

## REFERENCES

1. H. Föll, M. Christophersen, J. Carstensen, and G. Hasse, *Mater. Sci. Eng. R*, **39**, 93 (2002).
2. V. Lehmann, *Electrochemistry of Silicon*, (Wiley-VCH Verlag GmbH, Weinheim, Germany, 2002) p. 7.
3. A.G. Cullis, L.T. Canham, and P.D.J. Calcott, *J. Appl. Phys.*, **82** (3), 909, (1997).
4. X.G. Zhang, *Electrochemistry of Silicon and its Oxide*, (Kluwer Academic/Plenum Publishers, New York, 2001) p. 294.
5. H. Seidel, L. Csepregi, A. Heuberger, and H. Baumgärtel, *J. Electrochem. Soc.*, **137** (11), 3612, (1990).
6. M.P. Stewart, E.G. Robins, T.W. Geders, M.J. Allen, H. Cheul Choi, and J.M. Buriak, *Phys. Stat. Sol. a*, **182**, 109 (2000).
7. A. Grosman and C. Ortega, in *Properties of Porous Silicon,* edited by L. Canham (INSPEC, The Institution of Electrical Engineers, London, 1997), pg. 150.
8. V. Lehmann, *Electrochemistry of Silicon*, (Wiley-VCH Verlag GmbH, Weinheim, Germany, 2002) p. 28.
9. J. Rappich, V. Yu. Timoshenko, R. Würz, and Th. Dittrich, *Electrochim. Acta*, **45**, 4629 (2000).
10. P. Allongue, V. Kieling, and H. Gerischer, *Electrochim. Acta*, **40** (10), 1353, (1995).

# AUTHOR INDEX

Aarts, I.M.P., 111
Acciarri, M., 565
Adriaenssens, Guy J., 99
Agarwal, S.C., 773
Agashe, Chitra, 405
Águas, H., 217, 589
Albenze, Erik J., 681
Al-Jassim, M.M., 443, 539, 571
Allen, T., 649, 661
An, Ilsin, 515
Anderson, Wayne A., 735
Anelli, G., 205
Aneva, Zdravka, 143
Appenzeller, Wolfgang, 285
Arcari, A., 565
Arya, R.R., 393, 467
Asano, A., 583
Asl Soleimani, E., 675
Atwater, H.A., 551
Aydil, Eray S., 521
Aziz, Abdul, 33

Bailat, J., 503
Ballif, Cristoffer, 753
Banerjee, Arindam, 309
Barrell, Y., 431
Beaucage, Philippe, 3
Belich, T.J., 509
Ben-Hander, Fathi A., 753, 759
Bennett, M.S., 393, 467
Bettotti, Paolo, 767
Binetti, S., 565
Biswas, R., 15
Blanc, N., 205
Blecher, Frank, 211
Bochem, Hans P., 405
Böhm, Markus, 211
Bollani, M., 565
Borchert, Dietmar, 753
Brandt, Martin S., 473
Branz, Howard M., 75, 157, 413
Brendel, K., 69, 693
Brinza, Monica, 99
Brüggemann, Rudi, 137
Bu, Ian Y.Y., 533

Carius, Reinhard, 81, 321, 327, 497
Carlson, David E., 27, 381, 393, 467
Carlson, Eric D., 625
Carter, C.B., 509
Cavalier, L., 205
Cerny, G., 265
Chambouleyron, I.E., 717
Chan, Isaac, 193
Chan, Mansun, 705
Chen, C.M., 551
Chen, Chi, 443, 515
Chen, X.Y., 87
Chen, Yousu, 223
Cheng, Ju-Yin, 45
Cheng, Lawrence C.F., 705
Chi, Linfei, 607
Cho, B.J., 87
Choi, J.H., 265
Chu, V., 151, 259
Chung, Kyuha, 265, 711
Clancy, Paulette, 681
Cohen, J. David, 51
Collins, R.W., 303, 351, 443, 515, 539
Conde, J.P., 151, 259
Constant, Alan P., 655
Cui, R.Q., 357

Dalakos, George T., 595
Dalal, Vikram L., 33, 375, 619
Daldosso, Nicola, 767
Dal Negro, Luca, 767
Deal, Michael D., 669
Delahoy, Alan E., 545
Deng, J., 303, 351
Derakhshandeh, J., 675
Despeisse, M., 205
Dexheimer, S.L., 333
Dinca, S., 345
Ding, Z.M., 357
Dissertori, G., 205
Droz, C., 503
Dunand, S., 205
Durny, R., 39
Dylla, T., 81

Ehara, Takashi, 117
Eickhoff, Martin, 473
Engeln, R., 111

Fantoni, A., 199
Feitknecht, L., 503
Ferlauto, A.S., 443, 515, 539
Fernandes, M., 199
Ferreira, G.M., 351, 443, 515, 539
Finger, F., 81, 321, 327, 497
Fixe, F., 259
Flewitt, A.J., 533
Fonseca, Luis F., 779
Fortunato, E., 217, 589
Fuhs, W., 69, 125
Fujikake, Shinji, 521
Fumagalli, L., 565
Fynan, Nick F., 681

Gaburro, Zeno, 767
Ganguly, Gautam, 27, 345, 381, 393, 467, 515
Gaspar, J., 151
Gaspari, F., 649, 661
Gedvilas, L.N., 387, 643
Gedvilas, Lynn M., 455
Geng, Xinhua, 223, 613
Ghulinyan, Mher, 767
Gielis, J.J.H., 111
Ginley, Dave, 413
Golshani, N., 675
Gordijn, A., 637
Goullet, A., 217
Guadalupe, Ana R., 779
Guerrero-Lemus, Ricardo, 753, 759
Guha, Subhendu, 21, 297, 309, 339, 363
Guillet, J., 503

Hadi, G.M., 357
Hamada, Shiro, 687
Han, Daxing, 339, 399, 413
Han, Min-Koo, 723, 729, 741
Hanna, Jun-ichi, 229, 253
Hardeman, A.J., 631
Harikai, Atsushi, 437

Hawley, Frank, 699
Hernández-Rodríguez, Cecilio, 753, 759
Higashimine, K., 163
Hillebrand, Matthias, 211
Hoefnagels, J.P.M., 431
Hoex, B., 111
Hong, J., 169
Hong, W.S., 265
Hou, Guofu, 613
Hu, H., 87
Hu, Jian, 75, 157
Huang, Tiao-Yuan, 181
Huh, Hwang, 747
Hüpkes, Jürgen, 285, 405
Hwang, B.K., 265

Islam, Md.N., 105
Issaq, Farid, 699
Itoh, Masaya, 491
Iwaniczko, E., 387, 643
Izumi, A., 163

Jackson, Warren B., 75, 157
Jarron, P., 205
Ji, Chunhai, 735
Jiang, C.-S., 571
Jin, Cheng-Guo, 687
Johnson, Todd R., 381, 625
Jones, K.M., 443, 539, 571
Jordan, William B., 577, 625
Jung, K.W., 265
Jung, Sang-Hoon, 723, 741

Kakalios, J., 509
Kang, Myung-Koo, 711
Kang, Sook Young, 711
Kang, Su-Hyuk, 729, 741
Karim, K.S., 247
Katouno, Kouichi, 491
Kaufmann, R., 205
Keblinski, P.J., 45
Kenanoglu, Ali, 753
Kessels, W.M.M., 99, 111, 169, 431, 461
Kharchenko, A.V., 559

Kherani, N.P., 649, 661
Kikkawa, A., 163
Kim, Chang-Dong, 241
Kim, Chi-Woo, 711
Kim, Hyun Jae, 711
Kluth, Oliver, 285, 405
Koga, Kazunori, 437
Koh, Joohyun, 443
Komori, Hirohito, 687
Kortshagen, U., 509
Kosteski, T., 649, 661
Kosuge, T., 583
Koval, R.J., 303, 351, 443, 539
Kudo, Toshio, 687
Kwak, Joong Hwan, 515

Lambertson, Roy, 699
Lebib, S., 559
Lee, Czang-Ho, 193
Lee, Jeong-Woo, 253
Lee, Joon-Yong, 93
Lee, Seok-Woo, 241
Lee, Su-Kyung, 711
Leewis, C.M., 111
Lehner, Andrea, 473
Lei, Tan-Fu, 181
Leong, K., 649
Leung, Vincent T.C., 705
Levy, Roland A., 545
Li, G., 357
Li, Juan, 223
Li, Liwei, 545
Li, Yuan-Min, 545
Liang, J., 297
Lim, Koeng Su, 515
Lim, Kyoung Moon, 241
Lin, Horng-Chih, 181
Lin, Kuixun, 607
Lin, Xuanying, 607
Lips, K., 69
Liu, Shi, 607
Liu, Yaocheng, 669
Longeaud, C., 559
Lord, Kenneth, 21, 339
Louro, P., 199
Lu, Y.F., 87

Lu, Z., 345
Lussky, Th., 125

Mahan, A.H., 387, 643
Main, Charlie, 131, 137, 143, 327
Mandal, N.P., 773
Martínez-Duart, José M., 753, 759
Martins, R., 217, 589
Mason, M.S., 551
Masuda, Atsushi, 491
Matejik, Laura A., 681
Matsumura, Hideki, 163, 491
McCollum, John, 699
Meng, Zhiguo, 223
Merazga, Amar, 131
Merdzhanova, T., 321
Metselaar, J.W., 39
Miazza, C., 205
Milch, O., 125
Milne, W.I., 533
Milostnaya, I., 649, 661
Mizukami, Hiroyuki, 601
Mohajerzadeh, S., 675
Mohapatra, Y.N., 105, 175
Moraes, D., 205
Mousseau, Normand, 3
Moutinho, Helio R., 413, 571
Mulato, M., 717
Müller, Joachim, 285, 405
Myasnikov, Alexandre M., 705

Nádazdy, V., 39
Nakano, Y., 583
Nam, Dae Hyun, 241
Nam, Woo-Jin, 723, 729
Nampoothiri, A.V.V., 333
Nathan, Arokia, 187, 193, 247, 277
Nelson, Brent P., 333, 387, 455, 571, 643
Nesheva, Diana, 143
Nesládek, M., 111
Nickel, N.H., 69, 693
Niquille, X., 503
Niu, X.J., 619
Noack, Max, 619

Ogata, Takanori, 437
Ohto, Tokio, 521
Okamoto, H., 583
Oton, Claudio J., 767
Owens, Jessica M., 339

Pan, B.C., 15
Pancheri, Lucio, 767
Park, Dong-Hyun, 93
Park, Kee-Chan, 723
Park, Soo-Jeong, 723
Pavesi, Lorenzo, 767
Peacock, P.W., 425
Pearce, J.M., 303, 351, 443, 539
Pereira, L., 217, 589
Perkins, C.L., 75
Perkins, John, 413, 571
Perrey, C.R., 509
Persans, Peter D., 595
Petit, A.M.H.N., 369
Pizzini, S., 565
Plawsky, Joel L., 595
Plummer, James D., 669
Poon, Vincent M.C., 705
Prazeres, D.M.F., 259

Ranaweera, Jeewika, 699
Raniero, L., 589
Rath, J.K., 631, 637
Rech, Bernd, 285, 405
Reedy, Robert C., 455
Ren, Huizhi, 613
Repmann, Tobias, 285
Resto, Oscar, 779
Reynolds, Steve, 131, 137, 143, 327
Robertson, J., 425, 533
Robertson, M.D., 675
Roca i Cabarrocas, P., 105, 175, 559, 589
Rodrigues, I., 199
Rodríguez, Tomás, 753
Roes, W.L., 661
Romero, M.J., 571
Rosario-Canales, Mariem, 779
Roschek, Tobias, 285
Ross, Christoph, 497

Rovira, P.I., 515
Rusche, M.Y.S., 631

Saadane, O., 559
Sakariya, K., 247
Schiff, E.A., 297, 345
Schmidt, M., 125
Schoepke, A., 125
Schropp, R.E.I., 479, 631, 637
Seike, Koji, 687
Selvan, J.A. Anna, 545
Seo, Hyun Sik, 241
Servati, Peyman, 187
Shah, A., 205, 503
Sharma, Puneet, 33
Shimizu, Kousaku, 229, 253
Shin, Hee-Sun, 741
Shin, Jung H., 747
Shinar, Ruth, 655
Shiratani, Masaharu, 437
Siekmann, Hilde, 405
Silva, R., 217
Sirvent, A.G., 205
Smets, Arno H.M., 99, 111, 461
Smirnov, Vladimir, 327
Smit, C., 527
Song, In-Hyuk, 729
Steinhoff, Georg, 473
Sterzel, Jürgen, 211
Stiebig, Helmut, 285
Stradins, Pauls, 75, 157, 393
Striakhilev, Denis, 187
Stutzmann, Martin, 473
Su, T., 27
Suendo, V., 105, 559
Sultana, Mahmooda, 669
Sun, Jian, 613

Tabuchi, Toshihiro, 601
Takano, Akihiro, 521
Takashiri, Masayuki, 601
Tao, S., 277
Taylor, P.C., 27, 63
Tessler, Leandro R., 413
Theil, Jeremy A., 271
Thompson, S., 509

To, Bobby, 413, 571
Toet, D., 717
Toyama, T., 583
Toyoshima, Yasumasa, 601
Treacy, M.M.J., 45
Tripathi, V., 105, 175

Valiquette, Francis, 3
van den Oever, P.J., 431
van der Werf, C.H.M., 631
van de Sanden, M.C.M., 99, 111, 169, 369, 431, 461, 527
van Swaaij, R.A.C.M.M., 39, 369, 527
van Veenendaal, P.A.T.T., 631
Vieira, M., 199
Viertel, G., 205
Vigneron, P., 425
Vlahos, V., 303, 345, 351
von Känel, H., 565
Vygranenko, Y., 277

Wada, Takehito, 521
Wagner, Sigurd, 577, 625
Wang, Keda, 399
Wang, Qi, 75, 157, 223, 399, 413
Wang, W., 297
Ward, Scott, 157
Watanabe, Yukio, 437
Whitaker, J., 63
Williamson, D.L., 387, 455, 527, 643
Willing, F., 393
Wills, Matthew R., 655
Wood, George S., 381
Wronski, C.R., 303, 345, 351, 443, 515, 539
Wu, Chunya, 223
Wu, Y.H., 87
Wuttig, Matthias, 405

Wyrsch, N., 205

Xiong, K., 425
Xiong, Shaozheng, 223
Xu, Yueqin, 455, 571, 613
Xue, Junming, 613

Yamazaki, Kazunori, 687
Yan, Baojie, 21, 309, 339, 363, 387, 643
Yang, Guanghua, 223
Yang, Huidong, 223
Yang, Jeffrey, 21, 297, 309, 339, 363
Yang, S.H., 265
Yawaka, Sachi, 687
Yeghikyan, D., 649, 661
Yoon, Jin Mo, 241
Yoon, Jong-Hwan, 93
Yoshida, Takashi, 521
Yu, Cheng-Ming, 181
Yu, Chuying, 607
Yu, Yunpeng, 607
Yuan, Q., 345
Yue, Guozhen, 21, 309, 363

Zanatta, A.R., 717
Zeman, M., 39, 369
Zhang, Dekun, 613
Zhang, Jian Jun, 253
Zhang, Lizhu, 223
Zhang, Xiaodan, 613
Zhao, Ying, 223, 613
Zhou, Z.B., 357
Zhou, Zhenhua, 223
Zhu, Jianhua, 375
Zhu, K., 297
Zrinscak, Ivica, 131, 143
Zukotynski, S., 649, 661

# SUBJECT INDEX

absorption spectroscopy, 111
activation-relaxation technique, 3
AFM, 545
all solid state green lasers, 687
amorphous(-)
    hydrogenated silicon, 75
    microcrystalline silicon, 649, 661
    silicon, 3, 21, 33, 39, 125, 131,
        143, 151, 157, 175, 187,
        205, 217, 259, 277, 303,
        339, 345, 351, 363, 369,
        381, 431, 437, 443, 473,
        515, 539, 595, 699
            germanium, 455
            solar cells, 467
            TFT, 247
anisotropy, 767
area dependent, 157
a-SiCO:H, 265
a-Si:H, 99, 111, 193, 211, 461
a-Si PIN/OLED coupling, 223
atomic force microscopy, 571

buffer layer, 369

capping layer, 711
carbon content, 583
catalytic-CVD, 163
Cat-CVD (Hot-Wire CVD), 491
cavity ringdown spectroscopy, 111
charge deep-level transient
    spectroscopy, 39
cluster, 437
combinatorial approach, 413
conductivity, 399
constant photocurrent method, 131
crystallization, 3

dangling bond, 117
defect(s), 52, 93, 131
    density, 559
    spectroscopy, 137
degradation, 399
density of states, 99
deposition rate, 387

device modeling, 297
diffraction gratings, 717
diffusion, 425
diode edge, 271
DNA chip, 259
doped, 637
    with boron, 577
    with phosphorus, 577
double-pulsed laser annealing, 687

ELA, 729
electrical
    and numerical simulation, 199
    and optical properties, 661
    probe, 607
electron spin resonance, 63, 69, 117
electronic properties, 509
ellipsometry, 497
epitaxy, 551
ethane, 583
excimer laser annealing, 723
expanding thermal plasma, 143, 169

far-infrared, 333
fast deposition, 601
femtosecond, 333
field-assisted, 675
filament regime, 643
floating active structure, 729

gap state density, 125
gate oxide integrity, 241
germanium(-), 655, 681
    carbide, 655
    carbon, 619
    nitride, 717
    seed, 675
grain size, 625
growth, 533, 595
    mechanism, 425, 431

HEEPT, 601
high-rate low-temperature deposition,
    169
hole drift mobility, 297, 345

hot wire
    chemical vapor deposition, 479, 551, 631
    CVD, 387, 455, 643
HRTEM, 163, 649
HW a-Si:H, 399
hydrogen, 27
    bond, 559
    chloride, 521
    density-of-states, 693
    dilution, 363
hydrogenated amorphous silicon, 271, 333
hydrosilylation, 473

image
    fluctuations, 45
    sensor, 205
infrared spectroscopy, 521
instabilities and metastable effects, 81
instability, 33
interface response function, 681

junction defect, 723

laser
    annealing, 87
    crystallization, 693, 717
    deposition, 87
    scanned photodiode image sensor, 199
lateral grain growth, 741
layer-by-layer, 637
LEPECVD, 565
lifetime, 753
light
    induced changes, 393
    soaking, 21, 39
low-k, 265
low temperature, 533, 595
    poly Si (LTPS), 711
luminescence, 773

measurement(s), 705
    reliability, 607
mechanical stress, 193

MEMS, 151
metal induced
    crystallization, 669
    growth, 735
metallic filaments, 75
metal-to-metal antifuse, 699
metamict silica, 45
metastability, 51, 137, 247, 773
microcrystalline, 339, 503
    Si solar cells, 375
    SiC, 583
    silicon, 81, 93, 117, 157, 309, 321, 327, 387, 479, 491, 497, 527, 533, 637
        germanium, 375
    thin films, 601
MILC, 705
MIS, 217
    photodetector, 277
mixed phase, 15
molecular dynamics, 15
    simulation, 681
multi-junction, 309

nanocrystalline, 565, 619
    germanium, 577, 625
    inclusions, 509
    silicon, 15, 357, 571
narrow bandgap alloy, 455
nickel, 669
n-i-p solar cells, 503
NMR, 27
numerical and circuit derivation, 607

open circuit voltage, 21, 363
optical band gap, 521
optoelectronic detection, 259
oxidation and adsorption, 81

parameter extraction, 187
particle sensor, 205
PECVD, 661
phase transition, 613
photoconductivity spectroscopy, 143
photoconductor, 211
photoelectron spectroscopy, 125

photoluminescence, 93, 321, 759
photovoltaic(s), 357, 551, 565
plasma
    CVD, 437
    deposition, 527
    impedance, 589
    monitoring, 467
    passivation, 181
p-layer, 351
polycrystalline silicon, 69, 229, 253, 693
    germanium, 229
    thin films, 687
polymorphous silicon, 559, 589
poly-Si, 729, 735, 741
polysilicon, 631, 773
    thin film transistor, 241
porous silicon, 753, 759, 767
    sodium fluoride, 779
post-anneal, 241
post-transit photocurrent, 327
process control, 497
pulsed laser deposition, 655

Raman, 105, 339, 545
    spectroscopy, 571, 649
real time spectroscopic ellipsometry, 443, 539
recrystallization, 711
refractory metal, 699
resonators, 151
reverse bias, 271
room temperature annealing, 393
roughness, 425

scanning electron microscopy, 779
scattering, 767
secondary ion mass spectroscopy, 779
sensor, 223
shunts, 381
silicon(-), 27, 105, 503, 669
    crystallization, 675
    germanium alloys, 51
    nanocrystals, 87
    nitride, 163, 169

simulation(s), 223, 369
single ion beam sputtering, 357
solar cell(s), 285, 297, 303, 321, 351, 375, 381, 393, 443, 479, 491, 515, 527, 539, 619, 631, 643, 735
spectroscopic ellipsometry, 589
spectroscopy, 431
sputtering, 405
SRP, 705
stability, 467
stable hydrogenated amorphous silicon, 613
stain
    etched, 753
    etching, 759
sub-gate, 181
superposition, 303
surface
    diffusion, 461
    roughness, 625
switch, 175
switching, 75, 157

tandem n-i-p-n-i-p structures, 199
TFTs, 187, 253, 723
TFT-LCD, 265
thermal CVD, 253
thin film
    silicon, 285, 413
    solar cells, 405
    transistor(s), 105, 175, 181, 229
threshold voltage, 247
time-of-flight, 99, 345
topological order, 45
transient photoconductivity, 137, 327
transmission electron microscopy, 509
transparent conductor oxide, 515
traps, 33
trench, 741

UV
    detector, 211
    sensitivity, 277

very high frequency (VHF), 309
    PECVD, 613
voids/vacancies, 461
voltage-current characterization, 217

XANES, 413

x-ray
    detection, 193
    diffraction, 571
    photoelectron spectroscopy, 473
XRD, 545

zinc oxide, 285, 405